Cornerstones

Series Editors
Charles L. Epstein, *University of Pennsylvania, Philadelphia*
Steven G. Krantz, *Washington University, St. Louis*

Advisory Board
Anthony W. Knapp, *State University of New York at Stony Brook, Emeritus*

Anthony W. Knapp

Advanced Algebra

Along with a companion volume
Basic Algebra

Birkhäuser
Boston • Basel • Berlin

Anthony W. Knapp
81 Upper Sheep Pasture Road
East Setauket, NY 11733-1729
U.S.A.
e-mail to: aknapp@math.sunysb.edu
http://www.math.sunysb.edu/~aknapp/books/a-alg.html

Cover design by Mary Burgess.

Mathematics Subject Classicification (2000): 11-01, 13-01, 14-01, 16-01, 18G99, 55U99, 11R04, 11S15, 12F99, 14A05, 14H05, 12Y05, 14A10, 14Q99

Library of Congress Control Number: 2007936880

ISBN-13: 978-0-8176-4522-9   eISBN-13: 978-0-8176-4613-4

Basic Algebra   ISBN-13: 978-0-8176-3248-9
Basic Algebra and *Advanced Algebra* (Set)   ISBN-13: 978-0-8176-4533-5

Printed on acid-free paper.

©2007 Anthony W. Knapp
All rights reserved. This work may not be translated or copied in whole or in part without the written permission of the publisher (Birkhäuser Boston, c/o Springer Science+Business Media LLC, 233 Spring Street, New York, NY 10013, USA) and the author, except for brief excerpts in connection with reviews or scholarly analysis. Use in connection with any form of information storage and retrieval, electronic adaptation, computer software, or by similar or dissimilar methodology now known or hereafter developed is forbidden.
The use in this publication of trade names, trademarks, service marks and similar terms, even if they are not identified as such, is not to be taken as an expression of opinion as to whether or not they are subject to proprietary rights.

9 8 7 6 5 4 3 2 1

www.birkhauser.com   (MP)

To Susan

and

To My Algebra Teachers:

Ralph Fox, John Fraleigh, Robert Gunning,
John Kemeny, Bertram Kostant, Robert Langlands,
Goro Shimura, Hale Trotter, Richard Williamson

CONTENTS

Contents of Basic Algebra	x
Preface	xi
List of Figures	xv
Dependence among Chapters	xvi
Guide for the Reader	xvii
Notation and Terminology	xxi

I. TRANSITION TO MODERN NUMBER THEORY — 1

1. Historical Background — 1
2. Quadratic Reciprocity — 8
3. Equivalence and Reduction of Quadratic Forms — 12
4. Composition of Forms, Class Group — 24
5. Genera — 31
6. Quadratic Number Fields and Their Units — 35
7. Relationship of Quadratic Forms to Ideals — 38
8. Primes in the Progressions $4n+1$ and $4n+3$ — 50
9. Dirichlet Series and Euler Products — 56
10. Dirichlet's Theorem on Primes in Arithmetic Progressions — 61
11. Problems — 67

II. WEDDERBURN–ARTIN RING THEORY — 76

1. Historical Motivation — 77
2. Semisimple Rings and Wedderburn's Theorem — 81
3. Rings with Chain Condition and Artin's Theorem — 87
4. Wedderburn–Artin Radical — 89
5. Wedderburn's Main Theorem — 94
6. Semisimplicity and Tensor Products — 104
7. Skolem–Noether Theorem — 111
8. Double Centralizer Theorem — 114
9. Wedderburn's Theorem about Finite Division Rings — 117
10. Frobenius's Theorem about Division Algebras over the Reals — 118
11. Problems — 120

III. BRAUER GROUP 123

1. Definition and Examples, Relative Brauer Group 124
2. Factor Sets 132
3. Crossed Products 135
4. Hilbert's Theorem 90 145
5. Digression on Cohomology of Groups 147
6. Relative Brauer Group when the Galois Group Is Cyclic 158
7. Problems 162

IV. HOMOLOGICAL ALGEBRA 166

1. Overview 167
2. Complexes and Additive Functors 171
3. Long Exact Sequences 184
4. Projectives and Injectives 192
5. Derived Functors 202
6. Long Exact Sequences of Derived Functors 210
7. Ext and Tor 223
8. Abelian Categories 232
9. Problems 250

V. THREE THEOREMS IN ALGEBRAIC NUMBER THEORY 262

1. Setting 262
2. Discriminant 266
3. Dedekind Discriminant Theorem 274
4. Cubic Number Fields as Examples 279
5. Dirichlet Unit Theorem 288
6. Finiteness of the Class Number 298
7. Problems 307

VI. REINTERPRETATION WITH ADELES AND IDELES 313

1. p-adic Numbers 314
2. Discrete Valuations 320
3. Absolute Values 331
4. Completions 342
5. Hensel's Lemma 349
6. Ramification Indices and Residue Class Degrees 353
7. Special Features of Galois Extensions 368
8. Different and Discriminant 371
9. Global and Local Fields 382
10. Adeles and Ideles 388
11. Problems 397

VII. INFINITE FIELD EXTENSIONS — 403
1. Nullstellensatz — 404
2. Transcendence Degree — 408
3. Separable and Purely Inseparable Extensions — 414
4. Krull Dimension — 423
5. Nonsingular and Singular Points — 428
6. Infinite Galois Groups — 434
7. Problems — 445

VIII. BACKGROUND FOR ALGEBRAIC GEOMETRY — 447
1. Historical Origins and Overview — 448
2. Resultant and Bezout's Theorem — 451
3. Projective Plane Curves — 456
4. Intersection Multiplicity for a Line with a Curve — 466
5. Intersection Multiplicity for Two Curves — 473
6. General Form of Bezout's Theorem for Plane Curves — 488
7. Gröbner Bases — 491
8. Constructive Existence — 499
9. Uniqueness of Reduced Gröbner Bases — 508
10. Simultaneous Systems of Polynomial Equations — 510
11. Problems — 516

IX. THE NUMBER THEORY OF ALGEBRAIC CURVES — 520
1. Historical Origins and Overview — 520
2. Divisors — 531
3. Genus — 534
4. Riemann–Roch Theorem — 540
5. Applications of the Riemann–Roch Theorem — 552
6. Problems — 554

X. METHODS OF ALGEBRAIC GEOMETRY — 558
1. Affine Algebraic Sets and Affine Varieties — 559
2. Geometric Dimension — 563
3. Projective Algebraic Sets and Projective Varieties — 570
4. Rational Functions and Regular Functions — 579
5. Morphisms — 590
6. Rational Maps — 595
7. Zariski's Theorem about Nonsingular Points — 600
8. Classification Questions about Irreducible Curves — 604
9. Affine Algebraic Sets for Monomial Ideals — 618
10. Hilbert Polynomial in the Affine Case — 626

X. METHODS OF ALGEBRAIC GEOMETRY (Continued)

11.	Hilbert Polynomial in the Projective Case	633
12.	Intersections in Projective Space	635
13.	Schemes	638
14.	Problems	644

Hints for Solutions of Problems	649
Selected References	713
Index of Notation	717
Index	721

CONTENTS OF *BASIC ALGEBRA*

I. Preliminaries about the Integers, Polynomials, and Matrices
II. Vector Spaces over $\mathbb{Q}$, $\mathbb{R}$, and $\mathbb{C}$
III. Inner-Product Spaces
IV. Groups and Group Actions
V. Theory of a Single Linear Transformation
VI. Multilinear Algebra
VII. Advanced Group Theory
VIII. Commutative Rings and Their Modules
IX. Fields and Galois Theory
X. Modules over Noncommutative Rings

PREFACE

Advanced Algebra and its companion volume *Basic Algebra* systematically develop concepts and tools in algebra that are vital to every mathematician, whether pure or applied, aspiring or established. The two books together aim to give the reader a global view of algebra, its use, and its role in mathematics as a whole. The idea is to explain what the young mathematician needs to know about algebra in order to communicate well with colleagues in all branches of mathematics.

The books are written as textbooks, and their primary audience is students who are learning the material for the first time and who are planning a career in which they will use advanced mathematics professionally. Much of the material in the two books, including nearly all of *Basic Algebra* and some of *Advanced Algebra*, corresponds to normal course work, with the proportions depending on the university. The books include further topics that may be skipped in required courses but that the professional mathematician will ultimately want to learn by self-study. The test of each topic for inclusion is whether it is something that a plenary lecturer at a broad international or national meeting is likely to take as known by the audience.

Key topics and features of *Advanced Algebra* are as follows:

- Topics build on the linear algebra, group theory, factorization of ideals, structure of fields, Galois theory, and elementary theory of modules developed in *Basic Algebra*.
- Individual chapters treat various topics in commutative and noncommutative algebra, together providing introductions to the theory of associative algebras, homological algebra, algebraic number theory, and algebraic geometry.
- The text emphasizes connections between algebra and other branches of mathematics, particularly topology and complex analysis. All the while, it carries along two themes from *Basic Algebra*: the analogy between integers and polynomials in one variable over a field, and the relationship between number theory and geometry.
- Several sections in two chapters introduce the subject of Gröbner bases, which is the modern gateway toward handling simultaneous polynomial equations in applications.
- The development proceeds from the particular to the general, often introducing examples well before a theory that incorporates them.

- More than 250 problems at the ends of chapters illuminate aspects of the text, develop related topics, and point to additional applications. A separate section "Hints for Solutions of Problems" at the end of the book gives detailed hints for most of the problems, complete solutions for many.

It is assumed that the reader is already familiar with linear algebra, group theory, rings and modules, unique factorization domains, Dedekind domains, fields and algebraic extension fields, and Galois theory at the level discussed in *Basic Algebra*. Not all of this material is needed for each chapter of *Advanced Algebra*, and chapter-by-chapter information about prerequisites appears in the Guide for the Reader beginning on page xvii.

Historically the subjects of algebraic number theory and algebraic geometry have influenced each other as they have developed, and the present book tries to bring out this interaction to some extent. It is easy to see that there must be a close connection. In fact, one number-theory problem already solved by Fermat and Euler was to find all pairs (x, y) of integers satisfying $x^2 + y^2 = n$, where n is a given positive integer. More generally one can consider higher-order equations of this kind, such as $y^2 = x^3 + 8x$. Even this simple change of degree has a great effect on the difficulty, so much so that one is inclined first to solve an easier problem: find the *rational* pairs satisfying the equation. Is the search for rational solutions a problem in number theory or a problem about a curve in the plane? The answer is that really it is both. We can carry this kind of question further. Instead of considering solutions of a single polynomial equation in two variables, we can consider solutions of a system of polynomial equations in several variables. Within the system no individual equation is an intrinsic feature of the problem because one of the equations can always be replaced by its sum with another of the equations; if we regard each equation as an expression set equal to 0, then the intrinsic problem is to study the locus of common zeros of the equations. This formulation of the problem sounds much more like algebraic geometry than number theory.

A doubter might draw a distinction between integer solutions and rational solutions, saying that finding integer solutions is number theory while finding rational solutions is algebraic geometry. Experience shows that this is an artificial distinction. Although algebraic geometry was initially developed as a subject that studies solutions for which the variables take values in a field, particularly in an algebraically closed field, the insistence on working only with fields imposed artificial limitations on how problems could be approached. In the late 1950s and early 1960s the foundations of the subject were transformed by allowing variables to take values in an arbitrary commutative ring with identity. The very end of this book aims to give some idea of what those new foundations are.

Along the way we shall observe parallels between number theory and algebraic geometry, even as we nominally study one subject at a time. The book begins with a chapter on those aspects of number theory that mark the historical transition from classical number theory to modern algebraic number theory. Chapter I deals with three celebrated advances of Gauss and Dirichlet in classical number theory that one might wish to generalize by means of algebraic number theory. The detailed level of knowledge that one gains about those topics can be regarded as a goal for the desired level of understanding about more complicated problems. Chapter I thus establishes a framework for the whole book.

Associative algebras are the topic of Chapters II and III. The tools for studying such algebras provide methods for classifying noncommutative division rings. One such tool, known as the Brauer group, has a cohomological interpretation that ties the subject to algebraic number theory.

Because of other work done in the 1950s, homology and cohomology can be abstracted in such a way that the theory impacts several fields simultaneously, including topology and complex analysis. The resulting subject is called homological algebra and is the topic of Chapter IV. Having cohomology available at this point of the present book means that one is prepared to use it both in algebraic number theory and in situations in algebraic geometry that have grown out of complex analysis.

The last six chapters are about algebraic number theory, algebraic geometry, and the relationship between them. Chapters V–VI concern the three main foundational theorems in algebraic number theory. Chapter V goes at these results in a direct fashion but falls short of giving a complete proof in one case. Chapter VI goes at matters more indirectly. It explores the parallel between number theory and the theory of algebraic curves, makes use of tools from analysis concerning compactness and completeness, succeeds in giving full proofs of the three theorems of Chapter V, and introduces the modern approach via adeles and ideles to deeper questions in these subject areas.

Chapters VII–X are about algebraic geometry. Chapter VII fills in some prerequisites from the theories of fields and commutative rings that are needed to set up the foundations of algebraic geometry. Chapters VIII–X concern algebraic geometry itself. They come at the subject successively from three points of view—from the algebraic point of view of simultaneous systems of polynomial equations in several variables, from the number-theoretic point of view suggested by the classical theory of Riemann surfaces, and from the geometric point of view.

The topics most likely to be included in normal course work include the Wedderburn theory of semisimple algebras in Chapter II, homological algebra in Chapter IV, and some of the advanced material on fields in Chapter VII. A chart on page xvi tells the dependence of chapters on earlier chapters, and, as mentioned above, the section Guide for the Reader tells what knowledge of *Basic Algebra* is assumed for each chapter.

The problems at the ends of chapters are intended to play a more important role than is normal for problems in a mathematics book. Almost all problems are solved in the section of hints at the end of the book. This being so, some blocks of problems form additional topics that could have been included in the text but were not; these blocks may be regarded as optional topics, or they may be treated as challenges for the reader. The optional topics of this kind usually either carry out further development of the theory or introduce significant applications to other branches of mathematics. For example a number of applications to topology are treated in this way.

Not all problems are of this kind, of course. Some of the problems are really pure or applied theorems, some are examples showing the degree to which hypotheses can be stretched, and a few are just exercises. The reader gets no indication which problems are of which type, nor of which ones are relatively easy. Each problem can be solved with tools developed up to that point in the book, plus any additional prerequisites that are noted.

The theorems, propositions, lemmas, and corollaries within each chapter are indexed by a single number stream. Figures have their own number stream, and one can find the page reference for each figure from the table on page xv. Labels on displayed lines occur only within proofs and examples, and they are local to the particular proof or example in progress. Each occurrence of the word "PROOF" or "PROOF" is matched by an occurrence at the right margin of the symbol $\Box$ to mark the end of that proof.

I am grateful to Ann Kostant and Steven Krantz for encouraging this project and for making many suggestions about pursuing it, and I am indebted to David Kramer, who did the copyediting. The typesetting was by $A_\mathcal{M}S$-TEX, and the figures were drawn with Mathematica.

I invite corrections and other comments from readers. I plan to maintain a list of known corrections on my own Web page.

A. W. KNAPP
August 2007

LIST OF FIGURES

3.1.	A cochain map	154
4.1.	Snake diagram	185
4.2.	Enlarged snake diagram	185
4.3.	Defining property of a projective	192
4.4.	Defining property of an injective	195
4.5.	Formation of derived functors	205
4.6.	Universal mapping property of a kernel of a morphism	235
4.7.	Universal mapping property of a cokernel of a morphism	236
4.8.	The pullback of a pair of morphisms	243
6.1.	Commutativity of completion and extension as field mappings	356
6.2.	Commutativity of completion and extension as homomorphisms of valued fields	360

DEPENDENCE AMONG CHAPTERS

Below is a chart of the main lines of dependence of chapters on prior chapters. The dashed lines indicate helpful motivation but no logical dependence. Apart from that, particular examples may make use of information from earlier chapters that is not indicated by the chart.

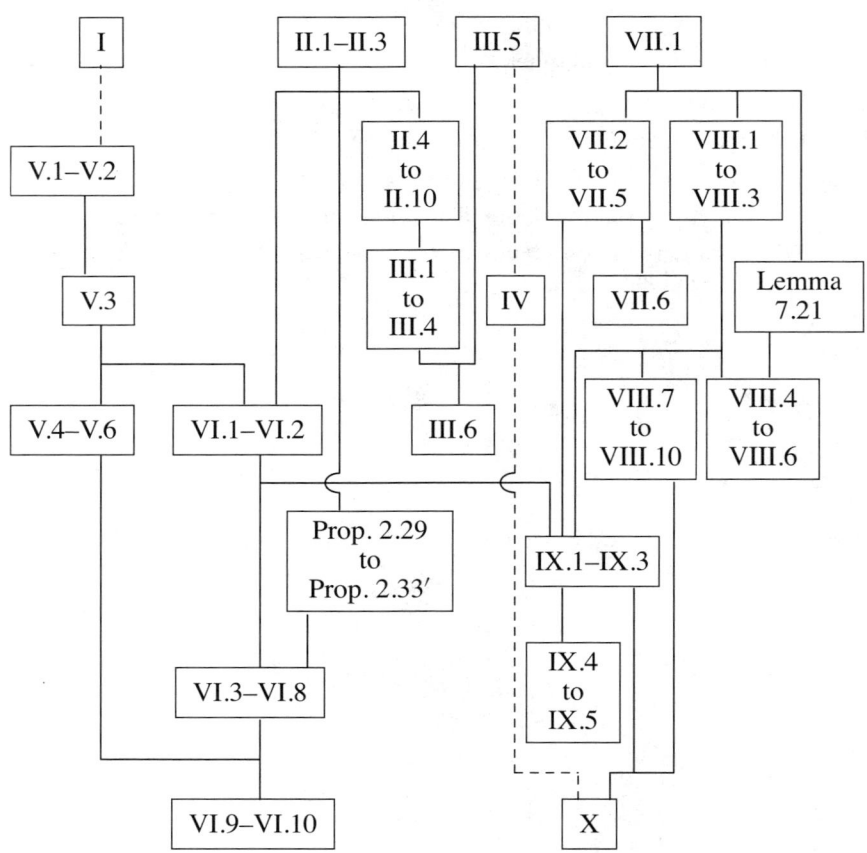

GUIDE FOR THE READER

This section is intended to help the reader find out what parts of each chapter are most important and how the chapters are interrelated. Further information of this kind is contained in the abstracts that begin each of the chapters.

The book treats its subject material as pointing toward algebraic number theory and algebraic geometry, with emphasis on aspects of these subjects that impact fields of mathematics other than algebra. Two chapters treat the theory of associative algebras, not necessarily commutative, and one chapter treats homological algebra; both these topics play a role in algebraic number theory and algebraic geometry, and homological algebra plays an important role in topology and complex analysis. The constant theme is a relationship between number theory and geometry, and this theme recurs throughout the book on different levels.

The book assumes knowledge of most of the content of *Basic Algebra*, either from that book itself or from some comparable source. Some of the less standard results that are needed from *Basic Algebra* are summarized in the section Notation and Terminology beginning on page xxi. The assumed knowledge of algebra includes facility with using the Axiom of Choice, Zorn's Lemma, and elementary properties of cardinality. All chapters of the present book but the first assume knowledge of Chapters I–IV of *Basic Algebra* other than the Sylow Theorems, facts from Chapter V about determinants and characteristic polynomials and minimal polynomials, simple properties of multilinear forms from Chapter VI, the definitions and elementary properties of ideals and modules from Chapter VIII, the Chinese Remainder Theorem and the theory of unique factorization domains from Chapter VIII, and the theory of algebraic field extensions and separability and Galois groups from Chapter IX. Additional knowledge of parts of *Basic Algebra* that is needed for particular chapters is discussed below. In addition, some sections of the book, as indicated below, make use of some real or complex analysis. The real analysis in question generally consists in the use of infinite series, uniform convergence, differential calculus in several variables, and some point-set topology. The complex analysis generally consists in the fundamentals of the one-variable theory of analytic functions, including the Cauchy Integral Formula, expansions in convergent power series, and analytic continuation.

The remainder of this section is an overview of individual chapters and groups of chapters.

Chapter I concerns three results of Gauss and Dirichlet that marked a transition from the classical number theory of Fermat, Euler, and Lagrange to the algebraic number theory of Kummer, Dedekind, Kronecker, Hermite, and Eisenstein. These results are Gauss's Law of Quadratic Reciprocity, the theory of binary quadratic forms begun by Gauss and continued by Dirichlet, and Dirichlet's Theorem on primes in arithmetic progressions. Quadratic reciprocity was a necessary preliminary for the theory of binary quadratic forms. When viewed as giving information about a certain class of Diophantine equations, the theory of binary quadratic forms gives a gauge of what to hope for more generally. The theory anticipates the definition of abstract abelian groups, which occurred later historically, and it anticipates the definition of the class number of an algebraic number field, at least in the quadratic case. Dirichlet obtained formulas for the class numbers that arise from binary quadratic forms, and these formulas led to the method by which he proved his theorem on primes in arithmetic progressions. Much of the chapter uses only elementary results from *Basic Algebra*. However, Sections 6–7 use facts about quadratic number fields, including the multiplication of ideals in their rings of integers, and Section 10 uses the Fourier inversion formula for finite abelian groups, which is in Section VII.4 of *Basic Algebra*. Sections 8–10 make use of a certain amount of real and complex analysis concerning uniform convergence and properties of analytic functions.

Chapters II–III introduce the theory of associative algebras over fields. Chapter II includes the original theory of Wedderburn, including an amplification by E. Artin, while Chapter III introduces the Brauer group and connects the theory with the cohomology of groups. The basic material on simple and semisimple associative algebras is in Sections 1–3 of Chapter II, which assumes familiarity with commutative Noetherian rings as in Chapter VIII of *Basic Algebra*, plus the material in Chapter X on semisimple modules, chain conditions for modules, and the Jordan–Hölder Theorem. Sections 4–6 contain the statement and proof of Wedderburn's Main Theorem, telling the structure of general finite-dimensional associative algebras in characteristic 0. These sections include a relatively self-contained segment from Proposition 2.29 through Proposition 2.33' on the role of separability in the structure of tensor products of algebras. This material is the part of Sections 4–6 that is used in the remainder of the chapter to analyze finite-dimensional associative division algebras over fields. Two easy consequences of this analysis are Wedderburn's Theorem that every finite division ring is commutative and Frobenius's Theorem that the only finite-dimensional associative division algebras over $\mathbb{R}$ are $\mathbb{R}$, $\mathbb{C}$, and the algebra $\mathbb{H}$ of quaternions, up to $\mathbb{R}$ isomorphism.

Chapter III introduces the Brauer group to parametrize the isomorphism classes of finite-dimensional associative division algebras whose center is a given field. Sections 2–3 exhibit an isomorphism of a relative Brauer group with what turns out to be a cohomology group in degree 2. This development runs parallel to the theory of factor sets for groups as in Chapter VII of *Basic Algebra*, and some familiarity with that theory can be helpful as motivation. The case that the relative Brauer group is cyclic is of special importance, and the theory is used in the problems to construct examples of division rings that would not have been otherwise available. The chapter makes use of material from Chapter X of *Basic Algebra* on the tensor product of algebras and on complexes and exact sequences.

Chapter IV is about homological algebra, with emphasis on connecting homomorphisms, long exact sequences, and derived functors. All but the last section is done in the context of "good" categories of unital left R modules, R being a ring with identity, where it is possible to work with individual elements in each object. The reader is expected to be familiar with some example for motivation; this can be knowledge of cohomology of groups at the level of Section III.5, or it can be some experience from topology or from the cohomology of Lie algebras as treated in other books. Knowledge of complexes and exact sequences from Chapter X of *Basic Algebra* is prerequisite. Homological algebra properly belongs in this book because it is fundamental in topology and complex analysis; in algebra its role becomes significant just beyond the level of the current book. Important applications are not limited in practice to "good" categories; "sheaf" cohomology is an example with significant applications that does not fit this mold. Section 8 sketches the theory of homological algebra in the context of "abelian" categories. In this case one does not have individual elements at hand, but some substitute is still possible; sheaf cohomology can be treated in this context.

Chapters V and VI are an introduction to algebraic number theory. The theory of Dedekind domains from Chapters VIII and IX of *Basic Algebra* is taken as known, along with knowledge of the ingredients of the theory—Noetherian rings, integral closure, and localization. Both chapters deal with three theorems—the Dedekind Discriminant Theorem, the Dirichlet Unit Theorem, and the finiteness of the class number. Chapter V attacks these directly, using no additional tools, and it comes up a little short in the case of the Dedekind Discriminant Theorem. Chapter VI introduces tools to get around the weakness of the development in Chapter V. These tools are valuations, completions, and decompositions of tensor products of fields with complete fields. Chapter VI makes extensive use of metric spaces and completeness, and compactness plays an important role in Sections 9–10. As noted in remarks with Proposition 6.7, Section VI.2 takes for granted that Theorem 8.54 of *Basic Algebra* about extensions of Dedekind domains does not need separability as a hypothesis; the actual proof of the improved theorem without a hypothesis of separability is deferred to Section VII.3.

Chapter VII supplies additional background needed for algebraic geometry, partly from field theory and partly from the theory of commutative rings. Knowledge of Noetherian rings is needed throughout the chapter. Sections 4–5 assume knowledge of localizations, and the indispensable Corollary 7.14 in Section 3 concerns Dedekind domains. The most important result is the Nullstellensatz in Section 1. Transcendence degree and Krull dimension in Sections 2 and 4 are tied to the notion of dimension in algebraic geometry. Zariski's Theorem in Section 5 is tied to the notion of singularities; part of its proof is deferred to Chapter X. The material on infinite Galois groups in Section 6 has applications to algebraic number theory and algebraic geometry but is not used in this book after Chapter VII.

Chapters VIII–X introduce algebraic geometry from three points of view. Chapter VIII approaches it as an attempt to understand solutions of simultaneous polynomial equations in several variables using module-theoretic tools. Chapter IX approaches the subject of curves as an outgrowth of the complex-analysis theory of compact Riemann surfaces and uses number-theoretic methods. Chapter X approaches its subject matter geometrically, using the field-theoretic and ring-theoretic tools developed in Chapter VII. All three chapters assume knowledge of Section VII.1 on the Nullstellensatz.

Chapter VIII is in three parts. Sections 1–4 are relatively elementary and concern the resultant and preliminary forms of Bezout's Theorem. Sections 5–6 concern intersection multiplicity for curves and make extensive use of localizations; the goal is a better form of Bezout's Theorem. Sections 7–10 are independent of Sections 5–6 and introduce the theory of Gröbner bases. This subject was developed comparatively recently and lies behind many of the symbolic manipulations of polynomials that are possible with computers.

Chapter IX concerns irreducible curves and is in two parts. Sections 1–3 define divisors and the genus of such a curve, while Sections 4–5 prove the Riemann–Roch Theorem and give applications of it. The tool for the development is discrete valuations as in Section VI.2, and the parallel between the theory in Chapter VI for algebraic number fields and the theory in Chapter IX for curves becomes more evident than ever. Some complex analysis is needed to understand the motivation in Sections 1 and 4.

Chapter X largely concerns algebraic sets defined as zero loci over an algebraically closed field. The irreducible such sets are called varieties. Sections 1–3 are concerned with algebraic sets and their dimension, Sections 4–6 treat maps between varieties, and Sections 7–8 deal with finer questions. Sections 9–12 are independent of Sections 6–8 and do two things simultaneously: they tie the theoretical work on dimension to the theory of Gröbner bases in Chapter VIII, making dimension computable, and they show how the dimension of a zero locus is affected by adding one equation to the defining system. The chapter concludes with an introductory section about schemes, in which the underlying algebraically closed field is replaced by a commutative ring with identity. The entire chapter assumes knowledge of elementary point-set topology.

NOTATION AND TERMINOLOGY

This section contains some items of notation and terminology from *Basic Algebra* that are not necessarily reviewed when they occur in the present book. A few results are mentioned as well. The items are grouped by topic.

Set theory

$\in$	membership symbol
$\#S$ or $\|S\|$	number of elements in S
$\varnothing$	empty set
$\{x \in E \mid P\}$	the set of x in E such that P holds
E^c	complement of the set E
$E \cup F,\ E \cap F,\ E - F$	union, intersection, difference of sets
$\bigcup_\alpha E_\alpha,\ \bigcap_\alpha E_\alpha$	union, intersection of the sets E_α
$E \subseteq F,\ E \supseteq F$	containment
$E \subsetneq F,\ E \supsetneq F$	proper containment
$(a_1, \ldots, a_n)$	ordered n-tuple
$\{a_1, \ldots, a_n\}$	unordered n-tuple
$f : E \to F,\ x \mapsto f(x)$	function, effect of function
$f \circ g$ or $fg,\ f\vert_E$	composition of f following g, restriction to E
$f(\,\cdot\,, y)$	the function $x \mapsto f(x, y)$
$f(E),\ f^{-1}(E)$	direct and inverse image of a set
in one-one correspondence	matched by a one-one onto function
countable	finite or in one-one correspondence with integers
2^A	set of all subsets of A

Number systems

δ_{ij}	Kronecker delta: 1 if $i = j$, 0 if $i \neq j$
$\binom{n}{k}$	binomial coefficient
n positive, n negative	$n > 0,\ n < 0$
$\mathbb{Z}, \mathbb{Q}, \mathbb{R}, \mathbb{C}$	integers, rationals, reals, complex numbers
max, min	maximum/minimum of finite subset of reals
$[x]$	greatest integer $\leq x$ if x is real
Re z, Im z	real and imaginary parts of complex z
$\bar{z}$	complex conjugate of z
$\|z\|$	absolute value of z

Linear algebra and elementary number theory

$\mathbb{F}^n$	space of n-dimensional column vectors
e_j	j^{th} standard basis vector of $\mathbb{F}^n$
V'	dual vector space of vector space V
$\dim_{\mathbb{F}} V$ or $\dim V$	dimension of vector space V over field $\mathbb{F}$
0	zero vector, matrix, or linear mapping
1 or I	identity matrix or linear mapping
A^t	transpose of A
$\det A$	determinant of A
$[M_{ij}]$	matrix with $(i,j)^{\text{th}}$ entry M_{ij}
$\binom{L}{\Delta\Gamma}$	matrix of L relative to domain ordered basis Γ and range ordered basis Δ
$x \cdot y$	dot product
$\cong$	is isomorphic to, is equivalent to
$\mathbb{F}_p$	integers modulo a prime p, as a field
GCD	greatest common divisor
$\equiv$	is congruent to
φ	Euler's φ function

Groups, rings, modules, and categories

0	additive identity in an abelian group
1	multiplicative identity in a group or ring
$\cong$	is isomorphic to, is equivalent to
C_m	cyclic group of order m
unit	invertible element in ring R with identity
$R^\times$	group of units in ring R with identity
R^n	space of column vectors with entries in ring R
R^o	opposite ring to R with $a \circ b = ba$
$M_{mn}(R)$	m-by-n matrices with entries in R
$M_n(R)$	n-by-n matrices with entries in R
unital left R module	left R module M with $1m = m$ for all $m \in M$
$\text{Hom}_R(M, N)$	group of R homomorphisms from M into N
$\text{End}_R(M)$	ring of R homomorphisms from M into M
$\ker \varphi$, image φ	kernel and image of φ
$H^n(G, N)$	n^{th} cohomology of group G with coefficients in abelian group N
simple left R module	nonzero unital left R module with no proper nonzero R submodules
semisimple left R module	sum (= direct sum) of simple left R modules
$\text{Obj}(\mathcal{C})$	class of objects for category $\mathcal{C}$
$\text{Morph}_\mathcal{C}(A, B)$	set of morphisms from object A to object B

Groups, rings, modules, and categories, continued

1_A	identity morphism on A
$\mathcal{C}^S$	category of S-tuples of objects from $\text{Obj}(\mathcal{C})$
product of $\{X_s\}_{s \in S}$	$(X, \{p_s\}_{s \in S})$ such that if A in $\text{Obj}(\mathcal{C})$ and $\{\varphi_s \in \text{Morph}_{\mathcal{C}}(A, X_s)\}$ are given, then there exists a unique $\varphi \in \text{Morph}_{\mathcal{C}}(A, X)$ with $p_s \varphi = \varphi_s$ for all s
coproduct of $\{X_s\}_{s \in S}$	$(X, \{i_s\}_{s \in S})$ such that if A in $\text{Obj}(\mathcal{C})$ and $\{\varphi_s \in \text{Morph}_{\mathcal{C}}(X_s, A)\}$ are given, then there exists a unique $\varphi \in \text{Morph}_{\mathcal{C}}(X, A)$ with $\varphi i_s = \varphi_s$ for all s
$\mathcal{C}^{\text{opp}}$	category opposite to $\mathcal{C}$

Commutative rings R with identity and factorization of elements

identity	denoted by 1, allowed to equal 0
ideal $I = (r_1, \ldots, r_n)$	ideal generated by $r_1, \ldots, r_n$
prime ideal I	*proper* ideal with $ab \in I$ implying $a \in I$ or $b \in I$
integral domain	R with no zero divisors and with $1 \neq 0$
R/I with I prime	always an integral domain
$GL(n, R)$	group of invertible n-by-n matrices, entries in R
Chinese Remainder Theorem	$I_1, \ldots, I_n$ given ideals with $I_i + I_j = R$ for $i \neq j$. Then the natural map $\varphi : R \to \prod_{j=1}^n R/I_j$ yields isomorphism $R / \bigcap_{j=1}^n I_j \cong R/I_1 \times \cdots \times R/I_n$ of rings. Also $\bigcap_{j=1}^n I_j = I_1 \cdots I_n$.
Nakayama's Lemma	If I is an ideal contained in all maximal ideals and M is a finitely generated unital R module with $IM = M$, then $M = 0$.
algebra A over R	unital R module with an R bilinear multiplication $A \times A \to A$. In this book nonassociative algebras appear only in Chapter II, and each associative algebra has an identity.
RG	group algebra over R for group G
$R[X_1, \ldots, X_n]$	polynomial algebra over R with n indeterminates
$R[x_1, \ldots, x_n]$	R algebra generated by $x_1, \ldots, x_n$
irreducible element $r \neq 0$	$r \notin R^\times$ such that $r = ab$ implies $a \in R^\times$ or $b \in R^\times$
prime element $r \neq 0$	$r \notin R^\times$ such that whenever r divides ab, then r divides a or r divides b
irreducible vs. prime	prime implies irreducible; in any unique factorization domain, irreducible implies prime
GCD	greatest common divisor in unique factorization domain

Fields

$\mathbb{F}_q$	a finite field with $q = p^n$ elements, p prime
K/F	an extension field K of a field F
$[K : F]$	degree of extension K/F, i.e., $\dim_F K$
$K(X_1, \ldots, X_n)$	field of fractions of $K[X_1, \ldots, X_n]$
$K(x_1, \ldots, x_n)$	field generated by K and $x_1, \ldots, x_n$
number field	finite-dimensional field extension of $\mathbb{Q}$
$\mathrm{Gal}(K/F)$	Galois group, automorphisms of K fixing F
$N_{K/F}(\,\cdot\,)$ and $\mathrm{Tr}_{K/F}(\,\cdot\,)$	norm and trace functions from K to F

Tools for algebraic number theory and algebraic geometry

Noetherian R	commutative ring with identity whose ideals satisfy the ascending chain condition; has the property that any R submodule of a finitely generated unital R module is finitely generated.
Hilbert Basis Theorem	R nonzero Noetherian implies $R[X]$ Noetherian

Integral closure
Situation: $R =$ integral domain, $F =$ field of fractions, $K/F =$ extension field.

$x \in K$ integral over R	x is a root of a monic polynomial in $R[X]$
integral closure of R in K	set of $x \in K$ integral over R, is a ring
R integrally closed	R equals its integral closure in F

Localization
Situation: $R =$ commutative ring with identity, $S =$ multiplicative system in R.

$S^{-1}R$	localization, pairs (r, s) with $r \in R$ and $s \in S$, modulo $(r, s) \sim (r', s')$ if $t(rs' - sr') = 0$ for some $t \in S$
property of $S^{-1}R$	$I \mapsto S^{-1}I$ is one-one from set of ideals I in R of form $I = R \cap J$ onto set of ideals in $S^{-1}R$
local ring	commutative ring with identity having a unique maximal ideal
R_P for prime ideal P	localization with $S =$ complement of P in R
Dedekind domain	Noetherian integrally closed integral domain in which every nonzero prime ideal is maximal, has unique factorization of nonzero ideals as product of prime ideals
Dedekind domain extension	R Dedekind, F field of fractions, K/F finite separable extension, T integral closure of R in K. Then T is Dedekind, and any nonzero prime ideal $\wp$ in R has $\wp R = \prod_{i=1}^{g} P_i^{e_i}$ for distinct prime ideals P_i with $P_i \cap R = \wp$. These have $\sum_{i=1}^{g} e_i f_i = [K : F]$, where $f_i = [T/P_i : R/\wp]$.

CHAPTER I

Transition to Modern Number Theory

Abstract. This chapter establishes Gauss's Law of Quadratic Reciprocity, the theory of binary quadratic forms, and Dirichlet's Theorem on primes in arithmetic progressions.

Section 1 outlines how the three topics of the chapter occurred in natural sequence and marked a transition as the subject of number theory developed a coherence and moved toward the kind of algebraic number theory that is studied today.

Section 2 establishes quadratic reciprocity, which is a reduction formula providing a rapid method for deciding solvability of congruences $x^2 \equiv m$ mod p for the unknown x when p is prime.

Sections 3–5 develop the theory of binary quadratic forms $ax^2 + bxy + cy^2$, where a, b, c are integers. The basic tool is that of proper equivalence of two such forms, which occurs when the two forms are related by an invertible linear substitution with integer coefficients and determinant 1. The theorems establish the finiteness of the number of proper equivalence classes for given discriminant, conditions for the representability of primes by forms of a given discriminant, canonical representatives of the finitely many proper equivalence classes of a given discriminant, a group law for proper equivalence classes of forms of the same discriminant that respects representability of integers by the classes, and a theory of genera that takes into account inequivalent forms whose values cannot be distinguished by linear congruences.

Sections 6–7 digress to leap forward historically and interpret the group law for proper equivalence classes of binary quadratic forms in terms of an equivalence relation on the nonzero ideals in the ring of integers of an associated quadratic number field.

Sections 8–10 concern Dirichlet's Theorem on primes in arithmetic progressions. Section 8 discusses Euler's product formula for $\sum_{n=1}^{\infty} n^{-s}$ and shows how Euler was able to modify it to prove that there are infinitely many primes $4k + 1$ and infinitely many primes $4k + 3$. Section 9 develops Dirichlet series as a tool to be used in the generalization, and Section 10 contains the proof of Dirichlet's Theorem. Section 8 uses some elementary real analysis, and Sections 9–10 use both elementary real analysis and elementary complex analysis.

1. Historical Background

The period 1800 to 1840 saw great advances in number theory as the subject developed a coherence and moved toward the kind of algebraic number theory that is studied today. The groundwork had been laid chiefly by Euclid, Diophantus, Fermat, Euler, Lagrange, and Legendre. Some of what those people did was remarkably insightful for its time, but what collectively had come out of their labors was more a collection of miscellaneous results than an organized theory. It was Gauss who first gave direction and depth to the subject, beginning with his book *Disquisitiones Arithmeticae* in 1801. Dirichlet built on Gauss's work, clarifying the deeper parts and adding analytic techniques that pointed toward the integrated subject of the future. This chapter concentrates on three jewels of classical number theory—largely the work of Gauss and Dirichlet—that seem on the surface to be only peripherally related but are actually a natural succession of developments leading from earlier results toward modern algebraic number theory. To understand the context, it is necessary to back up for a moment.

Diophantine equations in two or more variables have always lain at the heart of number theory. Fundamental examples that have played an important role in the development of the subject are $ax^2 + bxy + cy^2 = m$ for unknown integers x and y; $x_1^2 + x_2^2 + x_3^2 + x_4^2 = m$ for unknown integers x_1, x_2, x_3, x_4; $y^2 = x(x-1)(x+1)$ for unknown integers x and y; and $x^n + y^n = z^n$ for unknown integers x, y, z.

In every case one can get an immediate necessary condition on a solution by writing the equation modulo some integer n. The necessary condition is that the corresponding congruence modulo n have a solution. For example take the equation $x^2 + y^2 = p$, where p is a prime, and let us allow ourselves to use the more elementary results of *Basic Algebra*. Writing the equation modulo p leads to $x^2 + y^2 \equiv 0 \bmod p$. Certainly x cannot be divisible by p, since otherwise y would be divisible by p, x^2 and y^2 would be divisible by p^2, and $x^2 + y^2 = p$ would be divisible by p^2, contradiction. Thus we can divide, obtaining $1 + (yx^{-1})^2 \equiv 0 \bmod p$. Hence $z^2 \equiv -1 \bmod p$ for $z \equiv xy^{-1}$. If p is an odd prime, then -1 has order 2, and the necessary condition is that there exist some z in $\mathbb{F}_p^\times$ whose order is exactly 4. Since $\mathbb{F}_p^\times$ is cyclic of order $p-1$, the necessary condition is that 4 divide $p-1$.

Using a slightly more complicated argument, we can establish conversely that the divisibility of $p-1$ by 4 implies that $x^2 + y^2 = p$ is solvable for integers x and y. In fact, we know from the solvability of $z^2 \equiv -1 \bmod p$ that there exists an integer r such that p divides $r^2 + 1$. Consider the possibilities in the integral domain $\mathbb{Z}[i]$ of Gaussian integers, where $i = \sqrt{-1}$. It was shown in Chapter VIII of *Basic Algebra* that $\mathbb{Z}[i]$ is Euclidean. Hence $\mathbb{Z}[i]$ is a principal ideal domain, and its elements have unique factorization. If p remains prime in $\mathbb{Z}[i]$, then the fact that p divides $(r+i)(r-i)$ implies that p divides $r+i$ or $r-i$ in $\mathbb{Z}[i]$. Then at least one of $\frac{r}{p} + i\frac{1}{p}$ and $\frac{r}{p} - i\frac{1}{p}$ would have to be in $\mathbb{Z}[i]$. Since $i\frac{1}{p}$ is not in $\mathbb{Z}[i]$, this divisibility does not hold, and we conclude that p does not remain prime in $\mathbb{Z}[i]$. If we write $p = (a+bi)(c+di)$ nontrivially, then $p^2 = |a+bi|^2|c+di|^2 = (a^2+b^2)(c^2+d^2)$ as an equality in $\mathbb{Z}$, and we readily conclude that $a^2 + b^2 = p$.

This much argument solves the Diophantine equation $x^2 + y^2 = p$ for p prime. For p replaced by a general integer m, we use the identity

$$(x_1^2 + y_1^2)(x_2^2 + y_2^2) = (x_1 x_2 - y_1 y_2)^2 + (x_1 y_2 + x_2 y_1)^2,$$

which has been known since antiquity, and we see that $x^2 + y^2 = m$ is solvable if m is a product of odd primes of the form $4k + 1$. It is solvable also if $m = 2$ and if $m = p^2$ for any prime p. Thus $x^2 + y^2 = m$ is solvable whenever m is a positive integer such that each prime of the form $4k + 3$ dividing m divides m an even number of times. Using congruences modulo prime powers, we see that this condition is also necessary, and we arrive at the following result; historically it had already been asserted as a theorem by Fermat and was subsequently proved by Euler, albeit by more classical methods than we have used.

Proposition 1.1. The Diophantine equation $x^2 + y^2 = m$ is solvable in integers x and y for a given positive integer m if and only if every prime number $p = 4k+3$ dividing m occurs an even number of times in the prime factorization of m.

The first step in the above argument used congruence information; we had to know the primes p for which $z^2 \equiv -1 \bmod p$ is solvable. The second step was in two parts—both rather special. First we used specific information about the nature of factorization in a particular ring of algebraic integers, namely $\mathbb{Z}[i]$. Second we used that the norm of a product is the product of the norms in that same ring of algebraic integers.

It is too much to hope that some recognizable generalization of these steps with $x^2 + y^2 = m$ can handle all or most Diophantine equations. At least the first step is available in complete generality, and indeed number theory—both classical and modern—deduces many helpful conclusions by passing to congruences. There is the matter of deducing something useful from a given congruence, but doing so is a finite problem for each prime. Like some others before him, Gauss set about studying congruences systematically. Linear congruences are easy and had been handled before. Quadratic congruences are logically the next step. The first jewel of classical number theory to be discussed in this chapter is the Law of Quadratic Reciprocity of Gauss, which appears below as Theorem 1.2 and which makes useful deductions possible in the case of quadratic congruences. In effect quadratic reciprocity allows one to decide easily which integers are squares modulo a prime p. Euler had earlier come close to finding the statement of this result, and Legendre had found the exact statement without finding a complete proof. Gauss was the one who gave the first complete proof.

Part of the utility of quadratic reciprocity is that it helps one to attack quadratic Diophantine equations more systematically. The second jewel of classical number theory to be discussed in this chapter is the body of results concerning representing integers by binary quadratic forms $ax^2 + bxy + cy^2 = m$ that do not degenerate in some way. Lagrange and Legendre had already made advances in this theory, but Gauss's own discoveries were decisive. Dirichlet simplified the more advanced parts of the theory and investigated an aspect of it that Gauss had not addressed and that would lead Dirichlet to his celebrated theorem on primes in arithmetic progressions.[1]

Lagrange had introduced the notion of the discriminant of a quadratic form and a notion of equivalence of such forms—two forms of the same discriminant being equivalent if one can be obtained from the other by a linear invertible substitution with integer entries. Equivalence is important because equivalent forms represent the same numbers. He established also a theory of reduced forms that specifies representatives of each equivalence class. For an odd prime p, $ax^2 + bxy + cy^2 = p$ is solvable only if the discriminant $b^2 - 4ac$ is a square modulo p, and Lagrange was hampered by not knowing quadratic reciprocity. But he did know some special cases, such as when 5 is a square modulo p, and he was able to deal completely with discriminant -20. For this discriminant, there are two equivalence classes, represented by $x^2 + 5y^2$ and $2x^2 + 2xy + 3y^2$, and Lagrange showed for primes p other than 2 and 5 that

$$x^2 + 5y^2 = p \quad \text{is solvable if and only if} \quad p \equiv 1 \text{ or } 9 \bmod 20,$$
$$2x^2 + 2xy + 3y^2 = p \quad \text{is solvable if and only if} \quad p \equiv 3 \text{ or } 7 \bmod 20;$$

the fact about $x^2 + 5y^2 = p$ had been conjectured earlier by Euler. Lagrange observed further that

$$(2x_1^2 + 2x_1 y_1 + 3y_1^2)(2x_2^2 + 2x_2 y_2 + 3y_2^2)$$
$$= (2x_1 x_2 + x_1 y_2 + y_1 x_2 + 3y_1 y_2)^2 + 5(x_1 x_2 - y_1 y_2)^2,$$

from which it follows that the product of two primes congruent to 3 or 7 modulo 20 is representable as $x^2 + 5y^2$; this fact had been conjectured by Fermat.

Legendre added to this investigation the correct formula for quadratic reciprocity, which he incorrectly believed he had proved, and many of its consequences for representability of primes by binary quadratic forms. In addition, he tried to develop a theory of composition of forms that generalizes Lagrange's identity above, but he had only limited success.

In addition to establishing quadratic reciprocity, Gauss introduced the vital notion of "proper equivalence" for forms $ax^2 + bxy + cy^2$ of the same discriminant— two forms of the same discriminant being properly equivalent if one can be obtained from the other by a linear invertible substitution with integer entries and determinant $+1$. In terms of this definition, he settled the representability of primes by binary quadratic forms, he showed that there are only finitely many proper equivalences classes for each discriminant, and he gave an algorithm for

---

[1] These matters are affirmed in Dirichlet's *Lectures on Number Theory*. The aspect that Gauss had not addressed and that provided motivation for Dirichlet is the value of the "Dirichlet class number" $h(D)$ defined below.

deciding whether two forms are properly equivalent. The main results of Gauss in this direction appear as Theorems 1.6 and 1.8 below. In addition, Gauss showed, without the benefit of having a definition of "group," in effect that the set of proper equivalence classes of forms with a given discriminant becomes a finite abelian group in a way that controls representability of nonprime integers; by contrast, Lagrange's definition of equivalence does not lead to a group structure. Gauss's main results in this direction, as recast by Dirichlet, appear as Theorem 1.12 below.

The story does not stop here, but let us pause for a moment to say what Lagrange's theory, as amended by Gauss, says for the above example, first rephrasing the context in more modern terminology. We saw earlier that unique factorization in the ring $\mathbb{Z}[i]$ of Gaussian integers is the key to the representation of integers by the quadratic form $x^2 + y^2$. For a general quadratic form $ax^2 + bxy + cy^2$ with discriminant $D = b^2 - 4ac$, properties of the ring R of algebraic integers in the field $\mathbb{Q}(\sqrt{D})$ are relevant for the questions that Gauss investigated. It turns out that R is a principal ideal domain if Gauss's finite abelian group of proper equivalence classes is trivial and that when D is "fundamental," there is a suitable converse.[2]

With the context rephrased we come back to the example. Consider the equation $x^2 + 5y^2 = p$ for primes p. The discriminant of $x^2 + 5y^2$ is -20, and the relevant ring of algebraic integers is $\mathbb{Z}[\sqrt{-5}]$, which is not a unique factorization domain. Thus the argument used with $x^2 + y^2 = p$ does not apply, and we have no reason to expect that solvability of $x^2 + 5y^2 \equiv 0 \bmod p$ is sufficient for solvability of $x^2 + 5y^2 = p$. Let us look more closely. The congruence condition is that -20 is a square modulo p. Thus -5 is to be a square modulo p. If we leave aside the primes $p = 2$ and $p = 5$ that divide 20, the Law of Quadratic Reciprocity will tell us that the necessary congruence resulting from solvability of $x^2 + 5y^2 = p$ is that p be congruent to 1, 3, 7, or 9 modulo 20. However, we can compute all residues n of $x^2 + 5y^2$ modulo 20 for n with $\text{GCD}(n, 20) = 1$ to see that

$$x^2 + 5y^2 \equiv 1 \text{ or } 9 \bmod 20 \qquad \text{if } \text{GCD}(x^2 + 5y^2, 20) = 1.$$

Meanwhile, the form $2x^2 + 2xy + 3y^2$ has discriminant -20, and we can check that solvability of $2x^2 + 2xy + 3y^2 = p$ leads to the conclusion that

$$2x^2 + 2xy + 3y^2 \equiv 3 \text{ or } 7 \bmod 20 \qquad \text{if } \text{GCD}(2x^2 + 2xy + 3y^2, 20) = 1.$$

Lagrange's theory easily shows that representability of integers by a form depends only on the equivalence class of the form and that all primes congruent to 1, 3,

---

[2] In each of the situations (a) and (b) of Proposition 1.17 below, R is a principal ideal domain only if Gauss's group is trivial. In all other cases, Gauss's group is nontrivial, and R is a principal ideal domain only if the group has order 2.

7, or 9 modulo 20 are representable by some form. This example is special in that equivalence and proper equivalence come to the same thing. Gauss's multiplication rule for proper equivalence classes of forms with discriminant -20 produces a group of order 2, with $x^2 + 5y^2$ representing the identity class and $2x^2 + 2xy + 3y^2$ representing the other class. Consequently

$p \equiv 1$ or $9 \bmod 20$ implies $x^2 + 5y^2 = p$ solvable,

$p \equiv 3$ or $7 \bmod 20$ implies $2x^2 + 2xy + 3y^2 = p$ solvable.

In addition, the multiplication rule has the property that if m is representable by all forms in the class of $a_1 x^2 + b_1 xy + c_1 y^2$ and n is representable by all forms in the class of $a_2 x^2 + b_2 xy + c_2 y^2$, then mn is representable by all forms in the class of the product form. It is not necessary to have an explicit identity for the multiplication. Thus, for example, it follows without further argument that if p and q are primes congruent to 3 or 7 modulo 20, then $x^2 + 5y^2 = pq$ is solvable.

Let us elaborate a little about the rephrased context for Gauss's theory. We let D be the discriminant of the binary quadratic forms in question, and we assume that D is "fundamental." Let R be the ring of algebraic integers that lie in the field $\mathbb{Q}(\sqrt{D})$. It turns out to be possible to define a notion of "strict equivalence" on the set of ideals of R in such a way that multiplication of ideals descends to a multiplication of strict equivalence classes. The strict equivalence classes of ideals then form a group, and this group is isomorphic to Gauss's group. In particular, one obtains the nonobvious conclusion that the set of strict equivalence classes of ideals is finite. The main result giving this isomorphism is Theorem 1.20. This rephrasing of the theory points to a generalization to algebraic number fields of degree higher than 2 and is a starting point for modern algebraic number theory.

Now we return to the work of Gauss. Even the example with $D = -20$ that was described above does not give an idea of how complicated matters can become. For discriminant -56, for example, the two forms $x^2 + 14y^2$ and $2x^2 + 7y^2$ take on the same residues modulo 56 that are prime to 56, but no prime can be represented by both forms. These two forms and the forms $3x^2 \pm 2xy + 5y^2$ represent the four proper equivalence classes. By contrast, there are only three equivalence classes in Lagrange's sense, and we thus get some insight into why Legendre encountered difficulties in defining a useful multiplication even for $D = -56$. Gauss's theory goes on to address the problem that $x^2 + 14y^2$ and $2x^2 + 7y^2$ take on one set of residues modulo 56 and prime to 56 while $3x^2 \pm 2xy + 5y^2$ take on a disjoint set of such residues. Gauss defined a "genus" (plural: "genera") to consist of proper equivalence classes like these that cannot be distinguished by linear congruences, and he obtained some results about this notion. Gauss's set of genera inherits a group structure from the group structure on the proper equivalence classes of forms, and the group structure for the genera enables one to work with genera easily.

The third jewel of classical number theory to be discussed in this chapter is Dirichlet's celebrated theorem on primes in arithmetic progressions, given below as Theorem 1.21. The statement is that if m and b are positive relatively prime integers, then there are infinitely many primes of the form $km + b$ with k a positive integer. The proof mixes algebra, a little real analysis, and some complex analysis.

What is not immediately apparent is how this theorem fits into a natural historical sequence with Gauss's theory of binary quadratic forms. In fact, the statement about primes in arithmetic progressions was thrust upon Dirichlet in at least two ways. Dirichlet thoroughly studied the work of those who came before him. One aspect of that work was Legendre's progress toward obtaining quadratic reciprocity; in fact, Legendre actually had a proof of quadratic reciprocity except that he assumed the unproved result about primes in arithmetic progressions for part of it and argued in circular fashion for another part of it. Another aspect of the work Dirichlet studied was Gauss's theory of multiplication of proper equivalence classes of forms, which Dirichlet saw a need to simplify and explain; indeed, a complete answer to the representability of composite numbers requires establishing theorems about genera beyond what Gauss obtained and has to make use of the theorem about primes in arithmetic progressions.

In addition, Dirichlet asked and settled a question about proper equivalence classes for which Gauss had published nothing and for which Jacobi had conjectured an answer: How many such classes are there for each discriminant D? Let us call this number the "Dirichlet class number," denoting it by $h(D)$. Dirichlet's answer has several cases to it. When D is fundamental, even, negative, and not equal to -4, the answer is

$$h(D) = \frac{2\sqrt{|D/4|}}{\pi} \sum_{\substack{n \geq 1, \\ \mathrm{GCD}(n,D)=1}} \left(\frac{D/4}{n}\right) \frac{1}{n},$$

with the sum taken over positive integers prime to D. Here when p is a prime not dividing D, $\left(\frac{D/4}{p}\right)$ is $+1$ if $D/4$ is a square modulo p and is -1 if not. For general $n = \prod p^k$ prime to D, $\left(\frac{D/4}{n}\right)$ is the product of the expressions $\left(\frac{D/4}{p}\right)^k$ corresponding to the factorization[3] of n. When $D = -4$, the quantity on the right side has to be doubled to give the correct result, and thus the formula becomes

$$h(-4) = \frac{4}{\pi} \sum_{n \text{ odd} \geq 1} \left(\frac{-1}{n}\right) \frac{1}{n} = \frac{4}{\pi} \sum_{n \text{ odd} \geq 1} \frac{(-1)^{(n-1)/2}}{n}.$$

The adjusted formula correctly gives $h(-4) = -1$, since Leibniz had shown more than a century earlier that $1 - \frac{1}{3} + \frac{1}{5} - \frac{1}{7} + \cdots = \frac{\pi}{4}$. Dirichlet was able to

---

[3]The expression $\left(\frac{D/4}{n}\right)$ is called a "Jacobi symbol." See Problems 9–11 at the end of the chapter.

evaluate the displayed infinite series for general D as a finite sum, but that further step does not concern us here. The important thing to observe is that the infinite series is always an instance of a series $\sum_{n=1}^{\infty} \chi(n)/n$ with χ a periodic function on the positive integers satisfying $\chi(m+n) = \chi(m)\chi(n)$. Dirichlet's derivation of a series expansion for his class numbers required care because the series is only conditionally convergent. To be able to work with absolutely convergent series, he initially replaced $\frac{1}{n}$ by $\frac{1}{n^s}$ for $s > 1$, thus initially treating series he denoted by $L(s, \chi) = \sum_{n=1}^{\infty} \chi(n)/n^s$.

As a consequence of this work, Dirichlet was familiar with series $L(s, \chi)$ and was aware of the importance of expressions $L(1, \chi)$, knowing that at least when $\chi(n) = \left(\frac{D}{n}\right)$, $L(1, \chi)$ is not 0 because it is essentially a class number. This nonvanishing turns out to be the core of the proof of the theorem on primes in arithmetic progressions. Dirichlet would have known about Euler's proof that the progressions $4n + 1$ and $4n + 3$ contain infinitely many primes, a proof that we give in Section 8, and he would have recognized Euler's expression $\sum_{n=1}^{\infty} (-1)^n/(2n+1)$ as something that occurs in his formula for $h(-4)$. Thus he was well equipped with tools and motivation for a proof of his theorem on primes in arithmetic progressions.

2. Quadratic Reciprocity

If p is an odd prime number and a is an integer with $a \not\equiv 0 \bmod p$, the **Legendre symbol** $\left(\frac{a}{p}\right)$ is defined by

$$\left(\frac{a}{p}\right) = \begin{cases} +1 & \text{if } a \text{ is a square modulo } p, \\ -1 & \text{if } a \text{ is not a square modulo } p. \end{cases}$$

Since $\mathbb{F}_p^\times$ is a cyclic group of even order, the squares form a subgroup of index 2. Therefore $a \mapsto \left(\frac{a}{p}\right)$ is a group homomorphism of $\mathbb{F}_p^\times$ into $\{\pm 1\}$, and we have $\left(\frac{a}{p}\right)\left(\frac{b}{p}\right) = \left(\frac{ab}{p}\right)$ whenever a and b are not divisible by p.

Theorem 1.2 (Law of Quadratic Reciprocity). *If p and q are distinct odd prime numbers, then*

(a) $\left(\dfrac{-1}{p}\right) = (-1)^{\frac{1}{2}(p-1)}$,

(b) $\left(\dfrac{2}{p}\right) = (-1)^{\frac{1}{8}(p^2-1)}$,

(c) $\left(\dfrac{p}{q}\right)\left(\dfrac{q}{p}\right) = (-1)^{[\frac{1}{2}(p-1)][\frac{1}{2}(q-1)]}$.

2. Quadratic Reciprocity

REMARKS. Conclusion (a) is due to Fermat and says that -1 is a square modulo p if and only if $p = 4n + 1$. We proved this result already in Section 1 and will not re-prove it here. Conclusion (b) is due to Euler and says that 2 is a square modulo p if and only if $p = 8n \pm 1$. Conclusion (c) is due to Gauss and says that if p or q is $4n + 1$, then $\left(\frac{p}{q}\right) = \left(\frac{q}{p}\right)$ and otherwise $\left(\frac{p}{q}\right) = -\left(\frac{q}{p}\right)$. The proofs of (b) and (c) will occupy the remainder of this section.

EXAMPLES.

(1) This example illustrates how quickly iterated use of the theorem decides whether a given integer is a square. We compute $\left(\frac{17}{79}\right)$. We have

$$\left(\frac{17}{79}\right) = \left(\frac{79}{17}\right) = \left(\frac{11}{17}\right) = \left(\frac{17}{11}\right) = \left(\frac{6}{11}\right) = -\left(\frac{3}{11}\right) = +\left(\frac{11}{3}\right) = \left(\frac{2}{3}\right) = -1,$$

the successive equalities being justified by using (c), the formula $\left(\frac{a+kp}{p}\right) = \left(\frac{a}{p}\right)$, (c) again, $\left(\frac{a+kp}{p}\right) = \left(\frac{a}{p}\right)$ again, the formula $\left(\frac{a}{p}\right)\left(\frac{b}{p}\right) = \left(\frac{ab}{p}\right)$ and (b), (c) once more, $\left(\frac{a+kp}{p}\right) = \left(\frac{a}{p}\right)$ once more, and an explicit evaluation of $\left(\frac{2}{3}\right)$.

(2) Lemma 9.46 of *Basic Algebra* asserts that 3 is a generator of the cyclic group $\mathbb{F}_n^\times$ when n is prime of the form $2^{2^N} + 1$ with $N > 0$, and Theorem 1.2 enables us to give a proof. In fact, this n has $n \equiv 2 \bmod 3$ and $n \equiv 1 \bmod 4$. Thus $\left(\frac{3}{n}\right) = \left(\frac{n}{3}\right) = \left(\frac{2}{3}\right) = -1$. Since $\mathbb{F}_n^\times$ is a cyclic group whose order is a power of 2, every nonsquare is a generator. Thus 3 is a generator.

We prove two lemmas, give the proof of (b), prove a third lemma, and then give the proof of (c).

Lemma 1.3. If p is an odd prime and a is any integer such that p does not divide a, then $a^{\frac{1}{2}(p-1)} \equiv \left(\frac{a}{p}\right) \bmod p$.

PROOF. The multiplicative group $\mathbb{F}_p^\times$ being cyclic, let b be a generator. Write $a \equiv b^r \bmod p$ for some integer r. Since $\left(\frac{a}{p}\right) = (-1)^r$ and $a^{\frac{1}{2}(p-1)} \equiv (b^r)^{\frac{1}{2}(p-1)} = (b^{\frac{1}{2}(p-1)})^r \equiv (-1)^r \bmod p$, the lemma follows. □

Lemma 1.4 (Gauss). Let p be an odd prime, and let a be any integer such that p does not divide a. Among the least positive residues modulo p of the integers $a, 2a, 3a, \ldots, \frac{1}{2}(p-1)a$, let n denote the number of residues that exceed $p/2$. Then $\left(\frac{a}{p}\right) = (-1)^n$.

PROOF. Let $r_1, \ldots, r_n$ be the least positive residues exceeding $p/2$, and let $s_1, \ldots, s_k$ be those less than $p/2$, so that $n + k = \frac{1}{2}(p-1)$. The residues $r_1, \ldots, r_n, s_1, \ldots, s_k$ are distinct, since no two of $a, 2a, 3a, \ldots, \frac{1}{2}(p-1)a$ differ by a multiple of p. Each integer $p - r_i$ is strictly between 0 and $p/2$, and we cannot have any equality $p - r_i = s_j$, since $r_i + s_j = p$ would mean that $(u+v)a$ is divisible by p for some integers u and v with $1 \le u, v \le \frac{1}{2}(p-1)$. Hence

$$p - r_1, \ldots, p - r_n, s_1, \ldots, s_k$$

is a permutation of $1, \ldots, \frac{1}{2}(p-1)$. Modulo p, we therefore have

$$\begin{aligned} 1 \cdot 2 \cdots \tfrac{1}{2}(p-1) &\equiv (-1)^n r_1 \cdots r_n s_1 \cdots s_k \\ &\equiv (-1)^n a \cdot 2a \cdots \tfrac{1}{2}(p-1)a \\ &\equiv (-1)^n a^{\frac{1}{2}(p-1)} 1 \cdot 2 \cdots \tfrac{1}{2}(p-1), \end{aligned}$$

and cancellation yields $a^{\frac{1}{2}(p-1)} \equiv (-1)^n \bmod p$. The result follows by combining this congruence with the conclusion of Lemma 1.3. □

PROOF OF (b) IN THEOREM 1.2. We shall apply Lemma 1.4 with $a = 2$ after investigating the least positive residues of $2, 4, 6, \ldots, p-1$. We can list explicitly those residues that exceed $p/2$ for each odd value of p mod 8 as follows:

$$\begin{aligned} p &= 8k+1, & 4k+2, 4k+4, \ldots, 8k, \\ p &= 8k+3, & 4k+2, 4k+4, \ldots, 8k+2, \\ p &= 8k+5, & 4k+4, \ldots, 8k+2, 8k+4, \\ p &= 8k+7, & 4k+4, \ldots, 8k+4, 8k+6. \end{aligned}$$

If n denotes the number of such residues for a given p, a count of each line of the above table shows that

$$\begin{aligned} n &= 2k & \text{and} & \quad (-1)^n = +1 & \text{for} & \quad p = 8k+1, \\ n &= 2k+1 & \text{and} & \quad (-1)^n = -1 & \text{for} & \quad p = 8k+3, \\ n &= 2k+1 & \text{and} & \quad (-1)^n = -1 & \text{for} & \quad p = 8k+5, \\ n &= 2k+2 & \text{and} & \quad (-1)^n = +1 & \text{for} & \quad p = 8k+7. \end{aligned}$$

Thus Lemma 1.4 shows that $\left(\frac{2}{p}\right) = +1$ for $p = 8k \pm 1$ and $\left(\frac{2}{p}\right) = -1$ for $p = 8k \pm 3$. This completes the proof of (b). □

2. Quadratic Reciprocity

Lemma 1.5. *If p is an odd prime and a is a positive odd integer such that p does not divide a, then $\left(\frac{a}{p}\right) = (-1)^t$, where $t = \sum_{u=1}^{\frac{1}{2}(p-1)} [ua/p]$. Here $[\,\cdot\,]$ denotes the greatest-integer function.*

REMARKS. When $a = 2$, the equality $\left(\frac{a}{p}\right) = (-1)^t$ fails for $p = 3$, since $t = [2/3] = 0$.

PROOF. With notation as in Lemma 1.4 and its proof, we form each ua for $1 \leq u \leq \frac{1}{2}(p-1)$ and reduce modulo p, obtaining as least positive residue either some r_i for $i \leq n$ or some s_j for $j \leq k$. Then $ua/p = [ua/p] + p^{-1}(\text{some } r_i \text{ or } s_j)$. Hence

$$\sum_{u=1}^{\frac{1}{2}(p-1)} ua = \sum_{u=1}^{\frac{1}{2}(p-1)} p[ua/p] + \sum_{i=1}^{n} r_i + \sum_{j=1}^{k} s_j. \qquad (*)$$

The proof of Lemma 1.4 showed that $p-r_1, \ldots, p-r_n, s_1, \ldots, s_k$ is a permutation of $1, \ldots, \frac{1}{2}(p-1)$, and thus the sum is the same in the two cases:

$$\sum_{u=1}^{\frac{1}{2}(p-1)} u = \sum_{i=1}^{n} (p - r_i) + \sum_{j=1}^{k} s_j = np - \sum_{i=1}^{n} r_i + \sum_{j=1}^{k} s_j.$$

Subtracting this equation from $(*)$, we obtain

$$(a-1) \sum_{u=1}^{\frac{1}{2}(p-1)} u = p\left(\sum_{u=1}^{\frac{1}{2}(p-1)} [ua/p] - n \right) + 2 \sum_{i=1}^{n} r_i.$$

Replacing $\sum_{u=1}^{\frac{1}{2}(p-1)} u$ on the left side by its value $\frac{1}{8}(p^2 - 1)$ and taking into account that p is odd, we obtain the following congruence modulo 2:

$$(a-1)\tfrac{1}{8}(p^2 - 1) \equiv \sum_{u=1}^{\frac{1}{2}(p-1)} [ua/p] - n \mod 2.$$

Since a is odd, the left side is congruent to 0 modulo 2. Therefore $n \equiv \sum_{u=1}^{\frac{1}{2}(p-1)} [ua/p] \equiv t \mod 2$, and Lemma 1.4 allows us to conclude that $(-1)^t = (-1)^n = \left(\frac{a}{p}\right)$. $\square$

PROOF OF (c) IN THEOREM 1.2. Let

$$S = \{(x, y) \in \mathbb{Z} \times \mathbb{Z} \mid 1 \leq x \leq \tfrac{1}{2}(p-1) \text{ and } 1 \leq y \leq \tfrac{1}{2}(q-1)\},$$

the number of elements in question being $|S| = \frac{1}{4}(p-1)(q-1)$. We can write $S = S_1 \cup S_2$ disjointly with

$$S_1 = \{(x, y) \mid qx > py\} \quad \text{and} \quad S_2 = \{(x, y) \mid qx < py\};$$

the exhaustion of S by S_1 and S_2 follows because $qx = py$ would imply that p divides qx and hence that p divides x, contradiction. We can describe S_1 alternatively as

$$S_1 = \{(x, y) \mid 1 \leq x \leq \tfrac{1}{2}(p-1) \text{ and } 1 \leq y < qx/p\},$$

and therefore $|S_1| = \sum_{x=1}^{\frac{1}{2}(p-1)} [qx/p]$, which is the integer t in Lemma 1.5 such that $(-1)^t = \left(\frac{q}{p}\right)$. Similarly we have $|S_2| = \sum_{y=1}^{\frac{1}{2}(q-1)} [py/q]$, which is the integer t in Lemma 1.5 such that $(-1)^t = \left(\frac{p}{q}\right)$. Therefore

$$(-1)^{\frac{1}{4}(p-1)(q-1)} = (-1)^{|S|} = (-1)^{|S_1|}(-1)^{|S_2|} = \left(\tfrac{q}{p}\right)\left(\tfrac{p}{q}\right),$$

and the proof is complete. $\square$

3. Equivalence and Reduction of Quadratic Forms

A **binary quadratic form** over $\mathbb{Z}$ is a function $F(x, y) = ax^2 + bxy + cy^2$ from $\mathbb{Z} \times \mathbb{Z}$ to $\mathbb{Z}$ with a, b, c in $\mathbb{Z}$. Following Gauss,[4] we abbreviate this F as (a, b, c). We shall always assume, without explicitly saying so, that the **discriminant** $D = b^2 - 4ac$ is not the square of an integer and that F is **primitive** in the sense that $\mathrm{GCD}(a, b, c) = 1$. When there is no possible ambiguity, we may say "form" or "quadratic form" in place of "binary quadratic form."

Let $\begin{pmatrix} \alpha & \beta \\ \gamma & \delta \end{pmatrix}$ be a member of the group $\mathrm{GL}(2, \mathbb{Z})$ of integer matrices whose inverse is an integer matrix. The determinant of such a matrix is ± 1. We can use this matrix to change variables, writing

$$\begin{pmatrix} x \\ y \end{pmatrix} = \begin{pmatrix} \alpha & \beta \\ \gamma & \delta \end{pmatrix}\begin{pmatrix} x' \\ y' \end{pmatrix} = \begin{pmatrix} \alpha x' + \beta y' \\ \gamma x' + \delta y' \end{pmatrix}.$$

Then $ax^2 + bxy + cy^2$ becomes

$$a(\alpha x' + \beta y')^2 + b(\alpha x' + \beta y')(\gamma x' + \delta y') + c(\gamma x' + \delta y')^2$$
$$= (a\alpha^2 + b\alpha\gamma + c\gamma^2)x'^2 + (2a\alpha\beta + b\alpha\delta + b\beta\gamma + 2c\gamma\delta)x'y' + (a\beta^2 + b\beta\delta + c\delta^2)y'^2.$$

---

[4]*Disquisitiones Arithmeticae*, Article 153. Actually, Gauss always assumed that the coefficient of xy is even and consequently wrote (a, b, c) for $ax^2 + 2bxy + cy^2$. To study $x^2 + xy + y^2$, for example, he took $a = 2, b = 1, c = 2$. The convention of working with $ax^2 + bxy + cy^2$ is due to Eisenstein.

If we associate the triple (a, b, c) of $F(x, y)$ to the matrix $\begin{pmatrix} 2a & b \\ b & 2c \end{pmatrix}$, then this formula shows that the triple (a', b', c') of the new form $F'(x', y')$ is associated to the matrix

$$\begin{pmatrix} 2a' & b' \\ b' & 2c' \end{pmatrix} = \begin{pmatrix} \alpha & \gamma \\ \beta & \delta \end{pmatrix} \begin{pmatrix} 2a & b \\ b & 2c \end{pmatrix} \begin{pmatrix} \alpha & \beta \\ \gamma & \delta \end{pmatrix}.$$

From this equality of matrices, we see that

(i) the member $\begin{pmatrix} 1 & 0 \\ 0 & 1 \end{pmatrix}$ of $\mathrm{GL}(2, \mathbb{Z})$ has the effect of the identity transformation,

(ii) the member $\begin{pmatrix} \alpha & \beta \\ \gamma & \delta \end{pmatrix} \begin{pmatrix} \alpha' & \beta' \\ \gamma' & \delta' \end{pmatrix}$ of $\mathrm{GL}(2, \mathbb{Z})$ has the effect of applying first $\begin{pmatrix} \alpha & \beta \\ \gamma & \delta \end{pmatrix}$ and then $\begin{pmatrix} \alpha' & \beta' \\ \gamma' & \delta' \end{pmatrix}$.

These two facts say that we do not quite have the expected group action on forms on the left. Instead, we can say either that we have a group action on the right or that gF is obtained from F by operating by g^t. Anyway, there are orbits, and they are what we really need. The discriminant $D = b^2 - 4ac$ of the form F is evidently minus the determinant of the associated matrix $\begin{pmatrix} 2a & b \\ b & 2c \end{pmatrix}$, and the displayed equality of matrices thus implies that the discriminant of the form F' is $D(\alpha\delta - \beta\gamma)^2$. Since $(\alpha\delta - \beta\gamma)^2 = 1$ for matrices in $\mathrm{GL}(2, \mathbb{Z})$, we conclude that

(iii) each member of $\mathrm{GL}(2, \mathbb{Z})$ preserves the discriminant of the form.

Hence the group $\mathrm{GL}(2, \mathbb{Z})$ acts on the forms of discriminant D.

Forms in the same orbit under $\mathrm{GL}(2, \mathbb{Z})$ are said to be **equivalent**. Forms in the same orbit under the subgroup $\mathrm{SL}(2, \mathbb{Z})$ are said to be **properly equivalent**. A **proper equivalence class** of forms will refer to the latter relation. This notion is due to Gauss. Equivalence under $\mathrm{GL}(2, \mathbb{Z})$ is an earlier notion due to Lagrange, and we shall refer to its classes as **ordinary equivalence classes** on the infrequent occasions when the notion arises. Proper equivalence is necessary later in order to get a group operation on classes of forms. If one form can be carried to another form by a member of $\mathrm{GL}(2, \mathbb{Z})$ of determinant -1, we say that the two forms are **improperly equivalent**. Use of the matrix $\begin{pmatrix} 1 & 0 \\ 0 & -1 \end{pmatrix}$ shows that the form (a, b, c) is improperly equivalent to the form $(a, -b, c)$. In particular, $(a, 0, c)$ is improperly equivalent to itself.

The discriminant D is congruent to b^2 modulo 4 and hence is congruent to 0 or 1 modulo 4. All nonsquare integers D that are congruent to 0 or 1 modulo 4 arise as discriminants; in fact, we can always achieve such a D with $a = 1$ and with b equal either to 0 or to 1.

The discriminant is minus the determinant of the matrix $\begin{pmatrix} 2a & b \\ b & 2c \end{pmatrix}$ associated to

(a, b, c), and this matrix is real symmetric with trace $2(a+c)$. Since $D = b^2 - 4ac$ is assumed not to be the square of an integer, neither a nor c can be 0.

If $D > 0$, the symmetric matrix $\begin{pmatrix} 2a & b \\ b & 2c \end{pmatrix}$ is indefinite, having eigenvalues of opposite sign. In this case the **Dirichlet class number** of D, denoted by $h(D)$, is defined to be the number[5] of all proper equivalence classes of forms of discriminant D.

If $D < 0$, then a and c have the same sign. The matrix $\begin{pmatrix} 2a & b \\ b & 2c \end{pmatrix}$ is positive definite if a and c are positive, and it is negative definite if a and c are negative. Correspondingly we refer to the form (a, b, c) as **positive definite** or **negative definite** in the two cases. Since $g^t \begin{pmatrix} 2a & b \\ b & 2c \end{pmatrix} g$ is positive definite whenever $\begin{pmatrix} 2a & b \\ b & 2c \end{pmatrix}$ is positive definite, any form equivalent to a positive definite form is again positive definite. A similar remark applies to negative definite forms. Thus "positive definite" and "negative definite" are class properties. For any given discriminant $D < 0$, the **Dirichlet class number** of D, denoted by $h(D)$, is the number[6] of proper equivalence classes of *positive definite* forms of discriminant D.

The form (a, b, c) **represents** an integer m if $ax^2 + bxy + cy^2 = m$ is solvable for some integers x and y. The form **primitively represents** m if the x and y with $ax^2 + bxy + cy^2 = m$ can be chosen to be relatively prime. In any event, $\mathrm{GCD}(x, y)$ divides m, and thus whenever a form represents a prime p, it primitively represents p.

Theorem 1.6. Fix a nonsquare discriminant D.

(a) The Dirichlet class number $h(D)$ is finite. In fact, any form of discriminant D is properly equivalent to a form (a, b, c) with $|b| \leq |a| \leq |c|$ and therefore has $3|ac| \leq |D|$, and the number of forms of discriminant D satisfying all these inequalities is finite.

(b) An odd prime p with $\mathrm{GCD}(D, p) = 1$ is primitively representable by some form (a, b, c) of discriminant D if and only if $\left(\frac{D}{p}\right) = +1$. In this case the number of proper equivalence classes of forms primitively representing p is either 1 or 2, and these classes are carried to one another by $\mathrm{GL}(2, \mathbb{Z})$. In fact, if $\left(\frac{D}{p}\right) = +1$, then $b^2 \equiv D \bmod 4p$ for some integer b, and representatives of these classes may be taken to be $\left(p, \pm b, \frac{b^2 - D}{4p}\right)$.

---

[5]This number was studied by Dirichlet. According to Theorem 1.20 below, it counts the "strict equivalence classes" of ideals in a sense that is introduced in Section 7. This number either equals or is twice the number of equivalence classes of ideals in the other sense that is introduced in Section 7. The latter is what is generalized in Chapter V in the subject of algebraic number theory, and the latter is how "class number" is usually defined in modern books in algebraic number theory. Consequently Dirichlet class numbers sometimes are twice what modern class numbers are. We use "Dirichlet class numbers" in this chapter and change to the modern "class numbers" in Chapter V.

[6]This number was studied by Dirichlet. See the previous footnote for further information.

We come to the proof after some preliminary remarks and examples. The argument for (a) is constructive, and thus the forms given explicitly in (b) can be transformed constructively into properly equivalent forms satisfying the conditions of (a). Hence we are led to explicit forms as in (a) representing p. A generalization of (b) concerning how a composite integer m can be represented if $\mathrm{GCD}(D, m) = 1$ appears in Problem 2 at the end of the chapter. What is missing in all this is a description of proper equivalences among the forms as in (a). We shall solve this question readily in Proposition 1.7 when $D < 0$. For $D > 0$, the answer is more complicated; we shall say what it is in Theorem 1.8, but we shall omit some of the proof of that theorem.

EXAMPLES.

(1) $D = -4$. Theorem 1.2a shows that the odd primes with $\left(\frac{D}{p}\right) = +1$ are those of the form $4k + 1$. Theorem 1.6a says that each proper equivalence class of forms of discriminant -4 has a representative (a, b, c) with $3|ac| \leq 4$. Since $D < 0$, we are interested only in positive definite forms, which necessarily have a and c positive. Thus $a = c = 1$, and we must have $b = 0$. So there is only one class of (positive definite) forms of discriminant -4, namely $x^2 + y^2$, and Theorem 1.6b allows us to conclude that $x^2 + y^2 = p$ is solvable for each prime $p = 4k + 1$. In other words, we recover the conclusion of Proposition 1.1 as far as representability of primes is concerned.

(2) $D = -20$. To have $\left(\frac{D}{p}\right) = +1$ for an odd prime p, we must have either $\left(\frac{-1}{p}\right) = \left(\frac{5}{p}\right) = +1$ or $\left(\frac{-1}{p}\right) = \left(\frac{5}{p}\right) = -1$. Theorem 1.2 shows in the first case that $p \equiv 1 \bmod 4$ and $p \equiv \pm 1 \bmod 5$, while in the second case $p \equiv 3 \bmod 4$ and $p \equiv \pm 3 \bmod 5$. That is, p is congruent to one of 1 and 9 modulo 20 in the first case and to one of 3 and 7 modulo 20 in the second case. Let us consider the forms as in Theorem 1.6a. We know that $a > 0$ and $c > 0$. The inequality $3ac \leq |D|$ forces $ac \leq 6$. Since $|b| \leq a \leq c$, we obtain $a^2 \leq 6$ and $a \leq 2$. Since 4 divides D, b is even. Then $b = 0$ or $b = \pm 2$. So the only possibilities are $(1, 0, 5)$ and $(2, \pm 2, 3)$. Because of Theorem 1.6b, any prime congruent to one of 1, 3, 7, 9 modulo 20 is representable either by $(1, 0, 5)$ and not $(2, \pm 2, 3)$, or by $(2, \pm 2, 3)$ and not $(1, 0, 5)$. We can write down all residues modulo 20 for $x^2 + 5y^2$ and $2x^2 \pm 2xy + 3y^2$, and we find that the possible residues prime to 20 are 1 and 9 in the first case, and they are 3 and 7 in the second case. The conclusion for odd primes p with $\mathrm{GCD}(20, p) = 1$ is that

$p \equiv 1$ or 9 mod 20   implies p is representable as $x^2 + 5y^2$,

$p \equiv 3$ or 7 mod 20   implies p is representable as $2x^2 \pm 2xy + 3y^2$.

The residues modulo 20 have shown that $x^2 + 5y^2$ is not equivalent to either of $2x^2 \pm 2xy + 3y^2$, but they do not show whether $2x^2 \pm 2xy + 3y^2$ are properly equivalent to one another. Hence the Dirichlet class number $h(-20)$ is either 2 or 3. It will turn out to be 2.

(3) $D = -56$. To have $\left(\frac{D}{p}\right) = +1$ for an odd prime p, we must have an odd number of the Legendre symbols $\left(\frac{-1}{p}\right)$, $\left(\frac{2}{p}\right)$, and $\left(\frac{7}{p}\right)$ equal to $+1$ and the rest equal to -1. We readily find from Theorem 1.2 that the possibilities with $\text{GCD}(56, p) = 1$ are

$$p \equiv 1, 3, 5, 9, 13, 15, 19, 23, 25, 27, 39, 45 \bmod 56.$$

Applying Theorem 1.6a as in the previous example, we find that $x^2 + 14y^2$, $2x^2 + 7y^2$, and $3x^2 \pm 2xy + 5y^2$ are representatives of all proper equivalence classes of forms of discriminant -56. Taking into account Theorem 1.6b and the residue classes of these forms modulo 56, we conclude for odd primes p that

if $p \equiv$ any of $1, 9, 15, 23, 25, 39 \bmod 56$, then

p is representable as $x^2 + 14y^2$ or $2x^2 + 7y^2$,

if $p \equiv$ any of $3, 5, 13, 19, 27, 45 \bmod 56$, then

p is representable as both of $3x^2 \pm 2xy + 5y^2$.

The question left unsettled by the argument so far is whether $x^2 + 14y^2$ is properly equivalent to $2x^2 + 7y^2$. Equivalent forms represent the same integers, and the integer 1 is representable by $x^2 + 14y^2$ but not by $2x^2 + 7y^2$. Hence the two forms are not equivalent and cannot be properly equivalent. According to Theorem 1.6b, the primes of the first line are therefore representable by either $x^2 + 14y^2$ or $2x^2 + 7y^2$ but *never* by both. Hence the Dirichlet class number $h(-56)$ is either 3 or 4. It will turn out to be 4.

(4) $D = 5$. The forms of discriminant 5 are indefinite. Applying Theorem 1.6a, we obtain $3|ac| \leq 5$. Hence $|a| = |c| = 1$. Since D is odd, b is odd. The inequality $|b| \leq |a|$ thus forces $|b| = 1$. Then $D = 1 - 4ac$ shows that $ac < 0$. The possibilities are therefore $(1, \pm 1, -1)$ and $(-1, \pm 1, 1)$. The Dirichlet class number $h(5)$ is at most 4. It will turn out to be 1. Let us take this fact as known. The odd primes p with $\left(\frac{D}{p}\right) = +1$ are $p = 5k \pm 1$. Under the assumption that the class number is 1, Theorem 1.6b shows that every such prime is representable as $x^2 + xy - y^2$.

PROOF OF THEOREM 1.6a. We consider the effect of two transformations in $\text{SL}(2, \mathbb{Z})$, one via $\begin{pmatrix} 0 & -1 \\ 1 & 0 \end{pmatrix}$ and the other via $\begin{pmatrix} 1 & n \\ 0 & 1 \end{pmatrix}$. Under these, the matrix associated to (a, b, c) becomes

$$\begin{pmatrix} 0 & 1 \\ -1 & 0 \end{pmatrix} \begin{pmatrix} 2a & b \\ b & 2c \end{pmatrix} \begin{pmatrix} 0 & -1 \\ 1 & 0 \end{pmatrix} = \begin{pmatrix} 2c & -b \\ -b & 2a \end{pmatrix}$$

and $\begin{pmatrix} 1 & 0 \\ n & 1 \end{pmatrix} \begin{pmatrix} 2a & b \\ b & 2c \end{pmatrix} \begin{pmatrix} 1 & n \\ 0 & 1 \end{pmatrix} = \begin{pmatrix} 2a & 2an+b \\ 2an+b & 2an^2+2bn+2c \end{pmatrix}$,

respectively. Thus the transformations are

$$(a,b,c) \longmapsto (c,-b,a), \qquad (*)$$

$$(a,b,c) \longmapsto (a, 2an+b, c'). \qquad (**)$$

Possibly applying (∗) allows us to make $|a| \leq |c|$ while leaving $|b|$ alone. Since $a \neq 0$, we can apply (∗∗) with n the closest integer to $-\frac{b}{2a}$ to make $|b| \leq |a|$. This step possibly changes c. Thus after this step, we again apply (∗) if necessary to make $|a| \leq |c|$, and we apply (∗∗) again. In each pair of steps, we may assume that $|b|$ strictly decreases or else that $n = 0$. We cannot always be in the former case, since $|b|$ is bounded below by 0. Thus at some point we obtain $n = 0$. At this point, c does not change, and thus we have $|b| \leq |a| \leq |c|$, as required.

The inequalities $|b| \leq |a| \leq |c|$ imply that

$$4|ac| = |D - b^2| \leq |D| + |b|^2 \leq |D| + |ac|,$$

and hence $3|ac| \leq |D|$. Since neither a nor c is 0, it follows that the inequalities $|b| \leq |a| \leq |c|$ imply that $|a|, |b|, |c|$ are all bounded by $|D|$. Therefore the Dirichlet class number $h(D)$ is finite. □

PROOF OF NECESSITY IN THEOREM 1.6b. Suppose x and y are integers with $\mathrm{GCD}(x, y) = 1$ and $ax^2 + bxy + cy^2 = p$. Then $ax^2 + bxy + cy^2 \equiv 0 \bmod p$. Choose u and v with $ux + vy = 1$. Routine computation shows that

$$4(ax^2+bxy+cy^2)(av^2 - buv + cu^2)$$
$$= [u(xb+2yc) - v(2xa+yb)]^2 - (b^2 - 4ac)(xu+yv)^2$$
$$= [u(xb+2yc) - v(2xa+yb)]^2 - (b^2 - 4ac),$$

and hence

$$0 \equiv [u(xb+2yc) - v(2xa+yb)]^2 - (b^2 - 4ac) \bmod p.$$

Consequently $D \equiv [u(xb+2yc) - v(2xa+yb)]^2 \bmod p$, and D is exhibited as a square modulo p. □

PROOF OF SUFFICIENCY IN THEOREM 1.6b. Choose an integer solution b of $b^2 \equiv D \bmod p$. Since $b + p$ is another solution and has the opposite parity, we may assume that b and D have the same parity. Then $b^2 \equiv D \bmod p$ and $b^2 \equiv D \bmod 4$, so that $b^2 \equiv D \bmod 4p$. Since $\mathrm{GCD}(D, p) = 1$, p does not divide b, and the forms $\left(p, \pm b, \frac{b^2-D}{4p}\right)$ are primitive. They have discriminant $b^2 - 4p\frac{b^2-D}{4p} = D$, they take the value p for $(x, y) = (1, 0)$, and they are improperly equivalent via $\begin{pmatrix} 1 & 0 \\ 0 & -1 \end{pmatrix}$. Thus the forms in the statement of the theorem exist.

For the uniqueness suppose that a form (a, b, c) of discriminant D represents p, say with $ax_0^2 + bx_0y_0 + cy_0^2 = p$. Since this representation has to be primitive, we know that $\mathrm{GCD}(x_0, y_0) = 1$. Put $\begin{pmatrix} \alpha \\ \gamma \end{pmatrix} = \begin{pmatrix} x_0 \\ y_0 \end{pmatrix}$, and choose integers β and δ such that $\alpha\delta - \beta\gamma = 1$. Then $\begin{pmatrix} \alpha & \beta \\ \gamma & \delta \end{pmatrix}$ has determinant 1 and satisfies $\begin{pmatrix} \alpha & \beta \\ \gamma & \delta \end{pmatrix}\begin{pmatrix} 1 \\ 0 \end{pmatrix} = \begin{pmatrix} x_0 \\ y_0 \end{pmatrix}$. The equality $ax_0^2 + bx_0y_0 + cy_0^2 = \frac{1}{2}(x_0 \; y_0)\begin{pmatrix} 2a & b \\ b & 2c \end{pmatrix}\begin{pmatrix} x_0 \\ y_0 \end{pmatrix}$ therefore yields

$$p = \tfrac{1}{2}(1 \; 0)\begin{pmatrix} \alpha & \gamma \\ \beta & \delta \end{pmatrix}\begin{pmatrix} 2a & b \\ b & 2c \end{pmatrix}\begin{pmatrix} \alpha & \beta \\ \gamma & \delta \end{pmatrix}\begin{pmatrix} 1 \\ 0 \end{pmatrix}.$$

Consequently the form (a', b', c') associated to the matrix $\begin{pmatrix} \alpha & \gamma \\ \beta & \delta \end{pmatrix}\begin{pmatrix} 2a & b \\ b & 2c \end{pmatrix}\begin{pmatrix} \alpha & \beta \\ \gamma & \delta \end{pmatrix}$ takes on the value p at $(x, y) = (1, 0)$ and is properly equivalent to (a, b, c). In particular, it is a form (p, b', c') for some b' and c' such that $b'^2 - 4pc' = D$.

Thus in the proof of uniqueness, we may assume that we have two forms (p, b', c') and (p, b'', c'') of discriminant D. Then $b''^2 \equiv D \equiv b'^2 \bmod 4p$. The conditions $b''^2 \equiv b'^2 \bmod p$ and $b''^2 \equiv b'^2 \bmod 4$ imply that $b'' \equiv \pm b' \bmod p$ and $b'' \equiv b' \bmod 2$ for one of the choices of sign. Thus $b'' \equiv \pm b' \bmod 2p$ for that choice of sign. Let us write $b'' = \pm b' + 2np$ for some integer n. The matrix equality

$$\begin{pmatrix} 1 & 0 \\ n & 1 \end{pmatrix}\begin{pmatrix} 2p & \pm b' \\ \pm b' & 2c' \end{pmatrix}\begin{pmatrix} 1 & n \\ 0 & 1 \end{pmatrix} = \begin{pmatrix} 2p & 2pn \pm b' \\ 2pn \pm b' & 2(*) \end{pmatrix}$$

shows that $(p, \pm b', c')$ is properly equivalent to $(p, b'', *)$. Since the discriminant has to be D, we conclude that $* = c''$. That is, (p, b'', c'') is properly equivalent to $(p, \pm b', c')$ for that same choice of sign. Since (p, b', c') is improperly equivalent to $(p, -b', c')$, the proof of the theorem is complete. $\square$

Our discussion of representability of primes p by binary quadratic forms of discriminant D when $\mathrm{GCD}(D, p) = 1$ will be complete once we have a set of representatives of proper equivalence classes with no redundancy. For discriminant $D < 0$, this step is not difficult and amounts, according to Theorem 1.6a, to sorting out proper equivalences among forms (a, b, c) with $b^2 - 4ac = D$ and $|b| \leq |a| \leq |c|$. Let us call a form with $D < 0$ **reduced** when it satisfies these conditions.

There are two redundancies that are easy to spot, namely

(a, b, a) is properly equivalent to $(a, -b, a)$ via $\begin{pmatrix} 0 & 1 \\ -1 & 0 \end{pmatrix}$,

(a, a, c) is properly equivalent to $(a, -a, c)$ via $\begin{pmatrix} 1 & -1 \\ 0 & 1 \end{pmatrix}$.

The result for $D < 0$ is that there are no other redundancies among reduced forms.

Proposition 1.7. Fix a negative discriminant D. With the exception of the proper equivalences of

and
$$(a, b, a) \text{ to } (a, -b, a)$$
$$(a, a, c) \text{ to } (a, -a, c),$$

no two distinct reduced positive definite forms of discriminant D are properly equivalent.

PROOF. Suppose that (a, b, c) is properly equivalent to (a', b', c'), that both are reduced, and that $a \geq a' > 0$. For some $\begin{pmatrix} \alpha & \beta \\ \gamma & \delta \end{pmatrix}$ in $SL(2, \mathbb{Z})$, we have $a' = a\alpha^2 + b\alpha\gamma + c\gamma^2$. Hence the inequalities $c \geq a$ and $|b| \geq -a$ imply that

$$a \geq a\alpha^2 + b\alpha\gamma + c\gamma^2 \geq a(\alpha^2 + \gamma^2) + b\alpha\gamma \geq a(\alpha^2 + \gamma^2) - a|\alpha\gamma| \geq a|\alpha\gamma|, \quad (*)$$

and $\alpha\gamma$ equals 0 or ± 1. Thus the ordered pair (α, γ) is one of $(0, \pm 1)$, $(\pm 1, 0)$, $(\pm 1, 1)$, $(\pm 1, -1)$. Multiplying $\begin{pmatrix} \alpha & \beta \\ \gamma & \delta \end{pmatrix}$ if necessary by $\begin{pmatrix} -1 & 0 \\ 0 & -1 \end{pmatrix}$, which acts trivially on quadratic forms, we may assume that (α, γ) is one of $(0, 1)$, $(1, 0)$, $(1, \pm 1)$. We treat these three cases separately.

Case 1. $(\alpha, \gamma) = (0, 1)$. The condition $\alpha\delta - \beta\gamma = 1$ forces $\beta\gamma = -1$, and the formula $b' = 2a\alpha\beta + b\alpha\delta + b\beta\gamma + 2c\gamma\delta$ gives $(a', b', c') = (c, -b + 2c\delta, *)$. Since $|b| \leq c$ and $|b - 2c\delta| \leq c$, we must have $|\delta| \leq 1$. If $\delta = 0$, we are led to $(a', b', c') = (c, -b, a)$, which is reduced only if $c = a$, and this is the first of the two allowable exceptions. If $|\delta| = 1$, the triangle inequality gives $2c = |2c\delta| \leq |b| + |2c\delta - b| \leq c + c = 2c$, and therefore $|b| = c = |b - 2c\delta|$. Then $b = -(b - 2c\delta)$, and $b = c\delta = \pm c$. Since $|b| \leq a \leq c$, $b = \pm a$ also. Hence $(a', b', c') = (a, -b, a)$, and this is again the first of the two allowable exceptions.

Case 2. $(\alpha, \gamma) = (1, 0)$. The condition $\alpha\delta - \beta\gamma = 1$ forces $\alpha\delta = 1$, and thus $(a', b', c') = (a, b + 2a\beta, *)$. Since $|b| \leq a$ and $|b + 2a\beta| \leq a$, we must have $|\beta| \leq 1$. If $\beta = 0$, then $(a', b', c') = (a, b, c)$, and there is nothing to prove. If $|\beta| = 1$, the triangle inequality gives $2a = |2a\beta| \leq |-b| + |2a\beta + b|$, and therefore $|b| = a = |b + 2\beta a|$. Then $b = -(b + 2\beta a)$, and we conclude that $b = -a\beta = \pm a$ and $b + 2\beta a = \mp a$. Hence the proper equivalence in question is of (a, a, c) to $(a, -a, c)$, which is the second of the two allowable exceptions.

Case 3. $(\alpha, \gamma) = (1, \pm 1)$. From (*) and the assumption that $a \geq a'$, we have $a \geq a' \geq a|\alpha\gamma| = a$. Thus $a = a'$, and the definition of a' shows that $a = a + b\gamma + c$. Hence $c = -b\gamma$, and $c = |b|$. Since $|b| \leq a \leq c$, we obtain $-b\gamma = a = c$. The formula $b' = 2a\alpha\beta + b\alpha\delta + b\beta\gamma + 2c\gamma\delta$ then simplifies to $b' = 2a\beta + b\delta + b\beta\gamma + 2a\gamma\delta = (2a + b\gamma)(\beta + \gamma\delta)$. From $\alpha\delta - \beta\gamma = 1$, we have $\delta - \beta\gamma = 1$ and thus also $\gamma\delta = \gamma + \beta$. Therefore $\beta + \gamma\delta = 2\beta + \gamma$, and this cannot be 0. So $|b'| \geq |2a + b\gamma| = |2a - a| = a = a'$. Since (a', b', c') is reduced, $|b'| = a' = a = c = |b|$, and the proper equivalence is of (a, a, a) to $(a, -a, a)$. This is an instance of both allowable exceptions, and the proof is complete. □

EXAMPLES, CONTINUED.

(2) $D = -20$. We saw earlier that the reduced positive definite forms with $D = -20$ are $x^2 + 5y^2$ and $2x^2 \pm 2xy + 3y^2$, i.e., $(1, 0, 5)$ and $(2, \pm 2, 3)$. The remarks preceding Proposition 1.7 show that $(2, 2, 3)$ is properly equivalent to $(2, -2, 3)$, and the proposition shows that $(1, 0, 5)$ is not properly equivalent to $(2, 2, 3)$. (We saw this latter conclusion for this example earlier by considering residues.) Consequently $h(-20) = 2$.

(3) $D = -56$. We saw earlier that the reduced positive definite forms with $D = -56$ are $x^2 + 14y^2, 2x^2 + 7y^2$, and $3x^2 \pm 2xy + 5y^2$, i.e., $(1, 0, 14), (2, 0, 7), (3, 2, 5)$, and $(3, -2, 5)$. Proposition 1.7 shows that no two of these four forms are properly equivalent. Consequently $h(-56) = 4$.

Let us turn our attention to $D > 0$. We still have the proper equivalences of (a, b, a) to $(a, -b, a)$ and (a, a, c) to $(a, -a, c)$ as in the remarks before Proposition 1.7. But there can be others, and the question is subtle. Here are some simple examples.

EXAMPLES WITH POSITIVE DISCRIMINANT.

(1) $D = 5$. The forms with $D = 5$ satisfying the inequalities $|b| \leq |a| \leq |c|$ of Theorem 1.6a are $(1, \pm 1, -1)$ and $(-1, \pm 1, 1)$. The second standard equivalence allows us to discard one form from each pair, and we are left with $(1, 1, -1)$ and $(-1, -1, 1)$. The first of these two is equivalent to the second via $\begin{pmatrix} \alpha & \beta \\ \gamma & \delta \end{pmatrix} = \begin{pmatrix} 0 & 1 \\ -1 & 0 \end{pmatrix}$. Thus $h(5) = 1$, as was announced without proof in Example 4 earlier in this section.

(2) $D = 13$. The forms with $D = 13$ satisfying the inequalities $|b| \leq |a| \leq |c|$ of Theorem 1.6a are $(1, \pm 1, -3)$ and $(-1, \pm 1, 3)$. The second standard

equivalence allows us to discard one form from each pair, and we are left with $(1, 1, -3)$ and $(-1, -1, 3)$. The first of these two is equivalent to the second via $\begin{pmatrix} \alpha & \beta \\ \gamma & \delta \end{pmatrix} = \begin{pmatrix} 1 & -2 \\ 1 & -1 \end{pmatrix}$. Thus $h(13) = 1$.

(3) $D = 21$. The forms with $D = 21$ satisfying the inequalities $|b| \leq |a| \leq |c|$ of Theorem 1.6a are $(1, \pm 1, -5)$ and $(-1, \pm 1, 5)$. The second standard equivalence allows us to discard one form from each pair, and we are left with $(1, 1, -5)$ and $(-1, -1, 5)$. These are not properly equivalent. In fact, the form $-x^2 - xy + 5y^2$ is -1 for $(x, y) = (1, 0)$, but $x^2 + xy - 5y^2 = -1$ is not even solvable modulo 3. Thus $h(21) = 2$.

Although the starting data for these three examples are similar, the outcomes are strikingly different. The idea for what to do involves starting afresh with the reduction question that was addressed in Theorem 1.6a. For discriminant $D > 0$, a different reduction is to be used. The reduction in question appears in Theorem 1.8a below, but some preliminary remarks are needed to explain the proof.

Two forms (a, b, c) and (a', b', c') of discriminant $D > 0$ will be said to be **neighbors** if $c = a'$ and $b + b' \equiv 0$ mod $2c$. More precisely we say in this case that (a', b', c') is a **neighbor on the right** of (a, b, c) and that (a, b, c) is a **neighbor on the left** of (a', b', c'). A key observation is that neighbors are properly equivalent to one another. In fact, if (a', b', c') is a neighbor on the right of (a, b, c), define $\begin{pmatrix} \alpha & \beta \\ \gamma & \delta \end{pmatrix} = \begin{pmatrix} 0 & -1 \\ 1 & (b+b')/(2c) \end{pmatrix}$. Then computation gives

$$\begin{pmatrix} \alpha & \gamma \\ \beta & \delta \end{pmatrix} \begin{pmatrix} 2a & b \\ b & 2c \end{pmatrix} \begin{pmatrix} \alpha & \beta \\ \gamma & \delta \end{pmatrix} = \begin{pmatrix} 2c & b' \\ b' & (b-b')\frac{b+b'}{2c} \end{pmatrix}.$$

The lower right entry of this matrix is an even integer, since $b + b' \equiv 0$ mod $2c$ and since, as a consequence, $b + b' \equiv 0$ mod 2. Hence (a, b, c) is transformed into (c, b', c'), where $c' = \frac{1}{2}(b - b')\frac{b+b'}{2c}$.

Let us call a primitive form (a, b, c) of discriminant $D > 0$ **reduced** when it satisfies the conditions

$$0 < b < \sqrt{D} \quad \text{and} \quad \sqrt{D} - b < 2|a| < \sqrt{D} + b.$$

The first inequality shows that b is bounded if D is fixed, and the equality $-4ac = D^2 - b^2$ shows that there are only finitely many possibilities for a and c. Consequently there are only finitely many reduced forms for given D.

From $|b| < \sqrt{D}$, we see that $b^2 < D = b^2 - 4ac$ and $ac < 0$; thus any reduced form has a and c of opposite sign. Then $D - b^2 = -4ac = (2|a|)(2|c|)$, and it follows that $2|a| > \sqrt{D} - b$ implies $2|c| < \sqrt{D} + b$ and that $2|a| < \sqrt{D} + b$ implies $2|c| < \sqrt{D} - b$. Consequently

$$\sqrt{D} - b < 2|c| < \sqrt{D} + b.$$

Theorem 1.8. Fix a positive nonsquare discriminant D.

(a) Each form of discriminant D is properly equivalent to some reduced form of discriminant D.

(b) Each reduced form of discriminant D is a neighbor on the left of one and only one reduced form of discriminant D and is a neighbor on the right of one and only one reduced form of discriminant D.

(c) The reduced forms of discriminant D occur in uniquely determined cycles, each one of even length, such that each member of a cycle is an iterated neighbor on the right to all members of the cycle and consequently is properly equivalent to all other members of the cycle.

(d) Two reduced forms of discriminant D are properly equivalent if and only if they lie in the same cycle in the sense of (c).

REMARKS. Conclusion (d) is the deepest part of the theorem, involving a subtle argument that in essence uses the periodic continued-fraction expansion of the roots z of the polynomial $az^2 + bz + c$ if (a, b, c) is a form under consideration. We shall prove (a) through (c), omitting the proof of (d), and then we shall return to the three examples $D = 5, 13, 29$ begun just above.

PROOF OF THEOREM 1.8a. If (a, b, c) is given and is not reduced, let m be the unique integer such that

$$\sqrt{D} - 2|c| < -b + 2cm < \sqrt{D}, \qquad (*)$$

and define $(a', b', c') = (c, -b + 2cm, a - bm + cm^2)$. Then

$$b'^2 - 4a'c' = (-b + 2cm)^2 - 4c(a - bm + cm^2)$$
$$= b^2 - 4bcm + 4c^2m^2 - 4ac + 4bcm - 4c^2m^2 = b^2 - 4ac = D,$$

and we observe that $a' = c$ and that $b + b' = 2cm \equiv 0 \bmod 2c$. Consequently (a', b', c') is a form of discriminant D and is a right neighbor to (a, b, c). By the remarks before the theorem, (a, b, c) is properly equivalent to (a', b', c').

We repeat this process at least once, obtaining (a'', b'', c''). If $|a''| < |a'|$, we repeat it again, obtaining (a''', b''', c'''), and we continue in this way. Eventually the strict decrease of the magnitude of the first entry must stop. To keep the notation simple, we may assume without loss of generality that $|a''| \geq |a'|$. The claim is that (a', b', c') is then reduced.

Put $u = \sqrt{D} - b'$ and $v = b' - (\sqrt{D} - 2|a'|)$. The inequalities $(*)$ show that $u > 0$ and $v > 0$. Therefore

$$0 < v^2 + 2uv + 2u\sqrt{D} = (u + v)^2 - u^2 + 2u\sqrt{D}$$
$$= 4a'^2 - (D - 2b'\sqrt{D} + b'^2) + 2D - 2b'\sqrt{D}$$
$$= 4a'^2 + D - b'^2 = 4a'^2 - 4a'c'.$$

Since $|c'| = |a''| \geq |a'|$, this inequality shows that $a'c' < 0$. Therefore $b'^2 = D + 4a'c' < D$, and $|b'| < \sqrt{D}$.

From $a'c' < 0$ and $|a'| \leq |c'|$, we see that $4|a'|^2 \leq 4|a'c'| = -4a'c' = D - b'^2 \leq D$. Therefore $2|a'| < \sqrt{D}$. The inequality $\sqrt{D} - 2|c| < b'$ implies that $\sqrt{D} - b' < 2|c| = 2|a'|$. The right side has just been shown to be $< \sqrt{D}$, and therefore $b' > 0$. Hence $\sqrt{D} - b' < 2|a'| < \sqrt{D} < \sqrt{D} + b'$. □

PROOF OF THEOREM 1.8b. Suppose that (a, b, c) is reduced and that (a', b', c') is a reduced neighbor on the right of (a, b, c). Then we must have $a' = c$ and $b + b' \equiv 0 \bmod 2c$. Since $D - b' < 2|a'|$ and $b' < \sqrt{D}$, we have $\sqrt{D} - 2|a'| < b' < \sqrt{D}$. That is, $\sqrt{D} - 2|c| < b' < \sqrt{D}$. These inequalities in combination with the congruence $b + b' \equiv 0 \bmod 2c$ show that (a, b, c) uniquely determines b'. Since (a', b', c') is to have discriminant D, c' is uniquely determined also.

We turn this construction around to prove existence of a right neighbor. Define (a', b', c') in terms of (a, b, c) as in the proof of Theorem 1.8a. Then $a' = c$, and b' is the unique integer such that $b + b' \equiv 0 \bmod 2c$ and

$$\sqrt{D} - 2|c| < b' < \sqrt{D}.$$

The form (a', b', c') is a right neighbor of (a, b, c), and we are to show that (a', b', c') is reduced.

Since (a, b, c) is reduced, we have $\sqrt{D} - b < 2|c| < \sqrt{D} + b$ and $b < \sqrt{D}$. Let m be the integer such that $b + b' = 2m|c|$. Addition of the inequalities $b' - (\sqrt{D} - 2|c|) > 0$ and $\sqrt{D} + b - 2|c| > 0$ gives $2m|c| = b + b' > 0$, and thus $m > 0$. Hence $m - 1 \geq 0$. Addition of the inequalities $\sqrt{D} - b > 0$ and $b' - (\sqrt{D} - 2|c|) > 0$ gives $0 < b' - b + 2|c| = 2b' - (b + b') + 2|c| = 2b' - 2(m - 1)|c|$. Hence $2b' > 2(m - 1)|c| \geq 0$, and we see that $b' > 0$. Therefore $0 < b' < \sqrt{D}$.

The definition of b' gives $\sqrt{D} - b' < 2|c| = 2|a'|$. Addition of the inequalities $2(m - 1)|c| \geq 0$ and $\sqrt{D} - b > 0$ gives $b + b' - 2|c| + \sqrt{D} - b > 0$, which says that $2|a'| < \sqrt{D} + b'$. Therefore (a', b', c') is reduced.

Let R be the operation of passing from a reduced form (a, b, c) to its unique reduced right neighbor (a', b', c'). What we have just shown implies that R acts as a permutation of the finite set of reduced forms of discriminant D. This set being finite, let n be the order of R. Then the set $\{R^k \mid 0 \leq k \leq n - 1\}$ is a cyclic group of permutations of the set of reduced forms of discriminant D. The existence of a two-sided inverse of R as a permutation implies that each reduced form of discriminant D has exactly one left neighbor. Thus the existence and uniqueness of neighbors on one side for reduced forms, in the presence of the finiteness of the set, implies existence and uniqueness on the other side. □

PROOF OF THEOREM 1.8c. We continue with R as the operation of passing from a reduced form to its unique reduced right neighbor, letting $\{R^k \mid 0 \leq k \leq n-1\}$ be the finite cyclic group of powers of R. This group acts on the set of reduced forms of discriminant D, and the cycles in question are the orbits under this action. To see that each orbit has an even number of members, we recall that a reduced form (a, b, c) has a and c of opposite sign. Thus if, for example, a is positive, then $R^l(a, b, c) = (a', b', c')$ has $(-1)^l a'$ positive. If the orbit of (a, b, c) has k members, then $R^k(a, b, c) = (a, b, c)$. Consequently $(-1)^k a$ has to have the same sign as a, and k has to be even. Finally the members of each orbit are properly equivalent to one another because, as we observed before the statement of the theorem, a form is properly equivalent to each of its neighbors. □

EXAMPLES WITH POSITIVE DISCRIMINANT, CONTINUED.

(1) $D = 5$. The forms with $D = 5$ satisfying the inequalities of Theorem 1.8a are $(1, 1, -1)$ and $(-1, 1, 1)$, and these consequently represent all proper equivalence classes. They form a single cycle and are properly equivalent by Theorem 1.8c. Thus again we obtain the easy conclusion that $h(5) = 1$.

(2) $D = 13$. The forms with $D = 13$ satisfying the inequalities of Theorem 1.8a are $(1, 3, -1)$ and $(-1, 3, 1)$, which make up a single cycle. Thus $h(13) = 1$.

(3) $D = 21$. The forms with $D = 21$ satisfying the inequalities of Theorem 1.8a are $(1, 3, -2)$ and $(-2, 3, 1)$, which make up one cycle, and $(-1, 3, 2)$ and $(2, 3, -1)$, which make up another cycle. Thus $h(21) = 2$.

4. Composition of Forms, Class Group

The identity $(x_1^2 + y_1^2)(x_2^2 + y_2^2) = (x_1 x_2 - y_1 y_2)^2 + (x_1 y_2 + x_2 y_1)^2$, which can be derived by factoring the left side in $\mathbb{Q}(\sqrt{-1})[x_1, y_1, x_2, y_2]$ and rearranging the factors, readily generalizes to an identity involving any form $x^2 + bxy + cy^2$ of nonsquare discriminant $D = b^2 - 4c$. We complete the square, writing the form as $(x - \frac{1}{2}by)^2 - \frac{1}{4}y^2 D$ and factoring it as $\left(x - \frac{1}{2}by + \frac{1}{2}y\sqrt{D}\right)\left(x - \frac{1}{2}by - \frac{1}{2}y\sqrt{D}\right)$, and we obtain

$$(x_1^2 + bx_1 y_1 + cy_1^2)(x_2^2 + bx_2 y_2 + cy_2^2)$$
$$= (x_1 x_2 - cy_1 y_2)^2 + b(x_1 x_2 - cy_1 y_2)(x_1 y_2 + x_2 y_1 + by_1 y_2)$$
$$+ c(x_1 y_2 + x_2 y_1 + by_1 y_2)^2.$$

Improving on an earlier attempt by Legendre, Gauss made a thorough investigation of how one might multiply two distinct forms of the same nonsquare discriminant, not necessarily with first coefficient 1, and Dirichlet reworked the theory and simplified it. Out of this work comes the following **composition formula**, of which the above formula is manifestly a special case.

Proposition 1.9. Let (a_1, b, c_1) and (a_2, b, c_2) be two primitive forms with the same middle coefficient b and with the same nonsquare discriminant D, hence with $a_1 c_1 = a_2 c_2 \neq 0$. Suppose that $j = c_1 a_2^{-1} = c_2 a_1^{-1}$ is an integer. Then the form $(a_1 a_2, b, j)$ is primitive of discriminant D, and it has the property that

$$(a_1 x_1^2 + b x_1 y_1 + c y_1^2)(a_2 x_2^2 + b x_2 y_2 + c y_2^2)$$
$$= a_1 a_2 (x_1 x_2 - j y_1 y_2)^2 + b(x_1 x_2 - j y_1 y_2)(a_1 x_1 y_2 + a_2 x_2 y_1 + b y_1 y_2)$$
$$+ j (a_1 x_1 y_2 + a_2 x_2 y_1 + b y_1 y_2)^2.$$

REMARKS. Consequently if an integer m is represented by the form (a_1, b, c_1) and an integer n is represented by the form (a_2, b, c_2), then mn is represented by the form $(a_1 a_2, b, j)$. For example we saw in an example with $D = -20$ immediately following the statement of Theorem 1.6 that any prime that is congruent to 3 or 7 modulo 20 is representable as $2x^2 + 2xy + 3y^2$. If we have two such primes p and q, then p is representable by $(2, 2, 3)$ and q is representable by $(3, 2, 2)$. The proposition is applicable with $j = 1$ and shows that pq is representable by $(6, 2, 1)$. In turn, substitution using $\begin{pmatrix} \alpha & \beta \\ \gamma & \delta \end{pmatrix} = \begin{pmatrix} 1 & 0 \\ -1 & 1 \end{pmatrix}$ changes this form to the properly equivalent form $(5, 0, 1)$. Thus pq is representable as $x^2 + 5y^2$.

PROOF. The form $(a_1 a_2, b, j)$ is primitive because any prime that divides $\mathrm{GCD}(a_1 a_2, b, j)$ has to divide either $\mathrm{GCD}(a_1, b, j)$ or $\mathrm{GCD}(a_2, b, j)$ and then certainly has to divide $\mathrm{GCD}(a_1, b, c_1)$ or $\mathrm{GCD}(a_2, b, c_2)$. No such prime exists, and hence $(a_1 a_2, b, j)$ is primitive. The discriminant of $(a_1 a_2, b, j)$ is $b^2 - 4 j a_1 a_2 = D + 4 a_1 c_1 - 4 j a_1 a_2 = D + 4 a_1 c_1 - 4(c_1 a_2^{-1}) a_1 a_2 = D$, as asserted, and the verification of the displayed identity is a routine computation. □

Let us say that two primitive forms (a_1, b_1, c_1) and (a_2, b_2, c_2) of the same nonsquare discriminant are **aligned** if $b_1 = b_2$ and if $j = c_1 a_2^{-1} = c_2 a_1^{-1}$ is an integer. In the presence of equal nonsquare discriminants D and the equal middle entries b, the rational number j is automatically an integer if $\mathrm{GCD}(a_1, a_2) = 1$. In fact, the equality $D - b^2 = -4 a_1 c_1 = -4 a_2 c_2$ shows that $D - b^2$ is divisible by $4 a_1$ and by $4 a_2$; since $\mathrm{GCD}(a_1, a_2) = 1$, $D - b^2$ is divisible by $4 a_1 a_2$, and the quotient $-j$ is an integer.

The idea is that each pair of classes of properly equivalent primitive forms of discriminant D has a pair of aligned representatives, and a multiplication of proper equivalence classes is well defined if the product is defined as the class of the composition of these aligned representatives in the sense of Proposition 1.9. This multiplication for proper equivalence classes will make the set of classes into a finite abelian group. This group will be defined as the "form class group" for the discriminant D, except that we use only the positive definite classes in the case that $D < 0$. Before phrasing these statements as a theorem, we make some remarks and then state and prove two lemmas.

Let (a, b, c) be a form of nonsquare discriminant D, and let b' be an integer with $b' \equiv b \bmod 2a$. In this case the number $c' = (b'^2 - D)/(4a)$ is an integer; in fact, we certainly have the congruences $b'^2 \equiv b^2 \bmod 2a$ and $b'^2 \equiv b^2 \bmod 4$, and thus we obtain the automatic[7] consequence $b'^2 \equiv b^2 \bmod 4a$, the rewritten congruence $b'^2 \equiv D + 4ac \bmod 4a$, and the desired result $b'^2 - D \equiv 0 \bmod 4a$. Hence (a, b', c') is another form of discriminant D. We call (a, b', c') a **translate** of (a, b, c). The key observation about translates is that the translate (a, b', c') is properly equivalent to (a, b, c). This fact follows from the computation

$$\begin{pmatrix} 1 & 0 \\ l & 1 \end{pmatrix} \begin{pmatrix} 2a & b \\ b & 2c \end{pmatrix} \begin{pmatrix} 1 & l \\ 0 & 1 \end{pmatrix} = \begin{pmatrix} 2a & b + 2al \\ b + 2al & 2(al^2 + bl + c) \end{pmatrix} = \begin{pmatrix} 2a & b' \\ b' & 2c' \end{pmatrix},$$

valid for any integer l.

Lemma 1.10. If (a, b, c) is a primitive form of nonsquare discriminant and if $m \neq 0$ is an integer, then (a, b, c) primitively represents some integer relatively prime to m.

PROOF. Let

$w_0 =$ product of all primes dividing a, c, and m,

$x_0 =$ product of all primes dividing a and m but not c,

$y_0 =$ product of all primes dividing m but not a.

Referring to the definitions, we see that any prime dividing m divides exactly one of w_0, x_0, and y_0. In particular, $\gcd(x_0, y_0) = 1$. We shall show that $\gcd(m, ax_0^2 + bx_0 y_0 + cy_0^2) = 1$, and the proof will be complete. Arguing by contradiction, suppose that a prime p divides $\gcd(m, ax_0^2 + bx_0 y_0 + cy_0^2)$. There are three cases for p, as follows.

Case 1. If p divides x_0, then the fact that p divides $ax_0^2 + bx_0 y_0 + cy_0^2$ implies that p divides cy_0^2. Since p does not divide y_0, p divides c, in contradiction to the definition of x_0.

Case 2. If p divides y_0, then similarly p divides ax_0^2. Since p does not divide x_0, p divides a, in contradiction to the definition of y_0.

Case 3. If p divides w_0, then the fact that p divides a and c implies that p divides $bx_0 y_0$. Since p divides neither x_0 nor y_0, p divides b, in contradiction to the fact that (a, b, c) is primitive. □

---

[7]The argument being used here—that a congruence modulo $2a$ implies the congruence of the squares modulo $4a$—will be used again later in this section without detailed comment.

4. Composition of Forms, Class Group

Lemma 1.11. Suppose that (a_1, b, c_1) and (a_2, b, c_2) are properly equivalent forms of nonsquare discriminant. If l is an integer such that $\mathrm{GCD}(a_1, a_2, l) = 1$ and such that l divides $\mathrm{GCD}(c_1, c_2)$, then $(la_1, b, l^{-1}c_1)$ and $(la_2, b, l^{-1}c_2)$ are properly equivalent forms.

REMARK. Even if (a_1, b, c_1) and (a_2, b, c_2) are primitive, it does not follow that $(la_1, b, l^{-1}c_1)$ and $(la_2, b, l^{-1}c_2)$ are primitive. In fact, one need only take $l = 2$ and $(a_1, b, c_1) = (a_2, b, c_2) = (1, 2, 4)$.

PROOF. Since (a_1, b, c_1) and (a_2, b, c_2) are properly equivalent, there exists $\begin{pmatrix} \alpha & \beta \\ \gamma & \delta \end{pmatrix}$ with

$$\begin{pmatrix} \alpha & \gamma \\ \beta & \delta \end{pmatrix} \begin{pmatrix} 2a_1 & b \\ b & 2c_1 \end{pmatrix} \begin{pmatrix} \alpha & \beta \\ \gamma & \delta \end{pmatrix} = \begin{pmatrix} 2a_2 & b \\ b & 2c_2 \end{pmatrix}.$$

We multiply both sides on the right by $\begin{pmatrix} \alpha & \beta \\ \gamma & \delta \end{pmatrix}^{-1}$, and the result is the system of four scalar equations

$$2a_1\alpha + b\gamma = 2a_2\delta - b\gamma,$$
$$2a_1\beta + b\delta = b\delta - 2c_2\gamma,$$
$$b\alpha + 2c_1\gamma = -2a_2\beta + b\alpha,$$
$$b\beta + 2c_1\delta = -b\beta + 2c_2\alpha.$$

The second and third equations simplify to $a_1\beta + c_2\gamma = 0$ and $a_2\beta + c_1\gamma = 0$. Since l divides c_1 and c_2, these two simplified equations show that l divides $a_1\beta$ and $a_2\beta$. Since $\mathrm{GCD}(a_1, a_2, l) = 1$, it follows that l divides β.

Therefore the matrix $\begin{pmatrix} \alpha & l^{-1}\beta \\ l\gamma & \delta \end{pmatrix}$ of determinant 1 has integer entries. Direct computation shows that

$$\begin{pmatrix} \alpha & l\gamma \\ l^{-1}\beta & \delta \end{pmatrix} \begin{pmatrix} 2la_1 & b \\ b & 2l^{-1}c_1 \end{pmatrix} \begin{pmatrix} \alpha & l^{-1}\beta \\ l\gamma & \delta \end{pmatrix} = \begin{pmatrix} 2la_2 & b \\ b & 2l^{-1}c_2 \end{pmatrix}.$$

Consequently the forms $(la_1, b, l^{-1}c_1)$ and $(la_2, b, l^{-1}c_2)$ are properly equivalent. □

Theorem 1.12. Let D be a nonsquare discriminant, and let $\mathcal{C}_1$ and $\mathcal{C}_2$ be proper equivalence classes of primitive forms of discriminant D.

(a) There exist aligned forms $(a_1, b, c_1) \in \mathcal{C}_1$ and $(a_2, b, c_2) \in \mathcal{C}_2$, and these may be chosen in such a way that a_1 and a_2 are relatively prime to each other and to any integer $m \neq 0$ given in advance.

(b) If the product of C_1 and C_2 is defined to be the proper equivalence class of the composition of any aligned representatives of C_1 and C_2, as for example the ones in (a), then the resulting product operation is well defined on proper equivalence classes of primitive forms of discriminant D.

(c) Under the product operation in (b), the set of proper equivalence classes of primitive forms of discriminant D is a finite abelian group. The identity is the class of $(1, 0, -D/4)$ if $D \equiv 0$ mod 4 and is the class of $(1, 1, -(D-1)/4)$ if $D \equiv 1$ mod 4. The group inverse of the class of (a, b, c) is the class of $(a, -b, c)$.

REMARK. When $D < 0$, the proper equivalence classes of positive definite forms are a subgroup. In fact, if (a_1, b, c_1) and (a_2, b, c_2) are positive definite and are aligned, then a_1 and a_2 are positive, and therefore their composition $(a_1 a_2, b, j)$ has $a_1 a_2$ positive and is positive definite. As was indicated in the discussion before Lemma 1.10, the **form class group** for discriminant D is defined to be the group in (c) if $D > 0$, and it is defined to be the subgroup of classes of positive definite forms if $D < 0$.

PROOF OF THEOREM 1.12a. By two applications of Lemma 1.10, C_1 primitively represents some integer a_1 prime to m, and C_2 primitively represents some integer a_2 prime to $a_1 m$. Arguing as in the last part of the proof of Theorem 1.6b, we may assume without loss of generality that $(x, y) = (1, 0)$ yields these values in each case. Then C_1 contains a form $(a_1, b_1, *)$ for some b_1, and C_2 contains a form $(a_2, b_2, *)$ for some b_2. By the remarks before Lemma 1.10, C_1 contains every translate $(a_1, b_1 + 2a_1 l_1, *)$, and C_2 contains every translate $(a_2, b_2 + 2a_2 l_2, *)$.

Let us make specific choices of l_1 and l_2. We know that $b_1 \equiv D \equiv b_2$ mod 2, so that $b_2 - b_1$ is even. The construction of a_1 and a_2 was arranged to make $\mathrm{GCD}(a_1, a_2) = 1$, and therefore $\mathrm{GCD}(2a_1, 2a_2) = 2$. Since $b_2 - b_1$ is even, we can choose l_1 and l_2 such that $2a_1 l_1 - 2a_2 l_2 = b_2 - b_1$. Then $b_1 + 2a_1 l_1 = b_2 + 2a_2 l_2$, and we take the common value as b.

For this b, C_1 contains the form $(a_1, b, *)$, and C_2 contains the form $(a_2, b, *)$. Since we have arranged that $\mathrm{GCD}(a_1, a_2) = 1$, the remark immediately following the definition of "aligned" shows that these forms are aligned. □

PROOF OF THEOREM 1.12b. Suppose that

$(a_1', b', *)$ is properly equivalent to $(a_1'', b'', *)$,
$(a_2', b', *)$ is properly equivalent to $(a_2'', b'', *)$,

with the vertical pairs aligned. We are to show that

$(a_1' a_2', b', *)$ is properly equivalent to $(a_1'' a_2'', b'', *)$.   (*)

Theorem 1.12a applied to the integer $m = a_1' a_2' a_1'' a_2''$ gives us an aligned pair of forms $(a_1, b, *)$ and $(a_2, b, *)$ in the respective proper equivalence classes such that $\text{GCD}(a_1, a_2) = 1$ and $\text{GCD}(a_1 a_2, m) = 1$. If we can show that

$$(a_1' a_2', b', *) \quad \text{is properly equivalent to} \quad (a_1 a_2, b, *), \tag{**}$$

then we will have symmetrically that

$$(a_1'' a_2'', b'', *) \quad \text{is properly equivalent to} \quad (a_1 a_2, b, *),$$

and (*) will follow from this fact and (**) by transitivity of proper equivalence.

We can now argue as in the proof of Theorem 1.12a. We know that $b \equiv D \equiv b' \bmod 2$, so that $b' - b$ is even. The construction of a_1 and a_2 was arranged to make $\text{GCD}(a_1 a_2, a_1' a_2') = 1$, and therefore $\text{GCD}(2a_1 a_2, 2a_1' a_2') = 2$. Since $b_2 - b_1$ is even, we can choose l and l' such that $2a_1 a_2 l - 2a_1' a_2' l' = b' - b$. Then $b + 2a_1 a_2 l = b' + 2a_1' a_2' l'$, and we take the common value as B. This B has

$$B \equiv b \bmod 2a_1 a_2 \quad \text{and} \quad B \equiv b' \bmod 2a_1' a_2'.$$

Thus

$$\begin{aligned}
(a_1, b, *) &\quad \text{is properly equivalent to} \quad (a_1, B, *), \\
(a_2, b, *) &\quad \text{is properly equivalent to} \quad (a_2, B, *), \\
(a_1 a_2, b, *) &\quad \text{is properly equivalent to} \quad (a_1 a_2, B, *),
\end{aligned} \tag{†}$$

and similarly

$$\begin{aligned}
(a_1', b', *) &\quad \text{is properly equivalent to} \quad (a_1', B, *), \\
(a_2', b', *) &\quad \text{is properly equivalent to} \quad (a_2', B, *), \\
(a_1' a_2', b', *) &\quad \text{is properly equivalent to} \quad (a_1' a_2', B, *).
\end{aligned} \tag{††}$$

By construction of b, $(a_1, b, *)$ is properly equivalent to $(a_1', b', *)$. This equivalence, in combination with the first line of (†) and the first line of (††), shows that

$$(a_1, B, *) \quad \text{is properly equivalent to} \quad (a_1', B, *). \tag{‡}$$

Let us check that Lemma 1.11 is applicable to the two properly equivalent forms of (‡) and to the integer $l = a_2'$. In fact, $\text{GCD}(a_1, a_2, l) = 1$ follows from $\text{GCD}(a_1 a_2, a_1' a_2') = 1$, and the problem is to show that $l = a_2'$ divides $(D - B^2)/(4a_1)$ and $(D - B^2)/(4a_1')$. To see this divisibility, we observe that $D - b'^2$ is divisible by $4a_1' a_2'$ because $(a_1', b', *)$ and $(a_2', b', *)$ are given as aligned; the congruence $b' \equiv B \bmod 2a_1' a_2'$ implies that $b'^2 \equiv B^2 \bmod 4a_1' a_2'$, and addition gives $D - B^2 \equiv 0 \bmod 4a_1' a_2'$. Meanwhile, $D - B^2$ is divisible by $4a_1$ because the third member of $(a_1, B, *)$ is an integer. Since $D - B^2$ is divisible also by $4a_1' a_2'$ and since $\text{GCD}(a_1, a_1' a_2') = 1$, $D - B^2$ is divisible by

$4a_1a_1'a_2'$. Therefore $(D-B^2)/(4a_1)$ and $(D-B^2)/(4a_1')$ are divisible by a_2', and Lemma 1.11 is indeed applicable.

The application of Lemma 1.11 to (‡) with $l = a_2'$ shows that

$$(a_1a_2', B, *) \quad \text{is properly equivalent to} \quad (a_1'a_2', B, *).$$

Similarly $(a_2, B, *)$ is properly equivalent to $(a_2', B, *)$, and an application of Lemma 1.11 to this equivalence with $l = a_1$ shows that

$$(a_1a_2, B, *) \quad \text{is properly equivalent to} \quad (a_1a_2', B, *).$$

The two results together show that

$$(a_1a_2, B, *) \quad \text{is properly equivalent to} \quad (a_1'a_2', B, *).$$

Combining this equivalence with the third line of (†) and the third line of (††), we obtain (∗∗), and the proof of (b) is complete. □

PROOF OF THEOREM 1.12c. The set of proper equivalence classes is finite by Theorem 1.6a, and commutativity of multiplication is clear. Define δ to be 0 if $D \equiv 0 \bmod 4$ and to be 1 if $D \equiv 1 \bmod 4$. Let us see that the class of $(1, \delta, *)$ is the identity. If (a, b, c) has discriminant D, then $b \equiv \delta \bmod 2$, and hence $(1, b, *) = (1, \delta + 2 \cdot 1 \cdot \frac{1}{2}(b - \delta))$ is a translate of $(1, \delta, *)$. Consequently $(1, b, *)$ and $(1, \delta, *)$ are properly equivalent. Since Proposition 1.9 shows that the composition of (a, b, c) and $(1, b, *)$ is $(a, b, *)$, Theorem 1.12b allows us to conclude that the class of $(1, \delta, *)$ is the identity.

For inverses Theorem 1.12b shows that the product of the classes of (a, b, c) and $(a, -b, c)$ is the product of the classes of (a, b, c) and (c, b, a), which is the class of the composition $(a, b, c)(c, b, a)$. Proposition 1.9 shows that this composition is $(ac, b, 1)$. Since $(ac, b, 1)$ is properly equivalent to $(1, -b, ac)$ and since the latter is properly equivalent to $(1, \delta, *)$, the class of the composition $(a, b, c)(c, b, a)$ is the identity.

To complete the proof, we need to verify associativity. Let $\mathcal{C}_1, \mathcal{C}_2$, and $\mathcal{C}_3$ be three proper equivalence classes of primitive forms of discriminant D. Let (a_1, b_1, c_1) be a form in the class $\mathcal{C}_1$. Lemma 1.10 shows that $\mathcal{C}_2$ represents an integer a_2 prime to a_1, and then it follows that the form (a_2, b_2, c_2) is in $\mathcal{C}_2$ for some integers b_2 and c_2. A second application of Lemma 1.10 shows that $\mathcal{C}_3$ represents an integer a_3 prime to a_1a_2, and then it follows that the form (a_3, b_3, c_3) is in $\mathcal{C}_3$ for some integers b_3 and c_3. The middle components have $b_1 \equiv b_2 \equiv b_3 \equiv \delta \bmod 2$, and thus $\frac{1}{2}(b_j - \delta)$ is an integer for $j = 1, 2, 3$. Since a_1, a_2, a_3 are relatively prime in pairs, the Chinese Remainder Theorem shows that the congruences $x \equiv \frac{1}{2}(b_j - \delta) \bmod a_j$ have a common integer solution x for $j = 1, 2, 3$. Define

$b = 2x + \delta$. Then b is a solution of $b \equiv b_j \bmod 2a_j$ for $j = 1, 2, 3$. Write $b = b_j + 2a_j n_j$ for suitable integers n_j. Then $(a_j, b, *) = (a_j, b_j + 2a_j n_j, *)$ is a translate of (a_j, b_j, c_j) and consequently is properly equivalent to it. Thus $(a_j, b, *)$ lies in $\mathcal{C}_j$. Taking into account Theorem 1.12b and using Proposition 1.9, we see that $\mathcal{C}_1(\mathcal{C}_2\mathcal{C}_3)$ and $(\mathcal{C}_1\mathcal{C}_2)\mathcal{C}_3$ are both represented by the form $(a_1 a_2 a_3, b, *)$ and hence are equal. $\square$

5. Genera

The theory of genera lumps proper equivalence classes of forms of a given discriminant according to their values in some way. There are at least two possible definitions of "genus," and it is a deep result that they lead to the same thing in all cases of interest. By way of background, we saw in Sections 2 and 3 for discriminant $D = -56$ that the number of proper equivalence classes of binary quadratic forms is exactly 4, representatives being $x^2 + 14y^2$, $2x^2 + 7y^2$, and $3x^2 \pm 2xy + 5y^2$. The last two are improperly equivalent and take the same values at integer points (x, y), and there are no other improper equivalences. Thus the first two take on a disjoint set of prime values from the values of $3x^2 \pm 2xy + 5y^2$ for integer points (x, y), and the sets of prime values taken on by $x^2 + 14y^2$ and $2x^2 + 7y^2$ at integer points are disjoint from one another.

Two possible lumpings of proper equivalence classes arise for this discriminant. One is to identify forms when their values modulo 56 include the same residues prime to 56. It is just a finite computation to see that

$x^2 + 14y^2$ and $2x^2 + 7y^2$ take on the residues $1, 9, 15, 23, 25, 39$,

$3x^2 \pm 2xy + 5y^2$ take on the residues $3, 5, 13, 19, 27, 45$.

Thus the first kind of lumping treats $x^2 + 14y^2$ and $2x^2 + 7y^2$ together because of the residues they take on, and it treats $3x^2 + 2xy + 5y^2$ and $3x^2 - 2xy + 5y^2$ together. Gauss proceeded by using this kind of lumping to define "genus."

The other lumping is to identify integer forms that take on the same *rational* values at rational points. Here $2x^2 + 7y^2 = 1$ for $(x, y) = (\frac{1}{3}, \frac{1}{3})$, and of course $x^2 + 14y^2 = 1$ for $(x, y) = (1, 0)$. Hence the sets of values of $x^2 + 14y^2$ and $2x^2 + 7y^2$ for x and y rational have a nonzero value in common. Lemma 1.13 below implies that the sets of rational values taken on by the two forms are identical. The second kind of lumping treats $x^2 + 14y^2$ and $2x^2 + 7y^2$ together because they take on the same rational values. We shall use this latter kind of lumping because, as Theorem 1.14 below shows, this is the definition that more quickly identifies the genus group once the form class group is known.

Problems 25–40 at the end of the chapter show that the two definitions of genus lead to the same thing for discriminants that are "fundamental" in a sense that we define in a moment.

We have defined two forms (a, b, c) and (a', b', c') with integer entries to be "properly equivalent" if there is a matrix $\begin{pmatrix} \alpha & \beta \\ \gamma & \delta \end{pmatrix}$ in SL(2, $\mathbb{Z}$) with

$$\begin{pmatrix} \alpha & \gamma \\ \beta & \delta \end{pmatrix} \begin{pmatrix} 2a & b \\ b & 2c \end{pmatrix} \begin{pmatrix} \alpha & \beta \\ \gamma & \delta \end{pmatrix} = \begin{pmatrix} 2a' & b' \\ b' & 2c' \end{pmatrix}.$$

We say that two forms (a, b, c) and (a', b', c') with rational entries are **properly equivalent over** $\mathbb{Q}$ if there is a matrix $\begin{pmatrix} \alpha & \beta \\ \gamma & \delta \end{pmatrix}$ in SL(2, $\mathbb{Q}$) such that the displayed equality holds. For emphasis we can refer to the original notion as "proper equivalence over $\mathbb{Z}$" when it is advisable to be more specific. It is evident that if two forms with rational entries are properly equivalent over $\mathbb{Q}$, then their sets of values at points (x, y) in $\mathbb{Q} \times \mathbb{Q}$ are the same.

Lemma 1.13. If (a, b, c) is a form with rational coefficients and with non-square discriminant D that takes on a nonzero value $q \in \mathbb{Q}$ for some (x_0, y_0) in $\mathbb{Q} \times \mathbb{Q}$, then (a, b, c) is properly equivalent over $\mathbb{Q}$ to $(q, 0, -D/(4q))$. Consequently two forms over $\mathbb{Q}$ of the same discriminant that take on a nonzero value in common over $\mathbb{Q}$ are properly equivalent over $\mathbb{Q}$.

PROOF. Suppose that $ax_0^2 + bx_0 y_0 + cy_0^2 = q$. Put $\begin{pmatrix} \alpha \\ \gamma \end{pmatrix} = \begin{pmatrix} x_0 \\ y_0 \end{pmatrix}$. Since x_0 and y_0 cannot both be 0, we can choose rationals β and δ such that $\alpha\delta - \beta\gamma = 1$. Then $\begin{pmatrix} \alpha & \beta \\ \gamma & \delta \end{pmatrix}$ has determinant 1 and satisfies $\begin{pmatrix} \alpha & \beta \\ \gamma & \delta \end{pmatrix} \begin{pmatrix} 1 \\ 0 \end{pmatrix} = \begin{pmatrix} x_0 \\ y_0 \end{pmatrix}$. The equality $ax_0^2 + bx_0 y_0 + cy_0^2 = \frac{1}{2}(x_0 \ y_0) \begin{pmatrix} 2a & b \\ b & 2c \end{pmatrix} \begin{pmatrix} x_0 \\ y_0 \end{pmatrix}$ therefore yields

$$q = \tfrac{1}{2}(1 \ 0) \begin{pmatrix} \alpha & \gamma \\ \beta & \delta \end{pmatrix} \begin{pmatrix} 2a & b \\ b & 2c \end{pmatrix} \begin{pmatrix} \alpha & \beta \\ \gamma & \delta \end{pmatrix} \begin{pmatrix} 1 \\ 0 \end{pmatrix}.$$

It follows that (a, b, c) is properly equivalent over $\mathbb{Q}$ to some form (q, b', c') with b' and c' rational. Using a translation with a rational parameter, we see that (q, b', c') is properly equivalent over $\mathbb{Q}$ to a form $(q, 0, *)$. Inspection of the discriminant shows that this last form must be $(q, 0, -D/(4q))$. □

Two primitive integer forms having the same discriminant are said to be in the same **genus** (plural: *genera*) if they are properly equivalent over $\mathbb{Q}$. In view of Lemma 1.13 the condition is that they are primitive and take on a common nonzero value over $\mathbb{Q}$, or equivalently that they are primitive and take on the same set of values over $\mathbb{Q}$. Thus $x^2 + 14y^2$ and $2x^2 + 7y^2$ furnish an example of two forms in distinct classes that are in the same genus. Two primitive integer forms that are in the same proper equivalence class over $\mathbb{Z}$ are in the same genus. The genus of the class $\mathcal{C}$ will be denoted by $[\mathcal{C}]$. The identity class will be denoted by $\mathcal{E}$, and $P = [\mathcal{E}]$ is called the **principal genus**. If (a, b, c) is an integer form representing a class $\mathcal{C}$, then Theorem 1.12c shows that $(a, -b, c)$ represents $\mathcal{C}^{-1}$. On the other hand, $\mathcal{C}$ and $\mathcal{C}^{-1}$ take on the same values over $\mathbb{Z}$, as we see by replacing (x, y) by $(x, -y)$, and it follows that $[\mathcal{C}] = [\mathcal{C}^{-1}]$.

For the main theorem about genera, we shall introduce an extra hypothesis on the discriminant D. A nonsquare integer D will be said to be a **fundamental discriminant** if D is not divisible by the square of any odd prime and if when D is even, $D/4$ is congruent to 2 or 3 modulo 4. It will be seen later that this condition is equivalent to the requirement that D be the "field discriminant" of some quadratic number field. Examples of discriminants that are not fundamental are $D = -12, -44, -108$.

With this condition imposed on D, any integer form (a, b, c) of discriminant D is automatically primitive. In fact, no odd prime p can divide $\text{GCD}(a, b, c)$, since then p^2 would divide D. If 2 were to divide $\text{GCD}(a, b, c)$, then $(a/2, b/2, c/2)$ would be an integer form, and $D/4 = (b/2)^2 - 4(a/2)(c/2)$ would be an integer congruent to 1 or 4 modulo 4.

Theorem 1.14. For a fundamental discriminant D, the principal genus P of primitive integer forms[8] is a subgroup of the form class group H, and the cosets of P are the various genera. Thus the set G of genera is exactly the set of cosets H/P and inherits a group structure from class multiplication. The subgroup P coincides with the subgroup of squares in H, and consequently every nontrivial element of G has order 2.

REMARKS. The group G is called the **genus group** of discriminant D. The hypothesis that D is fundamental is needed only for the conclusion that every member of P is a square in H. Since every nontrivial element of G has order 2 when D is fundamental, application of the Fundamental Theorem of Finitely Generated Abelian Groups or use of vector-space theory over a 2-element field shows that G is the direct sum of cyclic groups of order 2; in particular, the order of G is a power of 2. Problems 25–29 at the end of the chapter show that the order of G is 2^g, where $g + 1$ is the number of distinct prime factors of D.

PROOF. Let $V(\mathcal{C})$ denote the set of $\mathbb{Q}$ values assumed by forms in the class $\mathcal{C}$ at points (x, y) in $\mathbb{Q} \times \mathbb{Q}$. If S and S' are two genera and if $\mathcal{C}$ is a class in S and $\mathcal{C}'$ is a class in S', we define $S \cdot S' = [\mathcal{C}\mathcal{C}']$.

---

[8]As usual, we exclude the negative definite classes in the discussion.

To see that this product operation is well defined on the set G of genera, let $\mathcal{C}''$ be in S' also. Then $V(\mathcal{C}') = V(\mathcal{C}'')$. If q is in $V(\mathcal{C})$ and q' is in $V(\mathcal{C}') = V(\mathcal{C}'')$, then the prescription for multiplying classes shows that qq' is in $V(\mathcal{CC}')$ and $V(\mathcal{CC}'')$. Hence $V(\mathcal{CC}') = V(\mathcal{CC}'')$, and $[\mathcal{CC}'] = [\mathcal{CC}'']$. Therefore multiplication of genera is well defined. Define a function $\varphi : H \to G$ by $\varphi(\mathcal{C}) = [\mathcal{C}]$. Then the computation

$$\varphi(\mathcal{CC}') = [\mathcal{CC}'] = [\mathcal{C}][\mathcal{C}'] = \varphi(\mathcal{C})\varphi(\mathcal{C}')$$

shows that φ is a homomorphism of H onto G. The kernel of φ is $[\mathcal{C}] = P$, which is therefore a subgroup, and the image of φ, which is the set G of genera with its product operation, has to be a group.

For any class $\mathcal{C}$, the equality $[\mathcal{C}] = [\mathcal{C}^{-1}]$ implies that $[\mathcal{C}^2] = [\mathcal{C}][\mathcal{C}] = [\mathcal{C}][\mathcal{C}^{-1}] = [\mathcal{CC}^{-1}] = [\mathcal{E}] = P$. Hence P contains all squares. Conversely let $\mathcal{C}$ be in P. Then $\mathcal{C}$ takes on the value 1 over $\mathbb{Q}$. If (a, b, c) is a form in the class $\mathcal{C}$, then there exist rationals r and s with $ar^2 + brs + cs^2 = 1$. Clearing fractions, we see that there exist integers x and y such that $ax^2 + bxy + cy^2 = n^2$ for some integer $n \neq 0$. Without loss of generality, we may assume that n is positive. Since (a, b, c) is primitive, a familiar argument allows us to make a substitution for which the value n^2 is taken on at $(x, y) = (1, 0)$. In other words, (a, b, c) is properly equivalent over $\mathbb{Z}$ to a form (n^2, b', c') for suitable integers b' and c'. The composition formula in Proposition 1.9 shows that the composition of $(n, b', c'n)$ with itself is (n^2, b', c'), and hence $\mathcal{C}$ is exhibited as the square of the class of $(n, b', c'n)$. Since $(n, b', c'n)$ has the same discriminant D as (n^2, b', c') and therefore as (a, b, c) and since D is fundamental, $(n, b', c'n)$ is primitive. Therefore $\mathcal{C}$ is the square of a class of primitive forms. If $\mathcal{C}$ is positive definite, then the above choice of the sign of n as positive makes $(n, b', c'n)$ positive definite. Hence the class of $(n, b', c'n)$ is in H. $\square$

EXAMPLE. The discriminant $D = -56$ is fundamental, and we have seen that the form class group is of order 4 with representatives $x^2 + 14y^2$, $2x^2 + 7y^2$, and $3x^2 \pm 2xy + 5y^2$. We have seen also that $x^2 + 14y^2$ and $2x^2 + 7y^2$ both lie in the principal genus P. A group of order 4 must be isomorphic to the cyclic group C_4 or to $C_2 \times C_2$. In the first case the subgroup of squares has order 2, and in the second case the subgroup of squares has order 1. Since we have already found two elements in P, P has order exactly 2. By the theorem we must be in the first case. Hence H is of type C_4, and the genus group G is of type C_2. It is possible to check directly that $3x^2 + 2xy + 5y^2$ has order 4 by making computations similar to those for Problem 4d at the end of the chapter.

6. Quadratic Number Fields and Their Units

In this section we review material about quadratic number fields that appears in various places in *Basic Algebra*, and we determine the units in the ring of integers of such a number field.

Quadratic number fields are extension fields K of $\mathbb{Q}$ with $[K : \mathbb{Q}] = 2$. Such a field is necessarily of the form $K = \mathbb{Q}(\sqrt{m})$, where m is a uniquely determined square-free integer not equal to 0 or 1. The set $\{1, \sqrt{m}\}$ is a vector-space basis of K over $\mathbb{Q}$.

The extension $K/\mathbb{Q}$ is a Galois extension, and the Galois group $\mathrm{Gal}(K/\mathbb{Q})$ of automorphisms of K fixing $\mathbb{Q}$ has two elements. We denote the nontrivial element of the Galois group by σ; its values on the members of the vector-space basis are $\sigma(1) = 1$ and $\sigma(\sqrt{m}) = -\sqrt{m}$.

The norm $N = N_{K/\mathbb{Q}}$ and trace $\mathrm{Tr} = \mathrm{Tr}_{K/\mathbb{Q}}$ are given by $N(\alpha) = \alpha \cdot \sigma(\alpha)$ and $\mathrm{Tr}(\alpha) = \alpha + \sigma(\alpha)$. Thus $N(a + b\sqrt{m}) = a^2 - mb^2$ and $\mathrm{Tr}(a + b\sqrt{m}) = 2a$. These values are members of $\mathbb{Q}$. The norm is multiplicative in the sense that $N(\alpha\beta) = N(\alpha)N(\beta)$, and $N(1) = 1$.

The ring R of algebraic integers in K is the integral closure of $\mathbb{Z}$ in K. It works out to be

$$R = \begin{cases} \mathbb{Z}[\sqrt{m}] & \text{if } m \equiv 2 \text{ or } 3 \bmod 4, \\ \mathbb{Z}[\tfrac{1}{2}(\sqrt{m} - 1)] & \text{if } m \equiv 1 \bmod 4 \end{cases}$$

and is therefore a free abelian group of rank 2. The automorphism σ carries R to itself. The norm and trace of any member of R are in $\mathbb{Z}$; conversely any member of K whose norm and trace are in $\mathbb{Z}$ is in R. We define the algebraic integer δ to be given by

$$\delta = \begin{cases} -\sqrt{m} & \text{if } m \equiv 2 \text{ or } 3 \bmod 4, \\ \tfrac{1}{2}(1 - \sqrt{m}) & \text{if } m \equiv 1 \bmod 4. \end{cases}$$

Then $\{1, \delta\}$ is a $\mathbb{Z}$ basis of R. The norm and trace of δ are given by

$$N(\delta) = \delta \cdot \sigma(\delta) = \begin{cases} -m & \text{if } m \equiv 2 \text{ or } 3 \bmod 4, \\ \tfrac{1}{4}(1 - m) & \text{if } m \equiv 1 \bmod 4, \end{cases}$$

$$\mathrm{Tr}(\delta) = \delta + \sigma(\delta) = \begin{cases} 0 & \text{if } m \equiv 2 \text{ or } 3 \bmod 4, \\ 1 & \text{if } m \equiv 1 \bmod 4. \end{cases}$$

There is a general notion of **field discriminant** D, or **absolute discriminant**, for an algebraic number field, whose definition will be given in Chapter V. We shall not give that definition in general now but will be content to give the formula for D in the quadratic number field $\mathbb{Q}(\sqrt{m})$, namely

$$D = \begin{cases} 4m & \text{if } m \equiv 2 \text{ or } 3 \bmod 4, \\ m & \text{if } m \equiv 1 \bmod 4. \end{cases}$$

The **units** of K are understood to be the members of the group $R^\times$ of units in the ring R. These are the members ε of R with $N(\varepsilon) = \pm 1$. In fact, if ε is a unit, then the equality $\varepsilon\varepsilon^{-1} = 1$ implies that $1 = N(1) = N(\varepsilon\varepsilon^{-1}) = N(\varepsilon)N(\varepsilon^{-1})$ and shows that $N(\varepsilon)$ is a unit in $\mathbb{Z}$. Thus $N(\varepsilon) = \pm 1$. Conversely if $N(\varepsilon) = \pm 1$, then $\pm \varepsilon \sigma(\varepsilon) = 1$ shows that $\sigma(\varepsilon) = \pm \varepsilon^{-1}$; since $\sigma(\varepsilon)$ is in R, ε is exhibited as in $R^\times$ and is therefore a unit.

For $m < 0$, the units of $\mathbb{Q}(\sqrt{m})$ are easily determined. In fact, if $\varepsilon = a + b\delta$ with a and b in $\mathbb{Z}$, then $N(\varepsilon) = (a + b\delta)(a + b\sigma(\delta)) = a^2 + b\,\mathrm{Tr}\,\delta + b^2 N(\delta)$ with each term equal to an integer and with the end terms ≥ 0. Sorting out the possibilities, we see that

$$R^\times = \begin{cases} \{\pm 1, \pm\sqrt{-1}\} & \text{if } m = -1, \\ \{\pm 1, \tfrac{1}{2}(\pm 1 \pm \sqrt{-3})\} & \text{if } m = -3, \\ \{\pm 1\} & \text{for all other } m < 0. \end{cases}$$

The respective orders of $R^\times$ are 4, 6, and 2.

Determination of the units when $m > 0$ is more delicate. We require a lemma.

Lemma 1.15. If α is a real irrational number and if $N > 0$ is an integer, then there exist integers A and B with

$$|B\alpha - A| < \frac{1}{N} \qquad \text{and} \qquad 0 < B \leq N.$$

For this A and this B,

$$\left|\alpha - \frac{A}{B}\right| < \frac{1}{B^2}.$$

PROOF. Put $\alpha_n = n\alpha - [n\alpha]$, where $[\cdot]$ denotes the greatest-integer function. Then $0 \leq \alpha_n < 1$. We partition the half-open interval $[0, 1)$ into N subintervals $\left[\frac{t-1}{N}, \frac{t}{N}\right)$ with $1 \leq t \leq N$. For $0 \leq n \leq N$, the expression α_n takes on $N+1$ distinct values because $\alpha_n = \alpha_m$ would imply that $(n-m)\alpha$ is in $\mathbb{Z}$. Hence there exist α_n and α_m with $n > m$ that lie in the same subinterval $\left[\frac{t-1}{N}, \frac{t}{N}\right)$. Then $|\alpha_n - \alpha_m| < \frac{1}{N}$. If we take $B = n - m$ and $A = [n\alpha] - [m\alpha]$, then $|B\alpha - A| = |\alpha_n - \alpha_m|$, and the inequality $|B\alpha - A| < \frac{1}{N}$ follows. Dividing this inequality by B gives $|\alpha - \frac{A}{B}| < \frac{1}{BN}$, and this is $\leq \frac{1}{B^2}$ because $N \geq B$. $\square$

Proposition 1.16. For $K = \mathbb{Q}(\sqrt{m})$ with $m > 0$, the units are the members of the infinite group

$$R^\times = \{(\pm 1)\varepsilon_1^n \mid n \in \mathbb{Z}\} \cong \mathbb{Z} \times C_2,$$

where ε_1 is the **fundamental unit**, defined as the least unit > 1.

6. Quadratic Number Fields and Their Units

REMARK. For example, when $m = 2$, the fundamental unit is $\varepsilon_1 = 1 + \sqrt{2}$.

PROOF. The units ω with $|\omega| = 1$ are ± 1, since the members of K are real numbers. We shall show shortly that there exists a unit ω with $|\omega| \neq 1$. Then ω or ω^{-1} has absolute value > 1. Let us say that $|\omega| > 1$. Then one of ω and $-\omega$ is > 1. Let us say that $\omega > 1$. Write $\omega = a + b\sqrt{m}$, so that $\sigma(\omega) = a - b\sqrt{m} = \pm \omega^{-1}$ has $|\sigma(\omega)| < 1$. Then

$$|2a| = |\omega + \sigma(\omega)| \leq |\omega| + |\sigma(\omega)| \leq |\omega| + 1$$

and

$$|2b\sqrt{m}| = |\omega - \sigma(\omega)| \leq |\omega| + |\sigma(\omega)| \leq |\omega| + 1$$

together show that there are only finitely many units ω' with $1 < |\omega'| < |\omega|$. Hence the existence of a unit ω with $|\omega| \neq 1$ implies the existence of a fundamental unit ε_1.

If ω' is any unit > 1, then we can choose a power ε_1^n of ε_1 with $\varepsilon_1^{n+1} > \omega' \geq \varepsilon_1^n$, by the archimedean property of $\mathbb{R}$. Then $\omega' \varepsilon_1^{-n}$ is a unit ≥ 1 with $|\omega' \varepsilon_1^{-n}| < \varepsilon_1$. Since ε_1 is fundamental, $\omega' \varepsilon_1^{-n}$ is 1, and thus $\omega' = \varepsilon_1^n$. Then it follows that the group of units has the asserted form.

Thus we need to exhibit some unit ω with $|\omega| \neq 1$. We apply Lemma 1.15 with $\alpha = \sqrt{m}$ and with N arbitrary. Then we obtain infinitely many pairs (A, B) of integers with $\left|\sqrt{m} - \frac{A}{B}\right| < \frac{1}{B^2} \leq 1$, hence with $|A/B| < 1 + \sqrt{m}$. For each such pair (A, B), the member $r = A - B\sqrt{m}$ of R has

$$|N(r)| = |(A + B\sqrt{m})(A - B\sqrt{m})| = \left|\frac{A}{B} - \sqrt{m}\right| |B^2| \left|\frac{A}{B} + \sqrt{m}\right|$$
$$\leq \frac{1}{B^2} B^2 (1 + 2\sqrt{m}) = 1 + 2\sqrt{m}.$$

Thus there are infinitely many r in R with $|N(r)| \leq 1 + 2\sqrt{m}$. Since the norm of an algebraic integer is in $\mathbb{Z}$, there is some integer n such that infinitely many $r \in R$ have $N(r) = n$. Among the elements $r \in R$ with $N(r) = n$, which we write as $r = A + B\sqrt{m}$ with A and B in $\frac{1}{2}\mathbb{Z}$, we consider the finitely many congruence classes of (A, B) modulo n, saying that two such (A, B) and (A', B') are congruent if $A - A'$ and $B - B'$ are integers divisible by n. Since infinitely many $r \in R$ have $N(r) = n$, there must be infinitely many of these in some particular congruence class. Take three such, say α_1, α_2, and α_3. Then

$$N(\alpha_1) = N(\alpha_2) = N(\alpha_3) = n$$

with

$$\frac{\alpha_1 - \alpha_2}{n} \text{ in } R \quad \text{and} \quad \frac{\alpha_1 - \alpha_3}{n} \text{ in } R.$$

Since $n = N(\alpha_2) = \alpha_2 \sigma(\alpha_2)$, we see that

$$\frac{\alpha_1}{\alpha_2} = 1 + \left(\frac{\alpha_1 - \alpha_2}{n}\right) \sigma(\alpha_2).$$

Thus α_1/α_2 is exhibited as in R, and it has $N(\alpha_1/\alpha_2) = N(\alpha_1)/N(\alpha_2) = n/n = 1$. Hence α_1/α_2 is a unit different from $+1$. Arguing similarly with α_1/α_3, we see that α_1/α_3 is a unit different from $+1$ and not equal to α_1/α_2. Hence one of α_1/α_2 and α_1/α_3 is a unit whose absolute value is not 1. □

7. Relationship of Quadratic Forms to Ideals

We continue with K as the quadratic number field $\mathbb{Q}(\sqrt{m})$ and R as the ring of algebraic integers in K. Here $R = \mathbb{Z}[\delta]$, where $\delta = -\sqrt{m}$ if $m \equiv 2$ or $3 \bmod 4$ and $\delta = \frac{1}{2}(1 - \sqrt{m})$ if $m \equiv 1 \bmod 4$. Let D be the field discriminant of $\mathbb{Q}(\sqrt{m})$ as defined in Section 6.

The topic of this section is a relationship between nonzero ideals in R and binary quadratic forms with discriminant D. Binary quadratic forms with D as discriminant are automatically primitive.

The relationship is not a one-one correspondence of ideals to forms but a one-one correspondence of a certain kind of equivalence class of ideals to proper equivalence classes of forms. We saw in Theorem 1.12 that the latter collection has the structure of a finite abelian group, and we shall see in this section that the former collection has the natural structure of a finite abelian group as well. The correspondence is a group isomorphism, according to Theorem 1.20 below.

Consider nonzero ideals I in R. The first observation is that I is additively a free abelian group of rank 2. In fact, R itself is additively a free abelian group of rank 2, and the additive subgroup I has to be free abelian of rank ≤ 2. If r is a nonzero element in I, then $N(r) = r\sigma(r)$ is in I, and thus I contains a nonzero integer. If n is an integer in I, then $n\sqrt{m}$ is in I, and thus I contains a noninteger. Therefore I is a free abelian group of rank exactly 2, as asserted.

Certainly I can then be generated as an ideal by two elements, and our customary notation has been to write $I = (r_1, r_2)$ in this case. However, without an extra condition on them, the two ideal generators need not together be a $\mathbb{Z}$ basis for I because they need not generate all of I additively. It will be helpful to have separate notation when the generators are known to give a $\mathbb{Z}$ basis. Accordingly we shall write $I = \langle r_1, r_2 \rangle$ when r_1, r_2 give a $\mathbb{Z}$ basis of I. In this case it will be helpful also to regard the set $\{r_1, r_2\}$ as *ordered* with r_1 preceding r_2, and we shall often do so.

Now suppose that $I = \langle r_1, r_2 \rangle$ is a nonzero ideal, and consider the expression

$$r_1 \sigma(r_2) - \sigma(r_1) r_2 = \det \begin{pmatrix} r_1 & \sigma(r_1) \\ r_2 & \sigma(r_2) \end{pmatrix}.$$

If I is written in terms of a second ordered $\mathbb{Z}$ basis as $I = \langle s_1, s_2 \rangle$, then the two ordered bases are related by a matrix $\begin{pmatrix} \alpha & \beta \\ \gamma & \delta \end{pmatrix}$ in GL(2, $\mathbb{Z}$), the relationship being

$$\begin{pmatrix} r_1 \\ r_2 \end{pmatrix} = \begin{pmatrix} \alpha & \beta \\ \gamma & \delta \end{pmatrix} \begin{pmatrix} s_1 \\ s_2 \end{pmatrix}.$$

Hence

$$\begin{pmatrix} r_1 & \sigma(r_1) \\ r_2 & \sigma(r_2) \end{pmatrix} = \begin{pmatrix} \alpha & \beta \\ \gamma & \delta \end{pmatrix} \begin{pmatrix} s_1 & \sigma(s_1) \\ s_2 & \sigma(s_2) \end{pmatrix},$$

and therefore

$$\det \begin{pmatrix} r_1 & \sigma(r_1) \\ r_2 & \sigma(r_2) \end{pmatrix} = \pm \det \begin{pmatrix} s_1 & \sigma(s_1) \\ s_2 & \sigma(s_2) \end{pmatrix},$$

where ± 1 is the determinant of $\begin{pmatrix} \alpha & \beta \\ \gamma & \delta \end{pmatrix}$. Consequently the expression

$$N(I) = \frac{|r_1 \sigma(r_2) - \sigma(r_1) r_2|}{|\sqrt{D}|},$$

where D is the field discriminant of K, is independent of the choice of $\mathbb{Z}$ basis. It is called the **norm** of the ideal I. The factor of $\sqrt{D}$ in the denominator is a normalization factor that arranges for the norm of the ideal $I = R$ to be 1; in fact, we can write $R = \langle 1, \delta \rangle$ with δ as in the first paragraph of this section, and then

$$N(R) = \frac{|\sigma(\delta) - \delta|}{|\sqrt{D}|} = \begin{cases} \frac{|\sqrt{m}+\sqrt{m}|}{|\sqrt{4m}|} & \text{if } m \equiv 2 \text{ or } 3 \bmod 4 \\ \frac{|\frac{1}{2}(1+\sqrt{m})-\frac{1}{2}(1-\sqrt{m})|}{|\sqrt{m}|} & \text{if } m \equiv 1 \bmod 4 \end{cases} = 1.$$

Since the norm of an element of R is given by $N(r) = r\sigma(r)$, it is immediate from the definition that

$$N(rI) = |N(r)| N(I) \quad \text{for } r \in R.$$

Consequently the norm of the principal ideal (r) is given by

$$N((r)) = |N(r)| N(R) = |N(r)| 1 = |N(r)| \quad \text{for } r \in R.$$

Still with $I = \langle r_1, r_2 \rangle$, let us observe that

$$\sigma\big(r_1 \sigma(r_2) - \sigma(r_1) r_2\big) = -\big(r_1 \sigma(r_2) - \sigma(r_1) r_2\big).$$

It follows that

$$r_1 \sigma(r_2) - \sigma(r_1) r_2 \quad \text{is} \quad \begin{cases} \text{real} & \text{if } m > 0, \\ \text{imaginary} & \text{if } m < 0. \end{cases}$$

Since $r_1\sigma(r_2) - \sigma(r_1)r_2$ changes sign when r_1 and r_2 are interchanged, let us say that the expression $I = \langle r_1, r_2\rangle$ for I is **positively oriented** if $r_1\sigma(r_2) - \sigma(r_1)r_2$ is positive or positive imaginary,[9] **negatively oriented** if $r_1\sigma(r_2) - \sigma(r_1)r_2$ is negative or negative imaginary. If $I = \langle r_1, r_2\rangle$, then exactly one of the expressions $I = \langle r_1, r_2\rangle$ and $I = \langle r_2, r_1\rangle$ is positively oriented. The notion of orientation will be critical to setting up the correspondence between classes of ideals and classes of forms.

The set of nonzero ideals of R has a commutative associative multiplication that was introduced in *Basic Algebra*: if I and J are nonzero ideals, then IJ is defined to be the set of sums of products from the two ideals, the product IJ again being an ideal. Later in this section we shall recall some properties of this multiplication that were proved in *Basic Algebra*.

We define two equivalence relations on the set of nonzero ideals of I. We say that I and J are **equivalent** if there exist nonzero r and s in R with $(r)I = (s)J$. Here (r) and (s) are understood to be principal ideals. The ideals I and J are **strictly equivalent**, or **narrowly equivalent**, if equivalence occurs and if r and s can be chosen with $N(rs^{-1}) > 0$. Both relations are certainly reflexive and symmetric. To see transitivity, let $(r_1)I_1 = (r_2)I_2$ and $(s_2)I_2 = (s_3)I_3$. Then $(r_1s_2)I_1 = (r_2s_2)I_2 = (r_2s_3)I_3$, and I_1 is equivalent to I_3. If also $N(r_1r_2^{-1}) > 0$ and $N(s_2s_3^{-1}) > 0$, then the product $N((r_1s_2)(r_2s_3)^{-1})$ is positive, and I_1 is strictly equivalent to I_3. In other words, "equivalent" and "strictly equivalent" are equivalence relations.

The principal ideals form one full equivalence class under "equivalent." First of all, (r) is equivalent to (s) because $(s)(r) = (rs) = (r)(s)$. In the reverse direction, if I and (1) are equivalent, let $(r)I = (s)$. Then there exists $x \in I$ with $rx = s$. Hence sr^{-1} is in I, and $(sr^{-1}) \subseteq I$. In fact, equality holds: if y is in I, then the equality $ry = sz$ with z in R says that $y = (sr^{-1})z$, and y is in (sr^{-1}). In other words, $I = (sr^{-1})$.

In a sense, therefore, equivalence of ideals measures the extent to which nonprincipal ideals exist.

Multiplication is a class property of ideals relative to equivalence and to strict equivalence. In fact, if $(r)I = (r')I'$ and $(s)J = (s')J'$, then $(rs)IJ = (r's')I'J'$, and the assertion follows.

The theorem will be that multiplication of *strict* equivalence classes of ideals of R makes the set of such classes into an abelian group that is isomorphic to the finite abelian form class group of discriminant D. This result is not as beautiful as one might hope, since the identity class of ideals under strict equivalence need not match the set of all principal ideals. However, we can quantify the discrepancy. The relevant result is as follows.

---

[9]If $m < 0$, we adopt the convention that $\sqrt{m}$ is positive imaginary.

7. Relationship of Quadratic Forms to Ideals

Proposition 1.17. Equivalence and strict equivalence are the same for ideals of R if and only if either

(a) $m > 0$ and the fundamental unit ε_1 has $N(\varepsilon_1) = -1$ or
(b) $m < 0$.

In the contrary case when $m > 0$ and the fundamental unit ε_1 has $N(\varepsilon) = +1$, a nonzero principal ideal (r) is strictly equivalent to (1) if and only if $N(r) > 0$; in particular, the principal ideal $(\sqrt{m})$ is not strictly equivalent to (1).

REMARKS. When $m > 0$, there are examples with $N(\varepsilon_1) = +1$ and examples with $N(\varepsilon_1) = -1$. Specifically when $m = 2$, $\varepsilon_1 = 1 + \sqrt{2}$, and this has $N(\varepsilon_1) = -1$. When $m > 0$ and m has any odd prime divisor p with $p \equiv 3 \bmod 4$, then $N(\varepsilon_1) = +1$; in fact, otherwise $\varepsilon_1 = x + y\sqrt{m}$ would imply that $-1 = N(\varepsilon_1) = x^2 - my^2$ and therefore that $-1 \equiv x^2 \bmod p$, but this congruence has no solutions by Theorem 1.2a.

PROOF. Suppose that $m > 0$ and $N(\varepsilon_1) = -1$. If $(r)I = (s)J$ with $N(rs^{-1}) < 0$, then $(\varepsilon_1 r)I = (s)J$ with $N(\varepsilon_1 rs^{-1}) > 0$. Thus equivalence implies strict equivalence in this case.

Suppose that $m < 0$. Then all norms of nonzero elements are > 0. Hence $N(rs^{-1}) > 0$ is an empty condition, and equivalence implies strict equivalence.

Conversely suppose that $m > 0$ and $N(\varepsilon_1) = +1$. Proposition 1.16 shows that the most general unit is $\varepsilon = \pm \varepsilon_1^n$, and consequently $N(\varepsilon) = N(\pm 1)N(\varepsilon_1)^n = +1$ for every unit. The element $\sqrt{m}$ is in R, and $N(\sqrt{m}) = -m < 0$. We know that the principal ideals (1) and $(\sqrt{m})$ are equivalent. Arguing by contradiction, suppose that they are strictly equivalent. Then $(r) = (r)(1) = (s)(\sqrt{m}) = (s\sqrt{m})$ for some r and s with $N(rs^{-1}) > 0$. Since the principal ideals generated by r and $s\sqrt{m}$ are the same, these elements must be related by $r = \varepsilon s \sqrt{m}$ for some unit ε. Then $N(rs^{-1}) = N(\varepsilon \sqrt{m}) = N(\varepsilon)N(\sqrt{m}) = -m < 0$, contradiction. The proposition follows. □

Once we have introduced group structures on the set of equivalence classes of ideals and the set of strict equivalence classes of ideals, it follows that the map that carries a strict equivalence class to the equivalence class containing it is a group homomorphism onto. If either of the conditions (a) and (b) in Proposition 1.17 is satisfied, then this homomorphism is one-one. Otherwise its kernel consists of the two strict equivalence classes of principal ideals—those whose generator has positive norm and those whose generator has negative norm.

At this point we could establish that the set of strict equivalence classes of ideals is a finite abelian group. The finiteness of the set of strict equivalence classes could be established directly by a geometric argument we give in Chapter V, and the group structure could be derived from the group structure on the set of "fractional ideals" of K that were introduced in Problems 48–53 at the end of Chapter VIII of *Basic Algebra*.

Although we could proceed with proofs along these lines, it is instructive to proceed in a different way. Rather than give a stand-alone proof of the finiteness of the number of strict equivalence classes of ideals, we prefer to derive this finiteness as part of the correspondence with proper equivalence classes of binary quadratic forms, since the number of such classes of binary quadratic forms has already been proved to be finite in Theorem 1.6a. The group structure then readily follows from this finiteness and the fact that R is a Dedekind domain.

Let us pause for a moment, therefore, to use results we already know in order to show how the group structure on the set of strict equivalence classes follows once it is known that there are only finitely many such classes. We know from Theorems 8.54 and 8.55 of *Basic Algebra* that R is a Dedekind domain and that R has unique factorization for its nonzero ideals. In other words, in terms of the already-defined multiplication of ideals, each nonzero ideal I in R is of the form $I = \prod_{j=1}^{k} P_j^{n_j}$, where the P_j are distinct nonzero prime ideals, the n_j are positive integers, and k is ≥ 0; moreover, this product expansion is unique up to the order of the factors.

Lemma 1.18. Let $\mathcal{H}$ be the set of strict equivalence classes of nonzero ideals in R, with its inherited commutative associative multiplication. If $\mathcal{H}$ is finite, then $\mathcal{H}$ is a group under this multiplication.

REMARKS. The group $\mathcal{H}$ will be seen in Theorem 1.20 to be isomorphic to the form class group of D. The set of ordinary equivalence classes is a quotient and is called the **ideal class group** of K. It will be generalized in Chapter V.

PROOF. The identity element of $\mathcal{H}$ is the strict equivalence class of the ideal $R = (1)$, and we are to prove the existence of inverses. Thus let I be given. For the sequence of ideals $I, I^2, I^3, \ldots$, the finiteness of $\mathcal{H}$ shows that two of these ideals must be strictly equivalent. Suppose that I^k is equivalent to I^{k+l} for some $k > 0$ and $l > 0$. Then there exist nonzero principal ideals (r) and (s) such that $(r)I^k = (s)I^{k+l}$. The uniqueness of factorization of ideals implies that we can cancel I^k from both sides of this equality, thereby obtaining $(r) = (s)I^l$. Let us define an element t in R. If $N(rs^{-1}) > 0$, we take t to be 1. Otherwise m must be positive, and we let $t = \sqrt{m}$, so that $N(t) < 0$. In both cases we then have $(rt)(1) = (s)(t)I^l$ with $N(rts^{-1}) > 0$, and the ideal $(t)I^l$ is strictly equivalent to (1). Hence the strict equivalence class of $(t)I^{l-1}$ is an inverse to the strict equivalence class of I, and $\mathcal{H}$ is a group. $\square$

Now we define the mappings $\mathcal{F}$ and $\mathcal{I}$ that we shall use to establish the main result of this section. Let I be a nonzero ideal in R, and suppose that I is given by an expression $I = \langle r_1, r_2 \rangle$ that is positively oriented. We regard x and y as integer variables. To I, we associate the binary quadratic form

$$\mathcal{F}(I, r_1, r_2) = N(I)^{-1} N(r_1 x + r_2 y) = N(I)^{-1} (r_1 x + r_2 y)(\sigma(r_1) x + \sigma(r_2) y).$$

7. Relationship of Quadratic Forms to Ideals

The associated 2-by-2 matrix for this form is

$$\frac{1}{N(I)}\begin{pmatrix} 2r_1\sigma(r_1) & r_1\sigma(r_2)+r_2\sigma(r_1) \\ r_1\sigma(r_2)+r_2\sigma(r_1) & 2r_2\sigma(r_2) \end{pmatrix}$$

$$= \frac{1}{N(I)}\begin{pmatrix} r_1 & \sigma(r_1) \\ r_2 & \sigma(r_2) \end{pmatrix}\begin{pmatrix} \sigma(r_1) & \sigma(r_2) \\ r_1 & r_2 \end{pmatrix},$$

and the discriminant of the quadratic form is therefore

$$-\det\left[\frac{1}{N(I)}\begin{pmatrix} r_1 & \sigma(r_1) \\ r_2 & \sigma(r_2) \end{pmatrix}\begin{pmatrix} \sigma(r_1) & \sigma(r_2) \\ r_1 & r_2 \end{pmatrix}\right] = N(I)^{-2}\bigl(r_1\sigma(r_2) - \sigma(r_1)r_2\bigr)^2$$

$$= |D|\frac{\bigl(r_1\sigma(r_2) - \sigma(r_1)r_2\bigr)^2}{\bigl|r_1\sigma(r_2) - \sigma(r_1)r_2\bigr|^2}$$

$$= |D|(\operatorname{sgn} m) = D.$$

Thus we have associated a quadratic form $\mathcal{F}(I, r_1, r_2)$ of discriminant D to an ideal I when I is given by a positively oriented expression $I = \langle r_1, r_2\rangle$. If $m < 0$, this quadratic form is positive definite because the coefficient of x^2, namely $N(I)^{-1}r_1\sigma(r_1) = N(I)^{-1}N(r_1)$, is positive when $m < 0$.

In the reverse direction we associate to an arbitrary form (a, b, c) of discriminant D an ideal $I = \mathcal{I}(a, b, c)$ given by a positively oriented expression $\langle r_1, r_2\rangle$. To begin with, if b is an integer with $b \equiv D \bmod 2$, let us define b' to be $\tfrac{1}{2}b$ if $D \equiv 0 \bmod 4$ and to be $\tfrac{1}{2}(b-1)$ if $D \equiv 1 \bmod 4$; in other words, $b' = \tfrac{1}{2}(b - \operatorname{Tr}(\delta))$ in both cases. The definition of $\mathcal{I}$ is to be

$$\mathcal{I}(a,b,c) = \begin{cases} \langle a, b'+\delta\rangle & \text{if } a > 0, \\ \langle \delta a, \delta(b'+\delta)\rangle & \text{if } a < 0. \end{cases}$$

The right sides in the above display make sense as ideals if the angular brackets are replaced by parentheses. To see that the definitions make sense, we thus need to check that $(a, b' + \delta) = \langle a, b' + \delta\rangle$ for all a and that the orientations are positive. Lemma 1.19a below shows that $(a, b' + \delta) = \langle a, b' + \delta\rangle$ if it is proved that a divides $N(b' + \delta)$, and the computation that verifies this equality is

$$N(b'+\delta) = b'^2 + b'(\delta + \sigma(\delta)) + \delta\sigma(\delta)$$

$$= \begin{cases} b'^2 + b' + \tfrac{1}{4}(1-m) & \text{if } D \equiv 1 \bmod 4, \\ b'^2 - m & \text{if } D \equiv 0 \bmod 4, \end{cases}$$

$$= \begin{cases} \tfrac{1}{4}(b-1)^2 + \tfrac{1}{2}(b-1) + \tfrac{1}{4}(1-D) & \text{if } D \equiv 1 \bmod 4, \\ \tfrac{1}{4}b^2 - \tfrac{1}{4}D & \text{if } D \equiv 0 \bmod 4, \end{cases}$$

$$= \tfrac{1}{4}(b^2 - D)$$

$$= ac.$$

44   I. Transition to Modern Number Theory

From the definitions near the beginning of this section, the orientation of $\langle r_1, r_2 \rangle$ is given by the sign of $(\sqrt{m})^{-1}(r_1 \sigma(r_2) - \sigma(r_1) r_2)$. Thus

$$\text{orientation}\langle a, b' + \delta \rangle = \text{sgn}\left((\sqrt{m})^{-1} a(\sigma(\delta) - \delta)\right) = \text{sgn}\, a,$$

$$\text{orientation}\langle \delta a, \delta(b' + \delta) \rangle = \text{sgn}\left((\sqrt{m})^{-1}\left(\delta a \sigma(\delta b' + \delta^2) - \sigma(\delta) a \delta(b' + \delta)\right)\right)$$
$$= \text{sgn}\left((\sqrt{m})^{-1} N(\delta) a (\sigma(\delta) - \delta)\right) = -\text{sgn}\, a,$$

and the orientations are positive in both cases.

Lemma 1.19.

(a) If $a \neq 0$ and b' are integers such that a divides $N(b' + \delta)$ in $\mathbb{Z}$, then $(a, b' + \delta) = \langle a, b' + \delta \rangle$ in the sense that the free abelian subgroup of R generated by a and $b' + \delta$ coincides with the ideal generated by a and $b' + \delta$.

(b) If I is any nonzero ideal in R, then I is of the form $I = \langle a, r \rangle$ for some integer $a > 0$ and some r in R.

PROOF. For (a), we are to show that $I' = \mathbb{Z} a + \mathbb{Z}(b' + \delta)$ is closed under multiplication by the generators 1 and δ of R. Closure of I' under multiplication by 1 is evident, and the formula $\delta a = -b' a + a(b' + \delta)$ shows that $\delta(\mathbb{Z} a) \subseteq I'$. Addition of $\delta b'$ to the sum of the two formulas $\delta^2 = \delta(\delta + \sigma(\delta)) - \delta \sigma(\delta) = \delta \text{Tr}(\delta) - N(\delta)$ and $N(b' + \delta) = b'^2 + b' \text{Tr}(\delta) + N(\delta)$ yields

$$\delta(b' + \delta) = -N(b' + \delta) + (b' + \text{Tr}(\delta))(b' + \delta),$$

which shows that $\delta(b' + \delta) \subseteq I'$ because $N(b' + \delta)$ is by assumption an integer multiple of a.

For (b), we start from any $\mathbb{Z}$ basis $\{r_1, r_2\}$ of I, say with $r_1 = a_1 + b_1 \delta$ and $r_2 = a_2 + b_2 \delta$, and let $d = \text{GCD}(b_1, b_2)$. Choose integers n_1 and n_2 with $n_1 b_1 + n_2 b_2 = d$. Then $\text{GCD}(n_1, n_2) = 1$, and we can therefore find integers k_1 and k_2 with $\det \begin{pmatrix} k_1 & k_2 \\ n_1 & n_2 \end{pmatrix} = 1$. Consequently $\begin{pmatrix} s_1 \\ s_2 \end{pmatrix} = \begin{pmatrix} k_1 & k_2 \\ n_1 & n_2 \end{pmatrix} \begin{pmatrix} r_1 \\ r_2 \end{pmatrix}$ is a new $\mathbb{Z}$ basis of I of the form

$$s_1 = c_1 + kd\delta,$$
$$s_2 = c_2 + d\delta.$$

If we put $a = s_1 - k s_2$ and possibly replace a by its negative, then $\{a, s_2\}$ is a $\mathbb{Z}$ basis of I of the required form. □

Theorem 1.20. The set $\mathcal{H}$ of strict equivalence classes of nonzero ideals relative to the field $K = \mathbb{Q}(\sqrt{m})$ is a finite abelian group. Moreover, the mapping $\mathcal{F}$ that carries a positively oriented expression $I = \langle r_1, r_2 \rangle$ for a nonzero ideal of R to a binary quadratic form depends only on I, not the ordered $\mathbb{Z}$ basis, and descends to an isomorphism of the group $\mathcal{H}$ onto the form class group H for the discriminant D of the field K, i.e., the group of proper equivalence classes of binary quadratic forms of discriminant D, subject to the remark below. Moreover, the mapping $\mathcal{I}$ with domain all binary quadratic forms whose discriminant equals the field discriminant of K, sending such a form to a positively oriented expression for a nonzero ideal of R, descends to be defined from H to $\mathcal{H}$, and the descended map is the two-sided inverse of the isomorphism induced by $\mathcal{F}$.

REMARK. If $m < 0$, H is understood as usual to include only the classes of the positive definite forms.

PROOF. The proof proceeds in six steps.

Step 1. We show that the proper equivalence class of the quadratic form $\mathcal{F}(I, r_1, r_2)$ depends only on the ideal I, not the positively oriented expression $I = \langle r_1, r_2 \rangle$ for it. Thus the class of the form can be abbreviated as $\mathcal{F}(I)$.

Suppose that $I = \langle s_1, s_2 \rangle$ is another positively oriented expression for I. Then we can write $\begin{pmatrix} r_1 \\ r_2 \end{pmatrix} = \begin{pmatrix} \alpha & \beta \\ \gamma & \delta \end{pmatrix} \begin{pmatrix} s_1 \\ s_2 \end{pmatrix}$ for a matrix $\begin{pmatrix} \alpha & \beta \\ \gamma & \delta \end{pmatrix}$ in $\mathrm{GL}(2, \mathbb{Z})$, and we have seen that

$$\begin{pmatrix} r_1 & \sigma(r_1) \\ r_2 & \sigma(r_2) \end{pmatrix} = \begin{pmatrix} \alpha & \beta \\ \gamma & \delta \end{pmatrix} \begin{pmatrix} s_1 & \sigma(s_1) \\ s_2 & \sigma(s_2) \end{pmatrix}, \tag{*}$$

and that

$$\det \begin{pmatrix} r_1 & \sigma(r_1) \\ r_2 & \sigma(r_2) \end{pmatrix} = \pm \det \begin{pmatrix} s_1 & \sigma(s_1) \\ s_2 & \sigma(s_2) \end{pmatrix},$$

where ± 1 is the determinant of $\begin{pmatrix} \alpha & \beta \\ \gamma & \delta \end{pmatrix}$. Since both expressions $I = \langle r_1, r_2 \rangle$ and $I = \langle s_1, s_2 \rangle$ are positively oriented, it follows that the sign in the determinant equation is plus, hence that $\begin{pmatrix} \alpha & \beta \\ \gamma & \delta \end{pmatrix}$ is in $\mathrm{SL}(2, \mathbb{Z})$. Substituting from (*) into the formula for the matrix associated to the binary quadratic form $\mathcal{F}(I, r_1, r_2)$, we obtain the matrix

$$N(I)^{-1} \begin{pmatrix} \alpha & \beta \\ \gamma & \delta \end{pmatrix} \begin{pmatrix} s_1 & \sigma(s_1) \\ s_2 & \sigma(s_2) \end{pmatrix} \begin{pmatrix} \sigma(s_1) & \sigma(s_2) \\ s_1 & s_2 \end{pmatrix} \begin{pmatrix} \alpha & \gamma \\ \beta & \delta \end{pmatrix}. \tag{**}$$

The product of the coefficient $N(I)^{-1}$ and the middle two matrices is the matrix associated to the quadratic form $\mathcal{F}(I, s_1, s_2)$, and (**) therefore exhibits the two quadratic forms as properly equivalent.

Step 2. We show that the proper equivalence class $\mathcal{F}(I)$ does not change when we replace I by a strictly equivalent ideal.

Thus let $I = \langle r_1, r_2 \rangle$ and $J = \langle s_1, s_2 \rangle$ be expressions for I and J, and suppose that (r) and (s) are nonzero principal ideals such that $(r)I = (s)J$ and $N(s/r) > 0$. The formula

$$\det \begin{pmatrix} rr_1 & \sigma(rr_1) \\ rr_2 & \sigma(rr_2) \end{pmatrix} = r\sigma(r) \det \begin{pmatrix} r_1 & \sigma(r_1) \\ r_2 & \sigma(r_2) \end{pmatrix} = N(r) \det \begin{pmatrix} r_1 & \sigma(r_1) \\ r_2 & \sigma(r_2) \end{pmatrix}$$

shows that the expression $(r)I = \langle rr_1, rr_2 \rangle$ is positively oriented if $N(r) > 0$ and is negatively oriented if $N(r) < 0$. Similarly $(s)J = \langle ss_1, ss_2 \rangle$ is positively oriented if $N(s) > 0$ and is negatively oriented if $N(s) < 0$. Since $N(r/s) > 0$, $N(r)$ and $N(s)$ are both positive or both negative. Possibly replacing r and s by $r\sqrt{m}$ and $s\sqrt{m}$, we may assume that $N(r)$ and $N(s)$ are both positive. Then the matrix associated to the quadratic form $\mathcal{F}((r)I, rr_1, rr_2)$ is

$$N(rI)^{-1} \begin{pmatrix} rr_1 & \sigma(rr_1) \\ rr_2 & \sigma(rr_2) \end{pmatrix} \begin{pmatrix} \sigma(rr_1) & \sigma(rr_2) \\ rr_1 & rr_2 \end{pmatrix}$$

$$= N(rI)^{-1} \begin{pmatrix} r_1 & \sigma(r_1) \\ r_2 & \sigma(r_2) \end{pmatrix} \begin{pmatrix} r & 0 \\ 0 & \sigma(r) \end{pmatrix} \begin{pmatrix} \sigma(r) & 0 \\ 0 & r \end{pmatrix} \begin{pmatrix} \sigma(r_1) & \sigma(r_2) \\ r_1 & r_2 \end{pmatrix}$$

$$= N(rI)^{-1} N(r) \begin{pmatrix} r_1 & \sigma(r_1) \\ r_2 & \sigma(r_2) \end{pmatrix} \begin{pmatrix} \sigma(r_1) & \sigma(r_2) \\ r_1 & r_2 \end{pmatrix}$$

$$= |N(r)|^{-1} N(I)^{-1} N(r) \begin{pmatrix} r_1 & \sigma(r_1) \\ r_2 & \sigma(r_2) \end{pmatrix} \begin{pmatrix} \sigma(r_1) & \sigma(r_2) \\ r_1 & r_2 \end{pmatrix}$$

$$= N(I)^{-1} \begin{pmatrix} r_1 & \sigma(r_1) \\ r_2 & \sigma(r_2) \end{pmatrix} \begin{pmatrix} \sigma(r_1) & \sigma(r_2) \\ r_1 & r_2 \end{pmatrix},$$

while the matrix associated to $\mathcal{F}((s)J, ss_1, ss_2)$, by a similar computation, is

$$N(J)^{-1} \begin{pmatrix} s_1 & \sigma(s_1) \\ s_2 & \sigma(s_2) \end{pmatrix} \begin{pmatrix} \sigma(s_1) & \sigma(s_2) \\ s_1 & s_2 \end{pmatrix}.$$

Since $(r)I = (s)J$, Step 1 shows that $\mathcal{F}((r)I, rr_1, rr_2)$ is properly equivalent to $\mathcal{F}((s)J, ss_1, ss_2)$.

Step 3. We show that $\mathcal{I}(a, b, c)$ depends only on the proper equivalence class of the binary quadratic form (a, b, c).

Problem 37 at the end of Chapter VII of *Basic Algebra* shows that $SL(2, \mathbb{Z})$ is generated by $\alpha = \begin{pmatrix} 0 & -1 \\ 1 & 0 \end{pmatrix}$ and $\beta = \begin{pmatrix} 0 & 1 \\ -1 & -1 \end{pmatrix}$, hence by $\alpha\beta = \begin{pmatrix} 1 & 1 \\ 0 & 1 \end{pmatrix}$ and $\alpha^{-1} = \begin{pmatrix} 0 & 1 \\ -1 & 0 \end{pmatrix}$. Thus it is enough to handle $\alpha\beta$ and α^{-1}.

The operation of $\alpha\beta = \begin{pmatrix} 1 & 1 \\ 0 & 1 \end{pmatrix}$ on forms sends (a, b, c) into the translate $(a, b+2a, *)$. Define $b' = \frac{1}{2}(b - \text{Tr}(\delta))$ in the same way as when $\mathcal{I}$ was defined. If $a > 0$, then $\mathcal{I}(a, b, c) = (a, b'+\delta)$, and $\mathcal{I}(a, b+2a, *) = (a, (b+2a)'+\delta) = (a, b'+a+\delta)$; thus the two image ideals are the same. If $a < 0$, then the respective images are $(\delta)(a, b' + \delta)$ and $(\delta)(a, b' + a + \delta)$, and again the image ideals are the same.

To handle $\alpha^{-1} = \begin{pmatrix} 0 & 1 \\ -1 & 0 \end{pmatrix}$, we are to show that the ideals $\mathcal{I}(a, b, c)$ and $\mathcal{I}(c, -b, a)$ are strictly equivalent. We saw just after the definition of $\mathcal{I}$ that $N(b' + \delta) = ac$. There are four cases to the proof of the strict equivalence according to the signs of a and c. Let us use the symbol $\sim$ to denote "is strictly equivalent to."

7. Relationship of Quadratic Forms to Ideals

Suppose that $a > 0$ and $c > 0$, so that $N(b' + \delta) > 0$. Then

$$\mathcal{I}(a, b, c) = (a, b' + \delta) \sim (b' + \sigma(\delta))(a, b' + \delta) = (a(b' + \sigma(\delta)), N(b' + \delta))$$
$$= (a(b' + \sigma(\delta)), ac) = (a)(b' + \sigma(\delta), c)$$
$$\sim (c, b' + \sigma(\delta)) = (c, -b' - \sigma(\delta)) = (c, (-b)' + \delta),$$

the last equality holding because $b' + (-b)' = -\operatorname{Tr}\delta = -\delta - \sigma(\delta)$. The right side equals $\mathcal{I}(c, -b, a)$, and the strict equivalence is proved in this case.

Suppose that $a < 0$ and $c < 0$, so that $N(b' + \delta) > 0$. Then

$$\mathcal{I}(a, b, c) = (\delta)(a, b' + \delta) \sim (b' + \sigma(\delta))(\delta)(a, b' + \delta)$$
$$= (\delta)(a(b' + \sigma(\delta)), N(b' + \delta)) = (\delta)(a(b' + \sigma(\delta)), ac)$$
$$= (a)(\delta)(b' + \sigma(\delta), c) \sim (\delta)(c, b' + \sigma(\delta))$$
$$= (\delta)(c, -b' - \sigma(\delta)) = (\delta)(c, (-b)' + \delta) = \mathcal{I}(c, -b, a),$$

and the strict equivalence is proved in this case.

Suppose that $a > 0$ and $c < 0$, so that $N(b' + \delta) < 0$. Then $N(\delta)N(b' + \delta)$ is positive, and

$$\mathcal{I}(a, b, c) = (a, b' + \delta) \sim (\delta)(b' + \sigma(\delta))(a, b' + \delta)$$
$$= (\delta)(a(b' + \sigma(\delta)), N(b' + \delta)) = (\delta)(a(b' + \sigma(\delta)), ac)$$
$$= (a)(\delta)(b' + \sigma(\delta), c) \sim (\delta)(c, b' + \sigma(\delta)) = (\delta)(c, -b' - \sigma(\delta))$$
$$= (\delta)(c, (-b)' + \delta) = \mathcal{I}(c, -b, a),$$

and the strict equivalence is proved in this case.

Suppose that $a < 0$ and $c > 0$, so that $N(b' + \delta) < 0$. Then $N(\delta)^{-1}N(b' + \delta)$ is positive, and

$$\mathcal{I}(a, b, c) = (\delta)(a, b' + \delta) \sim (b' + \sigma(\delta))(a, b' + \delta)$$
$$= (a(b' + \sigma(\delta)), N(b' + \delta)) = (a(b' + \sigma(\delta)), ac) = (a)(b' + \sigma(\delta), c)$$
$$\sim (c, -b' - \sigma(\delta)) = (c, (-b)' + \delta) = \mathcal{I}(c, -b, a),$$

and the strict equivalence is proved in this case.

Step 4. We show that the mapping of the set H of proper equivalence classes of forms to itself induced by $\mathcal{F}\mathcal{I}$ is the identity.

Let the given form be (a, b, c). With b' defined to be $\frac{1}{2}(b - \operatorname{Tr}(\delta))$ as usual, we have seen that $N(b' + \delta) = ac$. Therefore a divides $N(b' + \delta)$, and Lemma 1.19a shows that $(a, b' + \delta) = \langle a, b' + \delta \rangle$ in the sense that the ideal generated by a and $b' + \delta$ matches the free abelian group generated by these two elements.

First suppose that $a > 0$. Then $\mathcal{I}(a, b, c) = (a, b' + \delta) = \langle a, b' + \delta \rangle$, and we know that this expression is positively oriented. Calculation gives

$$N(I) = |\sqrt{D}|^{-1} \left| \det \begin{pmatrix} a & a \\ b'+\delta & b'+\sigma(\delta) \end{pmatrix} \right|$$

$$= a|\sqrt{D}|^{-1}|\sigma(\delta) - \delta|$$

$$= a \times \begin{cases} |\sqrt{m}|/|\sqrt{m}| & \text{if } D \equiv 1 \bmod 4, \\ 2|\sqrt{m}|/|\sqrt{4m}| & \text{if } D \equiv 0 \bmod 4, \end{cases}$$

$$= a. \qquad (\dagger)$$

Therefore the quadratic form $\mathcal{FI}(a, b, c)$ is

$$N(I)^{-1}(ax + (b' + \delta)y)(ax + (b' + \sigma(\delta))y)$$

$$= a^{-1}\big(a^2 x^2 + a(2b' + (\delta + \sigma(\delta)))xy + N(b' + \delta)y^2\big)$$

$$= ax^2 + \big(2b' + \mathrm{Tr}(\delta)\big)xy + cy^2$$

$$= ax^2 + bxy + cy^2,$$

and we see that $\mathcal{FI}(a, b, c) = (a, b, c)$ when $a > 0$.

Next suppose that $a < 0$. Then $\mathcal{I}(a, b, c) = (\delta a, \delta(b' + \delta)) = \langle \delta a, \delta(b' + \delta) \rangle$, and we know that this expression is positively oriented. Since $a < 0$ cannot occur for $m < 0$, $N(\delta)$ is negative. Thus calculation gives

$$N(I) = N((\delta)(a, b' + \delta)) = N((\delta)(-a, b' + \delta)) = |N(\delta)|N((-a, b' + \delta))$$
$$= |N(\delta)||a| = N(\delta)a,$$

the next-to-last equality following from the calculation that gives $(\dagger)$. Therefore the quadratic form $\mathcal{FI}(a, b, c)$ is

$$N(I)^{-1}(a\delta x + (b' + \delta)\delta y)(a\sigma(\delta)x + (b' + \sigma(\delta))\sigma(\delta)y)$$

$$= N(I)^{-1}N(\delta)(ax + (b' + \delta)y)(ax + (b' + \sigma(\delta))y)$$

$$= a^{-1}\big(a^2 x^2 + a(2b' + (\delta + \sigma(\delta)))xy + N(b' + \delta)y^2\big)$$

$$= ax^2 + \big(2b' + \mathrm{Tr}(\delta)\big)xy + cy^2$$

$$= ax^2 + bxy + cy^2,$$

and we see that $\mathcal{FI}(a, b, c) = (a, b, c)$ when $a < 0$.

Step 5. We show that the mapping of the set $\mathcal{H}$ of strict equivalence classes of ideals to itself induced by $\mathcal{IF}$ is the identity. In view of Step 4, it follows that $\mathcal{F}$ and $\mathcal{I}$ are both one-one onto. Since Theorem 1.6a shows H to be finite, $\mathcal{H}$ has to be finite, and Lemma 1.18 shows that the multiplication on $\mathcal{H}$ makes $\mathcal{H}$ into an abelian group.

Let an ideal I be given, and apply Lemma 1.19b to write $I = \langle \tilde{a}, r \rangle$ with $\tilde{a} > 0$ an integer. The expression deciding orientation is $\tilde{a}\sigma(r) - \sigma(\tilde{a})r = \tilde{a}(\sigma(r) - r)$, and this is multiplied by -1 if r is replaced by $-r$. Possibly changing r to $-r$ in the expression for I, we may therefore assume that the expression $I = \langle \tilde{a}, r \rangle$ is positively oriented. Write $r = c + d\delta$. Then

$$\sigma(r) - r = d(\sigma(\delta) - \delta) = \begin{cases} 2d\sqrt{m} & \text{if } m \equiv 2 \text{ or } 3 \bmod 4 \\ d\sqrt{m} & \text{if } m \equiv 1 \bmod 4 \end{cases} = d\sqrt{D}.$$

The orientation of I is given by $\tilde{a}(\sigma(r) - r) = \tilde{a}d\sqrt{D}$, and we deduce that $d > 0$ and that

$$N(I) = |\sqrt{D}|^{-1}\tilde{a}|\sigma(r) - r| = \tilde{a}d.$$

The definition of $\mathcal{F}$ gives $\mathcal{F}(I, \tilde{a}, r) = N(I)^{-1} N(\tilde{a}x + ry)$, which is a quadratic form whose x^2 coefficient is $a = N(I)^{-1}\tilde{a}^2 = d^{-1}\tilde{a}$ and whose xy coefficient is

$$b = N(I)^{-1}\tilde{a}\operatorname{Tr}(r) = d^{-1}\operatorname{Tr}(r) = d^{-1}(2c + d\operatorname{Tr}(\delta)) = 2d^{-1}c + \operatorname{Tr}(\delta).$$

With b' defined as usual to be $b' = \frac{1}{2}(b - \operatorname{Tr}(\delta))$, we see that $b' = d^{-1}c$. Consequently $\mathcal{IF}(I, \tilde{a}, r) = (a, b' + \delta) = (d^{-1}\tilde{a}, d^{-1}c + \delta)$. The product of this ideal with (d) is $(\tilde{a}, c + d\delta) = (\tilde{a}, r) = I$, and thus $\mathcal{IF}(I, \tilde{a}, r)$ is strictly equivalent to I.

Step 6. We show that the mapping induced by $\mathcal{I}$ from the set H of proper equivalence classes of forms to the set $\mathcal{H}$ of strict equivalence classes of ideals respects the group operations in H and $\mathcal{H}$ and hence is an isomorphism.

Let two proper equivalence classes of forms with discriminant D be given, and use Theorem 1.12a to choose representatives (a, b, c) and $(\tilde{a}, b, \tilde{c})$ with $\operatorname{GCD}(a, \tilde{a}) = 1$. The composition of the forms is well defined and is $(a\tilde{a}, b, *)$ for a suitable third entry in $\mathbb{Z}$. Let b' be $\frac{1}{2}(b - \operatorname{Tr}(\delta))$ as usual. We divide matters into cases according to the signs of a and $\tilde{a}$.

Suppose that $a > 0$ and $\tilde{a} > 0$. The definition of $\mathcal{I}$ shows that the ideals corresponding to the three quadratic forms in question are

$$(a, b' + \delta), \quad (\tilde{a}, b' + \delta), \quad \text{and} \quad (a\tilde{a}, b' + \delta).$$

The product of the first two ideals is $\big(a\tilde{a}, a(b' + \delta), \tilde{a}(b' + \delta), (b' + \delta)^2\big)$, and we are to show that this equals $(a\tilde{a}, b' + \delta)$. In fact, the inclusion

$$\big(a\tilde{a}, a(b' + \delta), \tilde{a}(b' + \delta), (b' + \delta)^2\big) \subseteq (a\tilde{a}, b' + \delta)$$

is clear. For the reverse inclusion we use the fact that $\mathrm{GCD}(a,\widetilde{a}) = 1$ to write $k_1 a + k_2 \widetilde{a} = 1$ for suitable integers k_1 and k_2. Then we see that $b' + \delta = k_1(a(b'+\delta)) + k_2(\widetilde{a}(b'+\delta))$, and the reverse inclusion follows.

Suppose that a and $\widetilde{a}$ are of opposite sign. By symmetry we may assume that $a > 0$ and $\widetilde{a} < 0$. The three ideals are then

$$(a, b' + \delta), \quad (\widetilde{a}\delta, (b'+\delta)\delta), \quad \text{and} \quad (a\widetilde{a}\delta, (b'+\delta)\delta),$$

while the product of the first two ideals is $\big(a\widetilde{a}\delta, a(b'+\delta)\delta, \widetilde{a}(b'+\delta)\delta, (b'+\delta)^2\delta\big) = (\delta)\big(a\widetilde{a}, a(b'+\delta), \widetilde{a}(b'+\delta), (b'+\delta)^2\big)$. From the previous paragraph this last ideal equals $(\delta)(a\widetilde{a}, b'+\delta) = (a\widetilde{a}\delta, (b'+\delta)\delta)$, and we have the required match.

Suppose that $a < 0$ and $\widetilde{a} < 0$. This time the product ideal is given by $(a\delta, (b'+\delta)\delta)(\widetilde{a}\delta, (b'+\delta)\delta) = (\delta^2)\big(a\widetilde{a}, a(b'+\delta), \widetilde{a}(b'+\delta), (b'+\delta)^2\big) = (\delta^2)(a\widetilde{a}, b'+\delta)$, the second equality following from the computation in the paragraph for a and $\widetilde{a}$ both positive. The ideal $(\delta^2)(a\widetilde{a}, b'+\delta)$ is strictly equivalent to $(a\widetilde{a}, b'+\delta)$ because $N(\delta^2) = N(\delta)^2$ is positive. Thus we have the required match on the level of strict equivalence classes. We conclude that the mapping of H to $\mathcal{H}$ is a group isomorphism. □

8. Primes in the Progressions $4n + 1$ and $4n + 3$

This section is the first of three sections about Dirichlet's Theorem on primes in arithmetic progressions, whose statement is as follows.

Theorem 1.21 (Dirichlet's Theorem). *If m and b are relatively prime integers with $m > 0$, then there exist infinitely many primes of the form $km + b$ with k a positive integer.*

We begin with the earlier treatment of the arithmetic progressions $4n + 1$ and $4n + 3$ by Euler. In 1737 Euler made the stunning discovery of the formula

$$\sum_{n=1}^{\infty} \frac{1}{n^s} = \prod_{p \text{ prime}} \frac{1}{1 - p^{-s}},$$

valid for $s > 1$. Actually, the formula is valid for complex s with $\operatorname{Re} s > 1$, but Euler had not considered powers n^s with s complex by this time and did not need them for his purpose. Euler's formula is a consequence of unique factorization of integers. In fact, the product for $p \leq N$ is

$$\prod_{p \leq N} \frac{1}{1 - p^{-s}} = \prod_{p \leq N} \left(1 + \frac{1}{p^s} + \frac{1}{p^{2s}} + \cdots \right) = \sum_{\substack{n \text{ with} \\ \text{no prime} \\ \text{divisors} > N}} \frac{1}{n^s}.$$

Letting $N \to \infty$, we obtain the desired formula.

Built into the formula is the result of Euclid's that there are infinitely many primes, i.e., infinitely many primes in the arithmetic progression n. There are two ways to see this. In both cases one starts from the observation that the sum $\sum_{n=1}^{\infty} 1/n^s$ is $\geq \int_1^{\infty} (1/x^s)\,dx = 1/(s-1)$, from which it follows that the sum tends to infinity as s decreases to 1. In one case the argument continues with the observation that if there were only finitely many primes, then $\prod_{p \text{ prime}} \frac{1}{1-p^{-s}}$ would certainly have finite limit as s decreases to 1, and we arrive at a contradiction. In the other case the argument continues with the observation that the logarithm of $\frac{1}{1-p^{-s}}$ is comparable in size to $1/p^s$, hence that $\log \sum_{n=1}^{\infty} 1/n^s$ is comparable to $\sum_{p \text{ prime}} 1/p^s$. Since $\sum_{n=1}^{\infty} 1/n^s$ tends to infinity, $\sum_{p \text{ prime}} 1/p^s$ must tend to infinity, and we conclude that there are infinitely many primes. We shall return to this observation shortly in order to justify it more rigorously.[10]

Euclid's proof was much simpler: if there were only finitely many primes, then the sum of 1 and the product of all the primes would be divisible by none of the primes and would give a contradiction. The difficulty with Euclid's argument is that there is no apparent way to adapt it to treat primes of the form $4n + 1$. Euler's argument, by contrast, does adapt to treat primes $4n + 1$.

Before continuing, let us make rigorous the notion of comparing sizes of factors of an infinite product with terms of an infinite series. An infinite product $\prod_{n=1}^{\infty} c_n$ with $c_n \in \mathbb{C}$ and with no factor 0 is said to **converge** if the sequence of partial products converges to a finite limit and the limit is not 0. A necessary condition for convergence is that c_n tend to 1.

Proposition 1.22. If $|a_n| < 1$ for all n, then the following conditions are equivalent:

(a) $\prod_{n=1}^{\infty} (1 + |a_n|)$ converges,

(b) $\sum_{n=1}^{\infty} |a_n|$ converges,

(c) $\prod_{n=1}^{\infty} (1 - |a_n|)$ converges.

In this case, $\prod_{n=1}^{\infty} (1 + a_n)$ converges.

PROOF. Condition (c) is equivalent to

(c') $\prod_{n=1}^{\infty} (1 - |a_n|)^{-1}$ converges.

For each of (a), (b), and (c'), convergence is equivalent to boundedness above. Since

$$1 + \sum_{n=1}^{N} |a_n| \leq \prod_{n=1}^{N} (1 + |a_n|) \leq \prod_{n=1}^{N} \frac{1}{1-|a_n|},$$

---

[10] In fact, this argument is showing that $\sum 1/p$ diverges, which says something more than just that there are infinitely many primes.

we see that (c′) implies (a) and that (a) implies (b). To see that (b) implies (c′), we may assume, without loss of generality, that $|a_n| \leq \frac{1}{2}$ for all n. Since $|x| \leq \frac{1}{2}$ implies that

$$\log \tfrac{1}{1-x} \leq |x| \sup_{|t| \leq |x| \leq \frac{1}{2}} \left| \tfrac{d}{dt} \log \tfrac{1}{1-t} \right| = |x| \sup_{|t| \leq |x| \leq \frac{1}{2}} \left(\tfrac{1}{1-t} \right) \leq 2|x|,$$

we have

$$\log \left(\prod_{n=1}^{N} \tfrac{1}{1-|a_n|} \right) = \sum_{n=1}^{N} \log \left(\tfrac{1}{1-|a_n|} \right) \leq 2 \sum_{n=1}^{N} |a_n|.$$

Thus (b) implies (c′).

Now suppose that (a) holds. To prove that $\prod_{n=1}^{\infty}(1+a_n)$ converges, it is enough to show that $\prod_{n=M}^{N}(1+a_n)$ tends to 1 as M and N tend to ∞. In the expression

$$\left| \prod_{n=M}^{N} (1+a_n) - 1 \right|,$$

we expand out the product, move the absolute values in for each term, and reassemble the product. The result is the inequality

$$\left| \prod_{n=M}^{N} (1+a_n) - 1 \right| \leq \prod_{n=M}^{N} (1+|a_n|) - 1.$$

By (a), the right side tends to 0 as M and N tend to ∞. Therefore so does the left side. This proves the proposition. □

Using this proposition and its proof, we can give a more rigorous justification for the comparison of $\log \sum_{n=1}^{\infty} n^{-s}$ and $\sum_{p \text{ prime}} p^{-s}$ in Euler's argument. Anticipating the notation that Riemann was to use for the function a century later, we introduce

$$\zeta(s) = \sum_{n=1}^{\infty} \frac{1}{n^s},$$

at the moment just for real s with $s > 1$. (This function subsequently was named the **Riemann zeta function** and is defined and analytic for complex s with $\operatorname{Re} s > 1$. We postpone a more serious discussion of $\zeta(s)$ to Proposition 1.24 below.) We begin from the formula

$$\log \zeta(s) = \sum_{p \text{ prime}} \log \tfrac{1}{1-p^{-s}} = \sum_{p \text{ prime}} \left(\tfrac{1}{p^s} + \tfrac{1}{2p^{2s}} + \tfrac{1}{3p^{3s}} + \cdots \right).$$

Let us see that this expression equals

$$\sum_{p \text{ prime}} \tfrac{1}{p^s} + \text{bounded term} \qquad \text{as } s \downarrow 1.$$

Going over the second displayed line in the proof of Proposition 1.22, which applied when $|x| \leq \frac{1}{2}$, we have

$$\left|\log \tfrac{1}{1-x} - x\right| \leq |x| \sup_{|t| \leq |x| \leq \frac{1}{2}} \left|\tfrac{d}{dt}\left(\log \tfrac{1}{1-t} - t\right)\right|$$

$$= |x| \sup_{|t| \leq |x| \leq \frac{1}{2}} \left|\tfrac{1}{1-t} - 1\right| = |x| \sup_{|t| \leq |x| \leq \frac{1}{2}} \left|\tfrac{t}{1-t}\right| \leq 2|x|^2.$$

For $x = p^{-s}$ with $s > 1$, this inequality becomes

$$\left|\log \tfrac{1}{1-p^{-s}} - \tfrac{1}{p^s}\right| \leq 2p^{-2s}.$$

Consequently

$$\left|\log \zeta(s) - \sum_{p \text{ prime}} \tfrac{1}{p^s}\right| = \left|\sum_{p \text{ prime}} \left[\log \tfrac{1}{1-p^{-s}} - \tfrac{1}{p^s}\right]\right|$$

$$\leq \sum_{p \text{ prime}} \left|\log \tfrac{1}{1-p^{-s}} - \tfrac{1}{p^s}\right| \leq 2 \sum_{p \text{ prime}} p^{-2s}.$$

The right side is $\leq 2 \sum_{n=1}^{\infty} n^{-2}$ for all $s > 1$, and we arrive at the desired formula

$$\log \zeta(s) = \sum_{p \text{ prime}} \frac{1}{p^s} + \text{bounded term} \qquad \text{as } s \downarrow 1.$$

Since we know that $\log \zeta(s)$ increases without bound as s decreases to 1, we can immediately conclude that there are infinitely many primes in the arithmetic progression n.

With this argument well understood as a prototype, let us modify it to treat primes $4k + 1$ separately from primes $4k + 3$. Euler needed one further key idea to succeed. It is tempting to replace the sum over all primes of p^{-s} in the above argument by

$$\sum_{\substack{p \text{ prime},\\ p \equiv 1 \bmod 4}} \frac{1}{p^s} \qquad \text{or} \qquad \sum_{\substack{p \text{ prime},\\ p \equiv 3 \bmod 4}} \frac{1}{p^s},$$

trace backward, and see what happens. What happens is that the expansion of the corresponding product of $(1 - p^{-s})^{-1}$ as a sum does not yield anything very manageable. For example, with the first of the two sums, we are led to the logarithm of the series $\sum_{n=1}^{\infty} c(n) n^{-s}$, where $c(n)$ is 1 if n is a product of primes $4k + 1$ and is 0 otherwise, and we have no direct way of deciding whether this diverges or converges as s decreases to 1.

Euler's key additional idea was to work with the sum and difference of the displayed series, rather than the two terms separately, and then to recover the two displayed series at the end. Let us see what this idea accomplishes. Tracing backward in the derivation of the formula $\log \zeta(s) = \sum_{p \text{ prime}} p^{-s} + \text{bounded term}$, we want to obtain a series $\sum_{p \text{ prime}} a_p p^{-s}$ from the logarithm of a product $\prod_p (1 - a_p p^{-s})^{-1}$ and be able to recognize this product as equal to a manageable series $\sum_{n=1}^{\infty} b_n n^{-s}$. Guided by what happens for $\zeta(s)$, we can hope that b_n will be readily computable from the a_p's and the unique factorization of n. The relevant identities, which we shall verify below, are as follows:

$$\sum_{n \text{ odd}} \frac{1}{n^s} = \prod_{\substack{p \text{ prime,} \\ p \text{ odd}}} \frac{1}{1 - p^{-s}},$$

$$\sum_{n \text{ odd}} \frac{(-1)^{\frac{1}{2}(n-1)}}{n^s} = \left(\prod_{\substack{p \text{ prime,} \\ p=4k+1}} \frac{1}{1 - p^{-s}} \right) \left(\prod_{\substack{p \text{ prime,} \\ p=4k+3}} \frac{1}{1 + p^{-s}} \right).$$

In more detail let us write

$$\chi_0(n) = \begin{cases} 0 & \text{if } n \equiv 0 \mod 2, \\ 1 & \text{if } n \equiv 1 \mod 2, \end{cases}$$

$$\chi_1(n) = \begin{cases} 0 & \text{if } n \equiv 0 \mod 2, \\ 1 & \text{if } n \equiv 1 \mod 4, \\ -1 & \text{if } n \equiv 3 \mod 4. \end{cases}$$

With χ equal to χ_0 or χ_1, we have $\chi(mn) = \chi(m)\chi(n)$ for all m and n. Consequently the two expressions $\sum_{n \text{ odd}} \frac{1}{n^s}$ and $\sum_{n \text{ odd}} \frac{(-1)^{\frac{1}{2}(n-1)}}{n^s}$ are both of the form

$$L(s, \chi) = \sum_{n=1}^{\infty} \frac{\chi(n)}{n^s},$$

the function χ being χ_0 for the first series and being χ_1 for the second series. As we shall verify rigorously in the next section, the same argument via unique factorization that yields Euler's identity $\sum_{n=1}^{\infty} n^{-s} = \sum_{p \text{ prime}} \frac{1}{1-p^{-s}}$ gives a factorization

$$L(s, \chi) = \sum_{n=1}^{\infty} \frac{\chi(n)}{n^s} = \prod_{p \text{ prime}} \frac{1}{1 - \chi(p)p^{-s}}$$

because of the identity $\chi(mn) = \chi(m)\chi(n)$. Going over the argument that $\log \zeta(s)$ is the sum of $\sum_{p \text{ prime}} p^{-s}$ and a bounded term, we find that

$$\log L(s, \chi) = \sum_{p \text{ prime}} \frac{\chi(p)}{p^s} + g(s, \chi).$$

with $g(s, \chi)$ bounded as $s \downarrow 1$. The sum and difference for the two choices of $\chi(n)$ gives

$$\log(L(s, \chi_0)L(s, \chi_1)) = 2 \sum_{\substack{p \text{ prime} \\ p=4k+1}} \frac{1}{p^s} + (g(s, \chi_0) + g(s, \chi_1))$$

and

$$\log(L(s, \chi_0)L(s, \chi_1)^{-1}) = 2 \sum_{\substack{p \text{ prime} \\ p=4k+3}} \frac{1}{p^s} + (g(s, \chi_0) - g(s, \chi_1)).$$

The function $L(s, \chi_0)$ is the product of $\zeta(s)$ and an elementary factor. In fact, a change of index of summation in the formula defining $\zeta(s)$ gives $2^{-s}\zeta(s) = \sum_{n \text{ even}} n^{-s}$. Subtracting this formula from the definition of $\zeta(s)$ gives

$$L(s, \chi_0) = \sum_{n \text{ odd}} \frac{1}{n^s} = (1 - 2^{-s})\zeta(s).$$

Therefore

$$\lim_{s \downarrow 1} L(s, \chi_0) = +\infty.$$

Meanwhile, the series $L(s, \chi_1) = \sum_{n \text{ odd}} \frac{(-1)^{\frac{1}{2}(n-1)}}{n^s}$ is alternating and converges for $s > 0$ by the Leibniz test. The convergence is uniform on compact sets, and the sum $L(s, \chi_1)$ is continuous for $s > 0$. Grouping the terms of this series in pairs, we see that $L(1, \chi_1)$ is positive.[11] Hence we have

$$0 < \lim_{s \downarrow 1} L(s, \chi_1) < +\infty.$$

Putting together the two limit relations for $L(s, \chi_0)$ and $L(s, \chi_1)$ as s decreases to 1, we see that

$$\log\left(L(s, \chi_0)L(s, \chi_1)\right) \quad \text{and} \quad \log\left(L(s, \chi_0)L(s, \chi_1)^{-1}\right)$$

both tend to $+\infty$ as $s \downarrow 1$. Referring to the values computed above for these expressions and taking into account that $\sum 1/p$ exceeds $\sum 1/p^s$ when $s > 1$, we see that

$$\sum_{\substack{p \text{ prime} \\ p=4k+1}} \frac{1}{p} \quad \text{and} \quad \sum_{\substack{p \text{ prime} \\ p=4k+3}} \frac{1}{p}$$

---

[11] We can even recognize the value of $L(1, \chi_1)$ as $\pi/4$ from the Taylor series of arctan x, but the explicit value is not needed in the argument.

are both infinite. Hence there are infinitely many primes $4k + 1$, and there are infinitely many primes $4k + 3$.

The proof of the general case of Dirichlet's Theorem (Theorem 1.21) will proceed in similar fashion. We return to it in Section 10 after a brief but systematic investigation of the kinds of series and products that we have encountered in the present section.

9. Dirichlet Series and Euler Products

A series $\sum_{n=1}^{\infty} a_n n^{-s}$ with a_n and s complex is called a **Dirichlet series**. The first result below shows that the region of convergence and the region of absolute convergence for such a series are each right half-planes in $\mathbb{C}$ unless they are equal to the empty set or to all of $\mathbb{C}$. These half-planes may not be the same: for example, $\sum_{n=1}^{\infty} (-1)^n n^{-s}$ is convergent for $\operatorname{Re} s > 0$ and absolutely convergent for $\operatorname{Re} s > 1$.

Proposition 1.23. Let $\sum_{n=1}^{\infty} a_n n^{-s}$ be a Dirichlet series.

(a) If the series is convergent for $s = s_0$, then it is convergent uniformly on compact sets for $\operatorname{Re} s > \operatorname{Re} s_0$, and the sum of the series is analytic in this region.

(b) If the series is absolutely convergent for $s = s_0$, then it is uniformly absolutely convergent for $\operatorname{Re} s \geq \operatorname{Re} s_0$.

(c) If the series is convergent for $s = s_0$, then it is absolutely convergent for $\operatorname{Re} s > \operatorname{Re} s_0 + 1$.

(d) If the series is convergent at some s_0 and sums to 0 in a right half-plane, then all the coefficients are 0.

REMARK. The proof of (a) will use the **summation by parts** formula. Namely if $\{u_n\}$ and $\{v_n\}$ are sequences and if $U_n = \sum_{k=1}^{n} u_k$ for $n \geq 0$, then $1 \leq M \leq N$ implies

$$\sum_{n=M}^{N} u_n v_n = \sum_{n=M}^{N-1} U_n (v_n - v_{n+1}) + U_N v_N - U_{M-1} v_M. \qquad (*)$$

PROOF. For (a), we write $a_n n^{-s} = a_n n^{-s_0} \cdot n^{-(s-s_0)} = u_n v_n$ and then apply the summation by parts formula $(*)$. The given convergence means that the sequence $\{U_n\}$ is convergent, and certainly v_n tends to 0 uniformly on any proper half-plane of $\operatorname{Re} s > \operatorname{Re} s_0$. Thus the second and third terms on the right side of $(*)$ tend to 0 with the required uniformity as M and N tend to ∞. For the first term, the sequence $\{U_n\}$ is bounded, and we shall show that

$$\sum_{n=1}^{\infty} |v_n - v_{n+1}| = \sum_{n=1}^{\infty} \left| \frac{1}{n^{s-s_0}} - \frac{1}{(n+1)^{s-s_0}} \right|$$

9. Dirichlet Series and Euler Products   57

is convergent uniformly on compact sets for which $\operatorname{Re} s > \operatorname{Re} s_0$. Use of (*) and the Cauchy criterion will complete the proof of convergence. For $n \leq t \leq n+1$, we have

$$\left| n^{-(s-s_0)} - t^{-(s-s_0)} \right| \leq \sup_{n \leq t \leq n+1} \left| \tfrac{d}{dt}\left(n^{-(s-s_0)} - t^{-(s-s_0)}\right) \right|$$

$$= \sup_{n \leq t \leq n+1} \left| \tfrac{s-s_0}{t^{s-s_0+1}} \right| \leq \tfrac{|s-s_0|}{n^{1+\operatorname{Re}(s-s_0)}}.$$

Thus

$$|v_n - v_{n+1}| = \left| n^{-(s-s_0)} - (n+1)^{-(s-s_0)} \right| \leq \tfrac{|s-s_0|}{n^{1+\operatorname{Re}(s-s_0)}},$$

and $\sum_{n=1}^{\infty} |v_n - v_{n+1}|$ is uniformly convergent on compact sets with $\operatorname{Re} s > \operatorname{Re} s_0$, by the Weierstrass M-test. It follows that the given Dirichlet series is uniformly convergent on compact sets for which $\operatorname{Re} s > \operatorname{Re} s_0$. Since each term is analytic in this region, the sum is analytic.

For (b), we have

$$\left| \tfrac{a_n}{n^s} \right| = \left| \tfrac{a_n}{n^{s_0}} \right| \cdot \left| \tfrac{1}{n^{s-s_0}} \right| \leq \left| \tfrac{a_n}{n^{s_0}} \right|.$$

Since the sum of the right side is convergent, the desired uniform convergence follows from the Weierstrass M-test.

For (c), let $\epsilon > 0$ be given. Then

$$\left| \tfrac{a_n}{n^{s_0+1+\epsilon}} \right| = \left| \tfrac{a_n}{n^{s_0}} \right| n^{-(1+\epsilon)}$$

with the first factor on the right bounded and the second factor contributing to a finite sum. Therefore we have absolute convergence at $s_0 + 1 + \epsilon$, and (c) follows from (b).

For (d), we may assume by (c) that there is absolute convergence at s_0. Suppose that $a_1 = \cdots = a_{N-1} = 0$. By (b), $\sum_{n=N}^{\infty} a_n n^{-s} = 0$ for $\operatorname{Re} s > \operatorname{Re} s_0$. The series

$$\sum_{n=N}^{\infty} a_n (n/N)^{-s} \tag{**}$$

is by assumption absolutely convergent at s_0, and $\operatorname{Re} s > \operatorname{Re} s_0$ implies

$$\left| a_n (n/N)^{-s} \right| \leq \left| a_n (n/N)^{-s_0} \right|.$$

By dominated convergence we can take the limit of (**) term by term as $s \to +\infty$. The only term that survives is a_N. Since (**) has sum 0 for all s, we conclude that $a_N = 0$. This completes the proof. $\square$

Proposition 1.24. The Riemann zeta function $\zeta(s) = \sum_{n=1}^{\infty} n^{-s}$, initially defined and analytic for $\operatorname{Re} s > 1$, extends to be meromorphic for $\operatorname{Re} s > 0$. Its only pole is at $s = 1$, and the pole is simple.

REMARK. Actually, $\zeta(s)$ extends to be meromorphic in $\mathbb{C}$ with no additional poles, but we do not need this additional information.

PROOF. For $\operatorname{Re} s > 1$, we have

$$\tfrac{1}{s-1} = \int_1^{\infty} t^{-s}\,dt = \sum_{n=1}^{\infty} \int_n^{n+1} t^{-s}\,dt.$$

Thus $\operatorname{Re} s > 1$ implies

$$\zeta(s) = \tfrac{1}{s-1} + \sum_{n=1}^{\infty}\left(\tfrac{1}{n^s} - \int_n^{n+1} t^{-s}\,dt\right) = \tfrac{1}{s-1} + \sum_{n=1}^{\infty} \int_n^{n+1}(n^{-s} - t^{-s})\,dt.$$

It is enough to show that the series on the right side converges uniformly on compact sets for $\operatorname{Re} s > 0$. Thus suppose that $\operatorname{Re} s \geq \sigma > 0$ and $|s| \leq C$. The proof of Proposition 1.23a showed that $|n^{-s} - t^{-s}| \leq |s|\,n^{-(1+\operatorname{Re} s)}$. Hence

$$\left|\int_n^{n+1}(n^{-s} - t^{-s})\,dt\right| \leq \int_n^{n+1} |n^{-s} - t^{-s}|\,dt \leq |s|\,n^{-(1+\operatorname{Re} s)} \leq C n^{-(1+\sigma)}.$$

Since $\sum_{n=1}^{\infty} n^{-(1+\sigma)} < \infty$, the desired uniform convergence follows from the Weierstrass M-test. $\square$

Proposition 1.25. Let $Z(s) = \sum_{n=1}^{\infty} a_n n^{-s}$ be a Dirichlet series with all $a_n \geq 0$. Suppose that the series is convergent in some half-plane and that the sum extends to be analytic for $\operatorname{Re} s > 0$. Then the series converges for $\operatorname{Re} s > 0$.

PROOF. By assumption the series converges somewhere, and therefore $s_0 = \inf\{s \geq 0 \mid \sum_{n=1}^{\infty} a_n n^{-s} \text{ converges}\}$ is a well-defined real number ≥ 0. Arguing by contradiction, suppose that $s_0 > 0$. Since $\sum a_n n^{-s}$ converges uniformly on compact sets for $\operatorname{Re} s > s_0$ by Proposition 1.23a and since the terms of the series are analytic, we can compute the derivatives of the series term by term. Thus

$$Z^{(N)}(s_0 + 1) = \sum_{n=1}^{\infty} \frac{a_n(-\log n)^N}{n^{s_0+1}}. \qquad (*)$$

The Taylor series of $Z(s)$ about $s_0 + 1$ is

$$Z(s) = \sum_{N=0}^{\infty} \tfrac{1}{N!}(s - s_0 - 1)^N Z^{(N)}(s_0 + 1)$$

and is convergent at $s = \frac{1}{2}s_0$, since $Z(s)$ is analytic in the open disk centered at $s_0 + 1$ and having radius $s_0 + 1$. Thus

$$Z(\tfrac{1}{2}s_0) = \sum_{N=0}^{\infty} \tfrac{1}{N!}(1 + \tfrac{1}{2}s_0)^N (-1)^N Z^{(N)}(s_0 + 1),$$

with the series convergent. Substituting from (∗), we have

$$Z(\tfrac{1}{2}s_0) = \sum_{N=0}^{\infty} \sum_{n=1}^{\infty} \tfrac{a_n (\log n)^N}{N! n^{s_0+1}} (1 + \tfrac{1}{2}s_0)^N.$$

This is a series with terms ≥ 0, and Fubini's Theorem allows us to interchange the order of summation and obtain

$$Z(\tfrac{1}{2}s_0) = \sum_{n=1}^{\infty} \sum_{N=0}^{\infty} \tfrac{a_n}{n^{s_0+1}} \tfrac{(\log n)^N (1+\tfrac{1}{2}s_0)^N}{N!} = \sum_{n=1}^{\infty} \tfrac{a_n}{n^{s_0+1}} e^{(\log n)(1+\tfrac{1}{2}s_0)} = \sum_{n=1}^{\infty} a_n n^{-\tfrac{1}{2}s_0}.$$

In other words, the assumption $s_0 > 0$ led to a point between 0 and s_0 (namely $\tfrac{1}{2}s_0$) for which there is convergence. This contradiction proves that $s_0 = 0$. Therefore $\sum_{n=1}^{\infty} a_n n^{-s}$ converges for $\operatorname{Re} s > 0$. □

We shall now examine special features of Dirichlet series that allow the series to have product expansions like the one for $\zeta(s)$, namely $\sum_{n=1}^{\infty} n^{-s} = \prod_{p \text{ prime}} \frac{1}{1-p^{-s}}$. Consider a formal product

$$\prod_{p \text{ prime}} (1 + a_p p^{-s} + \cdots + a_{p^m} p^{-ms} + \cdots).$$

If this product is expanded without regard to convergence, the result is the Dirichlet series $\sum_{n=1}^{\infty} a_n n^{-s}$, where $a_1 = 1$ and a_n is given by

$$a_n = a_{p_1^{r_1}} \cdots a_{p_k^{r_k}} \qquad \text{if } n = p_1^{r_1} \cdots p_k^{r_k}.$$

Suppose that the Dirichlet series $\sum_{n=1}^{\infty} a_n n^{-s}$ is in fact absolutely convergent in some right half-plane. Then every rearrangement is absolutely convergent to the same sum, and the same conclusion is valid for subseries. If E is a finite set of primes and if $\mathbb{N}(E)$ denotes the set of positive integers requiring only members of E for their factorization, then we have

$$\prod_{p \in E} (1 + a_p p^{-s} + \cdots + a_{p^m} p^{-ms} + \cdots) = \sum_{n \in \mathbb{N}(E)} a_n n^{-s}.$$

Letting E swell to the whole set of positive integers, we see that the infinite product has a limit in the half-plane of absolute convergence of the Dirichlet series, and the limit of the infinite product equals the sum of the series. The sum of the series is 0 only if one of the factors on the left side is 0. In particular, the sum of the series cannot be identically 0, by Proposition 1.23d. Thus the limit of the infinite product can can be given by only this one Dirichlet series.

Conversely if an absolutely convergent Dirichlet series $\sum_{n=1}^{\infty} a_n n^{-s}$ has the property that its coefficients are **multiplicative**, i.e.,

$$a_1 = 1 \quad \text{and} \quad a_{mn} = a_m a_n \quad \text{whenever GCD}(m, n) = 1,$$

then we can form the above infinite product and recover the given series by expanding the product and using the formula $a_n = a_{p_1^{r_1}} \cdots a_{p_k^{r_k}}$ when $n = p_1^{r_1} \cdots p_k^{r_k}$. In this case we say that the Dirichlet series $\sum_{n=1}^{\infty} a_n n^{-s}$ has the infinite product as an **Euler product**. Many functions in elementary number theory give rise to multiplicative sequences; an example is $a_n = \varphi(n)$, where φ is the Euler φ function.

If the coefficients are **strictly multiplicative**, i.e., if

$$a_1 = 1 \quad \text{and} \quad a_{mn} = a_m a_n \quad \text{for all } m \text{ and } n,$$

then the pth factor of the infinite product simplifies to

$$1 + a_p p^{-s} + \cdots + (a_p p^{-s})^m + \cdots = \frac{1}{1 - a_p p^{-s}}.$$

As a consequence we obtain the following proposition.

Proposition 1.26. If the coefficients of the Dirichlet series $\sum_{n=1}^{\infty} a_n n^{-s}$ are strictly multiplicative, then the Dirichlet series has an Euler product of the form

$$\sum_{n=1}^{\infty} \frac{a_n}{n^s} = \prod_{p \text{ prime}} \frac{1}{1 - a_p p^{-s}},$$

valid in its region of absolute convergence.

REMARK. We refer to the kind of Euler product in this proposition as a **first-degree Euler product**.

This is what happens with $\zeta(s)$, for which all the coefficients are 1, and with $a_n = \chi_0(n)$ and $a_n = \chi_1(n)$ as in the previous section. Conversely an Euler product expansion of the form in the proposition forces the coefficients of the Dirichlet series to be strictly multiplicative.

A Dirichlet series $\sum_{n=1}^{\infty} a_n n^{-s}$ with $|a_n| \leq n^c$ for some real c is absolutely convergent for $\operatorname{Re} s > c + 1$. This fact leads us to a convergence criterion for first-degree Euler products.

Proposition 1.27. A first-degree Euler product $\prod (1 - a_p p^{-s})^{-1}$ with $|a_p| \leq p^c$ for some real c and all primes p defines an absolutely convergent Dirichlet series for $\operatorname{Re} s > c + 1$ and hence a valid identity $\sum_{n=1}^{\infty} a_n n^{-s} = \prod_{p \text{ prime}} (1 - a_p p^{-s})^{-1}$ in that region.

PROOF. The coefficients a_n are strictly multiplicative, and thus $|a_n| \leq n^c$ for all n. The absolute convergence follows. $\square$

10. Dirichlet's Theorem on Primes in Arithmetic Progressions

In this section we shall prove Dirichlet's Theorem as stated in Theorem 1.21. Recall from Section 8 that the proof of Dirichlet's Theorem for the progressions $4n + 1$ and $4n + 3$ required taking the sum and difference of two expressions, working with them, and then passing back to the original expressions. Generalizing this step involves recognizing this process as Fourier analysis on the 2-element group $(\mathbb{Z}/4\mathbb{Z})^\times$. This kind of Fourier analysis was discussed in Section VII.4 of *Basic Algebra*. Let us begin by reviewing what is needed from that section of *Basic Algebra* and then pinpoint the Fourier analysis that was the key to the argument in Section 8.

Let G be a finite abelian group, such as $(\mathbb{Z}/m\mathbb{Z})^\times$. A **multiplicative character** of G is a homomorphism of G into the circle group $S^1 \subseteq \mathbb{C}^\times$. The multiplicative characters of G form a finite abelian group $\widehat{G}$ under pointwise multiplication:

$$(\chi \chi')(g) = \chi(g)\chi'(g).$$

In this setting we recall the statement of the Fourier inversion formula.

THEOREM 7.17 OF *Basic Algebra* (Fourier inversion formula). Let G be a finite abelian group, and introduce an inner product on the complex vector space $C(G, \mathbb{C})$ of all functions from G to $\mathbb{C}$ by the formula

$$\langle F, F' \rangle = \sum_{g \in G} F(g)\overline{F'(g)},$$

the corresponding norm being $\|F\| = \langle F, F \rangle^{1/2}$. Then the members of $\widehat{G}$ form an orthogonal basis of $C(G, \mathbb{C})$, each χ in $\widehat{G}$ satisfying $\|\chi\|^2 = |G|$. Consequently $|\widehat{G}| = |G|$, and any function $F : G \to C$ is given by the "sum of its Fourier series":

$$F(g) = \frac{1}{|G|} \sum_{\chi \in \widehat{G}} \left(\sum_{h \in G} F(h)\overline{\chi(h)} \right) \chi(g).$$

EXAMPLE. With the two-element group $G = \{\pm 1\}$, there are two multiplicative characters, with $\chi_0(+1) = \chi_0(-1) = 1$, $\chi_1(+1) = 1$, and $\chi_1(-1) = -1$. We can think of the Fourier-coefficient mapping as carrying any complex-valued function F on G to the function $\widehat{F}$ on $\widehat{G}$ given by $\widehat{F}(\chi) = \sum_{h \in G} F(h)\overline{\chi(h)}$. The inversion formula says that F is recovered as $F = \frac{1}{2}\bigl(\widehat{F}(\chi_0)\chi_0 + \widehat{F}(\chi_1)\chi_1\bigr)$. A basis for the 2-dimensional space of complex-valued functions on G consists of the two functions F^+ and F^-, with F^+ equal to 1 at $+1$ and 0 at -1 and with F^- equal to 0 at $+1$ and 1 at -1. The multiplicative characters are given by $\chi_0 = F^+ + F^-$ and $\chi_1 = F^+ - F^-$. For these two functions the inversion formula reads $F^+ = \frac{1}{2}(\chi_0 + \chi_1)$ and $F^- = \frac{1}{2}(\chi_0 - \chi_1)$. In Section 8 the roles of F^+ and F^- are played by functions of s, not by scalars, with F^+ corresponding to $\sum_{p \equiv 1 \bmod 4} p^{-s}$ and F^- corresponding to $\sum_{p \equiv 3 \bmod 4} p^{-s}$. We are to consider the functions of s corresponding to their sum χ_0 and to their difference χ_1. The results of Section 9 show that these are the series that come from Euler products. The role of the Fourier inversion formula is to ensure that we can reconstruct $\sum_{p \equiv 1 \bmod 4} p^{-s}$ and $\sum_{p \equiv 3 \bmod 4} p^{-s}$ from the sum and difference. The general proof of Dirichlet's Theorem is a direct generalization of this argument for $m = 4$.

Fix an integer $m > 1$. A **Dirichlet character modulo** m is a function $\chi : \mathbb{Z} \to S^1 \cup \{0\}$ such that

(i) $\chi(j) = 0$ if and only if $\mathrm{GCD}(j, m) > 1$,
(ii) $\chi(j)$ depends only on the residue class j mod m,
(iii) when regarded as a function on the residue classes modulo m, χ is a multiplicative character of $(\mathbb{Z}/m\mathbb{Z})^\times$.

In particular, a Dirichlet character modulo m determines a multiplicative character of $(\mathbb{Z}/m\mathbb{Z})^\times$. Conversely each multiplicative character of $(\mathbb{Z}/m\mathbb{Z})^\times$ defines a unique Dirichlet character modulo m as the lift of the multiplicative character on the set $\{j \in \mathbb{Z} \mid \mathrm{GCD}(j, m) = 1\}$ and as 0 on the rest of $\mathbb{Z}$. For example the multiplicative character on $(\mathbb{Z}/4\mathbb{Z})^\times$ that is 1 at 1 mod 4 and is -1 at 3 mod 4 lifts to the Dirichlet character that is 1 at integers congruent to 1 modulo 4, is -1 at integers congruent to 3 modulo 4, and is 0 at even integers. It will often be notationally helpful to use the same symbol for the Dirichlet character and the multiplicative character of $(\mathbb{Z}/m\mathbb{Z})^\times$. Because of this correspondence, the number of Dirichlet characters modulo m matches the order of $\widehat{G}$ for $G = (\mathbb{Z}/m\mathbb{Z})^\times$, which matches the order of G and is $\varphi(m)$, where φ is the Euler φ function. The **principal** Dirichlet character modulo m, denoted by χ_0, is the one built from the trivial character of $(\mathbb{Z}/m\mathbb{Z})^\times$:

$$\chi_0(j) = \begin{cases} 1 & \text{if } \mathrm{GCD}(j, m) = 1, \\ 0 & \text{if } \mathrm{GCD}(j, m) > 1. \end{cases}$$

10. Dirichlet's Theorem on Primes in Arithmetic Progressions

Each Dirichlet character modulo m is strictly multiplicative, in the sense of the previous section. We assemble each as the coefficients of a Dirichlet series, the associated **Dirichlet L function**, by the definition

$$L(s, \chi) = \sum_{n=1}^{\infty} \frac{\chi(n)}{n^s}.$$

Proposition 1.28. Fix m, and let χ be a Dirichlet character modulo m.

(a) The Dirichlet series $L(s, \chi)$ is absolutely convergent for $\operatorname{Re} s > 1$ and is given in that region by a first-degree Euler product

$$L(s, \chi) = \prod_{p \text{ prime}} \frac{1}{1 - \chi(p) p^{-s}}.$$

(b) If χ is not principal, then the series for $L(s, \chi)$ is convergent for $\operatorname{Re} s > 0$, and the sum is analytic for $\operatorname{Re} s > 0$.

(c) For the principal Dirichlet character χ_0 modulo m, $L(s, \chi_0)$ extends to be meromorphic for $\operatorname{Re} s > 0$. Its only pole for $\operatorname{Re} s > 0$ is at $s = 1$, and the pole is simple. It is given in terms of the Riemann zeta function by

$$L(s, \chi_0) = \zeta(s) \prod_{\substack{p \text{ prime}, \\ p \text{ dividing } m}} (1 - p^{-s}).$$

PROOF. For (a), the boundedness of χ implies that the series is absolutely convergent for $\operatorname{Re} s > 1$. Since χ is strictly multiplicative, $L(s, \chi)$ has a first-degree Euler product by Proposition 1.26, and the product is convergent in the same region.

For (b), let us notice that $\chi \neq \chi_0$ implies the equality

$$\sum_{n=1}^{m} \chi(n + b) = 0 \qquad \text{for any } b, \tag{$*$}$$

since the member of $(\mathbb{Z}/m\mathbb{Z})^\times$ that corresponds to χ is orthogonal to the trivial character, by the Fourier inversion formula as quoted above from *Basic Algebra*. For s real and positive, let us write

$$\tfrac{\chi(n)}{n^s} = \chi(n) \cdot \tfrac{1}{n^s} = u_n v_n$$

in the notation of the summation by parts formula that follows the statement of Proposition 1.23, and let us put $U_n = \sum_{k=1}^{n} u_k$. Equation $(*)$ implies that $\{U_n\}$ is bounded, say with $|U_n| \leq C$. Summation by parts then gives

$$\left| \sum_{n=M}^{N} \tfrac{\chi(n)}{n^s} \right| \leq \sum_{n=M}^{N-1} C \left(\tfrac{1}{n^s} - \tfrac{1}{(n+1)^s} \right) + \tfrac{C}{N^s} + \tfrac{C}{M^s} = \tfrac{2C}{M^s}.$$

This expression tends to 0 as M and N tend to ∞. Therefore the series $L(s, \chi) = \sum_{n=1}^{\infty} \frac{\chi(n)}{n^s}$ is convergent for s real and positive. By Proposition 1.23a the series is convergent for $\operatorname{Re} s > 0$, and the sum is analytic in this region.

For (c), let $\operatorname{Re} s > 1$. From the product formula in (a) with χ set equal to χ_0, we have

$$L(s, \chi_0) = \prod_{\substack{p \text{ prime,} \\ p \text{ not dividing } m}} \frac{1}{1 - p^{-s}}.$$

Using the Euler product expansion of $\zeta(s)$, we obtain the displayed formula of (c). The remaining statements in (c) follow from Proposition 1.24, since the product over primes p not dividing m is a finite product. $\square$

By Proposition 1.28b, $L(s, \chi)$ is well defined and finite at $s = 1$ if χ is not principal. The main step in the proof of Dirichlet's Theorem is the following lemma.

Lemma 1.29. $L(1, \chi) \neq 0$ if χ is not principal.

PROOF. Let $Z(s) = \prod_\chi L(s, \chi)$. Exactly one factor of $Z(s)$ has a pole at $s = 1$, according to Proposition 1.28. If any factor has a zero at $s = 1$, then $Z(s)$ is analytic for $\operatorname{Re} s > 0$. Assuming that $Z(s)$ is indeed analytic, we shall derive a contradiction.

Being the finite product of absolutely convergent Dirichlet series for $\operatorname{Re} s > 1$, $Z(s)$ is given by an absolutely convergent Dirichlet series. We shall prove that the coefficients of this series are ≥ 0. More precisely we shall prove for $\operatorname{Re} s > 1$ that

$$Z(s) = \prod_{p \text{ with GCD}(p,m)=1} \frac{1}{\left(1 - p^{-f(p)s}\right)^{g(p)}}, \quad (*)$$

where $f(p)$ is the order of p in $(\mathbb{Z}/m\mathbb{Z})^\times$ and where $g(p) = \varphi(m)/f(p)$, φ being Euler's φ function. The factor $(1 - p^{-f(p)s})^{-1}$ is given by a Dirichlet series with all coefficients ≥ 0. Hence so is the $g(p)^{\text{th}}$ power, and so is the product over p of the result. Thus $(*)$ will prove that all coefficients of $Z(s)$ are ≥ 0.

To prove $(*)$, we write, for $\operatorname{Re} s > 1$,

$$Z(s) = \prod_\chi L(s, \chi) = \prod_p \left(\prod_\chi \frac{1}{1 - \chi(p)p^{-s}} \right) = \prod_{\substack{p \text{ with} \\ \text{GCD}(p,m)=1}} \left(\prod_\chi \frac{1}{1 - \chi(p)p^{-s}} \right).$$

Fix p not dividing m. We shall show that

$$\prod_\chi \left(1 - \chi(p)p^{-s}\right) = \left(1 - p^{-fs}\right)^g, \quad (**)$$

where f is the order of p in $(\mathbb{Z}/m\mathbb{Z})^\times$ and where $g = \varphi(m)/f$; then $(*)$ will follow.

The function $\chi \to \chi(p)$ is a homomorphism of $(\mathbb{Z}/m\mathbb{Z})^\times$ into the subgroup $\{e^{2\pi i k/f}\}$ of S^1 and is onto some cyclic subgroup $\{e^{2\pi i k/f'}\}$ with f' dividing f. Let us see that $f' = f$. In fact, if $f' < f$, then $p^{f'} \not\equiv 1 \bmod m$, while $\chi(p^{f'}) = \chi(p)^{f'} = 1$ for all χ; since $\chi(p^{f'}) = \chi(1)$ for all χ, the χ's cannot span all functions on $(\mathbb{Z}/m\mathbb{Z})^\times$, in contradiction to the Fourier inversion formula (Theorem 7.17 of *Basic Algebra*).

Thus $\chi \to \chi(p)$ is onto $\{e^{2\pi i k/f}\}$. In other words, $\chi(p)$ takes on all f^{th} roots of unity as values, and the homomorphism property ensures that each is taken on the same number of times, namely $g = \varphi(m)/f$ times. If X is an indeterminate, we then have

$$\prod_\chi (1 - \chi(p)X) = \Big(\prod_{k=0}^{f-1} (1 - e^{2\pi i k/f} X) \Big)^g = (1 - X^f)^g.$$

Then $(**)$ follows and so does $(*)$. Hence all the coefficients of the Dirichlet series of $Z(s)$ are ≥ 0. We have already observed that this series, as the finite product of absolutely convergent series for $\operatorname{Re} s > 1$, is absolutely convergent for $\operatorname{Re} s > 1$. Thus Proposition 1.25 applies and shows that the Dirichlet series of $Z(s)$ converges for $\operatorname{Re} s > 0$.

Since the coefficients of the series are positive, the convergence is absolute for s real and positive. By Proposition 1.23b the convergence is absolute for $\operatorname{Re} s > 0$. Therefore the Euler product expansion $(*)$ is valid for $\operatorname{Re} s > 0$.

For primes p not dividing m and for real $s > 0$, we have

$$\frac{1}{(1 - p^{-fs})^g} = (1 + p^{-fs} + p^{-2fs} + \cdots)^g \geq 1 + p^{-fgs} + p^{-2fgs} + \cdots$$

$$= 1 + p^{-\varphi(m)s} + p^{-2\varphi(m)s} + \cdots = \frac{1}{1 - p^{-\varphi(m)s}}.$$

In combination with $(*)$, this inequality gives

$$Z(s)\Big(\prod_{p \text{ dividing } m} \frac{1}{1 - p^{-\varphi(m)s}} \Big)$$

$$= \Big(\prod_{p \text{ with GCD}(p,m)=1} \frac{1}{(1 - p^{-fs})^g} \Big) \Big(\prod_{p \text{ dividing } m} \frac{1}{1 - p^{-\varphi(m)s}} \Big)$$

$$\geq \prod_{p \text{ prime}} \frac{1}{1 - p^{-\varphi(m)s}} = \sum_{n=1}^\infty \frac{1}{n^{\varphi(m)s}}.$$

The sum on the right is $+\infty$ for $s = 1/\varphi(m)$, while the left side is finite for that s. This contradiction completes the proof of the lemma. $\square$

PROOF OF THEOREM 1.21. First we show for each Dirichlet character χ modulo m that

$$\log L(s, \chi) = \sum_{p \text{ prime}} \frac{\chi(p)}{p^s} + g(s, \chi) \qquad (*)$$

for *real* numbers $s > 1$, with $g(s, \chi)$ remaining bounded as $s \downarrow 1$. In this statement we have not yet specified a branch of the logarithm, and we shall choose it presently. Fix p and define, for $s \geq 1$, a value of the logarithm of the pth factor of the Euler product of $L(s, \chi)$ in Proposition 1.28a by

$$\log\left(\frac{1}{1-\chi(p)p^{-s}}\right) = \frac{\chi(p)}{p^s} + \frac{1}{2}\frac{\chi(p^2)}{p^{2s}} + \frac{1}{3}\frac{\chi(p^3)}{p^{3s}} + \cdots = \frac{\chi(p)}{p^s} + g(s, p, \chi). \qquad (**)$$

In Section 8 we obtained the inequality $|\log(1-x)^{-1} - x| \leq 2|x|^2$ for real x with $|x| \leq \frac{1}{2}$, but the proof remains valid for complex x with $|x| \leq \frac{1}{2}$. Since $x = \chi(p)p^{-s}$ is complex with $|\chi(p)p^{-s}| \leq \frac{1}{2}$, we obtain

$$|g(s, p, \chi)| = \left|\log\left(\frac{1}{1-\chi(p)p^{-s}}\right) - \chi(p)p^{-s}\right| \leq 2|\chi(p)p^{-s}|^2 \leq 2p^{-2}.$$

Since $\sum_{p \text{ prime}} p^{-2} \leq \sum_{n=1}^{\infty} n^{-2} < \infty$, the series $\sum_p g(s, p, \chi)$ is uniformly convergent for $s \geq 1$. Let $g(s, \chi)$ be the continuous function $\sum_p g(s, p, \chi)$. Summing $(**)$ over primes p, we obtain

$$\sum_p \log\left(\frac{1}{1-\chi(p)p^{-s}}\right) = \sum_p \frac{\chi(p)}{p^s} + g(s, \chi).$$

Because of the validity of the Euler product expansion of $L(s, \chi)$ in Proposition 1.28a, the left side represents a branch of $\log L(s, \chi)$. This proves $(*)$.

For each b prime to m, define a function F_b on the positive integers by

$$F_b(n) = \begin{cases} 1 & \text{if } n \equiv b \bmod m, \\ 0 & \text{otherwise.} \end{cases}$$

The Fourier inversion formula (Theorem 7.17 of *Basic Algebra*) gives

$$\sum_\chi \overline{\chi(b)}\chi(n) = \varphi(m) F_b(n). \qquad (\dagger)$$

Multiplying $(*)$ by $\overline{\chi(b)}$, summing on χ, and using $(\dagger)$ to handle the term that is summed over p prime, we obtain

$$\varphi(m) \sum_{\substack{p \text{ prime,} \\ p = km+b}} p^{-s} = \sum_\chi \overline{\chi(b)} \log L(s, \chi) - \sum_\chi \overline{\chi(b)} g(s, \chi). \qquad (\dagger\dagger)$$

The term $\sum_\chi \overline{\chi(b)} g(s, \chi)$ is bounded as $s \downarrow 1$, according to $(*)$. The term $\overline{\chi_0(b)} \log L(s, \chi_0)$ is unbounded as $s \downarrow 1$, by Proposition 1.28c. For χ nonprincipal, the term $\overline{\chi(b)} \log L(s, \chi)$ is bounded as $s \downarrow 1$, by Proposition 1.28b and Lemma 1.29. Therefore the left side of $(\dagger\dagger)$ is unbounded as $s \downarrow 1$. Hence the number of primes contributing to the sum is infinite. □

11. Problems

1. Fix an odd integer $m > 1$. Let P be the set of odd primes $p > 0$ such that $x^2 \equiv m \bmod p$ is solvable and such that p does not divide m. Show that P is nonempty and that there is a finite set S of arithmetic progressions such that the members of P are the odd primes > 0 that lie in at least one member of S.

2. Let D be a nonsquare integer, and let m be an odd integer with $\gcd(D, m) = 1$. By suitably adapting the proof of Theorem 1.6,
   (a) prove that if m is primitively representable by some binary quadratic form of discriminant D, then $x^2 \equiv D \bmod m$ is solvable,
   (b) prove that if $x^2 \equiv D \bmod m$ is solvable and m is odd, then m is primitively representable by some binary quadratic form of discriminant D.

3. For a fixed discriminant D, let H be the group of proper equivalence classes of binary quadratic forms of discriminant D, and let H' be the set of ordinary equivalence classes of discriminant D. Inclusion of a proper equivalence class into the ordinary equivalence class that contains it gives a map f of H onto H'. Give an example in which H' can admit no group structure for which f is a group homomorphism.

4. (a) Show that if (a, b, c) has order 3 in the form class group, then the product of any two integers of the form $ax^2 + bxy + cy^2$ is again of that form.
   (b) Show that $h(-23) = 3$.
   (c) Using the general theory, show that the class of $2x^2 + xy + 3y^2$ has order 3.
   (d) Find an explicit formula for (X, Y) in terms of (x_1, y_1) and (x_2, y_2) such that $(2x_1^2 + x_1 y_1 + 3y_1^2)(2x_2^2 + x_2 y_2 + 3y_2^2) = 2X^2 + XY + 3Y^2$.

5. If two integer forms are improperly equivalent over $\mathbb{Z}$, prove that they are properly equivalent over $\mathbb{Q}$.

6. Verify for the fundamental discriminant $D = -67$ that $h(D) = 1$. (Educational note: It is known that the only negative fundamental discriminants D with $h(D) = 1$ are $-3, -4, -7, -8, -11, -19, -43, -67, -163$. It is known also that the only other nonsquare $D < 0$ for which $h(D) = 1$ are $-12, -16, -28, -27$.)

7. This problem carries out the algorithm suggested by Theorem 1.8 to find representatives of all proper equivalence classes of binary quadratic forms (a, b, c) of discriminant $316 = 4 \cdot 79$. For each of these, b will be even.
   (a) For each even positive b with $b < \sqrt{4 \cdot 79}$, factor $(b^2 - 4 \cdot 79)/4$ as a product ac in all possible ways such that $a > 0$ and such that both $|a|$ and $|c|$ lie between $\sqrt{79} - b/2$ and $\sqrt{79} + b/2$, obtaining 16 forms (a, b, c). Expand the list by adjoining each form $(-a, b, -c)$, so that the expanded list has 32 members.

(b) Arrange the 32 members of the expanded list of (a) into 6 cycles, obtaining 2 cycles of length 4 and 4 cycles of length 6.
(c) Conclude that $h(4 \cdot 79) = 6$.

8. For discriminant $D = -47$, the class number is $h(-47) = 5$, and the reduced binary quadratic forms are $(1, 1, 12), (2, 1, 6), (2, -1, 6), (3, 1, 4), (3, -1, 4)$. Show what the multiplication table is for the proper equivalence classes of these forms.

Problems 9–11 concern the Jacobi symbol, which is a generalization of the Legendre symbol. Let m and n be integers with $n > 0$ odd, and let $n = p_1^{k_1} \cdots p_r^{k_r}$ be the prime factorization of n. The **Jacobi symbol** $\left(\frac{m}{n}\right)$ is defined to be 0 if $\mathrm{GCD}(m, n) > 1$ and is defined to be $\prod_{j=1}^{r} \left(\frac{m}{p_j}\right)^{k_j}$ if $\mathrm{GCD}(m, n) = 1$, where $\left(\frac{m}{p_j}\right)$ is a Legendre symbol. The Jacobi symbol therefore extends the domain of the Legendre symbol, and it depends only on the residue m mod n. Even when $\mathrm{GCD}(m, n) = 1$, the Jacobi symbol does *not* encode whether m is a square modulo n, however, since $\left(\frac{-1}{21}\right) = +1$ and since the residue -1 is not a square modulo 21.

9. Suppose that n and n' are odd positive integers and that m and m' are integers. Verify that
    (a) $\left(\frac{mm'}{nn'}\right) = \left(\frac{m}{n}\right)\left(\frac{m'}{n'}\right)$,
    (b) $\left(\frac{m^2}{n}\right) = \left(\frac{m}{n^2}\right) = 1$ if $\mathrm{GCD}(m, n) = 1$.

10. Prove for all odd positive integers n that
    (a) $\left(\frac{-1}{n}\right) = (-1)^{\frac{1}{2}(n-1)}$,
    (b) $\left(\frac{2}{n}\right) = (-1)^{\frac{1}{8}(n^2-1)}$.

11. **(Quadratic reciprocity)** Prove for all odd positive integers m and n satisfying $\mathrm{GCD}(m, n) = 1$ that $\left(\frac{m}{n}\right) = (-1)^{[\frac{1}{2}(m-1)][\frac{1}{2}(n-1)]}\left(\frac{n}{m}\right)$.

Problems 12–13 indicate, without spelling out what the group G is, two uses of Dirichlet's Theorem in the subject of "elliptic curves." No knowledge of the subject of elliptic curves is assumed, however.

12. Suppose that G is a finite abelian group whose order $|G|$ divides $p + 1$ for all sufficiently large primes p with $p \equiv 3 \bmod 4$. It is to be shown that $|G|$ divides 4 by means of multiple applications of Dirichlet's Theorem.
    (a) Deduce that 8 does not divide $|G|$ by considering the arithmetic progression $8k + 3$.
    (b) Deduce that 3 does not divide $|G|$ by considering the arithmetic progression $12k + 7$.
    (c) Deduce that no odd prime $q > 3$ divides $|G|$ by considering the arithmetic progression $4qk + 3$.

11. Problems   69

13. Suppose that G is a finite abelian group whose order $|G|$ divides $p + 1$ for all sufficiently large primes p with $p \equiv 2 \bmod 3$. It is to be shown that $|G|$ divides 6 by means of multiple applications of Dirichlet's Theorem.
    (a) Deduce that 4 does not divide $|G|$ by considering the arithmetic progression $12k + 5$.
    (b) Deduce that 9 does not divide $|G|$ by considering the arithmetic progression $9k + 2$.
    (c) Deduce that no odd prime $q > 3$ divides $|G|$ by considering the arithmetic progression $3qk + 2$.

Problems 14–19 develop some elementary properties of ideals and their norms in quadratic number fields. Notation is as in Sections 6–7. In particular, the number field is $K = \mathbb{Q}(\sqrt{m})$, the ring R of algebraic integers in it has $\mathbb{Z}$ basis $\{1, \delta\}$, and σ is the nontrivial automorphism of K fixing $\mathbb{Q}$.

14. Prove that if $I = \langle a, r \rangle$ is a nonzero ideal in R with $a \in \mathbb{Z}$ and $r \in R$, then a divides $N(s)$ for every s in I.

15. Prove that any nonzero ideal I in R can be written as $I = \langle a, b + g\delta \rangle$ with a, b, and g in $\mathbb{Z}$ and with $a > 0$, $0 \leq b < a$, and $0 < g \leq a$. Prove also that the $\mathbb{Z}$ basis with these properties is unique, and it has the properties that g divides a and b and that ag divides $N(b + g\delta)$.

16. Let a, b, and g be integers satisfying $a > 0$, $0 \leq b < a$, and $0 < g \leq a$ with g dividing a and b and with ag dividing $N(b + g\delta)$. Prove that the ideal $I = (a, b + g\delta)$ in R has $\{a, b + g\delta\}$ as a $\mathbb{Z}$ basis.

17. Prove that if $I = \langle a, r \rangle$ is a nonzero ideal in R with $a \in \mathbb{Z}$, $r \in R$, and $r = c + d\delta$ for integers c and d, then $N(I) = |ad|$.

18. (a) Prove that if I is a nonzero ideal in R, then $N(I)$ is the number of elements in R/I.
    (b) Deduce that if $I \subseteq J$ are nonzero ideals in R, then $N(J)$ divides $N(I)$, and $I = J$ if and only if $N(J) = N(I)$.

19. (a) Using the Chinese Remainder Theorem, prove that if I and J are nonzero ideals in R with $I + J = R$, then $N(IJ) = N(I)N(J)$.
    (b) Let P be a nonzero prime ideal in R, and let $p > 0$ be the prime number such that $P \cap \mathbb{Z} = (p)\mathbb{Z}$. Then R/P is a vector space over $\mathbb{Z}/p\mathbb{Z}$, and its order is of the form p^f for some integer $f > 0$. Show by induction on the integer $e > 0$ that R/P^e has order p^{ef}.
    (c) Using unique factorization of ideals, deduce that if I and J are any two nonzero ideals in R, then $N(IJ) = N(I)N(J)$.
    (d) Prove that any nonzero ideal I of R has $I\sigma(I) = (N(I))$.

Problems 20–24 concern the splitting of prime ideals when extended to quadratic number fields. Fix a quadratic number field $\mathbb{Q}(\sqrt{m})$, and let R, D, δ, and σ be as in Sections 6–7. Let $p > 0$ be a prime in $\mathbb{Z}$. According to Theorem 9.62 of *Basic Algebra*, the unique factorization of the ideal $(p)R$ in R is one of the following: $(p)R = (p)$ is already prime in R, $(p)R = P_1 P_2$ is the product of two distinct prime ideals, or $(p)R = P^2$ is the square of a prime ideal.

20. Deduce from the formula $N((p)R) = p^2$ that if P is a nontrivial factor in the unique factorization of the ideal $(p)R$, then $N(P) = p$.

21. This problem concerns the prime $p = 2$.
    (a) Use Problem 15 to prove that if $D \equiv 5 \bmod 8$, then $(2)R$ is a prime ideal in R.
    (b) Prove that if $D \equiv 1 \bmod 8$, then $(2)R$ factors into the product of two distinct prime factors as $(2)R = \langle 2, \delta \rangle \langle 2, 1 + \delta \rangle$.
    (c) Prove that if D is even and $D/4 \equiv 3 \bmod 4$, then $(2)R = (2, 1+\delta)^2$ exhibits $(2)R$ as the square of a prime ideal.
    (d) Prove that if D is even and $D/4 \equiv 2 \bmod 4$, then $(2)R = (2, \delta)^2$ exhibits $(2)R$ as the square of a prime ideal.

22. Let p be an odd prime.
    (a) Prove that if D is odd, then $(p)R$ has a nontrivial factorization into prime ideals if and only if $x^2 + x + \frac{1}{4}(1 - D) \equiv 0 \bmod p$ has a solution, and in this case a factorization of $(p)R$ is as $(p)R = (p, x + \delta)(p, x + \sigma(\delta))$.
    (b) Prove that if D is even, then $(p)R$ has a nontrivial factorization into prime ideals if and only if $x^2 \equiv 0 \bmod (D/4)$ has a solution, and in this case a factorization of $(p)R$ is as $(p)R = (p, x + \delta)(p, x + \sigma(\delta))$.
    (c) Deduce from (a) and (b) that $(p)R$ has a nontrivial factorization into prime ideals if and only if D is a square modulo p.

23. Let p be an odd prime such that D is a square modulo p, so that Problem 22c gives a nontrivial factorization of $(p)R$ into prime ideals of the form $(p)R = (p, x + \delta)(p, x + \sigma(\delta))$ for some integer x. Let $I = (p, x + \delta)$.
    (a) Prove that if D is odd, then $\sigma(I) = I$ if and only if the integer x is $\frac{1}{2}(p - 1)$.
    (b) Prove that if D is even, then $\sigma(I) = I$ if and only if the integer x is 0.

24. Let p be an odd prime such that D is a square modulo p, so that Problem 22c gives a nontrivial factorization of $(p)R$ into prime ideals of the form $(p)R = (p, x + \delta)(p, x + \sigma(\delta))$ for some integer x. Using the previous problem, show that the two factors on the right are the same ideal if and only if p divides D.

Problems 25–29 seek to identify the genus group explicitly for fundamental discriminants D. Let $K = \mathbb{Q}(\sqrt{m})$ be the corresponding quadratic number field, let R be the ring of algebraic integers in K, and let σ be the nontrivial automorphism of K fixing $\mathbb{Q}$. Let $E = \{p_1, \ldots, p_{g+1}\}$ with $g \geq 0$ be the set of distinct prime divisors of D. The goal of this set of problems is to prove that the order of the genus group is 2^g and to exhibit ideals in R representing each genus. Recall from Theorem 1.20

that strict equivalence classes of ideals correspond to proper equivalence classes of binary quadratic forms and therefore that each genus corresponds to a set of proper equivalence classes of binary quadratic forms.

25. Let the form class group H for discriminant D be isomorphic to a product of cyclic groups of orders $2^{k_1}, \ldots, 2^{k_r}, q_1^{l_1}, \ldots, q_s^{l_s}$, where $k_1, \ldots, k_r$ and $l_1, \ldots, l_s$ are positive integers and $q_1, \ldots, q_s$ are odd primes that are not necessarily distinct. Prove that the genus group has order 2^r and is abstractly isomorphic to the subgroup of H of elements whose order divides 2. (Educational note: Thus a goal of the present set of problems is to show that $r = g$.)

26. According to Problems 20–24, the nonzero prime ideals of R are of three kinds:
    (i) unique distinct ideals $I = (p, b+\delta)$ and $\sigma(I) = (p, b+\sigma(\delta))$ with product $(p)R$ if p is an odd prime not dividing D such that $x^2 \equiv D \bmod p$ is solvable, or if $p = 2$ and $D \equiv 1 \bmod 8$,
    (ii) the ideal $(p)R$ if p is an odd prime not dividing D such that $x^2 \equiv D \bmod p$ is not solvable, or if $p = 2$ and $D \equiv 5 \bmod 8$,
    (iii) a unique ideal I_p with $I_p^2 = (p)R$ if p divides D.

    For each subset $S \subseteq E$ of the $g + 1$ distinct prime divisors of D, define $J_S = \prod_{p \in S} I_p$.
    (a) Using unique factorization of ideals in R, show that any nonzero proper ideal I in R with $\sigma(I) = I$ is of the form $(a)J_S$ for some $a \in \mathbb{Z}$ and some subset $S \subseteq E$.
    (b) By considering norms of ideals, show that I uniquely determines S in (a).

27. (a) The element $x = -1$ of K has $N(x) = 1$ and factors as $x = \sigma(y)y^{-1}$ for the element $y = \sqrt{m}$ of K. For all other elements x of K with norm 1, verify the formula
    $$\frac{1+x}{1+\sigma(x)} = \frac{(1+x)x}{(1+\sigma(x))x} = \frac{(1+x)x}{x+x\sigma(x)} = \frac{(1+x)x}{1+x} = x,$$
    and explain why it shows that x is of the form $\sigma(y)y^{-1}$ for some $y \neq 0$ in K. (Educational note: This result is a special case of **Hilbert's Theorem 90**, which is a theorem in the cohomology of groups and appears in Chapter III. The general theorem says for a finite Galois extension K/k with Galois group Γ that the cohomology H^1 of the group Γ with coefficients in the abelian group $K^\times$ is 0.)
    (b) Show that the element y in (a) can be taken to be in R and that all such y's in R are $\mathbb{Z}$ multiples of one of them y_0, which is unique up to a factor of -1.

28. Let I be a nonzero ideal in R whose class in the ideal class group $\mathcal{H}$ has order 2, i.e., an ideal I such that $I^2 = (x)$ for some element $x \in R$.
    (a) Show that the element $xN(I)^{-1}$ of K has norm 1.

(b) Show that the corresponding element y_0 of R from the previous problem has the property that $\sigma((y_0)I) = (y_0)I$.

(c) Using either y_0 or $y_0\sqrt{m}$ from (b), deduce that for any nonzero ideal I in R with I^2 principal, there is a *strictly* equivalent ideal J_S for some subset $S \subseteq E$ of the $g+1$ prime divisors of E. Consequently the order of the genus group is a power of 2 equal to at most 2^{g+1}.

29. This problem shows that the number of ideals J_S in the previous problem that are mutually strictly inequivalent is exactly 2^g. To get at this fact, the problem investigates properties of principal ideals $I = (x)$ in R with the properties that $\sigma(I) = I$ and $N(x) > 0$. Since $\sigma(I) = I$, it must be true that $\sigma(x) = \varepsilon x$ for some unit ε in R, and then $N(\sigma(x)) = N(x)$ implies that $N(\varepsilon) = +1$. Matters now split into cases along the lines of the hypotheses of Proposition 1.17.

(a) Under the assumption that $m < 0$ and that m is neither -1 nor -3, show that if a principal ideal $I = (x)$ in R has $\sigma(I) = I$, then x is in $\mathbb{Z}$ or in $\mathbb{Z}\sqrt{m}$.

(b) Under the assumption that $m < 0$, show that the only subsets S of E for which the ideal J_S is principal are $S = \varnothing$ and S equal to the set of all prime divisors of m, i.e., S equal to E for D odd and for D even with $D/4 \equiv 2 \bmod 4$ and S equal to $E - \{2\}$ for D even with $D/4 \equiv 2 \bmod 4$.

(c) Under the assumption that $m < 0$, Proposition 1.17 says that strict equivalence for ideals coincides with equivalence. Show how to conclude from this fact and the results of (a) and (b) that the order of the genus group is 2^g when $m < 0$.

(d) Under the assumption that $m > 0$ and that the fundamental unit ε_1 has norm -1, Proposition 1.17 says that strict equivalence for ideals coincides with equivalence. With I, x, and ε as in the statement of the problem, show that $\varepsilon = \pm\varepsilon_1^{2n}$ for some integer $n \geq 0$. Deduce that $\sigma(\varepsilon_1^n x) = s\varepsilon_1^n x$ for a suitable choice of sign s, and show as a consequence that J_S is principal for the same S's as in (b) and that the order of the genus group is 2^g.

(e) Under the assumption that $m > 0$ and that the fundamental unit ε_1 has norm $+1$, Proposition 1.17 says that strict equivalence for ideals is distinct from equivalence; in particular, there are two strict equivalence classes of principal ideals: those with a generator of positive norm and those with a generator of negative norm. Let y_0^+ and y_0^- be the elements produced by Problem 27 that satisfy $\varepsilon_1 = \sigma(y_0^+)(y_0^+)^{-1}$ and $-\varepsilon_1 = \sigma(y_0^-)(y_0^-)^{-1}$. Prove that exactly one of y_0^+ and y_0^- has positive norm, so that two of the principal ideals (1), (y_0^+), (y_0^-), $(\sqrt{m})$ are strictly equivalent to (1), and two are not. Prove that all four of these principal ideals are of the form J_S and that they are distinct. By expressing elements arising from Problem 27 for the most general unit in R in terms of y_0 and ε_1, show that no other J_S is a principal ideal. Show as a consequence that the number of strict equivalence classes of ideals among the J_S's is 2^g.

Problems 30–34 show that proper equivalence over $\mathbb{Q}$ for two integer forms of fundamental discriminant D implies proper equivalence over $\mathbb{Z}/D\mathbb{Z}$. Consequently the order of the genus group is at most the number of classes of integer forms of discriminant D under proper equivalence over $\mathbb{Z}/D\mathbb{Z}$. It will follow from the next set of problems, concerning "genus characters," that the number of such classes is at least 2^g, where $g + 1$ is the number of distinct prime divisors of D. In combination with Problem 29, this result shows that the number of genera equals 2^g. Throughout this set of problems, let D be a fundamental discriminant.

30. Let (a_1, b_1, c_1) be a binary quadratic form over $\mathbb{Z}$ of discriminant D. Using Lemma 1.10, prove that (a_1, b_1, c_1) is properly equivalent over $\mathbb{Z}$ to a form (a, b, c) of discriminant D such that $\mathrm{GCD}(a, D) = 1$.

31. Suppose that (a, b, c) is a binary quadratic form over $\mathbb{Z}$ of discriminant D such that $\mathrm{GCD}(a, D) = 1$.
    (a) Prove that if D is odd, then (a, b, c) is properly equivalent over $\mathbb{Z}$ to a form (a, kD, lD) for some integers k and l.
    (b) Prove that if D is even, then (a, b, c) is properly equivalent over $\mathbb{Z}$ to a form $(a, 2kD, -a(D/4) + lD)$ for some integers k and l.

32. Suppose that (a, kD, lD) is a form over $\mathbb{Z}$ having odd discriminant D, satisfying $\mathrm{GCD}(a, D) = 1$, and taking on an integer value r relatively prime to D for some rational (x, y). Write x and y as fractions with a positive common denominator as small as possible: $x = u/w$ and $y = v/w$.
    (a) Prove that $\mathrm{GCD}(w, D) = 1$, and conclude that $a \equiv d^2 r \bmod D$ for some integer d relatively prime to D.
    (b) Suppose that $(a', k'D, l'D)$ is a second form over $\mathbb{Z}$ having discriminant D, satisfying $\mathrm{GCD}(a', D) = 1$, and taking on the value r at some rational point. Prove that $a' \equiv as^2 \bmod D$ for some s relatively prime to D.
    (c) Suppose that (a, b, c) and (a', b', c') are forms over $\mathbb{Z}$ of the same odd discriminant with $\mathrm{GCD}(a, D) = \mathrm{GCD}(a', D) = 1$, and suppose that these forms are properly equivalent over $\mathbb{Q}$. Deduce that (a, b, c) and (a', b', c') are properly equivalent over $\mathbb{Z}/D\mathbb{Z}$ in the sense that there exists a matrix $\begin{pmatrix} \alpha & \beta \\ \gamma & \delta \end{pmatrix}$ in $\mathrm{SL}(2, \mathbb{Z}/D\mathbb{Z})$ such that substitution of $x = \alpha x' + \beta y'$ and $y = \gamma x' + \delta y'$ leads from $ax^2 + bxy + cy^2$ modulo D to $a'x'^2 + b'x'y' + c'y'^2$ modulo D.

33. Suppose that $(a, 2kD, -a(D/4) + lD)$ is a form over $\mathbb{Z}$ having even discriminant D, satisfying $\mathrm{GCD}(a, D) = 1$, and taking on an integer value r relatively prime to D for some rational (x, y). Write x and y as fractions with a positive common denominator as small as possible: $x = u/w$ and $y = v/w$.
    (a) Prove that $\mathrm{GCD}(w, D) = 1$, and obtain a congruence relating a and r modulo D.

(b) Suppose that $(a', 2k'D, -a'(D/4) + l'D)$ is a second form over $\mathbb{Z}$ having discriminant D, satisfying $\operatorname{GCD}(a', D) = 1$, and taking on the value r at some rational point. Prove that $\left(\frac{a}{p}\right) = \left(\frac{a'}{p}\right)$ for every odd prime p dividing D.

(c) In the setting of (b), suppose in addition that $D/4 \equiv 3 \bmod 4$. Prove that $a \equiv a' \bmod 4$.

(d) In the setting of (b), suppose in addition that $D/4 \equiv 2 \bmod 4$. Prove for $D/4 \equiv 2 \bmod 8$ that $a' \equiv \pm a \bmod 8$, and prove for $D/4 \equiv 6 \bmod 8$ that either $a' \equiv a \bmod 8$ or $a' \equiv 3a \bmod 8$.

(e) Suppose that (a, b, c) and (a', b', c') are forms over $\mathbb{Z}$ of the same even discriminant with $\operatorname{GCD}(a, D) = \operatorname{GCD}(a', D) = 1$, and suppose that these forms are properly equivalent over $\mathbb{Q}$. Deduce that (a, b, c) and (a', b', c') are properly equivalent over $\mathbb{Z}/D\mathbb{Z}$.

34. Why does it follow from Problems 30–33 that the order of the genus group for discriminant D is at least as large as the number of proper equivalence classes under $\operatorname{SL}(2, \mathbb{Z}/D\mathbb{Z})$ of integer forms of discriminant D?

Problems 35–40 introduce "genus characters." In fact, genus characters are already implicit in Problems 32 and 33. Throughout this set of problems, let D be a fundamental discriminant, and suppose that D has exactly $g + 1$ distinct prime factors. The content of these problems will be summarized in Problem 40. Call two binary quadratic forms over $\mathbb{Z}$ of discriminant D **similar modulo** D if they take on the same residues r modulo D that are relatively prime to D. Proper equivalence over $\mathbb{Z}$ via $\operatorname{SL}(2, \mathbb{Z})$ implies proper equivalence modulo D via $\operatorname{SL}(2, \mathbb{Z}/D\mathbb{Z})$, and this in turn implies similarity modulo D in the sense that was just defined. Problems 30–31 show that it is enough to study forms $ax^2 \bmod D$ for D odd, where $\operatorname{GCD}(a, D) = 1$, and to study forms $a(x^2 - (D/4)y^2)$ for D even, again where $\operatorname{GCD}(a, D) = 1$. Initially the **genus characters** are functions of pairs (similarity class, r), where r is a residue modulo D with $\operatorname{GCD}(r, D) = 1$ such that r is represented by the form modulo D. The values of these functions are $\left(\frac{r}{p}\right)$ for each odd prime $p > 0$ dividing D, as well as the indicated one of the following for $p = 2$ if D is even:

$$\xi(r) = \left(\frac{-1}{r}\right) = (-1)^{\frac{1}{2}(r-1)} \qquad \text{if } D \text{ is even and } D/4 \equiv 3 \bmod 4,$$

$$\eta(r) = \left(\frac{2}{r}\right) = (-1)^{\frac{1}{8}(r^2-1)} \qquad \text{if } D \text{ is even and } D/4 \equiv 2 \bmod 8,$$

$$\xi(r)\eta(r) = \left(\frac{-2}{r}\right) = (-1)^{\frac{1}{2}(r-1)+\frac{1}{8}(r^2-1)} \qquad \text{if } D \text{ is even and } D/4 \equiv 6 \bmod 8.$$

Thus $g + 1$ expressions have been defined for each ordered pair (similarity class, r).

35. Using Problems 32 and 33, show that the genus characters are independent of the residue r modulo D with $\operatorname{GCD}(r, D) = 1$ such that r is represented by the form modulo D. Therefore the residue a in the quadratic form, either $ax^2 \bmod D$ for D odd or $a(x^2 - (D/4)y^2)$ for D even, can be used as r, and the genus characters are $g + 1$ functions defined on the set of similarity classes modulo D.

36. Prove that the genus characters respect the operation of multiplication of proper equivalence classes of forms over $\mathbb{Z}$.

37. The product of all $g+1$ genus characters is 1 in every case. A sketch of the argument for D odd is as follows: Since $D \equiv 1 \mod 4$, D has an even number $2t$ of prime factors $4k+3$. Use of the Jacobi symbol with a odd and p varying over the (odd) prime divisors of D gives

$$\prod_p \left(\tfrac{a}{p}\right) = \prod_{p=4k+1}\left(\tfrac{a}{p}\right) \prod_{p=4k+3}\left(\tfrac{a}{p}\right) = \xi(a)^{2t} \prod_{p=4k+1}\left(\tfrac{p}{a}\right) \prod_{p=4k+3}\left(\tfrac{p}{a}\right) = \left(\tfrac{D}{a}\right),$$

and the right side is $+1$ by Problem 2a. Using this sketch as a guide, show that the product of all $g+1$ genus characters is 1 for the cases that D is even and
   (a) $D/4 \equiv 3 \mod 4$,
   (b) $D/4 \equiv 2 \mod 8$,
   (c) $D/4 \equiv 6 \mod 8$.

38. If D is even, let α be ξ if $D/4 \equiv 3 \mod 4$, η if $D/4 \equiv 2 \mod 8$, and $\xi\eta$ if $D/4 \equiv 6 \mod 8$. Let $p \mapsto s_p$ be any function to $\{\pm 1\}$ from the set of distinct prime divisors of D. Using Dirichlet's Theorem on primes in arithmetic progressions, prove that there exists a prime q such that $\left(\tfrac{q}{p}\right) = s_p$ for each odd prime divisor p of D and $\alpha(q) = s_2$ in case D is even.

39. With α as in the previous problem, let $p \mapsto s_p$ be any function to $\{\pm 1\}$ from the set of distinct prime divisors of D such that $\prod_p s_p = +1$, and choose a prime q as in the previous problem. Prove that q is primitively representable by some integer binary quadratic form of discriminant D and that the values of the genus characters on this form are the numbers s_p. Conclude that the number of distinct similarity classes modulo D is at least 2^g.

40. For the quadratic number field $K = \mathbb{Q}(\sqrt{m})$ with discriminant D, suppose that D has $g+1$ distinct prime divisors. Conclude that the following equivalence classes of binary quadratic forms over $\mathbb{Z}$ of discriminant D coincide and that the number of such classes is 2^g:
   (i) classes relative to proper equivalence over $\mathbb{Q}$, i.e., genera,
   (ii) classes relative to proper equivalence over $\mathbb{Z}/D\mathbb{Z}$,
   (iii) classes relative to similarity modulo D.

CHAPTER II

Wedderburn–Artin Ring Theory

Abstract. This chapter studies finite-dimensional associative division algebras, as well as other finite-dimensional associative algebras and closely related rings. The chapter is in two parts that overlap slightly in Section 6. The first part gives the structure theory of the rings in question, and the second part aims at understanding limitations imposed by the structure of a division ring.

Section 1 briefly summarizes the structure theory for finite-dimensional (nonassociative) Lie algebras that was the primary historical motivation for structure theory in the associative case. All the algebras in this chapter except those explicitly called Lie algebras are understood to be associative.

Section 2 introduces left semisimple rings, defined as rings R with identity such that the left R module R is semisimple. Wedderburn's Theorem says that such a ring is the finite product of full matrix rings over division rings. The number of factors, the size of each matrix ring, and the isomorphism class of each division ring are uniquely determined. It follows that left semisimple and right semisimple are the same. If the ring is a finite-dimensional algebra over a field F, then the various division rings are finite-dimensional division algebras over F. The factors of semisimple rings are simple, i.e., are nonzero and have no nontrivial two-sided ideals, but an example is given to show that a simple ring need not be semisimple. Every finite-dimensional simple algebra is semisimple.

Section 3 introduces chain conditions into the discussion as a useful generalization of finite dimensionality. A ring R with identity is left Artinian if the left ideals of the ring satisfy the descending chain condition. Artin's Theorem for simple rings is that left Artinian is equivalent to semisimplicity, hence to the condition that the given ring be a full matrix ring over a division ring.

Sections 4–6 concern what happens when the assumption of semisimplicity is dropped but some finiteness condition is maintained. Section 4 introduces the Wedderburn–Artin radical rad R of a left Artinian ring R as the sum of all nilpotent left ideals. The radical is a two-sided nilpotent ideal. It is 0 if and only if the ring is semisimple. More generally $R/\text{rad } R$ is always semisimple if R is left Artinian. Sections 5–6 state and prove Wedderburn's Main Theorem—that a finite-dimensional algebra R with identity over a field F of characteristic 0 has a semisimple subalgebra S such that R is isomorphic as a vector space to $S \oplus \text{rad } R$. The semisimple algebra S is isomorphic to $R/\text{rad } R$. Section 5 gives the hard part of the proof, which handles the special case that $R/\text{rad } R$ is isomorphic to a product of full matrix algebras over F. The remainder of the proof, which appears in Section 6, follows relatively quickly from the special case in Section 5 and an investigation of circumstances under which the tensor product over F of two semisimple algebras is semisimple. Such a tensor product is not always semisimple, but it is semisimple in characteristic 0.

The results about tensor products in Section 6, but with other hypotheses in place of the condition of characteristic 0, play a role in the remainder of the chapter, which is aimed at identifying certain division rings. Sections 7–8 provide general tools. Section 7 begins with further results about tensor products. Then the Skolem–Noether Theorem gives a relationship between any two homomorphisms of a simple subalgebra into a simple algebra whose center coincides with the underlying field of

1. Historical Motivation

scalars. Section 8 proves the Double Centralizer Theorem, which says for this situation that the centralizer of the simple subalgebra in the whole algebra is simple and that the product of the dimensions of the subalgebra and the centralizer is the dimension of the whole algebra.

Sections 9–10 apply the results of Sections 6–8 to obtain two celebrated theorems—Wedderburn's Theorem about finite division rings and Frobenius's Theorem classifying the finite-dimensional associative division algebras over the reals.

1. Historical Motivation

Elementary ring theory came from several sources historically and was already in place by 1880. Some of the sources are field theory (studied by Galois and others), rings of algebraic integers (studied by Gauss, Dirichlet, Kummer, Kronecker, Dedekind, and others), and matrices (studied by Cayley, Hamilton, and others). More advanced general ring theory arose initially not on its own but as an effort to imitate the theory of "Lie algebras," which began about 1880.

A brief summary of some early theorems about Lie algebras will put matters in perspective. The term "algebra" in connection with a field F refers at least to an F vector space with a multiplication that is F bilinear. This chapter will deal only with two kinds of such algebras, the Lie algebras and those algebras whose multiplication is associative. If the modifier "Lie" is absent, the understanding is that the algebra is associative.

Lie algebras arose originally from "Lie groups"—which we can regard for current purposes as connected groups with finitely many smooth parameters—by a process of taking derivatives along curves at the identity element of the group. Precise knowledge of that process will be unnecessary in our treatment, but we describe one example: The vector space $M_n(\mathbb{R})$ of all n-by-n matrices over $\mathbb{R}$ becomes a Lie algebra with multiplication defined by the "bracket product" $[X, Y] = XY - YX$. If G is a closed subgroup of the matrix group $GL(n, \mathbb{R})$ and $\mathfrak{g}$ is the set of all members of $M_n(\mathbb{R})$ of the form $X = c'(0)$, where c is a smooth curve in G with $c(0)$ equal to the identity, then it turns out that the vector space $\mathfrak{g}$ is closed under the bracket product and is a Lie algebra. Although one might expect the Lie algebra $\mathfrak{g}$ to give information about the Lie group G only infinitesimally at the identity, it turns out that $\mathfrak{g}$ determines the multiplication rule for G in a whole open neighborhood of the identity. Thus the Lie group and Lie algebra are much more closely related than one might at first expect.

We turn to the underlying definitions and early main theorems about Lie algebras. Let F be a field. A vector space A over F with an F bilinear multiplication $(X, Y) \mapsto [X, Y]$ is a **Lie algebra** if the multiplication has the two properties
   (i) $[X, X] = 0$ for all $X \in A$,
   (ii) (**Jacobi identity**) $[X, [Y, Z]] + [Y, [Z, X]] + [Z, [X, Y]] = 0$ for all $X, Y, Z \in A$.

Multiplication is often referred to as **bracket**. It is usually not associative. The vector space $M_n(F)$ with $[X, Y] = XY - YX$ is a Lie algebra, as one easily checks by expanding out the various brackets that are involved; it is denoted by $\mathfrak{gl}(n, F)$.

The elementary structural definitions with Lie algebras run parallel to those with rings. A **Lie subalgebra** S of A is a vector subspace closed under brackets, an **ideal** I of A is a vector subspace such that $[X, Y]$ is in I for $X \in I$ and $Y \in A$, a **homomorphism** $\varphi : A_1 \to A_2$ of Lie algebras is a linear mapping respecting brackets in the sense that $\varphi[X, Y] = [\varphi(X), \varphi(Y)]$ for all $X, Y \in A_1$, and an **isomorphism** is an invertible homomorphism. Every ideal is a Lie subalgebra. In contrast to the case of rings, there is no distinction between "left ideals" and "right ideals" because the bracket product is skew symmetric. Under the passage from Lie groups to Lie algebras, abelian Lie groups yield Lie algebras with all brackets 0, and thus one says that a Lie algebra is **abelian** if all its brackets are 0.

Examples of Lie subalgebras of $\mathfrak{gl}(n, F)$ are the subalgebra $\mathfrak{sl}(n, F)$ of all matrices of trace 0, the subalgebra $\mathfrak{so}(n, F)$ of all skew-symmetric matrices, and the subalgebra of all upper-triangular matrices.

The elementary properties of subalgebras, homomorphisms, and so on for Lie algebras mimic what is true for rings: The kernel of a homomorphism is an ideal. Any ideal is the kernel of a quotient homomorphism. If I is an ideal in A, then the ideals of A/I correspond to the ideals of A containing I, just as in the First Isomorphism Theorem for rings. If I and J are ideals in A, then $(I + J)/I \cong J/(I \cap J)$, just as in the Second Isomorphism Theorem for rings.

The connection of Lie algebras to Lie groups makes one want to introduce definitions that lead toward classifying all Lie algebras that are finite-dimensional. We therefore assume for the remainder of this section that all Lie algebras under discussion are finite-dimensional over F. Some of the steps require conditions on F, and we shall assume that F has characteristic 0.

Group theory already had a notion of "solvable group" from Galois, and this leads to the notion of solvable Lie algebra. In A, let $[A, A]$ denote the linear span of all $[X, Y]$ with $X, Y \in A$; $[A, A]$ is called the **commutator ideal** of A, and $A/[A, A]$ is abelian. In fact, $[A, A]$ is the smallest ideal I in A such that A/I is abelian. Starting from A, let us form successive commutator ideals. Thus put $A_0 = A$, $A_1 = [A_0, A_0], \ldots, A_n = [A_{n-1}, A_{n-1}]$, so that

$$A = A_0 \supseteq A_1 \supseteq \cdots \supseteq A_n \supseteq \cdots.$$

The terms of this sequence are all the same from some point on, by finite dimensionality, and we say that A is **solvable** if the terms are ultimately 0. One easily checks that the sum $I + J$ of two solvable ideals in A, i.e., the set of sums, is a solvable ideal. By finite dimensionality, there exists a unique largest solvable ideal. This is called the **radical** of A and is denoted by rad A. The Lie algebra

A is said to be **semisimple** if rad $A = 0$. It is easy to use the First Isomorphism Theorem to check that $A/\operatorname{rad} A$ is always semisimple.

In the direction of classifying Lie algebras, one might therefore want to see how all solvable Lie algebras can be constructed by successive extensions, identify all semisimple Lie algebras, and determine how a general Lie algebra can be constructed from a semisimple Lie algebra and a solvable Lie algebra by an extension.

The first step in this direction historically concerned identifying semisimple Lie algebras. We say that the Lie algebra A is **simple** if dim $A > 1$ and if A contains no nonzero proper ideals.

Working with the field $\mathbb{C}$ but in a way that applies to other fields of characteristic 0, W. Killing proved in 1888 that A is semisimple if and only if A is the (internal) direct sum of simple ideals. In this case the direct summands are unique, and the only ideals in A are the partial direct sums.

This result is strikingly different from what happens for abelian Lie algebras, for which the theory reduces to the theory of vector spaces. A 2-dimensional vector space is the internal direct sum of two 1-dimensional subspaces in many ways. But Killing's theorem says that the decomposition of semisimple Lie algebras into simple ideals is unique, not just unique up to some isomorphism.

É. Cartan in his 1894 thesis classified the simple Lie algebras, up to isomorphism, for the case that the field is $\mathbb{C}$. The Lie algebras $\mathfrak{sl}(n, \mathbb{C})$ for $n \geq 2$ and $\mathfrak{so}(n, \mathbb{C})$ for $n = 3$ and $n \geq 5$ were in his list, and there were others. Killing had come close to this classification in his 1888 work, but he had made a number of errors in both his statements and his proofs.

E. E. Levi in 1905 addressed the extension problem for obtaining all finite-dimensional Lie algebras over $\mathbb{C}$ from semisimple ones and solvable ones. His theorem is that for any Lie algebra A, there exists a subalgebra S isomorphic to $A/\operatorname{rad} A$ such that $A = S \oplus \operatorname{rad} A$ as vector spaces. In essence, this result says that the extension defining A is given by a semidirect product.

The final theorem in this vein at this time in history was a 1914 result of Cartan classifying the simple Lie algebras when the field F is $\mathbb{R}$. This classification is a good bit more complicated than the classification when F is $\mathbb{C}$.

With this background in mind, we can put into context the corresponding developments for associative algebras. Although others had done some earlier work, J. H. M. Wedderburn made the first big advance for associative algebras in 1905. Wedderburn's theory in a certain sense is more complicated than the theory for Lie algebras because left ideals in the associative case are not necessarily two-sided ideals. Let us sketch this theory.

For the remainder of this section until the last paragraph, A will denote a finite-dimensional associative algebra over a field F of characteristic 0, possibly the 0 algebra. We shall always assume that A has an identity. Although we shall make some definitions here, we shall repeat them later in the chapter at the appropriate times. For many results later in the chapter, the field F will not be assumed to be of characteristic 0.

As in Chapter X of *Basic Algebra*, a unital left A module M is said to be simple if it is nonzero and it has no proper nonzero A submodules, semisimple if it is the sum (or equivalently the direct sum) of simple A submodules. The algebra A is **semisimple** if the left A module A is a semisimple module, i.e., if A is the direct sum of simple left ideals; A is **simple** if it is nonzero and has no nontrivial two-sided ideals. In contrast to the setting of Lie algebras, we make no exception for the 1-dimensional case; this distinction is necessary and is continually responsible for subtle differences between the two theories.

Wedderburn's first theorem has two parts to it, the first one modeled on Killing's theorem for Lie algebras and the second one modeled on Cartan's thesis:

(i) The algebra A is semisimple if and only if it is the (internal) direct sum of simple two-sided ideals. In this case the direct summands are unique, and the only two-sided ideals of A are the partial direct sums.

(ii) The algebra A is simple if and only if $A \cong M_n(D)$ for some integer $n \geq 1$ and some division algebra D over F. In particular, if F is algebraically closed, then $A \cong M_n(F)$ for some n.

E. Artin generalized the Wedderburn theory to a suitable kind of "semisimple ring." For part of the theory, he introduced a notion of "radical" for the associative case—the **radical** of a finite-dimensional associative algebra A being the sum of the "nilpotent" left ideals of A. Here a left ideal I is called **nilpotent** if $I^k = 0$ for some k. The radical rad A is a two-sided ideal, and $A/\operatorname{rad} A$ is a semisimple ring.

Wedderburn's Main Theorem, proved later in time and definitely assuming characteristic 0, is an analog for associative algebras of Levi's result about Lie algebras. The result for associative algebras is that A decomposes as a vector-space direct sum $A = S \oplus \operatorname{rad} A$, where S is a semisimple subalgebra isomorphic to $A/\operatorname{rad} A$.

The remaining structural question for finite-dimensional associative algebras is to say something about simple algebras when the field is not algebraically closed. Such a result may be regarded as an analog of the 1914 work by Cartan. In the associative case one then wants to know what the F isomorphism classes of finite-dimensional associative division algebras D are for a given field F. We now drop the assumption that the field F has characteristic 0. In asking this question, one does not want to repeat the theory of field extensions. Consequently one looks only for classes of division algebras whose center is F. If F is algebraically closed, the only such D is F itself, as we shall observe in more detail in Section 2.

If F is a finite field, one is led to another theorem of Wedderburn's, saying that D has to be commutative and hence that $D = F$; this theorem appears in Section 9. If F is $\mathbb{R}$, one is led to a theorem of Frobenius saying that there are just two such D's up to $\mathbb{R}$ isomorphism, namely $\mathbb{R}$ itself and the quaternions $\mathbb{H}$; this theorem appears in Section 10. For a general field F, it turns out that the set of classes of finite-dimensional division algebras with center F forms an abelian group. The group is called the "Brauer group" of F. Its multiplication is defined by the condition that the class of D_1 times D_2 is the class of a division algebra D_3 such that $D_1 \otimes_F D_2 \cong M_n(D_3)$ for some n; the inverse of the class of D is the class of the opposite algebra D^o, and the identity is the class of F. The study of the Brauer group is postponed to Chapter III. This group has an interpretation in terms of cohomology of groups, and it has applications to algebraic number theory.

2. Semisimple Rings and Wedderburn's Theorem

We now begin our detailed investigation of associative algebras over a field. In this section we shall address the first theorem of Wedderburn's that is mentioned in the previous section. It has two parts, one dealing with semisimple algebras and one dealing with finite-dimensional simple algebras. The first part does not need the finite dimensionality as a hypothesis, and we begin with that one.

Let R be a ring with identity. The ring R is **left semisimple** if the left R module R is a semisimple module, i.e., if R is the direct sum of minimal left ideals.[1] In this case $R = \bigoplus_{i \in S} I_i$ for some set S and suitable minimal left ideals I_i. Since R has an identity, we can decompose the identity according to the direct sum as $1 = 1_{i_1} + \cdots + 1_{i_n}$ for some finite subset $\{i_1, \ldots, i_n\}$ of S, where 1_{i_k} is the component of 1 in I_{i_k}. Multiplying by $r \in R$ on the left, we see that $R \subseteq \bigoplus_{k=1}^n I_{i_k}$. Consequently R has to be a *finite* sum of minimal left ideals. A ring R with identity is **right semisimple** if the right R module R is a semisimple module. We shall see later in this section that left semisimple and right semisimple are equivalent.

EXAMPLES OF SEMISIMPLE RINGS.

(1) If D is a division ring, then we saw in Example 4 in Section X.1 of *Basic Algebra* that the ring $R = M_n(D)$ is left semisimple in the sense of the above definition. Actually, that example showed more. It showed that R as a left R module is given by $M_n(D) \cong D^n \oplus \cdots \oplus D^n$, where each D^n is a simple left R module and the jth summand D^n corresponds to the matrices whose only nonzero entries are in the jth column. The left R module $M_n(D)$ has a composition series whose terms are the partial sums of the n summands D^n. If M is any simple left $M_n(D)$ module and if $x \neq 0$ is in M, then $M = M_n(D)x$. If we set $I = \{r \in M_n(D) \mid rx = 0\}$, then I is a left ideal in $M_n(D)$ and $M \cong M_n(D)/I$

---
[1] By convention, a "minimal left ideal" always means a "minimal nonzero left ideal."

as a left $M_n(D)$ module. In other words, M is an irreducible quotient module of the left $M_n(D)$ module $M_n(D)$. By the Jordan–Hölder Theorem (Corollary 10.7 of *Basic Algebra*), M occurs as a composition factor. Hence $M \cong D^n$ as a left $M_n(D)$ module. Hence every simple left $M_n(D)$ module is isomorphic to D^n. We shall use this style of argument repeatedly but will ordinarily include less detail.

(2) If $R_1, \ldots, R_n$ are left semisimple rings, then the direct product $R = \prod_{i=1}^n R_i$ is left semisimple.[2] In fact, each minimal left ideal of R_i, when included into R, is a minimal left ideal of R. Hence R is the sum of minimal left ideals and is left semisimple. By the same kind of argument as for Example 1, every simple left R module is isomorphic to one of these minimal left ideals.

Lemma 2.1. Let D be a division ring, let $R = M_n(D)$, and let D^n be the simple left R module of column vectors. Each member of D acts on D^n by scalar multiplication on the *right* side, yielding a member of $\mathrm{End}_R(D^n)$. In turn, $\mathrm{End}_R(D^n)$ is a ring, and this identification therefore is an inclusion of the members of D into the right D module $\mathrm{End}_R(D^n)$. The inclusion is in fact an isomorphism of rings: $D^o \cong \mathrm{End}_R(D^n)$, where D^o is the opposite ring of D.

PROOF. Let $\varphi : D \to \mathrm{End}_R(D^n)$ be the function given by $\varphi(d)(v) = vd$. Then $\varphi(dd')(v) = v(dd') = (vd)d' = \varphi(d')(vd) = \varphi(d')(\varphi(d)(v))$. Since the order of multiplication in D is reversed by φ, φ is a ring homomorphism of D^o into $\mathrm{End}_R(D^n)$. It is one-one because D^o is a division ring and has no nontrivial two-sided ideals. To see that it is onto $\mathrm{End}_R(D^n)$, let f be in $\mathrm{End}_R(D^n)$. Put

$$f\begin{pmatrix} 1 \\ 0 \\ \vdots \\ 0 \end{pmatrix} = \begin{pmatrix} d \\ d_2 \\ \vdots \\ d_n \end{pmatrix}.$$ Since f is an R module homomorphism,

$$f\begin{pmatrix} a_1 \\ a_2 \\ \vdots \\ a_n \end{pmatrix} = f\left(\begin{pmatrix} a_1 & 0 & \cdots & 0 \\ a_2 & 0 & \cdots & 0 \\ \vdots & & & \\ a_n & 0 & \cdots & 0 \end{pmatrix}\begin{pmatrix} 1 \\ 0 \\ \vdots \\ 0 \end{pmatrix}\right) = \begin{pmatrix} a_1 & 0 & \cdots & 0 \\ a_2 & 0 & \cdots & 0 \\ \vdots & & & \\ a_n & 0 & \cdots & 0 \end{pmatrix} f\begin{pmatrix} 1 \\ 0 \\ \vdots \\ 0 \end{pmatrix}$$

$$= \begin{pmatrix} a_1 & 0 & \cdots & 0 \\ a_2 & 0 & \cdots & 0 \\ \vdots & & & \\ a_n & 0 & \cdots & 0 \end{pmatrix}\begin{pmatrix} d \\ d_2 \\ \vdots \\ d_n \end{pmatrix} = \begin{pmatrix} a_1 d \\ a_2 d \\ \vdots \\ a_n d \end{pmatrix} = \varphi(d)\begin{pmatrix} a_1 \\ a_2 \\ \vdots \\ a_n \end{pmatrix}.$$

Therefore $\varphi(d) = f$, and φ is onto. □

---

[2] Some comment is appropriate about the notation $R = \prod_{i=1}^n R_i$ and the terminology "direct product." Indeed, $\prod_{i=1}^n R_i$ is a product in the sense of category theory within the category of rings or the category of rings with identity. Sometimes one views R alternatively as built from n two-sided ideals, each corresponding to one of the n coordinates; in this case, one may say that R is the "direct sum" of these ideals. This direct sum is to be regarded as a direct sum of abelian groups, or perhaps vector spaces or R modules, but it is not a coproduct within the category of rings with identity.

2. Semisimple Rings and Wedderburn's Theorem

Theorem 2.2 (Wedderburn). If R is any left semisimple ring, then

$$R \cong M_{n_1}(D_1) \times \cdots \times M_{n_r}(D_r)$$

for suitable division rings $D_1, \ldots, D_n$ and positive integers $n_1, \ldots, n_r$. The number r is uniquely determined by R, and the ordered pairs $(n_1, D_1), \ldots, (n_r, D_r)$ are determined up to a permutation of $\{1, \ldots, r\}$ and an isomorphism of each D_j. There are exactly r mutually nonisomorphic simple left R modules, namely $(D_1)^{n_1}, \ldots, (D_r)^{n_r}$.

PROOF. Write R as the direct sum of minimal left ideals, and then regroup the summands according to their R isomorphism type as $R \cong \bigoplus_{j=1}^{r} n_j V_j$, where $n_j V_j$ is the direct sum of n_j submodules R isomorphic to V_j and where $V_i \not\cong V_j$ for $i \neq j$. The isomorphism is one of unital left R modules. Put $D_i^o = \mathrm{End}_R(V_i)$. This is a division ring by Schur's Lemma (Proposition 10.4b of *Basic Algebra*). Using Proposition 10.14 of *Basic Algebra*, we obtain an isomorphism of rings

$$R^o \cong \mathrm{End}_R R \cong \mathrm{Hom}_R\left(\bigoplus_{i=1}^{r} n_i V_i, \bigoplus_{j=1}^{r} n_j V_j\right). \qquad (*)$$

Define $p_i : \bigoplus_{j=1}^{r} n_j V_j \to n_i V_i$ to be the ith projection and $q_i : n_i V_i \to \bigoplus_{j=1}^{r} n_j V_j$ to be the ith inclusion. Let us see that the right side of $(*)$ is isomorphic as a ring to $\prod_i \mathrm{End}_R(n_i V_i)$ via the mapping $f \mapsto (p_1 f q_1, \ldots, p_r f q_r)$. What is to be shown is that $p_j f q_i = 0$ for $i \neq j$. Here $p_j f q_i$ is a member of $\mathrm{Hom}_R(n_i V_i, n_j V_j)$. The abelian group $\mathrm{Hom}_R(n_i V_i, n_j V_j)$ is the direct sum of abelian groups isomorphic to $\mathrm{Hom}_R(V_i, V_j)$ by Proposition 10.12, and each $\mathrm{Hom}_R(V_i, V_j)$ is 0 by Schur's Lemma (Proposition 10.4a).

Referring to $(*)$, we therefore obtain ring isomorphisms

$$R^o \cong \prod_{i=1}^{r} \mathrm{Hom}_R(n_i V_i, n_i V_i) = \prod_{i=1}^{r} \mathrm{End}_R(n_i V_i)$$

$$\cong \prod_{i=1}^{r} M_{n_i}(\mathrm{End}_R(V_i)) \qquad \text{by Corollary 10.13}$$

$$\cong \prod_{i=1}^{r} M_{n_i}(D_i^o) \qquad \text{by definition of } D_i^o.$$

Reversing the order of multiplication in R^o and using the transpose map to reverse the order of multiplication in each $M_{n_i}(D_i^o)$, we conclude that $R \cong \prod_{i=1}^{r} M_{n_i}(D_i)$. This proves existence of the decomposition in the theorem.

We still have to identify the simple left R modules and prove an appropriate uniqueness statement. As we recalled in Example 1, we have a decomposition

$M_{n_i}(D_i) \cong D_i^{n_i} \oplus \cdots \oplus D_i^{n_i}$ of left $M_{n_i}(D_i)$ modules, and each term $D_i^{n_i}$ is a simple left $M_{n_i}(D_i)$ module. The decomposition just proved allows us to regard each term $D_i^{n_i}$ as a simple left R module, $1 \le i \le r$. Each of these modules is acted upon by a different coordinate of R, and hence we have produced at least r nonisomorphic simple left R modules. Any simple left R module must be a quotient of R by a maximal left ideal, as we observed in Example 2, hence a composition factor as a consequence of the Jordan–Hölder Theorem. Thus it must be one of the V_j's in the previous part of the proof. There are only r nonisomorphic such V_j's, and we conclude that the number of simple left R modules, up to isomorphism, is exactly r.

For uniqueness suppose that $R \cong M_{n'_1}(D'_1) \times \cdots \times M_{n'_s}(D'_s)$ as rings. Let $V'_j = (D'_j)^{n'_j}$ be the unique simple left $M_{n'_j}(D'_j)$ module up to isomorphism, and regard V'_j as a simple left R module. Then we have $R \cong \bigoplus_{j=1}^s n'_j V'_j$ as left R modules. By the Jordan–Hölder Theorem we must have $r = s$ and, after a suitable renumbering, $n_i = n'_i$ and $V_i \cong V'_i$ for $1 \le i \le r$. Thus we have ring isomorphisms

$$(D'_i)^o \cong \mathrm{End}_{M_{n'_i}(D'_i)}(V'_i) \qquad \text{by Lemma 2.1}$$
$$\cong \mathrm{End}_R(V'_i)$$
$$\cong \mathrm{End}_R(V_i) \qquad \text{since } V_i \cong V'_i$$
$$\cong D_i^o.$$

Reversing the order of multiplication gives $D'_i \cong D_i$, and the proof is complete. □

Corollary 2.3. For a ring R, left semisimple coincides with right semisimple.

REMARK. Therefore we can henceforth refer to left semisimple rings unambiguously as **semisimple**.

PROOF. The theorem gives the form of any left semisimple ring, and each ring of this form is certainly right semisimple. □

Wedderburn's original formulation of Theorem 2.2 was for algebras over a field F, and he assumed finite dimensionality. The theorem in this case gives

$$R \cong M_{n_1}(D_1) \times \cdots \times M_{n_r}(D_r),$$

and the proof shows that $D_i^o \cong \mathrm{End}_R(V_i)$, where V_i is a minimal left ideal of R of the ith isomorphism type. The field F lies inside $\mathrm{End}_R(V_i)$, each member of F yielding a scalar mapping, and hence each D_i is a division algebra over F. Each D_i is necessarily finite-dimensional over F, since R was assumed to be finite-dimensional.

We shall make occasional use in this chapter of the fact that if D is a finite-dimensional division algebra over an algebraically closed field F, then $D = F$. To see this equality, suppose that x is a member of D but not of F, i.e., is not an F multiple of the identity. Then x and F together generate a subfield $F(x)$ of D that is a nontrivial algebraic extension of F, contradiction. Consequently every finite-dimensional semisimple algebra R over an algebraically closed field F is of the form

$$R \cong M_{n_1}(F) \times \cdots \times M_{n_r}(F),$$

for suitable integers $n_1, \ldots, n_r$.

As we saw, the finite dimensionality plays no role in decomposing semisimple rings as the finite product of rings that we shall call "simple." The place where finite dimensionality enters the discussion is in identifying simple rings as semisimple, hence in establishing a converse theorem that every finite direct product of simple rings, each equal to an ideal of the given ring, is necessarily semisimple. We say that a nonzero ring R with identity is **simple** if its only two-sided ideals are 0 and R.

EXAMPLES OF SIMPLE RINGS.

(1) If D is a division ring, then $M_n(D)$ is a simple ring. In fact, let J be a two-sided ideal in $M_n(D)$, fix an ordered pair (i, j) of indices, and let

$$I = \{x \in D \mid \text{some member } X \text{ of } J \text{ has } X_{ij} = x\}.$$

Multiplying X in this definition on each side by scalar matrices with entries in D, we see that I is a two-sided ideal in D. If $I = 0$ for all (i, j), then $J = 0$. So assume for some (i, j) that $I \neq 0$. Then $I = D$ for that (i, j), and we may suppose that some X in J has $X_{ij} = 1$. If E_{kl} denotes the matrix that is 1 in the $(k, l)^{\text{th}}$ place and is 0 elsewhere, then $E_{ii} X E_{jj} = E_{ij}$ has to be in J. Hence $E_{kl} = E_{ki} E_{ij} E_{jl}$ has to be in J, and $J = M_n(D)$.

(2) Let R be the **Weyl algebra** over $\mathbb{C}$ in one variable, namely

$$R = \left\{ \sum_{n \geq 0} P_n(x) \left(\frac{d}{dx}\right)^n \,\bigg|\, \text{each } P_n \text{ is in } \mathbb{C}[x], \text{ and the sum is finite} \right\}.$$

To give a more abstract construction of R, we can view R as $\mathbb{C}[x, \frac{d}{dx}]$ subject to the relation $\frac{d}{dx} x = x \frac{d}{dx} + 1$; this is not to be a quotient of a polynomial algebra in two variables but a quotient of a tensor algebra in two variables. We omit the details. We shall now prove that the ring R is simple but not semisimple.

To see that R is a simple ring, we easily check the two identities

(i) $\frac{d}{dx}\left(x^m \frac{d^n}{dx^n}\right) = m x^{m-1} \frac{d^n}{dx^n} + x^m \frac{d^{n+1}}{dx^{n+1}}$ by the product rule,

(ii) $\frac{d^n}{dx^n} x = n \frac{d^{n-1}}{dx^{n-1}} + x \frac{d^n}{dx^n}$ by induction when applied to a polynomial $f(x)$.

Let I be a nonzero two-sided ideal in R, and fix an element $X \neq 0$ in I. Let x^m be the highest power of x appearing in X, and let $\frac{d^n}{dx^n}$ be the highest power of $\frac{d}{dx}$ appearing in terms of X involving x^m. Let l and r denote "left multiplication by" and "right multiplication by," and apply $\left(l\left(\frac{d}{dx}\right) - r\left(\frac{d}{dx}\right)\right)^m$ to X. Since (i) shows that

$$\left(l\left(\tfrac{d}{dx}\right) - r\left(\tfrac{d}{dx}\right)\right) x^k \left(\tfrac{d}{dx}\right)^l = kx^{k-1}\left(\tfrac{d}{dx}\right)^l,$$

the result of computing $\left(l\left(\frac{d}{dx}\right) - r\left(\frac{d}{dx}\right)\right)^m X$ is a polynomial in $\frac{d}{dx}$ of degree exactly n with no x's. Application of $(r(x) - l(x))^n$ to the result, using (ii), yields a nonzero constant. We conclude that 1 is in I and therefore that $I = R$. Hence R is simple.

To show that R is not semisimple, first note that $\mathbb{C}[x]$ is a natural unital left R module. We shall show that R has infinite length as a left R module, in the sense of the length of finite filtrations. In fact,

$$R \supseteq R\left(\tfrac{d}{dx}\right) \supseteq R\left(\tfrac{d}{dx}\right)^2 \supseteq \cdots \supseteq R\left(\tfrac{d}{dx}\right)^n \qquad (*)$$

is a finite filtration of left R submodules of R. If $R\left(\frac{d}{dx}\right)^k = R\left(\frac{d}{dx}\right)^{k+1}$, then $\left(\frac{d}{dx}\right)^k = r\left(\frac{d}{dx}\right)^{k+1}$ for some $r \in R$. Applying these two equal expressions for a member of R to the member x^k of the left R module $\mathbb{C}[x]$, we arrive at a contradiction and conclude that every inclusion in $(*)$ is strict. Therefore R has infinite length and is not semisimple.

The extra hypothesis that Wedderburn imposed so that simple rings would turn out to be semisimple is finite dimensionality. Wedderburn's result in this direction is Theorem 2.4 below. This hypothesis is quite natural to the extent that the subject was originally motivated by the theory of Lie algebras. E. Artin found a substitute for the assumption of finite dimensionality that takes the result beyond the realm of algebras, and we take up Artin's idea in the next section.

Theorem 2.4 (Wedderburn). Let R be a finite-dimensional algebra with identity over a field F. If R is a simple ring, then R is semisimple and hence is isomorphic to $M_n(D)$ for some integer $n \geq 1$ and some finite-dimensional division algebra D over F. The integer n is uniquely determined by R, and D is unique up to isomorphism.

PROOF. By finite dimensionality, R has a minimal left ideal V. For r in R, form the set Vr. This is a left ideal, and we claim that it is minimal or is 0. In fact, the function $v \mapsto vr$ is R linear from V onto Vr. Since V is simple as a left R module, Vr is simple or 0. The sum $I = \sum_{r \text{ with } Vr \neq 0} Vr$ is a two-sided ideal in R, and it is not 0 because $V1 \neq 0$. Since R is simple, $I = R$. Then the left R module R is exhibited as the sum of simple left R modules and is therefore semisimple. The isomorphism with $M_n(D)$ and the uniqueness now follow from Theorem 2.2. □

3. Rings with Chain Condition and Artin's Theorem

Parts of Chapters VIII and IX of *Basic Algebra* made considerable use of a hypothesis that certain commutative rings are "Noetherian," and we now extend this notion to noncommutative rings. A ring R with identity is **left Noetherian** if the left R module R satisfies the ascending chain condition for its left ideals. It is **left Artinian** if the left R module R satisfies the descending chain condition for its left ideals. The notions of **right Noetherian** and **right Artinian** are defined similarly.

We saw many examples of Noetherian rings in the commutative case in *Basic Algebra*. The ring of integers $\mathbb{Z}$ is Noetherian, and so is the ring of polynomials $R[X]$ in an indeterminate over a nonzero Noetherian ring R. It follows from the latter example that the ring $F[X_1, \ldots, X_n]$ in finitely many indeterminates over a field is a Noetherian ring. Other examples arose in connection with extensions of Dedekind domains.

Any finite direct product of fields is Noetherian and Artinian because it has a composition series and because its ideals therefore satisfy both chain conditions. If p is any prime, the ring $\mathbb{Z}/p^2\mathbb{Z}$ is Noetherian and Artinian for the same reason, and it is not a direct product of fields.

In the noncommutative setting, any semisimple ring is necessarily left Noetherian and left Artinian because it has a composition series for its left ideals and the left ideals therefore satisfy both chain conditions.

Proposition 2.5. Let R be a ring with identity, and let M be a finitely generated unital left R module. If R is left Noetherian, then M satisfies the ascending chain condition for its R submodules; if R is left Artinian, then M satisfies the descending chain condition for its R submodules.

PROOF. We prove the first conclusion by induction on the number of generators, and the proof of the second conclusion is completely similar. The result is trivial if M has 0 generators. If $M = Rx$, then M is a quotient of the left R module R and satisfies the ascending chain condition for its R submodules, according to Proposition 10.10 of *Basic Algebra*. For the inductive step with ≥ 2 generators, write $M = Rx_1 + \cdots + Rx_n$ and $N = Rx_1 + \cdots + Rx_{n-1}$. Then N satisfies the ascending chain condition for its R submodules by the inductive hypothesis, and M/N is isomorphic to $Rx_n/(N \cap Rx_n)$, which satisfies the ascending chain condition for its R submodules by the inductive hypothesis. Therefore M satisfies the ascending chain condition for its R submodules by application of the converse direction of Proposition 10.10. □

Artin's theorem (Theorem 2.6 below) will make use of the hypothesis "left Artinian" in identifying those simple rings that are semisimple. The hypothesis left Artinian may therefore be regarded as a useful generalization of finite dimensionality. Before we come to that theorem, we give a construction that produces large numbers of nontrivial examples of such rings.

EXAMPLE (triangular rings). Let R and S be nonzero rings with identity, and let M be an (R, S) bimodule.[3] Define a set A and operations of addition and multiplication symbolically by

$$A = \begin{pmatrix} R & M \\ 0 & S \end{pmatrix} = \left\{ \begin{pmatrix} r & m \\ 0 & s \end{pmatrix} \,\bigg|\, r \in R,\, m \in M,\, s \in S \right\}$$

with
$$\begin{pmatrix} r & m \\ 0 & s \end{pmatrix} \begin{pmatrix} r' & m' \\ 0 & s' \end{pmatrix} = \begin{pmatrix} rr' & rm' + ms' \\ 0 & ss' \end{pmatrix}.$$

Then A is a ring with identity, the bimodule property entering the proof of associativity of multiplication in A. We can identify R, M, and S with the additive subgroups of A given by $\begin{pmatrix} R & 0 \\ 0 & 0 \end{pmatrix}$, $\begin{pmatrix} 0 & M \\ 0 & 0 \end{pmatrix}$, and $\begin{pmatrix} 0 & 0 \\ 0 & S \end{pmatrix}$. Problems 8–11 at the end of the chapter ask one to check the following facts:

(i) The left ideals in A are of the form $I_1 \oplus I_2$, where I_2 is a left ideal in S and I_1 is a left R submodule of $R \oplus M$ containing MI_2.

(ii) The right ideals in A are of the form $J_1 \oplus J_2$, where J_1 is a right ideal in R and J_2 is a right S submodule of $M \oplus S$ containing $J_1 M$.

(iii) The ring A is left Noetherian if and only if R and S are left Noetherian and M satisfies the ascending chain condition for its left R submodules. The ring A is right Noetherian if and only if R and S are right Noetherian and M satisfies the ascending chain condition for its right S submodules.

(iv) The previous item remains valid if "Noetherian" is replaced by "Artinian" and "ascending" is replaced by "descending."

(v) If $A = \begin{pmatrix} R & R \\ 0 & S \end{pmatrix}$ is a ring such as $\begin{pmatrix} \mathbb{Q} & \mathbb{Q} \\ 0 & \mathbb{Z} \end{pmatrix}$ in which S is a (commutative) Noetherian integral domain with field of fractions R and if $S \neq R$, then A is left Noetherian and not right Noetherian, and A is neither left nor right Artinian.

(vi) If $A = \begin{pmatrix} R & R \\ 0 & S \end{pmatrix}$ is a ring such as $\begin{pmatrix} \mathbb{Q}(x) & \mathbb{Q}(x) \\ 0 & \mathbb{Q} \end{pmatrix}$ in which R and S are fields with $S \subseteq R$ and $\dim_S R$ infinite, then A is left Noetherian and left Artinian, and A is neither right Noetherian nor right Artinian.

From these examples we see, among other things, that "left" and "right" are somewhat independent for both the Noetherian and the Artinian conditions. We

---

[3]This means that M is an abelian group with the structure of a unital left R module and the structure of a unital right S module in such a way that $(rm)s = r(ms)$ for all $r \in R, m \in M$, and $s \in S$.

already know from the commutative case that Noetherian does not imply Artinian, $\mathbb{Z}$ being a counterexample. We shall see in Theorem 2.15 later that left Artinian implies left Noetherian and that right Artinian implies right Noetherian.

Theorem 2.6 (E. Artin). If R is a simple ring, then the following conditions are equivalent:
   (a) R is left Artinian,
   (b) R is semisimple,
   (c) R has a minimal left ideal,
   (d) $R \cong M_n(D)$ for some integer $n \geq 1$ and some division ring D.

In particular, a left Artinian simple ring is right Artinian.

REMARK. Theorem 2.4 is a special case of the assertion that (a) implies (d). In fact, if R is a finite-dimensional algebra over a field F, then the finite dimensionality forces R to be left Artinian.

PROOF. It is evident from Wedderburn's Theorem (Theorem 2.2) that (b) and (d) are equivalent. For the rest we prove that (a) implies (c), that (c) implies (b), and that (b) implies (a).

Suppose that (a) holds. Applying the minimum condition for left ideals in R, we obtain a minimal left ideal. Thus (c) holds.

Suppose that (c) holds. Let V be a minimal left ideal. Then the sum $I = \sum_{r \in R} Vr$ is a two-sided ideal in R, and it is nonzero because the term for $r = 1$ is nonzero. Since R is simple, $I = R$. Then the left R module R is spanned by the simple left R modules Vr, and R is semisimple. Thus (b) holds.

Suppose that (b) holds. Since R is semisimple, the left R module R has a composition series. Then the left ideals in R satisfy both chain conditions, and it follows that R is left Artinian. Thus (a) holds. □

4. Wedderburn–Artin Radical

In this section we introduce one notion of "radical" for certain rings with identity, and we show how it is related to semisimplicity. This notion, the "Wedderburn–Artin radical," is defined under the hypothesis that the ring is left Artinian. It is not the only notion of radical studied by ring theorists, however. There is a useful generalization, known as the "Jacobson radical," that is defined for arbitrary rings with identity. We shall not define and use the Jacobson radical in this text.

Fix a ring R with identity. A **nilpotent element** in R is an element a with $a^n = 0$ for some integer $n \geq 1$. A **nil left ideal** is a left ideal in which every element is nilpotent; nil right ideals and nil two-sided ideals are defined similarly.

A **nilpotent left ideal** is a left ideal I such that $I^n = 0$ for some integer $n \geq 1$, i.e., for which $a_1 \cdots a_n = 0$ for all n-fold products of elements from I; nilpotent right ideals and nilpotent two-sided ideals are defined similarly.

Lemma 2.7. If I_1 and I_2 are nilpotent left ideals in a ring R with identity, then $I_1 + I_2$ is nilpotent.

PROOF. Let $I_1^r = 0$ and $I_2^s = 0$. Expand $(I_1 + I_2)^k$ as $\sum I_{i_1} I_{i_2} \cdots I_{i_k}$ with each i_j equal to 1 or 2. Take $k = r + s$. In any term of the sum, there are $\geq r$ indices 1 or $\geq s$ indices 2. In the first case let there be t indices 2 at the right end. Since $I_2 I_1 \subseteq I_1$, we can absorb all other indices 2, and the term of the sum is contained in $I_1^r I_2^t = 0$. Similarly in the second case if there are t' indices 1 at the right end, then the term is contained in $I_2^s I_1^{t'} = 0$. $\square$

Lemma 2.8. If I is a nilpotent left ideal in a ring R with identity, then I is contained in a nilpotent two-sided ideal J.

PROOF. Put $J = \sum_{r \in R} Ir$. This is a two-sided ideal. For any integer $k \geq 0$, $J^k = \left(\sum_{r \in R} Ir\right)^k \subseteq \sum_{r_1, \ldots, r_k} Ir_1 Ir_2 \cdots Ir_k \subseteq \sum_{r_k} I^k r_k$. If $I^k = 0$, then $J^k = 0$. $\square$

Lemma 2.9. If R is a ring with identity, then the sum of all nilpotent left ideals in a nil two-sided ideal.

PROOF. Let K be the sum of all nilpotent left ideals in R, and let a be a member of K. Write $a = a_1 + \cdots + a_n$ with $a_i \in I_i$ for a nilpotent left ideal I_i. Lemma 2.7 shows that $I = \sum_{i=1}^n I_i$ is a nilpotent left ideal. Since a is in I, a is a nilpotent element.

The set K is certainly a left ideal, and we need to see that aR is in K in order to see that K is a two-sided ideal. Lemma 2.8 shows that $I \subseteq J$ for some nilpotent two-sided ideal J. Then $J \subseteq K$ because J is one of the nilpotent left ideals whose sum is K. Since a is in I and therefore in J and since J is a two-sided ideal, aR is contained in J. Therefore aR is contained in K, and K is a two-sided ideal. $\square$

Theorem 2.10. If R is a left Artinian ring, then any nil left ideal in R is nilpotent.

REMARK. Readers familiar with a little structure theory for finite-dimensional Lie algebras will recognize this theorem as an analog for associative algebras of Engel's Theorem.

PROOF. Let I be a nil left ideal of R, and form the filtration

$$I \supseteq I^2 \supseteq I^3 \supseteq \cdots.$$

4. Wedderburn–Artin Radical

Since R is left Artinian, this filtration is constant from some point on, and we have $I^k = I^{k+1} = I^{k+2} = \cdots$ for some $k \geq 1$. Put $J = I^k$. We shall show that $J = 0$, and then we shall have proved that I is a nilpotent ideal.

Suppose that $J \neq 0$. Since $J^2 = I^{2k} = I^k = J$, we have $J^2 = J$. Thus the left ideal J has the property that $JJ \neq 0$. Since R is left Artinian, the set of left ideals $K \subseteq J$ with $JK \neq 0$ has a minimal element K_0. Choose $a \in K_0$ with $Ja \neq 0$. Since $Ja \subseteq JK_0 \subseteq K_0$ and $J(Ja) = J^2a = Ja \neq 0$, the minimality of K_0 implies that $Ja = K_0$. Thus there exists $x \in J$ with $xa = a$. Applying powers of x, we obtain $x^n a = a$ for every integer $n \geq 1$. But x is a nilpotent element, being in I, and thus we have a contradiction. □

Corollary 2.11. If R is a left Artinian ring, then there exists a unique largest nilpotent two-sided ideal I in R. This ideal is the sum of all nilpotent left ideals and also is the sum of all nilpotent right ideals.

REMARKS. The two-sided ideal I of the corollary is called the **Wedderburn–Artin radical** of R and will be denoted by rad R. This exists under the hypothesis that R is left Artinian.

PROOF. By Lemma 2.9 and Theorem 2.10 the sum of all nilpotent left ideals in R is a two-sided nilpotent ideal I. Lemma 2.8 shows that any nilpotent right ideal is contained in a nilpotent two-sided ideal J. Since J is in particular a nilpotent left ideal, the definition of I forces $J \subseteq I$. Hence the sum of all nilpotent right ideals is contained in I. But I itself is a nilpotent right ideal and hence equals the sum of all the nilpotent right ideals. □

Lemma 2.12 (Brauer's Lemma). If R is any ring with identity and if V is a minimal left ideal in R, then either $V^2 = 0$ or $V = Re$ for some element e of V with $e^2 = e$.

REMARK. An element e with the property that $e^2 = e$ is said to be **idempotent**.

PROOF. Being a minimal left ideal, V is a simple left R module. Schur's Lemma (Proposition 10.4b of *Basic Algebra*) shows that $\text{End}_R V$ is a division ring. If a is in V, then the map $v \mapsto va$ of V into itself lies in $\text{End}_R V$ and hence is the 0 map or is one-one onto. If it is the 0 map for all $a \in V$, then $V^2 = 0$. Otherwise suppose that a is an element for which $v \mapsto va$ is one-one onto. Then there exists $e \in V$ with $ea = a$. Multiplying on the left by e gives $e^2 a = ea$ and therefore $(e^2 - e)a = 0$. Since the map $v \mapsto va$ is assumed to be one-one onto, we must have $e^2 - e = 0$ and $e^2 = e$. □

Theorem 2.13. If R is a left Artinian ring and if the Wedderburn–Artin radical of R is 0, then R is a semisimple ring.

REMARKS. Conversely semisimple rings are left Artinian and have radical 0. In fact, we already know that semisimple rings have a composition series for their left ideals and hence are left Artinian. To see that the radical is 0, apply Theorem 2.2 and write the ring as $R = M_{n_1}(D_1) \times \cdots \times M_{n_r}(D_r)$. The two-sided ideals of R are the various subproducts, with 0 in the missing coordinates. Such a subproduct cannot be nilpotent as an ideal unless it is 0, since the identity element in any factor is not a nilpotent element in R.

PROOF. Let us see that any minimal left ideal I of R is a direct summand as a left R submodule. Since rad $R = 0$, I is not nilpotent. Thus $I^2 \neq 0$, and Lemma 2.12 shows that I contains an idempotent e. This element satisfies $I = Re$. Put $I' = \{r \in R \mid re = 0\}$. Then I' is a left ideal in R. Since $I' \cap I \subseteq I$ and e is not in I', the minimality of I forces $I' \cap I = 0$. Writing $r = re + (r - re)$ with $re \in I$ and $r - re \in I'$, we see that $R = I + I'$. Therefore $R = I \oplus I'$.

Now put $I_1 = I$. If I' is not 0, choose a minimal left ideal $I_2 \subseteq I'$ by the minimum condition for left ideals in R. Arguing as in the previous paragraph, we have $I_2 = Re_2$ for some element e_2 with $e_2^2 = e_2$. The argument in the previous paragraph shows that $R = I_2 \oplus I_2'$, where $I_2' = \{r \in R \mid re_2 = 0\}$. Define $I'' = \{r \in R \mid re_1 = re_2 = 0\} = I' \cap I_2'$. Since I_2 is contained in I', we can intersect $R = I_2 \oplus I_2'$ with I' and obtain $I' = I_2 \oplus I''$. Then $R = I_1 \oplus I' = I_1 \oplus I_2 \oplus I''$. Continuing in this way, we obtain $R = I_1 \oplus I_2 \oplus I_3 \oplus I'''$, etc. As this construction continues, we have $I' \supseteq I'' \supseteq I''' \supseteq \cdots$. Since R is left Artinian, this sequence must terminate, evidently in 0. Then R is exhibited as the sum of simple left R modules and is semisimple. □

Corollary 2.14. If R is a left Artinian ring, then $R/\operatorname{rad} R$ is a semisimple ring.

PROOF. Let $I = \operatorname{rad} R$, and let $\varphi : R \to R/I$ be the quotient homomorphism. Arguing by contradiction, let $\overline{J}$ be a nonzero nilpotent left ideal in R/I, and let $J = \varphi^{-1}(\overline{J}) \subseteq R$. Since $\overline{J}$ is nilpotent, $J^k \subseteq I$ for some integer $k \geq 1$. But I, being the radical, is nilpotent, say with $I^l = 0$, and hence $J^{k+l} \subseteq I^l = 0$. Therefore J is a nilpotent left ideal in R strictly containing I, in contradiction to the maximality of I. We conclude that no such $\overline{J}$ exists. Then $\operatorname{rad}(R/\operatorname{rad} R) = 0$. Since $R/\operatorname{rad} R$ is left Artinian as a quotient of a left Artinian ring, Theorem 2.13 shows that $R/\operatorname{rad} R$ is a semisimple ring. □

We shall use this corollary to prove that left Artinian rings are left Noetherian. We state the theorem, state and prove a lemma, and then prove the theorem.

Theorem 2.15 (Hopkins). If R is a left Artinian ring, then R is left Noetherian.

Lemma 2.16. If R is a semisimple ring, then every unital left R module M is semisimple. Consequently any unital left R module satisfying the descending chain condition has a composition series and therefore satisfies the ascending chain condition.

PROOF. For each $m \in M$, let R_m be a copy of the left R module R, and define $\widetilde{M} = \bigoplus_{m \in M} R_m$ as a left R module. Since each R_m is semisimple, $\widetilde{M}$ is semisimple. Define a function $\varphi : \widetilde{M} \to M$ as follows: if $r_{m_1} + \cdots + r_{m_k}$ is given with r_{m_j} in R_{m_j} for each j, let $\varphi(r_{m_1} + \cdots + r_{m_k}) = \sum_{j=1}^{k} r_{m_j} m_j$. Then φ is an R module map with the property that $\varphi(1_m) = m$, and consequently φ carries $\widetilde{M}$ onto M. As the image of a semisimple R module under an R module map, M is semisimple.

Now suppose that M is a unital left R module satisfying the descending chain condition. We have just seen that M is semisimple, and thus we can write $M = \bigoplus_{i \in S} M_i$ as a direct sum over a set S of simple left R modules M_i. Let us see that S is a finite set. If S were not a finite set, then we could choose an infinite sequence $i_1, i_2, \ldots$ of distinct members of S, and we would obtain

$$M \supsetneq \bigoplus_{i \neq i_1} M_i \supsetneq \bigoplus_{i \neq i_1, i_2} M_i \supsetneq \cdots,$$

in contradiction to the fact that the R submodules of M satisfy the descending chain condition. □

PROOF OF THEOREM 2.15. Let $I = \operatorname{rad} R$. Since I is nilpotent, $I^n = 0$ for some n. Each I^k for $k \geq 0$ is a left R submodule of R. Since R is left Artinian, its left R submodules satisfy the descending chain condition, and the same thing is true of the R submodules of each I^k. Consequently the R submodules of each I^k/I^{k+1} satisfy the descending chain condition.

In the action of R on I^k/I^{k+1} on the left, I acts as 0. Hence I^k/I^{k+1} becomes a left R/I module, and the R/I submodules of this left R/I module must satisfy the descending chain condition. Corollary 2.14 shows that $R/I = R/\operatorname{rad} R$ is a semisimple ring. Since the R/I submodules of I^k/I^{k+1} satisfy the descending chain condition, Lemma 2.16 shows that these R/I submodules satisfy the ascending chain condition. Therefore the R submodules of each left R module I^k/I^{k+1} satisfy the ascending chain condition.

We shall show inductively for $k \geq 0$ that the R submodules of R/I^{k+1} satisfy the ascending chain condition. Since $I^n = 0$, this conclusion will establish that R is left Noetherian, as required. The case $k = 0$ was shown in the previous paragraph. Assume inductively that the R submodules of R/I^k satisfy the ascending chain condition. Since $R/I^k \cong (R/I^{k+1})/(I^k/I^{k+1})$ and since the R submodules of R/I^k and of I^k/I^{k+1} satisfy the ascending chain condition, the same is true for R/I^{k+1}. This completes the proof. □

5. Wedderburn's Main Theorem

Wedderburn's Main Theorem is an analog for finite-dimensional associative algebras over a field of characteristic 0 of the Levi decomposition of a finite-dimensional Lie algebra over a field of characteristic 0. Each of these results says that the given algebra is a "semidirect product" of the radical and a semisimple subalgebra isomorphic to the quotient of the given algebra by the radical. In other words, the whole algebra, as a vector space, is the direct sum of the radical and a vector subspace that is closed under multiplication.

An example of this phenomenon occurs with a block upper-triangular subalgebra A of $M_n(D)$ whenever D is a finite-dimensional division algebra over the given field. Let the diagonal blocks be of sizes $n_1, \ldots, n_r$ with $n_1 + \cdots + n_r = n$. The radical rad A is the nilpotent ideal of all matrices whose only nonzero entries are above and to the right of the diagonal blocks, and the semisimple subalgebra consists of all matrices whose only nonzero entries lie within the diagonal blocks.

Theorem 2.17 (Wedderburn's Main Theorem). Let A be a finite-dimensional associative algebra with identity over a field F of characteristic 0, and let rad A be the Wedderburn–Artin radical. Then there exists a subalgebra S of A isomorphic as an F algebra to $A/\operatorname{rad} A$ such that $A = S \oplus \operatorname{rad} A$ as vector spaces.

REMARKS. The finite dimensionality implies that A is left Artinian, and Corollary 2.14 shows that $A/\operatorname{rad} A$ is a semisimple algebra. The decomposition $A = S \oplus \operatorname{rad} A$ is different in nature from the one in Theorem 2.2, which involves complementary ideals. When there are complementary ideals, the identity of A decomposes as the sum of the identities for each summand. Here the identity of A is the identity of S and has 0 component in rad A. To see this, write $1 = a + b$ with $a \in S$ and $b \in \operatorname{rad} A$. Multiplying $1 = a + b$ on the left and right by $s \in S$, we see that $as = s = sa$ and that $bs = sb = 0$. Hence $a = 1_S$ is the identity of S. Then $b^2 = (1 - 1_S)^2 = 1 - 2 \cdot 1_S + 1_S^2 = 1 - 2 \cdot 1_S + 1_S = 1 - 1_S = b$, and $b^n = b$ for all $n \geq 1$. Since rad A is nilpotent, $b^n = 0$ for some n. Thus $b = 0$, and $1 = 1_S$ as asserted.

Theorem 2.17 is a deep result, and the proof will occupy all of the present section and the next. The key special case to understand occurs when $A/\operatorname{rad} A \cong M_{n_1}(F) \times \cdots \times M_{n_r}(F)$. We shall handle this case by means of Theorem 2.18 below, whose proof will be the main goal of the present section. Corollary 2.27 (of Theorem 2.18) near the end of this section will show that Theorem 2.18 implies this special case of Theorem 2.17 for $r = 1$, and Corollary 2.28 will deduce this special case of Theorem 2.17 for general r from Corollary 2.27.

5. Wedderburn's Main Theorem

Theorem 2.18. Let A be a left Artinian ring with Wedderburn–Artin radical rad A, and suppose that $A/\operatorname{rad} A$ is simple, i.e., is of the form $A/\operatorname{rad} A \cong M_n(D)$ for some division ring D. Then A is isomorphic as a ring to $M_n(R)$ for some left Artinian ring R such that $R/\operatorname{rad} R \cong D$.

The idea behind the proof of Theorem 2.18 is to give an abstract characterization of a ring of matrices in terms of the elements E_{ij} that are 1 in the $(i, j)^{\text{th}}$ place and are 0 elsewhere. In turn, these elements arise from the diagonal such elements E_{ii}, which are idempotents, i.e., have $E_{ii}^2 = E_{ii}$. The critical issue in the proof of Theorem 2.18 is to show that each idempotent of $A/\operatorname{rad} A$, which is assumed to be a full matrix ring $M_n(D)$, has an idempotent in its preimage in A. The lifted idempotents then point to $M_n(R)$ for a certain R.

Thus we begin with some discussion of idempotents. We shall intersperse facts about general rings with facts about left Artinian rings as we go along. For the moment let R be any ring with identity, and let e be an idempotent. Then $1 - e$ is an idempotent, and we have the three **Peirce**[4] **decompositions**

$$R = Re \oplus R(1-e),$$
$$R = eR \oplus (1-e)R,$$
$$R = eRe \oplus eR(1-e) \oplus (1-e)Re \oplus (1-e)R(1-e).$$

All the direct sums may be regarded as direct sums of abelian groups. The two members of the right side in the first case are left ideals, and the two members of the right side in the second case are right ideals. If $r \in R$ is given, then the first decomposition is as $r = re + r(1-e)$; the decomposition is direct because if $r_1 e = r_2(1-e)$, then right multiplication by e gives $r_1 e = 0$ since $e^2 = e$. The second decomposition is proved similarly, and the third decomposition follows by combining the first two. In the third decomposition, eRe is a ring with e as identity, and $(1-e)R(1-e)$ is a ring with $1-e$ as identity.

EXAMPLE. Let $R = M_n(F)$, and let

$$e = \begin{pmatrix} 1 & & & & \\ & \ddots & & & \\ & & 1 & & \\ & & & 0 & \\ & & & & \ddots \\ & & & & & 0 \end{pmatrix}, \quad \text{so that} \quad 1 - e = \begin{pmatrix} 0 & & & & \\ & \ddots & & & \\ & & 0 & & \\ & & & 1 & \\ & & & & \ddots \\ & & & & & 1 \end{pmatrix}.$$

---
[4]Pronounced "purse." Charles Sanders Peirce (1839–1914).

In block form we then have

$$eRe = \begin{pmatrix} * & 0 \\ 0 & 0 \end{pmatrix}, \qquad eR(1-e) = \begin{pmatrix} 0 & * \\ 0 & 0 \end{pmatrix},$$

$$(1-e)Re = \begin{pmatrix} 0 & 0 \\ * & 0 \end{pmatrix}, \qquad (1-e)R(1-e) = \begin{pmatrix} 0 & 0 \\ 0 & * \end{pmatrix}.$$

Proposition 2.19. In a ring R with identity, let e be an element of R with $e^2 = e$.

(a) If I is a left ideal in eRe, then $eRI = I$. Hence $I \mapsto RI$ is a one-one inclusion-preserving map of the left ideals of eRe to those of R.

(b) If J is a two-sided ideal of eRe, then $e(RJR)e = J$. Hence $J \mapsto RJR$ is a one-one inclusion-preserving map of the two-sided ideals of eRe to those of R. This map respects multiplication of ideals.

(c) If $\widetilde{J}$ is a two-sided ideal of R, then $e\widetilde{J}e$ is a two-sided ideal of eRe, and $eRe \cap \widetilde{J} = e\widetilde{J}e$.

PROOF. For (a), we have $eRI = eR(eI) = (eRe)I = I$, the first equality holding because e is the identity in eRe and the third equality holding because eRe contains its identity e. The rest of (a) then follows.

For (b), J satisfies $J = eJe$, since $ej = je = j$ for every $j \in eRe$, and therefore $eRJRe = eReJeRe = (eRe)J(eRe) = J$, the last equality holding because eRe contains its identity e. To see that $J \mapsto RJR$ respects multiplication, we compute that $(RJR)(RJ'R) = RJRJ'R = R(Je)R(eJ')R = RJ(eRe)J'R = RJJ'R$.

For (c), $eRe \cap \widetilde{J} \supseteq e\widetilde{J}e$ certainly. In the reverse direction, let j be in $eRe \cap \widetilde{J}$. Then $j = ere$ for some $r \in R$, and hence $eje = e^2re^2 = ere = j$ shows that j is in $e\widetilde{J}e$. □

Corollary 2.20. In a left Artinian ring R, let e be an element with $e^2 = e$. Then the ring eRe is left Artinian, and

$$\mathrm{rad}(eRe) = eRe \cap \mathrm{rad}\,R = e(\mathrm{rad}\,R)e.$$

If $\overline{R}$ denotes the quotient ring $R/\mathrm{rad}\,R$ and $\bar{e}$ denotes the element $e + \mathrm{rad}\,R$ of the quotient, then the quotient map carries eRe onto $\bar{e}\overline{R}\bar{e}$ and has kernel $\mathrm{rad}(eRe)$. Consequently

$$eRe/\mathrm{rad}(eRe) \cong \bar{e}\overline{R}\bar{e}.$$

PROOF. The ring eRe is left Artinian as an immediate consequence of Proposition 2.19a. For the first display we may assume that R and eRe are both left Artinian. Then $eRe \cap \mathrm{rad}\,R$ is a two-sided ideal of eRe, and $(eRe \cap \mathrm{rad}\,R)^n \subseteq$

$(\operatorname{rad} R)^n$ for every n. Since $(\operatorname{rad} R)^N = 0$ for some N, $eRe \cap \operatorname{rad} R$ is nilpotent, and $eRe \cap \operatorname{rad} R \subseteq \operatorname{rad}(eRe)$. Since the reverse inclusion is evident, we obtain $\operatorname{rad}(eRe) = eRe \cap \operatorname{rad} R$. The equality $eRe \cap \operatorname{rad} R = e(\operatorname{rad} R)e$ is the special case of Proposition 2.19c in which $J = \operatorname{rad} R$. This proves the equalities in the first display.

For the isomorphism in the second display, the quotient mapping carries ere to $ere + \operatorname{rad} R = (e + \operatorname{rad} R)(r + \operatorname{rad} R)(e + \operatorname{rad} R) = \bar{e}(r + \operatorname{rad} R)\bar{e}$. Thus the quotient map $R \to \overline{R}$ carries eRe onto $\bar{e}\overline{R}\bar{e}$. The kernel is $eRe \cap \operatorname{rad} R$, which we have just proved is $\operatorname{rad}(eRe)$. Therefore the quotient map exhibits an isomorphism of rings $eRe/\operatorname{rad}(eRe) \cong \bar{e}\overline{R}\bar{e}$. □

Proposition 2.21. In a ring R with identity, let e_1 and e_2 be idempotents. Then the unital left R modules Re_1 and Re_2 are isomorphic as left R modules if and only if there exist elements e_{12} and e_{21} in R such that

$$e_1 e_{12} e_2 = e_{12}, \qquad e_2 e_{21} e_1 = e_{21},$$
$$e_{12} e_{21} = e_1, \qquad e_{21} e_{12} = e_2.$$

REMARK. In this case we shall say that e_1 and e_2 are **isomorphic idempotents**, and we shall write $e_1 \cong e_2$.

PROOF. Let $\varphi : Re_1 \to Re_2$ be an R isomorphism. Define $e_{12} = \varphi(e_1)$ and $e_{21} = \varphi^{-1}(e_2)$. Every element s of Re_2 has the property that $se_2 = s$ because $e_2^2 = e_2$; since e_{12} lies in Re_2, $e_{12}e_2 = e_{12}$. Meanwhile, $e_{12} = \varphi(e_1) = \varphi(e_1^2) = e_1\varphi(e_1) = e_1 e_{12}$. Putting these two facts together gives $e_{12} = e_{12}e_2 = e_1 e_{12} e_2$. This proves the first equality in the display, and the equality $e_{21} = e_2 e_{21} e_1$ is proved similarly. Also, $e_1 = \varphi^{-1}(\varphi(e_1)) = \varphi^{-1}(e_{12}) = \varphi^{-1}(e_{12}e_2) = e_{12}\varphi^{-1}(e_2) = e_{12}e_{21}$, and similarly $e_{21}e_{12} = e_2$. This completes the proof that an R isomorphism $Re_1 \cong Re_2$ leads to elements e_{12} and e_{21} such that the four displayed identities hold.

For the converse, suppose that e_{12} and e_{21} exist and satisfy the four displayed identities. Define $\varphi : Re_1 \to R$ by $\varphi(re_1) = re_{12}$. To see that this map is well defined, suppose that $re_1 = 0$; then $re_{12} = r(e_1 e_{12} e_2) = (re_1)e_{12}e_2 = 0$, as required. Similarly we can define $\psi : Re_2 \to R$ by $\psi(re_2) = re_{21}$. Then

$$\psi\varphi(e_1) = \psi(e_{12}) = \psi(e_{12}e_2) = e_{12}\psi(e_2) = e_{12}e_{21} = e_1,$$

and similarly $\varphi\psi(e_2) = e_2$. Since $\psi\varphi$ and $\varphi\psi$ are R module homomorphisms, each is the identity on its domain. □

Corollary 2.22. Let R be a left Artinian ring. For each r in R, let $\bar{r}$ be the coset $r + \operatorname{rad} R$ in $R/\operatorname{rad} R$. If e_1 and e_2 are idempotents in R, then e_1 and e_2 are isomorphic if and only if $\bar{e}_1$ and $\bar{e}_2$ are isomorphic.

PROOF. If e_1 and e_2 are given as isomorphic in R, let e_{12} and e_{21} be as in Proposition 2.21, and pass to $R/\operatorname{rad} R$ by the quotient homomorphism to obtain elements $\bar{e}_{12}$ and $\bar{e}_{21}$ that exhibit $\bar{e}_1$ and $\bar{e}_2$ as isomorphic idempotents.

Conversely let $\bar{e}_1$ and $\bar{e}_2$ be isomorphic idempotents in $R/\operatorname{rad} R$, and use Proposition 2.21 to produce elements $\bar{u}_{12}$ and $\bar{u}_{21}$ in $R/\operatorname{rad} R$ such that

$$\bar{e}_1 \bar{u}_{12} \bar{e}_2 = \bar{u}_{12}, \quad \bar{e}_2 \bar{u}_{21} \bar{e}_1 = \bar{u}_{21}, \quad \bar{u}_{12} \bar{u}_{21} = \bar{e}_1, \quad \bar{u}_{21} \bar{u}_{12} = \bar{e}_2.$$

Let u_{12} and u_{21} be preimages of $\bar{u}_{12}$ and $\bar{u}_{21}$ in R. Possibly replacing u_{12} by $e_1 u_{12} e_2$ and u_{21} by $e_2 u_{21} e_1$, we may assume that $e_1 u_{12} e_2 = u_{12}$ and $e_2 u_{21} e_1 = u_{21}$. Our construction is such that $u_{12} u_{21} = e_1 - z_1$ with z_1 in $\operatorname{rad} R$ and $e_1 z_1 = z_1 = z_1 e_1$. Since z_1 is a nilpotent element,

$$(e_1 - z_1)(e_1 + z_1 + z_1^2 + \cdots + z_1^n) = e_1$$

as soon as $z_1^{n+1} = 0$. Thus we have $u_{12} u_{21}(e_1 + z_1 + z_1^2 + \cdots + z_1^n) = e_1$. Define $e_{12} = u_{12}$ and $e_{21} = u_{21}(e_1 + z_1 + z_1^2 + \cdots + z_1^n)$. Then it is immediate that $\bar{e}_{12} = \bar{u}_{12}$, $\bar{e}_{21} = \bar{u}_{21}$, and $e_{12} e_{21} = e_1$. Also, the equality $e_1 u_{12} e_2 = u_{12}$ implies that $e_1 e_{12} e_2 = e_{12}$, and the equality $e_2 u_{21} e_1 (e_1 + z_1 + z_1^2 + \cdots + z_1^n) = u_{21}(e_1 + z_1 + z_1^2 + \cdots + z_1^n)$ implies that $e_2 e_{21} e_1 = e_{21}$ since $e_1 z_1 = z_1 = z_1 e_1$.

In view of Proposition 2.21, we are left with checking the value of $e_{21} e_{12}$. We know that $\bar{e}_{21} \bar{e}_{12} = \bar{u}_{21} \bar{u}_{12} = \bar{e}_2$, and hence $e_{21} e_{12} = e_2 - z_2$ for some z_2 in $\operatorname{rad} R$. Multiplying by e_2 on both sides, we see that

$$e_2 z_2 = z_2 = z_2 e_2. \qquad (*)$$

Now $(e_{21} e_{12})(e_{21} e_{12}) = e_{21} e_1 e_{12} = e_{21} e_{12}$, and thus $(e_2 - z_2)^2 = e_2 - z_2$. Expanding out this equality and using $(*)$ gives $e_2 - 2z_2 + z_2^2 = e_2 - z_2$ and therefore gives $z_2^2 = z_2$. Hence $z_2^n = z_2$ for every $n \geq 1$. But z_2 is in $\operatorname{rad} R$, and every element of $\operatorname{rad} R$ is nilpotent. Thus $z_2 = 0$, and $e_{12} e_{21} = e_1$ as required. □

The proof of Corollary 2.22 shows a little more than the statement asserts, and we shall use this little extra conclusion when we finally get to the proof of Theorem 2.18. The extra fact is that any elements $\bar{u}_{12}$ and $\bar{u}_{21}$ exhibiting $\bar{e}_1$ and $\bar{e}_2$ have lifts to elements e_{12} and e_{21} exhibiting e_1 and e_2 as isomorphic.

The critical step of lifting a single idempotent from $A/\operatorname{rad} A$ to A is accomplished by the following proposition.

Proposition 2.23. Let R be a left Artinian ring. For each r in R, let $\bar{r}$ be the element $r + \operatorname{rad} R$ of $R/\operatorname{rad} R$. If a is an element of R such that $\bar{a}$ is idempotent in $R/\operatorname{rad} R$, then there exists an idempotent e in R such that $\bar{e} = \bar{a}$.

PROOF. Set $b = 1 - a$. The elements a and b commute, and $ab = a(1 - a)$ maps to $\bar{a} - \bar{a}^2 = 0$ in $R/\operatorname{rad} R$, since $\bar{a}$ is idempotent. Therefore ab lies in rad R and must satisfy $(ab)^n = 0$ for some n. Since a and b commute, we can apply the Binomial Theorem to obtain

$$1 = (a+b)^{2n} = \sum_{k=0}^{2n} \binom{2n}{k} a^{2n-k} b^k.$$

Define $\qquad e = \sum_{k=0}^{n} \binom{2n}{k} a^{2n-k} b^k \quad \text{and} \quad f = \sum_{k=n+1}^{2n} \binom{2n}{k} a^{2n-k} b^k.$

Each term of e contains at least the nth power of a, and each term of b contains at least the nth power of b. Thus each term of ef contains at least a factor $a^n b^n = (ab)^n = 0$, and we see that $ef = 0$. Therefore $e = e1 = e(e + f) = e^2 + 0 = e^2$, and e is an idempotent. Each term of e except the one for $k = 0$ contains a factor ab, and thus $e \equiv a^{2n}$ mod rad R. Since $\bar{a}$ is idempotent, $a^{2n} \equiv a$ mod rad R, and therefore $\bar{e} = \bar{a}$. $\qquad\square$

For the proof of Theorem 2.18, we need to lift an entire matrix ring to obtain a matrix ring, and this involves lifting more than a single idempotent. In effect, we have to lift compatibly an entire system $\bar{e}_{ij}$ that behaves like the usual system of E_{ij} for matrices. The idea is that if $R/\operatorname{rad} R$ is a matrix ring $M_n(K)$ with some ring of coefficients K, then the ith and j^{th} columns of $M_n(K)$ may be described compatibly as $M_n(K)\bar{e}_{ii}$ and $M_n(K)\bar{e}_{jj}$. Proposition 2.23 allows us to lift $\bar{e}_{ii}$ and $\bar{e}_{jj}$ to idempotents e_{ii} and e_{jj}, and Corollary 2.22 shows that an isomorphism $\bar{e}_{ii} \cong \bar{e}_{jj}$ implies an isomorphism $e_{ii} \cong e_{jj}$. The isomorphism gives us elements e_{ij} and e_{ji}, and then we can piece these together to form matrices.

Two idempotents e and f in a ring R with identity are said to be **orthogonal** if $ef = 0 = fe$. Suppose that $e_1, \ldots, e_n$ are mutually orthogonal idempotents such that $\sum_{i=1}^n e_i = 1$. Let us see in this case that

$$R = Re_1 \oplus \cdots \oplus Re_n$$

as left R modules. In fact, the condition $\sum_{i=1}^n e_i = 1$ shows that $r = \sum_{i=1}^n re_i$ for each $r \in R$, and thus $R = Re_1 + \cdots + Re_n$. If r lies in $Re_j \cap \sum_{i \neq j} Re_i$, then $r = se_j$ and $r = \sum_{i \neq j} r_i e_i$. Multiplying the first of these equalities on the right by e_j gives $re_j = se_j^2 = se_j = r$. Hence the second of these equalities, upon multiplication by e_j, yields $r = re_j = \sum_{i \neq j} r_i e_i e_j = 0$. In other words, the sum is direct, as asserted.

Corollary 2.24. Let R be a left Artinian ring. For each r in R, let $\bar{r}$ be the coset $r + \operatorname{rad} R$ in $R/\operatorname{rad} R$. If x and y are orthogonal idempotents in $\bar{R} = R/\operatorname{rad} R$ and if e is an idempotent in R with $\bar{e} = x$, then there exists an idempotent f in R with $\bar{f} = y$ and $ef = fe = 0$.

PROOF. By Proposition 2.23 choose an idempotent f_0 in R with $\bar{f_0} = y$. Then $f_0 e$ has $\overline{f_0 e} = yx = 0$. Hence $f_0 e$ is in rad R, and $(f_0 e)^{n+1} = 0$ for some n. Consequently $1 + f_0 e + (f_0 e)^2 + \cdots + (f_0 e)^n$ is a two-sided inverse to $1 - f_0 e$. Define

$$f = (1-e)\bigl(1 + f_0 e + (f_0 e)^2 + \cdots + (f_0 e)^n\bigr) f_0 (1 - f_0 e).$$

Then $\bar{f} = (1-x)(y+0+\cdots+0)y(1-0) = (1-x)y = y - xy = y$. Moreover,

$$fe = (1-e)\bigl(1 + f_0 e + (f_0 e)^2 + \cdots + (f_0 e)^n\bigr)(f_0 e - f_0^2 e^2) = 0$$

since $f_0 e - f_0^2 e^2 = f_0 e - f_0 e = 0$, and

$$ef = e(1-e)\bigl(1 + f_0 e + (f_0 e)^2 + \cdots + (f_0 e)^n\bigr) f_0 (1 - f_0 e) = 0$$

since $e(1-e) = 0$.

We still need to see that $f^2 = 0$. Since $f_0(1 - f_0 e) = f_0(1 - e)$, we can write $f = (1-e)(1 + f_0 e + \cdots) f_0 (1-e)$ and

$$\begin{aligned}
f^2 &= (1-e)(1 + f_0 e + \cdots) f_0 (1-e)(1 + f_0 e + \cdots) f_0 (1-e) \\
&= (1-e)(1 + f_0 e + \cdots) f_0 (1 - f_0 e)(1 + f_0 e + \cdots) f_0 (1-e) \\
&= (1-e)(1 + f_0 e + \cdots) f_0 \cdot 1 \cdot f_0 (1-e) \\
&= (1-e)(1 + f_0 e + \cdots) f_0 (1 - f_0 e) \\
&= f,
\end{aligned}$$

as required. □

Corollary 2.25. Let R be a left Artinian ring. For each r in R, let $\bar{r}$ be the coset $r + \operatorname{rad} R$ in $R/\operatorname{rad} R$. If $\{x_1, \ldots, x_N\}$ is a finite set of mutually orthogonal idempotents in $\bar{R} = R/\operatorname{rad} R$, then there exists a set of mutually orthogonal idempotents $\{e_1, \ldots, e_N\}$ in R such that $\bar{e}_i = x_i$ for all i. If $\sum_{i=1}^{N} x_i = 1$, then $\sum_{i=1}^{N} e_i = 1$.

PROOF. For the existence of $\{x_1, \ldots, x_N\}$, we proceed by induction on N, the case $N = 1$ being Proposition 2.23. Suppose we have found $e_1, \ldots, e_n$ and we want to find e_{n+1}. Let e be the idempotent $e_1 + \cdots + e_n$, and apply Corollary 2.24 to the idempotent e in R and the idempotent x_{n+1} in $R/\operatorname{rad} R$. The corollary gives us e_{n+1} orthogonal to e with $\bar{e}_{n+1} = x_{n+1}$. Since $e_i = e_i e = e e_i$ for $i \leq n$, we obtain $e_{n+1} e_i = e_{n+1}(e e_i) = (e_{n+1} e) e_i = 0$ and similarly $e_i e_{n+1} = 0$ for those i's, and the induction is complete.

Finally $\sum_i x_i = 1$ implies that $\sum_i e_i = 1 + r$ for some r in rad R. Then the idempotent $1 - \sum_i e_i$ is exhibited as in rad R and must be 0 because every element of rad R is nilpotent. □

In a nonzero ring R with identity, a finite subset $\{e_{ij} \mid i,j \in \{1,\ldots,n\}\}$ is called a set of **matrix units** in R if $\sum_{i=1}^n e_{ii} = 1$ and $e_{ij}e_{kl} = \delta_{jk}e_{il}$ for all i,j,k,l. It follows from these conditions that the e_{ii} are mutually orthogonal idempotents with sum 1, since $e_{ii}e_{jj} = \delta_{ij}e_{ij} = \delta_{ij}e_{ii}$. In view of the remarks before Corollary 2.24, we automatically have $R = \bigoplus_{i=1}^n Re_{ii}$. In addition, the product rule gives $e_{ii}e_{ij}e_{jj} = e_{ij}$, $e_{jj}e_{ji}e_{ii} = e_{ji}$, $e_{ij}e_{ji} = e_{ii}$, and $e_{ji}e_{ij} = e_{jj}$; by Proposition 2.21 the idempotents e_{ii} and e_{jj} are isomorphic in the sense that there is a left R module isomorphism $Re_{ii} \cong Re_{jj}$.

If $A = M_n(R)$, define E_{ij} to be the matrix that is 1 in the $(i,j)^{\text{th}}$ place and is 0 elsewhere. Then it is immediate that $\{E_{ij}\}$ is a set of matrix units in A. To recognize matrix rings, we prove the following converse.

Proposition 2.26. For a nonzero ring A with identity, suppose that

$$\{e_{ij} \mid i,j \in \{1,\ldots,n\}\}$$

is a set of matrix units in A. Let R be the subring of A of all elements of A commuting with all e_{ij}. Then every element of A can be written in one and only one way as $\sum_{i,j} r_{ij}e_{ij}$ with $r_{ij} \in R$ for all i and j, and the map $A \to M_n(R)$ given by $a \mapsto [r_{ij}]$ is a ring isomorphism. The ring R can be recovered from A by means of the isomorphism $R \cong e_{11}Ae_{11}$.

PROOF. To each $a \in A$, associate the matrix $[r_{ij}]$ in $M_n(A)$ whose entries are given by $r_{ij} = \sum_k e_{ki}ae_{jk}$. Then

$$r_{ij}e_{lm} = \sum_k e_{ki}ae_{jk}e_{lm} = \sum_k e_{ki}a\delta_{kl}e_{jm} = e_{li}ae_{jm}, \qquad (*)$$

and

$$e_{lm}r_{ij} = \sum_k e_{lm}e_{ki}ae_{jk} = \sum_k \delta_{mk}e_{li}ae_{jk} = e_{li}ae_{jm}.$$

Thus $r_{ij}e_{lm} = e_{li}ae_{jm} = e_{lm}r_{ij}$. Because of the definition of R, this equality shows that r_{ij} is in R. In particular, $[r_{ij}]$ is in $M_n(R)$. A special case of $(*)$ is that $r_{ij}e_{ij} = e_{ii}ae_{jj}$. Hence

$$\sum_{i,j} r_{ij}e_{ij} = \sum_{i,j} e_{ii}ae_{jj} = 1a1 = a.$$

This proves that a can be expanded as $a = \sum_{i,j} r_{ij}$.

For uniqueness, suppose that $a = \sum_{i,j} s_{ij}e_{ij}$ is given with each s_{ij} in R. Multiplication on the left by e_{kp} and right by e_{qk}, followed by addition, gives

$$r_{pq} = \sum_k e_{kp}ae_{qk} = \sum_k e_{kp}\Big(\sum_{i,j} s_{ij}e_{ij}\Big)e_{qk} = \sum_{i,j,k} s_{ij}e_{kp}e_{ij}e_{qk} = \sum_k s_{pq}e_{kk} = s_{pq}.$$

This proves that the map $A \to M_n(R)$ is one-one onto.

To see that the map $A \to M_n(R)$ respects multiplication, let a and a' be in A, and let the effect of the map on a, a', and aa' be $a \mapsto [r_{ij}]$, $a' \mapsto [r'_{ij}]$, and $aa' \mapsto [s_{ij}]$. Then we have

$$\sum_l r_{il} r'_{lj} = \sum_{l,k,k'} e_{ki} a e_{lk} e_{k'l} a' e_{jk'} = \sum_{l,k} e_{ki} a e_{ll} a' e_{jk} = \sum_k e_{ki} a a' e_{jk} = s_{ij},$$

and the matrix product of the images of a and a' coincides with the image of aa'.

Finally consider the image $E_{11} = [r_{ij}]$ of the element $a = e_{11}$ of A. It has $r_{ij} = \sum_k e_{ki} e_{11} e_{jk} = \delta_{i1}\delta_{1j} \sum_k e_{kk} = \delta_{i1}\delta_{1j}$. If a is a general element of A and its image is $[r_{ij}]$, then the result of the previous paragraph shows that $e_{11} a e_{11}$ maps to $E_{11}[r_{ij}]E_{11} = r_{11} E_{11}$. Hence the map $e_{11} a e_{11} \mapsto r_{11}$ is an isomorphism of $e_{11} A e_{11}$ with R. □

PROOF OF THEOREM 2.18. Let $\{x_{ij} \mid i, j \in \{1, \ldots, n\}\}$ be a set of matrix units for the matrix ring $A/\operatorname{rad} A \cong M_n(D)$. Then $x_{11}, \ldots, x_{nn}$ are mutually orthogonal idempotents in $A/\operatorname{rad} A$ with sum 1. By Corollary 2.25 we can choose mutually orthogonal idempotents $e_{11}, \ldots, e_{nn}$ in A with $\sum_{i=1}^n e_{ii} = 1$ and with $\bar{e}_{ii} = x_{ii}$.

We observed at the time of defining matrix units that $x_{11}, \ldots, x_{nn}$ are isomorphic as idempotents. Corollary 2.22 shows as a consequence that $e_{11}, \ldots, e_{nn}$ are isomorphic as idempotents. The remarks following Corollary 2.22 show that the isomorphism of Re_{11} with Re_{ii} can be exhibited by elements e_{1i} and e_{i1} in A satisfying the usual properties

$$e_{11} e_{1i} e_{ii} = e_{1i}, \quad e_{ii} e_{i1} e_{11} = e_{i1}, \quad e_{1i} e_{i1} = e_{11}, \quad e_{i1} e_{1i} = e_{ii}$$

and also the properties $\bar{e}_{1i} = x_{1i}$ and $\bar{e}_{i1} = x_{i1}$. Here $\bar{a}$ is shorthand for $a + \operatorname{rad} A$. Define $e_{ij} = e_{i1} e_{1j}$. Then $\bar{e}_{ij} = \bar{e}_{i1}\bar{e}_{1j} = x_{i1} x_{1j} = x_{ij}$, and we readily check that $\{e_{ij}\}$ is a set of matrix units for A.

By Proposition 2.26, $A \cong M_n(R)$ with $R \cong e_{11} A e_{11}$. From Corollary 2.20 we know that $e_{11} A e_{11}/\operatorname{rad}(e_{11} A e_{11}) \cong \bar{e}_{11}(A/\operatorname{rad} A)\bar{e}_{11}$, where $\bar{e}_{11}$ denotes the element $e_{11} + \operatorname{rad} A$ of $A/\operatorname{rad} A$. Hence

$$R/\operatorname{rad} R \cong \bar{e}_{11}(A/\operatorname{rad} A)\bar{e}_{11} \cong \bar{e}_{11} M_n(D) \bar{e}_{11} \cong D,$$

and the proof is complete. □

Corollary 2.27. If A is a finite-dimensional algebra with identity over a field F and if $A/\operatorname{rad} A \cong M_n(F)$ as algebras, then there is a subalgebra S isomorphic to $M_n(F)$ such that $A \cong S \oplus \operatorname{rad} A$ as vector spaces.

REMARKS. This corollary shows that Theorem 2.18 implies Theorem 2.17 under the additional assumption that the algebra A of Theorem 2.17 satisfies $A/\operatorname{rad} A \cong M_n(F)$. It is not necessary to assume characteristic 0.

PROOF. Suppose that A is a finite-dimensional algebra with identity over F such that $A/\operatorname{rad} A \cong M_n(F)$. Then A is left Artinian, and Theorem 2.18 produces a certain ring R with $A \cong M_n(R)$. Here Proposition 2.26 shows that R is isomorphic as a ring to $e_{11} A e_{11}$ for a certain idempotent e_{11} in A. It follows that R is an algebra with identity over F, necessarily finite-dimensional because A is finite-dimensional. The algebra R, according to Theorem 2.18, has $R/\operatorname{rad} R \cong F$. Therefore $R \cong F \oplus \operatorname{rad} R$ as F vector spaces. If we allow $M_n(\,\cdot\,)$ to be defined even for rings without identity, then we have F algebra isomorphisms

$$A \cong M_n(R) \cong M_n(F \oplus \operatorname{rad} R) \cong M_n(F) \oplus M_n(\operatorname{rad} R)$$

in which the direct sums are understood to be direct sums of vector spaces. We shall show that

$$\operatorname{rad}(M_n(R)) = M_n(\operatorname{rad} R), \qquad (*)$$

and then the decomposition $A = S \oplus \operatorname{rad} A$ will have been proved with $S \cong M_n(F)$.

To prove $(*)$, let E_{ij} be the member of $M_n(R)$ that is 1 in the $(i, j)^{\text{th}}$ place and is 0 elsewhere. Suppose that J is a two-sided ideal in $M_n(R)$. Let $I \subseteq R$ be the set of all elements x_{11} for $x \in J$. If r is in R, then rE_{11} is a member of $M_n(R)$, and the $(1, 1)^{\text{th}}$ entry of the element $(rE_{11})x$ of J is rx_{11}. Thus rx_{11} is in I. Similarly $x_{11}r$ is in I, and I is a two-sided ideal in R. Let us see that

$$J = M_n(I). \qquad (**)$$

If x is in J, then so is $E_{i1} x E_{1j} = x_{11} E_{ij}$, and hence $I E_{ij}$ is in J; taking sums over i and j shows that $M_n(I) \subseteq J$. In the reverse direction if x is in J, then so is $E_{1i} x E_{j1} = x_{ij} E_{11}$, and hence x_{ij} is in I; therefore $J \subseteq M_n(I)$. This proves $(**)$. Let us apply $(**)$ with $J = \operatorname{rad}(M_n(R))$. The corresponding ideal I of R consists of all entries x_{11} of members x of J. Using Corollary 2.20, we obtain

$$I E_{11} = E_{11} J E_{11} = E_{11} \operatorname{rad}(M_n(R)) E_{11} = \operatorname{rad}(E_{11} M_n(R) E_{11}) = \operatorname{rad}(R E_{11}).$$

Thus $I = \operatorname{rad} R$. Taking $M_n(\,\cdot\,)$ of both sides and applying $(**)$, we arrive at $(*)$. This completes the proof. □

Corollary 2.28. If A is a finite-dimensional associative algebra with identity over a field F and if $A/\operatorname{rad} A \cong M_{n_1}(F) \times \cdots \times M_{n_r}(F)$, then there is a subalgebra S of A isomorphic as an algebra to $A/\operatorname{rad} A$ such that $A \cong S \oplus \operatorname{rad} A$ as vector spaces.

REMARKS. This corollary gives the conclusion of Theorem 2.17 under the additional assumption that the semisimple algebra $A/\operatorname{rad} A$ over F is of the form $A/\operatorname{rad} A \cong M_{n_1}(F) \times \cdots \times M_{n_r}(F)$. If F is algebraically closed, then the division rings D_k in Theorem 2.2 are finite-dimensional division algebras over F and necessarily equal F, as was observed in the discussion after Corollary 2.3. Thus Theorem 2.2 shows that the additional assumption about the form of $A/\operatorname{rad} A$ is automatically satisfied if F is algebraically closed. In other words, Corollary 2.28 completes the proof of Theorem 2.17 if F is algebraically closed.

PROOF. For $1 \leq j \leq r$, let x_j be the identity matrix of $M_{n_j}(F)$ when $M_{n_j}(F)$ is regarded as a subalgebra of $A/\operatorname{rad} A$. The elements x_j are orthogonal idempotents in $A/\operatorname{rad} A$ with sum 1, and Corollary 2.25 shows that they lift to orthogonal idempotents e_j of A with sum 1. For each j, Corollary 2.20 shows that $e_j A e_j / \operatorname{rad}(e_j A e_j) = x_j (A/\operatorname{rad} A) x_j \cong M_{n_j}(F)$. By Corollary 2.27, $e_j A e_j$ has a subalgebra $S_j \cong M_{n_j}(F)$ with $e_j A e_j = S_j \oplus \operatorname{rad}(e_j A e_j)$ as vector spaces. Put $S = \bigoplus_{j=1}^r S_j$, the direct sum being understood in the sense of vector spaces. The subalgebra S_j has identity e_j, and the product of e_j with any other S_i is 0 because $e_i e_j = e_j e_i = 0$ when $i \neq j$. If $s = \sum_j s_j$ and $s' = \sum_j s'_j$ are two elements of S, then $ss' = \left(\sum_i s_i e_i\right)\left(\sum_j e_j s'_j\right) = \sum_{i,j} s_i e_i e_j s'_j = \sum_j s_j e_j s'_j = \sum_j s_j s'_j$. Hence S is a subalgebra. The element $\sum_{j=1}^r e_j$ is a two-sided identity in S.

Let us prove that $S \cap \operatorname{rad} A = 0$. If $s = \sum_j s_j$ is in $S \cap \operatorname{rad} A$, then $s_j = e_j s e_j$ is in $S_j = e_j S e_j$ and is in $e_j (\operatorname{rad} A) e_j$, which equals $\operatorname{rad}(e_j A e_j)$ by Corollary 2.20. Since $S_j \cap \operatorname{rad}(e_j A e_j) = 0$ by construction, $s_j = 0$. Thus $s = \sum_j s_j = 0$.

Consequently $S \cap \operatorname{rad} A = 0$. A count of dimensions gives $\dim S = \sum_j \dim S_j = \sum_j n_j^2 = \dim(A/\operatorname{rad} A)$. Thus $\dim A = \dim S + \dim(\operatorname{rad} A)$, and we conclude that $A = S \oplus \operatorname{rad} A$ as vector spaces. $\square$

6. Semisimplicity and Tensor Products

In this section we shall complete the proof of Wedderburn's Main Theorem (Theorem 2.17). In the previous section we proved in Corollary 2.28 the special case in which $A/\operatorname{rad} A$ is isomorphic to a product of full matrix rings over the base field F. This special case includes all cases of Theorem 2.17 in which F is algebraically closed.

The idea for the general case is to make a change of rings by tensoring A with the algebraic closure of the underlying field F, or at least with a large enough finite extension K of F for Corollary 2.28 to be applicable. That is, we first consider $A_K = A \otimes_F K$ and $(A/\operatorname{rad} A) \otimes_F K$ in place of A and $A/\operatorname{rad} A$. Inside A_K we can recognize $(\operatorname{rad} A) \otimes_F K$ as a subalgebra defined over K, and we expect that it is $\operatorname{rad} A_K$ and that we can find a complementary subalgebra S over

K; then the question is one of showing that S is of the form $S_0 \otimes_F K$ for some semisimple subalgebra S_0 of A defined over F. The trouble with this style of argument is that the tensor product $(A/\operatorname{rad} A) \otimes_F K$ need not be semisimple and there need not be a candidate for S. Some question about separability of field extensions plays a role, as the following example shows, and the assumption of characteristic 0 will ensure this separability.

EXAMPLE. We exhibit two extension fields K and L of a base field F such that $K \otimes_F L$ is not a semisimple algebra over F. The field extensions are each 1-by-1 matrix algebras over an extension field of F and hence are simple algebras, yet the tensor product is not semisimple. Fix a prime field $\mathbb{F}_p$, and let $F = \mathbb{F}_p(x^p)$ be a simple transcendental extension of $\mathbb{F}_p$. Define $K = L = \mathbb{F}_p(x) = F(\sqrt[p]{x^p})$. Both K and L are field extensions of F of degree p. Thus $K \otimes_F L$ is a finite-dimensional commutative algebra with identity over F, by the construction in Proposition 10.24 of *Basic Algebra*. The element $z = x \otimes 1 - 1 \otimes x$ in $K \otimes_F L$ is nonzero but has $z^p = x^p \otimes 1 - 1 \otimes x^p = x^p \otimes 1 - x^p \otimes 1 = 0$, the next-to-last equality following because x^p lies in the base field F. Consequently $K \otimes_F L$ has a nonzero nilpotent element. If $K \otimes_F L$ were semisimple, Theorem 2.2 would show that it was the direct product of fields, and it could not have any nonzero nilpotent elements. We conclude that $K \otimes_F L$ is not a semisimple algebra.

Proposition 2.29. Let F be a field, let $K = F(\alpha)$ be a simple algebraic extension, let $g(X)$ be the minimal polynomial of α over F, and let L be another field extension of F. Then

(a) $K \otimes_F L \cong L[X]/(g(X))$ as associative algebras over L,
(b) $K \otimes_F L$ is a semisimple algebra if the polynomial $g(X)$ is separable.

REMARKS. Proposition 10.24 of *Basic Algebra* shows that the tensor product $A \otimes_F B$ of two associative algebras with identity over F has a unique associative algebra structure such that $(a_1 \otimes b_1)(a_2 \otimes b_2) = a_1 a_2 \otimes b_1 b_2$. Problem 8 at the end of Chapter X shows that if B is an extension field of F, then $A \otimes_F B$ is in fact an associative algebra with identity over B, the multiplication by $b \in B$ being given by the mapping $1 \otimes$ (left by b).

PROOF. For (a), let $n = [K : F]$. Form the F bilinear mapping of $F[X] \times L$ into $L[X]$ given by $(P(X), \ell) \mapsto \ell P(X)$. Corresponding to this F bilinear mapping is a unique F linear map $\varphi : F[X] \otimes_F L \to L[X]$ carrying $P(X) \otimes \ell$ to $\ell P(X)$ for $P(X) \in F[X]$ and $\ell \in L$. The F vector space $F[X] \otimes_F L$ is an L vector space with multiplication by $\ell_0 \in L$ given by the linear mapping $1 \otimes$ (left by ℓ_0). Since $\varphi\big((1 \otimes (\text{left by } \ell_0))(P(X) \otimes \ell)\big) = \ell_0 \ell P(X) = \ell_0 \varphi(P(X) \otimes \ell)), \varphi$ is L linear. In addition, $\varphi((P(X) \otimes \ell)(Q(X) \otimes \ell')) = \varphi(P(X)Q(X) \otimes \ell\ell') = \ell\ell' P(X)Q(X) = \varphi(P(X) \otimes \ell)\varphi(Q(X) \otimes \ell')$, and therefore φ is an algebra homomorphism.

We follow φ with the quotient homomorphism $\psi : L[X] \to L[X]/(g(X))$, and the composition $\psi\varphi$ is 0 on the ideal $(g(x)) \otimes_F L$ of $F[X] \otimes_F L$. Therefore $\psi\varphi$ descends to a homomorphism $(F[X]/(g(X))) \otimes_F L \to L[X]/(g(X))$, hence to a homomorphism $\eta : K \otimes_F L \to L[X]/(g(X))$. Since φ and ψ are onto, so is η.

It is enough to prove that η is one-one. Thus suppose that $\eta\bigl(\sum_i k_i \otimes \ell_i\bigr) = 0$ with all k_i in K, all ℓ_i in L, and the ℓ_i linearly independent over F. Write $k_i = P_i(X) + (g(X))$ with $\deg P_i(X) < n$ whenever $P_i \neq 0$. Then $\sum_i \ell_i P_i(X) \equiv 0 \bmod g(X)$. Since $g(X)$ has degree n and each nonzero $P_i(X)$ has degree at most n, $\sum_i \ell_i P_i(X) = 0$. Write $P_i(X) = \sum_j c_{ij} X^j$ with each c_{ij} in F. Then $\sum_j \bigl(\sum_i \ell_i c_{ij}\bigr) X^j = 0$, and $\sum_i \ell_i c_{ij} = 0$ for all j. Since the ℓ_i are linearly independent over F, $c_{ij} = 0$ for all i and j. Thus $k_i = 0$ for all i, $\sum_i k_i \otimes \ell_i = 0$, and η is one-one. This proves (a).

For (b), factor $g(X)$ over L as $g_1(X) \cdots g_m(X)$ for polynomials $g_j(X)$ irreducible over L. Since the separability of g forces $g_1, \ldots, g_m$ to be relatively prime in pairs, the Chinese Remainder Theorem implies that

$$L[X]/(g_1(X) \cdots g_m(X)) \cong L[X]/(g_1(X)) \times \cdots \times L[X]/(g_m(X)).$$

Each $L[X]/(g_j(X))$ is a field, and thus $L[X]/(g(X))$ is exhibited as a product of fields and is semisimple. $\square$

Corollary 2.30. Let F be a field, let K be a finite separable algebraic extension of F, and let L be another field extension of F. Then the algebra $K \otimes_F L$ is semisimple.

REMARKS. The condition of separability of the extension K/F is automatic in characteristic 0. The two field extensions K and L in the example before Proposition 2.29 both failed to be separable extensions of the base field F.

PROOF. The Theorem of the Primitive Element (Theorem 9.34 of *Basic Algebra*) shows that K/F is a simple extension, say with $K = F(\alpha)$. Since this extension is assumed separable, the minimal polynomial over F of any element of K is a separable polynomial. The hypotheses of Proposition 2.29b are therefore satisfied, and $K \otimes_F L$ is semisimple. $\square$

Proposition 2.31. Suppose that A and B are algebras with identity over a field F, that B is simple, and that B has center F. Then the two-sided ideals of the tensor-product algebra $A \otimes_F B$ are all subsets $I \otimes_F B$ such that I is a two-sided ideal of A.

PROOF. The set $I \otimes_F B$ is a two-sided ideal of $A \otimes_F B$, since $(a \otimes b)(i \otimes b') = ai \otimes bb'$ and since a similar identity applies to multiplication in the other order.

Conversely suppose that J is an ideal in $A \otimes_F B$. Let 1_B be the identity of B, and define $I = \{a \in A \mid a \otimes 1_B \in J\}$. Then I is a two-sided ideal of A, and we shall prove that $J = I \otimes_F B$. The easy inclusion is $I \otimes_F B \subseteq J$. For this, let i be in I and b be in B. Then $i \otimes 1_B$ is in J and $1_A \otimes b$ is in $A \otimes_F B$. Their product $i \otimes b$ has to be in J, and thus $I \otimes_F B \subseteq J$.

For the reverse inclusion, take a basis $\{x_i\}$ of I over F and extend it to a basis of A by adjoining some vectors $\{y_j\}$. It is enough to show that any finite sum $\sum_j y_j \otimes b_j$ in J necessarily has all b_j equal to 0. Arguing by contradiction, suppose that $\sum_{k=1}^m y_{j_k} \otimes b_{j_k}$ is a nonzero sum in J with m as small as possible and in particular with all b_{j_k} nonzero. Let H be the subset of B defined by

$$H = \left\{ c_{j_1} \,\middle|\, \sum_{k=1}^m y_{j_k} \otimes c_{j_k} \in J \text{ for some } m\text{-tuple } \{c_{j_k}\} \subseteq B \right\}.$$

The set H is a two-sided ideal of B containing the nonzero element b_{j_1} of B. Since B is simple by assumption, $H = B$. Thus 1_B is in H. Therefore some element

$$y_{j_1} \otimes 1_B + \sum_{k=2}^m y_{j_k} \otimes c_{j_k}$$

is in J. Let $b \in B$ be arbitrary. Multiplying the displayed element on the left and right by $1_A \otimes b$ and subtracting the results shows that

$$y_{j_2} \otimes (bc_{j_2} - c_{j_2}b) + \cdots + y_{j_m} \otimes (bc_{j_m} - c_{j_m}b)$$

is in J. Since m was chosen to be minimal, this element must be 0 for all choices of b. Then all coefficients are 0, and the conclusion is that all coefficients c_{j_k} are in the center of B, which is F by assumption. Consequently we can rewrite our element of J as

$$y_{j_1} \otimes 1_B + \sum_{k=2}^m y_{j_k} \otimes c_{j_k} = y_{j_1} \otimes 1_B + \sum_{k=2}^m c_{j_k} y_{j_k} \otimes 1_B = \left(y_{j_1} + c_{j_2} y_{j_2} + \cdots + c_{j_m} y_{j_m} \right) \otimes 1_B.$$

The definition of I shows that the factor $y_{j_1} + c_{j_2} y_{j_2} + \cdots + c_{j_m} y_{j_m}$ in the pure tensor on the right is in I. Since the y_j's form a basis of a vector-space complement to I, this vector must be 0. The linear independence of the y_j's over F forces each coefficient to be 0, and we have arrived at a contradiction because the coefficient of y_{j_1} is 1, not 0. $\square$

Lemma 2.32. The center of a finite-dimensional simple algebra A over a field F is a field that is a finite extension of F.

PROOF. By Theorem 2.4, $A \cong M_n(D)$ for some finite-dimensional division algebra D over F. Let Z be the center of A. By inspection this consists of the scalar matrices whose entries lie in the center of D. The center of D is a field. Hence Z is a field, necessarily a finite extension of F. $\square$

Proposition 2.33. Let A be a finite-dimensional semisimple algebra over a field F of characteristic 0, and suppose that K is a field containing F. Then the algebra $A \otimes_F K$ over K is semisimple.

PROOF. Since the tensor product of a finite direct sum is the direct sum of tensor products, we may assume without loss of generality that A is simple. Lemma 2.32 shows that the center Z of A is a finite extension field of F. By Corollary 2.30 and the assumption that F has characteristic 0, the algebra $Z \otimes_F K$ is semisimple. Being commutative, it must be of the form $K_1 \oplus \cdots \oplus K_s$ with each ideal K_i equal to a field, by Theorem 2.2.

Each ideal K_i is a unital $Z \otimes_F K$ module, hence is both a unital Z module and a unital K module. Thus we can regard each K_i as an extension field of Z or of K, whichever we choose. First let us regard K_i as an extension field of Z. Since K_i has no nontrivial ideals and A has center Z, Proposition 2.31 shows that the Z algebra $A \otimes_Z K_i$ is simple as a ring.

Next let us regard K_i as an extension field of K; since A is finite-dimensional over F, so is Z. Therefore $Z \otimes_F K$ is finite-dimensional over K, and K_i is a finite extension of K. Hence $A \otimes_Z K_i$ is a finite-dimensional algebra over K, and it is left Artinian as a ring.

By Theorem 2.6, any left Artinian simple ring such as $A \otimes_Z K_i$ is necessarily semisimple. Using the associativity formula for tensor products given in Proposition 10.22 of *Basic Algebra*, we obtain an isomorphism of rings

$$A \otimes_F K \cong (A \otimes_Z Z) \otimes_F K \cong A \otimes_Z (Z \otimes_F K)$$
$$\cong A \otimes_Z (K_1 \oplus \cdots \oplus K_s) \cong \bigoplus_{j=1}^{s} (A \otimes_Z K_j),$$

the summands being two-sided ideals in each case. Since each $A \otimes_Z K_j$ is a finite-dimensional simple algebra over K, $A \otimes_F K$ is a semisimple algebra over K by Theorem 2.4. □

Let us digress for a moment, returning in Lemma 2.34 to the argument that leads to the proof of Theorem 2.17. In the next section we shall want to know circumstances under which we can draw the same conclusion as in Proposition 2.33 without assuming that the characteristic is 0. Write the finite-dimensional semisimple algebra A as $A = M_{n_1}(D_1) \times \cdots \times M_{n_r}(D_r)$, where each D_r is a division algebra over F. Let $Z_1, \ldots, Z_r$ be the respective centers of the simple factors of A. Lemma 2.32 observes that each Z_j is a finite extension field of F. The proof of Proposition 2.33 appealed to Corollary 2.30 to conclude from the condition characteristic 0 that $Z_j \otimes_F K$ is semisimple. Instead, by rereading the statement of Corollary 2.30, we see that it would have been enough for each Z_j to be a finite *separable* field extension of F, even if F did not have characteristic 0.

6. Semisimplicity and Tensor Products   109

Then the rest of the above proof goes through without change. Accordingly we define a finite-dimensional semisimple algebra A over a field F to be a **separable** semisimple algebra if the center of each simple component of A is a separable extension field of F. In terms of this definition, we obtain the following improved version of Proposition 2.33.

Proposition 2.33′. Let A be a finite-dimensional separable semisimple algebra over a field F, and suppose that K is a field containing F. Then the algebra $A \otimes_F K$ over K is semisimple.

Lemma 2.34. Suppose that A is a finite-dimensional algebra with identity over a field F, and suppose that N is a nilpotent two-sided ideal of A such that the algebra A/N is semisimple. Then $N = \operatorname{rad} A$.

PROOF. The algebra A is left Artinian, being finite-dimensional. Since N is nilpotent, we must have $N \subseteq \operatorname{rad} A$. The two-sided ideal $(\operatorname{rad} A)/N$ of the semisimple algebra A/N is nilpotent and hence must be 0. Therefore $N = \operatorname{rad} A$. □

PROOF OF THEOREM 2.17. Let A be the given finite-dimensional algebra of the field F of characteristic 0, and write N for rad A and $\overline{A}$ for A/N. For any extension field K of F, we write $A_K = A \otimes_F K$, $N_K = N \otimes_F K$, and $\overline{A}_K = \overline{A} \otimes_F K$.

For most of the proof, we shall treat the special case that $N^2 = 0$. Let $\overline{F}$ be an algebraic closure of F. Then $\overline{A}_{\overline{F}} = \overline{A} \otimes_F \overline{F} = (A/N) \otimes_F \overline{F} \cong (A \otimes_F \overline{F})/(N \otimes_F \overline{F}) = A_{\overline{F}}/N_{\overline{F}}$. Proposition 2.33 shows that $\overline{A}_{\overline{F}} = \overline{A} \otimes_F \overline{F}$ is a semisimple algebra over $\overline{F}$, and the claim is that the two-sided ideal $N_{\overline{F}}$ of $A_{\overline{F}}$ is nilpotent. In fact, any element of $N_{\overline{F}}$ is a finite sum of the form $\sum_i (a_i \otimes c_i)$ with each a_i in N and each c_i in $\overline{F}$. The product of this element with $\sum_j (a'_j \otimes c'_j)$ is $\sum_{i,j} (a_i a'_j \otimes c_i c'_j)$, and this is 0 because the assumption $N^2 = 0$ implies that $a_i a'_j = 0$ for all i and j. Thus $N_{\overline{F}}^2 = 0$, and $N_{\overline{F}}$ is nilpotent.

Since $A_{\overline{F}}/N_{\overline{F}}$ is semisimple and $N_{\overline{F}}$ is nilpotent, Lemma 2.34 shows that $N_{\overline{F}} = \operatorname{rad}(A_{\overline{F}})$. Corollary 2.28 (a special case of Theorem 2.17) is applicable to $A_{\overline{F}}$ because $\overline{F}$ is algebraically closed, and it follows that there exists a subalgebra $\widetilde{S}$ of $A_{\overline{F}}$ such that $A_{\overline{F}} = \widetilde{S} \oplus N_{\overline{F}}$ as vector spaces. Here $\widetilde{S}$ is a product of finitely many algebras $M_{n_j}(\overline{F})$. The embedded matrix units e_{ij} of $\widetilde{S}$ obtained from each $M_{n_j}(\overline{F})$ are members of $A_{\overline{F}} = A \otimes_F \overline{F}$ and hence are of the form $\sum_l x_l \otimes c_l$, where $\{x_l\}_{l=1}^n$ is a vector-space basis of A over F and each c_l is in $\overline{F}$. Only finitely many such c_l's are needed to handle all e_{ij}'s, and we let K be a finite extension of F within $\overline{F}$ containing all of them. Let $\rho_0 = 1, \rho_1, \ldots, \rho_s$ be a vector-space basis of K over F.

Relative to this K, we form A_K, N_K, and $\overline{A}_K$ as in the first paragraph of the proof. The same argument as with $\overline{F}$ shows that $\overline{A}_K \cong A_K/N_K$ is semisimple and that N_K is nilpotent. By Lemma 2.34, $N_K = \operatorname{rad} A_K$. The formulas for the e_{ij}'s in the previous paragraph are valid in A_K and give us a system of matrix units. As in the previous paragraph, Corollary 2.28 produces a subalgebra S of A_K isomorphic to some $M_{n_1}(K) \times \cdots \times M_{n_r}(K)$ such that $A_K = S \oplus N_K$ as vector spaces.

In the basis $\{x_i\}_{i=1}^n$ of A over F, we may assume that the first t vectors form a basis of $N = \operatorname{rad} A$ and the remaining vectors form a basis of a vector-space complement to N. We identify members a of A with members $a \otimes 1$ of A_K. With this identification in force, we decompose each basis vector x_i for $i > t$ according to $A_K = S \oplus N_K$ as $x_i = y_i - z_i$ with $y_i \in S$ and $z_i \in N_K$. Since the x_i's for $i \leq t$ are in $N \subseteq N_K$, the vectors y_i with $i > t$ form a vector-space basis of S over K. For $i > t$, write $z_i = \sum_{j=0}^s z_{ij} \otimes \rho_j$ with z_{ij} in N. Then we have

$$y_i = x_i + z_i = (x_i + z_{i0}) + \sum_{j=1}^s z_{ij} \otimes \rho_j \qquad \text{for } i > t.$$

Put

$$x_i' = x_i + z_{i0} \qquad \text{and} \qquad z_i' = \sum_{j=1}^s z_{ij} \otimes \rho_j \qquad \text{for } i > t.$$

Then $\{x_i\}_{i=1}^t \cup \{x_i'\}_{i=t+1}^n$ is a basis of A over F. We shall show that $S_0 = \sum_{i=t+1}^n F x_i'$ is a subalgebra of A, and then $A = S_0 \oplus N$ will be the required decomposition.

Let x_i' and x_j' be given with $i > t$ and $j > t$, and write

$$x_i' x_j' = \sum_k \gamma_{kij} x_k' + v_{ij} \qquad \text{with } \gamma_{kij} \in F \text{ and } v_{ij} \in N.$$

Substituting $x_i' = y_i - z_i'$ and taking into account that N_K is an ideal in A_K, we have

$$y_i y_j \equiv \sum_k \gamma_{kij} x_k' \mod N_K \equiv \sum_k \gamma_{kij} y_k \mod N_K.$$

Then $y_i y_j = \sum_k \gamma_{kij} y_k + u_{ij}$ with each $u_{ij} \in N_K$. Since the y_i are in S and S is a subalgebra, $u_{ij} = 0$. Thus $y_i y_j = \sum_k \gamma_{kij} y_k$. Let us resubstitute into this equality from $y_i = x_i' + z_i'$. Taking into account that $z_i' z_j' = 0$ because $N_K^2 = 0$, we obtain

$$x_i' x_j' + x_i' z_j' + z_i' x_j' = \sum_k \gamma_{kij} x_k' + \sum_k \gamma_{kij} z_k'.$$

Substituting from $z_i' = \sum_{j=1}^s z_{ij} \otimes \rho_j$ gives

$$x_i' x_j' \otimes 1 + \sum_{l=1}^s x_i' z_{jl} \otimes \rho_l + \sum_{l=1}^s z_{il} x_j' \otimes \rho_l = \sum_k \gamma_{kij} x_k' \otimes 1 + \sum_k \sum_{l=1}^s \gamma_{kij} z_{kl} \otimes \rho_l.$$

The coefficients of $\rho_0 = 1$ must be equal, and therefore

$$x'_i x'_j = \sum_k \gamma_{kij} x'_k.$$

This equation shows that S_0 is a subalgebra and completes the proof under the hypothesis that $N^2 = 0$.

Now we drop the assumption that $N^2 = 0$. We shall prove the theorem by induction on $\dim_F A$, the base cases of the induction being $\dim_F A = 0$ and $\dim_F A = 1$, for which the theorem is immediate by inspection. For the inductive case, let A be given, and assume the theorem to be known for algebras of dimension $< \dim_F A$. If $N^2 = 0$, then we are done. Thus we may assume that the product ideal N^2 is nonzero and therefore that $\dim_F(A/N^2) < \dim_F A$. The First Isomorphism Theorem shows that $(A/N^2)/(N/N^2) \cong A/N = \overline{A}$. The quotient A/N is semisimple, and N/N^2 is a nilpotent ideal in A/N^2. By Lemma 2.34, $N/N^2 = \text{rad}(A/N^2)$. The inductive hypothesis gives $A/N = S_1/N^2 \oplus N/N^2$ for a subalgebra S_1 of A with $S_1 \supseteq N^2$. This means that $A = S_1 + N$ and $S_1 \cap N = N^2$. Here

$$\dim_F A = \dim_F(S_1 + N) = \dim_F S_1 + \dim_F N - \dim_F(S_1 \cap N)$$
$$= \dim_F S_1 + \dim_F N - \dim_F N^2 = \dim_F S_1 + \dim_F(N/N^2),$$

and $N/N^2 \neq 0$ implies $\dim_F S_1 < \dim_F A$. The Second Isomorphism Theorem gives $A/N = (S_1 + N)/N \cong S_1/(S_1 \cap N) = S_1/N^2$. Thus S_1/N^2 is semisimple. Since N^2 is nilpotent, Lemma 2.34 shows that $N^2 = \text{rad}\, S_1$. The inductive hypothesis gives $S_1 = S \oplus N^2$ for a semisimple subalgebra S. Substituting into $A = S_1 + N$, we obtain $A = (S \oplus N^2) + N = S + N$. Meanwhile, $S \cap N = (S \cap S_1) \cap N = S \cap (S_1 \cap N) = S \cap N^2 = 0$. Therefore $A = S \oplus N$, and the induction is complete. $\square$

7. Skolem–Noether Theorem

In this section we begin an investigation of division algebras that are finite-dimensional over a given field F. A nonzero algebra A with identity over a field F will be called **central** if the center of A consists exactly of the scalar multiples of the identity, i.e., if $\text{center}(A) = F$. Of special interest will be algebras with identity that are **central simple**, i.e., are both central and simple.

Lemma 2.35. Let A and B be algebras with identity over a field F, and suppose that B is central. Then

(a) the members of $A \otimes_F B$ commuting with $1 \otimes B$ are the members of $A \otimes 1$,
(b) $\text{center}(A \otimes_F B) = (\text{center}\, A) \otimes_F 1$.

PROOF. For (a), suppose that $z = \sum_i a_i \otimes b_i$ commutes with $1 \otimes B$ and that the a_i are linearly independent over F. If b is in B, then

$$0 = (1 \otimes b)z - z(1 \otimes b) = \sum_i a_i \otimes (bb_i - b_i b),$$

from which it follows that $bb_i - b_i b = 0$ for all b and all i. Since B is central, each b_i is in F, and we can write z as

$$z = \sum_i a_i \otimes b_i = \sum_i (a_i b_i \otimes 1) = \left(\sum_i a_i b_i\right) \otimes 1.$$

In other words, z is of the form $z = a \otimes 1$.

For (b), we need to prove the inclusion $\subseteq$. Thus let z be in center$(A \otimes_F B)$. By (a), z is of the form $z = a \otimes 1$ for some $a \in A$. Now suppose that a' is in A. Then $0 = (a' \otimes 1)z - z(a' \otimes 1) = (a'a - aa') \otimes 1$. Hence $a'a = aa'$, and we conclude that a is in center(A). $\square$

Proposition 2.36. Let A and B be algebras with identity over a field F, and suppose that B is central simple. Then

  (a) A simple implies $A \otimes_F B$ simple,
  (b) A central simple implies $A \otimes_F B$ central simple.

PROOF. For (a), Proposition 2.31 shows that any two-sided ideal of $A \otimes_F B$ is of the form $I \otimes_F B$ for some two-sided ideal I of A. Since A is assumed simple, the only I's are 0 and A. Thus the only ideals in $A \otimes_F B$ are 0 and $A \otimes_F B$, and $A \otimes_F B$ is simple.

For (b), conclusion (a) shows that $A \otimes_F B$ is simple. By Lemma 2.35b the center of $A \otimes_F B$ is (center $A) \otimes 1 = F1 \otimes 1 = F(1 \otimes 1)$, and hence $A \otimes_F B$ is central. $\square$

Corollary 2.37. If A and B are finite-dimensional semisimple algebras over a field F and at least one of them is separable over F, then $A \otimes_F B$ is semisimple.

REMARK. The definition of separability of A or B appears between Proposition 2.33 and Proposition 2.33'.

PROOF. Without loss of generality, we may assume that A and B are simple. For definiteness let us say that A is the given separable algebra over F. Let $K = \text{center}(B)$. Lemma 2.32 shows that K is a field, and associativity of tensor products allows us to write

$$A \otimes_F B \cong A \otimes_F (K \otimes_K B) \cong (A \otimes_F K) \otimes_K B.$$

Here $A \otimes_F K$ is semisimple by Proposition 2.33', and B is central simple over K. Thus Proposition 2.36a applies and shows that $(A \otimes_F K) \otimes_K B$ is simple. $\square$

7. Skolem–Noether Theorem

Corollary 2.38. Let A be a central simple algebra of finite dimension n over a field F, and let A^o be the opposite algebra. Then $A \otimes_F A^o \cong M_n(F)$.

EXAMPLE. Take $F = \mathbb{R}$ and $A = \mathbb{H}$, the algebra of quaternions. Then conjugation, with $1 \mapsto 1$ and $i, j, k \mapsto -i, -j, -k$, is an antiautomorphism of $\mathbb{H}$. Consequently $H^o \cong H$. The corollary says in this case that $\mathbb{H} \otimes_\mathbb{R} \mathbb{H} \cong M_4(\mathbb{R})$.

PROOF. Let V be A considered as a vector space. For each $a_0 \in A$, we associate the members $l(a_0)$ and $r(a_0)$ of $\text{End}_F(V)$ given by $l(a_0)a = a_0 a$ and $r(a_0)a = aa_0$. Then $l(a_0 a'_0) = l(a_0)l(a'_0)$ and $r(a_0 a'_0) = r(a'_0)r(a_0)$, and it follows that $l : A \to \text{End}_F(V)$ and $r : A^o \to \text{End}_F(V)$ are algebra homomorphisms sending 1 to 1.

Meanwhile, the map $A \times A^o \to \text{End}_F(V)$ given by $(a, a') \mapsto l(a)r(a')$ is F bilinear and extends to an F linear map $\varphi : A \otimes_F A^o \to \text{End}_F(V)$. Because of the homomorphism properties of l and r, the mapping φ is an algebra homomorphism sending 1 to 1. Proposition 2.36 shows that $A \otimes_F A^o$ is simple, and it follows that φ is one-one. Since $\dim_F(A \otimes_F A^o) = (\dim_F A)^2 = \dim_F \text{End}_F(V)$, φ is onto. □

Corollary 2.39. Let A be a central simple algebra of finite dimension d over a field F. Then d is the square of an integer.

PROOF. Let $\overline{F}$ be an algebraic closure of F. Proposition 2.36a shows that the algebra $\overline{F} \otimes_F A$ is simple, and its dimension over $\overline{F}$ is d. A simple finite-dimensional algebra over an algebraically closed field is a full matrix algebra over that field, and thus $\overline{F} \otimes_F A \cong M_n(\overline{F})$. Comparing dimensions over $\overline{F}$, we see that $d = n^2$. □

Corollary 2.40. If D is a division algebra finite-dimensional over its center F, then $\dim_F D$ is the square of an integer.

PROOF. The algebra D is central simple over its center F, and the result is immediate from Corollary 2.39. □

Theorem 2.41 (Skolem–Noether Theorem). Let A be a finite-dimensional central simple algebra over the field F, and let B be any simple algebra over F. Suppose that f and g are F algebra homomorphisms of B into A carrying the identity to the identity. Then there exists an $x \in A$ with $f(b) = xg(b)x^{-1}$ for all b in B.

PROOF. Let us observe that the homomorphisms f and g are one-one because B is simple, and the finite dimensionality of A therefore forces B to be finite-dimensional.

We consider first the special case that $A = M_n(F)$ for some n. The homomorphism f makes the space F^n of column vectors into a unital left B module by the definition $bv = f(b)v$, and similarly the homomorphism g makes F^n into a unital left B module. Since B is finite-dimensional and simple, an argument given with Example 1 of semisimple rings in Section 2 shows that there is only one simple left B module up to isomorphism and that every unital left B module is a direct sum of copies of this simple left B module. Consequently the isomorphism classes of the B modules determined by f and g depend only on their dimension. The dimension is n in both cases, and hence there exists an invertible F linear map $L : F^n \to F^n$ such that $Lf(b)v = g(b)Lv$ for all $v \in F^n$. If L is given by the matrix x^{-1} in $M_n(F)$, then $x^{-1}f(b) = g(b)x^{-1}$, and the theorem is therefore proved in this special case.

For the general case we form the tensor products $B \otimes_F A^o$ and $A \otimes_F A^o$. The maps $f \otimes 1$ and $g \otimes 1$ are F algebra homomorphisms between these algebras, $B \otimes_F A^o$ is simple by Proposition 2.36a, and Corollary 2.38 shows that $A \otimes_F A^o$ is isomorphic to $M_n(F)$ for the integer $n = \dim A$. The special case is applicable, and we obtain an invertible element X of $A \otimes_F A^o$ such that

$$(f \otimes 1)(b \otimes a^o) = X(g \otimes 1)(b \otimes a^o)X^{-1} \quad \text{for all } b \in B \text{ and } a^o \in A^o. \quad (*)$$

Taking $b = 1$, we see that $1 \otimes a^o = X(1 \otimes a^o)X^{-1}$ for all $a^o \in A^o$. By Lemma 2.35a, X lies in $A \otimes 1$, hence is of the form $X = x \otimes 1$ for some x in A. Substituting for X in $(*)$, we obtain $f(b) = xg(b)x^{-1}$ as required. $\square$

Corollary 2.42. If A is a finite-dimensional central simple algebra over the field F, then every F automorphism of A is inner in the sense of being given by conjugation by an invertible element of A.

PROOF. This is the special cse of Theorem 2.41 in which $B = A$ and g is the identity map on B. $\square$

8. Double Centralizer Theorem

We saw in Corollary 2.40 that if D is a division algebra finite-dimensional over its center F, then $\dim_F D$ is the square of an integer. In this section we shall prove a theorem from which we can conclude that the positive integer of which $\dim_F D$ is the square is the dimension of any maximal subfield of D. We state the theorem, establish two lemmas, prove the theorem, and then derive two corollaries concerning maximal subfields of division algebras.

If A is an algebra with identity and B is a subalgebra containing the identity, then the **centralizer** of B in A is the subalgebra of all members of A commuting with every element of B.

Theorem 2.43 (Double Centralizer Theorem). Let A be a finite-dimensional central simple algebra over a field F, let B be a simple subalgebra of A, and let C be the centralizer of B in A. Then C is simple, B is the centralizer of C in A, and $(\dim_F B)(\dim_F C) = \dim_F A$.

Lemma 2.44. Let A and A' be algebras with identity over a field F, let B and B' be subalgebras of them, and let C and C' be the centralizers of B and B' in A and A', respectively. Then the centralizer of $B \otimes_F B'$ in $A \otimes_F A'$ is $C \otimes_F C'$.

PROOF. Expand an element of $A \otimes_F A'$ for the moment as $x = \sum_i a_i \otimes a'_i$ with the elements a'_i linearly independent over F. If x satisfies $x(b \otimes 1) = (b \otimes 1)x$ for all b in B, then $\sum_i (a_i b - b a_i) \otimes a'_i = 0$. Since the a'_i's are independent, $a_i b - b a_i = 0$ for all i, and each a_i is in C. Thus the centralizer of $B \otimes_F 1$ is $C \otimes_F A'$.

Rewriting x with the a_i's assumed independent, we see similarly that the centralizer of $1 \otimes_F B'$ is $A \otimes_F C'$. Putting these conclusions together, we see that

$$\text{centralizer}(B \otimes_F B') \subseteq \text{centralizer}(B \otimes_F 1) \cap \text{centralizer}(1 \otimes_F B')$$
$$= (C \otimes_F A') \cap (A \otimes_F C') = C \otimes_F C'.$$

The reverse inclusion, namely $\text{centralizer}(B \otimes_F B') \supseteq C \otimes_F C'$, is immediate, and the lemma follows. □

Lemma 2.45. Let B be a finite-dimensional simple algebra over a field F, and write V for the algebra B considered as a vector space. For b in B and v in V, define members $l(b)$ and $r(b)$ of $\text{End}_F(V)$ by $l(b)v = bv$ and $r(b)v = vb$. Then the centralizer in $\text{End}_F(V)$ of $l(B)$ is $r(B)$.

PROOF. Let K be the center of B. This is an extension field of F by Lemma 2.32, and B is central simple over K. Let us see that any member a of $\text{End}_F(V)$ that centralizes $l(B)$ is actually in $\text{End}_K(V)$. If c is in K, then c is in particular in B, and therefore $al(c) = l(c)a$. Applying this equality to $v \in V$ yields $a(cv) = ca(v)$, and this equality for all $c \in K$ says that a is in $\text{End}_K(V)$.

Thus it is enough to show that the centralizer of $l(B)$ in $\text{End}_K(V)$ is $r(B)$. We argue as in the proof of Corollary 2.38: The definitions of l and r make V into a unital left B module and a unital right B module, and the members of K operate consistently on either side of V because K lies in the center of B. The function $(b, b') \mapsto l(b)r(b')$ is therefore K bilinear, and it extends to the tensor product $B \otimes_K B^o$ as an algebra homomorphism $\varphi : B \otimes_K B^o \mapsto \text{End}_K(V)$. The homomorphism φ is one-one, since Proposition 2.36a shows $B \otimes_K B^o$ to be simple. The dimensional equality $\dim_K(B \otimes_K B^o) = (\dim_K B)^2 = \dim_K(\text{End}_K(V))$ allows us to conclude that φ is onto, hence is an isomorphism.

Lemma 2.35a shows that the centralizer of $B \otimes_K 1$ in $B \otimes_K B^o$ is $1 \otimes_K B^o$. If this statement is translated from the context of $B \otimes_K B^o$ into the isomorphic context of $\text{End}_K(V)$, then the centralizer of $l(B)$ in $\text{End}_K(V)$ is $r(B)$, and we saw that this fact is sufficient to imply the lemma. $\square$

PROOF OF THEOREM 2.43. Let V be the algebra B considered as a vector space over F, and let $l(B)$ and $r(B)$ be the sets of those members of $\text{End}_F(V)$ that are given by left multiplication and right multiplication by members of B. The algebra A is central simple by assumption, and $\text{End}_F(V)$ is central simple, being isomorphic to $M_n(F)$ for the integer $n = \dim_F(V)$. By Proposition 2.36b, $A \otimes_F \text{End}_F(V)$ is central simple. We define two algebra homomorphisms f and g of B into $A \otimes_F \text{End}_F(V)$ by $f(b) = l(b) \otimes 1$ and $g(b) = 1 \otimes l(b)$.

The Skolem–Noether Theorem (Theorem 2.41) produces an element x of $A \otimes_F \text{End}_F(V)$ with $f(b) = xg(b)x^{-1}$ for all $b \in B$. Hence

$$B \otimes_F 1 = x(1 \otimes_F l(B))x^{-1}. \qquad (*)$$

Lemma 2.44 shows that the centralizer of $B \otimes_F 1$ in $A \otimes_F \text{End}_F(V)$ is $C \otimes_F \text{End}_F(V)$ and that the centralizer of $1 \otimes_F l(B)$ is $A \otimes_F r(B)$. From the latter identification the centralizer of $x(1 \otimes_F l(B))x^{-1}$ is $x(A \otimes_F r(B))x^{-1}$. Combining $(*)$ with these computations of centralizers, we see that

$$C \otimes_F \text{End}_F(V) = x(A \otimes_F r(B))x^{-1}. \qquad (**)$$

The algebra $A \otimes_F r(B)$ is isomorphic to $A \otimes_F B^o$, which is simple by Proposition 2.36a. Therefore $C \otimes_F \text{End}_F(V)$ is simple, and C has to be simple.

Equating the dimensions of the two sides of $(**)$ gives

$$(\dim_F C)(\dim_F B)^2 = (\dim_F C)(\dim_F \text{End}_F(V)) = \dim_F(C \otimes_F \text{End}_F(V))$$
$$= \dim_F(A \otimes_F r(B)) = (\dim_F A)(\dim_F B),$$

and hence
$$(\dim_F C)(\dim_F B) = \dim_F A.$$

Finally the centralizer D of C contains B, and two applications of the dimensional equality gives

$$(\dim_F D)(\dim_F C) = \dim_F A = (\dim_F C)(\dim_F B).$$

Thus $\dim_F D = \dim_F B$, and we must have $D = B$. In other words, B is the centralizer of C. $\square$

Corollary 2.46. Let D be a central finite-dimensional division algebra over the field F. If K is any maximal subfield of D, then $\dim_F D = (\dim_F K)^2$.

PROOF. Apply the Double Centralizer Theorem (Theorem 2.43) with $A = D$. Let $Z(K)$ be the centralizer of the simple subalgebra K in D. Since K is commutative, $K \subseteq Z(K)$. If a is in $Z(K)$ but not K, then $K(a)$ is a field in D properly containing K, in contradiction to the assumption that K is a maximal subfield of D. Hence $K = Z(K)$. The dimensional equality in the theorem therefore gives $\dim_F D = (\dim_F K)(\dim_F Z(K)) = (\dim_F K)^2$. $\square$

Corollary 2.47. Let A be a finite-dimensional central simple algebra over a field F, and let K be a subfield of A. Then the following are equivalent:
(a) K is its own centralizer,
(b) $\dim_F A = (\dim_F K)^2$,
(c) K is a maximal commutative subalgebra of A.

PROOF. Let $Z(K)$ be the centralizer of K in A. The Double Centralizer Theorem (Theorem 2.43) gives the equality

$$\dim_F A = (\dim_F K)(\dim_F Z(K)). \qquad (*)$$

If (a) holds, then $Z(K) = K$, and $(*)$ yields (b).

If (b) holds, then $(*)$ and the equality $\dim_F A = (\dim_F K)^2$ together imply that $\dim_F Z(K) = \dim_F K$. Since K is commutative, $Z(K) \supseteq K$. The equality of dimensions implies that $Z(K) = K$, and then (c) follows.

If (c) holds, we start from the inclusion $K \subseteq Z(K)$. If x is in $Z(K)$ but not K, then $K(x)$ is a field strictly larger than K, in contradiction to (c). Thus $K = Z(K)$, and (a) holds. $\square$

9. Wedderburn's Theorem about Finite Division Rings

The theorem of this section is as follows.

Theorem 2.48 (Wedderburn). Every finite division ring is a field.

The proof will be preceded by a lemma.

Lemma 2.49. If G is a finite group and H is a proper subgroup, then $\bigcup_{g \in G} g H g^{-1}$ does not exhaust G.

PROOF. In the union $\bigcup_{g \in G} gHg^{-1}$, the terms corresponding to g and to gh, for h in H, are the same because $(gh)H(gh)^{-1} = g(hHh^{-1})g^{-1} = gHg^{-1}$. Thus the union can be rewritten as $\bigcup_{gH} gHg^{-1}$, it being understood that only one g is used from each coset gH. From this rewritten form of the union, we see that the number of elements other than the identity in the union is

$$\leq [G:H](|H|-1) = [G:H]|H| - [G:H] = |G| - [G:H] < |G| - 1,$$

and the lemma follows. □

PROOF OF THEOREM 2.48. Let D be a finite division ring, and let F be the center. Then F is a field, say of q elements. Maximal subfields of D certainly exist. Any such subfield K has $\dim_F D = (\dim_F K)^2$ by Corollary 2.46, and hence any two such subfields K and K' are isomorphic. The Skolem–Noether Theorem (Theorem 2.41) shows that $K' = xKx^{-1}$ for some invertible x in the group $D^\times$ of invertible elements of D.

On the other hand, F and any element of D generate a subfield of D, and this subfield is contained in a maximal subfield. Consequently any element of D is contained in some such K', and $D = \bigcup_{x \in D^\times} xKx^{-1}$. Discarding the element 0 from both sides, we obtain $D^\times = \bigcup_{x \in D^\times} xK^\times x^{-1}$. Applying Lemma 2.49 to the group $G = D^\times$ and the subgroup $H = K^\times$, we see that $K^\times$ cannot be a proper subgroup of $D^\times$. Therefore $D = K$, and D is commutative. □

10. Frobenius's Theorem about Division Algebras over the Reals

We conclude this chapter by bringing together our results to prove the following celebrated theorem of Frobenius.

Theorem 2.50 (Frobenius). Up to $\mathbb{R}$ isomorphism the only finite-dimensional associative division algebras over $\mathbb{R}$ are the algebras $\mathbb{R}$ of reals, $\mathbb{C}$ of complex numbers, and $\mathbb{H}$ of quaternions.

REMARKS. The text of this chapter has not produced any concrete examples of noncommutative division rings other than the quaternions. Problems 12–16 at the end of the chapter produce generalized quaternion algebras in which $\mathbb{R}$ can be replaced by many other fields; there are infinitely many nonisomorphic such examples when the field is $\mathbb{Q}$. In addition, Problems 17–19 produce examples of central division algebras of dimension 9 over suitable base fields. The next chapter will give further insight into the construction of division algebras.

10. Frobenius's Theorem about Division Algebras over the Reals

PROOF. Let D be such a division algebra, and let F be the center. Then F is a finite extension field of $\mathbb{R}$ and must be $\mathbb{R}$ or $\mathbb{C}$, since $\mathbb{C}$ is algebraically closed. If $F = \mathbb{C}$, then we have seen that $D = \mathbb{C}$. Thus we may assume that center$(D) = \mathbb{R}$.

Let K be a maximal subfield of D (existence by finite dimensionality), and let $n = \dim_{\mathbb{R}} K$. Corollary 2.46 shows that $\dim_{\mathbb{R}} D = n^2$. Since K has to be $\mathbb{R}$ or $\mathbb{C}$, n has to be 1 or 2. If $n = 1$, we obtain $D \cong \mathbb{R}$. Thus we may assume that $n = 2$, $K = \mathbb{C}$, and $\dim_{\mathbb{R}} D = 4$.

The map $f : K \to D$ given by $f(a + bi) = a - bi$, where i is the member of K corresponding to $\sqrt{-1}$ in $\mathbb{C}$, is an algebra homomorphism into a central simple algebra over $\mathbb{R}$, and so is the map $g : K \to D$ given by $g(a+bi) = a+bi$. By the Skolem–Noether Theorem (Theorem 2.41), there exists some x in D with $x(a + bi)x^{-1} = a - bi$ for all a and b in $\mathbb{R}$.

This element x has the property that x^2 commutes with every element of K and must lie in K, by Corollary 2.47. Let us see that x^2 lies in center$(D) = \mathbb{R}$. In fact, otherwise 1 and x^2 would generate K as an $\mathbb{R}$ algebra, and every member of D commuting with 1 and x^2 would commute with all of K; since x commutes with 1 and x^2, x would have to commute with K, contradiction. Thus x^2 lies in $\mathbb{R}$.

If $x^2 > 0$, then $x^2 = r^2$ for some $r \in \mathbb{R}$. The elements x and r together lie in some subfield K' of D, and K' has no zero divisors. Since $(x - r)(x + r) = 0$ within K', we conclude that $x = \pm r$. Then x commutes with the maximal subfield K above, and we arrive at a contradiction.

Thus $x^2 < 0$. Write $x^2 = -y^2$ for some $y \in \mathbb{R}$, and put $j = y^{-1}x$. The equation $x(a+bi)x^{-1} = a-bi$ says that $j(a+bi)j^{-1} = a-bi$ and in particular that $jij^{-1} = -i$. Define $k = ij$.

We have $j^2 = y^{-2}x^2 = -1$. Hence $k^2 = ijij = i(jij^{-1})j^2 = i(-i)(-1) = i^2 = -1$. Then $ijk = -1$, and $k = -1(j^{-1})(i^{-1}) = -1(-j)(-i) = -ji$; hence $ij + ji = 0$.

Let us show that $\{1, i, j, k\}$ is a linearly independent set over $\mathbb{R}$. Certainly j is not an $\mathbb{R}$ linear combination of 1 and i. If $k = a + bi + cj$ for some $a, b, c \in \mathbb{R}$, then squaring gives

$$-1 = k^2 = a^2 + b^2 i^2 + c^2 j^2 + 2abi + 2acj + bc(ij + ji)$$
$$= a^2 - b^2 - c^2 + 2abi + 2acj.$$

Equating coefficients of 1, i, and j, we obtain $-1 = a^2 - b^2 - c^2$, $ab = 0$, and $ac = 0$. We cannot have $-1 = a^2$, and thus at least one of b and c is nonzero. Then $a = 0$, and $ij = k = bi + cj$. Left multiplication by i gives $-j = -b + cij = -b + c(bi + cj)$; equating coefficients shows that $b = 0$. Hence $ij = cj$, and we arrive at the contradiction $i = c \in \mathbb{R}$. We conclude that $\{1, i, j, k\}$ is linearly independent over $\mathbb{R}$.

To complete the proof that D is isomorphic to $\mathbb{H}$, we have only to verify that $\{1, i, j, k\}$ satisfies the usual multiplication table for $\mathbb{H}$. We know that $i^2 = j^2 = k^2 = -1$, that $k = ij$, and that $k = -ji$. The last of these says that $ji = -k$. The other verifications are

$$jk = jij = (jij^{-1})j^2 = (-i)(-1) = i,$$
$$kj = ijj = i(-1) = -i,$$
$$ki = iji = i(jij^{-1})j = i(-i)j = j,$$
$$ik = iij = (-1)j = -j,$$

and the proof is complete. □

11. Problems

In all the problems below, all algebras are assumed to be associative.

1. Let G be a finite group, and let $\mathbb{C}G$ be its complex group algebra. Prove that $\mathbb{C}G$ is a semisimple ring, and identify the constituent matrix algebras that arise for $\mathbb{C}G$ in Theorem 2.2 in terms of the irreducible representations of G.

2. Wedderburn's Main Theorem (Theorem 2.17) decomposes finite-dimensional algebras A in characteristic 0 as $A = S \oplus \mathrm{rad}\, A$ for some subalgebra S.
   (a) What explicitly is a decomposition $A = S \oplus \mathrm{rad}\, A$ for the complex algebra $\mathbb{C}[X]/(X^2 + 1)^2$?
   (b) Is the subalgebra S in (a) unique? Prove that it is, or give a counterexample.
   (c) Answer the same questions as for (a) and (b) in the case of the real algebra $\mathbb{R}[X]/(X^2 + 1)^2$.

3. Let A and B be finite-dimensional algebras with identity over a field F, and suppose that B is central simple. Prove that $\mathrm{rad}(A \otimes_F B) = (\mathrm{rad}\, A) \otimes_F B$.

Problems 4–7 concern commutative Artinian rings. Let R be such a ring.

4. Prove that
   (a) R has only finitely many maximal ideals,
   (b) $\mathrm{rad}\, R$ is the set of all nilpotent elements in R,
   (c) R is semisimple if and only if it has no nonzero nilpotent elements,
   (d) R semisimple implies that R is the direct product of fields.

5. Let $\bar{e}$ be an idempotent in $R/\mathrm{rad}\, R$. Prove that the idempotent $e \in R$ in Proposition 2.23 with $\bar{e} = e + \mathrm{rad}\, R$ is unique.

6. Problem 4a shows that R has only finitely many maximal ideals. Let N be their product. Use Nakayama's Lemma (Lemma 8.51 of *Basic Algebra*, restated in the present book on page xxiii) to prove that N is a nilpotent ideal in R.

7. Deduce from the previous problem that any prime ideal in R contains one of the finitely many maximal ideals, hence that every prime ideal in R is maximal.

Problems 8–11 concern triangular rings, which were introduced in an example after Proposition 2.5. The problems ask for verifications for some assertions that were made in that example without proof. The notation is as follows: R and S are rings with identity, and M is a unital (R, S) bimodule. Define a set A and operations of addition and multiplication symbolically by

$$A = \begin{pmatrix} R & M \\ 0 & S \end{pmatrix} = \left\{ \begin{pmatrix} r & m \\ 0 & s \end{pmatrix} \middle| r \in R, \, m \in M, \, s \in S \right\}$$

with

$$\begin{pmatrix} r & m \\ 0 & s \end{pmatrix} \begin{pmatrix} r' & m' \\ 0 & s' \end{pmatrix} = \begin{pmatrix} rr' & rm' + ms' \\ 0 & ss' \end{pmatrix}.$$

8. Prove that the left ideals in A are of the form $I_1 \oplus I_2$, where I_2 is a left ideal in S and I_1 is a left R submodule of $R \oplus M$ containing MI_2. (Educational note: Then similarly the right ideals in A are of the form $J_1 \oplus J_2$, where J_1 is a right ideal in R and J_2 is a right S submodule of $M \oplus S$ containing $J_1 M$.)

9. (a) Prove that the ring A is left Noetherian if and only if R and S are left Noetherian and M satisfies the ascending chain condition for its left R submodules.

   (b) Prove that the ring A is right Noetherian if and only if R and S are right Noetherian and M satisfies the ascending chain condition for its right S submodules. (Educational note: By similar arguments the conclusions of (a) and (b) remain valid if "Noetherian" is replaced by "Artinian" and "ascending" is replaced by "descending.")

10. If $A = \begin{pmatrix} R & R \\ 0 & S \end{pmatrix}$ is any ring such as $\begin{pmatrix} \mathbb{Q} & \mathbb{Q} \\ 0 & \mathbb{Z} \end{pmatrix}$ in which S is a (commutative) Noetherian integral domain with field of fractions R and if $S \neq R$, prove that A is left Noetherian and not right Noetherian, and A is neither left nor right Artinian.

11. If $A = \begin{pmatrix} R & R \\ 0 & S \end{pmatrix}$ is a ring such as $\begin{pmatrix} \mathbb{Q}(x) & \mathbb{Q}(x) \\ 0 & \mathbb{Q} \end{pmatrix}$ in which R and S are fields with $S \subseteq R$ and $\dim_S R$ is infinite, prove that A is left Noetherian and left Artinian, and A is neither right Noetherian nor right Artinian.

Problems 12–16 concern **generalized quaternion algebras**. Let F be a field of characteristic other than 2, let K be a quadratic extension field, and let σ be the nontrivial element in the Galois group. The field K is necessarily of the form $K = F(\sqrt{m})$ for some nonsquare $m \in F$, and the elements c of K for which $\sigma(c) = -c$ are the F multiples of $\sqrt{m}$. Fix an element $r \neq 0$ of F, and let A be the subset of $M_2(K)$ given by $\begin{pmatrix} a & b \\ r\sigma(b) & \sigma(a) \end{pmatrix}$.

12. (a) Prove that A is a 4-dimensional algebra over F.

    (b) Prove that A is central simple by examining $cx - xc$ for $c = \begin{pmatrix} \sqrt{m} & 0 \\ 0 & -\sqrt{m} \end{pmatrix}$ when $x \neq 0$ is in a two-sided ideal I and is not in $K \cong \left\{ \begin{pmatrix} a & 0 \\ 0 & \sigma(a) \end{pmatrix} \right\}$.

13. Prove that A is a division algebra if and only if r is not of the form $N_{K/F}(c)$ for some $c \in K$. Why must A be isomorphic to $M_2(F)$ when A is not a division algebra?

14. Prove that if r and r' are two members of F such that $r = r' N_{K/F}(c)$ for some c in K, then the algebra A associated to r is isomorphic to the algebra associated to r'.

15. Let $\{1, i, j, k\}$ be the F basis of A consisting of the matrices
$$1 = \begin{pmatrix} 1 & 0 \\ 0 & 1 \end{pmatrix}, \quad i = \begin{pmatrix} \sqrt{m} & 0 \\ 0 & -\sqrt{m} \end{pmatrix}, \quad j = \begin{pmatrix} 0 & 1 \\ r & 0 \end{pmatrix}, \quad k = \begin{pmatrix} 0 & \sqrt{m} \\ -r\sqrt{m} & 0 \end{pmatrix}.$$
Prove that these satisfy $i^2 = m1$, $j^2 = r1$, $k^2 = -rm1$, $ij = k = -ji$, $jk = -ri = -kj$, and $ki = -mj = -ik$.

16. By going over the proof of Theorem 2.50 and using the relations of the previous problem, prove that every central simple algebra of dimension 4 over F is of the same kind as A for some quadratic extension $K = F(\sqrt{m})$ and some member $r \neq 0$ of F.

Problems 17–19 concern **cyclic algebras**, which were introduced by L. E. Dickson. These extend the theory of generalized quaternion algebras to other sizes of matrices. The analogy with the theory in Problems 12–16 is tightest when the size is a prime. For notational simplicity this set of problems asks about size 3. Let F be any field, and let K be a finite Galois extension of F with cyclic Galois group. It is assumed in these problems that K has degree 3 over F and that $\{1, \sigma, \sigma^2\}$ is the Galois group. Fix an element $r \neq 0$ of F, and let A be the subset of $M_3(K)$ given by $\begin{pmatrix} a & b & c \\ r\sigma(c) & \sigma(a) & \sigma(b) \\ r\sigma^2(b) & r\sigma^2(c) & \sigma^2(a) \end{pmatrix}$.
Identifying $a \in K$ with the member $\begin{pmatrix} a & 0 & 0 \\ 0 & \sigma(a) & 0 \\ 0 & 0 & \sigma^2(a) \end{pmatrix}$ of A and letting j be the member $\begin{pmatrix} 0 & 1 & 0 \\ 0 & 0 & 1 \\ r & 0 & 0 \end{pmatrix}$ of A allows one to view A as the set of all matrices $a + bj + cj^2$ with $a, b, c \in K$. The element j satisfies $jaj^{-1} = \sigma(a)$ for $a \in K$ and $j^3 = r$.

17. Arguing as for Problem 12, show that A is an algebra over F and that it is central simple of dimension 9.

18. Using the general theory, prove that A either is a division algebra over F or is isomorphic to $M_3(F)$, and that $A \cong M_3(F)$ if and only if there is a 3-dimensional vector subspace of A that is a left A submodule of A. (Educational note: This problem makes crucial use of the fact that the size 3 is a prime.)

19. (a) Prove that if $r = N_{K/F}(d)$ for some $d \in K$, then the 3-dimensional vector subspace $K(1 + d^{-1}j + d^{-1}\sigma(d)^{-1}j^2)$ of A is a left A submodule.
    (b) Prove that any 3-dimensional left K submodule of A is necessarily of the form $K(a_0 + b_0 j + c_0 j^2)$ for some nonzero $a_0 + b_0 j + c_0 j^2$ in A and that this left K submodule is a left A submodule only if there exists an element $d \in K$ with $N_{K/F}(d) = r$, $da_0 = r\sigma(c_0)$, $db_0 = \sigma(a_0)$, and $dc_0 = \sigma(b_0)$.

CHAPTER III

Brauer Group

Abstract. This chapter continues the study of finite-dimensional associative division algebras over a field F, with particular attention to those that are simple and have center F. Section 5 is a self-contained digression on cohomology of groups that is preparation for an application in Section 6 and for a general treatment of homological algebra in Chapter IV.

Section 1 introduces the Brauer group of F and the relative Brauer group of K/F, K being any finite extension field. The Brauer group $\mathcal{B}(F)$ is the abelian group of equivalence classes of finite-dimensional central simple algebras over F under a relation called Brauer equivalence. The inclusion $F \subseteq K$ induces a group homomorphism $\mathcal{B}(F) \to \mathcal{B}(K)$, and the relative Brauer group $\mathcal{B}(K/F)$ is the kernel of this homomorphism. The members of the kernel are those classes such that the tensor product with K of any member of the class is isomorphic to some full matrix algebra $M_n(K)$; such a class always has a representative A with $\dim_F A = (\dim_F K)^2$. One proves that $\mathcal{B}(F)$ is the union of all $\mathcal{B}(K/F)$ as K ranges over all finite Galois extensions of F.

Sections 2–3 establish a group isomorphism $\mathcal{B}(K/F) \cong H^2(\mathrm{Gal}(K/F), K^\times)$ when K is a finite Galois extension of F. With these hypotheses on K and F, Section 2 introduces data called a factor set for each member of $\mathcal{B}(K/F)$. The data depend on some choices, and the effect of making different choices is to multiply the factor set by a "trivial factor set." Passage to factor sets thereby yields a function from $\mathcal{B}(K/F)$ to the cohomology group $H^2(\mathrm{Gal}(K/F), K^\times)$. Section 3 shows how to construct a concrete central simple algebra over F from a factor set, and this construction is used to show that the function from $\mathcal{B}(K/F)$ to $H^2(\mathrm{Gal}(K/F), K^\times)$ constructed in Section 2 is one-one onto. An additional argument shows that this function in fact is a group isomorphism.

Section 4 proves under the same hypotheses that $H^1(\mathrm{Gal}(K/F), K^\times) = 0$, and a corollary makes this result concrete when the Galois group is cyclic. This result and the corollary are known as Hilbert's Theorem 90.

Section 5 is a self-contained digression on the cohomology of groups. If G is a group and $\mathbb{Z}G$ is its integral group ring, a standard resolution of $\mathbb{Z}$ by free $\mathbb{Z}G$ modules is constructed in the category of all unital left $\mathbb{Z}G$ modules. This has the property that if M is an abelian group on which G acts by automorphisms, then the groups $H^n(G, M)$ result from applying the functor $\mathrm{Hom}_{\mathbb{Z}G}(\,\cdot\,, M)$ to the members of this resolution, dropping the term $\mathrm{Hom}_{\mathbb{Z}G}(\mathbb{Z}, M)$, and taking the cohomology of the resulting complex. Section 5 goes on to show that the groups $H^n(G, M)$ arise whenever this construction is applied to any free resolution of $\mathbb{Z}$, not necessarily the standard one. This section serves as a prerequisite for Section 6 and as motivational background for Chapter IV.

Section 6 applies the result of Section 5 in the case that G is finite cyclic, producing a nonstandard free resolution of $\mathbb{Z}$ in this case. From this alternative free resolution, one obtains a rather explicit formula for $H^2(G, M)$ whenever G is finite cyclic. Application to the case that G is the Galois group $\mathrm{Gal}(K/F)$ for a finite Galois extension gives the explicit formula $\mathcal{B}(K/F) \cong F^\times / N_{K/F}(K^\times)$ for the relative Brauer group when the Galois group is cyclic.

1. Definition and Examples, Relative Brauer Group

The "Brauer group" of a field allows one to work with the set of all isomorphism classes of finite-dimensional central division algebras over the field. The basic theory in principle reduces the study of all such division algebras to questions in the cohomology theory of groups. The latter theory was introduced in Chapter VII of *Basic Algebra* and will be developed further in the present chapter and the next.

Let F be a field. Theorem 2.4 shows that every finite-dimensional central simple algebra A over F is of the form $A \cong M_n(D)$ for some uniquely determined integer $n \geq 1$ and some finite-dimensional central division algebra D over F that is uniquely determined up to F isomorphism. We can introduce an equivalence relation for finite-dimensional central division algebras over F that exactly mirrors the relation of F isomorphism of the underlying finite-dimensional central division algebras. Specifically if $A \cong M_n(D)$ and $A' \cong M_{n'}(D')$ are two such central simple algebras for the same F such that $D \cong D'$, then we say that A is **Brauer equivalent** to A', and we write $A \sim A'$. It is immediate from the definition that "Brauer equivalent" is an equivalence relation. We shall introduce an abelian-group structure into the set of Brauer equivalence classes, hence into the set of isomorphism classes of central finite-dimensional division algebras over F.

Proposition 10.24 of *Basic Algebra* gives the definition of the tensor product of two F algebras[1] over F, and this operation is associative, up to canonical isomorphism, by Proposition 10.22. It is also commutative, up to canonical isomorphism. In fact, if A and B are given algebras over F, then the canonical vector-space isomorphism $\varphi : A \otimes_F B \to B \otimes_F A$ is given by $\varphi(a \otimes b) = b \otimes a$. If $a_1 \otimes b_1$ and $a_2 \otimes b_2$ are given, then the computation

$$\varphi(a_1 \otimes b_1)\varphi(a_2 \otimes b_2) = (b_1 \otimes a_1)(b_2 \otimes a_2) = b_1 b_2 \otimes a_1 a_2$$
$$= \varphi(a_1 a_2 \otimes b_1 b_2) = \varphi\big((a_1 \otimes b_1)(a_2 \otimes b_2)\big)$$

shows that φ respects multiplication. Hence tensor product is commutative for algebras, up to canonical isomorphism.

Lemma 3.1. If F is a field, then

(a) $M_n(R) \cong R \otimes_F M_n(F)$ for any algebra R with identity over F,
(b) $M_m(F) \otimes_F M_n(F) \cong M_{(mn)}(F)$.

PROOF. For (a), the F bilinear map $(r, [a_{ij}]) \mapsto [r a_{ij}]$ of $R \times M_n(F)$ into

---

[1] All algebras in this chapter are understood to be associative.

$M_n(R)$ has a unique linear extension φ to an F linear map of $R \otimes_F M_n(F)$ into $M_n(R)$. The map φ has

$$\varphi\big((r \otimes [a_{ij}])(r' \otimes [a'_{ij}])\big) = \varphi(rr' \otimes [a_{ij}][a'_{ij}])$$
$$= rr'[a_{ij}][a'_{ij}]$$
$$= r[a_{ij}]r'[a'_{ij}] \qquad \text{since each } a_{ij} \text{ is in } F$$
$$= \varphi(r \otimes_F [a_{ij}])\varphi(r' \otimes [a'_{ij}]),$$

and hence φ is an F algebra homomorphism. If $\{r_k\}$ is a vector-space basis of R over F and if $\{E_{ij}\}$ is the usual basis of $M_n(F)$, then $\varphi(r_k \otimes E_{ij}) = r_k E_{ij}$, and it follows that φ carries a vector-space basis onto a vector-space basis. Hence φ is one-one and onto.

For (b), the result of (a) gives $M_m(F) \otimes_F M_n(F) \cong M_n(M_m(F))$, and the algebra on the right is isomorphic to the algebra $M_{(mn)}(F)$ of matrices of size mn by the multiplication-in-blocks isomorphism. □

Proposition 3.2. For the field F, the operation of tensor product on finite-dimensional central simple algebras over F descends to an operation on the set of Brauer equivalence classes of such algebras and makes this set into an abelian group.

PROOF. The tensor product of two finite-dimensional algebras over F is again a finite-dimensional algebra, and Proposition 2.36 shows that the tensor product of two central simple algebras is again central simple. Hence tensor product is well defined as an operation on finite-dimensional central simple algebras over F. Let us see that tensor product is a Brauer class property. Thus suppose that $A \sim A'$ and $B \sim B'$, say with $A = M_m(D)$, $A' \cong M_{m'}(D)$, $B = M_n(E)$, and $B' = M_{n'}(E)$. Since the tensor product of some $M_r(F)$ with an algebra over F, up to isomorphism, does not depend on the order of the two factors and since tensor product is associative up to isomorphism, Lemma 3.1 gives

$$A \otimes_F B = M_m(D) \otimes_F M_n(E) \cong D \otimes_F M_m(F) \otimes_F M_n(F) \otimes_F E$$
$$\cong D \otimes_F M_{(mn)}(F) \otimes_F E \cong M_{(mn)}(F) \otimes_F D \otimes_F E$$
$$\cong M_{(mn)}(D \otimes_F E).$$

Similarly $A' \otimes_F B' \cong M_{(m'n')}(D \otimes_F E)$. Thus $A \otimes_F B \sim A' \otimes_F B'$.

We have observed that the tensor product operation on algebras over F is associative and commutative, up to canonical isomorphisms, and hence so is the product operation on Brauer equivalence classes. The class of the 1-dimensional algebra F is the identity, and the class of the opposite algebra A^o is an inverse to the class of A because of the isomorphism $A \otimes_F A^o \cong M_n(F)$ given in Corollary 2.38. □

The abelian group of Brauer equivalence classes of finite-dimensional central simple algebras over F is called the **Brauer group** of F and is denoted by $\mathcal{B}(F)$. We use additive notation for its product operation.

EXAMPLES ALREADY SETTLED IN CHAPTER II.

(1) If F is algebraically closed, then $\mathcal{B}(F) = 0$.

(2) If $F = \mathbb{R}$, then $\mathcal{B}(F) = \mathbb{Z}/2\mathbb{Z}$ by Frobenius's Theorem (Theorem 2.50).

(3) If F is a finite field, then $\mathcal{B}(F) = 0$ by Wedderburn's Theorem about finite division rings (Theorem 2.48).

The group structure for $\mathcal{B}(F)$ given in Proposition 3.2 offers little help by itself in identifying the finite-dimensional division algebras over a particular field. Instead, the usual procedure for understanding $\mathcal{B}(F)$ is to isolate certain special subgroups of $\mathcal{B}(F)$, known as "relative Brauer groups" and denoted by $\mathcal{B}(K/F)$, K being any finite extension of F. Under the assumption that K is a finite Galois extension of F, Theorem 3.14 below says that $\mathcal{B}(K/F)$ is isomorphic to the cohomology group $H^2(G, N)$, where G is the finite group $G = \text{Gal}(K/F)$ and N is the (abelian) multiplicative group $K^\times$ of the field K. This cohomology group is in principle manageable. Corollary 3.9 below says that $\mathcal{B}(F)$ is the union over all finite Galois extensions K/F of $\mathcal{B}(K/F)$, and we therefore obtain a handle on $\mathcal{B}(F)$.

If A is any finite-dimensional central simple algebra over F and if K/F is any field extension, then Proposition 2.36a shows that $A \otimes_F K$ is simple as a ring, and Lemma 2.35b shows that $A \otimes_F K$ has center K. Therefore $A \otimes_F K$ is a central simple algebra over K, and its Brauer equivalence class is a member of $\mathcal{B}(K)$.

Let us see that this map of algebras A into $\mathcal{B}(K)$ depends only on the Brauer equivalence class of A in $\mathcal{B}(F)$. Thus suppose that $A = M_m(D)$ and $A' = M_n(D)$ for some finite-dimensional central division algebra D over F. Lemma 3.1a gives us isomorphisms of F algebras

$$A \otimes_F K \cong M_m(D) \otimes_F K \cong (M_m(F) \otimes_F D) \otimes_F K$$
$$\cong M_m(F) \otimes_F (D \otimes_F K) \cong M_m(D \otimes_F K),$$

and similarly $A' \otimes_F K \cong M_n(D \otimes_F K)$. In each case the left member of the isomorphism is a K algebra, with K contained in the center. Thus we can view each of our isomorphisms as isomorphisms of central simple K algebras. Since $D \otimes_F K$ is a finite-dimensional central simple K algebra, we know that $D \otimes_F K \cong M_r(E)$ for some finite-dimensional central division algebra E over K. Application of Lemma 3.1b allows us to continue the displayed isomorphisms as

$$A \otimes_F K \cong M_m(D \otimes_F K) \cong M_m(M_r(E)) \cong M_{(mr)}(E).$$

Similarly we have $A' \otimes_F K \cong M_{(nr)}(E)$. Thus $A \otimes_F K$ and $A' \otimes_F K$ yield the same member of $\mathcal{B}(K)$, and $(\,\cdot\,) \otimes_F K$ induces a well-defined function from $\mathcal{B}(F)$ into $\mathcal{B}(K)$.

The function from $\mathcal{B}(F)$ into $\mathcal{B}(K)$ is a group homomorphism. In fact, if A and B are finite-dimensional central simple over F, then we have K isomorphisms

$$(A \otimes_F K) \otimes_K (B \otimes_F K) \cong A \otimes_F (K \otimes_K (B \otimes_F K))$$
$$\cong A \otimes_F (B \otimes_F K) \cong (A \otimes_F B) \otimes_F K,$$

and the map is indeed a group homomorphism.

In addition, the resulting homomorphism satisfies the expected compatibility condition with respect to compositions. In more detail, if we have nested fields $F \subseteq K \subseteq L$, then the L isomorphisms

$$(A \otimes_F K) \otimes_K L \cong A \otimes_F (K \otimes_K L) \cong A \otimes_F L$$

show that the composition of tensoring with K over F, followed by tensoring with L over K, yields the same result as tensoring directly with L over F.

We define the **relative Brauer group** $\mathcal{B}(K/F)$ to be the kernel of the homomorphism of $\mathcal{B}(F)$ into $\mathcal{B}(K)$. The members of the group $\mathcal{B}(K/F)$ are the Brauer equivalence classes of finite-dimensional central simple F algebras A such that $A \otimes_F K$ is F isomorphic to $M_n(K)$ for some n. We say that such algebras are **split** over K, that K **splits** such algebras, and that K is a **splitting field** for these algebras and their Brauer equivalence classes.

Theorem 3.3. Let K/F be a finite extension of fields. Then K is a splitting field for a given member X of $\mathcal{B}(K/F)$ if and only if there exists an algebra A over F in the Brauer equivalence class X containing a subfield K' isomorphic to K such that $\dim_F A = (\dim_F K')^2$.

REMARKS.

(1) The theory of the Brauer group makes repeated use of this result. Corollary 2.47 shows that the subfield K' of A is a maximal commutative subalgebra of A and in particular is a maximal subfield of A.

(2) Observe that the field K is given in the theorem, and hence the integer $n = \dim_F K$ is known. Then A must have dimension n^2. The equality $\dim_F A = n^2$ determines A up to F isomorphism. In fact, Theorem 2.4 shows that $A \cong M_r(D)$ for a central division algebra whose isomorphism class is determined by the class X. Then $n^2 = \dim_F A = r^2 \dim_F D$, and $r^2 = n^2/\dim_F(D)$. So A is indeed determined up to F isomorphism.

(3) In view of the previous remark, any class X in $\mathcal{B}(K/F)$ has a distinguished representative that is unique up to F isomorphism; the distinguished representatives of the members of $\mathcal{B}(K/F)$ for fixed K all have the same dimension.

PROOF. Suppose that A is a central simple algebra in the Brauer equivalence class X containing a subfield K' isomorphic to K such that $\dim_F A = (\dim_F K')^2$. We are to prove that K' splits A. Write n for $\dim_F K'$, so that $\dim_F A = n^2$. Regard A as an n-dimensional K' vector space with K' acting by right multiplication on A. Define an F bilinear mapping $f : A \times K' \to \operatorname{End}_{K'}(A)$ by $f(a, c)(a') = aa'c$; the image $f(a, c)$ is in $\operatorname{End}_{K'}(A)$ because

$$f(a, c)(a'c') = aa'c'c = (aa'c)c' = \big(f(a, c)(a')\big)c'.$$

Extend f without changing its name to an F linear mapping $f : A \otimes_F K' \to \operatorname{End}_{K'}(A)$ such that $f(a \otimes c)(a') = aa'c$. The mapping f is actually K' linear because

$$f((a \otimes c)c')(a') = f(a \otimes cc')(a') = aa'cc' = \big(f(a \otimes c)(a')\big)c'.$$

Also, it respects multiplication, since

$$\begin{aligned}f(a \otimes c)\big(f(a' \otimes c')(a'')\big) &= f(a \otimes c)(a'a''c') = aa'a''c'c = aa'a''cc' \\ &= f(aa' \otimes cc')(a'') = f\big((a \otimes c)(a' \otimes c')\big)(a'').\end{aligned}$$

Thus f is a homomorphism of K' algebras. The domain $A \otimes_F K'$ is central simple over K', as we saw when setting up the homomorphism $\mathcal{B}(F) \to \mathcal{B}(K)$, and therefore f is one-one. Since $A \otimes_F K'$ and $\operatorname{End}_{K'}(A)$ both have K' dimension n^2, f has to be onto. Thus f exhibits $A \otimes_F K'$ as isomorphic to a full matrix ring over K', and K' splits A.

Conversely suppose that K is a splitting field for the members of the class X in $\mathcal{B}(F)$. Let D be a division algebra in the class X. Since $\mathcal{B}(K/F)$ is a group and therefore contains the inverse class D^o, we must have $D^o \otimes_F K \cong M_m(K)$ for the integer m such that $\dim_F D^o = m^2$. Let us rewrite this K isomorphism as $D^o \otimes_F K \cong \operatorname{End}_K(K^m)$. The algebra $\operatorname{End}_F(K^m)$ is central simple over F, and up to an isomorphism, it contains the K algebra $D^o \otimes_F K$ and hence also the F algebra $D^o \otimes_F F \cong D^o$. Let A be the centralizer of D^o in $\operatorname{End}_F(K^m)$. We shall prove that A has the required properties.

The algebra A contains (center $D^o) \otimes_F K$, which is a subfield K' isomorphic to K because D^o is central over F, and A is simple by the Double Centralizer Theorem (Theorem 2.43). The center of A matches the center of the centralizer of A, which is the center of D^o by Theorem 2.43, which in turn is F. Thus A is central simple over F. Yet another application of Theorem 2.43 gives

$$(\dim_F A)(\dim_F D^o) = \dim_F \operatorname{End}_F(K^m) = m^2 (\dim_F K)^2. \qquad (*)$$

Since $\dim_F D^o = m^2$, we see that $\dim_F A = (\dim_F K)^2$. Thus the subfield K' of A isomorphic to K has the required dimension.

To see that A is in the Brauer equivalence class X, start from the F bilinear map $A \times (D^o \otimes_F F) \to \operatorname{End}_F(K^m)$ given by $(a, d \otimes 1) \mapsto ad$, and form its F linear extension $\varphi : A \otimes_F (D^o \otimes_F F) \to \operatorname{End}_F(K^m)$. The map φ respects multiplication because the members of A commute with the members of $D^o \otimes_F F$:

$$\varphi(a \otimes (d \otimes 1))\big(\varphi(a' \otimes (d' \otimes 1))(v)\big) = \varphi(a \otimes (d \otimes 1))(a'd'v) = ada'd'v$$
$$= aa'dd'v = \varphi(aa' \otimes (dd' \otimes 1))(v).$$

Since $A \otimes_F (D^o \otimes_F F)$ is simple by Proposition 2.36, φ is one-one. A look at (∗) shows that

$$\dim_F(A \otimes_F (D^o \otimes_F F)) = (\dim_F A)(\dim_F D^o) = \dim_F \operatorname{End}_F(K^m)$$

and allows us to conclude that φ is onto. Therefore $A \otimes_F D^o \cong \operatorname{End}_F(K^m)$. Since $\operatorname{End}_F(K^m)$ is Brauer equivalent to F, the Brauer equivalence class of A is the inverse of the class of D^o. Hence the class of A equals the class of D, which is X. □

Corollary 3.4. If D is a finite-dimensional central division algebra over the field F, then any maximal subfield K of D splits D.

PROOF. This is the special case of Theorem 3.3 in which $A = D$. The formula for the dimensions holds by Corollary 2.47. □

Corollary 3.5. If F is a field, then the Brauer group $\mathcal{B}(F)$ is the union of all relative Brauer groups $\mathcal{B}(K/F)$ as K ranges over all finite extensions of F.

REMARKS. This result is all very tidy but is not very useful, since we have no indication how to identify $\mathcal{B}(K/F)$ for a general finite extension F. In Corollary 3.9 below, we sharpen this result to make K range only over the finite *Galois* extensions of F, and we shall see in Section 3 that $\mathcal{B}(K/F)$ can be realized for such fields K in terms of the cohomology of groups.

PROOF. Any member of $\mathcal{B}(F)$ has some central division algebra D as a representative, and Corollary 3.4 identifies an extension field K of F that splits D, namely any maximal subfield of D. □

Corollary 3.6. Let D be a finite-dimensional central division algebra over a field F, and let $\dim_F D = n^2$. If K is a splitting field for D, then $\dim_F K$ is a multiple of n.

PROOF. If K is a splitting field for D, then Theorem 3.3 says that there exists an integer r such that $M_r(D)$ contains a subfield K' isomorphic to K with $\dim_F M_r(D) = (\dim_F K')^2$. Thus $r^2 n^2 = (\dim_F K)^2$, and $rn = \dim_F K$. □

Theorem 3.7 (Noether–Jacobson Theorem). If D is a noncommutative finite-dimensional central division algebra over the field F, then there exists a member of D that is not in F and is separable over F.

REMARKS. Within a field extension K/F, we know from Corollary 9.31 of *Basic Algebra* that the subset of all elements of K that are separable over F is a subfield of K containing F. Consequently an equivalent formulation of the theorem is that D contains a nontrivial separable extension field of F.

PROOF (Herstein). Arguing by contradiction, suppose that no element of D outside F is separable over F. Let the characteristic of F be p, necessarily nonzero. If a is any element of D not in F, then the assumed nonseparability implies that the minimal polynomial $f(X)$ of a over F has $f'(X) = 0$, according to Proposition 9.27 of *Basic Algebra*. Hence $f(X) = f_1(X^p)$ for some polynomial $f_1(X)$ in $F[X]$. In turn, the minimal polynomial of a^p is $f_1(X)$, and if a^p is not in F, then $f_1(X) = f_2(X^p)$ for some polynomial $f_2(X)$ in $F[X]$. Since the degree decreases at each step as we pass from f to f_1, from f_1 to f_2, and so on, we conclude that a^{p^e} is in F for some e. In short, each a in D has the property that there is some integer $e \geq 0$ depending on a such that a^{p^e} is in F.

In view of the assumption that $D \neq F$ and the argument that we have just seen, there exists an element a in D outside F such that a^p is in F. Define a function $d : D \to D$ by $d(x) = xa - ax$. The function d is F linear, and it is not identically 0 because a is not in the center F of D. If r and l denote right and left multiplication, we can rewrite d as $d(x) = (r(a) - l(a))(x)$. The linear maps $r(a)$ and $l(a)$ commute with each other, and thus the Binomial Theorem is applicable in computing $d^p(x)$ as

$$d^p(x) = (r(a) - l(a))^p(x) = (r(a)^p - l(a)^p)(x) = xa^p - p^a x = 0,$$

the last equality holding because a^p is in F and is therefore central. Since d^p is the zero function and d is not, there exist an integer s with $2 \leq s \leq p$ and an element y in D with $d^{s-1}y \neq 0$ and $d^s y = 0$. Put $x = d^{s-1}y$. Since $x = d(d^{s-2}y)$, the element $w = d^{s-2}y$ has the property that $x = wa - aw$. The condition $dx = 0$ says that $xa = ax$. Put $x = au$. The elements a and u commute because a and x commute. If we set $c = wu^{-1}$, then $x = wa - aw = cua - acu$, and hence $a = xu^{-1} = cuau^{-1} - ac$. Since a and u commute, we obtain $a = ca - ac$. Right multiplying by a^{-1} gives $1 = c - aca^{-1}$ and therefore $c = 1 + aca^{-1}$. Raising both sides to the $p^{e'}$ power gives $c^{p^{e'}} = 1 + ac^{p^{e'}}a^{-1}$. The first paragraph of the proof shows that there is some $e' \geq 0$ for which $c^{p^{e'}}$ is in F, and for this integer e', we obtain the contradictory equation $c^{p^{e'}} = 1 + c^{p^{e'}}$ from the commutativity of a with F. This completes the proof. □

Corollary 3.8. If D is a noncommutative finite-dimensional central division algebra over the field F and if K is a subfield of D that is separable over F, then there exists a maximal subfield L of D containing K such that L is separable over F.

PROOF. Because of the finite dimensionality, we may assume without loss of generality that K is not properly contained in any larger subfield of D that is separable over F. Arguing by contradiction, we may assume that K is not a maximal subfield of D. Let E be the centralizer of K in D. This is a division algebra over F. It is simple by the Double Centralizer Theorem (Theorem 2.43), and it contains K because K is commutative. Moreover, we know from Theorem 2.43 that

$$\dim_F D = (\dim_F K)(\dim_F E)$$

and that K is the centralizer of E. The latter condition shows that the division algebra E is central simple over K. Since K is not a maximal subfield of D, Corollary 2.46 gives $\dim_F D > (\dim_F K)^2$. Thus $\dim_F K < \dim_F E$. Since E is central over K, E is noncommutative.

Application of Theorem 3.7 produces an element x in E outside K that is separable over K. Let L be the subfield $K(x)$ of E. Since K is a separable extension of F, the Theorem of the Primitive Element gives an element α of K such that $K = F(\alpha)$. Then $L = F(\alpha, x)$. The implication (b) implies (c) in Corollary 9.29 of *Basic Algebra* shows that if α is separable over F and x is separable over $F(\alpha)$, then α and x are both separable over F. The elements of L that are separable over F form a subfield of L, and we have just proved that this subfield properly contains K. This conclusion contradicts the assumption that K is a maximal separable extension of F within D, and the proof is complete. □

Corollary 3.9. If F is a field, then the Brauer group $\mathcal{B}(F)$ is the union of all relative Brauer groups $\mathcal{B}(K/F)$ as K ranges over all finite *Galois* extensions of F.

REMARKS. This is the result of interest. Each $\mathcal{B}(K/F)$ with K as in the corollary will be seen to be given as an H^2 in the cohomology of groups, and this group is in principle manageable. Thus we obtain a handle on $\mathcal{B}(F)$.

PROOF. If D is a central division algebra over F, then Corollaries 3.4 and 3.8 together show that some finite separable extension K' of F splits D. That is, the Brauer equivalence class of D lies in $\mathcal{B}(K'/F)$. Let us write $K' = F(\alpha)$ by the Theorem of the Primitive Element. If $f(X)$ is the minimal polynomial of α over F, then every root of $f(X)$ in an algebraic closure $\overline{F}$ of F containing K' is separable over F. Let K be the subfield of $\overline{F}$ generated by all the roots. This is a finite normal extension, and Corollary 9.30 of *Basic Algebra* shows that it is a separable extension. We have seen that the composition of the homomorphisms $\mathcal{B}(F) \to \mathcal{B}(K')$ and $\mathcal{B}(K') \to \mathcal{B}(K)$ is $\mathcal{B}(F) \to \mathcal{B}(K)$, and consequently $\mathcal{B}(K'/F) \subseteq \mathcal{B}(K/F)$. Therefore the Brauer equivalence class of D lies in $\mathcal{B}(K/F)$. □

2. Factor Sets

Throughout this section let K/F be a finite Galois extension of fields. Our objective is to construct a function from the relative Brauer group $\mathcal{B}(K/F)$ into the cohomology group $H^2(\text{Gal}(K/F), K^\times)$. In Section 3 we shall prove that this function is a group isomorphism.

We take as known the material in Chapter VII of *Basic Algebra* on cohomology of groups. For convenient reference we list the relevant formulas for cohomology in degree 2. If G is a group and N is an abelian group on which G acts by automorphisms, the group $C^2(G, N)$ of 2-cochains is the group of all functions $a : G \times G \to N$, the group $Z^2(G, N)$ of 2-cocycles is the set of members f of $C^2(G, N)$ such that

$$u(f(v, w)) + f(u, vw) = f(uv, w) + f(u, v) \qquad \text{for all } u, v, w \in G,$$

the group $B^2(G, N)$ of 2-coboundaries is the set of members f of $C^2(G, N)$ of the form

$$f(u, v) = u(\alpha(v)) - \alpha(uv) + \alpha(u) \qquad \text{for some } \alpha : G \to N,$$

and the cohomology group $H^2(G, N)$ is the quotient

$$H^2(G, N) = Z^2(G, N)/B^2(G, N).$$

Here it is understood that we are using additive notation for the group operation in N and that the action of $u \in G$ on a member n of N is denoted by $u(n)$.

In constructing the function from $\mathcal{B}(K/F)$ into $H^2(\text{Gal}(K/F), K^\times)$, we shall proceed in somewhat the same fashion as for the identification of group extensions with an H^2 that was carried out in Chapter VII of *Basic Algebra*. Namely we shall associate a "factor set" to some choices concerning a given finite-dimensional central simple algebra and see that this factor set is a cocyle. Then we shall show that the factor set for any set of choices for any Brauer-equivalent central simple algebra differs from this cocyle by a coboundary. The result will be the desired function from $\mathcal{B}(K/F)$ into $H^2(\text{Gal}(K/F), K^\times)$.

Thus write G for $\text{Gal}(K/F)$, fix a Brauer equivalence class X in $\mathcal{B}(K/F)$, and let A be a central simple algebra in the class X meeting the conditions of Theorem 3.3: A contains a subfield K' isomorphic to K, and $\dim_F A = (\dim_F K')^2$. Write $c \mapsto c'$ for the isomorphism $K \to K'$.

2. Factor Sets

Let σ be an element of the Galois group G. Then $c \mapsto c'$ and $c \mapsto \sigma(c)'$ are two algebra homomorphisms of the simple algebra K into the central simple algebra A, and the Skolem–Noether Theorem (Theorem 2.41) says that they are related by an inner automorphism:

$$\boxed{\sigma(c)' = x_\sigma c' x_\sigma^{-1}} \qquad \text{for some } x_\sigma \in A.$$

Some choice is involved in selecting x_σ, but the element x_σ is unique up to a factor from K' on the right. In fact, if x_σ and y_σ both behave as in the boxed formula, then $y_\sigma^{-1} x_\sigma$ commutes with K' and hence is in K'. Thus $x_\sigma = y_\sigma c_0'$ with c_0' in K'.

The nonuniqueness can be expressed also in terms of a factor from K' on the left. In fact, the boxed formula for $c = c_0$ implies that $x_\sigma = (x_\sigma c_0' x_\sigma^{-1})(x_\sigma c_0'^{-1}) = \sigma(c_0)' y_\sigma$.

At any rate, fix a choice of x_σ for all $\sigma \in G$, and let us examine the effect of composition. If σ and τ are in G, then

$$x_{\sigma\tau} c' x_{\sigma\tau}^{-1} = (\sigma\tau)(c)' = \sigma(\tau(c))' = x_\sigma \tau(c)' x_\sigma^{-1} = x_\sigma x_\tau c' x_\tau^{-1} x_\sigma^{-1}.$$

Using the result of the previous paragraph, we see that $x_{\sigma\tau}$ and $x_\sigma x_\tau$ are related by a factor from K' on the *left*. Hence we can write

$$\boxed{x_\sigma x_\tau = a(\sigma, \tau)' x_{\sigma\tau}} \qquad \text{with } a(\sigma, \tau) \in K^\times.$$

If we examine the effect of composing three elements of G, we obtain a consistency condition that the function $a : G \times G \to K^\times$ must satisfy. Namely, let ρ, σ, and τ be in G, and let us compute $x_\rho x_\sigma x_\tau$ in two ways, taking advantage of the associativity in A. With one grouping, we obtain

$$x_\rho x_\sigma x_\tau = (x_\rho x_\sigma) x_\tau = a(\rho, \sigma)' x_{\rho\sigma} x_\tau = a(\rho, \sigma)' a(\rho\sigma, \tau)' x_{\rho\sigma\tau},$$

and with the other grouping, we have

$$x_\rho x_\sigma x_\tau = x_\rho (x_\sigma x_\tau) = x_\rho a(\sigma, \tau)' x_{\sigma\tau}$$
$$= \rho(a(\sigma, \tau))' x_\rho x_{\sigma\tau} = \rho(a(\sigma, \tau))' a(\rho, \sigma\tau)' x_{\rho\sigma\tau}.$$

Therefore the function $a : G \times G \to K^\times$ satisfies

$$\boxed{\rho(a(\sigma, \tau)) a(\rho, \sigma\tau) = a(\rho, \sigma) a(\rho\sigma, \tau).}$$

A function $a : G \times G \to K^\times$ satisfying the above boxed formula is called a **factor set**. From A, an isomorphism $K \to K'$, and a choice of the elements x_σ for $\sigma \in G$, we have obtained a factor set.

Comparing this boxed formula with the formulas in the second paragraph of this section, we see that a factor set is exactly a member of $Z^2(\mathrm{Gal}(K/F), K^\times)$ except that the boxed formula uses multiplicative notation for $K^\times$ and the definition of 2-cocycle uses additive notation. Thus we have associated a member of $Z^2(\mathrm{Gal}(K/F), K^\times)$ to the triple consisting of A, an isomorphism $K \to K'$, and a choice of the elements x_σ for $\sigma \in G$.

With the extension K/F and the class $X \in \mathcal{B}(K/F)$ fixed, let us see the effect on the factor set of making different choices. The algebra A lies in the Brauer equivalence class X and has $\dim_F A = (\dim_F K)^2$. As we saw in the remarks with Theorem 3.3, A is determined up to isomorphism by these properties.

Thus let us start from a different system of choices: an algebra B in the class X, an isomorphism $K \to K''$, and elements y_σ for $\sigma \in G$ such that $\sigma(c)'' = y_\sigma c'' y_\sigma^{-1}$. Define the corresponding factor set $b : G \times G \to K^\times$ by

$$y_\sigma y_\tau = b(\sigma, \tau)'' y_{\sigma\tau}.$$

We wish to relate $a(\sigma, \tau)$ and $b(\sigma, \tau)$. We have just seen that A and B are isomorphic as algebras. Let $\varphi : A \to B$ be an isomorphism. Then $c \mapsto c' \mapsto \varphi(c')$ and $c \mapsto c''$ are two algebra homomorphisms of K into B, and the Skolem–Noether Theorem (Theorem 2.41) produces an element $t \in B$ with

$$c'' = t\varphi(c')t^{-1} \qquad \text{for all } c \in K.$$

Starting from the formula $\sigma(c)' = x_\sigma c' x_\sigma^{-1}$, apply φ and conjugate by t to obtain

$$\sigma(c)'' = t\varphi(\sigma(c)')t^{-1} = \bigl(t\varphi(x_\sigma)t^{-1}\bigr)c''\bigl(t\varphi(x_\sigma)t^{-1}\bigr)^{-1}.$$

This equation says that $t\varphi(x_\sigma)t^{-1}$ serves the same purpose as y_σ, and therefore

$$y_\sigma = c_\sigma'' t\varphi(x_\sigma)t^{-1}$$

for some member c_σ'' of K'' placed on the left. Substitution into the formula $y_\sigma y_\tau = b(\sigma, \tau)'' y_{\sigma\tau}$ gives

$$c_\sigma'' t\varphi(x_\sigma)t^{-1} c_\tau'' t\varphi(x_\tau)t^{-1} = b(\sigma, \tau)'' c_{\sigma\tau}'' t\varphi(x_{\sigma\tau})t^{-1}.$$

If we substitute from the formula $c'' = t\varphi(c')t^{-1}$ for all members of K'' and then conjugate by t^{-1} and apply φ^{-1}, we obtain

$$c_\sigma' x_\sigma c_\tau' x_\tau = b(\sigma, \tau)' c_{\sigma\tau}' x_{\sigma\tau}.$$

The left side equals

$$c_\sigma' \sigma(c_\tau)' x_\sigma x_\tau = c_\sigma' \sigma(c_\tau)' a(\sigma, \tau)' x_{\sigma\tau},$$

and comparison of this expression with the right side gives

$$b(\sigma, \tau)' c'_{\sigma\tau} = c'_\sigma \sigma(c_\tau)' a(\sigma, \tau)'.$$

Passing from K' back to K, we conclude that

$$\boxed{b(\sigma, \tau) c_{\sigma\tau} = c_\sigma \sigma(c_\tau) a(\sigma, \tau).}$$

This formula says that b is the product of a and the **trivial factor set** $c : G \times G \to K^\times$ given by

$$c(\sigma, \tau) = c_\sigma \sigma(c_\tau) c_{\sigma\tau}^{-1},$$

where $\sigma \mapsto c_\sigma$ is some function from G to $K^\times$. Again referring to the second paragraph of this section and remembering that we are using multiplicative notation for $K^\times$, we see that the trivial factor sets are the 2-coboundaries, lying in $B^2(\mathrm{Gal}(K/F), K^\times)$, in the same way that the general factor sets are the 2-cocycles, lying in $Z^2(\mathrm{Gal}(K/F), K^\times)$. We have thus proved the following proposition.

Proposition 3.10. Let K be a finite Galois extension of the field F. For X in $\mathcal{B}(K/F)$, let A be an algebra in the Brauer equivalence class X with $\dim_F A = (\dim_F K)^2$, let $K \to K'$ be an isomorphism of K into A, and let $\{x_\sigma \mid \sigma \in \mathrm{Gal}(K/F)\} \subseteq A^\times$ be a set of elements such that $\sigma(c)' = x_\sigma c' x_\sigma^{-1}$. Then the passage from X to the factor set determined by the triple of data $(A, K \to K', \{x_\sigma\})$ descends to a well-defined function from the abelian group $\mathcal{B}(K/F)$ to the abelian group $H^2(\mathrm{Gal}(K/F), K^\times)$.

3. Crossed Products

In this section we continue to assume that K/F is a finite Galois extension of fields. We are going to show that the function $\mathcal{B}(K/F) \to H^2(\mathrm{Gal}(K/F), K^\times)$ given in Proposition 3.10 is an isomorphism of groups. The homomorphism property comes last and is the hard part of the argument. In the meantime, we construct the inverse function by associating an algebra to each member of $Z^2(\mathrm{Gal}(K/F), K^\times)$ and showing in Corollary 3.13 that the resulting function on $Z^2(\mathrm{Gal}(K/F), K^\times)$ descends to an inverse function from $H^2(\mathrm{Gal}(K/F), K^\times)$ into $\mathcal{B}(K/F)$. The algebra is called a "crossed product" and is produced in Proposition 3.12 below. Before either of these steps, we establish one more property of the system $\{x_\sigma \mid \sigma \in \mathrm{Gal}(K/F)\}$ of the previous section that has not needed mentioning until now.

Thus let a central simple algebra A be given with $\dim_F A = (\dim_F K)^2$, along with an isomorphism $K \to K'$ denoted by $c \mapsto c'$. As in the previous section we choose $x_\sigma \in A^\times$ with

$$\sigma(c)' = x_\sigma c' x_\sigma^{-1} \qquad \text{for all } c \in K.$$

The corresponding factor set $a(\sigma, \tau)$ has

$$x_\sigma x_\tau = a(\sigma, \tau)' x_{\sigma\tau}.$$

We regard A as a vector space over K' with K' acting by multiplication on the left.

Lemma 3.11. With hypotheses as above, the set $\{x_\sigma \mid \sigma \in \text{Gal}(K/F)\}$ is a vector-space basis of A over K'.

PROOF. Let $G = \text{Gal}(K/F)$. Since $|G| = \dim_F K = \dim_F K' = \dim_{K'} A$, it is enough to prove linear independence. Arguing by contradiction, assume that the set $\{x_\sigma \mid \sigma \in G\}$ is linearly dependent. Choose a maximal subset J of G such that $\{x_\tau \mid \tau \in J\}$ is linearly independent. For σ not in J, we then have

$$x_\sigma = \sum_{\tau \in J} a'_\tau x_\tau \qquad \text{with } a_\tau \in K. \qquad (*)$$

Every c in K satisfies

$$\sigma(c)' x_\sigma = x_\sigma c' = \sum_{\tau \in J} a'_\tau x_\tau c' = \sum_{\tau \in J} a'_\tau \tau(c)' x_\tau,$$

and thus $x_\sigma = \sum_{\tau \in J} \sigma(c)'^{-1} a'_\tau \tau(c)' x_\tau$. Comparing this expansion with $(*)$ shows that

$$\sigma(c)'^{-1} a'_\tau \tau(c)' = a'_\tau \qquad \text{for } \tau \in J. \qquad (**)$$

Since $x_\sigma \neq 0$, some a'_τ in the expansion $(*)$ is nonzero. For this τ, we can cancel a'_τ in $(**)$ and obtain $\sigma(c)' = \tau(c)'$ for all $c \in K$. Then $\sigma = \tau$, in contradiction to the fact that σ is not in J. $\square$

The linear independence in Lemma 3.11 allows us to read off the structure of A: as a K' vector space, the algebra A is given by $A = \bigoplus_{\sigma \in \text{Gal}(K/F)} K' x_\sigma$, and the elements x_σ have the properties that

$$x_\sigma c' = \sigma(c)' x_\sigma \text{ for } c \in K \qquad \text{and} \qquad x_\sigma x_\tau = a(\sigma, \tau)' x_{\sigma\tau}.$$

Proposition 3.12 is motivated by these formulas, saying that we can reconstruct A from a given 2-cocycle $a(\sigma, \tau)$ in such a way that these formulas hold.

3. Crossed Products

Proposition 3.12. Let K/F be a finite Galois extension, and let $a = a(\sigma, \tau)$ be in $Z^2(\text{Gal}(K/F), K^\times)$. Then there exist a central simple algebra A over F with $\dim_F A = (\dim_F K)^2$, an isomorphism $K \to K'$ of K onto a subfield K' of A, and a subset $\{x_\sigma \in A \mid \sigma \in \text{Gal}(K/F)\}$ such that

(a) $A = \bigoplus_{\sigma \in \text{Gal}(K/F)} K' x_\sigma$,
(b) $x_\sigma c' x_\sigma^{-1} = \sigma(c)'$ for all c in K, with $c \mapsto c'$ denoting the isomorphism of K onto K',
(c) $x_\sigma x_\tau = a(\sigma, \tau)' x_{\sigma\tau}$.

REMARKS. We write $A = \mathcal{A}(K, \text{Gal}(K/F), a)$ and call A the **crossed-product algebra** corresponding to the factor set a. The algebra A is completely determined by the given conditions, up to canonical isomorphism, since (a), (b), and (c) determine the entire multiplication table of A.

PROOF. Let $G = \text{Gal}(K/F)$, form a set $\{x_\sigma \mid \sigma \in G\}$, and let A be the K vector space (free K module) with basis $\{x_\sigma\}$. Then $A = \bigoplus_{\sigma \in G} K x_\sigma$. Define a multiplication on K basis vectors in A by

$$(cx_\sigma)(dx_\tau) = c\sigma(d)a(\sigma, \tau)x_{\sigma\tau}, \qquad (*)$$

and extend it to a multiplication on A by additivity.

First we shall check that A is an associative F algebra with $a(1, 1)^{-1}x_1$ as identity by making use of the cocycle property

$$\rho(a(\sigma, \tau))a(\rho, \sigma\tau) = a(\rho, \sigma)a(\rho\sigma, \tau). \qquad (**)$$

For associativity, $(*)$ gives

$$\begin{aligned}(bx_\rho)\big((cx_\sigma)(dx_\tau)\big) &= (bx_\rho)(c\sigma(d)a(\sigma, \tau)x_{\sigma\tau}) \\ &= b\rho(c)(\rho\sigma(d))\rho(a(\sigma, \tau))a(\rho, \sigma\tau)x_{\rho\sigma\tau}\end{aligned}$$

and

$$\begin{aligned}\big((bx_\rho)(cx_\sigma)\big)(dx_\tau) &= (b\rho(c)a(\rho, \sigma)x_{\rho\sigma})(dx_\tau) \\ &= b\rho(c)a(\rho, \sigma)\rho\sigma(d))a(\rho\sigma, \tau)x_{\rho\sigma\tau},\end{aligned}$$

and the right sides are equal by $(**)$. To see that $a(1, 1)^{-1}x_1$ is a two-sided identity, take $\rho = \sigma = 1$ in $(**)$ to get $1(a(1, \tau))a(1, \tau) = a(1, 1)a(1, \tau)$. Since a takes values in $K^\times$, we can cancel and obtain

$$a(1, \tau) = a(1, 1). \qquad (\dagger)$$

Thus $(*)$ gives

$$\big(a(1, 1)^{-1}x_1\big)(dx_\tau) = a(1, 1)^{-1}1(d)a(1, \tau)x_\tau = dx_\tau.$$

Similarly another specialization of (∗∗) is $\sigma(a(1,1))a(\sigma,1) = a(\sigma,1)a(\sigma,1)$, from which we obtain
$$\sigma(a(1,1)) = a(\sigma,1). \qquad (\dagger\dagger)$$

Thus (∗) gives
$$(cx_\sigma)\bigl(a(1,1)^{-1}x_1\bigr) = c\sigma(a(1,1))^{-1}a(\sigma,1)x_\sigma = cx_\sigma,$$

and $a(1,1)^{-1}x_1$ is indeed a two-sided identity. We denote it by 1. Scalar multiplication by $r \in F$ is understood to be the additive extension of $r(cx_\sigma) = (rc)x_\sigma$ for $c \in K$, and the identities

$$\bigl(r(cx_\sigma)\bigr)(dx_\tau) = rc\sigma(d)a(\sigma,\tau)x_{\sigma\tau},$$
$$(cx_\sigma)\bigl(r(dx_\tau)\bigr) = c\sigma(rd)a(\sigma,\tau)x_{\sigma\tau} = rc\sigma(d)a(\sigma,\tau)x_{\sigma\tau},$$
$$r\bigl((cx_\sigma)(dx_\tau)\bigr) = rc\sigma(d)a(\sigma,\tau)x_{\sigma\tau}$$

show that multiplication in A is F linear with respect to scalars, hence show that A is an algebra over F.

Second we define $K' \subseteq A$ and an isomorphism $K \to K'$. For $b \in K$, we let b' be the member of A given by $b' = b1 = b(a(1,1)^{-1}x_1)$, and we let K' be the image of K under $b \mapsto b'$. The map $b \mapsto b'$ certainly respects addition, and it respects multiplication because the identity

$$(b_1 a(1,1)^{-1}x_1)(b_2 a(1,1)^{-1}x_1) = b_1 b_2 a(1,1)^{-1}x_1$$

is immediate from (∗). Hence K' is a subfield of A.

Third we prove properties (a), (b), and (c). For (a), we use (∗) and (†) to obtain the identity

$$b'x_\sigma = (ba(1,1)^{-1}x_1)x_\sigma = ba(1,1)^{-1}a(1,\sigma)x_\sigma = bx_\sigma. \qquad (\ddagger)$$

This identity shows that $K'x_\sigma = Kx_\sigma$, and (a) follows. From (‡), we see also that $x_\sigma(bx_{\sigma^{-1}}) = (1x_\sigma)(bx_{\sigma^{-1}}) = 1\sigma(b)a(\sigma,\sigma^{-1})x_1$ and that $(bx_{\sigma^{-1}})x_\sigma = b\sigma(1)a(\sigma^{-1},\sigma)x_1$; thus x_σ has a right inverse in A and also a left inverse, hence a two-sided inverse. Consequently the statement of (b) is meaningful; for its proof we have only to observe that

$$x_\sigma c' x_\sigma^{-1} = \bigl(x_\sigma(ca(1,1)^{-1}x_1)\bigr)x_\sigma^{-1} = \bigl(\sigma(c)\sigma(a(1,1))^{-1}a(\sigma,1)x_\sigma\bigr) \cdot x_\sigma^{-1}$$
$$= \sigma(c)x_\sigma \cdot x_\sigma^{-1} = \sigma(c)'x_\sigma x_\sigma^{-1} = \sigma(c)',$$

the last three equalities following from (††), (‡), and the identity $x_\sigma x_\sigma^{-1} = 1$. For (c), we have
$$x_\sigma x_\tau = a(\sigma,\tau)x_{\sigma\tau} = a(\sigma,\tau)'x_{\sigma\tau},$$

the second equality following from (‡).

Fourth we show that A is simple. Let I be a proper two-sided ideal in A, and let $\varphi : A \to A/I$ be the quotient homomorphism. Since 1 is not in I and since K' is a subfield of A, we know that $\ker(\varphi|_{K'}) = 0$ and that $\varphi(K')$ is a subfield of A/I. The field $\varphi(K')$ acts on A/I by left multiplication and makes A/I into a $\varphi(K')$ vector space. The members $\varphi(x_\sigma)$ of A/I certainly span A/I over $\varphi(K')$ because of (a), and the claim is that they are linearly independent. If so, then φ is one-one, I equals 0, and A is simple. For the linear independence, we argue by contradiction in the same way as for Lemma 3.11. Suppose that $J \subseteq G$ is a maximal subset such that $\{\varphi(x_\tau) \mid \tau \in J\}$ is linearly independent over $\varphi(K')$. For σ not in J, we then have

$$\varphi(x_\sigma) = \sum_{\tau \in J} \varphi(a'_\tau)\varphi(x_\tau) \qquad \text{with } a_\tau \in K. \tag{‡‡}$$

Every c in K satisfies

$$\varphi(\sigma(c)')\varphi(x_\sigma) = \varphi(x_\sigma)\varphi(c') = \sum_{\tau \in J} \varphi(a'_\tau)\varphi(x_\tau)\varphi(c') = \sum_{\tau \in J} \varphi(a'_\tau)\varphi(\tau(c)')\varphi(x_\tau),$$

and thus

$$\varphi(x_\sigma) = \sum_{\tau \in J} \varphi(\sigma(c)')^{-1}\varphi(a'_\tau)\varphi(\tau(c)')\varphi(x_\tau).$$

Comparing this expansion with (‡‡) shows that

$$\varphi(\sigma(c)')^{-1}\varphi(a'_\tau)\varphi(\tau(c)') = \varphi(a'_\tau) \qquad \text{for } \tau \in J. \tag{§}$$

Since x_σ is invertible in A, $\varphi(x_\sigma)$ is invertible in A/I and cannot be 0. Therefore some $\varphi(a'_\tau)$ in the expansion (‡‡) is nonzero. For this τ, we can cancel $\varphi(a'_\tau)$ in (§) and obtain $\varphi(\sigma(c)') = \varphi(\tau(c)')$ for all $c \in K$. Since φ is one-one on K', we conclude that $\sigma = \tau$, in contradiction to the fact that σ is not in J. Therefore A is simple.

Fifth we show that A has center F. Thus suppose that $\sum_\sigma c'_\sigma x_\sigma$ is central. Commutativity with $d' x_\tau$ forces the two expressions

$$\Big(\sum_\sigma c'_\sigma x_\sigma\Big) d' x_\tau = \sum_\sigma c'_\sigma \sigma(d)' x_\sigma x_\tau = \sum_\sigma c'_\sigma \sigma(d)' a(\sigma, \tau)' x_{\sigma\tau}$$

and

$$d' x_\tau \Big(\sum_\sigma c'_\sigma x_\sigma\Big) = \sum_\sigma (d' x_\tau)(c'_\sigma x_\sigma) = \sum_\sigma d' \tau(c_\sigma)' a(\tau, \sigma)' x_{\tau\sigma}$$
$$= \sum_\sigma d' \tau(c_{\tau^{-1}\sigma\tau})' a(\tau, \tau^{-1}\sigma\tau)' x_{\sigma\tau}$$

to be equal. Hence

$$d\tau(c_{\tau^{-1}\sigma\tau})a(\tau, \tau^{-1}\sigma\tau) = c_\sigma \sigma(d) a(\sigma, \tau) \qquad \text{for all } d, \sigma, \tau. \tag{§§}$$

Putting $d = 1$ in (§§) shows that $\tau(c_{\tau^{-1}\sigma\tau})a(\tau, \tau^{-1}\sigma\tau) = c_\sigma a(\sigma, \tau)$. Substituting from this equation into the left side of (§§) gives

$$dc_\sigma a(\sigma, \tau) = c_\sigma \sigma(d) a(\sigma, \tau) \quad \text{for all } d, \sigma, \tau.$$

If $c_\sigma \neq 0$, we see that $\sigma(d) = d$ for all $d \in K$; thus $c_\sigma \neq 0$ only for $\sigma = 1$. For $\sigma = 1$ and $d = 1$, (§§) reduces to

$$\tau(c_1)a(\tau, 1) = c_1 a(1, \tau).$$

Taking into account (†) and (††), we obtain

$$\tau(c_1 a(1, 1)) = c_1 a(1, 1).$$

Since τ is arbitrary, this says that $c_1 a(1, 1)$ is in F. Thus the central element is $c'_1 x_1 = c_1 x_1 = c_1 a(1, 1) a(1, 1)^{-1} x_1 = (c_1 a(1, 1))1$ and is an F multiple of the identity.

Since $\{x_\sigma\}$ by definition is a basis of A over K, we have $\dim_K A = |G| = \dim_F K$. Multiplying this equation by $\dim_F K$ yields $\dim_F A = (\dim_F K)^2$. This completes the proof. $\square$

Corollary 3.13. If K is a finite Galois extension of the field F, then the map $\mathcal{B}(K/F) \to H^2(\mathrm{Gal}(K/F), K^\times)$ defined via factor sets is one-one onto.

PROOF. Put $G = \mathrm{Gal}(K/F)$. If $a : G \times G \to K^\times$ is in $Z^2(G, K^\times)$, then we can construct an algebra A via Proposition 3.12, and the claim is that the map $a \mapsto A$ descends to $H^2(G, K^\times)$ and is a two-sided inverse to the map from $\mathcal{B}(K/F)$ into $H^2(G, K^\times)$ given in Proposition 3.10.

First we show that $a \mapsto A$ descends to $H^2(G, K^\times)$. Thus suppose that b is a second cocycle and is of the form $b(\sigma, \tau) = a(\sigma, \tau) c_\sigma \sigma(c_\tau) c_{\sigma\tau}^{-1}$, i.e., represents the same member of $H^2(G, K^\times)$. Let B be the algebra constructed from b by Proposition 3.12, say with K mapping to $K'' \subseteq B$ via $c \mapsto c''$ and with

(a') $B = \bigoplus_{\sigma \in G} K'' y_\sigma$ for a subset $\{y_\sigma\}$ of B,
(b') $y_\sigma c'' y_\sigma^{-1} = \sigma(c)''$,
(c') $y_\sigma y_\tau = b(\sigma, \tau)'' y_{\sigma\tau}$.

Define $\varphi : A \to B$ to be the additive extension of the function with $\varphi(c' x_\sigma) = c'' c''^{-1}_\sigma y_\sigma$. To check that φ is an algebra homomorphism, we start from the formula $(c' x_\sigma)(d' x_\tau) = c'\sigma(d)' a(\sigma, \tau)' x_{\sigma\tau}$ and apply φ to obtain

$$\varphi\big((c' x_\sigma)(d' x_\tau)\big) = c'' \sigma(d)'' a(\sigma, \tau)'' c''^{-1}_{\sigma\tau} y_{\sigma\tau}.$$

3. Crossed Products

Meanwhile,

$$\begin{aligned}\varphi(c'x_\sigma)\varphi(d'x_\tau) &= (c''c_\sigma''^{-1}y_\sigma)(d''c_\tau''^{-1}y_\tau) \\ &= c''c_\sigma''^{-1}\sigma(d)''\sigma(c_\tau)''^{-1}b(\sigma,\tau)''y_{\sigma\tau} \\ &= c''c_\sigma''^{-1}\sigma(d)''\sigma(c_\tau)''^{-1}a(\sigma,\tau)''c_\sigma''\sigma(c_\tau)''c_{\sigma\tau}''^{-1}y_{\sigma\tau}.\end{aligned}$$

Hence $\varphi((c'x_\sigma)(d'x_\tau)) = \varphi(c'x_\sigma)\varphi(d'x_\tau)$, and φ is an algebra homomorphism. Since φ carries K basis to K basis, φ is an algebra isomorphism.

Thus the map $a \mapsto A$ descends to a map from $H^2(G, K^\times)$ into $\mathcal{B}(K/F)$. Starting from a cocycle a in $Z^2(G, K^\times)$, we can construct A and elements x_σ by Proposition 3.12, we can apply Propositions 3.12b and 3.10 to the x_σ's to obtain another cocycle $\bar{a}$ in $Z^2(G, K^\times)$, and we can use Proposition 3.12c to see that $\bar{a} = a$. In the reverse direction if we start from an algebra A, make a set of choices, and form a factor set a by means of Proposition 3.10, then Proposition 3.12 constructs an algebra $\overline{A}$ that has to be isomorphic to A because conditions (a) through (c) in Proposition 3.12 determine the same multiplication table for an algebra as was used in constructing the cocycle a. $\square$

Theorem 3.14. If K is a finite Galois extension of the field F, then the map $\mathcal{B}(K/F) \to H^2(\text{Gal}(K/F), K^\times)$ defined via factor sets is a group isomorphism.

REMARKS. Put $G = \text{Gal}(K/F)$. In view of Corollary 3.13, is enough to prove that the mapping $Z^2(G, K^\times) \to \mathcal{B}(K/F)$ of Proposition 3.12 is a group homomorphism. Thus let A, B, and C be the crossed-product algebras $A = \mathcal{A}(K, G, a)$, $B = \mathcal{A}(K, G, b)$, and $C = \mathcal{A}(K, G, ab)$. We are to prove that $A \otimes_F B$ is Brauer equivalent to C. Each of A, B, and C has F dimension $(\dim_F K)^2$, and hence $A \otimes_F B$ will not be isomorphic to C. Consequently we need to prove Brauer equivalence of two specific nonisomorphic algebras. This is the circumstance that makes the proof complicated.

PROOF (Chase). Let G, a, b, A, B, and C be as in the remarks. We can regard A and B as vector spaces over K with K acting on the left in each case. We define an F vector space M to be the quotient of $A \otimes_F B$ by the F vector subspace I generated by all vectors $ca \otimes b - a \otimes cb$ with $a \in A$, $b \in B$, and $c \in K$. We write $M = A \otimes_K B$ for this quotient, even though more standard notation for it might be $A^o \otimes_K B$ with A^o as a right K module and B as a left K module.

The subspace I is carried to itself by right multiplication by any member of the algebra $A \otimes_F B$ and hence is a right ideal. The quotient M is therefore a unital right $A \otimes_F B$ module with $(a \otimes_K b)(a' \otimes_F b') = aa' \otimes_K bb'$ for $a \otimes_K b$ in M and $a' \otimes_F b'$ in $A \otimes_F B$.

We shall make the unital right $A \otimes_F B$ module M into a unital $(C, A \otimes_F B)$ bimodule by introducing an action by C on the left. For this purpose let $\{u_\sigma\}, \{v_\sigma\}$, and $\{w_\sigma\}$ be the distinguished K bases of the algebras A, B, and C indexed by G and used to form A, B, and C from the 2-cocycles a, b, and ab. Given an element xw_σ in C with $x \in K$, define xw_σ on $A \otimes_F B$ to be (left by xu_σ) $\otimes$ (left by v_σ). Let us see that this operation carries the generators of I into I. We have

$$(xw_\sigma)(ca \otimes_F b) - (xw_\sigma)(a \otimes_F cb) = xu_\sigma ca \otimes_F v_\sigma b - xu_\sigma a \otimes_F v_\sigma cb$$
$$= x\sigma(c)u_\sigma a \otimes_F v_\sigma b - xu_\sigma a \otimes_F \sigma(c)v_\sigma b$$
$$= \sigma(c)(xu_\sigma a) \otimes_F (v_\sigma b)$$
$$- (xu_\sigma a) \otimes_F \sigma(c)(v_\sigma b),$$

and the right side is indeed in I. Thus we obtain an operation of xw_σ on the left for $A \otimes_K B$ such that

$$(xw_\sigma)(a \otimes_K b) = xu_\sigma a \otimes_K v_\sigma b \quad \text{for } x \in K, \ \sigma \in G, \ a \in A, \ b \in B. \quad (*)$$

We extend this definition by additivity in such a way that all of C operates on the left for $A \otimes_K B$.

The claim is that the additive extension $(*)$ to C makes $M = A \otimes_K B$ into a unital left C module. What needs proof is that 1 acts as 1 and that

$$((xw_\sigma)(yw_\tau))(a \otimes_K b) = (xw_\sigma)((yw_\tau)(a \otimes_K b)). \quad (**)$$

The element 1 in C is $a(1, 1)^{-1}b(1, 1)^{-1}w_1$, and we have

$$(a(1, 1)^{-1}b(1, 1)^{-1}w_1)(a \otimes_K b) = a(1, 1)^{-1}b(1, 1)^{-1}u_1 a \otimes_K v_1 b$$
$$= a(1, 1)^{-1}u_1 a \otimes_K b(1, 1)^{-1}v_1 b = a \otimes_K b.$$

Thus 1 acts as 1. For $(**)$, the left side is

$$(x\sigma(y)a(\sigma, \tau)b(\sigma, \tau)w_{\sigma\tau})(a \otimes_K b) = x\sigma(y)a(\sigma, \tau)b(\sigma, \tau)u_{\sigma\tau} a \otimes_K v_{\sigma\tau} b,$$

while the right side is

$$(xw_\sigma)(yu_\tau a \otimes_K v_\tau b) = xu_\sigma yu_\tau a \otimes_K v_\sigma v_\tau b = x\sigma(y)u_\sigma u_\tau a \otimes_K v_\sigma v_\tau b$$
$$= x\sigma(y)a(\sigma, \tau)u_{\sigma\tau} a \otimes_K b(\sigma, \tau)v_{\sigma\tau} b.$$

These are equal, since $b(\sigma, \tau)$ is in K and therefore moves across the tensor-product sign.

Thus M is a unital left C module. The left action by C certainly commutes with the right action by $A \otimes_F B$, and M is consequently a unital $(C, A \otimes_F B)$ bimodule. Each member of $A \otimes_F B$ therefore yields by its right action a member of the ring $\text{End}_C(M)$, and we obtain a ring homomorphism of $(A \otimes_F B)^o$ into $\text{End}_C(M)$. Since $A \otimes_F B$ is a simple ring, this homomorphism is one-one. If we can prove that this homomorphism is onto, then we will have a ring isomorphism $(A \otimes_F B)^o \cong \text{End}_C(M)$, and the rest will be easy.

To see that the homomorphism is onto, we shall calculate dimensions. Let $n = \dim_F K$. Then each of A, B, and C has F dimension n^2, and we have

$$\dim_F M = (\dim_F A)(\dim_F B)/(\dim_F K) = n^2 n^2/n = n^3 = (\dim_F C)n.$$

Since the algebra C is simple, every unital left C module is semisimple and is in fact isomorphic to a multiple of a simple left C module V. The above dimensional equality says that if r is the integer such that C is isomorphic to rV as a left C module, then M is isomorphic to nrV.

Let D^o be the division algebra $\text{End}_C(V)$. As in the proof of Wedderburn's Theorem (Theorem 2.2), we know for each integer m that

$$\text{End}_C(mV) \cong M_m(\text{End}_C(V)) \cong M_m(D^o). \tag{$\dagger$}$$

Taking $m = r$ in ($\dagger$) gives $C^o \cong \text{End}_C(rV) \cong M_r(D^o)$. Hence

$$C \cong M_r(D), \tag{$\dagger\dagger$}$$

and $\dim_F C = r^2 \dim_F D$. Since $\dim_F C = (\dim_F K)^2 = n^2$, we obtain $\dim_F D = n^2/r^2$. Taking $m = nr$ in ($\dagger$) gives

$$\text{End}_C(M) \cong \text{End}_C(nrV) \cong M_{nr}(D^o), \tag{$\ddagger$}$$

and we therefore obtain

$$\dim_F \text{End}_C(M) = n^2 r^2 \dim_F D = (n^2 r^2)(n^2/r^2) = n^4.$$

Since $\dim_F(A \otimes_F B) = n^4$, we obtain $\dim_F(A \otimes_F B)^o = \dim_F \text{End}_C(M)$, and we conclude that the algebra homomorphism $(A \otimes_F B)^o \to \text{End}_C(M)$ is onto. Thus it is an isomorphism, and $A \otimes_F B \cong (\text{End}_C(M))^o$.

Combining this isomorphism with ($\ddagger$) shows that $A \otimes_F B \cong M_{nr}(D)$. In view of ($\dagger\dagger$), $A \otimes_F B$ is therefore Brauer equivalent to C. $\square$

Corollary 3.15. If D is a finite-dimensional central division algebra of dimension m^2 over a field F, then the m-fold tensor product of D with itself over F is a full matrix algebra over F.

PROOF. Corollary 3.9 produces a finite Galois extension K of F such that K splits D. Write G for $\text{Gal}(K/F)$. In view of Theorems 3.3 and 2.4, there exists an integer l such that $A = M_l(D)$ contains a subfield K' isomorphic to K with $\dim_F A = (\dim_F K')^2$. Changing notation, we may redefine $K = K'$. Let

$n = \dim_F K$. Then $n^2 = \dim_F A = l^2 \dim_F D = (lm)^2$, and $n = lm$. Following the construction of factor sets in Section 2 and using Lemma 3.11, we form a vector-space basis $\{x_\sigma \mid \sigma \in G\}$ of A over K and a factor set $\{a(\sigma, \tau)\}$ such that $x_\sigma x_\tau = a(\sigma, \tau) x_{\sigma\tau}$ and $\sigma(c) = x_\sigma c x_\sigma^{-1}$ for all c in K.

Example 1 of semisimple rings in Section II.2 shows that the left A module A is the direct sum of l isomorphic simple left A modules. Let V be one of these. Restricting the module structure of V from A to K makes V into a unital left K module, hence into a vector space over K. Then we have

$$n^2 = \dim_F A = l \dim_F V = l(\dim_K V)(\dim_F K) = ln \dim_K V,$$

and $\dim_K V = m$. Let $v_1, \ldots, v_m$ be a K basis of V. For each $x \in A$, define a matrix $C(x)$ in $M_m(K)$ by

$$x v_j = \sum_{i=1}^m C(x)_{ij} v_i.$$

For σ and τ in G, we compute $x_\sigma x_\tau v_j$ in two ways as

$$x_\sigma x_\tau v_j = a(\sigma, \tau) x_{\sigma\tau} v_j = a(\sigma, \tau) \sum_{i=1}^m C(x_{\sigma\tau})_{ij} v_i \qquad (*)$$

and as

$$x_\sigma x_\tau v_j = x_\sigma \sum_{k=1}^m C(x_\tau)_{kj} v_k = \sum_{k=1}^m \sigma(C(x_\tau)_{kj}) x_\sigma v_k = \sum_{i,k=1}^m \sigma(C(x_\tau)_{kj}) C(x_\sigma)_{ik} v_i.$$

If we write $\sigma(C(x_\tau))$ for the result of applying σ to each entry of $C(x_\tau)$, then we obtain

$$x_\sigma x_\tau v_j = \sum_{i=1}^m (C(x_\sigma) \sigma(C(x_\tau)))_{ij} v_i. \qquad (**)$$

Comparing $(*)$ and $(**)$ leads to the matrix equation in $M_m(K)$ given by

$$a(\sigma, \tau) C(x_{\sigma\tau}) = C(x_\sigma) \sigma(C(x_\tau)).$$

Putting $c_\sigma = \det C(x_\sigma)$ and taking the determinant of both sides yields

$$a(\sigma, \tau)^m c_{\sigma\tau} = c_\sigma \sigma(c_\tau).$$

This equation shows that $a(\sigma, \tau)^m$ is a trivial factor set. Applying Theorem 3.14, we see that the mth power of the Brauer equivalence class of A is trivial. Since A is Brauer equivalent to D, the corollary follows. $\square$

Corollary 3.16. If F is any field, then every element of $\mathcal{B}(F)$ has finite order.

PROOF. If A is any central simple algebra over F, then Theorem 2.4 shows that $A \cong M_l(D)$ for some integer $l \geq 1$ and some central division algebra D over F. Corollary 3.15 shows that the Brauer equivalence class of D has finite order in $\mathcal{B}(F)$. Since A is Brauer equivalent to D, the same thing is true for A. □

4. Hilbert's Theorem 90

Let K/F be a finite Galois extension of fields. Our interest in this section will be in the cohomology groups $H^q(\text{Gal}(K/F), K^\times)$ with q possibly different from 2. For $q = 0$, $H^0(G, N)$ is always the subgroup of elements of N fixed by every element of G. In the case of a Galois extension, the members of $K^\times$ fixed by the Galois group are the nonzero elements of the base field F. Thus we have

$$H^0(\text{Gal}(K/F), K^\times) \cong F^\times.$$

In addition, Theorem 3.14 has established an isomorphism

$$H^2(\text{Gal}(K/F), K^\times) \cong \mathcal{B}(K/F),$$

and thus we have already obtained some understanding of this group for $q = 2$.

We shall examine H^1 in a moment, but first we take note of another fact about H^2. Problem 16b at the end of Chapter VII of *Basic Algebra* shows that if G is a finite group and N is an abelian group on which G acts by automorphisms, then every element of $H^q(G, N)$ for $q > 0$ has order dividing $|G|$. In particular, every element of $H^2(\text{Gal}(K/F), K^\times)$ has order dividing $\dim_F K$ whenever K is a finite Galois extension of F. Applying Theorem 3.14, we see that every member of $\mathcal{B}(K/F)$ has order dividing $\dim_F K$. In view of Corollary 3.9, this argument gives a new and shorter proof of the result of Corollary 3.16 that every member of $\mathcal{B}(F)$ has finite order. The estimate of the order via Corollary 3.15, however, is sharper than the estimate obtained via the shorter proof, and this distinction makes all the difference in Problem 12 at the end of the chapter.

The result concerning H^1 and its important special case given as Corollary 3.18 below are known as **Hilbert's Theorem 90**.

Theorem 3.17. If K/F is any finite Galois extension of fields, then $H^1(\text{Gal}(K/F), K^\times) = 0$.

PROOF. Let $G = \text{Gal}(K/F)$, put $n = \dim_F K$, and enumerate G as $\sigma_1, \ldots, \sigma_n$. By the Theorem of the Primitive Element, we can write $K = F(\alpha)$ for some α in K, and then $\{1, \alpha, \alpha^2, \ldots, \alpha^{n-1}\}$ is a basis of K over F. Form the n-by-n matrix M

with entries in K whose $(i, j)^{\text{th}}$ entry is $\sigma_j(\alpha^{i-1})$. This is a Vandermonde matrix, and Corollary 5.3 of *Basic Algebra* gives its determinant as $\prod_{j>i} [\sigma_j(\alpha) - \sigma_i(\alpha)]$. This determinant cannot be 0, since $\sigma_j(\alpha) = \sigma_i(\alpha)$ implies $\sigma_j(\alpha^k) = \sigma_j(\alpha)^k = \sigma_i(\alpha)^k = \sigma_i(\alpha^k)$ for all k and then $\sigma_j(x) = \sigma_i(x)$ for all x. Hence the matrix M is nonsingular.

Let f be a nonzero element in $Z^1(G, K^\times)$. Such a function $f : G \to K$ is nowhere vanishing and has $f(\sigma\tau) = f(\sigma)\sigma(f(\tau))$ for all σ and τ in G. Since the matrix M is nonsingular, the nontrivial linear combination $\sum_{\sigma \in G} f(\sigma)\sigma$ cannot be 0 on all members of the basis $\{1, \alpha, \alpha^2, \ldots, \alpha^{n-1}\}$. Choose k with $\sum_{\sigma \in G} f(\sigma)\sigma(\alpha^k) = y \neq 0$. Applying $\tau \in G$ to this equation, we obtain

$$\tau(y) = \sum_{\sigma \in G} \tau(f(\sigma))\tau\sigma(\alpha^k) = \sum_{\sigma \in G} f(\tau\sigma)f(\tau)^{-1}\tau\sigma(\alpha^k)$$
$$= f(\tau^{-1}) \sum_{\sigma \in G} f(\sigma)\sigma(\alpha^k) = f(\tau)^{-1} y.$$

The equation $f(\tau)^{-1} = \tau(y)y^{-1}$ shows that f^{-1} is a coboundary, hence that f is a coboundary. $\square$

Corollary 3.18. If K/F is a finite Galois extension with cyclic Galois group and if σ is a generator of the Galois group, then every member x of K with $N_{K/F}(x) = 1$ is of the form $x = \sigma(y)y^{-1}$ for some $y \in K^\times$.

REMARKS. The instance of this corollary in which K is a quadratic number field and F is the field $\mathbb{Q}$ appears as Problem 27 at the end of Chapter I. In subsequent problems at the end of that chapter, Problem 27 plays a crucial role in showing that various possible definitions of genera are equivalent.

PROOF. Let $G = \{1, \sigma, \sigma^2, \ldots, \sigma^{n-1}\}$ be the Galois group, and define a function $F : \mathbb{Z} \to K^\times$ by $F(0) = 1$ and

$$F(k) = x\sigma(x)\sigma^2(x)\cdots\sigma^{k-1}(x) \qquad \text{for } k \geq 1.$$

Then we have

$$F(k+l) = x\sigma(x)\sigma^2(x)\cdots\sigma^{k+l-1}(x)$$
$$= \bigl(x\sigma(x)\sigma^2(x)\cdots\sigma^{k-1}(x)\bigr)\sigma^k\bigl(x\sigma(x)\sigma^2(x)\cdots\sigma^{l-1}(x)\bigr)$$
$$= F(k)\sigma^k(F(l)), \qquad (*)$$

The condition that $N_{K/F}(x) = 1$ is exactly the condition that $F(n) = 1$. Then $F(k+n) = F(k)\sigma^n(F(1)) = F(k)$ for all k, and it is meaningful to define a 1-cochain $f : G \to K^\times$ in $C^1(G, K^\times)$ by $f(\sigma^k) = F(k)$. Condition $(*)$ implies that $f(\sigma^k\sigma^l) = f(\sigma^k)\sigma^k(f(\sigma^l))$, and hence f is a cocycle in $Z^1(G, K^\times)$. Theorem 3.17 shows that f is a coboundary in $B^1(G, K^\times)$, necessarily satisfying $f(\tau) = \tau(y)y^{-1}$ for some $y \in K^\times$ and all $\tau \in G$. Taking $\tau = \sigma$, we obtain $x = f(\sigma) = \sigma(y)y^{-1}$, as required. $\square$

Our final result concerning $H^q(\text{Gal}(K/F), K^\times)$ for this chapter gives further information about the special case in which $\text{Gal}(K/F)$ is cyclic, but now for general q. In combination with the study of crossed-product algebras, the case $q = 2$ of this result provides a way of constructing new examples of noncommutative division algebras. A key step in the proof makes use of a fundamental general property concerning cohomology of groups, and we therefore digress in Section 5 to establish this property.

5. Digression on Cohomology of Groups

This section develops general material about cohomology of groups. Although the earlier sections of this chapter are helpful for motivation, the results that we discuss in this section do not rely on any previous material in this volume. It will be assumed that the reader is familiar with the definitions of complexes and exact sequences in Chapter X of *Basic Algebra*, as well as with the application of tensor-product functors and Hom functors to exact sequences and complexes. The material in Chapter VII of *Basic Algebra* on cohomology of groups will be helpful as background, but it is unnecessary from a logical point of view. If R is a ring with identity, we denote by $\mathcal{C}_R$ the category of all unital left R modules.

Let G be a group, not necessarily finite. We shall work with the integral group ring $\mathbb{Z}G$ of G. It has the universal mapping property that whenever G acts by automorphisms on an abelian group M, then the action by G on M extends to $\mathbb{Z}G$ in a unique way that makes M into a unital left $\mathbb{Z}G$ module.

Here is a brief overview of what is to happen in this section: If G acts on the abelian group M by automorphisms, then the abelian group $C^n(G, M)$ of n-cochains is the set of functions into M from the n-fold product of G with itself, the operation being given by addition of the values of the functions. To define the cohomology group $H^n(G, M)$, one introduces suitable homomorphisms known as "coboundary maps" $\delta_n : C^n(G, M) \to C^{n+1}(G, M)$ and shows that the sequence

$$0 \longrightarrow C_0(G, M) \xrightarrow{\delta_0} \cdots \xrightarrow{\delta_{n-1}} C_n(G, M) \xrightarrow{\delta_n} C_{n+1}(G, M) \longrightarrow \cdots$$

of abelian groups and homomorphisms is a complex in the category $\mathcal{C}_\mathbb{Z}$. Then it is meaningful to define $H^n(G, M) = (\ker \delta_n)/(\text{image } \delta_{n-1})$ for $n \geq 0$ if we adopt the convention that image $\delta_{-1} = 0$. The first thing that we shall do in this section is to exhibit a certain exact sequence in the category $\mathcal{C}_{\mathbb{Z}G}$ such that the above complex is obtained from it by application of the functor $\text{Hom}_{\mathbb{Z}G}(\,\cdot\,, M)$ and the dropping of one term of the form $\text{Hom}_{\mathbb{Z}G}(\mathbb{Z}, M)$. Except for a single term $\mathbb{Z}$, the members of this exact sequence will all be free $\mathbb{Z}G$ modules, and the exact sequence will be called the "standard resolution of $\mathbb{Z}$ in the category $\mathcal{C}_{\mathbb{Z}G}$." The exactness is proved in Theorem 3.20, and the application of $\text{Hom}_{\mathbb{Z}G}(\,\cdot\,, M)$ to it appears after the proof of the theorem.

The next thing that we shall do is show that if the standard resolution of $\mathbb{Z}$ is changed to any exact sequence in $\mathcal{C}_{\mathbb{Z}G}$ in such a way that the free $\mathbb{Z}G$ modules are replaced by other free $\mathbb{Z}G$ modules and the module $\mathbb{Z}$ is left unchanged, then application of $\text{Hom}_{\mathbb{Z}G}(\,\cdot\,, M)$ to the new exact sequence leads to canonically isomorphic cohomology groups. This result appears below as Theorem 3.31. In brief, the cohomology groups $H^n(G, M)$ can be computed starting from any "free resolution of $\mathbb{Z}$" in the category $\mathcal{C}_{\mathbb{Z}G}$ in place of the standard resolution.

We begin by constructing the "standard resolution of $\mathbb{Z}$." For $n \geq 0$, let F_n be the free abelian group with $\mathbb{Z}$ basis the set of all $(n+1)$-tuples $(g_0, \ldots, g_n)$ with all $g_j \in G$. The group G acts on F_n by automorphisms, the action on the members of the $\mathbb{Z}$ basis being

$$g(g_0, \ldots, g_n) = (gg_0, \ldots, gg_n).$$

The universal mapping property of $\mathbb{Z}G$ then allows us to regard each F_n as a unital left $\mathbb{Z}G$ module.

Lemma 3.19. For $n \geq 0$, the left $\mathbb{Z}G$ module F_n is a free $\mathbb{Z}G$ module with $\mathbb{Z}G$ basis consisting of all $(n+1)$-tuples $(1, g_1, \ldots, g_n)$, i.e., all $\mathbb{Z}$ basis elements with $g_0 = 1$.

PROOF. The formula $g_0(1, g_0^{-1}g_1, \ldots, g_0^{-1}g_n) = (g_0, g_1, \ldots, g_n)$ shows that all members of the $\mathbb{Z}$ basis defining F_n are $\mathbb{Z}G$ images of the asserted $\mathbb{Z}G$ basis; hence the asserted $\mathbb{Z}G$ basis is a spanning set of F_n relative to $\mathbb{Z}G$. Suppose that there are finitely many distinct members h_j of G and finitely many distinct $(n+1)$-tuples $(1, g_{i,1}, \ldots, g_{i,n})$, and members $\sum_j n_{ij} h_j$ of $\mathbb{Z}G$ such that

$$\sum_i \left(\sum_j n_{ij} h_j \right)(1, g_{i,1}, \ldots, g_{i,n}) = 0.$$

Then
$$\sum_{i,j} n_{ij}(h_j, h_j g_{i,1}, \ldots, h_j g_{i,n}) = 0.$$

Since the h_j's are distinct as j varies and the n-tuples $(g_{i,1}, \ldots, g_{i,n})$ are distinct as i varies, the $(n+1)$-tuples $(h_j, h_j g_{i,1}, \ldots, h_j g_{i,n})$ are distinct as the pair (i, j) varies. Thus the $\mathbb{Z}$ independence implies that $n_{ij} = 0$ for all i and j. This proves the lemma. □

For $n \geq 1$, we define $\partial_{n-1} : F_n \to F_{n-1}$ as a function from the $\mathbb{Z}$ basis into F_{n-1} by

$$\partial_{n-1}(g_0, \ldots, g_n) = \sum_{i=0}^{n} (-1)^i (g_0, \ldots, \widehat{g_i}, \ldots, g_n),$$

where the symbol $\widehat{}$ indicates an expression to be omitted. We extend ∂_{n-1} to all of F_n by the universal mapping property of free abelian groups. For g in G and for any $\mathbb{Z}$ generator x of F_n, it is evident that $\partial_{n-1}(gx) = g(\partial_{n-1}(x))$. Since ∂_{n-1} is a homomorphism of abelian groups, the formula $\partial_{n-1}(gx) = g(\partial_{n-1}(x))$ extends to all x's in F_n. Since G and $\mathbb{Z}$ generate $\mathbb{Z}G$, we obtain $\partial_{n-1}(rx) = r(\partial_{n-1}(x))$ for all $r \in \mathbb{Z}G$ and all $x \in F_n$. In other words, each ∂_{n-1} is a $\mathbb{Z}G$ homomorphism.

We shall make use of one additional $\mathbb{Z}G$ homomorphism. According to Lemma 3.19, the $\mathbb{Z}G$ module F_0 is free on the $\mathbb{Z}G$ basis $\{(1)\}$. Let us think of the group G as acting trivially by automorphisms on the abelian group $\mathbb{Z}$. Under this action, $\mathbb{Z}$ becomes a $\mathbb{Z}G$ module. Define $\varepsilon : F_0 \to \mathbb{Z}$ to be the $\mathbb{Z}G$ homomorphism with $\varepsilon((1)) = 1$. Then $\varepsilon((g_0)) = g_0(\varepsilon((1))) = g_0 \cdot 1 = 1$ for all $g_0 \in G$. The $\mathbb{Z}G$ homomorphism ε is called the **augmentation map**.

Theorem 3.20. If G is any group, then the sequence

$$\cdots \xrightarrow{\partial_{n+1}} F_{n+1} \xrightarrow{\partial_n} F_n \xrightarrow{\partial_{n-1}} \cdots \xrightarrow{\partial_0} F_0 \xrightarrow{\varepsilon} \mathbb{Z} \longrightarrow 0$$

of left unital $\mathbb{Z}G$ modules and $\mathbb{Z}G$ homomorphisms is exact.

REMARKS. The displayed sequence is called the **standard resolution of** $\mathbb{Z}$ in the category $\mathcal{C}_{\mathbb{Z}G}$. The proof will be preceded by two lemmas.

Lemma 3.21. The sequence

$$\cdots \xrightarrow{\partial_{n+1}} F_{n+1} \xrightarrow{\partial_n} F_n \xrightarrow{\partial_{n-1}} \cdots \xrightarrow{\partial_0} F_0 \xrightarrow{\varepsilon} \mathbb{Z} \longrightarrow 0$$

in $\mathcal{C}_{\mathbb{Z}G}$ is a complex, i.e., $\partial_{n-1}\partial_n = 0$ for $n \geq 1$ and also $\varepsilon \partial_0 = 0$.

PROOF. With the understanding that the symbol $\widehat{}$ indicates an expression to be omitted, we have

$$\partial_{n-1}\partial_n(g_0,\ldots,g_n) = \sum_{i=0}^{n}(-1)^i\partial_{n-1}(g_0,\ldots,\widehat{g_i},\ldots,g_n)$$

$$= \sum_{i=0}^{n}(-1)^i\sum_{j=0}^{i-1}(-1)^j(g_0,\ldots,\widehat{g_j},\ldots,\widehat{g_i},\ldots,g_n)$$

$$+ \sum_{i=0}^{n}(-1)^i\sum_{j=i+1}^{n}(-1)^{j+1}(g_0,\ldots,\widehat{g_i},\ldots,\widehat{g_j},\ldots,g_n)$$

$$= \sum_{i=0}^{n}\sum_{j=0}^{i-1}(-1)^{i+j}(g_0,\ldots,\widehat{g_j},\ldots,\widehat{g_i},\ldots,g_n)$$

$$- \sum_{i=0}^{n}\sum_{j=i+1}^{n}(-1)^{i+j}(g_0,\ldots,\widehat{g_i},\ldots,\widehat{g_j},\ldots,g_n).$$

If we interchange the order of summation in the second double sum on the right, we see that the result equals the first double sum on the right. Thus the difference is 0.

This handles all the consecutive compositions except for $\varepsilon\partial_0$. For this we have $\varepsilon\partial_0(g_0,g_1) = \varepsilon(g_1) - \varepsilon(g_0) = 1 - 1 = 0$. □

Lemma 3.22. Fix s in G. For $n \geq 0$, define a homomorphism $h_n : F_n \to F_{n+1}$ of abelian groups to be the additive extension of the function with

$$h_n(g_0,\ldots,g_n) = (s,g_0,\ldots,g_n),$$

and define $h_{-1} : \mathbb{Z} \to F_0$ by $h_{-1}(k) = k(s)$. Then $\partial_n h_n + h_{n-1}\partial_{n-1} = 1$ for $n \geq 1$, and also $\partial_0 h_0 + h_{-1}\varepsilon = 1$.

PROOF. On the $\mathbb{Z}$ basis of $(n+1)$-tuples in F_n, we have

$$\partial_n h_n(g_0,\ldots,g_n) = \partial_n(s,g_0,\ldots,g_n)$$

$$= (g_0,\ldots,g_n) + \sum_{i=0}^{n}(-1)^{i+1}(s,g_0,\ldots,\widehat{g_i},\ldots,g_n)$$

and also

$$h_{n-1}\partial_{n-1}(g_0,\ldots,g_n) = \sum_{i=0}^{n}(-1)^i(s,g_0,\ldots,\widehat{g_i},\ldots,g_n).$$

The sum of these is $(g_0,\ldots,g_n)$, as required. Also,

$$\partial_0 h_0(g_0) = \partial_0(s,g_0) = (g_0) - (s) \quad \text{and} \quad h_{-1}\varepsilon(g_0) = h_{-1}1 = (s).$$

Thus $\partial_0 h_0(g_0) + h_{-1}\varepsilon(g_0) = (g_0)$, and $\partial_0 h_0 + h_{-1}\varepsilon = 1$. □

5. Digression on Cohomology of Groups

PROOF OF THEOREM 3.20. Lemma 3.21 gives image $\partial_n \subseteq \ker \partial_{n-1}$ and image $\partial_0 \subseteq \ker \varepsilon$. For the reverse of the first inclusion, let $x \in F_n$ be given with $\partial_{n-1} x = 0$ and $n \geq 1$. Then Lemma 3.22 gives $x = \partial_n h_n x + h_{n-1} \partial_{n-1} x$. The second term on the right side is 0, and therefore $x = \partial_n(h_n x)$ is in image ∂_n.

For the reverse of the inclusion image $\partial_0 \subseteq \ker \varepsilon$, let $x \in F_0$ be given with $\varepsilon x = 0$. Then Lemma 3.22 gives $x = \partial_0 h_0 x + h_{-1} \varepsilon x$. The second term on the right side is 0, and therefore $x = \partial_0(h_0 x)$ is in image ∂_0. □

With the standard resolution of $\mathbb{Z}$ in $\mathcal{C}_{\mathbb{Z}G}$ now known to be exact, we examine the effect of applying the functor $\operatorname{Hom}_{\mathbb{Z}G}(\,\cdot\,, M)$ to it. This functor is contravariant and carries $\mathcal{C}_{\mathbb{Z}G}$ to the category $\mathcal{C}_{\mathbb{Z}}$ of all abelian groups. On a unital left $\mathbb{Z}G$ module F, this functor yields the abelian group $\operatorname{Hom}_{\mathbb{Z}G}(F, M)$. On a $\mathbb{Z}$ module homomorphism $\varphi : F \to F'$, it yields the homomorphism

$$\operatorname{Hom}(\varphi, 1) : \operatorname{Hom}_{\mathbb{Z}G}(F', M) \to \operatorname{Hom}_{\mathbb{Z}G}(F, M)$$

of abelian groups given by $\operatorname{Hom}(\varphi, 1)(\psi) = \psi \circ \varphi$ for $\psi \in \operatorname{Hom}_{\mathbb{Z}G}(F', M)$. We know from Chapter X of *Basic Algebra* that this functor carries complexes to complexes but does not necessarily preserve exactness.

Before applying $\operatorname{Hom}_{\mathbb{Z}G}(\,\cdot\,, M)$ to the standard resolution of $\mathbb{Z}$, it is customary to drop the term $\mathbb{Z}$ and the augmentation map, obtaining a modified sequence

$$\cdots \xrightarrow{\partial_{n+1}} F_{n+1} \xrightarrow{\partial_n} F_n \xrightarrow{\partial_{n-1}} \cdots \xrightarrow{\partial_0} F_0 \longrightarrow 0$$

that is still a complex in $\mathcal{C}_{\mathbb{Z}G}$. Let us define $d_n = \operatorname{Hom}(\partial_n, 1)$. Then the result of applying $\operatorname{Hom}_{\mathbb{Z}G}(\,\cdot\,, M)$ to the modified complex is the complex

$$0 \longrightarrow \operatorname{Hom}_{\mathbb{Z}G}(F_0, M) \xrightarrow{d_0} \cdots \operatorname{Hom}_{\mathbb{Z}G}(F_n, M) \xrightarrow{d_n} \operatorname{Hom}_{\mathbb{Z}G}(F_{n+1}, M) \xrightarrow{d_{n+1}}$$

in $\mathcal{C}_{\mathbb{Z}}$. To each φ in $\operatorname{Hom}_{\mathbb{Z}G}(F_n, M)$, we associate $f = \Phi(\varphi)$ in $C^n(G, M)$ by the definition

$$f(g_1, \ldots, g_n) = \varphi(1, g_1, g_1 g_2, \ldots, g_1 \cdots g_n).$$

Any member φ of $\operatorname{Hom}_{\mathbb{Z}G}(F_n, M)$ is determined by its values on $(n+1)$-tuples $(1, g_1, \ldots, g_n)$, since we can factor out the first entry of the argument of φ and commute it past φ, and it follows that the system of group homomorphisms

$$\Phi_n : \operatorname{Hom}_{\mathbb{Z}G}(F_n, M) \to C^n(G, M)$$

is a system of isomorphisms of abelian groups. Let

$$\delta_n : C^n(G, M) \to C^{n+1}(G, M)$$

be the map corresponding to $d_n : \text{Hom}_G(F_n, M) \to \text{Hom}_G(F_{n+1}, M)$ under this system of isomorphisms, namely $\delta_n = \Phi_{n+1} \circ d_n \circ \Phi_n^{-1}$. We can calculate δ_n explicitly as follows: If $f = \Phi_n(\varphi)$, then $\delta_n f = (\Phi_{n+1} d_n \Phi_n^{-1})(\Phi_n)(\varphi) = \Phi_{n+1} d_n \varphi$, and therefore

$$\begin{aligned}(\delta_n f)(g_1, \ldots, g_{n+1}) &= (d_n\varphi)(1, g_1, g_1 g_2, \ldots, g_1 \cdots g_{n+1}) \\ &= \varphi(\partial_n(1, g_1, g_1 g_2, \ldots, g_1 \cdots g_{n+1})) \\ &= \varphi(g_1, g_1 g_2, \ldots, g_1 \cdots g_{n+1}) \\ &\quad + \sum_{i=1}^{n} (-1)^i \varphi(1, g_1, \ldots, \widehat{g_1 \cdots g_i}, \ldots, g_1 \cdots g_{n+1}) \\ &\quad + (-1)^{n+1} \varphi(1, g_1, \ldots, g_1 \cdots g_n) \\ &= g_1(f(g_2, g_3, \ldots, g_{n+1})) \\ &\quad + \sum_{i=1}^{n} (-1)^i f(g_1, \ldots, \widehat{g_i}, \ldots, g_{n+1}) \\ &\quad + (-1)^{n+1} f(g_1, \ldots, g_n).\end{aligned}$$

Comparing this formula with the original formula defining δ_n in Chapter VII of *Basic Algebra*, we get a match. That is, we have obtained the complex in $\mathcal{C}_{\mathbb{Z}}$ defining the usual groups $H^n(G, M)$ by applying $\text{Hom}_{\mathbb{Z}G}(\,\cdot\,, M)$ to the standard resolution of $\mathbb{Z}$ in $\mathcal{C}_{\mathbb{Z}G}$ and implementing the system of isomorphisms Φ_n. In particular, we obtain a more conceptual proof than in *Basic Algebra* of the fact that the sequence

$$0 \longrightarrow C_0(G, M) \xrightarrow{\delta_0} \cdots \xrightarrow{\delta_{n-1}} C_n(G, M) \xrightarrow{\delta_n} C_{n+1}(G, M) \longrightarrow \cdots$$

is a complex and that cohomology groups are therefore well defined.

This completes the discussion of the first main point of the section as outlined in the overview at the beginning. Next, any exact sequence

$$\cdots \xrightarrow{\partial'_{n+1}} F'_{n+1} \xrightarrow{\partial'_n} F'_n \xrightarrow{\partial'_{n-1}} \cdots \xrightarrow{\partial'_0} F'_0 \xrightarrow{\varepsilon'} \mathbb{Z} \longrightarrow 0$$

in the category $\mathcal{C}_{\mathbb{Z}G}$ in which all $\mathbb{Z}G$ modules F'_n for $n \geq 0$ are free $\mathbb{Z}G$ modules is called a **free resolution of** $\mathbb{Z}$ in the category $\mathcal{C}_{\mathbb{Z}G}$. The second main point of the section is that if we apply the functor $\text{Hom}_{\mathbb{Z}G}(\,\cdot\,, M)$ to this sequence with $\mathbb{Z}$ dropped, then the consecutive quotients of kernels modulo images are canonically isomorphic to the cohomology groups $H^n(G, M)$ obtained above. Thus $H^n(G, M)$ can be computed from *any* free resolution of $\mathbb{Z}$, and we are not obliged to use the standard free resolution. This result is stated precisely as Theorem 3.31 below.

By way of preparation, let us establish a slightly more general setting and work with it for a moment. Let $\mathcal{C}_R$ be the category of all unital left R modules, where R is any ring with identity. According to circumstances, a complex X in $\mathcal{C}_R$ might be written with decreasing indices as

$$X: \quad \cdots \xrightarrow{\partial_{n+1}} X_{n+1} \xrightarrow{\partial_n} X_n \xrightarrow{\partial_{n-1}} X_{n-1} \xrightarrow{\partial_{n-2}} \cdots$$

or with increasing indices as

$$X: \quad \cdots \xrightarrow{d_{n-2}} X_{n-1} \xrightarrow{d_{n-1}} X_n \xrightarrow{d_n} X_{n+1} \xrightarrow{d_{n+1}} \cdots .$$

Mathematically these complexes amount to the same thing: if we rename each X_k in the second complex as X_{-k} and rename each d_k as ∂_{-k-1}, then we obtain the first complex. However, it is convenient to allow both systems of indexing because of applications.

For the first complex, which has decreasing indices, we define the nth **homology** of X, written $H_n(X)$, by

$$H_n(X) = (\ker \partial_{n-1})/(\text{image } \partial_n).$$

For the second complex, which has increasing indices, we define the nth **cohomology** of X, written $H^n(X)$, by

$$H^n(X) = (\ker d_n)/(\text{image } d_{n-1}).$$

In both cases the integer n is called the **degree**. In either case the homology or cohomology is again a module in $\mathcal{C}_R$. The condition that X be a complex is equivalent to the condition that the image of each incoming map be contained in the kernel of the corresponding outgoing map, and this is precisely the condition that the homology or cohomology be meaningful. Exactness at a particular module in one of the complexes is the statement that the image of the incoming map equals the kernel of the outgoing map. Thus the homology or cohomology of X measures the extent to which the complex X fails to be exact.

Because the nature of the indexing of a complex is not mathematically significant, we will treat only the case of increasing indices for a while, and the modules associated to our complexes will therefore be cohomology modules. A

cochain map[2] between two complexes X and Y in the same category $\mathcal{C}_R$ is a system $f = \{f_n\}$ of R homomorphisms $f_n : X_n \to Y_n$ such that the various squares commute in Figure 3.1.

$$X : \quad \cdots \xrightarrow{d_{n-2}} X_{n-1} \xrightarrow{d_{n-1}} X_n \xrightarrow{d_n} X_{n+1} \xrightarrow{d_{n+1}} \cdots$$
$$\downarrow f_{n-1} \qquad \downarrow f_n \qquad \downarrow f_{n+1}$$
$$Y : \quad \cdots \xrightarrow{d'_{n-2}} Y_{n-1} \xrightarrow{d'_{n-1}} Y_n \xrightarrow{d'_n} Y_{n+1} \xrightarrow{d'_{n+1}} \cdots$$

FIGURE 3.1. A cochain map $f : X \to Y$.

Proposition 3.23. A cochain map $f : X \to Y$ as in Figure 3.1 induces an R homomorphism on cohomology $H^n(X) \to H^n(Y)$ in each degree.

PROOF. Suppose that x_n is in $\ker d_n$, i.e., that $d_n(x_n) = 0$. The commutativity of the right square gives $d'_n(f_n(x_n)) = f_{n+1}(d_n(x_n)) = 0$, and hence $f(x_n)$ is in $\ker d'_n$. Suppose that x_n is in image d_{n-1}, i.e., that $x_n = d_{n-1}(x_{n-1})$ for some x_{n-1}. The commutativity of the left square gives $f_n(x_n) = f_n d_{n-1}(x_{n-1}) = d'_{n-1}(f_{n-1}(x_{n-1}))$, and hence $f_n(x_n)$ is in image d'_{n-1}. Then it follows that $f_n|_{\ker d_n}$ descends to the quotient $(\ker d_n)/(\text{image } d_{n-1})$, yielding a map of $H^n(X)$ into $H^n(Y)$. □

Suppose in the situation of Figure 3.1 that $g = \{g_n\}$ is a second cochain map of X into Y. We say that f is **homotopic**[3] to g, written $f \simeq g$, if there is a system $h = \{h_n\}$ of maps $h_n : X_n \to Y_{n-1}$ in $\mathcal{C}_R$ such that $d'h + hd = f - g$, i.e., if $d'_{n-1}h_n + h_{n+1}d_n = f_n - g_n$ for all n.

Proposition 3.24. In the situation of Figure 3.1 if $f = \{f_n\}$ and $g = \{g_n\}$ are two cochain maps of X into Y and if f and g are homotopic, then f and g induce *identical* maps $H^n(X) \to H^n(Y)$ in each degree.

PROOF. Suppose that $d_n(x_n) = 0$. Then $f_n(x_n) - g_n(x_n) = d'_{n-1}(h_n(x_n)) + h_{n+1}(d_n(x_n)) = d'_{n-1}(h_n(x_n)) + 0$ shows that the images of x_n under f_n and g_n in Y_n differ by a member of image d'_{n-1}. □

Now we bring free R modules into the discussion.

---

[2] The analogous kind of system in which the complexes have decreasing indices is called a **chain map**.
[3] An analogous definition is to be made in the case of two chain maps. If the maps of X are $\partial_n : X_{n+1} \to X_n$ and the maps of Y are $\partial'_n : Y_{n+1} \to Y_n$, then we are to have $h_n : X_n \to Y_{n+1}$ with $\partial'_n h_n + h_{n-1}\partial_{n-1} = f_n - g_n$.

5. Digression on Cohomology of Groups

Proposition 3.25. For the diagram

$$
\begin{array}{ccccc}
F & \xrightarrow{\partial} & M & \xrightarrow{\partial_1} & N \\
\bigg\downarrow \widetilde{f} & & \bigg\downarrow f & & \bigg\downarrow f_1 \\
F' & \xrightarrow{\partial'} & M' & \xrightarrow{\partial'_1} & N'
\end{array}
$$

in $\mathcal{C}_R$, suppose that the top and bottom rows are exact at M and M', suppose that the square on the right commutes, and suppose that F is a free R module. Then there exists an R homomorphism $\widetilde{f} : F \to F'$ that makes the left square commute.

PROOF. If x is a free generator of F, then $0 = f_1 \partial_1 \partial(x) = \partial'_1(f \partial x)$. By exactness at M', $f \partial x$ lies in image(∂'). Choose any $y \in F'$ with $\partial' y = f \partial x$, and define $\widetilde{f}(x)$ to be this y. Then $f \partial x = \partial' \widetilde{f} x$, and the left square commutes at x. The universal mapping property of free R modules says that $\widetilde{f}$ extends to an R homomorphism of F into F', and the extension has $f \partial = \partial' \widetilde{f}$, as required. □

Corollary 3.26. In the category $\mathcal{C}_{\mathbb{Z}G}$, if the rows of the diagram

$$
\begin{array}{ccccccccccc}
\xrightarrow{\partial'_{n+1}} & X_{n+1} & \xrightarrow{\partial'_n} & X_n & \xrightarrow{\partial'_{n-1}} & \cdots & \xrightarrow{\partial'_0} & X_0 & \xrightarrow{\varepsilon'} & \mathbb{Z} & \longrightarrow 0 \\
& \bigg\downarrow f_{n+1} & & \bigg\downarrow f_n & & & & \bigg\downarrow f_0 & & \bigg\downarrow 1 & \\
\xrightarrow{\partial''_{n+1}} & Y_{n+1} & \xrightarrow{\partial''_n} & Y_n & \xrightarrow{\partial''_{n-1}} & \cdots & \xrightarrow{\partial''_0} & Y_0 & \xrightarrow{\varepsilon'} & \mathbb{Z} & \longrightarrow 0
\end{array}
$$

are free resolutions and the vertical identity map $1 : \mathbb{Z} \to \mathbb{Z}$ is given, then the remaining vertical maps,

$$f_0 : X_0 \to Y_0, \quad \ldots, \quad f_n : X_n \to Y_n, \quad f_{n+1} : X_{n+1} \to Y_{n+1}, \quad \ldots,$$

can be constructed inductively from the right to make all the squares commute.

REMARK. The resulting system $f = \{f_n\}$ is called a **chain map over** the identity map $1 : \mathbb{Z} \to \mathbb{Z}$.

PROOF. There is no harm in including a vertical 0 map at the right between the two 0 modules. Certainly the square whose verticals are the identity map $1 : \mathbb{Z} \to \mathbb{Z}$ and the 0 map commutes. Proposition 3.25 is to be applied first to this square and the second square from the right (with vertical f_0 to be constructed and vertical $1 : \mathbb{Z} \to \mathbb{Z}$ given) to construct f_0, then to the second and third squares from the right to construct f_1, and so on, inductively. □

Proposition 3.27. For the diagram

$$\begin{array}{ccccc}
\widetilde{F} & \xrightarrow{\partial} & F & \xrightarrow{\partial_1} & N \\
{\scriptstyle \widetilde{f}}\downarrow & \swarrow_{h} & {\scriptstyle f}\downarrow & \swarrow_{h_1} & \downarrow{\scriptstyle f_1} \\
\widetilde{F} & \xrightarrow[\partial]{} & F & \xrightarrow[\partial_1]{} & N
\end{array}$$

in $\mathcal{C}_R$, suppose that the top and bottom rows are exact at F, that the left and right squares commute, that $\widetilde{F}$ and F are free R modules, and that $h_1 : N \to F$ exists with $f_1 - \partial_1 h_1$ vanishing on image(∂_1). Then there exists $h : F \to \widetilde{F}$ such that $\partial h + h_1 \partial_1 = f$, and this property implies that $f - \partial h$ vanishes on image(∂).

PROOF. If x is a free generator of F, then $f(x) - h_1(\partial_1(x))$ is in ker(∂_1) because $\partial_1(fx - h_1 \partial_1 x) = f_1 \partial_1 x - \partial_1 h_1 \partial_1 x = (f_1 - \partial_1 h_1)(\partial_1 x)$ and because $f_1 - \partial_1 h_1$ vanishes on image(∂_1) by assumption. Therefore $f(x) - h_1(\partial_1(x))$ is in image(∂), and we can write $f(x) - h_1(\partial_1(x)) = \partial a$ for some $a \in \widetilde{F}$. Put $h(x) = a$. Then $\partial h x = \partial a = fx - h_1 \partial_1 x$, and h has the required property on the generator x. The universal mapping property of the free R module F allows us to extend h to an R homomorphism $h : F \to \widetilde{F}$, and the extension satisfies $\partial h = f - h_1 \partial_1$. Once h has this property, then necessarily $(f - \partial h)\partial = (h_1 \partial_1)\partial = h_1(\partial_1 \partial) = 0$. □

Corollary 3.28. In the category $\mathcal{C}_{\mathbb{Z}G}$, if a free resolution $X = \{X_n\}$ of $\mathbb{Z}$ and a chain map $f = \{f_n\}$ of X with itself are given such that the map from $\mathbb{Z}$ to itself is 0, then the chain map f is homotopic to the zero chain map $g = \{g_n\}$ with $g_n = 0$ for all n.

PROOF. We are given the diagram

$$\begin{array}{ccccccccccc}
\longrightarrow & X_n & \longrightarrow & \cdots & \xrightarrow{\partial_1'} & X_1 & \xrightarrow{\partial_0'} & X_0 & \xrightarrow{\varepsilon'} & \mathbb{Z} & \longrightarrow 0 \\
& {\scriptstyle f_n}\downarrow & & & \swarrow_{h_1} & {\scriptstyle f_1}\downarrow & \swarrow_{h_0} & {\scriptstyle f_0}\downarrow & \swarrow_{h_{-1}} & \downarrow{\scriptstyle 0} & \\
\longrightarrow & X_n & \longrightarrow & \cdots & \xrightarrow[\partial_1']{} & X_1 & \xrightarrow[\partial_0']{} & X_0 & \xrightarrow[\varepsilon']{} & \mathbb{Z} & \longrightarrow 0
\end{array}$$

in the category $\mathcal{C}_{\mathbb{Z}G}$ with the two rows as free resolutions and all squares commuting. We are to construct maps $h_n : X_n \to X_{n+1}$ with $\partial_n' h_n + h_{n-1} \partial_{n-1}' = f_n$. Let h_{-2} be the 0 map from the top 0 module to the bottom $\mathbb{Z}$, and let h_{-1} be the 0 map from the top $\mathbb{Z}$ to the bottom X_0. Then $\partial_n' h_n + h_{n-1} \partial_{n-1}' = f_n$ is satisfied for $n = -1$ because the map f_{-1} is the 0 map from $\mathbb{Z}$ to itself. Proposition 3.27 then allows us to construct inductively first h_0, then h_1, then h_2, and so on. □

Corollary 3.29. In the category $\mathcal{C}_{\mathbb{Z}G}$, if a free resolution $X = \{X_n\}$ of $\mathbb{Z}$ and a chain map $f = \{f_n\}$ of X with itself are given such that the map from $\mathbb{Z}$ to itself is the identity 1, then the chain map f is homotopic to the identity chain map $g = \{g_n\}$ with $g_n = 1$ for all n.

PROOF. Apply Corollary 3.28 to $f - 1$. □

Corollary 3.30. In the category $\mathcal{C}_{\mathbb{Z}G}$, if two free resolutions $X = \{X_n\}$ of $\mathbb{Z}$ and $Y = \{Y_n\}$ of $\mathbb{Z}$ are given and if two chain maps $f : X \to Y$ and $g : Y \to X$ are given such that the map from $\mathbb{Z}$ to itself in each case is the identity 1, then gf is homotopic to 1 and fg is homotopic to 1.

PROOF. Apply Corollary 3.29 to fg and then to gf. □

Theorem 3.31. If
$$\cdots \xrightarrow{\partial'_{n+1}} F'_{n+1} \xrightarrow{\partial'_n} F'_n \xrightarrow{\partial'_{n-1}} \cdots \xrightarrow{\partial'_0} F'_0 \xrightarrow{\varepsilon} \mathbb{Z} \longrightarrow 0$$
is any free resolution of $\mathbb{Z}$ in the category $\mathcal{C}_{\mathbb{Z}G}$ and M is a unital left $\mathbb{Z}G$ module, then $H^n(G, M)$ is canonically isomorphic to the nth cohomology group of the complex in $\mathcal{C}_{\mathbb{Z}}$ given by
$$0 \longrightarrow \operatorname{Hom}_{\mathbb{Z}G}(F'_0, M) \xrightarrow{d_0} \cdots \operatorname{Hom}_{\mathbb{Z}G}(F'_n, M) \xrightarrow{d_n} \operatorname{Hom}_{\mathbb{Z}G}(F'_{n+1}, M) \xrightarrow{d_{n+1}}$$
with $d_n = \operatorname{Hom}(\partial'_n, 1)$ for $n \geq 0$.

PROOF. Let the resolution in the statement of the theorem be Y, and let X be the standard free resolution of $\mathbb{Z}$ in the category $\mathcal{C}_{\mathbb{Z}G}$. Two applications of Corollary 3.26 produce chain maps $f : X \to Y$ and $g : Y \to X$ over $1 : \mathbb{Z} \to \mathbb{Z}$. Corollary 3.30 shows that gf is homotopic to $1 = 1_X$ and fg is homotopic to $1 = 1_Y$. Apply the functor $\operatorname{Hom}_{\mathbb{Z}G}(\,\cdot\,, M)$ throughout, including to the members of the homotopies. Then we obtain chain maps

$$\operatorname{Hom}_{\mathbb{Z}G}(f, 1) : \operatorname{Hom}_{\mathbb{Z}G}(Y, M) \to \operatorname{Hom}_{\mathbb{Z}G}(X, M)$$

and

$$\operatorname{Hom}_{\mathbb{Z}G}(g, 1) : \operatorname{Hom}_{\mathbb{Z}G}(X, M) \to \operatorname{Hom}_{\mathbb{Z}G}(Y, M)$$

with

$\operatorname{Hom}_{\mathbb{Z}G}(f, 1) \circ \operatorname{Hom}_{\mathbb{Z}G}(g, 1)$ homotopic to 1

and

$\operatorname{Hom}_{\mathbb{Z}G}(g, 1) \circ \operatorname{Hom}_{\mathbb{Z}G}(f, 1)$ homotopic to 1.

Proposition 3.24 allows us to conclude that

$\operatorname{Hom}_{\mathbb{Z}G}(f, 1) \circ \operatorname{Hom}_{\mathbb{Z}G}(g, 1)$ induces the identity on $H^*(\operatorname{Hom}_{\mathbb{Z}G}(X, M))$

and

$\operatorname{Hom}_{\mathbb{Z}G}(g, 1) \circ \operatorname{Hom}_{\mathbb{Z}G}(f, 1)$ induces the identity on $H^*(\operatorname{Hom}_{\mathbb{Z}G}(Y, M))$.

Thus $\operatorname{Hom}_{\mathbb{Z}G}(g, 1)$ induces an isomorphism of each group $H^n(\operatorname{Hom}_{\mathbb{Z}G}(X, M))$ onto $H^n(\operatorname{Hom}_{\mathbb{Z}G}(Y, M))$. □

6. Relative Brauer Group when the Galois Group Is Cyclic

This section has two parts to it. The first part specializes Theorem 3.31 to compute group cohomology when the group in question is cyclic of finite order. The second part applies this computation to $H^2(\text{Gal}(K/F), K^\times)$ and obtains information about Brauer groups. As a consequence we obtain new information about the classification of noncommutative division algebras.

Let G be a finite cyclic group of order n. Theorem 3.31 says that if G acts by automorphisms on an abelian group M, then $H^2(G, M)$ can be computed from any free resolution of $\mathbb{Z}$ in the category $\mathcal{C}_{\mathbb{Z}G}$. The standard resolution of $\mathbb{Z}$ is one such resolution. We shall construct another such resolution that is special to the case of G cyclic and that makes the cohomology more transparent.

Let $G = \{1, s, s^2, \ldots, s^{n-1}\}$. Lemma 3.19 notes that the free abelian group on the 1-tuples $(1), (s), (s^2), \ldots, (s^{n-1})$ is a free $\mathbb{Z}G$ module with $\mathbb{Z}G$ basis (1). In other words, the elements of the left $\mathbb{Z}G$ module $\mathbb{Z}G$ may be identified with the integer linear combinations of these 1-tuples. Define two operators T and N from the left $\mathbb{Z}G$ module $\mathbb{Z}G$ into itself by

$$T = \text{multiplication by } (s) - (1),$$
$$N = \text{multiplication by } (1) + (s) + \cdots + (s^{n-1}).$$

Each of these respects addition and commutes with multiplication by (s), hence is a $\mathbb{Z}G$ module homomorphism. We shall compute the kernel and image of each.

The kernel of T consists of all elements for which left multiplication by (s) fixes the element. The elements of $\mathbb{Z}G$ are of the form $\sum_{j=0}^{n-1} c_j(s^j)$, and (s) times this gives $c_{n-1}(1) + \sum_{j=1}^{n-1} c_{j-1}(s^j)$. Since $(1), (s), \ldots, (s^{n-1})$ form a $\mathbb{Z}$ basis, the condition to be in the kernel of T is that $c_{n-1} = c_0 = c_1 = \cdots = c_{n-2}$. Thus

$$\ker T = \{c((1) + (s) + \cdots + (s^{n-1})) \mid c \in \mathbb{Z}\}.$$

Also,

$$\text{image } T = \{\text{integer polynomials in } (s) \text{ divisible by } (s) - (1)\}$$
$$= \{\text{integer polynomials equal to 0 when } s \text{ is set equal to 1}\}$$
$$= \left\{ \sum_{j=0}^{n-1} c_j(s^j) \,\Big|\, \sum_{j=0}^{n-1} c_j = 0 \right\}.$$

In the case of the operator N, we have $N(s^j) = (1) + (s) + \cdots + (s^{n-1})$, and therefore $N\left(\sum_j c_j(s^j)\right) = \sum_j c_j((1) + (s) + \cdots + (s^{n-1}))$. Hence

$$\ker N = \left\{ \sum_{j=0}^{n-1} c_j(s^j) \,\Big|\, \sum_{j=0}^{n-1} c_j = 0 \right\} = \text{image } T,$$
$$\text{image } N = \{c((1) + (s) + \cdots + (s^{n-1})) \mid c \in \mathbb{Z}\} = \ker T.$$

An immediate consequence of this and a supplementary argument concerning the augmentation map is the following proposition.

Proposition 3.32. If G is a finite cyclic group, then the sequence

$$\cdots \xrightarrow{T} \mathbb{Z}G \xrightarrow{N} \mathbb{Z}G \xrightarrow{T} \cdots \xrightarrow{T} \mathbb{Z}G \xrightarrow{N} \mathbb{Z}G \xrightarrow{T} \mathbb{Z}G \xrightarrow{\varepsilon} \mathbb{Z} \longrightarrow 0$$

is a free resolution of $\mathbb{Z}$ in the category $C_{\mathbb{Z}G}$.

PROOF. We still need to check exactness at the first $\mathbb{Z}G$ from the right. The map ε is the $\mathbb{Z}G$ homomorphism with $\varepsilon((1)) = 1$. Hence $\varepsilon((s^j)) = 1$ for all j, and $\varepsilon\left(\sum_{j=0}^{n-1} c_j(s^j)\right) = \sum_{j=0}^{n-1} c_j$. Thus $\ker \varepsilon = \ker N = \operatorname{image} T$, and exactness is proved. □

Corollary 3.33. If G is a finite cyclic group and M is an abelian group on which G acts by automorphisms, then

$$H^2(G, M) \cong M^G / \left((1) + (s) + \cdots + (s^{n-1})\right)M,$$

where M^G is the subgroup of all elements of M fixed by G.

PROOF. Let us number the terms $\mathbb{Z}G$ in the resolution of Proposition 3.32 starting with index 0 from the right. Combining Proposition 3.32 with Theorem 3.31, we see that we may compute $H^2(G, M)$ as the cohomology of the complex obtained by applying the functor $\operatorname{Hom}_{\mathbb{Z}G}(\,\cdot\,, M)$ to the terms with indices 1, 2, 3 in the resolution in Proposition 3.32. Thus $H^2(G, M)$ is the cohomology at the middle of the complex

$$\operatorname{Hom}_{\mathbb{Z}G}(\mathbb{Z}G, M) \xrightarrow{(\cdot)\circ N} \operatorname{Hom}_{\mathbb{Z}G}(\mathbb{Z}G, M) \xrightarrow{(\cdot)\circ T} \operatorname{Hom}_{\mathbb{Z}G}(\mathbb{Z}G, M).$$

The mapping $\alpha \mapsto \alpha((1))$ of $\operatorname{Hom}_{\mathbb{Z}G}(\mathbb{Z}G, M)$ into M is one-one and onto, and we can identify members α of $\operatorname{Hom}_{\mathbb{Z}G}(\mathbb{Z}G, M)$ with the corresponding elements $\alpha((1))$ accordingly. If α is in $\ker\left((\cdot)\circ T\right)$, then $\alpha(T((1))) = 0$, and we thus have $\alpha((s)) = \alpha((1))$ and $(s)\alpha((1)) = \alpha((1))$. Hence $\alpha((1))$ is in M^G. These steps can be reversed, and thus $\ker\left((\cdot)\circ T\right) = M^G$. If β is in image $\left((\cdot)\circ N\right)$, then $\beta = \alpha \circ N$ for some $\alpha \in \operatorname{Hom}_{\mathbb{Z}G}(\mathbb{Z}G, M)$, and thus

$$\beta((1)) = \alpha\left((1) + (s) + \cdots + (s^{n-1})\right) = \alpha((1)) + (s)\alpha((1)) + \cdots + (s^{n-1})\alpha((1)).$$

Since $\alpha((1))$ is a completely arbitrary element of M, we see that image $\left((\cdot)\circ N\right) = \left((1) + (s) + \cdots + (s^{n-1})\right)M$, and the result follows. □

Now we specialize to the Galois case that has occupied our attention in this chapter. Let K/F be a finite Galois extension of fields. We are going to set $G = \text{Gal}(K/F)$, $n = \dim_F K$, and $M = K^\times$. To take advantage of Corollary 3.33, we suppose that $\text{Gal}(K/F)$ is cyclic. Then $M^G = (K^\times)^G = F^\times$. If x is an element of $K^\times$, then the orbit Gx is $\{x, sx, s^2x, \ldots, s^{n-1}x\}$. Remembering that we are using additive notation in working with cohomology of groups and multiplicative notation in working with $K^\times$, we see that the element $\big((1) + (s) + \cdots + (s^{n-1})\big)$ of $\mathbb{Z}G$ is to be regarded as operating by giving the product of the members of an orbit in $K^\times$. This product for the orbit of $x \in K^\times$ is $N_{K/F}(x)$, and Corollary 3.33 thus specializes to the following result.

Corollary 3.34. If K/F is a finite Galois extension of fields such that $\text{Gal}(K/F)$ is cyclic, then

$$H^2(\text{Gal}(K/F), K^\times) \cong F^\times / N_{K/F}(K^\times).$$

Corollary 3.34 considerably simplifies the proofs of Frobenius's Theorem about division algebras over the reals (Theorem 2.50) and Wedderburn's Theorem about finite division rings (Theorem 2.48), and thus the theory in Chapter III has added something to the theory of Chapter II even in these very special situations. In the case of the Frobenius theorem, the only nontrivial algebraic extension of $\mathbb{R}$ is $\mathbb{C}$, and thus Theorem 3.14 and Corollary 3.34 give

$$\mathcal{B}(\mathbb{R}) = \mathcal{B}(\mathbb{C}/\mathbb{R}) \cong H^2(\text{Gal}(\mathbb{C}/\mathbb{R}), \mathbb{R}^\times)$$
$$\cong \mathbb{R}^\times / N_{\mathbb{C}/\mathbb{R}}(\mathbb{C}^\times) = \mathbb{R}^\times / (\mathbb{R}^\times)^+ \cong \mathbb{Z}/2\mathbb{Z}.$$

Hence the reals and the quaternions are the only finite-dimensional central simple division algebras over $\mathbb{R}$.

In the case of the Wedderburn theorem, suppose that a finite field K splits a central division algebra over a field F with q elements. Say that $|K| = q^n$. For finite fields the Galois groups are always cyclic, and thus $\text{Gal}(K/F)$ is cyclic of order n, generated by the map $x \mapsto x^q$. In view of Corollary 3.34, the Wedderburn theorem follows if $F^\times/N_{K/F}(K^\times)$ is shown to be trivial, i.e., if the norm map $N_{K/F} : K^\times \to F^\times$ is onto. The group $K^\times$ is cyclic, say with a generator x_0 of order $q^n - 1$. Since the norm of an element is the product of the images under the Galois group, the norm of x_0 is given by

$$N_{K/F}(x_0) = x_0 x_0^q x_0^{q^2} \cdots x_0^{q^{n-1}} = x_0^{1+q+\cdots+q^{n-1}} = x_0^{\frac{q^n-1}{q-1}}.$$

This has order $q - 1$, not less, and thus is a generator of $F^\times$. Thus the norm map is onto $F^\times$.

6. Relative Brauer Group when the Galois Group Is Cyclic

For a more difficult example that we can settle completely, consider the case that $F = \mathbb{Q}$ and $K = \mathbb{Q}(\sqrt{m})$ for a square-free integer m other than 1. The Galois group in this case is a 2-element group and is in particular cyclic. Thus Corollary 3.34 applies. The norm of the member $x + y\sqrt{m}$ of K, where x and y are in $\mathbb{Q}$, is $x^2 - my^2$. The problem of determining the quotient group $F^\times / N_{K/\mathbb{Q}}(K^\times)$ may be rephrased in terms of genera as in Section I.5. Specifically the field discriminant D is defined to be m if $m \equiv 1 \bmod 4$ and to be $4m$ if $m \not\equiv 1 \bmod 4$. A genus for $\mathbb{Q}(\sqrt{m})$ is an equivalence class of primitive quadratic forms $ax^2 + bxy + cy^2$ whose discriminant matches the field discriminant D, except that the theory of Chapter I discards all negative definite forms. Equivalence is determined by the action of $\mathrm{SL}(2, \mathbb{Q})$. Lemma 1.13 shows for $D > 0$ that each nonzero rational number is a value taken on by the members of one and only one genus at points $(x, y) \neq (0, 0)$ with x and y both rational; for $D < 0$, Lemma 1.13 applies to positive definite forms and positive rational numbers. Let us now enlarge the definition of genera to include negative definite forms and negative rational numbers when $D < 0$.

The definition of the multiplication of classes of forms is set up so as to be compatible with multiplication of the values of the quadratic forms, and the genera define a group, the identity element being the principal genus. Since a representative of the principal genus is $x^2 - my^2$, the nonzero rational values corresponding to the principal genus are exactly the members of the group $N_{K/\mathbb{Q}}(K^\times)$. Consequently the quotient group $F^\times / N_{K/\mathbb{Q}}(K^\times)$ is isomorphic to the group of genera.[4] The easy result concerning the group of genera is Theorem 1.14, which says that this group is finite abelian and that every nontrivial element has order 2; since $\mathcal{B}(K/F) \cong F^\times / N_{K/\mathbb{Q}}(K^\times)$, Corollary 3.15 gives another way of seeing that every nontrivial element has order 2. The hard result, which appears in Problems 25–29 at the end of Chapter I, identifies the order of the group of genera explicitly.[5] If $D > 0$, then the order of the group of genera is $2^{g'}$, where $g' + 1$ is the number of distinct prime divisors of D; if $D < 0$, then the order of the group of genera is $2^{g'+1}$.

Consequently if m has $g + 1$ distinct prime divisors, then the relative Brauer group is a product of 2-element groups whose order is given by

$$|\mathcal{B}(\mathbb{Q}(\sqrt{m})/\mathbb{Q})| = \begin{cases} 2^g & \text{if } m > 0 \text{ and } m \not\equiv 3 \bmod 4, \\ 2^{g+1} & \text{if } m > 0 \text{ and } m \equiv 3 \bmod 4, \\ 2^{g+1} & \text{if } m < 0 \text{ and } m \not\equiv 3 \bmod 4, \\ 2^{g+2} & \text{if } m < 0 \text{ and } m \equiv 3 \bmod 4. \end{cases}$$

---

[4]With the understanding that genera from negative definite forms are to be allowed if $D < 0$.
[5]In quoting this result, we are now making allowances for genera corresponding to negative definite forms.

The example with $K/\mathbb{Q}$ quadratic shows the kind of information that has to go into a complete determination of the relative Brauer group when $K/\mathbb{Q}$ is Galois. Showing that a relative Brauer group is nontrivial in a case with $\mathrm{Gal}(K/\mathbb{Q})$ cyclic is considerably easier. According to Corollary 3.34, all one needs to know is that the norm function does not carry $K^\times$ onto $\mathbb{Q}^\times$, and congruence conditions can be used as a first step in addressing this question; Problem 4 at the end of the chapter illustrates this principle. Problems 15–17 at the end of Chapter II give a construction in this situation of nontrivial central simple algebras over $\mathbb{Q}$ that are split by K, and such algebras whose dimension is the square of a prime are necessarily division algebras. Problems 6–12 at the end of the present chapter give a sufficient condition for obtaining a division algebra when the dimension is not the square of a prime.

7. Problems

1. Let A be a finite-dimensional central simple algebra over a field F, let K be a subfield of A, and let B be the centralizer of K in A.
   (a) Arguing as in the proof of Theorem 3.3, exhibit a one-one algebra homomorphism $A \otimes_F K \to \mathrm{End}_{B^o} A$.
   (b) Referring to the proof of Theorem 2.2 and counting dimensions with the aid of the Double Centralizer Theorem, prove that the mapping in (a) is onto $\mathrm{End}_{B^o} A$.
   (c) Deduce that $A \otimes_F K$ and B yield the same member of $\mathcal{B}(K)$.

2. Let $a = a(\sigma, \tau)$ be a 2-cocycle in $Z^2(\mathrm{Gal}(K/F), K^\times)$, where K/F is a finite Galois extension of fields. Prove for each τ that $\prod_{\sigma \in \mathrm{Gal}(K/F)} a(\sigma, \tau)$ lies in $F^\times$.

3. Let K/F be a finite Galois extension of fields with $\mathrm{Gal}(K/F)$ cyclic. Corollary 3.34 identifies $H^q(\mathrm{Gal}(K/F), K^\times)$ for $q = 2$. Identify this group for all other values of $q \geq 0$.

Problems 4–5 amplify the discussion of cyclic algebras that was begun in Problems 17–19 at the end of Chapter II. Problem 4 in effect produces an explicit division algebra of dimension 9 over $\mathbb{Q}$, and Problem 5 hints at the existence of an explicit division algebra of dimension n^2 over $\mathbb{Q}$ for each integer $n \geq 1$.

4. Let $\zeta = e^{2\pi i/7}$, and let $K = \mathbb{Q}(\zeta) \cap \mathbb{R}$.
   (a) Show that $K/\mathbb{Q}$ is a Galois extension of degree 3, that a basis for K over $\mathbb{Q}$ consists of $\tau_1 = \zeta + \zeta^{-1}$, $\tau_2 = \zeta^2 + \zeta^{-2}$, $\tau_3 = \zeta^3 + \zeta^{-3}$, and that the Galois group permutes τ_1, τ_2, τ_3 cyclically.
   (b) Show that if a, b, c are in $\mathbb{Q}$, then

$$N_{K/\mathbb{Q}}(a\tau_1 + b\tau_2 + c\tau_3) = abc(\tau_1^3 + \tau_2^3 + \tau_3^3)$$
$$+ (a^3 + b^3 + c^3 + 3abc)\tau_1\tau_2\tau_3$$
$$+ (a^2b + ac^2 + b^2c)(\tau_1^2\tau_2 + \tau_2^2\tau_3 + \tau_3^2\tau_1)$$
$$+ (a^2c + ab^2 + bc^2)(\tau_1\tau_2^2 + \tau_2\tau_3^2 + \tau_3\tau_1^2).$$

(c) Verify the following identities:
$$\tau_1 + \tau_2 + \tau_3 = -1,$$
$$\tau_1\tau_2 = \tau_1 + \tau_3, \quad \tau_1\tau_3 = \tau_2 + \tau_3, \quad \tau_2\tau_3 = \tau_1 + \tau_2,$$
$$\tau_1^2 = \tau_2 + 2, \quad \tau_2^2 = \tau_3 + 2, \quad \tau_3^2 = \tau_1 + 2.$$

(d) Combine (b) and (c) to show that
$$N_{K/\mathbb{Q}}(a\tau_1 + b\tau_2 + c\tau_3) = (a^3 + b^3 + c^3) - abc$$
$$+ 3(a^2b + ac^2 + b^2c) - 4(a^2c + ab^2 + bc^2).$$

(e) Under the assumption that a, b, c are integers with $\gcd(a, b, c) = 1$, show that $N_{K/\mathbb{Q}}(a\tau_1 + b\tau_2 + c\tau_3) \not\equiv 0 \bmod 3$.

(f) Deduce from (e) that $r = 3$ is not in $N_{K/\mathbb{Q}}(K^\times)$. (Educational note: Consequently Problems 18–19 at the end of Chapter II produce an explicit division algebra over $\mathbb{Q}$ of dimension 9.)

5. (a) Show for each integer $n \geq 1$ that there exists a prime p such that n divides $p - 1$.

(b) Deduce for this p that there exists a field L with $\mathbb{Q} \subseteq L \subseteq \mathbb{Q}(e^{2\pi i/p})$ such that the field extension $L/\mathbb{Q}$ is a Galois extension whose Galois group is cyclic of order n.

Problems 6–12 continue the discussion of cyclic algebras that was begun in Problems 17–19 at the end of Chapter II and continued in Problems 4–5 above. Let F be any field, and let K be a finite Galois extension of F whose Galois group $G = \text{Gal}(K/F)$ is cyclic of order n. Let σ be a generator of G, fix an element $r \neq 0$ in F, and let A be the subset of matrices in $M_n(K)$ of the form

$$\begin{pmatrix} c_1 & c_2 & c_3 & \cdots & c_n \\ r\sigma(c_n) & \sigma(c_1) & \sigma(c_2) & \cdots & \sigma(c_{n-1}) \\ r\sigma^2(c_{n-1}) & r\sigma^2(c_n) & \sigma^2(c_1) & \cdots & \sigma^2(c_{n-2}) \\ \vdots & \vdots & \vdots & \ddots & \vdots \\ r\sigma^{n-1}(c_2) & r\sigma^{n-1}(c_3) & r\sigma^{n-1}(c_4) & \cdots & \sigma^{n-1}(c_1) \end{pmatrix}.$$

Identify $c \in K$ with the diagonal member of A for which $c_1 = c$ and $c_2 = \cdots = c_n = 0$, and let j be the member of A for which $c_1 = 0$, $c_2 = 1$, and $c_3 = \cdots = c_n = 0$. Under this identification every member of A has a unique expansion as $\sum_{k=1}^n c_k j^{k-1}$ with all c_k in K, and the element j satisfies $j^n = r$ and $jcj^{-1} = \sigma(c)$ for $c \in K$. Take it as known that A is a central simple algebra over F of dimension n^2. This series of problems leads in part to another theorem due to Wedderburn. (However, a more direct proof of the theorem of Wedderburn without the other results is possible.)

6. In the construction of factor sets in Section 2, use $x_{\sigma^k} = j^k$ for $0 \leq k \leq n-1$. Show that the algebra A above corresponds to the 2-cocycle a with

$$a(\sigma^k, \sigma^l) = \begin{cases} 1 & \text{if } k+l < n, \\ r & \text{if } k+l \geq n. \end{cases}$$

7. Under the assumption that $r = N_{K/F}(x)$ with $x \in K^\times$, show that the choice $c_{\sigma^k} = x\sigma(x)\sigma^2(x)\cdots\sigma^{k-1}(x)$ exhibits the factor set of the previous problem as a trivial factor set and hence shows that $A \cong M_n(F)$.

8. Let $F = \{F_k\}$ be the standard free resolution of $\mathbb{Z}$ in $C_{\mathbb{Z}G}$, and let $X = \{X_k\}$ be the free resolution of Proposition 3.32. The latter has $X_k = \mathbb{Z}G$ for every $k \geq 0$. Trace through the proof of Corollary 3.26, and show that the proof allows a chain map $f = \{f_k\}$ to be defined in such a way that the values of f_0, f_1, f_2 on standard $\mathbb{Z}G$ basis elements of F_0, F_1, F_2 are $f_0(1) = 1$, $f_1(1, \sigma^k) = -(1 + \sigma + \cdots + \sigma^{k-1})$ for $0 \leq k < n$, and

$$f_2(1, \sigma^k, \sigma^l) = \begin{cases} 0 & \text{if } 0 \leq k \leq l < n, \\ -\sigma^l & \text{if } 0 \leq l < k < n. \end{cases}$$

9. Let $\Phi_2 : \operatorname{Hom}_{\mathbb{Z}G}(F_2, K^\times) \to C^2(G, K^\times)$ be the isomorphism of Section 5, and let ψ be in $\operatorname{Hom}_{\mathbb{Z}G}(\mathbb{Z}G, K^\times)$. Show that the member of $C^2(G, K^\times)$ that corresponds to ψ is $\Phi_2(\psi \circ f_2)$ and that

$$\Phi_2(\psi \circ f_2)(\sigma^k, \sigma^l) = \begin{cases} \psi(0) & \text{if } k+l < n, \\ \psi(\sigma^{k+l-n})^{-1} & \text{if } k+l \geq n. \end{cases}$$

10. Let y be a member of $K^\times$, and let ψ be the unique element of $\operatorname{Hom}_{\mathbb{Z}G}(\mathbb{Z}G, K^\times)$ with $\psi(1) = y$. Why in the context of Proposition 3.32 is ψ a 2-cocycle if and only if y is in $F^\times$?

11. Take ψ as in the previous problem with $\psi(1) = r^{-1}$, and show that the member of $C^2(G, K^\times)$ that corresponds to it under Problem 9 is the factor set a of Problem 6.

12. Deduce from the previous problem that the order of the Brauer equivalence class in $\mathcal{B}(K/F)$ is the order of the coset of r in $F^\times / N_{K/F}(K^\times)$. Why does it follow that A is a division algebra over F if the coset of r in $F^\times / N_{K/F}(K^\times)$ has exact order n? (Educational note: This result is a theorem of Wedderburn except that it is here dressed in more modern language. The special case that n is prime was already handled by Problems 18–19 at the end of Chapter II. Although the converse was seen in those problems to be valid for n prime, the converse is known to fail for $n = 4$.)

Problems 13–20 introduce the reduced norm of a central simple algebra and give an application. Let A be a central simple algebra over a field F with $\dim_F A = n^2$. For a in A, the **algebra polynomial** of a is defined to be the characteristic polynomial

$\det(X1 - A)$ of the F linear mapping $L(a) : A \to A$ given by the left multiplication $x \mapsto ax$. This monic polynomial lies in $F[X]$ and has degree n^2. The ordinary **norm** $N_{A/F}(a)$ is defined to be $(-1)^{n^2}$ times the constant term, and the ordinary **trace** $\text{Tr}_{A/F}(a)$ is defined to be minus the coefficient of X^{n^2-1}; these functions of a take values in F. Choose a finite Galois extension K of F that splits A, and fix an isomorphism $\varphi : A \otimes_F K \to M_n(K)$. The **reduced polynomial** of a is defined to be the monic polynomial $\det(\varphi(X1 - a \otimes 1))$. This polynomial lies in $K[X]$ and has degree n. The **reduced norm** $\text{Nrd}_{A/F}(a)$ is defined to be $(-1)^n$ times the constant term, and the **reduced trace** $\text{Trrd}_{A/F}(a)$ is defined to be minus the coefficient of X^{n-1}; these functions of a initially take values in K.

13. Prove that the reduced polynomial of a does not depend on the choice of the isomorphism φ.

14. Prove that $\det(X1 - a) = \det(\varphi(X1 - a \otimes 1))^n$.

15. Using Galois theory and unique factorization, prove that any monic polynomial $P(X)$ in $K[X]$ such that $P(X)^n$ lies in $F[X]$ already lies in $F[X]$. Conclude that the reduced polynomial of any element of A is in $F[X]$.

16. Prove that $\det(\varphi(X1 - a \otimes 1))$ does not depend on the choice of the Galois extension K of F that splits A.

17. Deduce that $\text{Nrd}_{A/F}$ is a function from A to F such that $\text{Nrd}_{A/F}(ab) = \text{Nrd}_{A/F}(a)\text{Nrd}_{A/F}(b)$ for all a and b in A, $\text{Nrd}_{A/F}(1) = 1$, and $\text{Nrd}_{A/F}(a)^n = N_{A/F}(a)$ for all a in A. How does it follow that
    (a) an element $a \in A$ is invertible if and only if $\text{Nrd}_{A/F}(a) \neq 0$ and
    (b) A is a division algebra if and only if $\text{Nrd}_{A/F}(a) = 0$ only for $a = 0$?

18. Let K/F be a finite Galois extension of fields, put $G = \text{Gal}(K/F)$, and suppose that a crossed-product algebra $A = \mathcal{A}(K, G, a)$ is given as in Proposition 3.12 with $K \subseteq A$ and with $\dim_F A = (\dim_F K)^2 = n^2$. Let $\{x_\sigma \mid \sigma \in G\}$ be the system in the proposition such that $A = \bigoplus_{\sigma \in G} Kx_\sigma$. Associate a matrix $m(v)$ in $M_n(K)$ to each $v \in A$ as follows. The rows and columns of the matrices are indexed by G, and $E_{\sigma,\tau}$ denotes the matrix that is 1 in the (σ, τ) entry and is 0 elsewhere. Let $m(cx_\tau) = \sum_\sigma \sigma(c)a(\sigma, \tau)E_{\sigma,\sigma\tau}$ for $c \in K$, and extend additively to handle all $v \in A$. Check that $v \mapsto m(v)$ is a one-one F algebra homomorphism of A into $M_n(K)$, and prove that $\text{Nrd}_{A/F}(v) = \det m(v)$. (Educational note: Thus by Proposition 3.12 the matrix algebra in Problems 6–12 is central simple.)

19. Identify the norm and the reduced norm for the real algebra $\mathbb{H}$ of quaternions.

20. A field F is said to satisfy **condition (C1)** if every homogeneous polynomial of degree d in n variables with $d < n$ has a nontrivial zero. Using the reduced norm for a central division algebra over F, prove that condition (C1) implies that $\mathcal{B}(F) = 0$. (Educational note: Algebraically closed fields and finite fields satisfy (C1), the latter by a theorem of Chevalley. A deeper fact is that a simple transcendental extension of an algebraically closed field satisfies (C1).)

CHAPTER IV

Homological Algebra

Abstract. This chapter develops the rudiments of the subject of homological algebra, which is an abstraction of various ideas concerning manipulations with homology and cohomology. Sections 1–7 work in the context of good categories of modules for a ring, and Section 8 extends the discussion to abelian categories.

Section 1 gives a historical overview, defines the good categories and additive functors used in most of the chapter, and gives a more detailed outline than appears in this abstract.

Section 2 introduces some notions that recur throughout the chapter—complexes, chain maps, homotopies, induced maps on homology and cohomology, exact sequences, and additive functors. Additive functors that are exact or left exact or right exact play a special role in the theory.

Section 3 contains the first main theorem, saying that a short exact sequence of chain or cochain complexes leads to a long exact sequence in homology or cohomology. This theorem sees repeated use throughout the chapter. Its proof is based on the Snake Lemma, which associates a connecting homomorphism to a certain kind of diagram of modules and maps and which establishes the exactness of a certain 6-term sequence of modules and maps. The section concludes with proofs of the crucial fact that the Snake Lemma and the first main theorem are functorial.

Section 4 introduces projectives and injectives and proves the second main theorem, which concerns extensions of partial chain and cochain maps and also construction of homotopies for them when the complexes in question satisfy appropriate hypotheses concerning exactness and the presence of projectives or injectives. The notion of a resolution is defined in this section, and the section concludes with a discussion of split exact sequences.

Section 5 introduces derived functors, which are the basic mathematical tool that takes advantage of the theory of homological algebra. Derived functors of all integer orders ≥ 0 are defined for any left exact or right exact additive functor when enough projectives or injectives are present, and they generalize homology and cohomology functors in topology, group theory, and Lie algebra theory.

Section 6 implements the two theorems of Section 3 in the situation in which a left exact or right exact additive functor is applied to an exact sequence. The result is a long exact sequence of derived functor modules. It is proved that the passage from short exact sequences to long exact sequences of derived functor modules is functorial.

Section 7 studies the derived functors of Hom and tensor product in each variable. These are called Ext and Tor, and the theorem is that one obtains the same result by using the derived functor mechanism in the first variable as by using the derived functor mechanism in the second variable.

Section 8 discusses the generalization of the preceding sections to abelian categories, which are abstract categories satisfying some strong axioms about the structure of morphisms and the presence of kernels and cokernels. Some generalization is needed because the theory for good categories is insufficient for the theory for sheaves, which is an essential tool in the theory of several complex variables and in algebraic geometry. Two-thirds of the section concerns the foundations, which involve unfamiliar manipulations that need to be internalized. The remaining one-third introduces an artificial definition of "member" for each object and shows that familiar manipulations with members can be used to verify equality of morphisms, commutativity of square diagrams, and exactness of sequences of objects and morphisms. The consequence is that general results for categories of modules in homological algebra requiring such verifications can readily be translated into results for general abelian categories. The method with members, however, does not provide for constructions of morphisms member by member. Thus the construction of the connecting homomorphism in the Snake Lemma needs a new proof, and that is given in a concluding example.

1. Overview

This chapter develops the rudiments of the subject of homological algebra. The only prerequisite within the present volume is the self-contained Section III.5 entitled "Digression on Cohomology of Groups," which is helpful primarily as motivation. The definitions of category, functor, object, morphism, natural transformation, product, and coproduct as in Chapters IV and VI of *Basic Algebra* will be taken as known, and it will be helpful as motivation to know also the material from Chapter VII of *Basic Algebra* on group extensions and cohomology of groups. The present chapter will make some allusions to notions from algebraic topology, particularly in this first section, and the reader is encouraged to skip lightly over anything of this kind that might be an impediment to continuing with the remainder of the chapter.

Homology and cohomology have their origins in attempts to assign algebraic invariants to topological obstructions. One example historically was the holes in a domain of the Euclidean plane that can make line integrals that are locally independent of the path fail to be globally independent of the path. Another was the handles on 2-dimensional closed surfaces. These obstructions were originally viewed as numbers (Betti numbers for example) and later viewed as algebraic objects such as abelian groups or vector spaces. A big advance was to regard them not just as objects attached to geometric configurations but as functors that attach objects to geometric configurations and also attach functions between such objects to reflect the behavior of functions between geometric configurations.

Hints of connections with algebra on a deeper level and hints that homology and cohomology could be computed quite flexibly began with work of W. Hurewicz in 1936 and H. Hopf in 1942. Hurewicz considered the following situation: M is a finite connected simplicial complex, U is its universal cover, and G is the fundamental group of M. Suppose that U is contractible. The group G acts freely on the group $C_*(U)$ of simplicial chains of U (with integer coefficients). The boundary operator then gives us an exact sequence

$$0 \leftarrow \mathbb{Z} \leftarrow C_0(U) \leftarrow C_1(U) \leftarrow C_2(U) \leftarrow \cdots$$

of abelian groups with an action of G on each $C_j(U)$ by automorphisms in such a way that each $C_j(U)$ in effect is a free $\mathbb{Z}G$ module. Applying $(\,\cdot\,) \otimes_{\mathbb{Z}G} \mathbb{Z}$, we

obtain the complex
$$0 \leftarrow C_0(M) \leftarrow C_1(M) \leftarrow C_2(M) \leftarrow \cdots.$$
The homology $H_0(M)$ is just $\mathbb{Z}$ because M is connected, and $H_1(M)$ is just the quotient of G by its commutator subgroup; thus $H_0(M)$ and $H_1(M)$ depend only on G. What Hurewicz showed is that all higher $H_i(M)$ depend only on G; he did not address existence of such spaces M and U for G.

Hopf clarified the situation and drew attention to it by making an explicit calculation: Dropping all assumptions on U other than its simple connectivity, he gave a formula for the quotient of $H_2(M)$ modulo the subgroup of "spherical homology classes" in terms of G. Later he obtained a result for higher-degree homology. In effect, Hopf was giving formulas for $H_n(G, \mathbb{Z})$ by discovering and applying the homology analog of the cohomology result given as Theorem 3.31 in Section III.5.

Meanwhile, S. Eilenberg in 1944 made an adjustment to Lefschetz's singular homology theory and showed for locally finite polyhedra that his adjusted theory gives the same groups as the more traditional simplicial theory. His method was to introduce a third complex, to exhibit chain maps from this to each of the complexes under study, and show that the chain maps possess inverses in a suitable sense.

In addition to the people mentioned above, some others who pursued these matters in the mid 1940s were R. Baer, B. Eckmann, H. Freudenthal, and S. Mac Lane. One thing that mathematicians gradually realized was that homology and cohomology in various situations can be calculated from suitable kinds of abstract resolutions, a fact that lies at the heart of the subject of homological algebra. Another was that the subject of cohomology of groups made sense on an abstract level without any reference to topology and that the theory of factor sets for group extensions, as had been introduced by O. Schreier in the 1920s, was actually one aspect of this theory.

With a great leap of generality, H. Cartan and Eilenberg set down such a theory in their celebrated book *Homological Algebra*, whose publication was delayed until 1956. Homology and cohomology became things attached to complexes, no longer dependent on topology, and the book developed enormous machinery for working with such complexes and homology/cohomology. By the time that Cartan and Eilenberg had published their book, other special cases of homological algebra had already arisen. One was the cohomology theory of Lie algebras, developed by C. Chevalley in the 1940s and by J.-L. Koszul in 1950. Another was the cohomology theory of sheaves, used in the subject of several complex variables starting about 1950 by K. Oka and H. Cartan; sheaves themselves had been introduced in 1946 by J. Leray in connection with partial differential equations.

In the eventual theory the fundamental notion is that of a "derived functor": homology or cohomology is obtained by starting from some kind of resolution, or exact complex, passing to another complex by means of a functor with some special properties, and then extracting the homology or cohomology of the image complex. Two categories are thus involved, one for the resolution and one for the values of the functor. From an expository point of view, it seems wise to start with concrete categories and not to try to identify the most general categories for which the theory makes sense. For much of the chapter, we shall work with a category not much more general than the category $\mathcal{C}_R$ of all unital left R modules, where R is a ring with identity, and our functors will pass from one such category to another. Use of categories $\mathcal{C}_R$ subsumes the following applications:

(i) manipulations with basic homology and cohomology in topology, in which one begins with the ring $R = \mathbb{Z}$ of integers. For more advanced applications in topology, one moves from $\mathbb{Z}$ to more general rings.

(ii) homology and cohomology of groups, in which one initially uses group rings of the form $\mathbb{Z}G$, where G is any group and $\mathbb{Z}$ is the ring of integers.

(iii) homology and cohomology of Lie algebras. If $\mathfrak{g}$ is a Lie algebra over a field such as $\mathbb{C}$, then $\mathfrak{g}$ has a "universal enveloping algebra" $U(\mathfrak{g})$ and a canonical mapping $\iota : \mathfrak{g} \to U(\mathfrak{g})$. Here $U(\mathfrak{g})$ is a complex associative algebra with identity, ι is a Lie algebra homomorphism, and the pair $(U(\mathfrak{g}), \iota)$ has the following universal mapping property: whenever $\varphi : \mathfrak{g} \to A$ is a Lie algebra homomorphism into a complex associative algebra A with identity, then there is a unique homomorphism $\Phi : U(\mathfrak{g}) \to A$ of associative algebras with identity such that $\varphi = \Phi \circ \iota$. Lie algebra homology and cohomology are the theory for the set-up in which the initial underlying rings are $U(\mathfrak{g})$ and $\mathbb{C}$.

In other words, in each of the three applications above, many derived functors of importance pass from the category $\mathcal{C}_R$ for a ring R with identity to the category $\mathcal{C}_S$ for another ring S with identity.

The slight generalization of categories $\mathcal{C}_R$ that we shall use for much of the chapter is as follows: Let R be a ring with identity. A **good category** $\mathcal{C}$ of R modules consists of

(i) some nonempty class of unital left R modules closed under passage to submodules, quotients, and finite direct sums (the **modules** of the category),

(ii) the full sets $\text{Hom}_R(A, B)$ of all R linear homomorphisms from A to B for each A and B as in (i) (the **morphisms**, or **maps**, of the category).

For example the collection of all finitely generated abelian groups, as a subcategory of $\mathcal{C}_{\mathbb{Z}}$, is a good category.[1] So is the collection of all **torsion abelian groups**,

---

[1] One reason for working with this slight generalization is to emphasize that a certain property of categories $\mathcal{C}_R$, namely that they have "enough projectives" and "enough injectives" in a sense to be made precise below in Section 5, does not necessarily persist for slight variants of $\mathcal{C}_R$.

i.e., abelian groups whose elements all have finite order, as a subcategory of $\mathcal{C}_\mathbb{Z}$.

The definition of "good category" specifies *left* R modules that are unital. However, the theory applies equally well to right R modules that are unital, since a unital right R module becomes a unital left module for the opposite ring R^o, i.e., the ring whose underlying abelian group is the same as for R and whose multiplication is given by $a \circ b = ba$.

The special property of a functor $F : \mathcal{C} \to \mathcal{C}'$ used for passing from a complex in one good category to a complex in another good category is that it is **additive**, namely that $F(\varphi_1 + \varphi_2) = F(\varphi_1) + F(\varphi_2)$ whenever φ_1 and φ_2 are in the same $\text{Hom}_R(A, B)$. The initial examples of additive functors are tensor product $M \otimes_R (\cdot)$, which passes from $\mathcal{C}_R$ to $\mathcal{C}_\mathbb{Z}$ if M is a right R module, and Hom in each variable: $\text{Hom}_R(\cdot, M)$ and $\text{Hom}_R(M, \cdot)$, both of which pass from $\mathcal{C}_R$ to $\mathcal{C}_\mathbb{Z}$ if M is a left R module. In Section 2 we shall consider additive functors in more detail.

The set-up with good categories does not subsume the cohomology of sheaves, nor some other applications of interest, such as the cohomology of vector bundles with a fixed base. The cohomology of sheaves is an important tool in algebraic geometry and several complex variables, and it cannot be ignored. Consequently one ultimately wants the theory to extend to other categories than good categories of modules. In addition, it is quite useful to have the theory work for the categories opposite to two given categories if it works for two given categories, and this feature means that the general theory should not insist that the objects be sets of elements and the morphisms be functions on such elements. Accordingly the abstract theory is carried out for "abelian categories," which will be defined in Section 8. The idea for creating the abstract theory is to take the theory for good categories of modules and rephrase all of the results for all abelian categories. In many instances the proofs will translate easily to the general setting, but in other instances it will be necessary to eliminate individual elements from arguments and obtain new arguments that rely only on complexes, exact sequences, and commutative diagrams. Some of this detail will be carried out in Section 8.

Sections 2–3 establish the framework of homology and cohomology in the context of good categories of modules. Section 2 discusses complexes and exact sequences at length, and Section 3 shows how a short exact sequence of complexes leads to a long exact sequence in homology or cohomology. This is the first main result of the theory and finds multiple uses later in the chapter.

Section 4 contains a discussion of "projectives and injectives" that expands and systematizes Theorem 3.31, which concerned the flexible role of resolutions in computing the cohomology of groups. Once that flexibility is in place in the more general setting of good categories, Sections 5–6 introduce derived functors and some of their properties. The main examples of derived functors at this stage are functors $\text{Ext}(\cdot, \cdot)$ and $\text{Tor}(\cdot, \cdot)$ obtained from Hom and tensor product; these are examined more closely in Section 7. The example given in Section III.5 and now being used as motivation requires some subtlety to be regarded as a derived functor. That example was the system of functors $H^n(G, \cdot)$ yielding cohomology of the group G with coefficients in the module $(\cdot)$; these were obtained in Section III.5 by applying the functor $\text{Hom}_{\mathbb{Z}G}(\cdot, M)$ to any free resolution of $\mathbb{Z}$ in the category $\mathcal{C}_{\mathbb{Z}G}$. It is seen in examples in Section 5 that the effect of using the free resolution was to compute $H^n(G, M)$ as $\text{Ext}^n_{\mathbb{Z}G}(\cdot, M)$ when the variable is set equal to $\mathbb{Z}$; realizing this result as a derived functor in the M variable requires knowing that one gets the same result from $\text{Ext}^n_{\mathbb{Z}G}(\mathbb{Z}, \cdot)$ when its variable is set equal to M. This conclusion is part of Theorem 4.31, which is proved in Section 7.

The first seven sections complete the treatment of the rudiments of homological algebra in the setting of good categories. One more central technique beyond that of derived functors is the mechanism of spectral sequences, but we shall omit this topic to save space.[2]

The chapter concludes with some discussion of abelian categories in Section 8. The foundations of homological algebra have to be redone completely when objects are no longer necessarily sets of elements. After this step, one introduces a substitute notion of "member" for elements, establishes its properties, and immediately obtains extensions of much of the theory to all abelian categories. A supplementary argument is needed whenever the theory for good categories uses an element-by-element construction of a homomorphism.

Sheaves are introduced in the last section of text in Chapter X, and their cohomology is mentioned very briefly there.

2. Complexes and Additive Functors

Let $\mathcal{C}$ be a good category of R modules in the sense of Section 1. A **complex** in $\mathcal{C}$ is a finite or infinite sequence of modules and maps in $\mathcal{C}$ such that the consecutive compositions are all 0. There is no harm in assuming that the indexing for the sequence is done by all of $\mathbb{Z}$, since we can always adjoin 0 modules and 0 maps as necessary to fill out the indexing. The indices may be increasing or decreasing, and, as we saw in Section III.5, this distinction is only a formality. However, the distinction is very convenient when it comes to applications, since homology is normally associated with decreasing indices and cohomology is normally associated with increasing indices.

Thus let us be more precise about the indexing. A **chain complex** in $\mathcal{C}$ is a sequence of pairs $X = \{(X_n, \partial_n)\}_{n=-\infty}^{\infty}$ in which each X_n is a module in $\mathcal{C}$,

---

[2]For the reader who is interested in learning about spectral sequences, this author is partial to the explanation of the topic in Appendix D of the book by Knapp and Vogan in the Selected References. The setting in that appendix is limited to good categories of modules, and some important applications are included.

each ∂_n is a map in $\operatorname{Hom}_R(X_{n+1}, X_n)$, and $\partial_n \partial_{n+1} = 0$ for all n. The maps ∂_n are sometimes called **boundary maps**, or **boundary operators**. We define the **homology** of X, written $H_*(X) = \{H_n(X)\}_{n=-\infty}^{\infty}$ with subscripts, to be the sequence of modules in $\mathcal{C}$ given by

$$H_n(X) = (\ker \partial_{n-1})/(\operatorname{image} \partial_n).$$

The members of the space $\ker \partial_{n-1}$ are called n-**cycles**, and the members of the space image ∂_n are called n-**boundaries**.

EXAMPLES OF CHAIN COMPLEXES.

(1) Simplicial homology. Let S be a simplicial complex of dimension N, and number its vertices. For each integer n, the group $C_n(S)$ of simplicial n-chains is the free abelian group on the set of simplices of dimension n. This is 0 for $n < 0$ and $n > N$. In elementary topology one defines the boundary of each n-simplex to be the member of $C_{n-1}(S)$ equal to an integer combination of its faces, the coefficient of the face being $(-1)^i$ if the missing vertex for the face is the ith of the $n+1$ vertices of the given n-simplex. This definition is extended additively to the boundary map $\partial_{n-1} : C_n(S) \to C_{n-1}(S)$, and a combinatorial argument gives $\partial_n \partial_{n-1} = 0$ for all n. Thus $X = \{(C_n(S), \partial_{n-1})\}$ is a complex. The associated homology $H_n(X)$ is the nth (integral simplicial) homology of the simplicial complex S and is usually denoted by $H_n(S)$.

(2) Cubical singular homology. Let S be a topological space. For $n \geq 0$, a **singular n-cube** in S is a continuous function $T : I^n \to S$, where I^n denotes the n-fold product of the closed interval $[0, 1]$ with itself. The free abelian group on the set of n-cubes is denoted by $Q_n(S)$. A singular n-cube T is **degenerate** if its values are independent of one of the n variables. The subgroup of $Q_n(S)$ generated by the degenerate singular n-cubes is denoted by $D_n(S)$, and the quotient $C_n(S) = Q_n(S)/D_n(S)$ is the group of **cubical singular n-chains**. One defines a boundary operator from $Q_n(S)$ to $Q_{n-1}(S)$ for each n in analogy with the definition in the previous example and shows that it carries $D_n(S)$ into $D_{n-1}(S)$. Consequently the boundary operator descends to a homomorphism of abelian groups $\partial_{n-1} : C_n(S) \to C_{n-1}(S)$. A combinatorial argument shows that $\partial_n \partial_{n-1} = 0$; thus we get a complex. The associated homology is the nth (integral singular) homology of S and is usually denoted by $H_n(S)$.

(3) Free resolution of $\mathbb{Z}$ in $\mathcal{C}_{\mathbb{Z}G}$. Let G be a group. Then the standard resolution of $\mathbb{Z}$ in the category $\mathcal{C}_{\mathbb{Z}G}$, as given in Theorem 3.20, is a chain complex in that category.

Let us make the class of chain complexes for the good category $\mathcal{C}$ into a category. Each chain complex is to be an object. If $X = \{(X_n, \partial_n)\}$ and $X' = \{(X'_n, \partial'_n)\}$

are two chain complexes in $\mathcal{C}$, a morphism in $\mathrm{Morph}(X, X')$ is any **chain map** $f = \{f_n\}$, defined as a sequence of maps $f_n \in \mathrm{Hom}_R(X_n, X'_n)$ such that the diagram

$$\begin{array}{ccc} X_n & \xrightarrow{\partial_{n-1}} & X_{n-1} \\ f_n \downarrow & & \downarrow f_{n-1} \\ X'_n & \xrightarrow{\partial'_{n-1}} & X'_{n-1} \end{array}$$

commutes for all n. Briefly $f\partial = \partial' f$. Since the f_n's are functions, it is customary to use function notation $f : X \to X'$ for chain maps. The system $\{1_{X_n}\}$ of identity maps serves as an identity morphism, and coordinate-by-coordinate composition is associative. Thus the result is a category.

The next step is to observe that homology H_*, as applied to chain maps for the category $\mathcal{C}$, is a covariant functor from the category of chain maps to itself. The effect of the functor on objects is to send X to $H_*(X) = \{(H_n(X), 0)\}$. If $f : X \to X'$ is a chain map, then the formula $\partial'_{n-1}(f_n(x_n)) = f_{n-1}(\partial_{n-1}(x_n))$ shows that $f_n(\ker \partial_{n-1}) \subseteq \ker \partial'_{n-1}$, and the formula $\partial'_n(f_{n+1}(x_{n+1})) = f_n(\partial_n(x_{n+1}))$ shows that $f_n(\mathrm{image}\, \partial_n) \subseteq \mathrm{image}\, \partial'_n$. Therefore f_n descends to the quotient, giving a map $H(f_n) : H_n(X) \to H_n(X')$. The assembled collection of maps $H_*(f) : H_*(X) \to H_*(X')$ is manifestly a chain map. Instead of writing $H(f_n)$ for the map induced by f_n on the nth homology, we shall often write $(f_n)_*$ or $\bar{f}_n$, especially in diagrams, to make the notation less cumbersome. Since the identity chain map yields the identity on $H_*(X)$ and since compositions go to compositions in the same order, homology H_* is a covariant functor.

If $f : X \to X'$ and $g : X \to X'$ are two chain maps, then a **homotopy** h of f to g is a system of maps $h = \{h_n\}$ increasing degrees by 1, i.e., having h_n carry X_n into X'_{n+1}, such that $h_{n-1}\partial_{n-1} + \partial'_n h_n = f_n - g_n$ for all n. Briefly $h\partial + \partial' h = f - g$. When such an h exists, we say that f and g are **homotopic**, and we write $f \simeq g$. This relation is an equivalence relation.

Proposition 4.1. If $f : X \to X'$ and $g : X \to X'$ are homotopic chain maps in the good category $\mathcal{C}$, then f and g induce the *same* maps $H_*(f)$ and $H_*(g)$ on homology, i.e., $H_n(f)$ and $H_n(g)$ are the same map of $H_n(X)$ into $H_n(X')$ for each n.

PROOF. Let h be a homotopy, and suppose that $\partial_{n-1}(x_n) = 0$. Then the computation $f_n(x_n) - g_n(x_n) = h_{n-1}\partial_{n-1}(x_n) + \partial'_n h_n(x_n) = 0 + \partial'_n h_n(x_n)$ shows that the images of x_n under f_n and g_n in X'_n differ by a member of image ∂'_n. □

Briefly let us translate all of these definitions and conclusions into statements when the complexes have increasing indices. A **cochain complex** in $\mathcal{C}$ is a sequence of pairs $X = \{(X_n, d_n)\}_{n=-\infty}^{\infty}$ in which each X_n is a module in $\mathcal{C}$, each d_n is a map in $\operatorname{Hom}_R(X_n, X_{n+1})$, and $d_{n+1}d_n = 0$ for all n. The maps d_n are sometimes called **coboundary maps**, or **coboundary operators**. We define the **cohomology** of X, written $H^*(X) = \{H^n(X)\}_{n=-\infty}^{\infty}$ with superscripts, to be the sequence of modules in $\mathcal{C}$ given by $H^n(X) = (\ker d_n)/(\operatorname{image} d_{n-1})$. The members of the space $\ker d_n$ are called n-**cocycles**, and the members of image d_{n-1} are called n-**coboundaries**.

EXAMPLES OF COCHAIN COMPLEXES.

(1) Singular cohomology. Let S be a topological space, let $X = \{(C_n(S), \partial_{n-1})\}$ be its complex of cubical singular n-chains, and let M be any abelian group. If $C^n(S, M) = \operatorname{Hom}_{\mathbb{Z}}(C_n(S), M)$ and if $d_n : C^n(S, M) \to C^{n+1}(S, M)$ is the map $d_n = \operatorname{Hom}(\partial_{n+1}, 1)$, then $Y = \{(C^n(S, M)), d_n)\}$ is a cochain complex, and its cohomology, written $H^*(Y) = \{H^n(S, M)\}$, is the (integral singular) cohomology of S with coefficients in M.

(2) Cohomology of groups. Let G be a group, and let M be an abelian group on which G acts by automorphisms. Let $C^n(G, M)$ be the abelian group of functions from the n-fold product of G with itself into M, the functions being added pointwise. Define $\delta_n : C^n(G, M) \to C^{n+1}(G, M)$ as in Section III.5. Then $X = \{(C^n(G, M), \delta_n)\}$ is a cochain complex, and its cohomology $H^*(X) = \{H^n(G, M)\}$ is the cohomology of G with coefficients in M.

The cochain complexes for the good category $\mathcal{C}$ form a category for which the morphisms from $X = \{(X_n, d_n)\}$ to $X' = \{(X'_n, d'_n)\}$ are **cochain maps** $f = \{f_n\}$; the latter are defined by the conditions that f_n carry X_n to X'_n and $fd = df$, i.e., $f_{n+1}d_n = d_n f_n$ for all n. Cohomology H^*, as applied to cochain maps for the category $\mathcal{C}$, is a covariant functor from the category of cochain maps to itself. The effect of the functor on objects is to send X to $H^*(X) = \{(H^n(X), 0)\}$, and the argument that a cochain map $f : X \to X'$ carries $H^*(X)$ to $H^*(X')$ via a cochain map $H^*(f)$ is the same as for chain maps. Instead of writing $H(f_n)$ for the map induced by f_n on the nth cohomology, we shall often write $(f_n)^*$ or $\bar{f}_n$, especially in diagrams, to make the notation less cumbersome.[3]

If $f : X \to X'$ and $g : X \to X'$ are two cochain maps, then a **homotopy** h of f to g is a system of maps $h = \{h_n\}$ decreasing degrees by 1, i.e., having h_n carry X_n into X'_{n-1}, such that $h_{n+1}d_n + d'_{n-1}h_n = f_n - g_n$ for all n. Briefly $hd + d'h = f - g$. When such an h exists, we say that f and g are **homotopic**, and we write $f \simeq g$. This relation is an equivalence relation.

---

[3]The notation with the bar is to be avoided when there might be some ambiguity about which of homology and cohomology is involved.

Proposition 4.1'. If $f : X \to X'$ and $g : X \to X'$ are homotopic cochain maps in the good category $\mathcal{C}$, then f and g induce the *same* maps $H^*(f)$ and $H^*(g)$ on cohomology, i.e., $H^n(f)$ and $H^n(g)$ are the same map of $H^n(X)$ into $H^n(X')$ for each n.

PROOF. Let h be a homotopy, and suppose that $d_n(x_n) = 0$. Then the computation $f_n(x_n) - g_n(x_n) = h_{n+1}d_n(x_n) + d'_{n-1}h_n(x_n) = 0 + d'_{n-1}h_n(x_n)$ shows that the images of x_n under f_n and g_n in X'_n differ by a member of image d'_{n-1}. □

A chain or cochain complex written neutrally as $X = \{X(n)\}$ is **exact** at $X(n)$ if the kernel of the outgoing map at $X(n)$ equals the image of the incoming map at $X(n)$ (as opposed to merely containing the image). The complex is **exact**, or is an **exact sequence**, if it is exact at every $X(n)$. A **short exact sequence** is an exact sequence of the form

$$0 \to A \xrightarrow{\varphi} B \xrightarrow{\psi} C \to 0,$$

understood to have 0's at all positions beyond each end. The conditions on the 5-term complex above for it to be exact are that φ be one-one, ψ be onto C, and that ψ exhibit C as isomorphic to $B/\,\text{image}\,\varphi$. To make the terminology more symmetric, it is customary to introduce a name for the quotient of the range of a homomorphism η by the image of η; this quotient is defined to be the **cokernel** of the homomorphism and is denoted by coker η. The conditions for exactness above can then be restated more symmetrically as

$$\ker \varphi = \operatorname{coker} \psi = 0 \quad \text{and} \quad \operatorname{image} \varphi = \ker \psi.$$

An exact sequence can always be broken into short exact sequences by stretching each link

$$\cdots \to A \xrightarrow{\varphi} B \xrightarrow{\psi} \cdots$$

into

$$\cdots \to A \xrightarrow{\varphi} \operatorname{image} \varphi \to 0 \to 0 \to \ker \psi \xrightarrow{\operatorname{inc}} B \xrightarrow{\psi} \cdots$$

and breaking it between the 0's; here "inc" denotes the inclusion mapping of ker ψ into B. This stretching process does not take us outside our good category, since good categories are assumed to be closed under passage to submodules and quotients. Conversely if we have two exact sequences

$$\cdots \to A \xrightarrow{\varphi} C \to 0 \quad \text{and} \quad 0 \to C \xrightarrow{i} B \xrightarrow{\psi} \cdots,$$

then we can combine them into an exact sequence

$$\cdots \to A \xrightarrow{i\varphi} B \xrightarrow{\psi} \cdots.$$

Exactness at A of the merged sequence follows because $\ker(i\varphi) = \ker\varphi$, and exactness at B follows because $\ker\psi = \operatorname{image} i = \operatorname{image}(i\varphi)$.

Any map $\varphi : A \to B$ in our good category can be expressed in terms of an exact sequence by including the kernel and cokernel:

$$0 \to \ker\varphi \xrightarrow{i} A \xrightarrow{\varphi} B \xrightarrow{q} \operatorname{coker}\varphi \to 0;$$

here $i : \ker\varphi \to A$ is the inclusion, and $q : B \to \operatorname{coker}\varphi$ is the quotient mapping. All the modules and maps in the exact sequence are in the category, since good categories are assumed to be closed under passage to submodules and quotients. We shall use the following special case of this observation in Section 3.

Proposition 4.2. Let $X = \{(X_n, \partial_n)\}_{n=-\infty}^{\infty}$ be a chain complex in a good category with ∂_n in $\operatorname{Hom}_R(X_{n+1}, X_n)$ for each n. Then the boundary operator ∂_{n-1} on X_n descends to the quotient as a mapping $\bar{\partial}_{n-1} : \operatorname{coker}\partial_n \to \ker\partial_{n-2}$ and yields an exact sequence

$$0 \to H_n(X) \xrightarrow{i} \operatorname{coker}\partial_n \xrightarrow{\bar{\partial}_{n-1}} \ker\partial_{n-2} \xrightarrow{q} H_{n-1}(X) \to 0.$$

Here i is the inclusion $i : \ker\partial_{n-1}/\operatorname{image}\partial_n \to X_n/\operatorname{image}\partial_n$, and q is the quotient $q : \ker\partial_{n-2} \to \ker\partial_{n-2}/\operatorname{image}\partial_{n-1}$. This association of a six-term exact sequence to X for each n is functorial in the sense that if $X' = \{(X'_n, \partial'_n)\}_{n=-\infty}^{\infty}$ is a second chain complex and if $f : X \to X'$ is a chain complex, then the diagram

$$\begin{array}{ccccccc}
H_n(X) & \xrightarrow{i} & \operatorname{coker}\partial_n & \xrightarrow{\bar{\partial}_{n-1}} & \ker\partial_{n-2} & \xrightarrow{q} & H_{n-1}(X) \\
\downarrow & & \downarrow & & \downarrow & & \downarrow \\
H_n(X') & \xrightarrow{i'} & \operatorname{coker}\partial'_n & \xrightarrow{\bar{\partial}'_{n-1}} & \ker\partial'_{n-2} & \xrightarrow{q'} & H_{n-1}(X')
\end{array}$$

commutes; here the vertical maps are those induced by f_{n-1} and f_n.

REMARKS.

(1) The term "functorial" in the statement has a precise meaning in this and other contexts. Each chain complex is being carried to a 6-term exact sequence for each n. The chain complexes and the 6-term exact sequences both form categories, the morphisms in each case being chain maps. To say that the passage

from the objects of one category to the other is **functorial** is to say that the passage between the categories is actually a functor, i.e., chain maps for the chain complexes are sent to chain maps for the 6-term exact sequences, the identity goes to the identity, and compositions go to compositions. The latter two conditions are evident, and what needs proof is that chain maps are carried to chain maps.[4]

(2) For a cochain complex $X = \{(X_n, d_n)\}_{n=-\infty}^{\infty}$ with d_n in $\mathrm{Hom}_R(X_n, X_{n+1})$, the corresponding exact sequence is

$$0 \longrightarrow H_{n-1}(X) \xrightarrow{i} \mathrm{coker}\, d_{n-2} \xrightarrow{\bar{d}_{n-1}} \ker d_n \xrightarrow{q} H_n(X) \longrightarrow 0,$$

and it is functorial with respect to cochain maps.

PROOF. To see that the map $\bar{\partial}_{n-1}$ carries $\mathrm{coker}\, \partial_n$ to $\ker \partial_{n-2}$, we write it as a composition

$$\mathrm{coker}\, \partial_n = X_n/\mathrm{image}\, \partial_n \to X_n/\ker \partial_{n-1} \cong \mathrm{image}\, \partial_{n-1} \subseteq \ker \partial_{n-2},$$

with the arrow induced by the inclusion $\mathrm{image}\, \partial_n \subseteq \ker \partial_{n-1}$ and with the isomorphism induced by applying ∂_{n-1} to X_n and passing to the quotient. Then we have $\ker \bar{\partial}_{n-1} = \ker \partial_{n-1}/\mathrm{image}\, \partial_n = H_n(X)$ and

$$\mathrm{coker}\, \bar{\partial}_{n-1} = \ker \partial_{n-2}/\bar{\partial}_{n-1}(X_n/\mathrm{image}\, \partial_n) = \ker \partial_{n-2}/\partial_{n-1} X_n$$
$$= \ker \partial_{n-2}/\mathrm{image}\, \partial_{n-1} = H_{n-1}(X),$$

and the exactness of the sequence is a special case of the exactness noted in the paragraph before the proposition.

For the assertion that the association is functorial, the left square commutes because the verticals are both induced by the same map f_n, and the right square commutes because the verticals are both induced by the same map f_{n-1}. For the middle square the commutativity follows from the fact that $f_{n-1}\partial_{n-1} = \partial'_{n-1} f_n$. □

---

[4]Some authors use the word "natural" instead of the word "functorial" in this situation. Authors who do this may have the notion of "natural transformation" between two functors in mind, or they may not. For those who do not, it seems advisable to use a different term like "functorial" to avoid confusion. For those who do, the allusion to a natural transformation is at best tortured in this instance. A natural transformation refers to two categories C and C', and the most intuitive choice for C here is the category of chain complexes X. There are to be two functors from C to C' and the natural transformation relates the values of those functors on X, for each X; no second complex X' enters into matters. To have X' involved in a natural transformation would mean including at least two chain complexes in each object of C. In other instances, however, some additional structure may be present. Then the distinction between "functorial" and "natural" may be one of emphasis concerning the data. The statements of Propositions 4.29 and 4.30 below provide examples.

As was mentioned in Section 1, our interest will be in functors $F : \mathcal{C} \to \mathcal{C}'$ between two good categories, not necessarily involving the same ring, with the property of being **additive**. This means that $F(\varphi_1 + \varphi_2) = F(\varphi_1) + F(\varphi_2)$ when φ_1 and φ_2 are in the same $\operatorname{Hom}_R(A, B)$.

An additive functor sends any 0 map to the corresponding 0 map. Consequently it always sends complexes to complexes. Moreover, since any functor carries the identity map of each $\operatorname{Hom}_R(A, A)$ to an identity map, an additive functor has to send any module A for which the 0 map and the identity coincide to another such module. The 0 module is the unique module A with this property, and thus an additive functor has to send the 0 module to a 0 module.

Moreover, additive functors carry finite direct sums to finite direct sums. (Recall that good categories are closed under finite direct sums.) This fact needs proper formulation, and we need first to express direct sums in terms of modules and maps. From the point of view of category theory, we shall take advantage of the fact that for left R modules, product and coproduct coincide and are given by direct sum. If $C \cong A \oplus B$, then there are thus projections $p_A : C \to A$ and $p_B : C \to B$ and injections $\iota_A : A \to C$ and $\iota_B : B \to C$ such that

$$p_A \iota_A = 1_A \quad \text{and} \quad p_B \iota_B = 1_B,$$
$$p_B \iota_A = 0 \quad \text{and} \quad p_A \iota_B = 0,$$

and

$$\iota_A p_A + \iota_B p_B = 1_C.$$

Conversely if we have maps $p_A, \iota_A, p_B,$ and ι_B with these properties, then the modules $A = \operatorname{image} p_A$ and $B = \operatorname{image} p_B$ have the property that C is the internal direct sum $C = \iota_A A \oplus \iota_B B$, and ι_A and ι_B are one-one. In fact, the equation $\iota_A p_A + \iota_B p_B = 1_C$ shows that $\iota_A A + \iota_B B = C$. To see that $\iota_A A \cap \iota_B B = 0$, let x be in the intersection. Then $p_B x$ lies in $p_B \iota_A A$, which is 0, and $p_A x$ lies in $p_A \iota_B B$, which is 0. Thus $\iota_A p_A + \iota_B p_B = 1_C$ gives $0 = \iota_A p_A x + \iota_B p_B x = x$. Hence $\iota_A A \cap \iota_B B = 0$ and $C = \iota_A A \oplus \iota_B B$. Finally the equations $p_A \iota_A = 1_A$ and $p_B \iota_B = 1_B$ imply that ι_A and ι_B are one-one.

With direct sum now expressed in terms of modules and maps, let us return to the effect of additive functors on direct sums. Let $C \cong A \oplus B$, and let $p_A, p_B, \iota_A,$ and ι_B be as above. Suppose that the additive functor F is covariant. Applying F to the displayed identities in the previous paragraph and using that F is additive, we see that $F(p_A), F(p_B), F(\iota_A),$ and $F(\iota_B)$ have the properties that allow us to recognize a direct sum. Hence

$$F(C) = F(\iota_A)F(A) \oplus F(\iota_B)F(B)$$

with $F(\iota_A)$ and $F(\iota_B)$ one-one. Thus

$$F(C) \cong F(A) \oplus F(B).$$

2. Complexes and Additive Functors

If instead F is contravariant, then the roles of the projections and the injections get interchanged, but we still obtain $F(C) \cong F(A) \oplus F(B)$.

An additive functor $F : \mathcal{C} \to \mathcal{C}'$ between two good categories is **exact** if it transforms exact sequences into exact sequences. Proposition 4.3 below will show that exact covariant functors preserve kernels, images, cokernels, submodules, quotients, and more. However, exact functors occur only infrequently; we shall see a few examples of them in Section 4. For examples of failures at exactness, it was shown in Section X.6 of *Basic Algebra* that if

$$0 \to M \xrightarrow{\varphi} N \xrightarrow{\psi} P \to 0$$

is a short exact sequence in the category $\mathcal{C}_R$, if E is a unital left R module, and if E' is a unital right R module, then the following sequences in $\mathcal{C}_\mathbb{Z}$ are exact:

$$E' \otimes_R M \xrightarrow{1 \otimes \varphi} E' \otimes_R N \xrightarrow{1 \otimes \psi} E' \otimes_R P \longrightarrow 0,$$

$$0 \longrightarrow \mathrm{Hom}_R(E, M) \xrightarrow{\mathrm{Hom}(1,\varphi)} \mathrm{Hom}_R(E, N) \xrightarrow{\mathrm{Hom}(1,\psi)} \mathrm{Hom}_R(E, P),$$

$$\mathrm{Hom}_R(M, E) \xleftarrow{\mathrm{Hom}(\varphi,1)} \mathrm{Hom}_R(N, E) \xleftarrow{\mathrm{Hom}(\psi,1)} \mathrm{Hom}_R(P, E) \longleftarrow 0;$$

on the other hand, the extensions of these complexes to 5-term complexes by the adjoining of a 0 need not be exact, and thus the functors $E' \otimes_R (\cdot)$, $\mathrm{Hom}_R(E, \cdot)$, and $\mathrm{Hom}_R(\cdot, E)$ are not exact for suitable choices of R, E, and E'.

Proposition 4.3. An additive functor $F : \mathcal{C} \to \mathcal{C}'$ between two good categories is exact if and only if it carries all short exact sequences into short exact sequences.

REMARK. This proposition makes it a little easier to test concrete additive functors for exactness than it would be from the definition.

PROOF. Necessity is obvious. For sufficiency, let

$$A \xrightarrow{\varphi} B \xrightarrow{\psi} C$$

be exact, and let the additive functor F be covariant, the contravariant case being completely analogous. Put $A_1 = \ker \varphi$, $B_1 = \ker \psi$, and $C_1 = \mathrm{image}\, \psi$. Since $\psi \varphi = 0$, we can factor φ as $\varphi = \varphi_2 \varphi_1$, where $\varphi_1 : A \to B_1$ is φ with its range space reduced and where $\varphi_2 : B_1 \to B$ is the inclusion. Similarly we can factor ψ as $\psi = \psi_2 \psi_1$, where $\psi_1 : B \to C_1$ is ψ with its range space reduced and where $\psi_2 : C_1 \to C$ is the inclusion. Of the sequences

$$0 \longrightarrow A_1 \longrightarrow A \xrightarrow{\varphi_1} B_1 \longrightarrow 0,$$

$$0 \longrightarrow B_1 \xrightarrow{\varphi_2} B \xrightarrow{\psi_1} C_1 \longrightarrow 0,$$

$$0 \longrightarrow C_1 \xrightarrow{\psi_2} C \longrightarrow C/C_1 \longrightarrow 0,$$

the first and the third are trivially exact, and the second is exact because $\ker \psi_1 = \ker \psi = \text{image } \varphi = \text{image } \varphi_2$. The hypothesis that F carries short exact sequences to short exact sequences thus implies that the three sequences

$$F(A) \xrightarrow{F(\varphi_1)} F(B_1) \longrightarrow 0,$$

$$F(B_1) \xrightarrow{F(\varphi_2)} F(B) \xrightarrow{F(\psi_1)} F(C_1),$$

$$0 \longrightarrow F(C_1) \xrightarrow{F(\psi_2)} F(C)$$

are exact. From these, $\ker F(\psi_1) = \text{image } F(\varphi_2)$. Also, $F(\psi_2)$ is one-one, so that

$$\ker F(\psi_1) = \ker \big(F(\psi_2) F(\psi_1)\big) = \ker F(\psi),$$

and $F(\varphi_1)$ is onto, so that

$$\text{image } F(\varphi_2) = \text{image }\big(F(\varphi_2) F(\varphi_1)\big) = \text{image } F(\varphi).$$

Hence $\ker F(\psi) = \text{image } F(\varphi)$, and

$$F(A) \xrightarrow{F(\varphi)} F(B) \xrightarrow{F(\psi)} F(C)$$

is exact, as required. □

Proposition 4.4. Let $F : \mathcal{C} \to \mathcal{C}'$ be an additive functor between good categories, let X be a complex in $\mathcal{C}$, and let $F(X)$ be the corresponding complex in $\mathcal{C}'$. If F is exact, then F carries the homology or cohomology of X to the homology or cohomology of $F(X)$.

REMARKS. Our convention is to refer to homology when the indexing goes down and cohomology when the indexing goes up. If F is covariant, it preserves the indexing, while if F is contravariant, it reverses it. For the proof we shall use notation A, B, C for modules that is neutral with respect to the indexing. The arguments are qualitatively different in the covariant and contravariant cases, and we shall give both of them.

PROOF IN THE COVARIANT CASE. Let

$$A \xrightarrow{\varphi} B \xrightarrow{\psi} C$$

be a given complex, thus having $\psi\varphi = 0$, and form the image complex

$$F(A) \xrightarrow{F(\varphi)} F(B) \xrightarrow{F(\psi)} F(C).$$

2. Complexes and Additive Functors 181

We are to exhibit an isomorphism

$$F(\ker\psi/\operatorname{image}\varphi) \cong \ker F(\psi)/\operatorname{image} F(\varphi). \tag{$*$}$$

Let $i : \operatorname{image}\varphi \to \ker\psi$ and $j : \ker\psi \to B$ be the inclusions, and let $q : \ker\psi \to \ker\psi/\operatorname{image}\varphi$ be the quotient map. Applying F to the exact sequence

$$0 \longrightarrow \operatorname{image}\varphi \xrightarrow{i} \ker\psi \xrightarrow{q} \ker\psi/\operatorname{image}\varphi \longrightarrow 0$$

and using exactness, we obtain an isomorphism via $F(q)$:

$$F(\ker\psi/\operatorname{image}\varphi) \cong F(\ker\psi)\big/F(i)F(\operatorname{image}\varphi). \tag{$**$}$$

Since j is one-one and F is exact, $F(j)$ is one-one. Thus application of $F(j)$ to the right side of $(**)$ gives

$$F(\ker\psi/\operatorname{image}\varphi) \cong F(j)F(\ker\psi)\big/F(ji)F(\operatorname{image}\varphi). \tag{$\dagger$}$$

If $\overline{\varphi}$ denotes φ with its range reduced to its image, then $\varphi = ji\overline{\varphi}$. Applying F to the two exact sequences

$$\ker\psi \xrightarrow{j} B \xrightarrow{\psi} C,$$

$$A \xrightarrow{\overline{\varphi}} \operatorname{image}\varphi \to 0$$

gives us $F(j)F(\ker\psi) = \ker F(\psi)$ and $F(\operatorname{image}\varphi) = F(\overline{\varphi})F(A)$. Applying $F(ji)$ to the second of these and substituting both into the right side of $(\dagger)$ transforms $(\dagger)$ into $(*)$ and gives the required isomorphism. □

PROOF IN THE CONTRAVARIANT CASE. Let

$$A \xrightarrow{\varphi} B \xrightarrow{\psi} C$$

be given with $\psi\varphi = 0$, and form the image complex

$$F(A) \xleftarrow{F(\varphi)} F(B) \xleftarrow{F(\psi)} F(C).$$

We are to exhibit an isomorphism

$$F(\ker\psi/\operatorname{image}\varphi) \cong \ker F(\varphi)/\operatorname{image} F(\psi). \tag{$*$}$$

Let $j : \ker\psi \to B$ be the inclusion, let $\bar{j} : \ker\psi/\operatorname{image}\varphi \to B/\operatorname{image}\psi$ be the induced map between quotients, and let q, q', q'' be the quotient maps

$$q : B \to B/\ker\psi,$$
$$q' : B \to B/\operatorname{image}\varphi,$$
$$q'' : B/\operatorname{image}\varphi \to B/\ker\psi.$$

These satisfy $q = q''q'$. Applying F to the exact sequence

$$0 \longrightarrow \ker\psi/\operatorname{image}\varphi \xrightarrow{\bar{j}} B/\operatorname{image}\varphi \xrightarrow{q''} B/\ker\psi \longrightarrow 0$$

and using exactness, we obtain an isomorphism via $F(\bar{j})$:

$$F(\ker\psi/\operatorname{image}\varphi) \cong F(B/\operatorname{image}\varphi)/F(q'')F(B/\ker\psi). \qquad (**)$$

Since q' is onto and F is exact, $F(q')$ is one-one. Thus application of $F(q')$ to the right side of $(**)$ gives

$$F(\ker\psi/\operatorname{image}\varphi) \cong F(q')F(B/\operatorname{image}\varphi)/F(q)F(B/\ker\psi). \qquad (\dagger)$$

Applying F to the three exact sequences

$$A \xrightarrow{\varphi} B \xrightarrow{q'} B/\operatorname{image}\varphi,$$

$$\ker\psi \xrightarrow{j} B \xrightarrow{\psi} C,$$

$$\ker\psi \xrightarrow{j} B \xrightarrow{q} B/\ker\psi$$

gives us $F(q')F(B/\operatorname{image}\varphi) = \ker F(\varphi)$ and $F(q)F(B/\ker\psi) = \ker F(j) = \operatorname{image} F(\psi)$. Substituting both these equalities into the right side of $(\dagger)$ transforms $(\dagger)$ into $(*)$ and gives the required isomorphism. □

We were reminded before Proposition 4.3 that Hom_R and $\otimes_R$ need not yield exact functors. The partial exactness that they exhibit, as opposed to exactness itself, is more typical of additive functors, and we incorporate this behavior into two definitions. We shall define left and right exactness in such a way that Hom_R is left exact in each variable and $\otimes_R$ is right exact. An additive functor F is **left exact** if the exactness of

$$0 \to A \xrightarrow{\varphi} B \xrightarrow{\psi} C \to 0$$

implies the exactness of

$$0 \longrightarrow F(A) \xrightarrow{F(\varphi)} F(B) \xrightarrow{F(\psi)} F(C) \qquad (F \text{ covariant}),$$

$$0 \longrightarrow F(C) \xrightarrow{F(\psi)} F(B) \xrightarrow{F(\varphi)} F(A) \qquad (F \text{ contravariant}).$$

2. Complexes and Additive Functors

We say that F is **right exact** if the exactness of the sequence with $0, A, B, C, 0$ above implies the exactness of

$$F(A) \xrightarrow{F(\varphi)} F(B) \xrightarrow{F(\psi)} F(C) \longrightarrow 0 \qquad (F \text{ covariant}),$$

$$F(C) \xrightarrow{F(\psi)} F(B) \xrightarrow{F(\varphi)} F(A) \longrightarrow 0 \qquad (F \text{ contravariant}).$$

The words "left" and "right" refer to the part of the *target* sequence that is exact when the arrows are arranged to point to the right. A consequence (but not the full content) of these definitions in each case is an assertion about one-one or onto maps. For example a left exact covariant F carries one-one maps to one-one maps; we have only to start from a one-one map $\varphi : A \to B$ and set up a short exact sequence with $C = B/\operatorname{image} \varphi$, and the definition shows that $F(\varphi)$ is one-one.

Proposition 4.5. If F is a covariant left exact functor, then F carries an exact sequence

$$0 \to A \xrightarrow{\varphi} B \xrightarrow{\psi} C$$

into an exact sequence

$$0 \longrightarrow F(A) \xrightarrow{F(\varphi)} F(B) \xrightarrow{F(\psi)} F(C).$$

REMARK. The expected analogs of this result are valid if F is contravariant or if F is right exact or both.

PROOF. Starting from the given exact sequence, let $i : \operatorname{image} \psi \to C$ be the inclusion, and let $\overline{\psi} : B \to \operatorname{image} \psi$ be ψ with its range space reduced. Then $\psi = i\overline{\psi}$, and the sequences

$$0 \longrightarrow A \xrightarrow{\varphi} B \xrightarrow{\overline{\psi}} \operatorname{image} \psi \longrightarrow 0$$

and

$$0 \longrightarrow \operatorname{image} \psi \xrightarrow{i} C \longrightarrow C/\operatorname{image} \psi \longrightarrow 0$$

are exact. Applying F and using its left exactness, we see that

$$0 \longrightarrow F(A) \xrightarrow{F(\varphi)} F(B) \xrightarrow{F(\overline{\psi})} F(\operatorname{image} \psi)$$

and

$$0 \longrightarrow F(\operatorname{image} \psi) \xrightarrow{F(i)} F(C)$$

are exact. Thus $F(i)$ is one-one, and $F(\psi) = F(i\overline{\psi}) = F(i)F(\overline{\psi})$ has the same kernel as $F(\overline{\psi})$. The exactness of the first image complex shows that $\ker F(\overline{\psi}) = \operatorname{image} F(\varphi)$, and the proof of the required exactness is complete. □

3. Long Exact Sequences

As in Section 2, let $\mathcal{C}$ be a good category. We have seen that chain complexes in $\mathcal{C}$ themselves form a category whose morphisms are chain maps. If we have several chain maps in succession, each with an index $n \in \mathbb{Z}$, we can say that they form an "exact sequence" of chain maps if for each n, the sequences of modules and maps having index n form an exact sequence in $\mathcal{C}$. Our objective in this section is to show that any short exact sequence of complexes of this kind yields a "long exact sequence" of modules and maps in $\mathcal{C}$ involving all the indices. More precisely we are able to construct for each n a "connecting homomorphism" relating[5] what happens with each index n to what happens for index $n+1$ or $n-1$ and incorporating modules and maps for all indices into a single exact sequence of infinite length.

By way of preparation for the construction of connecting homomorphisms, let us be more explicit about the discussion in Section 2 of how a chain map carries the homology of one complex to the homology of another complex. Let

$$\begin{array}{ccc} A & \xrightarrow{\varphi} & B \\ \downarrow{\alpha} & & \downarrow{\beta} \\ A' & \xrightarrow{\varphi'} & B' \end{array}$$

be a commutative diagram in the good category $\mathcal{C}$. Let us observe that $\varphi(\ker \alpha) \subseteq \ker \beta$; in fact, any $a \in \ker \alpha$ has $0 = \varphi'\alpha(a) = \beta\varphi(a)$, and thus $\varphi(a)$ is in $\ker \beta$. Let us observe further that $\varphi'(\alpha(A)) = \beta(\varphi(A)) \subseteq \beta(B)$; since φ' carries A' into B', it follows that φ' descends to a mapping $\overline{\varphi}'$ defined on $A'/\alpha(A) = \operatorname{coker} \alpha$ and taking values in $B'/\beta(B) = \operatorname{coker} \beta$. We can summarize these remarks by the inclusions

$$\varphi(\ker \alpha) \subseteq \ker \beta \quad \text{and} \quad \overline{\varphi}'(\operatorname{coker} \alpha) \subseteq \operatorname{coker} \beta.$$

Using these remarks, we can now construct a "connecting homomorphism" whenever we have a diagram as in Figure 4.1 below.

---

[5] For readers familiar with the use of homology in topology, connecting homomorphisms arise when one works with the homology of a topological space, the homology of a subspace, and the relative homology of the space and the subspace; the construction in this section may be regarded as an abstract version of that construction.

3. Long Exact Sequences

$$A \xrightarrow{\varphi} B \xrightarrow{\psi} C \longrightarrow 0$$
$$\downarrow \alpha \quad\quad \downarrow \beta \quad\quad \downarrow \gamma$$
$$0 \longrightarrow A' \xrightarrow{\varphi'} B' \xrightarrow{\psi'} C'$$

FIGURE 4.1. Snake diagram. The rows are assumed exact, and the squares commute. In this situation the Snake Lemma constructs a connecting homomorphism $\omega : \ker \gamma \to \coker \alpha$.

Lemma 4.6 (Snake Lemma). In a good category $\mathcal{C}$, a snake diagram as in Figure 4.1 induces a homomorphism $\omega : \ker \gamma \to \coker \alpha$ with

$$\ker \omega = \psi(\ker \beta) \quad \text{and} \quad \image \omega = \varphi'^{-1}(\image \beta)/\image \alpha,$$

and with $\omega(c) = \varphi'^{-1}(\beta(\psi^{-1}(c))) + \image \alpha$ for $c \in \ker \gamma$, and then

$$\ker \alpha \xrightarrow{\overline{\varphi}} \ker \beta \xrightarrow{\overline{\psi}} \ker \gamma \xrightarrow{\omega} \coker \alpha \xrightarrow{\overline{\varphi}'} \coker \beta \xrightarrow{\overline{\psi}'} \coker \gamma$$

is an exact sequence. Here $\overline{\varphi}$ and $\overline{\psi}$ are restrictions of φ and ψ, and $\overline{\varphi}'$ and $\overline{\psi}'$ are descended versions of φ and ψ. If φ is one-one, then $\overline{\varphi}$ is one-one. If ψ' is onto C', then $\overline{\psi}'$ is onto $\coker \gamma$.

REMARKS. The homomorphism ω is called a **connecting homomorphism**. The name "Snake Lemma" comes from the pattern that the six-term exact sequence makes when superimposed on the enlarged version of Figure 4.1 shown in Figure 4.2.

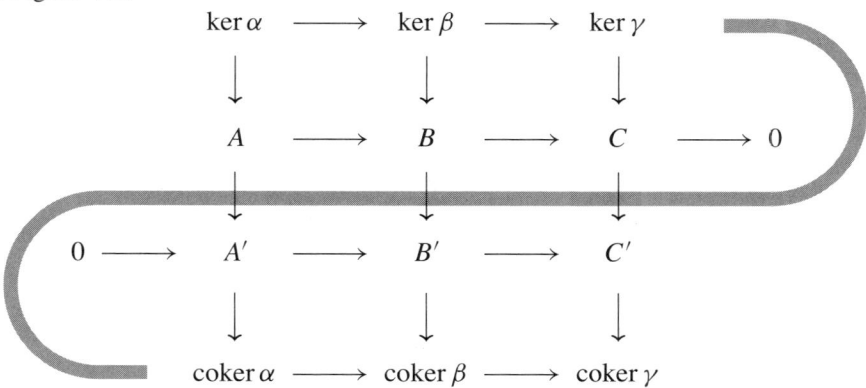

FIGURE 4.2. Enlarged snake diagram.

PROOF. First let us construct ω and see that it is well defined. Let c be in $\ker \gamma$. Since ψ is onto C, write $c = \psi(b)$ for some $b \in B$. The commutativity of the second square in Figure 4.1 gives $0 = \gamma(c) = \gamma\psi(b) = \psi'(\beta b)$. Thus $\beta(b)$ is in $\ker \psi' = \text{image }\varphi'$, and $\beta(b) = \varphi'(a')$ for some $a' \in A'$; the element a' is uniquely determined, since φ' is one-one. Define $\omega(c) = a' + \alpha(A)$.

The only choice in this definition is that of b, and we are to show that any other choice leads to the same member of $\operatorname{coker}\alpha$. If $\bar{b}$ is another choice and if $\beta(\bar{b}) = \varphi'(\bar{a}')$ with $a' \in A'$, then $\psi(\bar{b} - b) = c - c = 0$ shows that $\bar{b} - b = \varphi(a)$ for some $a \in A$. Thus $\varphi'(\bar{a}' - a') = \beta(\bar{b} - b) = \beta\varphi(a) = \varphi'(\alpha(a))$. Since φ' is one-one, $\bar{a}' - a' = \alpha(a)$, and $\bar{a}'$ and a' are exhibited as in the same coset of A' modulo $\alpha(A)$.

Let us compute $\ker \omega$. Suppose that $\omega(c) = 0$, i.e., that $\omega(c)$ is in $\alpha(A)$. Say $\omega(c) = \alpha(a)$. By construction of ω, $\omega(c) = a' + \alpha(A)$ for an element $a' \in A'$ such that $\beta(b) = \varphi'(a')$ and $c = \psi(b)$. In this case, $a' = \alpha(a)$. So $\beta(b) = \varphi'\alpha(a) = \beta\varphi(a)$, and thus $b - \varphi(a)$ is in $\ker \beta$. Consequently $c = \psi(b) = \psi(b) - \psi\varphi(a)$ is in $\psi(\ker \beta)$, and $\ker \omega \subseteq \psi(\ker \beta)$. For the reverse inclusion, if c is in $\psi(\ker \beta)$, choose $b \in \ker \beta$ with $\psi(b) = c$. Then $\gamma(c) = \gamma\psi(b) = \psi'\beta(b) = 0$ shows that $\omega(c)$ is defined. Since $c = \psi(b)$, the construction of ω shows that $\beta(b) = \varphi'(a')$ for some $a' \in A'$. Since b is in $\ker \beta$ and since φ' is one-one, this a' must be 0. Then $\omega(c) = a' + \alpha(A) = 0 + \alpha(A)$, c is in $\ker \omega$, and $\psi(\ker \beta) \subseteq \ker \omega$.

Now we compute image ω. Our step-by-step definition of ω shows that image $\omega \subseteq \varphi'^{-1}(\text{image }\beta)/\alpha(A)$. For the reverse inclusion, suppose that $a' \in A'$ is in $\varphi'^{-1}(\text{image }\beta)$, i.e., has $\varphi'(a') = \beta(b)$ for some $b \in B$. Then the element $c = \varphi(b)$ of C has $\gamma(c) = \gamma\psi(b) = \psi'\beta(b) = \psi'\varphi'(a') = 0$, and $\omega(c)$ is therefore defined. Our definition of ω makes $\omega(c) = a' + \alpha(A)$, and thus $\varphi'^{-1}(\text{image }\beta)/\alpha(A) \subseteq \text{image }\omega$.

We are left with establishing the exactness of the displayed sequence of six terms at the four positions other than the ends and with proving the two assertions in the last sentence of the lemma.

The condition of exactness at $\ker \beta$ is that $\varphi(\ker \alpha) = \ker \psi \cap \ker \beta$. The inclusion $\subseteq$ follows from the equalities $0 = \psi\varphi$ and $\beta\varphi(\ker \alpha) = \varphi'\alpha(\ker \alpha) = 0$. For the inclusion $\supseteq$, let $b \in B$ satisfy $\psi(b) = \beta(b) = 0$. Exactness at B gives $b = \varphi(a)$ with $a \in A$. Then $0 = \beta(b) = \beta\varphi(a) = \varphi'\alpha(a)$ with φ' one-one implies that $\alpha(a) = 0$, and a is in $\ker \alpha$. Thus b is in $\varphi(\ker \alpha)$, and exactness at $\ker \beta$ is proved. If φ is one-one, then certainly its restriction $\overline{\varphi}$ is one-one.

The condition of exactness at $\ker \gamma$ is that $\ker \omega = \psi(\ker \beta)$, and this was proved in the third paragraph of the proof.

By the result of the fourth paragraph, the condition of exactness at $\operatorname{coker}\alpha$ is that $\varphi'^{-1}(\beta(B))/\alpha(A)$ equal $\ker \overline{\varphi}'$, where $\overline{\varphi}' : A'/\alpha(A) \to B'/\beta(B)$ is the map induced by φ'. The members of $\ker \overline{\varphi}'$ are those cosets $a' + \alpha(A)$ with

$\varphi'(a' + \alpha(A)) \subseteq \beta(B)$. Since $\varphi'\alpha(A) = \beta\varphi(A) \subseteq \beta(B)$, the condition on $a' + \alpha(A)$ is that $\varphi'(a')$ be in $\beta(B)$, hence that a' be in $\varphi'^{-1}(\beta(B))$, hence that the coset $a' + \alpha(A)$ be in $\varphi'^{-1}(\beta(B))/\alpha(A)$. Thus we have exactness at coker α.

At coker β, we know that the descended map $\overline{\varphi}'$ maps coker α into coker β, and we are to show that $\overline{\varphi}'(\operatorname{coker}\alpha) = \ker \overline{\psi}'$. Inclusion $\subseteq$ follows because $\psi'\varphi' = 0$ implies $\overline{\psi}'\overline{\varphi}'(a' + \alpha(A)) = \overline{\psi}'(\varphi'(a') + \beta(B)) = \psi'\varphi'(a') + \gamma(C) = \gamma(C)$. For the reverse inclusion let $b' \in B'$ have $\overline{\psi}'(b' + \beta(B)) = \gamma(C)$. Then $\psi'(b')$ is in $\gamma(C)$. Since $\psi : B \to C$ is onto, we can find $b \in B$ with $\psi'(b') = \gamma\psi(b) = \psi'\beta(b)$. Hence $b' - \beta(b)$ is in ker ψ' = image φ', and $b' - \beta(b) = \varphi'(a')$ for some $a' \in A'$. Consequently $b' + \beta(B) = \varphi'(a') + \beta(b) + \beta(B) = \varphi'(a') + \beta(B) = (\varphi')_*(a' + \alpha(A))$, and $b' + \beta(B)$ is exhibited as in $(\varphi')_*(a' + \alpha(A))$, i.e., in $(\varphi')_*(\operatorname{coker}\alpha)$. Thus we have exactness at coker β. Finally if ψ' is onto C', then certainly its descended map $\overline{\psi}'$ is onto coker γ. This completes the proof. □

Theorem 4.7. Let $A = \{(A_n, \alpha_n)\}$, $B = \{(B_n, \beta_n)\}$, and $C = \{(C_n, \gamma_n)\}$ be chain complexes in a good category $\mathcal{C}$, and suppose that $\varphi = \{\varphi_n\} : A \to B$ and $\psi = \{\psi_n\} : B \to C$ are chain maps such that the sequence

$$0 \to A \xrightarrow{\varphi} B \xrightarrow{\psi} C \to 0$$

of chain complexes is exact. Then this exact sequence of chain complexes induces an exact sequence in homology of the form

$$\cdots \to H_{n+1}(C) \xrightarrow{\overline{\omega}_n} H_n(A) \xrightarrow{\overline{\varphi}_n} H_n(B) \xrightarrow{\overline{\psi}_n} H_n(C) \xrightarrow{\overline{\omega}_{n-1}} H_{n-1}(A) \to \cdots.$$

Here the map $\overline{\omega}_n : H_{n+1}(C) \to H_n(A)$ has descended from the connecting homomorphism ω_n defined on ker γ_n in C_{n+1} and having range coker $\alpha_n = A_n/\operatorname{image}\alpha_n$.

REMARKS.
(1) The exact sequence in homology is called the **long exact sequence** in homology corresponding to the short exact sequence of chain complexes, and the maps ω_n are called **connecting homomorphisms**. As the proof will show, these connecting homomorphisms arise by *two* applications of the Snake Lemma, not just one.

(2) In more detail the diagram of the short exact sequence of chain complexes is of the form

$$
\begin{array}{ccccccccc}
& & \vdots & & \vdots & & \vdots & & \\
& & \downarrow & & \downarrow & & \downarrow & & \\
0 & \longrightarrow & A_{n+1} & \xrightarrow{\varphi_{n+1}} & B_{n+1} & \xrightarrow{\psi_{n+1}} & C_{n+1} & \longrightarrow & 0 \\
& & \downarrow \alpha_n & & \downarrow \beta_n & & \downarrow \gamma_n & & \\
0 & \longrightarrow & A_n & \xrightarrow{\varphi_n} & B_n & \xrightarrow{\psi_n} & C_n & \longrightarrow & 0 \\
& & \downarrow \alpha_{n-1} & & \downarrow \beta_{n-1} & & \downarrow \gamma_{n-1} & & \\
0 & \longrightarrow & A_{n-1} & \xrightarrow{\varphi_{n-1}} & B_{n-1} & \xrightarrow{\psi_{n-1}} & C_{n-1} & \longrightarrow & 0 \\
& & \downarrow & & \downarrow & & \downarrow & & \\
& & \vdots & & \vdots & & \vdots & &
\end{array}
$$

The rows are exact, the columns are chain complexes, and the squares commute.

(3) The corresponding result for cochain complexes involves the diagram

$$
\begin{array}{ccccccccc}
& & \vdots & & \vdots & & \vdots & & \\
& & \uparrow & & \uparrow & & \uparrow & & \\
0 & \longrightarrow & A_{n+1} & \xrightarrow{\varphi_{n+1}} & B_{n+1} & \xrightarrow{\psi_{n+1}} & C_{n+1} & \longrightarrow & 0 \\
& & \uparrow \alpha_n & & \uparrow \beta_n & & \uparrow \gamma_n & & \\
0 & \longrightarrow & A_n & \xrightarrow{\varphi_n} & B_n & \xrightarrow{\psi_n} & C_n & \longrightarrow & 0 \\
& & \uparrow \alpha_{n-1} & & \uparrow \beta_{n-1} & & \uparrow \gamma_{n-1} & & \\
0 & \longrightarrow & A_{n-1} & \xrightarrow{\varphi_{n-1}} & B_{n-1} & \xrightarrow{\psi_{n-1}} & C_{n-1} & \longrightarrow & 0 \\
& & \uparrow & & \uparrow & & \uparrow & & \\
& & \vdots & & \vdots & & \vdots & &
\end{array}
$$

and the corresponding **long exact sequence** in cohomology is

$$\cdots \to H^{n-1}(C) \xrightarrow{\overline{\omega}_n} H^n(A) \xrightarrow{\overline{\varphi}_n} H^n(B) \xrightarrow{\overline{\psi}_n} H^n(C) \xrightarrow{\overline{\omega}_{n+1}} H^{n+1}(A) \to \cdots .$$

The result for cochain complexes is a consequence of the result for chain complexes and follows by making adjustments in the notation.

3. Long Exact Sequences

PROOF. We regard the top two displayed rows of the diagram in Remark 2 as a snake diagram. Applying the Snake Lemma (Lemma 4.6), we obtain a connecting homomorphism ω_n and an exact sequence

$$\ker \alpha_n \xrightarrow{\overline{\varphi}_{n+1}} \ker \beta_n \xrightarrow{\overline{\psi}_{n+1}} \ker \gamma_n \xrightarrow{\omega_n} \operatorname{coker} \alpha_n \xrightarrow{\overline{\varphi}'_n} \operatorname{coker} \beta_n \xrightarrow{\overline{\psi}'_n} \operatorname{coker} \gamma_n.$$

Using Proposition 4.2 for each of the chain complexes $A = \{(A_n, \alpha_n)\}$, $B = \{(B_n, \beta_n)\}$, and $C = \{(C_n, \gamma_n)\}$, we see that we obtain a diagram

$$\begin{array}{ccccc}
0 & & 0 & & 0 \\
\downarrow & & \downarrow & & \downarrow \\
H_n(A) & & H_n(B) & & H_n(C) \\
\downarrow & & \downarrow & & \downarrow \\
\operatorname{coker} \alpha_n & \xrightarrow{\overline{\varphi}_n} & \operatorname{coker} \beta_n & \xrightarrow{\overline{\psi}_n} & \operatorname{coker} \gamma_n \longrightarrow 0 \\
\downarrow \overline{\alpha}_{n-1} & & \downarrow \overline{\beta}_{n-1} & & \downarrow \overline{\gamma}_{n-1} \\
0 \longrightarrow \ker \alpha_{n-2} & \xrightarrow{\overline{\varphi}_{n-1}} & \ker \beta_{n-2} & \xrightarrow{\overline{\psi}_{n-1}} & \ker \gamma_{n-2} \\
\downarrow & & \downarrow & & \downarrow \\
H_{n-1}(A) & & H_{n-1}(B) & & H_{n-1}(C) \\
\downarrow & & \downarrow & & \downarrow \\
0 & & 0 & & 0
\end{array}$$

in which the rows and columns are exact and the squares commute. The third and fourth rows form a snake diagram, and the second and fifth rows identify the kernels and cokernels. Thus the Snake Lemma gives us an exact sequence

$$H_n(A) \xrightarrow{\overline{\varphi}_n} H_n(B) \xrightarrow{\overline{\psi}_n} H_n(C) \xrightarrow{\Omega} H_{n-1}(A) \xrightarrow{\overline{\varphi}_{n-1}} H_{n-1}(B) \xrightarrow{\overline{\psi}_{n-1}} H_{n-1}(C)$$

for a suitable connecting homomorphism Ω. Repeating this argument for all n proves exactness at all modules of the long exact sequence.

To complete the proof, we have only to identify Ω. Reference to the statement of the Snake Lemma shows that the formula for Ω is

$$\Omega(\bar{c}) = (\overline{\varphi}'_{n-1})^{-1}(\overline{\beta}_{n-1}(\overline{\psi}_n^{-1}(\bar{c}))) + \operatorname{image} \overline{\alpha}_{n-1}$$

for $\bar{c} \in H_n(C)$. Meanwhile, the connecting homomorphism from the first application of the Snake Lemma is $\omega_{n-1}(c) = (\varphi'_{n-1})^{-1}(\beta_{n-1}(\psi_n^{-1}(c))) + \operatorname{image} \alpha_{n-1}$ for $c \in \ker \gamma_{n-1}$. Thus $\Omega(c + \operatorname{image} \gamma_n) = \omega_{n-1}(c) + \operatorname{image} \alpha_{n-1}$ as asserted. □

Corollary 4.8. If
$$0 \to A \xrightarrow{\varphi} B \xrightarrow{\psi} C \to 0$$
is an exact sequence of chain complexes in a good category and if A is exact, then $H_n(B) \cong H_n(C)$ for all n; if instead C is exact, then $H_n(A) \cong H_n(B)$ for all n. Consequently if any two of the three chain complexes are exact, then the third one is exact.

PROOF. Theorem 4.7 gives the long exact sequence
$$\cdots \longrightarrow H_{n+1}(C) \longrightarrow H_n(A) \longrightarrow H_n(B) \longrightarrow H_n(C) \longrightarrow H_{n-1}(A) \longrightarrow \cdots.$$
If $H_n(A) = 0$ and $H_{n-1}(A) = 0$, then we see that $H_n(B) \cong H_n(C)$. If $H_{n+1}(C) = 0$ and $H_n(C) = 0$, then we see that $H_n(A) \cong H_n(B)$.

If two of the three chain complexes are exact, then one of the two is A or C, and the result in the previous paragraph applies. Then the other two complexes (B and C, or A and B) have isomorphic homology. The hypothesis says that one of these two sequences of homology groups is 0. Therefore the other one is 0. □

To conclude the discussion, we shall prove results saying that the exact sequences produced by Lemma 4.6 and Theorem 4.7 are functorial.

Lemma 4.9. In a good category $\mathcal{C}$, the six-term exact sequence that is obtained from a snake diagram as in Figure 4.1 is functorial in the following sense: If there are two horizontal planar snake diagrams, one with tildes ($\sim$) over all modules and maps and the other as is, and if there are vertical maps f_A, etc., in three dimensions from the tilde version of the snake diagram to the original version such that all vertical squares commute, then the squares of the diagram

$$\begin{array}{ccccccccccc}
\ker\widetilde{\alpha} & \xrightarrow{\widetilde{\widetilde{\varphi}}} & \ker\widetilde{\beta} & \xrightarrow{\widetilde{\widetilde{\psi}}} & \ker\widetilde{\gamma} & \xrightarrow{\widetilde{\omega}} & \operatorname{coker}\widetilde{\alpha} & \xrightarrow{\widetilde{\widetilde{\varphi'}}} & \operatorname{coker}\widetilde{\beta} & \xrightarrow{\widetilde{\widetilde{\psi'}}} & \operatorname{coker}\widetilde{\gamma} \\
\downarrow \tilde{f}_A & & \downarrow \tilde{f}_B & & \downarrow \tilde{f}_C & & \downarrow \tilde{f}_{A'} & & \downarrow \tilde{f}_{B'} & & \downarrow \tilde{f}_{C'} \\
\ker\alpha & \xrightarrow{\overline{\varphi}} & \ker\beta & \xrightarrow{\overline{\psi}} & \ker\gamma & \xrightarrow{\omega} & \operatorname{coker}\alpha & \xrightarrow{\overline{\varphi'}} & \operatorname{coker}\beta & \xrightarrow{\overline{\psi'}} & \operatorname{coker}\gamma
\end{array}$$

all commute.

PROOF. For the first square from the left, the assumed commutativity shows that $f_{A'}\widetilde{\alpha} = \alpha f_A$, and thus $x \in \ker\widetilde{\alpha}$ implies $f_A(x) \in \ker\alpha$; similarly $x \in \ker\widetilde{\beta}$ implies $f_B(x) \in \ker\beta$. Thus the maps of the square are well defined. We are given also that $\varphi f_A = f_B \widetilde{\varphi}$, and this proves that the square commutes. The second square from the left is handled similarly.

For the fourth square from the left, the equation $f_{A'}\widetilde{\alpha} = \alpha f_A$ shows that $y = \widetilde{\alpha}(x)$ implies $f_{A'}(y) = \alpha(f_A(x))$, and thus $y \in$ image $\widetilde{\alpha}$ implies $f_{A'}(y) \in$ image α; this means that $f_{A'}$ descends to a map $\bar{f}_{A'}$ of coker $\widetilde{\alpha}$ to coker α. Similarly $f_{B'}$ descends to a map $\bar{f}_{B'}$ of coker $\widetilde{\beta}$ to coker β. Thus the maps of the square are well defined. We are given also that $\varphi' f_{A'} = f_{B'}\widetilde{\varphi}'$, and this proves that the square commutes. The fifth square from the left is handled similarly.

We are left with the third square from the left. The map at the left side of this square was shown to be well defined in the first paragraph of the proof, and the map at the right side of this square was shown to be meaningful in the second paragraph of the proof. We are to prove that the square commutes. Referring to the construction of $\widetilde{\omega}$, let $\widetilde{c}$ be in ker $\widetilde{\gamma}$, choose $\widetilde{b}$ in $\widetilde{B}$ with $\widetilde{\psi}(\widetilde{b}) = \widetilde{c}$, and write $\widetilde{\beta}(\widetilde{b}) = \widetilde{\varphi}'(\widetilde{a}')$. Then $\widetilde{\omega}(\widetilde{c})$ is defined to be the coset of $\widetilde{a}'$. Using the assumed commutativity, we compute that $\psi f_B(\widetilde{b}) = f_C \widetilde{\psi}(\widetilde{b}) = f_C(\widetilde{c})$ and that

$$\varphi' f_{A'}(\widetilde{a}') = f_{B'}\widetilde{\varphi}'(\widetilde{a}') = f_{B'}\widetilde{\beta}(\widetilde{b}) = \beta f_B(\widetilde{b}).$$

Thus $f_B(\widetilde{b})$ is an element whose image under ψ is $f_C(\widetilde{c})$, and β of this element is $\varphi' f_{A'}(\widetilde{a}')$. Consequently the coset of $\omega(f_C(\widetilde{c}))$ is to be the coset of $f_{A'}(\widetilde{a}') = f_{A'}\widetilde{\omega}(c)$. This proves the desired commutativity. □

Theorem 4.10. In a good category $\mathcal{C}$, the long exact sequence that is obtained from a short exact sequence of chain complexes as in Theorem 4.7 is functorial in the following sense: if there are two short exact sequences of chain complexes as in the theorem, one with tildes ($\sim$) over all modules and maps and the other as is, each viewed as lying in a horizontal plane, and if there are vertical maps f_{A_n}, etc., from the tilde version of the exact sequence of chain complexes to the original version such that all vertical squares commute, then the squares of the diagram

$$\begin{array}{ccccccccc}
\longrightarrow & H_{n+1}(\widetilde{C}) & \xrightarrow{\widetilde{\omega}_n} & H_n(\widetilde{A}) & \xrightarrow{\widetilde{\varphi}_n} & H_n(\widetilde{B}) & \xrightarrow{\widetilde{\psi}_n} & H_n(\widetilde{C}) & \xrightarrow{\widetilde{\omega}_{n-1}} & H_{n-1}(\widetilde{A}) & \longrightarrow \\
& \downarrow f_{C_{n+1}} & & \downarrow f_{A_n} & & \downarrow f_{B_n} & & \downarrow f_{C_n} & & \downarrow f_{A_{n-1}} & \\
\longrightarrow & H_{n+1}(C) & \xrightarrow{\omega_n} & H_n(A) & \xrightarrow{\varphi_n} & H_n(B) & \xrightarrow{\psi_n} & H_n(C) & \xrightarrow{\omega_{n-1}} & H_{n-1}(A) & \longrightarrow
\end{array}$$

all commute.

PROOF. Theorem 4.7 was proved by three applications of Proposition 4.2, which includes its own assertion of functoriality, and two applications of Lemma 4.6, whose functoriality is addressed in Lemma 4.9. The argument involved only manipulations with diagrams, and functoriality is in place for every step. Hence functoriality is in place for the end result, and passage to the long exact sequence is functorial. □

4. Projectives and Injectives

In Section III.5 we exploited the fact that certain complexes were exact and involved free modules in order to obtain chain maps and homotopies. The hypothesis "free" entered the arguments through Propositions 3.25 and 3.27; in both cases an R homomorphism was to be constructed from a free R module to some other R module, and a computation revealed how the R homomorphism should be defined on free generators. The universal mapping property of free modules allowed the R homomorphism to be extended from the generators to the whole free module. Examination of those arguments shows that it is enough to assume that the domain on which this R homomorphism is to be constructed is a "projective" R module, in the sense to be defined below, and we begin with that notion.

Let $\mathcal{C}$ be a good category of unital left R modules. We say that a module P in this category is **projective** in $\mathcal{C}$ or is a **projective** in $\mathcal{C}$ if whenever a diagram in the category is given as in Figure 4.3 with ψ mapping onto B, then there exists $\sigma : P \to C$ in $\mathcal{C}$ such that the diagram commutes.

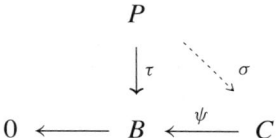

FIGURE 4.3. Defining property of a projective.

If P is a free R module in $\mathcal{C}$, then P is projective in $\mathcal{C}$. In fact, for each free generator x of P, we choose an element c_x in C with $\psi(c_x) = \tau(x)$. Then we define $\sigma(x) = c_x$ and extend σ to a homomorphism. We give further examples of projectives shortly. First let us establish in Lemma 4.11 an ostensibly stronger property that projectives automatically satisfy.

Lemma 4.11. If P is projective in the good category $\mathcal{C}$ and if the diagram

$$\begin{array}{ccccc} & & P & & \\ & & \downarrow{\tau} \searrow{\sigma} & & \\ A' & \xleftarrow{\varphi} & A & \xleftarrow{\psi} & A'' \end{array}$$

in $\mathcal{C}$ has $\ker \varphi = \operatorname{image} \psi$ and $\varphi\tau = 0$, then there exists a map $\sigma : P \to A''$ in $\mathcal{C}$ such that the diagram commutes.

PROOF. The hypotheses force $\operatorname{image} \tau \subseteq \ker \varphi = \operatorname{image} \psi$. Thus if we put $B = \operatorname{image} \psi$ and $C = A''$, then the above diagram leads to the diagram in Figure

4.3. The hypothesis "projective" therefore gives us the map σ in Figure 4.3 with $\tau = \psi\sigma$, and the same σ is the required map here. □

EXAMPLES OF PROJECTIVES.

(1) If R is a field F and if $\mathcal{C}$ is the category of all vector spaces over F, then every module is free, hence projective, since every vector space has a basis.

(2) For general R, if $\mathcal{C} = \mathcal{C}_R$ is the category of all unital left R modules, then the projectives are the direct summands of free modules. This fact is easily verified from Figure 4.3 as follows: In one direction if $F = P \oplus P'$ is a free R module and the diagram in Figure 4.3 is given, extend τ to F as 0 on P', find σ from the fact that the free module F is projective, and restrict σ to P. In the other direction if P is projective, find a free R module F mapping onto P by a map ψ, and put $B = P$, $C = F$, and $\tau = 1$ in Figure 4.3. Then the equality $1_P = \tau = \psi\sigma$ forces σ to be one-one, and it follows that $P \cong \text{image}\,\sigma$. Consequently $F = \text{image}\,\sigma \oplus \ker\psi$.

(3) For $R = \mathbb{Z}$, the category $\mathcal{C} = \mathcal{C}_\mathbb{Z}$ of all unital R modules is the category of all abelian groups. Then the projective modules are the free abelian groups by (2), since any subgroup of a free abelian group is free abelian.

(4) For R equal to any (commutative) principal ideal domain, the projective modules in the category $\mathcal{C}_R$ of all unital R modules are the free modules, by the same argument as in (3) in combination with the Fundamental Theorem of Finitely Generated Modules (Theorem 8.25 of *Basic Algebra*).

(5) For $R = \mathbb{Z}$, two good categories that were listed in Section 2 were the category of all finitely generated abelian groups and the category of all torsion abelian groups. With the first of these, the projectives are the free abelian groups of finite rank, by the same argument as in (3). With the second of these, Problem 1 at the end of the chapter asks for a verification that some module in the category fails to be the image of any projective in the category.

We come to the main result concerning flexibility in setting up chain complexes. This result generalizes Proposition 3.25 through Corollary 3.30 in Section III.5.

Theorem 4.12. Let $X = \{(X_n, \partial_n)\}_{n=-\infty}^{\infty}$ and $X' = \{(X'_n, \partial'_n)\}_{n=-\infty}^{\infty}$ be chain complexes in the good category $\mathcal{C}$, and let r be an integer. Let $\{f_n : X_n \to X'_n\}_{n \leq r}$ be a family of maps in $\mathcal{C}$ such that $\partial'_{n-1} f_n = f_{n-1} \partial_{n-1}$ for $n \leq r$. If X_n is projective for $n > r$ and X' is exact at each X'_n with $n \geq r$, then $\{f_n : X_n \to X'_n\}_{n \leq r}$ extends to a chain map $f : X \to X'$, and f is unique up to homotopy. More precisely any two extensions are homotopic by a homotopy h such that $h_n = 0$ for $n \leq r$.

REMARKS. The diagrams in question are

$$\cdots \xrightarrow{\partial_{n+1}} X_{n+1} \xrightarrow{\partial_n} X_n \xrightarrow{\partial_{n-1}} X_{n-1} \xrightarrow{\partial_{n-2}} \cdots$$
$$\downarrow f_{n+1} \qquad \downarrow f_n \qquad \downarrow f_{n-1}$$
$$\cdots \xrightarrow{\partial'_{n+1}} X'_{n+1} \xrightarrow{\partial'_n} X'_n \xrightarrow{\partial'_{n-1}} X'_{n-1} \xrightarrow{\partial'_{n-2}} \cdots$$

for the construction of the chain map and

$$\cdots \to X_{n+2} \xrightarrow{\partial_{n+1}} X_{n+1} \xrightarrow{\partial_n} X_n \xrightarrow{\partial_{n-1}} X_{n-1} \to \cdots$$
$$\downarrow f_{n+2} \;\swarrow h_{n+1}\; \downarrow f_{n+1} \;\swarrow h_n\; \downarrow f_n \;\swarrow h_{n-1}\; \downarrow f_{n-1}$$
$$\cdots \to X'_{n+2} \xrightarrow{\partial'_{n+1}} X'_{n+1} \xrightarrow{\partial'_n} X'_n \xrightarrow{\partial'_{n-1}} X'_{n-1} \to \cdots$$

for the construction of the homotopy.

PROOF. For the existence of the chain map, it is enough by induction to construct f_{r+1}. Matters are therefore as in the first of the above diagrams with $n = r$. Since X' is exact at X'_r and X_{r+1} is projective, we are in the situation of Lemma 4.11 with $P = X_{r+1}$, $A'' = X'_{r+1}$, $A = X'_r$, $A' = X'_{r-1}$, $\psi = \partial'_r$, $\varphi = \partial'_{r-1}$, and $\tau = f_r \partial_r$. The lemma gives a map $\sigma : P \to A''$ with $\psi \sigma = \tau$. If we take $f_{r+1} = \sigma$, then $\psi \sigma = \tau$ says that $\partial'_r f_{r+1} = f_r \partial_r$, and the inductive construction of the chain map is complete.

For the uniqueness up to homotopy, let $f : X \to X'$ and $g : X \to X'$ be two chain maps such that $f_n = g_n$ for $n \leq r$. Define $h_n : X_n \to X'_{n+1}$ to be 0 for $n \leq r$, and observe that the system of functions $\{h_n\}_{n \leq r}$ satisfies $h_{n-1} \partial_{n-1} + \partial'_n h_n = f_n - g_n$ for $n \leq r$ because $f_n = g_n$ for $n \leq r$. Proceeding inductively, suppose that $s \geq r$ and that h_n has been constructed for $n \leq s$ such that $h_{n-1} \partial_{n-1} + \partial'_n h_n = f_n - g_n$ for $n \leq s$. We are to construct $h_{s+1} : X_{s+1} \to X'_{s+2}$. This is the situation of the second diagram above with $n = s$. Since $s \geq r$, X' is exact at X'_{s+1} and X_{s+1} is projective. Thus we are in the situation of Lemma 4.11 with $P = X_{s+1}$, $A'' = X'_{s+2}$, $A = X'_{s+1}$, $A' = X'_s$, $\psi = \partial'_{s+1}$, $\varphi = \partial'_s$, and $\tau = (f_{s+1} - g_{s+1}) - h_s \partial_s$. The lemma gives a map $\sigma : P \to A''$ with $\psi \sigma = \tau$. If we take $h_{s+1} = \sigma$, then $\psi \sigma = \tau$ says that $\partial'_{s+1} h_{s+1} = (f_{s+1} - g_{s+1}) - h_s \partial_s$, and the inductive construction of the homotopy is complete. □

A **resolution** in the category $\mathcal{C}$ is an exact chain complex $X = \{(X_n, \partial_n)\}_{n=-\infty}^{\infty}$ or cochain complex $X = \{(X_n, d_n)\}_{n=-\infty}^{\infty}$ such that $X_n = 0$ for $n \leq -2$. We say that the complex is a **resolution of** X_{-1}, and we abbreviate it as

$$X = (X^+ \xrightarrow{\partial_{-1}} X_{-1}) \quad \text{or} \quad X = (X^+ \xleftarrow{d_{-1}} X_{-1}),$$

with X^+ referring to

$$X^+ : \quad \cdots \xrightarrow{\partial_2} X_2 \xrightarrow{\partial_1} X_1 \xrightarrow{\partial_0} X_0$$

or

$$X^+ : \quad \cdots \xleftarrow{d_2} X_2 \xleftarrow{d_1} X_1 \xleftarrow{d_0} X_0$$

in the respective cases. A chain complex $X = (X^+ \xrightarrow{\varepsilon} M)$ that forms a resolution is called a **free resolution** of M if every X_n for $n \geq 0$ is a free module. It is called a **projective resolution** of M if every X_n for $n \geq 0$ is projective.

Corollary 4.13. Let M be a module in a good category $\mathcal{C}$ and let

$$X = (X^+ \xrightarrow{\varepsilon} M) \quad \text{and} \quad X' = (X'^+ \xrightarrow{\varepsilon'} M)$$

be two projective resolutions of M. Then there exist chain maps $f : X \to X'$ and $g : X' \to X$ with $f_{-1} = 1_M$ and $g_{-1} = 1_M$, and any two such chain maps f and g have the property that $gf : X \to X$ is homotopic to 1_X and $fg : X' \to X'$ is homotopic to $1_{X'}$.

PROOF. The existence of f extending $f_{-1} = 1_M$ is immediate by applying the first part of Theorem 4.12 with $r = -1$. The hypotheses apply because X_n is projective for $n > -1$ and X' is exact at X'_n for $n \geq -1$. A similar argument shows the existence of g.

If we have f and g, then $gf : X \to X$ and $1_X : X \to X$ are chain maps that extend the partial chain map given for $n \leq -1$ by 1_M for $n = -1$ and by 0 for $n \leq -2$. Since again X_n is projective for $n > -1$ and X' is exact at X'_n for $n \geq -1$, the second part of the theorem shows that gf and 1_X are homotopic. A similar argument shows that fg and $1_{X'}$ are homotopic. □

There is an analogous sequence of results that ends with resolutions that are cochain maps. They will be equally as useful as the above results when we introduce derived functors in the next section. For the results below, the notion of a projective is replaced by that of an injective. We say that a module I in the good category $\mathcal{C}$ is **injective** in $\mathcal{C}$ or is an **injective** in $\mathcal{C}$ if whenever a diagram in the category is given as in Figure 4.4 with φ mapping one-one from B into C, then there exists $\sigma : B \to I$ in $\mathcal{C}$ such that the diagram commutes.

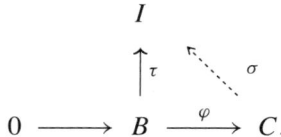

FIGURE 4.4. Defining property of an injective.

196   IV. Homological Algebra

We can think of the condition as saying that we can always extend such a τ from B to C, the extension being σ. In any event, we give some examples after proving an analog of Lemma 4.11.

Lemma 4.14. If I is injective in the good category $\mathcal{C}$ and if the diagram

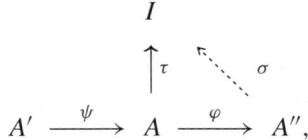

in $\mathcal{C}$ has $\ker \varphi = \operatorname{image} \psi$ and $\tau \psi = 0$, then there exists a map $\sigma : A'' \to I$ in $\mathcal{C}$ such that the diagram commutes.

PROOF. The hypotheses force $\ker \tau \supseteq \operatorname{image} \psi = \ker \varphi$. Thus $\tau : A \to I$ and $\varphi : A \to A''$ descend to maps $\overline{\tau} : A/\ker \varphi \to I$ and $\overline{\varphi} : A/\ker \varphi \to A''$. If we put $B = A/\ker \varphi$ and $C = A''$, then the above diagram leads to Figure 4.4 with $\overline{\tau}$ and $\overline{\varphi}$ in place of τ and φ. The hypothesis "injective" gives us σ in Figure 4.4 with $\overline{\tau} = \sigma \overline{\varphi}$, and the same σ is the required map in the diagram above. □

EXAMPLES OF INJECTIVES.

(1) If R is a field F and if $\mathcal{C}$ is the category of all vector spaces over F, then every module is injective. In fact, in Figure 4.4 we write $C = \operatorname{image} \varphi \oplus B'$, and we let $\eta : \operatorname{image} \varphi \to B$ be the inverse of $\varphi : B \to \operatorname{image} \varphi$. Then we can define σ to be 0 on B' and to be $\tau \eta$ on $\operatorname{image} \varphi$.

(2) Let $\mathcal{C}$ be the category of all abelian groups (unital $\mathbb{Z}$ modules). An abelian group G is said to be **divisible** if for each integer $n \neq 0$ and each $x \in G$, there exists $y \in G$ with $ny = x$. Two examples of divisible abelian groups are the additive group of rationals and the additive group of rationals modulo 1. It is easy to see that any quotient of a divisible group is divisible and that direct sums of divisible groups are divisible. Let us see for abelian groups that injective is equivalent to divisible.

The argument that injective implies divisible is easy: Let I be injective. Given $x \in I$ and $n \neq 0$, let $B = C = \mathbb{Z}$, let $\tau : \mathbb{Z} \to I$ have $\tau(k) = kx$, and let $\varphi : \mathbb{Z} \to \mathbb{Z}$ have $\varphi(k) = kn$. Setting up Figure 4.4, we obtain $\sigma : \mathbb{Z} \to I$ with $\tau = \sigma \varphi$. If we put $y = \sigma(1)$ and evaluate both sides at 1, then we obtain $x = \tau(1) = \sigma(\varphi(1)) = \sigma(n) = n\sigma(1) = ny$, as required.

The argument that divisible implies injective uses Zorn's Lemma. Let I be injective, and suppose that $B, C, \varphi,$ and τ are given as in Figure 4.4. Consider the set $\mathcal{S}$ of abelian-group homomorphisms σ' having domain a subgroup of C containing $\varphi(B)$, having range I, and having $\sigma' \varphi = \tau$. Order $\mathcal{S}$ by inclusion upward of the corresponding sets of ordered pairs. The set $\mathcal{S}$ is nonempty because the homomorphism σ' with domain $\varphi(B)$ and values $\sigma'(\varphi(b)) = \tau(b)$ lies in $\mathcal{S}$; σ' is well defined because φ is assumed one-one. Zorn's Lemma yields a maximal element σ in $\mathcal{S}$, say with domain $\overline{C}$. We show that $\overline{C} = C$. Arguing by contradiction, suppose that $\overline{C}$ is a proper subgroup. Let c be in C but not $\overline{C}$. The set of integers k with kc in $\overline{C}$ is an ideal in $\mathbb{Z}$, and we let n be a generator. Since I is divisible, there exists an element a in I with $na = \sigma(nc)$. Define $\widetilde{\sigma}$ on the subgroup generated by c and $\overline{C}$ by the formula $\widetilde{\sigma}(kc + \bar{c}) = ka + \sigma(\bar{c})$ for $k \in \mathbb{Z}$ and $\bar{c} \in \overline{C}$. We need to check that $\widetilde{\sigma}$ is well defined. If $kc + \bar{c} = k'c + \bar{c}'$, then $(k - k')c = \bar{c}' - \bar{c}$ is in $\overline{C}$, and thus $k - k' = qn$ for some integer q. Hence $\widetilde{\sigma}(kc + \bar{c}) - \widetilde{\sigma}(k'c + \bar{c}) = (k - k')a + \sigma(\bar{c} - \bar{c}') = qna + \sigma(\bar{c} - \bar{c}') = q\sigma(nc) + \sigma(\bar{c} - \bar{c}') = q\sigma(nc) - \sigma((k-k')c) = q\sigma(nc) - q\sigma(nc) = 0$. Therefore $\widetilde{\sigma}$ is a nontrivial additive extension of σ, in contradiction to maximality of σ, and the proof is complete.

(3) For $R = \mathbb{Z}$, two good categories that were listed in Section 2 were the category of all finitely generated abelian groups and the category of all torsion abelian groups. With the first of these, Problem 1 at the end of the chapter asks for a verification that some module in the category fails to be a submodule of any injective. With the second of these, the injectives are the torsion divisible groups.

The next proposition extends Example 2 and its proof to general R. Although the condition in the proposition is not very intuitive for general R, it has a simple interpretation for (commutative) principal ideal domains; see Problem 4 at the end of the chapter.

Proposition 4.15. A unital left R module I is injective for the good category of all unital left R modules if and only if every R homomorphism of a left ideal J of R into I extends to an R homomorphism $R \to I$.

PROOF. The necessity is immediate from Figure 4.4 and the definition of "injective" if we take $B = J$, $C = R$ and write τ for the given R homomorphism of J into I.

For the sufficiency, suppose that I and a diagram as in Figure 4.4 are given. Consider the set $\mathcal{S}$ of R module homomorphisms σ' having domain an R submodule of C containing $\varphi(B)$ and having range I such that $\sigma'\varphi = \tau$, and order $\mathcal{S}$ by inclusion upward of the corresponding sets of ordered pairs. The set $\mathcal{S}$ is nonempty because the homomorphism σ' with domain $\varphi(B)$ and values $\sigma'(\varphi(b)) = \tau(b)$ lies in $\mathcal{S}$; σ' is well defined because φ is assumed one-one. Zorn's Lemma yields a maximal element σ in $\mathcal{S}$, say with domain $\overline{C}$. We show that $\overline{C} = C$. Arguing by contradiction, suppose that $\overline{C}$ is a proper R submodule of C. Let c be in C but not $\overline{C}$. The set of elements $r \in R$ with rc in $\overline{C}$ is a left ideal J in R, and the mapping $\psi(r) = \sigma(rc)$ is a well-defined R homomorphism of J into I. By hypothesis, ψ extends to an R homomorphism

$\Psi : R \to I$. Define $\widetilde{\sigma}$ on the subgroup generated by c and $\overline{C}$ by the formula $\widetilde{\sigma}(rc + \bar{c}) = \Psi(r) + \sigma(\bar{c})$ for $r \in R$ and $\bar{c} \in \overline{C}$. We need to check that $\widetilde{\sigma}$ is well defined. If $rc + \bar{c} = r'c + \bar{c}'$, then $(r - r')c = \bar{c}' - \bar{c}$ is in $\overline{C}$, and thus $r - r'$ is in J. Consequently $\Psi(r) - \Psi(r') = \psi(r - r') = \sigma((r - r')c)$. Hence $\widetilde{\sigma}(rc+\bar{c}) - \widetilde{\sigma}(r'c+\bar{c}) = (\Psi(r) - \Psi(r')) + \sigma(\bar{c}-\bar{c}') = \sigma((r-r')c) + \sigma(\bar{c}-\bar{c}') = \sigma((r - r')c) - \sigma((r - r')c) = 0$. Therefore $\widetilde{\sigma}$ is a nontrivial extension of σ, in contradiction to maximality of σ, and the proof is complete. $\square$

Now we can prove an analog of Theorem 4.12 for cochain complexes. This result had no counterpart in Chapter III.

Theorem 4.16. Let $X = \{(X_n, d_n)\}_{n=-\infty}^{\infty}$ and $X' = \{(X'_n, d'_n)\}_{n=-\infty}^{\infty}$ be cochain complexes in the good category $\mathcal{C}$, and let r be an integer. Let $\{f_n : X_n \to X'_n\}_{n \leq r}$ be a family of maps in $\mathcal{C}$ such that $d'_{n-1} f_{n-1} = f_n d_{n-1}$ for $n \leq r$. If X is exact at each X_n with $n \geq r$ and X'_n is injective for $n > r$, then $\{f_n : X_n \to X'_n\}_{n \leq r}$ extends to a cochain map $f : X \to X'$, and f is unique up to homotopy. More precisely any two extensions are homotopic by a homotopy h such that $h_n = 0$ for $n \leq r$.

REMARKS. The diagrams in question are

$$\cdots \xrightarrow{d_{n-2}} X_{n-1} \xrightarrow{d_{n-1}} X_n \xrightarrow{d_n} X_{n+1} \xrightarrow{d_{n+1}} \cdots$$
$$\downarrow f_{n-1} \qquad \downarrow f_n \qquad \downarrow f_{n+1}$$
$$\cdots \xrightarrow{d'_{n-2}} X'_{n-1} \xrightarrow{d'_{n-1}} X'_n \xrightarrow{d'_n} X'_{n+1} \xrightarrow{d'_{n+1}} \cdots$$

for the construction of the cochain map and

$$\cdots \longrightarrow X_{n-1} \xrightarrow{d_{n-1}} X_n \xrightarrow{d_n} X_{n+1} \xrightarrow{d_{n+1}} X_{n+2} \longrightarrow \cdots$$
$$\downarrow f_{n-1} \;\swarrow h_n \;\downarrow f_n \;\swarrow h_{n+1} \;\downarrow f_{n+1} \;\swarrow h_{n+2} \;\downarrow f_{n+2}$$
$$\cdots \longrightarrow X'_{n-1} \xrightarrow{d'_{n-1}} X'_n \xrightarrow{d'_n} X'_{n+1} \xrightarrow{d'_{n+1}} X'_{n+2} \longrightarrow \cdots$$

for the construction of the homotopy.

PROOF. For the existence of the cochain map, it is enough by induction to construct f_{r+1}. Matters are therefore as in the first of the above diagrams with $n = r$. Since X is exact at X_r and X'_{r+1} is injective, we are in the situation of Lemma 4.14 with $I = X'_{r+1}$, $A'' = X_{r+1}$, $A = X_r$, $A' = X_{r-1}$, $\psi = d_{r-1}$, $\varphi = d_r$, and $\tau = d'_r f_r$. The lemma gives a map $\sigma : A'' \to I$ with $\sigma \varphi = \tau$. If we take $f_{r+1} = \sigma$, then $\sigma \varphi = \tau$ says that $f_{r+1} d_r = d'_r f_r$, and the inductive construction of the cochain map is complete.

For the uniqueness up to homotopy, let $f : X \to X'$ and $g : X \to X'$ be two cochain maps such that $f_n = g_n$ for $n \leq r$. Define $h_n : X_n \to X'_{n-1}$ to be 0 for $n \leq r+1$, and observe that the system of functions $\{h_n\}_{n \leq r}$ satisfies $h_{n+1}d_n + d'_{n-1}h_n = f_n - g_n$ for $n \leq r$ because $f_n = g_n$ for $n \leq r$. Proceeding inductively, suppose that $s \geq r$ and that h_n has been constructed for $n \leq s+1$ such that $h_{n+1}d_n + d'_{n-1}h_n = f_n - g_n$ for $n \leq s$. We are to construct $h_{s+2} : X_{s+2} \to X'_{s+1}$. This is the situation of the second diagram with $n = s$. Since $s \geq r$, X is exact at X_{s+1} and X'_{s+1} is injective. Thus we are in the situation of Lemma 4.14 with $I = X'_{s+1}$, $A'' = X_{s+2}$, $A = X_{s+1}$, $A' = X_s$, $\psi = d_s$, $\varphi = d_{s+1}$, and $\tau = (f_{s+1} - g_{s+1}) - d'_s h_{s+1}$. The lemma gives a map $\sigma : A'' \to I$ with $\sigma \varphi = \tau$. If we take $h_{s+2} = \sigma$, then $\sigma \varphi = \tau$ says that $h_{s+2}d_{s+1} = (f_{s+1} - g_{s+1}) - d'_s h_{s+1}$, and the inductive construction of the homotopy is complete. □

A cochain complex $X = (X^+ \xleftarrow{\varepsilon} M)$ that forms a resolution is called an **injective resolution** of M if every X_n for $n \geq 0$ is an injective.

Corollary 4.17. Let M be a module in a good category $\mathcal{C}$ and let

$$X = (X^+ \xleftarrow{\varepsilon} M) \quad \text{and} \quad X' = (X'^+ \xleftarrow{\varepsilon'} M)$$

be two injective resolutions of M. Then there exist cochain maps $f : X \to X'$ and $g : X' \to X$ with $f_{-1} = 1_M$ and $g_{-1} = 1_M$, and any two such cochain maps f and g have the property that $gf : X \to X$ is homotopic to 1_X and $fg : X' \to X'$ is homotopic to $1_{X'}$.

PROOF. The existence of f extending $f_{-1} = 1_M$ is immediate by applying the first part of Theorem 4.16 with $r = -1$. The hypotheses apply because X is exact at X_n for $n \geq -1$ and X'_n is injective for $n > -1$. A similar argument shows the existence of g.

If we have f and g, then $gf : X \to X$ and $1_X : X \to X$ are cochain maps that extend the partial cochain map given for $n \leq -1$ by 1_M for $n = -1$ and by 0 for $n \leq -2$. Since again X is exact at X_n for $n \geq -1$ and X'_n is injective for $n > -1$, the second part of the theorem shows that gf and 1_X are homotopic. A similar argument shows that fg and $1_{X'}$ are homotopic. □

We conclude with elementary characterizations of projectives and injectives that will turn out to be quite useful in the next two sections. We begin with a lemma[6] that will be useful now and will be helpful as motivation in the next section.

---

[6]The lemma is a slight variant of Problem 5 at the end of Chapter X of *Basic Algebra*.

Lemma 4.18. Let $\mathcal{C}$ be a good category of unital left R modules, and let

$$0 \longrightarrow A \xrightarrow{\varphi} B \xrightarrow{\psi} C \longrightarrow 0$$

be an exact sequence in $\mathcal{C}$. Then the following conditions are equivalent:
  (a) B is a direct sum $B = B' \oplus \ker \psi$ of modules in $\mathcal{C}$,
  (b) there exists an R homomorphism $\sigma : C \to B$ such that $\psi \sigma = 1_C$,
  (c) there exists an R homomorphism $\tau : B \to A$ such that $\tau \varphi = 1_A$.

REMARK. When the equivalent conditions of this lemma are satisfied, one says that the exact sequence is **split**.

PROOF. If (a) holds, then $\psi\big|_{B'}$ is one-one from B' onto C. Let σ be its inverse. Then $\sigma : C \to B'$ is one-one with $\psi \sigma = 1_C$. So (b) holds.

If (b) holds, then any b in B has the property that $b - \sigma \psi(b)$ has $\psi(b - \sigma\psi(b)) = \psi(b) - 1_C \psi(b) = 0$ and is therefore in image φ. Write $b - \sigma \psi(b) = \varphi(a)$ for some a depending on b; a is unique because φ is one-one. If $\tau : B \to A$ is defined by $\tau(b) = a$, then τ is an R homomorphism by the uniqueness of a. Consider $\tau(\varphi(a))$ for a in A. The element $b = \varphi(a)$ has $b - \sigma\psi(b) = \varphi(a) - \sigma\psi\varphi(a) = \varphi(a) - \sigma(0) = \varphi(a)$, and the definition of τ therefore says that $\tau(\varphi(a)) = a$. Hence $\tau \varphi = 1_A$, and (c) holds.

If (c) holds, then $B' = \ker \tau$ is an R submodule of B. If b is in $B' \cap \text{image } \varphi$, then $b = \varphi(a)$ for some $a \in A$ and also $0 = \tau(b) = \tau\varphi(a) = 1_A(a) = a$. So $b = 0$, and $B' \cap \text{image } \varphi = 0$. If $b \in B$ is given, write $b = (b - \varphi\tau(b)) + \varphi\tau(b)$. Then $\varphi\tau(b)$ is certainly in image φ, and $\tau(b - \varphi\tau(b)) = \tau(b) - 1_A\tau(b) = 0$ shows that $b - \varphi\tau(b)$ is in B'. Therefore $B = B' \oplus \text{image } \varphi$. Since image $\varphi = \ker \psi$, we see that $B = B' \oplus \ker \psi$ and that (a) holds. $\square$

Proposition 4.19. If $\mathcal{C}$ is a good category of unital left R modules, then
  (a) a module P in $\mathcal{C}$ is projective if and only if $\text{Hom}_R(P, \cdot)$ is an exact functor from $\mathcal{C}$ into $\mathcal{C}_\mathbb{Z}$, if and only if every exact sequence

$$0 \longrightarrow A \xrightarrow{\varphi} B \xrightarrow{\psi} C \longrightarrow 0$$

  in $\mathcal{C}$ splits when its third nonzero member C equals P, and
  (b) a module I in $\mathcal{C}$ is injective if and only if $\text{Hom}_R(\cdot, I)$ is an exact functor from $\mathcal{C}$ into $\mathcal{C}_\mathbb{Z}$, if and only if every exact sequence

$$0 \longrightarrow A \xrightarrow{\varphi} B \xrightarrow{\psi} C \longrightarrow 0$$

  in $\mathcal{C}$ splits when its first nonzero member A equals I.

PROOF. For (a), suppose that P is given. The functor $\operatorname{Hom}_R(P, \cdot)$ is covariant and left exact, no matter what P is. Proposition 4.3 shows it is exact if and only if it carries short exact sequences into short exact sequences, and the left exactness means that the functor is exact if and only if it carries onto maps from B to C to onto maps from $\operatorname{Hom}_R(P, B)$ to $\operatorname{Hom}_R(P, C)$. If $\psi : B \to C$ is given, then $\operatorname{Hom}(1, \psi) : \operatorname{Hom}_R(P, B) \to \operatorname{Hom}_R(P, C)$ operates on a map σ in $\operatorname{Hom}_R(P, B)$ by $\operatorname{Hom}(1, \psi)(\sigma) = \psi\sigma$. The statement that the equation $\psi\sigma = \tau$ is solvable for σ for each τ in $\operatorname{Hom}_R(P, C)$ whenever ψ is onto is precisely the statement that Figure 4.3 is solvable for σ for all possible τ's whenever $B \longrightarrow C \longrightarrow 0$ is exact, and thus P is projective if and only if the functor is exact.

If P is projective and an exact sequence with $C = P$ is given, take $\tau = 1_P$ in Figure 4.3. The projective property yields a map $\sigma : P \to B$ with $\psi\sigma = 1_P$, and Lemma 4.18b shows that the exact sequence splits.

Conversely suppose that every short exact sequence with P as its third nonzero member splits. Suppose that a diagram as in Figure 4.3 is given with $\psi : C \to B$ onto and with τ mapping P into B. Let $S = C \oplus P$, and let T be the R submodule $\{(c, x) \in C \oplus P \mid \psi(c) = \tau(x)\}$ of S. Denote the projections of S to C and P by p_C and p_P, and let $j : T \to S$ be the inclusion. The map[7] $p_P j$ carries T onto P; in fact, if $x \in P$ is given, then $\psi : C \to B$ onto implies that there exists $c_x \in C$ with $\psi(c_x) = \tau(x)$. Then (c_x, x) lies in T, and $p_P j(c_x, x) = p_P(c_x, x) = x$. Consequently we have a 5-term exact sequence with terms 0, $\ker(p_P j)$, T, P, 0, and this must split by hypothesis. Thus there exists a map $q : P \to T$ with $p_P jq = 1_P$. Define $\sigma = p_C jq$. For $x \in P$, $jq(x)$ is some member of S of the form (c, x) with $\psi(c) = \tau(x)$. Hence $\psi\sigma(x) = \psi p_C jq(x) = \psi p_C(c, x) = \psi(c) = \tau(x)$. Thus $\psi\sigma = \tau$, and $\sigma : P \to C$ is the required map that exhibits P as projective.

For (b), suppose that I is given. The functor $\operatorname{Hom}_R(\cdot, I)$ is contravariant and left exact, no matter what I is. It is exact if and only if it carries one-one maps from A to B to onto maps from $\operatorname{Hom}_R(B, I)$ to $\operatorname{Hom}_R(A, I)$. If $\varphi : A \to B$ is given, then $\operatorname{Hom}(\varphi, 1) : \operatorname{Hom}_R(B, I) \to \operatorname{Hom}_R(A, I)$ operates on a map σ in $\operatorname{Hom}_R(B, I)$ by $\operatorname{Hom}(\varphi, 1)(\sigma) = \sigma\varphi$. The statement that the equation $\sigma\varphi = \tau$ is solvable for σ for each τ in $\operatorname{Hom}_R(A, I)$ whenever φ is one-one is precisely the statement that Figure 4.4 is solvable for σ for all possible τ's whenever $0 \longrightarrow A \longrightarrow B$ is exact, and thus I is injective if and only if the functor is exact.

If I is injective and an exact sequence with $A = I$ is given, take $\tau = 1_I$ in Figure 4.4. The injective property yields a map $\sigma : B \to I$ with $\sigma\varphi = 1_I$, and Lemma 4.18c shows that the exact sequence splits.

Conversely suppose that every short exact sequence with I as its first nonzero member splits. Suppose that a diagram as in Figure 4.4 is given with $\varphi : A \to B$ one-one and with τ mapping A into I. Let $S = B \oplus I$, and let T be the quotient of

---

[7]The pair $(p_C j, p_P j)$ is called the **pullback** of (τ, ψ). See Problem 35 at the end of the chapter.

S by the R submodule $\{(\varphi(a), -\tau(a)) \mid a \in A\}$. Denote the inclusions of B and I into S by i_B and i_I, and let $k : S \to T$ be the quotient mapping. The composition[8] ki_I is one-one from I into T. In fact, if $ki_I(x) = 0$ for some $x \in I$, then $(0, x)$ is a member of S of the form $(\varphi(a), -\tau(a))$ for some $a \in A$; thus $\varphi(a) = 0$, and the fact that φ is one-one implies that $a = 0$ and hence that $x = -\tau(a) = 0$. Consequently we have a 5-term exact sequence with terms $0, I, T, T/I, 0$, and this must split by hypothesis. Thus there exists a map $r : T \to I$ with $rki_I = 1_I$. Define $\sigma = rki_B$. For $a \in A$, $i_B\varphi(a) - i_I\tau(a) = (\varphi(a), -\tau(a))$ is in $\ker k$. Thus $ki_B\varphi(a) = ki_I\tau(a)$, and $\sigma\varphi(a) = rki_B\varphi(a) = rki_I\tau(a) = 1_I\tau(a) = \tau(a)$ for $a \in A$. Therefore $\sigma\varphi = \tau$, and $\sigma : A \to I$ is the required map that exhibits I as injective. □

5. Derived Functors

Now we shall undertake the main construction of the chapter, that of "derived functors." Let $\mathcal{C}$ be a good category of unital left R modules. Arranging for derived functors to be defined on every module in $\mathcal{C}$ requires that each module M in $\mathcal{C}$ have either a projective resolution or an injective resolution, and thus $\mathcal{C}$ must have either many projectives or many injectives in a suitable sense. Let us make the condition precise.

We say that $\mathcal{C}$ has **enough projectives** if every module in $\mathcal{C}$ is a quotient of a projective in $\mathcal{C}$. Suppose that this condition is satisfied. Let M be a module in $\mathcal{C}$, and let X_0 be a projective that maps onto M, say by a map ε. Then $\ker \varepsilon$ is in $\mathcal{C}$, since good categories are closed under the passage to submodules, and we let X_1 be a projective in $\mathcal{C}$ that maps onto $\ker \varepsilon$, say by a map ∂_0. Similarly let X_2 be a projective that maps onto $\ker \partial_0$ in X_1, say by a map ∂_1, and so on. The result is that we obtain a projective resolution of the form $X^+ \xrightarrow{\varepsilon} M$ with X^+ given by

$$X^+ : \quad \cdots \longrightarrow X_2 \xrightarrow{\partial_1} X_1 \xrightarrow{\partial_0} X_0.$$

Consequently the condition "enough projectives" implies that every module in $\mathcal{C}$ has a projective resolution in $\mathcal{C}$.

Similarly we say that $\mathcal{C}$ has **enough injectives** if every module in $\mathcal{C}$ is a submodule of an injective in $\mathcal{C}$. Suppose that this condition is satisfied. Let M be a module in $\mathcal{C}$, and let X_0 be an injective into which M embeds, say by a map ε. Then $X_0/\operatorname{image} \varepsilon$ is in $\mathcal{C}$, since good categories are closed under the passage to quotient modules, and we let X_1 be an injective into which $X_0/\operatorname{image} \varepsilon$ embeds, say by a map $d_0^\#$. Let d_0 be the composition of the quotient map from X_0 to $X_0/\operatorname{image} \varepsilon$, followed by $d_0^\#$; then d_0 maps X_0 into X_1 with $\ker d_0 = \operatorname{image} \varepsilon$.

---

[8]The pair (ki_B, ki_I) is called the **pushout** of (τ, φ). See Problem 35 at the end of the chapter.

We let X_2 be an injective into which $X_1/\operatorname{image} d_0$ embeds, say by $d_1^\#$, and we let d_1 be the composition of the quotient map from X_1 to $X_1/\operatorname{image} d_0$, followed by $d_1^\#$; then d_1 maps X_1 into X_2 with $\ker d_1 = \operatorname{image} d_0$. Continuing in this way, we obtain an injective resolution of the form $X^+ \xleftarrow{\varepsilon} M$ with X^+ given by

$$X^+ : \quad \cdots \xleftarrow{d_2} X_2 \xleftarrow{d_1} X_1 \xleftarrow{d_0} X_0.$$

Consequently the condition "enough injectives" implies that every module in $\mathcal{C}$ has an injective resolution in $\mathcal{C}$.

The category $\mathcal{C}_R$ of all unital left R modules certainly has enough projectives. In fact, every module in $\mathcal{C}_R$ is the quotient of a free R module, and free R modules are projective in $\mathcal{C}_R$. It is less trivial but still true that $\mathcal{C}_R$ has enough injectives. Let us pause for a moment to prove this result in Proposition 4.20 below.

As is shown in Problems 1–2 at the end of the chapter, other good categories of unital left R modules may or may not have enough projectives or enough injectives, and a good category may have the one without the other.

Proposition 4.20. If R is any ring with identity, then the category of all unital left R modules has enough injectives.

PROOF. We treat first the case that $R = \mathbb{Z}$. In view of Example 2 of injectives, we are to exhibit an arbitrary abelian group A as isomorphic to a subgroup of a divisible group. We know that A is isomorphic to a quotient of some free abelian group. Write $A \cong F/S$ with F a direct sum of copies of $\mathbb{Z}$ and S equal to some subgroup of F. Taking a $\mathbb{Z}$ basis for F and forming a $\mathbb{Q}$ vector space with that same basis, we can regard F as a subgroup of the additive group D of a rational vector space. The group D is divisible, and A is isomorphic to a subgroup of D/S. Any quotient of a divisible group is divisible, and thus D/S is divisible.

Now we allow R to be any ring with identity. We shall make use of various results from Chapter X of *Basic Algebra*. If M is any unital left R module, let us denote by $\mathcal{F}M$ the underlying abelian group[9] of M. If we regard R as an $(\mathbb{Z}, R)$ bimodule, then Proposition 10.17 makes $\operatorname{Hom}_\mathbb{Z}(R, \mathcal{F}M)$ into a left R module, with $r\varphi(r') = \varphi(r'r)$ for r and r' in R. The mapping $m \mapsto \varphi_m$ with $\varphi_m(r) = rm$ is a one-one R homomorphism of M into $\operatorname{Hom}_\mathbb{Z}(R, \mathcal{F}M)$. From the previous paragraph we can find a divisible abelian group with $\mathcal{F}M \subseteq D$, and we can then regard the left R module $\operatorname{Hom}_\mathbb{Z}(R, \mathcal{F}M)$ as an R submodule of $\operatorname{Hom}_\mathbb{Z}(R, D)$. Consequently we can regard M as an R submodule of $\operatorname{Hom}_\mathbb{Z}(R, D)$. We are going to prove that $I = \operatorname{Hom}_\mathbb{Z}(R, D)$ is injective in $\mathcal{C}_R$.

We digress for a moment to make a side calculation. With D fixed and N equal to any unital left R module, we make use of the isomorphism

$$\operatorname{Hom}_R(N, \operatorname{Hom}_\mathbb{Z}(R, D)) \cong \operatorname{Hom}_\mathbb{Z}(R \otimes_R N, D)$$

---

[9] $\mathcal{F}$ is called the **forgetful functor** from $\mathcal{C}_R$ to $\mathcal{C}_\mathbb{Z}$.

given in Proposition 10.23 of *Basic Algebra*; in the expression $R \otimes_R N$, the left factor of R is to be regarded as a right R module (and not also a left R module), and then $R \otimes_R N$ is really $\mathcal{F}(R \otimes_R N)$ in the sense that the tensor product retains only the structure of an abelian group. Meanwhile, Corollary 10.19a gives us

$$\operatorname{Hom}_{\mathbb{Z}}(R \otimes_R N, D) \cong \operatorname{Hom}_{\mathbb{Z}}(N, D);$$

here the R on the left is an (R, R) bimodule, and the isomorphism is one of left R modules. However, there is no harm in applying $\mathcal{F}$ to both sides and obtaining

$$\operatorname{Hom}_{\mathbb{Z}}(\mathcal{F}(R \otimes_R N, D)) \cong \operatorname{Hom}_{\mathbb{Z}}(\mathcal{F}N, D).$$

Thus
$$\operatorname{Hom}_R(N, \operatorname{Hom}_{\mathbb{Z}}(R, D)) \cong \operatorname{Hom}_{\mathbb{Z}}(\mathcal{F}N, D). \qquad (*)$$

If we track down the isomorphisms in the results of Chapter X, we see that the map from left to right sends $\varphi \in \operatorname{Hom}_R(N, \operatorname{Hom}_{\mathbb{Z}}(R, D))$ to the map $\Phi \in \operatorname{Hom}_{\mathbb{Z}}(\mathcal{F}N, D)$ with $\Phi(x) = \varphi(x)(1)$ for $x \in N$, and the inverse sends Φ to φ with $\varphi(x)(r) = \Phi(rn)$.

Now we return to $I = \operatorname{Hom}_{\mathbb{Z}}(R, D)$. By Proposition 4.19b, I will be injective if and only if $\operatorname{Hom}_R(\cdot, I)$ is an exact functor. Since this functor is contravariant and left exact, it is enough to prove that if $0 \longrightarrow A \xrightarrow{\psi} B$ is exact in $\mathcal{C}_R$, then

$$\operatorname{Hom}_R(B, I) \xrightarrow{\operatorname{Hom}(\psi, 1)} \operatorname{Hom}_R(A, I) \longrightarrow 0 \qquad (**)$$

is exact in $\mathcal{C}_{\mathbb{Z}}$. Let us reinterpret $(**)$ in the light of the isomorphism $(*)$ when $N = B$ and $N = A$. If φ is in $\operatorname{Hom}_R(B, \operatorname{Hom}_{\mathbb{Z}}(R, D))$, then $\operatorname{Hom}(\psi, 1)(\varphi)$ is the member $\varphi\psi$ of $\operatorname{Hom}_R(A, \operatorname{Hom}_{\mathbb{Z}}(R, D))$. The corresponding members of $\operatorname{Hom}_{\mathbb{Z}}(\mathcal{F}B, D)$ and $\operatorname{Hom}_{\mathbb{Z}}(\mathcal{F}A, D)$ are Φ with $\Phi(b) = \varphi(b)(1)$ and a member Φ' of $\operatorname{Hom}_{\mathbb{Z}}(\mathcal{F}A, D)$ with $\Phi'(a) = \varphi\psi(a)(1)$. Thus $\Phi' = \Phi(\mathcal{F}\psi)$, and the mapping $\operatorname{Hom}(\psi, 1)$ in $(**)$ translates under the isomorphisms $(*)$ into the mapping $\operatorname{Hom}(\mathcal{F}\psi, 1)$ of $\operatorname{Hom}_{\mathbb{Z}}(\mathcal{F}B, D)$ into $\operatorname{Hom}_{\mathbb{Z}}(\mathcal{F}A, D)$. The group D is divisible, hence injective in $\mathcal{C}_{\mathbb{Z}}$. Since $\mathcal{F}\psi : \mathcal{F}A \to \mathcal{F}B$ is one-one and D is injective in $\mathcal{C}_{\mathbb{Z}}$, Proposition 4.19b shows that $\operatorname{Hom}(\mathcal{F}\psi, 1)$ carries $\operatorname{Hom}_{\mathbb{Z}}(\mathcal{F}B, D)$ onto $\operatorname{Hom}_{\mathbb{Z}}(\mathcal{F}A, D)$. Therefore $(**)$ is exact, and we conclude that I is injective in $\mathcal{C}_R$. □

Derived functors of an additive functor F from one good category to another will be useful when F is left exact or right exact, and there will be one derived functor for each integer $n \geq 0$. The value of the nth derived functor on a module M is obtained by taking a projective or injective resolution of M according to the rule in Figure 4.5, applying F to the resolution, dropping the term $F(M)$

that occurs in degree -1, and forming the nth homology or cohomology of the resulting complex. The full traditional notation for the derived functor in question appears in Figure 4.5, along with an abbreviated notation that we shall tend to use.

The choice of projective or injective resolution at the start is made in such a way that the 0th derived functor is naturally isomorphic to F; this condition will be clarified in Proposition 4.21 below. If a projective resolution is to be used, one makes the assumption that the domain category has enough projectives; if an injective resolution is to be used, one makes the assumption that the domain category has enough injectives.

If the resulting complex obtained by applying F to the resolution is a chain complex, the abbreviated notation is F_n for the nth derived functor; otherwise it is F^n. The full traditional notation involves using an L or R in front of F to denote the one-sided exactness, left or right, that F is *not* assumed to have, and the subscript or superscript n is moved from F to the L or R.

Exactness	—variant	Resolution	—ology	Notation	Example
right	co—	projective	hom—	F_n, $L_n F$	$M \otimes_R (\cdot)$
right	contra—	injective	hom—	F_n, $L_n F$	$M \otimes_{\mathbb{Z}} \text{Hom}_{\mathbb{Z}}(\cdot, I)$, I injective
left	co—	injective	cohom—	F^n, $R^n F$	$\text{Hom}_R(M, \cdot)$
left	contra—	projective	cohom—	F^n, $R^n F$	$\text{Hom}_R(\cdot, M)$

FIGURE 4.5. Formation of derived functors.

There are several things that need elaboration in this definition, and we take them up right away.

First there is the fact that $F_n(M)$ or $F^n(M)$ is well defined. Suppose that we start with two resolutions X and X' of M (projective or injective by the rules in Figure 4.5). Corollary 4.13 or 4.17 gives us chain or cochain maps $f : X \to X'$ and $g : X' \to X$ with $f_{-1} = 1_M$ and $g_{-1} = 1_M$ and shows that $gf : X \to X$ is homotopic to 1_X and that $fg : X' \to X'$ is homotopic to $1_{X'}$. For definiteness let us suppose that F is covariant and right exact; then chain maps are involved and the derived functors of F are to be denoted by F_n. Applying F to our chain maps, we obtain chain maps $F(f) : F(X) \to F(X')$, $F(g) : F(X') \to F(X)$, $F(gf) : F(X) \to F(X)$, and $F(fg) : F(X') \to F(X')$. The last two of these are homotopic to $1_{F(X)} : F(X) \to F(X)$ and to $1_{F(X')} : F(X') \to F(X')$, respectively, by F of the respective homotopies. Proposition 4.1 shows that $F(g)F(f) = F(gf)$ induces the identity on $H_*(F(X))$ and that $F(f)F(g) = F(fg)$ induces the identity on $H_*(F(X'))$. Consequently the mappings induced on homology by $F(f)$ and $F(g)$ are two-sided inverses of one another. Thus $F_n(M)$ as computed from X is isomorphic to $F_n(M)$ as computed from X'.

Moreover, this isomorphism is canonical. If $f' : X \to X'$ is another chain map, then the same calculation shows that $F(f')$ and $F(g)$ induce two-sided inverses of each other on homology, and hence $F(f) = F(f')$ on homology. Thus $F_n(M)$ is well defined up to canonical isomorphism when F is covariant and right exact. The other three situations in Figure 4.5 are handled in similar fashion and lead to analogous conclusions.

Next we make F_n or F^n into a functor. To do do, let $\varphi : M \to M'$ be given. For definiteness, again let us suppose that F is covariant and right exact. Let X and X' be projective resolutions of M and M', respectively, and apply Theorem 4.12 to produce a chain map $\Phi : X \to X'$ with $\Phi_{-1} = \varphi$. Then $F(\Phi) : F(X) \to F(X')$ is a chain map and induces maps on homology that we denote by $F_n(\varphi)$. Here $F_n(\varphi)$ maps $F_n(M)$ into $F_n(M')$.

Let us see that $F_n(\varphi)$ is well defined. If X is replaced by $\overline{X}$, Corollary 4.13 produces chain maps $f : X \to \overline{X}$ and $g : \overline{X} \to X$ with $f_{-1} = 1_M$ and $g_{-1} = 1_M$, and Theorem 4.12 produces a chain map $\overline{\Phi} : \overline{X} \to X'$ with $\overline{\Phi}_{-1} = \varphi$. Since $\overline{\Phi} \circ f$ and Φ are both chain maps from X to X' that equal φ in degree -1, Theorem 4.12 shows that $\overline{\Phi} \circ f$ is homotopic to Φ. Similarly $\Phi \circ g$ and $\overline{\Phi}$ are chain maps from $\overline{X}$ to X' and are homotopic. By Proposition 4.1, $F(\overline{\Phi} \circ f) = F(\Phi)$ on homology, and $F(\Phi \circ g) = F(\overline{\Phi})$ on homology. Thus on homology $F(\overline{\Phi})$ corresponds to $F(\Phi)$ under the canonical isomorphism $F(f)$, whose inverse on homology is $F(g)$. In short, $F_n(\varphi)$ is well defined up to the previously obtained canonical isomorphisms. The other three situations in Figure 4.5 are handled in similar fashion and lead to analogous conclusions.

Tracing through the definition of how derived functors affect maps, we see that the map 1 goes to the map 1 and that compositions go to compositions, in the same order as for F. Thus the derived functors are indeed functors. The derived functors of a covariant functor are covariant, and the derived functors of a contravariant functor are contravariant.

We need to check that the derived functors are additive. If $\varphi : M \to M'$ and $\varphi' : M \to M'$ are given, then we can proceed as above and use a single resolution of M and a single resolution of M' to investigate φ, φ', and $\varphi + \varphi'$. Then it is apparent that the chain or cochain maps built from maps of M to M' add in the same way as the maps, and the result is that each F_n or F^n is additive with particular choices of the resolutions in place. Allowing the resolutions to vary means that we have to take canonical isomorphisms into account, and after doing so, we still get additivity.

If two functors F and G from $\mathcal{C}$ to $\mathcal{C}'$ of the same type in Figure 4.5 are naturally isomorphic, then F_n and G_n (or else F^n and G^n) are naturally isomorphic for all n. In fact, if T is the natural isomorphism, then T associates a member T_A

of $\text{Hom}(F(A), G(A))$ to each module A in $\mathcal{C}$. Take a projective or injective resolution $X = \{X_n\}$ of A, as appropriate, and form the two complexes $F(X)$ and $G(X)$. The system $\{T_{X_n}\}$ is then a chain map from $F(X)$ to $G(X)$, with inverse $\{T_{X_n}^{-1}\}$, and the homology or cohomology of $F(X)$ is exhibited as isomorphic to the homology or cohomology of $G(X)$. This much shows that $F_n(A) \cong G_n(A)$ (or $F^n(A) \cong G^n(A)$) for all n. We omit the details of verifying the naturality of this isomorphism in the A variable for each n.

Proposition 4.21. In the four situations of derived functors in Figure 4.5, under the assumption that the domain category for F has enough projectives or enough injectives as appropriate, the 0th derived functor of F is naturally isomorphic to F.

PROOF IF F IS COVARIANT AND RIGHT EXACT. Let

$$X_1 \xrightarrow{\partial_0} X_0 \xrightarrow{\varepsilon} M \longrightarrow 0$$

be the terms in degree $1, 0, -1, -2$ of a projective resolution of M. By Proposition 4.5 and its remark, the right exactness and covariance of F imply that

$$F(X_1) \xrightarrow{F(\partial_0)} F(X_0) \xrightarrow{F(\varepsilon)} F(M) \longrightarrow 0$$

is exact. The derived-functor module $F_0(M)$ is computed as the 0th homology of

$$F(X_1) \xrightarrow{F(\partial_0)} F(X_0) \longrightarrow 0.$$

Thus
$$F_0(M) = F(X_0)/\text{image } F(\partial_0) = F(X_0)/\ker F(\varepsilon).$$

Since $F(\varepsilon)$ is onto $F(M)$, the right side here is $\cong F(M)$ via $F(\varepsilon)$.

This establishes the isomorphism. Let us prove that it is natural in the variable M. If $\varphi : M \to M'$ is given, we are to prove that the diagram

$$\begin{array}{ccc} F_0(M) & \xrightarrow{\text{via } F(\varepsilon)} & F(M) \\ {\scriptstyle F_0(\varphi)}\downarrow & & \downarrow{\scriptstyle F(\varphi)} \\ F_0(M') & \xrightarrow{\text{via } F(\varepsilon')} & F(M') \end{array} \qquad (*)$$

commutes. Using Theorem 4.12, we form the part of a chain map that is indicated:

$$\begin{array}{ccccccc} X_1 & \xrightarrow{\partial_0} & X_0 & \xrightarrow{\varepsilon} & M & \longrightarrow & 0 \\ {\scriptstyle f_1}\downarrow & & {\scriptstyle f_0}\downarrow & & {\scriptstyle \varphi}\downarrow & & \\ X'_1 & \xrightarrow{\partial'_0} & X'_0 & \xrightarrow{\varepsilon'} & M' & \longrightarrow & 0 \end{array}$$

Application of F gives a commutative diagram

$$\begin{array}{ccc} F(X_0) & \xrightarrow{F(\varepsilon)} & F(M) \\ {\scriptstyle F(f_0)}\downarrow & & \downarrow{\scriptstyle F(\varphi)} \\ F(X_0') & \xrightarrow{F(\varepsilon')} & F(M') \end{array}$$

and this becomes $(*)$ upon passage to the quotients $F(X_0)/\ker F(\varepsilon)$ and $F(X_0')/\ker F(\varepsilon')$. This completes the proof. $\square$

EXAMPLES.

(1) The invariants functor $F(M) = M^G$ for a group G. Suppose that a group G acts on an abelian group M by automorphisms. This situation is completely equivalent to considering M as a unital left $\mathbb{Z}G$ module, where $\mathbb{Z}G$ is the integer group ring of G. The subgroup of **invariants** of M is

$$M^G = \{m \in M \mid gm = m \text{ for all } g \in G\}.$$

The formulas $F(M) = M^G$ for such a module M and $F(h) = h\big|_{M^G}$ for h in $\text{Hom}_{\mathbb{Z}G}(M, M')$ define a covariant additive functor called the **invariants functor**; we can think of F as carrying $\mathcal{C}_{\mathbb{Z}G}$ into itself, but it is preferable to think of it as carrying $\mathcal{C}_{\mathbb{Z}G}$ into the category $\mathcal{C}_{\mathbb{Z}}$ of abelian groups. The functor F is naturally isomorphic to the functor $H = \text{Hom}_{\mathbb{Z}G}(\mathbb{Z}, \cdot)$, where $\mathbb{Z}$ is made into a $\mathbb{Z}G$ module with trivial G action; as with F, we consider H as a functor from $\mathcal{C}_{\mathbb{Z}G}$ to $\mathcal{C}_{\mathbb{Z}}$. To see the isomorphism, we associate to each module M the abelian-group homomorphism $T_M : M^G \to \text{Hom}_{\mathbb{Z}}(\mathbb{Z}, M)$ defined by $T_M(m) = \varphi_m$ with $\varphi_m(k) = m$ for all $k \in \mathbb{Z}$. If h is in $\text{Hom}_{\mathbb{Z}G}(M, M')$, then the two maps $T_{M'} \circ F(h)$ and $H(h) \circ T_M$ of $F(M)$ into $H(M')$ are equal, since at each $m \in M^G$ we have

$$H(h)T_M(m) = H(h)(\varphi_m) = \text{Hom}(1, h)(\varphi_m) = h\varphi_m = \varphi_{h(m)} = T_{M'}F(h)(m).$$

This identity means that $\{T_M\}$ is a natural transformation; we readily check for each M that T_M carries M^G one-one onto $\text{Hom}_{\mathbb{Z}}(\mathbb{Z}, M)$, and thus $\{T_M\}$ is a natural isomorphism.

Because of this natural isomorphism, the invariants functor is covariant and left exact. Its derived functors F^n or H^n are obtained by using an injective resolution $I \leftarrow M \leftarrow 0$, applying the functor $(\cdot)^G$ or $\text{Hom}_{\mathbb{Z}G}(\mathbb{Z}, \cdot)$, dropping the term in degree -1, and forming cohomology. Briefly

$$F^n(M) \cong H^n(I^G) \cong H^n(\text{Hom}_{\mathbb{Z}G}(\mathbb{Z}, I))$$

for an injective resolution $I \leftarrow M \leftarrow 0$.

It turns out that the result is given also by the cohomology-of-groups functors $H^n(G, M)$ even though this was not the procedure by which we obtained group cohomology in Section III.5. In fact, what Section III.5 said to do was to start from a free resolution (a projective resolution would have been good enough) such as $P \longrightarrow M \longrightarrow 0$ of $\mathbb{Z}$ in $\mathcal{C}_\mathbb{Z}$, apply the contravariant left exact functor $\text{Hom}_{\mathbb{Z}G}(\,\cdot\,, M)$, drop the term in degree -1, and form cohomology. Briefly then, Section III.5 said that

$$H^n(G, M) \cong H^n(\text{Hom}_{\mathbb{Z}G}(P, M)) \quad \text{for a projective resolution } P \to \mathbb{Z} \to 0.$$

The fact that $H^n(G, M)$ can be computed in either of these ways is not particularly obvious from what we have done so far, but it will be a special case of the natural isomorphism of functors Ext^n and ext^n that is proved as Theorem 4.31 in Section 7. With either formula for $H^n(G, M)$, we obtain $H^0(G, M) \cong M^G$ in agreement with Proposition 4.21.

(2) The co-invariants functor $F(M) = M_G$ for a group G. In the same setting as in Example 1, the subgroup of **co-invariants** of M is

$$M_G = M\big/(\text{subgroup generated by all } gm - m \text{ for } g \in G, \ m \in M).$$

The functor F can be seen to be naturally isomorphic to the functor H with $H(M) = \mathbb{Z} \otimes_{\mathbb{Z}G} M$. It is therefore covariant and right exact. Its derived functors are given by

$$F_n(M) \cong H_n(P_G) \cong H_n(\mathbb{Z} \otimes_{\mathbb{Z}G} P) \quad \text{for a projective resolution } P \to M \to 0.$$

These are by definition the **homology-of-groups** functors $H_n(G, M)$. Although the equality is not particularly obvious, $H_n(G, M)$ can be computed also from

$$H_n(G, M) \cong H_n(P \otimes_{\mathbb{Z}G} M) \quad \text{for a projective resolution } P \to \mathbb{Z} \to 0.$$

This isomorphism is a special case of the natural isomorphism of functors Tor_n and tor_n that is mentioned just before Proposition 4.29 in Section 7; the proof is completely analogous to the proof of Theorem 4.31. With either formula for $H_n(G, M)$, we obtain $H_0(G, M) \cong M_G$ in agreement with Proposition 4.21.

(3) Derived functors with $R = \mathbb{Z}$. For the ring $\mathbb{Z}$ and the category $\mathcal{C}_\mathbb{Z}$ (or more generally for $\mathcal{C}_R$ for any principal ideal domain R), projective resolutions and injective resolutions can be fairly short, and derived functors in degree ≥ 2 are all 0. Let M be a given unital $\mathbb{Z}$ module, i.e., an abelian group. We know that M is the quotient of some free abelian group X_0, say with a quotient map ε, and then $X_1 = \ker \varepsilon$ is a subgroup of a free abelian group and hence is free abelian. Thus a projective resolution of M is

$$0 \longrightarrow X_1 \xrightarrow{\text{inc}} X_0 \xrightarrow{\varepsilon} M \longrightarrow 0.$$

The kinds of derived functors that make use of projective resolutions are the covariant right exact ones and the contravariant left exact ones. If F is such a functor, then we are led to the complexes

$$0 \longrightarrow F(X_1) \xrightarrow{F(\text{inc})} F(X_0) \xrightarrow{F(\varepsilon)} 0$$

and
$$0 \longleftarrow F(X_1) \xleftarrow{F(\text{inc})} F(X_0) \xleftarrow{F(\varepsilon)} 0$$

in the two cases. Thus the values of the derived functors are $F_0(M) \cong M$ and $F_1(M) = \ker F(\varepsilon)$ in the first case, and $F^0(M) \cong M$ and $F^1(M) = \operatorname{coker} F(\varepsilon)$ in the second case. Higher derived functors are 0. Similar remarks apply to injective resolutions and the remaining two cases for derived functors in Figure 4.5. Every abelian group is a subgroup of a divisible group, which is injective in $\mathcal{C}_{\mathbb{Z}}$, and the quotient of the divisible group by the given abelian group is divisible, hence injective. Thus we can arrange for all terms of an injective resolution to be 0 beyond the X_1 term, and an analysis of the results similar to the one above is possible.

6. Long Exact Sequences of Derived Functors

The first four theorems of this section say that a short exact sequence of modules leads to a long exact sequence of derived functor modules and that it does so in a functorial way. Let us suppose that $F : \mathcal{C} \to \mathcal{C}'$ is an additive functor between good categories. For the first of the theorems, suppose further that $\mathcal{C}$ has enough projectives and that F is one of the types of functors in Figure 4.5 making use of projective resolutions in the definition of its derived functors. The last of these conditions means that F is to be covariant right exact or contravariant left exact.

To prove such a theorem, we shall want to apply Theorem 4.7, which produces a long exact sequence from a short exact sequence of complexes. To each of the modules in the given short exact sequence, we attach a projective resolution. If these projective resolutions can somehow be related by chain maps so as to give a short exact sequence of projectives in each degree, then we can apply F to the entire diagram, invoke Theorem 4.7, and obtain the desired long exact sequence. Application of Theorem 4.10, in combination with some further checking, will show that the passage from the given short exact sequence of modules to the long exact sequence of derived functor modules is functorial in the modules of the short exact sequence.

Thus the problem is to obtain the compatible projective resolutions. Proposition 4.19a gives us a clue about what to look for: any short exact sequence of projectives has to be split. Here is the statement of the first theorem.

Theorem 4.22. Let $F : \mathcal{C} \to \mathcal{C}'$ be an additive functor between two good categories. Suppose that F either is covariant right exact or is contravariant left exact, and suppose that $\mathcal{C}$ has enough projectives. Whenever there are three modules and two maps in $\mathcal{C}$ forming a short exact sequence

$$0 \longrightarrow A \overset{\varphi}{\longrightarrow} B \overset{\psi}{\longrightarrow} C \longrightarrow 0,$$

then the derived functors of F on the three modules form a long exact sequence in $\mathcal{C}'$ as follows:

(a) If F is covariant and right exact, then the long exact sequence is

$$0 \longleftarrow F(C) \longleftarrow F(B) \longleftarrow F(A) \longleftarrow F_1(C) \longleftarrow F_1(B) \longleftarrow F_1(A)$$
$$\longleftarrow F_2(C) \longleftarrow F_2(B) \longleftarrow F_2(A) \longleftarrow F_3(C) \longleftarrow \cdots.$$

(b) If F is contravariant and left exact, then the long exact sequence is

$$0 \longrightarrow F(C) \longrightarrow F(B) \longrightarrow F(A) \longrightarrow F^1(C) \longrightarrow F^1(B) \longrightarrow F^1(A)$$
$$\longrightarrow F^2(C) \longrightarrow F^2(B) \longrightarrow F^2(A) \longrightarrow F^3(C) \longrightarrow \cdots.$$

We begin with a lemma.

Lemma 4.23. In the good category $\mathcal{C}$, suppose that the diagram

$$
\begin{array}{ccccccc}
 & & 0 & & 0 & & 0 \\
 & & \downarrow & & \downarrow & & \downarrow \\
0 \longleftarrow & A & \overset{\varepsilon_A}{\longleftarrow} & P_A & \overset{\psi_A}{\longleftarrow} & M_A & \longleftarrow 0 \\
 & \varphi\downarrow & & i_A\downarrow & & \varphi_1\dashv\downarrow & \\
0 \longleftarrow & B & \overset{\varepsilon_B}{\dashleftarrow} & P_A \oplus P_C & \overset{\psi_B}{\dashleftarrow} & M_B & \dashleftarrow 0 \\
 & \psi\downarrow & & p_C\downarrow & & \psi_1\dashv\downarrow & \\
0 \longleftarrow & C & \overset{\varepsilon_C}{\longleftarrow} & P_C & \overset{\psi_C}{\longleftarrow} & M_C & \longleftarrow 0 \\
 & \downarrow & & \downarrow & & \downarrow & \\
 & 0 & & 0 & & 0 &
\end{array}
$$

has the first two columns and the two rows with solid arrows exact and has P_A and P_C projective. Here i_A is the inclusion into the first component of $P_A \oplus P_C$, and p_C is the projection onto the second component. Then there exist a module M_B and maps ε_B, ψ_B, φ_1, and ψ_1 such that the whole diagram, including the dashed arrows, has exact rows and columns and has all squares commuting.

PROOF. The module $P_A \oplus P_C$ is in $\mathcal{C}$ because $\mathcal{C}$ is good, and it is easy to see that $P_A \oplus P_C$ is projective. Let us define ε_B. Since P_C is projective, there exists $h : P_C \to B$ such that $\psi h = \varepsilon_C$, and we put $\varepsilon_B(x_A, x_C) = \varphi \varepsilon_A x_A + h x_C$. Then the equation

$$\varphi \varepsilon_A x_A = \varepsilon_B(x_A, 0) = \varepsilon_B i_A x_A$$

says that the upper left square commutes, and the equation

$$\psi \varepsilon_B(x_A, x_C) = \psi \varphi \varepsilon_A x_A + \psi h x_C = 0 + \varepsilon_C x_C = \varepsilon_C p_C(x_A, x_C)$$

says that the lower left square commutes.

To see that ε_B is onto B, let $b \in B$ be given. Since p_C and ε_C are onto, so is $\varepsilon_C p_C = \psi \varepsilon_B$. Thus we can choose (x_A, x_C) in $P_A \oplus P_C$ with $\psi(b) = \psi \varepsilon_B(x_A, x_C)$. Hence $b - \varepsilon_B(x_a, x_C)$ lies in $\ker \psi = \operatorname{image} \varphi$, and we can write

$$b - \varepsilon_B(x_A, x_C) = \varphi(a) = \varphi \varepsilon_A(x'_A) = \varepsilon_B i_A(x'_A) = \varepsilon_B(x'_A, 0)$$

for some $x'_A \in P_A$. Then $b = \varepsilon_B(x_A + x'_A, x_C)$, and ε_B is onto.

Let $M_B = \ker \varepsilon_B$, and let $\psi_B : M_B \to P_A \oplus P_C$ be the inclusion. For m_A in M_A, let $\varphi_1(m_A) = (\psi_A m_A, 0)$. Then $\varphi_1(m_A)$ is in M_B because

$$\varepsilon_B(\psi_A m_A, 0) = \varphi \varepsilon_A \psi_A m_A + h0 = \varphi 0 + h0 = 0.$$

Moreover, this definition of φ_1 makes the upper right square commute.

To define ψ_1, let (x_A, x_C) be in M_B, so that $\varepsilon_B(x_A, x_C) = 0$. Then $0 = \psi \varepsilon_B(x_A, x_C) = \varepsilon_C p_C(x_A, x_C) = \varepsilon_C(x_C)$, x_C lies in $\ker \varepsilon_C = \operatorname{image} \psi_C$, and $x_C = \psi_C(m_C)$ for a unique m_C in M_C. We put $\psi_1(x_A, x_C) = m_C$. Then the equation

$$\psi_C \psi_1(x_A, x_C) = \psi_C(m_C) = x_C = p_C(x_A, x_C) = p_C \psi_B(x_A, x_C)$$

shows that the lower right square commutes.

Now all the squares commute, and all the rows and columns are exact except possibly the third column. Corollary 4.8 allows us to conclude that the third column is exact, and the proof of the lemma is complete. □

PROOF OF THEOREM 4.22. The main step is to construct projective resolutions of A, B, and C by an inductive process in such a way that the three resolutions together form an exact sequence of chain complexes. We start by forming projective resolutions

$$0 \longleftarrow A \xleftarrow{\varepsilon_A} X_0 \xleftarrow{\alpha_0} X_1 \xleftarrow{\alpha_1} \cdots$$

and

$$0 \longleftarrow C \xleftarrow{\varepsilon_C} Z_0 \xleftarrow{\gamma_0} Z_1 \xleftarrow{\gamma_1} \cdots.$$

Replacing X_1 by $M_A = \ker \alpha_0$ and Z_1 by $M_C = \ker \gamma_0$, we are led to the starting diagram in Lemma 4.23. Application of the lemma produces a short exact sequence
$$0 \longleftarrow B \xleftarrow{\varepsilon_B} X_0 \oplus Z_0 \xleftarrow{\text{inc}} M_B \longleftarrow 0$$
and the vertical maps φ_1 and ψ_1 that make the squares commute in the lemma. Next we move everything one step to the right, applying the lemma to a diagram as in the lemma with first and third rows

$$0 \longleftarrow \ker \varepsilon_A \xleftarrow{\alpha_0} X_1 \xleftarrow{\text{inc}} \ker \alpha_1 \longleftarrow 0$$

and

$$0 \longleftarrow \ker \varepsilon_C \xleftarrow{\gamma_0} Z_1 \xleftarrow{\text{inc}} \ker \gamma_1 \longleftarrow 0$$

and with an exact sequence in the first column involving the maps φ_1 and ψ_1. Application of the lemma produces a short exact sequence
$$0 \longleftarrow \ker \varepsilon_B \xleftarrow{\beta_0} X_1 \oplus Z_1 \xleftarrow{\text{inc}} \ker \beta_0 \longleftarrow 0$$
and the vertical maps φ_2 and ψ_2 that make the squares commute in the lemma. We can put these steps together to form the following diagram with exact rows and columns and with commuting squares:

$$\begin{array}{ccccccccccc}
& & 0 & & 0 & & 0 & & 0 & & \\
& & \downarrow & & \downarrow & & \downarrow & & \downarrow & & \\
0 & \longleftarrow & A & \xleftarrow{\varepsilon_A} & X_0 & \xleftarrow{\alpha_0} & X_1 & \xleftarrow{\text{inc}} & \ker \alpha_1 & \longleftarrow & 0 \\
& & \varphi \downarrow & & i_{X_0} \downarrow & & i_{X_1} \downarrow & & \varphi_2 \downarrow & & \\
0 & \longleftarrow & B & \xleftarrow{\varepsilon_B} & X_0 \oplus Z_0 & \xleftarrow{\beta_0} & X_1 \oplus Z_1 & \xleftarrow{\text{inc}} & \ker \beta_1 & \longleftarrow & 0 \\
& & \psi \downarrow & & p_{Z_0} \downarrow & & p_{Z_1} \downarrow & & \psi_2 \downarrow & & \\
0 & \longleftarrow & C & \xleftarrow{\varepsilon_C} & Z_0 & \xleftarrow{\gamma_0} & Z_1 & \xleftarrow{\text{inc}} & \ker \gamma_1 & \longleftarrow & 0 \\
& & \downarrow & & \downarrow & & \downarrow & & \downarrow & & \\
& & 0 & & 0 & & 0 & & 0 & &
\end{array}$$

We can repeat the use of Lemma 4.23, starting from the last column of the above diagram and more of the projective resolutions of A and C, and then we can merge the new result with the diagram above to obtain a diagram with one additional column. Continuing in this way, we arrive at three projective resolutions and vertical maps that together form an exact sequence of chain complexes.

To obtain a long exact sequence for our derived functors, we apply the functor F to the final diagram above, except that we drop the left column of 0's and the column containing A, B, C. After the application of F, the remaining columns are still exact because the columns in $\mathcal{C}$ are split and because F sends split exact sequences to split exact sequences.[10] Then we apply Theorem 4.7, taking Proposition 4.21 into account, and the long exact sequence results except for the one detail of the 0 at the end. In other words, we still have to prove exactness at $F(C)$. But exactness at this point is immediate from the assumed one-sided exactness of F. This completes the proof. □

Before addressing the functoriality of the association in Theorem 4.22, let us record the corresponding result when the derived functor makes use of injective resolutions.

Theorem 4.24. Let $F : \mathcal{C} \to \mathcal{C}'$ be an additive functor between two good categories. Suppose that F either is contravariant right exact or is covariant left exact, and suppose that $\mathcal{C}$ has enough injectives. Whenever there are three modules and two maps in $\mathcal{C}$ forming a short exact sequence

$$0 \longrightarrow A \xrightarrow{\varphi} B \xrightarrow{\psi} C \longrightarrow 0,$$

then the derived functors of F on the three modules form a long exact sequence in $\mathcal{C}'$ as follows:

(a) If F is contravariant and right exact, then the long exact sequence is

$$0 \longleftarrow F(A) \longleftarrow F(B) \longleftarrow F(C) \longleftarrow F_1(A) \longleftarrow F_1(B) \longleftarrow F_1(C)$$
$$\longleftarrow F_2(A) \longleftarrow F_2(B) \longleftarrow F_2(C) \longleftarrow F_3(A) \longleftarrow \cdots.$$

(b) If F is covariant and left exact, then the long exact sequence is

$$0 \longrightarrow F(A) \longrightarrow F(B) \longrightarrow F(C) \longrightarrow F^1(A) \longrightarrow F^1(B) \longrightarrow F^1(C)$$
$$\longrightarrow F^2(A) \longrightarrow F^2(B) \longrightarrow F^2(C) \longrightarrow F^3(A) \longrightarrow \cdots.$$

PROOF. The necessary modifications to the proof of Theorem 4.22 are fairly straightforward, but some comments are in order concerning how Lemma 4.23 is to be modified. In the diagram in the statement of Lemma 4.23, all the horizontal arrows are to be reversed, the projectives P_A and P_C are to be replaced by injectives

---

[10] A split exact sequence is the union of two four-term exact sequences from each end, and F is exact on each of these. In addition, we saw in Section 2 that F respects direct sums. It follows that F carries split exact sequences to split exact sequences.

6. Long Exact Sequences of Derived Functors

I_A and I_C, and M_A and M_C are the quotients $M_A = I_A/\varepsilon_A(A)$ and $M_C = I_C/\varepsilon_C(C)$. Let us define ε_B. Since I_A is injective, choose $h : B \to I_A$ with $h\varphi = \varepsilon_A$, and put $\varepsilon_B(b) = (h(b), \varepsilon_C \psi(b))$. Then the equation

$$\varepsilon_B \varphi(a) = (h\varphi a, \varepsilon_C \psi \varphi a) = (\varepsilon_A(a), 0) = i_A \varepsilon_A(a)$$

says that the upper left square commutes, and the equation

$$\varepsilon_C \psi(b) = p_C(h(b), \varepsilon_C \psi(b)) = p_C \varepsilon_B(b)$$

says that the lower left square commutes.

To see that ε_B is one-one, let $\varepsilon_B(b) = 0$. Then $0 = p_C \varepsilon_B(b) = \varepsilon_C \psi(b)$. Since ε_C is one-one, $\psi(b) = 0$, b lies in ker ψ = image φ, and $b = \varphi(a)$. Then $0 = \varepsilon_B(b) = \varepsilon_B \varphi(a) = i_A \varepsilon_A(a)$, and $a = 0$ because i_A and ε_A are one-one. Hence $b = \varphi(a) = 0$, and ε_B is one-one.

Let $M_B = (I_A \oplus I_C)/\varepsilon_B(B)$, and let $\psi_B : I_A \oplus I_C \to M_B$ be the quotient map. To define φ_1, we let $\varphi_1(m_A) = \psi_B(x_A, 0)$ if $m_A = \psi_A x_A$ with $x_A \in I_A$. If x'_A is another preimage of m_A under ψ_A^{-1}, then $x'_A - x_A = \varepsilon_A(a)$ for some $a \in A$, and $\psi_B(x_A, 0) - \psi_B(x'_A, 0) = \psi_B i_A \varepsilon_A(a) = \psi_B \varepsilon_B \varphi(a) = 0$; hence φ_1 is well defined. Since $\psi_B i_A x_A = \psi_B(x_A, 0) = \varphi_1 m_A = \varphi_1 \psi_A x_A$, the upper right square commutes. To define ψ_1, let $m_B \in M_B$ be $\psi_B(x_A, x_C)$, and define $\psi_1(m_B) = \psi_C(x_C)$. If (x'_A, x'_C) is another preimage of m_B under ψ_B^{-1}, then $(x'_A, x'_C) - (x_A, x_C) = \varepsilon_B(b)$ for some $b \in B$, and $\psi_C(x'_C) - \psi_C(x_C) = \psi_C p_C(x'_A, x'_C) - \psi_C p_C(x_A, x_C) = \psi_C p_C \varepsilon_B(b) = \psi_C \varepsilon_C \psi(b) = 0$; hence ψ_1 is well defined. Since $\psi_C p_C(x_A, x_C) = \psi_C(x_C) = \psi_1(m_B) = \psi_1 \psi_B(x_A, x_C)$, the lower right square commutes.

Now all the squares commute, and all the rows and columns are exact except possibly the third column. Corollary 4.8 allows us to conclude that the third column is exact, and the proof of the analog of Lemma 4.23 for injectives is complete. Theorem 4.24 then follows routinely. □

Theorem 4.25. Let $F : \mathcal{C} \to \mathcal{C}'$ be an additive functor between two good categories. Suppose that F either is covariant right exact or is contravariant left exact, and suppose that $\mathcal{C}$ has enough projectives. Then the passage as in Theorem 4.22 from short exact sequences in $\mathcal{C}$ to long exact sequences of derived functor modules in $\mathcal{C}'$ is functorial in the following sense: whenever

$$\begin{array}{ccccccccc}
0 & \longrightarrow & \widetilde{A} & \xrightarrow{\widetilde{\varphi}} & \widetilde{B} & \xrightarrow{\widetilde{\psi}} & \widetilde{C} & \longrightarrow & 0 \\
& & f_A \downarrow & & f_B \downarrow & & f_C \downarrow & & \\
0 & \longrightarrow & A & \xrightarrow{\varphi} & B & \xrightarrow{\psi} & C & \longrightarrow & 0
\end{array}$$

is a diagram in $\mathcal{C}$ with exact rows and commuting squares, then the long exact sequences of derived functors of F on $\widetilde{A}, \widetilde{B}, \widetilde{C}$ and A, B, C make commutative squares with the maps induced by the derived functors on f_A, f_B, f_C.

PROOF. The proof of Theorem 4.22 involved constructing a diagram

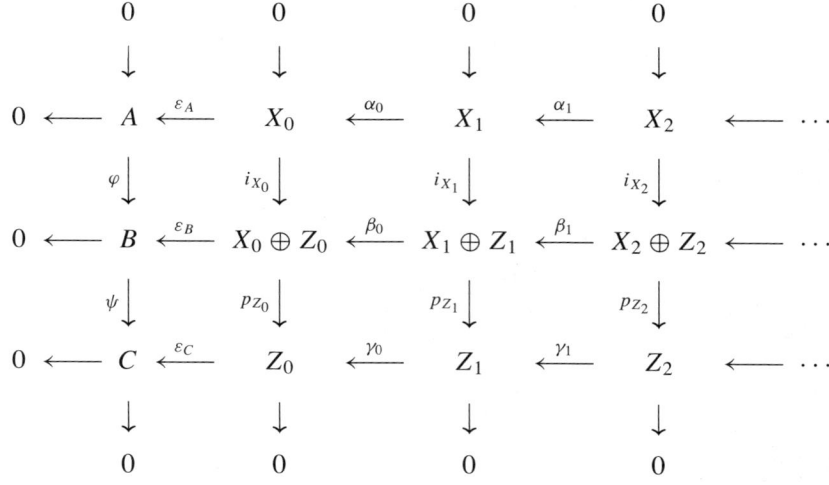

with exact rows and commuting squares in which each X_n and Z_n is projective, and a similar diagram corresponds to the given short exact sequence with tildes on it. The present theorem will follow from the functoriality in Theorem 4.10 if we can arrange that these two diagrams can be embedded in a 3-dimensional diagram with each of these diagrams in a horizontal plane and with vertical maps from the one diagram to the other such that all vertical squares commute.

We are given vertical maps f_A, f_B, and f_C, which we can regard as extending from the diagram with tildes to the other diagram. In addition, Theorem 4.12 gives us chain maps $\{f_{X_n}\}$ and $\{f_{Z_n}\}$ with $f_{X_{-1}} = f_A$ and $f_{X_{-1}} = f_C$, and all the completed vertical squares in the 3-dimensional diagram commute. To complete the proof, we construct by induction for $n \geq 0$ a map $f_n : \widetilde{X}_n \oplus \widetilde{Z}_n \to X_n \oplus Z_n$ such that

$$p_{Z_n} f_n = f_{Z_n} p_{\widetilde{Z}_n}, \qquad f_n i_{\widetilde{X}_n} = i_{X_n} f_{X_n}, \qquad \beta_{n-1} f_n = f_{n-1} \widetilde{\beta}_{n-1}, \qquad (*)$$

with the understanding that $\beta_{-1} = \varepsilon_B$. To make it possible for the inductive step to include the starting step of the induction, let us write $X_{-1} = A$, $Z_{-1} = B$, $i_{X_{-1}} = \varphi$, $p_{Z_{-1}} = \psi$, $\alpha_{-1} = \varepsilon_A$, $\gamma_{-1} = \varepsilon_C$, and $f_{-1} = f_B$. Also, let us understand any module or map with subscript -2 to be 0.

We shall construct f_n. For $\tilde{z} \in \tilde{Z}_n$, we apply $p_{Z_{n-1}}$ to the difference $\beta_{n-1}(0, f_{Z_n}\tilde{z}) - f_{n-1}\tilde{\beta}_{n-1}(0, \tilde{z})$ and get

$$p_{Z_{n-1}}\beta_{n-1}(0, f_{Z_n}\tilde{z}) - p_{Z_{n-1}}f_{n-1}\tilde{\beta}_{n-1}(0, \tilde{z})$$
$$= \gamma_{n-1}p_{Z_n}(0, f_{Z_n}\tilde{z}) - f_{Z_{n-1}}p_{\tilde{Z}_{n-1}}\tilde{\beta}_{n-1}(0, \tilde{z})$$
$$= \gamma_{n-1}f_{Z_n}\tilde{z} - f_{Z_{n-1}}\tilde{\gamma}_{n-1}p_{\tilde{Z}_n}(0, \tilde{z})$$
$$= f_{Z_{n-1}}\tilde{\gamma}_{n-1}\tilde{z} - f_{Z_{n-1}}\tilde{\gamma}_{n-1}\tilde{z} = 0.$$

Thus $\beta_{n-1}(0, f_{Z_n}\tilde{z}) - f_{n-1}\tilde{\beta}_{n-1}(0, \tilde{z}) = i_{X_{n-1}}(x)$ for a unique $x \in X_{n-1}$, and we define $\tau : \tilde{Z}_n \to X_{n-1}$ by saying that $\tau(\tilde{z})$ should be this x. This makes

$$i_{X_{n-1}}\tau(\tilde{z}) = \beta_{n-1}(0, f_{Z_n}\tilde{z}) - f_{n-1}\tilde{\beta}_{n-1}(0, \tilde{z}).$$

Setting up the diagram

$$\tilde{Z}_n$$
$$\downarrow \tau \quad \searrow \sigma$$
$$X_{n-2} \xleftarrow{\alpha_{n-2}} X_{n-1} \xleftarrow{\alpha_{n-1}} X_n$$

we prepare to invoke Lemma 4.11. We have

$$i_{X_{n-2}}\alpha_{n-2}\tau(\tilde{z}) = \beta_{n-2}i_{X_{n-1}}\tau(\tilde{z}) = \beta_{n-2}\beta_{n-1}(0, f_{Z_n}\tilde{z}) - \beta_{n-2}f_{n-1}\tilde{\beta}_{n-1}(0, \tilde{z})$$
$$= 0 - f_{n-2}\tilde{\beta}_{n-2}\tilde{\beta}_{n-1}(0, \tilde{z}) = 0.$$

Since $i_{X_{n-2}}$ is one-one, $\alpha_{n-2}\tau = 0$, and Lemma 4.11 applies. Thus we obtain $\sigma : \tilde{Z}_n \to X_n$ with $\alpha_{n-1}\sigma = \tau$, and σ satisfies

$$i_{X_{n-1}}\alpha_{n-1}\sigma(\tilde{z}) = \beta_{n-1}(0, f_{Z_n}\tilde{z}) - f_{n-1}\tilde{\beta}_{n-1}(0, \tilde{z}). \quad (**)$$

Define
$$f_n(\tilde{x}, \tilde{z}) = (f_{X_n}(\tilde{x}) - \sigma(\tilde{z}), f_{Z_n}(\tilde{z})). \quad (\dagger)$$

With f_n defined, we are to prove the three formulas $(*)$. For the first formula in $(*)$, we apply p_{Z_n} to both sides of $(\dagger)$ and obtain $p_{Z_n}f_n(\tilde{x}, \tilde{z}) = f_{Z_n}(\tilde{z}) = f_{Z_n}p_{\tilde{Z}_n}(\tilde{x}, \tilde{z})$, which is the desired formula. The second formula in $(*)$ at $\tilde{x}$ is just $(\dagger)$ with $\tilde{z} = 0$.

We are left with proving the third formula in $(*)$. Using the second formula in $(*)$, we have

$$\beta_{n-1}f_n(\tilde{x}, 0) = \beta_{n-1}f_n i_{\tilde{X}_n}(\tilde{x}) = \beta_{n-1}i_{X_n}f_{X_n}(\tilde{x})$$
$$= i_{X_{n-1}}\alpha_{n-1}f_{X_n}(\tilde{x}) = i_{X_{n-1}}f_{X_{n-1}}\tilde{\alpha}_{n-1}(\tilde{x})$$
$$= f_{n-1}i_{\tilde{X}_{n-1}}\tilde{\alpha}_{n-1}(\tilde{x}) = f_{n-1}\tilde{\beta}_{n-1}i_{\tilde{X}_n}(\tilde{x})$$
$$= f_{n-1}\tilde{\beta}_{n-1}(\tilde{x}, 0). \quad (\dagger\dagger)$$

Also,

$$\beta_{n-1}f_n(0,\widetilde{z}) = -\beta_{n-1}i_{X_n}\sigma(\widetilde{z}) + \beta_{n-1}(0, f_{Z_n}(\widetilde{z})) \qquad \text{by (†)}$$
$$= -i_{X_{n-1}}\alpha_{n-1}\sigma(\widetilde{z}) + \beta_{n-1}(0, f_{Z_n}(\widetilde{z})) \qquad \text{by commutativity}$$
$$= f_{n-1}\widetilde{\beta}_{n-1}(0,\widetilde{z}) \qquad \text{by (**)}.$$

Adding this equality and (††), we obtain the third formula of (*). This completes the proof. $\square$

The version of Theorem 4.25 appropriate for Theorem 4.24 is the following, and its proof is similar.

Theorem 4.26. Let $F : \mathcal{C} \to \mathcal{C}'$ be an additive functor between two good categories. Suppose that F either is contravariant right exact or is covariant left exact, and suppose that $\mathcal{C}$ has enough injectives. Then the passage as in Theorem 4.24 from short exact sequences in $\mathcal{C}$ to long exact sequences of derived functor modules in $\mathcal{C}'$ is functorial in the following sense: whenever

$$\begin{array}{ccccccccc} 0 & \longrightarrow & \widetilde{A} & \stackrel{\widetilde{\varphi}}{\longrightarrow} & \widetilde{B} & \stackrel{\widetilde{\psi}}{\longrightarrow} & \widetilde{C} & \longrightarrow & 0 \\ & & \downarrow f_A & & \downarrow f_B & & \downarrow f_C & & \\ 0 & \longrightarrow & A & \stackrel{\varphi}{\longrightarrow} & B & \stackrel{\psi}{\longrightarrow} & C & \longrightarrow & 0 \end{array}$$

is a diagram in $\mathcal{C}$ with exact rows and commuting squares, then the long exact sequences of derived functors of F on $\widetilde{A}, \widetilde{B}, \widetilde{C}$ and A, B, C make commutative squares with the maps induced by the derived functors on f_A, f_B, f_C.

We come to an important application of the long exact sequences in Theorems 4.22 and 4.24. Projective and injective resolutions make it easy to work with derived functors theoretically, but in practice any computations with them are likely to be difficult. It is therefore convenient to be able to compute derived functors from other resolutions than projective and injective ones.[11] For definiteness let us work with the case of a covariant *left* exact functor in a good category with

---

[11]The case of sheaf cohomology illustrates this point well. The present theory extends from good categories of modules to arbitrary abelian categories along the lines of Section 8 below, and the cohomology theory of sheaves fits into this more general framework. One additive functor of interest with sheaves is the "global-sections" functor. Its derived functors can be formed with injective resolutions, built from "flabby" sheaves, but flabby sheaves as a practical matter are too big to be useful in computations. In the theory of several complex variables for example, one approach is to substitute "fine" sheaves in resolutions; these permit computations and fall under the abelian-category generalization of Theorem 4.27 below.

enough injectives; this is the most important case in applications, and the other three cases in Figure 4.5 can be handled in similar fashion. Let $F : C \to C'$ be an additive functor between good categories that is covariant left exact. A module M in C is said to be F-acyclic if $F^n(M) = 0$ for all $n \geq 1$. Every module M that is injective in C is F-acyclic, since $0 \longrightarrow M \longrightarrow M \longrightarrow 0$ is an injective resolution of M from which we can see that $F^n(M) = 0$ for $n \geq 1$. An F-**acyclic resolution** of a module A in C is a resolution $X = (A \longrightarrow X^+)$ in which X_n is an F-acyclic module for all $n \geq 0$.

Theorem 4.27. Let C and C' be two good categories, let F be an additive functor from C to C' that is covariant and left exact, and suppose that C has enough injectives. If a module A in C has an F-acyclic resolution $X = (A \longrightarrow X^+)$ and if $I = (A \longrightarrow I^+)$ is any injective resolution of A, then any cochain map $f : X \to I$ with $f_{-1} = 1_A$ induces an isomorphism $F^n(A) \cong H^n(F(X))$ for each $n \geq 0$.

REMARKS. Such a cochain map always exists and is unique up to homotopy, according to Theorem 4.16. Theorem 4.27 says that the derived functors of F on any module A can be computed from any F-acyclic resolution of A; it is not necessary to work only with injective resolutions. The same result as in the theorem holds with $F_n(A) \cong H_n(F(A))$ if F is contravariant and right exact. If F is covariant right exact or contravariant left exact and if C has enough projectives, then any chain map from a projective resolution of A to an F-acyclic resolution[12] induces an isomorphism of the derived functors of A with the homology or cohomology of F of the F-acyclic resolution.

PROOF. The injective resolution is at our disposal, according to Corollary 4.17. Using the hypothesis that C has enough injectives, choose for each n an injective J_n containing X_n, let $g_n : X_n \to J_n$ be the inclusion, and make $\{J_n\}$ into an injective resolution of 0 with coboundary maps 0. Then replace I in the assumptions by $I \oplus J$ and f by (f, g). The result is that we have reduced the theorem to the case that f is one-one. Changing notation, we may assume from the outset that the injective resolution is $I = (A \longrightarrow I^+)$ and that the chain map $f : X \to I$ is one-one in each degree.

Put $Y_n = I_n/f_n(X_n) = \operatorname{coker} f_n$. The sequence

$$0 \longrightarrow X_n \xrightarrow{f_n} I_n \longrightarrow Y_n \longrightarrow 0 \qquad (*)$$

is exact, and Theorem 4.24a shows that the sequence

$$F^k(I_n) \longrightarrow F^k(Y_n) \longrightarrow F^{k+1}(X_n)$$

---

[12]For this situation, F-acyclic resolutions are understood to be chain complexes rather than cochain complexes.

is exact for every $k \geq 0$. Since I_n and X_n are F-acyclic for $n \geq 0$, the end terms are 0 for all $k \geq 1$. Consequently Y_n is F-acyclic for all $n \geq 0$.

Referring to $(*)$ for n and for $n+1$, we see that the coboundary map from I_n to I_{n+1} induces a compatible coboundary map from Y_n to Y_{n+1}. Thus we may consider $Y = (0 \longrightarrow Y^+)$ as a cochain complex with $Y^+ = \{Y_n\}_{n \geq 0}$. Then the equations $(*)$ for all $n \geq 0$, together with the coboundary maps, make

$$0 \longrightarrow X \xrightarrow{f} I \longrightarrow Y \longrightarrow 0 \qquad (**)$$

into a short exact sequence of complexes. Since X and I are exact, Corollary 4.8 shows that Y is exact.

If we apply F to the short exact sequence of complexes $(**)$, we obtain a planar diagram

$$0 \longrightarrow F(X) \xrightarrow{F(f)} F(I) \longrightarrow F(Y) \longrightarrow 0 \qquad (\dagger)$$

whose rows are the result of applying F to $(*)$, whose columns are complexes, and whose squares commutes. As usual we drop the row for $n = -1$, replacing it with a row of 0's. Let us prove that $(\dagger)$ is in fact a short exact sequence of complexes. In fact, the result of applying F to $(*)$ is the long exact sequence that begins

$$0 \longrightarrow F(X_n) \longrightarrow F(I_n) \longrightarrow F(Y_n) \longrightarrow F^1(X_n).$$

For $n \geq 0$, X_n is F-acyclic. Thus $F^1(X_n) = 0$, and the exactness for $n \geq 0$ follows. For $n \leq -1$, the rows of the diagram $(\dagger)$ are 0 and hence are exact. Thus $(\dagger)$ is a short exact sequence of complexes.

We shall now prove that $F(Y) = (0 \longrightarrow F(Y^+))$ is exact. Combining this fact with the exactness of the rows of $(\dagger)$ and applying Corollary 4.8 will then yield $H^n(F(X)) \cong H^n(F(I))$ for all $n \geq 0$. Since $H^n(F(I)) = F^n(A)$, this step will complete the proof.

To prove that $F(Y) = (0 \longrightarrow F(Y^+))$ is exact, define $Z_0 = Y_0$ and $Z_n = \mathrm{coker}(Y_{n-1} \to Y_n)$ for $n \geq 1$. Let $d_n : Y_n \to Y_{n+1}$ be the coboundary map. For each $n \geq 0$, the complex

$$0 \longrightarrow Y_n / \ker d_n \longrightarrow Y_{n+1} \longrightarrow Z_{n+1} \longrightarrow 0$$

is exact. Since $\ker d_n = \mathrm{image}\, d_{n-1}$ by exactness of Y, we have $Y_n / \ker d_n = Y_n / \mathrm{image}\, d_{n-1} = Z_n$, and thus

$$0 \longrightarrow Z_n \longrightarrow Y_{n+1} \longrightarrow Z_{n+1} \longrightarrow 0 \qquad (\dagger\dagger)$$

is exact for all $n \geq 0$.

Let us use $(\dagger\dagger)$ to prove the preliminary result that Z_n is F-acyclic for all $n \geq 0$. For $n = 0$, $Z_0 = Y_0$, and Y_0 is known to be F-acyclic. Proceeding

6. Long Exact Sequences of Derived Functors

inductively, suppose that Z_n is known to be F-acyclic. Applying Theorem 4.24a to (††), we see that

$$F^k(Y_{n+1}) \longrightarrow F^k(Z_{n+1}) \longrightarrow F^{k+1}(Z_n)$$

is exact for all $n \geq 0$ and all $k \geq 0$. For $n \geq 0$ and $k \geq 1$, the left end is 0 because Y_{n+1} is F-acyclic, and the right end is 0 because Z_n is F-acyclic by the inductive hypothesis. Therefore the middle term is 0, Z_{n+1} is F-acyclic, and the induction is complete.

Theorem 4.24a when applied to (††) shows that

$$0 \longrightarrow F(Z_n) \longrightarrow F(Y_{n+1}) \longrightarrow F(Z_{n+1}) \longrightarrow F^1(Z_n)$$

is exact for all $n \geq 0$, and we now know that the term at the right end is 0. Therefore

$$0 \longrightarrow F(Z_n) \longrightarrow F(Y_{n+1}) \longrightarrow F(Z_{n+1}) \longrightarrow 0 \qquad (\ddagger)$$

is exact for all $n \geq 0$.

Now we can prove that the complex

$$0 \longrightarrow F(Y_0) \longrightarrow F(Y_1) \longrightarrow F(Y_2) \longrightarrow F(Y_3) \longrightarrow \cdots \qquad (\ddagger\ddagger)$$

is exact at each module $F(Y_n)$. We know from Section 2 that we can merge two exact sequences

$$\cdots \to F(Y_{n+1}) \to F(Z_{n+1}) \to 0 \quad \text{and} \quad 0 \to F(Z_{n+1}) \to F(Y_{n+2}) \to \cdots$$

into a single exact sequence

$$\cdots \longrightarrow F(Y_{n+1}) \longrightarrow F(Y_{n+2}) \longrightarrow \cdots .$$

Consequently inductive application of ($\ddagger$) shows that the sequence

$$0 \to F(Z_0) \longrightarrow F(Y_1) \longrightarrow F(Y_2) \longrightarrow \cdots \longrightarrow F(Y_{n+1}) \longrightarrow F(Z_{n+1}) \to 0$$

is exact for each $n \geq 0$. In addition, we know that $Z_0 = Y_0$ by definition. Therefore ($\ddagger\ddagger$) is exact at $F(Y_n)$ for each $n \geq 0$, and the proof is complete. $\square$

Theorems 4.22 and 4.24 produce a long exact sequence from one additive functor and a short exact sequence of modules. Although it may at first seem odd to do so, we can obtain a different long exact sequence by varying the functor and fixing the module. This result, given as Proposition 4.28 below, will be used in the next section in analyzing the Ext and Tor functors.

Let C and C' be two good categories, and let F, G, H be three additive functors from C to C'. For definiteness, suppose that F, G, H are covariant and right exact. Suppose that there is a natural transformation S of F into G and there is a natural transformation T of G into H. We say that the sequence

$$F \xrightarrow{S} G \xrightarrow{T} H$$

is **exact on projectives** if for every projective P in C, the sequence

$$0 \longrightarrow F(P) \xrightarrow{S_P} G(P) \xrightarrow{T_P} H(P) \longrightarrow 0$$

is exact. Analogous definitions are to be made with projectives or injectives for the three other kinds of derived functors as in Figure 4.5.

Proposition 4.28. Let C and C' be two good categories, let F, G, H be three additive functors from C to C', suppose that F, G, H are covariant and right exact, and suppose that C has enough projectives. If there are natural transformations $S : F \to G$ and $T : G \to H$ such that the sequence $F \xrightarrow{S} G \xrightarrow{T} H$ is exact on projectives, then the derived functors of F, G, H on each module A in C form a long exact sequence

$$0 \longleftarrow H(A) \longleftarrow G(A) \longleftarrow F(A) \longleftarrow H_1(A) \longleftarrow G_1(A) \longleftarrow F_1(A)$$
$$\longleftarrow H_2(A) \longleftarrow G_2(A) \longleftarrow F_2(A) \longleftarrow H_3(A) \longleftarrow \cdots .$$

The passage from A to the long exact sequence is functorial in A.

REMARKS. The same long exact sequence and functoriality hold with the arrows reversed and F and H interchanged if the three functors are contravariant and left exact. If F, G, H are contravariant and right exact or are covariant and left exact, then analogous conclusions are valid provided C has enough injectives and the natural transformations S and T are exact on injectives.

PROOF. If $P = (P^+ \longrightarrow A)$ is a projective resolution of A, then the natural transformations S and T give us a planar diagram

$$\begin{array}{ccccccccc}
& & 0 & & 0 & & 0 & & \\
& & \uparrow & & \uparrow & & \uparrow & & \\
0 & \longrightarrow & F(P_0) & \xrightarrow{S_{P_0}} & G(P_0) & \xrightarrow{T_{P_0}} & H(P_0) & \longrightarrow & 0 \\
& & \uparrow & & \uparrow & & \uparrow & & \\
0 & \longrightarrow & F(P_1) & \xrightarrow{S_{P_1}} & G(P_1) & \xrightarrow{T_{P_1}} & H(P_1) & \longrightarrow & 0 \\
& & \uparrow & & \uparrow & & \uparrow & & \\
& & \vdots & & \vdots & & \vdots & &
\end{array}$$

in which the columns are complexes, the rows are exact because the sequence $F \xrightarrow{S} G \xrightarrow{T} H$ is exact on projectives, and the squares commute because S and T are natural transformations. The construction of the long exact sequence then follows from Theorem 4.7.

For the functoriality, suppose that $\varphi : A \to A'$ is a map between two modules of $\mathcal{C}$. Let $P = (P^+ \longrightarrow A)$ and $P' = (P'^+ \longrightarrow A)$ be projective resolutions of A and A', and use Theorem 4.12 to extend φ to a chain map $\{\varphi_n\}$ of P to P'. Then the planar diagrams as above for P and P' can be embedded in a 3-dimensional diagram in such a way that the various maps $F(\varphi_n)$, $G(\varphi_n)$, and $H(\varphi_n)$ connecting the diagram for P to the diagram for P' make all squares commute. The functoriality now follows immediately from Theorem 4.10. $\square$

7. Ext and Tor

In this section we study the derived functors of Hom and tensor product. Although we shall treat each as carrying unital left R modules, where R is a ring with identity, to abelian groups, the theory applies also to more complicated versions of Hom and tensor product, such as when one of the R modules in question is actually a bimodule for the rings R and S and the result of Hom or tensor product is an S module. Problems 9–11 at the end of the chapter address the situation with bimodules.

We know that $\operatorname{Hom}_R(A, B)$ is a contravariant left exact functor of the A variable and a left exact covariant functor of the B variable. Thus we have two initial choices for inserting resolutions and creating derived functors, namely

$$\operatorname{Ext}_R^n(A, B) = H^n(\operatorname{Hom}_R(P, B)), \quad \text{with } P = (A \leftarrow P^+) \text{ projective,}$$

and

$$\operatorname{ext}_R^n(A, B) = H^n(\operatorname{Hom}_R(A, I)), \quad \text{with } I = (B \to I^+) \text{ injective.}$$

Existence of the first one depends on having enough projectives in the category of the A variable, and existence of the second one depends on having enough injectives in the category of the B variable. Each of these, just as with Hom, depends on two variables, one in contravariant fashion and the other in covariant fashion. Thus Ext and ext are not functors of two variables in the strict sense of our definitions. Instead, they are examples of "bifunctors," of which $\operatorname{Hom}_R(\,\cdot\,,\,\cdot\,)$ is the prototype, and the main result, Theorem 4.31 below, in essence says that Ext and ext are naturally isomorphic as bifunctors, provided the first domain category has enough projectives *and* the second has enough injectives. Among other things this natural isomorphism will justify and explain how we were able to define cohomology of groups in more than one way.[13]

In the case of tensor product $A \otimes_R B$, similar remarks apply. Here A is a unital right R module, and B is a unital left R module. The module A in a natural way is a unital left R^o module, where R^o is the opposite ring of R, and thus tensor product is to be regarded as defined on the product of two categories of left modules just as Hom is. We can regard tensor product as an actual functor in either variable, and the functor is covariant right exact in both cases. Again we have two initial choices for inserting resolutions and creating derived functors, namely

$$\operatorname{Tor}_n^R(A, B) = H^n(P \otimes_R B), \qquad \text{with } P = (A \leftarrow P^+) \text{ projective,}$$

and

$$\operatorname{tor}_n^R(A, B) = H^n(A \otimes_R P), \qquad \text{with } P' = (B \leftarrow P'^+) \text{ projective.}$$

These exist if the domain categories have enough projectives. Both Tor and tor can be considered as covariant functors of two variables, or else as "bifunctors," and one can show in the same way as for Ext and ext that Tor and tor are naturally isomorphic. There is no need to write out the details. It is customary to write Tor for the common value.

Proposition 4.29. Let $\mathcal{C}$ and $\mathcal{C}'$ be good categories of unital left R modules, and suppose that $\mathcal{C}$ has enough projectives. Then the contravariant left exact functors $\operatorname{Hom}_R(\,\cdot\,, B)$ from $\mathcal{C}$ to $\mathcal{C}_\mathbb{Z}$ and their derived functors $\operatorname{Ext}_R^n(\,\cdot\,, B)$ have the following properties:

(a) Whenever $0 \to A' \to A \to A'' \to 0$ is a short exact sequence in $\mathcal{C}$, then there is a corresponding long exact sequence

$$0 \longrightarrow \operatorname{Hom}_R(A'', B) \longrightarrow \operatorname{Hom}_R(A, B) \longrightarrow \operatorname{Hom}_R(A', B)$$
$$\longrightarrow \operatorname{Ext}_R^1(A'', B) \longrightarrow \operatorname{Ext}_R^1(A, B) \longrightarrow \operatorname{Ext}_R^1(A', B)$$
$$\longrightarrow \operatorname{Ext}_R^2(A'', B) \longrightarrow \operatorname{Ext}_R^2(A, B) \longrightarrow \operatorname{Ext}_R^2(A', B) \to \operatorname{Ext}_R^3(A'', B) \to \cdots$$

in $\mathcal{C}_\mathbb{Z}$ for each module B in $\mathcal{C}'$. The passage from short exact sequences in $\mathcal{C}$ to long exact sequences of derived functor modules in $\mathcal{C}_\mathbb{Z}$ is functorial in its dependence on the exact sequence in the first variable in the sense of Theorem 4.25 and is natural in the second variable in the sense that if a map $\eta : \widetilde{B} \to B$ is given, then $\operatorname{Hom}(1, \eta)$ defines a chain map from the long exact sequence for $\widetilde{B}$ to the long exact sequence for B.

---

[13]It would add only definitions to our discussion to say precisely what a general bifunctor is and what a general natural transformation between bifunctors is, and we shall skip that detail, in effect incorporating the definitions into the theorem.

(b) If P is a projective in $\mathcal{C}$ and I is an injective in $\mathcal{C}'$, then $\operatorname{Ext}_R^n(P, B) = 0 = \operatorname{Ext}_R^n(A, I)$ for all $n \geq 1$ and all modules A in $\mathcal{C}$ and B in $\mathcal{C}'$.

(c) Whenever $0 \to B' \to B \to B'' \to 0$ is a short exact sequence in $\mathcal{C}'$, then there is a corresponding long exact sequence

$$0 \longrightarrow \operatorname{Hom}_R(A, B') \longrightarrow \operatorname{Hom}_R(A, B) \longrightarrow \operatorname{Hom}_R(A, B'')$$
$$\longrightarrow \operatorname{Ext}_R^1(A, B') \longrightarrow \operatorname{Ext}_R^1(A, B) \longrightarrow \operatorname{Ext}_R^1(A, B'')$$
$$\longrightarrow \operatorname{Ext}_R^2(A, B') \longrightarrow \operatorname{Ext}_R^2(A, B) \longrightarrow \operatorname{Ext}_R^2(A, B'') \to \operatorname{Ext}_R^3(A, B') \to \cdots$$

in $\mathcal{C}_\mathbb{Z}$ for each module A in $\mathcal{C}$. The passage from short exact sequences in $\mathcal{C}'$ to long exact sequences of derived functor modules in $\mathcal{C}_\mathbb{Z}$ is functorial in the exact sequence in the second variable and is natural in the first variable in the sense that if a map $\eta : \widetilde{A} \to A$ is given, then $\operatorname{Hom}(\eta, 1)$ defines a chain map from the long exact sequence for A to the long exact sequence for $\widetilde{A}$.

REMARKS. The naturality in the B parameter of the construction of the long exact sequence in (a) implies that Ext_R^n is a covariant functor of the second variable for fixed argument of the first variable. It implies also that all maps $\operatorname{Ext}_R^n(\alpha, 1)$ commute with all maps $\operatorname{Ext}_R^n(1, \beta)$.

PROOF. For (a), Theorem 4.22b gives the exact sequence, and Theorem 4.25 proves the functoriality in the first variable. For the naturality in the second variable, let $\eta : \widetilde{B} \to B$ be given. The proof of Theorem 4.22 produces a short exact sequence of projective resolutions of A', A, A'' to which the functor in that theorem is then applied. We now have two such functors $\operatorname{Hom}_R(\,\cdot\,, \widetilde{B})$ and $\operatorname{Hom}_R(\,\cdot\,, B)$, and the maps within each image diagram are all of the form $\operatorname{Hom}(\alpha, 1)$. The two diagrams fit into a 3-dimensional diagram, and the maps between the two diagrams are of the form $\operatorname{Hom}(1, \eta)$. Since all maps $\operatorname{Hom}(\alpha, 1)$ commute with all maps $\operatorname{Hom}(1, \beta)$, the 3-dimensional diagram is commutative. The corresponding long exact sequences are then related by a cochain map according to Theorem 4.10.

For (b), $0 \leftarrow P \leftarrow P \leftarrow 0$ is a projective resolution of P, and hence any derived functor that is defined by projective resolutions is 0 in degree ≥ 1. In addition, Proposition 4.19b shows that $\operatorname{Hom}_R(\,\cdot\,, I)$ is an exact functor, and hence its derived functors are 0 in degree ≥ 1.

For (c), we shall apply Proposition 4.28 in its version for contravariant left exact functors. Let $\varphi : B' \to B$ and $\psi : B \to B''$ be the maps in the given short exact sequence, and let F, G, H be the functors with $F(A) = \operatorname{Hom}_R(A, B')$, $G(A) = \operatorname{Hom}_R(A, B)$, $H(A) = \operatorname{Hom}_R(A, B'')$. Then we have a natural transformation S of F into G given by $S_A = \operatorname{Hom}(1, \varphi)$ and a natural transformation T of G into H given by $T_A = \operatorname{Hom}(1, \psi)$. Since

$$0 \longrightarrow \operatorname{Hom}_R(P, B') \xrightarrow{S_P} \operatorname{Hom}_R(P, B) \xrightarrow{T_P} \operatorname{Hom}_R(P, B'') \longrightarrow 0$$

is exact by Proposition 4.19a, the sequence

$$F \xrightarrow{S} G \xrightarrow{T} H$$

is exact on projectives. Proposition 4.28 in its version for contravariant left exact functors then says that there is a long exact sequence

$$0 \longrightarrow F(A) \longrightarrow G(A) \longrightarrow H(A) \longrightarrow F_1(A) \longrightarrow G_1(A) \longrightarrow H_1(A)$$
$$\longrightarrow F_2(A) \longrightarrow G_2(A) \longrightarrow H_2(A) \longrightarrow F_3(A) \longrightarrow \cdots$$

and that the passage to this long exact sequence is functorial in A. This much establishes the long exact sequence in (c) and the naturality in the A variable. For the behavior in the second variable with A fixed, suppose that we have a second exact sequence $0 \to \widetilde{B}' \to \widetilde{B} \to \widetilde{B}'' \to 0$ that maps to the given one by a chain map f. Let F', G', H' be the functors $\text{Hom}_R(\cdot, \widetilde{B}')$, $\text{Hom}_R(\cdot, \widetilde{B})$, $\text{Hom}_R(\cdot, \widetilde{B}'')$. We then get two horizontal planar diagrams of the kind in the proof of Proposition 4.28, one for F', G', H' and one for F, G, H. The maps within each of the two diagrams are maps in the A variable. The two diagrams embed in a 3-dimensional diagram with vertical maps $\text{Hom}_R(1, f)$, and the 3-dimensional diagram is commutative because all maps $\text{Hom}(\alpha, 1)$ commute with all maps $\text{Hom}(1, \beta)$. Application of Theorem 4.10 then completes the proof of functoriality in the exact sequence in the second variable. □

Proposition 4.30. Let $\mathcal{C}$ and $\mathcal{C}'$ be good categories of unital left R modules, and suppose that $\mathcal{C}'$ has enough injectives. Then the covariant left exact functors $\text{Hom}_R(A, \cdot)$ from $\mathcal{C}'$ to $\mathcal{C}_{\mathbb{Z}}$ and their derived functors $\text{ext}_R^n(A, \cdot)$ have the following properties:

(a) Whenever $0 \to A' \to A \to A'' \to 0$ is a short exact sequence in $\mathcal{C}$, then there is a corresponding long exact sequence

$$0 \longrightarrow \text{Hom}_R(A'', B) \longrightarrow \text{Hom}_R(A, B) \longrightarrow \text{Hom}_R(A', B)$$
$$\longrightarrow \text{ext}_R^1(A'', B) \longrightarrow \text{ext}_R^1(A, B) \longrightarrow \text{ext}_R^1(A', B)$$
$$\longrightarrow \text{ext}_R^2(A'', B) \longrightarrow \text{ext}_R^2(A, B) \longrightarrow \text{ext}_R^2(A', B) \to \text{ext}_R^3(A'', B) \to \cdots$$

in $\mathcal{C}_{\mathbb{Z}}$ for each module B in $\mathcal{C}'$. The passage from short exact sequences in $\mathcal{C}$ to long exact sequences of derived functor modules in $\mathcal{C}_{\mathbb{Z}}$ is functorial in its dependence on the exact sequence in the first variable and is natural in the second variable in the sense that if a map $\eta : \widetilde{B} \to B$ is given, then $\text{Hom}(1, \eta)$ defines a chain map from the long exact sequence for $\widetilde{B}$ to the long exact sequence for B.

(b) If P is a projective in $\mathcal{C}$ and I is an injective in $\mathcal{C}'$, then $\text{ext}_R^n(P, B) = 0 = \text{ext}_R^n(A, I)$ for all $n \geq 1$ and all modules A in $\mathcal{C}$ and B in $\mathcal{C}'$.

(c) Whenever $0 \to B' \to B \to B'' \to 0$ is a short exact sequence in $\mathcal{C}'$, then there is a corresponding long exact sequence

$$0 \longrightarrow \text{Hom}_R(A, B') \longrightarrow \text{Hom}_R(A, B) \longrightarrow \text{Hom}_R(A, B'')$$
$$\longrightarrow \text{ext}_R^1(A, B') \longrightarrow \text{ext}_R^1(A, B) \longrightarrow \text{ext}_R^1(A, B'')$$
$$\longrightarrow \text{ext}_R^2(A, B') \longrightarrow \text{ext}_R^2(A, B) \longrightarrow \text{ext}_R^2(A, B'') \to \text{ext}_R^3(A, B') \to \cdots$$

in $\mathcal{C}_\mathbb{Z}$ for each module A in $\mathcal{C}$. The passage from short exact sequences in $\mathcal{C}'$ to long exact sequences of derived functor modules in $\mathcal{C}_\mathbb{Z}$ is functorial in the exact sequence in the second variable and is natural in the first variable in the sense that if a map $\eta : \widetilde{A} \to A$ is given, then $\text{Hom}(\eta, 1)$ defines a chain map from the long exact sequence for A to the long exact sequence for $\widetilde{A}$.

REMARKS. The naturality in the A parameter of the construction of the long exact sequence in (c) implies that ext_R^n is a contravariant functor of the first variable for fixed argument of the second variable. It implies also that all maps $\text{ext}_R^n(\alpha, 1)$ commute with all maps $\text{ext}_R^n(1, \beta)$.

PROOF. The proof of (c) is a simple variant of the proof of Proposition 4.29a, the proof of (b) is a simple variant of the proof of Proposition 4.29b, and the proof of (a) is a simple variant of the proof of Proposition 4.29c. □

Propositions 4.29 and 4.30 show that Ext and ext, as functors of the first variable and as functors of the second variable, generate the same long exact sequences, the first under the assumption that $\mathcal{C}$ has enough projectives and the second under the assumption that $\mathcal{C}'$ has enough injectives. Theorem 4.31 will show that Ext and ext may be treated as equal if both assumptions are satisfied. It is customary therefore to use Ext as the notation in both cases; thus Ext exists if either $\mathcal{C}$ has enough projectives or $\mathcal{C}'$ has enough injectives. In both cases, Ext has a long exact sequence in the first variable and another long exact sequence in the second variable.

Theorem 4.31. Let $\mathcal{C}$ and $\mathcal{C}'$ be good categories of unital left R modules, and suppose that $\mathcal{C}$ has enough projectives and $\mathcal{C}'$ has enough injectives. Then $\text{Ext}_R^n(\cdot, \cdot)$ and $\text{ext}_R^n(\cdot, \cdot)$ are naturally isomorphic from $\mathcal{C} \times \mathcal{C}'$ to $\mathcal{C}_\mathbb{Z}$ in the sense that for each $n \geq 0$ and each pair of modules (A, B) in $\mathcal{C} \times \mathcal{C}'$, there exists an isomorphism $T_{(n,A,B)}$ in $\text{Hom}_\mathbb{Z}(\text{Ext}_R^n(A, B), \text{ext}_R^n(A, B))$ such that if φ is in

$\operatorname{Hom}_R(A, A')$ and ψ is in $\operatorname{Hom}_R(B, B')$, then the diagrams

$$\begin{array}{ccc} \operatorname{Ext}_R^n(A, B) & \xrightarrow{T_{(n,A,B)}} & \operatorname{ext}_R^n(A, B) \\ {\scriptstyle \operatorname{Ext}^n(\varphi,1)} \uparrow & & \uparrow {\scriptstyle \operatorname{ext}^n(\varphi,1)} \\ \operatorname{Ext}_R^n(A', B) & \xrightarrow{T_{(n,A',B)}} & \operatorname{ext}_R^n(A', B) \end{array}$$

and

$$\begin{array}{ccc} \operatorname{Ext}_R^n(A, B) & \xrightarrow{T_{(n,A,B)}} & \operatorname{ext}_R^n(A, B) \\ {\scriptstyle \operatorname{Ext}^n(1,\psi)} \downarrow & & \downarrow {\scriptstyle \operatorname{ext}^n(1,\psi)} \\ \operatorname{Ext}_R^n(A, B') & \xrightarrow{T_{(n,A,B')}} & \operatorname{ext}_R^n(A, B') \end{array}$$

commute.

REMARKS. The reader will be able to observe that a certain part of this proof amounts to showing that 3-dimensional diagrams in the shape of a cube having 5 faces equal to commuting squares and having suitable hypotheses on the maps automatically have their sixth face equal to a commuting square. The hypotheses concerning the faces and the maps come from Propositions 4.29 and 4.30, as well as induction. We shall not try to abstract a general result of this kind, however.

PROOF. We induct on n for $n \geq 0$. Several steps are involved in the proof, and we complete all of them for a particular n before going on to $n + 1$. The steps for a particular n are

(i) to define $T_{(n,A,B)}$ in the presence of an injective I and a one-one map $\mu : B \to I$ and to observe that $T_{(n,A,B)}$ is an isomorphism,
(ii) to show that the same $T_{(n,A,B)}$ results independently of the choice of I,
(iii) to prove the commutativity of the second diagram in the statement of the theorem, and
(iv) to prove the commutativity of the first diagram in the statement of the theorem.

The first base case of the induction is $n = 0$, for which we take $T_{(0,A,B)}$ to be the identity on $\operatorname{Hom}_R(A, B)$. Then (i) through (iv) are immediate.

The other base case of the induction is $n = 1$. Let (A, B) be given. An injective I and a one-one map $\mu : B \to I$ exist as in (i) because $\mathcal{C}'$ has enough injectives. Then we have an exact sequence

$$0 \longrightarrow B \xrightarrow{\mu} I \xrightarrow{\nu} C \longrightarrow 0 \qquad (*)$$

in which $C = I/\mu(B)$ and ν is the quotient map. We know from Propositions 4.29b and 4.30b that $\operatorname{Ext}_R^1(A, I) = 0 = \operatorname{ext}_R^1(A, I)$. Therefore Propositions

4.29c and 4.30c give us exact sequences

$$\text{Hom}_R(A, I) \xrightarrow{\text{Hom}(1,\nu)} \text{Hom}_R(A, C) \xrightarrow{\omega_{E,0}} \text{Ext}_R^1(A, B) \longrightarrow 0$$

and

$$\text{Hom}_R(A, I) \xrightarrow{\text{Hom}(1,\nu)} \text{Hom}_R(A, C) \xrightarrow{\omega_{e,0}} \text{ext}_R^1(A, B) \longrightarrow 0$$

in which $\omega_{E,0}$ and $\omega_{e,0}$ are suitable connecting homomorphisms. We define $T_{(1,A,B)} = \omega_{e,0}(\omega_{E,0})^{-1}$. This definition is meaningful, since the exactness of the two sequences gives

$$(\omega_{E,0})^{-1}(0) = \ker \omega_{E,0} = \text{Hom}(1, \nu)(\text{Hom}_R(A, I)) = \ker \omega_{e,0};$$

by an analogous computation, $\omega_{E,0}(\omega_{e,0})^{-1}$ is a well-defined function, and it is evidently a two-sided inverse. Thus $T_{(1,A,B)}$ is an isomorphism. This completes step (i).

In order to be able to handle steps (ii) and (iii) without being repetitive, let a map $\psi : B \to B'$ be given. For (ii), B' will be B, and ψ will be the identity on B. For (iii), B' and ψ will be general. Given ψ and one-one maps $\mu : B \to I$ and $\mu' : B' \to I'$, we can form the exact rows and the first column of the diagram

$$\begin{array}{ccccccccc} 0 & \longrightarrow & B & \xrightarrow{\mu} & I & \xrightarrow{\nu} & C & \longrightarrow & 0 \\ & & \psi \downarrow & & f \downarrow & & \bar{f} \downarrow & & \\ 0 & \longrightarrow & B' & \xrightarrow{\mu'} & I' & \xrightarrow{\nu'} & C' & \longrightarrow & 0. \end{array} \quad (**)$$

If we think of I and I' as extended to injective resolutions, Theorem 4.16 allows us to fill in a cochain map from the one extension to the other, and the first new step of that cochain map is f. If we define $\bar{f} = \nu' f \nu^{-1}$, then $\bar{f}$ is well defined because

$$\nu' f \nu^{-1}(0) = \nu' f \ker \nu = \nu' f \text{ image } \mu$$
$$= \nu' f \mu(B) = \nu' \mu' \psi(B) = 0(\psi(B)) = 0,$$

and the squares of the diagram $(**)$ now commute. Continuing with the effort to cut down on repetitive arguments, let $k \geq 1$ be an integer that will be 1 when $n = 1$ and will be different later in the proof. Applying Proposition 4.29c to $(**)$ gives us a commuting square

$$\begin{array}{ccc} \text{Ext}_R^{k-1}(A, C) & \xrightarrow{\omega_{E,k-1}} & \text{Ext}_R^k(A, B) \\ \text{Ext}^{k-1}(1,\bar{f}) \downarrow & & \downarrow \text{Ext}^k(1,\psi) \\ \text{Ext}_R^{k-1}(A, C') & \xrightarrow{\omega'_{E,k-1}} & \text{Ext}_R^k(A, B') \end{array} \quad (\dagger)$$

for $k \geq 1$, and Proposition 4.30c gives us a similar commuting square for ext for $k \geq 1$.

For each module in the diagram with Ext when $k = 1$, there is a map to the corresponding module in the diagram with ext. These maps are $T_{(k-1,A,C)}$ for the upper left and $T_{(k-1,A,C')}$ for the lower left. The maps for the upper right and lower right depend on the step of the argument.

For step (ii), we are taking $B' = B$, and the maps at the right are the two versions of $T_{(k,A,B)}$, one for the injective I and one for the injective I'. Let us call them $T_{(k,A,B)}$ and $T'_{(k,A,B)}$. We are to prove that $T'_{(k,A,B)} \operatorname{Ext}^k(1, \psi) = \operatorname{ext}^k(1, \psi) T_{(k,A,B)}$ for $\psi = 1$. The relevant definitions are

$$T_{(k,A,B)} = \omega_{(e,k-1)} T_{(k-1,A,C)} \omega_{(E,k-1)}^{-1}$$

and

$$T'_{(k,A,B)} = \omega'_{(e,k-1)} T_{(k-1,A,C')} (\omega'_{(E,k-1)})^{-1},$$

or equivalently

$$T_{(k,A,B)} \omega_{(E,k-1)} = \omega_{(e,k-1)} T_{(k-1,A,C)}$$

and

$$T'_{(k,A,B)} \omega'_{(E,k-1)} = \omega'_{(e,k-1)} T_{(k-1,A,C')}.$$

Since $T_{(k-1,A,C)}$ and $T_{(k-1,A,C')}$ are known inductively to be well defined and to satisfy (iii), we have $\operatorname{ext}^{k-1}(1, \bar{f}) T_{(k-1,A,C)} = T_{(k-1,A,C')} \operatorname{Ext}^{k-1}(1, \bar{f})$. Thus

$$\operatorname{ext}^k(1, \psi) T_{(k,A,B)} \omega_{(E,k-1)} = \operatorname{ext}^k(1, \psi) \omega_{(e,k-1)} T_{(k-1,A,C)}$$
$$= \omega'_{(e,k-1)} \operatorname{ext}^{k-1}(1, \bar{f}) T_{(k-1,A,C)} = \omega'_{(e,k-1)} T_{(k-1,A,C')} \operatorname{Ext}^{k-1}(1, \bar{f})$$
$$= T'_{(k,A,B)} \omega'_{(E,k-1)} \operatorname{Ext}^{k-1}(1, \bar{f}) = T'_{(k,A,B)} \operatorname{Ext}^k(1, \psi) \omega_{(E,k-1)}.$$

Since $\operatorname{Ext}^k(1, \psi) = 1$ and $\operatorname{ext}^k(1, \psi) = 1$ when $\psi = 1$, step (ii) follows for $n = 1$, i.e., $T_{(k,A,B)}$ is well defined.

For step (iii), we are allowing general B', and the maps at the right between the two versions of (†) are the well-defined isomorphisms $T_{(k,A,B)}$ and $T_{(k,A,B')}$. We are to prove that $T_{(k,A,B')} \operatorname{Ext}^k(1, \psi) = \operatorname{ext}^k(1, \psi) T_{(k,A,B)}$. The argument in the previous paragraph applies if we change $T'_{(k,A,B)}$ systematically to $T_{(k,A,B')}$ and take into account that $\omega_{(E,k-1)}$ is onto, and step (iii) follows for $n = 1$.

For step (iv), let $\varphi : A \to A'$ be given. The conclusion of Proposition 4.29c that the dependence is natural in the first variable gives us a commuting square

$$\begin{array}{ccc} \operatorname{Ext}^{k-1}_R(A, C) & \xrightarrow{\omega_{E,k-1}} & \operatorname{Ext}^k_R(A, B) \\ {\scriptstyle \operatorname{Ext}^{k-1}(\varphi,1)} \uparrow & & \uparrow {\scriptstyle \operatorname{Ext}^k(\varphi,1)} \\ \operatorname{Ext}^{k-1}_R(A', C) & \xrightarrow{\omega'_{E,k-1}} & \operatorname{Ext}^k_R(A', B) \end{array} \qquad (\dagger\dagger)$$

for $k \geq 1$ and for suitable connecting homomorphisms $\omega_{E,k-1}$ and $\omega'_{E,k-1}$, and Proposition 4.30c gives a similar commuting square for ext for $k \geq 1$. For each module in the diagram with Ext when $k = 1$, there is a map to the corresponding module in the diagram with ext. These maps are $T_{(k-1,A,C)}$ for the upper left, $T_{(k-1,A',C)}$ for the lower left, $T_{(k,A,B)}$ for the upper right, and $T_{(k,A',B)}$ for the lower right. We are to prove that $T_{(k,A,B)} \operatorname{Ext}^k(\varphi, 1) = \operatorname{ext}^k(\varphi, 1) T_{(k,A',B)}$. The relevant definitions are

$$T_{(k,A,B)} \omega_{(E,k-1)} = \omega_{(e,k-1)} T_{(k-1,A,C)}$$

and
$$T_{(k,A',B)} \omega'_{(E,k-1)} = \omega'_{(e,k-1)} T_{(k-1,A',C)}.$$

Since $T_{(k-1,A,C)}$ and $T_{(k-1,A',C)}$ are known inductively to satisfy (iv), we have $\operatorname{ext}^{k-1}(\varphi, 1) T_{(k-1,A',C)} = T_{(k-1,A,C)} \operatorname{Ext}^{k-1}(\varphi, 1)$. Thus

$$\operatorname{ext}^k(\varphi, 1) T_{(k,A',B)} \omega'_{(E,k-1)} = \operatorname{ext}^k(\varphi, 1) \omega'_{(e,k-1)} T_{(k-1,A',C)}$$
$$= \omega_{(e,k-1)} \operatorname{ext}^{k-1}(\varphi, 1) T_{(k-1,A',C)} = \omega_{(e,k-1)} T_{(k-1,A,C)} \operatorname{Ext}^{k-1}(\varphi, 1)$$
$$= T_{(k,A,B)} \omega_{(E,k-1)} \operatorname{Ext}^{k-1}(\varphi, 1) = T_{(k,A,B)} \operatorname{Ext}^k(\varphi, 1) \omega'_{(E,k-1)}.$$

Since $\omega'_{(E,k-1)}$ is onto, step (iv) follows for $n = 1$. This completes the proof for $n = 1$.

For the inductive step, suppose that steps (i) through (iv) have been carried out for some $n \geq 1$. Let us carry out step (i) for stage $n + 1$. For a given B, we know from Propositions 4.29b and 4.30b that $\operatorname{Ext}_R^n(A, I) = 0 = \operatorname{ext}_R^n(A, I)$. Hence Propositions 4.29c and 4.30c give us exact sequences

$$0 \longrightarrow \operatorname{Ext}_R^n(A, C) \xrightarrow{\omega_{E,n}} \operatorname{Ext}_R^{n+1}(A, B) \longrightarrow 0$$

and
$$0 \longrightarrow \operatorname{ext}_R^n(A, C) \xrightarrow{\omega_{e,n}} \operatorname{ext}_R^{n+1}(A, B) \longrightarrow 0.$$

In other words, $\omega_{E,n}$ and $\omega_{e,n}$ are isomorphisms. If we put

$$T_{(n+1,A,B)} = \omega_{e,n} T_{(n,A,C)} \omega_{E,n}^{-1},$$

then $T_{(n+1,A,B)}$ is an isomorphism of $\operatorname{Ext}_R^{n+1}(A, B)$ onto $\operatorname{ext}_R^{n+1}(A, B)$. This completes step (i) for stage $n + 1$.

We now refer back to our argument for $n = 1$ and put $k = n + 1$ throughout. Tracing matters through, we see that the argument carries out steps (ii) through (iv) for stage $n + 1$. This completes the induction and the proof. □

8. Abelian Categories

Not all situations in which one wants to apply homological algebra are limited to good categories of unital left R modules for some ring R. We have mentioned sheaves as one example, and we shall develop some properties of sheaves in Chapter X. Implicitly we have carried along a second example: all chain complexes within a good category, with chain maps as morphisms, form a category in which short exact sequences have remarkable properties, such as those in Theorems 4.7 and 4.10.

A setting to which one can generalize well such basic parts of homological algebra is that of "abelian categories," which we define in this section. It is advisable not to require that the objects in an abelian category actually be sets of individual elements; otherwise there is little chance that the notion of abelian category could be self dual. The morphisms of the category are then effectively all we have to work with, since a morphism already determines its "domain" and "range." If X and Y are objects, then a morphism in Morph(X, Y) need not be a function, but at least Morph(X, Y) is a set with elements to it. Since objects no longer have elements, books usually suppress the objects in the discussion to the point of referring to things like kernels and cokernels as morphisms rather than objects. It is perhaps more comfortable to think of a kernel as a pair, consisting of an object and a morphism into another object, rather than just as the embedding morphism, and we shall follow the more comfortable convention temporarily.

We introduce the notion of "abelian category" in stages. We begin with some definitions and remarks that make sense in a general category. First of all, let us have names for X and Y when referring to morphisms in Morph(X, Y) that do not require us to think in terms of functions. The convention is that if u is in Morph(X, Y), then X is the **domain** of u and Y is the **codomain**. We allow ourselves to write compositions of morphisms as gf or as $g \circ f$.

Next, it is possible to generalize usefully the notions of "one-one" and "onto" to make them applicable in any category. The definitions are in terms of cancellation laws. In the category $\mathcal{C}$, a morphism $u \in$ Morph(X, Y) is a **monomorphism**[14] if for any f and g in the same set Morph(W, X) such that $uf = ug$, it follows that $f = g$. Any isomorphism is certainly a monomorphism. The composition of two monomorphisms is a monomorphism. In fact, if u and v are monomorphisms with $vuf = vug$, then $uf = ug$ because v is a monomorphism, and $f = g$ because u is a monomorphism. If m is a monomorphism in Morph(X, Y) and u is any morphism in Morph(Y, X) such that $mu = 1_Y$, then m is an isomorphism. In fact, $mu = 1_Y$ implies $mum = 1_Y m = m$, which implies $um = 1_X$, since m is a monomorphism; therefore u is a two-sided inverse to m.

---

[14]Some authors use the word "monic" or the word "mono" as an adjectival form of this noun.

The morphism $u \in \text{Morph}(X, Y)$ is an **epimorphism**[15] if for any f' and g' in the same set $\text{Morph}(Y, Z)$ such that $f'u = g'u$, it follows that $f' = g'$. Any isomorphism is an epimorphism. The composition of two epimorphisms is an epimorphism. If e is an epimorphism in $\text{Morph}(X, Y)$ and u is any morphism in $\text{Morph}(Y, X)$ such that $ue = 1_X$, then e is an isomorphism.

Finally a **zero object** 0 in a category $\mathcal{C}$ is an object such that for each X in $\text{Obj}(\mathcal{C})$, each of $\text{Morph}(X, 0)$ and $\text{Morph}(0, X)$ has exactly one member. It is immediate that any two zero objects are isomorphic: if 0 and $0'$ are zero objects, then $\text{Morph}(0, 0)$ and $\text{Morph}(0', 0')$ each have just one member, which must be 1_0 and $1_{0'}$ in the two cases; the composition of the member of $\text{Morph}(0, 0')$ followed by the member of $\text{Morph}(0', 0)$ must be 1_0, and the composition in the other order must be $1_{0'}$, and the isomorphism of 0 with $0'$ has been exhibited.

Suppose that a zero object exists. Since the composition law for morphisms in $\mathcal{C}$ insists that the composite of a member of $\text{Morph}(X, 0)$ and a member of $\text{Morph}(0, Y)$ be in $\text{Morph}(X, Y)$, it follows that $\text{Morph}(X, Y)$ has a distinguished member, which we denote by 0_{XY}. This is called the **zero morphism** of $\text{Morph}(X, Y)$. By associativity it satisfies $f 0_{XY} = 0_{XZ}$ for all $f \in \text{Morph}(Y, Z)$ and $0_{XY} g = 0_{WY}$ for all $g \in \text{Hom}(W, X)$. Since $\text{Morph}(0, 0)$ has just one element, we have $0_{00} = 1_0$. If X is any other object such that $\text{Morph}(X, X)$ has $0_{XX} = 1_X$, then X is a zero object; in fact, the equalities $0_{X0} 0_{0X} = 0_{00} = 1_0$ and $0_{0X} 0_{X0} = 0_{XX} = 1_X$ show that X and 0 are isomorphic.

An **additive category** $\mathcal{C}$ is a category with the following three properties:

(i) $\mathcal{C}$ has a zero object,
(ii) the product and the coproduct[16] of any two objects in $\mathcal{C}$ exists in $\mathcal{C}$,
(iii) each set $\text{Morph}(X, Y)$ is an abelian group with the property that the operation is $\mathbb{Z}$ bilinear in the sense that if the operation is $+$ and if f, f' are arbitrary in $\text{Morph}(X, Y)$ and g, g' are arbitrary in $\text{Morph}(Y, Z)$, then

$$(g + g') \circ (f + f') = g \circ f + g' \circ f + g \circ f' + g' \circ f'$$

and
$$g \circ (-f) = (-g) \circ f = -(g \circ f).$$

If $\mathcal{C}$ is an additive category, then so is the opposite category $\mathcal{C}^{\text{opp}}$; this fact will enable us to use duality arguments occasionally. We shall henceforth write $\text{Hom}(X, Y)$ in place of $\text{Morph}(X, Y)$ for additive categories.

The zero morphism 0_{XY} of $\text{Hom}(X, Y)$ is the additive identity 0 of the abelian group $\text{Hom}(X, Y)$. In fact, 0_{0Y} is the additive identity of $\text{Hom}(0, Y)$, since $\text{Hom}(0, Y)$ has just one element. Therefore $0_{XY} = 0_{0Y} 0_{X0} = (0_{0Y} + 0_{0Y}) 0_{X0} = 0_{0Y} 0_{X0} + 0_{0Y} 0_{X0} = 0_{XY} + 0_{XY}$, and we obtain $0 = 0_{XY}$.

---

[15] Some authors use the word "epi" as an adjectival form of this noun.

[16] These are defined in Section IV.11 of *Basic Algebra*. They are always unique up to canonical isomorphism when they exist.

In an additive category a morphism u in $\operatorname{Hom}(X, Y)$ is a monomorphism if whenever $uf = 0$ with f in some $\operatorname{Hom}(W, X)$, then $f = 0$; a morphism u in $\operatorname{Hom}(X, Y)$ is an epimorphism if whenever $f'u = 0$ with f' in some $\operatorname{Hom}(Y, Z)$, then $f' = 0$.

This much structure forces products and coproducts to amount to the same thing in an additive category. The precise result is as follows.

Proposition 4.32. In an additive category, let (C, p_A, p_B) be a product of two objects A and B. Then there exist unique $i_A \in \operatorname{Hom}(A, C)$ and $i_B \in \operatorname{Hom}(B, C)$ such that
$$p_A i_A = 1_A, \quad p_B i_B = 1_B, \quad i_A p_A + i_B p_B = 1_C.$$
These satisfy $p_A i_B = 0$ and $p_B i_A = 0$, and (C, i_A, i_B) is a coproduct of A and B.

REMARKS.

(1) Since the defining properties of an additive category are self dual, any coproduct has a similar structure and becomes a product. The proof in effect will show more—that whenever there are data $A, B, C, i_A, i_B, p_A, p_B$ satisfying the displayed identities, then (C, p_A, p_B) is a product of A and B, and (C, i_A, i_B) is a coproduct. Thus a product/coproduct can be recognized without reference to other objects in the category.

(2) To emphasize the analogy with modules or vector spaces, we write $A \oplus B$ for a product or coproduct of A and B in $\mathcal{C}$ and call it the **direct sum** of A and B. The notation is understood to carry the morphisms i_A, i_B, p_A, p_B along with it. The direct sum is unique up to an isomorphism that carries the one set of morphisms i_A, i_B, p_A, p_B to the other.

PROOF. To the pair $1_A \in \operatorname{Hom}(A, A)$ and $0 \in \operatorname{Hom}(A, B)$, the product C associates a unique $i_A \in \operatorname{Hom}(A, C)$ with $p_A i_A = 1_A$ and $p_B i_A = 0$. Similarly the coproduct associates a unique $i_B \in \operatorname{Hom}(B, C)$ with $p_A i_B = 0$ and $p_B i_B = 1_B$. Computing with the aid of the $\mathbb{Z}$ bilinearity and associativity, we have

$$p_A(i_A p_A + i_B p_B) = 1_A p_A + 0 p_B = p_A$$

and
$$p_B(i_A p_A + i_B p_B) = 0 p_A + 1_B p_B = p_B.$$

Therefore $h = i_A p_A + i_B p_B$ is a member of $\operatorname{Hom}(C, C)$ with the property that $p_A h = p_A$ and $p_B h = p_B$. Since 1_C is another member of $\operatorname{Hom}(C, C)$ with this property, the assumed uniqueness shows that $h = 1_C$. This proves the displayed formulas in the proposition and the formulas $p_A i_B = 0$ and $p_B i_A = 0$.

For uniqueness of i_A and i_B, suppose that i'_A and i'_B satisfy $i'_A p_A + i'_B p_B = 1_C$. Right multiplication by i_A gives $i_A = 1_C i_A = (i'_A p_A + i'_B p_B) i_A = i'_A 1_A + i'_B 0 = i'_A$, and similarly $i_B = i'_B$.

8. Abelian Categories

To see that (C, i_A, i_B) is a coproduct of A and B, let $f \in \text{Hom}(A, X)$ and $g \in \text{Hom}(B, X)$ be given, and define $h = fp_A + gp_B$. This is in $\text{Hom}(C, X)$, has $hi_A = fp_Ai_A + gp_Bi_A = f1_A = f$, and similarly has $hi_B = g$. For uniqueness suppose that k is in $\text{Hom}(C, X)$ with $ki_A = f$ and $ki_B = g$. Then $ki_A p_A = fp_A$ and $ki_B p_B = gp_B$. Addition gives

$$k = k1_C = k(i_A p_A + i_B p_B) = fp_A + gp_B = h,$$

and uniqueness is proved. □

For an additive category $\mathcal{C}$, the notions of the kernel and cokernel of a morphism are defined by universal mapping properties. Problems 18–22 at the end of Chapter VI of *Basic Algebra* discussed universal mapping properties abstractly, saying what they are in a general context. For current purposes it is enough to know that what a universal mapping property produces (if it produces anything at all) is a pair consisting of an object and a morphism, and moreover the pair is automatically unique (if it exists) up to canonical isomorphism.

We allow ourselves to write morphisms as arrows in any of the customary ways for functions. Thus a member u of $\text{Hom}(A, B)$ may be written as $A \xrightarrow{u} B$, and a composition of u followed by a morphism $v \in \text{Hom}(B, C)$, which has been written as $v \circ u$ or as vu, may be written as $A \xrightarrow{u} B \xrightarrow{v} C$.

If $A \xrightarrow{u} B$ is a morphism in the additive category $\mathcal{C}$, then the **kernel** of u, denoted by $\ker u$, is a pair (K, i) with $i \in \text{Hom}(K, A)$ such that the composition $K \xrightarrow{i} A \xrightarrow{u} B$ has $ui = 0$ and such that for any pair (K', i') with i' in $\text{Hom}(K', A)$ for which $ui' = 0$, there exists a unique $\varphi \in \text{Hom}(K', K)$ with $i\varphi = i'$. See Figure 4.6. It is customary to drop all mention of K in the definition of kernel, saying that the kernel is i, since any mention of i carries along K as the domain of i; we shall adopt this abbreviated terminology shortly but shall refer to the pair (K, i) as the kernel for the time being.

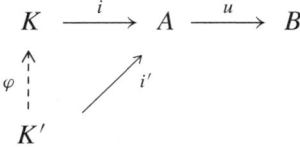

FIGURE 4.6. Universal mapping property of a kernel (K, i) of u.

The brief form of the definition of kernel is that $u \circ (\ker u) = 0$ and

$$ui' = 0 \quad \text{implies} \quad i' = (\ker u) \circ \varphi \quad \text{uniquely.}$$

The kernel of u is determined only up to an isomorphism applied to K; that is, i is determined only up to right multiplication by an isomorphism. The condition

for (K, i) to be a kernel is equivalent to the exactness of the sequence of abelian groups
$$0 \longrightarrow \mathrm{Hom}(K', K) \xrightarrow{i \circ (\cdot)} \mathrm{Hom}(K', A) \xrightarrow{u \circ (\cdot)} \mathrm{Hom}(K', B).$$
In fact, $ui = 0$ makes the sequence a complex, the existence of φ produces exactness at $\mathrm{Hom}(K', A)$, and the uniqueness of φ produces exactness at $\mathrm{Hom}(K', K)$.

Similarly the **cokernel** of u, denoted by $\mathrm{coker}\, u$, is a pair (C, p) with p in $\mathrm{Hom}(B, C)$ such that the composition $A \xrightarrow{u} B \xrightarrow{p} C$ has $pu = 0$ and such that for any pair (C', p') with p' in $\mathrm{Hom}(B, C')$ for which $p'u = 0$, there exists a unique $\psi \in \mathrm{Hom}(C, C')$ with $\psi p = p'$. See Figure 4.7. It is customary to drop all mention of the object C in the definition of cokernel, saying that the cokernel is p, since any mention of p carries along C as the *co*domain of p; we shall adopt this abbreviated terminology shortly but shall refer to the pair (C, p) as the cokernel for the time being.

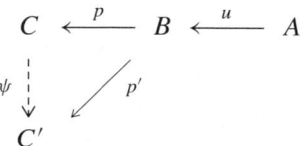

FIGURE 4.7. Universal mapping property of a cokernel (C, p) of u.

The brief form of the definition of cokernel is that $(\mathrm{coker}\, u) \circ u = 0$ and
$$p'u = 0 \quad \text{implies} \quad p' = \psi \circ (\mathrm{coker}\, u) \quad \text{uniquely}.$$
The cokernel of u is determined only up to an isomorphism applied to C; that is, p is determined only up to left multiplication by an isomorphism. The condition for (C, p) to be a cokernel is equivalent to the exactness of the sequence of abelian groups
$$0 \longrightarrow \mathrm{Hom}(C, C') \xrightarrow{(\cdot) \circ p} \mathrm{Hom}(B, C') \xrightarrow{(\cdot) \circ u} \mathrm{Hom}(A, C').$$
In fact, $pu = 0$ makes the sequence a complex, the existence of ψ produces exactness at $\mathrm{Hom}(B, C')$, and the uniqueness of ψ produces exactness at $\mathrm{Hom}(C, C')$.

Proposition 4.33. Let $\mathcal{C}$ be an additive category. If an element u of $\mathrm{Hom}(A, B)$ has a kernel (K, i) and if $m \in \mathrm{Hom}(B, B')$ is a monomorphism, then (K, i) is also a kernel of mu. If u has a cokernel (C, p) and if $e \in \mathrm{Hom}(A', A)$ is an epimorphism, then (C, p) is also a cokernel of ue. Briefly
$$\ker(mu) = \ker u \quad \text{and} \quad \mathrm{coker}(ue) = \mathrm{coker}\, u.$$

REMARK. We can safely omit the proof of any dual statement about additive categories, since the dual follows by expressing the original argument as a diagram, reversing all the arrows, and writing down the argument that the new diagram represents.

PROOF. We test whether $i = \ker u$ is a kernel of mu. We know that $(mu)i = m(ui) = 0$. Suppose that $mui' = 0$ with $i' \in \text{Morph}(K', A)$. Since m is a monomorphism, $ui' = 0$. Because i is a kernel of u, we obtain $i' = i\varphi$ for a unique $\varphi \in \text{Morph}(K', K)$. Hence i is a kernel of mu. The statement about cokernels is dual. □

Proposition 4.34. Let $\mathcal{C}$ be an additive category. If an element u of $\text{Hom}(A, B)$ has a kernel (K, i), then i is a monomorphism. Dually if u has a cokernel (C, p), then p is an epimorphism.

PROOF. Suppose that u has a kernel (K, i). For any object K', the zero morphism $i' = 0$ of $\text{Hom}(K', A)$ has the property that $ui' = 0$. The uniqueness property of the kernel says that the φ in $\text{Hom}(K', K)$ with $i\varphi = i'$ is unique. Evidently $\varphi = 0$ is one such choice and hence is the only such choice. Thus if f in $\text{Hom}(K', K)$ has $if = 0$, then $f = 0$. Therefore i is a monomorphism. □

Propositions 4.33 and 4.34 give a first hint that the notation (K, i) for the kernel, which we know is redundant, may also be inconvenient; it would be far simpler to refer to the kernel as i, and analogously for cokernels. Then Proposition 4.33 could truly be stated as the displayed formulas in its statement, and Proposition 4.34 would have the tidier statement that every kernel is a monomorphism and every cokernel is an epimorphism. Let us therefore now allow ourselves to regard kernels and cokernels as morphisms, rather than pairs consisting of an object and a morphism. With this convention in place, we always have $u \circ (\ker u) = 0$ and $(\text{coker } u) \circ u = 0$.

Proposition 4.35. Let $\mathcal{C}$ be an additive category, and let u be in $\text{Hom}(A, B)$. If u has a kernel and $\ker u$ has a cokernel, then $\text{coker}(\ker u)$ is a kernel of u. Briefly

$$\ker(\text{coker}(\ker u)) = \ker u.$$

Dually if u has a cokernel and $\text{coker } u$ has a kernel, then

$$\text{coker}(\ker(\text{coker } u)) = \text{coker } u.$$

PROOF. Let (K, i) be a kernel of u, and let (C, p) be a cokernel of i. We are to show that i is a kernel of p. For the existence step, suppose that i' in $\text{Hom}(K', A)$ has $pi' = 0$. We are to show that i' factors as $i' = i\varphi$ for some unique φ in $\text{Hom}(K', K)$. We know that $ui = 0$. Since $p = \text{coker } i$, u factors as $u = \psi p$ for some ψ in $\text{Hom}(C, B)$. Then $ui' = (\psi p)i' = \psi(pi') = 0$. Since $i = \ker u$, i' factors as $i' = i\varphi$ as required. This proves existence of φ.

For the uniqueness step, suppose that $pi' = 0$ for some i' in some $\text{Hom}(K', A)$. If i' were to have two distinct factorizations, say as $i' = i\varphi = i\overline{\varphi}$, then i could not be a monomorphism, in contradiction to Proposition 4.34 and the fact that $i = \ker u$. This proves uniqueness of φ. □

An **abelian category** $\mathcal{C}$ is an additive category with the following two properties:

(iv) every morphism has a kernel and a cokernel,
(v) every monomorphism is a kernel, and every epimorphism is a cokernel.

It is evident that the opposite category of any abelian category is abelian. Thus we can continue to use duality arguments.

Property (iv) is certainly desirable if one wants to have a theory involving homology and cohomology. Property (v) may be viewed as a converse to Proposition 4.34; some other authors use a different but equivalent formulation of this axiom. The objective is to have a generalization of the kind of factorization that one has with homomorphisms of abelian groups: any homomorphism factors canonically as the product of the canonical passage to the quotient by the kernel, followed by an isomorphism of this quotient onto the image of the homomorphism, followed by the inclusion of the image into the range.

Proposition 4.36. In any abelian category, every morphism that is both a monomorphism and an epimorphism is an isomorphism.

PROOF. If $f \in \text{Hom}(K, A)$ is a monomorphism, then $f = \ker g$ for some g in some $\text{Hom}(A, B)$ by (v). This fact implies that $gf = g \circ (\ker g) = 0$. If f is also an epimorphism, then the equality $gf = 0$ implies that $g = 0$. Hence $f = \ker 0_{AB}$. Taking $K' = A$ and $i' = 1_A$ in Figure 4.6, we have $0i' = 0$ and thus have $1_A = f\varphi$ for some φ in $\text{Hom}(A, K)$. Thus the monomorphism f has a right inverse and must be an isomorphism. □

Lemma 4.37. In an abelian category $\mathcal{C}$, every monomorphism is the kernel of its cokernel, and every epimorphism is the cokernel of its kernel.

PROOF. If m is a monomorphism, then (v) says that $m = \ker u$ for some u. Substituting into the first conclusion of Proposition 4.35, we obtain $\ker(\text{coker } m) = m$. If e is an epimorphism, then (v) says that $e = \text{coker } u$ for some u. Substituting into the second conclusion of Proposition 4.35, we obtain $\text{coker}(\ker e) = e$. □

Proposition 4.38. In an abelian category $\mathcal{C}$, any morphism f factors as $f = me$ for a monomorphism m and an epimorphism e. Here one such factorization is given by

$$m = \ker(\text{coker } f) \quad \text{and} \quad e = \text{coker}(\ker f).$$

Any other such factorization $f = m'e'$ has the property that there is some isomorphism x with $e' = xe$ and $m'x = m$.

PROOF. Put $m = \ker(\operatorname{coker} f)$. Since $(\operatorname{coker} f)f = 0$, the brief form of the definition of kernel gives $f = me$ for some e. We are going to prove that e is an epimorphism. Thus suppose that $re = 0$ for some morphism r. The brief form of the definition of kernel shows that $e = (\ker r)e'$ for some morphism e'. Then we have

$$f = me = m(\ker r)e' = m'e', \qquad \text{where } m' = m \ker r.$$

Being a kernel, $\ker r$ is a monomorphism. As the composition of two monomorphisms, m' is a monomorphism. Lemma 4.37 shows that $m' = \ker p'$, where $p' = \operatorname{coker} m'$.

Put $p = \operatorname{coker} m$. The definition of m and the second identity of Proposition 4.35 gives $p = \operatorname{coker}(\ker(\operatorname{coker} f)) = \operatorname{coker} f$. Since $m' = \ker p'$, we have $p'm' = 0$. Hence $p'f = p'm'e' = 0$. Since $p = \operatorname{coker} f$, the brief form of the definition of cokernel shows that $p' = sp$ for some s. Thus $p'm = spm = 0$, the latter equality holding because $p = \operatorname{coker} m$. Since $m' = \ker p'$, the brief form of the definition of kernel gives $m = m't$ for some t.

Resubstituting for m' gives $m = m't = m(\ker r)t$. Since m is a monomorphism, we can cancel and obtain $1_X = (\ker r)t$, where X is the codomain of $\ker r$. In other words, $\ker r$ has a right inverse. Being a monomorphism, it must be an isomorphism. Since any morphism v has $v \ker v = 0$, we obtain $r \ker r = 0$ and conclude that $r = 0$. Therefore e is an epimorphism, as asserted.

Since e is an epimorphism, Lemma 4.37 gives $e = \operatorname{coker}(\ker e)$, and Proposition 4.33 gives $\ker e = \ker(me) = \ker f$. Therefore $e = \operatorname{coker}(\ker f)$. This completes the proof of existence of the decomposition.

For uniqueness, suppose that $f = m'e'$ for a monomorphism m' and an epimorphism e'. Proposition 4.33 gives $\ker f = \ker(m'e') = \ker e'$, as well as $\ker f = \ker(me) = \ker e$, the understanding being that these equalities hold up to an isomorphism on the right. Set $u = \ker e$ and $u' = \ker e'$; then $u = u'w$ for some isomorphism w. Since e and e' are epimorphisms, Lemma 4.37 gives $e = \operatorname{coker} u$ and $e' = \operatorname{coker} u'$. Since m' is a monomorphism, the equality $0 = f(\ker f) = fu = m'e'u$ implies that $e'u = 0$; by the brief form of the definition of $\operatorname{coker} u$ as a cokernel, e' factors as $e' = xe$ for a unique x. Similarly the equality $0 = f \ker f = fu' = meu$ implies that $eu = 0$; by the brief form of the definition of $\operatorname{coker} u'$ as a cokernel, e factors as $e = x'e'$ for a unique x'. Then $e = x'e' = x'xe$; since e is an epimorphism, $x'x$ is the identity on its domain. Similarly $e' = xe = xx'e'$, and it follows that xx' is the identity on its domain. Consequently x is an isomorphism. Multiplying $e' = xe$ by m' on the left gives $me = f = m'e' = m'xe$; since e is an epimorphism, $m = m'x$. This completes the proof. □

With this canonical factorization in hand, we introduce two terms that will simplify the definition of "exact sequence." We define the **image** and **coimage** of $f = me$ in $\text{Hom}(A, B)$ by

$$m = \text{image } f \quad \text{and} \quad e = \text{coimage } f.$$

In words, the image of any morphism is its monomorphism factor, and the coimage is its epimorphism factor; in particular, a monomorphism is its own image, and an epimorphism is its own coimage.[17] Let us see what the factorization and these formulas say in terms of diagrams. We write (K, i) for the kernel of f and (C, p) for the cokernel of f. Let I be the codomain of e, which equals the domain of m. In terms of a diagram, the situation for f is then given by

$$K \xrightarrow[=\ker f]{i=\ker e} A \xrightarrow[=\text{coimage } f]{e=\text{coker } i} I \xrightarrow[=\text{image } f]{m=\ker p} B \xrightarrow[=\text{coker } f]{p=\text{coker } m} C.$$

The top row of labels explains the relationships among i, e, m, p, and the bottom row of labels relates i, e, m, p to f. The morphism f itself is the composition of the two morphisms in the center.

In a good category of modules, we can interpret this diagram in terms of the two short exact sequences

$$0 \longrightarrow K \xrightarrow{i} A \xrightarrow{e} A/\text{image } i \longrightarrow 0,$$

$$0 \longrightarrow A/\text{image } i \xrightarrow{m} B \xrightarrow{p} C \longrightarrow 0,$$

which we can merge into a single 6-term exact sequence

$$0 \longrightarrow K \xrightarrow{i} A \xrightarrow{me=f} B \xrightarrow{p} C \longrightarrow 0.$$

Now we can define complexes and exact sequences for abelian categories, and we can readily check that the new definitions are consistent with the definitions for good categories of modules. A **chain complex** is a doubly infinite sequence of morphisms with decreasing indexing such that the consecutive compositions are defined and are 0. If $f \in \text{Hom}(A, B)$ and $g \in \text{Hom}(B, C)$ are given morphisms, then the sequence

$$A \xrightarrow{f} B \xrightarrow{g} C$$

is **exact** at B if image f = ker g, or equivalently if coker f = coimage g. As usual in the subject of abelian categories, the equality sign here means "can be taken as." In more detail if f and g decompose as $f = me$ and $g = m'e'$, image f is defined to be m, and ker g equals ker e'. Thus the condition for exactness is

---

[17]The term "coimage" is not really needed for recognizing exact sequences, but it makes any implementation of duality more symmetric.

that m be a kernel of e'. Since $u(\ker u) = 0$ for any morphism u, exactness at B implies that $e'm = 0$. Then $gf = m'e'me = 0$, and we see that the given sequence (when extended by 0's at each end) is a complex.

Exactness of any finite or infinite sequence of morphisms whose consecutive compositions are defined means exactness at every object X in the sequence for which there is an incoming morphism in some $\mathrm{Hom}(W, X)$ and there is an outgoing morphism in some $\mathrm{Hom}(X, Y)$. With the kind of indexing used for a chain complex, a sequence

$$\cdots \longrightarrow X_{n+1} \xrightarrow{m_n e_n} X_n \xrightarrow{m_{n-1} e_{n-1}} X_{n-1} \longrightarrow \cdots$$

is exact if $m_n = \ker e_{n-1}$, or equivalently if $e_{n-1} = \mathrm{coker}\, m_n$, for all n.

For a sequence of four morphisms of the form

$$0 \longrightarrow K \xrightarrow{m} A \xrightarrow{e} C \longrightarrow 0,$$

exactness means exactness at K, A, and C. The conditions are that m is a monomorphism, e is an epimorphism, and $m = \ker e$ (or equivalently that $e = \mathrm{coker}\, m$). In this case the sequence is called a **short exact sequence**.

One can now proceed to define **projectives** and **injectives** for any abelian category as certain objects in the same way as in Figures 4.3 and 4.4, and extend all the results of earlier sections of this chapter to all abelian categories. We shall not carry out this detail.[18]

Instead, we shall indicate an approach to carrying out this detail that takes most of the difficulty out of translating results from the context of good categories to the context of abelian categories. It is to use the notion of "members." The word "members" in the present setting refers to something that substitutes for elements in situations in which objects need not necessarily be sets of elements. The idea is to recast elements, when they exist, in terms of morphisms and then to generalize the resulting definition. For orientation, consider the category $\mathcal{C}_R$ of all unital left R modules, R being a ring with identity. Let us write R_0 for the left R module R. The elements of a unital left R module X are then in one-one correspondence with the R homomorphisms of R_0 into X, the element x corresponding to the homomorphism that carries r to rx. Thus the category $\mathcal{C}_R$ has a distinguished object R_0 such that the elements of any object X are in one-one correspondence with $\mathrm{Hom}(R_0, X)$. Hence any argument about elements for this category immediately translates into an argument about morphisms.

The trouble is that a general abelian category has no distinguished object to play the role of R_0. The idea for getting around this difficulty is to take all possible

---

[18]The entire theory for abelian categories is carried out in detail in Freyd's book *Abelian Categories: An Introduction to the Theory of Functors*.

objects X_0 in place of R_0, consider the union on X_0 of *all* sets $\text{Hom}(X_0, X)$, introduce an equivalence relation, and hope for the best.

The definition is as follows. Let $\mathcal{C}$ be an abelian category, fix X in $\text{Obj}(\mathcal{C})$, and consider all morphisms with codomain X. Two such morphisms x and y are said to be **equivalent** morphisms for current purposes, written $x \equiv y$, if there exist epimorphisms u and v such that $xu = yv$. It is evident that "equivalent" is reflexive and symmetric. Transitivity requires proof, and we return to this matter in a moment. Once $\equiv$ has been shown to be an equivalence relation, an equivalence class of such morphisms is called a **member** of X. We write $x \in_m X$ to indicate that x is a morphism with codomain X, hence to indicate that x is a morphism whose equivalence class is a member of X. To avoid clumsy wording when there is really no possibility of confusion, we often simply say that x is a member of X. The question arises whether this definition presents any set-theoretic difficulties. As usual in category theory, one can answer the question painlessly by working when necessary only with subcategories for which the objects actually form a set; in this case, the union over all objects X and Y in the subcategory of all the groups $\text{Hom}(X, Y)$ of morphisms is a set, and there is no problem. Let us return to a special case of our example.

EXAMPLE OF MEMBERS. Let $\mathcal{C} = \mathcal{C}_\mathbb{Z}$ be the category of all abelian groups, and fix an abelian group X. If x is an abelian-group homomorphism with codomain X, let us use Proposition 4.38 to write $x = me$ for a monomorphism m and an epimorphism e. Then $x \equiv m$, and thus we might just as well consider only one-one homomorphisms into X. If H is the image of x, then we can view x as a composition $x = i_H y$ of a homomorphism y carrying the domain of x onto H, followed by the inclusion $i_H : H \to X$. The homomorphism y is an isomorphism, hence is an epimorphism. Thus $x \equiv i_H$. It is apparent that no two inclusions of subgroups of X into X are equivalent morphisms. Since every inclusion of a subgroup of X into X yields a member of X, the members of X are exactly the subgroups of X. Thus for example the set of members of $\mathbb{Z}$ corresponds to the set of integers ≥ 0, in which addition is lost, and does not correspond exactly to the set of elements of $\mathbb{Z}$. This fact is a little discouraging, but it turns out not to be as bad an omen as one might expect.

Returning to the setting of a general abelian category, we work toward a proof that $\equiv$ is an equivalence relation. We need the notion of the "pullback" of two morphisms, which we define by a universal mapping property momentarily. The appropriate construction establishing existence appears in the next proposition. Then we prove a proposition for using pullback as a tool, and afterward we prove the transitivity.

In an abelian category $\mathcal{C}$, let X, Y, Z be objects, and let $f \in \text{Hom}(Y, Z)$ and $g \in \text{Hom}(X, Z)$ be morphisms. A **pullback** of the pair (f, g) is a triple

$(W, \widetilde{f}, \widetilde{g})$ in which W is an object in $\mathcal{C}$, in which $\widetilde{f}$ and $\widetilde{g}$ are morphisms with $\widetilde{f} \in \operatorname{Hom}(W, Y)$ and $\widetilde{g} \in \operatorname{Hom}(W, X)$, and in which the following universal mapping property holds: whenever $(W', \widetilde{f}', \widetilde{g}')$ is a triple such that W' is an object in $\mathcal{C}$ and $\widetilde{f}'$ and $\widetilde{g}'$ are morphisms with $\widetilde{f}' \in \operatorname{Hom}(W', Y)$ and $\widetilde{g}' \in \operatorname{Hom}(W', X)$ and with $f\widetilde{g}' = g\widetilde{f}'$, then there exists a unique $\varphi \in \operatorname{Hom}(W', W)$ such that $\widetilde{f}' = \widetilde{f}\varphi$ and $\widetilde{g}' = \widetilde{g}\varphi$. See Figure 4.8.

$$\begin{array}{ccc} W & \xrightarrow{\widetilde{f}} & Y \\ \widetilde{g} \downarrow & & \downarrow g \\ X & \xrightarrow{f} & Z \end{array}$$

FIGURE 4.8. The pullback of a pair (f, g) of morphisms.

Proposition 4.39. In an abelian category $\mathcal{C}$, let X, Y, Z be objects, and let $f \in \operatorname{Hom}(X, Z)$ and $g \in \operatorname{Hom}(Y, Z)$ be morphisms. Let $X \oplus Y$ be the direct sum, let p_X and p_Y be the projections on the two factors, define $h = fp_X - gp_Y$ in $\operatorname{Hom}(X \oplus Y, Z)$, and let $m = \ker h$. Then a pullback $(W, \widetilde{f}, \widetilde{g})$ of (f, g) is given by $W = \operatorname{domain} m$, $\widetilde{f} = p_Y m$, and $\widetilde{g} = p_X m$.

REMARKS. The dual statement asserts the existence of a **pushout** of a pair of morphisms, and it is a consequence of Proposition 4.39. Problem 35 at the end of the chapter points out that the proof of Proposition 4.19a made use of a concretely constructed pullback, while the proof of Proposition 4.19b made use of a concretely constructed pushout.

PROOF. From $hm = h \ker h = 0$, we obtain $0 = fp_X m - gp_Y m = f\widetilde{g} - g\widetilde{f}$, and thus $f\widetilde{g} = g\widetilde{f}$. Now suppose that W', $\widetilde{f}'$, and $\widetilde{g}'$ are given with $f\widetilde{g}' = g\widetilde{f}'$. Then $m' = (\widetilde{g}', \widetilde{f}')$ is a morphism in $\operatorname{Hom}(W', X \oplus Y)$ such that $hm' = fp_X m' - gp_Y m' = f\widetilde{g}' - g\widetilde{f}' = 0$. Therefore m' factors through $m = \ker h$ as $(\widetilde{g}', \widetilde{f}') = m\varphi$ for a unique $\varphi \in \operatorname{Hom}(W', W)$. Application of p_X and p_Y to this equality gives $\widetilde{g}' = p_X m\varphi = \widetilde{g}\varphi$ and $\widetilde{f}' = p_Y m\varphi = \widetilde{f}\varphi$. $\square$

Proposition 4.40. In the notation of Figure 4.8 and Proposition 4.39 if f is a monomorphism, then so is $\widetilde{f}$. If f is an epimorphism, then so is $\widetilde{f}$; in the case of an epimorphism, $\ker f$ factors as $\ker f = \widetilde{g}(\ker \widetilde{f})$.

PROOF. Throughout the proof let i_X and i_Y be the injections associated with the direct sum $X \oplus Y$. Suppose that f is a monomorphism, and suppose that $\widetilde{f}w = 0$ for some morphism with codomain W. Since $\widetilde{f} = p_Y m$, $p_Y mw = 0$. Then $0 = (fp_X - gp_Y)mw = fp_X mw - 0 = fp_X mw$. Since f is a monomorphism, $p_X mw = 0$. Since also $\widetilde{f}w = p_Y mw = 0$, $mw = (i_X p_X + i_Y p_Y)mw = 0$. But m is a monomorphism, and therefore $w = 0$. Consequently $\widetilde{f}$ is a monomorphism.

For the remainder of the proof, assume that f is an epimorphism. Let us see that $h = fp_X - gp_Y$ is an epimorphism. In fact, if $zh = 0$, then $0 = z(fp_X - gp_Y)i_X = zfp_Xi_X = zf$. Since f is an epimorphism, $z = 0$. Thus h is an epimorphism.

It follows from Lemma 4.37 that $h = \mathrm{coker}(\ker h) = \mathrm{coker}\, m$. To prove that $\tilde{f}$ is an epimorphism, suppose that $v\tilde{f} = 0$ for some morphism v with domain Y. This means that $vp_Y m = 0$. Since h is the cokernel of m, vp_Y factors as $vp_Y = v'h$ for some morphism v'. Applying i_X on the right end of both sides gives $0 = vp_Y i_X = v'h i_X = v'(fp_X - gp_Y)i_X = v'fp_X i_X = v'f$. Since f is an epimorphism, $v' = 0$. Hence $vp_Y = v'h = 0$. Since p_Y is an epimorphism, $v = 0$. Therefore $\tilde{f}$ is an epimorphism.

Now set $k = \ker f$, and let K be its domain. The morphisms $k \in \mathrm{Hom}(K, X)$ and $0 \in \mathrm{Hom}(K, Y)$ have $fk = 0 = g0$. If we set $W' = K$, $\tilde{f}' = 0$, and $\tilde{g}' = k$, then $f\tilde{g}' = g\tilde{f}'$, and Proposition 4.39 produces a unique φ in $\mathrm{Hom}(K, W)$ with $0 = \tilde{f}\varphi$ and $k = \tilde{g}\varphi$. We shall show that φ is a kernel of $\tilde{f}$, and then the equation $k = \tilde{g}\varphi$ completes the proof.

We know that $\tilde{f}\varphi = 0$. Thus suppose that $\tilde{f}v = 0$ for some morphism v in some $\mathrm{Hom}(K', W)$. Since $f\tilde{g} = g\tilde{f}$, we have $f\tilde{g}v = g\tilde{f}v = 0$. Thus $\tilde{g}v$ factors through $k = \ker f$ as $\tilde{g}v = kv'$ for some v' in $\mathrm{Hom}(K', K)$.

Put $\Phi = v - \varphi v'$. Then $\tilde{f}\Phi = \tilde{f}v - \tilde{f}\varphi v' = 0 - 0 = 0$, and $\tilde{g}\Phi = \tilde{g}v - \tilde{g}\varphi v' = kv' - kv' = 0$. Consequently if we put $W'' = K'$, $\tilde{f}'' = 0$, and $\tilde{g}'' = 0$, then Φ and 0 are two morphisms in $\mathrm{Hom}(K', W)$ with $\tilde{f}'' = \tilde{f}\Phi = \tilde{f}0$ and $\tilde{g}'' = \tilde{g}\Phi = \tilde{g}0$. By uniqueness of the morphism in the universal mapping property for pullbacks, $\Phi = 0$. Therefore $v = \varphi v'$, and v has been exhibited as factoring through φ.

If v factors through φ also as $v = \varphi v''$, then $0 = \varphi(v' - v'')$, and we have $k(v' - v'') = \tilde{g}\varphi(v' - v'') = 0$. Since $k = \ker f$ is a monomorphism, $v' = v''$. Thus the factorization of v through φ is unique, and φ is a kernel of $\tilde{f}$. This completes the proof. □

Proposition 4.41. Let $\mathcal{C}$ be an abelian category, let X be an object in $\mathcal{C}$, and define $x \equiv y$ for two morphisms x and y with codomain X if there exist epimorphisms u and v with $xu = yv$. Then the relation $\equiv$ on the morphisms with codomain X is transitive and hence is an equivalence relation.

REMARK. A nontrivial special case is that the obvious equivalences $xu \equiv x$ and $x \equiv xv$ imply the nonobvious equivalence $xu \equiv xv$ when u and v are epimorphisms.

PROOF. Assuming that $x \equiv y$ and $y \equiv z$, write $xu = yv$ and $yr = zs$ for epimorphisms u, v, r, s. Since v and r have the same codomain, namely $\mathrm{domain}(y)$, the pullback $(\tilde{v}, \tilde{r})$ of (v, r) as in Proposition 4.39 is well defined, and Proposition 4.40 shows that $\tilde{v}$ and $\tilde{r}$ are epimorphisms. Since $r\tilde{v} = v\tilde{r}$, we obtain $xu\widetilde{r} = yv\widetilde{r} = yr\widetilde{v} = zs\widetilde{v}$. The morphisms $u\widetilde{r}$ and $s\widetilde{v}$ are epimorphisms as compositions of epimorphisms, and therefore $x \equiv z$. □

Fix an object X. Then 0_{0X} is a member of X called the **zero member**, denoted by 0. Every zero morphism 0_{YX} with codomain X is equivalent to 0_{0X}; in fact, $0_{YX} = 0_{0X} 0_{Y0}$. The morphism 0_{Y0} is an epimorphism because if $f \in \text{Hom}(0, Z)$ has $f 0_{Y0} = 0_{YZ}$, then f is the unique element 0_{0Z} of $\text{Hom}(0, Z)$. Conversely any nonzero morphism r in $\text{Hom}(Y, X)$ is inequivalent to 0_{YX}. In fact, an equality $ru = 0_{YX} v$ for epimorphisms u and v would imply that $r = 0_{YX}$, since we can cancel in the equality $ru = 0_{YX} v = 0_{YX} u$.

Each $x \in_m X$ has a "negative," namely the class of the negative of the representative x of the member; i.e., taking the negative of a morphism is respected in passing to classes. We write $-x \in_m X$ for the negative. (*Warning:* As the example with the category of abelian groups shows, one should use care in inferring any relationship between "negatives" and zero members.)

If f is a morphism in $\text{Hom}(X, Y)$, then each member $x \in_m X$ yields by composition a well-defined member $fx \in_m Y$. To see that this notion is indeed well defined, suppose that $x \equiv x'$, and choose epimorphisms u and v with $xu = x'v$. Then $(fx)u = f(xu) = f(x'v) = (fx')v$ shows that $fx \equiv fx'$.

The main result is Theorem 4.42 below, which gives a calculus for diagram chases using members in general abelian categories. After the proof we shall be content with one example of how the theorem allows all the diagram chases in earlier sections of this chapter to be extended to general abelian categories. The example is the proof of the part of the Snake Lemma that involves an explicit construction.[19] More examples appear in Problems 34–35 at the end of the chapter.

Theorem 4.42. The members of an abelian category satisfy the following properties:

(a) a morphism $f \in \text{Hom}(X, Y)$ is a monomorphism if and only if every $x \in_m X$ with $fx \equiv 0$ has $x \equiv 0$,

(b) a morphism $f \in \text{Hom}(X, Y)$ is a monomorphism if and only if every pair of members $x \in_m X$ and $x' \in_m X$ with $fx \equiv fx'$ has $x \equiv x'$,

(c) a morphism $g \in \text{Hom}(X, Y)$ is an epimorphism if and only if for each $y \in_m Y$, there exists some $x \in_m X$ with $gx \equiv y$,

(d) a morphism $h \in \text{Hom}(X, Y)$ is the 0 morphism if and only if every $x \in_m X$ has $hx \equiv 0$,

(e) a sequence $X \xrightarrow{f} Y \xrightarrow{g} Z$ is exact at Y if and only if $gf = 0$ and also each $y \in_m Y$ with $gy \equiv 0$ has some $x \in_m X$ with $fx \equiv y$,

---

[19]For more detail about this example and for further examples, see Mac Lane's *Categories for the Working Mathematician*.

(f) whenever x, y, z are members of an object X and $x \equiv yu + zv$ for some epimorphisms u and v, then $xu' - yv' \equiv z$ for some epimorphisms u' and v'.

REMARKS.
(1) The interpretations of (a) through (e) are straightforward enough and already give an indication that the notion of a member may be of some help in translating proofs for good categories into proofs for abelian categories. Application of (d) to the difference $f_1 - f_2$ of two morphisms in $\text{Hom}(X, Y)$ shows that $f_1 x \equiv f_2 x$ for all $x \in_m X$ implies $f_1 = f_2$.

(2) The interpretation of (f) is more subtle. As the example with the Snake Lemma below will show, conclusion (f) makes it possible to mirror in the theory of members the kind of subtraction that takes place with elements of a module to get their difference to be in the kernel of some homomorphism.

PROOF. For (a) and (b), if f is a monomorphism and $fx \equiv fx'$, then $fxu = fx'v$ for suitable epimorphisms u and v, and cancellation yields $xu = x'v$ and hence $x \equiv x'$. Conversely suppose $fx \equiv 0$ only for $x \equiv 0$. If f has $fx' = 0_{AY}$ for some x' in some $\text{Hom}(A, X)$, then $fx' \equiv 0$ and so $x' \equiv 0$ by hypothesis. In this case, $x' = 0_{AX}$ because we know that nonzero morphisms are not equivalent to 0.

For (c), suppose that g is an epimorphism. If $y \in_m Y$ is given, let y be in $\text{Hom}(X', Y)$, and let $(\widetilde{g}, \widetilde{y})$ be the pullback of (g, y), satisfying $y\widetilde{g} = g\widetilde{y}$. Proposition 4.40 shows that $\widetilde{g}$ is an epimorphism, and then $y \equiv gx$ for $x = \widetilde{y}$. Conversely if g fails to be an epimorphism, then there exists $r \neq 0$ in some $\text{Hom}(Y, Z)$ with $rg = 0_{XZ}$. If there is some x in some $\text{Hom}(A, X)$ with $gx \equiv 1_Y$, we can compose with r on the left of both sides and obtain $rgx \equiv r1_Y = r$. Since the left side equals 0_{AZ}, which is equivalent to 0_{YZ}, we obtain $0_{YZ} \equiv 0_{AZ} \equiv r$, which we know not to be true for nonzero members r of $\text{Hom}(Y, Z)$.

For (d), if $h = 0_{XY}$ and if x is in $\text{Hom}(Z, X)$, then $hx = 0_{XY}x = 0_{ZY} \equiv 0_{0Y}$. Conversely if every x in every $\text{Hom}(Z, X)$ has $hx \equiv 0_{0Y}$, we take $Z = X$ and $x = 1_X$. Then $hu = hxu = 0_{0Y}v$ for some epimorphisms $u \in \text{Hom}(A, X)$ and $v \in \text{Hom}(A, 0)$. This says that $hu = 0_{AY} = 0_{XY}u$. Since u is an epimorphism, $h = 0_{XY}$.

For (e), let $f = me$ be the decomposition of f as in Proposition 4.38. Then $m = \text{image } f$, and we define $k = \ker g$. If the sequence is exact at Y, then $gf = 0$ as part of the definition. Suppose $y \in_m Y$ has $gy \equiv 0$, i.e., $gy = 0$. Since $m = \ker g$ by exactness, the equality $gy = 0$ and the definition of kernel together imply that $y = my'$ for some y'. Using Proposition 4.39, let (e, y') have $(\widetilde{e}, \widetilde{y}')$ as pullback, satisfying $e\widetilde{y}' = y'\widetilde{e}$. Since e by construction is an epimorphism, Proposition 4.40 shows that $\widetilde{e}$ is an epimorphism. From the computation $f\widetilde{y}' = me\widetilde{y}' = my'\widetilde{e} = y\widetilde{e}$, we obtain $f\widetilde{y}' \equiv y$. Then $x = \widetilde{y}'$ has $x \in_m X$ and $fx \equiv y$.

Conversely suppose that $gf = 0$ and that the other condition holds. Since e is an epimorphism, the equality $gf = 0$ implies that $gm = 0$. The definition of $k = \ker g$ thus gives $m = k\varphi$ for some morphism φ. Meanwhile, the morphism $k = \ker g$ has $k \in_m Y$ and $gk = 0$. Thus $gk \equiv 0$. The hypothesis produces $x \in_m X$ with $fx \equiv k$, i.e., with $mexu = kv$ for suitable epimorphisms u and v. Write $ex = m'e'$ according to Proposition 4.38. Then $mm'e'u = kv$, and the uniqueness in Proposition 4.38 shows that $k = mm'\psi$ for some isomorphism ψ. Putting the results together gives $m = k\varphi = mm'\psi\varphi$ and $k = mm'\psi = k\varphi m'\psi$. Since m and k are monomorphisms, $1 = m'\psi\varphi$ and $1 = \varphi m'\psi$. These show that φ has a left inverse and a right inverse, hence is an isomorphism. Then m' too is an isomorphism, and $k = m$ except for a factor of an isomorphism on the right side. This means that we can take $\ker g = \text{image } f$ and that the given sequence is exact at Y.

For (f), let $x \equiv yu + zv$. Then $xu_1 \equiv (yu + zv)v_1$, and $xu_1 - y(uv_1) \equiv zvv_1$. Consequently $xu_1 - y(uv_1) \equiv zvv_1 \equiv z$, and (f) follows with $u' = u_1$ and $v' = uv_1$. $\square$

Theorem 4.42 enables us to use members to verify properties of morphisms in diagrams, but it does not by itself construct any morphisms. That is, just because we know what the equivalence class of fx should be for every $x \in_m X$ does not mean that we have a construction of f; it means only that we know how to work with f once f is known to exist. Specifically we know from Remark 1 with the theorem that there cannot be a different morphism g with $fx \equiv gx$ for all $x \in_m X$. Some tools that we have for constructing morphisms for a general abelian category are the existence of kernels and cokernels via Axiom (iv), Proposition 4.39 asserting the existence of pullbacks of pairs of morphisms, and the dual of Proposition 4.39 asserting the existence of pushouts of pairs of morphisms. For particular categories of interest, the hypotheses "enough projectives" and "enough injectives" provide additional constructions of morphisms.

The most complicated example of a constructed mapping that we encountered in the theory for good categories was the connecting homomorphism in the Snake Lemma. In the generalization to abelian categories, the construction of the connecting morphism has to go outside the usual diagram given in Figure 4.2. Problem 33 at the end of the chapter will compare the actual construction and Figure 4.2 for the chain map of exact sequences of abelian groups given below and observe that the two diagrams are different:

$$\begin{array}{ccccccccc}
0 & \longrightarrow & \mathbb{Z} & \xrightarrow{\times 8} & \mathbb{Z} & \xrightarrow{1 \mapsto 1 \bmod 8} & \mathbb{Z}/8\mathbb{Z} & \longrightarrow & 0 \\
& & \downarrow {\scriptstyle \times 4} & & \downarrow {\scriptstyle \times 2} & & \downarrow {\scriptstyle 1 \bmod 8 \mapsto 2 \bmod 4} & & \\
0 & \longrightarrow & \mathbb{Z} & \xrightarrow{\times 4} & \mathbb{Z} & \xrightarrow{1 \mapsto 1 \bmod 4} & \mathbb{Z}/4\mathbb{Z} & \longrightarrow & 0
\end{array}$$

The domain of the connecting homomorphism for this situation is the set of even members of $\mathbb{Z}/8\mathbb{Z}$, and the mapping carries $2 + 8\mathbb{Z}$ to $1 + 4\mathbb{Z}$ in $\mathbb{Z}/4\mathbb{Z}$.

EXAMPLE OF DIAGRAM CHASE. In the setting of the Snake Lemma (Lemma 4.6), we shall construct the connecting morphism ω and verify that its value on each member of its domain corresponds to what we expect on the basis of Lemma 4.6. The given snake diagram, partially enlarged toward Figure 4.2, is

$$\begin{array}{ccccccc}
& & & & C_0 & & \\
& & & & \downarrow k & & \\
A & \xrightarrow{\varphi} & B & \xrightarrow{\psi} & C & \longrightarrow & 0 \\
\downarrow \alpha & & \downarrow \beta & & \downarrow \gamma & & \\
0 & \longrightarrow & A' & \xrightarrow{\varphi'} & B' & \xrightarrow{\psi'} & C' \\
& & \downarrow p & & & & \\
& & A'_0 & & & &
\end{array} \qquad (*)$$

with the rows exact and the squares commuting. The added parts at the top and bottom are the kernel (C_0, k) of γ and the cokernel (A'_0, p) of α. Once the connecting homomorphism has been constructed, the proof of exactness will involve a diagram chase that makes rather straightforward use of Theorem 4.42, including conclusion (f). By contrast, the initial construction will involve a different sort of diagram, namely

$$\begin{array}{ccccccccc}
& & & & B_0 & \dashrightarrow^{\widetilde{\psi}} & C_0 & & \\
& & \widetilde{\varphi} \nearrow & & \downarrow \widetilde{k} & & \downarrow k & & \\
0 & \dashrightarrow & \overline{A} & \xrightarrow{\overline{\varphi}} & B & \xrightarrow{\psi} & C & \longrightarrow & 0 \\
& & \downarrow \overline{\alpha} & & \downarrow \beta & & \downarrow \gamma & & \\
0 & \longrightarrow & A' & \xrightarrow{\varphi'} & B' & \xrightarrow{\overline{\psi}'} & \overline{C}' & \dashrightarrow & 0 \\
& & \downarrow p & & \downarrow \widetilde{p} & \nearrow \widetilde{\psi}' & & & \\
& & A'_0 & \dashrightarrow^{\widetilde{\varphi}'} & B'_0 & & & &
\end{array}$$

8. Abelian Categories

In the construction we adjust the first row of $(*)$ to make it exact when a 0 is included at the left end. To do so, we factor φ according to Proposition 4.38 as $\varphi = me$, we let $\overline{A} = \text{domain } m = \text{codomain } e$, and we write $\overline{\varphi}$ for m. The commutativity of the left square of $(*)$ implies that $\varphi'\alpha(\ker \varphi) = \beta\varphi(\ker \varphi) = 0$. Since φ' is a monomorphism, $\alpha(\ker \varphi) = 0$. Then the fact $e = \text{coker}(\ker \varphi)$ implies that α factors through e as $\alpha = \overline{\alpha}e$ for some $\overline{\alpha}$ with domain $\overline{A}$. Consequently the left square in the adjusted diagram commutes, and the first row is exact with the 0 inserted at the left. Since e is an epimorphism, $p = \text{coker } \alpha = \text{coker}(\overline{\alpha}e) = \text{coker } \overline{\alpha}$, and the vertical line at the left is exact.

By a dual argument starting from a factorization of ψ', we can replace the triple (C', ψ', γ) in similar fashion by $(\overline{C}', \overline{\psi}', \overline{\gamma})$, see that $k = \ker \overline{\gamma}$, and add a 0 at the end of the second row to obtain an exact sequence.

Next, let $(B_0, \widetilde{\psi}, \widetilde{k})$ be a pullback of (ψ, k). Proposition 4.40 shows that $\widetilde{\psi}$ is an epimorphism and that $\ker \psi = \widetilde{k} \ker \widetilde{\psi}$. Since the first row is a short exact sequence, we know that $\overline{\varphi} = \ker \psi$, and the condition $\ker \psi = \widetilde{k} \ker \widetilde{\psi}$ shows that $\widetilde{\varphi} = \ker \widetilde{\psi}$ satisfies $\overline{\varphi} = \widetilde{k}\widetilde{\varphi}$. This completes the dashed arrows in the top part of the diagram. By a dual argument using $p = \text{coker } \overline{\alpha}$, we complete the dashed arrows in the bottom part of the diagram, deducing from $\overline{\psi}' = \text{coker } \varphi'$ the fact that $\widetilde{\psi}' = \text{coker } \widetilde{\varphi}'$ satisfies $\overline{\psi}' = \widetilde{\psi}'\widetilde{p}$.

Lemma 4.37 shows from $\widetilde{\varphi} = \ker \widetilde{\psi}$ that $\widetilde{\psi} = \text{coker } \widetilde{\varphi}$, and it shows from $\widetilde{\psi}' = \text{coker } \widetilde{\varphi}'$ that $\widetilde{\varphi}' = \ker \widetilde{\psi}'$. With these formulas in hand, we can construct the connecting homomorphism. Define $\omega_0 = \widetilde{p}\beta\widetilde{k}$ in $\text{Hom}(B_0, B_0')$ to be the composition down the center. Then $\omega_0\widetilde{\varphi} = \widetilde{p}\beta\widetilde{k}\widetilde{\varphi} = \widetilde{\varphi}'p\overline{\alpha} = 0$, the last equality holding because $p\overline{\alpha} = 0$. Therefore ω_0 factors through $\widetilde{\psi} = \text{coker } \widetilde{\varphi}$ as $\omega_0 = \omega_1\widetilde{\psi}$ for some $\omega_1 \in \text{Hom}(C_0, B_0')$. The morphism ω_1 satisfies $\widetilde{\psi}'\omega_1\widetilde{\psi} = \widetilde{\psi}'\widetilde{p}\beta\widetilde{k} = \overline{\gamma}\widetilde{k}\widetilde{\psi} = 0$, the last equality holding because $\overline{\gamma}k = 0$. Since $\widetilde{\psi}$ is an epimorphism, we can cancel it, obtaining $\widetilde{\psi}'\omega_1 = 0$. Therefore ω_1 factors through $\widetilde{\varphi}' = \ker \widetilde{\psi}'$ as $\omega_1 = \widetilde{\varphi}'\omega$ for some morphism $\omega \in \text{Hom}(C_0, A_0')$.

The construction of ω is now complete, and the assertion is that the value of ω on members corresponds to what we expect from the proof of Lemma 4.6. Since equivalences $\omega x \equiv \omega' x$ for some other candidate ω' for the connecting morphism and for all $x \in_m C_0$ would imply that $\omega = \omega'$, the argument will show that we have found the unique morphism taking the prescribed values on members.

During the verification we refer to $(*)$ to do the diagram chase. The member of C corresponding to $x \in_m C_0$ is $kx \in_m C$. Since ψ is an epimorphism, Theorem 4.42c produces $b \in_m B$ with $\psi b \equiv kx$. Then $\psi'\beta b \equiv \gamma \psi b \equiv \gamma kx \equiv 0$, since $\gamma k = 0$. Theorem 4.42e and exactness at B' imply that $\varphi'a' \equiv \beta b$ for some $a' \in A'$, and the class of a' is unique (for the b under consideration) by Theorem 4.42b because φ' is a monomorphism. We shall verify that $\omega x \equiv pa'$, and then the class of ωx matches what we expect from the proof of Lemma 4.6.

First let us show that a different choice of b, say b_1, leads to the same class pa'. We are given that $\psi b \equiv \psi b_1$. Let a' and a'_1 be the corresponding members of A' with $\varphi' a' \equiv \beta b$ and $\varphi' a'_1 \equiv \beta b_1$. We shall make repeated use of Theorem 4.42f, letting subscripted u's and v's denote suitable epimorphisms. From $\psi b \equiv \psi b_1$, Theorem 4.42f gives $\psi b u_1 - \psi b_1 v_1 \equiv 0$, i.e., $\psi(bu_1 - b_1 v_1) \equiv 0$. By Theorem 4.42e and exactness at B, $bu_1 - b_1 v_1 \equiv \varphi a$ for some $a \in_m A$. Hence $\beta b u_1 - \beta b_1 v_1 \equiv \beta \varphi a \equiv \varphi' \alpha a$. Two applications of Theorem 4.42f starting from $\beta b u_1 - \beta b_1 v_1 \equiv \varphi' \alpha a$ give

$$\varphi' a' \equiv \beta b \equiv \varphi' \alpha a u_2 + \beta b_1 v_2,$$

and then
$$\varphi' a' u_3 - \varphi' \alpha a v_3 \equiv \beta b_1 \equiv \varphi' a'_1.$$

Since φ' is a monomorphism, Theorem 4.42b says that

$$a' u_3 - \alpha a v_3 \equiv a'_1.$$

Applying p, we obtain $pa' u_3 - p\alpha a v_3 \equiv pa'_1$. Since $p\alpha = 0$, we can drop the term $p\alpha a v_3$, and we conclude that $pa' \equiv pa' u_3 \equiv pa'_1$.

We can now return to the verification that $\omega x \equiv pa'$, making use of the adjusted diagram as necessary.[20] Since $\widetilde{\psi}$ is an epimorphism, Theorem 4.42c produces $b_0 \in_m B_0$ with $\widetilde{\psi} b_0 \equiv x$. Then $\widetilde{k} b_0 \in_m B$ has $\psi \widetilde{k} b_0 \equiv k \widetilde{\psi} b_0 \equiv kx$. Hence $\widetilde{k} b_0$ is a member of B like b and b_1 in the previous paragraph. The above argument shows that $\beta \widetilde{k} b_0 \in_m B'$ has $\beta \widetilde{k} b_0 \equiv \varphi' a'$ for some $a' \in_m A'$ and that $pa' \in_m A'_0$ is what we should hope for as the value of ωx. So we compute that

$$\widetilde{\varphi}' \omega x \equiv \omega_1 x \equiv \omega_1 \widetilde{\psi} b_0 \equiv \omega_0 b_0 \equiv \widetilde{p} \beta \widetilde{k} b_0 \equiv \widetilde{p} \varphi' a' \equiv \widetilde{\varphi}' pa'.$$

Since $\widetilde{\varphi}'$ is a monomorphism by the dual of Proposition 4.40, Theorem 4.42b shows that $\omega x \equiv pa'$, which is the formula we were seeking.

9. Problems

1. (a) Prove that the good category of all finitely generated abelian groups has enough projectives but not enough injectives.
   (b) Prove that the good category of all torsion abelian groups has enough injectives but not enough projectives.

2. Let $\mathcal{C}_\mathbb{Z}$ be the category of all abelian groups. Give an example of a nonzero good category $\mathcal{C}$ of abelian groups that has enough projectives and enough injectives but for which no nonzero projective for $\mathcal{C}_\mathbb{Z}$ lies in $\mathcal{C}$ and no nonzero injective for $\mathcal{C}$ lies in $\mathcal{C}_\mathbb{Z}$.

---

[20]*Warning:* The construction of ω involves B_0 and B'_0, which are in the adjusted diagram but are not in (∗). These objects do not necessarily coincide with the domain of $\ker \beta$ and the codomain of $\operatorname{coker} \beta$.

9. Problems   251

3. Let R be a semisimple ring in the sense of Chapter II, and let $\mathcal{C}_R$ be the category of all unital left R modules. Prove that every module in $\mathcal{C}_R$ is projective and injective.

4. Let R be a (commutative) principal ideal domain, and let $\mathcal{C}_R$ be the category of all unital R modules. A module M in $\mathcal{C}_R$ is **divisible** if for each $a \neq 0$ in R and $x \in M$, there exists $y \in M$ with $ay = x$.
   (a) Referring to Example 2 of injectives in Section 4, prove that injective for $\mathcal{C}_R$ implies divisible.
   (b) Deduce from Proposition 4.15 that divisible implies injective for $\mathcal{C}_R$.

5. Let R be a (commutative) principal ideal domain, and let $\mathcal{C}_R$ be the category of all unital R modules. Prove that every module M in $\mathcal{C}_R$ has an injective resolution of the form $0 \to M \to I_0 \to I_1 \to 0$ with I_0 and I_1 injective.

6. Let $\mathcal{C}, \mathcal{C}', \mathcal{C}''$ be good categories of modules with enough projectives and enough injectives, let $G : \mathcal{C} \to \mathcal{C}'$ be a one-sided exact functor with derived functors G_n or G^n, and let $F : \mathcal{C}' \to \mathcal{C}''$ be an exact functor.
   (a) Prove that if F is covariant, then $F \circ G$ is one-sided exact, and its derived functors satisfy $(F \circ G)_n = F \circ G_n$ or $(F \circ G)^n = F \circ G^n$.
   (b) Prove that if F is contravariant, then $F \circ G$ is one-sided exact, and its derived functors satisfy $(F \circ G)^n = F \circ G_n$ or $(F \circ G)^n = F \circ G^n$.

7. Let $\mathcal{C}, \mathcal{C}', \mathcal{C}''$ be good categories of modules with enough projectives and enough injectives, let $F : \mathcal{C} \to \mathcal{C}'$ be an exact functor, and let $G : \mathcal{C}' \to \mathcal{C}''$ be a one-sided exact functor with derived functors G_n or G^n.
   (a) Suppose that F is covariant, that G_n or G^n is defined from projective resolutions, and that F carries projectives to projectives. Prove that $G \circ F$ is one-sided exact and that its derived functors satisfy $(G \circ F)_n = G_n \circ F$ or $(G \circ F)^n = G^n \circ F$.
   (b) Suppose that F is covariant, that G_n or G^n is defined from injective resolutions, and that F carries injectives to injectives. Prove that $G \circ F$ is one-sided exact and that its derived functors satisfy $(G \circ F)_n = G_n \circ F$ or $(G \circ F)^n = G^n \circ F$.
   (c) Suppose that F is contravariant, that G_n or G^n is defined from projective resolutions, and that F carries injectives to projectives. Prove that $G \circ F$ is one-sided exact and that its derived functors satisfy $(G \circ F)^n = G^n \circ F$ or $(G \circ F)_n = G_n \circ F$.
   (d) Suppose that F is contravariant, that G_n or G^n is defined from injective resolutions, and that F carries projectives to injectives. Prove that $G \circ F$ is one-sided exact and that its derived functors satisfy $(G \circ F)^n = G^n \circ F$ or $(G \circ F)_n = G_n \circ F$.

8. Let G be a group, and let $F = (F^+ \to \mathbb{Z})$ be a free resolution of the trivial $\mathbb{Z}G$ module $\mathbb{Z}$ in the category $\mathbb{Z}G$. If M is an abelian group on which G acts by automorphisms, then we know that the cohomology $H^n(G, M)$ is defined to be the nth cohomology of the cochain complex $\text{Hom}_{\mathbb{Z}G}(F^+, M)$ and the homology $H_n(G, M)$ is defined to be the nth homology of the chain complex $F^+ \otimes_{\mathbb{Z}G} M$. Take for granted the result of Proposition 3.32 that if G is a finite cyclic group with generator s, then

$$\cdots \xrightarrow{T} \mathbb{Z}G \xrightarrow{N} \mathbb{Z}G \xrightarrow{T} \cdots \xrightarrow{N} \mathbb{Z}G \xrightarrow{T} \mathbb{Z}G \xrightarrow{\varepsilon} \mathbb{Z} \longrightarrow 0$$

is a free resolution of $\mathbb{Z}G$, where T and N are the left $\mathbb{Z}G$ module homomorphisms defined by

$$T = \text{multiplication by } (s) - (1),$$
$$N = \text{multiplication by } (1) + (s) + \cdots + (s^{n-1}).$$

Prove that $H^n(G, M) \cong H^{n+2}(G, M)$ and $H_n(G, M) \cong H_{n+2}(G, M)$ for all $n \geq 1$ and all M when G is a finite cyclic group.

Problems 9–11 concern changes of rings. Fix a homomorphism $\rho : R \to S$ of rings with identity. This homomorphism determines three functors of interest, denoted by $\mathcal{F}_S^R : \mathcal{C}_S \to \mathcal{C}_R$, $P_R^S : \mathcal{C}_R \to \mathcal{C}_S$, and $I_R^S : \mathcal{C}_R \to \mathcal{C}_S$. The first takes an S module M and makes it into an R module $\mathcal{F}_S^R(M)$ by the definition $rm = \rho(r)m$ for $r \in R$ and $m \in M$; the effect on an S homomorphism is to leave the function unchanged and to regard it as an R homomorphism; this functor is manifestly exact. For the second, regard S as an (S, R) bimodule with right R action given by $sr = s\rho(r)$, and define $P_R^S(M) = S \otimes_R M$ for M in $\text{Obj}(\mathcal{C}_R)$ and $P_R^S(\varphi) = 1_S \otimes \varphi$ for φ in $\text{Hom}_R(M, N)$; this functor is covariant and right exact. For the third, regard S as an (R, S) bimodule with left R action given by $rs = \rho(r)s$, and define $I_R^S(M) = \text{Hom}_R(S, M)$ for M in $\text{Obj}(\mathcal{C}_R)$ and $I_R^S(\varphi) = \text{Hom}(1_S, \varphi)$ for φ in $\text{Hom}_R(M, N)$; this functor is covariant and left exact.

9. If $\mathcal{C}$ and $\mathcal{D}$ are good categories of modules and if $F : \mathcal{C} \to \mathcal{D}$ and $G : \mathcal{D} \to \mathcal{C}$ are covariant additive functors such that there exist isomorphisms of abelian groups

$$\text{Hom}(F(A), B) \cong \text{Hom}(A, G(B))$$

natural for A in $\text{Obj}(\mathcal{C})$ and for B in $\text{Obj}(\mathcal{D})$, then F is said to be **left adjoint** to G and G is said to be **right adjoint** to F.
(a) Prove that if G carries onto maps in $\mathcal{D}$ to onto maps in $\mathcal{C}$, then F carries projectives in $\mathcal{C}$ to projectives in $\mathcal{D}$.
(b) Prove that if F carries one-one maps in $\mathcal{C}$ to one-one maps in $\mathcal{D}$, then G carries injectives in $\mathcal{D}$ to injectives in $\mathcal{C}$. (Educational note: The conclusions in this problem extend to any abelian categories $\mathcal{C}$ and $\mathcal{D}$, and in this enlarged setting, (b) follows from (a) by duality.)

10. (a) Prove that P_R^S is left adjoint to $\mathcal{F}_S^R$.
   (b) Deduce from the previous problem that P_R^S sends projectives in $\mathcal{C}_R$ to projectives in $\mathcal{C}_S$.
   (c) Prove that if the right R module S is projective, then P_R^S is exact. (Educational note: In the subject of Lie algebra homology and cohomology, this hypothesis is satisfied when S is the universal enveloping algebra of a Lie algebra $\mathfrak{g}$ over a field $\mathbb{K}$, R is the universal enveloping algebra of a Lie subalgebra $\mathfrak{h}$ of $\mathfrak{g}$, and $\rho : R \to S$ is the inclusion. It is satisfied also in the subject of homology and cohomology of groups if S is the group algebra $\mathbb{K}G$ of a group G over a field $\mathbb{K}$ and if R is the group algebra $\mathbb{K}H$ of a subgroup H. See Problem 13c below.)
   (d) Using Problem 7, prove that if the right R module S is projective, then $\mathrm{Ext}_S^k(P_R^S M, N) \cong \mathrm{Ext}_R^k(M, \mathcal{F}_S^R N)$ naturally in each variable (M being in $\mathrm{Obj}(\mathcal{C}_R)$ and N being in $\mathrm{Obj}(\mathcal{C}_S)$).
   (e) Even without the assumption that the right R module S is projective, let $X = (X^+ \to M)$ be a projective resolution of a module M in $\mathcal{C}_R$, and let $Y = (Y^+ \to P_R^S M)$ be a projective resolution of $P_R^S M$ in $\mathcal{C}_S$. Construct a chain map from $P_R^S X$ to Y extending the identity map on $P_R^S M$, and use it to obtain the associated homomorphism $\mathrm{Ext}_S^k(P_R^S M, N) \to \mathrm{Ext}_R^k(M, \mathcal{F}_S^R N)$ natural in each variable.

11. (a) Prove that I_R^S is right adjoint to $\mathcal{F}_S^R$.
    (b) Deduce from Problem 9 that I_R^S sends injectives in $\mathcal{C}_R$ to injectives in $\mathcal{C}_S$.
    (c) Prove that if the right R module S is projective, then I_R^S is exact.
    (d) Using Problem 7, prove that if the right R module S is projective, then $\mathrm{Ext}_S^k(M, I_R^S N) \cong \mathrm{Ext}_R^k(\mathcal{F}_S^R M, N)$ naturally in each variable (M being in $\mathrm{Obj}(\mathcal{C}_S)$ and N being in $\mathrm{Obj}(\mathcal{C}_R)$).
    (e) Even without the assumption that the right R module S is projective, let $X = (X^+ \to N)$ be an injective resolution of a module N in $\mathcal{C}_R$, and let $Y = (Y^+ \to I_R^S N)$ be an injective resolution of $I_R^S N$ in $\mathcal{C}_S$. Construct a chain map from Y to $I_R^S N$ extending the identity map on $I_R^S N$, and use it to obtain the associated homomorphism $\mathrm{Ext}_S^k(M, I_R^S N) \to \mathrm{Ext}_R^k(\mathcal{F}_S^R M, N)$ natural in each variable.

Problems 12–13 concern the effect on cohomology of groups of changing the group. The main result is the exactness of the "inflation-restriction sequence"; this is applied particularly in algebraic number theory to relate Brauer groups (see Chapter III) for different field extensions. Let J and K be groups, and let $\rho : J \to K$ be a group homomorphism. By the universal mapping property of group rings, ρ extends to a ring homomorphism, also denoted by ρ, from $\mathbb{Z}J$ into $\mathbb{Z}K$. For any group G, we make use of the standard free resolution $F(G) = (F(G)^+ \xrightarrow{\varepsilon} \mathbb{Z})$ of $\mathbb{Z}$ in the category $\mathcal{C}_{\mathbb{Z}G}$, as described before Theorem 3.20. A $\mathbb{Z}$ basis of $F_n(G)$ consists of all tuples $(g_0, \ldots, g_n)$, and a $\mathbb{Z}G$ basis consists of those members of the $\mathbb{Z}$ basis with

$g_0 = 1$. In the context of the groups J and K, any $\mathbb{Z}K$ module M becomes a $\mathbb{Z}J$ module by the formula $xm = \rho(x)m$ for $x \in \mathbb{Z}J$ and $m \in M$. In particular, each free $\mathbb{Z}K$ module $F_n(K)$ can be regarded as a $\mathbb{Z}J$ module. Meanwhile, the homomorphism $\rho : J \to K$ induces a function from the $\mathbb{Z}J$ basis of $F_n(J)$ into $F_n(K)$ by the formula $\rho(1, j_1, \ldots, j_n) = (1, \rho(j_1), \ldots, \rho(j_n))$ for $j_1, \ldots, j_n \in J$, and this extends to a $\mathbb{Z}J$ homomorphism, still called ρ, of $F_n(J)$ into $F_n(K)$. A look at the formula for the boundary operators ∂_J and ∂_K in Section III.5 shows that ρ is a chain map in the sense that $\partial_K \rho = \rho \partial_J$. If M is any unital left $\mathbb{Z}K$ module, then it follows that $\text{Hom}(\rho, 1) : \text{Hom}(F(K), M) \to \text{Hom}(F(J), M)$ is a cochain map. Consequently we get maps on cohomology for each n of the form $H^n(\rho) : H^n(K, M) \to H^n(J, M)$. There are two cases of special interest:

(i) If $\rho : H \to G$ is the inclusion of a subgroup into a group, then the mapping on cohomology is called the **restriction homomorphism**

$$\text{Res} : H^n(G, M) \to H^n(H, M).$$

(ii) If H is a normal subgroup of G, let $\rho : G \to G/H$ be the quotient homomorphism. For any $\mathbb{Z}G$ module M, let M^H be the subgroup of H invariants. Then G/H acts on M^H. The above construction is applicable to the module M^H for the group ring $\mathbb{Z}(G/H)$ of G/H, and we form the mapping on cohomology $H^n(G/H, M^H) \to H^n(G, M^H)$. The inclusion of the $\mathbb{Z}G$ module M^H in M induces a mapping $H^n(G, M^H) \to H^n(G, M)$, and the composition is called the **inflation homomorphism**

$$\text{Inf} : H^n(G/H, M^H) \to H^n(G, M).$$

When H is a normal subgroup of G and M is a $\mathbb{Z}G$ module and $q \geq 1$ is an integer such that $H^k(H, M) = 0$ for $1 \leq k \leq q - 1$, the **inflation-restriction sequence** is the sequence of abelian groups and homomorphisms

$$0 \longrightarrow H^q(G/H, M^H) \xrightarrow{\text{Inf}} H^q(G, M) \xrightarrow{\text{Res}} H^q(H, M).$$

12. For $q = 1$, use direct arguments to prove the exactness of the inflation-restriction sequence by carrying out the following steps:
    (a) By sorting out the isomorphism $\Phi_q : \text{Hom}_{\mathbb{Z}G}(F_q, M) \to C^q(G, M)$ of Section III.5, show that the effect of a homomorphism $\rho : G \to G'$ on $C^q(G', M)$ is given by $(\rho^* f)(g_1, \ldots, g_q) = f(\rho(g_1), \ldots, \rho(g_q))$.
    (b) Verify that $\text{Res} \circ \text{Inf} = 0$ by looking at cocycles.
    (c) Show that Inf is one-one on $H^q(G/H, M^H)$ by showing that any cocycle $f : G/H \to M^H$ that is a coboundary when viewed as a function on G is itself a coboundary for G/H.
    (d) Show that every member of $\ker(\text{Res})$ lies in $\text{image}(\text{Inf})$ by showing that any cocycle $f : G \to M$ whose restriction to H is a coboundary may be adjusted to be 0 on H and that an examination of the equation $f(st) = f(s) + sf(t)$ in this case shows f to be a cocycle of G/H with values in M^H.

13. Assume inductively that $q > 1$, that $H^k(H, M) = 0$ for $1 \leq k \leq q-1$, and that the inflation-restriction sequence is exact for all N for degree $q-1$ whenever $H^k(H, N) = 0$ for $1 \leq k < q-1$. Form $B = I_{\mathbb{Z}}^{\mathbb{Z}G} \mathcal{F}_{\mathbb{Z}G}^{\mathbb{Z}} M = \mathrm{Hom}_{\mathbb{Z}}(\mathbb{Z}G, M)$ as in Problems 9–11. Elements of B can be identified with functions φ on G with values in M, and G acts by $(g_0\varphi)(g) = \varphi(gg_0)$.

   (a) For $m \in M$, show that the function $\varphi_m(t) = tm$ is a one-one $\mathbb{Z}G$ homomorphism of M into B. If $N = B/M$, then the sequence $0 \to M \to B \to N \to 0$ is therefore exact in $\mathcal{C}_{\mathbb{Z}G}$.

   (b) Use Problem 11 to verify that $H^k(G, B) \cong \mathrm{Ext}_{\mathbb{Z}}^k(\mathbb{Z}, \mathcal{F}_{\mathbb{Z}G}^{\mathbb{Z}} M)$, and deduce that $H^k(G, B) = 0$ for $k \geq 1$.

   (c) Verify the equality of right $\mathbb{Z}H$ modules $\mathbb{Z}G = A \otimes_{\mathbb{Z}} \mathbb{Z}H$ for some free abelian group A.

   (d) Using (c), show that $\mathcal{F}_{\mathbb{Z}G}^{\mathbb{Z}H} B \cong \mathrm{Hom}_{\mathbb{Z}}(\mathbb{Z}H, \mathrm{Hom}_{\mathbb{Z}}(A, M))$, and deduce that $H^k(H, B) = 0$ for $k \geq 1$.

   (e) Using the hypothesis that $H^1(H, M) = 0$ and a long exact sequence associated to the short exact sequence in (a), show that $0 \to M^H \to B^H \to N^H \to 0$ is exact.

   (f) Prove that $\mathbb{Z} \otimes_{\mathbb{Z}H} \mathbb{Z}G \cong \mathbb{Z}(G/H)$ as right $\mathbb{Z}G$ modules, where $\mathbb{Z}(G/H)$ is the integral group ring of G/H.

   (g) Show that $B^H = I_{\mathbb{Z}}^{\mathbb{Z}(G/H)} M$, and deduce that $H^k(G/H, B^H) = 0$ for $k \geq 1$.

   (h) Using the long exact sequences for G and for H associated to the short exact sequence of (a), as well as the long exact sequence for G/H associated to the short exact sequence of (e), establish isomorphisms of abelian groups
   $$H^{q-1}(G/H, N^H) \cong H^q(G/H, M^H),$$
   $$H^{q-1}(G, N) \cong H^q(G, M),$$
   $$H^{q-1}(H, N) \cong H^q(H, M).$$

   (i) Set up the diagram
   $$\begin{array}{ccccccccc} 0 & \longrightarrow & H^{q-1}(G/H, N^H) & \longrightarrow & H^{q-1}(G, N) & \longrightarrow & H^{q-1}(H, N) \\ & & \downarrow & & \downarrow & & \downarrow \\ 0 & \longrightarrow & H^q(G/H, M^H) & \longrightarrow & H^q(G, M) & \longrightarrow & H^q(H, M) \end{array}$$
   show that it is commutative, and deduce from the foregoing that the inflation-restriction sequence is exact for M in degree q. (Educational note: For an application to Brauer groups, let $F \subseteq K \subseteq L$ be fields, and assume that K/F, L/F, and L/K are all finite Galois extensions. The groups in question are $G = \mathrm{Gal}(L/F)$, $H = \mathrm{Gal}(L/K)$, and $G/H = \mathrm{Gal}(K/F)$, and the modules in question are $M = L^\times$ and $M^H = K^\times$. The index q is to be 2, and the vanishing of H^1 is by Hilbert's Theorem 90. The conclusion is that the sequence $0 \to \mathcal{B}(K/F) \to \mathcal{B}(L/F) \to \mathcal{B}(L/K)$ is exact.)

Problems 14–16 introduce the cup product in the cohomology of groups. This is a construction having applications to topology and algebraic number theory. Let G be a group, and form the standard free resolution $F = (F^+ \xrightarrow{\varepsilon} \mathbb{Z})$ of $\mathbb{Z}$ in the category $\mathcal{C}_{\mathbb{Z}G}$, as described before Theorem 3.20. A $\mathbb{Z}$ basis of F_n consists of all tuples $(g_0, \ldots, g_n)$, and a $\mathbb{Z}G$ basis consists of those members of the $\mathbb{Z}$ basis with $g_0 = 1$. Let ∂ denote the boundary operator, with the subscript dropped that indicates the degree. Define $\varphi_{p,q} : F_{p+q} \to F_p \otimes_{\mathbb{Z}} F_q$ by

$$\varphi_{p,q}(g_0, \ldots, g_{p+q}) = (g_0, \ldots, g_p) \otimes (g_p, \ldots, g_q).$$

14. Check that $(\varepsilon \otimes \varepsilon) \circ \varphi_{0,0} = \varepsilon$ and that each $\varphi_{p,q}$ with $p \geq 0$ and $q \geq 0$ is a $\mathbb{Z}G$ homomorphism satisfying

$$\varphi_{p,q} \circ \partial = (\partial \otimes 1) \circ \varphi_{p+1,q} + (-1)^p (1 \otimes \partial) \circ \varphi_{p,q+1}.$$

15. If A and B are abelian groups on which G acts by automorphisms, show that G acts by automorphisms on $A \otimes_{\mathbb{Z}} B$ in such a way that $g(a \otimes b) = ga \otimes gb$ for all $a \in A, b \in B, g \in G$. Thus whenever A and B are unital left $\mathbb{Z}G$ modules, then so is $A \otimes_{\mathbb{Z}} B$.

16. For any unital left $\mathbb{Z}G$ module M, we work with $\text{Hom}_{\mathbb{Z}G}(F_n, M)$ as the space of n-cochains. (Here it is not necessary to unravel the isomorphism given in Section III.5 that relates $\text{Hom}_{\mathbb{Z}G}(F_n, M)$ to the space $C^n(G, M)$ of cochains defined in Chapter VII of *Basic Algebra*.) Define the coboundary operator on the complex $\text{Hom}_{\mathbb{Z}G}(F^+, M)$ to be $d = \text{Hom}(\partial, 1)$. For any unital left $\mathbb{Z}G$ modules A and B, let $f \in \text{Hom}(F_p, A)$ and $g \in \text{Hom}(F_q, B)$ be given. The product cochain $f \cdot g$ is the member of $\text{Hom}_{\mathbb{Z}G}(F_{p+q}, A \otimes_{\mathbb{Z}} B)$ given by $f \cdot g = (f \otimes g) \circ \varphi_{p,q}$.
    (a) Check that $f \cdot g = (df) \cdot g + (-1)^p f \cdot (dg)$.
    (b) How does it follow that this product descends to a homomorphism of abelian groups $a \otimes b \mapsto a \cup b$ carrying the space $H^p(G, A) \otimes_{\mathbb{Z}} H^q(G, B)$ to $H^{p+q}(G, A \otimes_{\mathbb{Z}} B)$? The descended mapping is called the **cup product**.
    (c) Explain why the cup product is functorial in each variable A and B.
    (d) Explain why the cup product for $p = 0$ and $q = 0$ may be identified with the mapping on invariants given by $A^G \otimes B^G \to (A \otimes_{\mathbb{Z}} B)^G$.

Problems 17–20 introduce flat R modules, R being a ring with identity. These modules are of interest in topology and algebraic geometry. Let R^o be the opposite ring of R; right R modules may be identified with left R^o modules. Let $\mathcal{C}_R$ be the category of all unital left R modules; tensor product over R can be regarded as a functor in the second variable, carrying $\mathcal{C}_R$ to $\mathcal{C}_{\mathbb{Z}}$, or as a functor in the first variable, carrying $\mathcal{C}_{R^o}$ to $\mathcal{C}_{\mathbb{Z}}$. A unital right R module M (i.e., a unital left R^o module) is called **flat** if $M \otimes_R (\cdot)$ is an exact functor from $\mathcal{C}_R$ to $\mathcal{C}_{\mathbb{Z}}$. Since this functor is anyway right exact, M is flat if and only if tensoring with M carries one-one maps to one-one maps, i.e., if and only if whenever $f : A \to B$ is one-one, then $1_M \otimes f : M \otimes_R A \to M \otimes_R B$ is one-one. Take as known the analog for the functor Tor of all the facts about Ext proved in Section 7.

17. Prove for unital right R modules that
    (a) the right R module R is flat,
    (b) a direct sum $F = \bigoplus_{s \in S} F_s$ is flat if and only if each F_s is flat,
    (c) any projective in $\mathcal{C}_{R^o}$ is flat.

18. Let M be a unital right R module. For each finite subset F of M, let M_F be the right R submodule of M generated by the members of F. Prove that M is flat if and only if each M_F is flat.

19. Let B be in $\mathcal{C}_R$, write B as the R homomorphic image of a free left R module F, and form the exact sequence $0 \to K \to F \to B \to 0$ in which K is the kernel of $F \to B$. Prove for each unital right R module A that the sequence

$$0 \to \operatorname{Tor}_1^R(A, B) \to A \otimes_R K \to A \otimes_R F \to A \otimes_R B \to 0$$

is exact. Deduce that A is flat if and only if $\operatorname{Tor}_1^R(A, B) = 0$ for all B.

20. Suppose that R is a (commutative) principal ideal domain, so that in particular $R = R^o$. The **torsion submodule** $T(M)$ of a module M in $\mathcal{C}_R$ consists of all $m \in M$ with $rm = 0$ for some $r \neq 0$ in R.
    (a) Suppose that M is of the form $M = F \oplus T(M)$, where F is a free R module. Using the exact sequence $0 \to F \to M \to T(M) \to 0$, prove that $\operatorname{Tor}_1^R(M, B) = \operatorname{Tor}_1^R(T(M), B)$ for all modules B in $\mathcal{C}_R$.
    (b) Deduce from (a) and Problem 18 that a module M in $\mathcal{C}_R$ is flat if and only if $T(M)$ is flat. (Note that M is not assumed to be of the form $F \oplus T(M)$.)
    (c) By comparing the one-one inclusion $(a) \subseteq R$ for a nonzero $a \in R$ with the induced map from $(a) \otimes_R M$ to $R \otimes_R M$, prove that $T(M) \neq 0$ implies M not flat.
    (d) Deduce that a module M in $\mathcal{C}_R$ is flat if and only if M has 0 torsion, i.e., if and only if M is torsion free. (Educational note: In combination with the result of Problem 19, this condition explains the use of the notation "Tor" for the first derived functor of tensor product.)

Problems 21–25 deal with double chain complexes of abelian groups. A **double chain complex** is a system $\{E_{p,q}\}$ of abelian groups defined for all integers p and q and having boundary homomorphisms $\partial'_p : E_{p,q} \to E_{p-1,q}$ and $\partial''_q : E_{p,q} \to E_{p,q-1}$ such that $\partial'_{p-1,q} \partial'_{p,q} = 0$, $\partial''_{p,q-1} \partial''_{p,q} = 0$, and $\partial'_{p,q-1} \partial''_{p,q} + \partial''_{p-1,q} \partial'_{p,q} = 0$. This set of problems will assume that $E_{p,q} = 0$ if either p or q is sufficiently negative.

21. Let $\{E_{p,q}\}$ be a double complex of abelian groups with boundary homomorphisms as above, let $E_n = \bigoplus_{p+q=n} E_{p,q}$, and define $\partial_n : E_n \to E_{n-1}$ by $\partial_n\big|_{E_{p,q}} = \partial'_{p,q} + \partial''_{p,q}$. Show that the maps ∂_n make the system $\{E_n\}$ into a chain complex. (Note: The indexing on the boundary maps has been changed by 1 from earlier in the chapter in order to simplify the notation that occurs later in these problems.)

22. Let $\mathcal{C}_l$ be a good category of unital left R modules, and let $\mathcal{C}_r$ be a good category of unital left R^o modules; the latter modules are to be regarded as unital right R modules. Let $C = \{C_p\}_{p \geq -\infty}$ and $D = \{D_q\}_{q \geq -\infty}$ be chain complexes with boundary maps $\alpha_p : C_p \to C_{p-1}$ in $\mathcal{C}_r$ and $\beta_q : D_q \to D_{q-1}$ in $\mathcal{C}_l$. It is assumed that $C_p = 0$ for p sufficiently negative and that $D_q = 0$ for q sufficiently negative. Define $E_{p,q} = C_p \otimes_R D_q$, $\partial'_{p,q} = \alpha_p \otimes 1$, and $\partial''_{p,q} = (-1)^p (1 \otimes \beta_q)$. Prove that $\{E_{p,q}\}$ with these mappings is a double complex of abelian groups. (Educational note: Therefore the previous problem creates a chain complex $\{E_n\}$ with boundary maps $\partial_n : E_n \to E_{n-1}$ from this set of data. One writes $E = C \otimes_R D$ for this chain complex and calls it the **tensor product** of the two chain complexes.)

23. In the notation of the previous problem, suppose that $C_p = 0$ if $p < 0$ and $D_q = 0$ if $q < 0$. Let $Z_p = \ker \alpha_p$ and $\overline{Z}_q = \ker \beta_q$. Prove that if c is in Z_p and d is in $\overline{Z}_q$, then $c \otimes d$ is in the subgroup $\ker(\partial'_{p,q} + \partial''_{p,q})$ of $E_{p,q}$ and that as a consequence, there is a canonical homomorphism of $H^p(C) \otimes_R H^q(D)$ into $H^{p+q}(C \otimes_R D)$.

24. Suppose that a double complex E_{pq} of abelian groups has $E_{pq} = 0$ if $p < -1$ or $q < -1$ or $p = q = -1$. Suppose further that $E_{\cdot,q}$ is exact for each $q \geq 0$ and $E_{p,\cdot}$ is exact for each $p \geq 0$. Prove that the rth homology of $E_{-1,q}$ as q varies matches the rth homology of $E_{p,-1}$ as p varies. To do so, start from a cycle a under ∂'' in $E_{-1,k}$ with $k \geq 0$. It is mapped to 0 by ∂', hence has a preimage a' under ∂' in $E_{0,k}$. The element $\partial''a'$ in $E_{0,k-1}$ is mapped to 0 by ∂', hence has a preimage a'' in $E_{1,k-1}$. Continue in this way, and arrive at a cycle in $E_{k,0}$. Then sort out the details.

25. With notation as in Problem 22, let A be in $\mathcal{C}_r$, and let B be in $\mathcal{C}_l$. Let $C = (C^+ \to A)$ be a projective resolution of A, and let $D = (D^+ \to B)$ be a projective resolution of B. Form $E = C \otimes_R D$ as in Problem 22, and apply Problem 24 to give a direct proof (without the machinery of Section 7) that one gets the same result for $\text{Tor}_n^R(A, B)$ by using a projective resolution in the first variable as by using a projective resolution in the second variable.

Problems 26–31 concern the **Künneth Theorem** for homology and the Universal Coefficient Theorem for homology. Both these results have applications to topology. It will be assumed throughout that R is a (commutative) principal ideal domain.

STATEMENT OF KÜNNETH THEOREM. Let C and D be chain complexes over the principal ideal domain R, and assume that all modules in negative degrees are 0 and that C is flat. Then there is a natural short exact sequence

$$0 \to \bigoplus_{p+q=n} \left(H_p(C) \otimes_R H_q(D) \right) \xrightarrow{\alpha_n} H_n(C \otimes_R D)$$
$$\xrightarrow{\beta_{n-1}} \bigoplus_{p+q=n-1} \text{Tor}_1^R(H_p(C), H_q(D)) \to 0.$$

Moreover, the exact sequence splits, but not naturally.

The point of the theorem is to give circumstances under which the homology of each of two chain complexes C and D determines the homology of the tensor product $E = C \otimes_R D$, the tensor product complex being defined as in Problem 22. Problem 26 below shows that some further hypothesis is needed beyond the limitation on R. A sufficient condition is that one of C and D, say C, be **flat** in the sense that all the modules in it satisfy the condition of flatness defined in Problems 17–20. The problems in the set carry out some of the steps in proving the Künneth Theorem, and then they derive the Universal Coefficient Theorem for homology as a consequence. To keep the ideas in focus, the problems will suppress certain isomorphisms, writing them as equalities.

26. With $R = \mathbb{Z}$, let $C = D$ be the chain complex with $C_0 = \mathbb{Z}/2\mathbb{Z}$ and with $C_p = 0$ for $p \neq 0$. Let C' be the chain complex with $C'_0 = \mathbb{Z}$, with $C'_1 = \mathbb{Z}$, and with $C'_p = 0$ for $p > 1$ and for $p < 0$. Let the boundary map from C'_1 to C'_0 be $\times 2$. Compute the homology of $C, C', D, C \otimes_\mathbb{Z} D$, and $C' \otimes_\mathbb{Z} D$, and justify the conclusion that the homology of each of two chain complexes does not determine the homology of their tensor product.

27. Let ∂' be the boundary map for C. Show how to set up an exact sequence

$$0 \longrightarrow Z \xrightarrow{\iota} C \xrightarrow{\partial'} B' \longrightarrow 0$$

of complexes in which each module in Z is the submodule of cycles of the corresponding module in C, ι is the inclusion, B is the complex of boundaries, and B' is B with its indices shifted by 1. Why does it follow from the fact that C is flat that Z, B, and B' are flat?

28. Explain why

$$0 \longrightarrow Z \otimes_R D \xrightarrow{\iota \otimes 1} C \otimes_R D \xrightarrow{\partial' \otimes 1} B' \otimes_R D \longrightarrow 0$$

is exact even though D is not assumed to be flat.

29. The long exact sequence in homology corresponding to the short exact sequence in the previous problem has segments of the form

$$H_{n+1}(B' \otimes_R D) \xrightarrow{\omega_n} H_n(Z \otimes_R D) \xrightarrow{\iota_n \otimes 1} H_n(C \otimes_R D)$$
$$\xrightarrow{\partial'_n \otimes 1} H_n(B' \otimes_R D) \xrightarrow{\omega_{n-1}} H_{n-1}(Z \otimes_R D).$$

Let ∂'' be the boundary map for D, and let $\overline{Z}, \overline{B}$, and $\overline{B}'$ be the counterparts for D of the complexes Z, B, and B' for C. Show that
(a) the boundary map in $B' \otimes_R D$ may be regarded as $1 \otimes \partial''$ because the boundary map in B' is 0.

(b) $\ker(1 \otimes \partial'')_n = (B' \otimes_R \overline{Z})_n$ and $\text{image}(1 \otimes \partial'')_{n+1} = (B' \otimes_R \overline{B})_n$ because B' is flat.

(c) $H_n(B' \otimes_R D) \cong (B \otimes_R H(D))_{n-1}$ because B' is flat. (This isomorphism will be treated as an equality below.)

(d) similarly $H_n(Z \otimes_R D) \cong (Z \otimes_R H(D))_n$. (This isomorphism will be treated as an equality below.)

30. Form an exact sequence

$$0 \longrightarrow B \longrightarrow Z \longrightarrow H(C) \longrightarrow 0$$

of complexes, form the low-degree part of the long exact sequence corresponding to applying the functor $(\cdot) \otimes_R H(D)$, namely

$$0 \to \text{Tor}_1^R(H(C), H(D))_n \to (B \otimes_R H(D))_n$$
$$\to (Z \otimes_R H(D))_n \to (H(C) \otimes_R H(D))_n \to 0,$$

and rewrite it by (c) and (d) of Problem 29 as

$$0 \to \text{Tor}_1^R(H(C), H(D))_n \xrightarrow{\beta_n'} H_{n+1}(B' \otimes_R D)$$
$$\xrightarrow{\omega_{n-1}} H_n(Z \otimes_R D) \xrightarrow{\alpha_n'} (H(C) \otimes_R H(D))_n \to 0.$$

(a) Why is the term $\text{Tor}_1^R(Z, H(D))$ in the long exact sequence equal to 0?

(b) In the 5-term exact sequence of Problem 29, rewrite the part of the sequence centered at the map $\partial_n' \otimes 1$ in such a way that two exact sequences

$$\xrightarrow{\iota_n \otimes 1} H_n(C \otimes_R D) \xrightarrow{q} \text{coker}(\iota_n \otimes 1) \longrightarrow 0$$

and

$$0 \longrightarrow \ker \omega_{n-1} \xrightarrow{i} H_n(B' \otimes_R D) \xrightarrow{\omega_{n-1}} H_{n-1}(Z \otimes_R D)$$

result. Why can the group $\ker \omega_{n-1}$ and the homomorphism i be taken to be $\text{Tor}_1^R(H(C), H(D))_{n-1}$ and β_{n-1}'?

(c) Why in (b) can $\text{coker}(\iota_n \otimes 1)$ and q be taken to be $\text{Tor}_1^R(H(C), H(D))_{n-1}$ and some one-one homomorphism β_{n-1} such that $\beta_{n-1}'\beta_{n-1} = \partial_n' \otimes 1$?

(d) Arguing similarly with the map $\iota_n \otimes 1$ in Problem 29, obtain a factorization $\iota_n \otimes 1 = \alpha_n \alpha_n'$ in which $\alpha_n' : (Z \otimes_R H(D))_n \to (H(C) \otimes_R H(D))_n$ is onto and $\alpha_n : (H(C) \otimes_R H(D))_n \to H_n(C \otimes_R D)$ is one-one.

(e) The maps α_n and β_{n-1} having now been defined in the sequence in the statement of the Künneth Theorem, prove that the sequence is exact.

31. **(Universal Coefficient Theorem)** By specializing D in the statement of the Künneth Formula to a chain complex that is a module M in dimension 0 and is 0 in all other dimensions, obtain the natural short exact sequence

$$0 \longrightarrow H_n(C) \otimes_R M \longrightarrow H_n(C \otimes_R M) \longrightarrow \mathrm{Tor}_1^R(H_{n-1}(C), M) \longrightarrow 0,$$

valid whenever R is a principal ideal domain and C is a chain complex whose modules are all 0 in dimension < 0. (Educational note: The exact sequence splits, but not naturally.)

Problems 32–35 concern abelian categories.

32. Let $\mathcal{C}$ be an abelian category. Let $\mathcal{D}$ be the category for which $\mathrm{Obj}(\mathcal{D})$ consists of all chain complexes of objects and morphisms in $\mathcal{C}$ and for which $\mathrm{Morph}(X, Y)$ for any two objects X and Y in $\mathcal{D}$ consists of all chain maps from X to Y. Prove that $\mathcal{D}$ is an abelian category.

33. Consider the snake diagram in the category of all abelian groups consisting of the four rightmost groups in the first row and the four leftmost groups in the second row of the following commutative diagram:

$$\begin{array}{ccccccccc}
0 & \longrightarrow & \mathbb{Z} & \xrightarrow{\times 8} & \mathbb{Z} & \xrightarrow{1 \mapsto 1 \bmod 8} & \mathbb{Z}/8\mathbb{Z} & \longrightarrow & 0 \\
& & \downarrow{\scriptstyle \times 4} & & \downarrow{\scriptstyle \times 2} & & \downarrow{\scriptstyle \substack{1 \bmod 8 \\ \mapsto 2 \bmod 4}} & & \\
0 & \longrightarrow & \mathbb{Z} & \xrightarrow{\times 4} & \mathbb{Z} & \xrightarrow{1 \mapsto 1 \bmod 4} & \mathbb{Z}/4\mathbb{Z} & \longrightarrow & 0
\end{array}$$

Adjoin the 0's to make the diagram become what is displayed. Following the steps in the example of a diagram chase in Section 8, extend this diagram to the auxiliary diagram that appears in that discussion, and show that $(B_0, \widetilde{k})$ for the extended diagram is not a kernel of β.

34. For a general abelian category $\mathcal{C}$ and any M in $\mathrm{Obj}(\mathcal{C})$, verify that $\mathrm{Hom}(\,\cdot\,, M)$ is a left exact contravariant functor from $\mathcal{C}$ to $\mathcal{C}_\mathbb{Z}$ and $\mathrm{Hom}(M,\,\cdot\,)$ is a left exact covariant functor from $\mathcal{C}$ to $\mathcal{C}_\mathbb{Z}$.

35. Proposition 4.19 shows for any good category $\mathcal{C}$ of unital left R modules that a module P in $\mathcal{C}$ is projective for $\mathcal{C}$ if and only if $\mathrm{Hom}(P,\,\cdot\,)$ is an exact functor, if and only if every short exact sequence $0 \to X \to Y \to P \to 0$ splits. Rewrite this proof in such a way that it applies to arbitrary abelian categories $\mathcal{C}$. For the step in the argument that the splitting of every short exact sequence $0 \to X \to Y \to P \to 0$ implies that P is projective, use the notion of pullback that is developed in Section 8.

CHAPTER V

Three Theorems in Algebraic Number Theory

Abstract. This chapter establishes some essential foundational results in the subject of algebraic number theory beyond what was already in *Basic Algebra*.

Section 1 puts matters in perspective by examining what was proved in Chapter I for quadratic number fields and picking out questions that need to be addressed before one can hope to develop a comparable theory for number fields of degree greater than 2.

Sections 2–4 concern the field discriminant of a number field. Section 2 contains the definition of discriminant, as well as some formulas and examples. The main result of Section 3 is the Dedekind Discriminant Theorem. This concerns how prime ideals (p) in $\mathbb{Z}$ split when extended to the ideal $(p)R$ in the ring of integers R of a number field. The theorem says that ramification, i.e, the occurrence of some prime ideal factor in R to a power greater than 1, occurs if and only if p divides the field discriminant. The theorem is proved only in a very useful special case, the general case being deferred to Chapter VI. The useful special case is obtained as a consequence of Kummer's criterion, which relates the factorization modulo p of irreducible monic polynomials in $\mathbb{Z}[X]$ to the question of the splitting of the ideal $(p)R$. Section 4 gives a number of examples of the theory for number fields of degree 3.

Section 5 establishes the Dirichlet Unit Theorem, which describes the group of units in the ring of algebraic integers in a number field. The torsion subgroup is the subgroup of roots of unity, and it is finite. The quotient of the group of units by the torsion subgroup is a free abelian group of a certain finite rank. The proof is an application of the Minkowski Lattice-Point Theorem.

Section 6 concerns class numbers of algebraic number fields. Two nonzero ideals I and J in the ring of algebraic integers of a number field are equivalent if there are nonzero principal ideals (a) and (b) with $(a)I = (b)J$. It is relatively easy to prove that the set of equivalence classes has a group structure and that the order of this group, which is called the class number, is finite. The class number is 1 if and only if the ring is a principal ideal domain. One wants to be able to compute class numbers, and this easy proof of finiteness of class numbers is not helpful toward this end. Instead, one applies the Minkowski Lattice-Point Theorem a second time, obtaining a second proof of finiteness, one that has a sharp estimate for a finite set of ideals that need to be tested for equivalence. Some examples are provided. A by-product of the sharp estimate is Minkowski's theorem that the field discriminant of any number field other than $\mathbb{Q}$ is greater than 1. In combination with the Dedekind Discriminant Theorem, this result shows that there always exist ramified primes over $\mathbb{Q}$.

1. Setting

It is worth stepping back from the results of Chapter I to put matters into perspective. Chapter I studied three problems, all of which could be stated in terms of

1. Setting

elementary number theory. These were the questions of solvability of quadratic congruences, of representability of integers or rational numbers by primitive binary quadratic forms, and of the infinitude of primes in arithmetic progressions.

We had started from the more general problem of studying Diophantine equations, beginning with the observation that solvability in integers implies solvability modulo each prime.[1] Linear congruences being no problem, we began with quadratic congruences and were led to quadratic reciprocity. Then we sought to apply quadratic reciprocity to address representability of integers or rational numbers by binary quadratic forms. The reasons for studying the infinitude of primes in arithmetic progressions were more subtle; what we saw was that at various stages in dealing with binary quadratic forms, this question of infinitude kept arising, along with techniques that might be helpful in addressing it.

Work on at least the first two of the problems was helped to some extent by the use of algebraic integers, and we shall see momentarily that algebraic integers illuminate work on the third problem as well. In any event, it is apparent where to look for a natural generalization. We are to study higher-degree congruences, perhaps in more than one variable, and we are to use algebraic extensions of the rationals of degree greater than 2 to help in the study.

The situation studied in Section IX.17 of *Basic Algebra* will be general enough for now. Thus let $F(X)$ be a monic irreducible polynomial in $\mathbb{Z}[X]$. Section IX.17 began to look at the question of how $F(X)$ reduces modulo each prime p. We begin by reviewing the case of degree 2, the main results in this case having been obtained in Chapter I in the present volume. For the polynomial $F(X) = X^2 - m$ with $m \in \mathbb{Z}$, the assumed irreducibility means that m is not the square of an integer. For fixed m and most primes p, either $F(X)$ remains irreducible modulo p or $F(X)$ splits as the product of two distinct linear factors. The exceptional primes have the property that $F(X)$ modulo p is the square of a linear factor; these are the prime divisors of m and sometimes the prime 2. In short, they occur among the prime divisors of the discriminant $4m$ of $F(X)$. In terms of quadratic residues, the irreducibility of $F(X)$ modulo p means that m is not a quadratic residue modulo p, and the splitting into two distinct linear factors means that it is. The odd primes for which $F(X)$ modulo p is the square of a linear factor are the odd primes that divide m. Modulo 2, every integer is a square, and reduction modulo 2 was not helpful.

The number theory of quadratic number fields sheds additional light on this factorization. The relevant field is of course $\mathbb{Q}(\sqrt{m})$; this is a nontrivial extension of $\mathbb{Q}$, since m is not square. In working with this field in Chapter I, we imposed the additional condition that m be square free. Promising a general definition for

---

[1] Solvability modulo each prime power is also of interest but played a role in Chapter I only for powers of 2.

later, we defined the **field discriminant** of $\mathbb{Q}(\sqrt{m})$ in that chapter to be

$$D = \begin{cases} 4m & \text{if } m \equiv 2 \bmod 4 \text{ or } m \equiv 3 \bmod 4, \\ m & \text{if } m \equiv 1 \bmod 4. \end{cases}$$

Problems 20–24 in Chapter I implicitly related the splitting of $F(X)$ modulo p to the factorization of ideals. Let R be the ring of algebraic integers in $\mathbb{Q}(\sqrt{m})$. If p is an odd prime, those problems observed that $(p)R$ is a prime ideal in R if D is a nonsquare modulo p, is the product of two distinct prime ideals if D is a square modulo p but is not divisible by p, and is the square of a prime ideal if D is divisible by p. The factorization of $(2)R$ was more subtle and was addressed in Problem 21.

In any event, the pattern of reducibility modulo p of $X^2 - m$, at least when the prime p is odd, mirrors the pattern of factorization of the ideal generated by p in the ring of algebraic integers in the number field $\mathbb{Q}(\sqrt{m})$. The role of quadratic reciprocity was to explain this pattern. Problem 1 at the end of Chapter I showed that one qualitative consequence of quadratic reciprocity is that the odd primes p for which $X^2 - m$ remains irreducible are the ones in certain arithmetic progressions, and similarly for the odd primes not dividing p for which a factorization into two linear factors occurs.

One objective of a generalization is to produce a corresponding theory for an arbitrary monic irreducible polynomial $F(X)$ in $\mathbb{Z}[X]$, say of degree n. Let $\mathbb{K}$ be the extension of $\mathbb{Q}$ generated by a root of $F(X)$, and let R be the ring of algebraic integers in $\mathbb{K}$. Theorem 9.60 of *Basic Algebra* shows for each prime number p that the decomposition of the ideal $(p)R$ in R as a product of powers of distinct prime ideals takes the form $(p)R = \prod_{i=1}^{g} P_i^{e_i}$ with $f_i = [R/P_i : \mathbb{Z}/(p)]$ and $\sum_{i=1}^{g} e_i f_i = n$. Meanwhile, $F(X)$ factors modulo p as a product of powers of irreducible polynomials modulo p. Sections 2–3 will describe a theory begun by Kummer and Dedekind for how the factorization of the ideal $(p)R$ and the factorization of the polynomial $F(X)$ modulo p are related. One introduces a field discriminant for $\mathbb{K}$ that is closely related to the discriminant of the polynomial $F(X)$, and a key result, the Dedekind Discriminant Theorem, says that some e_i is > 1 if and only if p divides the field discriminant. The primes p for which some e_i is greater than 1 are said to **ramify** in the extension field $\mathbb{K}$. These primes are not as well behaved as the others, and one's first inclination might be to try to ignore them. However, Problems 25–40 at the end of Chapter I show that the ramified primes encode a great deal of information; in particular, they explain the theory of genera and the relationship between exact representability of rational numbers and representability of integers modulo the field discriminant.

Generalizations of quadratic reciprocity lie much deeper and are central results of the subject of class field theory, a subject that is beyond the scope of the present book. Suffice it to say that class field theory in its established form seeks to parametrize all finite Galois extensions of any number field having abelian Galois group; the parametrization is to refer only to data within the given number field. The reciprocity theorem in this setting goes under the name "Artin reciprocity," which includes quadratic reciprocity as a very special case. Class field theory for nonabelian finite Galois extensions is at present largely conjectural, and the conjectural reciprocity statement goes under the name "Langlands reciprocity."

Beginning in Section I.6, we translated some of the theory of binary quadratic forms into facts about quadratic number fields. One tool we needed was a description of the units in the ring of algebraic integers within the quadratic number field. It is to be expected that a similar description for an arbitrary number field will play a foundational role in number theory beyond the quadratic case. The description in question is captured in the Dirichlet Unit Theorem, which appears as Theorem 5.13 in Section 5.

The translation of the notion of proper equivalence class of binary quadratic forms into the language of quadratic field extensions led to a notion of strict equivalence of ideals, as well as a notion of ordinary equivalence. Because there are only finitely many proper equivalence classes of forms, there could be only finitely many strict equivalence classes of ideals, and this set of classes of ideals acquired the structure of a finite abelian group. Dirichlet studied the order of this group, which figures into formulas for the value of certain Dirichlet L functions $L(s, \chi)$ at $s = 1$. The ideal class group for ordinary equivalence is a quotient of this group by a subgroup of order at most 2.

Although we shall not be concerned with representability of integers by forms of degree greater than 2, the ideal class group and its order (the "class number" of the field) are of interest for general number fields when defined in terms of ordinary equivalence, not strict equivalence. Section 6 is devoted to proving that the class number is finite for any number field and to developing some tools for computing class numbers. Class number 1 is equivalent to having the ring of algebraic integers in question be a principal ideal domain. Apart from the appearance of class numbers in various limit formulas, here is one other indicator of the importance of the ideal class group: It is possible to extend the above theory of ramification in such a way that it applies to any extension $\mathbb{K}/\mathbb{F}$ of number fields, not just to finite extensions of $\mathbb{Q}$. Hilbert proved that for any $\mathbb{F}$, there is a finite Galois extension $\mathbb{K}/\mathbb{F}$ with abelian Galois group that is small enough for the extension to be unramified at every prime ideal of $\mathbb{F}$ and that is large enough for any unramified abelian extension of $\mathbb{F}$ to lie in $\mathbb{K}$. Artin reciprocity can be used to show that $\text{Gal}(\mathbb{K}/\mathbb{F})$ is isomorphic to the ideal class group[2] of $\mathbb{F}$ and thus gives some control over the nature of $\mathbb{K}$. In particular, $\mathbb{K} = \mathbb{F}$ if and only if every ideal in the ring of integers of $\mathbb{F}$ is principal. When $\mathbb{F}$ is quadratic over $\mathbb{Q}$, the

---

[2]The field $\mathbb{K}$ is called the **Hilbert class field** of $\mathbb{F}$. The name "class field" is meant to be a reminder of this isomorphism.

field $\mathbb{K}$ can be used to give more definitive results than in Chapter I concerning representability of integers by binary quadratic forms.

2. Discriminant

Let us recall some material about Dedekind domains from Chapters VIII and IX of *Basic Algebra*. A Dedekind domain is a Noetherian integral domain that is integrally closed and has the property that every nonzero prime ideal is maximal. Any principal ideal domain is an example. Any Dedekind domain has unique factorization for its ideals. Theorem 8.54 of the book gave a construction for extending certain Dedekind domains to larger Dedekind domains: if D is a Dedekind domain with field of fractions $\mathbb{F}$ and if $\mathbb{K}$ is a finite separable extension of $\mathbb{F}$, then the integral closure of D in $\mathbb{K}$ is a Dedekind domain R. The hard step in the proof, which was not carried out until Section IX.15, was to deduce from the separability that R is finitely generated over D. The role of separability was to force the bilinear form $(a, b) \mapsto \mathrm{Tr}_{\mathbb{K}/\mathbb{F}}(ab)$ to be nondegenerate, and this nondegeneracy in turn implied the desired result about finite generation.

In this section we introduce a tool that captures this last implication in quantitative fashion—that nondegeneracy of the trace form implies that the extended domain is finitely generated over the given domain. In a full-fledged treatment of algebraic number theory, one might well want to work in this full generality,[3] but we need less for our purposes: Throughout this section we assume that the given Dedekind domain is the ring $\mathbb{Z}$ of integers, that $\mathbb{K}$ is a number field, and that R is the integral closure of $\mathbb{Z}$ in $\mathbb{K}$, i.e., R is the ring of algebraic integers within $\mathbb{K}$. Let $n = [\mathbb{K} : \mathbb{Q}]$ be the degree of the field extension. Since $\mathbb{C}$ is algebraically closed, we can regard $\mathbb{K}$ as a subfield of $\mathbb{C}$.

The separability of $\mathbb{K}/\mathbb{Q}$ in combination with the fact that $\mathbb{C}$ is algebraically closed implies that there exist exactly n distinct field maps $\sigma_1, \ldots, \sigma_n$ of $\mathbb{K}$ into $\mathbb{C}$; one of them is the identity. Recall how $\sigma_1, \ldots, \sigma_n$ can be constructed: if ξ is a primitive element for $\mathbb{K}/\mathbb{Q}$, if $F(X)$ is the minimal polynomial of ξ over $\mathbb{Q}$, and if $\xi_1 = \xi, \xi_2, \ldots, \xi_n$ are the n distinct roots of $F(X)$ in $\mathbb{C}$, then σ_j can be defined by $\sigma_j\left(\sum_{i=0}^{n-1} c_i \xi^i\right) = \sum_{i=0}^{n-1} c_i \xi_j^i$ on any $\mathbb{Q}$ linear combination of powers of ξ. For any $\eta = \sum_{i=0}^{n-1} c_i \xi^i$ in $\mathbb{K}$, primitive or not, the n elements $\sigma_i(\eta)$ of $\mathbb{C}$ are called the **conjugates** of η relative to $\mathbb{K}$. They are the roots of the field polynomial of η over $\mathbb{K}$, and each occurs with multiplicity $[\mathbb{K} : \mathbb{Q}(\eta)]$.[4]

---

[3] For example this full level of generality would be appropriate if one planned ultimately to study class field theory.

[4] The field polynomial of an element of $\mathbb{K}$ is the characteristic polynomial of left multiplication on $\mathbb{K}$ by the element. This notion is discussed in Section IX.15 of *Basic Algebra*.

2. Discriminant

Let $\Gamma = (v_1, \ldots, v_n)$ be an ordered basis of $\mathbb{K}$ over $\mathbb{Q}$. The symmetric bilinear form $(u, v) \mapsto \text{Tr}_{\mathbb{K}/\mathbb{Q}}(uv)$ determines an n-by-n symmetric matrix $B_{ij} = \text{Tr}_{\mathbb{K}/\mathbb{Q}}(v_i v_j)$, and we can recover the form from the matrix B by the formula $\text{Tr}_{\mathbb{K}/\mathbb{Q}}(uv) = a^t Bb$ if $a = \binom{u}{\Gamma}$ and $b = \binom{v}{\Gamma}$ are the column vectors of u and v in the ordered basis Γ, i.e., if $u = \sum_{i=1}^n a_i v_i$ and $v = \sum_{j=1}^n b_j v_j$. From Section VI.1 of *Basic Algebra*, we know that the bilinear form determines a canonical $\mathbb{Q}$ linear map L from $\mathbb{K}$ to its vector space dual by the formula $L(u)(v) = \text{Tr}_{\mathbb{K}/\mathbb{Q}}(uv)$ and that the nondegeneracy of the form[5] implies that this linear map is one-one onto. Moreover, the matrix of L with respect to Γ and the dual basis of Γ is B. Thus the nondegeneracy implies that the matrix B is nonsingular. The **discriminant** $D(\Gamma)$ of the ordered basis Γ is given by

$$D(\Gamma) = \det B, \quad \text{where } B \text{ is the matrix of } (u, v) \mapsto \text{Tr}_{L/K}(uv) \text{ in the basis } \Gamma.$$

Because of the nonsingularity of B, this is a nonzero member of $\mathbb{Q}$.

Proposition 6.1 of *Basic Algebra* shows the effect on the matrix B of changing the basis. Specifically let $\Delta = (w_1, \ldots, w_n)$ be a second ordered basis, and let C be the matrix of the form in this basis, namely $C_{ij} = \text{Tr}_{\mathbb{K}/\mathbb{Q}}(w_i w_j)$. Let the two bases be related by $w_j = \sum_{i=1}^n a_{ij} v_i$, i.e., let $[a_{ij}] = \binom{I}{\Gamma\Delta}$. Then the proposition gives

$$C = \binom{I}{\Gamma\Delta}^t B \binom{I}{\Gamma\Delta}.$$

Taking determinants and using the fact that a matrix and its transpose have the same determinant, we obtain

$$D(\Delta) = D(\Gamma) \left(\det \binom{I}{\Gamma\Delta}\right)^2.$$

One consequence of this formula is that the sign of $D(\Gamma)$ is independent of Γ. Another is that the value of $D(\Gamma)$ does not depend on the ordering of the n members of Γ; it depends only on Γ as an unordered set.

Now suppose that the members of the ordered basis Γ are in the subring R of algebraic integers within $\mathbb{K}$. Bases of $\mathbb{K}$ over $\mathbb{Q}$ consisting of members of R always exist, since we can always multiply the members of a basis of $\mathbb{K}$ over $\mathbb{Q}$ by a suitable integer to get them to be in R. In this case the entries $B_{ij} = \text{Tr}_{\mathbb{K}/\mathbb{Q}}(v_i v_j)$ of the matrix of the bilinear form are in $\mathbb{Z}$, and $D(\Gamma)$ is therefore a nonzero member of $\mathbb{Z}$.

The **field discriminant**, or **absolute discriminant**, of $\mathbb{K}$, denoted by $D_{\mathbb{K}}$, is the value of $D(\Gamma)$ that minimizes $|D(\Gamma)|$ for all bases of $\mathbb{K}$ consisting of members

---

[5]The nondegeneracy of the trace form for a number field is a transparent result, not requiring anything deep from Section IX.15 of *Basic Algebra*, since any $u \neq 0$ in $\mathbb{K}$ has $\text{Tr}_{\mathbb{K}/\mathbb{Q}}(uu^{-1}) = \text{Tr}_{\mathbb{K}/\mathbb{Q}}(1) = n \neq 0$.

of R. This is a nonzero integer. The sign of $D_\mathbb{K}$ is well defined, since all values of $D(\Gamma)$ have the same sign.[6]

Fix an ordered basis $\Gamma = (v_1, \ldots, v_n)$ of $\mathbb{K}$, and consider the abelian group consisting of the $\mathbb{Z}$ span $\mathbb{Z}(\Gamma)$ of the members of Γ. This is evidently a free abelian group of rank n. If an ordered basis $\Delta = (w_1, \ldots, w_n)$ has the property that $\mathbb{Z}(\Delta) \subseteq \mathbb{Z}(\Gamma)$, then the theory in Section IV.9 of *Basic Algebra* that leads to the Fundamental Theorem of Finitely Generated Abelian Groups shows that if we write formally

$$\begin{pmatrix} w_1 \\ \vdots \\ w_n \end{pmatrix} = C \begin{pmatrix} v_1 \\ \vdots \\ v_n \end{pmatrix},$$

then there exist n-by-n integer matrices M_1 and M_2 of determinant ± 1 such that $D = M_1 C M_2$ is diagonal, and moreover the order of $\mathbb{Z}(\Gamma)/\mathbb{Z}(\Delta)$ is $|\det D| = |\det C|$. Examining the definition of C, we see that $C = \begin{pmatrix} I \\ \Gamma_\Delta \end{pmatrix}^t$. Consequently we obtain

$$|\mathbb{Z}(\Gamma)/\mathbb{Z}(\Delta)| = \left|\det \begin{pmatrix} I \\ \Gamma_\Delta \end{pmatrix}\right|,$$

a formula we shall use repeatedly in this chapter without specific reference.

Proposition 5.1. If Γ is a basis of $\mathbb{K}$ over $\mathbb{Q}$ whose members all lie in R, then $|R/\mathbb{Z}(\Gamma)|^2 = D(\Gamma)/D_\mathbb{K}$. In particular, Γ is a $\mathbb{Z}$ basis of R if and only if $D(\Gamma) = D_\mathbb{K}$.

REMARKS. We already know from *Basic Algebra* that R is a free abelian group of rank n. The second conclusion of this proposition, in combination with the transparent observation that the trace form is nonsingular for a number field, gives a more direct proof of this fact. Introductory treatments of algebraic number theory sometimes give this more direct proof, whose details are spelled out in the second paragraph below.

PROOF. Let Δ and Ω be two bases of $\mathbb{K}$ over $\mathbb{Q}$ whose members all lie in R, and suppose that $\mathbb{Z}(\Delta) \subseteq \mathbb{Z}(\Omega)$. Then the above discussion shows that

$$|D(\Delta)| = |D(\Omega)| \left(\det \begin{pmatrix} I \\ \Omega_\Delta \end{pmatrix}\right)^2$$

and that

$$|\mathbb{Z}(\Omega)/\mathbb{Z}(\Delta)|^2 = \left(\det \begin{pmatrix} I \\ \Omega_\Delta \end{pmatrix}\right)^2.$$

Since $D(\Delta)$ and $D(\Omega)$ are nonzero and have the same sign, we obtain

$$D(\Delta)/D(\Omega) = |\mathbb{Z}(\Omega)/\mathbb{Z}(\Delta)|^2. \qquad (*)$$

---

[6]As was observed above, any $D(\Delta)$ is the product of $D(\Gamma)$ and the square of a rational number. Hence $D(\Delta)$ and $D(\Gamma)$ have the same sign.

2. Discriminant

To prove the proposition, we prove the "if" part of the second conclusion first—without using the known fact that R is free abelian. Choose Δ such that $D(\Delta) = D_{\mathbb{K}}$ and such that Δ has all its members in R. Arguing by contradiction, suppose that Δ fails to be a $\mathbb{Z}$ basis of R. Let r be an element of R not in $\mathbb{Z}(\Delta)$. Then the $\mathbb{Z}$ span of $\mathbb{Z}(\Delta) \cup \{r\}$ is a finitely generated additive subgroup of $\mathbb{K}$ and must be free abelian of rank $\geq n$. Being a subgroup of the additive group of $\mathbb{K}$, it cannot have rank greater than n and hence has rank exactly n. Let Ω be an ordered $\mathbb{Z}$ basis of this subgroup. Since $\mathbb{Z}(\Delta) \subsetneq \mathbb{Z}(\Omega)$, the right side of $(*)$ is > 1, and thus $D_{\mathbb{K}} > D(\Omega)$. But this is a contradiction because the members of Ω lie in R, and hence Δ is a $\mathbb{Z}$ basis of R. In particular, a $\mathbb{Z}$ basis of R exists.

To prove the rest of the proposition, take Ω in $(*)$ to be a $\mathbb{Z}$ basis of R, and let $\Delta = \Gamma$ be any given basis of $\mathbb{K}$ over $\mathbb{Q}$ that lies in R. Then $(*)$ gives $|R/\mathbb{Z}(\Gamma)|^2 = D(\Gamma)/D(\Omega)$. Since $|R/\mathbb{Z}(\Gamma)|$ cannot be less than 1, $|D(\Gamma)|$ cannot be less than $|D(\Omega)|$. Thus $D_{\mathbb{K}} = D(\Omega)$, and $|R/\mathbb{Z}(\Gamma)|^2 = D(\Gamma)/D_{\mathbb{K}}$. This proves the first conclusion of the proposition, and the "only if" part of the second conclusion is immediate. □

EXAMPLE. Field discriminant of a quadratic number field. Let $\mathbb{K} = \mathbb{Q}(\sqrt{m})$, where m is a square-free integer other than 1. From Section I.6 a $\mathbb{Z}$ ordered basis Γ of R is given by

$$\Gamma = \begin{cases} \{1, \sqrt{m}\} & \text{if } m \equiv 2 \text{ or } 3 \bmod 4, \\ \{1, \tfrac{1}{2}(\sqrt{m} - 1)\} & \text{if } m \equiv 1 \bmod 4. \end{cases}$$

Proposition 5.1 allows us to compute $D_{\mathbb{K}}$ from this information. The matrix whose determinant is $D_{\mathbb{K}}$ in the two cases is $\begin{pmatrix} 2 & 0 \\ 0 & 2m \end{pmatrix}$ and $\begin{pmatrix} 2 & -1 \\ -1 & \tfrac{1}{2}(m+1) \end{pmatrix}$, respectively, and thus

$$D_{\mathbb{K}} = \begin{cases} 4m & \text{if } m \equiv 2 \text{ or } 3 \bmod 4, \\ m & \text{if } m \equiv 1 \bmod 4. \end{cases}$$

This is the formula that we took as a definition of field discriminant in Section I.6.

For a general number field $\mathbb{K}$ of degree n over $\mathbb{Q}$, there is no easy way to obtain a $\mathbb{Z}$ basis of R. Instead, one tries to compute $D_{\mathbb{K}}$ and find such a basis at the same time by successive refinements.

The first step is to use the special kind of $\mathbb{Q}$ basis of $\mathbb{K}$ whose existence is guaranteed by the Theorem of the Primitive Element. Specifically one can write $\mathbb{K} = \mathbb{Q}(\xi)$ for some ξ in $\mathbb{K}$, since $\mathbb{K}/\mathbb{Q}$ is a separable extension. Possibly after multiplying ξ by a suitably large integer, we may assume that ξ is in R. Then $\Gamma(\xi) = \{1, \xi, \xi^2, \ldots, \xi^{n-1}\}$ is a $\mathbb{Q}$ basis of $\mathbb{K}$ lying in R. We normally write $D(\xi)$ instead of $D(\Gamma(\xi))$ for the discriminant of $\Gamma(\xi)$. Write $\xi_i = \sigma_i(\xi)$ for the

i^{th} conjugate of ξ. Let $B = [B_{ij}]$ be the matrix whose determinant is $D(\xi)$. Since the trace of an element is the sum of its conjugates, B_{ij} is given by

$$B_{ij} = \text{Tr}_{\mathbb{K}/\mathbb{Q}}(\xi^{i-1}\xi^{j-1}) = \sum_{k=1}^{n} \sigma_k(\xi^{i-1}\xi^{j-1}) = \sum_{k=1}^{n} \xi_k^{i-1}\xi_k^{j-1},$$

and this is of the form $\sum_{k=1}^{n} V_{ik} V_{jk}^t$, where $V_{ik} = \xi_k^{i-1}$ is an entry of a Vandermonde matrix. Therefore

$$D(\xi) = \det B = (\det V)^2 = \Big(\prod_{i<j}(\xi_j - \xi_i)\Big)^2 = \prod_{i<j}(\xi_j - \xi_i)^2,$$

which coincides with the discriminant of the field polynomial of ξ over $\mathbb{Q}$.

EXAMPLES OF $D(\xi)$.

(1) $\mathbb{K} = \mathbb{Q}(\xi)$, where $\xi^5 - \xi - 1 = 0$. This field was studied in Example 1 of Section IX.17 of *Basic Algebra*. The discriminant of the polynomial $X^5 - X - 1$ is $2869 = 19 \cdot 151$, and thus $D(\xi) = 2869$. Proposition 5.1 shows that $D(\xi) = D_{\mathbb{K}} k^2$ for some nonzero integer k. Since 2869 is square free, we conclude that $D_{\mathbb{K}} = 2869$.

(2) $\mathbb{K} = \mathbb{Q}(\sqrt[3]{2})$. The minimal polynomial of $\xi = \sqrt[3]{2}$ is $X^3 - 2$, and its roots are $\xi, \xi\omega$, and $\xi\omega^2$, where $\omega = e^{2\pi i/3}$. Then

$$D(\xi) = (\xi - \xi\omega)^2(\xi - \xi\omega^2)^2(\xi\omega - \xi\omega^2)^2 = \xi^6(1-\omega)^2(1-\omega^2)^2(\omega - \omega^2)^2,$$

and this simplifies to $D(\xi) = -2^2 3^3$. This quantity is the product of $D_{\mathbb{K}}$ by the square of an integer. Thus $D_{\mathbb{K}}$ is one of $-3, -12, -27$, and -108.

What happens with Example 2 is typical: a second step is needed to decide among finitely many possibilities for $D_{\mathbb{K}}$. In the general case an induction is involved, and Proposition 5.2 below says what is to be done at each step. At the end of this section, we shall return to Example 2 and use the proposition to see that $D_{\mathbb{K}} = -108$ is the correct choice.

Before stating Proposition 5.2, let us interpolate a generalization of the computation of $D(\xi)$ that preceded the above examples. Suppose that $\Gamma = (\alpha_1, \ldots, \alpha_n)$ is any ordered $\mathbb{Q}$ basis of $\mathbb{K}$ lying in R. Let $B = [B_{ij}]$ be the matrix whose determinant is the discriminant of Γ. Then we have

$$B_{ij} = \text{Tr}_{\mathbb{K}/\mathbb{Q}}(\alpha_i \alpha_j) = \sum_{k=1}^{n} \sigma_k(\alpha_i \alpha_j) = \sum_{k=1}^{n} \sigma_k(\alpha_i)\sigma_k(\alpha_j) = \sum_{k=1}^{n} A_{ik}(A^t)_{kj},$$

where $A = [A_{ij}]$ is the matrix with $A_{ij} = \sigma_j(\alpha_i)$, and it follows that

$$D(\Gamma) = \big(\det[\sigma_j(\alpha_i)]\big)^2.$$

This formula can be useful for computing $D(\Gamma)$ when the conjugates of the α_i are readily available.

2. Discriminant

Proposition 5.2. Let $\Gamma = (v_1, \ldots, v_n)$ be an ordered $\mathbb{Q}$ basis of $\mathbb{K}$ lying in R. If the $\mathbb{Z}$ span $\mathbb{Z}(\Gamma)$ of Γ is a proper subgroup of R, then there exists a prime number p such that p^2 divides $D(\Gamma)$ and such that some member

$$v'_k = p^{-1}(c_1 v_1 + c_2 v_2 + \cdots + c_{k-1} v_{k-1} + v_k)$$

of $\mathbb{K}$ lies in R with $1 \leq k \leq n$ and $0 \leq c_j \leq p - 1$ for $j \leq k - 1$. If such an element v'_k is found, then $\Delta = (v_1, \ldots, v_{k-1}, v'_k, v_{k+1}, \ldots, v_n)$ has $\mathbb{Z}(\Delta)$ properly containing $\mathbb{Z}(\Gamma)$ with $D(\Delta) = p^{-2} D(\Gamma)$.

REMARKS. A finite computation is involved in finding p and k. On the one hand, for given p, at most $1 + p + p^2 + \cdots + p^{n-1}$ elements have to be checked for integrality. On the other hand, we in principle have to find the field polynomial of a certain element of $\mathbb{K}$ in each case and decide whether the coefficients are integers, and this computation may be lengthy. See Problem 2 at the end of the chapter for an easy example, Problem 16 for a harder example, and Problem 4b for a related computation.

PROOF. Let $\mathbb{Z}(\Gamma)$ be a proper subgroup of R, and put $m = |R/\mathbb{Z}(\Gamma)|$. Choose a $\mathbb{Z}$ basis $(w_1, \ldots, w_n)$ of R, and write $v_i = \sum_{j=1}^n c_{ij} w_j$ with all $c_{ij} \in \mathbb{Z}$. We know that $|\det[c_{ij}]| = m$, and we let p be any prime divisor of m. Reducing the c_{ij} modulo p, we see that the matrix $[c_{ij}]$ is singular modulo p, and thus there exist integers $a_1, \ldots, a_n$ not all divisible by p such that

$$\sum_{i=1}^n a_i c_{ij} \equiv 0 \bmod p \qquad \text{for } 1 \leq j \leq n.$$

Find k with $1 \leq k \leq n$ for which p divides all of $a_{k+1}, \ldots, a_n$ but not a_k, and write $\sum_{i=1}^n a_i c_{ij} = p l_j$ for integers l_j. Then

$$\sum_{i=1}^k a_i v_i = \sum_{j=1}^n \sum_{i=1}^k a_i c_{ij} w_j = \sum_{j=1}^n \left(p l_j - \sum_{i=k+1}^n a_i c_{ij} \right) w_j,$$

and the integer in parentheses on the right side is a multiple of p. Therefore $r = \sum_{i=1}^k a_i v_i$ is exhibited as ps for some $s \in R$. Choose a' and d_k in $\mathbb{Z}$ with $a' a_k - d_k p = 1$, and choose c_i and d_i in $\mathbb{Z}$ for each i with $i \leq k-1$ such that $0 \leq c_i \leq p - 1$ and $a' a_i - p d_i = c_i$. Then the computation

$$p a' s = a' r = \sum_{i=1}^k a' a_i v_i = \sum_{i=1}^{k-1}(c_i + p d_i) v_i + (1 + p d_k) v_k = \sum_{i=1}^{k-1} c_i v_i + v_k + p \sum_{i=1}^k d_i v_i$$

shows that $p^{-1}\left(\sum_{i=1}^{k-1} c_i v_i + v_k\right) = a' s - \sum_{i=1}^k d_i v_i$ lies in R. □

Proposition 5.1 shows that any primitive element ξ of $\mathbb{K}$ that lies in R has the property that $D(\xi)/D_\mathbb{K}$ is the square of a nonzero integer, and we write this quotient as $J(\xi)^2$ with $J(\xi) > 0$. One might hope that although some particular choice of ξ fails to have $J(\xi) = 1$, some other choice may be found for which equality holds. We shall see in Section 4 that for a class of integers m, $\mathbb{Q}(\sqrt[3]{m})$ has such an element ξ if and only if a certain nontrivial Diophantine equation in two variables has a solution. Both cases arise: for $m = 2$, such a ξ exists, while for $m = 175$, no such ξ exists.

But matters can be worse than this for a general $\mathbb{K}$. The quotient $J(\xi)^2 = D(\xi)/D_\mathbb{K}$ for a primitive element ξ of $\mathbb{K}$ lying in R is sometimes called the **index** of ξ. One might hope at least that each prime not dividing $D_\mathbb{K}$ fails to divide the index $J(\xi)^2$ for some ξ. However, Dedekind showed that there exist number fields $\mathbb{K}$ and primes p that are **common index divisors**[7] in the sense that p divides $J(\xi)$ for every primitive element ξ of $\mathbb{K}$ lying in R. Specifically he showed that $p = 2$ is such a prime when $\mathbb{K}$ is obtained by adjoining to $\mathbb{Q}$ a root of $X^3 + X^2 - 2X + 8$; here $D_\mathbb{K} = -503$. We shall study this example further in Section 4.

Let us now specialize our considerations from general additive subgroups of the form $\mathbb{Z}(\Gamma)$ to those that are ideals in R.

Proposition 5.3. If I is a nonzero ideal in R, then
(a) I contains a positive k in $\mathbb{Z}$ and
(b) I additively is of the form $I = \mathbb{Z}(\Gamma)$ for some $\mathbb{Q}$ basis Γ of $\mathbb{K}$ whose members lie in R.

Consequently R/I is a finite ring and satisfies $|R/I|^2 = D(\Gamma)/D_\mathbb{K}$.

PROOF. Let r be a nonzero member of I, and let $P(X)$ be the field polynomial of r. Then $P(X)$ is of the form $P(X) = X^n + a_{n-1}X^{n-1} + \cdots + a_1 X + (-1)^n N_{\mathbb{K}/\mathbb{Q}}(r)$, has integers for coefficients, and has r as one of its roots. Consequently the formula
$$(-1)^{n+1} N_{\mathbb{K}/\mathbb{Q}}(r) = r(r^{n-1} + a_{n-1}r^{n-2} + \cdots + a_1)$$
shows that the nonzero integer $N_{\mathbb{K}/\mathbb{Q}}(r)$ is the product of r by a member of R and hence lies in I. This proves (a) with $k = |N_{\mathbb{K}/\mathbb{Q}}(r)|$.

The ideal I additively is a subgroup of R and is thus free abelian of rank at most n. By (a), the integer $k = |N_{\mathbb{K}/\mathbb{Q}}(r)|$ has the property that $kR \subseteq I \subseteq R$. Since R/kR has k^n elements, R/I is finite. Therefore I has rank n as an additive group and must be of the asserted form $\mathbb{Z}(\Gamma)$. This proves (b). The formula $|R/I|^2 = D(\Gamma)/D_\mathbb{K}$ is immediate from Proposition 5.1. □

---

[7]Terminology varies for this notion. Such primes p are more usually called **common inessential discriminant divisors** or **essential discriminant divisors**. The very fact that these two more usual names appear to contradict each other is sufficient reason to avoid using either name.

2. Discriminant

The **absolute norm** $N(I)$ of a nonzero ideal I of R is defined to be $N(I) = |R/I|$. This is necessarily a positive integer by Proposition 5.3. To be able to work with this notion, we shall make use of the unique factorization of ideals of R as given in Theorem 8.55 of *Basic Algebra*. That theorem says that such an ideal I has a factorization of the form $\prod_{j=1}^{l} P_j^{e_j}$, where the P_j are distinct prime ideals of R, and that this factorization is unique except for the order of the factors.

Proposition 5.4. The absolute norms of nonzero ideals of R have the following properties:
  (a) $N(R) = 1$.
  (b) If $I \subseteq J$ are nonzero ideals in R, then $N(J)$ divides $N(I)$, and $I = J$ if and only if $N(J) = N(I)$.
  (c) If I and J are nonzero ideals in R, then $N(IJ) = N(I)N(J)$.
  (d) If (α) is a nonzero principal ideal in R, then $N((\alpha)) = |N_{\mathbb{K}/\mathbb{Q}}(\alpha)|$.

PROOF. Conclusion (a) is immediate, and so is most of (b). If $I \subseteq J$ and $N(J) = N(I)$, then the First Isomorphism Theorem for abelian groups yields $(R/I)/(J/I) \cong R/J$, and it follows that $N(I)/|J/I| = N(J)$. Since $N(I)$ and $N(J)$ are finite, $N(I) = N(J)$ if and only if $|J/I| = 1$, i.e., if and only if $I = J$.

For (c), we begin with the special case that I and J are powers of a nonzero prime ideal P. Inductively it is enough to show that $N(P^k) = N(P)N(P^{k-1})$ for $k \geq 1$. Since $(R/P^k)/(P^{k-1}/P^k) \cong R/P^{k-1}$ as abelian groups, it is enough to show that

$$|P^{k-1}/P^k| = |R/P|. \qquad (*)$$

The ring R operates on the ideal P^{k-1}, carrying P^k into itself, and P carries P^{k-1} into P^k. Thus P^{k-1}/P^k is a unital module for the ring R/P, which is a field because P is maximal. Hence P^{k-1}/P^k is a vector space over R/P. Corollary 8.60 of *Basic Algebra* shows that this vector space is 1-dimensional, and then $(*)$ is immediate.

For the general case in (c), Corollary 8.63 of *Basic Algebra* shows that if $I = \prod_{j=1}^{l} P_j^{e_j}$ is the unique factorization of the nonzero ideal I as the product of positive powers of distinct prime ideals P_j, then $R/I \cong \prod_{j=1}^{l} R/P_j^{e_j}$. Hence $N(I) = \prod_{j=1}^{l} N(P_j^{e_j})$. Because of the special case that is already proved, $N(I) = \prod_{j=1}^{l} N(P_j)^{e_j}$. Then (c) follows in the general case.

For (d), if $\Gamma = (u_1, \ldots, u_n)$ is an ordered $\mathbb{Z}$ basis of R, then the tuple $\alpha\Gamma = (\alpha u_1, \ldots, \alpha u_n)$ is an ordered $\mathbb{Z}$ basis of (α), and we know that $N((\alpha)) = |R/(\alpha)| = |\mathbb{Z}(\Gamma)/\mathbb{Z}(\alpha\Gamma)| = \left|\det\begin{pmatrix} I \\ \Gamma, \alpha\Gamma \end{pmatrix}\right|$. But $\begin{pmatrix} I \\ \Gamma, \alpha\Gamma \end{pmatrix}$ is just the matrix of the $\mathbb{Q}$ linear map left-by-α in the $\mathbb{Q}$ basis Γ, and the determinant of this linear map is $N_{\mathbb{K}/\mathbb{Q}}(\alpha)$ by definition of the norm of an element. $\square$

EXAMPLE 2 OF $D(\xi)$, CONTINUED. For $\mathbb{K} = \mathbb{Q}(\sqrt[3]{2})$, we have seen that the discriminant of the $\mathbb{K}$ basis $\Gamma(\sqrt[3]{2})$ is $D(\sqrt[3]{2}) = -3^3 2^2$. We are going to show that $(1, \sqrt[3]{2}, \sqrt[3]{4})$ is a $\mathbb{Z}$ basis of R, and then it follows that the field discriminant of $\mathbb{K}$ is $D_\mathbb{K} = -3^3 2^2$. We apply Proposition 5.2. The only primes that need testing in that proposition are the ones dividing $D(\sqrt[3]{2})$, and thus we consider $p = 2$ and $p = 3$. We want to see that no expression $p^{-1}(1)$ or $p^{-1}(c_1 + \sqrt[3]{2})$ or $p^{-1}(c_1 + c_2\sqrt[3]{2} + \sqrt[3]{4})$ is an algebraic integer for some coefficients c_0 and c_1 between 0 and $p - 1$. We can discard $p^{-1}(1)$ because the only rational numbers that are algebraic integers are the members of $\mathbb{Z}$. If the field polynomial over $\mathbb{Q}$ of some ξ in $\mathbb{K}$ is $X^3 + a_2 X^2 + a_1 X + a_0$, then the field polynomial of $p^{-1}\xi$ is $X^3 + p^{-1} a_2 X^2 + p^{-2} a_1 X + p^{-3} a_0$. So the question of integrality is one of divisibility of the coefficients of the field polynomials of certain algebraic integers ξ by suitable powers of p. These coefficients, up to sign, are the values of the elementary symmetric polynomials on the three conjugates of ξ.

In the case at hand, only the coefficient a_0 is needed. That is, it is enough to see that the norm of ξ is never divisible by 8 or 27 for ξ equal to $c_1 + \sqrt[3]{2}$ or $c_1 + c_2\sqrt[3]{2} + \sqrt[3]{4}$ as above. Let us write $\xi = c_1 + c_2\theta + c_3\theta^2$ with $\theta = \sqrt[3]{2}$ and with c_1, c_2, c_3 in $\mathbb{Z}$. Then $a_0 = -N_{\mathbb{K}/\mathbb{Q}}(\xi)$, and the norm is the product of the three conjugates of ξ. If $\omega = e^{2\pi i/3}$, we compute that

$$N_{\mathbb{K}/\mathbb{Q}}(\xi) = (c_1 + c_2\theta + c_3\theta^2)(c_1 + c_2\theta\omega + c_3\theta^2\omega^2)(c_1 + c_2\theta\omega^2 + c_3\theta^2\omega)$$
$$= (c_1^3 + 2c_2^3 + 4c_3^3) + 2c_1 c_2 c_3(2\omega + 3\omega^2 + \omega^4)$$
$$= (c_1^3 + 2c_2^3 + 4c_3^3) - 6c_1 c_2 c_3.$$

For $p = 2$, we consider this expression when c_1, c_2, c_3 are chosen from $\{0, 1\}$. To get divisibility by 8, we check this expression modulo 8. Each c_i^3 is c_i for $c_i \in \{0, 1\}$. Looking at the expression modulo 2, we see that c_1 must be even, i.e., $c_1 = 0$. Then 8 must divide $2c_2^3 + 4c_3^3$, and we obtain $c_2 = c_3 = 0$, in contradiction to the formulas for the ξ's under consideration.

For $p = 3$, it is enough to consider this expression when c_1, c_2, c_3 are chosen from $\{-1, 0, +1\}$. Since each c_i has $|c_i| \leq 1$, we see that $|N_{\mathbb{K}/\mathbb{Q}}(\xi)| \leq 13$, and divisibility by 27 can occur only if $N_{\mathbb{K}/\mathbb{Q}}(\xi) = 0$, which we know entails $\xi = 0$. Thus no ξ meets the test of Proposition 5.2, and the conclusion is that $(1, \sqrt[3]{3}, \sqrt[3]{4})$ is a $\mathbb{Z}$ basis of R in $\mathbb{Q}(\sqrt[3]{2})$.

3. Dedekind Discriminant Theorem

The field discriminant plays a role in determining how a prime ideal (p) in $\mathbb{Z}$, p being a prime number, splits when one extends (p) to an ideal $(p)R$ in the ring R of algebraic integers in a number field $\mathbb{K}$ of degree n over $\mathbb{Q}$. In this

3. Dedekind Discriminant Theorem

situation, recall from Theorem 9.60 of *Basic Algebra* that the prime factorization of the ideal $(p)R$ in R is of the form $(p)R = \prod_{i=1}^{g} P_i^{e_i}$ with $\sum_{i=1}^{g} e_i f_i = n$; here $n = [\mathbb{K} : \mathbb{Q}]$, the P_i are distinct, and $f_i = \dim_{\mathbb{F}_p}(R/P_i)$. The integers e_i are called **ramification indices**, and the integers f_i are called **residue class degrees**. The extension $\mathbb{K}/\mathbb{Q}$ is said to be **ramified at** p, and the prime p of $\mathbb{Z}$ is said to **ramify in** $\mathbb{K}$, if some e_i is > 1 in this decomposition.[8]

Theorem 5.5 (Dedekind Discriminant Theorem). The prime p of $\mathbb{Z}$ ramifies in a number field $\mathbb{K}$ if and only if p divides the field discriminant $D_\mathbb{K}$ of $\mathbb{K}$.

In this chapter we shall prove this theorem only in a useful special case, namely in the case that p is not a common index divisor. Only finitely many primes can divide the index $J(\xi) = (D(\xi)/D_\mathbb{K})^{1/2}$ for a single primitive element ξ of $\mathbb{K}$ lying in R, and thus there are only finitely many common index divisors.[9] Consequently the special case that we are proving implies that only finitely many primes of $\mathbb{Z}$ ramify in $\mathbb{K}$.

The difficulty in proving Theorem 5.5 in full generality is that we lack sufficient tools for addressing questions by localization. At the end of this section, we shall make some comments about how one can proceed with further tools.

As we shall see later in this section, Theorem 5.5 for primes that are not common index divisors is an easy consequence of the following theorem.

Theorem 5.6 (Kummer's criterion). Let $\mathbb{K}$ be a number field, and let R be its ring of algebraic integers. Suppose that $F(X)$ is a monic irreducible polynomial in $\mathbb{Z}[X]$, that ξ is a root of $F(X)$ in $\mathbb{C}$, and that p is a prime number that does not divide the integer $J(\xi)$ such that $J(\xi)^2 = D(\xi)/D_\mathbb{K}$. Write $\overline{F}(X)$ for the reduction of $F(X)$ modulo p, let

$$\overline{F}(X) = \overline{F}_1(X)^{e_1} \cdots \overline{F}_g(X)^{e_g}$$

be the unique factorization of $\overline{F}(X)$ in $\mathbb{F}_p[X]$ into a product of powers of distinct irreducible monic polynomials, and let $f_i = \deg(\overline{F}_i)$. For each i with $1 \leq i \leq g$, select a monic polynomial $F_i(X)$ in $\mathbb{Z}[X]$ whose reduction modulo p is $\overline{F}_i(X)$, and let P_i be the ideal in R defined by

$$P_i = pR + F_i(\xi)R.$$

Then the P_i's are distinct prime ideals of R with $\dim_{\mathbb{F}_p}(R/P_i) = f_i$, and the unique factorization of $(p)R$ into prime ideals is

$$(p)R = P_1^{e_1} \cdots P_g^{e_g}.$$

---

[8] More generally "relative discriminants," which we have not defined, play a role in the splitting of prime ideals in passing from a general number field to a finite extension. The cited Theorem 9.60 applies in this more general situation as well. This more general topic will be discussed further in Problems 5–9 at the end of this chapter and very briefly in Chapter VI.

[9] In fact, it can be shown that every common index divisor is less than $[\mathbb{K} : \mathbb{Q}]$.

REMARKS. The additive group $\mathbb{Z}(\Gamma(\xi))$ generated by the powers of ξ through ξ^{n-1} is a ring, since ξ^n is an integral combination of the lower powers of ξ, and this ring has index $J(\xi)$ as a subring of R. We divide the proof into two parts. The first part will give a complete proof in the special case that the subring $\mathbb{Z}(\Gamma(\xi))$ is all of R, but we shall retain notation that distinguishes the subring from the whole ring in order to see how much of the proof works for the general case. After the first part we pause for a lemma that will be used to tie results for the subring to results for all of R, and then we return to apply the lemma and complete the proof of Theorem 5.6.

FIRST PART OF PROOF. Let P_i' be the ideal $p\mathbb{Z}[X] + F_i(X)\mathbb{Z}[X]$ in $\mathbb{Z}[X]$. The passage from $\mathbb{Z}[X]$ to the quotient $\mathbb{Z}[X]/P_i'$ can be achieved in two steps, first using the substitution homomorphism carrying $\mathbb{Z}$ to $\mathbb{F}_p$ and X to itself and then taking the quotient by the principal ideal $(\overline{F}_i(X))$. Since $\overline{F}_i(X)$ is irreducible in $\mathbb{F}_p[X]$, the quotient is a field and P_i' has to be prime. The number of elements in $\mathbb{Z}[X]/P_i'$ is p^{f_i} because $\deg(\overline{F}_i(X)) = f_i$. The ideals P_i' are distinct because the polynomials $\overline{F}_i(X)$ are distinct.

Meanwhile, the substitution homomorphism of $\mathbb{Z}[X]$ leaving $\mathbb{Z}$ fixed and carrying X to ξ is a ring homomorphism of $\mathbb{Z}[X]$ onto $\mathbb{Z}(\Gamma(\xi))$. Let P_i'' be the image of P_i' under this homomorphism, i.e., let $P_i'' = p\mathbb{Z}(\Gamma(\xi)) + F_i(\xi)\mathbb{Z}(\Gamma(\xi))$. This is an ideal. The composite ring homomorphism of $\mathbb{Z}[X]$ onto $\mathbb{Z}(\Gamma(\xi))/P_i''$ factors through to a ring homomorphism of $\mathbb{Z}[X]/P_i'$ onto $\mathbb{Z}(\Gamma(\xi))/P_i''$. Since the domain is a field and the identity maps to the identity, the homomorphism is one-one and the image is a field. Thus P_i'' is a prime ideal, the order of $\mathbb{Z}(\Gamma(\xi))/P_i''$ is p^{f_i}, and and P_i' is the complete inverse image of P_i''. Since the ideals P_i' can be recovered from the P_i'' and since the P_i' are distinct, the P_i'' are distinct.

The next step is to compare the ideals $\prod_{i=1}^{g} P_i^{e_i}$ and $(p)R$. We shall use the fact that the polynomial $\prod_{i=1}^{g} F_i(X)^{e_i} - F(X)$ in $\mathbb{Z}[X]$ has coefficients divisible by p and therefore lies in $p\mathbb{Z}[X]$. The computation

$$\prod_{i=1}^{g} P_i^{e_i} = \prod_{i=1}^{g} (pR + F_i(\xi)R)^{e_i}$$

$$\subseteq pR + \prod_{i=1}^{g} F_i(\xi)^{e_i} R$$

$$\subseteq pR + \Big(\prod_{i=1}^{g} F_i^{e_i} - F\Big)(\xi) \quad \text{since } F(\xi) = 0$$

$$\subseteq pR + p\mathbb{Z}(\Gamma(\xi)) \quad \text{since } \prod_{i=1}^{g} F_i(X)^{e_i} - F(X) \text{ lies in } p\mathbb{Z}[X]$$

$$= pR$$

shows that $\prod_{i=1}^{g} P_i^{e_i} \subseteq (p)R$. If we can show that $N\big(\prod_{i=1}^{g} P_i^{e_i}\big) = N((p)R)$, then Proposition 5.4b will allow us to conclude that $\prod_{i=1}^{g} P_i^{e_i} = (p)R$.

At this point let us specialize to the case that $\mathbb{Z}(\Gamma(\xi)) = R$ and see how to complete the proof. Under this assumption the definitions of P_i and P_i'' exactly match. What we have shown about the P_i'' thus says that the P_i are distinct prime ideals in R with $|R/P_i| = p^{f_i}$, hence with $\dim_{\mathbb{F}_p}(R/P_i) = f_i$. Use of Proposition 5.4 and the fact that $|\mathbb{Z}(\Gamma(\xi))/P_i''| = p^{f_i}$ gives $N\left(\prod_{i=1}^g P_i^{e_i}\right) = \prod_{i=1}^g N(P_i)^{e_i} = \prod_{i=1}^g p^{e_i f_i} = p^{\sum_{i=1}^g e_i f_i} = p^n$, the last equality holding because $\deg \overline{F}(X) = \sum_{i=1}^g e_i \deg \overline{F}_i(X)$. Since p^n equals $N((p)R)$, the desired equality of norms has been proved. This completes the proof of the theorem when $\mathbb{Z}(\Gamma(\xi)) = R$. □

We interrupt the general proof for the promised lemma. When we apply the lemma to finish the proof of Theorem 5.6, we shall take $A = \mathbb{Z}(\Gamma(\xi))$, $J = J(\xi)$, and $m = p$. The hypotheses of Theorem 5.6 show that the condition $\mathrm{GCD}(p, J(\xi)) = 1$ is satisfied.

Lemma 5.7. Suppose that A is an additive subgroup of finite index J in R and that $m \geq 1$ is an integer relatively prime to J. Then for each $r \in R$, there exists $a \in A$ with $r - a$ in mR.

PROOF. Let $\{u_1, \ldots, u_n\}$ be a $\mathbb{Z}$ basis of R, and let $\{v_1, \ldots, v_n\}$ be a $\mathbb{Z}$ basis of A. We can write $v_j = \sum_{i=1}^n c_{ij} u_i$ for an integer matrix $[c_{ij}]$ with $|\det[c_{ij}]| = J$. Let $r = \sum_{i=1}^n b_i u_i$ be given, and let the unknown $a \in A$ be expanded as $a = \sum_{i=1}^n a_j v_j$. Then $a = \sum_{i,j} a_j c_{ij} u_i$, and we are to arrange that the element

$$r - a = \sum_{i=1}^n \left(b_i - \sum_{j=1}^n c_{ij} a_j\right) u_i$$

is in mR. Thus we are to arrange that each coefficient of a u_i is divisible by m. Since $|\det[c_{ij}]| = J$ is relatively prime to m, the system of linear equations

$$\sum_{j=1}^n c_{ij} a_j \equiv b_i \bmod m$$

with unknowns $a_1, \ldots, a_n$ has a nonsingular coefficient matrix modulo m and therefore has a solution. □

SECOND PART OF PROOF OF THEOREM 5.6. The ring homomorphism of $\mathbb{Z}(\Gamma(\xi))$ into $R/(pR + F_i(\xi)R)$ given by the composition of the inclusion followed by the quotient map descends to a ring homomorphism

$$\mathbb{Z}(\Gamma(\xi))\big/(p\mathbb{Z}(\Gamma(\xi)) + F_i(\xi)\mathbb{Z}(\Gamma(\xi))) \longrightarrow R/(pR + F_i(\xi)R). \qquad (*)$$

To see that $(*)$ is onto, let $r \in R$ be given. Take $A = pR$ in Lemma 5.7. Choose $z \in \mathbb{Z}(\Gamma(\xi))$ by the lemma in such a way that $z - r$ is in pR. Under the mapping

($*$), the coset of z goes to $r + (z-r) + pR + F_i(\xi)R = r + pR + F_i(\xi)R$, which is the coset of r. Hence ($*$) is onto.

To see that ($*$) is one-one, suppose that z maps to the 0 coset in the image. Then $z = pr_1 + F_i(\xi)r_2$ with r_1 and r_2 in R. Lemma 5.7 produces z_2 in $\mathbb{Z}(\Gamma(\xi))$ with $r_2 - z_2$ in pR. Hence the decomposition $z = pr_1 + F_i(\xi)(r_2 - z_2) + F_i(\xi)z_2$ exhibits z as in $pR + F_i(\xi)\mathbb{Z}(\Gamma(\xi))$. The product $F_i(\xi)\mathbb{Z}(\Gamma(\xi))$ is in $\mathbb{Z}(\Gamma(\xi))$, since $\mathbb{Z}(\Gamma(\xi))$ is a ring, and ($*$) will be one-one if we show that $pR \cap \mathbb{Z}(\Gamma(\xi)) \subseteq p\mathbb{Z}(\Gamma(\xi))$. Let $\{u_i\}$ be a $\mathbb{Z}$ basis of R, let $\{v_j\}$ be a $\mathbb{Z}$ basis of $\mathbb{Z}(\Gamma(\xi))$, and write $v_j = \sum_i c_{ij}u_i$ for integers c_{ij}. If z' is in $pR \cap \mathbb{Z}(\Gamma(\xi))$, let us write $z' = \sum_j a_j v_j$. Substitution gives $z' = \sum_i \left(\sum_j a_j c_{ij}\right)u_i$. Since z' is in pR, we see that $\sum_j c_{ij}a_j \equiv 0 \bmod p$ for all i. The determinant of $[c_{ij}]$ is the index $J(\xi)$, up to sign, and this by assumption is not divisible by p. Therefore $a_j \equiv 0 \bmod p$ for all j, and it follows that z' is in $p\mathbb{Z}(\Gamma(\xi))$. Hence ($*$) is one-one.

We have thus proved that ($*$) is a ring isomorphism, i.e., that $\mathbb{Z}(\Gamma(\xi))/P_i'' \cong R/P_i$ for all i. The left side is a field, and hence P_i is a prime ideal. From the isomorphism we obtain $N(P_i) = |\mathbb{Z}(\Gamma(\xi))/P_i''| = p^{f_i}$. The computation $N\left(\prod_{i=1}^g P_i^{e_i}\right) = \prod_{i=1}^g N(P_i)^{e_i} = \prod_{i=1}^g p^{e_i f_i} = p^{\sum_{i=1}^g e_i f_i} = p^n$ in the last paragraph of the first part of the proof is now fully justified, and we can therefore conclude as in the special case that $\prod_{i=1}^g P_i^{e_i} = (p)R$.

Finally we have to prove that the ideals P_i are distinct. If indices $i \neq j$ are given, we know that $P_i'' \neq P_j''$. Choose z in P_i'' but not P_j''. Then z is in P_i because $P_i'' \subseteq P_i$, and z is not in P_j because the proof above that ($*$) is one-one showed that $\mathbb{Z}(\Gamma(\xi)) \cap P_j \subseteq P_j''$. This completes the proof of Theorem 5.6. $\square$

PROOF OF THEOREM 5.5 WHEN p IS NOT A COMMON INDEX DIVISOR. If p is not a common index divisor, we can choose a primitive ξ for $\mathbb{K}/\mathbb{Q}$ such that ξ is in R and p does not divide $J(\xi) = |R/\mathbb{Z}(\Gamma(\xi))|$. Let $F(X)$ be the field polynomial of ξ over $\mathbb{Q}$. Since $D(\xi) = J(\xi)^2 D_\mathbb{K}$, p divides $D_\mathbb{K}$ if and only if p divides $D(\xi)$. Thus p divides $D_\mathbb{K}$ if and only if p divides the discriminant of $F(X)$. This happens if and only if the discriminant of $\overline{F}(X)$ is $\equiv 0 \bmod p$, if and only if $\overline{F}(X)$ has a root of multiplicity > 1 in an algebraic closure of $\mathbb{F}_p$, if and only if the factorization over $\mathbb{F}_p$ of $\overline{F}(X)$ as a product of powers of distinct irreducible monic polynomials has some factor with exponent > 1. Applying Theorem 5.6, we see that this last condition is satisfied if and only if the unique factorization of the ideal $(p)R$ in R as $\prod_{i=1}^g P_i^{e_i}$ has some $e_i > 1$. $\square$

As was mentioned earlier in this section, the difficulty in proving Theorem 5.5 in complete generality is that we lack sufficient tools for addressing questions by localization. The different prime numbers are interacting in some fashion, and the above proofs were unable to separate them. The usual technique of localization in our situation[10] suggests enlarging one or the other of the rings $\mathbb{Z}$ and R by adjoining inverses for all elements not in some prime ideal of interest. Then we piece together the results. If the localizing is done with respect to a prime ideal (p) of $\mathbb{Z}$, then $\mathbb{Z}$ gets replaced by the subring $S^{-1}\mathbb{Z}$ of all members of $\mathbb{Q}$ with no factors of p in the denominators, and R gets replaced by $S^{-1}R$. One advantage of this procedure is that $S^{-1}R$ is a principal ideal domain, whereas R is typically not such a domain.

Localization in that formulation does not by itself reveal a clear path to a proof of Theorem 5.5. Two additional ideas enter the argument to make a path seem natural; Dedekind succeeded without the second of them, and historically it is only with hindsight that one sees the benefit of the second idea. The first idea is to use a more fundamental object than the discriminant of $\mathbb{K}$, called the "relative different" of $\mathbb{K}/\mathbb{Q}$; this makes it possible to aim for a more precise description of the ramification indices when they are not equal to 1. The second idea is due to K. Hensel and involves forming a kind of completion of the localized rings; the ring $\mathbb{Z}$ gets replaced by the ring $\mathbb{Z}_p$ of "p-adic integers," and the field $\mathbb{Q}$ gets replaced by the field $\mathbb{Q}_p$ of "p-adic numbers." We return to these ideas in Chapter VI.

4. Cubic Number Fields as Examples

In treating examples of cubic fields, it will be convenient to have one further tool available for computing discriminants. Let $\mathbb{K}$ be a number field, let ξ be a primitive element of $\mathbb{K}/\mathbb{Q}$, and let $F(X)$ be its field polynomial over $\mathbb{Q}$. Let $\xi_i = \sigma_i(\xi)$ be the conjugates of ξ, and assume that $\xi_1 = \xi$. The conjugates are the roots of $F(X)$ in $\mathbb{C}$, and hence

$$F(X) = \prod_{i=1}^{n}(X - \xi_i).$$

The derivative is $F'(X) = \sum_{i=1}^{n}\prod_{j \neq i}(X - \xi_j)$, and therefore

$$F'(\xi) = \prod_{j=2}^{n}(\xi - \xi_j).$$

Observe that the form of the left side shows that this element lies in $\mathbb{K}$, and it lies in R if ξ lies in R. The **different** $\mathcal{D}(\xi)$ of the element ξ is defined to be this element of $\mathbb{K}$, namely[11]

---

[10]Localization was introduced in Section VIII.10 of *Basic Algebra*.

[11]The different of an element is related to the notion of relative different mentioned at the end of Section 3, but the nature of that relationship will not concern us at this time.

$$\mathcal{D}(\xi) = F'(\xi) = \prod_{j=2}^{n}(\xi - \xi_j).$$

Since $F'(X)$ has coefficients in $\mathbb{Q}$, the conjugates $\sigma_i(F'(\xi))$ of $F'(\xi)$ are the elements $F'(\sigma_i(\xi)) = F'(\xi_i)$ for $1 \leq i \leq n$. The formula for $F'(X)$ shows that $F'(\xi_i) = \prod_{j \neq i}(\xi_i - \xi_j)$. Therefore the norm of $\mathcal{D}(\xi)$ is

$$N_{\mathbb{K}/\mathbb{Q}}(\mathcal{D}(\xi)) = N_{\mathbb{K}/\mathbb{Q}}(F'(\xi)) = \prod_{i=1}^{n} F'(\xi_i) = \prod_{i=1}^{n}\prod_{j \neq i}(\xi_i - \xi_j)$$
$$= (-1)^{n(n-1)/2}\prod_{i<j}(\xi_i - \xi_j)^2 = (-1)^{n(n-1)/2}D(\xi).$$

In other words, the norm of the different of ξ is, up to sign, equal to the discriminant of $\Gamma(\xi)$, which in turn equals the discriminant of the field polynomial of the primitive element ξ. The definitions of $\mathcal{D}(\xi)$ and $D(\xi)$ and the formula connecting them make sense if ξ is allowed to be any element of $\mathbb{K}$, primitive or not. Both $\mathcal{D}(\xi)$ and $D(\xi)$ have the property of being nonzero if and only if ξ is primitive.

EXAMPLE. For the field $\mathbb{K} = \mathbb{Q}(\sqrt[3]{2})$, the different of $\xi = \sqrt[3]{2}$ is $3X^2|_{X=\sqrt[3]{2}} = 3\sqrt[3]{4}$, and the discriminant of $X^3 - 2$, up to the sign $(-1)^{3\cdot 2/2}$, is the norm of this, i.e.,

$$D(\sqrt[3]{2}) = -(3\sqrt[3]{4})(3\sqrt[3]{4}\,\omega)(3\sqrt[3]{4}\,\omega^2), \quad \text{where } \omega = e^{2\pi i/3},$$
$$= -3^3 2^2.$$

Alternatively, the norm can be computed from a field polynomial. Specifically the norm of $3\sqrt[3]{4}$ is the determinant of left multiplication by this element when considered as a $\mathbb{Q}$ linear mapping of $\mathbb{K}$ into itself.

We saw already in Example 2 of Section 2 that $D(\sqrt[3]{2}) = -3^3 2^2$, but the earlier method of computation was longer. At the end of Section 2, we saw in addition that $\{1, \sqrt[3]{2}, \sqrt[3]{4}\}$ is a $\mathbb{Z}$ basis of the ring of algebraic integers in the field $\mathbb{K} = \mathbb{Q}(\sqrt[3]{2})$. The use of differents does not simplify the proof of this latter fact.

In this section we consider further examples of cubic extensions of $\mathbb{Q}$. The first such fields that we study are the **pure cubic** extensions $\mathbb{K} = \mathbb{Q}(\sqrt[3]{m})$, where m is any cube-free positive integer > 1. Already with these fields $\mathbb{K}$, we shall see that $D_{\mathbb{K}}$ is not necessarily equal to $\mathcal{D}(\xi)$ for some algebraic integer ξ. However, all these fields have no common index divisors. Then we examine Dedekind's example of a cubic number field for which 2 is a common index divisor.

4. Cubic Number Fields as Examples

The correspondence of cube-free integers $m > 1$ to fields $\mathbb{Q}(\sqrt[3]{m})$ is many-to-one: if m is given and p is a prime dividing m, let $m' = m/p$ if p^2 divides m and $m' = mp$ if p^2 does not divide m; then $\mathbb{Q}(\sqrt[3]{m}) = \mathbb{Q}(\sqrt[3]{m'})$. In analyzing $\mathbb{Q}(\sqrt[3]{m})$, it will be convenient to normalize matters so as to resolve this ambiguity. We can write m uniquely as a product $m = ab^2$ for positive square-free integers a and b; these have $\text{GCD}(a, b) = 1$, b^2 is the largest square dividing m, and a is given by $a = m/b^2$. Then m and $m' = a^2 b$ lead to the same field.

Proposition 5.8. For a cube-free integer $m > 1$, let $\mathbb{K} = \mathbb{Q}(\sqrt[3]{m})$, and let R be the ring of algebraic integers in $\mathbb{K}$. Write $m = ab^2$ for positive square-free integers a and b with $\text{GCD}(a, b) = 1$, and define two members of R to be the real cube roots $\theta_1 = \sqrt[3]{ab^2}$ and $\theta_2 = \sqrt[3]{a^2 b}$. Then a $\mathbb{Z}$ basis of R consists of

(a) $\{1, \theta_1, \theta_2\}$ if $a \not\equiv \pm b \bmod 9$, i.e., if m is of **Type I**,
(b) $\{\frac{1}{3}(1 \pm \theta_1 \pm \theta_2), \theta_1, \theta_2\}$ for exactly one choice of the pair of signs if $a \equiv \pm b \bmod 9$, i.e., if m is of **Type II**.

In the respective cases the field discriminant is given by

$$D_\mathbb{K} = \begin{cases} -27 a^2 b^2 & \text{if } m \text{ is of Type I,} \\ -3 a^2 b^2 & \text{if } m \text{ is of Type II.} \end{cases}$$

REMARKS. More precisely in Type II, the congruence $a \equiv \pm b \bmod 9$ implies that a and b are prime to 3. Choose signs $s = \pm 1$ and $t = \pm 1$ such that $sa \equiv 1 \bmod 3$ and $tb \equiv 1 \bmod 3$. Then the first member of the $\mathbb{Z}$ basis is to be $\frac{1}{3}(1 + s\theta_1 + t\theta_2)$. The smallest m leading to Type I is $m = 2$, and this case was examined in Example 2 in Section 2. The smallest m leading to Type II is $m = 10$, and then the first member of the asserted $\mathbb{Z}$ basis of R is $\frac{1}{3}(1 + \sqrt[3]{10} + \sqrt[3]{100})$.

PROOF. Let $\omega = e^{2\pi i/3}$. The conjugates of θ_1 can be taken to be $\sigma_1(\theta_1) = \theta_1$, $\sigma_2(\theta_1) = \omega \theta_1$, and $\sigma_3(\theta_1) = \omega^2 \theta_1$. Since $\theta_1^2 = b \theta_2$, we have $\sigma_i(\theta_2) = b^{-1} \sigma_i(\theta_1)^2$, and therefore $\sigma_1(\theta_2) = \theta_2$, $\sigma_2(\theta_2) = \omega^2 \theta_2$, and $\sigma_3(\theta_2) = \omega \theta_2$. In view of the formula before Proposition 5.2, $D((1, \theta_1, \theta_2))$ is the square of

$$\det \begin{pmatrix} 1 & 1 & 1 \\ \theta_1 & \omega \theta_1 & \omega^2 \theta_1 \\ \theta_2 & \omega^2 \theta_2 & \omega \theta_2 \end{pmatrix},$$

and we calculate that $D((1, \theta_1, \theta_2)) = -27 a^2 b^2$.

Let us apply Proposition 5.2 to the triple $\{1, \theta_1, \theta_2\}$ of members of R. For each prime p dividing $27 a^2 b^2$, we are to check whether certain elements are integral. First suppose that p divides a but $p \neq 3$. It is enough to check the elements $p^{-1}(a_0 + \theta_1)$ or $p^{-1}(a_0 + a_1 \theta_1 + \theta_2)$ for integrality when a_0 and a_1 are integers from 0 to $p - 1$. Form the extension $\mathbb{L} = \mathbb{K}(\sqrt[3]{p}) = \mathbb{Q}(\sqrt[3]{m}, \sqrt[3]{p})$ of $\mathbb{K}$, and let T be its ring of algebraic integers. The degree $[\mathbb{L}:\mathbb{Q}]$ equals 9 if $\mathbb{L} \neq \mathbb{K}$ and equals 3 if $\mathbb{L} = \mathbb{K}$. If $p^{-1}(a_0 + \theta_1)$ is integral, then $a_0 + p^{1/3}((a/p)b^2)^{1/3} = pr$ with $r \in R$, and hence $a_0 = p^{1/3}c$ with $c \in T$. Applying $N_{\mathbb{L}/\mathbb{Q}}$ to both sides, we obtain $a_0^9 = p^3 N_{\mathbb{L}/\mathbb{Q}}(c)$ if $\mathbb{L} \neq \mathbb{K}$, and we obtain $a_0^3 = p N_{\mathbb{K}/\mathbb{Q}}(c)$ if $\mathbb{L} = \mathbb{K}$. In either case, p divides a_0, and $a_0 = 0$. So $p^{-1}\theta_1$ is integral, in contradiction to the facts that the field polynomial for $\mathbb{K}$ of $p^{-1}\theta_1$ is $X^3 - p^{-3}ab^2$ and that ab^2 contains p as a factor only once. We conclude that $p^{-1}(a_0 + \theta_1)$ is not integral.

Similarly if the element $p^{-1}(a_0 + a_1\theta_1 + \theta_2)$ is integral, then we see that $a_0 + a_1 p^{1/3}((a/p)b^2)^{1/3} + p^{2/3}((a/p)^2 b)^{1/3} = pr$ with $r \in R$. So $a_0 = p^{1/3}c$ with $c \in T$, and the same argument as above shows that $a_0 = 0$. Hence $a_1((a/p)b^2)^{1/3} + p^{1/3}((a/p)^2 b)^{1/3} = p^{2/3}r$, and $a_1((a/p)b^2)^{1/3} = p^{1/3}c'$ with $c' \in T$. Taking the norm gives $a_1^9((a/p)b^2)^3 = p^3 N_{\mathbb{L}/\mathbb{Q}}(c')$ if $\mathbb{L} \neq \mathbb{K}$ and $a_1^3(a/p)b^2 = p N_{\mathbb{K}/\mathbb{Q}}(c')$ if $\mathbb{L} = \mathbb{K}$. Since a/p and b are prime to p, we conclude that p divides a_1 in both cases. Therefore $a_1 = 0$, and $p^{-1}\theta_2$ is integral. The field polynomial for $\mathbb{K}$ of $p^{-1}\theta_2$ is $X^3 - p^{-3}a^2 b$, and $a^2 b$ contains p as a factor only twice. We conclude that $p^{-1}(a_0 + a_1\theta_1 + \theta_2)$ is not integral.

This disposes of the prime divisors of a other than $p = 3$, and we handle the prime divisors of b other than $p = 3$ in the same way, except that we start from the ordered triple $(1, \theta_2, \theta_1)$ and therefore need check only $p^{-1}(a_0 + \theta_2)$ and $p^{-1}(a_0 + a_1\theta_2 + \theta_1)$.

Now let us apply Proposition 5.2 to the ordered triple $(1, \theta_1, \theta_2)$ for the prime $p = 3$, except that we allow coefficients 0 and ± 1 instead of $0, 1, 2$. We check integrality for the elements $\frac{1}{3}(1 \pm \theta_1)$, $\frac{1}{3}(1 \pm \theta_2)$, $\frac{1}{3}(\theta_1 \pm \theta_2)$, and $\frac{1}{3}(1 \pm \theta_1 \pm \theta_2)$ by checking whether the coefficients of their field polynomials are in $\mathbb{Z}$. For the first two, let φ be $\pm\theta_1$ or $\pm\theta_2$. The coefficient of the first-degree term in the field polynomial of $\frac{1}{3}(1 + \varphi)$ is $\frac{1}{9}$ times

$$(1+\varphi)(1+\omega\varphi) + (1+\varphi)(1+\omega^2\varphi) + (1+\omega\varphi)(1+\omega^2\varphi)$$
$$= (1+\varphi)(2+\omega\varphi+\omega^2\varphi) + (1+\omega\varphi)(1+\omega^2\varphi)$$
$$= (1+\varphi)(2-\varphi) + (1-\varphi+\varphi^2) = 2+\varphi-\varphi^2 + 1-\varphi+\varphi^2 = 3,$$

hence is $\frac{1}{3}$. This is not an integer, and thus $\frac{1}{3}(1+\varphi)$ is not in R. If $\varphi = \pm\theta_1$ and $\psi = \pm\theta_2$, then the corresponding computation for $\varphi + \psi$ is

$$(\varphi+\psi)(\omega\varphi+\omega^2\psi) + (\varphi+\psi)(\omega^2\varphi+\omega\psi) + (\omega\varphi+\omega^2\psi)(\omega^2\varphi+\omega\psi)$$
$$= -(\varphi+\psi)(\varphi+\psi) + (\varphi^2-\varphi\psi+\psi^2)$$
$$= -3\varphi\psi = -3ab(\operatorname{sgn}\varphi)(\operatorname{sgn}\psi), \qquad (*)$$

and $\frac{1}{9}$ of this is an integer only if 3 divides ab. In this case our hypotheses show that 9 does not divide ab. The constant term in the field polynomial of $\frac{1}{3}(\varphi + \psi)$ is $-\frac{1}{27}$ times

$$(\varphi + \psi)(\omega\varphi + \omega^2\psi)(\omega^2\varphi + \omega\psi) = \varphi^3 + \psi^3$$
$$= (\operatorname{sgn}\varphi)ab^2 + (\operatorname{sgn}\psi)a^2b$$
$$= ab(b\operatorname{sgn}\varphi + a\operatorname{sgn}\psi). \qquad (**)$$

When 3 divides ab exactly once, 3 divides $(**)$ exactly once, and hence $-\frac{1}{27}$ of $(**)$ is not an integer. Thus $\frac{1}{3}(\varphi + \psi)$ is not in R.

It remains to check $\frac{1}{3}(1+\varphi+\psi)$ with $\varphi = \pm\theta_1$ and $\psi = \pm\theta_2$. The coefficient of the second-degree term in the field polynomial of $\frac{1}{3}(1+\varphi+\psi)$ is equal to $-\frac{1}{3}\operatorname{Tr}(1+\varphi+\psi) = -1$ and is an integer; thus it imposes no restrictions. The first-degree term of the field polynomial is $\frac{1}{9}$ of

$$(1+\varphi+\psi)(1+\omega\varphi+\omega^2\psi) + (1+\varphi+\psi)(1+\omega^2\varphi+\omega\psi)$$
$$+ (1+\omega\varphi+\omega^2\psi)(1+\omega^2\varphi+\omega\psi)$$
$$= (1+\varphi+\psi)(2-\varphi-\psi) + (1-\varphi-\psi+\varphi^2-\varphi\psi+\psi^2)$$
$$= 3 - 3\varphi\psi = 3(1 - ab(\operatorname{sgn}\varphi)(\operatorname{sgn}\psi)), \qquad (\dagger)$$

and $\frac{1}{9}$ of $(\dagger)$ is an integer if and only if $ab \equiv (\operatorname{sgn}\varphi)(\operatorname{sgn}\psi)$ mod 3. In particular, the proof is now complete unless $ab \equiv (\operatorname{sgn}\varphi)(\operatorname{sgn}\psi)$ mod 3. Thus we may assume from now on that neither a nor b is divisible by 3.

The constant term of the field polynomial of $\frac{1}{3}(1+\varphi+\psi)$ is $-\frac{1}{27}$ times

$$(1+\varphi+\psi)(1+\omega\varphi+\omega^2\psi)(1+\omega^2\varphi+\omega\psi)$$
$$= 1 + \operatorname{Tr}_{\mathbb{K}/\mathbb{Q}}(\varphi+\psi) + (*) + (**)$$
$$= 1 + 0 - 3ab(\operatorname{sgn}\varphi)(\operatorname{sgn}\psi) + ab(b\operatorname{sgn}\varphi + a\operatorname{sgn}\psi).$$

Put $\alpha = a\operatorname{sgn}\varphi$ and $\beta = b\operatorname{sgn}\psi$, so that $1 - 3\alpha\beta + \alpha\beta(\alpha+\beta)$ is to be divisible by 27. Since neither β nor α is divisible by 3, we can define l mod 27 by the congruence $\beta = l\alpha$ mod 27. Substituting shows that $1 - 3l\alpha^2 + l\alpha^2(\alpha+l\alpha) \equiv 0$ mod 27, hence that $l(l+1)\alpha^3 \equiv 3l\alpha^2 - 1$ mod 27, which we can rewrite as

$$\alpha^3 l^2 + (\alpha^3 - 3\alpha^2)l + 1 \equiv 0 \text{ mod } 27.$$

Completing the square in l allows us to write this congruence as

$$(l + \tfrac{1}{2}(1 - 3\alpha^{-1}))^2 \equiv \tfrac{1}{4}(1 - 3\alpha^{-1})^2 - \alpha^{-3} \text{ mod } 27.$$

Factoring the right side, we obtain

$$(l + \tfrac{1}{2}(1 - 3\alpha^{-1}))^2 \equiv \tfrac{1}{4}\alpha^{-4}[\alpha(\alpha-1)^2(\alpha-4)] \text{ mod } 27. \qquad (\dagger\dagger)$$

If $\alpha \equiv 1 \bmod 3$, the expression in square brackets on the right side is $\equiv 0 \bmod 27$, and 0 is the square of 0 and ± 9. If $\alpha \equiv 2 \bmod 3$, then the expression in square brackets is a square if and only if $\alpha(\alpha - 4) \equiv c^2 \bmod 27$. Considering the congruence only modulo 3 gives $2(-2) \equiv c^2 \bmod 3$ and therefore $c^2 \equiv 2 \bmod 3$, which has no solutions. Thus $\alpha \equiv 2 \bmod 3$ leads to no solutions of (††). We can summarize by saying that the solutions of (††) are given by $\alpha \equiv 1 \bmod 3$ and

$$l + \tfrac{1}{2}(1 - 3\alpha^{-1}) \equiv 0 \bmod 9.$$

One checks that the values $\alpha \equiv 1, 4, 7 \bmod 9$ all lead to $l = 1$.

Let us summarize. Let s and t be signs $\pm$. Then $\tfrac{1}{3}(1 + s\theta_1 + t\theta_2)$ is integral if and only if both of the following conditions are satisfied:

(i) $sa \equiv tb \equiv 1 \bmod 3$,
(ii) $sa \equiv tb \bmod 9$.

When these conditions are satisfied, we are in Type II; otherwise we are in Type I. This completes the proof. $\square$

In the setting of Type I in Proposition 5.8, let us form the discriminants of $\Gamma(\theta_1) = (1, \theta_1, \theta_1^2)$ and $\Gamma(\theta_2) = (1, \theta_2, \theta_2^2)$. Using the method of computation at the beginning of this section, we see that the differents in the two cases are $3\theta_1^2$ and $3\theta_2^2$. Therefore the discriminant of $\Gamma(\theta_1)$ is $D(\theta_1) = -N_{\mathbb{K}/\mathbb{Q}}(3\theta_1^2) = -3^3(\theta_1^2)^3 = -3^3(ab^2)^2 = -3^3 a^2 b^4$, and the discriminant of $\Gamma(\theta_2)$ similarly is $D(\theta_2) = -3^3 a^4 b^2$. The absolute value of the greatest common divisor of these two expressions is $3^3 a^2 b^2 = |D_{\mathbb{K}}|$, and therefore there are never any common index divisors in Type I.

On the other hand, there exist situations in Type I in which no primitive element ξ of $\mathbb{Q}(\sqrt[3]{m})$ lying in R has $\Gamma(\xi)$ as a $\mathbb{Z}$ basis. To prove this fact, we make use of the following proposition.

Proposition 5.9. For a pure cubic extension $\mathbb{K} = \mathbb{Q}(\sqrt[3]{ab^2})$ of Type I, an element $\xi = x + y\theta_1 + z\theta_2$ with $\mathbb{Z}$ coefficients has $D(\xi) = D_{\mathbb{K}}$ if and only if $y^3 b - z^3 a = \pm 1$.

PROOF. The matrix whose determinant is $D(\Gamma(\xi))$ is given by

$$M = \begin{pmatrix} 3 & \text{Tr}(\xi) & \text{Tr}(\xi^2) \\ \text{Tr}(\xi) & \text{Tr}(\xi^2) & \text{Tr}(\xi^3) \\ \text{Tr}(\xi^2) & \text{Tr}(\xi^3) & \text{Tr}(\xi^4) \end{pmatrix},$$

where Tr is short for $\text{Tr}_{\mathbb{K}/\mathbb{Q}}$. The element $\theta_1^i \theta_2^j$ has conjugates $\theta_1^i \theta_2^j$, $\omega^{i+2j} \theta_1^i \theta_2^j$, and $\omega^{2i+j} \theta_1^i \theta_2^j$, where $\omega = e^{2\pi i/3}$. Thus

$$\text{Tr}(\theta_1^i \theta_2^j) = (1 + \omega^{i+2j} + \omega^{2i+j})\theta_1^i \theta_2^j = (1 + \omega^{i+2j} + \omega^{2(i+2j)})\theta_1^i \theta_2^j.$$

This is 0 if $i + 2j$ is not divisible by 3 and is $3\theta_1^i \theta_2^j$ otherwise. We compute the trace of each power of ξ by applying the formula

$$\text{Tr}(\xi^l) = \sum_{k=0}^{n} \binom{l}{k} x^{l-k} \text{Tr}((y\theta_1 + z\theta_2)^k),$$

which comes from treating ξ as a binomial. The traces of the powers of $y\theta_1 + z\theta_2$ work out to be

$$\tfrac{1}{3} \text{Tr}(y\theta_1 + z\theta_2) = 0,$$
$$\tfrac{1}{3} \text{Tr}((y\theta_1 + z\theta_2)^2) = 2yz\theta_1\theta_2 = ab(2yz),$$
$$\tfrac{1}{3} \text{Tr}((y\theta_1 + z\theta_2)^3) = ab(y^3 b + z^3 a),$$
$$\tfrac{1}{3} \text{Tr}((y\theta_1 + z\theta_2)^4) = (ab)^2 6 y^2 z^2.$$

Substituting, we find the following formulas for the trace of each power of ξ:

$$\tfrac{1}{3} \text{Tr}(\xi) = x,$$
$$\tfrac{1}{3} \text{Tr}(\xi^2) = x^2 + 2(ab)yz,$$
$$\tfrac{1}{3} \text{Tr}(\xi^3) = x^3 + 3x(ab)2yz + (ab)(y^3 b + z^3 a),$$
$$\tfrac{1}{3} \text{Tr}(\xi^4) = x^4 + 6x^2(ab)2yz + 4x(ab)(y^3 b + z^3 a) + (ab)^2 6 y^2 z^2.$$

The matrix M is therefore of the form

$$\tfrac{1}{3} M = \begin{pmatrix} 1 & x & x^2 + A \\ x & x^2 + A & x^3 + B \\ x^2 + A & x^3 + B & x^4 + C \end{pmatrix},$$

where

$$A = 2(ab)yz,$$
$$B = 3x(ab)2yz + (ab)(y^3 b + z^3 a),$$
$$C = 6x^2(ab)2yz + 4x(ab)(y^3 b + z^3 a) + (ab)^2 6 y^2 z^2.$$

Expansion of $\det \tfrac{1}{3} M$ results in an expression that simplifies to

$$\det \tfrac{1}{3} M = AC + 2x AB - 3x^2 A^2 - A^3 - B^2.$$

Thus we have only to substitute. The resulting expression simplifies greatly, and we obtain $\det \tfrac{1}{3} M = -(ab)^2 (y^3 b - z^3 a)^2$. Consequently

$$D(\xi) = -3^3 (ab)^2 (y^3 b - z^3 a)^2.$$

Since Proposition 5.8 has shown that $D_\mathbb{K} = -3^3(ab)^2$, the result follows. $\square$

Thus in order to give an example of an m for which no ξ has $D(\xi) = D_{\mathbb{K}}$, we have only to select a and b for which the Diophantine equation $y^3 b - z^3 a = 1$ in y, z has no solution. Choose $a = 7$ and $b = 5$, so that $m = ab^2 = 175$. To verify that the Diophantine equation has no solution, take the equation modulo 7 and then modulo 5, obtaining $5y^3 \equiv 1 \bmod 7$ and $-7z^3 \equiv 1 \bmod 5$. These congruences say that $y^3 \equiv 3 \bmod 7$ and $z^3 \equiv 2 \bmod 5$. The only cubes modulo 7 are ± 1, and thus the congruence for y has no solution.

We turn to the question of the splitting of prime ideals in pure cubic extensions $\mathbb{K} = \mathbb{Q}(\sqrt[3]{m})$. In the notation of Proposition 5.8, we again write $m = ab^2$, and we shall assume that the extension is of Type I. We saw in Proposition 5.8 and the remarks afterward that $D_{\mathbb{K}}$ equals the greatest common divisor of $D(\sqrt[3]{ab^2})$ and $D(\sqrt[3]{a^2 b})$. Therefore the splitting of every prime ideal (p) in $\mathbb{Z}$ is described by Theorem 5.6. We have only to sort out the details.

Proposition 5.10. Let $\mathbb{K} = \mathbb{Q}(\sqrt[3]{m})$ be a pure cubic extension of Type I, and let R be its ring of algebraic integers. If p is a prime number, then the ideal $(p)R$ of R splits into prime ideals as follows:

(a) $(p)R = P_1 P_2$ with $N(P_1) = p$ and $N(P_2) = p^2$ if $p \equiv -1 \bmod 3$ and p does not divide $D_{\mathbb{K}}$,
(b) $(p)R = P_1 P_2 P_3$ with P_1, P_2, P_3 distinct of norm p if $p \equiv 1 \bmod 3$, $x^3 \equiv m \bmod p$ is solvable in $\mathbb{F}_p$, and p does not divide $D_{\mathbb{K}}$,
(c) $(p)R$ is prime of norm p^3 if $p \equiv 1 \bmod 3$, $x^3 \equiv m \bmod p$ is not solvable in $\mathbb{F}_p$, and p does not divide $D_{\mathbb{K}}$,
(d) $(p)R = P^3$ with $N(P) = p$ if p divides $D_{\mathbb{K}}$.

PROOF. The prime divisors of $D_{\mathbb{K}}$ are 3 and the prime divisors of a and b. For all other primes Theorem 5.6 shows that all ramification indices are 1. Let p be a prime of the form $6k \pm 1$ not dividing $D_{\mathbb{K}}$. The multiplicative group $\mathbb{F}_p^\times$ of $\mathbb{F}_p$ is cyclic of order $p - 1$ and hence has order divisible by 3 if and only if $p = 6k + 1$. Thus there are three cube roots of 1 when $p = 6k + 1$ but only 1 when $p = 6k - 1$. In the latter case the cubing map is one-one onto from $\mathbb{F}_p^\times$ to itself. Thus $X^3 - m$ factors modulo p as the product of a first-degree factor and an irreducible second-degree factor if $p = 6k - 1$, and (a) follows for such primes from Theorem 5.6. If $p = 6k + 1$, then $X^3 - m$ either factors modulo p as the product of three first-degree factors or is irreducible, since 1 has three cube roots. Thus (b) and (c) follow for such primes from Theorem 5.6.

For $p = 2$ if m is odd, then $X^3 - m \equiv X^3 - 1 \equiv (X - 1)(X^2 + X + 1) \bmod 2$, and we are in the situation of (a). This completes the discussion of primes that do not divide $D_{\mathbb{K}}$. If p divides m, then $X^3 - m \equiv X^3 \bmod p$ is the cube of a first-degree factor, and (d) follows in these cases. For $p = 3$ whether or not p divides m, we have $X^3 - m \equiv X^3 - m^3 \equiv (X - m)^3 \bmod 3$, and (d) follows in this case. $\square$

4. Cubic Number Fields as Examples

We conclude this section by discussing Dedekind's example of a common index divisor. The field in question is again of degree 3 over $\mathbb{Q}$ but is not of the form $\mathbb{Q}(\sqrt[3]{m})$. Instead, the field is $\mathbb{K} = \mathbb{Q}(\xi)$, where ξ is a root of $F(X) = X^3 + X^2 - 2X + 8$. The polynomial $F(X)$ is irreducible over $\mathbb{Q}$ because Gauss's Lemma shows that its only possible linear factors are $X - k$ with k dividing 8 and because routine computation rules out each such linear factor. As usual, let R be the ring of algebraic integers in $\mathbb{K}$.

The different of ξ is $\mathcal{D}(\xi) = F'(\xi) = 3\xi^2 + 2\xi - 2$, and the discriminant $D(\xi)$ therefore is given by $D(\xi) = -N_{\mathbb{K}/\mathbb{Q}}(3\xi^2 + 2\xi - 2)$. We calculate this norm as the determinant of left multiplication by $3\xi^2 + 2\xi - 2$ on $\mathbb{K}$, using the ordered basis $(1, \xi, \xi^2)$. Since $\xi^3 = -\xi^2 + 2\xi - 8$ and $\xi^4 = -\xi^3 + 2\xi^2 - 8\xi = 3\xi^2 - 10\xi + 8$, we have

$$(3\xi^2 + 2\xi - 2)(1) = -2 + 2\xi + 3\xi^2,$$
$$(3\xi^2 + 2\xi - 2)(\xi) = -2\xi + 2\xi^2 + 3\xi^3 = -24 + 4\xi - \xi^2,$$
$$(3\xi^2 + 2\xi - 2)(\xi^2) = -2\xi^2 + 2\xi^3 + 3\xi^4 = 8 - 26\xi + 5\xi^2.$$

Thus

$$N_{\mathbb{K}/\mathbb{Q}}(3\xi^2 + 2\xi - 2) = \det\begin{pmatrix} -2 & -24 & 8 \\ 2 & 4 & -26 \\ 3 & -1 & 5 \end{pmatrix} = 2^2 \cdot 503,$$

and $D(\xi) = -2^2 \cdot 503$. Thus either the index $J(\xi)$ of $\mathbb{Z}(\Gamma(\xi))$ in R is 1 with $D_{\mathbb{K}} = -2^2 \cdot 503$, or $J(\xi) = 2$ with $D_{\mathbb{K}} = -503$.

Problems 24–25 at the end of the chapter show that $\frac{1}{2}(\xi^2 + \xi)$ is in R and that consequently the correct choice is $J(\xi) = 2$ with $D_{\mathbb{K}} = -503$ and with $\{1, \xi, \frac{1}{2}(\xi^2 + \xi)\}$ as a $\mathbb{Z}$ basis of R. In fact, 2 divides $J(\eta)$ for every primitive element of $\mathbb{K}$ lying in R, and therefore 2 is a common index divisor in the sense of Section 2. One way to check this assertion would be to calculate $D(\eta)$ for every such η. The computation would be feasible because we can express η as a $\mathbb{Z}$ linear combination of the members of $\{1, \xi, \frac{1}{2}(\xi^2 + \xi)\}$ and calculate the field polynomial of η in the same way that $N_{\mathbb{K}/\mathbb{Q}}(\xi)$ was calculated above.

However, there is an easier way. Problem 28 at the end of the chapter shows that $(2)R$ splits as the product of three distinct prime ideals of R. If there were some η for which 2 did not divide $J(\eta)$, then Theorem 5.6 would show that the minimal polynomial of η when reduced modulo 2 splits as the product of three distinct first-degree factors. But $\mathbb{F}_2$ has only 2 elements, hence only two possible distinct linear factors to offer. Thus Theorem 5.6 must not be applicable to η and the prime 2, and we conclude that 2 divides $J(\eta)$. Going over this argument, we see that we have established the following more general result.

Proposition 5.11. Let $\mathbb{K}/\mathbb{Q}$ be a field extension of degree n, and let R be the ring of algebraic integers in $\mathbb{K}$. If p is a prime number with $2 \leq p \leq n-1$ such that $(p)R$ splits as the product of n distinct prime ideals of R, then p is a common index divisor for $\mathbb{K}$.

5. Dirichlet Unit Theorem

Let $\mathbb{K}$ be a number field of degree n over $\mathbb{Q}$, and let R be its ring of algebraic integers. We regard $\mathbb{K}$ as a subfield of $\mathbb{C}$. The **units** of $\mathbb{K}$ are understood to be the members of the group $R^\times$ of units of the ring R. As was observed in Section 2, there exist exactly n field mappings of $\mathbb{K}$ into $\mathbb{C}$, and we denote them by $\sigma_1, \ldots, \sigma_n$; one of these is the inclusion of $\mathbb{K}$ into $\mathbb{C}$. If x is in $\mathbb{K}$, then the images $\sigma_1(x), \ldots, \sigma_n(x)$ are called the **conjugates** of x.

In Section I.6 we studied the group of units in the quadratic case $n = 2$, and we found, particularly in the problems at the end of that chapter, that an understanding of this group was essential to working successfully on the number-theoretic problems studied in that chapter. When $n = 2$, we found that the qualitative nature of the group $R^\times$ depends on the sign of the field discriminant. The group turned out to be the finite subgroup of roots of unity in $\mathbb{K}$ if $D_\mathbb{K} < 0$, and it turned out to be isomorphic to the product of a copy of $\mathbb{Z}$ and a cyclic group of order 2 if $D_\mathbb{K} > 0$. The hard step in this analysis was constructing an element in the subgroup $\mathbb{Z}$ in the latter case.

Because of the importance of $R^\times$ in the quadratic case, we can expect that an understanding of $R^\times$ for our general number field $\mathbb{K}$ is important for higher-degree number-theoretic questions. In this section we shall obtain a structure theorem for $R^\times$ for general n analogous to the structure theorem for $n = 2$ mentioned in the previous paragraph. Such a theorem may not answer all important questions about $R^\times$, but it will be a good start.[12] The main theorem is Theorem 5.13 below, the Dirichlet Unit Theorem.

The units of R are the members ε of R with $N_{\mathbb{K}/\mathbb{Q}}(\varepsilon) = \pm 1$. This simple fact is verified for general $\mathbb{K}$ in the same way that it was verified for quadratic $\mathbb{K}$ in Section I.6.

Any element ε of finite order in $R^\times$ is a complex number with $\varepsilon^k = 1$ for some k and hence lies on the unit circle of $\mathbb{C}$. Since such an element ε is a root of $X^k - 1$, all its conjugates $\sigma_j(\varepsilon)$ lie on the unit circle of $\mathbb{C}$. We shall prove the following proposition about these elements.

---

[12]For example, when $n = 2$, we defined the **fundamental unit** ε_1 for the case $D_\mathbb{K} > 0$ to be the least unit > 1, and the sign of $N_{\mathbb{K}/\mathbb{Q}}(\varepsilon_1)$ was a thorny question that we did not answer fully but that affected results in the problems at the end of the chapter.

Proposition 5.12. The subgroup of $R^\times$ of elements of finite order consists of all l^{th} roots of unity in $\mathbb{C}$, where l is an integer depending on $\mathbb{K}$ that is bounded when the degree $n = [\mathbb{K} : \mathbb{Q}]$ is bounded.

PROOF. We are to bound the integers k for which primitive k^{th} roots of unity occur in $\mathbb{K}$. Let k have prime decomposition $k = p_1^{m_1} \cdots p_r^{m_r}$. From Section IX.9 of *Basic Algebra*, we know that the cyclotomic polynomial $\Phi_k(X)$ is a monic irreducible member of $\mathbb{Z}[X]$ whose roots in $\mathbb{C}$ are exactly all primitive k^{th} roots of unity; moreover, the degree of $\Phi_k(X)$ is given by the Euler φ function:

$$\varphi(k) = k \prod_{p \text{ divides } k} \left(1 - \tfrac{1}{p}\right).$$

If primitive k^{th} roots of unity occur in $\mathbb{K}$, then $\varphi(k) \leq n$ because $\Phi_k(X)$ is irreducible over $\mathbb{Q}$, and hence $(p_1 - 1) \cdots (p_r - 1) \leq n$. Allowing $p_1 = 2$ possibly, we see that each factor $p_j - 1$ with $j > 1$ is at least 2, and thus $2^{r-1} \leq n$. So r is bounded as a function of n by $\log_2 2n$, and we obtain

$$\varphi(k) \geq k \prod_{\text{first } \log_2 2n \text{ primes}} \left(1 - \tfrac{1}{2}\right) = 2^{-\log_2 2n} k = \tfrac{k}{2n}.$$

Consequently $k \leq 2n\varphi(k) \leq 2n^2$, as required. If $R^\times$ contains one primitive k^{th} root of unity in $\mathbb{C}$, then it contains them all, since the kth roots of unity form a cyclic group and any primitive such root is a generator. The result follows. □

We shall use the field mappings $\sigma_j : \mathbb{K} \to \mathbb{C}$ for $1 \leq j \leq n$ to introduce useful "absolute values" on $\mathbb{K}$. The mappings σ_j are of two types:

  (i) those carrying $\mathbb{K}$ into $\mathbb{R}$,
  (ii) those carrying $\mathbb{K}$ into $\mathbb{C}$ but not into $\mathbb{R}$; these come in pairs σ and $\overline{\sigma}$, where $\overline{\sigma}$ denotes the composition of σ followed by complex conjugation.

Suppose that there are r_1 mappings σ_j of the first kind and that there are r_2 pairs of the second kind. Then $r_1 + 2r_2 = n$. Renumbering $\sigma_1, \ldots, \sigma_n$ if necessary, let us arrange that $\sigma_1, \ldots, \sigma_{r_1}$ are of the first kind, that $\sigma_{r_1+1}, \ldots, \sigma_n$ are of the second kind, and that $\sigma_{r_1+r_2+i} = \overline{\sigma}_{r_1+i}$ for $1 \leq i \leq r_2$. We introduce $r_1 + r_2$ **absolute values**[13] on $\mathbb{K}$ by the definition

$$\|x\|_s = |\sigma_s(x)| \qquad \text{for } 1 \leq s \leq r_1 + r_2,$$

where $|\cdot|$ denotes the usual absolute value function on $\mathbb{C}$. Then the function $\text{Log} : \mathbb{K}^\times \to \mathbb{R}^{r_1+r_2}$ given by

$$\text{Log}(\varepsilon) = (\log \|\varepsilon\|_1, \ldots, \log \|\varepsilon\|_{r_1+r_2})$$

---

[13] These are called **archimedean absolute values** of $\mathbb{K}$ in the general theory. Some authors refer to them as **archimedean valuations**.

is evidently a group homomorphism.

A **lattice** in a Euclidean space $\mathbb{R}^l$ is an additive subgroup $\mathbb{Z}u_1 \oplus \cdots \oplus \mathbb{Z}u_l$ such that $\{u_1, \ldots, u_l\}$ is linearly independent over $\mathbb{R}$. Such a subgroup is discrete,[14] and the quotient is compact, by the Heine–Borel Theorem.

Theorem 5.13 (Dirichlet Unit Theorem). Let $\mathbb{K}$ be a number field of degree n with $r_1 + r_2$ absolute values, and let R be the ring of algebraic integers in $\mathbb{K}$. The kernel of the restriction to $R^\times$ of the function Log is the finite subgroup of roots of unity in $\mathbb{K}^\times$, and the image of this restriction of Log is a lattice in the vector subspace of elements $(x_1, \ldots, x_{r_1+r_2})$ in $\mathbb{R}^{r_1+r_2}$ satisfying

$$x_1 + \cdots + x_{r_1} + 2x_{r_1+1} + \cdots + 2x_{r_1+r_2} = 0.$$

Consequently $R^\times$ is a finitely generated abelian group of rank $r_1 + r_2 - 1$.

EXAMPLES.

(1) The theorem reduces when $n = 2$ to results known from Chapter I. Specifically if $\mathbb{K} = \mathbb{Q}(\sqrt{m})$, then $m > 0$ makes $r_1 = 2$ and $r_2 = 0$, while $m < 0$ makes $r_1 = 0$ and $r_2 = 1$.

(2) For $\mathbb{K} = \mathbb{Q}(\sqrt[3]{2})$, let $\omega = e^{2\pi i/3}$. The field mappings of $\mathbb{K}$ into $\mathbb{C}$ carry $\mathbb{K}$ into $\mathbb{R}$ or $\mathbb{R}\omega$ or $\mathbb{R}\omega^2$. Thus $r_1 = 1$ and $r_2 = 1$.

(3) The polynomial $F(X) = X^5 - 5X + 1$ in $\mathbb{Q}[X]$ was studied as an example in connection with Galois theory in Section IX.11 of *Basic Algebra*. The polynomial was shown to be irreducible over $\mathbb{Q}$ and to have three real roots and one pair of complex conjugate roots. For $\mathbb{K} = \mathbb{Q}[X]/(X^5 - 5X + 1)$, we therefore have $r_1 = 3$ and $r_2 = 1$. The primitive element ξ of $\mathbb{K}$ with $\xi^5 - 5\xi + 1 = 0$ lies in R; it is a nontrivial example of a member of $R^\times$ because $\xi(\xi^4 - 5) = -1$.

The proof of Theorem 5.13 will occupy the remainder of this section. We begin by clarifying in Lemma 5.14 the relationship between discrete subgroups and lattices in Euclidean space and by proving in Proposition 5.15 a weak version of Theorem 5.13 that addresses everything except the existence questions.

Lemma 5.14. A discrete subgroup of $\mathbb{R}^l$ is a free abelian group of rank $\leq l$ and is necessarily of the form $\mathbb{Z}u_1 \oplus \cdots \oplus \mathbb{Z}u_m$ for some set $\{u_1, \ldots, u_m\}$ that is linearly independent over $\mathbb{R}$. The discrete subgroup is a lattice if and only if the rank is l.

---

[14] A discrete subset of $\mathbb{R}^l$ is a subset S such that every one-point subset of S is open when S is given the relative topology. See Lemma 5.14 below for a converse assertion.

PROOF. We begin by proving that any discrete subgroup of $\mathbb{R}^l$ is topologically closed. Let G be the subgroup, and choose by discreteness an open ball $V = \{x \in \mathbb{R}^l \mid |x| < \epsilon\}$ V about 0 with $V \cap G = \{0\}$. The open ball $U = \{x \in \mathbb{R}^l \mid |x| < \epsilon/2\}$ has the property that $U + U \subseteq V$. If G is not closed, let x_0 be a limit point of G that is not in G. Then the open ball $x_0 - U$ about x_0 must contain a member g of G, and g cannot equal x_0. Write $x_0 - u = g$ with $u \in U$. Then $u = x_0 - g$ is a limit point of G that is not in G, and we can find $g' \neq 1$ in G such that g' is in $u + U$. But $u + U \subseteq U + U \subseteq V$, and so g' is in $G \cap V = \{0\}$, contradiction. We conclude that G contains all its limit points and is therefore closed.

From the fact that any discrete subgroup G of $\mathbb{R}^l$ is closed, let us see that any bounded subset of G is finite. It is enough to see that the intersection X of G with any (finite-radius) closed ball is finite. The set X is closed because G is closed, and it is therefore compact by the Heine–Borel Theorem. By discreteness, find for each $g \in G$ an open ball U_x centered at x that contains no member of G other than x. These open sets form an open cover of the compact set X, and a finite subcollection of them covers X. Each such open set contains only one member of X, and hence X is finite.

Returning to the statement of the lemma, we induct on the dimension of the $\mathbb{R}$ linear span of the discrete subgroup, the base case being that the $\mathbb{R}$ linear span is 0. Let G be the discrete subgroup, and let $\{v_1, \ldots, v_m\}$ in G be a maximal set that is linearly independent over $\mathbb{R}$. Let $G_0 = G \cap \left(\sum_{j=0}^{m-1} \mathbb{R} v_j\right)$. By induction we may assume that every $u \in G_0$ is a $\mathbb{Z}$ linear combination of $v_1, \ldots, v_{m-1}$. Let S be the set of $\mathbb{R}$ linear combinations of $\{v_1, \ldots, v_m\}$ of the form

$$S = \left\{ v = c_1 v_1 + \cdots + c_m v_m \in G \,\middle|\, \begin{array}{l} 0 \leq c_i < 1 \text{ for } 1 \leq i \leq m-1, \\ 0 \leq c_m \leq 1 \end{array} \right\}.$$

The set S is bounded, and we saw in the previous paragraph that any bounded subset of G is finite. So S is finite. Let v' be a member of S with the smallest positive coefficient for v_m, say

$$v' = a_1 v_1 + \cdots + a_m v_m.$$

If v is any member of S and its coefficient c_m is not a multiple of a_m, then $v - jv'$ for a suitable integer j has m^{th} coefficient positive but less than a_m; by subtracting from $v - jv'$ a suitable $\mathbb{Z}$ linear combination v'' of $v_1, \ldots, v_{m-1}$, we can make $v - jv' - v''$ be in S, and then we have a contradiction to the minimality of a_m. We conclude that c_m is always a multiple of a_m. Then $v - jv'$ is in G_0 for some integer j, and it follows that the $\mathbb{Z}$ linear combinations of $v_1, \ldots, v_{m-1}, v'$ span G. This completes the induction and the proof of the first conclusion of the lemma. The second conclusion is an immediate consequence of the first. $\square$

For the remainder of the section, we adopt the notation in the statement of Theorem 5.13, and we shall not repeat it in the statement of every intermediate result.

Proposition 5.15 (weak form of Dirichlet Unit Theorem). The kernel of the restriction to $R^\times$ of Log is the finite subgroup of roots of unity in $\mathbb{K}^\times$, and the image of this restriction of Log is a discrete additive subgroup in the vector subspace of elements $(x_1, \ldots, x_{r_1+r_2})$ in $\mathbb{R}^{r_1+r_2}$ satisfying

$$x_1 + \cdots + x_{r_1} + 2x_{r_1+1} + \cdots + 2x_{r_1+r_2} = 0.$$

Consequently $R^\times$ is a finitely generated abelian group of rank $\leq r_1 + r_2 - 1$.

PROOF. For α in $R^\times$, we calculate that

$$\log \|\alpha\|_1 + \cdots + \log \|\alpha\|_{r_1} + 2\log \|\alpha\|_{r_1+1} + \cdots + 2\log \|\alpha\|_{r_1+r_2}$$
$$= \log \left(|\sigma_1(\alpha)| \cdots |\sigma_{r_1}(\alpha)||\sigma_{r_1+1}(\alpha)|^2 \cdots |\sigma_{r_1+r_2}(\alpha)|^2\right)$$
$$= \log \Big| \prod_{j=1}^n \sigma_j(\alpha) \Big|$$
$$= \log |N_{\mathbb{K}/\mathbb{Q}}(\alpha)| = \log 1 = 0.$$

Hence the image lies in the vector subspace in the statement of the proposition.

Fix a (large) positive number M, and consider the set E_M of all members α of $R^\times$ for which all coordinates of $\mathrm{Log}(\alpha)$ are $\leq M$ in absolute value. Then the field polynomials

$$\det\left(XI - (\text{left by } \alpha)\right) = \prod_{j=1}^n (X - \sigma_j(\alpha))$$

of such elements α have all coefficients bounded by some M' depending on M, since each $|\sigma_j(\alpha)|$ is of the form $\|\alpha\|_j$ and is $\leq e^M$. Such a field polynomial is equal to $g(X)^r$, where $g(X)$ is the minimal polynomial of α and r is given by $r \deg(g(X)) = n$. Since α is in R, the coefficients of $g(X)$ are integers, and hence so are the coefficients of the corresponding field polynomial. There are only finitely many members of $\mathbb{Z}[X]$ of degree n whose coefficients are in a given bounded set, and hence there are only finitely many α's in E_M.

It follows that the image subgroup is discrete. Taking $M = 0$, we see also that the kernel of the restriction of Log to $R^\times$ is finite. Hence every element of this kernel has finite order and is therefore a root of unity. □

We come to the proof of Theorem 5.13. For quadratic extensions of $\mathbb{Q}$, which were handled in Section I.6, the crucial question of existence was addressed by means of an approximation result (Lemma 1.15) for irrational numbers. That result did not immediately establish the existence of units of infinite order, but it was applied infinitely many times in the course of proving Proposition 1.16, and the total effect was to produce a unit of infinite order.

We do something similar in general. In place of the approximation result in Lemma 1.15, we shall use a result known as the Minkowski Lattice-Point Theorem, which asserts the existence of lattice points in certain compact convex sets in Euclidean space. This result appears as Theorem 5.16 below. As was true in the quadratic case, it is not just a single application of this theorem that produces the desired units, but an infinite sequence of applications of it. The details will be more complicated here than in the quadratic case. Before describing how the argument is to proceed, let us establish the Minkowski theorem.

Let $\{v_1, \ldots, v_m\}$ be an $\mathbb{R}$ basis of $\mathbb{R}^m$, and let $L = \mathbb{Z}v_1 \oplus \cdots \oplus \mathbb{Z}v_m$ be the corresponding lattice. The **fundamental parallelotope** for L corresponding to this basis is the set

$$\{c_1 v_1 + \cdots + c_m v_m \mid 0 \leq c_j \leq 1 \text{ for } 1 \leq j \leq m\}.$$

The volume of this fundamental parallelotope is independent of the choice of the $\mathbb{Z}$ basis for L. In fact, any two such $\mathbb{Z}$ bases are carried from one to the other by an integer matrix of determinant ± 1, and any linear transformation from $\mathbb{R}^m$ to itself of determinant ± 1 is volume preserving. The one fundamental parallelotope is mapped to the other when the one basis is carried to the other, and hence the two fundamental parallelotopes have the same volume.

Theorem 5.16 (Minkowski Lattice-Point Theorem).[15] Let L be a lattice in $\mathbb{R}^m$, and let V_0 be the volume of a fundamental parallelotope. If E is any compact convex set in $\mathbb{R}^m$ containing 0, closed under negatives, and having volume(E) $\geq 2^m V_0$, then E contains a nonzero point of L.

REMARK. The constant 2^m in the statement is best possible, as is shown by taking L to be the standard lattice and E to be a cube oriented consistently with L, centered at 0, and having each side slightly less than 2. We need merely some constant, not the best possible one, in the application to Theorem 5.13, and the proof can be simplified a little for that purpose.[16] But the present theorem will be applied again in the next section, and this time the best possible constant yields the most useful information.

---

[15]The simple proof given here is due to H. Blichfeldt and is the standard one, so standard that Blichfeldt's name is sometimes attached to the theorem.

[16]In particular, the final paragraph of the proof can be omitted, and we can fix a value of M proportional to s in making the argument.

PROOF. Without loss of generality, L is the standard lattice of points with all coordinates in $\mathbb{Z}$, and V_0 is 1. Fix an arbitrarily small positive constant ϵ, and first assume that the given set E has volume$(E) \geq (2+\epsilon)^m V_0$. Arguing by contradiction, suppose that the only lattice point in E is 0. Since E is bounded, we can choose a number $s > 0$ in such a way that E is contained in the cube C_s centered at 0, oriented consistently with the lattice, and having side $2s$. Let us see that the sets $l + \frac{1}{2}E$ for $l \in L$ are disjoint. In fact, in obvious notation if $l_1 + \frac{1}{2}e_1 = l_2 + \frac{1}{2}e_2$ with $l_1 \neq l_2$, then $l_1 - l_2 = \frac{1}{2}(e_2 - e_1)$, and this is in E because e_2 and $-e_1$ are in E and E is convex. Thus the sets $l + \frac{1}{2}E$ are indeed disjoint.

Choose an integer M large enough to have $s/M < \epsilon$. Any lattice point l whose coordinates are all $\leq M$ in absolute value has $l + \frac{1}{2}E \subseteq C_{M+\frac{1}{2}s}$. Since the sets $l + \frac{1}{2}E$ for these l's are disjoint,

$$(2(M + \tfrac{1}{2}s))^m = \text{volume}(C_{M+\frac{1}{2}s}) \geq \sum_{\substack{\text{all } l \in L \text{ with} \\ \text{all coordinates } \leq M}} \text{volume}(l + \tfrac{1}{2}E)$$

$$\geq (2M)^m \text{volume}(\tfrac{1}{2}E) = M^m \text{volume}(E),$$

and therefore volume$(E) \leq (2 + s/M)^m$, in contradiction to our extra assumption that volume$(E) \geq (2+\epsilon)^m$.

Now suppose that volume$(E) = 2^m$. For each $\epsilon > 0$, let E_ϵ be the dilate $(1 + \frac{1}{2}\epsilon)E$. The sets E_ϵ satisfy the extra assumption made in the previous part of the proof, and therefore E_ϵ contains a nonzero lattice point. Since E_1 is bounded, there are only finitely many possibilities for this nonzero lattice point for each $\epsilon \leq 1$. Thus we can find a sequence of ϵ's tending to 0 for which this lattice point is the same. The convexity of the sets E_ϵ, in combination with the fact that the sets contain 0, implies that the sets are nested, and therefore this lattice point lies in E_ϵ for all $\epsilon > 0$. Since E is compact, $E = \bigcap_{\epsilon > 0} E_\epsilon$, and therefore this lattice point lies in E. $\square$

Let us describe the lattice to be used when the Minkowski Lattice-Point Theorem is applied to obtain the Dirichlet Unit Theorem. Let Ω be the real vector space $\Omega = \mathbb{R}^{r_1} \times \mathbb{C}^{r_2} \cong \mathbb{R}^n$, and let $|\omega|_s$ be the magnitude of the sth component of $\omega \in \Omega$ for $1 \leq s \leq r_1 + r_2$. We introduce a homomorphism Φ of the additive group of $\mathbb{K}$ into the additive group of Ω given by

$$\Phi(x) = \big(\sigma_1(x), \ldots, \sigma_{r_1}(x), \sigma_{r_1+1}(x), \ldots, \sigma_{r_1+r_2}(x)\big)$$

for $x \in \mathbb{K}$. We shall be mostly interested in the restriction of Φ to R, but the values on $\mathbb{K}$ will help a little with motivation when the Minkowski Lattice-Point Theorem is applied once again in the next section. Observe that our definitions make $\|x\|_s = |\sigma_s(x)| = |\Phi(x)|_s$ for $x \in \mathbb{K}$ and $1 \leq s \leq r_1 + r_2$.

5. Dirichlet Unit Theorem

Lemma 5.17. The image $\Phi(R)$ is a lattice in Ω.

PROOF. The homomorphism Φ is one-one on R because σ_1, being a field map, is one-one. Since R is a free abelian group of rank n and Φ is one-one, $\Phi(R)$ is free abelian of rank n. Lemma 5.14 therefore shows that it is sufficient to show that $\Phi(R)$ is discrete as an additive subgroup of Ω. It is enough to show that a bounded region of Ω contains only finitely many points of $\Phi(R)$.

The verification of this fact is similar to an argument in the proof of Proposition 5.15: A bound by some M on all $|\sigma_j(\alpha)|$ for certain elements $\alpha \in R$ implies that each field polynomial

$$\det \left(XI - (\text{left by } \alpha) \right) = \prod_{j=1}^{n} (X - \sigma_j(\alpha))$$

has all its coefficients bounded by some M' depending on M. These coefficients are integers when α is in R, and thus there are only finitely many such polynomials. Each polynomial has at most n distinct roots, and consequently only finitely many α's satisfy such a bound. □

We are now ready to prove Theorem 5.13, but we precede the proof by an outline. The proof has three steps to it:

(1) We apply the Minkowski Lattice-Point Theorem to the set $\Phi(R) \subseteq \Omega$, which we know is a lattice because of Lemma 5.17. For each s_0 with $1 \leq s_0 \leq r_1 + r_2$, let E_{s_0} be a set of ω's in Ω defined by the conditions that $|\omega|_s$ is to be small for $s \neq s_0$ and $|\omega|_{s_0}$ is allowed to be large—with the understanding that E_{s_0} is a bounded set and that E_{s_0} has volume $\geq 2^n V_0$, where V_0 is the volume of a fundamental parallelotope of $\Phi(R)$. Using a nonzero lattice point in $\Phi(R)$ obtained from applying Theorem 5.16 to E_{s_0} and squeezing E_{s_0} even more, we can obtain an infinite sequence of points α in R such that $|N_{\mathbb{K}/\mathbb{Q}}(\alpha)|$ remains bounded and such that the size of this norm is contributed to mostly by $\|\alpha\|_{s_0}$.

(2) Applying the same argument that was used for quadratic extensions of $\mathbb{Q}$ in the proof of Proposition 1.16, we obtain infinite sequences of units whose norm is contributed to mostly by $\| \cdot \|_{s_0}$. We can do this for $1 \leq s_0 \leq r_1 + r_2$.

(3) We pass to the Log map, proving and applying the following result from linear algebra: a real square matrix $[a_{ij}]$ with the property that $|a_{ii}| > \sum_{j \neq i} |a_{ij}|$ for all i is nonsingular. In the application of this result, we have $\log \|\varepsilon_{s_0}\|_{s_0} > 0$ for the s_0^{th} constructed unit, $\log \|\varepsilon_{s_0}\|_s < 0$ for $s \neq s_0$, and an equality that we can write either as $\sum_{s=1}^{n} \log \|\varepsilon_{s_0}\|_s = 0$ or as $\sum_{s=1}^{r_1} \log \|\varepsilon_{s_0}\|_s + 2\sum_{s=r_1+1}^{r_1+r_2} \log \|\varepsilon_{s_0}\|_s = 0$. If we drop all terms corresponding to the $(r_1+r_2)^{\text{th}}$ unit, then we are in a situation for which the result from linear algebra immediately implies the theorem.

296   V. Three Theorems in Algebraic Number Theory

PROOF OF THEOREM 5.13. The proof is carried out in three steps.

Step 1. For fixed s_0 with $1 \leq s_0 \leq r_1 + r_2$, we construct an infinite sequence $\alpha_j^{(s_0)}$ in R with

(i) $|N_{\mathbb{K}/\mathbb{Q}}(\alpha_j^{(s_0)})| \leq 2^n V_0$,
(ii) $\|\alpha_j^{(s_0)}\|_s$ tends to 0 for each $s \neq s_0$ as j tends to infinity,
(iii) $\|\alpha_j^{(s_0)}\|_{s_0}$ tends to infinity as j tends to infinity.

For the construction, form for each $j > 0$ the compact convex set in Ω closed under multiplication by -1 consisting of all ω such that

$$|\omega|_s \leq j^{-1} \qquad \text{for } s \neq s_0,$$

$$|\omega|_{s_0} \leq \begin{cases} 2^n j^{n-1} 2^{-r_1} \pi^{-r_2} V_0 & \text{if } 1 \leq s_0 \leq r_1, \\ (2^n j^{n-2} 2^{-r_1} \pi^{-r_2} V_0)^{1/2} & \text{if } r_1 + 1 \leq s_0 \leq r_1 + r_2. \end{cases}$$

This set has volume

$$\begin{cases} (2j^{-1})^{r_1-1} \cdot 2(2^n j^{n-1} 2^{-r_1} \pi^{-r_2} V_0)(\pi j^{-2})^{r_2} = 2^n V_0 & \text{if } s_0 \leq r_1, \\ (2j^{-1})^{r_1}(\pi j^{-2})^{r_2-1} \pi(2^n j^{n-2} 2^{-r_1} \pi^{-r_2} V_0) = 2^n V_0 & \text{if } s_0 > r_1. \end{cases}$$

Theorem 5.16 shows that the set contains a nonzero lattice point $\alpha_j^{(s_0)}$. Let us check that this point satisfies (i), (ii), and (iii). For (i), we have

$$|N_{\mathbb{K}/\mathbb{Q}}(\alpha_j^{(s_0)})| = \Big(\prod_{j=1}^{r_1} \|\alpha_j^{(s_0)}\|_s\Big)\Big(\prod_{s=r_1+1}^{r_1+r_2} \|\alpha_j^{(s_0)}\|_s\Big)^2$$

$$\leq \begin{cases} (j^{-1})^{r_1-1}(2^n j^{n-1} 2^{-r_1} \pi^{-r_2} V_0) j^{-2r_2} & \text{if } s_0 \leq r_1 \\ (j^{-1})^{r_1}(j^{-2})^{r_2-1}(2^n j^{n-2} 2^{-r_1} \pi^{-r_2} V_0) & \text{if } s_0 > r_1 \end{cases}$$

$$= 2^n V_0 2^{-r_1} \pi^{-r_2}$$

$$\leq 2^n V_0.$$

Property (ii) is immediate from the inequality $\|\alpha_j^{(s_0)}\|_s \leq j^{-1}$ for $s \neq s_0$. For (iii), we have

$$1 \leq |N_{\mathbb{K}/\mathbb{Q}}(\alpha_j^{(s_0)})| = \Big(\prod_{j=1}^{r_1} \|\alpha_j^{(s_0)}\|_s\Big)\Big(\prod_{s=r_1+1}^{r_1+r_2} \|\alpha_j^{(s_0)}\|_s\Big)^2;$$

thus (ii) implies (iii).

Step 2. For fixed s_0 with $1 \leq s_0 \leq r_1 + r_2$, we construct an infinite sequence of units $\varepsilon_j^{(s_0)}$ such that

(ii′) $\|\varepsilon_j^{(s_0)}\|_s$ tends to 0 for each $s \neq s_0$ as j tends to infinity,
(iii′) $\|\varepsilon_j^{(s_0)}\|_{s_0}$ tends to infinity as j tends to infinity.

For the construction, we pass to a subsequence from Step 1, still denoting it by $\alpha_j^{(s_0)}$, such that $N_{\mathbb{K}/\mathbb{Q}}(\alpha_j^{(s_0)})$ is a constant integer, say M. Since $R/(M)$ is finite, we can pass to a further subsequence, still with no change in notation, such that all $\alpha_j^{(s_0)}$ lie in the same residue class[17] modulo the principal ideal (M) of R. Put

$$\varepsilon_j^{(s_0)} = \alpha_j^{(s_0)} / \alpha_1^{(s_0)}.$$

Then $N_{\mathbb{K}/\mathbb{Q}}(\alpha_j^{(s_0)}) = N_{\mathbb{K}/\mathbb{Q}}(\alpha_1^{(s_0)})$, since $N_{\mathbb{K}/\mathbb{Q}}(\alpha_j^{(s_0)})$ is a constant integer, and $\frac{1}{M}(\alpha_j^{(s_0)} - \alpha_1^{(s_0)})$ is in R, since all $\alpha_j^{(s_0)}$ lie in the same residue class modulo (M). The computation

$$\varepsilon_j^{(s_0)} = 1 + \frac{\alpha_j^{(s_0)} - \alpha_1^{(s_0)}}{\alpha_1^{(s_0)}} = 1 + \frac{\alpha_j^{(s_0)} - \alpha_1^{(s_0)}}{M} \prod_{\sigma \neq 1} \sigma(\alpha_1^{(s_0)})$$

shows that $\varepsilon_j^{(s_0)}$ is an algebraic integer. Hence it is in R. We certainly have

$$N_{\mathbb{K}/\mathbb{Q}}(\varepsilon_j^{(s_0)}) = \frac{N_{\mathbb{K}/\mathbb{Q}}(\alpha_j^{(s_0)})}{N_{\mathbb{K}/\mathbb{Q}}(\alpha_1^{(s_0)})} = \frac{M}{M} = 1.$$

Therefore $\varepsilon_j^{(s_0)}$ is a unit. Also, the computation

$$\|\varepsilon_j^{(s_0)}\|_s = \frac{\|\alpha_j^{(s_0)}\|_s}{\|\alpha_1^{(s_0)}\|_s}$$

shows that (ii) and (iii) in Step 1 imply (ii′) and (iii′) here.

Step 3. For each s_0 with $1 \leq s_0 \leq r_1 + r_2$, choose j large enough for the unit $\varepsilon^{(s_0)} = \varepsilon_j^{(s_0)}$ in Step 2 to satisfy

(ii″) $\|\varepsilon^{(s_0)}\|_s < 1$ if $s \neq s_0$,
(iii″) $\|\varepsilon^{(s_0)}\|_{s_0} > 1$.

We assert that the vectors $\mathrm{Log}(\varepsilon^{(s_0)})$ for $1 \leq s_0 \leq r_1 + r_2 - 1$ are linearly independent over $\mathbb{R}$. Hence $\mathrm{Log}(R^\times)$ has rank $\geq r_1 + r_2 - 1$, and Proposition 5.15 therefore implies that $\mathrm{Log}(R^\times)$ has rank equal to $r_1 + r_2 - 1$.

To verify this assertion, form the square matrix $[a_{ij}]$ of size $r_1 + r_2$ given by

$$a_{ij} = \begin{cases} \log \|\varepsilon^{(i)}\|_j & \text{if } 1 \leq j \leq r_1, \\ 2\log \|\varepsilon^{(i)}\|_j & \text{if } r_1 + 1 \leq j \leq r_1 + r_2. \end{cases}$$

---

[17]This conclusion uses a result known as the **Dirichlet pigeonhole principle** or the **Dirichlet box principle**.

Then $a_{ii} > 0$ for each i by (iii''), $a_{ij} < 0$ for $i \neq j$ by (ii''), and $\sum_j a_{ij} = 0$ for each i because $N_{\mathbb{K}/\mathbb{Q}}(\varepsilon^{(i)}) = 1$. Let $[b_{ij}]$ be the upper left block of $[a_{ij}]$ of size $r_1 + r_2 - 1$. For each i, we then have $b_{ii} > 0$ and $\sum_{j \text{ with } j \neq i} |b_{ij}| < b_{ii}$. Let us prove that the matrix $[b_{ij}]$ is nonsingular. Assuming the contrary, let $[c_j]$ be a nonzero column vector with

$$\sum_j b_{ij} c_j = 0 \qquad \text{for all } i. \tag{$*$}$$

If i_0 is an index such that $|c_{i_0}| \geq |c_j|$ for all j, then setting $i = i_0$ leads to the strict inequality

$$|c_{i_0} b_{i_0 i_0}| = |c_{i_0}| |b_{i_0 i_0}| > |c_{i_0}| \sum_{j \neq i_0} |b_{i_0 j}| \geq \sum_{j \neq i_0} |b_{i_0 j} c_j| \geq \Big| \sum_{j \neq i_0} b_{i_0 j} c_j \Big|,$$

which contradicts ($*$). Thus $[b_{ij}]$ is nonsingular.

We conclude that $[b_{ij}]$ has rank $r_1 + r_2 - 1$. Thus its rows are linearly independent, and the first $r_1 + r_2 - 1$ rows of $[a_{ij}]$ must be linearly independent. Therefore the vectors

$$\big(\log \|\varepsilon^{(s_0)}\|_1, \ldots, \log \|\varepsilon^{(s_0)}\|_{r_1}, 2 \log \|\varepsilon^{(s_0)}\|_{r_1+1}, \ldots, 2 \log \|\varepsilon^{(s_0)}\|_{r_1+r_2} \big),$$

indexed by s_0 for $1 \leq s_0 \leq r_1 + r_2 - 1$, are linearly independent in $\mathbb{R}^{r_1+r_2}$. In other words, the vectors $\mathrm{Log}(\varepsilon^{(s_0)})$ are linearly independent for $1 \leq s_0 \leq r_1 + r_2 - 1$. □

6. Finiteness of the Class Number

As in Section 5, let $\mathbb{K}$ be a number field of degree n over $\mathbb{Q}$, and let R be its ring of algebraic integers. Let $\sigma_1, \ldots, \sigma_n$ be the distinct field maps of $\mathbb{K}$ into $\mathbb{C}$, and assume that the first r_1 of them have image in $\mathbb{R}$ and the remaining ones come in conjugate pairs with $\sigma_{r_1+r_2+k} = \overline{\sigma}_{r_1+k}$ for $1 \leq k \leq r_2$.

As in Section I.7, where we treated the case of quadratic extensions, we define two nonzero ideals I and J of R to be **equivalent** if $(r)I = (s)J$ for suitable nonzero elements r and s of R. The same argument as given in that section shows that the result is an equivalence relation. The principal ideals form a single equivalence class.[18]

---

[18]Section I.7 worked also with a notion of strict equivalence of ideals, but we shall not attempt to extend strict equivalence to the present setting.

Proposition 5.18. Multiplication of nonzero ideals in R descends to a well-defined multiplication of equivalence classes of ideals, and the resulting multiplication makes the set of equivalence classes into an abelian group. The identity element of this group is the class of principal ideals.

REMARKS. The proofs of this result and of Theorem 5.19 below will use the following fact proved in Problems 48–53 of Chapter VIII of *Basic Algebra*: if I is any nonzero ideal in R and if I^{-1} is defined by $I^{-1} = \{x \in \mathbb{K} \mid xI \subseteq R\}$, then $I^{-1}I = R$ and there exists $r \in R$ with rI^{-1} equal to an ideal of R. This fact can be made to look more beautiful by introducing the notion of "fractional ideal," but we shall not carry out that step at this time.[19]

PROOF. If I is a nonzero ideal, let $[I]$ denote its equivalence class, and define $[I][J] = [IJ]$. Suppose that $(r)I = (s)I'$ exhibits an equivalence. Then the equality $(s)I'J = (r)IJ$ shows that $[I'J] = [IJ]$. A similar argument applies in the J variable, and therefore multiplication of classes is well defined. It is immediate that multiplication of classes is associative and commutative and also that the class of principal ideals is an identity. If a class $[I]$ is given, let I^{-1} be as in the remarks above, and choose a nonzero $r \in R$ such that $rI^{-1} = J$ is an ideal in R. Multiplying by J gives $(r) = r(I^{-1}I) = (rI^{-1})I = JI$, and thus $[J][I]$ is the class of the principal ideals. So $[I]$ has an inverse. □

The group of equivalence classes of nonzero ideals as in Proposition 5.18 is called the **ideal class group** of $\mathbb{K}$. Its order is called the **class number** of $\mathbb{K}$ and will be denoted by $h_\mathbb{K}$. The main theorem of this section is as follows.

Theorem 5.19. The class number $h_\mathbb{K}$ of any number field is finite.

As we shall see in a moment, it is not too difficult at this stage to prove this finiteness. However, $h_\mathbb{K}$ is an important invariant of a number field that determines whether R is a principal ideal domain, that occurs in various limit formulas in the subject, and that occurs also in dimension formulas connected with "Hilbert class fields." It is therefore of considerable interest to be able to compute $h_\mathbb{K}$ in specific examples. For quadratic fields this computation can be carried out by the techniques of Chapter I because of the close connection between ideal classes and proper equivalence classes of binary quadratic forms. But no comparable theory is available as an aid in computation for number fields of degree greater than 2. As we shall see, the relatively easy proof of Theorem 5.19 that we give in a moment does not offer any helpful clues about the value of $h_\mathbb{K}$. The main

---

[19]The result of the beautification is that the fractional ideals form a group generated by the ideals, and the group of equivalence classes is a homomorphic image of the group of fractional ideals.

task of this section will therefore be to provide a better proof of Theorem 5.19 that helps us find the value of $h_{\mathbb{K}}$ in specific examples.

The two proofs have the following lemma in common. The lemma eliminates the notion of equivalence of ideals from the investigation and shows that the problem is really that of finding elements in each ideal of relatively small norm.

Lemma 5.20. For a particular number field $\mathbb{K}$, if there exists a real constant C with the property that each nonzero ideal J of R contains an element $s \neq 0$ with

$$|N_{\mathbb{K}/\mathbb{Q}}(s)| \leq C\, N(J),$$

then each equivalence class of ideals contains a member L whose absolute norm satisfies $N(L) \leq C$. Consequently the class number $h_{\mathbb{K}}$ is at most the number of nonzero ideals I in R with $N(I) \leq C$. This is a finite number.

PROOF. Let a nonzero ideal I in R be given. By the remarks with Proposition 5.18, choose a nonzero element r in R and an ideal J such that $rI^{-1} = J$. Multiplication by I and use of the remarks shows that $(r) = JI$. By hypothesis for the lemma, choose a nonzero $s \in J$ with $|N_{\mathbb{K}/\mathbb{Q}}(s)| \leq C\, N(J)$. Since s is in J, (s) is contained in J, and therefore $(s) = JL$ for some ideal L. Multiplying both sides of $(r) = JI$ by L gives $(r)L = LJI = (s)I$, and L is therefore equivalent to I. Applying Proposition 5.4, we obtain $N(J)N(L) = N(JL) = N((s)) = |N_{\mathbb{K}/\mathbb{Q}}(s)| \leq C\, N(J)$. Therefore $N(L) \leq C$ as required.

Let us now count the ideals I with $N(I) \leq C$. In terms of the unique factorization $I = \prod_{i=1}^{l} P_i^{e_i}$ of I, we have $N(I) \geq \prod_{i=1}^{l} p_i^{e_i}$, where p_i is the prime number such that $P_i \cap \mathbb{Z} = (p_i)$. In each case, $N(P_i) \geq p_i$. There are only finitely many primes p with $p \leq C$, each is associated with only finitely many prime ideals P of R with $P \cap \mathbb{Z} = (p)$, and P^e contributes at least 2^e toward $N(I)$. The inequality $N(I) \leq C$ shows that these p's and their associated P's are the only possible contributors to I and that each exponent is bounded by $\log_2 N(I)$. Hence there are only finitely many possibilities for I. □

Here is the relatively easy proof of Theorem 5.19.

FIRST PROOF OF THEOREM 5.19. Let $x_1, \ldots, x_n$ be a $\mathbb{Z}$ basis of R, and express members of R in terms of this basis as $r = \sum_{i=1}^{n} c_i x_i$ with all $c_i \in \mathbb{Z}$. The value of $N_{\mathbb{K}/\mathbb{Q}}(r)$ is the value of the determinant of left multiplication by r on $\mathbb{K}$, and this value, as a function of $c_1, \ldots, c_n$, is a homogeneous polynomial of degree n. Consequently we can find a constant C such that $\left|N_{\mathbb{K}/\mathbb{Q}}\left(\sum_{i=1}^{n} c_i x_i\right)\right| \leq C \max_{1 \leq i \leq n} |c_i|^n$.

It is enough to show that the condition of Lemma 5.20 is satisfied for this C. Thus let an ideal J be given. As each c_i runs through the integers from 0 to

$N(J)^{1/n}$, we obtain more than $N(J)$ members $r = \sum_{i=1}^{n} c_i x_i$ of R. Since there are only $N(J)$ cosets modulo J, at least two of these members of r, say r_1 and r_2, must lie in the same coset.[20] Then $r_1 - r_2$ is a nonzero member of J, it has all coefficients between $-N(J)^{1/n}$ and $+N(J)^{1/n}$, and our construction of C forces $|N_{\mathbb{K}/\mathbb{Q}}(r_1 - r_2)| \leq C(N(J)^{1/n})^n = C N(J)$. □

The second proof of Theorem 5.19 is to combine Lemma 5.20 with the deeper and more quantitative estimate given in the following theorem.

Theorem 5.21 (Minkowski). For any number field $\mathbb{K}$ of degree n, each nonzero ideal J of R contains an element $s \neq 0$ with

$$|N_{\mathbb{K}/\mathbb{Q}}(s)| \leq \left(\frac{4}{\pi}\right)^{r_2} \frac{n!}{n^n} |D_{\mathbb{K}}|^{1/2} N(J).$$

Here r_2 is half the number of nonreal embeddings of $\mathbb{K}$ in $\mathbb{C}$, and $D_{\mathbb{K}}$ is the field discriminant. Therefore every equivalence class of ideals contains a member L whose absolute norm satisfies

$$N(L) \leq \left(\frac{4}{\pi}\right)^{r_2} \frac{n!}{n^n} |D_{\mathbb{K}}|^{1/2}.$$

We shall prove Theorem 5.21 shortly by applying Minkowski's Lattice-Point Theorem to the lattice $\Phi(J)$ in $\Omega = \mathbb{R}^{r_1} \times \mathbb{C}^{r_2}$, where Φ is the mapping described after the proof of Theorem 5.16. The particular compact convex set in the application takes some time to describe, and we return to that matter shortly.

Meanwhile, let us see a little of the utility of Theorem 5.21. The techniques of Chapter I are more useful for computing class numbers for $n = 2$ than Theorem 5.21 is, and we therefore consider only $n \geq 3$. For $n = 3$, we must have $r_2 \leq 1$. Theorem 5.21 shows that every equivalence class of ideals in R has a representative L with

$$N(L) \leq \frac{4}{\pi} \frac{3!}{3^3} |D_{\mathbb{K}}|^{1/2} = \frac{8}{9\pi} |D_{\mathbb{K}}|^{1/2} < (0.283) |D_{\mathbb{K}}|^{1/2}.$$

Problems 1–2 at the end of the chapter give examples of cubic extensions of $\mathbb{Q}$ whose discriminants are $-23, -31,$ and -44. Since these have $(0.283)|D_{\mathbb{K}}|^{1/2} \leq (0.283)7 < 2$, the representative ideal in each case must have norm 1 and must be R. Thus for all three of these cubic fields, R is a principal ideal domain.

---
[20]Again we are applying the Dirichlet pigeonhole principle.

For the cubic field $\mathbb{K} = \mathbb{Q}(\sqrt[3]{2})$, we know from Section 2 that the discriminant is $D_\mathbb{K} = -108$. Consequently the estimate shows that every class of ideals has a representative with norm ≤ 2. If an ideal J has $N(J) = 2$, then 2 has to be a member, and J divides $(2)R$. Proposition 5.10d shows that the factorization of $(2)R$ is as P^3 for a certain unique prime ideal P. Thus R and P represent all equivalence classes, and $h_\mathbb{K}$ is 1 or 2. If there is some $r \in R$ with $N_{\mathbb{K}/\mathbb{Q}}(r) = 2$, then $P = (r)$, and the class number is 1; otherwise it is 2. The element $\sqrt[3]{2}$ has $|N_{\mathbb{K}/\mathbb{Q}}(\sqrt[3]{2})| = 2$, and thus $P = (\sqrt[3]{2})$. Therefore R is a principal ideal domain when $\mathbb{K} = \mathbb{Q}(\sqrt[3]{2})$.

For Dedekind's example, namely the cubic number field $\mathbb{K}$ built from $X^3 + X^2 - 2X + 8$, we saw in Section 4 that the discriminant is $D_\mathbb{K} = -503$. Then the constant in the estimate is $< (0.283)\sqrt{503} < 6.35$. So the interest is in ideals of norm ≤ 6. In ruling out ideals that are principal, we need consider only prime ideals with norm ≤ 6. Problems 24–32 at the end of the chapter identify all the prime ideals of this form and show that they are all principal ideals! We conclude that $h_\mathbb{K} = 1$, i.e., that the R in Dedekind's example is a principal ideal domain. Not every cubic number field has class number 1, however; Problem 4 gives an example.

Before turning to the proof of Theorem 5.21, let us observe the following striking consequence.

Corollary 5.22 (Minkowski). For any number field $\mathbb{K}$ of degree n,

$$|D_\mathbb{K}|^{1/2} \geq \left(\frac{\pi}{4}\right)^{r_2} \frac{n^n}{n!}.$$

Therefore $D_\mathbb{K} > 1$ if $n \geq 2$, and there exists at least one prime number that ramifies in $\mathbb{K}$.

REMARKS. With a more general number field $\mathbb{F}$ than $\mathbb{Q}$ as base field, it can happen that no prime ideal ramifies in a certain nontrivial extension field $\mathbb{K}/\mathbb{F}$. See Problems 5–9 at the end of the chapter.

PROOF. Set $J = R$ in Theorem 5.21, so that $N(J) = 1$. The nonzero element s must have $|N_{\mathbb{K}/\mathbb{Q}}(s)| \geq 1$. The theorem says that $(4/\pi)^{r_2}(n!/n^n)|D_\mathbb{K}|^{1/2} \geq 1$, and this is the displayed inequality of the corollary. Since $r_2 \leq \frac{1}{2}n$, $(\pi/4)^{r_2} \geq (\pi/4)^{n/2}$, and thus $|D_\mathbb{K}|^{1/2} \geq 2^{-n}\pi^{n/2}n^n/n!$. Denote the right side of this inequality by a_n. For $n = 2$, we have $a_2 = \pi/2 > 1$. Also, $a_{n+1}/a_n = \frac{1}{2}\pi^{1/2}(1 + \frac{1}{n})^n \geq \pi^{1/2}$, since $(1 + \frac{1}{n})^n$ is monotone increasing[21] with n and is ≥ 2 for $n = 2$. Hence $a_n > 1$ for all $n \geq 2$. By Theorem 5.5 some prime number ramifies in $\mathbb{K}$. □

---

[21]To see this monotonicity, expand $a_{n+1} = (1 + \frac{1}{n+1})^{n+1}$ and $a_n = (1 + \frac{1}{n})^n$ by the Binomial Theorem, and observe that the asserted inequality holds term by term.

We turn to the proof of Theorem 5.21. We again make use of the map $\Phi : \mathbb{K} \to \Omega = \mathbb{R}^{r_1} \times \mathbb{C}^{r_2} \cong \mathbb{R}^n$ of the previous section. Lemma 5.17 shows that $\Phi(R)$ is a lattice in Ω, and our interest will be in the sublattice $\Phi(J)$, J being the nonzero ideal under study. The idea is to consider the set of $\omega \in \Omega$ for which the function

$$N(\omega) = \Big(\prod_{i=1}^{r_1} |\omega|_i\Big)\Big(\prod_{i=r_1+1}^{r_1+r_2} |\omega|_i^2\Big)$$

has $N(\omega) \leq c$, c being a positive number. Since $N(\Phi(x)) = |N_{\mathbb{K}/\mathbb{Q}}(x)|$ for $x \in \mathbb{K}$, the question of finding a member s of J with $|N_{\mathbb{K}/\mathbb{Q}}(s)| \leq c$ is the same as the question of finding a nonzero lattice point in the set for which $N(\omega) \leq c$. Once we sort out how large c has to be for the answer to be affirmative, then the inequality of the theorem will result. The tool will again be the Minkowski Lattice-Point Theorem (Theorem 5.16), but the difficulty is that the set for which $N(\omega) \leq c$ is not necessarily convex.

The nature of the set for which $N(\omega) \leq c$ becomes clearer by considering the case of $\mathbb{K} = \mathbb{Q}(\sqrt{m})$ with $m > 0$. The map Φ carries $x + y\sqrt{m}$ for x and y in $\mathbb{Q}$ to the pair $(x + y\sqrt{m}, x - y\sqrt{m})$ in $\mathbb{R}^2$, and if we parametrize ω by the pair (x, y), then the set for which $N(\omega) \leq c$ is the part of the (x, y) plane containing the origin and bounded by the two hyperbolas $x^2 - my^2 = c$ and $x^2 - my^2 = -c$. This set is not convex, and it is not even bounded.

Briefly, an individual coordinate of our $\Omega = \mathbb{R}^{r_1} \times \mathbb{C}^{r_2}$, whether a factor of type $\mathbb{R}$ or a factor of type $\mathbb{C}$, contributes something compact convex to the set for which $N(\omega) \leq c$ as long as the other coordinates are fixed, but as soon as we allow more than one coordinate to vary, then the product formula defining $N(\omega)$ produces sets that are neither convex nor bounded. To use Theorem 5.16, we want to inscribe a compact convex set within the set for which $N(\omega) \leq c$, making the inscribed set contain the origin, be closed under negatives, and have volume as large as possible.

If we were trying to inscribe such a compact convex set in a region cut out by two hyperbolas as above, then the best possible set to use would be a rectangle with sides parallel to the axes. However, the description above in terms of those two hyperbolas used a noncanonical parametrization of elements of $\mathbb{Q}(\sqrt{m})$ as all rational combinations $x + y\sqrt{m}$.

Let us proceed for the general case by using only the structure that is given to us, without using any noncanonical parametrization. The things that are canonical are the factors $\mathbb{R}$ and $\mathbb{C}$, the functions $\|\cdot\|_i$ defined on them, and functions of these. For the example above, the function $N(\omega)$ is given by $N(\omega) = |\omega|_1 |\omega|_2$. The geometric set in $\mathbb{R}^2 = \{(\omega_1, \omega_2)\}$ to consider is changed from above; it is still the set toward the origin from two hyperbolas, but the hyperbolas are changed to be $\omega_1 \omega_2 = \pm c$, having the axes as asymptotes. The inscribed convex set becomes the set with $|\omega_1| + |\omega_2| \leq 2c^{1/2}$. The containment of the latter set in the set toward the origin from the two hyperbolas follows from the inequality $|\omega_1\omega_2|^{1/2} \leq \frac{1}{2}(|\omega_1| + |\omega_2|)$, which is a consequence of the inequality $\frac{1}{4}(|\omega_1| - |\omega_2|)^2 \geq 0$.

In the general case the inscribed convex set is described in terms of the function

$$T(\omega) = \sum_{i=1}^{r_1} |\omega|_i + 2 \sum_{i=r_1+1}^{r_1+r_2} |\omega|_i.$$

The set of ω with $T(\omega) \leq t$, t being a positive constant, is evidently a compact convex set containing 0 and closed under negatives, and the functions $T(\omega)$ and $N(\omega)$ are connected by the arithmetic–geometric mean inequality, which says that

$$N(\omega)^{1/n} \leq \frac{1}{n} T(\omega).$$

Because of this inequality the set with $T(\omega) \leq t$ is contained in the set with $N(\omega) \leq t^n/n^n$.

Since the absolute value in each $\mathbb{R}$ or $\mathbb{C}$ coordinate is canonical, so is the notion of volume, given on rectangular sets by taking products; as usual the understanding is that the set in a factor of $\mathbb{R}$ on which the absolute value is $\leq k$ contributes a factor of $2k$ to the volume, and the comparable set in a factor of $\mathbb{C}$ contributes a factor of πk^2. If V_0 denotes the volume of a fundamental parallelotope for the lattice $\Phi(J)$ in the n-dimensional Euclidean space Ω, then the Minkowski Lattice-Point Theorem says that the set with $T(\omega) \leq t$, and therefore also the set with $N(\omega) \leq t^n/n^n$, contains a nonzero lattice point as soon as the volume of the set with $T(\omega) \leq t$ is $\geq 2^n V_0$. In other words, as soon as the volume of the set with $T(\omega) \leq t$ is $\geq 2^n V_0$, there exists an $s \neq 0$ in J with $|N_{\mathbb{K}/\mathbb{Q}}(s)| \leq t^n/n^n$.

To prove Theorem 5.21, we therefore need to know two things—the volume V_0 of a fundamental parallelotope for $\Phi(J)$ and the volume of the set with $T(\omega) \leq t$. Then we can find the smallest t for which the set with $T(\omega) \leq t$ has volume $\geq 2^n V_0$, and we can sort out the details.

Let us compute the volume V_0. Let $\Gamma = (\alpha_1, \ldots, \alpha_n)$ be an ordered $\mathbb{Z}$ basis of the ideal J. The easy case in which to compute V_0 is that $r_1 = n$, i.e., that all the field embeddings of $\mathbb{K}$ into $\mathbb{C}$ are real. In this case the discriminant $D(\Gamma)$ is the determinant of the n-by-n matrix $[B_{ij}]$ with

$$B_{ij} = \mathrm{Tr}_{\mathbb{K}/\mathbb{Q}}(\alpha_i\alpha_j) = \sum_{k=1}^n \sigma_k(\alpha_i\alpha_j) = \sum_{k=1}^n \sigma_k(\alpha_i)\sigma_k(\alpha_j) = \sum_{k=1}^n A_{ik} A^t_{jk},$$

where $[A_{ij}]$ is the matrix with $A_{ij} = \sigma_j(\alpha_i)$. We recognize $|\det[A_{ij}]|$ as the volume of a fundamental parallelotope for $\Phi(J)$, and therefore $|D(\Gamma)| = V_0^2$. By Proposition 5.1, $D(\Gamma) = N(J)^2 D_\mathbb{K}$, and therefore $V_0 = N(J)|D_\mathbb{K}|^{1/2}$.

6. Finiteness of the Class Number

This answer for the value of V_0 is not correct if some of the embeddings of $\mathbb{K}$ into $\mathbb{C}$ are nonreal, since $|\det[\sigma_j(\alpha_i)]|$ no longer equals V_0. To see how to adjust matters, suppose that σ is a nonreal field mapping of $\mathbb{K}$ into $\mathbb{C}$. Then the n-by-n matrix $[\sigma_j(\alpha_i)]$ contains one column $z = \begin{pmatrix} z_1 \\ \vdots \\ z_n \end{pmatrix}$ corresponding to σ and another column $\bar{z} = \begin{pmatrix} \bar{z}_1 \\ \vdots \\ \bar{z}_n \end{pmatrix}$ corresponding to $\bar{\sigma}$. The entries in the kth row tell how α_k is embedded in Ω, namely at some point $z_k = x_k + i y_k$ for σ and at $\bar{z}_k = x_k - i y_k$. To compute V_0 properly, we should have x_k in one column and y_k in the other, instead of z_k and $\bar{z}_k$. We can transform from the matrix with columns containing z_k and $\bar{z}_k$ to one containing x_k and y_k by first replacing the first column by the sum of the two, which is $2x_k = z_k + \bar{z}_k$, and by then replacing the second column by the difference of the second column and half the new first column, which is $\frac{1}{2}(\bar{z}_k - z_k) = -i y_k$. These operations do not change the determinant. Repeating these steps for each of the r_2 pairs of nonreal field mappings, we obtain a matrix for which the absolute value of the determinant, apart from factors of 2 in r_2 of the columns, is V_0. Consequently $V_0 = 2^{-r_2} |\det[\sigma_j(\alpha_i)]|$. Then $V_0^2 = 2^{-2r_2} |D(\Gamma)|$, and we obtain
$$V_0 = 2^{-r_2} N(J) |D_{\mathbb{K}}|^{1/2}.$$

Now let us compute the volume of the set of ω in Ω for which $T(\omega) \leq t$. Write $\omega = (x_1, \ldots, x_{r_1}, z_{r_1+1}, \ldots, z_{r_1+r_2})$. The volume is the integral of 1 over the set on which $|x_1| + \cdots + |x_{r_1}| + 2|z_{r_1+1}| + \cdots + 2|z_{r_1+r_2}| \leq t$. The set for the integration is invariant under $x_i \mapsto -x_i$ and under rotation in any variable z_i, and hence the volume equals
$$2^{r_1}(2\pi)^{r_2} \int_E \rho_{r_1+1} \cdots \rho_{r_1+r_2} \, dx_1 \cdots dx_{r_1} \, d\rho_{r_1+1} \cdots d\rho_{r_1+r_2},$$
where E is the set on which all variables are ≥ 0 and
$$\sum_{i=1}^{r_1} x_i + 2 \sum_{i=r_1+1}^{r_1+r_2} \rho_i \leq t.$$

For $r_1 + 1 \leq i \leq r_1 + r_2$, introduce $x_i = 2\rho_i$, and make the change of variables. Then the volume becomes
$$2^{r_1 - r_2} \pi^{r_2} \int_{E'} x_{r_1+1} \cdots x_{r_1+r_2} \, dx_1 \cdots dx_{r_1+r_2},$$
where E' is the set of $(x_1, \ldots, x_n)$ in $\mathbb{R}^{r_1+r_2}$ with all $x_i \geq 0$ and with $\sum_{i=1}^{r_1+r_2} x_i \leq t$. Finally we make a change of variables that replaces each x_i by $t y_i$, and the result is that

$$\text{volume}(\{T(\omega) \leq t\}) = 2^{r_1-r_2}\pi^{r_2}t^n \int_S y_{r_1+1}\cdots y_{r_1+r_2}\,dy_1\cdots dy_{r_1+r_2},$$

where S is the standard simplex in $\mathbb{R}^{r_1+r_2}$ with all $y_i \geq 0$ and with $\sum_{i=1}^{r_1+r_2} y_i \leq 1$. This definite integral is of a standard type that is evaluated by the following lemma.

Lemma 5.23. In $\mathbb{R}^m$, let S be the standard simplex with all $x_i \geq 0$ and with $\sum_{i=1}^m x_i \leq 1$. If $a_1, \ldots, a_m$ are positive real numbers, then

$$\int_S x_1^{a_1-1} x_2^{a_2-1} \cdots x_m^{a_m-1}\,dx_1\cdots dx_m = \frac{\Gamma(a_1)\Gamma(a_2)\cdots\Gamma(a_m)}{\Gamma(a_1+\cdots+a_m+1)}.$$

REMARKS. The expression $\Gamma(\cdot)$ is understood to be the usual gamma function, whose value at positive integers is given by $\Gamma(n+1) = n!$. We merely sketch the proof; the details can be found in many books that treat changes of variables for multiple integrals.[22]

SKETCH OF PROOF. Let I be the unit cube, given by $0 \leq u_i \leq 1$ for $1 \leq i \leq m$. We make the change of variables $x = \varphi(u)$ that carries the points u of the cube I one-one onto the points x of the simplex S and that is given by

$$x_1 = u_1,$$
$$x_2 = (1-u_1)u_2,$$
$$\vdots$$
$$x_m = (1-u_1)\cdots(1-u_{m-1})u_m.$$

The volume element transforms by the absolute value of the Jacobian determinant, specifically by

$$dx = |\varphi'(u)|\,du = (1-u_1)^{m-1}(1-u_2)^{m-2}\cdots(1-u_{m-1})\,du,$$

and the result of the change of variables is that the given integral equals

$$\prod_{i=1}^m \int_0^1 u_i^{a_i-1}(1-u_i)^{\sum_{k=i+1}^m a_k}\,du_i.$$

The factors here can be evaluated by means of Euler's formula

$$\int_0^1 u^{a-1}(1-u)^{b-1} = \frac{\Gamma(a)\Gamma(b)}{\Gamma(a+b)},$$

and the lemma follows. $\square$

---

[22]One such is the author's *Basic Real Analysis*; the details appear in the problems at the end of Chapter VI of that book. Another such book is Rudin's *Principles of Mathematical Analysis*.

For the integral of interest to us, we have $m = r_1 + r_2$, $a_1 = \cdots = a_{r_1} = 1$, and $a_{r_1+1} = \cdots = a_{r_1+r_2} = 2$. Thus $a_1 + \cdots + a_m = r_1 + 2r_2 = n$, and we obtain

$$\text{volume}(\{T(\omega) \le t\}) = 2^{r_1-r_2} \pi^{r_2} t^n \frac{\Gamma(1)^{r_1} \Gamma(2)^{r_1+r_2}}{\Gamma(n+1)} = \frac{2^{r_1-r_2} \pi^{r_2} t^n}{n!}.$$

Finally we can put everything together. We are to solve for t such that this expression is equal to $2^n V_0$, and then there exists an element $s \ne 0$ in J with $|N_{\mathbb{K}/\mathbb{Q}}(s)| \le t^n/n^n$. Since $V_0 = 2^{-r_2} N(J) |D_\mathbb{K}|^{1/2}$, the equation to solve for t is

$$\frac{2^{r_1-r_2} \pi^{r_2} t^n}{n!} = 2^n 2^{-r_2} N(J) |D_\mathbb{K}|^{1/2}.$$

Thus $t^n = \left(\frac{4}{\pi}\right)^{r_2} n! N(J) |D_\mathbb{K}|^{1/2}$, and the element $s \ne 0$ in J satisfies

$$|N_{\mathbb{K}/\mathbb{Q}}(s)| \le \left(\frac{4}{\pi}\right)^{r_2} \frac{n!}{n^n} |D_\mathbb{K}|^{1/2} N(J).$$

This completes the proof of Theorem 5.21.

7. Problems

1. Take as known that the discriminant of a cubic polynomial $F(X) = X^3 + pX + q$ is $-(4p^3 + 27q^2)$. In each of the following cases, let $\mathbb{K} = \mathbb{Q}[X]/(F(X))$ with $F(X)$ as indicated, and verify that the field discriminant $D_\mathbb{K}$ is as indicated:
   (a) $F(X) = X^3 - X - 1$, $D_\mathbb{K} = -23$.
   (b) $F(X) = X^3 + X + 1$, $D_\mathbb{K} = -31$.

2. Let $\mathbb{K} = \mathbb{Q}[X]/(F(X))$, where $F(X) = X^3 - 2X^2 + 2$.
   (a) Use the formula of the previous problem to show that the discriminant of the polynomial $F(X)$ is -44.
   (b) Using Proposition 5.2, show that $D_\mathbb{K}$ cannot be -11, and conclude that $D_\mathbb{K} = -44$.

3. This problem computes the class number of $\mathbb{K} = \mathbb{Q}(\sqrt[3]{3})$.
   (a) Show that every equivalence class of nonzero ideals contains an ideal with norm ≤ 4.
   (b) Show that the prime ideals whose norm is a power of 2 are $P_1 = (2, \sqrt[3]{3}-1)$, whose norm is 2, and $P_2 = (2, \sqrt[3]{9} + \sqrt[3]{3} + 1)$, whose norm is 4.
   (c) Show for P_1 that 2 is a multiple of $\sqrt[3]{3} - 1$, and show for P_2 that 2 is a multiple of $\sqrt[3]{9} + \sqrt[3]{3} + 1$.
   (d) Show that the only prime ideal whose norm is 3 is $(\sqrt[3]{3})$.
   (e) Deduce that the class number of $\mathbb{K}$ is 1.

4. Let R be the ring of algebraic integers in the number field $\mathbb{K} = \mathbb{Q}(\sqrt[3]{7})$, and let I be the doubly generated ideal $I = (2, 1 + \sqrt[3]{7})$ in R.
   (a) Prove that $N(I) = 2$.
   (b) Prove that I is not a principal ideal.

Problems 5–9 give an example of a nontrivial finite extension $\mathbb{L}/\mathbb{K}$ of number fields in which no prime ideal for $\mathbb{K}$ ramifies in passing to $\mathbb{L}$. By contrast, Corollary 5.22 says that there always exists a prime that ramifies in passing from $\mathbb{Q}$ to a nontrivial finite extension. The example has $\mathbb{L} = \mathbb{Q}(\sqrt{-5}, \sqrt{-1})$ and $\mathbb{K} = \mathbb{Q}(\sqrt{-5})$. Let $\mathbb{K}' = \mathbb{Q}(\sqrt{5})$ and $\mathbb{K}'' = \mathbb{Q}(\sqrt{-1})$. Observe that $\mathbb{L}/\mathbb{Q}$ is a Galois extension, and so are all the various quadratic extensions of $\mathbb{L}$ over $\mathbb{K}$, $\mathbb{K}'$, and $\mathbb{K}''$, as well as of $\mathbb{K}$, $\mathbb{K}'$, and $\mathbb{K}''$ over $\mathbb{Q}$. The problems make use of the fact that ramification indices multiply in passing to an extension in stages, and so do residue class degrees.

5. Show that the minimal polynomial of $\sqrt{-1} + \sqrt{-5}$ over $\mathbb{Q}$ is $X^4 + 12X^2 + 16$, and deduce that the elements $\frac{1}{2}(\pm\sqrt{-1} \pm \sqrt{-5})$ are algebraic integers in $\mathbb{L}$.

6. By making use the formula for $D(\xi)$ in terms of $\mathcal{D}(\xi)$, where ξ is an element in $\mathbb{L}$, prove that $|D(\frac{1}{2}(\sqrt{-1} + \sqrt{-5}))| = 2^4 5^2$. Consequently $D_\mathbb{L}$ divides $2^4 5^2$.

7. Verify the following decompositions of the ideals (2) and (5) when extended from $\mathbb{Z}$ to the rings R, R', and R'' of algebraic integers in $\mathbb{K}$, $\mathbb{K}'$, and $\mathbb{K}''$:
   (a) $(2)R = \wp^2$ with $f = 1$, and $(5)R = \wp^2$ with $f = 1$.
   (b) $(2)R' = \wp$ with $f = 2$, and $(5)R' = \wp^2$ with $f = 1$.
   (c) $(2)R'' = \wp^2$ with $f = 1$, and $(5)R'' = \wp_1\wp_2$ with $f = 1$.

8. Let T be the ring of algebraic integers in $\mathbb{L}$. Since $\mathbb{L}/\mathbb{Q}$ is a Galois extension, the only possible decompositions of $(p)T$, when p is a prime number, have (e, f, g) equal to $(4, 1, 1)$ or $(2, 2, 1)$ or $(2, 1, 2)$ or $(1, 4, 1)$ or $(1, 2, 2)$ or $(1, 1, 4)$. Here e is the ramification index, f is the residue class degree, and g is the number of distinct prime factors. Using the product formulas for ramification degrees and comparing what happens for the passage $\mathbb{Q} \subseteq \mathbb{K}' \subseteq \mathbb{L}$ with what happens for the passage $\mathbb{Q} \subseteq \mathbb{K}'' \subseteq \mathbb{L}$, show that the only possibilities for $(p)T$ with $p = 2$ and $p = 5$ are
   (a) $(e, f, g) = (2, 2, 1)$ for $(2)T$, i.e., $(2)T = P^2$ with $\dim_{\mathbb{F}_2}(T/P) = 2$.
   (b) $(e, f, g) = (2, 1, 2)$ for $(5)T$, i.e., $(5)T = P_1^2 P_2^2$ with $\dim_{\mathbb{F}_5}(T/P_1) = \dim_{\mathbb{F}_5}(T/P_2) = 1$.

9. Return to the situation with $\mathbb{Q} \subseteq \mathbb{K} \subseteq \mathbb{L}$, where $\mathbb{K} = \mathbb{Q}(\sqrt{-5})$. According to Problem 7a, the prime decompositions of $(2)R$ and $(5)R$ are $(2)R = \wp_2^2$ and $(5)R = \wp_5^2$.
   (a) Using the results of Problem 8, show that $\wp_2 T = P$ and $\wp_5 T = P_1 P_2$, i.e., $\wp_2 T$ is prime, and $\wp_5 T$ is the product of two distinct prime ideals.

(b) Show how to conclude from these facts and from Theorem 5.6 that no prime ideal in R ramifies in T. (Educational note: The field $\mathbb{L}$ is the "Hilbert class field" of $\mathbb{K}$ in the sense of Section 1; the order of the Galois group $\text{Gal}(\mathbb{L}/\mathbb{K})$ matches the class number of $\mathbb{K}$.)

Problems 10–16 concern the cyclotomic field $\mathbb{K} = \mathbb{Q}(e^{2\pi i/p})$, where $p > 2$ is a prime number. They show that the discriminant is given by $D_\mathbb{K} = p^{p-2}$ and that a $\mathbb{Z}$ basis of the ring R of algebraic integers in $\mathbb{K}$ consists of $\{1, \zeta, \zeta^2, \ldots, \zeta^{p-2}\}$, where $\zeta = e^{2\pi i/p}$.

10. Show that $\mathbb{K}$ has no real-valued field mappings into $\mathbb{C}$, and deduce that $N_{\mathbb{K}/\mathbb{Q}}(x)$ is positive for every $x \neq 0$ in $\mathbb{K}$.

11. Let $F(X) = X^{p-1} + X^{p-2} + \cdots + 1$ be the minimal polynomial of ζ over $\mathbb{Q}$, and let $G(X) = F(X + 1)$. Suppose that k is an integer with $\text{GCD}(k, p) = 1$.
    (a) Prove that $G(X)$ is the minimal polynomial of $\zeta^k - 1$, and deduce that the norm of $\zeta^k - 1$ is given by $F(1) = p$.
    (b) Why does it follow that $N_{\mathbb{K}/\mathbb{Q}}(1 - \zeta^k) = p$?
    (c) Prove that $(1 - \zeta^k)/(1 - \zeta)$ is a unit of R.

12. With notation as in the previous problem, prove that the different $\mathcal{D}(\zeta^k)$ of ζ^k has $|\mathcal{D}(\zeta^k)| = p/|\zeta^k - 1|$.

13. Deduce from the previous problem that $D(\zeta) = (-1)^{(p-1)(p-2)/2} p^{p-2}$.

14. Let $\lambda = 1 - \zeta$. Problem 11b shows that $N_{\mathbb{K}/\mathbb{Q}}(\lambda) = p$. Prove that
    (a) the $\mathbb{Z}$ span of $\{1, \zeta, \zeta^2, \ldots, \zeta^{p-2}\}$ equals the $\mathbb{Z}$ span of $\{1, \lambda, \lambda^2, \ldots, \lambda^{p-2}\}$.
    (b) an equality $p = \prod_{k=1}^{p-1}(1 - \zeta^k)$ holds.
    (c) there exists a unit ε of R such that $p = \varepsilon(1 - \zeta)^{p-1} = \varepsilon \lambda^{p-1}$.

15. Using Problem 14c, prove that the principal ideals $(p)R$ and (λ) in R are related by $(p)R = (\lambda)^{p-1}$, and deduce from this fact that (λ) is a prime ideal.

16. Apply Proposition 5.2 to the $\mathbb{Q}$ basis $\{1, \lambda, \lambda^2, \ldots, \lambda^{p-2}\}$ of $\mathbb{K}$ lying in R to show that no factor of p^2 can be eliminated from $D(\lambda) = D(\zeta)$; take into account the highest powers of λ that divide each term. Conclude that $D_\mathbb{K} = D(\zeta)$ and that $\{1, \zeta, \zeta^2, \ldots, \zeta^{p-2}\}$ is a $\mathbb{Z}$ basis of R.

Problems 17–18 use the same notation as in the text of the chapter: $\mathbb{K}$ is a number field of degree n over $\mathbb{Q}$, R is its ring of algebraic integers, $D_\mathbb{K}$ is its field discriminant, the field mappings of $\mathbb{K}$ into $\mathbb{C}$ are denoted by σ_i for $1 \leq i \leq n$, r_1 of the σ_i's are real-valued, and r_2 complex-conjugate pairs of the σ_i's are nonreal.

17. Prove that the sign of $D_\mathbb{K}$ is $(-1)^{r_2}$.

18. **(Stickelberger's condition)** Let $\Gamma = (\alpha_1, \ldots, \alpha_n)$ be an ordered n-tuple of members of R linearly independent over $\mathbb{Q}$, and suppose that $\mathbb{K}/\mathbb{Q}$ is a Galois extension. Write $\det[\sigma_j(\alpha_i)] = P - N$, where P is the sum of all the terms of

the determinant corresponding to even permutations and N is the sum corresponding to even permutations. Using Galois theory, prove that $P + N$ and PN are in $\mathbb{Z}$. Then write $D(\Gamma) = (\det[\sigma_j(\alpha_i)])^2 = (P+N)^2 - 4PN$, and deduce that the integer $D(\Gamma)$ is congruent to 1 or 0 modulo 4. (Educational note: A variant of this argument proves the same conclusion about $D(\Gamma)$ without the assumption that $\mathbb{K}/\mathbb{Q}$ is a Galois extension. One makes use of the smallest normal extension of $\mathbb{Q}$ containing $\mathbb{K}$; this is the splitting field of the minimal polynomial of any primitive element of $\mathbb{K}$.)

Problems 19–23 continue with the notation of Problems 17–18. It is to be proved that a suitable localization $S^{-1}R$ of R is a principal ideal domain for which the group of units is finitely generated as an abelian group. Let h be the class number of $\mathbb{K}$.

19. Let $I_1, \ldots, I_h$ be ideals representing all the equivalence classes of ideals in R. For each I_j, let u_j be a nonzero element of I_j, and put $u = u_1 \cdots u_h$. Define $S = \{1, u, u^2, \ldots\}$. Prove that $S^{-1}R$ is a principal ideal domain.

20. (a) Prove that if a member a of R divides u^k within R for some $k \geq 0$, then a is a unit in $S^{-1}R$, i.e., a^{-1} is in $S^{-1}R$.
    (b) Prove conversely that if a member a of R has the property that au^{-m} is a unit in $S^{-1}R$ for some $m \geq 0$, then a divides u^k within R for some integer $k \geq 0$.

21. Let $P_1, \ldots, P_l$ be the distinct prime ideals appearing in the unique factorization of (u), and suppose that $P_j^h = (b_j)$ for $1 \leq j \leq l$. Let au^{-m} and k be as in Problem 20b, and write $u^k = ab$ with $b \in R$.
    (a) Why must each b_j necessarily be a unit in $S^{-1}R$?
    (b) Prove that there exist integers $n_j \geq 0$ for $1 \leq j \leq l$ such that the element $d = \prod_j b_j^{n_j}$ has $(a) = (d) P_1^{t_1} \cdots P_l^{t_l}$ for some integers t_j with $0 \leq t_j \leq h-1$.
    (c) In this case, why must $P_1^{e_1} \cdots P_l^{e_l}$ be a principal ideal?

22. Suppose that there are N tuples $(e_1, \ldots, e_l)$ with $0 \leq e_j \leq h - 1$ for all j such that $P_1^{e_1} \cdots P_l^{e_l}$ is a principal ideal. For the ith such tuple, let the principal ideal be denoted by (c_i), $1 \leq i \leq N$. Prove that if k, a, and b are as in the previous problem and if the principal ideal in (c) of that problem is (c_i), then $a = bc_i\varepsilon$ for some ε in $R^\times$.

23. Conclude from the three previous problems that the group of units of $S^{-1}R$ is finitely generated as an abelian group.

Problems 24–32 complete the discussion in Section 4 of Dedekind's example of a cubic extension of $\mathbb{Q}$ with a common index divisor. The field is $\mathbb{K} = \mathbb{Q}(\xi)$, where ξ is a root of $F(X) = X^3 + X^2 - 2X + 8$, and it was shown in Section 4 that $D(\xi) = -2^2 \cdot 503$. Let R be the ring of algebraic integers in $\mathbb{K}$. It will be shown that R is a principal ideal domain.

7. Problems 311

24. Show that $\eta = 4/\xi$ is a root of the polynomial $G(X) = X^3 - X^2 + 2X + 8$, and conclude that η is in R.

25. (a) By rewriting $F(\xi)/\xi$ in terms of ξ and η, show that $\xi^2 + \xi - 2 + 2\eta = 0$.
   (b) By rewriting $G(\eta)/\eta$ in terms of ξ and η, show that $2\xi + 2 - \eta + \eta^2 = 0$. Conclude from this formula and (a) that products of ξ and η may be simplified according to the table

   $$\xi^2 = -\xi + 2 - 2\eta, \qquad \eta^2 = -2\xi - 2 + \eta, \qquad \xi\eta = 4.$$

   (c) Using the first formula in (b), deduce the containment of abelian groups given by $\mathbb{Z}(\{1, \xi, \xi^2\}) \subseteq \mathbb{Z}(\{1, \xi, \eta\})$.
   (d) Using the first formula in (b), deduce that η does not lie in $\mathbb{Z}(\{1, \xi, \xi^2\})$.
   (e) Conclude from the above facts that $\{1, \xi, \eta\}$ and $\{1, \xi, \frac{1}{2}(\xi^2 + \xi)\}$ are $\mathbb{Z}$ bases of R.

26. Let P be a prime ideal in R containing $(2)R$, write $\mathbb{F}$ for the field R/P, let $\varphi : R \to \mathbb{F}$ be the quotient homomorphism, and let $\bar{\xi} = \varphi(\xi)$ and $\bar{\eta} = \varphi(\eta)$. By applying φ to the table in Problem 25b and using the fact that the additive group generated by $\{1, \xi, \eta\}$ is all of R, prove that $\mathbb{F}$ has only two elements, i.e., that the residue class degree is $f = 1$, and that the only possibilities for φ are the following:

   $$\varphi = \varphi_{0,0} \quad \text{with} \quad \varphi_{0,0}(\xi) = 0, \quad \varphi_{0,0}(\eta) = 0,$$
   $$\varphi = \varphi_{1,0} \quad \text{with} \quad \varphi_{1,0}(\xi) = 1, \quad \varphi_{1,0}(\eta) = 0,$$
   $$\varphi = \varphi_{0,1} \quad \text{with} \quad \varphi_{0,1}(\xi) = 0, \quad \varphi_{0,1}(\eta) = 1.$$

27. Conversely show that the three functions $\varphi_{0,0}, \varphi_{1,0}, \varphi_{0,1}$ defined on ξ and η in the previous problem extend to well-defined ring homomorphisms of R onto $\mathbb{F}_2$.

28. Let $P_{0,0}, P_{1,0}$, and $P_{0,1}$ be the kernels of the ring homomorphisms in the previous problem. Prove that these ideals all have norm 2 and that $(2)R = P_{0,0} P_{1,0} P_{0,1}$.

29. (a) Prove that $P_{0,0} = (2, \xi, \eta)$, $P_{1,0} = (2, \xi + 1, \eta)$, and $P_{0,1} = (2, \xi, \eta + 1)$.
   (b) Exhibit η as a member of the ideal $(2, \xi + 1)$, and show therefore that $P_{1,0} = (2, \xi + 1)$.
   (c) Similarly show that $P_{0,1} = (2, \eta + 1)$ and that $P_{0,0} = (2, \xi - \eta)$.

30. The previous problem exhibited $P_{0,0}, P_{1,0}$, and $P_{0,1}$ explicitly as doubly generated. In fact, use of the norm map $N_{\mathbb{K}/\mathbb{Q}}$ will ultimately show them to be principal ideals.
   (a) Show that if $H(X)$ is the field polynomial over $\mathbb{Q}$ of an element θ in $\mathbb{K}$, then $N_{\mathbb{K}/\mathbb{Q}}(\theta) = -H(0)$ and $N_{\mathbb{K}/\mathbb{Q}}(\theta - q) = -H(q)$ for every $q \in \mathbb{Q}$.
   (b) Prove that $N_{\mathbb{K}/\mathbb{Q}}(\xi) = N_{\mathbb{K}/\mathbb{Q}}(\eta) = -8 = -2^3$, that $|N_{\mathbb{K}/\mathbb{Q}}(\xi + 3)| = 2^2$, that $|N_{\mathbb{K}/\mathbb{Q}}(\xi - 1)| = |N_{\mathbb{K}/\mathbb{Q}}(\xi + 2)| = 2^3$, and that $|N_{\mathbb{K}/\mathbb{Q}}(\xi - 2)| = 2^4$.

(c) Prove that $(\xi) = P_{0,0}^a P_{1,0}^b P_{0,1}^c$ for unique exponents ≥ 0 whose sum is 3, and that $(\eta) = P_{0,0}^\alpha P_{1,0}^\beta P_{0,1}^\gamma$ for unique exponents ≥ 0 whose sum is 3.

(d) Using the fact that $\xi\eta = 4$, prove that $a + \alpha = b + \beta = c + \gamma = 2$.

(e) Using the definitions of $P_{0,0}$, $P_{1,0}$, and $P_{0,1}$ as kernels, prove that $b = 0$ and $\gamma = 0$.

(f) Conclude that $(\xi) = P_{0,0} P_{0,1}^2$ and that $(\eta) = P_{0,0} P_{1,0}^2$.

31. This problem uses the norm computations in Problem 30b.
    (a) Using the defining homomorphisms, show that if l is an odd integer, then $P_{1,0}$ contains $(\xi + l)$, but $P_{0,0}$ and $P_{0,1}$ do not.
    (b) Show that $(\xi + 3) = P_{1,0}^2$ and that $(\xi - 1) = P_{1,0}^3$.
    (c) Using the defining homomorphisms, show that if l is an even integer, then $P_{0,1}$ contains $(\xi + l)$, but $P_{1,0}$ does not.
    (d) Show that $(2, \xi) = P_{0,0} P_{0,1}$.
    (e) Show that if l is an even integer not divisible by 4, then $P_{0,1}^2$ does not contain $(\xi + l)$.
    (f) Show that $(\xi + 2) = P_{0,0}^2 P_{0,1}$ and that $(\xi - 2) = P_{0,0}^3 P_{0,1}$.

32. (a) From the identity $(\xi + 2) P_{0,0} = (\xi - 2)$ that results from Problem 31f, deduce that $r_{0,0} = \frac{\xi - 2}{\xi + 2}$ is in R and that $P_{0,0} = (r_{0,0})$.
    (b) Deduce similarly that $P_{1,0}$ and $P_{0,1}$ are principal ideals.
    (c) Using Theorem 5.6, show that R contains no ideals of norm 3.
    (d) Using Theorem 5.6, show that the only ideal in R of norm 5 is $(5, 1 + \xi)$.
    (e) Show that $|N_{\mathbb{K}/\mathbb{Q}}(1 + \xi)| = 10$, and deduce that $(1 + \xi) = (5, 1 + \xi) P$, where P is one of the three ideals $P_{0,0}$, $P_{1,0}$, and $P_{0,1}$.
    (f) Why does it follow that $(5, 1 + \xi)$ is a principal ideal?
    (g) Prove that R is a principal ideal domain.

CHAPTER VI

Reinterpretation with Adeles and Ideles

Abstract. This chapter develops tools for a more penetrating study of algebraic number theory than was possible in Chapter V and concludes by formulating two of the main three theorems of Chapter V in the modern setting of "adeles" and "ideles" commonly used in the subject.

Sections 1–5 introduce discrete valuations, absolute values, and completions for fields, always paying attention to implications for number fields and for certain kinds of function fields. Section 1 contains a prototype for all these notions in the construction of the field $\mathbb{Q}_p$ of p-adic numbers formed out of the rationals. Discrete valuations in Section 2 are a generalization of the order-of-vanishing function about a point in the theory of one complex variable. Absolute values in Section 3 are real-valued multiplicative functions that give a metric on a field, and the pair consisting of a field and an absolute value is called a valued field. Inequivalent absolute values have a certain independence property that is captured by the Weak Approximation Theorem. Completions in Section 4 are functions mapping valued fields into their metric-space completions. Section 5 concerns Hensel's Lemma, which in its simplest form allows one to lift roots of polynomials over finite prime fields $\mathbb{F}_p$ to roots of corresponding polynomials over p-adic fields $\mathbb{Q}_p$.

Section 6 contains the main theorem for investigating the fundamental question of how prime ideals split in extensions. Let K be a finite separable extension of a field F, let R be a Dedekind domain with field of fractions F, and let T be the integral closure of R in K. The question concerns the factorization of an ideal $\mathfrak{p}T$ in T when $\mathfrak{p}$ is a nonzero prime ideal in R. If $F_\mathfrak{p}$ denotes the completion of F with respect to $\mathfrak{p}$, the theorem explains how the tensor product $K \otimes_F F_\mathfrak{p}$ splits uniquely as a direct sum of completions of valued fields. The theorem in effect reduces the question of the splitting of $\mathfrak{p}T$ in T to the splitting of $F_\mathfrak{p}$ in a complete field in which only one of the prime factors of $\mathfrak{p}T$ plays a role.

Section 7 is a brief aside mentioning additional conclusions one can draw when the extension K/F is a Galois extension.

Section 8 applies the main theorem of Section 6 to an analysis of the different of K/F and ultimately to the absolute discriminant of a number field. With the new sharp tools developed in the present chapter, including a Strong Approximation Theorem that is proved in Section 8, a complete proof is given for the Dedekind Discriminant Theorem; only a partial proof had been accessible in Chapter V.

Sections 9–10 specialize to the case of number fields and to function fields that are finite separable extensions of $\mathbb{F}_q(X)$, where $\mathbb{F}_q$ is a finite field. The adele ring and the idele group are introduced for each of these kinds of fields, and it is shown how the original field embeds discretely in the adeles and how the multiplicative group embeds discretely in the ideles. The main theorems are compactness theorems about the quotient of the adeles by the embedded field and about the quotient of the normalized ideles by the embedded multiplicative group. Proofs are given only for number fields. In the first case the compactness encodes the Strong Approximation Theorem of Section 8 and the Artin product formula of Section 9. In the second case the compactness encodes both the finiteness of the class number and the Dirichlet Unit Theorem.

1. p-adic Numbers

This chapter will sharpen some of the number-theoretic techniques used in Chapter V, finally arriving at the setting of "adeles" and "ideles" in which many of the more recent results in number theory have tidy formulations. Although Chapter V dealt only with number fields, the present chapter will allow a greater degree of generality that includes results in the algebraic geometry of curves. This greater degree of generality will not require much extra effort, and it will allow us to use each of the subjects of number theory and algebraic geometry to motivate the other.

The first section of Chapter V returned to the idea that one can get some information about the integer solutions of a Diophantine equation by considering the equation as a system of congruences modulo each prime number. However, we lose information by considering only primes for the modulus, and this fact lies behind the failure of Chapter V to give a complete proof of the Dedekind Discriminant Theorem (Theorem 5.5). The proof that we did give was of a related result, Kummer's criterion (Theorem 5.6), which concerns a field $\mathbb{Q}(\xi)$, where ξ is a root of an irreducible monic polynomial $F(X)$ in $\mathbb{Z}[X]$. The statement of Theorem 5.6 involves the reduction of $F(X)$ modulo certain prime numbers p and no other congruences.

The Chinese Remainder Theorem tells us that a congruence modulo any integer can be solved by means of congruences modulo prime powers, and the formulation of Theorem 5.6 uses only congruences modulo primes raised to the first power. Let us strip away the complicated setting from such congruences and see some examples of how the use of prime powers can make a difference.

EXAMPLES.

(1) Consider the problem of finding a square root of 5 modulo powers of 2. For the first power, we have

$$x^2 - 5 = (x-1)^2 + 2x - 6 \equiv (x-1)^2 \bmod 2,$$

i.e., $x^2 - 5$ is the square of a linear factor modulo 2. For the second power, the computation is

$$x^2 - 5 = (x-1)(x+1) - 4 \equiv (x-1)(x+1) \bmod 4,$$

and $x^2 - 5$ is the product of two distinct linear factors modulo 4. For the third power, $x^2 - 5$ is irreducible modulo 8 because the only odd squares modulo 8 are ± 1. Thus the polynomial $x^2 - 5$ exhibits a third kind of behavior when considered modulo 8. For higher powers of 2, the irreducibility persists because a nontrivial factorization modulo 2^k with $k > 3$ would imply a nontrivial factorization modulo 8.

(2) Consider the problem of finding a square root of 17 modulo powers of 2. We readily compute that

$$x^2 - 17 = (x-1)^2 + 2x - 18 \equiv (x-1)^2 \bmod 2,$$
$$x^2 - 17 = (x-1)(x+1) - 16 \equiv (x-1)(x+1) \bmod 4,$$
$$x^2 - 17 = (x-1)(x+1) - 16 \equiv (x-1)(x+1) \bmod 8,$$
$$x^2 - 17 = (x-1)(x+1) - 16 \equiv (x-1)(x+1) \bmod 16,$$
$$x^2 - 17 = (x-7)(x+7) + 32 \equiv (x-7)(x+7) \bmod 32,$$
$$x^2 - 17 = (x-9)(x+9) + 64 \equiv (x-9)(x+9) \bmod 64,$$

i.e., that the factorization of $x^2 - 17$ begins in the same way as for $x^2 - 5$ but that $x^2 - 17$ continues to factor as the product of two distinct linear factors modulo $2^3, 2^4, 2^5$, and 2^6. We can argue inductively that this pattern persists through all higher powers. In fact, suppose that $x^2 - 17 = (x-m)(x+m) \bmod 2^k$ for an integer $k \geq 3$. Then

$$x^2 - 17 = x^2 - m^2 + a2^k,$$

and m must be odd. Then we can write

$$x^2 - 17 = x^2 - (m - a2^{k-1})^2 + a2^k(1 - m + a2^{k-2}).$$

The factor $(1 - m + a2^{k-2})$ is even, and this equality shows that $x^2 - 17$ is the product of two distinct linear factors modulo 2^{k+1}. This completes the induction.

One immediate observation from the two examples is that the factorizations of $x^2 - 5$ and $x^2 - 17$ are the same modulo 2 and modulo 2^2 but are qualitatively distinct modulo higher powers of 2. Another observation is the nature of the data produced by the inductive argument in Example 2: For each k, we obtain an odd integer m_k such that $m_k^2 \equiv 17 \bmod 2^k$, and the m_k's are constructed in such a way that $m_{k+1} = m_k - a_k 2^{k-1}$ if $m_k^2 = 17 + a_k 2^k$. It follows that if $l \geq k$, then $m_k - m_l$ is divisible by 2^{k-1}, i.e., by higher and higher powers of 2 as k increases.

A first conclusion is that we get additional information by using congruences modulo prime powers. A second and more subtle conclusion is that it would be desirable to regard the sequence $\{m_k\}$ as stabilizing in some sense; then we could regard the system of congruences modulo all powers 2^k as having a single pair of solutions that we can consider as square roots of 17. In this case we would not have to think about infinitely many solutions to infinitely many unrelated congruences.

The construction that is to follow in this section, which is due to K. Hensel, will capture this information as a single "2-adic number." Conversely the 2-adic number carries with it the congruence information modulo 2^k for all positive integers k.

Thus the revised method of considering congruences prime by prime will be a two-step process, first a step of "localization" and then a step of "completion." In our application in Chapter V, we did not explicitly make use of localization in the sense of Chapter VIII of *Basic Algebra*, but it was there implicitly—in Proposition 5.2 for example and in the proof of Theorem 5.6. Carrying out the details of setting up the theory behind the two-stage process will take some work and will occupy the first four sections of this chapter. Let us get started.

Let p be a prime number. We define a real-valued function $|\cdot|_p$ on the field $\mathbb{Q}$ of rationals as follows: we take $|0|_p = 0$, and for any rational $r = p^m ab^{-1}$ with a and b equal to integers relatively prime to p, we define $|r|_p = p^{-m}$. The function $|\cdot|_p$ is called the *p*-**adic absolute value** on $\mathbb{Q}$. It has the following properties:

(i) $|x|_p \geq 0$ with equality if and only if $x = 0$,
(ii) $|x + y|_p \leq \max(|x|_p, |y|_p)$,
(iii) $|xy|_p = |x|_p |y|_p$,
(iv) $|-1|_p = |1|_p = 1$, and
(v) $|-x|_p = |x|_p$.

In fact, with (ii), equality holds if $|x|_p \neq |y|_p$, and the case with $|x|_p = |y|_p$ comes down to the observation that $\frac{a}{b} + \frac{c}{d} = \frac{ad+bc}{bd}$ has no factor of p in its denominator if b and d are relatively prime to p. Property (iii) comes down to the fact that if a, b, c, d are relatively prime to p, then so are ac and bd. The other properties follow from the first three: To see that $|1|_p = 1$ in (iv), we observe from (iii) that $|1|_p$ is a nonzero solution of $x^2 = x$ and thus has to be 1. This conclusion and (iii) together show that $|-1|_p$ is a positive solution of $x^2 = 1$ and thus has to be 1. Property (v) follows immediately by combining (iii) and (iv).

Inequality (ii) is called the **ultrametric inequality**. It implies that $|x + y|_p \leq |x|_p + |y|_p$, and consequently the function $d(x, y) = |x - y|_p$ satisfies the triangle inequality

$$d(x, y) \leq d(x, z) + d(z, y).$$

Since (i) shows that $d(x, y) \geq 0$ with equality exactly when $x = y$ and since (v) implies that $d(x, y) = |x - y|_p = d(y, x)$, the function d on $\mathbb{Q} \times \mathbb{Q}$ is a metric. It is called the *p*-**adic metric** on $\mathbb{Q}$.

The field $\mathbb{Q}_p$ of *p*-**adic numbers** will be obtained by completing this metric and extending the field operations to the completion. Let us see to the details. Regard the space $\prod_{j=1}^{\infty} \mathbb{Q}$ of sequences $\{q_j\}_{j=1}^{\infty}$ of rational numbers as the direct product of copies of the ring $\mathbb{Q}$, the operations being taken coordinate by coordinate. Then $\prod_{j=1}^{\infty} \mathbb{Q}$ is a commutative ring with identity, the identity being the sequence whose terms are all equal to 1.

1. p-adic Numbers

As is usual for metric spaces, we say that a sequence of rationals, i.e., a member $\{q_j\}$ of $\prod_{j=1}^{\infty} \mathbb{Q}$, is **convergent** to $q \in \mathbb{Q}$ in the p-adic metric if for any real $\epsilon > 0$, there exists an integer N such that $|q_n - q|_p < \epsilon$ for all $n \geq N$. Convergence in this metric is quite different from what one might expect; for example the sequence $\{2^j\}_{j=1}^{\infty}$ is convergent to 0 when $p = 2$. The sequence $\{q_j\}$ is a **Cauchy sequence** in the p-adic metric if for any real $\epsilon > 0$, there exists an integer N such that $|q_m - q_n|_p < \epsilon$ for all $m \geq N$ and all $n \geq N$. Convergent sequences are Cauchy, as follows from the inequality $|q_m - q_n|_p \leq |q_m - q|_p + |q - q_n|_p$. Cauchy sequences need not be convergent, but every Cauchy sequence $\{q_n\}$ is **bounded** in the sense that there is some real C with $|q_n|_p \leq C$ for all n.

EXAMPLE 2, CONTINUED. We obtained a sequence $\{m_k\}$ of odd integers such that $l \geq k$ implies that $m_k - m_l$ is divisible by 2^{k-1} and $m_k^2 - 17$ is divisible by 2^k. In terms of the 2-adic absolute value, $|m_k - m_l|_p \leq 2^{-(k-1)}$ and $|m_k^2 - 17|_p \leq 2^{-k}$. The sequence $\{m_k\}$ is therefore a Cauchy sequence in the 2-adic metric, and the sequence $\{m_k^2\}$ is convergent in the 2-adic metric to 17.

It follows from the ultrametric inequality that the sum and difference of Cauchy sequences is bounded, and (ii) and the boundedness of Cauchy sequences implies that the product of two Cauchy sequences is Cauchy. Therefore the subset $\mathcal{R}$ of Cauchy sequences is a subring with identity within $\prod_{j=1}^{\infty} \mathbb{Q}$.

In the theory of metric spaces, one defines a suitable notion of equivalence of Cauchy sequences, and the set of equivalence classes becomes a complete metric space,[1] any member q of $\mathbb{Q}$ being identified with the constant Cauchy sequence whose terms all equal q. With the p-adic metric, one can then prove that the field operations extend to the completion, and the completion is the field of p-adic numbers. This verification is a little tedious when done directly, and we can proceed more expeditiously by using some elementary ring theory.

Since convergent sequences are Cauchy, the set $\mathcal{I}$ of sequences convergent to 0 is a subset of the ring $\mathcal{R}$. The sum or difference of two such sequences is again convergent to 0, and $\mathcal{I}$ is an additive subgroup. We shall show that $\mathcal{I}$ is in fact an ideal in $\mathcal{R}$. Thus let $\{z_n\}$ be convergent to 0, and let $\{q_n\}$ be Cauchy. Since $\{q_n\}$ is Cauchy, it is bounded, say with $|q_n|_p \leq M$ for all n. If $\epsilon > 0$ is given, choose N such that $n \geq N$ implies $|z_n|_p \leq \epsilon/M$. Then $n \geq N$ implies that $|z_n q_n|_p = |z_n|_p |q_n|_p \leq (\epsilon/M)M = \epsilon$. Hence $\{z_n q_n\}$ is convergent to 0, and $\mathcal{I}$ is an ideal in $\mathcal{R}$.

Proposition 6.1. With the p-adic absolute value imposed on $\mathbb{Q}$, let $\mathcal{R}$ be the subring of $\prod_{j=1}^{\infty} \mathbb{Q}$ consisting of all Cauchy sequences, and let $\mathcal{I}$ be the ideal in

---

[1]This construction is carried out in detail in Section II.11 of the author's *Basic Real Analysis*.

$\mathcal{R}$ consisting of all sequences convergent to 0. Then $\mathcal{I}$ is a maximal ideal in $\mathcal{R}$, and the quotient $\mathcal{R}/\mathcal{I}$ is a field. Consequently the Cauchy completion of $\mathbb{Q}$ in the p-adic metric is a topological field $\mathbb{Q}_p$ into which $\mathbb{Q}$ embeds via a field mapping. If $|\cdot|_p$ denotes the function $d(\cdot, 0)$ on $\mathbb{Q}_p$, then $|\cdot|_p$ is a continuous extension of the p-adic absolute value from $\mathbb{Q}$ to $\mathbb{Q}_p$, and it satisfies

(a) $|x|_p \geq 0$ with equality if and only if $x = 0$,
(b) $|x + y|_p \leq \max(|x|_p, |y|_p)$, and
(c) $|xy|_p = |x|_p |y|_p$.

The subset $\mathbb{Z}_p = \{x \in \mathbb{Q}_p \mid |x|_p \leq 1\}$ is an open closed subring of $\mathbb{Q}_p$ in which $\mathbb{Z}$ is dense, and $\mathbb{Z}_p$ is compact. Consequently the topological field $\mathbb{Q}_p$ is locally compact.

REMARKS. The field $\mathbb{Q}_p$ is called the field of p-**adic numbers**, and the ring $\mathbb{Z}_p$ is called the ring of p-**adic integers**. The ring $\mathbb{Z}_p$ contains the identity of $\mathbb{Q}_p$.

PROOF. First let us prove that $\mathcal{I}$ is a maximal ideal. Arguing by contradiction, let $\{q_n\}$ be a Cauchy sequence that is not in $\mathcal{I}$, i.e., is not convergent to 0. Then there exists an $\epsilon_0 > 0$ such that $|q_n|_p \geq \epsilon_0$ for infinitely many n. Choose N such that $|q_n - q_m| < \epsilon_0/2$ whenever $n \geq N$ and $m \geq N$, and find some $n_0 \geq N$ with $|q_{n_0}|_p \geq \epsilon_0$. Then $n \geq N$ implies that $|q_n|_p \geq \epsilon_0/2$ because otherwise we would have $\epsilon_0 \leq |q_{n_0}|_p \leq |q_n - q_{n_0}|_p + |q_n|_p < \epsilon_0/2 + \epsilon_0/2 = \epsilon_0$, contradiction. Let $\{r_n\}$ be the sequence with $r_n = 0$ for $n < N$ and $r_n = q_n^{-1}$ for $n \geq N$. For $n \geq N$ and $m \geq N$, we have

$$|r_n - r_m|_p = |q_n^{-1} - q_m^{-1}|_p = |(q_m - q_n)/(q_m q_n)|_p$$
$$= |q_m - q_n|_p |q_m|_p^{-1} |q_n|_p^{-1} \leq 4\epsilon_0^{-2} |q_m - q_n|_p,$$

and it follows that $\{r_n\}_p$ is Cauchy and hence lies in $\mathcal{R}$. Since $\mathcal{I}$ is an ideal in $\mathcal{R}$, $\{r_n q_n\}$ is Cauchy. The terms of the sequence $\{r_n q_n\}$ are all equal to 1 for $n \geq N$, and hence $\{r_n q_n\}$ differs from the identity of $\mathcal{R}$ by a member of $\mathcal{I}$. Consequently the identity is in $\mathcal{I}$. This is a contradiction, since the members of the constant sequence $\{1\}$ are at distance $|1 - 0|_p = 1$ from 0. Hence $\mathcal{I}$ is a maximal ideal, and $\mathcal{R}/\mathcal{I}$ is necessarily a field.

Meanwhile, the Cauchy completion $\mathbb{Q}_p$ of $\mathbb{Q}$ is the set of equivalence classes from $\mathcal{R}$, two members of $\mathcal{R}$ being equivalent if they differ by a sequence convergent to 0. Consequently the Cauchy completion $\mathbb{Q}_p$ is precisely $\mathcal{R}/\mathcal{I}$ as a set. The mapping $\mathbb{Q} \to \mathcal{R} \to \mathcal{R}/\mathcal{I}$ carrying a member q of $\mathbb{Q}$ to the constant sequence $\{q_n\}$ with all $q_n = q$ and then from $\mathcal{R}$ to the quotient $\mathcal{R}/\mathcal{I} = \mathbb{Q}_p$ evidently respects the operations and hence is a field mapping. This mapping identifies $\mathbb{Q}$ with a subset of $\mathbb{Q}_p$. The metric d on $\mathbb{Q}$ extends uniquely to a continuous function on

1. p-adic Numbers

the completion $\mathbb{Q}_p \times \mathbb{Q}_p$, and therefore the p-adic absolute value $|\cdot|_p = d(\cdot, 0)$ extends to a continuous function on $\mathbb{Q}_p$.

Property (a) for the function $|\cdot|_p$ on $\mathbb{Q}_p$ follows from the fact that the continuous extension of d is a metric on $\mathbb{Q}_p$. To see that (b) and (c) hold on $\mathbb{Q}_p$, let x and y be members of $\mathbb{Q}_p = \mathcal{R}/\mathcal{I}$, and let $\{q_n\}$ and $\{r_n\}$ be respective coset representatives of them in $\mathcal{R}$. Then $\{q_n + r_n\}$ and $\{q_n r_n\}$ are representatives of $x + y$ and xy by definition, and the continuity of the p-adic absolute value on $\mathbb{Q}_p$ implies that $\lim_n |q_n + r_n|_p = |x+y|_p$ and $\lim_n |q_n r_n|_p = |xy|_p$. From the first of these limit formulas and from (b) on $\mathbb{Q}$, we obtain

$$|x+y|_p = \limsup_n |q_n + r_n|_p \le \limsup_n \max(|q_n|_p, |r_n|_p) = \max(|x|_p, |y|_p),$$

since $\lim_n |q_n|_p = |x|_p$ and $\lim_n |r_n|_p = |y|_p$. This proves (b) on $\mathbb{Q}_p$. Similarly

$$|xy|_p = \lim_n |q_n r_n|_p = \lim_n |q_n|_p |r_n|_p = (\lim_n |q_n|_p)(\lim_n |r_n|_p) = |x|_p |y|_p,$$

and this proves (c) on $\mathbb{Q}_p$.

To see that addition, subtraction, and multiplication are continuous on $\mathbb{Q}_p \times \mathbb{Q}_p$, let $\{x_n\}$ and $\{y_n\}$ be convergent sequences in $\mathbb{Q}_p$ with respective limits x and y. Use of (b) on $\mathbb{Q}_p$ gives

$$|(x_n + y_n) - (x+y)|_p = |(x_n - x) + (y_n - y)|_p \le \max(|x_n - x|_p, |y_n - y|_p).$$

The right side has limit 0 in $\mathbb{R}$, and therefore $x_n + y_n$ has limit $x+y$ in $\mathbb{Q}_p$. A completely analogous argument, making use also of the equality $|-1|_p = |1|_p$, shows that subtraction is continuous. Consider multiplication. If M is an upper bound for the absolute values $|x_n|_p$ and $|y_n|_p$, then use of (c) on $\mathbb{Q}_p$ gives

$$\begin{aligned}|x_n y_n - xy|_p &= |x_n(y_n - y) + y(x_n - x)|_p \\ &\le \max(|x_n(y_n - y)|_p, |y(x_n - x)|_p) \\ &= \max(|x_n|_p |y_n - y|_p, |y|_p |x_n - x|_p) \\ &\le \max(M|y_n - y|_p, |y|_p |x_n - x|_p).\end{aligned}$$

The right side has limit 0 in $\mathbb{R}$, and therefore $x_n y_n$ has limit xy in $\mathbb{Q}_p$.

To see that inversion $x \mapsto x^{-1}$ is continuous on $\mathbb{Q}_p^\times$, let $\{x_n\}$ be a sequence in $\mathbb{Q}_p^\times$ with limit x in $\mathbb{Q}_p^\times$. Since $\lim_n |x_n|_p = |x|_p$, we can find an integer N such that $|x_n|_p \ge \tfrac{1}{2}|x|_p$ for $n \ge N$. The computation

$$|x_n^{-1} - x^{-1}|_p = |(x - x_n)/(x_n x)|_p = |x - x_n|_p / (|x_n|_p |x|_p) \le 2|x|_p^{-1}|x - x_n|_p,$$

valid for $n \geq N$, shows that $\lim x_n^{-1} = x^{-1}$, and inversion is continuous. Consequently $\mathbb{Q}_p$ is a topological field.

It follows immediately from properties (b) and (c) and from the equality $|-x|_p = |x|_p$ that $\mathbb{Z}_p$ is a subring of $\mathbb{Q}_p$. Since $\mathbb{Z}_p$ is defined in terms of a continuous function and an inequality, it is closed. It can also be defined as the subset with $|x|_p < p$ because the p-adic absolute value takes no values between 1 and p, and therefore $\mathbb{Z}_p$ is open. The most general nonzero member of $\mathbb{Q} \cap \mathbb{Z}_p$ is of the form $q = a/b$, where a and b are relatively prime nonzero integers with $|a/b|_p \leq 1$. Here $|b|_p = 1$, and p cannot divide b. If $k > 0$ is given, then it follows that there exists n with $bn - a \equiv 0 \bmod p^k$. This n has $|n - \frac{a}{b}|_p = |bn - a|_p \leq p^{-k}$. So q is in the closure of $\mathbb{Z}$ in $\mathbb{Q}_p$. In other words, the closure of $\mathbb{Z}$ contains $\mathbb{Q} \cap \mathbb{Z}_p$. Since $\mathbb{Q}$ is dense in $\mathbb{Q}_p$, $\mathbb{Z}$ is dense in $\mathbb{Z}_p$.

For each integer $n \geq 0$, the set $\mathbb{Z}_p$ is covered by the closed balls of radius p^{-n} centered at the integers $0, 1, 2, \ldots, p^n - 1$. In fact, every integer z has $z \equiv k \bmod p^n$ for some integer $k \in \{0, 1, 2, \ldots, p^n - 1\}$. For this k, $|z - k|_p \leq p^{-n}$. Thus $\mathbb{Z}$ is contained in the union of the closed balls of radius p^{-n} centered at $0, 1, 2, \ldots, p^n - 1$. This union is closed; since $\mathbb{Z}$ is dense in $\mathbb{Z}_p$, $\mathbb{Z}_p$ is contained in this union. In turn, these closed balls are contained in the open balls of radius p^{-n+1} centered at the integers $0, 1, 2, \ldots, p^n - 1$. Thus for any positive radius, there exists a finite collection of open balls of that radius or less such that the union of the open balls covers $\mathbb{Z}_p$. This means that $\mathbb{Z}_p$ is totally bounded in the metric space $\mathbb{Q}_p$. A totally bounded closed subset of a complete metric space is compact, and consequently $\mathbb{Z}_p$ is compact.

Thus the 0 element of $\mathbb{Q}_p$ has $\mathbb{Z}_p$ as a compact neighborhood. Since addition is continuous, $x + \mathbb{Z}_p$ is a compact neighborhood of x, and therefore $\mathbb{Q}_p$ is locally compact. $\square$

2. Discrete Valuations

The construction of the p-adic absolute value on $\mathbb{Q}$ seemingly made use of unique factorization of the members of $\mathbb{Z}$, but actually the unique factorization of the ideals in $\mathbb{Z}$ would have been sufficient. Thus we shall see in a moment that the construction extends to apply to any number field F as soon as we specify a nonzero prime ideal P in the ring R of algebraic integers of F. In fact, there is nothing special about a number field. If R is any Dedekind domain and F is its field of fractions, then the construction extends to F as soon as we specify a nonzero prime ideal P in R.

Before describing the extended construction, let us look at the definition of the p-adic absolute value on $\mathbb{Q}$ more closely. Recall that if $x = p^m ab^{-1}$ for integers a and b relatively prime to p, then $|x|_p = p^{-m}$. Actually, the base p in this exponential is not very important at this point, and we could have used any real number $r > 1$ in place of p in p^{-m}. With this adjustment the p-adic absolute value would have been given by $|x|_p = r^{-v_p(x)}$, where $v_p(x)$ is the exact net power of p that occurs when the prime factorizations of the numerator and denominator of x are used. The exponent $v_p(x)$ is what is important; the base r is unimportant.

The expression $v_p(x)$ for $\mathbb{Q}$ is analogous to the order of vanishing of a polynomial in one complex variable at a point, and Hensel was led to the p-adic absolute value by carrying the notion for $\mathbb{C}[X]$ to the setting with $\mathbb{Q}$. In setting up a generalization, we shall work first with the generalization of the order of vanishing $v_p(x)$, since it is the more primitive notion, and in Section 3 we shall exponentiate to obtain a generalization of the absolute value for which we can form a completion.

To make the definitions, it is convenient to make use of fractional ideals, which were the subject of a set of problems in Chapter VIII of *Basic Algebra*. Let us recall the definition and the relevant properties. Again let R be a Dedekind domain, and let F be its field of fractions. A **fractional ideal** of F is any finitely generated R module M. For such an R module, there exists some $a \in R$ with $aM \subseteq R$, and then aM is an ideal of R. If M is any nonzero fractional ideal, then $M^{-1} = \{x \in F \mid xM \in R\}$ is a nonzero fractional ideal, and $MM^{-1} = R$. With this definition and property, it readily follows from the unique factorization of ideals in R that any nonzero fractional ideal M of F is of the form

$$M = \prod_{j=1}^{l} P_j^{k_j},$$

for a suitable set $\{P_1, \ldots, P_l\}$ of distinct nonzero prime ideals of R and for suitable nonzero integer exponents k_j. This expansion is unique up to the order of the factors, and every such expression is a fractional ideal. It follows that the nonzero fractional ideals form a group under multiplication. At the end of this section, we shall mention how this group is related to the ideal class group of F as defined in Section V.6.

If $x \neq 0$ is in F, then the **principal fractional ideal** $(x) = xR$ has a factorization as above. If P is a nonzero prime ideal of R, we let $v_P(x)$ be the negative of the integer exponent of P in the prime factorization of (x). For example, if x is a nonzero element of R, then $v_P(x)$ is a nonnegative integer. To make $v_P(\cdot)$ be everywhere defined on F, we define $v_P(0) = +\infty$. Then $v_P(\cdot)$ is function from F onto $\mathbb{Z} \cup \{+\infty\}$ such that

(i) $v_P(x) = +\infty$ if and only if $x = 0$,
(ii) $v_P(x+y) \geq \min(v_P(x), v_P(y))$ for all x and y, and
(iii) $v_P(xy) = v_P(x) + v_P(y)$ for all x and y.

We shall see in Proposition 6.4 below that the effect of $v_P(\cdot)$ is to pick out from F the localization of R at P.

To proceed further, we abstract the above construction and see what information we can recover from it. Let F be any field. A **discrete valuation** of F is a function $v(\cdot)$ from F onto $\mathbb{Z} \cup \{\infty\}$ such that

(i) $v(x) = +\infty$ if and only if $x = 0$,
(ii) $v(x + y) \geq \min(v(x), v(y))$ for all x and y, and
(iii) $v(xy) = v(x) + v(y)$ for all x and y.

Observe as a consequence that

(iv) $v(-1) = v(1) = 0$,
(v) $v(-x) = v(x)$ for all x, and
(vi) $v(x + y) = v(x)$ if $v(y) > v(x)$.

In fact, $v(1) = 0$ follows by taking $x = y = 1$ in (iii), and then $v(-1) = 0$ follows by taking $x = y = -1$ in (iii). This proves (iv), and (v) follows by combining (iv) with (iii) for $x = -1$. For (vi), we have $v(x + y) \geq v(x)$ by (ii). In the reverse direction, $v(x) \geq \min(v(x + y), v(y))$ by (ii) and (v); since $v(y) > v(x)$, the minimum must be the first of the two, and thus $v(x) \geq v(x+y)$.

Define $R_v = \{x \in F \mid v(x) \geq 0\}$. Property (i) shows that 0 is in R_v, (ii) and (v) show that R_v is closed under addition and subtraction, (iii) shows that R_v is closed under multiplication, and (iv) shows that 1 is in R_v. Consequently R_v is an integral domain. The ring R_v is called the **valuation ring** of v in F.

If x is in F but not in R_v, then $v(x) < 0$. This inequality forces $v(x^{-1}) > 0$, and x^{-1} is in R_v. As a consequence, F can be regarded as the field of fractions of R_v.

Let $P_v = \{x \in F \mid v(x) > 0\}$. Arguing in similar fashion, we see that P_v is an ideal in R_v. Any x in R_v that is not in P_v has $v(x) = v(x^{-1}) = 0$ and is thus a unit in R_v. In other words, R_v is a local ring with P_v as its unique maximal ideal. The ideal P_v is called the **valuation ideal** of v in F. We write $\Bbbk_v$ for the field R_v/P_v; it is called the **residue class field** of v.

Proposition 6.2. Let v be a discrete valuation of a field F, let R_v be the valuation ring, and let P_v be the valuation ideal. Then

(a) R_v is a principal ideal domain,
(b) there exists an element π in P_v with $v(\pi) = 1$, and any such π has $P_v = (\pi)$,
(c) the nonzero ideals of R_v are exactly the nonnegative integer powers of P_v and are given by $P_v^n = (\pi^n) = \{x \in R_v \mid v(x) \geq n\}$ for $n \geq 0$,
(d) the nonzero fractional ideals of R_v are exactly the integer powers of P_v and are given by $P_v^n = (\pi^n) = \{x \in R_v \mid v(x) \geq n\}$ for $n \in \mathbb{Z}$.

REMARKS. When F equals $\mathbb{Q}$ and v counts the net power of a prime number p dividing a rational number, we see by inspection that the ring R_v is the localization of $\mathbb{Z}$ at p, consisting of all rational numbers with no factor of p in their denominators. The choices[2] for π in (b) are the elements rp, where r is any nonzero rational whose numerator and denominator are both prime to p, and the nonzero ideals are of the form (p^n) with $n \geq 0$.

PROOF. The ideal P_v contains an element π with $v(\pi) = 1$ because $v(\cdot)$ is assumed to be onto $\mathbb{Z} \cup \{+\infty\}$. Suppose that x is a nonzero member of P_v and that $v(x) = n > 0$. Then $v(\pi^{-n}x) = 0$, and the elements $\pi^{-n}x$ and $x^{-1}\pi^n$ lie in R_v. Hence $x = \pi^n(\pi^{-n}x)$ exhibits x as a member of (π^n), and $\pi^n = x(x^{-1}\pi^n)$ exhibits π^n as a member of (x). Consequently $(x) = (\pi^n)$. If I is a nonzero proper ideal in R_v, then it follows that $I = \pi^{n_0} R_v$, where n_0 is the smallest integer such that some element x_0 of I has $v(x_0) = n_0$. This proves (a), (b), and (c).

Since R_v is a principal ideal domain, it is a Dedekind domain, and the theory of fractional ideals is applicable. Since (c) shows the nonzero ideals to be all P_v^n with $n \geq 0$, it follows that the fractional ideals are all P_v^n with n an arbitrary integer. For any integer $n > 0$, we have $(\pi^{-n})P_v^n = \pi^{-n}R_v\pi^n R_v = R_v = P_v^{-n}P_v^n$, and thus $P_v^{-n} = (\pi^{-n})$. The latter ideal equals $\pi^{-n}R_v = \{x \in R_v \mid v(x) \geq -n\}$, and this proves (d). $\square$

From property (vi) it follows for $n > 0$ that the members x of the set $1 + P_v^n$ all have $v(x) = 0$. The product of two such elements is again in the set because P_v^n is an ideal. Let us see that the multiplicative inverse x^{-1} of a member x of the set is in the set. We calculate that $v(x^{-1} - 1) = v(x^{-1}) + v(1-x) = 0 + v(1-x) = v(1-x) \geq n$. Hence x^{-1} is in $1 + P_v^n$, and $1 + P_v^n$ is a group under multiplication. It is a subgroup of the group $R_v^\times$ of units in R_v.

EXAMPLE. When $F = \mathbb{Q}$ and v counts the net power of a prime number p dividing a rational number, the residue class field $\mathbb{k}_v$ has p elements, with the integers $0, 1, \ldots, p-1$ being coset representatives. The group $R_v^\times$ is the multiplicative group of rationals having numerators and denominators prime to p. The members of $1 + P_v^n$ are rationals of the form $1 + p^n ab^{-1}$, where a and b are integers and b is prime to p. If we write this as $b^{-1}(b + p^n a)$, we see that the condition on a rational to be in $1 + P_v^n$ is that its numerator and denominator be prime to p and be congruent to each other modulo p^n.

Now we return to our first example of a discrete valuation, which was constructed from a nonzero prime ideal P in a Dedekind domain R. We called the valuation $v_P(\cdot)$. We asserted earlier that the construction via $v_P(\cdot)$ picks out the localization of R at P and the associated data. This assertion will be proved in Proposition 6.4 below. We begin with a handy lemma.

---

[2] Some books use the term "uniformizer" or "uniformizing element" for any generator π of the principal ideal P_v. The generators are exactly the prime elements of the ring R_v.

Lemma 6.3. Let R be a Dedekind domain regarded as a subring of its field of fractions F, let P be a nonzero prime ideal in R, and let v_P be the valuation of F defined by P. Then any element x of F with $v_P(x) = 0$ is of the form $x = ab^{-1}$ with a and b in R and $v_P(a) = v_P(b) = 0$.

PROOF. If x is an element of F with $v_P(x) = 0$, write $x = a'b'^{-1}$ with $a' \in R$ and $b' \in R$. Then $v_P(a') = v_P(b') = n$ for some integer $n \geq 0$. Since a' and b' are in R, (a') and (b') are ordinary ideals, and their prime factorizations are into ordinary ideals. Let the factorizations be $(a') = P^n Q_1$ and $(b') = P^n Q_2$, where Q_1 and Q_2 are products of prime ideals not involving P. Since we are dealing with ordinary ideals, a' and b' lie in P^n. Choose an element z in the fractional ideal P^{-n} that is not in P^{-n+1}. By definition of P^{-n}, zP^n is contained in R. Hence za' and zb' lie in R. Write $(za') = P^m Q_3$ and $(zb') = P^{m'} Q_4$, where $m \geq 0$ and where Q_3 and Q_4 are ordinary ideals whose prime factorizations do not involve P. Substituting for (a'), we obtain $(z)P^n Q_1 = P^m Q_3$ and hence $(z)P^n = P^m Q_3 Q_1^{-1}$. From this expression we see that $Q_3 Q_1^{-1}$ is an ordinary ideal. By definition of P^{-n+1}, $(z)P^{n-1}$ is not contained in R. Since $(z)P^{n-1} = P^{m-1} Q_3 Q_1^{-1}$, it follows that $m = 0$. Similarly $m' = 0$. Consequently $v_P(za') = v_P(zb') = 0$, and the lemma follows with $a = za'$ and $b = zb'$. □

Proposition 6.4. Let R be a Dedekind domain regarded as a subring of its field of fractions F, let P be a nonzero prime ideal in R, and let $v_P(\cdot)$ be the corresponding valuation of F. If S denotes the multiplicative system in R consisting of the complement of P and if the localization $S^{-1}R$ is regarded as a subring of F, then the valuation ring R_{v_P} coincides with $S^{-1}R$ and the valuation ideal P_{v_P} coincides with $S^{-1}P$.

PROOF. The set S consists exactly of the members x of R with $v_P(x) \leq 0$. Since v_P is nonnegative on R, these are the members x of R with $v_P(x) = 0$. Thus each x in $S^{-1}R$ has $v_P(x) \geq 0$, and $S^{-1}R$ is a subset of R_{v_P}.

For the reverse inclusion, fix a member π of P that is not in P^2. This element has $v_P(\pi) = 1$. If x is given in R_{v_P} with $v_P(x) = n \geq 0$, then we can write $x = \pi^n u$ for some member u of F with $v_P(u) = 0$. By Lemma 6.3 we can decompose u as $u = ab^{-1}$ with a and b in R and $v_P(a) = v_P(b) = 0$. The members of R on which v_P takes the value 0 are exactly the members of S. Thus u is exhibited as the quotient of two members of S, and u is in $S^{-1}R$. Since π is in the ideal P of R, $x = \pi^n u$ is in $S^{-1}R$. Hence $R_{v_P} = S^{-1}R$.

The ideal $S^{-1}P$ is a maximal ideal of $S^{-1}R = R_{v_P}$, and we observed just before Proposition 6.2 that P_{v_P} is the unique maximal ideal of R_{v_P}. Therefore $S^{-1}P = P_{v_P}$. □

Let us investigate the nature of an arbitrary discrete valuation in various settings involving a Dedekind domain. The main general result of this section is as follows.

2. Discrete Valuations

Theorem 6.5. Let R be a Dedekind domain regarded as a subring of its field of fractions F, and let v be a discrete valuation of F such that $R \subseteq R_v$. Then
  (a) $P = R \cap P_v$ is a nonzero prime ideal of R,
  (b) the associated discrete valuation v_P defined by P coincides with v,
  (c) $PR_v = P_v$,
  (d) $R + P_v = R_v$, and in fact $R + P_v^n = R_v$ for every integer $n \geq 1$, and
  (e) the inclusion of R into R_v induces a field isomorphism $R/P \cong R_v/P_v$.

PROOF. Since 1 is not in P_v, the ideal P in (a) is proper. If a and b are members of R such that ab is in P, then ab is in P_v, one of a and b is in P_v as well as R, and $P = R \cap P_v$ is a prime ideal. The ideal P cannot be 0 because otherwise every nonzero element x of R would have $v(x) = 0$, in contradiction to the fact that F is the field of fractions of R. Thus P is a nonzero prime ideal of R. This proves (a).

For (b) and (c), let us begin by showing that $v_P(x) = 0$ implies $v(x) = 0$. By Lemma 6.3 we can write $x = ab^{-1}$ with a and b in R and with $v_P(a) = v_P(b) = 0$. The values of v_P show that the members a and b of R are not in P. Since $P = R \cap P_v$, neither a nor b is in P_v. Therefore $v(a) \leq 0$ and $v(b) \leq 0$. Since $R \subseteq R_v$ by assumption, $v(a) \geq 0$ and $v(b) \geq 0$. We conclude that $v(a) = v(b) = 0$ and that $v(x) = v(ab^{-1}) = v(a) - v(b) = 0$.

Now we can show that $v = v_P$ and that $PR_v = P_v$. The ideal PR_v of R_v has to be of the form P_v^e for some integer $e \geq 0$ by Proposition 6.2c, and the integer e has to be > 0 because 1 is not in PR_v. If a nonzero $x \in R$ has $v_P(x) = n$ for some integer $n \geq 0$, then $xR = P^n Q$, where Q is an ideal of R whose prime factorization does not involve P. The function v_P is 0 on Q, and the result of the previous paragraph shows that v is 0 on Q. Hence the members of Q are units in R_v, and $QR_v = R_v$. Therefore $xR_v = xRR_v = P^n QR_v = P^n R_v = (PR_v)^n = P_v^{en}$, and $v(x) = en = ev_P(x)$. Since F is the field of fractions of R, $v = ev_P$ everywhere. The image of v_P is $\mathbb{Z} \cup \{+\infty\}$, and we conclude that $e = 1$. In other words, $v = v_P$ and $PR_v = P_v$. This proves (b) and (c).

For the first conclusion in (d), we certainly have $R + P_v \subseteq R_v$. In the reverse direction, let $x \in R_v$ be given. If $v(x) > 0$, then x is in P_v, and there is nothing to prove. If $v(x) = 0$, then (b) and Lemma 6.3 together show that we can write $x = ab^{-1}$, where a and b are members of R but not P. Since R/P is a field, we can choose c in R with bc in $1 + P$. Then

$$x - ac = a(b^{-1} - c) = ab^{-1}(1 - bc) = x(1 - bc).$$

The right side is a member of $R_v P$, and (c) showed that $R_v P = P_v$. Therefore x is exhibited as the sum of the member ac of R and the member $x(1 - bc)$ of P_v, and we conclude that $R + P_v = R_v$. This proves the first conclusion in (d).

For the second conclusion in (d), we show inductively for $n \geq 1$ that $P^{n-1} + P_v^n = P_v^{n-1}$, the case $n = 1$ being what has already been proved in (d). Assume that case n has been proved. Multiplying the equality by P and using (c), we obtain $P^n + PP_v^n = (PR_v)P_v^{n-1} = P_v P_v^{n-1} = P_v^n$. Since $P \subseteq P_v$, the term PP_v^n is contained in P_v^{n+1}, but increasing the left side in this way does not increase the right side. Thus $P^n + P_v^{n+1} = P_v^n$. This completes the induction. Using a second induction, we show that $R + P_v^n = R_v$. We have already proved this equality for $n = 1$. If we assume it for n and substitute from what has just been proved, we obtain $R + (P^n + P_v^{n+1}) = R_v$, and this proves case $n + 1$ since $P^n \subseteq R$. The second conclusion of (d) thus follows by induction.

For (e), we are assuming that $R \subseteq R_v$, and we have defined $P = R \cap P_v$. Thus the inclusion $R \to R_v$, when followed by the passage to the quotient R_v/P_v, descends to the quotient as a field map $R/P \to R_v/P_v$. By (d), any member x of R_v is the sum of a member y of R and a member z of P_v; then $y + P$ is the member of R/P that maps to $x + P_v$ in R_v/P_v. Thus the field map $R/P \to R_v/P_v$ is onto, and (e) is proved. □

Corollary 6.6. Let R be a Dedekind domain regarded as a subring of its field of fractions F. If x is a member of $\mathbb{F}$ such that $v(x) \geq 0$ for every discrete valuation v of F satisfying $R \subseteq R_v$, then x lies in R.

PROOF. We may assume that $x \neq 0$. Write $x = ab^{-1}$ with a and b in R. Theorem 6.5 shows that the valuations in question are the ones determined by the nonzero prime ideals of R. If the principal ideals (a) and (b) factor as $(a) = P_1^{j_1} \cdots P_r^{j_r}$ and $(b) = P_1^{k_1} \cdots P_r^{k_r}$, then $0 \leq v_{P_i}(x) = v_{P_i}(ab^{-1}) = j_i - k_i$ for $1 \leq i \leq r$. Thus $j_i \geq k_i$ for all i, and the fractional ideal (ab^{-1}) equals the product $P_1^{j_1-k_1} \cdots P_r^{j_r-k_r}$, which is contained in R. Hence $x = ab^{-1}$ lies in R. □

A finite field has no discrete valuations because of the requirement that the image of a discrete valuation be $\mathbb{Z} \cup \{+\infty\}$. If we drop this requirement in the definition and let a be a multiplicative generator of a finite field, then any discrete valuation v would have $v(a^k) = kv(a)$ by property (ii). Taking k equal to the order of a and using that $v(1) = 0$, we obtain $v(a) = 0$. Thus if we drop the requirement about the image of a discrete valuation, the only possibility has $v(0) = +\infty$ and $v(x) = 0$ for all $x \neq 0$. Thus this setting is not very interesting.

The settings in which discrete valuations v are of most interest to us are the following:

(i) number fields,
(ii) "function fields in one variable" over a base field,[3]

---

[3] This notion has not been defined thus far in the book but will be treated in Chapter VII. The fields in question are finite algebraic extensions of a field $\Bbbk(X)$, where X is an indeterminate and $\Bbbk$

(iii) fields obtained from (i) or (ii) by a process of completion similar to that used in forming the field of p-adic numbers.

The first of these are the initial subject matter of algebraic number theory, and the second of these are the initial subject matter of algebraic geometry—the geometry of curves. The third of these are used as a tool in studying the other two. Section VIII.7 of *Basic Algebra* explained parts of the analogy between the first two kinds of fields, and that is why we treat them together. We shall use Proposition 6.7 below to determine their discrete valuations. In the case of (ii), the members of the base field $\Bbbk$ are regarded as constants, and the interest is only in valuations that are 0 on $\Bbbk^\times$.

Proposition 6.7. Let R be a Dedekind domain, let F be its field of fractions, let K be a finite algebraic extension of F, and let T be the integral closure of R in K. If a discrete valuation v of K is ≥ 0 on R, then it is ≥ 0 on T.

REMARKS. We make repeated use in this chapter of the fact that T is a Dedekind domain in this situation. This fact was proved as Theorem 8.54 of *Basic Algebra* for the case that K is a finite *separable* extension of F, but it is valid without the hypothesis of separability. The result without the hypothesis of separability will be proved in Chapter VII as part of an investigation of separable and "purely inseparable" extensions.

PROOF. If $x \neq 0$ is in T, then the minimal polynomial of x over R is a monic polynomial in $T[X]$, and thus there exist an integer n and coefficients $a_{n-1}, \ldots, a_0$ in R such that
$$x^n = a_{n-1}x^{n-1} + \cdots + a_1 x + a_0.$$
Properties (ii) and (iii) of discrete valuations show from this equation that
$$nv(x) \geq \min_{0 \leq j \leq n-1} \left(v(a_j) + jv(x)\right).$$
Since $v(a_j) \geq 0$, we obtain $nv(x) \geq \min_{0 \leq j \leq n-1} jv(x)$, and it follows that $v(x) \geq 0$. Thus v is nonnegative on T. $\square$

Corollary 6.8. The only discrete valuations of the field $\mathbb{Q}$ of rationals are the ones leading to the p-adic absolute value for each prime number p. If K is a number field and T is its ring of algebraic integers, then the only discrete valuations of K are the valuations v_P corresponding to each nonzero prime ideal P of T.

---

is a field called the **base field**. At times later in the chapter, we shall be interested only in the case that the algebraic extension is separable. It will be proved in Chapter VII that for perfect fields $\Bbbk$, this separability can always be arranged by adjusting the indeterminate X suitably.

PROOF. If v is an arbitrary discrete valuation of $\mathbb{Q}$, then property (iv) of discrete valuations shows that $v(-1) = v(1) = 0$, and property (ii) allows us to conclude that v is nonnegative on all of $\mathbb{Z}$. Thus $\mathbb{Z}$ is contained in the valuation ring of v, and Theorem 6.5 applies. By (a) in the theorem, the intersection of $\mathbb{Z}$ with the valuation ideal is a nonzero prime ideal of $\mathbb{Z}$, hence is $p\mathbb{Z}$ for some prime number p. Part (b) in the theorem then identifies v as the valuation corresponding to $p\mathbb{Z}$. This proves the first conclusion.

For the second conclusion, let v be a discrete valuation of K. The restriction to $\mathbb{Q}$ has to be a positive integral multiple of a discrete valuation of $\mathbb{Q}$ or else a function that is identically 0 on $\mathbb{Q}^\times$. In either case, v is ≥ 0 on $\mathbb{Z}$, and Proposition 6.7 shows that v is ≥ 0 on T. If R_v denotes the valuation ring of v and P_v denotes the valuation ideal, then this says that $T \subseteq R_v$. We can therefore apply Theorem 6.5. If P is defined by $P = T \cap P_v$, then (a) in the theorem shows that P is a nonzero prime ideal, and (b) shows that $v = v_P$. $\square$

Let us now consider the field $\mathbb{C}(X)$, regarding it as having some properties in common with the number field $\mathbb{Q}$. We want to know whether some analog of Corollary 6.8 is valid for $\mathbb{C}(X)$. The ring $\mathbb{C}[X]$ of polynomials is a principal ideal domain with $\mathbb{C}(X)$ as field of fractions, and the prime ideals of $\mathbb{C}[X]$ are all of the form $(X - c)$ with $c \in \mathbb{C}$ because $\mathbb{C}$ is algebraically closed. For each such c, we therefore obtain a discrete valuation $v_{(X-c)}$. Are there any other discrete valuations? If we think geometrically about this question, we can regard $\mathbb{C}(X)$ as the rational functions on the Riemann sphere, and each discrete valuation addresses the order of vanishing of rational functions at some point of the sphere. For the points of the sphere that correspond to points c of $\mathbb{C}$, such a valuation picks out the power of $(X - c)$ by which the rational function should be divided in order to be regular and nonvanishing at c. The point ∞ on the Riemann sphere behaves differently. The usual technique in complex-variable theory is to replace X by $1/X$ and examine the behavior at 0. Following that prescription, we are led to a discrete valuation v_∞ that is not of the form v_P for some prime ideal P of $\mathbb{C}[X]$. The definition of v_∞ on the quotient $f(X)/g(X)$ of nonzero polynomials is

$$v_\infty(f(X)/g(X)) = \deg g - \deg f$$

with $v_\infty(0) = +\infty$ as usual. The next proposition, which extends one of Liouville's theorems in complex-variable theory[4] from $\mathbb{C}$ to a general field $\Bbbk$, says that there are no other discrete valuations of interest for this example.

Proposition 6.9. Let $\Bbbk$ be any field, and let $F = \Bbbk(X)$ be the field of rational expressions in one indeterminate over $\Bbbk$. Regard F as the field of fractions of

---

[4]For a meromorphic function on the Riemann sphere, the sum of the orders of the poles equals the sum of the orders of the zeros.

the principal ideal domain $\Bbbk[X]$. Then the only discrete valuations of F that are 0 on the multiplicative group $\Bbbk^\times$ of nonzero constant polynomials are the various valuations $v_{(p)}$, where $p(X)$ is a monic prime polynomial in $\Bbbk[X]$, and the valuation v_∞ that is defined on nonzero elements of F by

$$v_\infty(f(X)/g(X)) = \deg g - \deg f$$

if f and g are polynomials. Moreover, any nonzero $h(X)$ in F has

$$v_\infty(h) + \sum_{\substack{p(X) \text{ monic} \\ \text{prime in } R}} (\deg p) v_{(p)}(h) = 0.$$

PROOF. Let v be a discrete valuation of F that is 0 on $\Bbbk^\times$. First suppose that $v(X) \geq 0$. Being 0 on the coefficients, v is nonnegative on all polynomials. Thus $\Bbbk[X]$ is contained in the valuation ring of v, and Theorem 6.5 applies. By (a) in the theorem, the intersection of $\Bbbk[X]$ with the valuation ideal is a nonzero prime ideal of $\Bbbk[X]$, hence is $(p(X))$ for some monic prime polynomial $p(X)$. Part (b) in the theorem then identifies v as the valuation corresponding to $(p(X))$.

Next suppose that $v(X) < 0$. Since $\Bbbk[X^{-1}]$ has $\Bbbk(X)$ as field of fractions, the argument in the previous paragraph is applicable, and we find that v is the valuation determined by the prime ideal (X^{-1}) in $\Bbbk[X^{-1}]$. In particular, $v(X) = -1$. To find $v(f)$ for a general polynomial $f(X) = a_n X^n + \cdots + a_1 X + a_0$ in $\Bbbk[X]$ under the assumption that $a_n \neq 0$, we write f as $X^n(a_n + \cdots + a_1 X^{1-n} + a_0 X^{-n})$. The member $a_n + \cdots + a_1 X^{1-n} + a_0 X^{-n}$ of $\Bbbk[X^{-1}]$ is not divisible by X^{-1}, and thus v is 0 on it. Consequently $v(f) = v(X^n) = nv(X) = -n = -\deg f$. If f and g are both nonzero in $\Bbbk[X]$, then it follows that $v(f/g) = v(f) - v(g) = -\deg f + \deg g = v_\infty(f/g)$. That is, $v = v_\infty$.

To prove the displayed formula, write a given nonzero member $h(X)$ of F as the quotient of two relatively prime polynomials, thus as $h(X) = f(X)/g(X)$. Factor the numerator as $f(X) = c \prod_{i=1}^m p_i(X)^{k_i}$ with $c \in \Bbbk^\times$, and factor the denominator similarly. If $p(X)$ is a monic prime polynomial, then inspection of the formula for $f(X)$ shows that $v_{(p)}(f)$ is k_i if $p = p_i$ and is 0 otherwise. Hence $\sum_p (\deg p) v_{(p)}(f) = \sum_{i=1}^m k_i \deg p_i = \deg f$. Subtracting this formula and a corresponding formula for g, we obtain

$$\sum_p (\deg p) v_{(p)}(f/g) = \deg f - \deg g = -v_\infty(h),$$

and the result follows. □

Corollary 6.10. Let $\Bbbk$ be a field, let $F = \Bbbk(X)$ be the field of rational expressions in one indeterminate over $\Bbbk$, let K be a finite algebraic extension of

$\Bbbk[X]$, let T be the integral closure of $\Bbbk[X]$ in K, and let v be a discrete valuation of K that is 0 on the multiplicative group $\Bbbk^\times$. Then the only possibilities for v are as follows:

(a) $v(X) \geq 0$, and there exists a unique nonzero prime ideal P in T such that $v = v_P$,

(b) $v(X) < 0$, and there exists a prime ideal P in the integral closure T' of $\Bbbk[X^{-1}]$ in K such that $P \cap \Bbbk[X^{-1}] = X^{-1}\Bbbk[X^{-1}]$ and such that v is the valuation of K determined by P.

REMARK. The ideals P that occur in (b) are the ones in the prime factorization of the ideal $X^{-1}T'$ in T'. There is at least one, and there are only finitely many.

PROOF. The argument is similar to the one for Corollary 6.8, except that we have to take into account what Proposition 6.9 says when $v(X) < 0$. The conclusion is that either v is ≥ 0 on $\Bbbk[X]$, and then Proposition 6.7 and Theorem 6.5 show that v is as in (a), or else $v(X) < 0$, and then Proposition 6.7 and Theorem 6.5 show that v is as in (b). □

To conclude, let us complete the remarks about fractional ideals begun early in this section. In the context that R is a Dedekind domain and F is its field of fractions, we mentioned that the nonzero fractional ideals of F form a group. We denote this group by $\mathcal{I}$. The nonzero principal fractional ideals form a subgroup $\mathcal{P}$, and $\mathcal{P}$ is isomorphic to the multiplicative group $F^\times$.

The point of the present discussion is that the group $\mathcal{I}/\mathcal{P}$ is isomorphic to the ideal class group of F as defined in the number-field setting in Section V.6. Recall the nature of this group. Two nonzero ideals I and J of R are equivalent if there exist nonzero members a and b of R with $aI = bJ$. Proposition 5.18 showed in the number-field setting that multiplication of such ideals descends to a multiplication on the set of equivalence classes and that the result is a group. This result holds for any Dedekind domain. The group is called the **ideal class group** of F; we denote it here by $\mathcal{C}$.

To verify that $\mathcal{C} \cong \mathcal{I}/\mathcal{P}$, we map each ideal I of R to its coset in $\mathcal{I}/\mathcal{P}$. If I and J are equivalent ideals of R and $aI = bJ$, then $(ab^{-1})I = J$, and I and J map to the same coset. Thus $\mathcal{C}$ maps homomorphically into $\mathcal{I}/\mathcal{P}$. If I maps into the identity coset, then $xI = R$ for some $x \in F^\times$. Writing x as ab^{-1} with a and b in R shows that $aI = bR = (b)$, hence that I is equivalent to a principal ideal. Thus the homomorphism $\mathcal{C} \to \mathcal{I}/\mathcal{P}$ is one-one. Finally if M is any nonzero fractional ideal of F, then we can find some $x \in F^\times$ with $xM \subseteq R$. Here xM is an ideal of R, and the equivalence of M and xM exhibits the class of M in $\mathcal{I}/\mathcal{P}$ as in the image of $\mathcal{C}$. Consequently $\mathcal{C} = \mathcal{I}/\mathcal{P}$, as asserted.

3. Absolute Values

The next step in analyzing and generalizing the construction of the p-adic absolute value is to pass from the valuation, which appears in the exponent, to the absolute value itself. If F is a field, an **absolute value** on F is a function $|\cdot|$ from F to $\mathbb{R}$ such that

(i) $|x| \geq 0$ with equality if and only if $x = 0$,
(ii) $|x + y| \leq |x| + |y|$ for all x and y in F,
(iii) $|xy| = |x||y|$ for all x and y in F.

It follows directly that

(iv) $|-1| = |1| = 1$ and that
(v) $|-x| = |x|$ for all x in F.

In fact, (iv) follows by combining (i) with (iii) for $x = y = 1$ and then with (iii) for $x = y = -1$; then (v) follows by combining (iii) and (iv). The absolute value $|\cdot|$ on F is said to be **nonarchimedean** if the following strong form of (ii) holds:[5]

(ii') $|x + y| \leq \max(|x|, |y|)$ for all x and y in F.

Otherwise it is called **archimedean**. The inequality in (ii') is called the **ultrametric inequality**. When the ultrametric inequality holds, then the following additional condition holds:

(vi) $|x + y| = |x|$ whenever x and y in F have $|y| < |x|$.

In fact, when $|y| < |x|$, (ii') immediately gives $|x + y| \leq |x|$. But also (ii') and (v) give $|x| \leq \max(|x + y|, |-y|) = \max(|x + y|, |y|)$. On the right side, the maximum cannot be $|y|$ because $|x| \leq |y|$ is false. Thus $|x| \leq |x + y|$, and (vi) holds.

Although it might seem counterintuitive, it turns out that the archimedean absolute values are easier to understand than the nonarchimedean ones in the number fields and function fields of interest to us.

Because of (iii), any absolute value of F when restricted to $F^\times$ is a multiplicative homomorphism into the positive real numbers. The image in the positive reals is therefore a group.

EXAMPLES OF NONARCHIMEDEAN ABSOLUTE VALUES.

(1) Let F be any field, and define $|x| = 0$ for $x = 0$ and $|x| = 1$ for $x \neq 0$. The result is a nonarchimedean absolute value called the **trivial absolute value**. It is of no interest, and we shall tend to exclude consideration of it from our results.

---

[5] Some authors refer to a nonarchimedean absolute value as a "valuation," using the same term as for the functions $v(\cdot)$ in Section 2. There is little danger of confusing the two notions, but we shall use the two distinct names anyway.

Any other absolute value will be said to be **nontrivial**. Observe for a finite field F that the fact that $x \mapsto |x|$ is a homomorphism from $F^\times$ to the positive reals implies that the only absolute value on a finite field is the trivial one.

(2) Let F be any field, let v be a discrete valuation on F, and fix a real number $r > 1$. Then $|x| = r^{-v(x)}$ defines a nonarchimedean absolute value on F. Property (i) of absolute values follows because $v(x)$ takes values in $\mathbb{Z} \cup \{+\infty\}$ and is infinite if and only if $x = 0$, property (ii') follows because $v(x+y) \geq \min(v(x), v(y))$, and property (iii) follows because $v(xy) = v(x) + v(y)$. In particular, the p-adic absolute value is obtained in this way when we take $r = p$, and we obtain corresponding examples for any number field F by taking $v = v_P$ and fixing $r > 1$, where P is any nonzero prime ideal in the ring of algebraic integers in F. For the function field $F = \Bbbk(X)$, we obtain corresponding examples by taking $v = v_{(p)}$ and fixing $r > 1$, where $p(X)$ is any monic prime polynomial in $\Bbbk(X)$. The choice $v = v_\infty$ gives us another example. In all of these cases, the image of $F^\times$ in $\mathbb{R}^\times$ under the absolute value is discrete in the sense that each one-point set of the image is open in the relative topology from the positive reals. Corollary 6.17 will show conversely that any absolute value for which the image in $\mathbb{R}^\times$ of the nonzero elements is discrete and nontrivial is obtained in this way from a discrete valuation. It is worth pausing to interpret some of the conclusions of Theorem 6.5 in terms of absolute values and metrics.

Proposition 6.11. Let R be a Dedekind domain regarded as a subring of its field of fractions F, suppose that $|\cdot|$ is an absolute value on F defined by means of a discrete valuation v, and suppose that the subset R_v of F for which $|x| \leq 1$ contains R. If P_v denotes the subset of F with $|x| < 1$, then $P = R \cap P_v$ is a nonzero prime ideal of R, and also

(a) R is dense in R_v,
(b) P^n is dense in P_v^n for every $n \geq 1$,
(c) $R/P \cong R_v/P_v$.

PROOF. In terms of v, the set R_v is the valuation ring, and the set P_v is the valuation ideal. The hypothesis $R \subseteq R_v$ is the hypothesis of Theorem 6.5. Part (a) of that theorem shows that $P = R \cap P_v$ is a prime ideal in R. Conclusions (a) and (b) here follow from Theorem 6.5d. In fact, let $|x| = r^{-v(x)}$ with $r > 1$. Suppose that x is given in P_v^n with $n \geq 0$ and that a positive number r^{-N} is specified. We may assume that $N \geq n$. The condition for x to be in P_v^n is that $|x| \leq r^{-n}$. Theorem 6.5d shows that we can find an x_0 in R such that $x_0 + y = x$ with y in P_v^N, hence with $|y| \leq r^{-N}$. Then x_0 is in R and has $|x_0 - x| = |y| \leq r^{-N}$. Hence x_0 is within r^{-N} of x. Since $|x_0| \leq \max(|x|, |y|) = \max(r^{-n}, r^{-N}) = r^{-n}$, x_0 is in $R \cap P_v^n = P^n$. Conclusion (c) is immediate from Theorem 6.5e. □

EXAMPLES OF ARCHIMEDEAN ABSOLUTE VALUES. If F is any subfield of $\mathbb{R}$ or $\mathbb{C}$ and if $|\cdot|$ is defined as the restriction to F of the ordinary absolute value function, then $|\cdot|$ is an archimedean absolute value. Remarkably it turns out that there are no other archimedean absolute values, apart from "equivalent" ones in the sense to be defined below. We return to this matter at the end of Section 4. Actually, we shall be interested in archimedean absolute values only when F is a number field or is all of $\mathbb{R}$ or all of $\mathbb{C}$, and we will not need to invoke any deep theorem for the cases of interest to us.

Properties (i), (ii), and (v) of absolute values show that the function d with $d(x, y) = |x - y|$ is a metric on F, and the next section will examine what happens when this metric is completed. The resulting fields will be generalizations of the field of p-adic numbers and will useful as tools in investigating number fields and function fields in one variable.

Two absolute values $|\cdot|_1$ and $|\cdot|_2$ on the same field are said to be **equivalent** if there is a positive number α such that $|\cdot|_1 = (|\cdot|_2)^\alpha$. In our passage from a discrete valuation v to a nonarchimedean absolute value $|\cdot|$, we fixed $r > 1$ and defined $|x| = r^{-v(x)}$. Changing r changes the absolute value to an equivalent absolute value. In the archimedean case a positive power of an absolute value need not be an absolute value, since the triangle inequality may fail. For example the ordinary absolute value on $\mathbb{R}$ satisfies the triangle inequality; so does its αth power for $\alpha < 1$ but not for $\alpha > 1$.

Equivalent absolute values yield the same topology on F and in fact the same Cauchy sequences.[6] Conversely two absolute values that yield the same topology are equivalent, according to the following proposition.

Proposition 6.12. Two nontrivial absolute values on a field F are equivalent if and only if
$$\{x \in F \mid |x|_1 > 1\} \subseteq \{x \in F \mid |x|_2 > 1\},$$
if and only if they induce the same topology on F.

REMARKS. If $|\cdot|_1$ is the trivial absolute value, then the stated inclusion holds for all $|\cdot|_2$, but the equivalence may fail; that is why the statement has to exclude this case. The statement of the proposition remains true if the inequalities $|x|_1 > 1$ and $|x|_2 > 1$ are replaced by $|x|_1 < 1$ and $|x|_2 < 1$, as we see by replacing x by x^{-1}.

PROOF. If the two absolute values are equivalent, then it is immediate from the definition of equivalent that equality holds in the stated inclusion. Conversely

---
[6]In many books an equivalence class of absolute values on a field is called a "place" of the field. We shall use this term in Sections 9 and 10 of this chapter,

suppose that the inclusion holds. Fix $x \in F$ with $|x|_1 > 1$. Such an x exists because $|\cdot|_1$ is nontrivial. Since $|x|_2 > 1$, there exists a real $s > 0$ with $|x|_1 = |x|_2^s$. We shall show that $|\cdot|_1 = |\cdot|_2^s$.

Let $y \in F$ be arbitrary with $|y|_1 \geq 1$. Find the number $r \geq 0$ depending on y such that $|y|_1 = |x|_1^r$. Let $\{a_n/b_n\}$ be a sequence of positive rationals strictly decreasing to r such that a_n and b_n are both positive. Then $|y|_1 = |x|_1^r < |x|_1^{a_n/b_n}$, from which we obtain $|y^{b_n}|_1 < |x^{a_n}|_1$ and $|x^{a_n}y^{-b_n}|_1 > 1$. By assumption, $|x^{a_n}y^{-b_n}|_2 > 1$, and therefore $|y|_2 < |x|_2^{a_n/b_n}$. Passing to the limit, we obtain $|y|_2 \leq |x|_2^r$.

Now suppose that $|y|_1 > 1$. Arguing similarly with a sequence of positive rationals strictly increasing to r, we obtain $|y|_2 \geq |x|_2^r$. Thus $|y|_2 = |x|_2^r$. Then we have

$$|y|_1 = |x|_1^r = |x|_2^{rs} = |y|_2^s \qquad \text{whenever } |y|_1 > 1. \qquad (*)$$

If instead $|y|_1 = 1$, then the number r in the second paragraph of the proof is 0, and we obtain $|y|_2 \leq |x|_2^r = 1$. Replacing y by y^{-1} shows also that $|y|_2 \geq 1$. Thus $|y|_1 = 1$ implies $|y|_2 = 1$.

The remaining case is that $|y|_1 < 1$. Then we apply $(*)$ to y^{-1} and conclude that $|y|_1 = |y|_2^s$ in this case as well. This completes the proof of the first conclusion of the proposition.

For the final statement we know that equivalent absolute values lead to the same topology. Conversely suppose that the absolute values are not equivalent. By what we have just shown, there exists $x \in F$ with $|x|_1 > 1$ and $|x|_2 \leq 1$. Then $\{x^{-n}\}$ is a sequence convergent to 0 in the topology from $|\cdot|_1$ but not convergent to 0 in the topology from $|\cdot|_2$. Therefore the topologies are different. □

Proposition 6.13. If $|\cdot|$ is an absolute value on the field F, then the topology on F induced by the associated metric makes F into a topological field.

REMARK. The proof is similar to part of the argument that proves Proposition 6.1 except that the general triangle inequality has to be used in place of the ultrametric inequality.

PROOF. To see that addition, subtraction, and multiplication are continuous on F, let $\{x_n\}$ and $\{y_n\}$ be convergent sequences in F with respective limits x and y. Use of the triangle inequality on F gives

$$|(x_n + y_n) - (x + y)| = |(x_n - x) + (y_n - y)| \leq |x_n - x| + |y_n - y|.$$

The right side has limit 0 in $\mathbb{R}$, and therefore $x_n + y_n$ has limit $x + y$ in F. A completely analogous argument, making use also of the equality $|-1| = |1|$, shows that subtraction is continuous. Consider multiplication. If M is an upper

3. Absolute Values

bound for the absolute values $|x_n|$, then use of the multiplicative property of the absolute value on F gives

$$|x_n y_n - xy| = |x_n(y_n - y) + y(x_n - x)| \le |x_n(y_n - y)| + |y(x_n - x)|$$
$$= |x_n||y_n - y| + |y||x_n - x| \le M|y_n - y| + |y||x_n - x|.$$

The right side has limit 0 in $\mathbb{R}$, and therefore $x_n y_n$ has limit xy in F.

To see that inversion $x \mapsto x^{-1}$ is continuous on $F^\times$, let $\{x_n\}$ be a sequence in $F^\times$ with limit x in $F^\times$. Since $\lim_n |x_n| = |x|$, we can find an integer N such that $|x_n| \ge \frac{1}{2}|x|$ for $n \ge N$. The computation

$$|x_n^{-1} - x^{-1}| = |(x - x_n)/(x_n x)| = |x - x_n|/(|x_n||x|) \le 2|x|^{-1}|x - x_n|,$$

valid for $n \ge N$, then shows that $\lim x_n^{-1} = x^{-1}$, and inversion is continuous. Consequently F is a topological field. □

We now give a few results that limit the kinds of absolute values that can arise in particular situations.

Proposition 6.14. If $|\cdot|$ is an absolute value on the field F for which there is some c with $|n| \le c$ for all integers $n \in \mathbb{Z}$, i.e., for all additive multiples of 1, then $|\cdot|$ is nonarchimedean. In particular, $|\cdot|$ is necessarily nonarchimedean if F has characteristic different from 0.

REMARK. When c exists, then c can be taken to be 1, since the image of $F^\times$ under the absolute value is a subgroup of the positive reals and the only bounded such subgroup is $\{1\}$.

PROOF. If x and y are in F and if n is any positive integer, then the Binomial Theorem gives $(x + y)^n = \sum_{j=0}^{n} \binom{n}{j} x^{n-j} y^j$. Therefore

$$|x + y|^n = \sum_{j=0}^{n} \left|\binom{n}{j}\right| |x|^{n-j} |y|^j$$
$$\le c \sum_{j=0}^{n} \max(|x|, |y|)^{n-j} \max(|x|, |y|)^j$$
$$= c(n + 1) \max(|x|, |y|)^n.$$

Extraction of the nth root gives $|x + y| \le c^{1/n}(n + 1)^{1/n} \max(|x|, |y|)$. Passing to the limit, we obtain $|x + y| \le \max(|x|, |y|)$. □

Theorem 6.15 (Ostrowski's Theorem). If $|\cdot|$ is a nontrivial absolute value on the field $\mathbb{Q}$, then $|\cdot|$ is equivalent either to the p-adic absolute value $|\cdot|_p$ for some prime number p or to the ordinary absolute value $|\cdot|_\mathbb{R}$.

REMARKS. No two of these are equivalent because $\{p^n\}$ tends to 0 relative to the p-adic absolute value, $\{p^{-n}\}$ tends to 0 relative to the ordinary absolute value, and p^n has absolute value 1 relative to the ℓ-adic absolute value for all prime numbers $\ell \neq p$.

PROOF. First suppose that every integer n has $|n| \leq 1$. Proposition 6.14 shows that $|\cdot|$ is nonarchimedean. Since $|\cdot|$ is nontrivial, we must have $|n| < 1$ for some n, and we may take n to be positive. Since $|n|$ is the product of $|p|$ over all primes dividing n, multiplicities included, some prime number p has $|p| < 1$. Let us see that p is unique. If, on the contrary, $|q| < 1$ for a second prime number q, choose integers a and b with $ap + bq = 1$. Then $1 = |1| = |ap + bq| \leq \max(|ap|, |bq|) = \max(|a||p|, |b||q|) \leq \max(|p|, |q|) < 1$, contradiction. If we now define a positive real α by $|p| = p^{-\alpha}$, then it follows that $|n| = (|n|_p)^\alpha$ for all integers n. Therefore $|\cdot| = (|\cdot|_p)^\alpha$ on all of $\mathbb{Q}$.

Now suppose that n is some integer with $|n| > 1$. We may assume that n is positive. For any positive integer m, the triangle inequality gives

$$|m| = |1 + \cdots + 1| \leq |1| + \cdots + |1| = m.$$

In particular we have $|n| = n^\alpha$ for some real α with $0 < \alpha \leq 1$.

We shall prove that

$$|m| \leq m^\alpha \qquad (*)$$

for all positive integers m. We start by expanding m to the base n, writing

$$m = c_0 + c_1 n + c_2 n^2 + \cdots + c_{k-1} n^{k-1},$$

where k is the integer such that $n^{k-1} \leq m < n^k$ and where each c_j satisfies $0 \leq c_j < n$. The triangle inequality gives

$$\begin{aligned}
|m| &\leq |c_0| + |c_1||n| + |c_2||n|^2 + \cdots + |c_{k-1}||n|^{k-1} \\
&\leq (n-1)(1 + n^\alpha + n^{2\alpha} + \cdots + n^{\alpha(k-1)}) \qquad \text{by definition of } \alpha \\
&= \frac{(n-1)n^{\alpha k}}{n^\alpha - 1} = \frac{(n-1)n^\alpha}{n^\alpha - 1} n^{\alpha(k-1)} \\
&\leq \frac{(n-1)n^\alpha}{n^\alpha - 1} m^\alpha \qquad \text{since } n^{k-1} \leq m.
\end{aligned}$$

In other words, there is a positive number C independent of m such that $|m| \leq Cm^\alpha$ for every positive integer m. For every positive integer N, we then have

$|m|^N = |m^N| \leq Cm^{\alpha N}$, and thus $|m| \leq C^{1/N} m^\alpha$. Letting N tend to infinity, we obtain $(*)$.

Let us now improve $(*)$ to the equality

$$|m| = m^\alpha \qquad \text{for every positive integer } m. \qquad (**)$$

The integer k above has $n^{k-1} \leq m < n^k$. Put $d = n^k - m$; this satisfies $0 < d \leq n^k - n^{k-1}$. Then

$$n^{\alpha k} = |n|^k = |n^k| \leq |m| + |d| \leq |m| + d^\alpha \leq |m| + (n^k - n^{k-1})^\alpha,$$

and consequently

$$|m| \geq n^{\alpha k} - (n^k - n^{k-1})^\alpha = n^{\alpha k}\left(1 - \left(1 - \tfrac{1}{n}\right)^\alpha\right) \geq m^\alpha \left(1 - \left(1 - \tfrac{1}{n}\right)^\alpha\right).$$

Thus $|m| \geq C'm^\alpha$ for some positive constant C' independent of m. For every positive integer N, we then have $|m|^N = |m^N| \geq C'm^{\alpha N}$ and hence $|m| \geq C'^{1/N} m^\alpha$. Letting N tend to infinity, we obtain $|m| \geq m^\alpha$. In combination with $(*)$, this proves $(**)$.

Since $|-m| = |m|$, the equality $(**)$ implies $|m| = (|m|_\mathbb{R})^\alpha$ for every integer m. Taking quotients, we obtain $|q| = (|q|_\mathbb{R})^\alpha$ for every rational q. $\square$

Corollary 6.16. If $|\cdot|$ is a nontrivial absolute value on a number field F, then the restriction of $|\cdot|$ to $\mathbb{Q}$ is nontrivial.

REMARK. In view of Ostrowski's Theorem (Theorem 6.15), the restriction to $\mathbb{Q}$ therefore has to be equivalent to the p-adic absolute value for some p or to the ordinary absolute value.

PROOF. Since $|\cdot|$ is nontrivial, there exists x with $|x| > 1$. Raising x to a power if necessary, we may assume that $|x| \geq 2$. Arguing by contradiction, suppose that $|q| = 1$ for all nonzero q in $\mathbb{Q}$. Since x is algebraic over $\mathbb{Q}$, there exist an integer $n \geq 1$ and rational coefficients $q_{n-1}, \ldots, q_0$ such that

$$x^n = q_{n-1}x^{n-1} + \cdots + q_1 x + q_0.$$

Applying $|\cdot|$ to both sides and using that $|q_j| \leq 1$ for all j gives

$$|x|^n \leq |x|^{n-1} + \cdots + |x| + 1 = \frac{|x|^n - 1}{|x| - 1} \leq |x|^n - 1,$$

the right-hand inequality holding because $|x| \geq 2$. We have thus obtained $|x|^n \leq |x|^n - 1$ and have arrived at a contradiction. $\square$

An absolute value $|\cdot|$ on a field F such that the image of $F^\times$ is discrete is called a **discrete absolute value**. The p-adic absolute values on $\mathbb{Q}$ and on $\mathbb{Q}_p$ furnish examples.

Corollary 6.17. If $|\cdot|$ is a nontrivial discrete absolute value on the field F, then $|\cdot|$ is nonarchimedean, and $|x| = r^{-v(x)}$ for some discrete valuation of F.

REMARKS. Example 1 of nonarchimedean absolute values shows that discrete valuations always lead to discrete absolute values. This corollary is a converse. The trivial absolute value is of course nonarchimedean, but it does not arise from a discrete valuation. We shall not be interested in any nonarchimedean absolute values that do not arise from discrete valuations.

PROOF. First we show that $|\cdot|$ is nonarchimedean. Proposition 6.14 immediately handles the case that F has nonzero characteristic, and we may therefore take the characteristic to be 0. Let D be the discrete image subgroup of $F^\times$. This D in particular must contain the image of $\mathbb{Q}^\times$. Meanwhile, Theorem 6.15 says that the restriction of $|\cdot|$ to $\mathbb{Q}$ has to be trivial, or equivalent to the p-adic absolute value for some p, or equivalent to the ordinary absolute value. Under the ordinary absolute value, the image of $\mathbb{Q}^\times$ cannot be contained in D, and the restriction must be one of the other kinds. For all of the other kinds, the image of $\mathbb{Z}$ is bounded, and Proposition 6.14 allows us to conclude that $|\cdot|$ is nonarchimedean.

Now that $|\cdot|$ is nonarchimedean, we set $v(0) = +\infty$ and $v(x) = -\log_r |x|$ for $x \neq 0$. Properties (i), (ii'), and (iii) of nonarchimedean absolute values immediately imply the three defining properties of a discrete valuation. $\square$

Corollary 6.18. If $|\cdot|$ is a nontrivial discrete absolute value on a field F, then the corresponding valuation ring $R = \{x \in F \mid |x| \leq 1\}$ and the valuation ideal $P = \{x \in F \mid |x| < 1\}$ are open and closed in F.

REMARK. Corollary 6.17 shows that $|\cdot|$ is defined by a discrete valuation.

PROOF. The definitions of R and P in the statement show that R is closed and P is open. Let D be the image of $F^\times$ under $|\cdot|$. A discrete subgroup of positive reals has to be equal[7] to $\{1\}$ or to the subgroup $r^\mathbb{Z}$ for a unique real $r > 1$. The nontriviality of $|\cdot|$ implies that the correct alternative is $r^\mathbb{Z}$. Then the equality $R = \{x \in F \mid |x| < r\}$ shows that R is open, and the equality $P = \{x \in F \mid |x| \leq r^{-1}\}$ shows that P is closed. $\square$

Next we prove a general result applicable to number fields and to function fields in one variable that yields the conclusion that nonarchimedean absolute values in these cases are automatically discrete. The general result is obtained in two parts, stated as Lemma 6.19 and Proposition 6.20.

---

[7]One can invoke Lemma 5.14, for example.

Lemma 6.19. If R is a Dedekind domain regarded as a subring of its field of fractions F, and if $|\cdot|$ is a nonarchimedean absolute value on F that is ≤ 1 on R, then $|\cdot|$ is discrete. Hence either $|\cdot|$ is trivial or else it is defined by the valuation relative to a nonzero prime ideal of R.

PROOF. The subset of $x \in R$ for which $|x| < 1$ is a proper ideal I in R, and we let P be a prime ideal containing I. Since R is a Dedekind domain, P defines a corresponding discrete valuation v_P. Let $|x|_P = 2^{-v_P(x)}$. Then

$$\{x \in R \mid |x| < 1\} = I \subseteq P = \{x \in R \mid |x|_P < 1\},$$

and hence

$$\{x \in R \mid |x|_P = 1\} \subseteq \{x \in R \mid |x| = 1\}. \tag{$*$}$$

Let π be an element of R with $|\pi|_P = \frac{1}{2}$. If x is an arbitrary nonzero member of F with $|x|_P < 1$, then Proposition 6.4 shows that we can write $x = \pi^k x'$ with $k > 0$, x' in R, and $|x'|_P = 1$. Then $|x'| = 1$ by $(*)$, and it follows that $|x| = |\pi|^k$. Since $|x|_P = |\pi|_P^k$ also, there are only two possibilities. One possibility is that $|x| = |\pi| = 1$ for all $x \neq 0$, and then $|\cdot|$ is trivial. The other possibility is that the subsets of F for which $|x| < 1$ and for which $|x|_P < 1$ coincide. In this case we apply Proposition 6.12 and conclude that $|\cdot|$ and $|\cdot|_P$ are equivalent. $\square$

Proposition 6.20. Let R be a Dedekind domain regarded as a subring of its field of fractions F, let K be a finite algebraic extension of F, and let T be the integral closure of R in K. If $|\cdot|$ is a nonarchimedean absolute value on K that is ≤ 1 on R, then it is ≤ 1 on T. Hence $|\cdot|$ is discrete, and either $|\cdot|$ is trivial or else it is defined by the valuation relative to a nonzero prime ideal of T.

PROOF. As with Proposition 6.7, T is a Dedekind domain. If $x \neq 0$ is in T, then the minimal polynomial of x over R is a monic polynomial in $R[X]$, and thus there exist an integer n and coefficients $a_{n-1}, \ldots, a_0$ in R such that

$$x^n = a_{n-1}x^{n-1} + \cdots + a_1 x + a_0.$$

Taking the absolute value of both sides and using the nonarchimedean property, we obtain

$$|x|^n \leq \max_{0 \leq j \leq n-1} (|a_j||x|^j) \leq \max_{0 \leq j \leq n-1} (|x|^j) = \max(1, |x|^{n-1}),$$

the inequality holding because $|\cdot|$ is assumed to be ≤ 1 on R. If we could have $|x| > 1$, then this inequality would read $|x|^n \leq |x|^{n-1}$, which is a contradiction. We conclude that $|x| \leq 1$ for all $x \in T$. The conclusions in the last sentence of the proposition now follow from Lemma 6.19. $\square$

Corollary 6.21. If K is a number field, then every nontrivial nonarchimedean absolute value $|\cdot|$ on K comes from the valuation v_P relative to some nonzero prime ideal P in the ring of algebraic integers in K.

REMARK. Proposition 6.27 below will classify the archimedean absolute values on a number field.

PROOF. Since $|\cdot|$ is nonarchimedean, its restriction to $\mathbb{Q}$ is nonarchimedean. By Ostrowski's Theorem (or by inspection), it is ≤ 1 on $\mathbb{Z}$. The result now follows from Proposition 6.20 if we take R to be $\mathbb{Z}$ and F to be $\mathbb{Q}$. □

Corollary 6.22. Let $\Bbbk$ be a field, let $F = \Bbbk(X)$ be the field of rational expressions in one indeterminate over $\Bbbk$, let K be a finite algebraic extension of $\Bbbk[X]$, let T be the integral closure of $\Bbbk[X]$ in K, and let $|\cdot|$ be a nontrivial nonarchimedean absolute value on K that is 1 on the multiplicative group $\Bbbk^\times$. Then $|\cdot|$ is discrete, and the only possibilities for it are as follows:
  (a) $|X| \leq 1$, and there exists a unique nonzero prime ideal P in T such that $|\cdot|$ comes from the valuation determined by P,
  (b) $|X| > 1$, and there exists a prime ideal P in the integral closure T' of $\Bbbk[X^{-1}]$ in K such that $P \cap \Bbbk[X^{-1}] = X^{-1}\Bbbk[X^{-1}]$ and such that $|\cdot|$ comes from the valuation of K determined by P.

REMARKS. As with Proposition 6.7, T and T' are Dedekind domains. If $\Bbbk$ has nonzero characteristic, then Proposition 6.14 shows that every absolute value is nonarchimedean. For the case that $\Bbbk$ has characteristic zero, remarks at the end of Section 4 will indicate why every absolute value that is 1 on $\Bbbk^\times$ is nonarchimedean; we shall not need to make use of this fact, however. In any event, just as with Corollary 6.10, the ideals P that occur in (b) are the ones in the prime factorization of the ideal $X^{-1}T'$ in T'; there is at least one, and there are only finitely many.

PROOF. The argument is similar to the one for Corollary 6.21, except that we have to take into account what happens when $|X| > 1$. We apply Proposition 6.20 either with $R = \Bbbk[X]$ or with $R = \Bbbk[X^{-1}]$.

Since $|\cdot|$ is 1 on $\Bbbk^\times$, an inequality $|X| \leq 1$ implies that $|\cdot|$ is ≤ 1 on $\Bbbk[X]$, $|\cdot|$ being assumed to be nonarchimedean. Then Proposition 6.20 and Corollary 6.10 show that (a) holds. Similarly an inequality $|X| > 1$ implies that $|\cdot|$ is ≤ 1 on $\Bbbk[X^{-1}]$ because $|\cdot|$ is assumed nonarchimedean. Then Proposition 6.20 and Corollary 6.10 show that (b) holds. □

Theorem 6.23 (Weak Approximation Theorem). Let $|\cdot|_1, \ldots, |\cdot|_n$ be inequivalent nontrivial absolute values on a field F. If $\epsilon > 0$ is a real number and $x_1, \ldots, x_n$ are elements of F, then there exists y in F such that

$$|y - x_j|_j < \epsilon \qquad \text{for } 1 \leq j \leq n.$$

REMARKS. The special case of this theorem in which F is a number field and the absolute values are defined by n distinct nonzero prime ideals in the ring of algebraic integers follows from the Chinese Remainder Theorem (Theorem 8.27 of *Basic Algebra*, restated in the present book on page xxiii). In fact, it is enough to handle the case that all the x_j's are algebraic integers in F. Let the prime ideals be $P_1, \ldots, P_n$, and let $|\cdot|_j = r_j^{-v_{P_j}(\cdot)}$ with $r_j > 1$. If we specify any positive integers $k_1, \ldots, k_n$, then the Chinese Remainder Theorem produces an algebraic integer y in F such that $y \equiv x_j \bmod P_j^{k_j}$ for $1 \leq j \leq n$. These congruences say that $v_{P_j}(y - x_j) \geq k_j$, hence that $|y - x_j|_j \leq r_j^{-k_j}$. Thus we have only to choose $k_1, \ldots, k_n$ large enough to make $r_j^{-k_j} < \epsilon$ for all j, and the inequalities of the theorem will hold.

PROOF. First let us prove that we can find an element z in F with

$$|z|_1 > 1 \quad \text{and} \quad |z|_j < 1 \text{ for } 2 \leq j \leq n. \qquad (*)$$

We do so by induction on n, the case $n = 2$ being Proposition 6.12. Assuming the result for $n - 1$, find u with $|u|_1 > 1$ and $|u|_j < 1$ for $2 \leq j \leq n - 1$. Then by the result for $n = 2$, find v with $|v|_1 > 1$ and $|v|_n < 1$. Let $k > 0$ be an integer to be specified, and put

$$z = \begin{cases} v & \text{if } |u|_n < 1, \\ u^k v & \text{if } |u|_n = 1, \\ \frac{u^k v}{1 + u^k} & \text{if } |u|_n > 1. \end{cases}$$

In the second case, k is to be chosen large enough to make $|u|_j^k |v|_j < 1$ for $2 \leq j \leq n - 1$. In the third case, k is to be chosen large enough to make $|u|_1^k (1 + |u|_1^k)^{-1} |v|_1 > 1$, $|u|_j^k (1 - |u|_j^k)^{-1} |v|_j < 1$ for $2 \leq j \leq n - 1$, and $|u|_n^k (|u|_n^k - 1)^{-1} |v|_n < 1$. Then z satisfies the conditions in $(*)$, and the inductive proof of $(*)$ is complete.

Applying $(*)$, find z_j such that $|z_j|_j > 1$ and $|z_j|_i < 1$ for $i \neq j$. Let l be a positive integer to be specified, and put

$$y = \sum_{i=0}^{n} \frac{x_i z_i^l}{1 + z_i^l}.$$

Since $y - x_j = -x_j (1 + z_j^l)^{-1} + \sum_{i \neq j} x_i z_i^l (1 + z_i^l)^{-1}$, we obtain

$$|y - x_j|_j \leq |x_j|_j (|z_j|_j^l - 1)^{-1} + \sum_{i \neq j} |x_i|_j (|z_i|_j^l (1 - |z_i|_j^l)^{-1}). \qquad (**)$$

For l large enough, the coefficients $(|z_j|_j^l - 1)^{-1}$ and $|z_i|_j^l (1 - |z_i|_j^l)^{-1}$ for $i \neq j$ can be made as small as we please, and thus the right side of $(**)$ can be made to be $< \epsilon$. □

4. Completions

In this section we finish our project of establishing an abstract theory that generalizes the construction of the field of p-adic numbers. A little care is appropriate in stating the results. Here is an example of the cost of imprecision: We know that the field $\mathbb{Q}_p$ is obtained by completing $\mathbb{Q}$ with respect to the p-adic absolute value. We shall see in Section 5 that $\mathbb{Q}_p$ for $p = 5$ is obtained also by completing the field $\mathbb{Q}(i)$ with respect to a certain absolute value and that in fact there are two distinct equivalence classes of absolute values on $\mathbb{Q}(i)$ for which $\mathbb{Q}_5$ results in this way. Thus a completion process is not well specified unless we include all the data—the original field, the absolute value on it (or at least the equivalence class of absolute values), and the mapping into the completed space.

For this reason we introduce the notions of a **valued field**, namely a pair $(F, |\cdot|_F)$ consisting of a field and an absolute value on it, and a **homomorphism of valued fields**. If $(F, |\cdot|_F)$ and $(K, |\cdot|_K)$ are the two valued fields in question, a homomorphism from the first to the second is a field map $\varphi : F \to K$ such that $|x|_F = |\varphi(x)|_K$ for all x in F. We write φ^* for the corresponding operation of restriction: $\varphi^*(|\cdot|_K) = |\cdot|_F$. If φ carries F onto K, then φ is called an **isomorphism of valued fields**.

A **completion** of a valued field $(F, |\cdot|_F)$ is defined to be a homomorphism of valued fields $\varphi : (F, |\cdot|_F) \to (K, |\cdot|_K)$ such that $(K, |\cdot|_K)$ is complete as a metric space and $\varphi(F)$ is dense in K. The first theorem establishes existence.

Theorem 6.24. Let F be a field with a nontrivial absolute value $|\cdot|_F$, let d be the associated metric on F, let $\mathcal{R}$ be the subring of $\prod_{j=1}^{\infty} F$ consisting of all Cauchy sequences relative to d, and let $\mathcal{I}$ be the ideal in $\mathcal{R}$ consisting of all sequences convergent to 0. Then $\mathcal{I}$ is a maximal ideal in $\mathcal{R}$, and the quotient $\mathcal{R}/\mathcal{I}$ is a field. Consequently the Cauchy completion of F relative to d is a topological field $\overline{F} = \mathcal{R}/\mathcal{I}$. Let $i : F \to \overline{F}$ be the natural map $F \to \mathcal{R} \to \mathcal{R}/\mathcal{I}$ of F into the Cauchy completion given by carrying members of F into constant sequences in $\mathcal{R}$, followed by passage to the quotient. The metric $\bar{d}$ on the Cauchy completion is the unique continuous function $\bar{d} : \overline{F} \times \overline{F} \to \mathbb{R}$ such that $\bar{d}(i(x), i(y)) = d(x, y)$. If a real-valued function $|\cdot|_{\overline{F}}$ is defined on $\overline{F}$ by $|x|_{\overline{F}} = \bar{d}(x, 0)$ for $x \in \overline{F}$, then $|\cdot|_{\overline{F}}$ is an absolute value on $\overline{F}$, and $i : (F, |\cdot|_F) \to (\overline{F}, |\cdot|_{\overline{F}})$ is a homomorphism of valued fields. Moreover, the absolute value on $\overline{F}$ is nonarchimedean if the absolute value on F is nonarchimedean.

REMARKS. The usual construction of the Cauchy completion embeds the original metric subspace as a dense subset of a complete metric space, and therefore this theorem is showing that $i : (F, |\cdot|_F) \to (\overline{F}, |\cdot|_{\overline{F}})$ is a completion of $(F, |\cdot|_F)$.

PROOF. The proof of this theorem is almost the same as the first part of the proof of Proposition 6.1, apart from notational changes. The differences occur in spots where the ultrametric inequality was invoked in the proof of Proposition 6.1 and only the triangle inequality is available here. The main such difference is the argument that the validity of the triangle inequality on F implies the validity of the triangle inequality on $\overline{F}$, and we give that argument in a moment. Correspondingly it is unnecessary for us to prove that the validity of the ultrametric inequality on F implies the validity of the ultrametric inequality on $\overline{F}$, because that argument does occur in the proof of Proposition 6.1.

The other places in the proof of Proposition 6.1 where the ultrametric inequality was used are in the proof that the completion is a topological field. It is not necessary to modify that proof here, however, since we can invoke Proposition 6.13.

Thus let us see that the validity of the triangle inequality on F implies the validity of the triangle inequality on $\overline{F}$. To proceed, let x and y be members of $\overline{F} = \mathcal{R}/\mathcal{I}$, and let $\{q_n\}$ and $\{r_n\}$ be respective coset representatives of them in $\mathcal{R}$. Then $\{q_n + r_n\}$ is a representative of $x + y$, by definition, and the continuity of $|\cdot|_{\overline{F}}$ on $\overline{F}$ implies that $\lim_n |q_n + r_n|_p = |x + y|_p$. From this limit formula and the triangle inequality for F, we obtain

$$|x + y|_{\overline{F}} = \lim_n |q_n + r_n|_{\overline{F}} \leq \limsup_n (|q_n|_{\overline{F}} + |r_n|_{\overline{F}})$$
$$\leq \limsup_n |q_n|_{\overline{F}} + \limsup_n |r_n|_{\overline{F}} = |x|_{\overline{F}} + |y|_{\overline{F}},$$

since $\lim_n |q_n|_{\overline{F}} = |x|_{\overline{F}}$ and $\lim_n |r_n|_{\overline{F}} = |y|_{\overline{F}}$. This proves the triangle inequality on $\overline{F}$. □

A valued field $(L, |\cdot|_L)$ is said to be **complete** if L is Cauchy complete in the metric defined by $|\cdot|_L$. In Section 6 we shall make crucial use of a universal mapping property of the completion of a valued field.

Theorem 6.25. If $\iota : (F, |\cdot|_F) \to (K, |\cdot|_K)$ is a completion of the valued field $(F, |\cdot|_F)$ and if $\varphi : (F, |\cdot|_F) \to (L, |\cdot|_L)$ is a homomorphism of valued fields with $(L, |\cdot|_L)$ complete, then there exists a unique homomorphism of valued fields $\Phi : (K, |\cdot|_K) \to (L, |\cdot|_L)$ such that $\varphi = \Phi \circ \iota$.

REMARKS. As usual with universal mapping properties, this theorem implies a uniqueness result: any two completions of a valued field are canonically isomorphic. It is not necessary to write out the details. Making a small adjustment to the proof below, we see also that if a field has two equivalent absolute values on it, then the corresponding two completions are canonically isomorphic by a field map that respects the topologies.

PROOF. The theory of completion of a metric space produces a unique continuous function $\Phi : K \to L$ such that $\varphi = \Phi \circ \iota$, and this continuous function respects the metrics. It is necessary to check only that Φ respects addition and multiplication.

The argument is the same for the two operations, and we check only addition. Let x and y be given in K, and choose sequences $\{x_n\}$ and $\{y_n\}$ in F with $\lim \iota(x_n) = x$, $\lim \iota(y_n) = y$. Since addition is continuous in K, $\lim \iota(x_n + y_n) = x + y$. Since Φ is a continuous function with $\varphi = \Phi \circ \iota$,

$$\Phi(x) + \Phi(y) = \Phi(\lim \iota(x_n)) + \Phi(\lim \iota(y_n))$$
$$= \lim(\Phi(\iota(x_n))) + \lim(\Phi(\iota(y_n))) = \lim(\varphi(x_n)) + \lim(\varphi(y_n))$$
$$= \lim(\varphi(x_n) + \varphi(y_n)) = \lim(\varphi(x_n + y_n))$$
$$= \lim(\Phi\iota(x_n + y_n)) = \Phi(\lim \iota(x_n + y_n)) = \Phi(x + y),$$

and Φ respects addition. $\square$

Theorem 6.24 generalizes the parts of Proposition 6.1 concerning $\mathbb{Q}_p$ but not those concerning $\mathbb{Z}_p$. The arguments concerning $\mathbb{Z}_p$ transparently made use of the ultrametric inequality, and they used a little more. The extra fact used is that the p-adic absolute value is defined from a discrete valuation. In view of Corollary 6.17 and Example 1 of nonarchimedean absolute values in the previous section, a necessary and sufficient condition for a nontrivial absolute value on a field F to be obtained from a discrete valuation is that the image of $F^\times$ under the valuation be a discrete subset of the positive reals. Such an absolute value is automatically nonarchimedean.

Theorem 6.26. Let $\iota : (F, |\cdot|_F) \to (\overline{F}, |\cdot|_{\overline{F}})$ be a completion of a valued field, and suppose that $|\cdot|_F$ is nontrivial and discrete. Let $v(\cdot)$ be the discrete valuation that defines $|\cdot|$ on F. Then

(a) the image $|\overline{F}^\times|_{\overline{F}}$ equals the image $|F^\times|_F$, and $|\cdot|_{\overline{F}}$ on $\overline{F}$ is therefore defined by a discrete valuation $\bar{v}(\cdot)$ on $\overline{F}$ such that $\bar{v} \circ \iota = v$,

(b) the image $\iota(R)$ of the valuation ring R of v is dense in the valuation ring $\overline{R}$ of $\bar{v}$,

(c) for every integer $n > 0$, the image $\iota(P^n)$ of the nth power P^n of the valuation ideal P of v is dense in the nth power $\overline{P}^n$ of the valuation ideal $\overline{P}$ of $\bar{v}$,

(d) the residue class fields of F and $\overline{F}$ coincide in the sense that the mapping $\iota : R \to \overline{R}$ descends to a field isomorphism of R/P onto $\overline{R}/\overline{P}$,

(e) for every integer $n > 0$, the mapping $\iota : R \to \overline{R}$ descends to a ring isomorphism of R/P^n onto $\overline{R}/\overline{P}^n$,

(f) $\overline{R}$ is compact if R/P is finite, and in this case the topological field $\overline{F}$ is locally compact.

4. Completions

REMARK. No assertion is made in (d) and (e) about whether the topologies match under the constructed isomorphisms. Our interest will be mostly in the case that R/P is finite, in which case the topologies match because they are discrete.

PROOF. Write $|F^\times|_F$ in the form $r^{\mathbb{Z}}$ for a unique real number $r > 1$. For (a), since $|\iota(x)|_{\overline{F}} = |x|_F$ and since $\iota(F)$ is dense in $\overline{F}$, the continuity of the absolute value $|\cdot|_{\overline{F}}$ implies that the image of $\overline{F}^\times$ is contained in the closure of $r^{\mathbb{Z}}$ within the positive reals, which is $r^{\mathbb{Z}}$. The formula $\bar{v} \circ \iota = v$ follows from the computation $r^{-v(x)} = |x|_F = |\iota(x)|_{\overline{F}} = r^{-\bar{v}(\iota(x))}$ by taking the logarithm to the base r.

For (b) and (c), we use that $\iota(F)$ is dense in $\overline{F}$, and we treat (b) as the case $n = 0$ of (c). Fix $n \geq 0$ and consider $\overline{P}^n$. Choose a sequence $\{x_k\}$ in F with $\{\iota(x_k)\}$ converging to a point x in $\overline{P}^n$. Since $|x|_{\overline{F}} \leq r^{-n}$, we must have $|x_k|_F < r^{-n+1}$ for all sufficiently large k. The elements x_k satisfying this condition are in P^n, and thus $\iota(P^n)$ is dense in $\overline{P}^n$.

For (d) and (e), the mapping $R \to \overline{R}/\overline{P}^n$ descends to R/P^n, since $\iota(P) \subseteq \overline{P}$. The descended map is one-one, since if $x \in R$ maps to the 0 coset, then x is in $\iota^{-1}(\overline{P}^n) = P^n$. To see that the descended map is onto, let a coset $\bar{x} + \overline{P}^n$ be given. Since $\iota(R)$ is dense in $\overline{R}$, we can choose $x \in R$ with $|\iota(x) - \bar{x}|_{\overline{F}} < r^{-n}$. Since $\overline{P}^n = \{y \in \overline{F} \mid |y| < r^{-n+1}\}$, $\iota(x) - \bar{x}$ is in $\overline{P}^n$. Hence $\iota(x)$ is exhibited as in $\bar{x} + \overline{P}^n$, and the coset $x + P^n$ maps to the coset $\bar{x} + \overline{P}^n$.

In (f), Corollary 8.60 of *Basic Algebra* shows that P^n/P^{n+1} is a 1-dimensional vector space over R/P. The First Isomorphism Theorem gives an R module isomorphism $(R/P^{n+1})/(R/P^n) \cong P^n/P^{n+1}$, and it follows by induction on n that the finiteness of R/P implies the finiteness of R/P^n. In view of (e), $\overline{R}/\overline{P}^n$ is finite for every $n > 0$.

For each $n > 0$, the set $\overline{R}$ is covered by the cosets of $\overline{P}^n$, which are closed balls in $\overline{F}$ of radius r^{-n} and open balls of radius r^{-n+1}. Thus for any positive radius, there exists a finite collection of open balls of that radius or less such that the union of the open balls covers $\overline{R}$. This means that $\overline{R}$ is totally bounded in the metric space $\overline{F}$. A totally bounded closed subset of a complete metric space is compact, and consequently $\overline{R}$ is compact.

Thus the 0 element of $\overline{F}$ has $\overline{R}$ as a compact neighborhood. Since addition is continuous, each member x of $\overline{F}$ has $x + \overline{R}$ as a compact neighborhood of x, and therefore $\overline{F}$ is locally compact. □

Let us review briefly. We start with an absolute value on a field F. The cases of initial interest are that F is a number field or is a function field in one variable, namely a finite algebraic extension of a field $\Bbbk(X)$, where $\Bbbk$ is a given base field; in the latter case we assume that the absolute value is identically 1 on $\Bbbk^\times$. A number field can have archimedean absolute values, and we come to them in a moment. In the function-field case we know that every absolute value is nonarchimedean if $\mathbb{k}$ has nonzero characteristic; this remains true for characteristic zero but we did not prove it. For our cases of interest the nonarchimedean nontrivial absolute values are always given by a discrete valuation.

Thus let us summarize what happens for a nonarchimedean nontrivial absolute value that is given by a discrete valuation. Within the given field F we have singled out a Dedekind domain R for which F is the field of fractions,[8] and the absolute value is ≤ 1 on R. For example, in the number-field case R is the ring of algebraic integers in F. In all cases the discrete valuation v is determined by a nonzero prime ideal $\mathfrak{p}$ of R, and the absolute value on F is given by $|x|_F = r^{-v(x)}$ for some number $r > 1$. Our two-step process consists in a step of localization and a step of completion. The step of localization passes to the principal ideal domain $S^{-1}R$ with maximal ideal $S^{-1}\mathfrak{p}$, where S is the complement of $\mathfrak{p}$ in R. The domain $S^{-1}R$ coincides with the valuation ring of v, and the ideal $S^{-1}\mathfrak{p}$ coincides with the valuation ideal of v. The absolute value on F does not change during this process of localization. The ideal $S^{-1}\mathfrak{p}$ is principal in $S^{-1}R$, say with π as a generator. The element π can be chosen to be in $\mathfrak{p}$, and it has $v(\pi) = 1$. Theorem 6.5 and Proposition 6.11 govern relationships between R and $S^{-1}R$. Briefly the powers of $\mathfrak{p}$ are dense in the powers of $S^{-1}\mathfrak{p}$, and the natural map of residue class fields $R/\mathfrak{p} \to S^{-1}R/S^{-1}\mathfrak{p}$ is a field isomorphism onto.

The second step is a step of completion with respect to the absolute value. The completion of a valued field $(F, |\cdot|_F)$ is a homomorphism of valued fields $\iota : (F, |\cdot|_F) \to (L, |\cdot|_L)$ such that $(L, |\cdot|_L)$ is complete as a metric space and ι carries F onto a dense subfield of L. This exists by Theorem 6.24. In the situation with a nonarchimedean nontrivial absolute value that is given by a discrete valuation, one often writes $F_\mathfrak{p}$ for the completed field L. The eventual interest is partly in what happens to R and $\mathfrak{p}$, but we first consider $S^{-1}R$ and $S^{-1}\mathfrak{p}$. The completed absolute value $|\cdot|_{F_\mathfrak{p}}$ is given by a discrete valuation $\bar{v}$ with $\bar{v} \circ \iota = v$. Let us write $R_\mathfrak{p}$ for its valuation ring and $\mathfrak{p}_\mathfrak{p}$ for its valuation ideal. Theorem 6.26 governs the relationships between $S^{-1}R$ and $R_\mathfrak{p}$. Briefly the images under ι of the powers of $S^{-1}\mathfrak{p}$ are dense in the powers of $\mathfrak{p}_\mathfrak{p}$, and the natural map of residue class fields $S^{-1}R/S^{-1}\mathfrak{p} \to R_\mathfrak{p}/\mathfrak{p}_\mathfrak{p}$ induced by ι is a field isomorphism onto.

The case of most interest for number theory is the case of a number field F and the absolute value determined by a nonzero prime ideal $\mathfrak{p}$ in the ring of algebraic integers of F. The field $F_\mathfrak{p}$ is called the field of **$\mathfrak{p}$-adic numbers**, and the ring $R_\mathfrak{p}$ is called the ring of **$\mathfrak{p}$-adic integers**. When $F = \mathbb{Q}$ and $\mathfrak{p} = p\mathbb{Z}$ for a prime number p, the element π can be taken to be p.

---

[8]The case $R = F$ is excluded; this is the case that produces the trivial absolute value, which does not interest us.

In the case of a function field in one variable that is most analogous to a number field, one starts from a field F that is a finite algebraic extension of $\mathbb{F}_q(X)$, where $\mathbb{F}_q$ is a finite field with q elements. According to Corollary 6.22, all but finitely many of the nonarchimedean absolute values are defined in terms of nonzero prime ideals in the integral closure of $\mathbb{F}_q[X]$ in F; the others are the prime constituents of the ideal $X^{-1}\mathbb{F}_q[X^{-1}]$ in $\mathbb{F}_q[X^{-1}]$. One can show that the ring in the completion analogous to $R_\mathfrak{p}$ is always a **ring of formal power series** $\mathbb{F}_{q'}[[X]]$ in one indeterminate X and with coefficients in a finite extension $\mathbb{F}_{q'}$ of $\mathbb{F}_q$. Elements of this ring are arbitrary formal power series of the form $\sum_{k=0}^{\infty} c_k X^k$ with all c_k in $\mathbb{F}_{q'}$. The field of fractions analogous to $F_\mathfrak{p}$ is always a **field of formal Laurent series** $\mathbb{F}_q((X))$ in one indeterminate; nonzero elements of this field are arbitrary expressions of the form $\sum_{k=-N}^{\infty} c_k X^k$ with all c_k in $\mathbb{F}_{q'}$, with $c_{-N} \neq 0$, and with N depending on the element.

Let us now examine archimedean completions. We shall discuss what happens when we start from a number field, and then we make some remarks without proof about the general case. Thus let F be a number field, and let an archimedean absolute value be given on it. To have notation parallel to the nonarchimedean case, it is customary to index the absolute value[9] by a symbol like v, writing $|\cdot|_v$ for it. Corollary 6.16 shows that the restriction of $|\cdot|_v$ to $\mathbb{Q}$ is nontrivial, and the combination of Proposition 6.14 and Ostrowski's Theorem (Theorem 6.15) shows that the restriction to $\mathbb{Q}$ is equivalent to the ordinary absolute value. Adjusting $|\cdot|_v$ within its equivalence class, we may assume that its restriction to $\mathbb{Q}$ matches the ordinary absolute value. Using Theorem 6.24, we form the completion of F with respect to $|\cdot|_v$, writing F_v for the completed space. The limits of Cauchy sequences from $\mathbb{Q}$ itself show that $\mathbb{R}$ lies in the completed space, since $|\cdot|_v$ matches the ordinary absolute value on $\mathbb{Q}$. Thus we can regard $\mathbb{R}$ as a subfield of F_v, and F is a subfield as well. Consequently the set $\mathbb{R}F$ of sums of products is a subring of F_v. The multiplication mapping of $\mathbb{R} \times F$ into F_v is $\mathbb{Q}$ bilinear and has a linear extension $\mathbb{R} \otimes_\mathbb{Q} F \to F_v$ whose image is $\mathbb{R}F$. The $\mathbb{R}$ dimension of $\mathbb{R} \otimes_\mathbb{Q} F$ is $[F:\mathbb{Q}]$, and consequently the $\mathbb{R}$ dimension of $\mathbb{R}F$ is $\leq [F:\mathbb{Q}]$, hence finite. Being a finite-dimensional $\mathbb{R}$ algebra embedded in a field, $\mathbb{R}F$ is a subfield[10] of F_v. It is therefore a finite algebraic extension of $\mathbb{R}$ and must be $\mathbb{R}$ or $\mathbb{C}$. Thus F lies in $\mathbb{R}$ or $\mathbb{C}$. The fields $\mathbb{R}$ and $\mathbb{C}$ are complete relative to the ordinary absolute value, and hence $\mathbb{R}F$ is a closed subset of F_v. Since F is dense, we conclude that F_v is $\mathbb{R}$ or $\mathbb{C}$.

Visualize having a standard copy of $\mathbb{C}$ available, with $\mathbb{R}$ embedded in it. From the above remarks, any archimedean absolute value of the number field F, after

---

[9]Or the equivalence class of the absolute value.

[10]Within a field if a nonzero element is algebraic over a base field, then the smallest ring containing the base field and the element contains also the inverse of the element.

adjustment within its equivalence class, yields a completion that takes one of the two forms

$$\sigma : (F, |\cdot|_v) \to (\mathbb{R}, |\cdot|) \quad \text{and} \quad \sigma : (F, |\cdot|_v) \to (\mathbb{C}, |\cdot|),$$

where $|\cdot|$ is ordinary absolute value on $\mathbb{R}$ or $\mathbb{C}$. Conversely any field mapping σ of F into $\mathbb{R}$ or $\mathbb{C}$ has dense image either in $\mathbb{R}$ or in $\mathbb{C}$ and defines an archimedean absolute value on F by $|\cdot|_v = \sigma^*(|\cdot|)$. Then $\sigma : (F, |\cdot|_v) \to (\mathbb{R} \text{ or } \mathbb{C}, |\cdot|)$ is a completion by Theorem 6.25.

To classify the archimedean absolute values up to equivalence, we recall from Section V.2 that the number of distinct field maps σ into $\mathbb{C}$ of a number field F of degree $[F : \mathbb{Q}] = n$ is exactly n, with a certain number r_1 of them having image in $\mathbb{R}$ and with the remainder $2r_2$ having image in $\mathbb{C}$ but not $\mathbb{R}$ and occurring in complex conjugate pairs. Each such field map σ gives us a completion. The members of a complex conjugate pair result in the same absolute value on F when the ordinary absolute value of $\mathbb{C}$ is restricted to F. We shall show that there are no other equivalences.

Proposition 6.27. Let F be a number field with $[F : \mathbb{Q}] = n$, and let there be r_1 distinct field maps of F into $\mathbb{R}$ and r_2 complex conjugate pairs of distinct field maps of F into $\mathbb{C}$, with $r_1 + 2r_2 = n$. Each such field map σ induces an archimedean absolute value on F by restriction from $\mathbb{R}$ or $\mathbb{C}$, the only equivalences are the ones from pairs of field maps related by complex conjugation, and the resulting collection of $r_1 + r_2$ absolute values exhausts the archimedean absolute values on F, up to equivalence.

PROOF. The remarks above show everything except that these $r_1 + r_2$ absolute values are mutually inequivalent. To prove this fact, suppose that σ and σ' are two field maps of F into the same field, $\mathbb{R}$ or $\mathbb{C}$, such that $x \mapsto |\sigma(x)|$ is equivalent to $x \mapsto |\sigma'(x)|$. Then $\varphi = \sigma' \sigma^{-1}$ is a field isomorphism from image σ onto image σ' that respects the absolute value, up to a power. It is therefore uniformly continuous from image σ onto image σ'. Consequently φ extends to all of $\mathbb{R}$ or $\mathbb{C}$, and the continuous extension respects the field operations. On $\mathbb{Q}$, φ is the identity, and hence its continuous extension to $\mathbb{R}$ must be the identity. Thus the continuous extension is an automorphism of $\mathbb{R}$ or $\mathbb{C}$ that fixes $\mathbb{R}$, and consequently it must be the identity or complex conjugation. □

It is of some interest to know what archimedean absolute values can occur in other situations, besides number fields, and Theorem 6.24 shows that it is enough to classify the complete ones. Ostrowski did so, and the result is that $\mathbb{R}$ and $\mathbb{C}$, with their ordinary absolute values, are the only complete archimedean fields up to equivalence.[11]

---

[11] A proof of the Ostrowski result may be found in Hasse's *Number Theory*, pp. 191–194. Gelfand

5. Hensel's Lemma

Hensel's Lemma is a device that in its simplest forms allows one to solve polynomial equations in the field $\mathbb{Q}_p$ of p-adic numbers by using congruence information modulo some power of p. It has a number of distinct formulations, all of which work within any complete nonarchimedean valued field, not limited to $\mathbb{Q}_p$. We shall give a fairly simple formulation and obtain a handy special case as a corollary, using an adaptation of Newton's method of iterations in calculus for finding roots of polynomials. At the end of the section, we shall state without proof a version of Hensel's Lemma that works to factor polynomials rather than to find their roots. Yet another formulation of Hensel's Lemma, whose precise statement we omit, applies to systems of polynomial equations in several variables.

No overarching result of this chapter actually makes use of any version of Hensel's Lemma. Instead, versions of Hensel's Lemma are indispensable in analyzing the fine structure of complete valued fields and in handling examples. Thus the applications of Hensel's Lemma in this book will occur in the examples of this section and the next and also in problems at the end of the chapter. Problem 16 is one such problem.

Theorem 6.28 (Hensel's Lemma). Let F be a field with a nontrivial discrete absolute value $|\cdot|$, necessarily nonarchimedean, and assume that F is complete. Let R be the valuation ring, and let $f(X)$ be a polynomial in $R[X]$. Suppose that a_0 is a member of R such that

$$|f(a_0)| < |f'(a_0)|^2.$$

Then the sequence $\{a_n\}$ recursively given by

$$a_{n+1} = a_n - \frac{f(a_n)}{f'(a_n)}$$

is well defined in R and converges to a root a of $f(X)$ that satisfies $|a - a_0| < 1$.

PROOF. Put $c = |f(a_0)|/|f'(a_0)|^2 < 1$. We prove the following three statements together by induction on n:

(i) a_n is well defined and is in R,
(ii) $|f'(a_n)| = |f'(a_0)| \neq 0$, and
(iii) $|f(a_n)|/|f'(a_n)| \leq c^{2^n}|f'(a_0)|$.

---

and Tornheim proved a more general result, with the same conclusion, that allows the multiplicative property of absolute values to be relaxed somewhat. A proof of this result appears in Artin's *Theory of Algebraic Numbers*, pp. 45–51.

The base case for the induction is the case $n = 0$, and the three statements are true by hypothesis in this case.

Assume that the three statements hold for n. From (ii), a_{n+1} is defined, and then (iii) shows that a_{n+1} satisfies

(iii') $|a_{n+1} - a_n| = |f(a_n)|/|f'(a_n)| \leq c^{2^n}|f'(a_0)|.$

The fact that a_n and $f'(a_0)$ are in R, in combination with (iii'), shows that a_{n+1} is in R. This proves (i) for $n + 1$.

For (ii) and (iii), we make use of the following Taylor expansions of $f(X)$ and $f'(X)$ about b:

$$f(X) = f(b) + (X - b)f'(b) + (X - b)^2 g(X) \quad \text{with } g(X) \in R[X]$$

and

$$f'(X) = f'(b) + (X - b)h(X) \quad \text{with } h(X) \in R[X].$$

To check that these expansions are valid in any characteristic, it is enough to check the first one, since the second one follows by differentiation. For the first one, it is enough to treat the special case X^k. Dividing $X^k - b^k$ by $X - b$, we see that we are to produce $g(X)$ such that

$$(X - b)g(X) = \sum_{j=0}^{k-1} b^{k-1-j} X^j - kb^{k-1} = \sum_{j=0}^{k-1} b^{k-1-j}(X^j - b^j).$$

Every term on the right side is divisible by $X - b$, and thus the quotient $g(X)$ is in $R[X]$.

Put $Q_n = a_{n+1} - a_n = -f(a_n)/f'(a_n)$. By (iii) for n, $|Q_n| \leq |f'(a_n)|c^{2^n}$; in particular, $|Q_n| < |f'(a_n)|$. In the expansion of $f'(X)$, we take $b = a_n$ and evaluate at $X = a_{n+1}$ to obtain

$$f'(a_{n+1}) = f'(a_n) + Q_n h(a_{n+1}).$$

Since $|Q_n| < |f'(a_n)|$ and $|h(a_{n+1})| \leq 1$, we see that $|f'(a_{n+1})| = |f'(a_n)|$. This proves (ii) for $n + 1$.

In the expansion of $f(X)$, we take $b = a_n$ and evaluate at $X = a_{n+1}$ to obtain

$$f(a_{n+1}) = f(a_n) + (a_{n+1} - a_n)f'(a_n) + (a_{n+1} - a_n)^2 g(a_{n+1}).$$

But $(a_{n+1} - a_n)f'(a_n) = -f(a_n)$, and hence this equation simplifies to

$$f(a_{n+1}) = Q_n^2 g(a_{n+1}).$$

Since $g(a_{n+1})$ is in R, application of (iii) for n and (ii) for $n + 1$ gives

$$\frac{|f(a_{n+1})|}{|f'(a_{n+1})|^2} = \frac{|Q_n|^2|g(a_{n+1})|}{|f'(a_n)|^2} \leq \left(\frac{|f(a_n)|}{|f'(a_n)|^2}\right)^2 \leq (c^{2^n})^2 = c^{2^{n+1}},$$

and this proves (iii) for $n + 1$. This completes the induction.

Now we can prove the theorem. If $n < m$, then (iii′) and the ultrametric inequality imply that

$$|a_m - a_n| \leq \max_{n \leq k < m} |a_{k+1} - a_k| \leq |f'(a_0)| \max_{n \leq k < m} c^{2^k} \leq |f'(a_0)|c^{2^n}. \qquad (*)$$

Consequently $\{a_n\}$ is a Cauchy sequence. Let a be its limit. Substituting into the definition of a_{n+1}, using (ii), and passing to the limit, we obtain $a = a - f(a)/f'(a)$. Thus $f(a) = 0$. Taking $n = 0$ in $(*)$ and letting m tend to infinity gives $|a - a_0| \leq |f'(a_0)|c$, and this is $\leq c < 1$ because $f'(a_0)$ is in R. $\square$

Corollary 6.29 (Hensel's Lemma). *Let F be a field with a nontrivial discrete absolute value, necessarily nonarchimedean, and assume that F is complete. Let R be the valuation ring, let $\mathfrak{p}$ be the unique maximal ideal, and let $f(X)$ be a polynomial in $R[X]$. If $\overline{f}(X)$ is the reduced polynomial with coefficients in $R/\mathfrak{p}$ and if $\bar{a}$ is a simple root of $\overline{f}(X)$, then $f(X)$ has a simple root $a \in R$ whose image in $R/\mathfrak{p}$ is $\bar{a}$.*

PROOF. Let a_0 be any member of R whose image in $R/\mathfrak{p}$ is $\bar{a}$. The assumptions imply that $f(a_0)$ is in $\mathfrak{p}$ and that $f'(a_0)$ is in R but not $\mathfrak{p}$. Thus the hypotheses of Theorem 6.28 are satisfied, and the theorem produces a root a of $f(X)$ with $a - a_0$ in $\mathfrak{p}$. $\square$

EXAMPLES WITH $F = \mathbb{Q}_p$ AND $R = \mathbb{Z}_p$.

(1) Suppose that p is an odd prime and that n is an integer for which the Legendre symbol $\left(\frac{n}{p}\right)$ is $+1$, i.e., for which $\gcd(n, p) = 1$ and n has a square root modulo p. Then n has a square root in $\mathbb{Z}_p$. This is immediate from Corollary 6.29 with $f(X) = X^2 - n$.

(2) Suppose that $p = 2$ and that n is an integer[12] having the form $8k + 1$. The maximal ideal in $\mathbb{Z}_2$ is (2). Corollary 6.29 is not applicable to $f(X) = X^2 - n$, since evaluation of the derivative $f'(X) = 2X$ at any point of $\mathbb{Z}_2$ leads to a member of the ideal (2). However, we can apply Theorem 6.28. Let $a_0 = 1$, so that $f(a_0) = 1 - n$ and $f'(a_0) = 2$. The theorem produces a root a in $\mathbb{Z}_2$ if $|1 - n|_2/|2|_2^2 < 1$, i.e., if $|1 - n|_2 < \frac{1}{4}$. Since $|1 - n|_2 = |-8k|_2 = \frac{1}{8}|k|_2 < \frac{1}{4}$, the theorem indeed applies. The resulting root a in $\mathbb{Z}_2$ has $a \equiv 1 \bmod (2)$.

---

[12]In fact, n could be a 2-adic integer in this argument.

(3) Suppose that $p > 3$. Every nonzero residue $\bar{a}$ in $\mathbb{Z}/p\mathbb{Z}$ has $\bar{a}^{p-1} \equiv 1 \bmod p$. Corollary 6.29 shows immediately that the polynomial $X^{p-1} - 1$ has a root a whose image in $\mathbb{Z}_p/p\mathbb{Z}_p$ is $\bar{a}$. Since the elements $\bar{a}$ are distinct, we conclude that $\mathbb{Z}_p$ contains all $p - 1$ of the $(p-1)^{\text{st}}$ root of unity.

(4) As promised at the beginning of Section 4, we show that $\mathbb{Q}_p$ for $p = 5$ is obtained also by completing the field $\mathbb{Q}(i)$ with respect to a certain absolute value and that in fact there are two distinct equivalence classes of absolute values on $\mathbb{Q}(i)$ for which $\mathbb{Q}_5$ results. Thus let $F = \mathbb{Q}$, $K = \mathbb{Q}(i)$, and $\mathfrak{p} = (5)$. The prime factorization of $(5)\mathbb{Z}[i]$ is as $(2+i)(2-i)$. If we put $P_1 = (2+i)$ and $P_2 = (2-i)$, then K_{P_1} and K_{P_2} are both equal to $\mathbb{Q}_5$ because Example 1 above shows that the square roots of -1 already appear in $\mathbb{Q}_5$. If a is one of the square roots, then $|2+a|_5 |2-a|_5 = |(2+a)(2-a)|_5 = |5|_5 = \frac{1}{5}$. Thus one of $|2+a|_5$ and $|2-a|_5$ equals $\frac{1}{5}$ and the other equals 1. What is happening is that there are two field mappings $\mathbb{Q}(i) \to \mathbb{Q}_5$. For each of them, the effect on the base field $\mathbb{Q}$ is the same; however, one field mapping sends i in $\mathbb{Q}(i)$ to a in $\mathbb{Q}_5$, and the other sends i to $-a$. For definiteness, let us say that $|2+a|_5 = \frac{1}{5}$. Then the valuation of $\mathbb{Q}(i)$ with respect to $P_1 = (2+i)$ is consistent with the 5-adic valuation of $\mathbb{Q}_5$, but the valuation of $P_2 = (2-i)$ is not. This example shows why the definition of completion insists on a mapping of valued fields (respecting absolute values), not merely a mapping of fields.

(5) Suppose that $p = 2$. The question is the prime factorization of $f(X) = X^3 + X^2 - 2X + 8$ in $\mathbb{Z}_2$. This polynomial was studied at length toward the end of Section V.4 in connection with common index divisors. It is irreducible over $\mathbb{Q}$, but we are to factor it over $\mathbb{Q}_2$. We shall show that it splits into first-degree factors. Considering the polynomial modulo 2, we find that $f(X) \equiv (X-1)X^2 \bmod 2$. Since 1 is a simple root modulo 2, Corollary 6.29 says that there exists an element θ_1 in $\mathbb{Z}_2$ such that $f(\theta_1) = 0$ and $\theta_1 \equiv 1 \bmod 2$. Dividing $f(X)$ by $X - \theta_1$, we obtain

$$f(X) = (X - \theta_1)(X^2 + (\theta_1 + 1)X + (\theta_1(\theta_1+1) - 2)).$$

To show that the quadratic factor splits over $\mathbb{Q}_2$, it is necessary and sufficient to show that its discriminant is a square, since $\mathbb{Q}_2$ has characteristic 0. The discriminant is

$$(\theta_1 + 1)^2 - 4(\theta_1(\theta_1+1) - 2) = 4\big((\tfrac{1}{2}(\theta_1+1))^2 - (\theta_1(\theta_1+1) - 2)\big),$$

and we can ignore the square factor of 4. We know that $\theta_1 \equiv 1 \bmod 2$. Let us compute θ_1 modulo $8\mathbb{Z}_2$ by writing $\theta_1 = 8\varphi + c$ with $\varphi \in \mathbb{Z}_2$ and with $c = \pm 1$ or ± 3. Substituting into $f(X)$ and computing modulo $8\mathbb{Z}_2$, we have

$$0 = f(\theta_1) \equiv c^3 + c^2 - 2c \bmod 8\mathbb{Z}_2.$$

Since c is odd, $c^3 \equiv c$ and $c^2 \equiv 1$ mod 8. Thus $0 \equiv c + 1 - 2c$ mod 8 and $c \equiv 1$ mod 8. Consequently

$$(\tfrac{1}{2}(\theta_1 + 1))^2 - (\theta_1(\theta_1 + 1) - 2) \equiv 1 \text{ mod } 8.$$

By Example 2 any 2-adic integer that is $\equiv 1$ mod $8\mathbb{Z}_2$ is a square in $\mathbb{Z}_2$, and thus $f(X)$ indeed factors over $\mathbb{Z}_2$ as the product of three first-degree factors.

We conclude this section with a version of Hensel's Lemma that we state without proof.[13] This version deals with factorizations rather than roots. Briefly it says that we can lift a *relatively prime* factorization modulo $\mathfrak{p}$ to a factorization in $R[X]$ if at least one of the two factors modulo $\mathfrak{p}$ has leading coefficient 1. This theorem certainly implies Corollary 6.29.

Theorem 6.30 (Hensel's Lemma). Let F be a field with a nontrivial discrete absolute value, necessarily nonarchimedean, and assume that F is complete. Let R be the valuation ring, let $\mathfrak{p}$ be the unique maximal ideal, let $\Bbbk$ be the residue class field, and let $f(X)$ be a polynomial in $R[X]$. Suppose that there exist polynomials $g_0(X)$ and $h_0(X)$ in $R[X]$ such that $g_0(X)$ mod $\mathfrak{p}$ and $h_0(X)$ mod $\mathfrak{p}$ are relatively prime in $\Bbbk[X]$, g_0 has leading coefficient 1, and $f(X)$ factors modulo $\mathfrak{p}$ as $f(X) \equiv g_0(X)h_0(X)$ mod $\mathfrak{p}$. Then there exist polynomials $g(X)$ and $h(X)$ in $R[X]$ such that $g(X)$ has leading coefficient 1, $g(X) \equiv g_0(X)$ mod $\mathfrak{p}$, $h(X) \equiv h_0(X)$ mod $\mathfrak{p}$, and $f(X)$ factors in $R(X)$ as $f(X) = g(X)h(X)$.

6. Ramification Indices and Residue Class Degrees

Sections 1–4 have presented the ingredients of a two-stage process for analyzing congruence information, and now it is time to use everything together. The goal is to have techniques for extracting information about a global number-theoretic problem by seeing what the problem says about ideals, for reducing the questions about ideals to questions about powers of prime ideals, and for then assembling the results.

We give one illustration of the utility of our constructions: With the techniques we had in Chapter V, we gave only a partial proof of the Dedekind Discriminant Theorem (Theorem 5.5). By contrast, we shall see in Section 8 that the present techniques lead naturally to a complete proof.

Although we might want to work just within one number field, it is helpful to change the context so that we are comparing a number field with a finite extension. There is no loss of generality in doing so; we can always take the base field to

---

[13] A proof may be found in Hasse's *Number Theory*, pp. 169–172.

be the rationals $\mathbb{Q}$, and the effect is that we consider only the finite set of prime ideals for the extension field that contain a given prime number p.

As long as we are going to consider finite extensions of fields in addressing number theory, we might as well treat also the case of function fields in one variable, at least to the extent that the two theories are quite analogous. Thus we are led to the following set-up.

Let R be a Dedekind domain considered as a subring of its field of fractions F, let K be a finite *separable*[14] extension of F with $[K : F] = n$, and let T be the integral closure of R in K. We shall work with F and K as valued fields, having some absolute value on them. The case of interest in this section will be that the absolute value is nonarchimedean and arises from a discrete valuation whose valuation ring contains R or T, respectively. Theorem 6.5 shows that the valuation is defined by means of some prime ideal $\wp$ of R or T, and the associated absolute value may thus be denoted by an expression[15] like $|\cdot|_\wp$.

We start from a prime ideal $\mathfrak{p}$ in R and form the corresponding absolute value on F as in Section 3, obtaining a valued field $(F, |\cdot|_\mathfrak{p})$. Then we complete as in Section 4, writing the completion as

$$\psi_0 : (F, |\cdot|_\mathfrak{p}) \to (F_\mathfrak{p}, |\cdot|_\mathfrak{p}).$$

We know that the ideal $\mathfrak{p}T$ in T has a prime factorization of the form $\mathfrak{p}T = P_1^{e_1} \cdots P_g^{e_g}$, where $P_1, \ldots, P_g$ are distinct prime ideals in T. The integers e_i are called **ramification indices** and the dimensions $f_i = \dim_{R/\mathfrak{p}}(T/P_i)$ are called **residue class degrees**. We are interested in saying everything we can about $P_1, \ldots, P_g$ and about the indices e_i and f_i. The fundamental relationship is given by Theorem 9.60 of *Basic Algebra*, namely

$$\sum_{i=1}^g e_i f_i = n.$$

We know that each P_i gives us a nonarchimedean absolute value $|\cdot|_{P_i}$ on K, unique up to equivalence, and then a completion

$$\psi_i : (K, |\cdot|_{P_i}) \to (K_{P_i}, |\cdot|_{P_i}).$$

---

[14]The role of separability will become apparent before the statement of Theorem 6.31 below.

[15]The number-theory case ultimately requires also a limited amount of analysis of archimedean absolute values, and that will be carried out in Section 9. In the context of passing from a Diophantine equation to congruence information, part of the role that archimedean absolute values play is in analyzing signs. Thus for example the simple-minded equation $x^2 + y^2 = -1$ has no solutions in integers; the reason for the absence of solutions is a constraint on signs, not some limitation from congruences with respect to powers of primes. Archimedean absolute values control signs.

The first important step is to establish an isomorphism involving fields such that the identity $\sum_{i=1}^{g} e_i f_i = n$ is a dimension formula that follows from the isomorphism. The identity in question concerns the ring $K \otimes_F F_{\mathfrak{p}}$, which is a commutative algebra over K or over $F_{\mathfrak{p}}$, whichever we like, and which is semisimple by Corollary 2.30 under our assumption that K is a finite separable extension of $\mathbb{F}$. The Wedderburn theory (Theorems 2.2 and 2.4) shows that $K \otimes_F F_{\mathfrak{p}}$ is isomorphic to a finite direct product of fields,[16] each of which is a finite extension of $F_{\mathfrak{p}}$. What we shall prove later in this section is the following theorem.

Theorem 6.31. Let R be a Dedekind domain considered as a subring of its field of fractions F, let K be a finite separable extension of F with $[K : F] = n$, and let T be the integral closure of R in K. If $\mathfrak{p}$ is a nonzero prime ideal of R and if the ideal $\mathfrak{p}T$ in T has a prime factorization of the form $\mathfrak{p}T = P_1^{e_1} \cdots P_g^{e_g}$, where $P_1, \ldots, P_g$ are distinct prime ideals in T and the e_j are positive integers, then

$$K \otimes_F F_{\mathfrak{p}} \cong \prod_{j=1}^{g} K_{P_j}.$$

When the formula $\sum_{j=1}^{g} e_j f_j = n$ is specialized to the field extension $K_{P_j}/F_{\mathfrak{p}}$, it becomes $e_j^* f_j^* = [K_{P_j} : F_{\mathfrak{p}}]$, where e_j^* and f_j^* are the ramification index and residue class degree associated to $K_{P_j}/F_{\mathfrak{p}}$. If we accept for the moment the result of Lemma 6.36 below that e_j^* and f_j^* coincide with the corresponding indices e_j and f_j for K/F, then $n = \sum_{j=1}^{g} e_j f_j = \sum_{j=1}^{n} e_j^* f_j^* = \sum_{j=1}^{g} [K_{P_j} : F_{\mathfrak{p}}]$ indeed counts the $F_{\mathfrak{p}}$ dimensions of both sides of the formula $K \otimes_F F_{\mathfrak{p}} \cong \prod_{j=1}^{g} K_{P_j}$ in the theorem. The theorem says much more than this, and we shall mine its consequences after giving the proof of the theorem.

For orientation, let us recall Example 4 from Section 5. In that example, we had $R = \mathbb{Z}$, $F = \mathbb{Q}$, $K = \mathbb{Q}(i)$, $T = \mathbb{Z}[i]$, $\mathfrak{p} = 5\mathbb{Z}$, and $F_{\mathfrak{p}} = \mathbb{Q}_5$. The factorization $\mathfrak{p}T = \prod P_j^{e_j}$ is $5\mathbb{Z}[i] = (2+i)(2-i)$, and the two completed versions of K are $K_{(2+i)} \cong \mathbb{Q}_5$ and $K_{(2-i)} \cong \mathbb{Q}_5$. Thus the identity in the theorem specializes to

$$\mathbb{Q}(i) \otimes_{\mathbb{Q}} \mathbb{Q}_5 \cong \mathbb{Q}_5 \times \mathbb{Q}_5.$$

Proving the identity on this level would be more challenging than necessary because the isomorphism cannot be unique; it can always be composed with the interchange of the two factors on the right side. For this reason the proof makes use of valued fields, and then in effect the desired isomorphism becomes a constructive one that we can write down rather explicitly.

---

[16]The words "direct product" in connection with finitely many fields refer to the direct sum of the additive structures, with multiplication given coordinate by coordinate.

Let us now work toward proving Theorem 6.31. Above, we mentioned the completion mapping ψ_0 for F relative to an absolute value in the equivalence class determined by $\mathfrak{p}$, as well as ψ_j for K relative to some absolute value in the class determined by P_j. In addition, we have inclusion mappings corresponding to the field extensions K/F and $K_{P_j}/F_\mathfrak{p}$. Figure 6.1 below is a square diagram that assigns the names φ_0 and φ_j to these as well.

$$\begin{array}{ccc} F & \xrightarrow{\psi_0} & F_\mathfrak{p} \\ \varphi_0 \downarrow & & \downarrow \varphi_j \\ K & \xrightarrow{\psi_j} & K_{P_j} \end{array}$$

FIGURE 6.1. Commutativity of completion and extension as field mappings.

The diagram in Figure 6.1 commutes. In fact, $\psi_j \varphi_0$ and $\varphi_j \psi_0$ are both F homomorphisms, being compositions of F homomorphisms, and hence $x \in F$ implies $\psi_j \varphi(x) = x(\psi_j \varphi(1)) = x(1) = x(\varphi_j \psi_0(1)) = \varphi_j \psi_0(x)$.

But more is true: we are going to impose absolute values on the four fields in the diagram in such a way that the four field mappings are homomorphisms of valued fields. We have already defined $|\cdot|_\mathfrak{p}$ on F as any absolute value corresponding to $\mathfrak{p}$, and then $|\cdot|_\mathfrak{p}$ is defined on $F_\mathfrak{p}$ in such a way that the completion mapping ψ_0 preserves absolute values. Theorem 6.33 below will enable us to define an absolute value in a unique fashion on K_{P_j} such that φ_j preserves absolute values. Proposition 6.34 will give us the definition of an absolute value on K, and we shall check in Lemma 6.35 that Figure 6.1 with these absolute values in place is a commutative diagram of valued fields. Finally we use this commutativity to prove in Lemma 6.36 that the ramification index e_j^* and residue class degree f_j^* for $K_{P_j}/F_\mathfrak{p}$ match the corresponding parameters e_j and f_j for K/F, and then we are ready for the main part of the proof of the theorem.

We begin our preliminary work by limiting the possibilities for a finite extension of a complete valued field $(F, |\cdot|_F)$. If K is a finite extension of F, a **norm** on the F vector space K relative to $|\cdot|_F$ is a function $\|\cdot\|$ from K to $\mathbb{R}$ having

(i) $\|x\| \geq 0$ on K with equality if and only if $x = 0$,
(ii) $\|cx\| = |c|_F \|x\|$ for $c \in F$ and $x \in K$,
(iii) $\|x + y\| \leq \|x\| + \|y\|$ for all x and y in K.

Lemma 6.32. If $(F, |\cdot|_F)$ is a complete valued field, if K is a finite extension of F, and if $\|\cdot\|_1$ and $\|\cdot\|_2$ are any two norms on K relative to $|\cdot|_F$, then there exist real constants C and C' such that

$$\|x\|_1 \leq C \|x\|_2 \quad \text{and} \quad \|x\|_2 \leq C' \|x\|_1 \qquad \text{for all } x \in K.$$

Consequently K is Cauchy complete in the metric induced by either norm.

REMARK. It is not important that K be a field in this lemma, only that it be a finite-dimensional vector space over F.

PROOF. Let $n = \dim_F K$. Fixing an ordered basis $(x_1, \ldots, x_n)$ of K over F, we may express any member x of K in the form $x = \sum_{i=1}^n c_i x_i$ with all c_i in F. With the c_i's defined this way, we define $\|x\|_{\sup} = \max_{1 \leq i \leq n} |c_i|_F$. To prove the displayed inequalities, it is enough to prove them for $\|\cdot\|_{\sup}$ and any other norm $\|\cdot\|$. For one direction of the inequality, we have

$$\|x\| = \big\|\sum_i c_i x_i\big\| \leq \sum_i \|c_i x_i\| = \sum_i |c_i|_F \|x_i\| \leq \Big(\sum_i \|x_i\|\Big)\|x\|_{\sup}.$$

This proves that $\|x\| \leq C\|x\|_{\sup}$ with $C = \sum_i \|x_i\|$.

For the reverse inequality we shall prove by induction on k that an inequality $\|x\|_{\sup} \leq C'_k \|x\|$ holds for all x in the F linear span of at most k of the vectors $x_1, \ldots, x_n$. The base case for the induction is $k = 1$, and then $\|x\|_{\sup} = \|x_i\|^{-1}\|x\|$ whenever x is a multiple of x_i. So $C'_1 = \max_{1 \leq i \leq n}(\|x_i\|^{-1})$.

Assume that $C'_1, \ldots, C'_k$ exist and that we are to produce C'_{k+1}. Arguing by contradiction, we may assume that there is some sequence $\{x^{(m)}\}$ in K, each term having at most $k+1$ nonzero coefficients, such that $\|x^{(m)}\| = 1$ for all m and $\|x^{(m)}\|_{\sup}$ tends to infinity. Possibly by passing to a subsequence, we may assume that the nonzero coefficients of $x^{(m)}$ all lie in a particular subset of $k+1$ of the coefficients, and there is no harm in assuming that this subset is $\{1, \ldots, k+1\}$. Passing to a further subsequence, we may assume that there is some index j such that the largest coefficient of each $x^{(m)}$, when measured by $|\cdot|_F$, is the jth, and there is no harm in assuming that $j = k+1$.

Let $c_1^{(m)}, \ldots, c_{k+1}^{(m)}$ be the coefficients of $x^{(m)}$, so that $x^{(m)} = \sum_{i=1}^{k+1} c_i^{(m)} x_i$. Put $y^{(m)} = (c_{k+1}^{(m)})^{-1} x^{(m)} = \sum_{i=1}^k d_i^{(m)} x_i + x_{k+1}$, where $d_i^{(m)} = (c_{k+1}^{(m)})^{-1} c_i^{(m)}$. Here $|d_i^{(m)}|_F \leq 1$ for $1 \leq i \leq k$ and for all m, and also $\|y^{(m)}\| = |c_{k+1}^{(m)}|_F^{-1}\|x^{(m)}\| = |c_{k+1}^{(m)}|_F^{-1}$ tends to 0.

For each vector $y^{(m)} - x_{k+1}$, only the first k coefficients can be nonzero, and the same thing is true of differences $y^{(m)} - y^{(m')}$ of two such vectors. The inductive hypothesis tells us that $\|y^{(m)} - y^{(m')}\|_{\sup} \leq C'_k \|y^{(m)} - y^{(m')}\|$, and the right side tends to 0 as m and m' tend to infinity because $\|y^{(m)}\|$ and $\|y^{(m')}\|$ tend to 0. Therefore the ith coordinate of $y^{(m)}$ forms a Cauchy sequence. Since F is given as complete, $\{y^{(m)}\}$ is convergent in the norm $\|\cdot\|_{\sup}$ to some $y = \sum_{i=1}^k d_i x_i + x_{k+1}$ in K.

By the easy direction of our inequality, $\|y^{(m)} - y\| \leq C\|y^{(m)} - y\|_{\sup}$. The right side tends to 0, and hence so does the left. We know that $\|y^{(m)}\|$ tends to 0, and hence $y = 0$. But this conclusion contradicts the form of y as $\sum_{i=1}^k d_i x_i + x_{k+1}$ with coefficient 1 for x_{k+1}. We conclude that C'_{k+1} exists as asserted, and the lemma follows. □

Theorem 6.33. If $(F, |\cdot|_F)$ is a complete valued field relative to a nontrivial nonarchimedean discrete absolute value and if K is a finite separable extension of F with $[K : F] = n$, then K has a unique absolute value $|\cdot|_K$ extending $|\cdot|_F$, K is complete and nonarchimedean, and the integral closure T in K of the valuation ring R of F is the valuation ring of K. The extension is given by $|x|_K = |N_{K/F}(x)|_F^{1/n}$.

REMARKS. Since T is the valuation ring, Proposition 6.2 shows that T has a unique nonzero prime ideal. It follows that if $\mathfrak{p}$ is a nonzero prime ideal of R, then $\mathfrak{p}T = P^e$ for a single prime ideal P of T. We shall make frequent use of this fact in applications without explicit mention.

PROOF. For uniqueness, suppose that $|\cdot|_1$ and $|\cdot|_2$ are two absolute values on K that extend $|\cdot|_F$. Let us see that each of these is a norm on K relative to $|\cdot|_F$. In fact, what needs checking for $|\cdot|_1$ is that the function respects scalars from F appropriately. If c is in F and x_0 is in K, then $|cx_0|_1 = |c|_1|x_0|_1 = |c|_F|x_0|_1$, the second equality following because $|\cdot|_1$ restricts to $|\cdot|_F$ on F. A similar argument applies to $|\cdot|_2$, and thus we are dealing with two norms.

If the two given absolute values are inequivalent, then Proposition 6.12 shows in the presence of the nontriviality of $|\cdot|_F$ that we can find an $x \in K$ with $|x|_1 > 1$ and $|x|_2 \leq 1$. Then $\lim_k |x^{-k}|_1 = 0$ while $|x^{-k}|_2 \geq 1$ for all k. Consequently there cannot exist a constant C such that $|y|_2 \leq C|y|_1$ for all $y \in F$, in contradiction to Lemma 6.32.

We conclude that $|\cdot|_1$ and $|\cdot|_2$ are equivalent, say that $|x|_1 = |x|_2^s$ for all $x \in K$ and some $s > 0$. Since $|\cdot|_F$ is nontrivial, there exists some $x_0 \in F$ with $|x_0|_1 > 1$. The equality $|x_0|_1 = |x_0|_2^s$ then implies that $s = 1$. This proves uniqueness.

We turn to existence. Proposition 6.2 shows that the valuation ring R in F for the discrete valuation v_F corresponding to $|\cdot|_F$ on F is a local principal ideal domain and that the valuation ideal $\mathfrak{p}$ is the unique maximal ideal of R. Theorem 6.5 shows that the valuation $v_\mathfrak{p}$ determined by $\mathfrak{p}$ is the same as the given valuation v_F. Hence $|\cdot|_F$ is given for all $a \in F$ by $|a|_F = r^{-v_\mathfrak{p}(a)}$ for some $r > 1$. Let π be a generator of the principal ideal $\mathfrak{p}$ of R.

Since K/F is finite and separable, Theorem 8.54 of *Basic Algebra* shows that the integral closure T of R in K is a Dedekind domain. Let $\mathfrak{p}T = P_1^{e_1} \cdots P_g^{e_g}$ be the factorization of the ideal $\mathfrak{p}T$ of T into the product of powers of distinct prime ideals of T. Each P_j defines a nonarchimedean valuation v_{P_j} of K. If a is any element of F, then we can write $a = \pi^k u$ for some $u \in R^\times$ and some integer k. The computation $aT = aRT = \pi^k uRT = \pi^k RT = \pi^k T = \mathfrak{p}^k T = P_1^{ke_1} \cdots P_g^{ke_g}$ shows that $v_\mathfrak{p}(a) = k$ and that $v_{P_j}(a) = ke_j$. Hence $v_{P_j} = e_j v_\mathfrak{p}$ on F, and therefore the formula $|x|_{P_j} = (r^{e_j^{-1}})^{-v_{P_j}(x)}$ for $x \in K$ defines an absolute

6. Ramification Indices and Residue Class Degrees

value on K that has $|a|_F = r^{-v_\mathfrak{p}(a)} = r^{-e_j^{-1} v_{P_j}(a)} = (r^{e_j^{-1}})^{-v_{P_j}(a)} = |a|_{P_j}$ for all a in F. This proves existence. The absolute value $|\cdot|_{P_j}$ on K is complete by Lemma 6.32 and is nonarchimedean because it is given by a discrete valuation.

Let us show that $g = 1$. Arguing by contradiction, suppose that there are at least two distinct prime ideals P_1 and P_2 of T that contain $\mathfrak{p}$. Since $P_1 + P_2 = T$, we can choose $x_1 \in P_1$ and $x_2 \in P_2$ with $x_1 + x_2 = 1$. Then $v_{P_1}(x_1) > 0$ and $v_{P_1}(1) = 0$, from which we see that $v_{P_1}(x_2) = 0$. Since $v_{P_2}(x_2) > 0$, we obtain a contradiction to the uniqueness part of the theorem. Thus the prime ideal of T is unique. Let us write P for this ideal.

We know that $v_P(T) \geq 0$, i.e., that T is contained in the valuation ring of v_P. Proposition 6.4 shows that the valuation ring of v_P equals $S^{-1}T$, where S is the complement of P in T. The uniqueness of P means that T is local, and hence every member of S is a unit in T. Thus $S^{-1}T = T$, and T is the valuation ring.

Write $|\cdot|_K$ in place of $|\cdot|_{P_j}$. To prove the explicit formula for $|\cdot|_K$ in the statement of the proposition, choose a finite Galois extension L of F that contains K; such a field L exists because K/F is separable.[17] By the existence just proved, let $|\cdot|_L$ be an extension of $|\cdot|_K$ to L. If σ is in $\mathrm{Gal}(L/F)$, then $x \mapsto |\sigma(x)|_L$ and $x \mapsto |x|_L$ are both absolute values on L that extend $|\cdot|_F$. By the uniqueness just proved, $|\sigma(x)|_L = |x|_L$. Applying $|\cdot|_L$ to both sides of the formula $N_{L/F}(x) = \prod_{\sigma \in \mathrm{Gal}(L/F)} \sigma(x)$ gives

$$|N_{L/F}(x)|_F = |N_{L/F}(x)|_L = \prod_{\sigma \in \mathrm{Gal}(L/F)} |\sigma(x)|_L = |x|_L^{[L:F]}. \qquad (*)$$

If x is in K, then the left side equals $(|N_{K/F}(x)|_F)^{[L:K]}$, and the right side equals $(|x|_K)^{[L:K][K:F]} = (|x|_K^{[K:F]})^{[L:K]}$. Thus the desired formula follows by extracting the positive $[L:K]^{\mathrm{th}}$ root of both sides of $(*)$. $\square$

Proposition 6.34. Under the hypotheses of Theorem 6.31, let $v_\mathfrak{p}$ be the valuation of F defined by $\mathfrak{p}$, and let v_{P_j} be the valuation of K defined by P_j, $1 \leq j \leq g$. Then $e_j v_\mathfrak{p} = v_{P_j}|_F$. Consequently if $|\cdot|_\mathfrak{p}$ is an absolute value on F defined by $\mathfrak{p}$, then for each j some member $|\cdot|_{P_j}$ of the equivalence class of absolute values defined on K by P_j is an extension of $|\cdot|_\mathfrak{p}$. In this case the inclusion of $(F, |\cdot|_\mathfrak{p})$ into $(K, |\cdot|_{P_j})$ is a homomorphism of valued fields.

PROOF. Let S be the multiplicative system in R given as the set-theoretic complement of $\mathfrak{p}$ in R. For the first conclusion Proposition 6.4 and Theorem 6.5 together show that it is enough to prove that

$$e_j v_{S^{-1}\mathfrak{p}} = v_{S^{-1}P_j}|_F. \qquad (*)$$

---

[17] The field L can be taken to be a splitting field of the minimal polynomial over F of an element ξ such that $K = F(\xi)$. The extension L/F is separable by Corollary 9.30 of *Basic Algebra*.

From the identity
$$\mathfrak{p}T = P_1^{e_1} \cdots P_g^{e_g},$$
we have
$$S^{-1}\mathfrak{p}T = (S^{-1}P_1)^{e_1} \cdots (S^{-1}P_g)^{e_g}. \quad (**)$$

Since S is the complement of $\mathfrak{p}$ in R, $v_\mathfrak{p}$ is 0 on S. Hence $v_{S^{-1}\mathfrak{p}}$ is 0 on S. From $R \cap P_j = \mathfrak{p}$, we have $S \cap P_j \subseteq S \cap \mathfrak{p} = \varnothing$. Thus the members of S lie in $R \subseteq T$ but in no P_j, and v_{P_j} is 0 on S. Hence $v_{S^{-1}P_j}$ is 0 on S.

Let π be a generator of the principal ideal $S^{-1}\mathfrak{p}$ in $S^{-1}R$, so that $v_{S^{-1}\mathfrak{p}}(\pi) = 1$. Since $\pi S^{-1}T = S^{-1}\mathfrak{p}T$, equation $(**)$ shows that $v_{S^{-1}P_j}(\pi) = e_j$. Each element y of F is of the form $y = \pi^k u$ for some integer k and some $u \in F$ with $v_{S^{-1}\mathfrak{p}}(u) = 0$. The element u must be in $S^{-1}R$ but not $S^{-1}\mathfrak{p}$ and hence is in S^{-1}. Thus $v_{S^{-1}P_j}(u) = 0$. We have now seen that $v_{S^{-1}P_j}(x) = e_j v_{S^{-1}\mathfrak{p}}(x)$ for the element $x = u$ above and also for $x = \pi$. Therefore $v_{S^{-1}P_j}(x) = e_j v_{S^{-1}\mathfrak{p}}(x)$ for all $x \in F$, and $(*)$ is proved.

Now that $e_j v_\mathfrak{p} = v_{P_j}\big|_F$, choose $r > 1$ such that $|x|_\mathfrak{p} = r^{-v_\mathfrak{p}(x)}$ for $x \in F$. If r' is defined by $r = (r')^{e_j}$, then the definition $|x|_{P_j} = (r')^{-v_{P_j}(x)}$ for $x \in K$ restricts for $x \in F$ to $|x|_{P_j} = (r')^{-v_{P_j}(x)} = (r')^{-e_j v_\mathfrak{p}(x)} = r^{-v_\mathfrak{p}(x)} = |x|_\mathfrak{p}$, and the inclusion is indeed a homomorphism of valued fields. $\square$

With these facts in place, let us make Figure 6.1 into a commutative diagram of valued fields. From $\mathfrak{p}$, we use any corresponding choice of $|\cdot|_\mathfrak{p}$ on F, and this uniquely determines an absolute value by the same name on $F_\mathfrak{p}$. Next we apply Theorem 6.33 to the inclusion $\varphi_j : F_\mathfrak{p} \to K_{P_j}$ to obtain a unique extension of $|\cdot|_\mathfrak{p}$ from $F_\mathfrak{p}$ to an absolute value $|\cdot|_{P_j}$ on K_{P_j}.

Meanwhile, with the index j specified, Proposition 6.34 gives us a unique absolute value $|\cdot|_{P_j}$ on K such that the inclusion $\varphi_0 : F \to K$ is a homomorphism of valued fields. The completion mapping $\psi_j : K \to K_{P_j}$ in turn gives us a second determination of $|\cdot|_{P_j}$ on K_{P_j}, and Lemma 6.35 below says that these two determinations match, i.e., that Figure 6.2 is a commutative diagram of homomorphisms of valued fields.

$$\begin{array}{ccc} (F, |\cdot|_\mathfrak{p}) & \xrightarrow{\psi_0} & (F_\mathfrak{p}, |\cdot|_\mathfrak{p}) \\ \varphi_0 \downarrow & & \downarrow \varphi_j \\ (K, |\cdot|_{P_j}) & \xrightarrow{\psi_j} & (K_{P_j}, |\cdot|_{P_j}) \end{array}$$

FIGURE 6.2. Commutativity of completion and extension as homomorphisms of valued fields.

Lemma 6.35. In the above notation the two determinations of $|\cdot|_{P_j}$ on K_{P_j} coincide—one by using Theorem 6.33 to insist that $\varphi_j \psi_0$ in Figure 6.2 be the composition of homomorphisms of valued fields, and the other by using Proposition 6.34 to insist that $\psi_j \varphi_0$ in Figure 6.2 be the composition of homomorphisms of valued fields.

REMARKS. The commutativity formula $\psi_j \varphi_0 = \varphi_j \psi_0$ for field mappings is known from the discussion concerning Figure 6.1.

PROOF. Let us give two different names to the two possible absolute values on K_{P_j}, writing $|\cdot|'$ for the one that makes $|\psi_j(k)|' = |k|_{P_j}$ for $k \in K$ and writing $|\cdot|''$ for the other, which makes $|\varphi_j(x)|'' = |x|_{\mathfrak{p}}$ for $x \in F_{\mathfrak{p}}$. Let y be in F. Then the equality $\varphi_j \psi_0 = \psi_j \varphi_0$ implies that

$$|\varphi_j \psi_0(y)|' = |\psi_j \varphi_0(y)|' = |\varphi_0(y)|_{P_j} = |y|_{\mathfrak{p}} = |\psi_0(y)|_{\mathfrak{p}}. \tag{$*$}$$

If x_0 is given in $F_{\mathfrak{p}}$, then we can choose a sequence $\{x_n\}$ in F with $\{\psi_0(x_n)\}$ convergent to x_0 in $F_{\mathfrak{p}}$. Then $\{\psi_0(x_n)\}$ is Cauchy in the metric on $F_{\mathfrak{p}}$, and it follows from $(*)$ applied with $y = x_n - x_{n'}$ that $\{\varphi_j \psi_0(x_n)\}$ is Cauchy in the metric from $|\cdot|'$ on K_{P_j}. If we have a second such sequence $\{x'_n\}$ in F with $\psi_0(x'_n)$ convergent to x_0 and if we alternate the terms of $\{x_n\}$ and $\{x'_n\}$ to produce a sequence $\{z_n\}$, then $\{\varphi_j \psi_0(z_n)\}$ remains Cauchy in the metric from $|\cdot|'$. Since $|\cdot|'$ is complete, it follows that $|\varphi_j(x_0)|'$ is given by a well-defined limit independently of the sequence in $\psi_0(F)$ used to approximate x_0. The formula $(*)$ shows that $|\varphi_j(x_0)|' = |x_0|_{\mathfrak{p}}$, and the definition of $|\cdot|''$ shows that this equals $|\varphi_j(x_0)|''$. By the uniqueness in Theorem 6.33, $|\cdot|' = |\cdot|''$ on K_{P_j}. □

Lemma 6.36. In the above notation and that of Theorem 6.31, the ramification index e_j^* corresponding to $K_{P_j}/F_{\mathfrak{p}}$ for the closure of the ideal $\psi_j(P_j)$ coincides with the ramification index e_j corresponding to K/F for the ideal P_j.

REMARK. In addition, the residue class degree f_j^* for $K_{P_j}/F_{\mathfrak{p}}$ coincides with the residue class degree f_j for K/F. In fact, the five paragraphs of review that follow Theorem 6.26 mention that residue class fields change neither during the localization step nor in the completion step of our two-step process. Thus $R/\mathfrak{p}$ remains the same during the two steps, and so does T/P_j. Hence the dimension of T/P_j as a vector space over $R/\mathfrak{p}$ remains the same.

PROOF. Let $v_{\mathfrak{p},F}$, $v_{P_j,K}$, $v_{\mathfrak{p},F_{\mathfrak{p}}}$, and $v_{P_j,K_{P_j}}$ be the valuations corresponding to the absolute values on F, K, $F_{\mathfrak{p}}$, and K_{P_j}, respectively. The last of these is well defined by Lemma 6.35. Proposition 6.34 shows that

$$e_j v_{\mathfrak{p},F} = v_{P_j,K} \varphi_0 \quad \text{and} \quad e_j^* v_{\mathfrak{p},F_{\mathfrak{p}}} = v_{P_j,K_{P_j}} \varphi_j. \tag{$*$}$$

Meanwhile, the completion mappings ψ_0 and ψ_j satisfy

$$v_{\mathfrak{p},F_{\mathfrak{p}}}\psi_0 = v_{\mathfrak{p},F} \quad\text{and}\quad v_{P_j,K_{P_j}}\psi_j = v_{P_j,K}. \tag{**}$$

Multiplying the second equation of $(*)$ on the right by ψ_0 and substituting from the first equation of $(**)$, we obtain

$$e_j^* v_{\mathfrak{p},F} = e_j^* v_{\mathfrak{p},F_{\mathfrak{p}}}\psi_0 = v_{P_j,K_{P_j}}\varphi_j\psi_0.$$

We substitute from the commutativity formula $\varphi_j\psi_0 = \psi_j\varphi_0$ and unwind the right side as

$$v_{P_j,K_{P_j}}\psi_j\varphi_0 = v_{P_j,K}\varphi_0 = e_j v_{\mathfrak{p},F}.$$

Thus $e_j^* v_{\mathfrak{p},F} = e_j v_{\mathfrak{p},F}$. Since $v_{\mathfrak{p},F}$ is not identically 0, we obtain $e_j^* = e_j$. □

PROOF OF THEOREM 6.31. As was mentioned before the statement of the theorem, it follows from Proposition 2.29 and the Wedderburn theory that $K \otimes_F F_{\mathfrak{p}}$ is isomorphic to a product $\prod_{i=1}^{g'} L_i$ of fields, each of which is a finite extension of $F_{\mathfrak{p}}$ and each of which has K embedded in it. The subfields L_i are uniquely determined within $K \otimes_F F_{\mathfrak{p}}$, and we let η_i be the projection of $K \otimes_F F_{\mathfrak{p}}$ onto L_i. Each η_i is a ring homomorphism and is given by multiplication by a specific element of $K \otimes_F F_{\mathfrak{p}}$, namely the element that is 1 in the ith position and is 0 in the other positions. When restricted to $K \otimes 1$, η_i gives a field map $\alpha_i : K \to L_i$; when restricted to $1 \otimes F_{\mathfrak{p}}$, it gives a field map $\beta_i : F_{\mathfrak{p}} \to L_i$.

We shall develop a small abstract theory about these field maps α_i and β_i. Suppose that M is a field containing F, that $\alpha : K \to M$ and $\beta : F_{\mathfrak{p}} \to M$ are F algebra homomorphisms, and that M is a finite separable extension of $\beta(F_{\mathfrak{p}})$. Theorem 6.33 says that M has a unique absolute value $|\cdot|_{\mathfrak{p},\beta}$ extending $|\cdot|_{\mathfrak{p}}$ and that the valued field $(M, |\cdot|_{\mathfrak{p},\beta})$ is complete. The extension property means that $\beta : (F_{\mathfrak{p}}, |\cdot|_{\mathfrak{p}}) \to (M, |\cdot|_{\mathfrak{p},\beta})$ is a homomorphism of valued fields. The restriction $\alpha^*(|\cdot|_{\mathfrak{p},\beta})$ to K makes $(K, \alpha^*(|\cdot|_{\mathfrak{p},\beta}))$ into a valued field in such a way that

$$\alpha : (K, \alpha^*(|\cdot|_{\mathfrak{p},\beta})) \to (M, |\cdot|_{\mathfrak{p},\beta}) \tag{*}$$

is a homomorphism of valued fields. Let us see that

$$\alpha^*(|\cdot|_{\mathfrak{p},\beta}) \text{ is one (and only one) of the absolute values } |\cdot|_{P_j} \text{ on } K \tag{**}$$

and that α in $(*)$ factors as the composition of the completion mapping

$$\psi_j : (K, |\cdot|_{P_j}) \to (K_{P_j}, |\cdot|_{P_j})$$

6. Ramification Indices and Residue Class Degrees

followed by some other homomorphism of valued fields

$$\iota : (K_{P_j}, |\cdot|_{P_j}) \to (M, |\cdot|_{\mathfrak{p},\beta}).$$

To get at (∗∗) and the factorization of α, let us show that the field mapping

$$\varphi_0 : (F, |\cdot|_{\mathfrak{p}}) \to (K, \alpha^*(|\cdot|_{\mathfrak{p},\beta})) \qquad (\dagger)$$

is a homomorphism of valued fields, i.e., that $\varphi_0^* \alpha^*(|\cdot|_{\mathfrak{p},\beta}) = |\cdot|_{\mathfrak{p}}$. The field mappings $\alpha \varphi_0$ and $\beta \psi_0$, which carry F into M via K and $F_{\mathfrak{p}}$, respectively, are compositions of F homomorphisms and hence are F homomorphisms. Therefore $x \in F$ implies that $\alpha \varphi_0(x) = x(\alpha \varphi_0(1)) = x(1) = x(\beta \psi_0(1)) = \beta \psi_0(x)$, and we see that $\alpha \varphi_0 = \beta \psi_0$ on F. For $x \in F$, this identity accounts for the third equality in the following computation proving (†):

$$|x|_{\mathfrak{p}} = |\psi_0 x|_{\mathfrak{p}} = |\beta \psi_0 x|_{\mathfrak{p},\beta}$$
$$= |\alpha \varphi_0 x|_{\mathfrak{p},\beta} = \alpha^*(|\cdot|_{\mathfrak{p},\beta})(\varphi_0 x) = \varphi_0^* \alpha^*(|\cdot|_{\mathfrak{p},\beta})(x).$$

Returning to (∗∗) and applying (†), we see that $\alpha^*(|\cdot|_{\mathfrak{p},\beta})$ is ≤ 1 on R. Since T is the integral closure of R, Proposition 6.20 shows that $\alpha^*(|\cdot|_{\mathfrak{p},\beta})$ is ≤ 1 on T and that it arises from some nonzero prime ideal of T, necessarily one of the ideals $P_1, \ldots, P_g$. This proves (∗∗). Then the factorization (∗) follows from (∗∗) and the universal mapping property of completions as given in Theorem 6.25, since $(M, |\cdot|_{\mathfrak{p},\beta})$ is complete.

Now let us specialize by taking $M = L_i$ with i fixed. As in the first paragraph of the proof, the projection $\eta_i : K \otimes_F F_{\mathfrak{p}} \to L_i$ gives us field mappings $\alpha_i : K \to L_i$ and $\beta_i : F_{\mathfrak{p}} \to L_i$ by composing η_i with $K \to K \otimes 1$ and with $F_{\mathfrak{p}} \to 1 \otimes F_{\mathfrak{p}}$. If $u_1, \ldots, u_n$ is a vector-space basis of K over F, then $u_1 \otimes 1, \ldots, u_n \otimes 1$ is a vector-space basis of $K \otimes_F F_{\mathfrak{p}}$ over $F_{\mathfrak{p}}$, and it follows that L_i is finite-dimensional over $F_{\mathfrak{p}}$. Let us check that L_i is separable over $F_{\mathfrak{p}}$. We are given that K is separable over F, hence that $K = F(\xi)$ for an element ξ whose minimal polynomial $g(X)$ over F is separable. Then $\xi \otimes 1$ is a root of $g(X)$ regarded as in $F_{\mathfrak{p}}[X]$, and so is $\eta_i(\xi \otimes 1)$. Therefore $L_i/F_{\mathfrak{p}}$ is separable, and the above theory is applicable. In the theory, L_i acquires an absolute value $|\cdot|_{\mathfrak{p},\beta_i}$ such that $\beta_i : (F_{\mathfrak{p}}, |\cdot|_{\mathfrak{p}}) \to (L_i, |\cdot|_{\mathfrak{p},\beta_i})$ is a homomorphism of valued fields, and then $(L_i, |\cdot|_{\mathfrak{p},\beta_i})$ is complete. The theory produces a unique index $j = j(i)$ making $\alpha_i : (K, |\cdot|_{P_j}) \to (L_i, |\cdot|_{\mathfrak{p},\beta_i})$ into a homomorphism of valued fields.

Let us see that $\alpha_i(K)$ is dense in L_i. Every member of L_i is the image under η_i of some member $\sum_{l=1}^n u_l \otimes c_l$ of $K \otimes_F F_{\mathfrak{p}}$ with each c_l in $F_{\mathfrak{p}}$. The computation

$$\eta_i(u_l \otimes c_l) = \eta_i(u_l \otimes 1)\eta_i(1 \otimes c_l) = \alpha_i(u_l)\beta_i(c_l)$$

shows that every member of L_i is of the form $\sum_{l=1}^{n} \alpha_i(u_l)\beta_i(c_l)$. Since F is dense in $F_\mathfrak{p}$, we can choose members c'_l of F as close as we please to c_l. Since β_i is isometric, $\sum_{l=1}^{n} \alpha_i(u_l)\beta_i(c_l)$ is then close to $\sum_{l=1}^{n} \alpha_i(u_l)\beta_i(c'_l) = \sum_{l=1}^{n} \alpha_i(c'_l u_l)$. Consequently $\alpha_i(K)$ is indeed dense in L_i.

Recall in connection with (∗) that $\alpha_i : K \to L_i$ factors as a composition of homomorphisms of valued fields, namely as $\psi_j : (K, |\cdot|_{P_j}) \to (K_{P_j}, |\cdot|_{P_j})$ followed by $\iota : (K_{P_j}, |\cdot|_{P_j}) \to (L_i, |\cdot|_{\mathfrak{p},\beta_i})$. Since K_{P_j} is complete, $\iota(K_{P_j})$ is closed in L_i. The dense image $\alpha_i(K) = \iota(\psi_j(K))$ in L_i is contained in the closed subset $\iota(K_{P_j})$, and it follows that ι is onto L_i. That is, the homomorphism of valued fields

$$\iota : (K_{P_j}, |\cdot|_{P_j}) \to (L_i, |\cdot|_{\mathfrak{p},\beta_i})$$

is an isomorphism. This identifies the valued field $(L_i, |\cdot|_{\mathfrak{p},\beta_i})$ as isomorphic to $(K_{P_j}, |\cdot|_{P_j})$.

As a consequence of the argument thus far, we have constructed a choice-free function $i \mapsto j(i)$ carrying $\{1, \ldots, g'\}$ into $\{1, \ldots, g\}$. The function has the property that $K_{P_{j(i)}}$ is isomorphic as a valued field to L_i for each i. We are going to show that $i \mapsto j(i)$ is onto $\{1, \ldots, g\}$. Thus let the completion homomorphism $\psi_j : (K, |\cdot|_{P_j}) \to (K_{P_j}, |\cdot|_{P_j})$ be given.

The F bilinear mapping $(\psi_j, \varphi_j) : K \times F_\mathfrak{p} \to K_{P_j}$ given by multiplication has a linear extension

$$\psi_j \otimes \varphi_j : K \otimes_F F_\mathfrak{p} \to K_{P_j}$$

that is a ring homomorphism. The range K_{P_j} is a field that is finite-dimensional over $\varphi_j(F_\mathfrak{p})$, and the image of $\psi_j \otimes \varphi_j$ is a $\varphi_j(F_\mathfrak{p})$ vector subspace of K_{P_j} that is closed under multiplication. Consequently the image of $\psi_j \otimes \varphi_j$ is closed under inverses[18] and is a field. The kernel of $\psi_j \otimes \varphi_j$ is therefore a maximal ideal, and it follows that there exists some i such that $\psi_j \otimes \varphi_j$ factors as a composition of $\eta_i : K \otimes_F F_\mathfrak{p} \to L_i$ followed by a field map $\gamma : L_i \to K_{P_j}$.

Having constructed a particular L_i, let us form α_i, β_i, and $P_{j(i)}$ as in the abstract theory with M. The map $\beta_i : (F_\mathfrak{p}, |\cdot|_\mathfrak{p}) \to (L_i, |\cdot|_{\mathfrak{p},\beta_i})$ is a homomorphism of valued fields such that $\gamma\beta_i = \varphi_j$, and the map $\alpha_i : (K, |\cdot|_{P_{j(i)}}) \to (L_i, |\cdot|_{\mathfrak{p},\beta_i})$ is a homomorphism of valued fields such that $\gamma\alpha_i = \psi_j$. The existence part of Theorem 6.33 shows that there exists an absolute value $|\cdot|_\gamma$ on K_{P_j} such that $\gamma : (L_i, |\cdot|_{\mathfrak{p},\beta_i}) \to (K_{P_j}, |\cdot|_\gamma)$ is a homomorphism of valued fields. Since $\varphi_j^*(|\cdot|_{P_j})) = |\cdot|_\mathfrak{p} = \beta_i^*(|\cdot|_{\mathfrak{p},\beta_i}) = \beta_i^*\gamma^*(|\cdot|_\gamma) = \varphi_j^*(|\cdot|_\gamma))$, the uniqueness

---

[18]The same argument applies here with $F_\mathfrak{p}$ as was used in Section 4 with $\mathbb{R}$: within a field if a nonzero element is algebraic over a base field, then the smallest ring containing the base field and the element contains also the inverse of the element.

part of Theorem 6.33 shows that $|\cdot|_\gamma = |\cdot|_{P_j}$ on K_{P_j}. Meanwhile, the equality $\psi_j = \gamma \alpha_i$ implies that $\psi_j^* = \alpha_i^* \gamma^*$. Then we have

$$\begin{aligned}
\left(|\cdot|_{P_j} \text{ on } K\right) &= \psi_j^*\left(|\cdot|_{P_j} \text{ on } K_{P_j}\right) \\
&= \alpha_i^* \gamma^*\left(|\cdot|_{P_j} \text{ on } K_{P_j}\right) && \text{since } \psi_j^* = \alpha_i^* \gamma^* \\
&= \alpha_i^* \gamma^*\left(|\cdot|_\gamma \text{ on } K_{P_j}\right) && \text{since } |\cdot|_\gamma = |\cdot|_{P_j} \\
&= \alpha_i^*\left(|\cdot|_{\mathfrak{p}, \beta_i} \text{ on } L_i\right) \\
&= \left(|,\cdot|_{P_{j(i)}}\right) \text{ on } K).
\end{aligned}$$

Therefore $j = j(i)$, and the map $i \mapsto j(i)$ is onto.

To complete the proof, let us compute dimensions relative to $F_\mathfrak{p}$, starting from the decomposition into fields L_i. The ramification index e_j^* and the residue class degree f_j^* for the valuation ring and ideal of K_{P_j} equal the corresponding parameters e_j and f_j for T and P_j, by Lemma 6.36. Thus we have

$$n = \sum_{i=1}^{g'} \dim_{F_\mathfrak{p}} L_i = \sum_{i=1}^{g'} \dim_{F_\mathfrak{p}} K_{P_{j(i)}} = \sum_{j=1}^{g} \sum_{j(i)=j} \dim_{F_\mathfrak{p}} K_{P_{j(i)}}$$
$$= \sum_{j=1}^{g} \sum_{j(i)=j} e_{j(i)}^* f_{j(i)}^* = \sum_{j=1}^{g} \sum_{j(i)=j} e_{j(i)} f_{j(i)} = \sum_{j=1}^{g} |\{i \mid j(i)=j\}| \, e_j f_j.$$

On the other hand, we know that $n = \sum_j e_j f_j$, and we have just proved that $|\{i \mid j(i) = j\}| \geq 1$ for each j. It follows that $|\{i \mid j(i) = j\}| = 1$ for each j, i.e., that the function $i \mapsto j(i)$ is one-one onto. In particular, $g' = g$. The theorem follows. □

Notationally what is happening in the proof of the theorem is that a function $i \mapsto j(i)$ is constructed such that $\alpha_i : K \to L_i$ factors as $\alpha_i = \iota \psi_{j(i)}$ for some canonical isomorphism $\iota : K_{P_{j(i)}} \to L_i$ of complete valued fields. Renumbering the factors and ignoring canonical isomorphisms, we find that $K \otimes_F F_\mathfrak{p}$ is the direct product of the factors K_{P_i} and that $\alpha_i = \psi_i$ carries K to $K \otimes 1$ and then to the ith factor K_{P_i}. Any linear mapping of the form $A \otimes 1$ in effect is therefore block diagonal with each block corresponding to the effect on some K_{P_i}.

Let us apply these considerations to operations "left-multiplication-by," which we write as $l(\cdot)$. If ξ is a member of K, the characteristic polynomial of $l(\xi)$ over F is $\det(X1 - l(\xi))$, and the characteristic polynomial of $l(\xi) \otimes 1$ over $F_\mathfrak{p}$ is still $\det(X1 - l(\xi))$, but now with its coefficients from F regarded as members of $F_\mathfrak{p}$ via the inclusion $\psi_0 : F \to F_\mathfrak{p}$.

The linear function $X(1 \otimes 1) - l(\xi) \otimes 1$ is block diagonal, equal to $X1 - l(\psi_i(\xi))$ on the ith block for $1 \leq i \leq g$. The characteristic polynomial

$\det(X1 - l(\xi))$, regarded as having coefficients in $F_\mathfrak{p}$, is therefore the product of the g characteristic polynomials $X1 - l(\psi_i(\xi))$, each with coefficients in $F_\mathfrak{p}$. In turn, this product formula yields a sum formula for the trace $\text{Tr}_{K/F}(\xi)$ and a product formula for the norm $N_{K/F}(\xi)$. If ξ is a primitive element for the extension K/F, then we can say even more. Let us write all these consequences as a corollary.

Corollary 6.37. Let R be a Dedekind domain regarded as a subring of its field of fractions F, let K be a finite separable extension of F with $[K:F] = n$, and let T be the integral closure of R in K. Let $\mathfrak{p}$ be a nonzero prime ideal of R, and let the ideal $\mathfrak{p}T$ in T have a prime factorization of the form $\mathfrak{p}T = P_1^{e_1} \cdots P_g^{e_g}$, where $P_1, \ldots, P_g$ are distinct prime ideals in T and $e_1, \ldots, e_g$ are positive. For $1 \le i \le g$, let $f_i = [T/P_i : R/\mathfrak{p}]$. If ξ is *any* element of K, then

(a) the F linear map $l(\xi)$ on K given by left multiplication by ξ has the property that its field polynomial $\det(X - l(\xi))$ over F, when reinterpreted as having coefficients in $F_\mathfrak{p}$, factors over $F_\mathfrak{p}$ as the product

$$\det(X - l(\xi)) = \prod_{i=1}^{g} \det(X - l(\xi_i))$$

of the g field polynomials of the images $\xi_i = \psi_i(\xi)$ under the completion map $\psi_i : K \to K_{P_i}$,

(b) $N_{K/F}(\xi) = \prod_{i=1}^{g} N_{K_{P_i}/F_\mathfrak{p}}(\xi_i)$,

(c) $\text{Tr}_{K/F}(\xi) = \sum_{i=1}^{g} \text{Tr}_{K_{P_i}/F_\mathfrak{p}}(\xi_i)$.

Furthermore, if ξ and F together generate K, if $m(X)$ is the minimal polynomial of ξ over F, and if $m(X) = \prod_{j=1}^{g'} m_j(X)$ expresses $m(X)$ as the product of distinct monic irreducible polynomials in $F_\mathfrak{p}[X]$, then

(d) $g' = g$,

(e) there is a one-one onto function $i \mapsto k(i)$ on the set $\{1, \ldots, g\}$ such that K_{P_i} is isomorphic as a field to $F_\mathfrak{p}[X]/(m_{k(i)}(X))$,

(f) $\deg m_{k(i)}(X) = e_i f_i$.

PROOF. Conclusion (a) was proved in the paragraph before the statement of the corollary, and (b) and (c) follow immediately from (a).

Under the assumption that $K = F(\xi)$, the minimal polynomial $m(X)$ of ξ and the characteristic polynomial $\det(X1 - l(\xi))$ are equal; thus $m(X) = \det(X - l(\xi))$ is irreducible over F. Applying Proposition 2.29a, we see that $K \otimes_F F_\mathfrak{p} \cong F_\mathfrak{p}[X]/(m(X))$ as an $F_\mathfrak{p}$ algebra. The assumed separability of K/F means that $m(X)$ is a separable polynomial, and $m(X)$ therefore factors over the extension field $F_\mathfrak{p}$ of F as a product of distinct monic irreducible polynomials

6. Ramification Indices and Residue Class Degrees

in $F_\mathfrak{p}[X]$, say as $m(X) = m_1(X) \cdots m_{g'}(X)$. The Chinese Remainder Theorem implies that

$$K \otimes_F F_\mathfrak{p} \cong \prod_{i=1}^{g'} F_\mathfrak{p}[X]/(m_i(X)),$$

and each $F_\mathfrak{p}[X]/(m_i(X))$ is a field. The factors on the right must coincide with the factors in Theorem 6.31, and it follows that $g' = g$ and that each K_{P_i} is of the form $F_\mathfrak{p}[X]/(m_k(X))$ for some $k = k(i)$. This proves (d) and (e). For (f), $\deg m_{k(i)}(X)$ is the product of the ramification index and the residue class degree for $K_{P_i}/F_\mathfrak{p}$, and this product equals $e_i f_i$ as a consequence of Lemma 6.36 and its remark. $\square$

A by-product of (d) is that we obtain a way of computing g for the extension: it is the number of irreducible factors into which $m(X)$ splits when it is factored over $F_\mathfrak{p}$ instead of F. Hensel's Lemma in the form of Theorem 6.30 can help with carrying out this factorization in favorable cases if ξ is chosen to be integral over R, i.e., to be in T. Namely we reduce the coefficients of $m(X)$ modulo $\mathfrak{p}$, obtaining a monic polynomial in $(R/\mathfrak{p})[X]$, and we factor this polynomial[19] as a product of powers of distinct primes in $(R/\mathfrak{p})[X]$. Since the powers of distinct primes are relatively prime and since everything is monic, Theorem 6.30 is applicable and allows us to lift the factorization to $F_\mathfrak{p}[X]$. The resulting monic factors in $F_\mathfrak{p}[X]$ may not be irreducible in unfavorable circumstances,[20] but we have at least made progress.

Theorem 6.31 has accomplished even more than is stated in Corollary 6.37. For each i, it has identified a field extension, namely $K_{P_i}/F_\mathfrak{p}$, in which the indices e_i and f_i are isolated from the other e_j's and f_j's. Under an additional hypothesis on the residue class field (it is enough to assume that the residue class field is finite), Proposition 6.38 below shows that it is possible to interpolate a unique intermediate field L with $F_\mathfrak{p} \subseteq L \subseteq K_{P_i}$ such that the residue class degree (the parameter f) of K_{P_i}/L is 1 and the ramification index (the parameter e) of $K/F_\mathfrak{p}$ is 1. Thus the proposition says that we can separate e_i and f_i from each other. One says that K_{P_i}/L is **totally ramified** and $L/F_\mathfrak{p}$ is **unramified**.

Proposition 6.38. Let F be a complete valued field under a nonarchimedean discrete valuation v, let R and $\mathfrak{p}$ be the valuation ring and valuation ideal for v, let K be a finite separable extension of F of degree n, let T be the integral closure of R in K, and let P be the unique maximal ideal in T as in Theorem 6.33. Suppose

---

[19]On a computer, for example, if $R/\mathfrak{p}$ is finite.
[20]In Example 5 in the previous section, the given polynomial in $\mathbb{Z}[X]$ is $m(X) = X^3+X^2-2X+8$, and the reduced polynomial in $\mathbb{F}_2[X]$ is $X^2(X+1)$. Theorem 6.30 exhibits a factorization of $m(X)$ over $\mathbb{Z}_2[X]$ as the product of a linear factor and a quadratic factor, and we saw in Example 5 of Section 5 that the quadratic factor is reducible over $\mathbb{Z}_2[X]$.

that $R/\mathfrak{p}$ is a finite field. Let e be the integer such that $\mathfrak{p}T = P^e$, and let f be the dimension of T/P over $R/\mathfrak{p}$. Then there exists a unique intermediate field L for which the integral closure U of R in L and the unique maximal ideal $\wp$ in U have the following properties:

(a) $\mathfrak{p}U = \wp$ and $\wp T = P^e$,
(b) $[U/\wp : R/\mathfrak{p}] = f$ and $[T/P : U/\wp] = 1$.

The proof is carried out in Problems 15–16 at the end of the chapter. We shall apply Proposition 6.38 in Section 8. The intermediate field L in the proposition is called the **inertia subfield** of K/F.

Once this separation of an extension of a complete valued field into a totally ramified extension and an unramified extension has been accomplished, one can go on to study each kind of extension separately, in order to find out what kind of ramification is possible. The results are stated as Lemmas 6.47 and 6.48, and proofs are carried out in Problems 17–19 at the end of the chapter.

7. Special Features of Galois Extensions

In this section we analyze what happens in the setting of Theorem 6.31 when the extension of fields is a Galois extension. For simplicity for the moment, let us work with the number-field setting, even though analogous results hold for function fields in one variable as well. Thus let K/F be a finite *Galois* extension of number fields, let T and R be the rings of algebraic integers in K and F respectively, and let $\mathfrak{p}$ be a nonzero prime ideal in R. Since the extension K/F is Galois, the Galois group $\text{Gal}(K/F)$ permutes transitively the nonzero prime ideals containing $\mathfrak{p}T$, and the factorization of $\mathfrak{p}T$ into powers of distinct prime ideals of T takes the special form $\mathfrak{p}T = P_1^e \cdots P_g^e$ with all the exponents the same.[21] In addition, the dimension of each finite field T/P_i over $R/\mathfrak{p}$ is an integer f independent of i, and we have $efg = [K : F]$.

Let us review Theorem 9.64 and its surrounding discussion in *Basic Algebra*. If we write P for one of the ideals P_i, then the subgroup G_P of $G = \text{Gal}(K/F)$ is called the **decomposition group** at P. Each $\sigma \in G_P$ descends to an automorphism $\bar{\sigma}$ of T/P that fixes $R/\mathfrak{p}$, thereby yielding a member of $\bar{G} = \text{Gal}((T/P)/(R/\mathfrak{p}))$. The map $G \to \bar{G}$ is certainly a homomorphism, and Theorem 9.64 of *Basic Algebra* says that it is onto. It follows that this homomorphism is e-to-1. In *Basic Algebra* this homomorphism was of interest when $F = \mathbb{Q}$ and $e = 1$, since it ensures the presence of certain kinds of permutations in G and makes it possible to determine G completely in certain circumstances.

---
[21]Lemma 9.61 and Theorem 9.62 of *Basic Algebra*.

Theorem 6.31 allows us to isolate each prime ideal P in such an analysis, reinterpreting everything in the context of a particular p-adic field. Carrying through this process gives insights into the decomposition group and the nature of the homomorphism $G_P \to \overline{G}$. The point of this section is to explain some of these insights.

We work within the setting of Theorem 6.31 except that we assume that the residue class fields are finite fields, as they are in the number-theory context. Thus let R be a Dedekind domain regarded as a subring of its field of fractions F, let K be a finite *Galois* extension of F with $[K : F] = n$, and let T be the integral closure of R in K. We suppose that $\mathfrak{p}$ is a nonzero prime ideal of R and that $R/\mathfrak{p}$ is a finite field. Let $\mathfrak{p}T = P_1^e \cdots P_g^e$ be the prime factorization of the ideal $\mathfrak{p}T$ in T; here $P_1, \ldots, P_g$ are assumed to be distinct prime ideals in T. Let f be the common value of the dimension of T/P_i over R/P.

In the decomposition $K \otimes_F F_\mathfrak{p} \cong \prod_{i=1}^g K_{P_i}$ of Theorem 6.31, the projection η_i to the ith factor on the right side is a member of $K \otimes_F F_\mathfrak{p}$; specifically it is the member of the direct product whose i^{th} coordinate is the multiplicative identity of K_{P_i} and whose other coordinates are 0. The element η_i is an **idempotent** in the sense that $\eta_i^2 = \eta_i$, and the η_i's are **orthogonal** in the sense that $\eta_i \eta_j = 0$ for $i \neq j$. The only idempotents of $K \otimes_F F_\mathfrak{p}$ are the sums of distinct elements η_i, and the η_i's are distinguished from the other idempotents in being **primitive**: η_i is not the sum of two nonzero orthogonal idempotents.

Recall the relationship derived in the proof of Theorem 6.31 between P_i and the element η_i: the mapping $\beta_i : F_\mathfrak{p} \to K_{P_i}$ given by $\beta_i(x) = (1 \otimes x)\eta_i$ for $x \in F_\mathfrak{p}$ is a homomorphism of valued fields, and so is the mapping $\alpha_i : K \to K_{P_i}$ given by $\alpha_i(k) = (k \otimes 1)\eta_i$ for $k \in K$. These facts uniquely determine P_i from among the ideals $P_1, \ldots, P_g$.

We extend the action by each member σ of $G = \text{Gal}(K/F)$ to $K \otimes_F F_\mathfrak{p}$ as the transformation $\sigma \otimes 1$. Then G acts on $K \otimes_F F_\mathfrak{p}$, manifestly keeping each element of $F_\mathfrak{p}$ fixed. Since the members of G respect multiplication and addition, they map idempotents to idempotents in $K \otimes_F F_\mathfrak{p}$, sending primitive idempotents to primitive idempotents. Thus G permutes the elements η_i. The elements x with $\eta_i x = x$ are exactly the members of K_{P_i}, and hence G permutes the fields K_{P_i}.

Lemma 6.39. In the above setting with K/F Galois, let P_i be one of the ideals $P_1, \ldots, P_g$. Then a member σ of the Galois group $G = \text{Gal}(K/F)$ extends to a field automorphism of K_{P_i} fixing $F_\mathfrak{p}$ if and only if it is an isometry of $(K, |\cdot|_{P_i})$, i.e., if and only if σ satisfies $|\sigma x|_{P_i} = |x|_{P_i}$ for all $x \in K$.

PROOF. If σ is an isometry from K into itself in the metric determined by $|\cdot|_{P_i}$, then σ is uniformly continuous as a function from K into the complete space K_{P_i} and therefore extends to a continuous function from the completion K_{P_i} into K_{P_i}.

It follows from the continuity of the extension and the fact that σ respects the operations on K that σ respects the operations on K_{P_i}. These remarks apply also to the extension of σ^{-1}, and the extension of σ^{-1} is a two-sided inverse to the extension of σ. Since σ is the identity on F, the continuity forces the extension of σ to be the identity on $F_\mathfrak{p}$.

Conversely suppose that σ extends to an automorphism of K_{P_i} fixing $F_\mathfrak{p}$. Let us use the name σ also for the extension. On K_{P_i}, the functions $x \mapsto |x|_{P_i}$ and $x \mapsto |\sigma(x)|_{P_i}$ are absolute values that extend $|\cdot|_\mathfrak{p}$ on $F_\mathfrak{p}$. Theorem 6.33 shows that they must be equal, and therefore σ is an isometry. $\square$

Proposition 6.40. In the above setting with K/F Galois, let P be one of the ideals $P_1, \ldots, P_g$, let $G = \text{Gal}(K/F)$ be the Galois group, and let G_P be the decomposition group at P. Then K_P is a Galois extension of $F_\mathfrak{p}$, the members of G_P extend to be isometries of K_P that fix $F_\mathfrak{p}$, and the resulting map $\varphi : G_P \to \text{Gal}(K_P/F_\mathfrak{p})$ exhibits G_P as isomorphic to $\text{Gal}(K_P/F_\mathfrak{p})$.

PROOF. Since K_P is generated by $F_\mathfrak{p}$ and K, it is obtained by adjoining to $F_\mathfrak{p}$ the same roots of the same polynomials over F that are used to generate K. Therefore $K_P/F_\mathfrak{p}$ is a Galois extension.

Lemma 6.39 gives us the map of G_P into $\text{Gal}(K_P/F_\mathfrak{p})$. The map φ is a homomorphism because the extension of each member of G_P is unique. It is one-one because the inclusion $K \subseteq K_P$ is one-one.

To see that it is onto, let σ be in $\text{Gal}(K_P/F_\mathfrak{p})$, and choose an element $\xi \in K$ such that $K = F(\xi)$. If $m(X)$ is the minimal polynomial of ξ over F, then $\sigma(\xi)$ is an element of K_P with $m(\sigma(\xi)) = 0$. Consequently $\sigma(\xi)$ is a root of $m(X)$. Since K/F is Galois and $m(X)$ has one root in K, all its roots are in K. Thus $\sigma(\xi)$ is in K. The most general member of K is of the form $q(\xi)$, where $q(X)$ is a polynomial of degree less than $\deg m(X)$, and $q(\sigma(\xi))$ has to be in K also. Thus σ is an automorphism of K fixing F. As such, σ must send T into itself and must send P into some ideal P_i of T containing $\mathfrak{p}T$. Meanwhile, Lemma 6.39 shows that σ is an isometry of K relative to $|\cdot|_P$. Thus σ must send P into itself. In other words, the restriction of σ to K is in the decomposition group G_P. $\square$

We know from Theorem 9.64 of *Basic Algebra* that every member σ of the decomposition group G_P yields a member $\overline{\sigma}$ of $\text{Gal}((T/P)/(R/\mathfrak{p}))$ and that the resulting map $\sigma \mapsto \overline{\sigma}$ is a homomorphism onto. Proposition 6.40 allows us to reinterpret this homomorphism as carrying the Galois group of K_P onto the Galois group of T/P. The order of $\text{Gal}(K_P/F_\mathfrak{p})$ is ef, and the order of $\text{Gal}((T/P)/(R/\mathfrak{p}))$ is f. Thus the kernel of this homomorphism, which is called the **inertia group** of $K_P/F_\mathfrak{p}$, has order e. By Galois theory the fixed field L of the inertia group has $[K_P : L] = e$, $L/F_\mathfrak{p}$ is a Galois extension, and $\text{Gal}(L/F_\mathfrak{p})$ has order f. This construction has been arranged to make $\text{Gal}(L/F_\mathfrak{p}) \cong \text{Gal}((T/P)/(R/F_\mathfrak{p}))$. As the Galois group of a finite extension of finite fields, the Galois group on the right is cyclic of order f. Therefore $\text{Gal}(L/F_\mathfrak{p})$ is cyclic of order f.

Referring back to the statement of Proposition 6.38, we might guess that the fixed field L of the inertia group is the unique intermediate field such that K/L is totally ramified and L/F is unramified. This guess is completely correct, but we omit the proof.

8. Different and Discriminant

Theorem 6.31 is the key to a "local/global" approach to handling certain kinds of problems in algebraic number theory and in its analog in algebraic geometry. To illustrate the approach and its power, we shall give in this section and in the problems at the end of the chapter a full proof for the Dedekind Discriminant Theorem (Theorem 5.5), which was left only partially proved in Chapter V. That theorem as stated in Chapter V says that the prime numbers p for which ramification occurs in passing from $\mathbb{Q}$ to a number field K are exactly the primes dividing the field discriminant. The result we obtain now[22] will in fact generalize Theorem 5.5 significantly. In giving the details, we leave the proofs of Proposition 6.38 and Lemmas 6.47 and 6.48 to Problems 15–19 at the end of the chapter.

In the approach used in Chapter V, we were unable to handle primes that are "common index divisors" in the sense of Section V.2. Section V.4 exhibited an example of a common index divisor. The difficulty with the approach in Chapter V is that localization by itself does not ostensibly separate the primes from one another sufficiently for us fully to handle them one at a time. The completion step is a tool powerful enough to complete the separation.

For part of this section, we shall work in the setting of Theorem 6.31, in which we compare two Dedekind domains whose fields of fractions are related by a separable field extension. The situation of eventual interest is that the two Dedekind domains are the rings of algebraic integers within two number fields, but we shall encounter also p-adic versions of this situation. Thus let R be a Dedekind domain regarded as a subring of its field of fractions F, let K be a finite separable extension of F with $[K : F] = n$, and let T be the integral closure of R in K. In this setting we shall introduce an ideal $\mathcal{D}(K/F)$ of T known as the "relative different" of the two fields, and we shall establish conditions under which the relative different captures fairly precisely what ramification occurs in passing from R to T. This is the generalized version of the Dedekind Discriminant Theorem and appears as Theorem 6.45 below.

---

[22]Dedekind's Theorem on Differents, given as Theorem 6.45.

In the special case that $F = \mathbb{Q}$, we shall see that the field discriminant D_K satisfies $|D_K| = N(\mathcal{D}(K/\mathbb{Q}))$. In words, the field discriminant is the absolute norm of the relative different $\mathcal{D}(K/\mathbb{Q})$ except possibly for a sign. Using the properties of $N(\,\cdot\,)$ listed in Proposition 5.4, we can read off the version of the Dedekind Discriminant Theorem stated in Theorem 5.5 from the results we establish about the relative different.

We work with fractional ideals in F and in K. If M is any nonzero fractional ideal of K, we define its (relative) **dual** as

$$\widehat{M} = \{x \in K \mid \mathrm{Tr}_{K/F}(xy) \text{ is in } R \text{ for all } y \in M\}.$$

Lemma 6.41. In the above setting, if M is a nonzero fractional ideal of K, then so is its dual $\widehat{M}$.

PROOF. Since T has K as its field of fractions, there exists an F vector space basis $\{t_1, \ldots, t_n\}$ of K consisting of members of T. If m_0 is a nonzero member of M and $m_j = t_j m_0$, then $\{m_1, \ldots, m_n\}$ is an F vector space basis of K lying in M. Form the R submodule $M_1 = \sum_{j=1}^{n} R m_j$ of M, and let $\{x_1, \ldots, x_n\}$ be the F vector space basis of K such that $\mathrm{Tr}_{K/F}(x_j m_j) = \delta_{ij}$. Let

$$\widehat{M_1} = \{x \in K \mid \mathrm{Tr}_{K/F}(xm) \text{ is in } R \text{ for all } m \in M_1\}.$$

If we expand a general element x of K as $x = \sum_{j=1}^{n} c_i x_i$, then a necessary condition for x to be in $\widehat{M_1}$ is that $c_j = \mathrm{Tr}_{K/F}(xm_j)$ be in R for all j. On the other hand, this condition is also sufficient because an element x with all $c_j \in R$ has $\mathrm{Tr}_{K/F}(xm) = \sum_{j=1}^{n} c_j r_j$ if $m = \sum_{j=1}^{n} r_j m_j$. Thus $\widehat{M_1}$ is a finitely generated R module with $x_1, \ldots, x_n$ as generators. Let S be the T submodule of K given by $S = \sum_{j=1}^{n} T x_j$. This is a finitely generated T submodule of K that contains $\widehat{M_1}$. The inclusion $M \supseteq M_1$ evidently implies that $\widehat{M} \subseteq \widehat{M_1}$, and hence $\widehat{M} \subseteq S$. In this way, $\widehat{M}$ is exhibited as a T submodule of the finitely generated T submodule S of K, and $\widehat{M}$ must itself be finitely generated because T is a Noetherian ring. $\square$

Proposition 6.42. In the above setting, the dual $\widehat{T}$ of T is of the form $\widehat{T} = \mathcal{D}(K/F)^{-1}$ for an ideal $\mathcal{D}(K/F)$ of T. This ideal $\mathcal{D}(K/F)$ has the property that

$$\widehat{M} = M^{-1} \mathcal{D}(K/F)^{-1}$$

for every nonzero fractional ideal M of K.

REMARK. The ideal $\mathcal{D}(K/F)$ in T is called the **relative different** of K with respect to F.

PROOF. From the definition, $\widehat{T}$ consists of all x in K for which $\text{Tr}_{K/F}(xt)$ is in R; any member x of T has this property, and thus $T \subseteq \widehat{T}$. Lemma 6.41 shows that $\widehat{T}$ is a fractional ideal of K. Since $\widehat{T}$ contains T, it is the inverse of an ideal of T. This ideal we define as $\mathcal{D}(K/F)$.

Let M be an arbitrary nonzero fractional ideal of K. Since $M^{-1}M = T$, we have $\text{Tr}_{K/F}(M^{-1}\mathcal{D}(K/F)^{-1} \cdot M) = \text{Tr}_{K/F}(\mathcal{D}(K/F)^{-1}) = \text{Tr}_{K/F}(\widehat{T}T) \subseteq R$, and it follows that $M^{-1}\mathcal{D}(K/F)^{-1} \subseteq \widehat{M}$. For the reverse inclusion, let x be in $\widehat{M}$. Then $\text{Tr}_{K/F}(xM \cdot t) \subseteq \text{Tr}_{K/F}(xM) \subseteq R$ for all $t \in T$, and hence $xM \subseteq \widehat{T} = \mathcal{D}(K/F)^{-1}$. This being true for all $x \in \widehat{M}$, we obtain $\widehat{M}M \subseteq \mathcal{D}(K/F)^{-1}$. Therefore $\widehat{M} \subseteq M^{-1}\mathcal{D}(K/F)^{-1}$. □

Proposition 6.43. In the above setting, if L is a field with $F \subseteq L \subseteq K$, then

$$\mathcal{D}(K/F) = \mathcal{D}(K/L)\mathcal{D}(L/F)$$

as an equality of fractional ideals in K.

REMARKS. Let U be the integral closure of R in L. In the displayed line of the proposition, $\mathcal{D}(L/F)$ is an ideal in U, and the right side amounts to the product in T given by $\mathcal{D}(K/L) \cdot \mathcal{D}(L/F)T$.

PROOF. We use the fact that traces can be computed in stages. An element x of K is in $\mathcal{D}(K/F)^{-1}$ if and only if $\text{Tr}_{K/F}(xT) \subseteq R$, if and only if $\text{Tr}_{L/F}\left(\text{Tr}_{K/L}(xT)\right) \subseteq R$, if and only if $\text{Tr}_{K/L}(xT) \subseteq \widehat{U} = \mathcal{D}(L/F)^{-1}$, if and only if $\text{Tr}_{K/L}(xT\mathcal{D}(L/F)) \subseteq U$, if and only if $xT\mathcal{D}(L/F) \subseteq \mathcal{D}(K/L)^{-1}$. Thus $\mathcal{D}(K/F)^{-1}\mathcal{D}(L/F) = \mathcal{D}(K/L)^{-1}$, and the result follows. □

The main result of this section, from which the Dedekind Discriminant Theorem will be derived as Corollary 6.49, is Theorem 6.45 below, Dedekind's Theorem on Differents. The proof requires some preparation. Two results will be used to reduce Theorem 6.45 to a statement about complete fields, for which only a single prime ideal is involved, both for R and for T. The first of these is Theorem 6.31, or more particularly its consequence for traces given in Corollary 6.37c. The other is the following strengthening of the Weak Approximation Theorem in the presence of additional hypotheses. The reduction step to a statement about complete fields then appears as Corollary 6.46.

Theorem 6.44 (Strong Approximation Theorem). Let F be a number field, let R be its ring of algebraic integers, let $P_1, \ldots, P_r$ be distinct nonzero prime ideals in R, and let v_{P_j} for each j be the valuation of F and of its completion that corresponds to P_j. If $l_1, \ldots, l_r$ are integers and if x_j for $1 \leq j \leq r$ is a member of the completed field F_{P_j}, then there exists y in F such that

$$v_{P_j}(y - x_j) \geq l_j \qquad \text{for } 1 \leq j \leq r$$

and such that $v_Q(y) \geq 0$ for all other nonzero prime ideals Q of R.

REMARKS.

(1) It will be helpful to have a name for the property in the conclusion of Theorem 6.44. Thus let T be a Dedekind domain regarded as a subring of its field of fractions K. We say that T has the **strong approximation property** if whenever distinct nonzero prime ideals $P_1, \ldots, P_r$ of T are given, along with integers $l_1, \ldots, l_r$ and members x_j of the completed field K_{P_j} for $1 \leq j \leq r$, then there exists y in K such that $v_{P_j}(y - x_j) \geq l_j$ for $1 \leq j \leq r$ and such that $v_Q(y) \geq 0$ for all other nonzero prime ideals Q of T. The content of Theorem 6.44 is that the ring of algebraic integers in any number field has the strong approximation property.

(2) More generally any principal ideal domain has the strong approximation property. In fact, if R is a principal ideal domain with field of fractions F, if K is a finite extension of F, and if T is the integral closure of R in K, then K is a Dedekind domain (according to the remarks with Proposition 6.7), and K has the strong approximation property. The proof is an easy adaptation of the proof below, with the principal ideal domain substituting for the ring $\mathbb{Z}$ of integers. As a consequence if $\Bbbk$ is a field and if T is the integral closure of $\Bbbk[X]$ in a finite extension of $\Bbbk(X)$, then T has the strong approximation property.

(3) Any Dedekind domain with only finitely many prime ideals has the strong approximation property as an immediate consequence of the Weak Approximation Theorem (Theorem 6.23). One does not need to make use of the fact that such a domain is always a principal ideal domain.

(4) For a number field the conclusion of the theorem as stated imposes a limitation on all the nonarchimedean absolute values. The conclusion cannot be strengthened to impose a limitation on *all* equivalence classes of absolute values, since the Artin product formula (Theorem 6.51 below) imposes a constraint on the set of all of them.

PROOF.[23] We may assume that each l_j satisfies $l_j \geq 0$. Recall that for each prime number p, there are only finitely many prime ideals P in R with $P \cap \mathbb{Z} = p\mathbb{Z}$. Possibly by moving some of the conditions $v_Q(y) \geq 0$ into the displayed hypothesis concerning the P_j's, we may assume that there is some finite set $\{p_1, \ldots, p_q\}$ of primes such that $\{P_1, \ldots, P_r\}$ consists exactly of all prime ideals P such that $P \cap \mathbb{Z} = p_i \mathbb{Z}$ for some i with $1 \leq i \leq q$.

Application of the Weak Approximation Theorem (Theorem 6.23) to the absolute values corresponding to $P_1, \ldots, P_r$ produces an element $z \in F$ with

---

[23] This proof is from Hasse's *Number Theory*, pp. 379–380. The argument for $R = \mathbb{Z}$ and all $l_j = 0$ is the key. After an application of the Weak Approximation Theorem, what has to be shown is that if $P_j = p_j \mathbb{Z}$ for $1 \leq j \leq r$ and if a rational ab^{-1} is given, then there exists a rational mn^{-1} with l prime to $p_1, \ldots, p_r$ such that the denominator of $ab^{-1} - mn^{-1}$ is divisible only by the primes $p_1, \ldots, p_r$. Another proof of Theorem 6.44, which appears in other books, uses the theory of adeles and ideles to be developed in the next two sections, and again the argument for $\mathbb{Z}$ is the key.

$$v_{P_j}(z - x_j) \geq l_j \qquad \text{for } 1 \leq j \leq r.$$

Form the fractional ideal zR in F, and let its unique factorization be $zR = P_1^{a_1} \cdots P_r^{a_r} Q_1 Q_2^{-1}$, where the a_j are in $\mathbb{Z}$ and where Q_1 and Q_2 are ideals of R whose prime factorizations involve no P_j. Let us see that Q_2 divides a nonzero principal ideal (N) of R whose generator N is in $\mathbb{Z}$ and that N can be chosen to be relatively prime to $p_1, \ldots, p_q$. In fact, it is enough to treat each prime factor of Q_2 separately and multiply the results. For a prime factor P, we know that $P \cap \mathbb{Z} = p\mathbb{Z}$ for some prime p in $\mathbb{Z}$, and we know that pR is the product of P and another ideal of R. This prime p is nonassociate to each of $p_1, \ldots, p_q$ because the only prime ideals whose intersection with $\mathbb{Z}$ is some $p_i \mathbb{Z}$ are $P_1, \ldots, P_r$ and because no such prime ideal divides Q_2. Therefore the prime factorization of (N) contains no factor $P_1, \ldots, P_r$.

Let b be a positive integer to be specified, and choose an integer l such that $lN \equiv 1 \bmod p_i^b$ for $1 \leq i \leq q$. If $p_i R$ factors as $\prod_k P_{i_k}^{m_{i_k}}$ with each P_{i_k} in $\{P_1, \ldots, P_r\}$, then l has the property that $lN - 1$ lies in $\left(\prod_k P_{i_k}^{m_{i_k}}\right)^b$, hence in each $P_{i_k}^b$. Consequently $lN - 1$ lies in P_j^b for $1 \leq j \leq r$.

We show that if b is sufficiently large, then the element $y = lNz$ is the element we seek. First consider nonzero prime ideals Q not in $\{P_1, \ldots, P_r\}$. Our factorizations of zR and (N) show that $yR = lQ_3 Q_1 P_1^{a_1} \cdots P_r^{a_r}$. The power of Q on the right side is ≥ 0 because Q_1 and Q_3 are ideals of R, and thus

$$v_Q(y) \geq 0. \qquad (*)$$

Now write $y - x_j = (lN - 1)z + (z - x_j)$, and apply the valuation v_{P_j}. Then we have

$$v_{P_j}(y - x_j) \geq \min\left(v_{P_j}((lN - 1)z), v_{P_j}(z - x_j)\right),$$

and it follows from $v_{P_j}(z - x_j)) \geq l_j$ that

$$v_{P_j}(y - x_j) \geq l_j \qquad (**)$$

if we can arrange that

$$v_{P_j}((lN - 1)z) \geq l_j. \qquad (\dagger)$$

Since $lN - 1$ lies in P_j^b and since $v_{P_j}(z) = a_j$, a sufficient condition for ($\dagger$) is that $b + a_j \geq l_j$. As j varies, we impose only finitely many conditions on b to get ($\dagger$) to hold for all j, and then the result is that ($**$) holds for all j. In combination with ($*$), this inequality shows that y has the required properties. $\square$

The preparation is all in place to prove Dedekind's Theorem on Differents, from which we shall easily derive the Dedekind Discriminant Theorem. The statement is as follows.

Theorem 6.45 (Dedekind's Theorem on Differents). Let R be a Dedekind domain regarded as a subring of its field of fractions F, let K be a finite separable extension of F with $[K : F] = n$, and let T be the integral closure of R in K. Suppose that T has the strong approximation property. Let $p > 0$ be the characteristic of the residue class field of $R/\mathfrak{p}$, let $\mathfrak{p}$ be a nonzero prime ideal in R, let $\mathfrak{p}T = P_1^{e_1} \cdots P_g^{e_g}$ be the factorization of $\mathfrak{p}T$ as the product of positive powers of distinct prime ideals in T, and let the relative different of K/F split as $\mathcal{D}(K/F) = P_1^{e'_1} \cdots P_g^{e'_g} Q$ for an ideal Q relatively prime to all P_j. Then for each j with $1 \leq j \leq g$, e'_j is given by

$$e'_j = \begin{cases} e_j - 1 & \text{if } p \text{ does not divide } e_j, \\ \bar{e}_j & \text{with } \bar{e}_j \geq e_j \text{ if } p \text{ divides } e_j. \end{cases}$$

Consequently $\mathcal{D}(K/F)$ has all $e'_j = 0$ if and only if $e_j = 1$ for all j.

The idea is to reduce Theorem 6.45 to the case of complete fields. In the notation in the statement of the theorem, the prime ideals $P_1, \ldots, P_g$ are exactly the prime ideals of T that divide $\mathfrak{p}T$, and it is customary to write $P_j \mid \mathfrak{p}$ for these prime ideals of T and only these. If M is a nonzero fractional ideal of K and if $M = P_1^{k_1} \cdots P_g^{k_g} Q$ with Q a fractional ideal whose factorization involves no P_j, we define the $\mathfrak{p}^{\text{th}}$ component of M to be

$$M_\mathfrak{p} = P_1^{k_1} \cdots P_g^{k_g}.$$

The understanding in the special case that all k_j are 0 is that $M_\mathfrak{p}$ is taken to be T. In all cases, M is then the product over all $\mathfrak{p}$ of its $\mathfrak{p}^{\text{th}}$ component, since the complete factorization of M has nonzero exponents for only finitely many nonzero prime ideals of T. For the two examples that appear in the statement of Theorem 6.45,

$$(\mathfrak{p}T)_\mathfrak{p} = \prod_{P_j \mid \mathfrak{p}} P_j^{e_j} \qquad \text{and} \qquad \mathcal{D}(K/F)_\mathfrak{p} = \prod_{P_j \mid \mathfrak{p}} P_j^{e'_j}.$$

The reduction of Theorem 6.45 to the case of complete fields results from the following proposition, which combines Theorem 6.31 and the strong approximation property (Theorem 6.44 in the case of number fields).

Proposition 6.46. Let R be a Dedekind domain regarded as a subring of its field of fractions F, let K be a finite separable extension of F with $[K : F] = n$, and let T be the integral closure of R in K. Suppose that T has the strong approximation property. If $\mathfrak{p}$ is any nonzero prime ideal in R, then the different $\mathcal{D}(K/F)$ has the property that

$$\mathcal{D}(K/F) = \prod_\mathfrak{p} \prod_{P \mid \mathfrak{p}} \mathcal{D}(K_P/F_\mathfrak{p}),$$

the outer product being taken over all nonzero prime ideals $\mathfrak{p}$ of R and the inner product being taken over all prime ideals P of T containing $\mathfrak{p}T$. Here the fields K_P and $F_\mathfrak{p}$ are the completions of K and F corresponding to P and $\mathfrak{p}$, respectively.

PROOF. We actually will show equality of the inverses of the two sides of the displayed formula. By the first conclusion of Proposition 6.42, we are to show that a member x of K has

$$\operatorname{Tr}_{K/F}(xT) \subseteq R \quad \text{if and only if} \quad \operatorname{Tr}_{K_P/F_\mathfrak{p}}((xT)_i) \subseteq R_\mathfrak{p} \quad (*)$$

for all $\mathfrak{p}$ and all P with $P \mid \mathfrak{p}$. Here $(\cdot)_i$ refers to the embedding $K \to K_{P_i}$ in Theorem 6.31 given by $\xi \mapsto \xi_i = \eta_i(1 \otimes \xi)$, where η_i is the ith projection. To prove $(*)$, we use the formula of Corollary 6.37c, namely

$$\operatorname{Tr}_{K/F}(\xi) = \sum_{i=1}^{g} \operatorname{Tr}_{K_{P_i}/F_\mathfrak{p}}(\xi_i) \quad \text{for all } \xi \in K. \quad (**)$$

This formula is valid for every $\mathfrak{p}$.

First suppose that $\operatorname{Tr}_{K_P/F_\mathfrak{p}}((xT)_i) \subseteq R_\mathfrak{p}$ for all $\mathfrak{p}$ and all P with $P \mid \mathfrak{p}$. Fix $\mathfrak{p}$, and put $\xi = xt$ with $t \in T$. Summing the traces over P with $P \mid \mathfrak{p}$ and applying $(**)$, we see that the valuation with respect to $\mathfrak{p}$ of the member $\operatorname{Tr}_{K/F}(\xi)$ of F is ≥ 0. That is, the factor $\mathfrak{p}^k$ that appears in the factorization of the principal fractional ideal $\operatorname{Tr}_{K/F}(\xi)R$ of F has $k \geq 0$. This being true for all $\mathfrak{p}$ means that $\operatorname{Tr}_{K/F}(\xi)R$ is an ordinary ideal. Hence $\operatorname{Tr}_{K/F}(\xi)$ is in R.

In the reverse direction, suppose that $\operatorname{Tr}_{K/F}(xT) \subseteq R$. For each nonzero prime ideal P in T, let v_P be the corresponding valuation. Fix $\mathfrak{p}$. Let $\{P_1, \ldots, P_g\}$ be the set of P's with $P \mid \mathfrak{p}$. Now fix i. By the assumed strong approximation property of K, there exists an element y in K with

$$v_{P_i}(y - x) \geq \max(v_{P_i}(x), 0),$$
$$v_{P_j}(y) \geq \max(v_{P_j}(x), 0) \quad \text{for } j \neq i,$$
$$v_Q(y) \geq 0 \quad \text{for all prime ideals } Q \notin \{P_1, \ldots, P_g\}.$$

Let us see that $v_{P_j}(yx^{-1}) \geq 0$ for all j. For $j \neq i$, this is immediate because $v_{P_j}(y) \geq v_{P_j}(x)$. For $j = i$, we compute that

$$v_{P_i}(yx^{-1} - 1) = v_{P_i}(y - x) - v_{P_i}(x) \geq \max(v_{P_i}(x), 0) - v_{P_i}(x)$$
$$= \max(0, -v_{P_i}(x)) \geq 0,$$

and then we see that $v_{P_i}(yx^{-1}) \geq \min(v_{P_i}(yx^{-1} - 1), v_{P_i}(1)) \geq 0$.

With y now fixed, we make use of the strong approximation property of K a second time, obtaining an element z in K with

$$v_{P_j}(z - yx^{-1}) \geq \max(v_{P_j}(x^{-1}), 0) \quad \text{for } 1 \leq j \leq g,$$
$$v_Q(z) \geq 0 \quad \text{for all prime ideals } Q \notin \{P_1, \ldots, P_g\}.$$

Since $v_{P_j}(yx^{-1}) \geq 0$ and $v_{P_j}(z - yx^{-1}) \geq 0$ for all j, we find that $v_{P_j}(z) \geq 0$ for all j. From $v_Q(z) \geq 0$ for all other Q, we conclude that z is in T. Since $\text{Tr}_{K/F}(xT) \subseteq R$, $\text{Tr}_{K/F}(xz)$ lies in R. The trace formula (∗∗) therefore shows that

$$\sum_{j=1}^{g} \text{Tr}_{K_{P_j}/F_{\mathfrak{p}}}(x_j z_j) \quad \text{lies in } R_{\mathfrak{p}}. \tag{†}$$

Meanwhile, we have

$$\text{Tr}_{K_{P_j}/F_{\mathfrak{p}}}(x_j z_j) = \text{Tr}_{K_{P_j}/F_{\mathfrak{p}}}(x_j(z_j - y_j x_j^{-1})) + \text{Tr}_{K_{P_j}/F_{\mathfrak{p}}}(y_j) \tag{††}$$

for $1 \leq j \leq g$. For all j, the first term on the right side of (††) lies in $R_{\mathfrak{p}}$ because the definition of z makes $v_{P_j}(x(z - yx^{-1})) \geq 0$. For $j \neq i$, the second term on the right side lies in $R_{\mathfrak{p}}$ because of the definition of y. Thus (††) shows that $\text{Tr}_{K_{P_j}/F_{\mathfrak{p}}}(x_j z_j)$ lies in $R_{\mathfrak{p}}$ for $j \neq i$. Comparing this conclusion with (†), we see that $\text{Tr}_{K_{P_i}/F_{\mathfrak{p}}}(x_i z_i)$ lies in $R_{\mathfrak{p}}$. Resubstituting into (††), we find that

$$\text{Tr}_{K_{P_i}/F_{\mathfrak{p}}}(y_i) \quad \text{lies in } R_{\mathfrak{p}}. \tag{‡}$$

Finally the definition of y shows that $v_{P_i}(y - x) \geq 0$. Hence $\text{Tr}_{K_{P_i}/F_{\mathfrak{p}}}(y_i - x_i)$ is in $R_{\mathfrak{p}}$. Combining this fact with (‡), we conclude that $\text{Tr}_{K_{P_i}/F_{\mathfrak{p}}}(x_i)$ is in $R_{\mathfrak{p}}$. Since i is arbitrary, $\text{Tr}_{K_{P_j}/F_{\mathfrak{p}}}(x_j)$ is in $R_{\mathfrak{p}}$ for $1 \leq j \leq g$. □

With the proof of Theorem 6.45 reduced to the case of complete valued fields by Proposition 6.46, we need to make use of Lemmas 6.47 and 6.48 below, whose proofs are carried out in Problems 17–19 at the end of the chapter.

Lemma 6.47. Let F be a complete valued field with respect to a discrete nonarchimedean valuation, let R be its valuation ring, let $\mathfrak{p}$ be its valuation ideal, let K be a finite separable extension of F with $[K : F] = n$, let T be the integral closure of R in K, and let P be the unique nonzero prime ideal in T. Suppose that K/F is totally ramified with $\mathfrak{p}T = P^e$ for an integer $e \geq 1$, and suppose that the isomorphic residue class fields $R/\mathfrak{p}$ and T/P are finite fields of characteristic p. Then the different $\mathcal{D}(K/F)$ is given by $\mathcal{D}(K/F) = P^{e'}$, where

$$e' = \begin{cases} e - 1 & \text{if } p \text{ does not divide } e, \\ \bar{e} & \text{with } \bar{e} \geq e \text{ if } p \text{ divides } e. \end{cases}$$

Lemma 6.48. Let F be a complete valued field with respect to a discrete nonarchimedean valuation, let R be its valuation ring, let $\mathfrak{p}$ be its valuation ideal, let K be a finite separable extension of F with $[K : F] = n$, let T be the integral closure of R in K, and let P be the unique nonzero prime ideal in T. Suppose that K/F is unramified, i.e., has $\mathfrak{p}T = P$, and suppose that the residue class fields $R/\mathfrak{p}$ and T/P are finite fields of characteristic p. Then the different $\mathcal{D}(K/F)$ equals T.

PROOF OF THEOREM 6.45. Proposition 6.46 shows that

$$\mathcal{D}(K/F)_\mathfrak{p} = \prod_{P|\mathfrak{p}} \mathcal{D}(K_P/F_\mathfrak{p}). \qquad (*)$$

Thus consider an extension $K_P/F_\mathfrak{p}$ of complete valued fields. Let L be the inertia subfield of $K_P/F_\mathfrak{p}$ as given by Proposition 6.38. The intermediate field L has the properties that K_P/L is totally ramified and that $L/F_\mathfrak{p}$ is unramified.

Let U be the integral closure of R in L, and let $\wp$ be the unique nonzero prime ideal in U. The properties of L make $\wp T = P^e$ for a suitable integer $e = e(P \mid \wp)$, $T/P \cong U/\wp$, and $\mathfrak{p}U = \wp$. Lemmas 6.47 and 6.48 tell us that $\mathcal{D}(L/F_\mathfrak{p}) = U$ and that $\mathcal{D}(K_P/L) = P^{e'}$, where

$$e' = \begin{cases} e - 1 & \text{if } p \text{ does not divide } e, \\ \bar{e} & \text{with } \bar{e} \geq e \text{ if } p \text{ divides } e. \end{cases} \qquad (**)$$

Problem 33 at the end of Chapter IX of *Basic Algebra* shows that ramification indices multiply for successive extensions. Thus $e(P \mid \mathfrak{p}) = e(P \mid \wp)e(\wp \mid \mathfrak{p}) = e \cdot 1 = e$. Proposition 6.43 shows that differents multiply in corresponding fashion. Therefore $\mathcal{D}(K_P/F_\mathfrak{p}) = \mathcal{D}(K_P/L)\mathcal{D}(L/F_\mathfrak{p}) = P^{e'}U = P^{e'}$. Substituting into $(*)$, we obtain

$$\mathcal{D}(K/F)_\mathfrak{p} = \bigoplus_{P|\mathfrak{p}} \mathcal{D}(K_P/F_\mathfrak{p}) = \bigoplus_{P|\mathfrak{p}} P^{e'(P|\mathfrak{p})},$$

where $e'(P \mid \mathfrak{p})$ is the integer e' of $(**)$ when $e = e(P \mid \mathfrak{p})$. This proves Theorem 6.45 for the $\mathfrak{p}^{\text{th}}$ component of $D(K/F)$. Since $\mathfrak{p}$ is arbitrary and only finitely many components can be unequal to T, the theorem follows. $\square$

Corollary 6.49 (= THEOREM 5.5, Dedekind Discriminant Theorem). Let K be a number field, let T be its ring of algebraic integers, let p be a prime number, and let $(p)T = P_1^{e_1} \cdots P_g^{e_g}$ be the factorization of $(p)T$ as the product of powers of distinct prime ideals in T. Then e_j is greater than 1 for some j if and only if p divides the field discriminant D_K.

PROOF. Let us observe first that the discriminant D_K is given up to sign by the index $|\widehat{T}/T|$. In fact, T is a torsion-free finitely generated abelian group and hence is free abelian of rank $n = [K : \mathbb{Q}]$, say with an ordered $\mathbb{Z}$ basis $\Gamma = (x_1, \ldots, x_n)$. Since the $\mathbb{Q}$ bilinear form $(x, y) \mapsto \operatorname{Tr}_{K/\mathbb{Q}}(xy)$ is nondegenerate on K, there exists an ordered basis $\Delta = (y_1, \ldots, y_n)$ of K with $\operatorname{Tr}_{K/\mathbb{Q}}(x_i y_j) = \delta_{ij}$. Let us write $x_j = \sum_i a_{ij} y_i$ with all a_{ij} in $\mathbb{Q}$. According to Proposition 5.1, D_K equals the discriminant $D(\Gamma)$ of Γ, defined in Section V.2 by $D(\Gamma) = \det[\operatorname{Tr}_{K/\mathbb{Q}}(x_i x_j)]_{ij}$. Substituting $x_j = \sum_i a_{ij} y_i$, we obtain

$$D_K = \det\Big[\sum_k a_{kj} \operatorname{Tr}_{K/\mathbb{Q}}(x_i y_k)\Big]_{ij} = \det\Big[\sum_k a_{kj} \delta_{ik}\Big]_{ij} = \det[a_{ij}]_{ij}.$$

Thus $|D_K| = |\widehat{T}/T| = \big|\mathcal{D}(K/\mathbb{Q})^{-1}/T\big|$, as asserted.

In a moment we shall show that

$$\big|\mathcal{D}(K/\mathbb{Q})^{-1}/T\big| = |T/\mathcal{D}(K/\mathbb{Q})|, \qquad (*)$$

from which we conclude that $|D_K| = N(\mathcal{D}(K/\mathbb{Q}))$. Assuming $(*)$, we continue. Unique factorization of ideals allows us to write $\mathcal{D}(K/\mathbb{Q}) = P_1^{e'_1} \cdots P_g^{e'_g} Q$, where Q is an ideal relatively prime to (p). Combining the equality $D_K = N(\mathcal{D}(K/\mathbb{Q}))$ with Proposition 5.4 shows that

$$D_K = N(\mathcal{D}(K/\mathbb{Q})) = N(Q) \prod_{j=1}^{g} N(P_j^{e'_j}) = N(Q) \prod_{j=1}^{g} p^{e'_j f_j},$$

where $N(Q)$ is an integer not divisible by p and where $f_j = \dim_{\mathbb{F}_p}(T/P_j)$ for $1 \leq j \leq g$. Consequently D_K is prime to p if and only if $e'_j = 0$ for all j. If we take into account that T has the strong approximation property as a consequence of Theorem 6.44, then application of Theorem 6.45 completes the proof of the present corollary except for the verification of $(*)$.

Thus we are left with proving that $\big|\mathcal{D}(K/\mathbb{Q})^{-1}/T\big| = |T/\mathcal{D}(K/\mathbb{Q})|$. More generally we shall show that

$$|I^{-1}/T| = |T/I| \qquad (**)$$

for every nonzero ideal I in T. In turn, we shall deduce $(**)$ after showing that

$$|M/PM| = N(P) \qquad (\dagger)$$

whenever M is a nonzero fractional ideal in K and P is a nonzero prime ideal in T. We do so by showing that M/PM is a vector space over the field T/P of dimension 1. It is evident that T carries M to itself and PM to itself, and that

P carries M to PM. Thus the action of T on M/PM descends to an action of T/P on M/PM. The vector space M/PM is not 0 because $M \neq PM$ by unique factorization of fractional ideals. To see that M/PM has dimension at most 1, fix an element x of M that does not lie in PM. Then $xT + PM$ is a fractional ideal of K that is contained in $M + PM = M$ and contains PM and a member of M that is not in PM. Hence it equals M. Accordingly, if $y \in M$ is given, we can choose $t \in T$ such that $xt - y$ is in PM. Then $(t + P)(x + PM) = y + PM$, and T/P carries $x + PM$ onto M/PM. So M/PM is 1-dimensional over T/P, and (†) follows.

Returning to (∗∗), let $I = Q_1 \cdots Q_k$ express I as the product of nonzero prime ideals. Iterated application of (∗∗) and the First Isomorphism Theorem gives

$$|I^{-1}/T| = |I^{-1}/Q_1 \cdots Q_k I^{-1}| = |I^{-1}/Q_1 \cdots Q_{k-1} I^{-1}| N(Q_k)$$
$$= |I^{-1}/Q_1 \cdots Q_{k-2} I^{-1}| N(Q_k) N(Q_{k-1})$$
$$= \cdots = |I^{-1}/I^{-1}| \prod_{j=1}^{k} N(Q_j) = N(I).$$

This proves (∗∗) and therefore also (∗). □

One more point needs explanation. The discussion in Section IX.17 of *Basic Algebra* concerned a monic irreducible polynomial $F(X)$ in $\mathbb{Z}[X]$ and its reduction $\overline{F}(X)$ modulo p, and the interest was in the Galois group G of the splitting field $\mathbb{K}'$ of $F(X)$ over $\mathbb{Q}$. Theorem 9.64 of that book dealt with the natural homomorphism from a decomposition subgroup G_P of G onto the Galois group $\overline{G}$ of the splitting field over $\mathbb{F}_p$ of $\overline{F}(X)$, and it was asserted without proof that this homomorphism is one-one if p does not divide the discriminant of $F(X)$. The order of the kernel of the homomorphism was identified as the common ramification index of the prime ideals P' containing $(p)R'$, R' being the ring of algebraic integers in $\mathbb{K}'$. Let $\mathbb{K} = \mathbb{Q}[X]/(F(X))$. Except in the quadratic case, the field $\mathbb{K}$ typically has much lower dimension over $\mathbb{Q}$ than $\mathbb{K}'$ does. The Dedekind Discriminant Theorem relates $D_\mathbb{K}$ to ramification relative to $\mathbb{K}$, as well as $D_{\mathbb{K}'}$ to ramification relative to $\mathbb{K}'$. We know that primes not dividing the discriminant of $F(X)$ do not divide $D_\mathbb{K}$, but we need a proof that primes not dividing the discriminant of $F(X)$ do not divide $D_{\mathbb{K}'}$.

To approach this question, one needs the notion of "relative discriminant" analogous to that of "relative different" for an extension $\mathbb{K}/\mathbb{F}$ of number fields. The relative different is defined so as to be an ideal for $\mathbb{K}$, and the relative discriminant is an ideal for $\mathbb{F}$. (The field discriminant is the generator of the relative discriminant for $\mathbb{K}/\mathbb{Q}$ with the appropriate sign attached.) One proves that the behavior of the relative discriminant under successive extension is reasonable, just as it is for degree of extension, ramification indices, residue class degrees, and relative differents. These results show that if $\mathbb{Q} \subseteq \mathbb{K} \subseteq \mathbb{L}$, then the field discriminant for $\mathbb{K}$ divides the field discriminant for $\mathbb{L}$. The next step is to extend the notion of field discriminant so that it applies to commutative semisimple algebras and to show that the discriminant of a tensor product over $\mathbb{Q}$ of finitely many number fields is a certain function of the field discriminants and dimensions of the factors. Finally we return to $F(X)$ and its splitting field $\mathbb{K}'$. Let ξ be a root of $F(X)$ in $\mathbb{K}'$, and let $\sigma_1(\xi), \ldots, \sigma_n(\xi)$ be the distinct conjugates of ξ. Then $\mathbb{K}'$ is generated by the subfields $\mathbb{Q}(\xi_1), \ldots, \mathbb{Q}(\xi_n)$, and the ($\mathbb{Q}$ multilinear) multiplication map extends to an algebra homomorphism of $\mathbb{Q}(\xi_1) \otimes_\mathbb{Q} \cdots \otimes_\mathbb{Q} \mathbb{Q}(\xi_n)$ onto $\mathbb{K}'$. As the tensor product of commutative semisimple algebras in characteristic 0, this is commutative semisimple (Corollary 2.37) and is therefore a direct sum of fields (Theorem 2.2). Thus we can regard $\mathbb{K}'$ as a subfield of the tensor product of fields isomorphic to $\mathbb{Q}[X]/(F(X))$, and the discriminant of $\mathbb{K}'$ divides the discriminant of the tensor product. Putting everything together, we see that the only possible primes dividing $D_{\mathbb{K}'}$ are the primes that divide $D_\mathbb{K}$. Therefore the primes that fail to divide the discriminant of $F(X)$ do not ramify in $\mathbb{K}'$.

9. Global and Local Fields

A **global field** K is either a number field, i.e., a finite extension of $\mathbb{Q}$, or a function field in one variable over a finite field, i.e., a finite extension of some $\mathbb{F}_q(X)$, where $\mathbb{F}_q$ is a finite field.[24] An example of the latter is

$$K = \mathbb{F}_p(x)[y]/(y^2 - (x^3 - x)) \cong \mathbb{F}_p(x)\left[\sqrt{x^3 - x}\right].$$

In this section we shall develop some machinery for working with global fields. Our interest at present is in number fields, but function fields in one variable are the object of study in Chapter IX. Consequently the results will be stated for all global fields as long as all global fields can readily be treated together, and thereafter we shall specialize to number fields.

The virtue of global fields for current purposes is that their completions with respect to nontrivial absolute values are always locally compact with a nontrivial topology. In the case of number fields, we know this for archimedean absolute values by Proposition 6.27, and it follows for nonarchimedean absolute values by Corollary 6.21 and Theorem 6.26. In the function-field case as above, the completions have to be nonarchimedean by Proposition 6.14, and their absolute values have to be discrete by Corollary 6.22; then the residue class fields are always

---

[24]It will be shown in Chapter VII that a function field in one variable over a finite field is always a finite *separable* extension of $\mathbb{F}_q(Y)$ for a suitable indeterminate Y.

finite, and Theorem 6.26 shows that the completions are all locally compact with a nontrivial topology.

To study a global field K in the style of this chapter, one studies simultaneously the completions[25] of K with respect to one absolute value from each equivalence class.[26] Two completions are said to be **equivalent completions** if the absolute values on the domains of the completion maps are equivalent in the sense of Section 3. An equivalence class of completions of nontrivial absolute values is called a **place** of K. A place is called **archimedean** or **nonarchimedean** according as the corresponding absolute values are archimedean or nonarchimedean; in the archimedean case it is called **real** or **complex** according as the locally compact completed field is $\mathbb{R}$ or $\mathbb{C}$.

Because of the special hypotheses for the situation with global fields, we shall see that to each place corresponds a distinguished choice of an absolute value on K from the equivalence class, called the **normalized** absolute value in the class.[27] These normalized completions are glued together[28] in a fashion to be described in the next section to form the ring of "adeles" of K and the group of "ideles" of K. Historically ideles preceded adeles, and ideles were introduced in order to reinterpret class field theory and improve upon it; convincing motivation is therefore not readily at hand without knowledge that extends beyond this book. However, we can get some advance insight into how adeles and ideles might be useful from the first part of the classical proof of the Dirichlet Unit Theorem (Theorem 5.13) as given in Section V.5.

That proof in effect handles archimedean places in a way similar to the way that adeles handle all places. In more detail let K be a number field of degree n over $\mathbb{Q}$, and let R be its ring of algebraic integers. In Chapter V we usually regarded K as a subfield of $\mathbb{C}$, but we shall not do so here. As was observed in Section V.2, there exist exactly n field mappings of K into $\mathbb{C}$, and we denote them by $\sigma_1, \ldots, \sigma_n$. If x is in K, then the images $\sigma_1(x), \ldots, \sigma_n(x)$ are called the **conjugates** of x. Among $\sigma_1, \ldots, \sigma_n$ are r_1 real-valued mappings and r_2 complex conjugate pairs, with $r_1 + 2r_2 = n$. Let us number the mappings so that $\sigma_1, \ldots, \sigma_{r_1}$ are real-valued and so that $\sigma_{r_1+1}, \ldots, \sigma_{r_1+r_2}$ pick out one from each complex conjugate pair. Proposition 6.27 shows that the functions $x \mapsto |\sigma_1(x)|$,

---

[25] It is important not to lose sight of the fact that a "completion" is a certain kind of homomorphism of valued fields and does not consist merely of the range space.

[26] The completion of the trivial absolute value is excluded.

[27] The range of each completion is a locally compact field whose topology is not the discrete topology. Such a field is often called a **local field** in books. Examples are $\mathbb{R}$, $\mathbb{C}$, p-adic fields, and fields $\mathbb{F}_q((X))$ of formal Laurent series. One can show that there are no other locally compact fields whose topology is not discrete. The definition of "local field" in some books is arranged to exclude $\mathbb{R}$ and $\mathbb{C}$.

[28] It is tempting to think in terms of the gluing as involving just the locally compact fields, but the completion mappings play a role and that description is thus an oversimplification.

..., $x \mapsto |\sigma_{r_1+r_2}(x)|$ are a complete set of representatives for the archimedean places of K; the first r_1 are real, and the last r_2 are complex.

Just before Lemma 5.17 we introduced the mapping $\Phi : K \to \mathbb{R}^{r_1} \times \mathbb{C}^{r_2}$ given by

$$\Phi(x) = \big(\sigma_1(x), \ldots, \sigma_{r_1}(x), \sigma_{r_1+1}(x), \ldots, \sigma_{r_1+r_2}(x)\big) \quad \text{for } x \in K.$$

Lemma 5.17 observed that the image $\Phi(R)$ of R is a lattice in $\mathbb{R}^{r_1} \times \mathbb{C}^{r_2} \cong \mathbb{R}^n$. The starting point for proving the Dirichlet Unit Theorem in Section V.5 was to apply the Minkowski Lattice-Point Theorem to this lattice $\Phi(R)$. Proposition 6.27 allows us to interpret the mapping Φ as the natural embedding of K into the product of its completions at all archimedean places.

The ring of adeles of K will be a corresponding space for dealing with completions with respect to all nontrivial absolute values, archimedean and nonarchimedean.

While we have the archimedean places of the number field K at hand, let us address the question of their normalized representatives. Since the field maps from K into $\mathbb{C}$ given by $\sigma_{r_1+1,\ldots,r_1+r_2}$ are equal to the complex conjugates of $\sigma_{r_1+r_2+1}, \ldots, \sigma_n$, every member x of K has

$$N_{F/\mathbb{Q}}(x) = \prod_{j=1}^{n} \sigma_j(x) = \Big(\prod_{j=1}^{r_1} \sigma_j(x) \Big) \Big(\prod_{j=r_1+1}^{r_1+r_2} |\sigma_j(x)|^2 \Big).$$

This formula can be viewed as an archimedean analog of the formula in Corollary 6.37b. The number field $\mathbb{Q}$ has one archimedean place, and ordinary absolute value is taken as its normalized representative. We denote this representative by $|\cdot|_\infty$. With $|\cdot|$ denoting ordinary absolute value on $\mathbb{R}$ and $\mathbb{C}$, we obtain

$$|N_{K/\mathbb{Q}}(x)|_\infty = \Big(\prod_{j=1}^{r_1} |\sigma_j(x)| \Big) \Big(\prod_{j=r_1+1}^{r_1+r_2} |\sigma_j(x)|^2 \Big).$$

It is customary to use letters like v and w as indices for places. The real places are the completions $x \mapsto \sigma_j(x)$, $1 \leq j \leq r_1$, of K into $\mathbb{R}$, and the **normalized absolute value** on K for a real place is the pullback from ordinary absolute value on $\mathbb{R}$. Thus if $|\cdot|_\mathbb{R}$ denotes ordinary absolute value on $\mathbb{R}$ and if v is a real place corresponding to σ_j, then we define $|x|_v = |\sigma_j(x)|_\mathbb{R}$ for $x \in K$. The normalization to use for the complex places is motivated by the formula above. If $r_1 + 1 \leq j \leq r_1 + r_2$, then σ_j in effect contributes twice to the above formula, once from j and once from $j + r_2$, and the notion of normalized absolute value is to take this double contribution into account. Thus we write $|\cdot|_\mathbb{C}$ for the *square* of the ordinary absolute value on $\mathbb{C}$; this quantity is not really an absolute value, since the triangle inequality fails for it, but it has too many desirable features to be ignored. We define the **normalized absolute value** on K for a complex place to be the pullback from this function $|\cdot|_\mathbb{C}$ on $\mathbb{C}$ even though the result fails to satisfy the triangle inequality. Thus if v is a complex place corresponding to σ_j with $r_1 + 1 \leq j \leq r_1 + r_2$, then we define $|x|_v = |\sigma_j(x)|_\mathbb{C} = |\sigma_j(x)|^2$ for $x \in K$. With these definitions of normalized absolute values for archimedean places, the formula above for $|N_{F/\mathbb{Q}}(x)|_\infty$ can be rewritten as

$$|N_{K/\mathbb{Q}}(x)|_\infty = \Big(\prod_{j=1}^{r_1} |\sigma_j(x)|_\mathbb{R}\Big)\Big(\prod_{j=r_1+1}^{r_1+r_2} |\sigma_j(x)|_\mathbb{C}\Big) = \Big(\prod_{v \text{ real}} |x|_v\Big)\Big(\prod_{v \text{ complex}} |x|_v\Big).$$

We summarize matters in the following proposition.

Proposition 6.50. If K is a number field, then

$$|N_{F/\mathbb{Q}}(x)|_\infty = \prod_{v \text{ archimedean}} |x|_v \quad \text{for } x \in K,$$

where $|\cdot|_v$ is the pullback of $|\cdot|_\mathbb{R}$, the ordinary absolute value, for real places and where $|\cdot|_v$ is the pullback of $|\cdot|_\mathbb{C}$, the ordinary absolute value *squared*, for complex places.

At this point we could give a definition of normalized absolute value corresponding to nonarchimedean places. But we shall digress in order to motivate the definition using concepts from measure theory that may be known to some readers and not to others. These concepts play a role within the text only in the next paragraph and in Example 4 of normalized discrete absolute values below, and the reader will not miss any results or proofs by skipping this material.

The digression begins. Any locally compact group has a nonzero measure on it that is invariant under left translation,[29] and this measure is unique up to multiplication by a scalar. Let a locally compact field L be given, and let μ be an invariant measure of this kind with respect to the additive group of L. Each nonzero element c of L has the property that $\mu(cE)$ is a multiple of $\mu(E)$ that is independent of E. If we write $|c|_L$ for this multiple and put $|0|_L = 0$, then it turns out that some power $|\cdot|_L^\alpha$ with $0 < \alpha \leq 1$ is necessarily an absolute value and that this power α can be taken to be 1 in all cases except when $L = \mathbb{C}$. In the case of $\mathbb{C}$, it is easy to check that $|c|_\mathbb{C} = |c|^2$, and the triangle inequality therefore

---

[29]Although the details will not be important for us, let us be more precise: The measure is on the σ-algebra of "Baire sets" on the group—the smallest σ-algebra containing those compact sets that are intersections of countably many open sets. The measure is not the 0 measure, it is finite on all the generating compact sets, and it takes the same value on a set as it does on any left translate of the set. It is called a left **Haar measure**. For more information, see the author's *Advanced Real Analysis*, Chapter VI.

fails for $\alpha = 1$. But in all other cases, $|\cdot|_L$ is a canonical choice for an absolute value on L. Now suppose that $\psi : K \to L$ is a field map of a global field K onto a dense subfield of a locally compact field. We impose this special absolute value $|\cdot|_L$ on L. Then a necessary and sufficient condition on an absolute value $|\cdot|_K$ for $\psi : (K, |\cdot|_K) \to (L, |\cdot|_L)$ to be a completion is that $|\cdot|_K = \psi^*(|\cdot|_L)$. In other words, the pullback of the special normalization of the absolute value on the locally compact field is the natural normalization to use for the absolute value on the global field.

With the digression now over, we want to associate to each nonarchimedean place of a global field a special normalization of an absolute value. (We handled the question of normalization at archimedean places earlier in the section.) We can be a bit more general. Suppose that F is an arbitrary field with a discrete valuation v and with corresponding nontrivial absolute value given by $|x|_v = r^{-v(x)}$ for some $r > 0$. Let R be the valuation ring and $\mathfrak{p}$ the valuation ideal; $\mathfrak{p}$ is a principal ideal of the form (π) for some $\pi \in R$. Suppose that the residue class field $R/\mathfrak{p}$ is finite. Then we say that $|\cdot|_v$ is **normalized** if $|\pi|_v = |R/\mathfrak{p}|^{-1}$. This definition is independent of the choice of π.

EXAMPLES OF NORMALIZED DISCRETE ABSOLUTE VALUES.

(1) The field $\mathbb{Q}$ and the p-adic absolute value given by $|ab^{-1}p^k|_p = p^{-k}$ when a and b are integers prime to p. The valuation ring R consists of all ab^{-1} with $a \in \mathbb{Z}, b \in \mathbb{Z}$, and b prime to p. The valuation ideal consists of all such ab^{-1} with a divisible by p, and the quotient $R/\mathfrak{p}$ is isomorphic to $\mathbb{F}_p$. The element π may be taken to be p, and $|p|_p$ equals p^{-1}, which equals $|R/\mathfrak{p}|^{-1}$. Thus the p-adic absolute value on $\mathbb{Q}$ is normalized.

(2) Let K be a number field of degree n over $\mathbb{Q}$, and let T be its ring of algebraic integers. Let $\mathfrak{p}$ be a nonzero prime ideal in T, and let v be the corresponding valuation of K. Let $q = |T/\mathfrak{p}|$, and define $|x|_\mathfrak{p} = q^{-v(x)}$. Then $|\cdot|_\mathfrak{p}$ is normalized because Theorem 6.5e shows that the residue class field obtained from the valuation is isomorphic to $T/\mathfrak{p}$.

(3) Let $K = \mathbb{F}_q(X)$, fix a prime polynomial $c(X)$ in $\mathbb{F}_q[X]$, and consider the absolute value on K defined by $|a(X)b(X)^{-1}c(X)^k| = q^{-k \deg c(X)}$ whenever $a(X)$ and $b(X)$ are polynomials relatively prime to $c(X)$. This example runs completely parallel to the two previous examples, and π may be taken to be $c(X)$. The residue class field has as representatives all polynomials $h(X)$ with $\deg h(X) < \deg c(X)$ and thus has order $q^{\deg c(X)}$. This order matches $|c(X)|^{-1}$, and hence $|\cdot|$ is normalized.

(4) If F is a locally compact field whose topology comes from some nontrivial discrete absolute value with finite residue class field, then the canonical absolute value $|\cdot|_F$ described in the digression above and obtained from an invariant

measure μ on the additive group of F is normalized. To see this, let R and $\mathfrak{p}$ be the valuation ring and valuation ideal, and write $\mathfrak{p} = (\pi)$. Put $m = |R/\mathfrak{p}|$, and let $x_1, \ldots, x_m$ be representatives of the m cosets of $R/\mathfrak{p}$ in R. Then $\mu(x_j + \mathfrak{p}) = \mu(\mathfrak{p})$ for $1 \leq j \leq m$ by translation invariance of μ, and hence $\mu(R) = \sum_{j=1}^m \mu(x_j + \mathfrak{p}) = m\mu(\mathfrak{p})$. Substituting and using the definition of $|\cdot|_F$ gives $\mu(\mathfrak{p}) = \mu(\pi R) = |\pi|_F \mu(R) = |\pi|_F m\mu(\mathfrak{p})$. The number $\mu(\mathfrak{p})$ is positive, since $\mathfrak{p}$ is a nonempty open subset of F, and we can cancel to get $|\pi|_F m = 1$. Thus $|\pi|_F = |R/\mathfrak{p}|^{-1}$, and $|\cdot|_F$ is normalized.

Theorem 6.51 (Artin product formula). If F is a number field and if normalized absolute values are used, then

$$\prod_v |x|_v = 1 \qquad \text{for all nonzero } x \in F,$$

the product being taken over all places v. In this product, only finitely many of the factors can be different from 1.

REMARKS. A version of this theorem is valid for function fields in one variable. As Corollary 6.22 permits, one can state this analogous theorem in terms of discrete valuations that are trivial on the base field, and absolute values need play no role. The precise statement and proof appear in Chapter IX. Corollary 6.9 in the present chapter is a special case.

PROOF. First we prove the result for $\mathbb{Q}$. Let a rational $y = \pm p_1^{k_1} \cdots p_r^{k_r}$ be given; here $p_1, \ldots, p_r$ are distinct primes. The product $\prod_v |y|_v$ is taken over all places, hence over all primes and the one archimedean place ∞. For this $y \in \mathbb{Q}$, we have $|y|_{p_j} = p_j^{-k_j}$ for $1 \leq j \leq r$ and $|y|_{p'} = 1$ for all other primes p'. So $\prod_{p \text{ prime}} |y|_p = p_1^{-k_1} \cdots p_r^{-k_r}$. Since $|y|_\infty = p_1^{k_1} \cdots p_r^{k_r}$, we obtain $\prod_{\text{all } v} |y|_v = 1$.

Let R be the ring of algebraic integers in F. Given x in F, factor the fractional ideal xR. The nonarchimedean places correspond to the nonzero prime ideals in R, and $|x|_v$ is 1 except for the v's corresponding to those prime ideals in the factorization. There are only finitely many of these. Since also there are only finitely many archimedean places, we see that $|x|_v = 1$ for all but finitely many v.

Let us consider the nonarchimedean places separately from the archimedean ones. The nonarchimedean places correspond to nonzero prime ideals $\wp$, and we group these according to the prime number p such that $\wp \cap \mathbb{Z} = p\mathbb{Z}$, writing $\wp \mid p\mathbb{Z}$ for this correspondence. For fixed p and for each $\wp$ with $\wp \mid p\mathbb{Z}$, let $x_\wp$ be the image of x under the local embedding in $F_\wp$. Corollary 6.37b gives $N_{F/\mathbb{Q}}(x) = \prod_{\wp \mid p\mathbb{Z}} N_{F_\wp/\mathbb{Q}_p}(x_\wp)$. Theorem 6.33 shows that $|x_\wp|_{F_\wp}$ is a power of $|N_{F_\wp/\mathbb{Q}_p}(x_\wp)|_{\mathbb{Q}_p}$. To determine the power, we observe from Example 2 that the canonical absolute values on $\mathbb{Q}_p$ and $F_\wp$ are normalized, and we specialize $|x_\wp|_{F_\wp}$ and $|N_{F_\wp/\mathbb{Q}_p}(x_\wp)|_{\mathbb{Q}_p}$ to $x_\wp$ in $\mathbb{Q}_p$. Making the comparison, we find that $|N_{F_\wp/\mathbb{Q}_p}(x_\wp)|_{\mathbb{Q}_p} = |x_\wp|_{F_\wp}$. We know that each local embedding respects absolute values; since Theorems 6.5e and 6.26e together show that the residue class fields of $F_\wp$ and $\mathbb{Q}_p$ have orders $|R/\wp|$ and $|\mathbb{Z}/p\mathbb{Z}|$, it follows that $|x_\wp|_{F_\wp} = |x|_\wp$. Therefore

$$|N_{F/\mathbb{Q}}(x)|_p = |N_{F/\mathbb{Q}}(x)|_{\mathbb{Q}_p} = \prod_{\wp | p\mathbb{Z}} |N_{F_\wp/\mathbb{Q}_p}(x_\wp)|_{\mathbb{Q}_p}$$
$$= \prod_{\wp | p\mathbb{Z}} |x_\wp|_{F_\wp} = \prod_{\wp | p\mathbb{Z}} |x|_\wp. \qquad (*)$$

For the finitely many archimedean places, Proposition 6.50 gives us the formula

$$|N_{F/\mathbb{Q}}(x)|_\infty = \prod_{v \text{ archimedean}} |x|_v, \qquad (**)$$

where $|\cdot|_\infty$ is the ordinary absolute value on $\mathbb{Q}$. Multiplying $(*)$ and $(**)$ and using the known identity $\prod_v |y|_v = 1$ for the element $y = N_{F/\mathbb{Q}}(x)$ of $\mathbb{Q}$, we obtain the theorem. □

10. Adeles and Ideles

In this section we do the gluing that creates the adeles and the ideles out of the places of a global field. We begin with a topological construction, and then we superimpose the algebraic structure. The general constructions and the two main theorems will be valid for all global fields, but we shall discuss proofs of the theorems only for number fields.

Suppose that $\{X_i \mid i \in I\}$ is a nonempty family of locally compact Hausdorff spaces. Assume that for all but finitely many $i \in I$ we are given a compact open subset Z_i of X_i. The **restricted direct product** of the X_i's relative to the Z_i's is the subset

$$\prod_{i \in I}{}' X_i \subseteq \prod_{i \in I} X_i$$

defined by

$$(x_i)_{i \in I} \in \prod_{i \in I}{}' X_i \qquad \text{if and only if} \qquad x_i \in Z_i \text{ for all but finitely many } i.$$

The restricted direct product is topologized as follows. Suppose that $S \subseteq I$ is a finite subset and that Z_i is defined for $i \notin S$. Put

$$X(S) = \prod_{i \in S} X_i \times \prod_{i \notin S} Z_i.$$

In their respective product topologies the first factor is locally compact, and the second factor is compact. Certainly $X(S)$ is a subset of the restricted direct product, and evidently the restricted direct product is the union of the subsets $X(S)$ over all finite subsets S for which Z_i is defined when $i \notin S$. We topologize $\prod'_{i \in I} X_i$ by insisting that each $X(S)$ be an open subset.[30] The resulting topology is locally compact Hausdorff. In fact, any two members of $\prod'_{i \in I} X_i$ lie in a common $X(S)$, and the open sets that separate them in $X(S)$ separate them in $\prod'_{i \in I} X_i$. Also, any $(x_i)_{i \in I}$ is in some $X(S)$, which is locally compact, and a compact neighborhood within $X(S)$ will be a compact neighborhood in $\prod'_{i \in I} X_i$.

Now we superimpose the algebraic structure. Let K be a global field. To each place v of K, we have associated a normalized absolute value $|\cdot|_v$ on K and a completion $\iota_v : (K, |\cdot|_v) \to (K_v, |\cdot|_{K_v})$. Each of the complete valued fields K_v is locally compact. Except at the finitely many archimedean places, which occur only in the number-field case, $|\cdot|_{K_v}$ arises from a discrete valuation. We take R_v to be the corresponding valuation ring, i.e., $R_v = \{x \in K_v \mid |x|_v \leq 1\}$. This is a compact open additive subgroup of K_v. Thus we can form a restricted direct product in which the index set I is the set of places of K, the vth locally compact Hausdorff space is K_v, and the vth compact open subset is R_v. This restricted direct product carries the structure of a commutative ring with identity, with its addition and multiplication defined in coordinate-by-coordinate fashion, and the operations are continuous. Thus we obtain a topological ring, known as the ring of **adeles** of K and denoted by $\mathbb{A}_K$ or simply by $\mathbb{A}$ when no ambiguity is possible.

If for each $x \in K_{v_0}$, we send x into the tuple $(a_v)_v$ that has $a_{v_0} = x$ and $a_v = 0$ for $v \neq v_0$, then the result is a one-one continuous ring homomorphism of K_v into $\mathbb{A}$. This homomorphism of course does not send the multiplicative identity of K_v to the multiplicative identity of $\mathbb{A}$.

The completion mappings $\iota_v : K \to K_v$ embed K into each K_v, and we can form a corresponding diagonal map $\iota : K \to \prod_v K_v$ into the full product of K_v's by defining $\iota(x) = (\iota_v(x))_v$. Actually, we shall check for $x \neq 0$ that only finitely many places have $|\iota_v(x)|_v = |x|_v$ unequal to 1, and therefore the image of the diagonal map is in the adeles. Thus we have a diagonal ring homomorphism

$$\iota : K \to \mathbb{A} \quad \text{given by} \quad \iota(x) = (\iota_v(x))_v \text{ for } x \in K.$$

The fact that in the number-field case, $|x|_v$ is unequal to 1 for only finitely many places appears as part of Theorem 6.51. For the function-field case, the field K is a finite separable extension of some field $\mathbb{F}_q(X)$, and all but finitely many places come from nonzero prime ideals in the integral closure R of $\mathbb{F}_q[X]$ in K. At the

---

[30]In other words, a set in $\prod'_{i \in I} X_i$ is open if and only if its intersection with each $X(S)$ is open in $X(S)$.

unexceptional such places the value of $|x|_v$ comes by treating xR as a fractional ideal and factoring it; only finitely many ideals are involved in the factorization, and only those among all the unexceptional places can have $|x|_v \neq 1$. The main structural theorem about the adeles is as follows.

Theorem 6.52. If K is a global field, then the image of K in the adeles $\mathbb{A}$ under the diagonal mapping $\iota : K \to \mathbb{A}$ is discrete, and the quotient $\mathbb{A}/\iota(K)$ of additive groups is compact.

For a number field the compactness in Theorem 6.52 encodes Lemma 5.17 and the Strong Approximation Theorem. The proof of the theorem is not hard, and we return to it in a moment. In the current discussion Theorem 6.52 is not something to appreciate for its own consequences but instead is a prototype for a corresponding theorem about "ideles" that encodes for number fields the finiteness of the class number and the Dirichlet Unit Theorem.

The construction of the "ideles" of K proceeds similarly to the construction of the adeles. Again we use a restricted direct product, with the set of places as index set. The locally compact Hausdorff space associated to the place v is the multiplicative group $K_v^\times$. For v nonarchimedean, we again let R_v be the valuation ring in K_v, and take the compact open subset of $K_v^\times$ to be the group $R_v^\times$ of units in R_v, i.e., $R_v^\times = \{x \in K_v \mid |x|_v = 1\}$. The group of ideles is the restricted direct product of the groups $K_v^\times$ relative to the compact subgroups $R_v^\times$. The result is a locally compact abelian group, known as the group of **ideles** of K and denoted by $\mathbb{I}_K$ or simply by $\mathbb{I}$.

Warning: As a set, $\mathbb{I}$ coincides with the group of units $\mathbb{A}^\times$. However, the topologies do not match. The topology for $\mathbb{I}$ is finer than the relative topology on $\mathbb{A}^\times$. See Problems 7–8 at the end of the chapter.

If for each $x \in K_{v_0}$, we send x into the tuple $(a_v)_v$ that has $a_{v_0} = x$ and $a_v = 1$ for $v \neq v_0$, then the result is a one-one continuous group homomorphism of $K_v^\times$ into $\mathbb{I}$. As with the ideles we also have a diagonal mapping $\iota : K^\times \to \mathbb{I}$ given by $\iota(x) = (\iota_v(x))_v$; the image is contained in I, since for a nonzero $x \in K$, $|x|_v$ can be unequal to 1 for only finitely many v.

The Artin product formula (Theorem 6.51) and the corresponding result for function fields in one variable over a finite field put a constraint on the image. We define the **absolute value** $|(a_v)_v|$ of an idele $(a_v)_v$ to be the product of the absolute values of the components: $|(a_v)_v| = \prod_v |a_v|_v$. This is well defined because only finitely many factors are allowed to be different from 1. If $\mathbb{I}^1$ denotes the group of ideles of absolute value 1, then $\mathbb{I}^1$ is a closed subgroup of $\mathbb{I}$. The Artin product formula and its function-field analog imply that the image of the diagonal mapping is contained in $\mathbb{I}^1$. The main structural theorem about the ideles is as follows.

Theorem 6.53. If K is a global field, then the image of $K^\times$ in the subgroup $\mathbb{I}^1$ of the ideles $\mathbb{I}$ under the diagonal mapping $\iota : K^\times \to \mathbb{I}$ is discrete, and the quotient group $\mathbb{I}^1/\iota(K^\times)$ is compact.

From now on, we suppose that the global field K is a number field. Let S_∞ be the set of archimedean places. We begin by supplying direct proofs of the discreteness in Theorems 6.52 and 6.53 and of the compactness of the quotient in Theorem 6.52. After some additional discussion we return to prove the compactness of the quotient in Theorem 6.53.

PROOF OF DISCRETENESS OF $\iota(K)$ IN THEOREM 6.52. It is enough to produce a neighborhood U of 0 in $\mathbb{A}$ such that $U \cap \iota(K) = \{0\}$. The set U of all $(x_v)_v \in \mathbb{A}$ such that $|x_v|_v < 1$ for all archimedean places and $|x_v|_v \leq 1$ for all nonarchimedean places is an open product set in $\mathbb{A}(S_\infty)$ and hence is an open neighborhood of 0 in $\mathbb{A}$. Since Theorem 6.51 shows that $\prod_v |\iota_v(y)|_v = 1$ for all $y \neq 0$ in K and since $\prod_v |x_v|_v < 1$ for all $(x_v)_v$ in U, $U \cap \iota(K) = \{0\}$. □

PROOF OF DISCRETENESS OF $\iota(K^\times)$ IN THEOREM 6.53. The set U of all $(x_v)_v \in \mathbb{I}$ such that $|x_v - 1|_v < 1$ for all archimedean places and $|x_v - 1|_v \leq 1$ for all nonarchimedean places is an open product set in $\mathbb{I}(S_\infty)$ and hence is an open neighborhood of 1 in $\mathbb{I}$. If $(x_v)_v = \iota(y)$ with $y \in K^\times$ and $y \neq 1$, then $x_v - 1 = \iota_v(y - 1)$ with $y - 1 \neq 0$, and Theorem 6.51 shows that $\prod_v |\iota_v(y) - 1|_v = \prod_v |\iota_v(y-1)|_v = 1$. The members $(x_v)_v$ of U all have $\prod_v |x_v - 1|_v < 1$, and thus $U \cap \iota(K^\times) = \{1\}$. □

PROOF OF COMPACTNESS OF $\mathbb{A}/\iota(K)$ IN THEOREM 6.52. We begin by observing that
$$\mathbb{A} = \iota(K) + \mathbb{A}(S_\infty), \qquad (*)$$
i.e., that the set of sums of a member of $\iota(K)$ and a member of $\mathbb{A}(S_\infty)$ exhausts $\mathbb{A}$. In fact, given $(x_v)_v$ in $\mathbb{A}$, we let $v_1, \ldots, v_r$ be the finitely many nonarchimedean places for which $|x_{v_j}|_{v_j} > 1$. The Strong Approximation Theorem (Theorem 6.44) applied to the elements $x_{v_1}, \ldots, x_{v_r}$ produces a member y of K such that $|\iota_{v_j}(y) - x_{v_j}|_{v_j} < 1$ for $1 \leq j \leq r$ and such that $|\iota_v(y)|_v \leq 1$ for all other nonarchimedean places v. Consequently $|\iota_v(y) - x_v|_v \leq 1$ for all nonarchimedean v. This inequality means exactly that $(x_v)_v - \iota(y)$ is in $\mathbb{A}(S_\infty)$. Hence
$$x = \iota(y) + ((x_v)_v - \iota(y))$$
is the required decomposition, and $(*)$ is proved.

In addition, we have
$$\iota(R) = \iota(K) \cap \mathbb{A}(S_\infty). \qquad (**)$$

In fact, the inclusion $\subseteq$ is clear. For the inclusion $\supseteq$, let y be a member of K such that $\iota(y)$ is in $\mathbb{A}(S_\infty)$. Then $|\iota_v(y)|_v \leq 1$ for all nonarchimedean v, and it follows that y is in R.

To prove the compactness, we use the identity $(M+N)/M \cong N/(M \cap N)$ given by the Second Isomorphism Theorem in the category of locally compact abelian groups, taking $M = \iota(K)$ and $N = \mathbb{A}(S_\infty)$. Then (*) shows that $M + N = \mathbb{A}$, and (**) shows that $M \cap N = \iota(R)$. Hence

$$\mathbb{A}/\iota(K) \cong \mathbb{A}(S_\infty)/\iota(R). \tag{$\dagger$}$$

Let us write $\mathbb{A}(S_\infty) = \Omega \times \Delta$, where $\Omega = \mathbb{R}^{r_1} \times \mathbb{C}^{r_2} = \prod_{v \text{ archimedean}} K_v$ and $\Delta = \prod_{v \text{ nonarchimedean}} R_v$. The mapping $\Phi : K \to \Omega$ defined near the beginning of Section 9 has the property that

$$\iota(R) + (\{0\} \times \Delta) = \Phi(R) \times \Delta.$$

From this equality we obtain

$$\mathbb{A}(S_\infty)/(\iota(R) + (\{0\} \times \Delta)) \cong (\Omega \times \Delta)/(\Phi(R) \times \Delta) \cong \Omega/\Phi(R),$$

and Lemma 5.17 shows that this is compact. Since $(\{0\} \times \Delta) \cap \iota(R) = \{0\}$, application of the First Isomorphism Theorem and then the Second Isomorphism Theorem gives

$$\big(\mathbb{A}(S_\infty)/\iota(R)\big)\big/\big(\mathbb{A}(S_\infty)/(\iota(R) + (\{0\} \times \Delta))\big) \cong \big(\iota(R) + (\{0\} \times \Delta)\big)/\iota(R)$$
$$\cong (\{0\} \times \Delta)/\big((\{0\} \times \Delta) \cap \iota(R)\big)$$
$$= \{0\} \times \Delta,$$

and this is compact also. So the closed subgroup $\mathbb{A}(S_\infty)/(\iota(R) + (\{0\} \times \Delta))$ of $\mathbb{A}(S_\infty)/\iota(R)$ and the quotient by this subgroup are both exhibited as compact, and it follows that $\mathbb{A}(S_\infty)/\iota(R)$ is compact. Application of ($\dagger$) shows that $\mathbb{A}/\iota(K)$ is compact. $\square$

A first approach to proving the compactness of $\mathbb{I}^1/\iota(K^\times)$ in Theorem 6.53 is to pursue an analogy with the above proof for $\mathbb{A}/\iota(K)$ by showing that multiplicative analogs of (*) and (**) from that proof are valid here:

$$\mathbb{I} \stackrel{?}{=} \iota(K^\times) \mathbb{I}(S_\infty),$$
$$\iota(R^\times) = \iota(K^\times) \cap \mathbb{I}(S_\infty).$$

The second of these formulas is fine and is easily proved: The inclusion $\iota(R^\times) \subseteq \iota(K^\times) \cap \mathbb{I}(S_\infty)$ is clear. For the inclusion $\iota(R^\times) \supseteq \iota(K^\times) \cap \mathbb{I}(S_\infty)$, let y be a member of $K^\times$ such that $\iota(y)$ is in $\mathbb{I}(S_\infty)$. Then $|\iota_v(y)|_v = 1$ for all nonarchimedean v, and it follows that y and y^{-1} are in R, hence that y is in $R^\times$.

The difficulty is that an equality $\mathbb{I} \stackrel{?}{=} \iota(K^\times)\mathbb{I}(S_\infty)$ holds if and only if the ring R of algebraic integers in K is a principal ideal domain. Let us elaborate on this point, since we will be led by it to the relationship between ideles and the ideal class group that makes ideles useful.

Let us enumerate the nonzero prime ideals of R as $P_1, P_2, \ldots$ in some fashion. As was mentioned in Section 2, each nonzero fractional ideal I in K has a finite unique factorization of the form $I = P_{i_1}^{k_{i_1}} \cdots P_{i_m}^{k_{i_m}}$, where $k_{i_1}, \ldots, k_{i_m}$ are integers. The mapping that carries I to the tuple $(a_j)_{j \geq 1}$ with $a_j = k_{i_l}$ when $j = i_l$ and $a_j = 0$ when j is not in $\{k_{i_1}, \ldots, k_{i_m}\}$ is a group isomorphism Ψ from the group $\mathcal{I}$ of fractional ideals onto a free abelian group $\bigoplus_{j=1}^\infty \mathbb{Z}$ of countably infinite rank. Some of these fractional ideals are of the form xR for some $x \in K^\times$, and they are the principal fractional ideals. They form a subgroup $\mathcal{P}$ of $\mathcal{I}$ that is isomorphic to $K^\times$, and the quotient $\mathcal{I}/\mathcal{P}$ is isomorphic to the ideal class group of K, as was shown at the end of Section 2. Theorem 5.19 says that the group $\mathcal{I}/\mathcal{P}$ is a finite group; its order is the **class number** of K.

Meanwhile, suppose that $(x_v)_v$ is a member of the group $\mathbb{I}$ of ideles. To each nonarchimedean place v, Corollary 6.8 associates a unique nonzero prime ideal, which we write as $P_{i(v)}$ for a function $i(\cdot)$. If $q_v = |R/P_{i(v)}|$, then the relationship between the valuation $\mathrm{ord}_v(\cdot)$ and the normalized absolute value associated to $P_{i(v)}$ is $|x_v|_v = q_v^{-\mathrm{ord}_v(x_v)}$. Since $(x_v)_v$ is an idele, there are only finitely many nonarchimedean v's for which $\mathrm{ord}_v(x_v)$ is not 0. We can therefore map $(x_v)_v$ into the tuple of integers $(\mathrm{ord}_v(x_v))_v$ and compose with Ψ^{-1} to obtain a homomorphism of the group $\mathbb{I}$ into the group $\mathcal{I}$ of fractional ideals. In more detail, the mapping from $\mathbb{I}$ to $\bigoplus_{j=1}^\infty \mathbb{Z}$ is given by $(x_v)_v \mapsto (a_j)_{j \geq 1}$ with $a_{i(v)} = \mathrm{ord}_v(x_v)$, and then Ψ^{-1} interprets this sequence of integers as the exponents of the appropriate prime ideals. Since any association of members of $K_v^\times$ at finitely many nonarchimedean places can be extended to an idele by making the idele be 1 at the remaining places, this homomorphism of $\mathbb{I}$ into $\mathcal{I}$ is onto $\mathcal{I}$.

Now suppose that the given idele $(x_v)_v$ is of form $\iota(x)$ for some x in $K^\times$. Then the procedure for mapping this idele to a product of powers of the nonzero prime ideals of R is the same as the procedure for decomposing the fractional ideal xR as a product of powers of nonzero prime ideals of R. Consequently our homomorphism descends to a homomorphism

$$\mathbb{I}/\iota(K^\times) \longrightarrow \mathcal{I}/\mathcal{P}$$

of the **idele class group** $\mathbb{I}/\iota(K^\times)$ onto the (finite) ideal class group $\mathcal{I}/\mathcal{P}$. This is the fundamental fact about the ideles; the displayed homomorphism in effect says that the idele class group refines the information in the ideal class group. The subject of class field theory shows that this refined information is useful.

Under the homomorphism of $\mathbb{I}$ onto $\mathcal{I}$, the kernel consists exactly of $\mathbb{I}(S_\infty)$, the ideles whose components at each nonarchimedean place v are in $R_v^\times$. Thus

$\mathbb{I}/\mathbb{I}(S_\infty) \to \mathcal{I}$ is an isomorphism. Taking into account the effect on $\iota(K^\times)$, we obtain an isomorphism

$$\mathbb{I}/\big(\iota(K^\times)\,\mathbb{I}(S_\infty)\big) \cong \mathcal{I}/\mathcal{P}.$$

Returning to our hoped-for equality $\mathbb{I} \stackrel{?}{=} \iota(K^\times)\,\mathbb{I}(S_\infty)$ and comparing with the displayed isomorphism, we see that $\mathbb{I}$ equals $\iota(K^\times)\,\mathbb{I}(S_\infty)$ if and only if $\mathcal{I} = \mathcal{P}$. Equality $\mathcal{I} = \mathcal{P}$ holds if and only if every fractional ideal of K is principal, if and only if every ordinary ideal of R is principal.

Thus we see why a direct analog of the proof of Theorem 6.52 does not work for Theorem 6.53. But at the same time we obtain information about how to give a correct proof. We saw that factoring $\mathbb{I}/\iota(K^\times)$ by $\mathbb{I}(S_\infty)$ leads to the finite group $\mathcal{I}/\mathcal{P}$. We shall see that if we factor $\mathbb{I}/\iota(K^\times)$ by a suitably larger group $\mathbb{I}(S)$ with S still finite, then the quotient is the trivial group. An indication of this fact was in Problems 19–23 at the end of Chapter V, which showed that if we localize R at a large enough finite set of nonzero prime ideals, then the result is a principal ideal domain. In adelic/idelic terms the corresponding procedure is to enlarge S_∞ to a suitable finite set S containing S_∞ and to replace $\mathbb{I}(S_\infty)$ by $\mathbb{I}(S)$; this enlargement has the effect of replacing $R_v^\times$ by $K_v^\times$ at finitely many places v in considering what happens to ideals, and this is exactly what the localization in those problems accomplishes. Thus for a suitable finite set S containing S_∞, we will have an isomorphism

$$\mathbb{I}/\big(\iota(K^\times)\,\mathbb{I}(S)\big) \cong \{1\};$$

in other words,

$$\mathbb{I} = \iota(K^\times)\,\mathbb{I}(S)$$

for a suitable finite set S containing S_∞.

One final remark is needed, and then we are ready to carry out the proof of the compactness of $\mathbb{I}^1/\iota(K^\times)$. The remark is that we always have at least one archimedean place, and adjusting an idele suitably at one archimedean place can change it from being in $\mathbb{I}$ to being in the subgroup $\mathbb{I}^1$ of ideles for which $\prod_v |x_v|_v = 1$. The members of $\iota(K^\times)$ are already in this subgroup, but the members of $\mathbb{I}(S)$ need not be. Thus we replace $\mathbb{I}(S)$ by $\mathbb{I}(S) \cap \mathbb{I}^1 = \mathbb{I}^1(S)$, and the above equality becomes

$$\mathbb{I}^1 = \iota(K^\times)\,\mathbb{I}^1(S)$$

for a suitable finite set S.

PROOF OF COMPACTNESS OF $\mathbb{I}^1/\iota(K^\times)$ IN THEOREM 6.53. Let S be as above. Since $\mathbb{I}^1 = \iota(K^\times)\,\mathbb{I}^1(S)$, the Second Isomorphism Theorem gives

$$\mathbb{I}^1/\iota(K^\times) \cong \mathbb{I}^1(S)/(\iota(K^\times)\,\mathbb{I}^1(S)). \qquad (*)$$

We shall prove that the right side is compact.

Let T be the complement of S_∞ in S, and define

$$\Omega_1^\times = \prod_{v \in S_\infty} K_v^\times, \quad \Omega_2^\times = \prod_{v \in T} K_v^\times, \quad \Delta_2^\times = \prod_{v \in T} R_v^\times, \quad \Delta_3^\times = \prod_{v \notin S} R_v^\times.$$

If E is any subset of $\mathbb{I}(S)$, E^1 will denote the set of members of E of total absolute value 1. Thus for example, $(\Omega_1^\times)^1$ is the set of tuples $(x_v)_{v \in S_\infty}$ with $\prod_{v \in S_\infty} |x_v|_v = 1$.

Let $\Phi : K^\times \to \Omega_1^\times$ be the mapping given in Section 9. Each member u of the group of units $R^\times$ has the property that $|u|_v = 1$ for every nonarchimedean place v. Then it follows from the Artin product formula (Theorem 6.51) that Φ carries $R^\times$ into $(\Omega_1^\times)^1$. One of the two key ingredients in the proof of Theorem 6.51 is the observation that

$$(\Omega_1^\times)^1 / \Phi(R^\times) \quad \text{is compact.} \tag{$**$}$$

In fact, $\Omega_1^\times$ is a product of r_1 copies of $\mathbb{R}^\times$ and r_2 copies of $\mathbb{C}^\times$. The function $\mathrm{Log} : \Omega_1^\times \to \mathbb{R}^{r_1+r_2}$ given by

$$\mathrm{Log}(x_1, \ldots, x_r, x_{r_1+1}, \ldots, x_{r_1+r_2+1})$$
$$= (\log|x_1|_\mathbb{R}, \ldots, \log|x_{r_1}|_\mathbb{R}, \log|x_{r_1+1}|_\mathbb{C}, \ldots, \log|x_{r_1+r_2}|_\mathbb{C})$$

is a continuous homomorphism of $\Omega_1^\times$ onto $\mathbb{R}^{r_1+r_2}$, and its kernel is compact, being the product of r_1 two-element groups and r_2 circles. The image of $(\Omega_1^\times)^1$ is a hyperplane, and the proof of the Dirichlet Unit Theorem (Theorem 5.13) shows that $\mathrm{Log}(\Omega_1^\times)^1 / \mathrm{Log}\Phi(R^\times)$ is compact. Then $(**)$ follows.

The other key ingredient is the finiteness of the class number of K, which was proved as Theorem 5.19. Let h be this class number. For each v in $T = (S_\infty)^c$, let P_v be the corresponding nonzero prime ideal in R. The ideal P_v^h in R is principal, and we let π_v be a generator. This element has the properties that $K_v^\times / \iota_v(\pi_v)^\mathbb{Z} R_v$ is compact and that $|\iota_{v'}(\pi_v)|_{v'} = |\pi_v|_{v'} = 1$ for all nonarchimedean v' with $v' \neq v$. Let

$$\Sigma_2 = \prod_{v \in T} \iota_v(\pi_v)^\mathbb{Z} R_v;$$

this is a subgroup between Δ_2 and Ω_2 such that Ω_2 / Σ_2 is compact. Let Π be the subgroup of $K^\times$ given by $\Pi = \prod_{v \in T} \pi_v^\mathbb{Z}$.

The group $\iota(\Pi)$ is certainly a subgroup of $\iota(K^\times)$, and the fact that $|\pi_v|_{v'} = 1$ for $v' \notin S$ implies that $\iota(\Pi)$ is contained in $\mathbb{I}^1(S)$. Each member of $\iota(R^\times)$ has all nonarchimedean absolute values equal to 1, and consequently we have an inclusion $\iota(R^\times)\iota(\Pi) \subseteq \iota(K^\times)\mathbb{I}^1(S)$. In view of $(*)$, $\mathbb{I}^1(S)/(\iota(K^\times)\mathbb{I}^1(S))$ is a homomorphic image of

$$\mathbb{I}^1(S) \big/ \big(\iota(R^\times)\iota(\Pi)(\{1\} \times \Delta_2^\times \times \Delta_3^\times)\big), \tag{$\dagger$}$$

and it is therefore enough to prove that (†) is compact.

The members of $\iota(R^\times)$ have all nonarchimedean absolute values equal to 1 and consequently

$$\iota(R^\times)(\{1\} \times \Delta_2^\times \times \Delta_3^\times) = \Phi(R^\times) \times \Delta_2^\times \times \Delta_3^\times.$$

Therefore the quotient of (†) by

$$\mathbb{I}^1(S)\big/\big(\iota(\Pi)((\Omega_1^\times)^1 \times \Delta_2^\times \times \Delta_3^\times)\big) \tag{††}$$

is isomorphic to

$$\mathbb{I}^1(S)\big/\big(\iota(\Pi)(\Phi(R^\times) \times \Delta_2^\times \times \Delta_3^\times)\big)\Big/\mathbb{I}^1(S)\big/\big(\iota(\Pi)((\Omega_1^\times)^1 \times \Delta_2^\times \times \Delta_3^\times)\big),$$

which in turn is isomorphic to

$$\big(\iota(\Pi)((\Omega_1^\times)^1 \times \Delta_2^\times \times \Delta_3^\times)\big)\big/\big(\iota(\Pi)(\Phi(R^\times) \times \Delta_2^\times \times \Delta_3^\times)\big),$$

which is a homomorphic image of

$$((\Omega_1^\times)^1 \times \Delta_2^\times \times \Delta_3^\times)/(\Phi(R^\times) \times \Delta_2^\times \times \Delta_3^\times) \cong (\Omega_1^\times)^1/\Phi(R^\times).$$

The right side is compact by (∗∗), and therefore it is enough to prove that (††) is compact.

Let us check that

$$\iota(\Pi)((\Omega_1^\times)^1 \times \Delta_2^\times \times \Delta_3^\times) = (\Omega_1^\times \times \Sigma_2 \times \Delta_3)^1. \tag{‡}$$

The inclusion ⊆ is immediate. Thus suppose that $((\omega_v)_{v \in S_\infty}, (\sigma_v)_{v \in T}, (\delta_v)_{v \notin S})$ lies in the right side of (‡). Since $(\sigma_v)_{v \in T}$ lies in Σ_2, there exists an element π_0 in Π such that $r_v = \iota_v(\pi_0)^{-1}\sigma_v$ lies in R_v for all $v \in T$. Define $(\omega'_v)_{v \in S_\infty}$ in $\Omega_1^\times$ by $\omega'_v = \iota_v(\pi_0)^{-1}\omega_v$. For a suitable $(\delta'_v)_{v \notin S}$, we then have $\iota(\pi_0)((\omega'_v)_{v \in S_\infty}, (r_v)_{v \in T}, (\delta'_v)_{v \notin S}) = ((\omega_v)_{v \in S_\infty}, (\sigma_v)_{v \in T}, (\delta_v)_{v \notin S})$, and (‡) is proved.

Combining (‡) and (††), we see that it is enough to prove that

$$\mathbb{I}^1(S)\big/(\Omega_1^\times \times \Sigma_2 \times \Delta_3)^1 \tag{‡‡}$$

is compact. The inclusion of $\mathbb{I}^1(S)$ into $\mathbb{I}(S)$ induces a homomorphism

$$\mathbb{I}^1(S)\big/(\Omega_1^\times \times \Sigma_2 \times \Delta_3)^1 \to \mathbb{I}(S)\big/(\Omega_1^\times \times \Sigma_2 \times \Delta_3) \tag{§}$$

that is evidently one-one. But it is also onto because if v_0 is an archimedean place and if $(x_v)_v$ is given in $\mathbb{I}(S)$, then we can adjust (x_{v_0}) in such a way that the replacement $(x_v)_v$ has absolute value 1. The adjustment is by a member of $\Omega_1^\times \times \{1\} \times \{1\}$, and thus (§) is onto. The right side of (§) is

$$(\Omega_1^\times \times \Omega_2 \times \Delta_3)/(\Omega_1^\times \times \Sigma_2 \times \Delta_3) \cong \Omega_2/\Sigma_2,$$

and we have arranged that this is compact. Consequently (‡‡) is compact, and the proof is complete. □

11. Problems

1. If F is a complete field with a nonarchimedean absolute value and if $\sum_{n=1}^{\infty} a_n$ is an infinite series whose terms a_n are in F, prove that the series converges in F if and only if $\lim_n a_n = 0$.

2. Let the 2-adic absolute value be imposed on $\mathbb{Q}$. Theorem 6.5 shows that $\mathbb{Z}$ is dense in the subring of $\mathbb{Q}$ consisting of all rationals with odd denominator.
   (a) Find a sequence of integers converging in this metric to $\frac{1}{3}$.
   (b) Generalize the result of (a) by finding an explicit sequence of integers converging in this metric to any given rational ab^{-1}, where a and b are nonzero integers with b odd.

3. For the Dedekind domain $R = \mathbb{Z}$ and its field of fractions $K = \mathbb{Q}$, the ring of units $R^\times$ is just $\{\pm 1\}$, and the set of archimedean places is just $S_\infty = \{\infty\}$. The formula $\iota(R^\times) = \iota(K^\times) \cap \mathbb{I}(S_\infty)$ of Section 10 therefore becomes $\{\iota(\pm 1)\} = \iota(\mathbb{Q}^\times) \cap (\mathbb{R}^\times \times \prod_p \mathbb{Z}_p^\times)$.
   (a) Verify this formula directly.
   (b) Since $\mathbb{Z}$ is a principal ideal domain, the theory of Section 10 and the above remarks show that $\mathbb{I} = \iota(\mathbb{Q}^\times)(\mathbb{R}^\times \times \prod_p \mathbb{Z}_p^\times)$. Prove this formula by an explicit construction whose only allowable choice, in view of (a), is a certain sign.

4. Let R be the Dedekind domain $\mathbb{Z}[\sqrt{-5}]$.
   (a) Verify for each choice of sign that the ideals $(1 \pm \sqrt{-5}, 3)$ and $(1 \pm \sqrt{-5}, 2)$ are prime and that $(1 + \sqrt{-5}, 2) = (1 - \sqrt{-5}, 2)$.
   (b) Find the prime factorizations of the principal ideals $(1 + \sqrt{-5})$ and (3).
   (c) Let P be the prime ideal $P = (1 + \sqrt{-5}, 3)$, and let v_P be the valuation of R determined by P. Prove that $v_P((1 + \sqrt{-5})/3) = 0$.
   (d) Lemma 6.3 shows that $(1 + \sqrt{-5})/3$ can be written as the quotient of two members a and b of R with $v_P(a) = v_P(b) = 0$. Find such a choice of a and b.

5. Let v be a discrete valuation of a field F, let R_v be the valuation ring, and let P_v be the valuation ideal. It was observed after Proposition 6.2 that $1 + P_v^n$ is a group under multiplication for any $n \geq 1$. Prove for $n \geq 1$ that the multiplicative group $(1 + P_v^n)/(1 + P_v^{n+1})$ is isomorphic to the additive group P_v^n/P_v^{n+1} under the mapping induced by $1 + x \mapsto x + P_v^{n+1}$.

6. Derive the finiteness of the class number of a number field K from the compactness of $\mathbb{I}_K^1/\iota(K^\times)$ given as Theorem 6.53.

Problems 7–8 compare the topology on the ideles $\mathbb{I} = \mathbb{I}_K$ of a number field K with the topology of the adeles $\mathbb{A} = \mathbb{A}_K$. The notation is as in Section 10.

7. For each finite set S of places containing the archimedean places, exhibit the mappings $\mathbb{I}(S) \to K_v$ for $v \in S$ and $\mathbb{I}(S) \to R_v$ for $v \notin S$ as continuous, and deduce that the inclusion $\mathbb{I} \to \mathbb{A}$ is continuous.

8. Let p_n be the nth positive prime in $\mathbb{Z}$, and let $x_n = (x_{n,v})_v$ be the adele in $\mathbb{A}_{\mathbb{Q}}$ with $x_{n,v} = p_n$ if $v = p_n$ and $x_{n,v} = 1$ if $v \neq p_n$. The result is a sequence $\{x_n\}$ of ideles in $\mathbb{I}_{\mathbb{Q}}$. Show that this sequence converges to the idele $(1)_v$ in the topology of the adeles but does not converge in the topology of the ideles.

Problems 9–10 below assume knowledge from measure theory of elementary properties of measures and of the existence–uniqueness theorem for translation-invariant measures (Haar measures) on locally compact abelian groups. The continuity in Problem 10a requires making estimates of integrals.

9. Let G be a locally compact abelian topological group with a Haar measure written as dx, and let Φ be an automorphism of G as a topological group, i.e., an automorphism of the group structure that is also a homeomorphism of G. Prove that there is a positive constant $a(\Phi)$ such that $d(\Phi(x)) = a(\Phi) \, dx$.

10. Let F be a locally compact topological field, and let $F^\times$ be the group of nonzero elements, the group operation being multiplication.
    (a) Let c be in $F^\times$, and define $|c|_F$ to be the constant $a(\Phi)$ from the previous problem when the measure is an additive Haar measure and Φ is multiplication by c. Define $|0|_F = 0$. Prove that $c \mapsto |c|_F$ is a continuous function from F into $[0, +\infty)$ such that $|c_1 c_2|_F = |c_1|_F |c_2|_F$.
    (b) If dx is a Haar measure for F as an additive locally compact group, prove that $dx/|x|_F$ is a Haar measure for $F^\times$ as a multiplicative locally compact group.
    (c) Let $F = \mathbb{R}$ be the locally compact field of real numbers. Compute the function $x \mapsto |x|_F$. Do the same thing for the locally compact field $F = \mathbb{C}$ of complex numbers.
    (d) Let $F = \mathbb{Q}_p$ be the locally compact field of p-adic numbers, where p is a prime. Compute the function $x \mapsto |x|_F$.
    (e) For the field $F = \mathbb{Q}_p$ of p-adic numbers, suppose that the ring $\mathbb{Z}_p$ of p-adic integers has additive Haar measure 1. What is the additive Haar measure of the maximal ideal I of $\mathbb{Z}_p$?

Problems 11–14 analyze the structure of complete valued fields whose residue class fields are finite, showing that the only kinds are p-adic fields and fields of formal Laurent series over a finite field. Let F be a complete valued field with a discrete nonarchimedean valuation, let v be the valuation, let R be the valuation ring, and let $\mathfrak{p}$ be the maximal ideal of R. Suppose that the residue class field $R/\mathfrak{p}$ is finite of order $q = p^m$ for a prime number p. Theorem 6.26 shows that the topology on F is locally compact. The normalized absolute value on F corresponding to v is $| \cdot |_F = q^{-v(\cdot)}$. For some purposes it is convenient to separate the **equal-characteristic case** for F and $R/\mathfrak{p}$ from the **unequal-characteristic case**.

11. Show in the unequal-characteristic case that F has characteristic 0.
12. (a) In both cases, use Hensel's Lemma to show that F has a full set of $(q-1)^{\text{st}}$ roots of unity and that coset representatives in F for $R/\mathfrak{p}$ can be taken to be these elements and 0. Denote this subset of q elements of F by E. The subset E is of course closed under multiplication.
    (b) Show in the equal-characteristic case that E is closed under addition and subtraction and is therefore a subfield of F isomorphic to $\mathbb{F}_q$.
13. In the equal-characteristic case, write $\mathbb{F}_q$ for the subfield of F constructed in Problem 12b, and let t be a generator of the principal ideal $\mathfrak{p}$, so that $v(t) = 1$.
    (a) Show that each nonzero element of R has a convergent infinite-series expansion of the form $\sum_{k=0}^{\infty} a_k t^k$ with all a_k in $\mathbb{F}_q$ and that the value of v on such an element is the smallest $k \geq 0$ such that $a_k \neq 0$.
    (b) Show conversely that every series $\sum_{k=0}^{\infty} a_k t^k$ with all a_k in $\mathbb{F}_q$ lies in R, and conclude that $R \cong \mathbb{F}_q[[t]]$.
    (c) Deduce that F is isomorphic to the field $\mathbb{F}_q((t))$ of formal Laurent series over $\mathbb{F}_q$, the understanding being that each such series involves only finitely many negative powers of t.
14. Let F be an arbitrary complete valued field in the unequal-characteristic case. Since Problem 11 shows F to be of characteristic 0, F contains a subgroup $\mathbb{Q}'$ isomorphic as a field to $\mathbb{Q}$.
    (a) Show that the integer $q = p^m$ in $\mathbb{Q}'$ lies in $\mathfrak{p}$.
    (b) Deduce that the number $v_0 = v(p)$ is positive.
    (c) For each nonzero member $ab^{-1}p^k$ of $\mathbb{Q}'$ for which a and b are integers relatively prime to p, show that $v(ab^{-1}p^k) = kv_0$.
    (d) Deduce that $(\mathbb{Q}', |\cdot|_F^{1/(mv_0)})$ is isomorphic as a valued field to $(\mathbb{Q}, |\cdot|_p)$.
    (e) Let $\overline{\mathbb{Q}'}$ be the closure of $\mathbb{Q}'$ in F, and explain why $(\overline{\mathbb{Q}'}, |\cdot|_F^{1/m})$ is isomorphic as a valued field to $(\mathbb{Q}_p, |\cdot|_p)$.
    (f) Let t be a generator of $\mathfrak{p}$. With E as in Problem 12a, show that each member of F has a unique series expansion $\sum_{k=-N}^{\infty} a_k t^k$ with each a_k in E and with N depending on the element, and show furthermore that every such series expansion converges to an element of F.
    (g) Let $c_1, \ldots, c_l$ with $l = q^{v_0}$ be an enumeration of the elements $\sum_{k=0}^{v_0-1} a_k t^k$ with all a_k in E. Show that to each element x in R corresponds some c_j such that $p^{-1}(x - c_j)$ lies in R. Deduce that every element of R is the sum of a convergent series of the form $\sum_{k=0}^{\infty} c_{j_k} p^k$.
    (h) Explain how it follows from the previous part that F is a finite-dimensional vector space over $\overline{\mathbb{Q}'}$, hence that F is a finite extension of the field $\mathbb{Q}_p$.

Problems 15–19 continue the analysis in Problems 11–14 by examining finite separable extensions of complete valued fields whose residue class fields are finite. The goal is to prove Proposition 6.38 and Lemmas 6.47 and 6.48. Let F be a complete valued field with a discrete nonarchimedean valuation, let R be the valuation ring, and let $\mathfrak{p}$ be the maximal ideal of R. Suppose that the residue class field $R/\mathfrak{p}$ is finite of order $q = p^m$ for a prime number p. Let K be a finite separable extension of F, put $n = [K : F]$, and let T be the integral closure of R in K. Theorem 6.33 shows that K is a valued field, that it has a unique nonzero prime ideal P, that the valuation ring of K is T, and that the valuation ideal is P. Write f for the dimension of T/P over $R/\mathfrak{p}$, so that T/P has order q^f. Also, write e for the power such that $\mathfrak{p}T = P^e$. It is known from Chapter IX of *Basic Algebra* that $n = ef$. In the equal-characteristic case, there is an especially transparent argument for proving Proposition 6.38, and Problem 15 gives that. Problem 16 gives a less transparent argument that handles both cases at once. The remaining problems address Lemmas 6.47 and 6.48.

15. In the equal-characteristic case, let E be the subset of q elements of F described in Problem 12, and let $\widetilde{E}$ be the corresponding subset of q^f elements of K. Problem 13 shows that E is a field isomorphic to $\mathbb{F}_q$ and that $\widetilde{E}$ is an extension field isomorphic to $\mathbb{F}_{q^f}$. Let t be a generator in R of $\mathfrak{p}$, and let $\widetilde{t}$ be a generator in T of P. Problem 13 shows that $F = \mathbb{F}_q((t))$ and that $K = \mathbb{F}_{q^f}((\widetilde{t}))$.
    (a) Show that the set L of formal Laurent series in t with coefficients from $\mathbb{F}_{q^f}$ is an intermediate field between F and K, so that $L = \mathbb{F}_{q^f}((t))$.
    (b) Why does it follow that the integral closure of R in L is $U = \mathbb{F}_{q^f}[[t]]$ and that the maximal ideal of U is $\wp = tU$?
    (c) Deduce that the residue class field of L is $\mathbb{F}_{q^f}$ of order q^f and that $\wp T = P^e$, so that the residue class degree of L/F is f and the ramification index of K/L is e.
    (d) How can one conclude that L/F is unramified and that K/L is totally ramified?

16. In this problem no distinction is made between the equal-characteristic case and the unequal-characteristic case. Let $\Bbbk_F$ and $\Bbbk_K$ be the residue class fields of F and K, and write $\Bbbk_K = \Bbbk_F(\overline{\alpha})$, where $\overline{\alpha}$ is a root of a monic irreducible polynomial $\overline{g}(X)$ in $\Bbbk_F[X]$. Let $g(X)$ be a monic polynomial in $R[X]$ that reduces modulo $\mathfrak{p}$ to $\overline{g}(X)$.
    (a) Prove that there exists $\alpha \in T$ with $\alpha + P = \overline{\alpha}$ and with $g(\alpha) = 0$.
    (b) With α as in (a), let L be the intermediate field between F and K given by $L = F(\alpha)$, let U be the integral closure of R in L, let $\wp$ be the maximal ideal of U, and let $\Bbbk_L = U/\wp$. Show that α lies in U and that the member $\overline{\alpha}$ of $\Bbbk_K$ is in the image of the natural field map $\Bbbk_L \to \Bbbk_K$.
    (c) Conclude from (b) that $\Bbbk_L = \Bbbk_K$.
    (d) By comparing $[L : K]$, the degrees of $g(X)$ and $\overline{g}(X)$, and the indices e and f for K/F and L/F, prove that L has the properties required by Proposition 6.38.

17. This problem applies to both the equal-characteristic case and the unequal-characteristic case. Let ξ be a member of T such that $K = F(\xi)$, and let $g(X) = X^n + c_1 X^{n-1} + \cdots + c_n$ be its minimal polynomial over F.
    (a) Let $N = \sum_{k=0}^{n-1} R\xi^k$. This is a free R submodule of T of rank n with $\{1, \xi, \ldots, \xi^{n-1}\}$ as an R basis. Define
    $$\widehat{N} = \{y \in K \mid \mathrm{Tr}_{K/F}(xy) \text{ is in } R \text{ for all } x \in M\}.$$
    Put $x_i = \xi^{i-1}$ for $1 \leq i \leq n$. Why is there a unique y_j in K with $\mathrm{Tr}_{K/F}(x_i y_j) = \delta_{ij}$? Show that $\widehat{N}$ is a free R module with $\{y_1, \ldots, y_n\}$ as R basis.
    (b) If A is a matrix in $M_n(R)$ with $\det A = \pm 1$ and if $z_k = \sum_j A_{jk} y_j$, why is $\sum_{k=1}^n R z_k = \sum_{k=1}^n R y_k$?
    (c) Let K' be a splitting field of $g(X)$ over F, and let $\xi_1, \ldots, \xi_n$ be the roots of $g(X)$ in K', with $\xi_1 = \xi$. It is known from *Basic Algebra* that $\xi_1, \ldots, \xi_n$ are distinct. Prove that
    $$\sum_{i=1}^n \frac{g(X)}{g'(\xi_i)(X - \xi_i)} = 1$$
    by observing that the difference of the two sides is a polynomial in X of degree at most $n - 1$ and all of $\xi_1, \ldots, \xi_n$ are roots.
    (d) Let σ_j be the field map that fixes F and carries $F(\xi)$ into K' in such a way that $\sigma_j(\xi) = \xi_j$. These mappings have the property that $\mathrm{Tr}_{K/F}(\xi) = \sum_{j=1}^n \sigma_j(\xi)$ for all $\xi \in K$. If $h(X)$ is in the ring $K[[X]]$ of formal power series over K, let $h^{\sigma_j}(X)$ be the polynomial obtained by applying σ_j to each coefficient, and extend $\mathrm{Tr}_{K/F} : K \to F$ to a mapping of $K[[X]]$ to $F[[X]]$ by letting $\mathrm{Tr}_{K/F} h(X) = \sum_{j=1}^n h^{\sigma_j}(X)$. By making the substitution $X \mapsto 1/X$ in (c) and using the extended trace function just defined, show that
    $$\frac{X^n}{1 + c_1 X + \cdots + c_n X^n} = \mathrm{Tr}_{K/F}\left(\frac{X}{g'(\xi)(1 - \xi X)}\right).$$
    (e) Write the identity in (d) out with power series, equate the coefficients of $X, X^2, \ldots, X^n$ on the two sides, and deduce that $\mathrm{Tr}_{K/F}\left(\xi^{k-1} g'(\xi)^{-1}\right)$ equals 0 for $1 \leq k < n$ and equals 1 for $k = n$.
    (f) Form the n-by-n matrix A with $A_{ij} = \mathrm{Tr}_{K/F}\left((\xi^{i-1} g'(\xi)^{-1})(\xi^{j-1})\right)$. The result of (e) shows that this matrix has all entries equal to 0 that lie above the off-diagonal $i + j = n + 1$ and all entries equal to 1 that lie on the off-diagonal. By writing $\xi^{i+j-2} = \xi^n \xi^{i+j-(n+1)-1}$ and by substituting for ξ^n, show that the remaining entries A_{ij} lie in R.
    (g) Combine the conclusions of (a), (b), and (f) to prove that $\widehat{N} = g'(\xi)^{-1} N$.

18. This problem continues with the notation of Problem 17 and assumes in addition that K/F is unramified, i.e., that $f = n$ and $e = 1$. The objective is to prove the assertion of Lemma 6.48 that $\mathcal{D}(K/F) = T$.

(a) Prove that the intermediate field L constructed in Problem 16 is K itself, that the polynomial $g(X)$ is the minimal polynomial of α over F, and that $K = F(\alpha)$.

(b) Let $N = \sum_{k=0}^{n-1} R\alpha^k$. Apply Problem 17 to obtain $\widehat{N} = g'(\alpha)^{-1}N$. Using the inclusion $N \subseteq T$, deduce that $\widehat{N} \supseteq \widehat{T}$, and conclude that $\mathcal{D}(K/F)^{-1} \subseteq g'(\alpha)^{-1}T$.

(c) Prove that $g'(\alpha)$ is a unit in T, and deduce that $\mathcal{D}(K/F) = T$.

19. This problem continues with the notation of Problem 17 and assumes in addition that K/F is totally ramified, i.e., that $e = n$ and $f = 1$. The objective is to prove the assertion of Lemma 6.47 that $\mathcal{D}(K/F) = P^{e'}$ with e' equal to $e-1$ if p does not divide e and with $e' \geq e$ if p divides e. Let E be the set of representatives in R of the members of $R/\mathfrak{p}$ as constructed in Problem 12. Since $f = 1$, the set E is also a set of representatives in T of the members of T/P. Let v_K and v_F be the respective discrete valuations of K and F, so that $v_F = nv_K\big|_F$ by Proposition 6.34. Let π and λ be respective generators of P and $\mathfrak{p}$.

(a) Prove that if M is a field with a discrete valuation w and if $x_1, \ldots, x_m$ are elements of M with $x_1 + \cdots + x_m = 0$ and $m \geq 2$, then the number of j's for which $w(x_j) = \min_{1 \leq i \leq m} w(x_i)$ is at least 2.

(b) Let $g(X) = c_0 X^n + c_1 X^{n-1} + \cdots + c_n$ with $c_0 = 1$ be the field polynomial of π over F. Why are all the coefficients c_j in R, and why is $v_K(c_j)$ divisible by n for each j?

(c) Taking into account that π is a root of its field polynomial and applying (a), show that there exist integers i and j with $0 \leq i < j \leq n$ such that $j - i = v_K(c_j) - v_K(c_i)$ and that all other integers k with $0 \leq k \leq n$ have $v_K(c_k \pi^{n-k}) \geq n$.

(d) Using the divisibility conclusion of (b), show that $g(X)$ is an **Eisenstein polynomial** relative to $\mathfrak{p}$ in the sense that $c_0 = 1$, that all of $c_1, \ldots, c_n$ lie in $\mathfrak{p}$, and that c_n does not lie in $\mathfrak{p}^2$.

(e) Conclude from (d) that $g(X)$ is irreducible over F, that $g(X)$ is the minimal polynomial of π over F, and that $K = F(\pi)$.

(f) For each $k \geq 0$, apply the division algorithm to write $k = ni + j$ with $0 \leq j < n = e$, and define $y_k = \lambda^i \pi^j$. Show that every member of T has a unique convergent series expansion as $\sum_{k=0}^{\infty} a_k y_k$ and that all such series expansions have sum in T.

(g) By rewriting the expansion in (f) suitably, show that $\{1, \pi, \ldots, \pi^{n-1}\}$ is an R basis for the free R module T.

(h) By applying Problem 17 with $N = \sum_{k=0}^{n-1} R\pi^k$, prove that $\widehat{T} = g'(\pi)^{-1}T$, and deduce that $\mathcal{D}(K/F) = (g'(\pi))$.

(i) Computing $g'(\pi)$ and applying the valuation v to it, show that $v(g'(\pi)) = e - 1$ if $v(e) = 0$ and that $v(g'(\pi)) \geq e$ if $v(e) > 0$. Explain how this conclusion proves Lemma 6.47.

CHAPTER VII

Infinite Field Extensions

Abstract. This chapter provides algebraic background for directly addressing some simple-sounding yet fundamental questions in algebraic geometry. All the questions relate to the set of simultaneous zeros of finitely many polynomials in n variables over a field.

Section 1 concerns existence of zeros. The main theorem is the Nullstellensatz, which in part says that there is always a zero if the finitely many polynomials generate a proper ideal and if the underlying field is algebraically closed.

Section 2 introduces the transcendence degree of a field extension. If L/K is a field extension, a subset of L is algebraically independent over K if no nonzero polynomial in finitely many of the members of the subset vanishes. A transcendence basis is a maximal subset of algebraically independent elements; a transcendence basis exists, and its cardinality is independent of the particular basis in question. This cardinality is the transcendence degree of the extension. Then L is algebraic over the subfield generated by a transcendence basis. Briefly any field extension can be obtained by a purely transcendental extension followed by an algebraic extension. The dimension of the set of common zeros of a prime ideal of polynomials over an algebraically closed field is defined to be the transcendence degree of the field of fractions of the quotient of the polynomial ring by the ideal.

Section 3 elaborates on the notion of separability of field extensions in characteristic p. Every algebraic extension L/K can be obtained by a separable extension followed by an extension that is purely inseparable in the sense that every element x of L has a power x^{p^e} for some integer $e \geq 0$ with x^{p^e} separable over K.

Section 4 introduces the Krull dimension of a commutative ring with identity. This number is one more than the maximum number of ideals occurring in a strictly increasing chain of prime ideals in the ring. For $K[X_1, \ldots, X_n]$ when K is a field, the Krull dimension in n. If P is a prime ideal in $K[X_1, \ldots, X_n]$, then the Krull dimension of the integral domain $R = K[X_1, \ldots, X_n]/P$ matches the transcendence degree over K of the field of fractions of R. Thus Krull dimension extends the notion of dimension that was defined in Section 2.

Section 5 concerns nonsingular and singular points of the set of common zeros of a prime ideal of polynomials in n variables over an algebraically closed field. According to Zariski's Theorem, nonsingularity of a point may be defined in either of two equivalent ways—in terms of the rank of a Jacobian matrix obtained from generators of the ideal, or in terms of the dimension of the quotient of the maximal ideal at the point in question factored by the square of this ideal. The point is nonsingular if the rank of the Jacobian matrix is n minus the dimension of the zero locus, or equivalently if the dimension of the quotient of the maximal ideal by its square equals the dimension of the zero locus. Nonsingular points always exist.

Section 6 extends Galois theory to certain infinite field extensions. In the algebraic case inverse limit topologies are imposed on Galois groups, and the generalization of the Fundamental Theorem of Galois Theory to an arbitrary separable normal extension L/K gives a one-one correspondence between the fields F with $K \subseteq F \subseteq L$ and the closed subgroups of $\text{Gal}(L/K)$.

1. Nullstellensatz

Algebraic geometry studies the geometric properties of sets defined by algebraic equations. In the simplest case some field K is specified, the equations are polynomial equations in several variables with coefficients in K, and one seeks solutions to the system of equations with the variables taking values in K or some larger field.

The nature of the subject is that even fairly simple-sounding geometric questions require algebraic background beyond what is in *Basic Algebra* and the first six chapters of the present book. This chapter addresses the necessary background, largely from the theory of fields, for addressing fundamental questions concerning existence of solutions, the dimension of the space of solutions, singularity of the solution set at a particular point, and effects of changing fields.

The present section supplies background for the question of existence. We have a system of polynomial equations in n variables with coefficients in K, and we are interested in simultaneous solutions in a given extension field L of K. A solution can be regarded as a column vector in L^n. Think of the equations as of the form $F_i(X_1, \ldots, X_n) = 0$ with each F_i a polynomial, and then the set of solutions is the locus of common zeros of the F_i's in L^n. The locus of common zeros is unaffected by enlarging the system of equations by allowing all equations of the form $\sum_i G_i F_i = 0$ with each G_i is arbitrary in $K[X_1, \ldots, X_n]$; thus we may as well regard the left sides as all members of some ideal I in $K[X_1, \ldots, X_n]$. The Hilbert Basis Theorem says that any ideal in $K[X_1, \ldots, X_n]$ is finitely generated, and hence studying the common zero locus for an ideal is always the same as studying the common zero locus for a finite set of polynomials.

A proper ideal need not have a nonempty locus of common zeros. For example, if $K = \mathbb{R}$, then the single equation $X^2 + Y^2 + 1 = 0$ has no solutions in $\mathbb{R}^2$. Hilbert's Nullstellensatz[1] is partly the affirmative statement that any proper ideal has a nonzero locus of common zeros under the additional assumption that K is algebraically closed.

Theorem 7.1 (Nullstellensatz). Let K be a field, let $\overline{K}$ be an algebraic closure, and let n be a positive integer. Then every maximal ideal J of $K[X_1, \ldots, X_n]$ has the property that $K[X_1, \ldots, X_n]/J$ is a finite algebraic extension of K, and in particular the maximal ideals of $\overline{K}[X_1, \ldots, X_n]$ are of the form

$$(X_1 - a_1, \ldots, X_n - a_n),$$

where $(a_1, \ldots, a_n)$ is an arbitrary member of $\overline{K}^n$. Consequently if I is any proper ideal in $K[X_1, \ldots, X_n]$, then

(a) the locus of common zeros of I in $\overline{K}^n$ is nonempty,

---
[1] German for "zero-locus theorem."

(b) any f in $K[X_1, \ldots, X_n]$ that vanishes on the locus of common zeros of I in $\overline{K}^n$ has the property that f^k is in I for some integer $k > 0$.

Before coming to the proof, we mention an important corollary.

Corollary 7.2. Let K be a field, let $\overline{K}$ be an algebraic closure, let n be a positive integer, and let I be a *prime* ideal in $K[X_1, \ldots, X_n]$. Then I contains every polynomial in $K[X_1, \ldots, X_n]$ that vanishes on the locus of common zeros of I in $K[X_1, \ldots, X_n]$.

PROOF. If f is a member of $K[X_1, \ldots, X_n]$ that vanishes on the locus of common zeros of I, then (b) in the theorem shows that f^k is in I for some k. Since I is prime, one of the factors of $f^k = f \cdots f$ lies in I. □

EXAMPLE FOR COROLLARY. Let $K = L = \mathbb{C}$, and let I be the principal ideal in $\mathbb{C}[X, Y]$ generated by $Y^2 - X(X+1)(X-1)$. Consider $\mathbb{C}[X, Y]$ as isomorphic to $\mathbb{C}[X][Y]$. As a polynomial in Y over $\mathbb{C}[X]$, $p(X, Y) = Y^2 - X(X+1)(X-1)$ is irreducible because $X(X+1)(X-1)$ is not the square of a polynomial in X. Since $\mathbb{C}[X, Y]$ is a unique factorization domain, $p(X, Y)$ is prime. Therefore $I = (p(X, Y))$ is a prime ideal. The corollary says that every polynomial vanishing on the locus of points $(x, y) \in \mathbb{C}^2$ for which $y^2 = x(x+1)(x-1)$ is the product of $Y^2 - X(X+1)(X-1)$ and a polynomial in (X, Y). Consequently the ring of restrictions of polynomials to the locus for which $y^2 = x(x+1)(x-1)$ is isomorphic to $\mathbb{C}[X, Y]/(Y^2 - X(X+1)(X-1))$.

Theorem 7.1b has a tidy formulation in terms of the "radical" of an ideal. If R is a commutative ring with identity and I is an ideal in R, then the **radical** of I, denoted by $\sqrt{I}$, is the set of all r in R such that r^k is in I for some $k \geq 1$. It is immediate that the radical of I is an ideal containing I and that $\sqrt{I}$ is proper if I is proper. If I is an ideal in $K[X_1, \ldots, X_n]$ and if f is in $\sqrt{I}$, then f^k is in I for some $k > 0$, and hence f vanishes on the locus of common zeros of I. Theorem 7.1b says conversely that any f vanishing on the locus of common zeros of I has f^k in I for some $k > 0$. This means that f is in $\sqrt{I}$. We can therefore rewrite (b) in the theorem as follows:

(b') the ideal of all f in $K[X_1, \ldots, X_n]$ that vanish on the locus of common zeros of I in $\overline{K}^n$ is exactly $\sqrt{I}$.

The proof of Theorem 7.1 will follow comparatively easily from the following two lemmas.

Lemma 7.3. If K is a field and L is an extension field that is generated as a K algebra by n elements $x_1, \ldots, x_n$, i.e., if $L = K[x_1, \ldots, x_n]$, then every x_j is algebraic over K.

REMARKS. Conversely if $x_1, \ldots, x_n$ are elements of an extension field L that are algebraic over K, then $K(x_1, \ldots, x_n) = K[x_1, \ldots, x_n]$. The reason is that
$$K(x_1, \ldots, x_n) = K(x_1, \ldots, x_{n-1})(x_n) = K(x_1, \ldots, x_{n-1})[x_n]$$
$$= K(x_1, \ldots, x_{n-2})(x_{n-1})[x_n] = K(x_1, \ldots, x_{n-2})[x_{n-1}][x_n]$$
$$= \cdots = K[x_1] \cdots [x_{n-1}][x_n] = K[x_1, \ldots, x_n].$$

PROOF. We proceed by induction on n. For $n = 1$, if $L = K[x_1]$, then we know from the elementary theory of fields that x_1 is algebraic over K.

For the inductive step, suppose that $L = K[x_1, \ldots, x_n]$. Since L is a field, $K(x_1) \subseteq L$, and hence $L = K(x_1)[x_2, \ldots, x_n]$. By the inductive hypothesis applied to L and $K(x_1)$, the elements $x_2, \ldots, x_n$ are algebraic over $K(x_1)$. To complete the proof, it is enough to show that x_1 is algebraic over K.

Fix $j \geq 2$. The element x_j, being algebraic over $K(x_1)$, satisfies a polynomial equation
$$X^m + a_{m-1}X^{m-1} + \cdots + a_0 = 0$$
with $a_{m-1}, \ldots, a_0$ in $K(x_1)$. Clearing fractions, we see that x_j satisfies an equation
$$b_m X^m + b_{m-1} X^{m-1} + \cdots + b_0 = 0$$
with $b_m, \ldots, b_0$ in $K[x_1]$ and $b_m \neq 0$. Multiplying through by b_m^{m-1} shows that x_j satisfies
$$(b_m X)^m + b_{m-1}(b_m X)^{m-1} + \cdots + b_0(b_m)^{m-1} = 0,$$
and we see that $b_m x_j$ is integral over the ring $K[x_1]$. Let us write c_j for the element $b_m \in K[x_1]$ that we have just produced for this j.

In the case of $j = 1$, we can use $m = 1$ and $a_0 = -x_1$ in the above argument, and we are then led to $c_1 = x_1$. If $x_1^{l_1} \cdots x_n^{l_n}$ is any monomial in $K[x_1, \ldots, x_n]$ and if l is defined as $l = \max(l_1, \ldots, l_n)$, then the fact that the integral elements over $K[x_1]$ form a ring implies that $(c_1 \cdots c_n)^l x_1^{l_1} \cdots x_n^{l_n}$ is integral over $K[x_1]$. Hence for any f in $K[x_1, \ldots, x_n]$, $(c_1 \cdots c_n)^l f$ is integral over $K[x_1]$ for a suitable integer $l = l(f)$. Since $K(x_1) \subseteq K[x_1, \ldots, x_n]$, this conclusion applies in particular to any member f of $K(x_1)$.

The ring $K[x_1]$ is a principal ideal domain and is therefore integrally closed in its field of fractions $K(x_1)$. For f in $K(x_1)$, we have seen that $(c_1 \cdots c_n)^l f$ is integral over $K[x_1]$ for some $l = l(f)$. The element $(c_1 \cdots c_n)^l f$ is in $K(x_1)$, and the integral-closure property therefore implies that $(c_1 \cdots c_n)^l f$ is in $K[x_1]$.

Consequently there exists a fixed element h of $K[x_1]$ such that every element f of $K(x_1)$ is of the form g/h^l for some g in $K[x_1]$ and some integer $l \geq 0$. We apply this observation to $f = q(x_1)^{-1}$ for each irreducible polynomial $q(X)$ in $K[X]$, and we obtain $q(x_1)g = h^l$ with g and l depending on $q(X)$. If x_1 is transcendental over K, this equality implies the polynomial identity $q(X)g(X) = h(X)^l$.

Consequently every irreducible polynomial $q(X)$ divides $h(X)$. If K is infinite, this is a contradiction because there are infinitely many distinct polynomials $X - a$ in $K[X]$; if K is finite, this is a contradiction because there exists at least one irreducible polynomial of each degree ≥ 1. We arrive at a contradiction in either case, and therefore x_1 is algebraic over K. This completes the induction and the proof. □

Lemma 7.4. Let K be a field, and let L be an algebraic extension of K. If I is a proper ideal in $K[X_1, \ldots, X_n]$, then $IL[X_1, \ldots, X_n]$ is a proper ideal in $L[X_1, \ldots, X_n]$.

REMARK. As usual, the notation $IL[X_1, \ldots, X_n]$ refers to the set of sums of products of elements of I and elements of $L[X_1, \ldots, X_n]$.

PROOF. First let us identify the integral closure of $K[X_1, \ldots, X_n]$ in the field $L(X_1, \ldots, X_n)$ as $L[X_1, \ldots, X_n]$. The ring $L[X_1, \ldots, X_n]$ is a unique factorization domain, and Proposition 8.41 of *Basic Algebra* shows that it is integrally closed. Consequently the integral closure of $K[X_1, \ldots, X_n]$ in $L(X_1, \ldots, X_n)$ is contained in $L[X_1, \ldots, X_n]$. On the other hand, the integral closure of $K[X_1, \ldots, X_n]$ in $L(X_1, \ldots, X_n)$ contains L because L/K is algebraic, and it contains each X_j. Therefore it contains $L[X_1, \ldots, X_n]$ and must equal $L[X_1, \ldots, X_n]$.

Now we apply Proposition 8.53 of *Basic Algebra* to the ring $K[X_1, \ldots, X_n]$, its field of fractions $K(X_1, \ldots, X_n)$, the extension field $L(X_1, \ldots, X_n)$, and the integral closure $L[X_1, \ldots, X_n]$ of $K[X_1, \ldots, X_n]$ in $L(X_1, \ldots, X_n)$. The proposition says that if P is any maximal ideal of $K[X_1, \ldots, X_n]$, then the ideal $PL[X_1, \ldots, X_n]$ is proper in $L[X_1, \ldots, X_n]$. This result is to be applied to any maximal ideal P of $K[X_1, \ldots, X_n]$ that contains I. □

PROOF OF THEOREM 7.1. Let J be a maximal ideal in $K[X_1, \ldots, X_n]$. Then $L = K[X_1, \ldots, X_n]/J$ is a field. Hence $L = K[x_1, \ldots, x_n]$ is a field if the x_i's are defined by $x_i = X_i + J$. Lemma 7.3 shows that each x_j is algebraic over K, and the first conclusion of the theorem follows.

When this conclusion is applied to $\overline{K}$ instead of K, then the fact that $\overline{K}$ is algebraically closed implies that each x_j lies in the cosets determined by $\overline{K}$, i.e., the cosets of the constant polynomials. Consequently for each j, there is an element a_j in $\overline{K}$ such that $x_j - a_j$ lies in J. Then it follows that $(X_1 - a_1, \ldots, X_n - a_n)$ is contained in J. Since the ideal $(X_1 - a_1, \ldots, X_n - a_n)$ is maximal, $J = (X_1 - a_1, \ldots, X_n - a_n)$. This proves that the maximal ideals are as in the displayed expression in the theorem.

To prove (a), we apply Lemma 7.4 to the ideal I in $K[X_1, \ldots, X_n]$ and to the algebraic extension $\overline{K}$ of K. The lemma produces a proper ideal of $\overline{K}[X_1, \ldots, X_n]$

containing I, and we extend it to a maximal ideal J of $\overline{K}[X_1, \ldots, X_n]$. From the previous paragraph of the proof, J is of the form $J = (X_1 - a_1, \ldots, X_n - a_n)$ for some $(a_1, \ldots, a_n)$ in $\overline{K}^n$. The ideal J is therefore identified as the kernel of the evaluation homomorphism of $\overline{K}[X_1, \ldots, X_n]$ at the point $(a_1, \ldots, a_n)$. Every member of J thus vanishes at $(a_1, \ldots, a_n)$, and the same thing is true of every member of I. This proves (a).

For (b), let I be a proper ideal in $K[X_1, \ldots, X_n]$, and let f be as in (b). Introduce an additional indeterminate Y, and let J be the ideal in $K[X_1, \ldots, X_n, Y]$ generated by I and $fY - 1$. If some point $(x_1, \ldots, x_n, y)$ lies on the locus of common zeros of J in $\overline{K}^{n+1}$, then $(x_1, \ldots, x_n)$ lies on the locus of common zeros of I in $\overline{K}^n$, since $I \subseteq J$; thus $f(x_1, \ldots, x_n) = 0$, since f is assumed to vanish on all common zeros of I in $\overline{K}^n$. Consequently $f(x_1, \ldots, x_n)y - 1 = -1 \neq 0$, and we find that $f(X_1, \ldots, X_n)Y - 1$ does not vanish on the locus of common zeros of J in $\overline{K}^{n+1}$, contradiction. We conclude that no point $(x_1, \ldots, x_n, y)$ lies on the locus of common zeros of J in $\overline{K}^{n+1}$. By (a), we see that

$$J = K[X_1, \ldots, X_n, Y]. \qquad (*)$$

Let us write X for the expression $X_1, \ldots, X_n$. Then $(*)$ implies that

$$1 = \sum_{i=1}^r p_i(X, Y) g_i(X) + q(X, Y)(f(X)Y - 1) \qquad (**)$$

for some $g_1, \ldots, g_r$ in I and some $p_1, \ldots, p_r$ and q in $K[X, Y]$. Let ψ be the substitution homomorphism of $K[X, Y]$ into $K(X)$ that carries K into itself, X into itself, and Y into $f(X)^{-1}$. Application of ψ to $(**)$ gives

$$1 = \sum_{i=1}^r p_i(X, f(X)^{-1}) g_i(X), \qquad (\dagger)$$

since $\psi(f(X)Y - 1) = 0$. If Y^k is the largest power of Y that appears in any of the polynomials $p_i(X, Y)$, then we can rewrite $(\dagger)$ as

$$f(X)^k = \sum_{i=1}^r \left(f(X)^k p_i(X, f(X)^{-1}) \right) g_i(X)$$

and exhibit $f(X)^k$ as the sum of products of the members g_i of I by members of $K[X]$. Thus $f(X)^k$ is in I, and (b) is proved. $\square$

2. Transcendence Degree

Let K be a field, and let L be an extension field. The algebraic construction in this section will show that L can be obtained from K in two steps, by a "purely transcendental" extension followed by an algebraic extension. The number of indeterminates in the first step (or the cardinality if the number is infinite) will be seen to be an invariant of the construction and will be called the "transcendence degree" of L/K.

Before coming to the details, let us mention what transcendence degree will mean geometrically. Suppose that the field K is algebraically closed, suppose that I is a prime ideal in $K[X_1, \ldots, X_n]$, and suppose that V is the locus of common zeros of I. Corollary 7.2 shows that I is the set of all polynomials vanishing on V, and thus the integral domain $K[X_1, \ldots, X_n]/I$ may be regarded as the set of all restrictions to V of polynomials. If L is the field of fractions of $K[X_1, \ldots, X_n]/I$, then the transcendence degree of L/K will be interpreted as the "number of independent variables" or "dimension" of the locus V.

Now we can make the precise definitions. Let K be a field, and let L be an extension field. A finite subset $x_1, \ldots, x_n$ of L is said to be **algebraically independent** over K if the ring homomorphism $K[X_1, \ldots, X_n] \to L$ given by $f \mapsto f(x_1, \ldots, x_n)$ is one-one.[2] Otherwise it is algebraically dependent.

EXAMPLE. Let $K = \mathbb{C}$, and let $p(X, Y) = Y^2 - X(X + 1)(X - 1)$. The principal ideal $I = (p(X, Y))$ was shown to be prime in $\mathbb{C}[X, Y]$ in the example with Corollary 7.2. Therefore $\mathbb{C}[X, Y]/I$ is an integral domain. Let x and y be the cosets $x = X + I$ and $y = Y + I$. If L denotes the field of fractions of $\mathbb{C}[X, Y]/I$, then we may regard x and y as members of L. The subset $\{x, y\}$ of L is algebraically dependent because the polynomial $p(X, Y)$ maps to 0 under the substitution homomorphism of $\mathbb{C}[X, Y]$ into L with $X \mapsto x$ and $Y \mapsto y$.

A subset S of L is called a **transcendence set** over K if each finite subset of S is algebraically independent over K. A maximal transcendence set over K is called a **transcendence basis** of L over K. For each transcendence set S of L over K, we write $K(S)$ for the smallest subfield of L containing K and S. If some transcendence basis S has the property that $K(S) = L$, then L is said to be a **purely transcendental** extension of K; in this case it follows from the definitions that S is a transcendence basis of L over K.

EXAMPLE, CONTINUED. With K and L as in the example above, the sets $S = \{x\}$ and $S = \{y\}$ are transcendence sets over $K = \mathbb{C}$. It is not hard to see that $\{x\}$ is a transcendence basis of L over K. Actually, if z is any member of L that is not in $\mathbb{C}$, then $\{z\}$ is a transcendence set over $\mathbb{C}$. The reason is that $\mathbb{C}$ is algebraically closed; hence either z is transcendental over $\mathbb{C}$ or else z lies in $\mathbb{C}$. Lemma 7.6 below shows that any transcendence set of L over $\mathbb{C}$ can be extended to a transcendence basis, and Theorem 7.9 shows that all transcendence bases of L over $\mathbb{C}$ have the same cardinality. It follows that if z is any member of L that is not in $\mathbb{C}$, then $\{z\}$ is a

---

[2] By convention the empty set is algebraically independent over K.

transcendence basis of L over $\mathbb{C}$ and that every transcendence basis of L over $\mathbb{C}$ is of this form. The two-element set $\{x, y\}$ cannot be a transcendence set by this reasoning, but we can see this conclusion more directly just by observing that $\{x, y\}$ was shown in the example above to be algebraically dependent.

Shortly we shall establish the existence of transcendence bases in general. If S is a transcendence basis and if K' is defined to be $K(S)$, then we shall show that L is algebraic over K'. The subfield K' of L depends on the choice of S, but there is a uniqueness theorem: the cardinality of a transcendence basis of L/K is independent of the particular transcendence basis.

Lemma 7.5. Let L/K be a field extension, let S be a transcendence set of L over K, let $K(S)$ be the subfield of L generated by K and S, and let x be an element of L not in S. Then $S' = S \cup \{x\}$ is a transcendence set of L over K if and only if x is transcendental over $K(S)$.

PROOF. Suppose that x is transcendental over $K(S)$ and is not in S. Let n distinct elements $x_1, \ldots, x_n$ of S' be given. If these are all in S, then $f \mapsto f(x_1, \ldots, x_n)$ is one-one because S is a transcendence set. Suppose that one of the n elements is x; say $x_n = x$. If f is in the kernel of the homomorphism $f \mapsto f(x_1, \ldots, x_n)$, i.e., if $f(x_1, \ldots, x_n) = 0$, then x is a root of the polynomial $g(X) = f(x_1, \ldots, x_{n-1}, X)$ in $K(x_1, \ldots, x_{n-1})[X]$. Since x is assumed to be transcendental over $K(S)$, the polynomial g must be 0. If we expand the polynomial f in powers of X as

$$f(X_1, \ldots, X_{n-1}, X) = c_l(X_1, \ldots, X_{n-1})X^l + \cdots + c_0(X_1, \ldots, X_{n-1}),$$

the condition that g be 0 says that $c_j(x_1, \ldots, x_{n-1}) = 0$ for all j. Since the set $\{x_1, \ldots, x_{n-1}\}$ is algebraically independent, we see that $c_j = 0$. Therefore $f = 0$. Hence $\{x_1, \ldots, x_n\}$ is algebraically independent, and S' is a transcendence set.

Conversely suppose that S' is a transcendence set of L over K. We are to show that the only polynomial $F(X)$ in $K(S)[X]$ such that $F(x) = 0$ is the 0 polynomial. Since only finitely many coefficients of F are in question, we may view F as in $K(\{x_1, \ldots, x_n\})[X]$ for some finite subset $\{x_1, \ldots, x_n\}$ of S. Clearing fractions, we can write F as

$$F(X) = d(x_1, \ldots, x_n)^{-1}\bigl(c_l(x_1, \ldots, x_n)X^l + \cdots + c_0(x_1, \ldots, x_n)\bigr)$$

for suitable polynomials $d, c_0, \ldots, c_l$ in $K[X_1, \ldots, X_n]$ with $d(x_1, \ldots, x_n) \neq 0$. Define

$$\widetilde{F}(X_1, \ldots, X_n, X) = c_l(X_1, \ldots, X_n)X^l + \cdots + c_0(X_1, \ldots, X_n).$$

The condition $F(x) = 0$ yields $\widetilde{F}(x_1, \ldots, x_n, x) = 0$. Since $\{x_1, \ldots, x_n, x\}$ is by assumption algebraically independent over K, we see that $\widetilde{F} = 0$. Thus $c_j(X_1, \ldots, X_n) = 0$ for all j, and consequently $c_j(x_1, \ldots, x_n) = 0$ for all j. Therefore $F = 0$, as required. □

2. Transcendence Degree 411

Lemma 7.6. If L/K is a field extension, then

(a) any transcendence set of L over K can be extended to a transcendence basis of L over K,
(b) any subset of L that generates L as a field over K has a subset that is a transcendence basis of L over K.

In particular, there exists a transcendence basis of L over K.

PROOF. For (a), order by inclusion upward the transcendence sets containing the given one. To apply Zorn's Lemma, we need only show that the union of a chain of transcendence sets in L over K is again a transcendence set. Thus let finitely many elements of the union of the sets in the chain be given. Since the sets in the chain are nested, all these elements lie in one member of the chain. Hence they are algebraically independent over K, and it follows from the definition that the union of the sets in the chain is a transcendence set. By Zorn's Lemma there exists a maximal transcendence set, and this is a transcendence basis by definition.

For (b), we argue in the same way as for (a). Let the given generating set be G. Order by inclusion upward the transcendence sets that are subsets of G. The empty set is such a transcendence set. As with (a), the union of a chain of transcendence sets in L over K is again a transcendence set, and the union is contained in G if each individual set is. By Zorn's Lemma there exists a maximal transcendence subset S of G. To complete the proof, it is enough to show that every member of G is algebraic over $K(S)$. Let x be in G. We may assume that x is not in S. By maximality, $S \cup \{x\}$ is not a transcendence set. Then Lemma 7.5 shows that x is algebraic over $K(S)$. Hence S is the required transcendence basis.

For the final conclusion we apply (a) to the empty set, which is a transcendence set of L over K. □

Theorem 7.7. If L/K is a field extension, then there exists an intermediate field K' such that K'/K is purely transcendental and L/K' is algebraic.

PROOF. Lemma 7.6 produces a transcendence basis S for L/K. Define K' to be the intermediate field $K(S)$ generated by K and S. Then K' is purely transcendental over K by definition. If x is a member of L that is not in K', then $S \cup \{x\}$ is not a transcendence set of L over K by maximality of S, and Lemma 7.5 shows that x is algebraic over $K(S) = K'$. Hence L is algebraic over K'. □

As was mentioned earlier in the section, the intermediate field K' with the properties stated in the theorem is not unique. In the example above with $\mathbb{K} = \mathbb{C}$ and with L equal to the field of fractions of $\mathbb{C}[X, Y]/(Y^2 - X(X + 1)(X - 1))$, K' can be any subfield $\mathbb{C}(z)$ with z not in the subfield $\mathbb{C}$. For an even simpler example, let K be arbitrary, and let $L = K(x)$ be any purely transcendental extension. Use of the transcendence basis $\{x\}$ of L over K leads to $K' = L$ in the proof of Theorem 7.7. But $\{x^2\}$ is another transcendence basis, and for it we have $K' = K(x^2)$. The extension L/K' is algebraic because x is a root of the polynomial $X^2 - x^2$ in $K(x^2)[X]$.

We turn to the matter of showing that any two transcendence bases of L over K have the same cardinality. We shall make use of the following result, which was proved at the end of the appendix of *Basic Algebra*:

> Let S and E be nonempty sets with S infinite, and suppose that to each element s of S is associated a countable subset E_x of E in such a way that $E = \bigcup_{s \in S} E_s$. Then card $E \leq$ card S.

In our application of this result, the sets E_x will all be finite sets.

Lemma 7.8 (Exchange Lemma). Let L/K be a field extension. If E is any subset of L, let $K(E)$ be the subfield of L generated by K and E, and let $\overline{K(E)}$ be the subfield of all elements in L that are algebraic over $K(E)$. If $E \cup \{x\}$ and $E \cup \{y\}$ are finite transcendence sets of L over K and if x lies in $\overline{K(E \cup \{y\})}$ but not $\overline{K(E)}$, then y lies in $\overline{K(E \cup \{x\})}$.

PROOF. The condition that x lie in $\overline{K(E \cup \{y\})}$ implies that there exist a finite subset $\{x_1, \ldots, x_n\}$ of E and a member f of $K(X_1, \ldots, X_n, Y)[Z]$ such that

$$f(x_1, \ldots, x_n, y, Z) \neq 0 \quad \text{but} \quad f(x_1, \ldots, x_n, y, x) = 0. \quad (*)$$

Clearing fractions, we may assume that f lies in $K[X_1, \ldots, X_n, Y, Z]$. Expand f in powers of Y as

$$f(X_1, \ldots, X_n, Y, Z) = \sum_{j=0}^{l} c_j(X_1, \ldots, X_n, Z) Y^j.$$

Since $f(x_1, \ldots, x_n, y, Z) \neq 0$ by $(*)$, at least one of the coefficients, say c_i, has to satisfy $c_i(x_1, \ldots, x_n, Z) \neq 0$. Lemma 7.5 shows that x is transcendental over $K(E)$, and therefore $c_i(x_1, \ldots, x_n, x) \neq 0$. Consequently $f(x_1, \ldots, x_n, Y, x)$ is nonzero. Since $f(x_1, \ldots, x_n, y, x) = 0$ by $(*)$, y is algebraic over $K(\{x_1, \ldots, x_n, x\})$. Therefore y lies in $\overline{K(E \cup \{x\})}$. □

The statement of Lemma 7.8 defines an operation $E \mapsto \overline{K(E)}$ on subsets of L. Because an algebraic extension of an algebraic extension is algebraic, applying this operation a second time does nothing new: $\overline{K(\overline{K(E)})} = \overline{K(E)}$. We shall make use of this fact in the proof of Theorem 7.9 below.

Theorem 7.9. If L/K is a field extension, then any two transcendence bases of L over K have the same cardinality.

2. Transcendence Degree

REMARKS. The cardinality is called the **transcendence degree** of L/K. For applications to algebraic geometry, the situation of interest is that this cardinality is finite, but we give a complete proof of the theorem anyway.

PROOF. First suppose that L/K has a finite transcendence basis B. Let $|B| = n$. Let B' be another transcendence basis, and let $m = |B \cap B'|$. We prove that $|B'| = |B|$ by induction downward on m. The base case of the induction is that $m = n$. Then $B \subseteq B'$, and we must have $B = B'$ by maximality of B.

For the inductive step, suppose that $m < n$ and that $|B'| = |B|$ whenever $|B \cap B'| \geq m + 1$. We write the elements of B in an order such that $B = \{x_1, \ldots, x_n\}$ and $B \cap B' = \{x_1, \ldots, x_m\}$. Lemma 7.5 shows that x_{m+1} is transcendental over $K(B - \{x_{m+1}\})$. Hence x_{m+1} does not lie in $\overline{K(B - \{x_{m+1}\})}$. A second application of Lemma 7.5 shows that $L = \overline{K(B')}$. The inclusion $B' \subseteq \overline{K(B - \{x_{m+1}\})}$ is impossible because otherwise we would have

$$L = \overline{K(B')} \subseteq \overline{K\big(\overline{K(B - \{x_{m+1}\})}\big)} = \overline{K(B - \{x_{m+1}\})}.$$

Hence there exists an element y of B' that does not lie in $\overline{K(B - \{x_{m+1}\})}$. A third application of Lemma 7.5 shows that $(B - \{x_{m+1}\}) \cup \{y\}$ is a transcendence set for L/K. Since y lies in $L = \overline{K(B)}$, the Exchange Lemma (Lemma 7.8) shows that x_{m+1} lies in $\overline{K((B - \{x_{m+1}\}) \cup \{y\})}$. Consequently B is contained in $\overline{K((B - \{x_{m+1}\}) \cup \{y\})}$, and $L = \overline{K((B - \{x_{m+1}\}) \cup \{y\})}$. A fourth application of Lemma 7.5 shows that the transcendence set $B_1 = (B - \{x_{m+1}\}) \cup \{y\}$ is a transcendence basis. The set B_1 has n elements, and the inclusion $B_1 \cap B' \supseteq \{x_1, \ldots, x_m, y\}$ shows that $|B_1 \cap B'| \geq m + 1$. The inductive hypothesis shows that $|B'| = |B_1|$, and therefore $|B'| = |B|$. This completes the proof under the assumption that L/K has a finite transcendence basis.

We may now suppose that L/K has no finite transcendence basis. Let B be a transcendence basis of L/K; existence is by Lemma 7.6. To each element x of L, we shall associate a canonical finite subset E_x of L.

Since the element x is algebraic over $K(B)$, use of the field polynomial of x over $K(B)$ shows that x is algebraic over $K(E)$ for some finite subset E of B. Let E_0 be such a finite set E with the smallest cardinality; the set E_0 will be the canonical finite subset E_x that we seek. To show that E_0 is canonical, we show that whenever x lies in $\overline{K(E)}$ for some finite subset E of B, then $E_0 \subseteq E$. Arguing by contradiction, suppose that y is a member of E_0 that is not in E, and define $E_1 = E_0 - \{y\}$. By minimality of $|E_0|$, x does not lie in $\overline{K(E_1)}$. However, x does lie in $\overline{K(E_1 \cup \{y\})}$. Application of the Exchange Lemma shows that y lies in $\overline{K(E_1 \cup \{x\})}$. Since

$$\overline{K(E_1 \cup \{x\})} \subseteq \overline{K(E_1 \cup \overline{K(E)})} \subseteq \overline{K\big(\overline{K(E_1 \cup E)}\big)} = \overline{K(E_1 \cup E)},$$

y lies in $\overline{K(E_1 \cup E)}$. Since y is in B but is not in $E_1 \cup E$, Lemma 7.5 shows that y is not algebraic over $K(E_1 \cup E)$, and we arrive at a contradiction. This completes the proof that whenever x lies in $\overline{K(E)}$ for some finite subset E of B, then $E_0 \subseteq E$. Hence E_0 is canonical.

For each element x of L, we let E_x be the finite subset of B constructed in the previous paragraph. Then we have a well-defined map of L to the set of all finite subsets of B given by $x \mapsto E_x \subseteq B$. Now let B' be a second transcendence basis of L/K, and restrict the map from L to B'. Taking $S = B'$ and $E = \bigcup_{x \in B'} E_x$ in the indented result quoted just before Lemma 7.8, we find that

$$\operatorname{card}\Big(\bigcup_{x \in B'} E_x\Big) \leq \operatorname{card}(B'). \tag{$*$}$$

On the other hand, any x in B' lies in $\overline{K(E_x)}$ by definition of E_x. Hence $B' \subseteq \overline{K\big(\bigcup_{x \in B'} E_x\big)}$. Applying the operation $\overline{K(\cdot)}$ to both sides gives

$$L = \overline{K(B')} \subseteq \overline{K\big(\overline{K(\bigcup_{x \in B'} E_x)}\big)} = \overline{K\big(\bigcup_{x \in B'} E_x\big)}.$$

Since $\bigcup_{x \in B'} E_x$ is a subset of B and since a proper subset of B cannot be a transcendence basis of L/K, we conclude that

$$B = \bigcup_{x \in B'} E_x.$$

Consequently

$$\operatorname{card} B = \operatorname{card}\Big(\bigcup_{x \in B'} E_x\Big).$$

In combination with $(*)$, this equality implies that $\operatorname{card} B \leq \operatorname{card} B'$. Reversing the roles of B and B' gives $\operatorname{card} B' \leq \operatorname{card} B$. Therefore $\operatorname{card} B = \operatorname{card} B'$ by the Schroeder–Bernstein Theorem.[3] □

3. Separable and Purely Inseparable Extensions

Thus far in this book, we have been interested in the detailed structure of algebraic field extensions only when they are separable. For applications to algebraic geometry, however, algebraic extensions that are not separable arise and even play a special role. Thus it is essential to have some understanding of their nature.

Let us review the material on separability in Section IX.6 of *Basic Algebra*. Let K be a field. A polynomial in $K[X]$ is defined to be **separable** if it splits into distinct first-degree factors in its splitting field over K. Let L/K be an algebraic extension of fields. An element of L is defined to be **separable** over K if its minimal polynomial over K is separable. Elements of L that fail to be separable over K are called **inseparable** over K. The prototype of an inseparable

---
[3] A proof of the Schroeder–Bernstein Theorem appears in the appendix of *Basic Algebra*.

3. Separable and Purely Inseparable Extensions

element is the element $a^{1/p}$ in the extension $\Bbbk(a^{1/p})$, where $\Bbbk = \mathbb{F}_p(a)$ is a simple transcendental extension of the finite field $\mathbb{F}_p$. Corollary 9.31 of *Basic Algebra* shows that the separable elements of L over K form a subfield, and L/K is defined to be separable if every every member of L is separable over K. As a consequence of Corollary 9.29 of *Basic Algebra*, we know that a separable extension of a separable extension is separable.

One further tool from *Basic Algebra* is needed in order to handle the failure of separability. This is Proposition 9.27, which says that an irreducible polynomial $f(X)$ in $K[X]$ is separable if and only if $f'(X)$ is not the zero polynomial. It is immediate that every irreducible polynomial is separable if K has characteristic 0. Thus we need discuss only characteristic p in the remainder of this section.

The consequence of Proposition 9.27 for characteristic p is that an irreducible polynomial $f(X)$ fails to be separable over K if and only if the only powers of X that appear with nonzero coefficient in $f(X)$ are the powers X^{kp}, i.e., if and only if $f(X) = g(X^p)$ for some g in $K[X]$.

In this case the polynomial $g(X)$ is certainly irreducible in $K[X]$, and we can repeat this process. The polynomial $g(X)$ fails to be separable over $K[X]$ if and only if $g(X) = h(X^p)$ for some h in $K[X]$. Then $f(X) = h(X^{p^2})$. Repeating this process as many times as possible, we see that to each irreducible polynomial $f(X)$ in $K[X]$ correspond a unique nonnegative integer e and a unique *separable* irreducible polynomial $g(X)$ such that $f(X) = g(X^{p^e})$. We call p^e the **degree of inseparability** of $f(X)$ over K. From the definitions an element of an algebraic extension of K is inseparable if and only if the degree of inseparability of its minimal polynomial over K is greater than 1.

If L/K is an algebraic field extension, then an element α of L is said to be **purely inseparable**[4] over K if α^{p^μ} lies in K for some integer $\mu \geq 0$. Let us see in this case that the minimal polynomial of α over K is of the form $X^{p^e} - \alpha^{p^e}$ for some $e \geq 0$.

Proposition 7.10. If K is a field of characteristic p and if α is a member of K such that $\sqrt[p]{\alpha}$ is not in K, then $X^{p^\mu} - \alpha$ is irreducible in $K[X]$ for every $\mu \geq 0$.

PROOF. Let L be a splitting field of $X^{p^\mu} - \alpha$ over K. If β is a root of $X^{p^\mu} - \alpha$, then $\beta^{p^\mu} = \alpha$, and hence $X^{p^\mu} - \alpha = X^{p^\mu} - \beta^{p^\mu} = (X - \beta)^{p^\mu}$.

Let $f(X)$ be a monic irreducible factor of $X^{p^\mu} - \alpha$ in $K[X]$. Let us see that $X^{p^\mu} - \alpha = f(X)^n$ for some n. In fact, if the contrary were true, then there would be a second monic irreducible factor $g(X)$ of $X^{p^\mu} - \alpha$ in $K[X]$ relatively prime to $f(X)$. Then we can write $u(X)f(X) + v(X)g(X) = 1$ for suitable

---

[4]*Warning*: Not every element of L that is purely inseparable over K is inseparable over K. The elements of K are counterexamples. Corollary 7.12 below shows that the elements of K are the only counterexamples.

polynomials $u(X)$ and $v(X)$ in $K[X]$. As members of $L[X]$, both $f(X)$ and $g(X)$ have to be powers of $X - \beta$ by unique factorization, and thus they both vanish at β. Substitution of β into $uf + vg = 1$ therefore yields a contradiction. Hence $X^{p^\mu} - \alpha = f(X)^n$.

Since $f(X)$ has to be $(X - \beta)^m$ for some m, we obtain $X^{p^\mu} - \alpha = f(X)^n = (X - \beta)^{mn}$. The integers m and n must divide p^μ. Thus $m = p^\nu$, and $f(X) = (X - \beta)^{p^\nu} = X^{p^\nu} - \beta^{p^\nu}$. Since $f(X)$ is assumed to be in $K[X]$, β^{p^ν} lies in K. An inequality $\nu < \mu$ would imply that $\gamma = (\beta^{p^\nu})^{p^{\mu-\nu-1}}$ lies in K; the pth power of γ is α, however, and the hypothesis of the proposition says that such an element γ cannot be in K. We conclude that $\nu = \mu$, and thus $f(X) = X^{p^\mu} - \alpha$. In other words, $X^{p^\mu} - \alpha$ is irreducible in $K[X]$. □

Corollary 7.11. If L/K is an algebraic extension in characteristic p, if α is a purely inseparable element of L over K, and if e is the smallest nonnegative integer such that α^{p^e} lies in K, then the minimal polynomial of α over K is $X^{p^e} - \alpha^{p^e}$.

PROOF. This is immediate from Proposition 7.10. □

Corollary 7.12. If L/K is an algebraic extension in characteristic p and if α is an element of L that is separable and purely inseparable over K, then α lies in K.

PROOF. Since α is purely inseparable over K, Corollary 7.11 says that the minimal polynomial of α over K is $X^{p^e} - \alpha^{p^e}$, where e is the smallest nonnegative integer such that α^{p^e} lies in K. The separability of α says that this polynomial is separable. Unless $p^e = 1$, the polynomial has derivative 0 and thus repeated roots. Therefore $p^e = 1$ and $e = 0$, and we conclude that α lies in K. □

An algebraic field extension L/K in characteristic p is said to be **purely inseparable** if every element of L is purely inseparable over K. Since purely inseparable elements α have minimal polynomials of the form $X^{p^e} - \alpha^{p^e}$, the degree of a purely inseparable extension has to be a power of p.

Theorem 7.13. If L/K is an algebraic field extension in characteristic p and if K_s is the subfield of all elements of L that are separable over K, then L/K_s is a purely inseparable extension.

PROOF. Let α be an element of L, and let $f(X)$ be the minimal polynomial of α over K. Then we can write $f(X) = g(X^{p^e})$, where p^e is the degree of inseparability of f. The polynomial $g(X)$ is irreducible over K, and it is separable. Since α^{p^e} is a root, α^{p^e} is a separable element. Therefore α^{p^e} lies in K_s. By definition of pure inseparability, α is purely inseparable over K_s. Since α is arbitrary in L, L is purely inseparable over K_s. □

3. Separable and Purely Inseparable Extensions

Corollary 7.14. Let R be a Dedekind domain, let F be its field of fractions, let K be a finite algebraic extension of F, and let T be the integral closure of R in K. Then T is a Dedekind domain.

REMARKS. This result is quite important. It was used extensively in Chapter VI, as was explained in the remarks with Proposition 6.7, and it plays a foundational role in the theory of algebraic curves as presented in Chapters IX and X. Theorem 8.54 of *Basic Algebra* proved this result under the assumption that K is a finite separable extension of F, and we are now dropping the hypothesis of separability. Since K/F is automatically a separable extension in characteristic 0, we may assume that the characteristic is not 0.

PROOF. Theorem 7.13 shows that K can be obtained in two steps from F, a separable extension followed by a purely inseparable extension. The integral closure of F in the separable extension field is a Dedekind domain D by Theorem 8.54 of *Basic Algebra*, and the integral closure of D in K equals T by the transitivity of integral closure. Consequently it is enough to prove the corollary under the additional hypothesis that K is a purely inseparable extension of F. What needs proof (in view of the statement of Theorem 8.54 of *Basic Algebra*) is that T is Noetherian, i.e., that each ideal of T is finitely generated.

Let p be the characteristic. Since K/F is finite and purely inseparable, there exists some power $q = p^m$ of p such that the field K^q is contained in F; specifically, the integer q is to be large enough for the qth power of each element of a vector-space basis of K over F to lie in F. We begin by proving that

$$T = \{b \in K \mid b^q \in R\}. \qquad (*)$$

The inclusion $\subseteq$ follows, since $b \in T$ implies that b^q is in $T \cap F = R$. For the inclusion $\supseteq$, let $b \neq 0$ be in K. Corollary 7.11 shows that the minimal polynomial of b over F is $X^{p^e} - b^{p^e}$, where e is the smallest integer ≥ 0 such that b^{p^e} lies in F. Since $K^{p^m} \subseteq F$, $e \leq m$. Thus b is a root of a polynomial $X^{p^m} - a$, where $a = b^{p^m}$ is a member of R. Consequently b is integral over R and must lie in T. This proves $(*)$.

Fix an algebraic closure K_{alg} of K, and let $H = F^{q^{-1}}$ denote the inverse image of F under the qth power isomorphism of K_{alg} onto itself. This is a subfield of K_{alg}, and it contains K because $K^q \subseteq F$. Let $S \subseteq H$ be the ring of all b in H with b^q in R. Since $x \mapsto x^q$ is a field isomorphism of H onto F, $x \mapsto x^q$ is a ring isomorphism of S onto R. Therefore S is a Dedekind domain. It contains T by $(*)$.

Let I be a nonzero ideal in T, and form the ideal $J = SI$ in S generated by I. Since S is Dedekind, J is invertible as a fractional ideal of H relative to S. If J^{-1} denotes the inverse, then J^{-1} is a finitely generated S module in H such that

$J^{-1}J = S$. Thus $S = J^{-1}J = J^{-1}SI = J^{-1}I$. Accordingly, choose finite sets $\{x_i\}$ in J^{-1} and $\{a_i\}$ in I such that $\sum x_i a_i = 1$.

We shall show that $\{a_i\}$ is a set of generators of I as an ideal in T. We apply the qth power mapping to $\sum x_i a_i = 1$, obtaining $\sum x_i^q a_i^q = 1$ with x_i^q in $H^q = F \subseteq K$ and with a_i^q in $S^q = R$. Put $b_i = a_i^{q-1} x_i^q$. Then $\sum x_i^q a_i^q = 1$ implies that $\sum a_i b_i = 1$; here a_i is in I and b_i is in $I^{q-1}K \subseteq K$. If a is in I, then $\sum (b_i a) a_i = a$, and it is enough to show that $b_i a$ is in T for each i, i.e., to show that $b_i I \subseteq T$ for each i.

The q-fold product $(x_i I) \cdots (x_i I)$ is contained in S because $x_i I \subseteq J^{-1}J = S$. Thus $b_i I = x_i^q a_i^{q-1} I \subseteq S$. So $b_i I \subseteq S \cap K$. If s is any element in $S \cap K$, then we know that $r = s^q$ is a member of R because $S^q = R$. Hence s is a root of $X^q - r$ with r in R. That is, s is integral over R. Since s also is in K, s lies in the integral closure of R in K, which is T. Thus $b_i I \subseteq T$, and the proof is complete. $\square$

A field K is **perfect** if either it has characteristic 0 or else it has characteristic p and the field map $x \mapsto x^p$ of K into itself is onto. Examples of perfect fields include all finite fields, all algebraically closed fields, and of course all fields of characteristic 0.

Proposition 7.15. A field K is perfect if and only if every algebraic extension of K is separable.

PROOF. We need to consider only the case that K has characteristic p. Suppose that $x \mapsto x^p$ fails to be onto K. Choose β in K such that $X^p - \beta$ has no root in K. Proposition 7.10 shows that $X^p - \beta$ is irreducible over K. Since this polynomial has derivative 0, it is not separable. Thus $X^p - \beta$ is a polynomial that is irreducible but not separable, and adjunction of a root of $X^p - \beta$ to K produces an extension L of K that is not separable.

Conversely suppose that the field map $x \mapsto x^p$ of K to itself is onto. Then $x \mapsto x^{p^e}$ is onto K for every $e \geq 0$. Let L be an algebraic extension of K, and let K_s be the subfield of elements separable over K. If α is given in L, then Theorem 7.13 shows that there exists a nonnegative integer e such that α^{p^e} is in K_s. Let $g(X)$ be the minimal polynomial of α^{p^e} over K, and write $g(X) = X^m + c_1 X^{m-1} + \cdots + c_m$. Since K is perfect, there exists b_j for each j with $1 \leq j \leq m$ such that $b_j^{p^e} = c_j$. Put $f(X) = X^m + b_1 X^{m-1} + \cdots + b_m$. Then

$$f(\alpha)^{p^e} = (\alpha^{p^e})^m + b_1^{p^e}(\alpha^{p^e})^{m-1} + \cdots + b_m^{p^e} = g(\alpha^{p^e}) = 0,$$

and therefore $f(\alpha) = 0$. Consequently $f(X)$ divides the minimal polynomial of α over K, and the fact that α^{p^e} lies in $K(\alpha)$ implies that

$$[K(\alpha) : K] \leq \deg f(X) = \deg g(X) = [K(\alpha^{p^e}) : K] \leq [K(\alpha) : K].$$

Equality must hold throughout, and therefore $K(\alpha) = K(\alpha^{p^e})$. Since $K(\alpha^{p^e})$ is contained in K_s, α lies in K_s. Therefore every member of L lies in K_s, and L is separable over K. $\square$

A **function field** in r variables over a field K is a field L that is finitely generated over K and has transcendence degree r over K. A transcendence basis $\{x_1, \ldots, x_r\}$ of such an extension L/K is called a **separating transcendence basis** of L/K if L is a separable algebraic extension of $K(x_1, \ldots, x_r)$. If the function field L in r variables over K has a separating transcendence basis, we say that L is **separably generated** over K.

The two kinds of fields of continual interest in Chapter VI were number fields and function fields in one variable over a base field. In the latter case some results beginning in Section VI.6 assumed in effect that the function field is separably generated over the base field. It was asserted at the beginning of Section VI.9 that function fields in one variable over finite fields are always separably generated; this assertion is a special case of Theorem 7.20 below.

Proposition 4.28 of *Basic Algebra* gave a version of the Factor Theorem valid for all commutative rings with identity. For the present investigation we need a version of the division algorithm that is valid in this wider context.

Lemma 7.16. Let R be a commutative ring with identity, let $f(X)$ and $g(X)$ be members of $R[X]$ of respective degrees m and n, and let a be the leading coefficient of $g(X)$. For the integer $k = \max(m-n+1, 0)$, there exist $q(X)$ and $r(X)$ in $R[X]$ such that

$$a^k f(X) = g(X)q(X) + r(X) \qquad \text{with } \deg r < n \text{ or } r = 0.$$

PROOF. If $m < n$, then $k = 0$, and the displayed formula holds with $q(X) = 0$ and $r(X) = f(X)$. For $m \geq n - 1$, we proceed by induction on m. The base case of the induction is $m = n - 1$, which we have already handled. For the inductive step, suppose that $m \geq n$. The integer k is $m - n + 1$. If b is the leading coefficient of $f(X)$, then $af(X) - bX^{m-n}g(X)$ is a polynomial that either is 0 or has degree less than m. The inductive hypothesis allows us to write

$$a^{(m-1)-n+1}\bigl(af(X) - bX^{m-n}g(X)\bigr) = g(X)q_1(X) + r_1(X)$$

with $\deg r_1 < n$ or $r_1 = 0$. If we set $q(X) = ba^{m-n}X^{m-n} + q_1(X)$ and $r(X) = r_1(X)$, then we obtain $a^k f(X) = g(X)q(X) + r(X)$, and the lemma follows. $\square$

Lemma 7.17. Let L/K be a field extension, let $x_1, \ldots, x_n, x_{n+1}$ be elements of L, and suppose that $x_1, \ldots, x_n$ are algebraically independent over K but that $x_1, \ldots, x_n, x_{n+1}$ are not algebraically independent. Then the ideal I of all polynomials in $K[X_1, \ldots, X_{n+1}]$ that vanish at $(x_1, \ldots, x_{n+1})$ is principal with a generator that is irreducible in $K[X_1, \ldots, X_{n+1}]$ and involves X_{n+1} nontrivially.

PROOF. The algebraic dependence implies that I contains nonzero polynomials. Let $g(X_1, \ldots, X_n, X_{n+1})$ be one whose degree in X_{n+1} is as small as possible, say l. Expand g as

$$g = c_0(X_1, \ldots, X_n) X_{n+1}^l + c_1(X_1, \ldots, X_n) X_{n+1}^{l-1} + \cdots + c_l(X_1, \ldots, X_n).$$

The algebraic independence of $X_1, \ldots, X_n$ implies that at least one of $c_0, \ldots, c_{l-1}$ is nonzero. Since $K[X_1, \ldots, X_n]$ is a unique factorization domain, we can factor out and discard the greatest common divisor of the coefficients $c_0, \ldots, c_l$. Thus we may assume that g is primitive as a polynomial in X_{n+1}. If f is any element in I, then Lemma 7.16 applied to the ring $K[X_1, \ldots, X_n]$ allows us to write $a^k f = gq + r$ with $r = 0$ or $\deg r < k$. Substituting $(x_1, \ldots, x_{n+1})$, we see that r is in I. The minimality of l implies that $r = 0$, and thus $a^k f = gq$. Write $c(h)$ for the greatest common divisor of the coefficients of a polynomial h. Taking the greatest common divisor of the coefficients on each side of $a^k f = gq$ and applying Gauss's Lemma, we obtain $a^k c(f) = c(q)$. Therefore a^k divides q, and we obtain $f = gq_0$ for some q_0. Consequently I is principal. If $g = g_1 g_2$, then the definition of I shows that at least one of g_1 and g_2 is in I, say g_1. The minimality of l implies that the degree of g_1 in X_{n+1} is l. Therefore g_2 is in $K[X_1, \ldots, X_n]$. Since g is primitive, g_2 divides 1. Hence g_2 lies in K. □

Theorem 7.18 (Mac Lane). *If L/K is a field extension that is finitely generated and separably generated, then any set of generators contains a subset that is a separating transcendence basis of L/K.*

PROOF. Let the characteristic be p. The proof is by induction on the transcendence degree of the extension. For transcendence degree 0, the required set is the empty set, and there is nothing to prove. The main step is transcendence degree 1.

Thus let $L = K(x_1, \ldots, x_n)$, and suppose that $\{z\}$ is a transcendence basis of L over K such that L is separable over $K(z)$. Since z is transcendental, z does not lie in $K(z^p)$. Thus Proposition 7.10 shows that $X^p - z^p$ is irreducible over $K(z^p)$, and z is inseparable over $K(z^p)$. The field L is algebraic over $K(z^p)$, and the subset of separable elements over $K(z^p)$ is a subfield. Since $L = K(x_1, \ldots, x_n)$ and since z is a member of L that is not separable over $K(z^p)$, it follows that some x_i, say x_1, is inseparable over $K(z^p)$. It will be proved that $\{x_1\}$ is a separating transcendence basis of L over K, i.e., that x_1 is transcendental over K and that L is separable algebraic over $K(x_1)$.

We apply Lemma 7.17 with $n = 2$ to the elements z, x_1. The lemma produces an irreducible polynomial $f(Z, X)$ in $K[Z, X]$ such that $f(z, x_1) = 0$. Gauss's Lemma shows that this polynomial remains irreducible when considered in $K(Z)[X]$, and we have a ring isomorphism $K(Z)[X] \cong K(z)[X]$ because z is transcendental over K. Up to a nonzero factor from $K(z)$, $f(z, X)$ is the minimal polynomial of x_1 over $K(z)$. Since L is separable over $K(z)$, the element x_1 is separable over $K(z)$, and its minimal polynomial over $K(z)$ involves some power of X that is not a power of X^p.

Let us prove that x_1 is transcendental over K. In the contrary case, let $g(X)$ be its minimal polynomial over K. Since g vanishes when $X = x_1$ and $Z = z$, $g(X)$ satisfies an identity $g(X) = q(Z, X)f(Z, X)$ in $K[Z, X]$. It therefore satisfies the same identity in $K(X)[Z]$. Since $g(X)$ is a unit in $K(X)[Z]$, so is $f(Z, X)$. Therefore $f(Z, X)$ is independent of Z. Since $g(X)$ is the minimal polynomial for x_1 over K, $g(X) = cf(Z, X)$ for some c in K. Since $f(Z, X)$ involves a power of X that is not a power of X^p, the same thing is true of $g(X)$, and consequently x_1 is separable over K. Therefore x_1 is separable over the larger field $K(z^p)$, in contradiction to the defining condition on x_1. We conclude that x_1 is transcendental over K.

Since L has transcendence degree 1 over K, it follows that z is algebraic over $K(x_1)$. Let us see that z is separable over $K(x_1)$. In fact, Gauss's Lemma shows that $f(Z, X)$ remains irreducible when considered in $K(X)[Z]$, and we have a ring isomorphism $K(X)[Z] \cong K(x_1)[Z]$ because x_1 is transcendental over K. Therefore $f(Z, x_1)$ is the product of a nonzero member of $K(x_1)$ and the minimal polynomial $m(Z)$ of z over $K(x_1)$. If z were inseparable over $K(x_1)$, then $m(Z)$ would be a polynomial in Z^p, and we would have $f(Z, X) = h(Z^p, X)$ with h in $K[Z, X]$. We know that $f(Z, X)$ involves some power of X that is not a power of X^p, and hence the same thing is true of $h(Z^p, X)$. Since $h(z^p, X)$ is irreducible in $K[X]$, x_1 is separable over $K(z^p)$, in contradiction to the defining property of x_1. Therefore z is separable over $K(x_1)$.

The defining property of z is that all x_j are separable over $K(z)$. Since z is separable over $K(x_1)$, all of $x_2, \ldots, x_n$ are separable over $K(x_1)$. Therefore L is separable over x_1, and $\{x_1\}$ is a separable transcendence basis of L/K. This completes the proof of the theorem for transcendence degree 1.

The inductive step is somewhat a formal consequence of what has just been proved. To see this, suppose that the theorem is known for transcendence degrees 1 and $r - 1$, and let $L = K(x_1, \ldots, x_n)$ have transcendence degree r. The assumption is that L has a transcendence basis $\{z_1, \ldots, z_r\}$ such that L is separable over $K(z_1, \ldots, z_r)$. Put $K_1 = K(z_1)$. Then the set $\{z_2, \ldots, z_r\}$ is a transcendence basis of L over K_1 consisting of $r - 1$ elements, and L is separable over $K_1(z_2, \ldots, z_r) = K(z_1, \ldots, z_r)$ by assumption. By the inductive hypothesis for the case of transcendence degree $r - 1$, some subset of $r - 1$ elements from among $x_1, \ldots, x_n$ forms a separating transcendence basis of L over K_1; let us say that this basis is $\{x_1, \ldots, x_{r-1}\}$. This implies that L is separable over $K_1(x_1, \ldots, x_{r-1}) = K(z_1, x_1, \ldots, x_{r-1})$. In other words, if $K' = K(x_1, \ldots, x_{r-1})$, then $L = K'(x_r, \ldots, x_n)$ is separable over $K'(z_1)$. Since

L/K' has transcendence degree 1, $\{z\}$ is a separating transcendence basis of L/K'. By the inductive hypothesis for transcendence degree 1, some x_j for $r \leq j \leq n$ forms a separating transcendence basis of L/K'. For this j, $\{x_1, \ldots, x_{r-1}, x_j\}$ is then a separating transcendence basis of L/K. □

Lemma 7.19. Suppose that L is a field extension of transcendence degree r over a field K and that L is not separably generated over K. If $x_1, \ldots, x_n$ are elements of L such that $L = K(x_1, \ldots, x_n)$, then for a suitable relabeling of the x_i's, the subfield $K(x_1, \ldots, x_{r+1})$ of L is of transcendence degree r and is not separably generated over K.

PROOF. We fix K and r, and we proceed by induction on n. The base case is that $n = r + 1$, and then there is nothing to prove. For the inductive step, suppose that the lemma has been proved for $n - 1$ when $n > r + 1$. We prove the lemma for n. Since $r < n$, we can renumber the x_i's and assume that $K(x_2, \ldots, x_n)$ has transcendence degree r over K. If this field is not separably generated over K, then we are in a situation with $n - 1$ elements. The inductive hypothesis is applicable, and the lemma follows in this case.

Thus suppose that $K(x_2, \ldots, x_n)$ is separably generated over K. Theorem 7.18 shows that after a renumbering of the indices, we may assume that $\{x_2, \ldots, x_{r+1}\}$ is a separating transcendence basis of $K(x_2, \ldots, x_n)$ over K. This implies that $K(x_2, \ldots, x_n)$ is a separable extension of $K(x_2, \ldots, x_{r+1})$. Since by assumption $L = K(x_1, \ldots, x_n)$ is not separably generated over K, $K(x_1, \ldots, x_n)$ is not separable over $K(x_2, \ldots, x_{r+1})$. A separable extension of a separable extension is separable, and we deduce that $K(x_1, \ldots, x_n)$ is not separable over $K(x_2, \ldots, x_n)$. Thus x_1 is inseparable over $K(x_2, \ldots, x_n)$ and is consequently inseparable over the subfield $K(x_2, \ldots, x_{r+1})$. Hence $K(x_1, \ldots, x_{r+1})$ is not separably generated over K. □

Theorem 7.20 (F. K. Schmidt). If K is a perfect field, then every finitely generated field extension of K is separably generated over K.

REMARK. In particular, the theorem applies if K is a finite field or is algebraically closed or has characteristic 0.

PROOF. Let K have characteristic p. We induct on the transcendence degree of the field extension of K. The base case of the induction is transcendence degree 0, and then the theorem is handled by Proposition 7.15. For the inductive step, assume that the theorem holds for all finitely generated field extensions of K having transcendence degree $r - 1$ over K. Let $L = K(x_1, \ldots, x_n)$ have transcendence degree r over K. Arguing by contradiction, suppose that L is not separably generated over K. Lemma 7.19 shows for a suitable renumbering of the

x_i's that $K' = K(x_1, \ldots, x_{r+1})$ has transcendence degree r and is not separably generated over K.

We divide matters into two cases. First suppose that the transcendence degree of $K'' = K(x_1, \ldots, x_r)$ is $r - 1$. The inductive hypothesis shows that K'' is separably generated over K, and then Theorem 7.18 shows that we may renumber the variables in such a way that $\{x_1, \ldots, x_{r-1}\}$ is a transcendence basis of K'' over K and K'' is separable algebraic over $K(x_1, \ldots, x_{r-1})$. Then $\{x_1, \ldots, x_{r-1}, x_{r+1}\}$ is a transcendence basis of K', and x_r is algebraic over $K(x_1, \ldots, x_{r-1}, x_{r+1})$. Since x_r is separable over $K(x_1, \ldots, x_{r-1})$, it is separable over the larger field $K(x_1, \ldots, x_{r-1}, x_{r+1})$. Therefore K' is separably generated over K, contradiction.

The remaining case is that every subset of r members of $\{x_1, \ldots, x_{r+1}\}$ is a transcendence basis of K' over K. Lemma 7.17 produces an irreducible polynomial f in $K[X_1, \ldots, X_{r+1}]$ such that $f(x_1, \ldots, x_{r+1}) = 0$. Since $\{x_1, \ldots, x_r\}$ is a transcendence basis of K', application of Gauss's Lemma shows that f is irreducible in $K(X_1, \ldots, X_r)[X_{r+1}] \cong K(x_1, \ldots, x_r)[X_{r+1}]$. Hence up to a nonzero factor from K, $f(x_1, \ldots, x_r, X_{r+1})$ is the minimal polynomial of x_{r+1} over $K(x_1, \ldots, x_r)$. The failure of K' to be separably generated over K implies that x_{r+1} is inseparable over $K(x_1, \ldots, x_r)$, and thus the only powers of X_{r+1} that appear in its minimal polynomial over $K(x_1, \ldots, x_r)$ are powers X_{r+1}^{pk}. In other words, f is in $K[X_1, \ldots, X_r, X_{r+1}^p]$. Since we are assuming that any r of the elements $x_1, \ldots, x_{r+1}$ form a transcendence basis of K' over K, there is nothing special about X_{r+1} in this argument. Consequently f is in $K[X_1^p, \ldots, X_r^p, X_{r+1}^p]$. Since K is perfect, any polynomial involving only p^{th} powers of each indeterminate is the pth power of some polynomial. Consequently f is reducible in $K[X_1, \ldots, X_{r+1}]$, in contradiction to the irreducibility guaranteed by Lemma 7.17. All cases thus lead to a contradiction, and the proof is complete. $\square$

4. Krull Dimension

In this section we develop the algebraic background necessary for a discussion of dimension. Suppose that K is an algebraically closed field, suppose that I is a prime ideal in $K[X_1, \ldots, X_n]$, and suppose that $V(I)$ is the locus of common zeros of I. Corollary 7.2 shows that I is the set of all polynomials vanishing on $V(I)$, and thus the integral domain $R = K[X_1, \ldots, X_n]/I$ may be regarded as the set of all restrictions to $V(I)$ of polynomials. If L is the field of fractions of R, then the transcendence degree of L/K is interpreted as the "number of independent variables" on the locus $V(I)$. We define it to be the **dimension** of $V(I)$. The elements $X_j + I$ of R for $1 \leq j \leq n$ generate R as a K algebra, and therefore they generate L over K as a field. We shall make critical use of the fact implied by Lemma 7.6b that some subset of $\{X_1 + I, \ldots, X_n + I\}$ is a transcendence basis of L. We shall speak of such a subset as a **transcendence basis** of R for economy of words. We denote its cardinality by tr. deg R.

EXAMPLE. We continue with the example from Sections 1–2. Let $K = \mathbb{C}$, let I be the principal ideal $(Y^2 - X(X+1)(X-1))$ in $\mathbb{C}[X, Y]$, and let L be the field of fractions of the integral domain $R = \mathbb{C}[X, Y]/I$. Corollary 7.2 shows that the ring R is the ring of restrictions of polynomials to the locus $V(I) = \{(x, y) \in \mathbb{C}^2 \mid y^2 = x(x+1)(x-1)\}$. According to the above definition, the dimension of $V(I)$ is the transcendence degree of L, which we have seen is 1. This is in accord with the intuition that the locus $V(I)$ is a "curve" in the sense of having one independent complex parameter.

The goal of this section is to produce an equivalent definition of dimension that does not depend on the fact that $K[X_1, \ldots, X_n]/I$ is an integral domain. The rephrased definition will extend to any commutative ring with identity and is essential for modern algebraic geometry.

Let R be any commutative ring with identity. The **Krull dimension** of R, denoted by dim R, is the supremum of the indices d of all strictly increasing chains

$$P_0 \subsetneq P_1 \subsetneq \cdots \subsetneq P_d$$

of prime ideals in R. We define dim $R = \infty$ if there is no finite supremum.

EXAMPLES OF KRULL DIMENSION.

(1) R equal to a field. The only prime ideal is 0. Thus the Krull dimension of any field is 0.

(2) $R = \mathbb{Z}$. The prime ideals are of the form $p\mathbb{Z}$ for each prime number p, together with 0. Each nonzero prime ideal is maximal. Consequently there is a strictly increasing chain $0 \subsetneq p\mathbb{Z}$ of prime ideals for each prime number p, but there are no longer such chains. Thus dim $\mathbb{Z} = 1$. More generally any principal ideal domain R that is not a field, or even any Dedekind domain R that is not a field, has dim $R = 1$ because every nonzero prime ideal is maximal.

(3) R commutative Artinian. In Chapter II a ring with identity was defined to be Artinian if its two-sided ideals satisfy the descending chain condition. Problem 8 at the end of that chapter showed that *every* prime ideal in such a ring is maximal. In other words, every commutative Artinian ring has Krull dimension 0.

(4) Polynomial ring $R = K[X_1, \ldots, X_n]$, where K is a field. In geometric terms for the case that K is algebraically closed, the relevant zero locus for this R is K^n, which we certainly want to have dimension equal to n, and the field of fractions of R is $K(X_1, \ldots, X_n)$, which indeed has transcendence degree n. Let

us examine the Krull dimension of R. If $0 \leq k \leq n$ and if we form the ideal $(X_1, \ldots, X_k)$, then the ring isomorphism

$$R \cong K[X_{k+1}, \ldots, X_n][X_1, \ldots, X_k]$$

shows that the quotient $R/(X_1, \ldots, X_k)$ is isomorphic to $K[X_{k+1}, \ldots, X_n]$, which is an integral domain. Therefore $(X_1, \ldots, X_k)$ is prime, and we have a strictly increasing chain

$$0 \subsetneqq (X_1) \subsetneqq \cdots \subsetneqq (X_1, \ldots, X_{n-1}) \subsetneqq (X_1, \ldots, X_n).$$

So $\dim K[X_1, \ldots, X_n] \geq n$. Actually, equality holds, as Theorem 7.22 will show.

Lemma 7.21. Let R be a commutative ring with identity, let $S^{-1}R$ be the localization relative to a multiplicative system S in R, let I be an ideal in R, and let $\overline{S}$ be the image of S in R/I. Then

$$S^{-1}R \big/ S^{-1}I \cong \overline{S}^{-1}(R/I)$$

via the mapping $s^{-1}r + S^{-1}I \mapsto (s+I)^{-1}(r+I)$.

PROOF. Let $q : R \to R/I$ and $\bar{q} : S^{-1}R \to S^{-1}R/S^{-1}I$ be the quotient homomorphisms, and let $\eta : R \to S^{-1}R$ and $\bar{\eta} : R/I \to \overline{S}^{-1}(R/I)$ be the canonical homomorphisms of R and R/I into their localizations. To each of the rings $X_1 = S^{-1}R\big/S^{-1}I$ and $X_2 = \overline{S}^{-1}(R/I)$ is associated a canonical map, namely $\eta_1 : R \to X_1$ and $\eta_2 : R \to X_2$ with $\eta_1 = \bar{q}\eta$ and $\eta_2 = \bar{\eta}q$. Let us see that the pairs (X_i, η_i) for $i = 1, 2$ have the following universal mapping property with respect to ring homomorphisms φ of R into a commutative ring T with identity such that $\varphi(1) = 1$, $\varphi(I) = 0$, and $\varphi(S) \subseteq T^\times$: there exists a unique homomorphism $\overline{\varphi}_i : X_i \to T$ such that $\varphi = \overline{\varphi}_i \eta_i$.

For $i = 1$, we first apply the universal mapping property of the localization $S^{-1}R$ to write $\varphi = \varphi_1 \eta$ and then apply the universal mapping property of the quotient to write $\varphi = \overline{\varphi}_1 \bar{q} \eta$. For $i = 2$, we first apply the universal mapping property of the quotient R/I to write $\varphi = \varphi_2 q$ and then apply the universal mapping property of the localization to write $\varphi = \overline{\varphi}_2 \bar{\eta} q$. From these constructions we deduce existence and uniqueness of $\overline{\varphi}_i$ in both cases. The asserted isomorphism then follows from the general fact that objects satisfying a universal mapping property are unique up to isomorphism; tracking down that isomorphism gives the explicit formula in the lemma. □

Theorem 7.22. Let K be a field, let R be an integral domain that is finitely generated as a K algebra, and let L be the field of fractions of R. Then the Krull dimension of R equals the transcendence degree of L over K.

PROOF. If $x_1, \ldots, x_n$ are generators of R as a K algebra, then $R \cong K[X_1, \ldots, X_n]/I$, where I is the ideal of all polynomials in $K[X_1, \ldots, X_n]$ that vanish at $(x_1, \ldots, x_n)$. The ideal I is prime, since R is assumed to be an integral domain. Let r be the transcendence degree of L over K. We know from Lemma 7.6b that some subset of $\{x_1, \ldots, x_n\}$ is a transcendence basis of L over K; therefore $r \leq n$. To prove the theorem, we shall prove that $r \geq \dim R$ and that $r \leq \dim R$.

Suppose that P and Q are prime ideals of R with $P \subseteq Q$. Then the identity map on R descends to a K algebra homomorphism $\varphi : R/P \to R/Q$. If $\alpha_j = x_j + P$ and $\beta_j = x_j + Q$ are the images of x_j under the respective quotient maps $R \to R/P$ and $R \to R/Q$, then $\{\alpha_1, \ldots, \alpha_n\}$ is a set of generators of R/P, $\{\beta_1, \ldots, \beta_n\}$ is a set of generators of R/Q, and $\varphi(\alpha_j) = \beta_j$ for $1 \leq j \leq n$. If $r' = \text{tr. deg } R/Q$, we may assume that $\{\beta_1, \ldots, \beta_{r'}\}$ is a transcendence basis of R/Q. Then $\{\alpha_1, \ldots, \alpha_{r'}\}$ is an algebraically independent subset of R/P over K because if f is a nonzero polynomial in $K[X_1, \ldots, X_{r'}]$ such that $f(\alpha_1, \ldots, \alpha_{r'}) = 0$, then application of φ and use of the fact that φ fixes each coefficient of f yields $f(\beta_1, \ldots, \beta_{r'}) = 0$; the latter equation contradicts the algebraic independence of $\{\beta_1, \ldots, \beta_{r'}\}$. We conclude that

$$P \subseteq Q \quad \text{implies} \quad \text{tr. deg}(R/P) \geq \text{tr. deg}(R/Q). \tag{$*$}$$

To prove the inequality $r \geq \dim R$, let a chain of prime ideals

$$0 \subseteq P_0 \subsetneq P_1 \subsetneq \cdots \subsetneq P_d$$

of R be given. We are to show that $r \geq d$. Abbreviate $K[X_1, \ldots, X_n]$ as A, so that $R = A/I$. Pull the chain of ideals of R back to a chain of ideals in A as

$$I \subseteq P'_0 \subsetneq P'_1 \subsetneq \cdots \subsetneq P'_d. \tag{$**$}$$

Inequality $(*)$ shows that

$$\text{tr. deg}(A/P'_0) \geq \text{tr. deg}(A/P'_1) \geq \cdots \geq \text{tr. deg}(A/P'_d). \tag{$\dagger$}$$

Since taking $P'_0 = I$ shows that $\text{tr. deg}(A/I) = \text{tr. deg}(R) = r$, every member of $(\dagger)$ is $\leq r$. It will follow from $(\dagger)$ that $r \geq d$ if we show that each inequality in $(\dagger)$ is strict, i.e., that for prime ideals P and Q in A,

$$P \subsetneq Q \quad \text{implies} \quad \text{tr. deg}(A/P) > \text{tr. deg}(A/Q). \tag{$\dagger\dagger$}$$

Since dim R is the supremum of the integers d as in (∗∗) and (†), proving (††) will prove that $r \geq \dim R$.

Thus let P and Q be prime ideals in $A = K[X_1, \ldots, X_n]$ with $P \subsetneq Q$. Put $\alpha_j = X_j + P$ and $\beta_j = X_j + Q$, so that the mappings of A to A/P and A/Q are $f(X_1, \ldots, X_n) \mapsto f(\alpha_1, \ldots, \alpha_n)$ and $f(X_1, \ldots, X_n) \mapsto f(\beta_1, \ldots, \beta_n)$. Then $A/P = K[\alpha_1, \ldots, \alpha_n]$ and $A/Q = K[\beta_1, \ldots, \beta_n]$. As above, if $r' = \text{tr.deg}\, A/Q$, then we may assume that $\{\beta_1, \ldots, \beta_{r'}\}$ is a transcendence basis of A/Q. Arguing by contradiction, we may assume that tr. deg A/P = tr. deg A/Q. Then it follows that $\{\alpha_1, \ldots, \alpha_{r'}\}$ is a transcendence basis of A/P. We localize A with respect to the multiplicative system S consisting of the complement of 0 in $K[X_1, \ldots, X_{r'}]$. Then $S^{-1}A = K(X_1, \ldots, X_{r'})[X_{r'+1}, \ldots, X_n]$. To understand $S^{-1}P$, we apply Lemma 7.21 to write

$$S^{-1}A/S^{-1}P \cong \overline{S}^{-1}(A/P), \qquad (\ddagger)$$

where $\overline{S}$ is the image of S in A/P. The restriction to $K[X_1, \ldots, X_{r'}]$ of the map $A \to A/P$ carries $f(X_1, \ldots, X_{r'})$ to $f(\alpha_1, \ldots, \alpha_{r'})$ and is one-one because $\{\alpha_1, \ldots, \alpha_{r'}\}$ is a transcendence set. Therefore $S \cap P = \varnothing$, and $S \to \overline{S}$ is one-one. Corollary 8.48d of *Basic Algebra* shows from $S \cap P = \varnothing$ that $S^{-1}P$ is a proper ideal of $S^{-1}A$. Since $S \to \overline{S}$ is one-one, let us view $\overline{S}$ as $\overline{S} = \{f(\alpha_1, \ldots, \alpha_{r'}) \mid f \neq 0\}$. Then

$$\overline{S}^{-1}(A/P) = K(\alpha_1, \ldots, \alpha_{r'})[\alpha_{r'+1}, \ldots, \alpha_n]. \qquad (\ddagger\ddagger)$$

Since $\alpha_{r'+1}, \ldots, \alpha_n$ are algebraic over $K(\alpha_1, \ldots, \alpha_{r'})$ because of the assumption tr. deg A/P = tr. deg $A/Q = r'$, the remark with Lemma 7.3 shows that ($\ddagger\ddagger$) is a field. By ($\ddagger$), $S^{-1}P$ is a maximal ideal. Arguing similarly with Q, we see that $S \cap Q = \varnothing$ and that $S^{-1}Q$ is a maximal ideal. From $P \subseteq Q$, we have $S^{-1}P \subseteq S^{-1}Q$. Because $S^{-1}P$ and $S^{-1}Q$ are maximal, $S^{-1}P = S^{-1}Q$. Therefore $Q \subseteq S^{-1}P$. Since Q properly contains P, we can choose g in Q that is not in P. This g is an element of $K[X_1, \ldots, X_n]$ such that $g(\alpha_1, \ldots, \alpha_n) \neq 0$ and $g(\beta_1, \ldots, \beta_n) = 0$. From the inclusion $Q \subseteq S^{-1}P$, there exist an f in P and a nonzero s in $K[X_1, \ldots, X_r]$ with $g = s^{-1}f$. Then $f = sg$. Since $f(\alpha_1, \ldots, \alpha_n) = 0$ and $s(\alpha_1, \ldots, \alpha_{r'})g(\alpha_1, \ldots, \alpha_n) \neq 0$, we obtain a contradiction. This contradiction proves (††) and shows that $r \geq \dim R$.

The argument that $r \leq \dim R$ will proceed by induction on r. If $r = 0$, then $R = K[x_1, \ldots, x_n]$ is a field by the remark with Lemma 7.3, and $\dim R = 0$ by Example 1 of Krull dimension. Now suppose inductively that $r > 0$ and that the inequality is known when tr. deg $R < r$. Put $A = K[X_1, \ldots, X_n]$, and suppose that $R = A/I = K[x_1, \ldots, x_n]$ with x_1 transcendental over K. We localize A with respect to the multiplicative system S consisting of the complement of 0

in $K[X_1]$. Then $S^{-1}A = K(X_1)[X_2, \ldots, X_n]$. To understand $S^{-1}I$, we apply Lemma 7.21 to write
$$S^{-1}A/S^{-1}I \cong \overline{S}^{-1}(A/I),$$
where $\overline{S}$ is the image of S in A/I. Arguing as in the previous paragraph, we see that
$$\overline{S}^{-1}(A/I) \cong K(x_1)[x_2, \ldots, x_n].$$
Combining these two isomorphisms, we see that $S^{-1}A/S^{-1}I$ has transcendence degree $r - 1$ over $K(x_1)$. By the inductive hypothesis, $S^{-1}A/S^{-1}I$ has Krull dimension $\geq r - 1$. Thus there exists a strictly increasing chain
$$S^{-1}I = Q_0 \subsetneqq Q_1 \subsetneqq \cdots \subsetneqq Q_{r-1}$$
of prime ideals in $S^{-1}A$. If we put $P_i = A \cap Q_i$ for each i, then each P_i is prime in A. From the theory of localization, we know that Q_i is recovered from P_i by $Q_i = S^{-1}P_i$, and thus we have a strictly increasing chain
$$I = P_0 \subsetneqq P_1 \subsetneqq \cdots \subsetneqq P_{r-1} \tag{§}$$
of prime ideals in A. The fact that P_{r-1} is proper implies that $S \cap P_{r-1} = \varnothing$. That is, no nonzero member of $K[X_1]$ lies in P_{r-1}. Consequently the image of X_1 in A/P_{r-1} is transcendental over K. The Nullstellensatz (Theorem 7.1) shows that P_{r-1} is not maximal in A. Hence the chain (§) can be extended by a strict inclusion in a maximal ideal P_r, and $r \leq \dim A/I = \dim R$. This completes the induction and the proof. □

5. Nonsingular and Singular Points

In this section we develop the initial algebraic background necessary for a discussion of nonsingular and singular points. Unlike what happened in previous sections, we shall not try to separate completely the algebra from the geometric setting, because the points to be investigated are the actual points of a zero locus.

The motivation comes from the Implicit Function Theorem in the calculus of several variables. In that setting, suppose that we have l numerical-valued smooth functions $f_1, \ldots, f_l$ of n variables. Let k be an integer with $1 < k < n$, and abbreviate $(x_1, \ldots, x_n)$ as (x, y), where $x = (x_1, \ldots, x_k)$ and $y = (x_{k+1}, \ldots, x_n)$. Suppose that (x_0, y_0) has the property that $f_i(x_0, y_0) = 0$ for $1 \leq i \leq l$. The hope is that there is a smooth vector-valued function $y = g(x)$ defined near $x = x_0$ such that $y_0 = g(x_0)$ and such that $f_i(x, y) = 0$ for $1 \leq i \leq l$ with (x, y) near (x_0, y_0) if and only if $y = g(x)$, i.e., that the locus of common zeros of $f_1, \ldots, f_l$ is locally the graph of a smooth function of k variables. According to the Implicit

Function Theorem, a sufficient condition for this to happen is that $k + l = n$ and that the (square) matrix of the first partial derivatives at (x_0, y_0) of the f_i's for $1 \leq i \leq l$ with respect to the y_j's for $k + 1 \leq j \leq n$ be invertible. A little more generally but still with $k + l = n$, the locus of common zeros is locally the graph of a smooth function of l of the variables in terms of the remaining k variables if the matrix of all the first partial derivatives of the f_i's has the maximum possible rank, namely l.

Let us describe the setting for a comparable situation in algebraic geometry. *Throughout this section we assume that K is an algebraically closed field.* Suppose that I is a prime ideal in $K[X_1, \ldots, X_n]$, and let $V(I)$ be the locus of common zeros[5] of I in K^n. The Hilbert Basis Theorem shows that I is finitely generated over K as an ideal, and we let $\{f_1, \ldots, f_l\}$ be a set of generators. Corollary 7.2 shows that I is the set of all polynomials vanishing on $V(I)$, and thus the integral domain $R = K[X_1, \ldots, X_n]/I$ may be regarded as the set of all restrictions to $V(I)$ of polynomials in the following sense: if $x = (x_1, \ldots, x_n)$ is a member of $V(I)$ and $f(X_1, \ldots, X_n)$ is in $K[X_1, \ldots, X_n]$, then every member of the coset $f + I$ has the same value at x, and it is consequently meaningful to write $f(x)$ for f in R.

From Theorem 7.22 the transcendence degree over K of the field of fractions of R equals the Krull dimension of the ring R, and these numbers are what is taken as the dimension of $V(I)$ over K. We write $\dim V(I)$ for this dimension. In this setting, a point x of $V(I)$ is called a **nonsingular point**, or **regular point**, if the matrix $\left[\frac{\partial f_i}{\partial X_j}(x)\right]$ has rank equal to $n - \dim V(I)$. Otherwise x is a **singular point**.

It is important to observe that these definitions do not depend on the choice of the set $\{f_1, \ldots, f_l\}$ of generators of I. In fact, it is enough to show that the row space of the matrix $\left[\frac{\partial f_i}{\partial X_j}(x)\right]$ is exactly the space of all row vectors

$$\left(\frac{\partial f}{\partial X_1}(x) \quad \cdots \quad \frac{\partial f}{\partial X_n}(x)\right) \qquad \text{for } f \in I,$$

since the latter space is manifestly independent of the choice of generators. To see that the displayed space equals the row space of the matrix whose rank appears in the definition of singular point, let $g_1, \ldots, g_n$ be arbitrary polynomials. Then $f = \sum_i g_i f_i$ is the most general member of I. Use of the product rule and the fact that $f_i(x) = 0$ for each i shows that $\frac{\partial f}{\partial X_j}(x) = \sum_i g_i(x) \frac{\partial f_i}{\partial X_j}(x)$. Since the g_i are arbitrary, we can arrange for $(g_1(x), \ldots, g_n(x))$ to be any given member of K^n. Thus the space of all row vectors $\left(\frac{\partial f}{\partial X_1}(x) \quad \cdots \quad \frac{\partial f}{\partial X_n}(x)\right)$ for $f \in I$ is the set of all K linear combinations of row vectors $\left(\frac{\partial f_i}{\partial X_1}(x) \quad \cdots \quad \frac{\partial f_i}{\partial X_n}(x)\right)$ for $1 \leq i \leq l$, as asserted.

---

[5] In terminology to be used in later chapters, one says that $V(I)$ is the **affine variety** corresponding to I.

EXAMPLES.

(1) **Irreducible affine curve**[6] in K^2. Suppose that $n = 2$ in the notation above and that I is nonzero and is generated by a single nonconstant polynomial $f(X, Y)$. The condition that I be prime is exactly the condition that $f(X, Y)$ be a prime polynomial. In turn, since $K[X, Y]$ is a unique factorization domain, the condition that $f(X, Y)$ be prime is exactly the condition that $f(X, Y)$ be irreducible. Let us specialize to a case for which the first partial derivatives take an especially simple form: suppose that

$$f(X, Y) = Y^2 - h(X).$$

The only possible factorization is $f(X, Y) = (Y + \sqrt{h(X)})(Y - \sqrt{h(X)})$, and thus $f(X, Y)$ is irreducible in $K[X, Y]$ if $h(X)$ is not the square of a member of $K[X]$. The relevant integral domain is $R = K[X, Y]/(f(X, Y))$, and we let $x = X + (f(X, Y))$ and $y = Y + (f(X, Y))$. Then x is transcendental over K, and the equation $y^2 = h(x)$ shows that y is algebraic over $K(x)$. Hence tr. deg $R = 1$, and the corresponding $V(I)$ has dim $V(I) = 1$. If (x_0, y_0) is a point of $V(I)$, then the matrix of first partial derivatives is

$$\left(\frac{\partial f}{\partial X} \quad \frac{\partial f}{\partial Y}\right)_{(x_0, y_0)} = (-h'(X) \quad 2Y)_{(x_0, y_0)}.$$

The rank of this matrix is ≤ 1, and nonsingularity of (x_0, y_0) means that the matrix has rank equal to 1. If the characteristic is $\neq 2$, then the condition for a singularity is that $y_0^2 = h(x_0)$, $y_0 = 0$, and $h'(x_0) = 0$ simultaneously. Hence $V(I)$ is everywhere nonsingular[7] if and only if h has no multiple roots in K.

(2) **Irreducible affine hypersurface**[8] in K^n. For general n, again suppose that I is a prime ideal generated by a single nonconstant polynomial $f(X_1, \ldots, X_n)$. The condition on f for I to be prime is that f be irreducible in $K[X_1, \ldots, X_n]$. The relevant ring is $R = K[X_1, \ldots, X_n]/(f(X_1, \ldots, X_n))$, and the image in R of a polynomial $g(X_1, \ldots, X_n)$ is 0 only if g is divisible by f, by Corollary 7.2. The polynomial f is nonconstant in some X_j, say for $j = n$. Then no nonzero polynomial $g(X_1, \ldots, X_{n-1})$ maps to 0 in R. Consequently the elements $x_i = X_i + (f(X_1, \ldots, X_n))$ have the property that $\{x_1, \ldots, x_{n-1}\}$ is a transcendence set in R. The equation $f(x_1, \ldots, x_n) = 0$ shows that x_n is algebraic over $K(X_1, \ldots, X_{n-1})$. Hence the corresponding $V(I)$ has dim $V(I) =$ tr. deg $R = n - 1$. The nonsingular points of $V(I)$ are the points of $V(I)$ for which some first partial derivative of f is nonzero.

---

[6]Some authors include irreducibility in the definition of "affine curve." This book does not.

[7]If K has characteristic 2 and if x_0 has the property that $h'(x_0) = 0$, then we can choose y_0 with $y_0^2 = h(x_0)$ because K is algebraically closed, and (x_0, y_0) will be a singular point. Hence $V(I)$ is everywhere nonsingular if and only if h has degree exactly 1.

[8]Some authors include irreducibility in the definition of "affine hypersurface." This book does not.

5. Nonsingular and Singular Points

Theorem 7.23 (Zariski's Theorem). With K algebraically closed, let I be a prime ideal in $K[X_1, \ldots, X_n]$, let $R = K[X_1, \ldots, X_n]/I$, and let $V(I)$ be the locus of common zeros of I in K^n. If $x = (x_1, \ldots, x_n)$ is a point of $V(I)$, define $\mathfrak{m}_x$ to be the maximal ideal

$$\mathfrak{m}_x = \{f \in R \mid f(x) = 0\}$$

of R, let R_x be the localization of R with respect to $\mathfrak{m}_x$, and let M_x be the maximal ideal of R_x. Then

$$\dim_K(M_x/M_x^2) = \dim_K(\mathfrak{m}_x/\mathfrak{m}_x^2) \geq \dim V(I),$$

and x is nonsingular if and only if equality holds. The set of nonsingular points of $V(I)$ is nonempty.

REMARKS. We are going to prove for each point x of $V(I)$ that

$$\dim_K(M_x/M_x^2) = \dim_K(\mathfrak{m}_x/\mathfrak{m}_x^2)$$

and that

$$\dim_K(\mathfrak{m}_x/\mathfrak{m}_x^2) + \operatorname{rank}\left[\tfrac{\partial f_i}{\partial X_j}\right] = n,$$

where $\{f_i\}$ is a finite set of generators of I. Since by definition x is nonsingular if and only if $\operatorname{rank}\left[\tfrac{\partial f_i}{\partial X_j}\right] = n - \dim V(I)$, it will follow that x is a nonsingular point if and only if $\dim_K(\mathfrak{m}_x/\mathfrak{m}_x^2) = \dim V(I)$. Only for the special case that $V(I)$ is an irreducible affine hypersurface do we prove that the inequality $\dim_K(\mathfrak{m}_x/\mathfrak{m}_x^2) \geq \dim V(I)$ always holds for all x and that equality always holds for some x. The general case will ultimately be reduced to the special case; we return to this matter in Chapter X. The partial proof that we give in the present section will be preceded by an example.

EXAMPLE 1, CONTINUED. Suppose that an affine variety V in K^2 is obtained from the irreducible polynomial $f(X, Y) = Y^2 - h(X)$. Let us assume that K has characteristic $\neq 2$ and that $(0, 0)$ lies in V. The latter condition means that $h(0) = 0$. Let $x = X + (f(X, Y))$ and $y = Y + (f(X, Y))$. Since $y^2 = h(x)$, any polynomial in (x, y) can be rewritten in such a way that the only powers of y that occur are 0 and 1. Thus $R = \{p(x) + yq(x) \mid p \in K[x], q \in K[x]\}$, and

$$\mathfrak{m}_{(0,0)} = \{xp(x) + yq(x) \mid p \in K[x],\ q \in K[x]\}.$$

The ideal $\mathfrak{m}_{(0,0)}^2$ consists of all sums of products of two elements of this kind. From two polynomials $xp(x)$, we can get any polynomial $x^2 a(x)$; from $xp(x)$

and $yq(x)$, we can get any $xyb(x)$; and from two polynomials $yq(x)$, we can get any $y^2c(x) = h(x)c(x)$. Thus

$$\mathfrak{m}_{(0,0)}^2 = \{x^2 a(x) + h(x)c(x) + yxb(x)\}.$$

What happens depends on the first-degree term in $h(x)$. Examining the possibilities, we see that

$$\mathfrak{m}_{(0,0)}^2 = \begin{cases} \{xa(x) + yxb(x)\} & \text{if } h'(0) \neq 0, \\ \{x^2 a(x) + yxb(x)\} & \text{if } h'(0) = 0. \end{cases}$$

Hence

$$\mathfrak{m}_{(0,0)}/\mathfrak{m}_{(0,0)}^2 \cong \begin{cases} Ky & \text{if } h'(0) \neq 0, \\ Kx + KY & \text{if } h'(0) = 0. \end{cases}$$

In other words, $\dim_K \mathfrak{m}_{(0,0)}/\mathfrak{m}_{(0,0)}^2$ equals 1 if $(0, 0)$ is nonsingular and equals 2 if $(0, 0)$ is singular. Since $\dim V(I) = 1$, this result is consistent with the statement of Theorem 7.23.

PARTIAL PROOF OF THEOREM 7.23. As mentioned in the remarks, one thing that we are going to prove for each point x of $V(I)$ is that

$$\dim_K (\mathfrak{m}_x/\mathfrak{m}_x^2) + \operatorname{rank}\left[\frac{\partial f_i}{\partial X_j}\right] = n, \qquad (*)$$

where $\{f_1, \ldots, f_l\}$ is a finite set of generators of I.

Let I_x be the pullback to $K[X_1, \ldots, X_n]$ of the ideal $\mathfrak{m}_x$, i.e., let

$$I_x = \{f \mid f + I \in \mathfrak{m}_x\} = \{f \in K[X_1, \ldots, X_n] \mid f(x_1, \ldots, x_n) = 0\}.$$

The K linear mapping $f \mapsto f + I$ carries I_x onto $\mathfrak{m}_x$; composing with the quotient mapping $\mathfrak{m}_x \to \mathfrak{m}_x/\mathfrak{m}_x^2$ gives a K linear mapping φ of I_x onto $\mathfrak{m}_x/\mathfrak{m}_x^2$. If f maps under φ to the 0 coset, then $f + I = \sum_j (g_j + I)(h_j + I)$ for suitable polynomials g_j and h_j with $g_j + I$ and $h_j + I$ in $\mathfrak{m}_x$. Then $f - \sum_j g_j h_j$ lies in I, and f is exhibited as a member of $I_x^2 + I$. Conversely φ does carry I_x^2 and I to the 0 coset. Thus the kernel of φ is exactly $I_x^2 + I$, and φ descends to a K linear isomorphism $I_x/(I_x^2 + I) \cong \mathfrak{m}_x/\mathfrak{m}_x^2$. Therefore

$$\dim_K \left(I_x/(I_x^2 + I)\right) \cong \dim_K (\mathfrak{m}_x/\mathfrak{m}_x^2). \qquad (**)$$

We define a K linear map θ of $K[X_1, \ldots, X_n]$ to the space $M_{1n}(K)$ of all n-dimensional row vectors over K by

$$\theta(f) = \left(\tfrac{\partial f}{\partial X_1}(x) \quad \cdots \quad \tfrac{\partial f}{\partial X_n}(x)\right).$$

The product rule for differentiation shows that $\theta(I_x^2) = 0$. The ideal I_x, considered as a K vector space, is spanned by I_x^2 and the various polynomials $X_j - x_j$. Since $\theta(X_j - x_j)$ is the jth standard basis vector of $M_{1n}(K)$, the vectors $\theta(X_j - x_j)$ form a basis of $M_{1n}(K)$. Therefore θ descends to a K linear isomorphism $\bar{\theta} : I_x/I_x^2 \to M_{1n}(K)$.

We observed just before Examples 1 and 2 that the vector space of all row vectors $\theta(f)$ for $f \in I$ equals the row space for the matrix $\left[\frac{\partial f_i}{\partial X_j}\right]$. Hence

$$\dim_K \theta(I) = \operatorname{rank}\left[\tfrac{\partial f_i}{\partial X_j}\right].$$

Since $\theta(I) = \bar{\theta}\big((I + I_x^2)/I_x^2\big)$ and since $\bar{\theta}$ is one-one, this equality shows that

$$\dim_K \big((I + I_x^2)/I_x^2\big) = \operatorname{rank}\left[\tfrac{\partial f_i}{\partial X_j}\right]. \tag{†}$$

Adding (∗∗) and (†) gives

$$\dim_K (I_x/I_x^2) = \dim_K(\mathfrak{m}_x/\mathfrak{m}_x^2) + \operatorname{rank}\left[\tfrac{\partial f_i}{\partial X_j}\right].$$

Since, as we have seen, I_x/I_x^2 is isomorphic to $M_{1n}(K)$ via $\bar{\theta}$, the left side is n, and (∗) is proved.

The second thing that we are going to prove now is that

$$\dim_K(\mathfrak{m}_x/\mathfrak{m}_x^2) = \dim_K(M_x/M_x^2). \tag{††}$$

If L is the field of fractions of the integral domain R, then the localization R_x is the subring of L of all quotients g/h with g and h in R and $h(x) \neq 0$. The inclusion $\mathfrak{m}_x \subseteq M_x$ induces a K linear ring homomorphism $\varphi : \mathfrak{m}_x/\mathfrak{m}_x^2 \to M_x/M_x^2$, and (††) will follow if φ is shown to be one-one onto.

If g/h is given in M_x with $g \in \mathfrak{m}_x$ and with $h \in R$ having $h(x) \neq 0$, then the decomposition

$$h(x)^{-1}g = \tfrac{g}{h} + \left(\tfrac{g}{h}\right)\left(\tfrac{h(x)^{-1}h-1}{1}\right)$$

exhibits $h(x)^{-1}g$ in $\mathfrak{m}_x$ as mapping to $g/h + M_x^2$. Therefore φ is onto.

If g in $\mathfrak{m}_x$ maps to $\sum_i \left(\tfrac{g_i}{h_i}\right)\left(\tfrac{g_i'}{h_i'}\right)$ in M_x^2, then we can clear fractions and write $hg = \sum_i g_i g_i' h_i''$ for an element h of R with $h(x) \neq 0$. Here $\sum_i g_i g_i' h_i''$ is in $\mathfrak{m}_x^2$. The set of elements f in R such that fg is in $\mathfrak{m}_x^2$ is an ideal in R that contains $\mathfrak{m}_x$ and that contains h. Since h is not in $\mathfrak{m}_x$ and since $\mathfrak{m}_x$ is maximal, this ideal in R contains $f = 1$, and it follows that g is in $\mathfrak{m}_x^2$. Consequently φ is an isomorphism, and (††) is proved. □

PROOF OF REMAINDER OF THEOREM 7.23 FOR IRREDUCIBLE AFFINE HYPERSURFACES. Let I be the principal ideal $(f(X_1, \ldots, X_n))$, where f is irreducible. We saw in Example 2 above that dim $V(I) = n - 1$. The matrix that appears in (∗) has only one row, corresponding to f, and hence it has rank 1 or rank 0. Substituting this fact into (∗), we see that $\dim_K(\mathfrak{m}_x/\mathfrak{m}_x^2) \geq n - 1 = \dim V(I)$.

Arguing by contradiction, suppose that strict inequality holds for every x in $V(I)$. Then $\frac{\partial f}{\partial X_j}(x) = 0$ for all $x \in V(I)$ and for all j. By Corollary 7.2, each $\frac{\partial f}{\partial X_j}$ is the product of f and a polynomial. Since the degree of $\frac{\partial f}{\partial X_j}$ in X_j is less than the degree of f in X_j, it follows that $\frac{\partial f}{\partial X_j} = 0$ for all j. In characteristic 0, this condition forces f to be constant and contradicts the assumption that f is an irreducible polynomial (and in particular the assumption that f is not a unit). In characteristic p, this condition forces each power of each X_j that occurs in f to be a multiple of p. That is, it says that $f(X_1, \ldots, X_n) = g(X_1^p, \ldots, X_n^p)$. Let Fr : $K \to K$ be the field map given by $a \mapsto a^p$. This is onto K, since K is algebraically closed. Hence there exists a polynomial $h(X_1, \ldots, X_n)$ such that $h^{\text{Fr}} = g$. Then $f(X_1, \ldots, X_n) = g(X_1^p, \ldots, X_n^p) = \big(h(X_1, \ldots, X_n)\big)^p$ exhibits f as reducible, contradiction. Hence strict inequality cannot hold for all $x \in V(I)$, and some point of $V(I)$ is nonsingular. □

6. Infinite Galois Groups

In this section, K denotes a field, and K_{alg} denotes a fixed algebraic closure of K. We define K_{sep} to be the subfield of all elements of K_{alg} that are separable over K. The field K_{sep} is called a **separable algebraic closure** of K. Theorem 7.13 shows that K_{alg} is a purely inseparable extension of K_{sep}. If F_1 and F_2 are any fields with $F_1 \subseteq F_2$, then the group of all field automorphisms of F_2 fixing F_1 is denoted by $\text{Gal}(F_2/F_1)$ and is called the **Galois group** of F_2 over F_1.

The purpose of this section is to extend the theory of Galois groups to handle infinite extensions. Such an extended theory has at least two important applications in the current context. A first application is to developments in algebraic number theory beyond what appears in Chapters V and VI. For example one way of viewing traditional class field theory for a number field F is that one forms $\text{Gal}(F_{\text{alg}}/F)$, defines the maximal abelian extension F_{ab} of F to be the fixed field of the closure of the commutator subgroup of $\text{Gal}(F_{\text{alg}}/F)$, and asks for a description of F_{ab} in terms of F. A second application is to the study of varieties over fields that are not algebraically closed. If a field K is given and a prime ideal I in $K_{\text{alg}}[X_1, \ldots, X_n]$ is specified by giving a finite set of generators, we can ask whether the same ideal can be defined via generators that lie in K. The given generators have coefficients in K_{alg}, and it is usually not obvious whether they can be adjusted to have coefficients in K. However, if Galois theory is available, then the question becomes whether the operation of each element of the Galois group $\text{Gal}(K_{\text{alg}}/K)$ carries each generator into a member of the ideal,[9] and this question is decidable by methods to be discussed in Chapter VIII. More generally algebraic geometry from before 1960 frequently worked with a field K and an algebraically closed field L that is larger than K_{alg}, for example with $K = \mathbb{Q}$ and $L = \mathbb{C}$. Under the assumption that K is perfect and L is algebraically closed, Theorem 7.34 below shows that $\text{Gal}(L/K)$ fixes only the elements of K, and thus Galois theory can still be used to decide in this situation whether a prime ideal in $L[X_1, \ldots, X_n]$ is generated by members of $K[X_1, \ldots, X_n]$.

The definition of "normal field extension" in *Basic Algebra* was limited to suitable finite separable algebraic extensions. We now drop both the finiteness assumption and the separability assumption: A field L with $K \subseteq L \subseteq K_{\text{alg}}$ is said to be a **normal extension** of K if there exists some nonempty family $\{f_i\}_{i \in S}$ of nonconstant polynomials in $K[X]$ such that L is generated by K and all the roots in K_{alg} of all the polynomials f_i. More specifically all the polynomials f_i split in K_{alg}, say as $f_i(X) = c_i \prod_{j=1}^{d(i)}(X - \alpha_{ij})$, and L is to be the subfield of K_{alg} generated by K and all the roots α_{ij}.

Proposition 7.24. The following conditions on a field L with $K \subseteq L \subseteq K_{\text{alg}}$ are equivalent:

(a) L is a normal extension of K,
(b) $\text{Gal}(K_{\text{alg}}/K)$ carries L to itself,
(c) any K isomorphism of L into K_{alg} carries L to itself,
(d) any polynomial f in $K[X]$ that is irreducible over K and has one root in L necessarily splits in L.

PROOF. If (a) holds, let L be generated by K and elements α_{ij} as in the paragraph before the proposition. If φ is in $\text{Gal}(K_{\text{alg}}/K)$, then $\varphi(\alpha_{ij})$ is a root of $f_i^\varphi = f_i$ because f_i has coefficients in K. Hence α_{ij} equals some $\alpha_{ij'}$. Thus φ permutes the generators of L over K, and $\varphi(L) = L$. Therefore (b) holds.

If (b) holds, then any K field map of L into K_{alg} extends to a K automorphism of K_{alg}, by Theorem 9.23 of *Basic Algebra*. By (b), the extended mapping carries L into itself. Thus (c) holds.

If (c) holds, let f in $K[X]$ be irreducible over K, and suppose that x_0 is a root of f in L. Let x_1 be another root of f in K_{alg}. By the uniqueness of simple extensions, we know that there exists a K isomorphism $\varphi_0 : K(x_0) \to K(x_1) \subseteq K_{\text{alg}}$, and we can regard φ_0 as a K field map of $K(x_0)$ into K_{alg}. The map φ_0 extends to a K field automorphism of K_{alg}, and we restrict the extension

---

[9]This condition is always necessary. For it to be sufficient, one has to show that the only members of K_{alg} fixed by all elements of $\text{Gal}(K_{\text{alg}}/K)$ are the members of K.

to a map $\varphi : L \to K_{\text{alg}}$. By (c), $\varphi(L) \subseteq L$. Since $K(x_0) \subseteq L$, we obtain $K(x_1) = \varphi(K(x_0)) \subseteq \varphi(L) \subseteq L$. Thus x_1 is in L, and (d) holds.

If (d) holds, then for each element x_i of L, let f_i be the minimal polynomial of x_i over K. Then certainly L is generated by K and the elements x_i. By (d), each f_i splits in L. Therefore L is generated over K by all the roots of the polynomials f_i and is normal. Thus (a) holds. □

Proposition 7.25. Every member of $\text{Gal}(K_{\text{alg}}/K)$ carries K_{sep} into itself, any two members of $\text{Gal}(K_{\text{alg}}/K)$ that agree on K_{sep} are equal on K_{alg}, and any field map of K_{sep} into K_{alg} extends to an automorphism of K_{alg}. Consequently the operation of restriction from K_{alg} to K_{sep} defines an isomorphism

$$\text{Gal}(K_{\text{alg}}/K) \cong \text{Gal}(K_{\text{sep}}/K).$$

PROOF. The first statement has three conclusions to it. For the first conclusion, if φ is in $\text{Gal}(K_{\text{alg}}/K)$ and if x_0 is in K_{sep}, let f be the minimal polynomial of x_0 over K. By separability, f is a separable polynomial over K. Since φ fixes f, φ carries x_0 to some root x_1 of f, and hence f is the minimal polynomial of x_1 over K. Since f is a separable polynomial over K, x_1 is separable over K and lies in K_{sep}.

For the second conclusion, let φ be a member of $\text{Gal}(K_{\text{alg}}/K)$ that is 1 on K_{sep}. If x is in K_{alg}, then the pure inseparability of $K_{\text{alg}}/K_{\text{sep}}$ implies that $x^{p^e} = a$ for some $a \in K_{\text{sep}}$ and some integer $e \geq 0$. The element x has $(X - x)^{p^e} = X^{p^e} - x^{p^e} = X^{p^e} - a$ and hence is the unique root of $X^{p^e} - a$. Since $\varphi(x)$ has to be a root of this polynomial, $\varphi(x) = x$.

The third conclusion is a special case of the extendability to all of K_{alg} of any field mapping of a subfield of K_{alg} into K_{alg}. For the displayed isomorphism the first conclusion shows that restriction carries $\text{Gal}(K_{\text{alg}}/K)$ into $\text{Gal}(K_{\text{sep}}/K)$, the second conclusion shows that restriction is one-one, and the third conclusion shows that restriction is onto. □

Corollary 7.26. Let L be a field with $K \subseteq L \subseteq K_{\text{sep}}$, form $\text{Gal}(L/K)$, and let $L^{\text{Gal}(L/K)}$ be the fixed field

$$L^{\text{Gal}(L/K)} = \{x \in L \mid \gamma x = x \text{ for all } x \in \text{Gal}(L/K)\}.$$

Then L is normal over K if and only if $L^{\text{Gal}(L/K)} = K$.

PROOF. Let L be normal over K, let x be in $L^{\text{Gal}(L/K)}$, and let f be the minimal polynomial of x over K. Since L is normal, f splits in L. Since $L \subseteq K_{\text{sep}}$, the roots of f in L all have multiplicity one. Arguing by contradiction, suppose that x is not in K. Then $\deg f > 1$, and f has another root x_1 besides x.

Hence we can find a K isomorphism $\varphi : K(x) \to K(x_1)$ with $\varphi(x) = x_1$. The mapping φ extends to a field automorphism of K_{alg}, and Proposition 7.24 shows that $\varphi(L) = L$, since L is normal. Thus φ defines by restriction a member of $\text{Gal}(L/K)$. Since $\varphi(x) = x_1$, we have a contradiction to the assumption that x is in $L^{\text{Gal}(L/K)} = K$.

Conversely let $L^{\text{Gal}(L/K)} = K$. Let x be in L, and let f be its minimal polynomial over K. Let $x_1 = x$ and $x_2, \ldots, x_r$ be the distinct images of x in L under members of $\text{Gal}(L/K)$. These are all roots of f, and the roots of f have multiplicity 1 because x lies in K_{sep}. Each member of $\text{Gal}(L/K)$ permutes $x_1, \ldots, x_r$ and hence acts via a permutation in the symmetric group $\mathfrak{S}_r$. Put $g(X) = \prod_{i=1}^{r}(X - x_i)$. Expanding g gives

$$g(X) = X^r - \Big(\sum_i x_i\Big) X^{r-1} + \Big(\sum_{i<j} x_i x_j\Big) X^{r-2} - \cdots \pm \Big(\prod_i x_i\Big).$$

Each permutation of $\{x_1, \ldots, x_r\}$ fixes the coefficients of $g(X)$, which are members of L, and hence the coefficients are in $L^{\text{Gal}(L/K)} = K$. Therefore $g(X)$ is in $K[X]$. Since $g(x) = 0$, $f(X)$ divides $g(X)$. Over L, $g(X)$ splits. By unique factorization in $L[X]$, $f(X)$ must split, too. By Proposition 7.24, L is normal over K. □

To obtain a version of the Fundamental Theorem of Galois Theory in the present context, it is necessary to introduce a topology on each Galois group. An example will illustrate.

EXAMPLE. Let K be the finite field $\mathbb{F}_q$, where $q = p^r$ for a prime p. If L_n is a finite extension of K of degree n, then Proposition 9.40 of *Basic Algebra* shows that $\text{Gal}(L_n/K)$ is cyclic of order n, a generator being the **Frobenius element** Fr_q defined by $\text{Fr}_q(x) = x^q$. The thing about the Frobenius element is that it really makes sense on all L_n's simultaneously. We know (from Proposition 7.15 for example) that every algebraic extension of K is separable, and hence $K_{\text{sep}} = K_{\text{alg}}$. Here we can view K_{sep} as an aligned union of the fields L_n for $n \geq 1$, and Fr_q really makes sense as a member of $\text{Gal}(K_{\text{sep}}/K)$ under the same definition: $\text{Fr}_q(x) = x^q$. On each L_n, some nonzero power of Fr_q is the identity, but this is no longer true on the infinite field K_{sep}. Thus the mapping $1 \mapsto \text{Fr}_q$ extends to a one-one homomorphism of $\mathbb{Z}$ into $\text{Gal}(K_{\text{sep}}/K)$. However, it is not onto. Any element γ of $\text{Gal}(K_{\text{sep}}/K)$ has the property that for each n, there is a unique integer k_n with $0 \leq k_n < n$ such that $\gamma\big|_{L_n} = \text{Fr}_q^{k_n}$, and the sequence $\{k_n\}$ determines γ; nevertheless Problem 3 at the end of the chapter shows that the sequence need not ultimately be constant, and therefore γ need not be in the image of $\mathbb{Z}$. The Galois group $\text{Gal}(K_{\text{sep}}/K)$ is instead a certain topological completion of $\mathbb{Z}$ that is usually denoted by $\widehat{\mathbb{Z}}$. Taking the topology into account will be essential to extending the Fundamental Theorem of Galois Theory, since $\mathbb{Z}$ and $\widehat{\mathbb{Z}}$ are distinct subgroups of $\text{Gal}(K_{\text{sep}}/K)$ that have the same fixed field, namely K itself.

If L is a normal extension of K with $L \subseteq K_{\text{sep}}$, we shall introduce a topology on $\text{Gal}(L/K)$ to make "close" mean "equal on a large finite-dimensional subspace." With this intuition as a guide, we could define a basic neighborhood of an element γ_0 of $\text{Gal}(L/K)$ by taking finitely many elements $\alpha_1, \ldots, \alpha_n$ in K and forming

$$\{\gamma \in \text{Gal}(L/K) \mid \gamma\alpha_i = \gamma_0\alpha_i \text{ for } 1 \leq i \leq n\}.$$

It is more useful, however, to define the topology in another way, and then it will turn out that we indeed would have obtained a neighborhood basis by the above definition. In any event, the topology turns out to be compact Hausdorff and to make $\text{Gal}(L/K)$ into a topological group.

The method we use will be to define the topology as an "inverse limit." Inverse limit is a general notion in category theory defined by a universal mapping property. As usual it consists of an object and a morphism; it need not exist in a general category, but when it does exist, it is unique up to canonical isomorphism. For the category of interest, the objects are the compact (Hausdorff) topological groups, and the morphisms are continuous group homomorphisms. If we wanted to emphasize the category-theory aspects of the construction, we would also need products of this category with itself, but we shall not belabor this point.

Let I be a **directed set**, i.e., a nonempty partially ordered set under an ordering $\leq$ such that for any a and b in I, there is an element c in I with $a \leq c$ and $b \leq c$. We allow ourselves to write $b \geq a$ in place of $a \leq b$ whenever convenient. Two examples of directed sets of particular interest both have $I = \{1, 2, 3, \ldots\}$; in one case the ordering is given by $a \leq b$ if a divides b, and in the other case the ordering is given by the usual notion of inequality.

An **inverse system** $(I, \{G_i\}, \{f_{ij}\})$ in the category of compact topological groups consists of a directed set I, a system of compact topological groups G_i, one for each $i \in I$, and a system of continuous homomorphisms $f_{ij} : G_j \to G_i$, defined whenever i and j are in I with $i \leq j$, such that

- $f_{ii} = 1$ for all $i \in I$,
- $f_{ij} \circ f_{jk} = f_{ik}$ whenever $i \leq j \leq k$.

EXAMPLES.

(1) Let $I = \{1, 2, 3, \ldots\}$ with $a \leq b$ meaning that a divides b. Let G_a be the cyclic group $\mathbb{Z}/a\mathbb{Z}$ of order a. Define $f_{ab} : G_b \to G_a$ to be the homomorphism such that $f_{ab}(1 + b\mathbb{Z}) = 1 + a\mathbb{Z}$.

(2) Let $I = \{1, 2, 3, \ldots\}$ with the usual ordering. Fix a prime number p, and define G_a to be the cyclic group $\mathbb{Z}/p^a\mathbb{Z}$ of order p^a. Define $f_{ab} : G_b \to G_a$ to be the homomorphism such that $f_{ab}(1 + p^b\mathbb{Z}) = 1 + p^a\mathbb{Z}$.

An **inverse limit** $(G, \{f_i\}_{i \in I})$ of the inverse system $(I, \{G_i\}, \{f_{ij}\})$, often written $G = \varprojlim G_i$ and sometimes also called the **projective limit**, consists of a compact topological group G and continuous homomorphisms $f_i : G \to G_i$ such that

(i) $f_{ij} \circ f_j = f_i$ whenever $i \leq j$,
(ii) whenever $(G', \{f_i'\}_{i \in I})$ is a pair consisting of a compact topological group G' and continuous homomorphisms $f_i' : G' \to G_i$ such that $i \leq j$ implies $f_{ij} \circ f_j' = f_i'$, then there exists a unique continuous homomorphism $F : G' \to G$ such that $f_i \circ F = f_i'$ for all i.

In the two examples the inverse limit group in the first case is $\widehat{\mathbb{Z}}$; in the second case the inverse limit is isomorphic to the additive group $\mathbb{Z}_p$ of p-adic integers. In the first case we omit a description of the homomorphisms $f_a : \widehat{\mathbb{Z}} \to \mathbb{Z}/a\mathbb{Z}$. In the second case the homomorphisms f_a are easy to describe: $f_a : \mathbb{Z}_p \to \mathbb{Z}/p^a\mathbb{Z}$ is given by the composition of the quotient homomorphism $\mathbb{Z}_p \to \mathbb{Z}_p/p^a\mathbb{Z}_p$ and the isomorphism $\mathbb{Z}_p/p^a\mathbb{Z}_p \to \mathbb{Z}/p^a\mathbb{Z}$ asserted by Theorem 6.26e.

Proposition 7.27. In the category of compact topological groups, an inverse system $(I, \{G_i\}, \{f_{ij}\})$ has at least one inverse limit, namely $(G, \{f_i\}_{i \in I})$ with

$$G = \left\{(g_i)_{i \in I} \in \prod_{i \in I} G_i \;\middle|\; f_{ij}(g_j) = g_i \text{ whenever } i \leq j\right\},$$

$f_i =$ restriction to G of the ith projection $\prod_j G_j \to G_i$.

REMARKS. It is to be understood from the statement that G gets the relative topology from $\prod_{i \in I} G_i$. We refer to this $(G, \{f_i\}_{i \in I})$ as the **standard inverse limit** of $(I, \{G_i\}, \{f_{ij}\})$.

PROOF. If $(g_i)_{i \in I}$ and $(g_i')_{i \in I}$ are in G, then the fact that each f_{ij} is a homomorphism implies that $f_{ij}(g_j g_j') = g_i g_i'$ and that $f_{ij}(g_j^{-1}) = g_i^{-1}$. Therefore $(g_i g_i')_{i \in I}$ and $(g_i^{-1})_{i \in I}$ are in G, and G is a group. The subset of $G_i \times G_j$ with $f_{ij}(x_j) = x_i$ is topologically closed, and it follows that G is the intersection of closed sets and hence is closed. Since $\prod_{j \in I} G_j$ is compact Hausdorff, G is compact Hausdorff. The continuity of the multiplication and inversion is a consequence of those properties for $\prod_{j \in I} G_j$. The ith projection of $\prod_{j \in I} G_j$ onto G_i is a continuous homomorphism, and hence so is the restriction of this projection to G.

Condition (i) in the definition of inverse limit is immediate, and we have to prove (ii). Let $(G', \{f_i'\}_{i \in I})$ be given with each $f_i' : G' \to G_i$ having the property that $i \leq j$ implies $f_{ij} \circ f_j' = f_i'$. For each g' in G', the I-tuple $(f_i'(g'))_{i \in I}$ is a member of $\prod_i G_i$, and the map $g' \mapsto (f_i'(g'))_{i \in I}$ is continuous into the product topology because each entry is continuous. If $i \leq j$, then the tuple $(f_i'(g'))_{i \in I}$ has the property that $f_{ij}(f_j'(g')) = f_i'(g')$ because of the given compatibility condition for the f_i''s. Therefore the map F given by $g' \mapsto (f_i'(g'))_{i \in I}$ has its image in the subset G of $\prod_i G_i$, and it is evidently a continuous group homomorphism. The map F proves the existence assertion in (ii) because $f_i \circ F(g') = f_i((f_j'(g'))_{j \in I}) = f_i'(g')$.

For uniqueness, suppose that $H : G' \to G$ is a continuous homomorphism such that $f_i \circ H = f_i'$ for all i. For each $g' \in G'$, we have $f_i(H(g')) = f_i'(g')$. Thus $H(g')$ is the member $(g_i)_{i \in I}$ of $\prod_{i \in I} G_i$ for which $g_i = f_i'(g')$ for all i. Hence H is uniquely determined. $\square$

Proposition 7.28. In the category of compact topological groups, any two inverse limits for an inverse system $(I, \{G_i\}, \{f_{ij}\})$ are canonically isomorphic.

PROOF. This is a special case of the uniqueness in category theory of objects having a specific universal mapping property, as established in *Basic Algebra*. $\square$

It is important in applications that the inverse limit of an inverse system of compact groups depend only on what happens far out in the directed set. We have not yet used that the indexing set is a directed set, rather than merely a partially ordered set, and we shall use this property now.

Corollary 7.29. Let I be a directed set, let j_0 be in I, and let I' be the set of members of I that are $\geq j_0$. If $(I, \{G_i\}, \{f_{ij}\})$ is an inverse system of compact groups, then the two inverse systems $(I, \{G_i\}, \{f_{ij}\})$ and $(I', \{G_i\}, \{f_{ij}\})$ have canonically isomorphic inverse limits, the isomorphism of the standard inverse limit $G \subseteq \prod_{i \in I} G_i$ onto the standard inverse limit $G' \subseteq \prod_{i \geq j_0} G_i$ being given by projection to the coordinates $\geq j_0$.

PROOF. Let $P : G \to G'$ be the projection, and let $f_i' : G' \to G_i$ for $i \geq j_0$ be the associated maps. Certainly $f_i' \circ P = f_i$ for $i \geq j_0$. We shall extend the definition of f_i' to apply to all $i \in I$. If $i \in I$ is given, we use the fact that I is directed to choose i' with $i' \geq i$ and $i' \geq j_0$. Define $f_i' = f_{ii'} \circ f_{i'}'$. Let us see that f_i' is well defined. Let i'' have $i'' \geq i$ and $i'' \geq j_0$. Choose i''' with $i''' \geq i'$ and $i''' \geq i''$. The computation

$$f_{ii'''} \circ f_{i'''}' = f_{ii'} \circ f_{i'i'''} \circ f_{i'''}' = f_{ii'} \circ f_{i'}'$$

shows that i' and i''' yield the same definition of f_i', and a similar argument shows that i'' and i''' yield the same definition. Therefore i' and i'' yield the same definition. Thus f_i' is now defined for all i in I.

We shall show that $(G', \{f_i'\}_{i \in I})$ is an inverse limit of $(I, \{G_i\}, \{f_{ij}\})$, and then the corollary follows from Proposition 7.28. Property (i) of inverse limits is built into the definition of the homomorphisms f_i'. For property (ii) of G', suppose that

$(\widetilde{G}, \{\widetilde{f}_i\}_{i \in I})$ is a pair consisting of a compact topological group $\widetilde{G}$ and continuous homomorphisms $\widetilde{f}_i : \widetilde{G} \to G_i$ such that $i \leq j$ implies $f_{ij} \circ \widetilde{f}_j = \widetilde{f}_i$. By (ii) for existence with G, find a continuous homomorphism $F : \widetilde{G} \to G$ with $f_i \circ F = \widetilde{f}_i$ for all i. Substituting from $f'_i \circ P = f_i$, we obtain $f'_i \circ (P \circ F) = \widetilde{f}_i$, and this says that $P \circ F : \widetilde{G} \to G'$ is the map we seek for the existence in (ii) for G'. For uniqueness in (ii), suppose that $F' : \widetilde{G} \to G'$ satisfies $f'_i \circ F' = \widetilde{f}_i$ for all i. Then $f'_i \circ F' = f'_i \circ (P \circ F)$ for $i \geq j_0$. By (ii) for uniqueness with G', $F' = P \circ F$. This says that the map from $\widetilde{G}$ to G' in (ii) is unique. □

Let us now apply these considerations to topologize Galois groups of infinite separable normal algebraic extensions. The topologized Galois group will be the inverse limit of finite Galois groups, each with the discrete topology.[10]

We return to our field K, its algebraic closure K_{alg}, and its separable algebraic closure K_{sep} within K_{alg}. Let L be a field with $K \subseteq L \subseteq K_{\text{sep}}$, and assume that L/K is a normal extension, not necessarily finite. We shall topologize $\text{Gal}(L/K)$. Let x be any element of L, and let F be the finite extension $F = K(x)$ of K. If f is the minimal polynomial of x over K, then f has a root in L and must split in L because L/K is normal. Let $x_1, \ldots, x_n$ be the roots of f, with $x_1 = x$. Then $E = K(x_1, \ldots, x_n)$ is a finite normal extension of K with $K \subseteq F \subseteq E \subseteq L$. Since x is arbitrary in L, L is the union of all the finite normal extensions of K lying within L.

For each pair (E, E') of normal extensions of K with $K \subseteq E \subseteq E' \subseteq L$, Proposition 7.24 gives us restriction homomorphisms $\varphi_{EE'} : \text{Gal}(E'/K) \to \text{Gal}(E/K)$. We write φ_E for the special case that $E' = L$, so that $\varphi_{EL} = \varphi_E$. If $K \subseteq E \subseteq E' \subseteq E'' \subseteq L$, then $\varphi_{EE'} \circ \varphi_{E'E''} = \varphi_{EE''}$, and consequently the system

$$\left(\left\{ \begin{matrix} E \text{ finite normal} \\ \text{extension of } K \\ \text{in } L \end{matrix} \right\}, \{\text{Gal}(E/K)\}, \{\varphi_{EE'}\} \right)$$

is an inverse system of (discrete finite) topological groups. Meanwhile, we can form the group $\text{Gal}(L/K)$ and the system $\{\varphi_E\}$ of homomorphisms with $\varphi_E = \varphi_{EL}$.

Proposition 7.30. With the above notation, the group $\text{Gal}(L/K)$ may be identified with the underlying abstract group of the inverse limit $\varprojlim_{L \leftarrow E} \text{Gal}(E/K)$, taken over finite normal extensions E/K with $E \subseteq L$, in such a way that the homomorphisms φ_E become the homomorphisms of the inverse limit.

---

[10]The inverse limit of a finite group is called a **profinite group**. Profinite groups have special properties by comparison with general compact groups, but it will not be necessary for us to undertake a study of them.

PROOF. Let $G = \varprojlim_{L \leftarrow E} \mathrm{Gal}(E/K)$, put $G_E = \mathrm{Gal}(E/K)$, and regard G as the standard inverse limit given as in Proposition 7.27:

$$G = \{(\gamma_E)_E \in \prod_E G_E \mid \varphi_{EE'}(\gamma_{E'}) = \gamma_E \text{ whenever } E \subseteq E'\}.$$

For each E, we have a homomorphism $\varphi_E : \mathrm{Gal}(L/K) \to G_E$, and the product of the values of these defines a homomorphism $\Phi : \mathrm{Gal}(L/K) \to \prod_E G_E$. The relations $\varphi_{EE'} \circ \varphi_{E'E''} = \varphi_{EE''}$ show that the image of Φ is contained in the subgroup G of $\prod_E G_E$. We shall show that $\Phi : \mathrm{Gal}(L/K) \to G$ is one-one onto.

Let us see that Φ is one-one. If $\gamma \neq 1$ is in $\mathrm{Gal}(L/K)$, then there exists $x \in K$ with $\gamma(x) \neq x$. Let E be a finite normal extension of K within L containing x. Then $\gamma|_E \neq 1$, and thus $\varphi_E(\gamma) \neq 1$. Hence $\Phi(\gamma) \neq 1$, and Φ is one-one.

Let us see that Φ is onto G. Let $(\gamma_E)_E \in G$ be given. For x in L, choose a finite normal E with $x \in E$ and $E \subseteq L$, and define $\gamma(x) = \gamma_E(x)$. The relations among the $\varphi_{EE'}$ show that this definition of $\gamma(x)$ is independent of the choice of E, and γ is therefore a field map of L into itself. Certainly γ fixes K, and we can construct an inverse to γ from the mappings γ_E^{-1}. Thus γ is in $\mathrm{Gal}(L/K)$. Application of Φ gives $\Phi(\gamma) = (\varphi_E(\gamma))_E = (\gamma_E)_E$, and Φ is onto. □

Using Proposition 7.30, we transfer the topology from $\varprojlim_{L \leftarrow E} \mathrm{Gal}(E/K)$ to $\mathrm{Gal}(L/K)$, and we can now regard $\mathrm{Gal}(L/K)$ as a compact topological group. For any finite normal extension F of K with $F \subseteq L$, consider the group $\mathrm{Gal}(L/F)$. The inverse-limit topology identifies $\mathrm{Gal}(L/K)$ with a subgroup of $\prod_{E \supseteq K} \mathrm{Gal}(E/K)$, the product being taken over all finite normal extensions E of K contained in L, and Corollary 7.29 allows us to identify $\mathrm{Gal}(L/K)$ with a subgroup of

$$\prod_{E \supseteq F} \mathrm{Gal}(E/K),$$

the product being taken over all finite normal extensions E of F contained in L. Under this identification $\mathrm{Gal}(L/F)$ is identified with the subgroup of elements γ of the image of $\mathrm{Gal}(L/K)$ for which $\varphi_F(\gamma) = 1$. Since φ_F is continuous, this is a closed set. In turn, this set equals the image of $\mathrm{Gal}(L/F)$ in the subset

$$\prod_{E \supseteq F} \mathrm{Gal}(E/F).$$

The latter gives the standard inverse limit topology on $\mathrm{Gal}(L/F)$. Except for some details, the conclusion is as follows.

6. Infinite Galois Groups

Corollary 7.31. With the notation of Proposition 7.30, give $\mathrm{Gal}(L/K)$ the inverse-limit topology. If F is a finite normal extension of K contained in L, then $\mathrm{Gal}(L/F)$ is a closed subgroup of $\mathrm{Gal}(L/K)$, and the relative topology on $\mathrm{Gal}(L/F)$ coincides with the inverse-limit topology of $\mathrm{Gal}(L/F)$. The subgroup $\mathrm{Gal}(L/F)$ of $\mathrm{Gal}(L/K)$ is a normal subgroup of finite index in $\mathrm{Gal}(L/K)$. Being a closed subgroup of finite index, it is an open subgroup.

PROOF. We still need to prove that $\mathrm{Gal}(L/F)$ has finite index in $\mathrm{Gal}(L/K)$. Proposition 7.24 shows that the restriction to F of any member of $\mathrm{Gal}(L/K)$ is an automorphism of F. Since F is a finite extension of K, there are only finitely many possibilities for this automorphism. If two elements γ and γ' of $\mathrm{Gal}(L/K)$ restrict to the same automorphism of F, then $\gamma^{-1}\gamma'$ is a member of $\mathrm{Gal}(L/K)$ fixing F, i.e., a member of $\mathrm{Gal}(L/F)$. Thus γ' lies in the coset $\gamma \,\mathrm{Gal}(L/F)$, and we conclude that there are only finitely many cosets. Since every member of $\mathrm{Gal}(L/K)$ restricts on F to an automorphism of F, the subgroup of members of $\mathrm{Gal}(L/K)$ restricting to the identity on F is a normal subgroup. Thus $\mathrm{Gal}(L/F)$ is normal in $\mathrm{Gal}(L/K)$. $\square$

Corollary 7.32. With the notation of Proposition 7.30, $\mathrm{Gal}(L/K)$ has a system of open normal subgroups with intersection $\{1\}$. Hence the same thing is true of any closed subgroup of T of $\mathrm{Gal}(L/K)$. Moreover, if U is any open neighborhood of 1 in T, then some open normal subgroup lies in U; consequently the open normal subgroups of T form a neighborhood base about the identity.

PROOF. The open normal subgroups in the first conclusion are the subgroups $\mathrm{Gal}(L/F)$ as in Corollary 7.31. Since every member of L lies in some finite normal extension of K within L, a member of $\mathrm{Gal}(L/K)$ cannot lie in every $\mathrm{Gal}(L/F)$ unless it is the identity on L.

Let U be an open neighborhood of 1 in the closed subgroup T of $\mathrm{Gal}(L/K)$. The set-theoretic complement U^c of U in T is a compact set, and the complements of the open normal subgroups of T are open sets whose union covers U^c, by the result of the previous paragraph. By compactness finitely many complements of open normal subgroups of T together cover U^c. The intersection of these open normal subgroups is then an open normal subgroup contained in U. $\square$

Theorem 7.33 (Fundamental Theorem of Galois Theory). Let K be a field, and let K_{alg} be an algebraic closure, so that $K \subseteq K_{\mathrm{sep}} \subseteq K_{\mathrm{alg}}$. Let L be a normal extension of K lying in K_{sep}. Let $\mathcal{S}$ be the set of all closed subgroups of $\mathrm{Gal}(L/K)$, and let $\mathcal{F}$ be the set of all intermediate fields between K and L. Then $F \mapsto \mathrm{Gal}(L/F)$ is a one-one mapping of $\mathcal{F}$ onto $\mathcal{S}$ with inverse $S \mapsto L^S$, L^S being the fixed field within L of the group S.

PROOF. First we show that $\mathrm{Gal}(L/F)$ is closed; Corollary 7.31 shows this only when F is a *normal* extension of K. Let $\{F_\alpha\}$ be the set of all finite extensions of K contained in F. Then $F = \bigcup_\alpha F_\alpha$, and thus $\text{Gal}(L/F) = \bigcap_\alpha \text{Gal}(L/F_\alpha)$. Each F_α is contained in a finite normal extension E_α of K lying in L, and hence $\text{Gal}(L/F_\alpha) \supseteq \text{Gal}(L/E_\alpha)$. Corollary 7.31 shows that $\text{Gal}(L/E_\alpha)$ is an open subgroup of $\text{Gal}(L/K)$, and hence the larger subgroup $\text{Gal}(L/F_\alpha)$ is open (as a union of cosets, each of which is open). Open subgroups are closed. Thus $\text{Gal}(L/F_\alpha)$ is closed, and so is $\text{Gal}(L/F) = \bigcap_\alpha \text{Gal}(L/F_\alpha)$.

Next if F is in $\mathcal{F}$, then the inclusion $L \supseteq F$ and the fact that L is normal over K together imply that L is normal over F. By Corollary 7.26, $F = L^{\text{Gal}(L/F)}$. Hence $F \mapsto \text{Gal}(L/F)$ is one-one, and $S \mapsto L^S$ is a left inverse of it.

Finally we show that $S \mapsto L^S$ is a right inverse by showing that $\text{Gal}(L/L^S) = S$ for any closed subgroup S of $\text{Gal}(L/K)$. Define $T = \text{Gal}(L/L^S)$. Certainly $S \subseteq T$. The previous step shows that $F = L^{\text{Gal}(L/F)}$ for all $F \in \mathcal{F}$. Taking $F = L^S$ gives $L^S = L^{\text{Gal}(L/L^S)} = L^T$. Let V be an arbitrary open normal subgroup of T, and put $E = L^V$. The members of T/V give well-defined automorphisms of E, and

$$E^{T/V} = (L^V)^{T/V} = L^T = L^S = (L^V)^{SV/V} = E^{SV/V}. \qquad (*)$$

The group T/V is a finite group of automorphisms of E fixing K, and Corollary 9.37 of *Basic Algebra*, when applied to the group T/V and the separable extension $E/E^{T/V}$, shows that $T/V = \text{Gal}(E/E^{T/V})$. Similarly it shows that $SV/V = \text{Gal}(E/E^{SV/V})$. By $(*)$, $T/V = SV/V$, i.e., $T = SV$. Corollary 7.32 shows that the open normal subgroups of T form a neighborhood base about the identity of T. From the equality $T = SV$ for arbitrary V, let us see that

$$S \text{ is dense in } T. \qquad (**)$$

Arguing by contradiction, let g be in T but not in the closure of S. Find V small enough so that $gV^{-1} \cap S = \emptyset$. From $T = SV$, we can write $g = sv$ with $s \in S$ and $v \in V$. Then $svV^{-1} \cap S = \emptyset$, and hence $vV^{-1} \cap S = \emptyset$. This last equality is a contradiction, since the identity lies in vV^{-1}, and $(**)$ is proved. Since S is closed, it follows from $(**)$ that $S = T$. But $T = \text{Gal}(L/L^S)$ by definition. Therefore $\text{Gal}(L/L^S) = S$, and the proof of the theorem is complete. □

Theorem 7.34. Let K be a perfect field, and L be an algebraically closed field containing K. Then the only members of L fixed by every element of $\text{Gal}(L/K)$ are the members of K.

PROOF. Proposition 7.15 shows that $K_{\text{sep}} = K_{\text{alg}}$, and Corollary 7.26 implies that the only members of K_{alg} fixed by $\text{Gal}(K_{\text{alg}}/K)$ are the members of K. Thus we are done unless L contains elements not in K_{alg}.

Let x and y be any two members of L not in K_{alg}, and let ψ be in $\text{Gal}(K_{\text{alg}}/K)$. The singleton sets $\{x\}$ and $\{y\}$ are transcendence sets over K_{alg}, and Lemma 7.6

shows that they can be extended to transcendence bases of L over K_{alg}. Call these transcendence bases E and F, respectively. Theorem 7.9 shows that E and F have the same cardinality. Therefore there exists a one-one function φ of E onto F such that $\varphi(x) = y$. This function φ extends uniquely to a field map Φ of $K_{\text{alg}}(E)$ onto $K_{\text{alg}}(F)$ that restricts to ψ on K_{alg}. Theorem 7.7 shows that L is an algebraic extension of $K_{\text{alg}}(E)$ and of $K_{\text{alg}}(F)$; hence L is an algebraic closure of $K_{\text{alg}}(E)$ and of $K_{\text{alg}}(F)$. The composition of Φ followed by inclusion is a field map of $K_{\text{alg}}(E)$ into L, and Theorem 9.23 of *Basic Algebra* shows that it can be extended to a field map $\widetilde{\Phi}$ of L into L. Since $\widetilde{\Phi}(L)$ is an algebraic closure of $K_{\text{alg}}(F)$, $\widetilde{\Phi}(L) = L$. Thus there exists a member $\widetilde{\Phi}$ of $\text{Gal}(L/K_{\text{alg}})$ with $\widetilde{\Phi}(x) = y$ such that $\widetilde{\Phi}\big|_{K_{\text{alg}}} = \psi$.

Taking ψ to be the identity shows that no element of L transcendental over K is fixed by $\text{Gal}(L/K)$. If an element z of K_{alg} is given that is not in K, then the first paragraph of the proof produces a member ψ of $\text{Gal}(K_{\text{alg}}/K)$ that moves z. Applying the result of the second paragraph to this ψ with x arbitrary and with $y = x$ shows that ψ extends to a member of $\text{Gal}(L/K)$ that moves z. □

7. Problems

1. Let L/K be a field extension in characteristic p. Prove that the set of elements of L that are purely inseparable over K is a subfield of L.

2. In characteristic p, let $K(\alpha)$ be an algebraic extension of a field K, and form the inclusions $K \subseteq K(\alpha^{p^e}) \subseteq K(\alpha)$, where α^{p^e} is the smallest power of α that is separable over K. Prove that the subfield of separable elements in the extension $K(\alpha)/K$ consists exactly of $K(\alpha^{p^e})$, i.e., that no separable elements of $K(\alpha)$ over K lie outside $K(\alpha^{p^e})$.

3. Partially order the positive integers by saying that $a \leq b$ if a divides b. Let $(\widehat{\mathbb{Z}}, \{f_a\}_{a \geq 1})$ be the inverse limit of the cyclic groups $\mathbb{Z}/a\mathbb{Z}$, with the homomorphism f_{ab} from $\mathbb{Z}/b\mathbb{Z}$ to $\mathbb{Z}/a\mathbb{Z}$ being given by $f_{ab}(1 + b\mathbb{Z}) = 1 + a\mathbb{Z}$ when a divides b. Each member c of $\mathbb{Z}$ defines a member z_c of $\widehat{\mathbb{Z}}$ such that $f_a(z_c) = c + a\mathbb{Z}$ for all a. Exhibit some other explicit member of $\widehat{\mathbb{Z}}$.

4. Prove that the only members of $\mathbb{C}$ fixed by all members of $\text{Gal}(\mathbb{C}/\mathbb{Q})$ are the members of $\mathbb{Q}$. What members of $\mathbb{R}$ are fixed by $\text{Gal}(\mathbb{R}/\mathbb{Q})$?

5. By making use of the field $K = \mathbb{Q}(\sqrt{2}, \sqrt{3}, \sqrt{5}, \sqrt{7}, \dots)$, show that there exist subgroups of $\text{Gal}(\mathbb{Q}_{\text{alg}}/\mathbb{Q})$ of index 2 that are not open.

Problems 6–14 concern primary ideals and make use of the notion of the radical $\sqrt{I}$ of an ideal I as defined in Section 1. Throughout, R will denote a commutative ring with identity. A proper ideal I of R is **primary** if whenever a and b are in R, ab is in I, and a is not in I, then b^m is in I for some integer $m > 0$. It is immediate that every prime ideal is primary.

6. Prove that an ideal I of R is primary if and only if every zero divisor in R/I is nilpotent (in the sense that some power of it is 0), if and only if 0 is primary in R/I.

7. (a) Prove that if I is a primary ideal, then $\sqrt{I}$ is a prime ideal. (Educational note: In this case the prime ideal $\sqrt{I}$ is called the **associated prime ideal** to I.)
    (b) Prove that if I is any ideal and if $I \subseteq J$ for a prime ideal J, then $\sqrt{I} \subseteq J$.

8. (a) Show that the primary ideals in $\mathbb{Z}$ are 0 and (p^n) for p prime and $n > 0$.
    (b) Let $R = \mathbb{C}[x, y]$ and $I = (x, y^2)$. Use Problem 6 to show that I is primary. Show that $P = \sqrt{I}$ is given by $P = (x, y)$. Deduce that $P^2 \subsetneq I \subsetneq P$ and that a primary ideal is not necessarily a power of a prime ideal.
    (c) Let K be a field, let $R = K[X, Y, Z]/(XY - Z^2)$, and let x, y, z be the images of X, Y, Z in R. Show that $P = (x, z)$ is prime by showing that R/P is an integral domain. Show that P^2 is not primary by starting from the fact that $xy = z^2$ lies in P^2.

9. Prove that if I is an ideal such that $\sqrt{I}$ is maximal, then I is primary. Deduce that the powers of a maximal ideal are primary.

10. An ideal is **reducible** if it is the finite intersection of ideals strictly containing it; otherwise it is **irreducible**.
    (a) Show that every prime ideal is irreducible.
    (b) Let $R = \mathbb{C}[x, y]$, and let I be the maximal ideal (x, y). Show that I^2 is primary and that the equality $I^2 = (Rx + I^2) \cap (Ry + I^2)$ exhibits I^2 as reducible.

11. Prove that if R is Noetherian, then every ideal is a finite intersection of proper irreducible ideals. (The ideal R is understood to be an empty intersection.)

12. Suppose that R is Noetherian and that Q is a proper irreducible ideal in R. Prove that 0 is primary in R/Q, and deduce that Q is primary in R.

13. Prove that if $Q_1, \ldots, Q_n$ are primary ideals in R that all have $\sqrt{Q_i} = P$, then $Q = \bigcap_{i=1}^n Q_i$ is primary with $\sqrt{Q} = P$.

14. (**Lasker–Noether Decomposition Theorem**) The expression $I = \bigcap_{i=1}^n Q_i$ of an ideal I as an intersection of primary ideals Q_i is said to be **irredundant** if
    (i) no Q_i contains the intersection of the other ones, and
    (ii) the Q_i have distinct associated prime ideals.

    Prove that if R is Noetherian, then every ideal is the irredundant intersection of finitely many primary ideals.

CHAPTER VIII

Background for Algebraic Geometry

Abstract. This chapter introduces aspects of the algebraic theory of systems of polynomial equations in several variables.

Section 1 gives a brief history of the subject, treating it as one of two early sources of questions to be addressed in algebraic geometry.

Section 2 introduces the resultant as a tool for eliminating one of the variables in a system of two such equations. A first form of Bezout's Theorem is an application, saying that if $f(X, Y)$ and $g(X, Y)$ are polynomials of respective degrees m and n whose locus of common zeros has more than mn points, then f and g have a nontrivial common factor. This version of the theorem may be regarded as pertaining to a pair of affine plane curves.

Section 3 passes to projective plane curves, which are nonconstant homogeneous polynomials in three variables, two such being regarded as the same if they are multiples of one another. Versions of the resultant and Bezout's Theorem are valid in this context, and two projective plane curves defined over an algebraically closed field always have a common zero.

Sections 4–5 introduce intersection multiplicity for projective plane curves. Section 4 treats a line and a curve, and Section 5 treats the general case of two curves. The theory in Section 4 is completely elementary, and a version of Bezout's Theorem is proved that says that a line and a curve of degree d have exactly d common zeros, provided the underlying field is algebraically closed, the zeros are counted as often as their intersection multiplicities, and the line does not divide the curve. Section 5 makes more serious use of algebraic background, particularly localizations and the Nullstellensatz. It gives an indication that ostensibly simple phenomena in the subject can require sophisticated tools to analyze.

Section 6 proves a version of Bezout's Theorem appropriate for the context of Section 5: if F and G are two projective plane curves of respective degrees m and n over an algebraically closed field, then either they have a nontrivial common factor or they have exactly mn common zeros when the intersection multiplicities of the zeros are taken into account.

Sections 7–10 concern Gröbner bases, which are finite generating sets of a special kind for ideals in a polynomial algebra over a field. Section 7 sets the stage, introducing monomial orders and defining Gröbner bases. Section 8 establishes a several-variable analog of the division algorithm for polynomials in one variable and derives from it a usable criterion for a finite set of generators to be a Gröbner basis. From this it is easy to give a constructive proof of the existence of Gröbner bases and to obtain as consequences solutions of the ideal-membership problem and the proper-ideal problem. Section 9 obtains a uniqueness theorem under the condition that the Gröbner basis be reduced. Adjusting a Gröbner basis to make it reduced is an easy matter. A consequence of the uniqueness result is a solution of the ideal-equality problem. Section 10 gives two theorems concerning solutions of systems of polynomial equations. The Elimination Theorem identifies in terms of Gröbner bases those members of the ideal that depend only on a certain subset of the variables. The Extension Theorem, proved under the additional assumption that the underlying field is algebraically closed, gives conditions under which a solution to the subsystem of equations that depend on all but one variable can be extended to a solution of the whole system. The latter theorem makes use of the theory of resultants.

1. Historical Origins and Overview

Modern algebraic geometry grew out of early attempts to solve simultaneous polynomial equations in several variables and out of the theory of Riemann surfaces. We shall discuss the first of these sources in the present chapter and the second of the sources in Chapter IX.

Serious consideration of simultaneous polynomial equations of degree > 2 dates to a 1750 book[1] by Gabriel Cramer (1704–1752), who may be better known for Cramer's rule in connection with determinants. Cramer was interested in various aspects of the zero loci of polynomials in two variables with real coefficients. Thinking of the zero locus, we refer to a nonconstant polynomial in two variables as a plane curve.

One of the problems of interest to Cramer was to find the number of points in the plane that would uniquely determine a plane curve of degree n up to a constant multiple. Cramer gave the answer $\frac{1}{2}n(n+3)$ to this problem. For example, when $n = 2$, if we normalize matters by taking the coefficient of x^2 to be 1, then the possible quadratic polynomials

$$f(x, y) = x^2 + bxy + cy^2 + dx + ey + f$$

involve five unknown coefficients. Each condition $f(x_i, y_i) = 0$ gives a linear condition on the coefficients, and Cramer was able to write down explicitly a plane curve through the given points in question by introducing determinants and applying his rule to solve the problem.

Already with this much description the reader will see a certain subtlety—that there will be special choices of the five points for which existence or uniqueness will fail. We could also ask about the effect of multiplicities: what does it mean geometrically to take two or more of the points to be equal, and how does such an occurrence affect the number of points that can be specified?

Cramer noticed a subtlety that is less easy to resolve, even in hindsight. If we are given any two plane curves of degree 3, then Cardan's formula says that we can solve one equation for y in terms of x, obtaining three expressions in x; then we can substitute for y in the other equation each of the three expressions in x and obtain a cubic equation in x each time. In other words, we should expect up to 9 points of intersection for two cubics, and 9 should sometimes occur. (The various

---

[1]G. Cramer, *Introduction à l'Analyse des Lignes courbes algébriques*, Chez les Frères Cramer & Cl. Philibert, Geneva, 1750.

forms of Bezout's Theorem, which came a little later, confirm this argument.) The number of points that determine a cubic completely is $\frac{1}{2}n(n+3)$ for $n = 3$, i.e., is 9. Thus we have 9 points determining a unique cubic, and yet the second cubic goes through these 9 points as well. What is happening? This question has come to be known as **Cramer's paradox**.

Explaining this kind of mystery became an early impetus for the development of algebraic geometry.

The question of the number of points of intersection had been the subject of conjecture for some time earlier, and it was expected that two plane curves of respective total degrees m and n in some sense had mn points of intersection. Étienne Bezout (1730–1783) took up this question and dealt with parts of it rigorously. The quadratic case can be solved by finding one variable in terms of the other and by substituting, but let us handle it by the method that Bezout used. If we view each polynomial as quadratic in y and having coefficients that depend on x, then we have a system

$$a_0 + a_1 y + a_2 y^2 = 0,$$
$$b_0 + b_1 y + b_2 y^2 = 0.$$

Instead of regarding this as a system of two equations for y, we regard it as a system of two homogeneous linear equations for variables x_0, x_1, x_2, where $x_0 = 1, x_1 = y, x_2 = y^2$. We can get two further equations by multiplying each equation by y:

$$a_0 y + a_1 y^2 + a_2 y^3 = 0,$$
$$b_0 y + b_1 y^2 + b_2 y^3 = 0,$$

and then we have four homogeneous linear equations for $x_0 = 1, x_1 = y, x_2 = y^2, x_3 = y^3$. Since the system has the nonzero solution $(1, y, y^2, y^3)$, the determinant of the coefficient matrix must be 0. Remembering that the coefficients depend on x, we see that we have eliminated the variable y and obtained a polynomial equation for x without using any solution formula for polynomials in one variable. The device that Bezout introduced for this purpose—the determinant of the coefficient matrix—is called the **resultant** of the system and is a fundamental tool in handling simultaneous polynomial equations. With it Bezout went on in 1779 to give a rigorous proof that when two polynomials in (x, y) are set equal to 0 simultaneously, one of degree m and the other of degree n, then there cannot be more than mn solutions unless the two polynomials have a common factor. This is a first form of Bezout's Theorem and is proved in Section 2.

In order to have a chance of obtaining a full complement of mn solutions, we make three adjustments—allow complex solutions instead of just real solutions (even in the case $(m, n) = (2, 1)$), consider "projective plane curves" instead of ordinary plane curves to allow for solutions at infinity (even in the case $(m, n) =$

(1, 1)), and introduce a suitable notion of intersection number of two plane curves at a point in order to take multiplicities into account (even in the case $(m, n) = (2, 1)$). We shall allow complex solutions already in Section 2, and we shall make an adjustment for projective plane curves in Section 3. The issue of intersection multiplicity is more complicated. The beginnings of a classical approach to it are indicated in Section 4, and a somewhat more modern approach appears in Section 5. With the full theory of intersection multiplicities of projective plane curves in place, we obtain a general form of Bezout's Theorem[2] in Section 6.

The theory of the resultant can be extended in various ways, but we shall largely not pursue this matter. Studies of zero loci of systems of equations took a more geometric turn in the first part of the nineteenth century through the work of Julius Plücker (1801–1858) and others, but these matters will be left for an implicit discussion in Chapter X. Instead, we skip to a development that began with the doctoral thesis of Bruno Buchberger in 1965. Buchberger was interested in being able to decide when a polynomial is a member of an ideal that is specified by a finite list of generators. For this purpose he learned that each ideal has a special finite set of generators that is unique once certain declarations are made. He devised an algorithm for determining such a set of generators,[3] and he gave the name "Gröbner basis" to the set, in honor of his thesis advisor.[4] The special unique such basis is called a "reduced Gröbner basis."

An unfortunate feature of the algorithm (and even of later improved algorithms) is that Gröbner bases are extraordinarily complicated to calculate. The timing of Buchberger's discovery was therefore especially fortuitous, coming when computers were becoming more common, more economical, and more powerful.

Buchberger was able to give a test for membership in an ideal in terms of a multivariable division algorithm involving any Gröbner basis. Other general problems involving ideals were solvable as well. Because of the uniqueness of the reduced Gröbner basis, two ideals are identical if and only if their reduced Gröbner bases are equal. When some of the theory of resultants was incorporated into the theory of Gröbner bases, these bases could also be used to address various questions of identifying zero loci. Other problems involving ideals could be addressed by similar methods. The theory has flowered tremendously since its initial discovery and by the present day has found many imaginative applications to applied problems. Sections 7–10 give an introductory account of this important theory.

---

[2] A correct proof of the general form of the theorem seems to have been published for the first time by Georges-Henri Halphen (1844–1889) in 1873.

[3] Devising the algorithm was Buchberger's real contribution, since the abstract existence of the special set of generators is an easy consequence of the Hilbert Basis Theorem and had already been used in papers of H. Hironaka in 1964.

[4] Wolfgang Gröbner (1899–1980). The name is often spelled out as "Groebner," particularly when it is used in connection with computer algorithms.

2. Resultant and Bezout's Theorem

Let A be a unique factorization domain. The case that $A = K[X_1, \ldots, X_r]$ for a field K will be the main case of interest for us. If f and g are polynomials in $A[X]$ of the form
$$f(X) = f_0 + f_1 X + \cdots + f_m X^m,$$
$$g(X) = g_0 + g_1 X + \cdots + g_n X^n,$$
with m and n both positive, then we let $\mathcal{R}(f, g)$ be the $(m+n)$-by-$(m+n)$ matrix

$$\begin{pmatrix} f_0 & f_1 & \cdots & f_{m-1} & f_m & 0 & 0 & 0 & \cdots & 0 \\ 0 & f_0 & \cdots & f_{m-2} & f_{m-1} & f_m & 0 & 0 & \cdots & 0 \\ \vdots & & \ddots & & & & \ddots & & & \vdots \\ 0 & \cdots & & & f_0 & & & & \cdots & f_m \\ g_0 & g_1 & \cdots & & g_{n-1} & g_n & 0 & \cdots & & 0 \\ 0 & g_0 & \cdots & & g_{n-2} & g_{n-1} & g_n & & \cdots & 0 \\ \vdots & & \ddots & & & & & \ddots & & \vdots \\ 0 & \cdots & & g_0 & g_1 & & \cdots & & & g_n \end{pmatrix},$$

in which there are n rows above the g_0 in the first column and there are m remaining rows. The **resultant** of f and g is the determinant
$$R(f, g) = \det \mathcal{R}(f, g).$$

Theorem 8.1. If A is a unique factorization domain and if f and g are nonzero members of $A[X]$ of the form $f(X) = \sum_{i=0}^{m} f_i X^i$ and $g(X) = \sum_{j=0}^{n} g_j X^j$ with $m > 0$ and $n > 0$ and with at least one of f_m and g_n nonzero, then the following are equivalent:
  (a) f and g have a common factor of degree > 0 in X,
  (b) $af + bg = 0$ for some nonzero a and b in $A[X]$ with $\deg a < n$ and $\deg b < m$.
  (c) $R(f, g) = 0$.

Regard $R(f, g)$ as a constant polynomial in X. When $R(f, g) \neq 0$, there exist unique a and b in $A[X]$ such that $a(X) f(X) + b(X) g(X) = R(f, g)$ with $\deg a < n$ and $\deg b < m$. Both the polynomials a and b are nonzero if both $f(X)$ and $g(X)$ are nonconstant.

REMARKS. The theorem says that $af + bg = R(f, g)$ holds in every case for which at least one of the coefficients f_m and g_n is nonzero. Sometimes the theorem appears in texts with the assumption that both coefficients are nonzero; in this connection, see Problem 5 at the end of the chapter. When $R(f, g) = 0$, the theorem does not point to a useful way to identify a common factor; the division algorithm can be used for this purpose in some circumstances, but the use of Gröbner bases as in Section 7 will be more helpful.

PROOF. Let us prove the equivalence of (a) and (b). Suppose that (a) holds. If u is a nonconstant polynomial in X that divides both f and g, let us write $f = bu$ and $g = -au$. Then $af + bg = 0$. Also, $\deg a + \deg u = \deg g$; since $\deg u > 0$, $\deg a < \deg g \leq n$. Similarly $\deg b < m$. Thus (b) holds. Conversely suppose that (b) holds, so that $af = -bg$ with a and b nonzero and with $\deg a < n$ and $\deg b < m$. Suppose that $f_m \neq 0$. The equality $af = -bg$ shows that f divides bg. Since $\deg b < m = \deg f$, f cannot divide b. But $A[X]$ is a unique factorization domain, and thus there is some prime factor p of f of positive degree such that p^k for some k divides f but not b. Then p divides both f and g, and (a) holds. A similar argument works if $g_n \neq 0$.

Now we prove the equivalence of (b) and (c). Let F be the field of fractions of A. We set up a one-one correspondence between polynomials $a(X)$ in $A[X]$ of degree at most $n - 1$ and n-dimensional row vectors $(\alpha_0 \quad \alpha_1 \quad \cdots \quad \alpha_{n-1})$ with entries in A by the formula

$$a(X) = \alpha_0 + \alpha_1 X + \cdots + \alpha_{n-1} X^{n-1},$$

and similarly we set up one-one correspondences for degrees at most $m - 1$ and at most $m + n - 1$ by the formulas

$$b(X) = \beta_0 + \beta_1 X + \cdots + \beta_{m-1} X^{m-1},$$
$$c(X) = \gamma_0 + \gamma_1 X + \cdots + \gamma_{m+n-1} X^{m+n-1}.$$

Examining the form of $\mathcal{R}(f, g)$, we see that the matrix equality

$$(\alpha_0 \quad \alpha_1 \quad \cdots \quad \alpha_{n-1} \quad \beta_0 \quad \cdots \quad \beta_{m-1}) \mathcal{R}(f, g)$$
$$= (\gamma_0 \quad \gamma_1 \quad \cdots \quad \gamma_{m+n-1}) \quad (*)$$

holds if and only if the polynomial equality

$$a(X) f(X) + b(X) g(X) = c(X). \quad (**)$$

holds. If (b) holds, then $af = -bg$, and $(**)$ shows that $c = 0$. That is, $(\gamma_0 \quad \gamma_1 \quad \cdots \quad \gamma_{m+n-1})$ is the 0 row vector. Interpreting $(*)$ as a matrix equality over F and assuming that a and b are not both 0, we see that the transpose of $\mathcal{R}(f, g)$ has a nontrivial null space. Therefore $R(f, g) = \det \mathcal{R}(f, g) = 0$. This proves (c). Conversely if (c) holds, then we can find row vectors $(\alpha_0 \quad \alpha_1 \quad \cdots \quad \alpha_{n-1})$ and $(\beta_0 \quad \beta_1 \quad \cdots \quad \beta_{m-1})$ not both 0, having entries in F, such that the left side of $(*)$ equals the 0 row vector. Clearing fractions, we may assume that $(\alpha_0 \quad \alpha_1 \quad \cdots \quad \alpha_{n-1})$ and $(\beta_0 \quad \beta_1 \quad \cdots \quad \beta_{m-1})$ have entries in A. Referring to $(*)$, we obtain $af + bg = 0$ with $\deg a$ at most $n - 1$ and $\deg b$ at most $m - 1$. We know that at least one of a and b is nonzero, and we have to see that both are nonzero. The situation is symmetric in a and b. If a were to equal 0, then we would have $bg = 0$ and we could conclude that $b = 0$ because $g \neq 0$. So we would obtain the contradiction $a = b = 0$. This proves (b).

For the last statement of the theorem, suppose that $R(f, g) \neq 0$. Then Cramer's rule applied over the field of fractions F of A shows that the matrix inverse of $\mathcal{R}(f, g)$ is of the form

$$\mathcal{R}(f, g)^{-1} = R(f, g)^{-1} \mathcal{S}(f, g),$$

where $\mathcal{S}(f, g)$ is a matrix with entries in A. Consequently the row vector

$$(\, R(f,g) \quad 0 \quad \cdots \quad 0 \,) \mathcal{R}(f, g)^{-1}$$

has entries in A, and we can define members $\alpha_0, \ldots, \alpha_{n-1}, \beta_0, \ldots, \beta_{m-1}$ of A by

$$(\alpha_0 \quad \alpha_1 \quad \cdots \quad \alpha_{n-1} \quad \beta_0 \quad \cdots \quad \beta_{m-1})$$
$$= (\, R(f,g) \quad 0 \quad \cdots \quad 0 \,) \mathcal{R}(f, g)^{-1}.$$

Then (*) holds with $(\gamma_0 \quad \gamma_1 \quad \cdots \quad \gamma_{m+n-1}) = (\, R(f,g) \quad 0 \quad \cdots \quad 0 \,)$, and the equality (**) shows that $a(X)f(X) + b(X)g(X) = R(f, g)$. If both f and g are nonconstant, then neither $a(X)$ nor $b(X)$ can be 0, since otherwise the equation would show that $R(f, g)$ is a nonconstant polynomial. □

Theorem 8.2 (Bezout's Theorem). *Let K be any field, and let $f(X, Y)$ and $g(X, Y)$ be nonconstant polynomials in $K[X, Y]$, of exact respective degrees m and n. If the locus of common zeros of f and g in K^2 has more than mn points, then f and g have a nonconstant common factor in $K[X, Y]$.*

PROOF. For most of the proof, we assume that K is infinite. Arguing by contradiction, suppose that f and g both vanish at distinct points (x_i, y_i) for $1 \leq i \leq mn + 1$, and suppose that f and g have no nonconstant common factor. Since there are only finitely many members c of K such that $y_i - y_j = c(x_i - x_j)$ for some i and j with $i \neq j$ and since K is assumed to be infinite, we can find c in K such that $y_i - y_j \neq c(x_i - x_j)$ for all i and j with $i \neq j$. For this c, $y_i - cx_i \neq y_j - cx_j$ when $i \neq j$, and therefore the second coordinates of the points $(x_i, y_i - cx_i)$ are distinct. The common zeros of $f(X, Y)$ and $g(X, Y)$ include the points (x_i, y_i), and thus the common zeros of $f(X, Y + cX)$ and $g(X, Y + cX)$ include the $mn + 1$ points $(x_i, y_i - cx_i)$ whose second coordinates are distinct.

In other words, there is no loss of generality in assuming that the given polynomials f and g vanish at $mn + 1$ points whose second coordinates are distinct. Regard $f(X, Y)$ and $g(X, Y)$ as members $f(X)$ and $g(X)$ of $A[X]$, where $A = K[Y]$, and write

$$f(X) = f_0 + f_1 X + \cdots + f_{m'} X^{m'},$$
$$g(X) = g_0 + g_1 X + \cdots + g_{n'} X^{n'},$$

with each f_i and g_i in A and with $f_{m'} \neq 0$ and $g_{n'} \neq 0$. Here $m' \leq m$ and $n' \leq n$.

Let us rule out the possibility that $m' = 0$ or $n' = 0$. Indeed, if we had $m' = 0$, then the polynomial f would be nonzero and would depend on Y alone. Since f is nonzero and has degree $m \geq 1$, it has at most m roots. But we are assuming that f and g vanish at $mn + 1$ points whose Y coordinates are distinct, and the inequalities $m \leq mn < mn + 1$ therefore give a contradiction. Thus $m' \neq 0$. Similarly $n' \neq 0$. So Theorem 8.1 is applicable.

Form the square matrix $\mathcal{R}(f, g)$ of size $m' + n'$ and its determinant $R(f, g)$. The latter is a member of $K[Y]$, and Theorem 8.1 shows that it cannot be 0, since f and g are assumed to have no nonconstant common factor in $K[X, Y]$.

Let us bound the degree of the member $R(f, g) = \det \mathcal{R}(f, g)$ of $K[Y]$. Each term in the expansion of the determinant is of the form

$$\pm \prod_{1 \leq i \leq m'+n'} \mathcal{R}(f, g)_{i, \sigma(i)} \qquad (*)$$

for some permutation σ of $\{1, \ldots, m' + n'\}$. Here $\mathcal{R}(f, g)_{ij}$ is given by

$$\mathcal{R}(f, g)_{ij} = \begin{cases} f_{j-i} & \text{for } 1 \leq i \leq n' \text{ and for } j \text{ with } i \leq j \leq m' + i, \\ 0 & \text{for } 1 \leq i \leq n' \text{ and for all other } j, \\ g_{j+n'-i} & \text{for } n' + 1 \leq i \leq n' + m' \text{ and for } j \\ & \text{with } i \leq n' + j \leq m' + i, \\ 0 & \text{for } n' + 1 \leq i \leq n' + m' \text{ and for all other } j. \end{cases}$$

In addition, the degree of f_{j-i} as a member of $K[Y]$ is at most $m' - (j - i)$, and the degree of $g_{j+n'-i}$ is at most $n' - (j + n' - i) = i - j$. Setting $j = \sigma(i)$, we see that the degree of $(*)$ is at most

$$\sum_{1 \leq i \leq n'} (m' - \sigma(i) + i) + \sum_{n'+1 \leq i \leq m'+n'} (i - \sigma(i))$$
$$= m'n' - \sum_{1 \leq i \leq m'+n'} \sigma(i) + \sum_{1 \leq i \leq m'+n'} i = m'n' \leq mn.$$

Thus $R(f, g)$ is a nonzero polynomial in $K[Y]$ of degree at most mn. Consequently it has at most mn roots.

Theorem 8.1 shows that $af + bg = R(f, g)$ for suitable members a and b of $K[X, Y]$. Recalling that f and g are assumed to vanish at $mn + 1$ points whose second coordinates are distinct, we see that $R(f, g)$ vanishes at each of these second coordinates, and we arrive at a contradiction.

2. Resultant and Bezout's Theorem

Now we can allow K to be finite. Let K' be an infinite extension. We have just seen that f and g have a nonconstant factor in $K'[X, Y]$. Without loss of generality, this factor depends nontrivially on X. Theorem 8.1 applied with $A = K'[Y]$ shows that $R[f, g] = 0$. The same theorem with $A = K[Y]$ then shows that f and g have a common factor in $A[X] = K[X, Y]$ depending nontrivially on X. □

Let us introduce some geometric language for the situation in Theorem 8.2. **Affine n-space** over a field K is the set of n-dimensional column vectors

$$\mathbb{A}^n = \mathbb{A}^n_{K_{\text{alg}}} = \{(x_1, \ldots, x_n) \in K^n_{\text{alg}}\}$$

with entries in a fixed algebraic closure K_{alg} of K. The set of K **rational points**, or K **points**, in $\mathbb{A}^n$ is the subset

$$\mathbb{A}^n_K = \{(x_1, \ldots, x_n) \in K^n\}.$$

We shall comment on the appearance of K_{alg} in these definitions shortly.

Members of $\mathbb{A}^n$ are called **points** in n-dimensional affine space, and the functions $P \mapsto x_j(P)$ give the **coordinates** of the points. If L is any field between K and K_{alg}, then any polynomial f in $K[X_1, \ldots, X_n]$ defines a corresponding polynomial function from $\mathbb{A}^n_L$ into L.

For algebraic geometry the case of interest for Sections 1–6 of this chapter is the case $n = 2$. The way of viewing a curve is influenced by Cramer's thinking as discussed in Section 1: the particular polynomial that defines a curve is important, not just the zero locus in the affine plane, but two curves are to be regarded as the same if each is a nonzero multiple of the other. We can incorporate this viewpoint into algebraic language by defining an **affine plane curve** C over the field K to be any nonzero proper principal ideal[5] in $K[X, Y]$. The curve is an **affine plane line** if the degree of any generator is 1.

In practice in studying affine plane curves, there is ordinarily no need to distinguish between a polynomial and the principal ideal that it generates, and we shall feel free to refer to an affine plane curve $C = (f)$ as f when there is no possibility of confusion.

The zero locus of a curve is the corresponding geometric notion, but it can readily be empty, as is the case with $X^2 + Y^2 + 1$ when $K = \mathbb{R}$. On the other hand, the Nullstellensatz (Theorem 7.1) ensures that the zero locus will be nonempty if the underlying field is algebraically closed. Thus we define the **zero locus** $V(C) = V((f))$ of the curve $C = (f(X, Y))$ by[6]

---

[5]*Warning:* This definition will be changed slightly in Chapter IX and again in Chapter X to reflect changed emphasis in those chapters.

[6]The letter "V" is the letter that is commonly used in the notation for a zero locus. It stands for "variety," a notion that we have not yet defined. But beware: not all objects labeled with a "V" are actually varieties the way the term is normally defined. An affine plane curve will turn out to be a variety exactly when the generating polynomial f is prime in $K_{\text{alg}}[X, Y]$.

$$V(C) = V_{K_{\text{alg}}}(C) = \{(x, y) \in K_{\text{alg}}^2 \mid f(x, y) = 0\}.$$

This is the same as the set of all (x, y) such that every member of the ideal C vanishes at (x, y). The set of K **rational points**, or K **points**, of C is

$$V_K(C) = V_K((f)) = \{(x, y) \in K \mid f(x, y) = 0\}.$$

When we are content to refer to an affine curve $C = (f)$ as f, we are content also to write $V(f)$ in place of $V(C) = V((f))$.

In Chapter X, under the assumption that K is algebraically closed, we shall extend these definitions from the case $n = 2$ and C as above to the case that n is general and C is replaced by any ideal I in $K[x_1, \ldots, X_n]$. The set $V(I)$ of common zeros of the members of I in $K^n = K_{\text{alg}}^n$ will be called an "affine algebraic set." The case of affine n-space itself arises when the ideal is 0.

For general K, not necessarily algebraically closed, it is meaningful to consider the set $V_K(I)$ of K rational points, i.e., the subset of common zeros lying in K^n. For $I = 0$ and $V(I) = \mathbb{A}^n$, the distinction between $V_K(I)$ and $V_{K_{\text{alg}}}(I)$ is hardly worth mentioning, but the distinction is well worth making for general I and is made for the case $V(I) = \mathbb{A}^n$ for consistency. Although the study of sets $V_K(I)$ is of importance in number theory, in geometry over $\mathbb{R}$, and in other areas, we shall not pursue it in Chapter X for lack of space.

Returning to Theorem 8.2, we see that the statement concerns $V_K(C) \cap V_K(D)$, where C and D are the principal ideals $C = (f)$ and $D = (g)$ in $K[X, Y]$. The theorem says that if $V_K(C) \cap V_K(D)$ contains more than mn points, then there is a nonzero principal ideal h with $(h) \subseteq (f) \cap (g)$.

3. Projective Plane Curves

Section 2 dealt with intersections of affine plane curves. Even over an algebraically closed field, two affine plane curves need not intersect. An example is the pair of straight lines $X + Y - 1$ and $X + Y - 2$, whose locus of common zeros is empty. To get these lines to intersect, we have to introduce "points at infinity." The projective plane is the device for including such points.

Let K be a field, and let K_{alg} be an algebraic closure. The **projective plane** over K is defined set theoretically as the quotient of $K_{\text{alg}}^3 - \{0\}$ by an equivalence relation:

$$\mathbb{P}^2 = \mathbb{P}^2_{K_{\text{alg}}} = \{(x, y, w) \in K_{\text{alg}}^3 - \{0\}\}/\sim,$$

where $(x', y', w') \sim (x, y, w)$ if $(x', y', w') = \lambda(x, y, w)$ for some $\lambda \in K_{\text{alg}}^\times$. The set of K **rational points**, or K **points**, of $\mathbb{P}^2$ is the quotient

$$\mathbb{P}^2_K = \{(x, y, w) \in K^3 - \{0\}\}/\sim,$$

where $(x', y', w') \sim (x, y, w)$ if $(x', y', w') = \lambda(x, y, w)$ for some $\lambda \in K$. When there is a need to be careful, we shall write $[x, y, w]$ for the member of $\mathbb{P}^2_K$ corresponding to (x, y, w) in $K^3 - \{0\}$. But often there will not be such a need, and we shall simply refer to (x, y, w) as a member of $\mathbb{P}^2_K$. Both $\mathbb{P}^2$ and $\mathbb{P}^2_K$ have additional structure on them, given by "affine local coordinates," and we come to that matter later in this section.

Let us record briefly the obvious generalization of the projective plane to other dimensions: **Projective n-space** over K is defined set theoretically as the quotient

$$\mathbb{P}^n = \mathbb{P}^n_{K_{\mathrm{alg}}} = \{(x_1, \ldots, x_{n+1}) \in K^{n+1}_{\mathrm{alg}} - \{0\}\} / \sim,$$

where $(x'_1, \ldots, x'_{n+1}) \sim (x_1, \ldots, x_{n+1})$ if $(x'_1, \ldots, x'_{n+1}) = \lambda(x_1, \ldots, x_{n+1})$ for some $\lambda \in K^\times_{\mathrm{alg}}$. The set $\mathbb{P}^n_K$ of K **rational points** of $\mathbb{P}^n$ is the set defined in similar fashion using just nonzero vectors in K^{n+1} and scalars in $K^\times$.

Scalar-valued functions on $\mathbb{P}^n_K$ are of little interest because they amount to scalar-valued functions of $K^n - \{0\}$ that are unchanged when $(x_1, \ldots, x_n)$ is replaced by a multiple of itself. A polynomial of this kind, for example, is necessarily constant. Instead, the polynomials of interest that are related to $\mathbb{P}^n_K$ are "homogeneous polynomials." A **monomial** in $K[X_1, \ldots, X_{n+1}]$ is a polynomial of the form $X_1^{j_1} \cdots X_{n+1}^{j_{n+1}}$; its **total degree** is $\sum_{i=1}^{n+1} j_i$. We say that a nonzero F in $K[X_1, \ldots, X_{n+1}]$ is **homogeneous** of degree $d \geq 0$ if every monomial appearing in F with nonzero coefficient has total degree d. By convention the 0 polynomial is homogeneous of every degree. We write $K[X_1, \ldots, X_{n+1}]_d$ for the set of homogeneous polynomials of degree d. Each such F satisfies

$$F(\lambda x_1, \ldots, \lambda x_{n+1}) = \lambda^d F(x_1, \ldots, x_{n+1})$$

for all $(x_1, \ldots, x_{n+1}) \in K^{n+1}$ and $\lambda \in K^\times$. Conversely the fact that the mapping of polynomials into polynomial functions is one-one for an infinite field implies that homogeneous polynomials over an infinite field can be detected by this property.

Let us assemble some further properties of homogeneous polynomials: The monomials of total degree d form a K basis of the vector space $K[X_1, \ldots, X_{n+1}]_d$; this fact follows from the definition of polynomials over K. To calculate the dimension of $K[X_1, \ldots, X_{n+1}]_d$, consider the problem of taking d factors X on which to place subscripts and using n dividers to separate the X_1's from the X_2's and so on. The number of monomials in question is just the number of ways of selecting the n dividers from among the $d + n$ symbols and dividers. Thus we obtain the important formula

$$\dim_K K[X_1, \ldots, X_{n+1}]_d = \binom{d+n}{n}.$$

Lemma 8.3. Any polynomial factor of a homogeneous polynomial over a field K is homogeneous.

PROOF. Write $F = F_1 F_2$ nontrivially. Let d_1 and e_1 be the highest and lowest total degrees of terms in F_1, and let d_2 and e_2 be the highest and lowest total degrees of terms in F_2. The product of the terms of total degree d_1 in F_1 and the terms of total degree d_2 in F_2 is nonzero and is the $d_1 d_2$ total-degree part of F. The product of the terms of total degree e_1 in F_1 and the terms of total degree e_2 in F_2 is nonzero and is the $e_1 e_2$ total-degree part of F. Since F is homogeneous, $d_1 d_2 = e_1 e_2$. It follows that $d_1 = e_1$ and $d_2 = e_2$; thus F_1 and F_2 are homogeneous. $\square$

An ideal I in $K[X_1, \ldots, X_{n+1}]$ is called a **homogeneous ideal** if it is the sum over $d \geq 0$ of its intersections with $K[X_1, \ldots, X_{n+1}]_d$:

$$I = \bigoplus_{d=0}^{\infty} (I \cap K[X_1, \ldots, X_{n+1}]_d).$$

The sum is to be regarded as a direct sum of vector spaces. For such an ideal, we can compute the quotient $K[X_1, \ldots, X_{n+1}]/I$ term by term:

$$K[X_1, \ldots, X_{n+1}]/I = \bigoplus_{d=0}^{\infty} K[X_1, \ldots, X_{n+1}]_d / (I \cap K[X_1, \ldots, X_{n+1}]_d).$$

We can often recognize a homogeneous ideal from its generators: an ideal with a set of generators that are all homogeneous is necessarily a homogeneous ideal. In fact, if an ideal I has homogeneous generators F_j, then the most general member of I is a finite sum of terms $A_j F_j$. The terms of total degree d in $A_j F_j$ are the product of F_j with the terms in A_j of total degree $d - \deg F_j$, and each such term is in I. Hence each member of I is a sum of homogeneous polynomials that lie in I, and the assertion follows.

In the setting of $\mathbb{P}^2$, projective plane curves over K are initially defined to be nonconstant *homogeneous* polynomials in $K[X, Y, W]$. Although such polynomials are not well defined on the projective plane, their zero loci are well defined subsets of $\mathbb{P}^2$. As in the affine case, the particular polynomial that defines a curve is important, not just the zero locus, but two curves are to be regarded as the same if each is a nonzero multiple of the other. We can incorporate this viewpoint into algebraic language by defining a **projective plane curve** of **degree** $d > 0$ over the field K to be any nonzero proper principal ideal in $K[X, Y, W]$ generated by a homogeneous polynomial of degree d. Such an ideal is necessarily homogeneous. In the special cases that $d = 1, 2, 3$, or 4, the curve is called a **projective line**, **conic**, **cubic**, or **quartic** respectively.

Just as in the affine case, in practice in studying projective plane curves, there is often no need to distinguish between a homogeneous polynomial and the homogeneous principal ideal that it generates, and we shall feel free to refer to a projective plane curve $C = (F) \subseteq K[X, Y, W]$ as F when there is no possibility of confusion.

3. Projective Plane Curves

If (F) is a projective plane curve of degree d, then its zero locus is denoted by

$$V((F)) = V_{K_{\text{alg}}}((F)) = \{[x, y, w] \in \mathbb{P}^2 \mid F(x, y, w) = 0\}.$$

The locus

$$V_K((F)) = \{[x, y, w] \in \mathbb{P}^2_K \mid F(x, y, w) = 0\}$$

is called the set of K **rational points**, or K **points**, of the curve. When we allow ourselves to refer to the curve simply as F, then we can write $V(F)$ in place of $V((F))$.

The affine plane $\mathbb{A}^2_K = \{(x, y)\}$ has a standard one-one embedding into the projective plane $\mathbb{P}^2_K$. Namely we map (x, y) into $[x, y, 1]$. The set that is missed by the image is the set with $w = 0$, which is the set of K rational points of the line L with $L(X, Y, W) = W$, a line called the **line at infinity**. We shall denote this line by W. The points of $V_K(W)$, i.e., those with $w = 0$, are called the **points at infinity**.

Except for the line at infinity, lines in $\mathbb{P}^2_K$ correspond under restriction exactly to lines in K^2. Namely the projective line $L(X, Y, W) = aX + bY + cW$ corresponds to the affine line $l(x, y) = aX + bY + c$, and vice versa. In certain ways the geometry of $\mathbb{P}^2_K$ is simpler than the geometry of $\mathbb{A}^2_K$:

(i) Two distinct lines in $\mathbb{P}^2_K$ intersect in a unique point. In fact, we set up the system of equations

$$\begin{pmatrix} a & b & c \\ a' & b' & c' \end{pmatrix} \begin{pmatrix} x \\ y \\ w \end{pmatrix} = \begin{pmatrix} 0 \\ 0 \end{pmatrix}.$$

Since the lines are distinct, the coefficient matrix has rank 2. Thus the kernel has dimension 1, and there is just one point $[x, y, w]$ in the intersection.

(ii) Two distinct points in $\mathbb{P}^2_K$ lie on a unique line. In fact, we set up the system of equations

$$\begin{pmatrix} x & y & w \\ x' & y' & w' \end{pmatrix} \begin{pmatrix} a \\ b \\ c \end{pmatrix} = \begin{pmatrix} 0 \\ 0 \end{pmatrix}$$

and argue in similar fashion.

Along with the embedding of $\mathbb{A}^2_K$ into $\mathbb{P}^2_K$ is a correspondence between projective curves and affine curves. Let us work with the polynomials themselves, without identifying each polynomial with every nonzero scalar multiple of itself. The passage from a nonzero homogeneous polynomial $F(X, Y, W)$ of degree

$d > 0$ to a polynomial $f(X, Y)$ is given by $f(X, Y) = F(X, Y, 1)$. The mapping $F \mapsto f$ is a substitution homomorphism, and it therefore respects products. However, the degree may drop in the process, and in particular $f(X, Y)$ is a constant if and only if $F(X, Y, W)$ is a multiple of W^d.

In the reverse direction if $f(X, Y)$ is a polynomial of degree e, then $f(X, Y)$ arises from a polynomial $F(X, Y, W)$, but we have to specify the degree d of F and we must have $d \geq e$. Operationally we obtain F by inserting a power of W into each term of f to make the total degree of the term become d. For example, with $f(X, Y) = Y^2 + XY + X^3$ if the desired degree is 3, then $F(X, Y, W) = Y^2 W + XYW + X^3$. On the other hand, if the desired degree is 4, then $F(X, Y, W) = Y^2 W^2 + XYW^2 + X^3 W$.

The formula for this reverse process is $F(X, Y, W) = W^d f(XW^{-1}, YW^{-1})$. That is, F is given by a substitution homomorphism, followed by multiplication by a power of W. From this fact, we can read off conclusions of the following kind:

> If polynomials $f(X, Y)$ and $g(X, Y)$ are obtained from homogeneous polynomials $F(X, Y, W)$ and $G(X, Y, W)$ by taking $W = 1$, then there exist integers r and s such that the polynomial $W^r F(X, Y, W) + W^s G(X, Y, W)$ is homogeneous and such that $f(X, Y) + g(X, Y)$ is obtained from it by taking $W = 1$.

As we mentioned above, $\mathbb{P}_K^2$ has more structure than simply the structure of a set. About any point in $\mathbb{P}_K^2$ we can introduce various systems of "affine local coordinates." The idea is to imitate what happens in the definition of a manifold: the whole manifold is covered by charts, each giving an invertible mapping of a set in the manifold to an open subset of Euclidean space. Here a single system of affine local coordinates plays the role of a chart; it puts $\mathbb{A}_K^2$ into one-one correspondence with the complement of the zero locus of a line in $\mathbb{P}_K^2$.

Let Φ be a member of the matrix group $\mathrm{GL}(3, K)$. Then Φ maps the set K^3 of column vectors in one-one fashion onto K^3 and passes to a one-one map of $\mathbb{P}_K^2$ onto $\mathbb{P}_K^2$ called the **projective transformation** corresponding to Φ. Two Φ's give the same map of $\mathbb{P}_K^2$ if and only if they are multiples of one another. The group action of $\mathrm{GL}(3, K)$ on $\mathbb{P}_K^2$ is transitive because $\mathrm{GL}(3, K)$ acts transitively on $K^3 - \{(0, 0, 0)\}$.

If L is the projective line whose coefficients are given by the row vector $(a \quad b \quad c)$ and if Φ is is in $\mathrm{GL}(3, K)$, then the row vector $(a \quad b \quad c)\Phi^{-1}$ defines a new projective line L^Φ, and the K rational points of L^Φ are given by

$$V_K(L^\Phi) = \Phi(V_K(L)).$$

In fact, let $\begin{pmatrix} x \\ y \\ w \end{pmatrix}$ be in $V_K(L)$. Then $\begin{pmatrix} x' \\ y' \\ w' \end{pmatrix} = \Phi \begin{pmatrix} x \\ y \\ w \end{pmatrix}$ is in $\Phi(V_K(L))$ and satisfies

3. Projective Plane Curves

$$(a \quad b \quad c)\Phi^{-1}\begin{pmatrix} x' \\ y' \\ w' \end{pmatrix} = 0;$$

hence it is in $V_K(L^\Phi)$. Conversely if $\begin{pmatrix} x' \\ y' \\ w' \end{pmatrix}$ is in $V_K(L^\Phi)$, then $\begin{pmatrix} x \\ y \\ w \end{pmatrix} = \Phi^{-1}\begin{pmatrix} x' \\ y' \\ w' \end{pmatrix}$ satisfies

$$(a \quad b \quad c)\begin{pmatrix} x \\ y \\ w \end{pmatrix} = (a \quad b \quad c)\Phi^{-1}\begin{pmatrix} x' \\ y' \\ w' \end{pmatrix} = 0,$$

and thus $\begin{pmatrix} x' \\ y' \\ w' \end{pmatrix}$ is Φ of something in $V_K(L)$.

To form the analog of a chart, fix $[x_0, y_0, w_0]$ in $\mathbb{P}^2_K$. Choose (by transitivity) some Φ in $GL(3, K)$ with $\Phi(x_0, y_0, w_0) = (0, 0, 1)$. Then we can define **affine local coordinates** on $\Phi^{-1}(K \times K \times \{1\})$ to K^2 by the one-one map

$$\varphi(\Phi^{-1}(x, y, 1)) = (x, y).$$

This definition generalizes the standard embedding of the affine plane K^2 into $\mathbb{P}^2_K$ earlier; that embedding was the case $\Phi = 1$.

EXAMPLES OF AFFINE LOCAL COORDINATES FOR $\mathbb{P}^2_K$.

(1) Suppose $(x_0, y_0, w_0) = (x_0, y_0, 1)$. We can choose $\Phi = \begin{pmatrix} 1 & 0 & -x_0 \\ 0 & 1 & -y_0 \\ 0 & 0 & 1 \end{pmatrix}$. Then

$$\Phi\begin{pmatrix} x \\ y \\ 1 \end{pmatrix} = \begin{pmatrix} 1 & 0 & -x_0 \\ 0 & 1 & -y_0 \\ 0 & 0 & 1 \end{pmatrix}\begin{pmatrix} x \\ y \\ 1 \end{pmatrix} = \begin{pmatrix} x - x_0 \\ y - y_0 \\ 1 \end{pmatrix}.$$

In this case, the local coordinates are defined on

$$\Phi^{-1}(K \times K \times 1) = K \times K \times 1$$

and are given by

$$\varphi(x, y, 1) = \varphi(\Phi^{-1}(\Phi(x, y, 1)))$$
$$= \varphi(\Phi^{-1}(x - x_0, y - y_0, 1)) = (x - x_0, y - y_0).$$

This Φ is handy for reducing behavior about $(x_0, y_0, 1)$ in $\mathbb{P}^2_K$ to behavior about $(0, 0)$ in K^2.

(2) Suppose $(x_0, y_0, w_0) = (0, 1, 0)$. We can choose $\Phi = \begin{pmatrix} 0 & 0 & 1 \\ 1 & 0 & 0 \\ 0 & 1 & 0 \end{pmatrix}$. Then

$$\Phi \begin{pmatrix} x \\ 1 \\ w \end{pmatrix} = \begin{pmatrix} 0 & 0 & 1 \\ 1 & 0 & 0 \\ 0 & 1 & 0 \end{pmatrix} \begin{pmatrix} x \\ 1 \\ w \end{pmatrix} = \begin{pmatrix} w \\ x \\ 1 \end{pmatrix}$$

and

$$\varphi(x, 1, w) = \varphi(\Phi^{-1}(\Phi(x, 1, w))) = \varphi(\Phi^{-1}(w, x, 1)) = (w, x).$$

This Φ is handy for studying behavior near one of the points at infinity in $\mathbb{P}^2_K$.

We can use affine local coordinates to examine the behavior of a projective plane curve "near a particular point," by which is meant "with that point as the center point in the analysis." To examine behavior near $(0, 0, 1)$, we use the correspondence $f(X, Y) = F(X, Y, 1)$ that we discussed earlier. For a general point, we make use of the fact that whenever F is a homogeneous polynomial of degree d, then so is $F \circ \Phi^{-1}$. To examine the behavior of F near a point (x_0, y_0, w_0) in $K^3 - \{(0, 0, 0)\}$, we choose Φ in GL$(3, K)$ with $\Phi(x_0, y_0, w_0) = (0, 0, 1)$, and we define

$$f(X, Y) = F(\Phi^{-1}(X, Y, 1)).$$

Under this correspondence the behavior of F at (x_0, y_0, w_0) is reflected in the behavior of f at $(0, 0)$. We call $f(X, Y)$ the **local expression** for F in the affine local coordinates determined by Φ. This local expression is a polynomial in $K[X, Y]$, and it is nonconstant unless F is a scalar multiple of $(W \circ \Phi)^d$ for some d.

EXAMPLES, CONTINUED.

(1) Suppose that $(x_0, y_0, w_0) = (x_0, y_0, 1)$ and that $\Phi = \begin{pmatrix} 1 & 0 & -x_0 \\ 0 & 1 & -y_0 \\ 0 & 0 & 1 \end{pmatrix}$. Computation gives

$$\Phi^{-1} \begin{pmatrix} x \\ y \\ 1 \end{pmatrix} = \begin{pmatrix} x + x_0 \\ y + y_0 \\ 1 \end{pmatrix},$$

and the corresponding local expression for a projective plane curve F is

$$f(X, Y) = F(X + x_0, Y + y_0, 1).$$

For the projective plane curve

$$F(X, Y, W) = X^2 Y + XYW + W^3$$

and the same Φ, the local expression $f(X, Y)$ splits into homogeneous terms as

3. Projective Plane Curves

$$f(X, Y) = (x_0^2 y_0 + x_0 y_0 + 1) + (x_0^2 Y + 2x_0 y_0 X + x_0 Y + y_0 X)$$
$$+ (y_0 X^2 + 2x_0 XY + XY) + (X^2 Y).$$

We shall use this splitting in the next section in the first example of intersection multiplicity.

(2) Suppose that $(x_0, y_0, w_0) = (0, 1, 0)$ and that $\Phi = \begin{pmatrix} 0 & 0 & 1 \\ 1 & 0 & 0 \\ 0 & 1 & 0 \end{pmatrix}$. Then

$$\Phi^{-1} \begin{pmatrix} x \\ y \\ 1 \end{pmatrix} = \begin{pmatrix} y \\ 1 \\ x \end{pmatrix},$$

and the local expression for a projective plane curve F relative to this Φ is

$$f(X, Y) = F(Y, 1, X).$$

For the same projective plane curve F as in Example 1, namely

$$F(X, Y, W) = X^2 Y + XYW + W^3,$$

we obtain

$$f(X, Y) = (Y^2 + XY) + (X^3).$$

We shall examine this example further in the next section.

In this way we have associated to each projective plane curve F and to the system of affine local coordinates determined by a member Φ of $\mathrm{GL}(3, K)$ a local expression that is a nonzero polynomial in $K[X, Y]$. Conversely if the degree d and the member Φ of $\mathrm{GL}(3, K)$ are given and if f in $K[X, Y]$ is nonzero of degree at most d, then we can reconstruct a projective plane curve F of degree d whose local expression relative to Φ is f. We have only to form the unique homogeneous polynomial G of degree d with $f(X, Y) = G(X, Y, 1)$ and then put $F = G \circ \Phi$.

With these preparations in place, we return to a consideration of resultants and Bezout's Theorem. Our objective is to rephrase Theorem 8.2 to take advantage of properties of the projective plane.

Lemma 8.4. Let K be a field, let A be the polynomial ring $A = K[x_1, \ldots, x_r]$, and let f and g be members of $A[X]$ of the form

$$f(X) = f_0 + f_1 X + \cdots + f_m X^m,$$
$$g(X) = g_0 + g_1 X + \cdots + g_n X^n,$$

where f_j is a member of A homogeneous of degree $m' - j$ and g_j is a member of A homogeneous of degree $n' - j$. Then the resultant $R(f, g)$ is a homogeneous member of A of degree $mn' + m'n - mn$.

REMARKS. In the application to proving Theorem 8.5, we will have $m' = m$ and $n' = n$, and then $R(f, g)$ is homogeneous of degree mn. Problem 8 at the end of the chapter concerns a situation for which $m' \neq m$ and $n' \neq n$.

PROOF. There is no loss of generality in assuming that K is algebraically closed, hence in particular is infinite. Each nonzero entry $\mathcal{R}(f, g)_{ij}$ of $\mathcal{R}(f, g)$ is a coefficient of f or of g. For each entry, define $p(i, j)$ such that $\mathcal{R}(f, g)_{ij}(tx_1, \ldots, tx_r) = t^{p(i,j)} \mathcal{R}(f, g)_{ij}(x_1, \ldots, x_r)$. The assembled matrix $\mathcal{R}$ with powers of t in place is

$$\begin{pmatrix} t^{m'} f_0 & t^{m'-1} f_1 & \cdots & t^{m'-m} f_m & \cdots \\ 0 & t^{m'} f_0 & & \cdots & \\ \vdots & & \ddots & & \\ t^{n'} g_0 & t^{n'-1} g_1 & \cdots & t^{n'-n} g_n & \cdots \\ 0 & t^{n'} g_0 & & \cdots & \end{pmatrix}. \tag{$*$}$$

It turns out that there is a function $q(i)$ such that $r(j) = q(i) + p(i, j)$ depends only on j. Here $t^{q(i)}$ is the ith entry of

$$(t^{n'}, t^{n'-1}, \ldots, t^{n'-n+1}; t^{m'}, t^{m'-1}, \ldots, t^{m'-m+1}).$$

The matrix $(*)$ with $t^{q(i)}$ multiplying every entry of the ith row is

$$\begin{pmatrix} t^{n'} t^{m'} f_0 & t^{n'} t^{m'-1} f_1 & \cdots & t^{n'} t^{m'-m} f_m & \cdots \\ 0 & t^{n'-1} t^{m'} f_0 & & \cdots & \\ \vdots & & \ddots & & \\ t^{m'} t^{n'} g_0 & t^{m'} t^{n'-1} g_1 & \cdots & t^{m'} t^{n'-n} g_n & \cdots \\ 0 & t^{m'-1} t^{n'} g_0 & & \cdots & \end{pmatrix}. \tag{$**$}$$

In $(**)$, $t^{r(j)}$ is the jth entry of $(t^{m'+n'}, t^{m'+n'-1}, \ldots, t^{m'+n'-m-n+1})$. Then we have

$$t^u R(f, g)(tx_1, \ldots, tx_r) = t^v R(f, g)(x_1, \ldots, x_r),$$

where $u = \sum_i q(i)$ and $v = \sum_j r(j)$. So

$$R(f, g)(tx_1, \ldots, tx_r) = t^{v-u} R(f, g)(x_1, \ldots, x_r).$$

In other words, $R(f, g)$ is a homogeneous function. Since K is infinite, $R(f, g)$ is homogeneous as a member of A. Computing u and v, we find that $u = mm' + nn' - \frac{1}{2}m(m-1) - \frac{1}{2}n(n-1)$ and $v = (m+n)(m'+n') - \frac{1}{2}(m+n)(m+n-1)$. Therefore $v - u = mn' + m'n - mn$, and the degree of homogeneity of $R(f, g)$ is $mn' + m'n - mn$. □

3. Projective Plane Curves

Theorem 8.5 (Bezout's Theorem). Let K be a field, let K_{alg} be an algebraic closure, and suppose that F in $K[X, Y, W]_m$ and G in $K[X, Y, W]_n$ are projective plane curves. Then their locus $V(F) \cap V(G)$ of common zeros in $\mathbb{P}^2_{K_{\text{alg}}}$ is nonempty. If this zero locus has more than mn points, then F and G have as a common factor some homogeneous polynomial in (X, Y, W) of positive degree.

REMARKS. For two polynomials $f(X, Y)$ and $g(X, Y)$ in affine space, application of Theorem 8.1 concerning the resultant in the Y variable involves checking that at least one of the polynomials has the expected degree in the Y variable, and doing so may not be so easy. In the projective setting, this problem disappears if we apply a projective transformation and arrange that $[0, 0, 1]$ not be on the zero locus of one of the given polynomials, say $F(X, Y, W)$. In fact, if F is in $K[X, Y, W]_m$, then the coefficient of W^m has to be a constant, and this term is the only term of F that contributes to the value of F at $(0, 0, 1)$. With the above adjustment the coefficient must be nonzero, and Theorem 8.1 is applicable.

PROOF. Without loss of generality, we may assume throughout that K is algebraically closed. Write F and G in the form

$$F(X, Y, W) = f_0 + f_1 W + \cdots + f_m W^m \quad \text{with } f_j \in K[X, Y]_{m-j},$$
$$G(X, Y, W) = g_0 + g_1 W + \cdots + g_n W^n \quad \text{with } g_j \in K[X, Y]_{n-j}.$$
(∗)

Pick a point (x, y, w) at which F is nonzero, and move it to $(0, 0, 1)$ by a projective transformation, so that $F(0, 0, 1) \neq 0$. Regarding F and G as polynomials in W, with coefficients in $A = K[X, Y]$, we form $R(F, G)$, which Lemma 8.4 identifies as a member of $K[X, Y]_{mn}$.

Since $R(F, G)$ is homogeneous as a member of $K[X, Y]$ and since K is algebraically closed, we can choose a point $(x_0, y_0) \neq (0, 0)$ with $R(F, G)(x_0, y_0) = 0$. Then the resultant of $F(x_0, y_0, W)$ and $G(x_0, y_0, W)$ is 0, and Theorem 8.1 applies because $F(x_0, y_0, W)$ has degree m in W. The theorem says that these two polynomials in W have a common factor. Since K is algebraically closed, this common factor vanishes at some w_0, and then we must have $F(x_0, y_0, w_0) = G(x_0, y_0, w_0) = 0$. This proves the first conclusion.

For the second conclusion, suppose that $V(F) \cap V(G)$ contains $mn + 1$ points. Join these points by lines, and pick a point of $\mathbb{P}^2_K$ that is not on any of the lines. We can do so because K, being algebraically closed, is infinite. Applying a projective transformation, we may assume that the point is $[0, 0, 1]$. Write F and G in the form (∗). Regarding F and G as polynomials in W, with coefficients in $A = K[X, Y]$, we again form $R(F, G)$, which Lemma 8.4 identifies as a member of $K[X, Y]_{mn}$. For fixed (x_0, y_0), Theorem 8.1 says that $R(F, G)(x_0, y_0) = 0$ if and only if $F(x_0, y_0, W)$ and $G(x_0, y_0, W)$ have a common factor (necessarily a common factor of the form $W - w_0$ because K is algebraically closed), if and only if $F(x_0, y_0, w_0) = G(x_0, y_0, w_0) = 0$ for some w_0. So at each of our $mn + 1$

points, say (x_i, y_i, w_i), we have $R(F, G)(cx_i, cy_i) = 0$ for all scalars c. Since $(x_i, y_i) \neq (0, 0)$, $R(F, G)$ vanishes on the line $y_i X - x_i Y = 0$. Consequently $y_i X - x_i Y$ divides $R(F, G)$ in $K[X, Y]$.

Suppose that (x_i, y_i) is a multiple of (x_j, y_j) with $i \neq j$. Then (x_i, y_i, w_i) and (x_j, y_j, w_j) both satisfy $y_i X - x_i Y = 0$. Since $(0, 0, 1)$ satisfies this also and since $(0, 0, 1)$ is not to be on any of the connecting lines, we obtain a contradiction.

Thus the $mn+1$ factors $y_i X - x_i Y$ are nonassociate primes in $K[X, Y]$ dividing $R(F, G)$. By unique factorization for $K[X, Y]$, their product divides $R(F, G)$. Since $\deg R(F, G) = mn$, we conclude that $R(F, G) = 0$. Then Theorem 8.1 shows that F and G have a nonconstant common factor in $K[X, Y][W] = K[X, Y, W]$. The common factor is homogeneous by Lemma 8.3, and the second conclusion is proved. $\square$

4. Intersection Multiplicity for a Line with a Curve

In this section we begin the topic of "intersection multiplicity" for projective plane curves. The idea is that the number of points in the intersection $V(F) \cap V(G)$ in Bezout's Theorem as formulated in Theorem 8.5 should actually equal mn, not merely be bounded above by mn, if the field is algebraically closed and the points are counted according to their "multiplicities," whatever that might mean.

The prototype is the factorization of a polynomial of degree n in one variable. The polynomial has at most n roots, and it has exactly n if the field is algebraically closed and each root is counted according to its multiplicity. In this case, as we well know, a root z_0 of $f(z)$ has multiplicity k if $(z - z_0)^k$ is the largest power of $z - z_0$ that divides $f(z)$.

Our objective in this section is to develop a notion of intersection multiplicity for the case of a line and a curve at a point; the case of two curves is less intuitive and is postponed to the next section. The main result is to be that the sum of the intersection multiplicities at all points for a line and a projective plane curve equals the degree of the curve, provided that the underlying field is algebraically closed and that the line does not divide the curve. The statement in the previous paragraph about polynomials in one variable will amount to a special case; for this special case the projective line is Y, the projective curve is of the form $W^{d-1}Y - F(X, W)$, where F is homogeneous of degree d and where $f(X) = F(X, 1)$, and the divisibility proviso is that F not be the 0 polynomial, i.e., that $f(z)$ not be identically 0.

Let K be a field, let L be in $K[X, Y, W]_1$, and let F be in $K[X, Y, W]_d$. The notation for intersection multiplicity will be $I(P, L \cap F)$, where $P = (x_0, y_0, w_0)$ is in $V_K(F) \cap V_K(L)$. To make the definition, we introduce affine local coordinates. Choose Φ in GL$(3, K)$ with $\Phi(x_0, y_0, w_0) = (0, 0, 1)$, and form the corresponding local expressions

$$f(X, Y) = F(\Phi^{-1}(X, Y, 1)) = f_1(X, Y) + \cdots + f_d(X, Y),$$

$$l(X, Y) = L(\Phi^{-1}(X, Y, 1)).$$

Here f_j is the part of f that is homogeneous of degree j. Since $l(0, 0) = 0$, we see that $l(X, Y) = bX - aY$ for some constants a and b not both 0. Then $\varphi(t) = \begin{pmatrix} at \\ bt \end{pmatrix}$, for $t \in K$, is a parametrization of the locus in $\mathbb{A}_K^2$ on which $l(x, y) = 0$. The composition $f(\varphi(t))$ is a polynomial in t with $f(\varphi(0)) = 0$. In fact,

$$f(\varphi(t)) = f_1(at, bt) + f_2(at, bt) + \cdots + f_d(at, bt)$$
$$= t f_1(a, b) + t^2 f_2(a, b) + \cdots + t^d f_d(a, b).$$

There are two possibilities. If $f \circ \varphi$ is not the 0 polynomial, then $f(\varphi(t))$ has a zero of some finite order at $t = 0$, and this order is defined to be the **intersection multiplicity**, or **intersection number**, $I(P, L \cap F)$. If $f \circ \varphi$ is the 0 polynomial, then we say that $I(P, L \cap F) = +\infty$. It will be convenient to define $I(P, L \cap F) = 0$ if P is not in $V_K(L) \cap V_K(F)$. We need to check that $I(P, L \cap F)$ does not depend on the choice of Φ, but we postpone this verification until after we consider two examples.

EXAMPLES OF INTERSECTION MULTIPLICITY.

(1) Example 1 in the previous section showed that relative to a suitable Φ in GL$(3, K)$, the projective plane curve

$$F(X, Y, W) = X^2 Y + XYW + W^3$$

has local expression $f(X, Y)$ about $P = (x_0, y_0, 1)$ given by

$$f(X, Y) = (x_0^2 y_0 + x_0 y_0 + 1) + (x_0^2 Y + 2x_0 y_0 X + x_0 Y + y_0 X)$$
$$+ (y_0 X^2 + 2x_0 XY + XY) + (X^2 Y)$$
$$= f_0 + f_1(X, Y) + f_2(X, Y) + f_3(X, Y).$$

For a line L, the intersection multiplicity $I(P, L \cap F)$ is 0 unless P lies in $V_K(F)$, i.e., unless $f_0 = x_0^2 y_0 + x_0 y_0 + 1 = 0$. Suppose that the line L is given by

$$L(X, Y, W) = \alpha X + \beta Y + \gamma W,$$

with local expression

$$l(X, Y) = L(X + x_0, Y + y_0, 1) = (\alpha x_0 + \beta y_0 + \gamma) + (\alpha X + \beta Y).$$

Here α and β are not both 0. The intersection multiplicity $I(P, L \cap F)$ is 0 unless P lies also in $V_K(L)$, i.e., unless $\alpha x_0 + \beta y_0 + \gamma = 0$. Thus suppose that P lies in $V_K(L) \cap V_K(F)$. Then we can parametrize the locus for which $l(x, y) = 0$ by $\binom{x}{y} = \varphi(t) = \binom{-\beta t}{\alpha t}$, and we obtain

$$f_1(\varphi(t)) = f_1(-\beta t, \alpha t) = t(x_0^2 \alpha - 2x_0 y_0 \beta + x_0 \alpha - y_0 \beta),$$
$$f_2(\varphi(t)) = f_2(-\beta t, \alpha t) = t^2(y_0 \beta^2 - 2x_0 \alpha \beta + \alpha \beta).$$

One point lying in $V_K(F)$ is $P = (x_0, y_0, 1) = \left(1, -\frac{1}{2}, 1\right)$, and P lies also in $V_K(L)$ if $\alpha - \frac{1}{2}\beta + \gamma = 0$, i.e., if γ satisfies $\gamma = \frac{1}{2}\beta - \alpha$. Then we have $f_1(\varphi(t)) = t(2\alpha + \frac{3}{2}\beta)$ and $f_2(\varphi(t)) = t^2(-\frac{1}{2}\beta^2 - \alpha \beta)$. Consequently, $I(P, L \cap F)$ is ≥ 1 if and only if $\gamma = \frac{1}{2}\beta - \alpha$. In this case, $I(P, L \cap F)$ is ≥ 2 if and only if $2\alpha + \frac{3}{2}\beta = 0$, i.e., if $\alpha = -\frac{3}{4}\beta$. When both conditions are satisfied, we have $f_2(\varphi(t)) = t^2(-\frac{1}{2}\beta^2 - \alpha \beta) = t^2(\frac{1}{4}\beta^2)$, and this is not the 0 function because under these conditions, $\beta = 0$ would imply that $(\alpha, \beta, \gamma) = (0, 0, 0)$; hence $I(P, L \cap F) = 2$.

(2) Example 2 in the previous section considered the point $P = (x_0, y_0, w_0) = (0, 1, 0)$ for the same F, namely $F(X, Y, W) = X^2Y + XYW + W^3$. This P lies in $V_K(F)$. For a suitable Φ, the earlier computations showed that the local expression for F is

$$f(X, Y) = (Y^2 + XY) + (X^3).$$

The most general line L for which P lies in $V_K(L)$ is $\alpha X + \gamma W = 0$, and the corresponding local expression is

$$l(X, Y) = L(Y, 1, X) = \alpha Y + \gamma X.$$

We use the parametrization $\varphi(t) = (-\alpha t, \gamma t)$ for L and obtain

$$f(\varphi(t)) = t^2(\gamma^2 - \alpha \gamma) + t^3(-\alpha^3).$$

By inspection we see that $I(P, L \cap F) \geq 2$ for all choices of α and γ, and that $I(P, L \cap F) \geq 3$ if and only if $\gamma = 0$ or $\gamma = \alpha$. If $\gamma = 0$ or $\gamma = \alpha$, then α^3 cannot be 0, and thus $I(P, L \cap F) = 3$.

Let us return to the verification that $I(P, L \cap F)$ does not depend on the choice of Φ. Thus suppose that Ψ is another member of $GL(3, K)$ with $\Psi(x_0, y_0, w_0) = (0, 0, 1)$. Write

$$\Psi \circ \Phi^{-1} = \begin{pmatrix} \alpha & \beta & 0 \\ \gamma & \delta & 0 \\ r & s & 1 \end{pmatrix},$$

4. Intersection Multiplicity for a Line with a Curve

form the local expressions

$$f'(X, Y) = F(\Psi^{-1}(X, Y, 1)) = f'_1(X, Y) + \cdots + f'_d(X, Y),$$
$$l'(X, Y) = L(\Psi^{-1}(X, Y, 1)) = b'X - a'Y,$$

and parametrize the locus in $\mathbb{A}_K^2$ with $l'(x, y) = 0$ by

$$\begin{pmatrix} x \\ y \end{pmatrix} = \varphi'(t) = \begin{pmatrix} a't \\ b't \end{pmatrix}.$$

We need a lemma.

Lemma 8.6. In the above notation, $f(X, Y)$ equals

$$(rX + sY + 1)^{d-1} f'_1(\alpha X + \beta Y, \gamma X + \delta Y)$$
$$+ (rX + sY + 1)^{d-2} f'_2(\alpha X + \beta Y, \gamma X + \delta Y)$$
$$+ \cdots + f'_d(\alpha X + \beta Y, \gamma X + \delta Y),$$

and therefore

$$f_1(X, Y) = f'_1(\alpha X + \beta Y, \gamma X + \delta Y).$$

PROOF. For the first conclusion, let us justify the following computation:

$$f(X, Y) = (F \circ \Psi^{-1})(\Psi \circ \Phi^{-1})(X, Y, 1)$$
$$= (F \circ \Psi^{-1})(\alpha X + \beta Y, \gamma X + \delta Y, rX + sY + 1)$$
$$= (F \circ \Psi^{-1})\left((rX + sY + 1)\left(\tfrac{\alpha X+\beta Y}{rX+sY+1}, \tfrac{\gamma X+\delta Y}{rX+sY+1}, 1\right)\right)$$
$$= (rX + sY + 1)^d f'\left(\tfrac{\alpha X+\beta Y}{rX+sY+1}, \tfrac{\gamma X+\delta Y}{rX+sY+1}\right)$$
$$= (rX + sY + 1)^d (f'_1 + \cdots + f'_d)\left(\tfrac{\alpha X+\beta Y}{rX+sY+1}, \tfrac{\gamma X+\delta Y}{rX+sY+1}\right)$$
$$= (rX + sY + 1)^{d-1} f'_1(\alpha X + \beta Y, \gamma X + \delta Y)$$
$$+ (rX + sY + 1)^{d-2} f'_2(\alpha X + \beta Y, \gamma X + \delta Y)$$
$$+ \cdots + f'_d(\alpha X + \beta Y, \gamma X + \delta Y).$$

In fact, the first three lines are valid if we make the computation in the field of fractions $K(X, Y)$, the fourth line uses the homogeneity of F and a substitution homomorphism that evaluates members of $K[X, Y, W]$ at points of $K(X, Y, W)$, and the remaining lines use the homogeneity of $f'_1, \ldots, f'_d$ and a substitution homomorphism that evaluates their arguments at points of $K(X, Y)$.

This proves the first conclusion. To derive the second conclusion from it, we expand each of the coefficients on the right side and group terms of the same degree of homogeneity under $(X, Y) \mapsto (\lambda X, \lambda Y)$. The only term whose degree of homogeneity is 1 is $f_1'(\alpha X + \beta Y, \gamma X + \delta Y)$ with a coefficient 1 coming from the expansion of $(rX + sY + 1)^{d-1}$; all other terms have higher degree of homogeneity. When $f(X, Y)$ on the left side is expanded as a sum of homogeneous polynomials, the term of degree 1 is $f_1(X, Y)$. The second conclusion follows. □

Continuing with the verification that $I(P, L \cap F)$ does not depend on the choice of Φ, we apply Lemma 8.6 to L in place of F, and we obtain

$$l(X, Y) = l'(\alpha X + \beta Y, \gamma X + \delta Y).$$

Since $l(X, Y) = bX - aY$ and $l'(X, Y) = b'X - a'Y$, this equation shows that

$$b = b'\alpha - a'\gamma \quad \text{and} \quad -a = b'\beta - a'\delta.$$

Putting $\Delta = \alpha\delta - \beta\gamma$, we solve for a' and b' and obtain

$$\alpha a + \beta b = \Delta a' \quad \text{and} \quad \gamma a + \delta b = \Delta b'.$$

When $x = at$ and $y = bt$, we thus have

$$\alpha x + \beta y = \alpha a t + \beta b t = t\Delta a' \quad \text{and} \quad \gamma x + \delta y = \gamma a t + \delta b t = t\Delta b'.$$

Substituting these formulas into the first conclusion of Lemma 8.6 and using the homogeneity of each f_j' gives

$$f(\varphi(t)) = (art + bst + 1)^{d-1} t \Delta f_1'(a', b')$$
$$+ (art + bst + 1)^{d-2} t^2 \Delta^2 f_2'(a', b') + \cdots + t^d \Delta^d f_d'(a', b').$$

If j is the smallest index for which $f_j'(a', b') \neq 0$, then the lowest power of t remaining on the right side after expansion of the coefficients is t^j, and its coefficient is $\Delta^j f_j'(a', b')$. Thus we can conclude that the lowest power of t with nonzero coefficient on the left side is t^j, and its coefficient $f_j(a, b)$ must equal $\Delta^j f_j'(a', b')$. The equality of the lowest power of t remaining on each side shows that $I(P, L \cap F)$ is the same when computed from f as when computed from f', and we obtain as a bonus the formula $f_j(a, b) = \Delta^j f_j'(a', b')$ if t^j is that power. This completes the verification that $I(P, L \cap F)$ does not depend on the choice of Φ.

Now we come back to the circle of ideas around Bezout's Theorem. The first task is to clarify the meaning of infinite intersection multiplicity.

4. Intersection Multiplicity for a Line with a Curve

Proposition 8.7. Over the field K if a projective line L and a projective plane curve F meet at a point P in $\mathbb{P}_K^2$, then $I(P, L \cap F) = +\infty$ if and only if L divides F.

PROOF. If L divides F, then in the above notation the local expression $l(X, Y)$ divides $f(X, Y)$. Since $l(\varphi(t))$ is the 0 polynomial, so is $f(\varphi(t))$.

Conversely suppose that $f(\varphi(t))$ is the 0 polynomial, so that $f_r(a, b) = 0$ for all r with $1 \leq r \leq d = \deg F$. Without loss of generality, suppose $b \neq 0$. The equality

$$0 = f_r(a, b) = c_0 a^r + c_1 a^{r-1} b + \cdots + c_r b^r$$
$$= b^r \left(c_0 (ab^{-1})^r + c_1 (ab^{-1})^{r-1} + \cdots + c_r \right)$$

says that $Z - ab^{-1}$ is a factor of $b^r(c_0 Z^r + c_1 Z^{r-1} + \cdots + c_r)$. If we write

$$b^r(c_0 Z^r + c_1 Z^{r-1} + \cdots + c_r) = (Z - ab^{-1}) u(Z)$$

and take $Z = XY^{-1}$, then

$$b^r f_r(X, Y) = b^r Y^r \left(c_0 (XY^{-1})^r + c_1 (XY^{-1})^{r-1} + \cdots + c_r \right)$$
$$= Y^r (XY^{-1} - ab^{-1}) u(XY^{-1}) = b^{-1} l(X, Y) \left(Y^{r-1} u(XY^{-1}) \right).$$

Hence $l(X, Y)$ divides $f_r(X, Y)$ for all r. It follows that $l(X, Y)$ divides $f(X, Y)$ and then that L divides F. □

The full-strength version of Bezout's Theorem says that two projective plane curves F and G of degrees m and n meet in at most mn points even when multiplicities are counted, and that the number is equal to mn if K is algebraically closed and multiplicities are counted. This theorem will be proved in Section 6. For the time being, we shall limit ourselves to the special case of the full-strength theorem in which one of the curves is a line.

Theorem 8.8 (Bezout's Theorem). Let K be an algebraically closed field. If F is a projective plane curve over K of degree d and if L is a projective line such that L does not divide F, then $\sum_P I(P, L \cap F) = d$.

PROOF. First we show that

$$\sum_P I(P, L \cap F) < +\infty. \tag{$*$}$$

Since L is assumed not to divide F, Proposition 8.7 shows that $I(P, L \cap F)$ is finite at every point of $V_K(L) \cap V_K(F)$. Thus $\sum_P I(P, L \cap F)$ is finite if there are only finitely many points in $V_K(L) \cap V_K(F)$. Bezout's Theorem in the form of Theorem 8.5 shows that either $V_K(L) \cap V_K(F)$ is finite or else L and F have as a common factor some homogeneous polynomial of positive degree. Since L has degree 1, L is prime, and thus L and F can have a common factor of positive degree only if L divides F. We are assuming the contrary, and therefore $V_K(L) \cap V_K(F)$ is finite. This proves (∗).

Possibly by applying a projective transformation, we may assume[7] that the given line L is the line at infinity W. Then the points P_j with $I(P_j, W \cap F) > 0$ are of the form $[x_j, y_j, 0]$. Taking into account that the algebraically closed field K is necessarily infinite, we can apply a second projective transformation, one that translates the Y variable, and assume that no y_j is 0. Then we can write $P_j = [r_j, 1, 0]$ with r_j in K. Let us see that

$$H(X) = F(X, 1, 0) \quad \text{is a nonzero polynomial of degree exactly } d. \quad (**)$$

In fact, $F(X, Y, W)$ is homogeneous of degree d, and we have arranged that $[1, 0, 0]$, which certainly lies in $V_K(W)$, is not in $V_K(F)$. Consequently the X^d term in $F(X, Y, W)$ has nonzero coefficient, and (∗∗) follows.

Next let us prove that

$$I\big((r, 1, 0), W \cap F\big) = \text{multiplicity of } r \text{ as a root of } H(X) = F(X, 1, 0). \quad (\dagger)$$

Then it will follow that $\sum_P I(P, W \cap F)$ equals the number of roots of $H(X) = F(X, 1, 0)$, each counted as many times as its multiplicity. In view of (∗∗) and the fact that K is algebraically closed, we will then have proved that $\sum_P I(P, W \cap F) = d$, as required.

To prove (†), we introduce affine local coordinates about $(r, 1, 0)$, using $\Phi^{-1} = \begin{pmatrix} 1 & 0 & r \\ 0 & 0 & 1 \\ 0 & 1 & 0 \end{pmatrix}$, so that $\Phi(r, 1, 0) = (0, 0, 1)$. The local versions f of F and l of W relative to this Φ are

$$f(X, Y) = F(\Phi^{-1}(X, Y, 1)) = F(X + r, 1, Y),$$
$$l(X, Y) = W(\Phi^{-1}(X, Y, 1)) = Y.$$

Hence $l(X, Y)$ is of the form $bX - aY$ with $a = -1$ and $b = 0$. If we parametrize l by $\varphi(t) = (at, bt) = (-t, 0)$, then

$$f(\varphi(t)) = f(-t, 0) = F(-t + r, 1, 0).$$

---

[7]If P and P' are distinct points in $\mathbb{P}_K^2$, then there exists a projective transformation carrying P to $[1, 0, 0]$ and P' to $[0, 1, 0]$. This transformation carries the unique line through P and P' to the line at infinity.

The order of vanishing of $f(\varphi(t))$ at $t = 0$, which is $I\bigl([r, 1, 0], W \cap F\bigr)$, thus equals the order of the zero of $F(-t + r, 1, 0)$ at $t = 0$, which equals the multiplicity of r as a root of $H(X) = F(X, 1, 0)$. This proves (†), and the theorem follows. □

5. Intersection Multiplicity for Two Curves

In this section we continue the topic of "intersection multiplicity" begun in Section 4. That section dealt with intersection multiplicity for the special case of a projective line and a projective plane curve, and the present section deals with the general case of two projective plane curves. The next section will use the general notion to address Bezout's Theorem in full generality. In this section and the next we shall make occasional use of material from Chapter VII, especially Lemma 7.21 and the results in Section VII.1.

It is worth reviewing qualitatively what happened in Section 4. What we did was refer the given line and curve to affine space, parametrize the line in a natural way, and substitute the parametrization into the formula for the curve to obtain a scalar-valued function of one variable. The order of vanishing of the resulting scalar-valued function of one variable was defined to be the intersection multiplicity. The classical approach[8] for handling two curves proceeds by trying to generalize this construction, in effect parametrizing one curve and substituting into the other. The fact that there need be no natural parametrization of either of the curves leads to a number of complications, and ultimately the argument involves a complicated ring of power series.

We shall follow a somewhat more modern approach[9] based on localizations.[10] The definition is not particularly intuitive, and it is necessary to study some examples to see its virtues. We give the definition, show that the definition is consistent with the definition in the special case of Section 4, check that the definition makes sense in general, state some properties that are useful in making computations, work out an example, and then verify the properties. Thus let F and G be homogeneous polynomials in (X, Y, W) of respective degrees m and n, and let $P = [x_0, y_0, w_0]$ be a point of the projective plane $\mathbb{P}^2_K$ over a field K. We refer matters back to affine space in the usual way by letting Φ be any member of $GL(3, K)$ such that $\Phi(x_0, y_0, w_0) = (0, 0, 1)$. The local expressions from Φ

---

[8] An account appears in Walker, Chapter IV.
[9] See Fulton, Chapter 3, for the present section and Fulton, Chapter 5, for the next section.
[10] For a still more modern and more general approach, see Serre's *Algèbre Locale*. Serre's opening sentence summarizes matters by saying, "Intersection multiplicities in algebraic geometry are equal to certain 'Euler–Poincaré characteristics' formed by means of the Tor functors of Cartan–Eilenberg."

about $(0, 0)$ corresponding to F and G are the polynomials f and g with

$$f(X, Y) = F(\Phi^{-1}(X, Y, 1)),$$
$$g(X, Y) = G(\Phi^{-1}(X, Y, 1)).$$

These polynomials break into homogeneous parts as

$$f(X, Y) = f_0 + f_1(X, Y) + \cdots + f_m(X, Y),$$
$$g(X, Y) = g_0 + g_1(X, Y) + \cdots + g_n(X, Y),$$

with f_j and g_j homogeneous of degree j in the pair (X, Y). We assume that P lies on the locus $V_K(F) \cap V_K(G)$ of common zeros of F and G, and the condition for this to happen is that $f_0 = g_0 = 0$. The **order of vanishing** $m_P(F)$ of F at P is the first j for which f_j is not the zero polynomial; we saw as a consequence of Lemma 8.6 that this quantity is well defined independently of the choice of Φ.

The **intersection multiplicity** $I(P, F \cap G)$ of F and G at P can be defined in either of two equivalent ways. The equivalence of the two definitions will be used repeatedly in the discussion and follows from the fact that localization commutes with passage to the quotient by an ideal, a fact that was proved as Lemma 7.21. One definition is

$$I(P, F \cap G) = \dim_K \big((K[X, Y]/(f, g))_{(0,0)} \big),$$

where $(K[X, Y]/(f, g))_{(0,0)}$ is the localization at $(0, 0)$ of the K algebra $K[X, Y]/(f, g)$. That is, we form the quotient ring of $K[X, Y]$ by the ideal generated over K by f and g, localize with respect to the maximal ideal of all members of the quotient vanishing at $(0, 0)$, and compute the dimension of this localization over K. The other definition is

$$I(P, F \cap G) = \dim_K \big(S^{-1} K[X, Y] / S^{-1}(f, g) \big),$$

where S is the multiplicative system in $K[X, Y]$ consisting of the complement of the maximal ideal (X, Y), i.e., consisting of all polynomials that are nonvanishing at $(0, 0)$. In either case all elements of the ring being localized have interpretations as functions, and the multiplicative system consists of all the functions that are nonzero at a certain point. Nevertheless, the matter is a little subtle because some members of the multiplicative system in the first case may be zero divisors. Here is a lower-dimensional example of that phenomenon that can also serve as a guiding example for Theorem 8.12 below.

EXAMPLE OF GEOMETRIC LOCALIZATION. $R = \bigl(K[X]/((X^2(X-1)^2))\bigr)_{(0)}$, with the subscript indicating localization at 0. Before passage to the localization, the quotient $Q = K[X]/((X^2(X-1)^2))$ has dimension 4, with a basis consisting of the cosets of 1, X, X^2, X^3. The multiplicative system S for localization at 0 consists of all members of the quotient that are nonzero at 0. The localization as a set consists of equivalence classes of pairs (r, s) with r in Q and s in S, two pairs (r, s) and (r', s') being equivalent if $t(rs' - r's) = 0$ for some t in S. Localization is a ring homomorphism, and we therefore consider the pairs (r, s) in the class of the additive identity. These have $t(r1 - 0s) = 0$ for some t. Then t and r have representatives $t(X)$ and $r(X)$ in $K[X]$ such that $t(X)r(X) = p(X)X^2(X-1)^2$ for some $p(X)$. Furthermore, $t(0) \neq 0$. Then X^2 must divide $r(X)$, and this condition is also sufficient for the choice $t(X) = (X-1)^2$. Thus the members $X^2 q(X)$ of $K[X]$ give 0 in the localization, and the localization is isomorphic to the 2-dimensional algebra $K[X]/(X^2)$.

Proposition 8.9 below will show that $I(P, F \cap G)$ is independent of the function Φ used to introduce affine local coordinates. Assuming this independence, we begin with an example that shows that the definition is consistent with the definition in Section 4.

EXAMPLE 1 OF INTERSECTION MULTIPLICITY. Case of a line L and a curve F homogeneous of degree d. Assuming that P lies in $V_K(L) \cap V_K(F)$, we introduce affine local coordinates by means of a member Φ of $GL(3, K)$ that carries a representative of P to $(0, 0, 1)$, and we let $l(X, Y)$ and $f(X, Y)$ be the corresponding local expressions for L and F. Let $f = f_1 + \cdots + f_d$ be the decomposition of f into its homogeneous parts. Since the intersection multiplicity is being assumed to be independent of the choice of Φ and since for any second point on a line through $(0, 0, 1)$, there exists a Φ that fixes $(0, 0, 1)$ and carries that second point to $(1, 0, 1)$, we may assume that $l(X, Y) = Y$. We introduce the parametrization $(x, y) = \varphi(t) = (t, 0)$ for the line $l(X, Y)$ and substitute into $f(X, Y)$, obtaining $f(\varphi(t)) = f_1(t, 0) + \cdots + f_d(t, 0)$. In the definition of Section 4, the intersection multiplicity is the least r such that $f_r(t, 0)$ is not identically 0, or else it is $+\infty$ if $f(\varphi(t))$ is identically 0. With the new definition we observe from the definition of r that f is of the form

$$f(X, Y) = (c_r X^r + \cdots + c_d X^d) + Y g(X, Y) = c_r X^r (1 + X h(X)) + Y g(X, Y)$$

with $c_r \neq 0$, $g(X, Y) \in K[X, Y]$, and $h(X) \in K[X]$. The ideal in $K[X, Y]$ generated by Y and f is the same as the ideal generated by Y and $X^r(1 + Xh(X))$. Hence

$$K[X, Y]/(Y, f) \cong K[X, Y]/(Y, X^r(1 + Xh)) \cong K[X]/(X^r(1 + Xh)).$$

The polynomial $1 + Xh(X)$ takes a nonzero value at 0 and hence is a member of the multiplicative system that we use to form the localization. Thus

$$\big(K[X, Y]/(Y, f)\big)_{(0,0)} \cong \big(K[X]/(X^r(1 + Xh))\big)_{(0)} \cong \big(K[X]/(X^r)\big)_{(0)}.$$

The dimension of the right side is r, and thus the new definition of intersection multiplicity matches the old one.

Proposition 8.9. The intersection multiplicity of two projective plane curves F and G at P is well defined independently of the member of Φ that moves a representative of P to $(0, 0, 1)$.

PROOF. It is enough to take $P = [0, 0, 1]$ and to compare the effect of passing to affine local coordinates determined by the identity with the effect of passing to the coordinates determined by a general element Φ of $GL(3, K)$ of the form $\Phi = \begin{pmatrix} \alpha & \beta & 0 \\ \gamma & \delta & 0 \\ r & s & 1 \end{pmatrix}$. Let $\deg F = m$ and $\deg G = n$. If $f(X, Y) = F(X, Y, 1)$ and $\widetilde{f}(X, Y) = F(\Phi^{-1}(X, Y, 1))$, then the computation in the proof of Lemma 8.6 shows that

$$f(X, Y) = (1 + rX + sY)^m \, \widetilde{f}\Big(\tfrac{\alpha X + \beta Y}{1 + rX + sY}, \tfrac{\gamma X + \delta Y}{1 + rX + sY}\Big). \tag{$*$}$$

Similarly if $g(X, Y) = G(X, Y, 1)$ and $\widetilde{g}(X, Y) = G(\Phi^{-1}(X, Y, 1))$, then

$$g(X, Y) = (1 + rX + sY)^n \, \widetilde{g}\Big(\tfrac{\alpha X + \beta Y}{1 + rX + sY}, \tfrac{\gamma X + \delta Y}{1 + rX + sY}\Big).$$

Let

$$X' = \tfrac{\alpha X + \beta Y}{1 + rX + sY}, \quad Y' = \tfrac{\gamma X + \delta Y}{1 + rX + sY}, \quad \text{and} \quad \Phi^{-1} = \begin{pmatrix} \alpha' & \beta' & 0 \\ \gamma' & \delta' & 0 \\ r' & s' & 1 \end{pmatrix}.$$

It is purely a formal matter that the mapping T defined by $(Th)(X, Y) = h(X', Y')$ is a field isomorphism of $K(X, Y)$ onto $K(X', Y')$. It sends $K[X, Y]$ onto $K[X', Y']$ and sends $\big(K[X, Y]\big)_{(0,0)}$ onto $\big(K[X', Y']\big)_{(0,0)}$. Referring to the formulas for X' and Y', we see that the image of $K[X, Y]$ is contained in the localization $\big(K[X, Y]\big)_{(0,0)}$; by the universal mapping property of localizations, the image of $\big(K[X, Y]\big)_{(0,0)}$ is contained in $\big(K[X, Y]\big)_{(0,0)}$. Comparing these two conclusions, we see that $\big(K[X', Y']\big)_{(0,0)} \subseteq \big(K[X, Y]\big)_{(0,0)}$.

Meanwhile, we can solve the equations defining X' and Y' for X and Y. If we compare the results with the formula for Φ^{-1}, we find that

$$X = \tfrac{\alpha' X' + \beta' Y'}{1 + r'X' + s'Y'} \quad \text{and} \quad Y = \tfrac{\gamma' X' + \delta' Y'}{1 + r'X' + s'Y'}.$$

Thus the situation is symmetric, and we have $(K[X, Y])_{(0,0)} \subseteq (K[X', Y'])_{(0,0)}$. Consequently the mapping

$$(Th)(X, Y) = h\left(\tfrac{\alpha X+\beta Y}{1+rX+sY}, \tfrac{\gamma X+\delta Y}{1+rX+sY}\right)$$

is an algebra automorphism of $(K[X, Y])_{(0,0)}$.

To prove the proposition, recall that localization commutes with passage to the quotient by an ideal. In view of (∗), it is therefore enough to show that

$$\dim_K \left((K[X, Y])_{(0,0)}/(f, g)\right)$$
$$\stackrel{?}{=} \dim_K \left((K[X, Y])_{(0,0)}/((1+rX+sY)^m Tf, (1+rX+sY)^n Tg)\right). \quad (**)$$

The factor $(1+rX+sY)$ is a unit in $(K[X, Y])_{(0,0)}$, and we can simplify the quotient algebra on the right side of (∗∗) to

$$(K[X, Y])_{(0,0)}/(Tf, Tg).$$

In turn, this algebra is K isomorphic to $(K[X, Y])_{(0,0)}/(f, g)$ because T is an automorphism of $(K[X, Y])_{(0,0)}$. The dimensional equality in (∗∗) follows. □

Let us extend the definition of intersection multiplicity to include the case that the point of interest does not lie in the locus of common zeros. We define $I(P, F \cap G) = 0$ if P is not in $V_K(F) \cap V_K(G)$. Assume now that K is algebraically closed. Below we compute a fairly typical example of intersection multiplicity. To do so, we shall make use of certain properties of $I(P, F \cap G)$ that we list in Theorem 8.10 below. In fact, there is an algorithm for computing $I(P, F \cap G)$ using only these properties,[11] but we shall not give it.

Before stating the properties, we need to make some definitions. Recall from earlier in the section that the order of vanishing $m_P(G)$ of G at P is computed using a suitable Φ in $GL(3, K)$ to refer G to affine local coordinates about P, defining $g(X, Y) = G(\Phi^{-1}(X, Y, 1))$, expanding $g(X, Y)$ as a sum of homogeneous terms $g(X, Y) = g_0 + g_1(X, Y) + \cdots + g_n(X, Y)$, and defining $m_P(G)$ to be the least j such that g_j is not the 0 polynomial. The homogeneous polynomial $g_j(X, Y)$ is X^j times a polynomial in the one variable YX^{-1}, and the fact that K is algebraically closed implies that g_j has a factorization of the form

$$g_j(X, Y) = c \prod_i (\alpha_i X + \beta_i Y)^{m_i}$$

---

[11]Fulton, p. 76.

with c in K. Here $j = \sum_i m_i$, and the pairs (α_i, β_i) correspond to distinct members of $\mathbb{P}^1_K$ that are uniquely determined up to indexing if $c \neq 0$. Let $l_i(X, Y) = \alpha_i X + \beta_i Y$, and let L_i be the corresponding projective line. We refer to all the lines L_i as the **tangent lines** to G at P, and we say that m_i is the **multiplicity** of L_i. The geometry of the situation is indicated in Problem 12 at the end of the chapter.

Theorem 8.10. Let K be an algebraically closed field, let P be in $\mathbb{P}^2_K$, and let F and G be projective plane curves over K. Then the intersection multiplicity $I(P, F \cap G)$ has the following properties:

(a) $I(P, F \cap G) = I(P, G \cap F)$,
(b) $I(P, F \cap G) = I(P, F \cap (G + HF))$ for any projective plane curve H with $\deg HF = \deg G$ such that $G + HF \neq 0$,
(c) $I(P, F \cap G) > 0$ if and only if P lies in $V_K(F) \cap V_K(G)$,
(d) $I(P, F \cap G) \leq I(P, AF \cap BG)$ for any projective plane curves A and B, with equality if A and B are nonvanishing at P,
(e) $I(P, F \cap G)$ is finite if and only if F and G have no common factor of degree ≥ 1 having P on its zero locus,
(f) $I(P, F \cap GH) = I(P, F \cap G) + I(P, F \cap H)$ and consequently if $F = \prod_i F_i^{r_i}$ and $G = \prod_j G_j^{s_j}$, then $I(P, F \cap G) = \sum_{i,j} r_i s_j I(P, F_i \cap G_j)$,
(g) $I(P, F \cap G) \geq m_P(F) m_P(G)$, with equality if F and G have no tangent lines in common at P.

REMARKS. Properties (a) and (b) are evident. Properties (c) and (d) are conversational and will be proved in these remarks. Properties (e), (f), and (g) require proofs, and we give those proofs after computing an example. For (c), if P lies in $V_K(F) \cap V_K(G)$, then the local expressions $f(X, Y)$ and $g(X, Y)$ vanish at 0, and so does every member of the ideal (f, g); therefore (f, g) is a proper ideal in $(K[X, Y])_{(0,0)}$, and the dimension of the quotient is positive. Conversely if P is not in $V_K(F)$, say, then $f(X, Y)$ lies in the multiplicative system S of nonvanishing polynomials at $(0, 0)$, and $S^{-1}(f, g) = (1)$; hence $S^{-1}K[X, Y]/S^{-1}(f, g) = 0$, and $I(P, F \cap G) = 0$. For (d), $S^{-1}(af, bg) \subseteq S^{-1}(f, g)$ with equality if a and b are nonvanishing at $(0, 0)$, and hence $S^{-1}K[X, Y]/S^{-1}(f, g)$ is a homomorphic image of $S^{-1}K[X, Y]/S^{-1}(af, bg)$ and is a one-one homomorphic image if a and b are nonvanishing at $(0, 0)$.

EXAMPLE 2 OF INTERSECTION MULTIPLICITY. Let $K = \mathbb{C}$, and let the two projective curves be the homogeneous versions of $Y^2 = X^3$ and $Y^2 = X^5$. In other words, let

$$F(X, Y, W) = Y^2 W - X^3 \quad \text{and} \quad G(X, Y, W) = Y^2 W^3 - X^5.$$

We compute $I(P, F \cap G)$ for all points P in $V_K(F) \cap V_K(G)$. In the affine plane the intersections (x, y) may be found by substituting the one equation into the other (or, with more effort in this case, by using the resultant). We obtain $x^5 - x^3 = 0$. This gives $x^3(x^2 - 1) = 0$. The factor $x^2 - 1$ has two distinct roots, and each gives two distinct y's. Thus we obtain the five affine solutions $(+1, \pm 1), (-1, \pm i), (0, 0)$. The fact that the first four occurred routinely with multiplicity 1 translates into intersection multiplicity 1 for each: In fact, (b) shows that $I(P, F \cap G) = I(P, F \cap (W^2 F - G))$, and $W^2 F - G$ restricts at $(X, Y, 1)$ to $X^5 - X^3 = X^3(X^2 - 1)$. At each of the points $(+1, \pm 1)$, $X^5 - X^3$ when viewed as equal to 0 has a vertical tangent $X - 1$ of multiplicity 1, while $Y^2 - X^3$ has a tangent that is not vertical. A similar argument applies at each of the points $(-1, \pm i)$. By (g), the intersection multiplicity is 1 at each of the four points $(+1, \pm 1)$ and $(-1, \pm i)$.

Next let us consider $(0, 0)$. The order of $X^5 - X^3$ is 3, and the homogeneous term of degree 3, namely $-X^3$, factors as the cube of a linear factor that gives the vertical line X. Meanwhile, $Y^2 - X^3$ has order 2 at $(0, 0)$, and Y^2 factors as the square of a linear factor that gives the horizontal line Y. The two curves have no tangents in common. Hence equality holds in (g), and the intersection multiplicity is 6 at $(0, 0)$.

Finally let us check points (x, y, w) on the line at infinity, i.e., those with $w = 0$. Putting $w = 0$ in the formula $F = G = 0$ shows that $x = 0$. Thus the only point of $V_K(F) \cap V_K(G)$ on the line at infinity is $P = [x_0, y_0, w_0] = [0, 1, 0]$. The local versions of F and G may be given in the variables X and W by restricting (X, Y, W) to $(X, 1, W)$ and considering the polynomials about $(x, w) = (0, 0)$. As above, (b) gives $I(P, F \cap G) = I(P, F \cap (W^2 F - G))$, but $F = Y^2 W - X^3$ restricts to $W - X^3$ and $W^2 F - G = -W^2 X^3 + X^5$ remains unchanged upon restriction. The respective lowest-order terms, in factored form, are W and $-X^3(X + W)(X - W)$. None of the factors of the first polynomial matches a factor of the second polynomial, and (g) says that the intersection multiplicity is $1 \cdot 5 = 5$.

The upshot is that we get multiplicity 6 from $(0, 0)$, multiplicity 1 apiece from four other points in the affine plane, and multiplicity 5 from $P = [0, 1, 0]$. The total is 15, the product of the degrees of the given curves, as it must be if we are to have any chance of obtaining the desired generalization of Bezout's Theorem.

To get at Theorem 8.10, we make use of a structure theorem about ideals I in $K[X_1, \ldots, X_n]$ for which $V(I)$ is a finite set. To prove the structure theorem, which appears as Theorem 8.12 below, we first prove a lemma about the radical $\sqrt{I}$ of an ideal I, a notion defined in Section VII.1.

Lemma 8.11. If R is a commutative Noetherian ring and I is an ideal in R, then $(\sqrt{I})^m \subseteq I$ for some integer $m \geq 1$.

PROOF. Since R is Noetherian, the ideal $\sqrt{I}$ is finitely generated. Let $\{a_1, \ldots, a_n\}$ be a set of generators for it. By definition of radical, choose integers $k_1, \ldots, k_n$ such that $a_j^{k_j}$ is in I for $1 \leq j \leq n$, and put $m = \sum_{j=1}^n k_j$. The most general element of $\sqrt{I}$ is of the form $\sum_{j=1}^n r_j a_j$ with all r_j in R. The mth power of this element is a sum of terms of the form $r a_1^{l_1} \cdots a_n^{l_n}$ with $\sum_{j=1}^n l_j = m$. In view of the definition of m, we must have $l_j \geq k_j$ for some j. Then the factor $a_j^{l_j}$ is in I, and hence the whole term $r a_1^{l_1} \cdots a_n^{l_n}$ is in I. □

Theorem 8.12. Let K be an algebraically closed field, and let I be an ideal in the polynomial ring $K[X_1, \ldots, X_n]$ whose locus of common zeros in K^n is a finite set $\{P_1, \ldots, P_k\}$. Then $K[X_1, \ldots, X_n]/I$ is isomorphic as a ring to the product of its localizations at the points P_j:

$$K[X_1, \ldots, X_n]/I \cong \prod_{j=1}^k \left(K[X_1, \ldots, X_n]/I\right)_{(P_j)}.$$

Consequently

$$\dim_K (K[X_1, \ldots, X_n]/I) = \sum_{j=1}^k \dim_K \left(K[X_1, \ldots, X_n]/I\right)_{(P_j)}.$$

REMARKS. The one-variable case is a guide: The ideal I is principal, and we can write $K[X]/I$ as $K[X]/(\prod_{j=1}^k (X-c_j)^{m_j})$. The points P_j of the theorem are the members c_j of K, and the same argument as for the first example of the section shows that $\left(K[X]/(\prod_j (X-c_j)^{m_j})\right)_{(c_j)} \cong K[X]/(X-c_j)^{m_j}$. The isomorphism of the theorem therefore reduces to an instance of the Chinese Remainder Theorem.

PROOF. Let $\varphi_j : K[X_1, \ldots, X_n]/I \to \left(K[X_1, \ldots, X_n]/I\right)_{(P_j)}$ be the canonical homomorphism, and let $\varphi = (\varphi_1, \ldots, \varphi_k)$. The mapping φ is a ring homomorphism into $\prod_{j=1}^k \left(K[X_1, \ldots, X_n]/I\right)_{(P_j)}$, and we shall prove that φ is one-one onto. Doing so requires some preparation.

Let I_j be the maximal ideal of all polynomials vanishing at P_j. The Nullstellensatz (Theorem 7.1) shows that $\sqrt{I}$ consists of all $f \in K[X, Y]$ such that f vanishes at each P_i, i.e., that $\sqrt{I} = \bigcap_{j=1}^k I_j$. Lemma 8.11 shows that $(\sqrt{I})^m \subseteq I$ for some m, and thus $\left(\bigcap_{j=1}^k I_j\right)^m \subseteq I$. For $i \neq j$, $I_i^m + I_j^m$ is an ideal whose locus of common zeros is empty, and the Nullstellensatz shows that $I_i^m + I_j^m = K[X_1, \ldots, X_n]$. The Chinese Remainder Theorem (Theorem 8.27 of *Basic Algebra*) therefore applies and shows that the intersection $\bigcap_{j=1}^k I_j^m$ and

5. Intersection Multiplicity for Two Curves

the product $\prod_{j=1}^{k} I_j^m$ coincide. Similarly $I_i + I_j = K[X_1, \ldots, X_n]$, and hence $\bigcap_{j=1}^{k} I_j = \prod_{j=1}^{k} I_j$. Putting these facts together, we conclude that

$$\bigcap_{j=1}^{k} I_j^m = \prod_{j=1}^{k} I_j^m = \left(\prod_{j=1}^{k} I_j\right)^m = \left(\bigcap_{j=1}^{k} I_j\right)^m \subseteq I. \quad (*)$$

Let us now denote members of $K[X_1, \ldots, X_n]$ by uppercase letters and their cosets modulo I by the corresponding lowercase letters. Let us observe for $1 \leq i \leq k$ that there exists $F_i \in K[X_1, \ldots, X_n]$ with $F_i(P_j) = \delta_{ij}$. In fact, we start from the special case that if $P \neq Q$, then there exists F with $F(P) = 1$ and $F(Q) = 0$. For the special case, P and Q differ in some coordinate; say that $x_l(P) \neq x_l(Q)$. Then the polynomial

$$F(X_1, \ldots, X_n) = (X_l - x_l(Q))(x_l(P) - x_l(Q))^{-1}$$

has the required properties. To construct F_1 with $F_1(P_j) = \delta_{1j}$, choose G_j with $G_j(P_1) = 1$ and $G_j(P_j) = 0$. Then $F_1 = \prod_{i \neq 1} G_i$ has $F_1(P_1) = 1$ and $F_1(P_j) = 0$ for $j \neq 1$. The polynomials $F_2, \ldots, F_k$ are constructed similarly.

With m as in the second paragraph of the proof, fix j and define $E_i = 1 - (1 - F_i^m)^m$. This is divisible by F_i^m and hence lies in I_j^m if $i \neq j$. In addition, $1 - F_j^m$ lies in I_j, and hence $1 - E_j = (1 - F_j^m)^m$ is in I_j^m. Therefore $1 - \sum_{i=1}^{k} E_i = (1 - E_j) - \sum_{i \neq j} E_i$ lies in I_j^m. Since the left side is independent of j, $1 - \sum_{i=1}^{k} E_i$ lies in $\bigcap_{j=1}^{k} I_j^m$, and we conclude from $(*)$ that

$$1 - \sum_{i=1}^{k} E_i \quad \text{lies in } I. \quad (**)$$

We just saw that E_i lies in $\bigcap_{j \neq i} I_j^m$. Hence if $i \neq j$, then $E_i E_j$ lies in $\bigcap_{l=1}^{k} I_l^m \subseteq I$. Passing to cosets modulo I, we find from this fact and from $(**)$ that

$$e_i e_j = 0 \text{ for } i \neq j, \quad \text{and that} \quad \sum_{i=1}^{k} e_i = 1. \quad (\dagger)$$

Multiplying the second equation by e_j and substituting from the first equation, we obtain

$$e_i^2 = e_i \quad \text{for all } i. \quad (\dagger\dagger)$$

Using $(\dagger)$ and $(\dagger\dagger)$, let us prove for each i that

to each $G \in K[X_1, \ldots, X_n]$ with $G(P_i) \neq 0$

corresponds a polynomial H with $hg = e_i$. $(\ddagger)$

In fact, we may assume that $G(P_i) = 1$. Let Q be the member of I_i given by $Q = 1 - G$. The element $Q^m E_i$ is in I_i^m because Q is in I_i, and it is in I_j^m for $j \neq i$ because E_i is in I_j^m for $j \neq i$. Thus $Q^m E_i$ is in $\bigcap_{j=1}^k I_j^m \subseteq I$, and $q^m e_i = 0$. Consequently

$$g(e_i + qe_i + \cdots + q^{m-1}e_i) = (1-q)e_i(1 + q + \cdots + q^{m-1}) = e_i(1 - q^m) = e_i,$$

and $H = E_i(1 + Q + \cdots + Q^{m-1})$ is a polynomial as in (‡).

Now we can prove that φ is one-one. If f is a member of $K[X_1, \ldots, X_n]/I$ such that $\varphi(f) = 0$, then $\varphi_i(f) = 0$ for all i. This means that there exists a member g_i of the multiplicative system for localization at P_i such that $g_i f = 0$. Any corresponding polynomial G_i has $G_i(P_i) \neq 0$. By (‡), there exists h_i with $h_i g_i = e_i$. Then (†) gives $f = \sum_{i=1}^k e_i f = \sum_{i=1}^k h_i g_i f = 0$. Thus φ is one-one.

For the proof that φ is onto, we recall that the multiplicative system used to obtain $\bigl(K[X_1, \ldots, X_n]/I\bigr)_{(P_j)}$ consists of the elements $K[X_1, \ldots, X_n]/I$ that are nonzero at P_j, and φ_j carries these to units in $\bigl(K[X_1, \ldots, X_n]/I\bigr)_{(P_j)}$. Since $E_j(P_j) = 1$, $\varphi_j(e_j)$ is a unit. For $i \neq j$, we have $\varphi_j(e_i)\varphi_j(e_j) = \varphi_j(e_i e_j) = 0$, and therefore $\varphi_j(e_i) = 0$. Consequently

$$\varphi_j(e_j) = \sum_{l=1}^k \varphi_j(e_l) = \varphi_j\Bigl(\sum_{l=1}^k e_l\Bigr) = \varphi_j(1) = 1,$$

and $\varphi_j(e_j)$ is the identity of $\bigl(K[X_1, \ldots, X_n]/I\bigr)_{(P_j)}$. The localization at P_j consists of the equivalence classes of all pairs (r_j, s_j) with r_j and s_j in $K[X_1, \ldots, X_n]/I$ and s_j in the multiplicative system for index j. Thus let such pairs (r_j, s_j) be given for $1 \leq j \leq k$. We are to produce an element a of $K[X_1, \ldots, X_n]/I$ such that $\varphi_j(a) = \varphi_j(r_j)(\varphi_j(s_j))^{-1}$ for all j. Use of (‡) produces h_j with $h_j s_j = e_j$ for all j, and this element has the property that $\varphi_j(h_j)\varphi_j(s_j) = \varphi_j(e_j) = 1$, hence that $\varphi_j(h_j) = \varphi_j(s_j)^{-1}$. Consequently the element $a = \sum_j r_j h_j e_j$ has the property that

$$\varphi_j(a) = \varphi_j\Bigl(\sum_i r_i h_i e_i\Bigr) = \sum_i \varphi_j(r_i)\varphi_j(h_i)\varphi_j(e_i) = \varphi_j(r_j)(\varphi_j(s_j))^{-1}$$

and exhibits φ as onto. □

Corollary 8.13. Let K be an algebraically closed field, and let I be an ideal in the polynomial ring $K[X_1, \ldots, X_n]$ whose locus of common zeros in K^n is a finite set $\{P_1, \ldots, P_k\}$. Then $K[X_1, \ldots, X_n]/I$ is finite-dimensional, and so is the localization $\bigl(K[X_1, \ldots, X_n]/I\bigr)_{(P_j)}$ for each j.

PROOF. This is a corollary partly of the statement of Theorem 8.12 and partly of the proof. Let m be as in the proof. If I_0 is the maximal ideal $(X_1, \ldots, X_n)$ of $K[X_1, \ldots, X_n]$, then I_0^m is the ideal generated by all monomials of degree m, and $K[X_1, \ldots, X_n]/I_0^m$ is finite-dimensional. Consequently the maximal ideal $I_j = (X_1 - x_1(P_j), \ldots, X_n - x_n(P_j))$ has the property that $K[X_1, \ldots, X_n]/I_j^m$ is finite-dimensional. Since $I_i^m + I_j^m = K[X_1, \ldots, X_n]$ for $i \neq j$, the Chinese Remainder Theorem shows that

$$K[X_1, \ldots, X_n] / \bigcap_{j=1}^{k} I_j^m \cong \prod_{j=1}^{k} K[X_1, \ldots, X_n]/I_j^m,$$

and the left side is therefore finite-dimensional. By $(*)$ in the proof of Theorem 8.12, $\bigcap_{j+1}^{k} I_j^m \subseteq I$, and hence $K[X_1, \ldots, X_n]/I$ is finite-dimensional. Then $\left(K[X_1, \ldots, X_n]/I\right)_{(P_j)}$ is finite-dimensional as a consequence of the statement of Theorem 8.12. □

PROOF OF THEOREM 8.10e. If F and G have a common factor H of degree ≥ 1 such that $H(P) = 0$, we may assume that H is irreducible. Introduce affine local coordinates about P. If f, g, h denote the local versions of F, G, H, then the ideal (f, g) of $K[X, Y]$ is contained in the principal ideal (h). The latter ideal is proper because $h(0, 0) = 0$, and the irreducibility of H thus implies that (h) is prime. If S denotes the multiplicative system in $K[X, Y]$ of polynomials that are nonvanishing at $(0, 0)$, then $S^{-1}(f, g) \subseteq S^{-1}(h)$, and we have a natural quotient homomorphism of $S^{-1}K[X, Y]/S^{-1}(f, g)$ onto $S^{-1}K[X, Y]/S^{-1}(h)$. The latter is isomorphic as a K algebra to $(K[X, Y]/(h))_{(0,0)}$, and the dimension of this localization is a lower bound for $I(P, F \cap G)$. Since $K[X, Y]/(h)$ is an integral domain, $K[X, Y]/(h)$ maps one-one into any localization of itself, and $\dim_K(K[X, Y]/(h))$ is a lower bound for $I(P, F \cap G)$. Since h is nonconstant, either X or Y actually occurs in it, say Y. Then h divides no member of $K[X]$, and the mapping of $K[X]$ into cosets modulo (h) is one-one. Therefore $K[X, Y]/(h)$ contains a subalgebra isomorphic to $K[X]$ and must be infinite-dimensional.

Conversely if F and G have no common factor of degree ≥ 1 with P on its locus, then (d) shows that we may assume F and G to have no common factor of degree ≥ 1 of any kind. In this case Theorem 8.5 shows that the locus of common zeros of F and G is finite, and Corollary 8.13 shows that $I(P, F \cap G)$ is finite. □

PROOF OF THEOREM 8.10f. We are to prove that

$$I(P, F \cap GH) = I(P, F \cap G) + I(P, F \cap H). \qquad (*)$$

If F and GH have a common factor of degree ≥ 1 that vanishes at P, then F and one of G and H have such a factor. By symmetry we may assume that F and G

have that common factor. Then the left side of (∗) and the first term on the right are infinite by (e), and (∗) is verified.

Thus we may assume that F and GH have no common factor that vanishes at P. If F has a prime factor that does not vanish at P, then (d) shows that we can drop that factor from all three appearances of F in (∗). In other words, it is enough to prove (f) under the assumption that F and GH have no common factor of degree ≥ 1 of any kind.

With this assumption in place, introduce affine local coordinates about P, let S denote the multiplicative system in $K[X, Y]$ of polynomials that are nonvanishing at $(0, 0)$, and let f, g, h be the local versions of the given curves F, G, H. The inclusion of ideals $(f, gh) \subseteq (f, g)$ induces an inclusion $S^{-1}(f, gh) \subseteq S^{-1}(f, g)$ and then an onto algebra homomorphism

$$\varphi : S^{-1}K[X, Y]/S^{-1}(f, gh) \to S^{-1}K[X, Y]/S^{-1}(f, g).$$

We shall exhibit a K vector-space isomorphism ψ of $S^{-1}K[X, Y]/S^{-1}(f, h)$ onto ker φ, and the resulting dimensional equality

$$\dim_K \left(S^{-1}K[X, Y]/S^{-1}(f, gh)\right)$$
$$= \dim_K \left(S^{-1}K[X, Y]/S^{-1}(f, g)\right) + \dim_K \left(S^{-1}K[X, Y]/S^{-1}(f, h)\right) \quad (**)$$

will prove (∗) and hence (f). We define

$$\Psi : S^{-1}K[X, Y] \to S^{-1}K[X, Y]/S^{-1}(f, gh)$$

as a K linear map by $\Psi(u) = gu + S^{-1}(f, gh)$. If $af + bh$ is in $S^{-1}(f, h)$, then $\Psi(af + bh) = afg + bgh + S^{-1}(f, gh) = S^{-1}(f, gh)$. Thus Ψ descends to a K linear map ψ of $S^{-1}K[X, Y]/S^{-1}(f, h)$ into $S^{-1}K[X, Y]/S^{-1}(f, gh)$. It is evident that $\varphi\Psi = 0$ and hence that $\varphi\psi = 0$, i.e., image $\psi \subseteq \ker \varphi$.

If any member $u + S^{-1}(f, gh)$ of ker φ is given, then $0 = \varphi(u + S^{-1}(f, gh)) = u + S^{-1}(f, g)$ shows that u is in $S^{-1}(f, g)$. Say that $u = af + bg$. Then $\psi(b + S^{-1}(f, h)) = bg + S^{-1}(f, gh) = bg + af + S^{-1}(f, gh) = u + S^{-1}(f, gh)$ shows that image $\psi \supseteq \ker \varphi$. Hence image $\psi = \ker \varphi$, i.e., ψ is onto.

To see that ψ is one-one, suppose that $\psi(u + S^{-1}(f, h))$ is the 0 coset, i.e., that $gu + S^{-1}(f, gh) = S^{-1}(f, gh)$. Then $gu = af + bgh$ with u, a, b in $S^{-1}K[X, Y]$. Clearing fractions, we may assume that u, a, b are in $K[X, Y]$. The formula $g(u - bh) = af$ in $K[X, Y]$, in the presence of the assumption that F and G have no common factor of degree ≥ 1, implies that f divides $u - bh$. Write $u - bh = cf$ with c in $K[X, Y]$. Then $u = cf + bh$, and u lies in the ideal (f, h). In other words, $u + S^{-1}(f, h)$ is the trivial coset, and ψ has been shown to be one-one. This proves (∗∗) and hence (f). □

5. Intersection Multiplicity for Two Curves

Lemma 8.14. For any field K, let $\{L_i\}_{i\geq 1}$ be a system of nonzero homogeneous polynomials in $K[X, Y]$ of the form $L_i = a_i X + b_i Y$, let $\{M_j\}_{j\geq 1}$ be another such system with $M_j = c_j X + d_j Y$, and suppose that no L_i is a scalar multiple of some M_j. For $n \geq 1$, let $B_0, \ldots, B_n$ be the system of homogeneous polynomials

$$B_k = L_1 \cdots L_k M_1 \cdots M_{n-k} \qquad \text{for } 0 \leq k \leq n.$$

Then $\{B_0, \ldots, B_n\}$ is a vector-space basis of the space $K[X, Y]_n$ of all homogeneous polynomials in (X, Y) of degree n.

PROOF. The set $\{B_0, \ldots, B_n\}$ has $n + 1$ elements, and $n + 1$ is the dimension of $K[X, Y]_n$ because $\{X^n, X^{n-1}Y, \ldots, Y^n\}$ is a basis. Thus it is enough to show that $\{B_0, \ldots, B_n\}$ is linearly independent. If we have a relation $\sum_{k=0}^{n} c_k B_k = 0$ for scalars c_k, then we observe that L_1 divides each B_k for $k \geq 0$, and L_1 does not divide B_0 because by assumption L_1 does not divide any factor M_j. Thus $c_0 = 0$. In effect, case n of the lemma has now been reduced to case $n - 1$, and the result readily follows by induction. $\square$

PROOF OF THEOREM 8.10g. Put $p = m_P(F)$ and $q = m_P(G)$. We pass to affine local coordinates about P, letting f and g be the members of $K[X, Y]$ corresponding to F and G. If I denotes the maximal ideal $I = (X, Y)$ in $K[X, Y]$, then f lies in I^p and g lies in I^q. We form the following sequence of K vector spaces and K linear mappings:

$$K[X,Y]/I^q \oplus K[X,Y]/I^p \xrightarrow{\psi} K[X,Y]/I^{p+q} \xrightarrow{\varphi} K[X,Y]/(I^{p+q}+(f,g)) \to 0.$$

Here the mapping φ is the algebra homomorphism induced by the inclusion $I^{p+q} \subseteq I^{p+q} + (f, g)$, and it is onto $K[X, Y]/(I^{p+q} + (f, g))$. The mapping ψ is defined by

$$\psi(a + I^q, b + I^p) = af + bg + I^{p+q}$$

and is merely K linear.

Let us see that the sequence is exact at $K[X, Y]/I^{p+q}$. Since

$$\varphi\psi(a + I^q, b + I^p) = \varphi(af + bg + I^{p+q}) = I^{p+q} + (f, g),$$

we obtain image $\psi \subseteq \ker \varphi$. If $h + I^{p+q}$ is in $\ker \varphi$, then h is in $I^{p+q} + (f, g)$, hence is of the form $u + af + bg$ with u in I^{p+q}. Then $h - u = af + bg$, and $\psi(a + I^q, b + I^p) = h - u + I^{p+q} = h + I^{p+q}$. So image $\psi \supseteq \ker \varphi$, and we have image $\psi = \ker \varphi$.

The mapping ψ descends to a one-one linear map of

$$M = (K[X, Y]/I^q \oplus K[X, Y]/I^p)\big/\ker \psi$$

into $K[X,Y]/I^{p+q}$. The vector space $K[X,Y]/I^q$ may be identified with the space of all polynomials of degree less than q, and that space is finite-dimensional. Similarly $K[X,Y]/I^p$ is finite-dimensional, and therefore

$$\dim_K M = \dim_K K[X,Y]/I^q + \dim_K K[X,Y]/I^p - \dim_K \ker \psi. \quad (*)$$

Meanwhile, φ exhibits $K[X,Y]/(I^{p+q} + (f,g))$ as isomorphic as a vector space to $(K[X,Y]/I^{p+q})/M$. Consequently

$$\dim_K K[X,Y]/I^{p+q} = \dim_K M + \dim_K K[X,Y]/(I^{p+q} + (f,g)). \quad (**)$$

Combining $(*)$ and $(**)$ with the simple vector-space isomorphism $K[X,Y]/I^d \cong K[X,Y,W]_{d-1}$ and with the fact from Section 3 that $\dim_K K[X,Y,W]_{d-1} = \binom{d+1}{2}$ gives

$$\dim_K K[X,Y]/(I^{p+q} + (f,g))$$
$$= \dim_K K[X,Y]/I^{p+q} - \dim_K K[X,Y]/I^q$$
$$\quad - \dim_K K[X,Y]/I^p + \dim_K \ker \psi$$
$$\geq \dim_K K[X,Y]/I^{p+q} - \dim_K K[X,Y]/I^q - \dim_K K[X,Y]/I^p$$
$$= \binom{p+q+1}{2} - \binom{q+1}{2} - \binom{p+1}{2}$$
$$= pq, \quad (\dagger)$$

with equality on the fourth line if and only if $\ker \psi = 0$.

The locus of common zeros of $I^{p+q} + (f,g)$ is just $\{0\}$, and Theorem 8.12 therefore shows that

$$\dim_K \left(K[X,Y]/(I^{p+q} + (f,g))\right)_{(0,0)} = \dim_K K[X,Y]/(I^{p+q} + (f,g)). \quad (\dagger\dagger)$$

The inclusion $(f,g) \subseteq I^{p+q} + (f,g)$ induces an algebra homomorphism of $\left(K[X,Y]/(f,g)\right)_{(0,0)}$ onto $\left(K[X,Y]/(I^{p+q} + (f,g))\right)_{(0,0)}$. Therefore

$$\dim_K \left(K[X,Y]/(f,g)\right)_{(0,0)} \geq \dim_K \left(K[X,Y]/(I^{p+q} + (f,g))\right)_{(0,0)}. \quad (\ddagger)$$

Let S be the set-theoretic complement of $I = (X,Y)$ in $K[X,Y]$. Because of the isomorphism $S^{-1}K[X,Y]/S^{-1}J \cong \left(K[X,Y]/J\right)_{(0,0)}$ for any ideal J, equality will hold in $(\ddagger)$ if $S^{-1}(f,g) = S^{-1}(I^{p+q} + (f,g))$. Combining $(\dagger)$, $(\dagger\dagger)$, and $(\ddagger)$, we find that

$$I(P, F \cap G) \geq pq, \quad (\ddagger\ddagger)$$

with equality if

$$I^{p+q} \subseteq S^{-1}(f, g) \quad \text{and} \quad \psi \text{ is one-one.} \tag{\S}$$

Inequality (‡‡) completes the proof of the inequality in (g) of the theorem. Because equality holds in (‡‡) if (§) holds, we can complete the proof of all of (g) by showing that (§) holds if F and G have no tangent line in common.

Thus for the remainder of the proof, we assume that F and G have no tangent line in common. Let the tangent lines of F, repeated according to their multiplicities, be $L_1, \ldots, L_p$, and let the tangent lines of G be $M_1, \ldots, M_q$. Define L_i for $i > p$ to be L_p, and define M_j for $j > q$ to be M_q.

In order to prove that the first conclusion of (§), namely that $I^{p+q} \subseteq S^{-1}(f, g)$, we shall prove that $I^t \subseteq S^{-1}(f, g)$ for t sufficiently large, and then we shall prove by induction downward on t that $I^t \subseteq S^{-1}(f, g)$ as long as $t \geq p + q$. If f and g were to have a nonconstant common factor, then a tangent line for that common factor would be a tangent line for both f and g, and no such tangent line exists according to our assumption. Therefore Bezout's Theorem (Theorem 8.2) applies to f and g and shows that their locus of common zeros is finite. Let it be $\{(0, 0), Q_1, \ldots, Q_l\}$. The third paragraph of the proof of Theorem 8.12 shows that there exists a polynomial h in $K[X, Y]$ such that $h(0, 0) = 1$ and $h(Q_i) = 0$ for $1 \leq i \leq l$. Then Xh and Yh vanish on $\{(0, 0), Q_1, \ldots, Q_l\}$, and the Nullstellensatz (Theorem 7.1) shows that there exists N such that $(Xh)^N$ and $(Yh)^N$ lie in (f, g). Since h is in the multiplicative system S, X^N and Y^N lie in $S^{-1}(f, g)$. Any monomial of degree $\geq 2N$ contains either a factor X^N or a factor Y^N, and consequently $I^{2N} \subseteq S^{-1}(f, g)$.

Proceeding inductively downward on t, suppose that $I^t \subseteq S^{-1}(f, g)$ and that $t - 1 \geq p + q$. As in Lemma 8.14, the polynomials defined by $B_k = L_1 \cdots L_k M_1 \cdots M_{t-1-k}$ for $0 \leq k \leq t-1$ form a vector-space basis of $K[X, Y]_{t-1}$. We show that each of these lies in $S^{-1}(f, g)$; then we can conclude that $I^{t-1} \subseteq S^{-1}(f, g)$, and our induction will be complete. Let $f = f_p + f_{p+1} + \cdots$ and $g = g_q + g_{q+1} + \cdots$ be the expansions of f and g as sums of homogeneous polynomials in (X, Y). If B_k is given, then an inequality $k \geq p$ would imply that B_k contains a factor $L_1 \cdots L_p$; this is f_p up to a constant factor. An inequality $t - 1 - k \geq q$ would imply that B_k contains a factor $M_1 \cdots M_q$; this is g_q up to a constant factor. Since $k < p$ and $t-1-k < q$ would together imply the inequality $t - 1 < p + q$ that we are assuming not to be the case, one of the alternatives $k \geq p$ and $t - 1 - k \geq q$ must occur. Say the first occurs. Except for a constant factor, we then have $B_k = f_p C$ for some homogeneous polynomial $C(X, Y)$ of degree $t - 1 - p$. Substituting for f_p gives $B_k = (f - f_{p+1} - \cdots)C$. Each term $f_{p+r}C$ with $r > 0$ is of degree $(p+r) + (t - 1 - p) > t - 1$ and therefore lies in $I^t \subseteq S^{-1}(f, g)$. Also, the term fC lies in $S^{-1}(f, g)$. Hence B_k lies in $S^{-1}(f, g)$. This completes the induction, and we conclude that $I^{p+q} \subseteq S^{-1}(f, g)$.

In order to prove the second conclusion of (§), namely that ψ is one-one, suppose that $0 = \psi(a + I^q, b + I^p) = af + bg + I^{p+q}$, i.e., that all terms of $af + bg$ are of order $\geq p+q$. Write $a = a_r + a_{r+1} + \cdots$ with $a_r \neq 0$ if a is not in I^q, and write $b = b_s + b_{s+1} + \cdots$ with $b_s \neq 0$ if b is not in I^p, so that

$$af + bg = a_r f_p + b_s g_q + \text{(higher-order terms)}.$$

The right side is assumed to be in I^{p+q}, which means that one of the following two conditions is satisfied:

(i) $r + p = s + q < p + q$ and $a_r f_p + b_s g_q = 0$,
(ii) $a_r f_p$ is in I^{p+q}, and $b_s g_q$ is in I^{p+q}.

If (i) holds, then the facts that $a_r f_p = -b_s g_q$ and that f and g have no tangent lines in common imply that f_p divides b_s. Since $s < p$, we must have $b_s = 0$. Therefore $a_r = 0$, and the conditions on a_r and b_s imply that a is in I^q and b is in I^p, which we are trying to show. If (ii) holds, then the fact that $a_r f_p$ is in I^{p+q} implies that $a_r = 0$ or $r \geq q$; in either case, a is in I^q. Similarly the fact that $b_s g_q = 0$ implies that $b_s = 0$ or $s \geq p$; in either case, b is in I^p. We conclude that ψ is one-one, as was to be shown. □

6. General Form of Bezout's Theorem for Plane Curves

With the discussion complete concerning intersection multiplicity for general projective plane curves, we arrive at the general form of Bezout's Theorem for plane curves.

Theorem 8.15 (Bezout's Theorem). Let K be an algebraically closed field, and let F and G be projective plane curves over K of respective degrees m and n. If F and G have no common factor of positive degree, then

$$\sum_{P \in \mathbb{P}^2_K} I(P, F \cap G) = mn.$$

REMARKS. The sum over P has only finitely many nonzero terms by Theorem 8.5, and each intersection multiplicity in the sum is finite by Theorem 8.10e.

PROOF. Theorem 8.5 shows that the locus of common zeros of F and G is a finite set. By applying a suitable Φ in $\text{GL}(3, K)$, we may assume that none of these zeros lies on the line at infinity, namely W. To do so, we choose a point P not in the finite set of common zeros. There are only finitely many lines passing through P and some member of the set of common zeros, and we choose a line through P different from all these. If Φ is chosen so as to move this line to the line at infinity W, then none of the common zeros will lie on the line W.

6. General Form of Bezout's Theorem for Plane Curves

With this normalization in place, let $\{P_1, \ldots, P_k\}$ be the set of common zeros of F and G. We introduce local versions f and g of F and G by the definitions $f(X, Y) = F(X, Y, 1)$ and $g(X, Y) = G(X, Y, 1)$. Application of Theorem 8.12 to the ideal $I = (f, g)$ in $K[X, Y]$ gives

$$\dim_K K[X, Y]/(f, g) = \sum_{j=1}^{k} \dim_K \left(K[X, Y]/(f, g)\right)_{(P_j)} = \sum_{j=1}^{k} I(P_j, F \cap G).$$

The theorem will therefore follow if we prove that

$$\dim_K K[X, Y]/(f, g) = mn. \qquad (*)$$

To prove $(*)$, we shall first prove a related equality concerning $K[X, Y, W]$ and the ideal (F, G) in it, and then we shall use the fact that F and G have no common zeros with W to transfer the conclusion to $K[X, Y]$.

Define K linear mappings $\varphi : K[X, Y, W] \oplus K[X, Y, W] \to K[X, Y, W]$ and $\psi : K[X, Y, W] \to K[X, Y, W] \oplus K[X, Y, W]$ by

$$\varphi(A, B) = AF + BG \qquad \text{and} \qquad \psi(C) = (CG, -CF),$$

and form the sequence of K vector spaces and K linear maps given by

$$0 \to K[X, Y, W] \xrightarrow{\psi} K[X, Y, W] \oplus K[X, Y, W] \xrightarrow{\varphi} K[X, Y, W]. \qquad (**)$$

It is evident that ψ is one-one, that $\varphi\psi = 0$, and that $\text{image}\,\varphi = (F, G)$. If (A, B) is in $\ker \varphi$, then $AF + BG = 0$. Since F and G have no common factor of positive degree, F divides B and G divides A. Setting $C = AG^{-1}$ therefore gives $A = CG$ and $B = -AG^{-1}F = -CF$. Hence (A, B) lies in image ψ. In other words, $(**)$ is exact, and image $\varphi = (F, G)$.

Let $d \geq m + n$. If we denote by ψ_d and φ_d the restrictions of ψ and φ to $K[X, Y, W]_{d-m-n}$ and $K[X, Y, W]_{d-n} \oplus K[X, Y, W]_{d-m}$, respectively, and if we go over the argument in the previous paragraph, then we see that the sequence

$$0 \to K[X,Y,W]_{d-m-n} \xrightarrow{\psi_d} K[X,Y,W]_{d-n} \oplus K[X,Y,W]_{d-m} \xrightarrow{\varphi_d} K[X,Y,W]_d$$

is exact and that image $\varphi_d = (F, G)_d$. The vector spaces in question here are all finite-dimensional, and thus we obtain

$$\dim_K (F, G)_d$$
$$= \dim_K K[X, Y, W]_{d-n} + \dim_K K[X, Y, W]_{d-m} - \dim_K K[X, Y, W]_{d-m-n}$$
$$= \binom{d-n+2}{2} + \binom{d-m+2}{2} - \binom{d-m-n+2}{2}$$
$$= -mn + \binom{d+2}{2}$$
$$= -mn + \dim_K K[X, Y, W]_d. \qquad (\dagger)$$

The ideal (F, G) is homogeneous, and thus we know from Section 3 that the image of $K[X, Y, W]_d$ in $K[X, Y, W]/(F, G)$ is $K[X, Y, W]_d/(F, G)_d$. If we write $\big(K[X, Y, W]/(F, G)\big)_d$ for this quotient, then (†) shows that

$$\dim_K \big(K[X, Y, W]/(F, G)\big)_d = mn \qquad (\dagger\dagger)$$

for all $d \geq m + n$.

To prove (∗) and the theorem, we shall translate (††) into a conclusion about $K[X, Y]/(f, g)$. Fix $d \geq m + n$, and let $\{V_1 + (F, G), \ldots, V_{mn} + (F, G)\}$ be a K basis of $\big(K[X, Y, W]/(F, G)\big)_d$. Define $v_j(X, Y) = V_j(X, Y, 1)$ for each j. We shall prove that the vectors

$$v_1 + (f, g), \ldots, v_{mn} + (f, g) \qquad (\ddagger)$$

form a K basis of $K[X, Y]/(f, g)$.

We need to make use of the fact that F and G have no common zeros on the line at infinity. Since $W(F, G) \subseteq (F, G)$, the K linear mapping of multiplication by W on $K[X, Y, W]$ descends to a K linear mapping L of $K[X, Y, W]/(F, G)$ to itself defined by $L(H + (F, G)) = WH + (F, G)$. Let us see that

$$L : K[X, Y, W]/(F, G) \to K[X, Y, W]/(F, G) \quad \text{is one-one.} \qquad (\ddagger\ddagger)$$

In fact, suppose that $WH = AF + BG$ for some H in $K[X, Y, W]$. For any U in $K[X, Y, W]$, let $U_0(X, Y) = U(X, Y, 0)$. If U is homogeneous, then so is U_0. In this notation we can write $F = F_0 + WM$ and $G = G_0 + WN$ for homogeneous members M and N of $K[X, Y, W]$. The polynomials F_0 and G_0 are relatively prime: in fact, if F_0 and G_0 have a nontrivial common factor D_0, then we can regard D_0 as a projective plane curve, and it must have a common zero Q with W, by Theorem 8.5; but then F, G, and W have Q as a common zero, in contradiction to the normalization in the first paragraph of the proof. Since $WH = AF + BG$ implies $A_0 F_0 = -B_0 G_0$, it follows that F_0 divides B_0 and that G_0 divides A_0. In other words, $B_0 = C_0 F_0$ and $A_0 = -C_0 G_0$ for some C_0 in $K[X, Y]$. If we define $A' = A + C_0 G$ and $B' = B - C_0 F$, then the formulas for A_0 and B_0 show that $A'_0 = B'_0 = 0$. Hence $A' = WA''$ and $B' = WB''$ for some homogeneous polynomials A'' and B''. Then $WH = AF + BG = (A' - C_0 G)F + (B' + C_0 F)G = A'F + B'G = W(A''F + B''G)$, and we obtain $H = A''F + B''G$. Thus H lies in (F, G), and (‡‡) is proved.

Left multiplication L by W carries $K[X, Y, W]_d$ into $K[X, Y, W]_{d+1}$ and carries $(F, G)_d$ into $(F, G)_{d+1}$. Therefore L is well defined as a mapping from $\big(K[X, Y, W]/(F, G)\big)_d$ into $\big(K[X, Y, W]/(F, G)\big)_{d+1}$. Since it is one-one by (‡‡) and since the spaces are finite-dimensional, it is onto. Therefore

$$\{W^r V_1 + (F, G), \ldots, W^r V_{mn} + (F, G)\} \qquad \text{is a basis} \qquad (\S)$$

of $\left(K[X, Y, W]/(F, G)\right)_{d+r}$ for every $r \geq 0$.

To prove that (‡) spans $K[X, Y]/(f, g)$, let h be in $K[X, Y]$. Let H be a homogeneous polynomial in $K[X, Y, W]$ with $h(X, Y) = H(X, Y, 1)$, and choose an integer s such that $W^s H$ lies in $K[X, Y, W]_{d+r}$ for some $r \geq 0$. Then we can write $W^s H = \sum_{j=1}^{mn} c_j W^r V_j + AF + BG$ for suitable scalars c_j and homogeneous polynomials A and B. Restricting the domain to points $(X, Y, 1)$ gives $h = \sum_{j=1}^{mn} c_j v_j + af + bg$, and therefore $h + (f, g) = \sum_{j=1}^{mn} c_j v_j + (f, g)$. This proves that (‡) spans $K[X, Y]/(f, g)$.

To prove that (‡) is linearly independent, suppose that $\sum_{j=1}^{mn} c_j v_j = af + bg$ with a and b in $K[X, Y]$. If A and B are homogeneous polynomials such that $a(X, Y) = A(X, Y, 1)$ and $b(X, Y) = B(X, Y, 1)$, then $W^r \sum_{j=1}^{mn} c_j V_j = W^s AF + W^t BG$, provided the exponents r, s, t are chosen to make the degrees of the terms $W^r \sum_{j=1}^{mn} c_j V_j$, $W^s AF$, and $W^t BG$ match. Consequently $W^r \sum_{j=1}^{mn} c_j V_j$ lies in $(F, G)_{d+r}$, and (§) shows that the coefficients are all 0. This proves that (‡) is linearly independent. □

7. Gröbner Bases

The remainder of the chapter returns to the main question introduced in Section 1, that of how to get information about the set of simultaneous solutions of polynomial equations in several variables. The resultant introduced in Section 2 gave us one tool, but the tool is of most use when there are only two equations. Beyond two equations the number of cases to check quickly grows, and the resultant is of limited usefulness.[12]

The tool to be introduced in this section is of a completely different nature. Historically it was introduced in order to have a way of deciding whether an ideal in $K[X_1, \ldots, X_n]$ contains a given polynomial. We know from the Hilbert Basis Theorem that every such ideal is finitely generated, and it is assumed that the ideal to be tested is specified by such a set of generators.

The proof of the Hilbert Basis Theorem gives a clue how to start studying an ideal of polynomials. In the statement of the theorem, R is a Noetherian integral domain, and I is a nonzero ideal in $R[X]$. It is to be proved that I is finitely generated. The proof by Hilbert is longer than the proof given in *Basic Algebra*, but the idea is clearer. To each nonzero member $f(X)$ of I, we associate the coefficient of the highest power of X appearing in $f(X)$. These coefficients, together with 0, form an ideal $L(I)$ in R, and $L(I)$ is finitely generated because R is Noetherian. Let $a_1, \ldots, a_r$ be generators, let $f_1(X), \ldots, f_r(X)$ be members

---

[12]The nature of the extended theory can be found in Van der Waerden, Volume II, Chapter XI. Theorem 8.31 below in effect reproduces some of this extended theory in a context that is manageable because of the theory of Gröbner bases.

of I with respective highest coefficients $a_1, \ldots, a_r$, and let q be the largest of the degrees of $f_1(X), \ldots, f_r(X)$. If a general $g(X)$ in I is given and if $a \in R$ is its highest coefficient, then we know that $a = \sum_i c_i a_i$ with $c_i \in R$. The polynomial $h(X)$ given by $h(X) = g(X) - \sum_i c_i f_i(X) X^{\deg g - \deg f_i}$ has degree lower than $\deg g$, and $g(X)$ will be in $(f_1, \ldots, f_r)$ if $h(X)$ is in $(f_1, \ldots, f_r)$. Iterating this construction, we see that it is enough to account for all the members of I of degree $\leq q - 1$. To handle these, one way to proceed is to enlarge the set $\{f_1, \ldots, f_r\}$ a little. For each k with $0 \leq k \leq q - 1$, let $L_k(I)$ be the union of $\{0\}$ and the set of coefficients of X^k in members of I of degree k. Each of these is an ideal of R and hence is finitely generated, and we adjoin to $\{f_1, \ldots, f_r\}$ a finite set of generators for each $L_k(I)$ with $0 \leq k \leq q - 1$. The result is a finite set $\{g_1, \ldots, g_s\}$ of generators of I, as one easily checks.

In fact, the set $\{g_1, \ldots, g_s\}$ is a special set of generators. For any member f of $R[X]$, let $\mathrm{LT}(f)$ be the complete term of $f(X)$ containing the highest power of X. What the argument shows is that $\{g_1, \ldots, g_s\}$ is a subset of I such that $\mathrm{LT}(I) = \bigl(\mathrm{LT}(g_1), \ldots, \mathrm{LT}(g_s)\bigr)$, where $\mathrm{LT}(I)$ denotes the ideal given as the linear span of all polynomials $\mathrm{LT}(g)$ for g in I. One can show that this property of $\{g_1, \ldots, g_s\}$ implies that $\{g_1, \ldots, g_s\}$ generates I. In essence this property will be the defining property of a "Gröbner basis" of I. It is not automatically satisfied for just any finite generating set $\{f_1, \ldots, f_r\}$, as the example below shows. We shall see that it is easy to use such a set of generators to test any polynomial in $R[X]$ for membership in I. Thus the original problem historically for introducing such sets is solved except for one little detail: the proof of the Hilbert Basis Theorem is not constructive, and we are left with no idea how actually to construct a Gröbner basis.[13]

EXAMPLE. Treat $K[X, Y]$ as an instance of the above setting by letting $R = K[Y]$ and regarding $K[X, Y]$ as $R[X]$. Consider the ideal $I = (f_1, f_2)$ in $R[X]$ with $f_1(X, Y) = X^2 + 2XY^2$ and $f_2(X, Y) = XY + 2Y^3 - 1$. Then $\bigl(\mathrm{LT}(f_1), \mathrm{LT}(f_2)\bigr) = (X^2, XY)$, and every monomial appearing with nonzero coefficient in a member of the latter ideal has total degree at least 2. On the other hand, I contains the polynomial

$$Y f_1(X, Y) - X f_2(X, Y) = Y(X^2 + 2XY) - X(XY + 2Y^3 - 1) = X,$$

and its leading term is X, whose total degree is 1. Thus $\mathrm{LT}(I)$ properly contains $\bigl(\mathrm{LT}(f_1), \mathrm{LT}(f_2)\bigr)$.

Because of the nonconstructive nature of the proof of the Hilbert Basis Theorem, it is necessary to start afresh. One message to glean from the abstract proof

---

[13]The exposition in this section and the next three is based partly on the book of Cox–Little–O'Shea and the Web tutorial of Fabrizio listed in the Selected References.

is that the leading terms of the members of I are important and somewhat control the nature of I. To handle $K[X_1, \ldots, X_n]$ when K is a field, it is of course necessary to use an additional induction that enumerates the variables. In the example above, we treated X as more significant than Y. For the inductive step for general $K[X_1, \ldots, X_n]$, the ring R in the above argument is K with some number m of the indeterminates included, and X is the $(m+1)^{st}$ indeterminate. Putting all the steps of the induction together, we see that the order in which the variables are processed appears to be important.

The theory of Gröbner bases as it has evolved allows a healthy extra measure of generality. Instead of defining leading terms by insisting on an ordering of the indeterminates, it defines them by using a suitable kind of ordering of monomials, and that is where we begin. Let $K[X_1, \ldots, X_n]$ be given, K being a field. Let $\mathcal{M}$ be the set of all monomials in $K[X_1, \ldots, X_n]$. A **monomial ordering** $\leq$ on $\mathcal{M}$ is a total ordering[14] with the two additional properties that

(i) $M_1 \leq M_2$ implies $M_1 M_3 \leq M_2 M_3$ for all M_1, M_2, M_3 in $\mathcal{M}$,
(ii) $1 \leq M$ for all M in $\mathcal{M}$.

We write $M_2 \geq M_1$ to mean $M_1 \leq M_2$. Also, $M_1 < M_2$ means $M_1 \leq M_2$ with $M_1 \neq M_2$, and $M_1 > M_2$ means $M_1 \geq M_2$ with $M_1 \neq M_2$.

EXAMPLES OF MONOMIAL ORDERINGS. Each ordering assumes that the variables are enumerated in some way. In these examples we take this enumeration to be $X_1, \ldots, X_n$. The first four examples all have the property that the largest X_j is X_1 and the smallest is X_n.

(1) **Lexicographic ordering**, abbreviated as "lex" by many authors and written as $\leq_{\text{LEX}}$ in this list of examples. This, the most important monomial ordering, is already suggested by the proof of the Hilbert Basis Theorem. In principle it can be used for all purposes in Sections 7–10, but one application in Chapter X will require a different monomial ordering. Its disadvantage is that it sometimes makes lengthy computations take longer than necessary; this matter will be discussed more in Section 9. The definition is that $X_1^{i_1} \cdots X_n^{i_n} \leq_{\text{LEX}} X_1^{j_1} \cdots X_n^{j_n}$ if either the two monomials are equal or else the first k for which $i_k \neq j_k$ has $i_k < j_k$. Thus for example, $X_1 X_2^2 X_3^3 \leq_{\text{LEX}} X_1^2$. The word "lexicographic" refers to the dictionary system for alphabetizing in which a first word comes before a second word if for the first position in which the two words differ, the letter of the first word in that position precedes alphabetically the letter of the second word in that position.

(2) **Graded lexicographic ordering**, abbreviated as "glex" or "grlex" by many authors. As in Section 3 the **total degree** of a monomial $X_1^{i_1} \cdots X_n^{i_n}$ is

---

[14]This means a partial ordering with the properties that each pair a, b has $a \leq b$ or $b \leq a$ and that both hold only if $a = b$.

$\deg(X_1^{i_1} \cdots X_n^{i_n}) = \sum_{k=1}^{n} i_k$. The definition of the ordering is that $M \leq_{\text{GLEX}} N$ if either $\deg M < \deg N$ or else if $\deg M = \deg N$ and $M \leq_{\text{LEX}} N$. Thus for example, $X_1^2 \leq_{\text{GLEX}} X_1 X_2^2 X_3^3$ because the total degree 2 of the first monomial is less than the total degree 6 of the second monomial. But $X_1 X_2^2 X_3^3 \leq_{\text{GLEX}} X_1^2 X_3^4$ because both monomials have the same total degree 6 and the second monomial involves a higher power of X_1 than does the first. This monomial ordering is not much used; more common is the variant of it in the next example.

(3) **Graded reverse lexicographic ordering**, abbreviated as "grevlex" by many authors. The definition is that $M \leq_{\text{GREVLEX}} N$ if either $\deg M < \deg N$ or else if $\deg M = \deg N$ and $N^t \leq_{\text{LEX}} M^t$, where M^t is M but with the exponents of X_j and X_{n-j} interchanged for each j, and where N^t is defined similarly. This ordering takes some getting used to. For example, $X_1^2 X_3^4 \leq_{\text{GREVLEX}} X_1 X_2^2 X_3^3$ when $n = 3$ because both monomials have the same total degree and $X_1^3 X_2^2 X_3 = (X_1 X_2^2 X_3^3)^t \leq_{\text{LEX}} (X_1^2 X_3^4)^t = X_1^4 X_3^2$. By contrast, $X_1 X_2^2 X_3^3 \leq_{\text{GLEX}} X_1^2 X_3^4$.

(4) Orderings of k-**elimination type**, where $1 \leq k \leq n - 1$. These are orderings such that any monomial containing one of $X_1, \ldots, X_k$ to a positive power exceeds any monomial in $X_{k+1}, \ldots, X_n$ alone. These will be discussed in Section 10. Of them, one of particular importance is the **Bayer–Stillman ordering** of k-elimination type. Here a monomial M is $\leq$ a monomial N if the sum of the exponents of $X_1, \ldots, X_k$ for M is less than the corresponding sum for N or else the two sums are equal and $M \leq_{\text{GREVLEX}} N$. This ordering is commonly used for making computations in the context of Section 10.

(5) Ordering from a tuple of **weight vectors**. For $1 \leq i \leq n$, let $w^{(i)}$ be a vector in $\mathbb{R}^n$ of the form $w^{(i)} = (w_1^{(i)}, \ldots, w_n^{(i)})$, and assume that $w^{(1)}, \ldots, w^{(n)}$ are linearly independent over $\mathbb{R}$. Identify the monomial X^α with the vector of individual exponents $\alpha = (\alpha_1, \ldots, \alpha_n)$. The ordering given by the weight vectors $w_j^{(i)}$ is defined by saying that $X^\alpha \leq X^\beta$ if $X^\alpha = X^\beta$ or if the first i such that $w^{(i)} \cdot \alpha \neq w^{(i)} \cdot \beta$ has $w^{(i)} \cdot \alpha < w^{(i)} \cdot \beta$. Here the dot refers to the ordinary dot product. A condition is needed on the $w^{(i)}$'s to ensure that $1 \leq X^\alpha$ for all α. (See Problem 14 at the end of the chapter.) Here are two specific examples for which the condition is satisfied. Let $e^{(i)}$ be the ith standard basis vector of $\mathbb{R}^n$. The lexicographic ordering in Example 1 is determined by the tuple of weight vectors $(e^{(1)}, \ldots, e^{(n)})$. The Bayer–Stillman ordering in Example 4 is determined by the tuple of weight vectors

$$\left(e^{(1)} + \cdots + e^{(k)}, e^{(k+1)} + \cdots + e^{(n)}, -e^{(n)}, \ldots, -e^{(k+2)}, -e^{(k)}, \ldots, -e^{(2)}\right).$$

Further discussion of monomial orderings determined by weight vectors occurs in Problems 14–15 at the end of the chapter.

Property (i) of monomial orderings insists that the ordering respect multiplication of monomials in the natural way. Property (ii), according to the next proposition, is a well-ordering property. The proof of the proposition will be preceded by a lemma.

Proposition 8.16. In any monomial ordering for $K[X_1, \ldots, X_n]$, any decreasing sequence $M_1 \geq M_2 \geq M_3 \geq \cdots$ is eventually constant. Consequently each nonempty subset of $\mathcal{M}$ has a smallest element in the ordering.

Lemma 8.17. If I is an ideal in $K[X_1, \ldots, X_n]$ generated by monomials and if $f(X_1, \ldots, X_n)$ is in I, then each monomial appearing in the expansion of f with nonzero coefficient lies in I. Consequently I has a finite set of monomials as generators. Moreover, if $\{M_1, \ldots, M_s\}$ is a set of monomials that generate I and if M is any monomial in I, then some M_j divides M.

PROOF. Let $\{M_\alpha\}$ be the set of monomials that generates I. If f is in I, then we can write $f = \sum_{j=1}^{k} h_j M_{\alpha_j}$ for polynomials h_j. Let $h_j = \sum_{i=1}^{l_j} c_{ij} M_{ij}$ be the expansion of h_j in terms of monomials. If M_0 is a monomial appearing in f with nonzero coefficient c, then the only possible monomial M_{ij} in h_j that can contribute toward c is one with $M_{ij} M_{\alpha_j} = M_0$ if such a monomial exists. For some j, such a monomial must exist, or c would be 0; thus M_0 lies in I.

For the second conclusion, write $\{f_1, \ldots, f_l\}$ by the Hilbert Basis Theorem. The first conclusion shows that each monomial contributing to each f_j lies in I, and the set of all these monomials, as j varies, is therefore a finite set of monomials generating I.

For the third conclusion, write $M = \sum_{i=1}^{s} a_i M_i$ for polynomials a_i. Expanding each a_i in terms of monomials, we see that some a_i contains with nonzero coefficient a monomial M' such that $M = M' M_i$. The divisibility follows. □

PROOF OF PROPOSITION 8.16. Let M be a monomial, and let I be the linear span of all monomials M' with $M' \geq M$. If M' is a such a monomial and N is any monomial, then $NM' \geq NM$ by (i), and $NM \geq 1M = M$ by (i) and (ii). Therefore NM' lies in I, and I is an ideal.

From such an ideal I, we can recover M as the unique monomial M_0 in I such that $M_0 \leq M'$ for every monomial M' in I, since any such M_0 has $M_0 \leq M$ as well as $M \leq M_0$.

With $M_1, M_2, \ldots$ given as in the proposition, let I_k be the linear span of all monomials $M' \geq M_k$. We have just seen that I_k is an ideal, and the I_k's are increasing in k. Then $I = \bigcup_{k=1}^{\infty} I_k$ is an ideal generated by monomials, and Lemma 8.17 shows that it has a finite set of monomials as a set of generators. Each such monomial generator lies in some I_k. Since the I_k's are nested, all the generators lie in some I_{k_0}, and we conclude that $I = I_{k_0}$. The previous paragraph of the proof shows that I_{k_0} determines M_{k_0}, and therefore $M_k = M_{k_0}$ for all $k \geq k_0$.

For the last statement of the proposition, if there were no least element, then for any element in the subset, we could always find a smaller element in the subset. In this way, we would be able to construct a strictly decreasing infinite sequence in $\mathcal{M}$, in contradiction to what has just been proved. $\square$

Fix a monomial ordering for $K[X_1, \ldots, X_n]$. If f is any nonzero member of $K[X_1, \ldots, X_n]$ and if f is expanded as a K linear combination of monomials, then we define the leading monomial, leading coefficient, and leading term of f by

$\text{LM}(f) = $ largest monomial with nonzero coefficient in expansion of f,

$\text{LC}(f) = $ coefficient of $\text{LM}(f)$ in f,

$\text{LT}(f) = \text{LC}(f) \text{LM}(f)$.

It will be convenient to be able to use these definitions without having to distinguish the cases $f \neq 0$ and $f = 0$. Accordingly, let us adjoin 0 to the set $\mathcal{M}$, agreeing that $0 < M$ and $0M = 0$ for every monomial M. We adopt the convention that $\text{LM}(0) = 0$, $\text{LT}(0) = 0$, and $\text{LC}(0) = 0$.

Since any monomial that occurs in a sum of two polynomials occurs in one or the other of them, it is immediate from the definition that

$$\text{LM}(f_1 + f_2) \leq \max(\text{LM}(f_1), \text{LM}(f_2))$$

if f_1, f_2, and $f_1 + f_2$ are nonzero. Checking the various cases, we see that this inequality persists if one or more of f_1, f_2, and $f_1 + f_2$ are 0.

The comparable results concerning multiplication are contained in the next proposition.

Proposition 8.18. If f_1 and f_2 are two nonzero members of $K[X_1, \ldots, X_n]$, then

$$\text{LM}(f_1 f_2) = \text{LM}(f_1) \text{LM}(f_2) \quad \text{and} \quad \text{LC}(f_1 f_2) = \text{LC}(f_1) \text{LC}(f_2);$$

hence

$$\text{LT}(f_1 f_2) = \text{LT}(f_1) \text{LT}(f_2).$$

These equalities persist if one or both of f_1 and f_2 are 0. Moreover, if f_1 and f_2 are nonzero and have $\text{LT}(f_1) = \text{LT}(f_2)$, then $\text{LM}(f_1 - f_2) < \text{LM}(f_1)$.

PROOF. For the first statement, let the expansions of f_1 and f_2 as linear combinations of distinct monomials be $f_1 = a_1 \text{LM}(f_1) + \sum_i c_i M_i$ and $f_2 = a_2 \text{LM}(f_2) + \sum_j d_j N_j$ with $M_i < \text{LM}(f_1)$ for all i and $N_j < \text{LM}(f_2)$ for all j. Then $f_1 f_2$ equals

$$a_1 a_2 \text{LM}(f_1) \text{LM}(f_2) + a_2 \sum_i c_i M_i \text{LM}(f_2) + a_1 \sum_j d_j \text{LM}(f_1) N_j + \sum_{i,j} c_i d_j M_i N_j,$$

and the conclusions in the first sentence of the proposition will follow if it is shown that $M_i \operatorname{LM}(f_2) < \operatorname{LM}(f_1) \operatorname{LM}(f_2)$, that $\operatorname{LM}(f_1) N_j < \operatorname{LM}(f_1) \operatorname{LM}(f_2)$, and that $M_i N_j < \operatorname{LM}(f_1) \operatorname{LM}(f_2)$. The first inequality follows from (i) because $M_i < \operatorname{LM}(f_1)$, and the second inequality is similar. For the third we apply (i) twice to obtain $M_i N_j \le M_i \operatorname{LM}(f_2) \le \operatorname{LM}(f_1) \operatorname{LM}(f_2)$ and observe that the end expressions can be equal only if equality holds in both instances. The latter is impossible because $K[X_1, \ldots, X_n]$ is an integral domain, and thus $M_i N_j < \operatorname{LM}(f_1) \operatorname{LM}(f_2)$.

The three displayed equalities persist if one or both of f_1 and f_2 are 0 because $\operatorname{LM}(f), \operatorname{LT}(f)$, and $\operatorname{LC}(f)$ can be 0 only if $f = 0$.

Finally if f_1 and f_2 are nonzero and have expansions as in the first paragraph of the proof with $\operatorname{LT}(f_1) = \operatorname{LT}(f_2)$, then $\operatorname{LC}(f_1) = a_1$ and $\operatorname{LC}(f_2) = a_2$. Hence $f_1 - f_2$ has an expansion involving only the monomials M_i and N_j. Consequently if $f_1 - f_2 \ne 0$, then the largest of the M_i's and N_j's is $< \operatorname{LM}(f_1)$. Thus $\operatorname{LM}(f_1 - f_2) < \operatorname{LM}(f_1)$. This inequality holds also if $f_1 - f_2 = 0$. □

If I is a nonzero ideal in $K[X_1, \ldots, X_n]$, we define $\operatorname{LT}(I)$ to be the vector space of all K linear combinations of polynomials $\operatorname{LT}(f)$ with f in I. It follows from Proposition 8.18 that $K[X_1, \ldots, X_n] \operatorname{LT}(I) \subseteq \operatorname{LT}(I)$, and therefore $\operatorname{LT}(I)$ is an ideal in $K[X_1, \ldots, X_n]$. A finite unordered subset $\{g_1, \ldots, g_k\}$ of nonzero elements of the ideal I is called a **Gröbner basis** of I if $\operatorname{LT}(I) = \big(\operatorname{LT}(g_1), \ldots, \operatorname{LT}(g_k)\big)$. The inclusion $\supseteq$ follows from the definition, and the question is whether $\operatorname{LT}(g_1), \ldots, \operatorname{LT}(g_k)$ generate $\operatorname{LT}(I)$.

Among the examples below, Example 3 is particularly suggestive of the utility of a Gröbner basis. The idea is that an ordinary set of generators may have the property that certain "small" elements of I can be expanded in terms of the generators only using "large" coefficients and that this property is reflected in the failure of $(\operatorname{LT}(g_1), \ldots, \operatorname{LT}(g_k))$ to exhaust $\operatorname{LT}(I)$.

EXAMPLES WITH LEXICOGRAPHIC ORDERING.

(1) Principal ideal. If $I = (f(X_1, \ldots, X_n))$, then $\{f\}$ is a Gröbner basis. In fact, the most general member of I is of the form hf with h in $K[X_1, \ldots, X_n]$, and Proposition 8.18 gives $\operatorname{LT}(hf) = \operatorname{LT}(h) \operatorname{LT}(f)$. Therefore $\operatorname{LT}(I) = (\operatorname{LT}(f))$, as required.

(2) Ideal generated by members of $K[X_1, \ldots, X_n]_1$. Suppose that $I = (L_1, \ldots, L_k)$, where each L_j is a homogeneous linear polynomial of degree 1. For example, I could be $(X_1 + X_2 + X_3, X_1 - X_3)$. Let us form the corresponding k-by-n coefficient matrix, specifically $\begin{pmatrix} 1 & 1 & 1 \\ 1 & 0 & -1 \end{pmatrix}$ in the 3-variable example. If we perform row operations to transform this matrix into reduced row-echelon form and let $L'_1, \ldots, L'_{k'}$ be the members of $K[X_1, \ldots, X_n]_1$ corresponding to the reduced matrix, specifically $X_1 - X_3$ and $X_2 + 2X_3$ for the reduced form

$\begin{pmatrix} 1 & 0 & -1 \\ 0 & 1 & 2 \end{pmatrix}$ of $\begin{pmatrix} 1 & 1 & 1 \\ 1 & 0 & -1 \end{pmatrix}$, then $I = (L'_1, \ldots, L'_{k'})$ and moreover $\{L'_1, \ldots, L'_{k'}\}$ is a Gröbner basis of I. This fact is not particularly obvious in the full generality of this example, but it will be shown to be an easy consequence of Theorem 8.23 in the next section.

(3) Earlier example in this section. In $K[X, Y]$, let $I = (f_1, f_2)$ with $f_1(X, Y) = X^2 + 2XY^2$ and $f_2(X, Y) = XY + 2Y^3 - 1$. Then $(\text{LT}(f_1), \text{LT}(f_2)) = (X^2, XY)$. We saw that X is a member of I and that $\text{LT}(X) = X$ is not in $(\text{LT}(f_1), \text{LT}(f_2))$. So $\{f_1, f_2\}$ is not a Gröbner basis. If we enlarge the set of generators of I to $\{f_1, f_2, X\}$, then we still do not have a Gröbner basis because $f_2 - YX = 2Y^3 - 1$ is in I and $\text{LT}(f_2 - YX) = 2Y^3$ does not lie in $(\text{LT}(f_1), \text{LT}(f_2), \text{LT}(X)) = (X^2, XY, X) = (X)$. We can enlarge the set of generators still further to $\{f_1, f_2, X, 2Y^3 - 1\}$. Is this a Gröbner basis? Here we have $(\text{LT}(f_1), \text{LT}(f_2), \text{LT}(X), \text{LT}(2Y^3 - 1)) = (X, Y^3)$, and it seems as if this equals $\text{LT}(I)$. But we need a way of checking easily. We shall obtain a way of checking in Theorem 8.23 in the next section.

The question of existence–uniqueness of a Gröbner basis will be addressed constructively in Sections 8–9; however, we did observe at the beginning of this section that Hilbert's proof of the Hilbert Basis Theorem essentially handles existence when the monomial ordering is the usual lexicographic ordering. Actually, the argument at the beginning of the section had two parts to it—a nonconstructive argument producing a certain finite set of leading terms and a verification that those leading terms lead to a set of generators of the ideal. The first part, being a nonconstructive existence proof, does not help us in our current efforts, and we defer to Problem 13 at the end of the chapter the question of adapting it to a general monomial order. The second part, on the other hand, is a useful kind of verification in our current efforts. It shows that a certain kind of finite subset of an ideal is necessarily a set of generators, and it generalizes as follows. The generalization will play a role in Section 9.

Proposition 8.19. If K is a field, if a monomial ordering is specified for $K[X_1, \ldots, X_n]$, and if $\{g_1, \ldots, g_k\}$ is a Gröbner basis for a nonzero ideal I of $K[X_1, \ldots, X_n]$, then $\{g_1, \ldots, g_k\}$ generates I.

PROOF. First we prove that if $f \neq 0$ is in I, then there exist a g_j, a monomial M_0, and a nonzero scalar c such that $\text{LM}(f - cM_0 g_j) < \text{LM}(f)$. To see this, we use the hypothesis that $\{g_1, \ldots, g_k\}$ is a Gröbner basis to find polynomials $h_1, \ldots, h_k$ such that $\text{LM}(f) = \sum_{i=1}^k h_i \text{LM}(g_i)$. Then it must be true for i equal to some index j that $\text{LM}(f) = M_0 \text{LM}(g_j)$ for one of the monomials M_0 that appears in h_j with nonzero coefficient. Since $M_0 \text{LM}(g_j) = \text{LM}(M_0) \text{LM}(g_j) = \text{LM}(M_0 g_j)$, we can rewrite this equality as $\text{LT}(f) = c \text{LT}(M_0 g_j)$ for some scalar $c \neq 0$. Then

$\operatorname{LT}(f) = \operatorname{LT}(cM_0 g_j)$, and Proposition 8.18 shows that $\operatorname{LM}(f - cM_0 g_j) < \operatorname{LM}(f)$, as asserted.

Iterating this construction and assuming that we never get 0, we can find successively nonzero scalars c_i, monomials M_i, and members g_{j_i} of the Gröbner basis such that the sequence $\operatorname{LM}\left(f - \sum_{i=1}^{l} c_j M_j g_{j_i}\right)$ indexed by l is strictly decreasing, in contradiction to Proposition 8.16. To avoid the contradiction, we must have $f - \sum_{i=1}^{l} c_j M_j g_{j_i} = 0$ for some l, and then f is exhibited as in the ideal $(g_1, \ldots, g_k)$. Hence the Gröbner basis generates I. □

8. Constructive Existence

Throughout this section, K denotes a field, and we work with a fixed monomial ordering on $K[X_1, \ldots, X_n]$. Ideals in $K[X_1, \ldots, X_n]$ will always be specified by giving finite sets of generators. Our objective is to obtain a constructive proof of the existence of a Gröbner basis for each nonzero ideal in $K[X_1, \ldots, X_n]$, along with a useful test procedure for deciding whether a given finite set of generators of I is a Gröbner basis. As is often the case with existence proofs, the motivation for the proof comes from a certain amount of deduction of properties that a Gröbner basis must satisfy if its exists. It was mentioned in the previous section that the failure of a set of generators to be a Gröbner basis has something to do with its failure to be able to represent all "small" elements of the ideal by means of expansions in terms of the generators that use "small" coefficients. The first part of this section will explore this idea, seeking to make it precise. The main step will be a checkable text for a set to be a Gröbner basis; this is Theorem 8.23. The existence argument will be an easy corollary. A by-product of the existence argument will be a way of testing a polynomial for membership in I.

In the one-variable case any ideal is principal, necessarily of the form $(g(X))$, and the test for membership of a polynomial f in the ideal is to apply the division algorithm, writing $f(X) = q(X)g(X) + r(X)$ with $r = 0$ or $\deg r < \deg g$. Then f is a member of the ideal if and only if $r = 0$. The starting point for the several-variable theory is to do the best we can to generalize the division algorithm to several variables, recognizing that we cannot expect too much because of the complicated ideal structure in several variables.

Proposition 8.20 (generalized division algorithm). Let $(f_1, \ldots, f_s)$ be a fixed enumeration of a set of nonzero members of $K[X_1, \ldots, X_n]$, and let f be an arbitrary nonzero member of $K[X_1, \ldots, X_n]$. Then there exist polynomials $a_1, \ldots, a_s$ and r such that

$$f = a_1 f_1 + \cdots + a_s f_s + r,$$

such that $\text{LM}(a_j f_j) \leq \text{LM}(f)$ for all j, and such that no monomial appearing in r with nonzero coefficient is divisible by $\text{LM}(f_j)$ for any j.

REMARK. The proof below will stop short of giving an algorithm, because omitting the details of the algorithm will make the invariant of the construction clearer. To make the proof into an algorithm, one merely needs to be systematic about the choices in the proof. There is no claim of any uniqueness of $a_1, \ldots, a_s$ or r in the statement; in fact, Problem 16 at the end of the chapter shows that more than one kind of nonuniqueness is possible. Corollary 8.21 below, however, will show that if the given $f_1, \ldots, f_s$ form a Gröbner basis of an ideal I, then r is independent of the enumeration of the Gröbner basis, even without the requirement that $\text{LM}(a_j f_j) \leq \text{LM}(f)$ for all j.

PROOF. We shall do a kind of induction involving decompositions of f of the form
$$f = (a_1 f_1 + \cdots + a_s f_s) + p + r, \qquad (*)$$
where $a_1, \ldots, a_s, p, r$ are polynomials with the properties that
(i) $\text{LM}(p) \leq \text{LM}(f)$,
(ii) $\text{LM}(a_i f_i) \leq \text{LM}(f)$ for all i,
(iii) no monomial M appearing in r with nonzero coefficient has M divisible by any $\text{LM}(f_i)$,

and we shall demonstrate that $\text{LM}(p)$ decreases at every step of the induction as long as $p \neq 0$. Initially we take all $a_i = 0$, $p = f$, and $r = 0$. Then $(*)$ and the three properties hold at the start. Let us describe the inductive step.

If $\text{LT}(f_j)$ divides $\text{LT}(p)$ for some j, then we replace a_j by $a_j + \text{LT}(p)/\text{LT}(f_j)$, we change p to $p - \bigl(\text{LT}(p)/\text{LT}(f_j)\bigr) f_j$, and we leave r alone. The equality $(*)$ is maintained, and (iii) continues to hold. Since
$$\text{LT}\bigl(\bigl(\text{LT}(p)/\text{LT}(f_j)\bigr) f_j\bigr) = \text{LT}\bigl(\text{LT}(p)/\text{LT}(f_j)\bigr) \text{LT}(f_j)$$
$$= \bigl(\text{LT}(p)/\text{LT}(f_j)\bigr) \text{LT}(f_j) = \text{LT}(p), \qquad (**)$$

Proposition 8.18 shows that $\text{LM}(p)$ strictly decreases. Consequently (i) continues to hold. By the same kind of computation as for $(**)$,
$$\text{LM}\bigl(\bigl(a_j + \text{LT}(p)/\text{LT}(f_j)\bigr) f_j\bigr) \leq \max\bigl(\text{LM}(a_j f_j), \text{LM}\bigl(\text{LT}(p)/\text{LT}(f_j)\bigr) f_j\bigr)$$
$$\leq \max(\text{LM}(f), \text{LM}(p)) = \text{LM}(f),$$

and therefore (ii) continues to hold. This completes the inductive step if $\text{LT}(f_j)$ divides $\text{LT}(p)$ for some j.

The contrary case is that $\text{LT}(p)$ is divisible by $\text{LT}(f_i)$ for no i. Then we replace p by $p - \text{LT}(p)$, we change r to $r + \text{LT}(p)$, and we leave all a_i alone. The equality (∗) is maintained, and (ii) continues to hold. Since $\text{LM}(p) = \text{LM}(\text{LT}(p))$, Proposition 8.18 shows that $\text{LM}(p)$ strictly decreases. Consequently (i) continues to hold. Also, (iii) continues to hold because of the assumption that $\text{LT}(p)$ is divisible by $\text{LT}(f_i)$ for no i. This completes the inductive step if $\text{LT}(p)$ is divisible by $\text{LT}(f_i)$ for no i.

Proposition 8.16 shows that the induction can continue for only finitely many steps. Since it must continue as long as $p \neq 0$, the conclusion is that $p = 0$ after some stage, and then the decomposition of the proposition has been proved. □

Corollary 8.21. If $\{g_1, \ldots, g_s\}$ is a Gröbner basis of a nonzero ideal I of $K[X_1, \ldots, X_n]$ and if f is any nonzero member of $K[X_1, \ldots, X_n]$, then there exist polynomials g and r such that $f = g + r$, g is in I, and no monomial appearing in r with nonzero coefficient is divisible by $\text{LM}(g_j)$ for any j. Moreover, r is uniquely determined by these properties, and g has an expansion $g = \sum_{i=1}^{s} a_i g_i$ with $\text{LM}(a_i g_i) \leq \text{LM}(f)$ for all i.

REMARKS. The uniqueness statement implies in particular that r is independent of the enumeration of the set $\{g_1, \ldots, g_s\}$. This corollary will give us some insight into the way a Gröbner basis can resolve cancellation. Shortly we shall introduce specific members of I that have cancellation built into their definition. Being in I, they have expansions with remainder term 0, according to this corollary. Since the remainder is unique, the corollary says that they can be rewritten in terms of the Gröbner basis in a way that eliminates the cancellation.

PROOF. For existence, let $\{g_1, \ldots, g_s\}$ be a Gröbner basis of I, and apply Proposition 8.20 to f and the ordered set $(g_1, \ldots, g_s)$. Then the existence follows immediately.

For uniqueness, suppose that $f = g_1 + r_1 = g_2 + r_2$. Then $r_1 - r_2 = g_2 - g_1$ exhibits $r_1 - r_2$ as in I. Arguing by contradiction, suppose that $r_1 \neq r_2$. The hypothesis on r_1 and r_2 shows that no monomial with nonzero coefficient in $r_1 - r_2$ is divisible by any $\text{LM}(g_j)$, and in particular $\text{LM}(r_1 - r_2)$ is not divisible by any of the generators of the monomial ideal $\big(\text{LM}(g_1), \ldots, \text{LM}(g_s)\big) = \text{LM}(I)$. Since $\text{LM}(r_1 - r_2)$ is a monomial in this ideal, this conclusion contradicts the last conclusion of Lemma 8.17. □

Suppose that $X^\alpha = X_1^{\alpha_1} \cdots X_n^{\alpha_n}$ and $X^\beta = X_1^{\beta_1} \cdots X_n^{\beta_n}$ are two monomials in $K[X_1, \ldots, X_n]$. Then we define their **least common multiple** $\text{LCM}(X^\alpha, X^\beta)$ to be

$$\text{LCM}(X^\alpha, X^\beta) = X^\gamma = X_1^{\gamma_1} \cdots X_n^{\gamma_n} \qquad \text{with } \gamma_j = \max(\alpha_j, \beta_j) \text{ for all } j.$$

This notion does not depend on the choice of a monomial ordering. Observe for any two monomials M and N that $\text{LCM}(M, N)/M$ and $\text{LCM}(M, N)/N$ are monomials.

If f_1 and f_2 are nonzero polynomials, then the expression

$$\frac{\text{LCM}\big(\text{LM}(f_1), \text{LM}(f_2)\big)}{\text{LT}(f_1)} f_1 = \frac{\text{LCM}\big(\text{LM}(f_1), \text{LM}(f_2)\big)}{\text{LM}(f_1)} \frac{f_1}{\text{LC}(f_1)}$$

is a polynomial whose leading monomial is $\text{LCM}\big(\text{LM}(f_1), \text{LM}(f_2)\big)$ and whose leading coefficient is 1. We define the *S*-**polynomial** of f_1 and f_2 to be

$$S(f_1, f_2) = \frac{\text{LCM}\big(\text{LM}(f_1), \text{LM}(f_2)\big)}{\text{LT}(f_1)} f_1 - \frac{\text{LCM}\big(\text{LM}(f_1), \text{LM}(f_2)\big)}{\text{LT}(f_2)} f_2.$$

This is the difference of two polynomials with the same leading monomial $\text{LCM}\big(\text{LM}(f_1), \text{LM}(f_2)\big)$ and with the same leading coefficient 1. Accordingly, Proposition 8.18 shows that

$$\text{LM}(S(f_1, f_2)) < \text{LCM}\big(\text{LM}(f_1), \text{LM}(f_2)\big).$$

The elements $S(f_1, f_2)$ are the elements mentioned in the remarks with Corollary 8.21; the above inequality is a precise formulation of their built-in cancellation.

Lemma 8.22 below says that whenever cancellation of this kind occurs in any sum of products with functions $f_1, \ldots, f_s$, then the sum of products can be rewritten in terms of the *S*-polynomials $S(f_j, f_k)$. In this way the nature of the cancellation has been made more transparent, partly being accounted for by the definitions of the individual polynomials $S(f_j, f_k)$.

Lemma 8.22. Let M and $M_1, \ldots, M_s$ be monomials, let $f_1, \ldots, f_s$ be nonzero polynomials, and suppose that $M_i \text{LM}(f_i) = M$ for all i. If $c_1, \ldots, c_s$ are constants such that $\text{LM}\left(\sum_{i=1}^{s} c_i M_i f_i\right) < M$, then the sum $\sum_{i=1}^{s} c_i M_i f_i$ can be rewritten in the form

$$\sum_{i=1}^{s} c_i M_i f_i = \sum_{j<k} \frac{d_{jk} M}{\text{LCM}\big(\text{LM}(f_j), \text{LM}(f_k)\big)} S(f_j, f_k)$$

for suitable constants d_{jk}. In the sum on the right side, each nonzero term has leading monomial $< M$.

PROOF. Let us write $L_{ij} = \text{LCM}\big(\text{LM}(f_i), \text{LM}(f_j)\big)$ for $i \neq j$. We may assume that all the c_i are nonzero, and we proceed by induction on s. There is nothing to prove for $s = 1$. The key step is $s = 2$, for which we are given that the M term of $c_1 M_1 f_1 + c_2 M_2 f_2$ is 0, i.e., that

$$c_1 \text{LC}(f_1) + c_2 \text{LC}(f_2) = 0. \qquad (*)$$

Substituting for $\text{LC}(f_2)$ from $(*)$ gives

$$\begin{aligned} ML_{12}^{-1}S(f_1, f_2) &= Mf_1/\text{LT}(f_1) - Mf_2/\text{LT}(f_2) \\ &= M_1 f_1/\text{LC}(f_1) - M_2 f_2/\text{LC}(f_2) \\ &= c_1^{-1}\text{LC}(f_1)^{-1}(c_1 M_1 f_1 + c_2 M_2 f_2), \end{aligned}$$

and this proves the displayed formula of the lemma with $d_{12} = c_1 \text{LC}(f_1)$.

Assume the result for $s - 1 \geq 2$. We are given that $\sum_{i=1}^{s} c_i \text{LC}(f_i) = 0$, which we break into two parts as

$$c_1 \text{LC}(f_1) - \tfrac{c_1 \text{LC}(f_1)}{\text{LC}(f_2)} \text{LC}(f_2) = 0,$$

$$\left(c_2 + \tfrac{c_1 \text{LC}(f_1)}{\text{LC}(f_2)}\right) \text{LC}(f_2) + \sum_{i=3}^{s} c_i \text{LC}(f_i) = 0.$$

The inductive hypothesis gives

$$c_1 M_1 f_1 - \tfrac{c_1 \text{LC}(f_1)}{\text{LC}(f_2)} M_2 f_2 = d_{12} M L_{12}^{-1} S(f_1, f_2),$$

$$\left(c_2 + \tfrac{c_1 \text{LC}(f_1)}{\text{LC}(f_2)}\right) M_2 f_2 + \sum_{i=3}^{s} c_i M_i f_i = \sum_{2 \leq j < k} d_{jk} M L_{jk}^{-1} S(f_j, f_k).$$

Adding these two formulas, we obtain the displayed formula of the lemma for the case s, and the induction is complete. $\square$

Theorem 8.23. Let $\{g_1, \ldots, g_s\}$ be a set of generators of a nonzero ideal I of $K[X_1, \ldots, X_n]$, and assume that $g_i \neq 0$ for all i. Then the following conditions on $\{g_1, \ldots, g_s\}$ are equivalent:
  (a) $\{g_1, \ldots, g_s\}$ is a Gröbner basis of I,
  (b) for each pair (g_j, g_k) with $S(g_j, g_k) \neq 0$, every expansion of $S(g_j, g_k)$ as $S(g_j, g_k) = \sum_{i=1}^{s} a_{ijk} g_i + r$ with the two properties that
      (i) $\text{LM}(a_{ijk} g_i) \leq \text{LM}(S(g_j, g_k))$ and
      (ii) no monomial appearing in r with nonzero coefficient is divisible by $\text{LM}(g_j)$ for any j
    has $r = 0$,
  (c) for each pair (g_j, g_k) with $S(g_j, g_k) \neq 0$, there is an expansion of the form $S(g_j, g_k) = \sum_{i=1}^{s} a_{ijk} g_i$ with $\text{LM}(a_{ijk} g_i) \leq \text{LM}(S(g_j, g_k))$.

REMARKS. Because of the equivalence of (b) and (c), the generalized division algorithm (Proposition 8.20) gives us a procedure for testing whether these conditions are satisfied by $\{g_1, \ldots, g_s\}$. Namely we follow through the steps in the proof of Proposition 8.20 in whatever fashion we please for each nonzero

$S(g_j, g_k)$. If we get remainder $r = 0$ for each pair (j, k), then the conditions are satisfied. If we get a nonzero remainder r for some pair, then the conditions are not satisfied. In view of the equivalence of (a) with these conditions, we have an effective (though somewhat tedious) way of checking whether $\{g_1, \ldots, g_s\}$ is a Gröbner basis.

PROOF. We prove that (a) implies (b) and that (c) implies (a). Since (b) certainly implies (c), the proof will be complete.

Let (a) hold, i.e., let $\{g_1, \ldots, g_s\}$ be a Gröbner basis. If $S(g_j, g_k) \neq 0$, then $S(g_j, g_k)$ is a nonzero member of I because each g_i lies in I, and $S(g_j, g_k)$ consequently has an expansion as $\sum_{i=1}^{s} a_i g_i + r$ with $r = 0$. By Corollary 8.21 it has a possibly different expansion with $r = 0$ and with $\mathrm{LM}(a_i g_i) \leq \mathrm{LM}(S(g_j, g_k))$ for each i. On the other hand, in any expansion of $S(g_j, g_k)$ as $\sum_{i=1}^{s} a_i g_i + r$ such that (ii) holds, whether or not $\mathrm{LM}(a_i g_i) \leq \mathrm{LM}(S(g_j, g_k))$, r must be 0 by Corollary 8.21. This proves (b).

To prove that (c) implies (a), we argue by contradiction. Among all expansions of members of I as $\sum_{i=1}^{s} b_i g_i$ such that $\mathrm{LT}\left(\sum_{i=1}^{s} b_i g_i\right)$ is not in the ideal $\left(\mathrm{LT}(g_1), \ldots, \mathrm{LT}(g_s)\right)$, choose one for which

$$M = \max_{1 \leq i \leq s} \mathrm{LM}(b_i g_i)$$

is as small as possible; this choice exists by Proposition 8.16. For this choice, let

$$f = \sum_{i=1}^{s} b_i g_i. \qquad (*)$$

Define $M_i = \mathrm{LM}(b_i)$ for each i with $b_i \neq 0$. If i_0 is an index with $M = \mathrm{LM}(b_{i_0} g_{i_0})$, then $M = M_{i_0} \mathrm{LM}(g_{i_0})$ by Proposition 8.18, and hence M lies in $\left(\mathrm{LT}(g_1), \ldots, \mathrm{LT}(g_s)\right)$. Since $\mathrm{LT}\left(\sum_{i=1}^{s} b_i g_i\right)$ is not in $\left(\mathrm{LT}(g_1), \ldots, \mathrm{LT}(g_s)\right)$, it follows that $\mathrm{LT}\left(\sum_i b_i g_i\right) < M$. Within the set $\{1, \ldots, s\}$, define a subset E to consist of those i for which $M_i \mathrm{LM}(g_i) = M$. This set contains i_0, and it has the property that all i not in E have $\mathrm{LM}(b_i g_i) < M$. We regroup f as

$$f = \sum_{i \in E} b_i g_i + \sum_{i \notin E} b_i g_i = \sum_{i \in E} \mathrm{LC}(b_i) M_i g_i + \sum_{i \in E} (b_i - \mathrm{LT}(b_i)) g_i + \sum_{i \notin E} b_i g_i.$$

Every term in the second and third sums on the right side has leading monomial $< M$, and so does f. Therefore $\mathrm{LM}\left(\sum_{i \in E} \mathrm{LC}(b_i) M_i g_i\right) < M$. It follows that the expression $\sum_{i \in E} \mathrm{LC}(b_i) M_i g_i$ is of the form considered in Lemma 8.22 with $c_i = \mathrm{LC}(b_i)$ for $i \in E$ (and $c_i = 0$ for $i \notin E$). The lemma tells us that

$$\sum_{i \in E} \mathrm{LC}(b_i) M_i g_i = \sum_{j,k} d_{jk} (M/L_{jk}) S(g_j, g_k)$$

for suitable scalars d_{jk}, where $L_{jk} = \text{LCM}(\text{LM}(g_j), \text{LM}(g_k))$.

Now we apply the hypothesis (c), expanding each $S(g_j, g_k)$ in some way as $S(g_j, g_k) = \sum_{i=1}^{s} a_{ijk} g_i$ with the a_{ijk} equal to polynomials such that

$$\text{LM}(a_{ijk} g_i) \leq \text{LM}(S(g_j, g_k)). \tag{**}$$

Substituting for $S(g_j, g_k)$, we obtain

$$f = \sum_{i,j,k} d_{jk} (M/L_{jk}) a_{ijk} g_i + \sum_{i \in E} (b_i - \text{LT}(b_i)) g_i + \sum_{i \notin E} b_i g_i. \tag{$\dagger$}$$

We know that every term in the second and third sums on the right side of ($\dagger$) has leading monomial $< M$, and we shall estimate the leading monomial of each term in the first sum. Multiplying the inequality

$$\text{LM}(S(g_j, g_k)) < \text{LCM}(\text{LM}(g_j), \text{LM}(g_k)) = L_{jk}$$

by the monomial M/L_{jk} yields

$$(M/L_{jk}) \text{LM}(S(g_j, g_k)) < M \tag{$\dagger\dagger$}$$

for every pair (j, k). Combining (**) and ($\dagger\dagger$) gives

$$\text{LM}\big((M/L_{jk}) a_{ijk} g_i\big) = (M/L_{jk}) \text{LM}(a_{ijk} g_i) \leq (M/L_{jk}) \text{LM}(S(g_j, g_k)) < M.$$

Since each d_{jk} is a scalar, every term in the first sum on the right side of ($\dagger$) has leading monomial $< M$. Thus ($\dagger$) is an expansion of a member of I that contradicts the minimality of $\max_i \text{LM}(b_i g_i)$ in the expansion (*). From this contradiction we conclude that (a) holds. $\square$

EXAMPLE OF A VERIFICATION THAT A SET IS A GRÖBNER BASIS. This example continues Example 2 of "Examples with lexicographic ordering" in the previous section. A nonzero ideal I is generated by members of $K[X_1, \ldots, X_n]_1$ of the form $(L_1, \ldots, L_s)$, where each L_j is a linear combination of $X_1, \ldots, X_n$. After initial manipulations we assume that the matrix of coefficients of $L_1, \ldots, L_s$ is in reduced row-echelon form. The assertion is that $\{L_1, \ldots, L_s\}$ is then a Gröbner basis of I. To prove this, we write $L_j = X_{n_j} + l_j$, where X_{n_j} is the associated corner variable and l_j is a linear combination of $X_{n_j+1}, \ldots, X_n$ such that the coefficient of each corner variable is 0. If $j < k$, then

$$S(L_j, L_k) = -l_k X_{n_j} + l_j X_{n_k} = -l_k (X_{n_j} + l_j) + l_j (X_{n_k} + l_k) = -l_k L_j + l_j L_k.$$

The second term on the right side contains no variable $X_1, \ldots, X_{n_j}$, but the first term on the right side contains X_{n_j}. Therefore, relative to the lexicographic ordering, we have $\text{LM}\big(S(L_j, L_k)\big) = \text{LM}(-l_k L_j) = \text{LM}(l_k) X_{n_j}$. Consequently $\text{LM}(l_j L_k) \leq \text{LM}\big(S(L_j, L_k)\big)$ (and actually strict inequality must hold). Thus the displayed formula shows that $S(L_j, L_k) = a_1 L_j + a_2 L_k$ in the form demanded by (c) of Theorem 8.23. Since (c) implies (a) in the theorem, $\{L_1, \ldots, L_s\}$ is a Gröbner basis of I.

Corollary 8.24 (Buchberger's algorithm).[15] Each nonzero ideal in the polynomial ring $K[X_1, \ldots, X_n]$ has a Gröbner basis. Such a basis can be obtained by the following procedure: Start from any set $\{f_1, \ldots, f_t\}$ of nonzero generators, apply the generalized division algorithm in some fashion to each $S(f_j, f_k)$ and to the generating set $\{f_1, \ldots, f_t\}$, and adjoin to the set of generators any nonzero remainders obtained from this process. Iterate this process for enlarging a set $\{f'_1, \ldots, f'_{t'}\}$ of generators as long as a nonzero remainder is obtained for some $S(f'_j, f'_k)$. This process must terminate at some point with all remainders equal to 0, and the resulting generating set is a Gröbner basis.

PROOF. At the stage of the iteration that works with the set $\{f'_1, \ldots, f'_{t'}\}$ of generators, any nonzero remainder r that arises has the property that no monomial occurring in r is divisible by any $\text{LM}(f'_j)$. By Lemma 8.17, $\text{LT}(r)$ is not a member of $(\text{LT}(f'_1), \ldots, \text{LT}(f'_t))$. However, at the next stage when r has been designated as one of the generators of I, $\text{LT}(r)$ has become one of the generators of this ideal. Therefore the ideal $(\text{LT}(f'_1), \ldots, \text{LT}(f'_t))$ strictly increases as we pass from one stage to the next. Since $K[X_1, \ldots, X_n]$ is Noetherian, its ideals satisfy the ascending chain condition, and this chain of ideals must stabilize. Consequently all the remainders must be 0 at some point, and then Theorem 8.23 shows that the set of generators is a Gröbner basis. □

EXAMPLE OF THE COMPUTATION OF A GRÖBNER BASIS. We return to Example 3 of "Examples with lexicographic ordering" in the previous section. In $K[X, Y]$, we let $f_1(X, Y) = X^2 + 2XY^2$ and $f_2(X, Y) = XY + 2Y^3 - 1$, and we define $I = (f_1, f_2)$. We seek a Gröbner basis of I, using the lexicographic ordering. Direct computation gives $S(f_1, f_2) = Y(X^2 + 2XY^2) - X(XY + 2Y^3 - 1) = X$. Since X is not divisible by $\text{LM}(f_1)$ or by $\text{LM}(f_2)$, $S(f_1, f_2) = 0f_1 + 0f_2 + X$ is an expansion of $S(f_1, f_2)$ as in Theorem 8.23c with $r = X$. The procedure of Corollary 8.24 says to adjoin $f_3 = X$ to the generating set and test again. Direct computation gives $S(f_1, f_3) = 1(X^2 + 2XY^2) - X \cdot X = 2XY$, and $S(f_1, f_3) = 0f_1 + 0F_2 + (2Y)f_3 + 0$ is an expansion of $S(f_1, f_3)$ as in (c), since $\text{LM}(2Yf_3) \leq \text{LM}(S(f_1, f_3))$. Thus $S(f_1, f_3)$ gives us a 0 remainder, hence nothing new to process. In addition, we have $S(f_2, f_3) = 1(XY + 2Y^3 - 1) - Y \cdot X = 2Y^3 - 1$. No term of this is divisible by any of the leading monomials of f_1, f_2, f_3, namely X^2, XY, X. Hence $2Y^3 - 1$ is a nonzero remainder.[16] Therefore we are to adjoin $f_4 = 2Y^3 - 1$ to our set. Computation gives $S(f_1, f_4) = 2XY^4 + X^2 = (2Y^4 + X)f_3$, $S(f_2, f_4) = 2Y^5 - Y^2 + \frac{1}{2}X = \frac{1}{2}f_3 + Y^2 f_4$,

---

[15]Computer programs typically use an improved version of this algorithm to compute Gröbner bases.

[16]It was not a bad choice of decomposition that led to a nonzero remainder when some other decomposition might have given us 0; the equivalence of (b) and (c) in Theorem 8.23 assures us of that fact.

and $S(f_3, f_4) = \frac{1}{2}X = \frac{1}{2}f_3$. In every case each term has leading monomial at most the leading monomial of the S-polynomial. Hence all remainders are 0, and Corollary 8.24 says that $\{f_1, f_2, f_3, f_4\}$ is a Gröbner basis of I.

Corollary 8.25 (solution of the ideal-membership problem). If I is a nonzero ideal in $K[X_1, \ldots, X_n]$ and f is a polynomial, then a procedure for deciding whether f lies in I is as follows: introduce a monomial ordering, construct a Gröbner basis $\{g_1, \ldots, g_s\}$ of I by means of Corollary 8.24, and apply the generalized division algorithm to write $f = \sum_{i=1}^{s} a_i g_i + r$ for polynomials $a_1, \ldots, a_r, r$ such that no monomial appearing in r with nonzero coefficient is divisible by $\text{LM}(g_j)$ for any j. Then f lies in I if and only if $r = 0$.

PROOF. Corollary 8.24 produces the Gröbner basis, and Corollary 8.21 affirms that this procedure decides whether f lies in I. □

Corollary 8.26 (solution of the proper-ideal problem). If I is a nonzero ideal in $K[X_1, \ldots, X_n]$, then a procedure for deciding whether $I = K[X_1, \ldots, X_n]$ is to compute a Gröbner basis for I and to see whether one of its members is a nonzero scalar c.

PROOF. If I has a nonzero scalar as one of its generators, then 1 lies in I, and hence I certainly equals $K[X_1, \ldots, X_n]$. Conversely if I is given, then Corollary 8.24 produces a Gröbner basis $\{g_1, \ldots, g_s\}$. Since $\text{LT}(1) = 1$ and since $\text{LT}(I) = (\text{LT}(g_1), \ldots, \text{LT}(g_s))$, the monomial 1 must lie in $(\text{LT}(g_1), \ldots, \text{LT}(g_s))$. Since 1 is a monomial, Lemma 8.17 shows that it must be divisible by $\text{LM}(g_j)$ for some j. Therefore $\text{LM}(g_j) = 1$. Since 1 is the smallest monomial in any monomial ordering, it is the only monomial appearing with a nonzero coefficient in g_j. Therefore g_j is a nonzero scalar. □

In many applications of Gröbner bases, there is some flexibility in what monomial ordering to impose in obtaining the Gröbner basis. In Corollaries 8.25 and 8.26, for example, absolutely any monomial ordering works fine. The actual calculation of Gröbner bases is often computationally demanding, and thus it is worthwhile to use such a basis that takes relatively little time to compute. According to computer scientists,[17] Gröbner bases are the most widely useful when computed relative to the lexicographic ordering, but they are then also the most time-consuming to compute. The monomial orderings that make the computation of Gröbner bases proceed quickly tend to be ones that first bound

---

[17]The Web essay "Representation and monomial orders," http://www.umich.edu/~gpcc/scs/magma/text835.htm, within the publication of the Statistics and Computation Service listed in the Selected References contains a discussion of various monomial orders and their uses and advantages.

the total degree in one or two steps. One of the reasons that this kind of monomial ordering works so efficiently is that once the total degree is bounded, there are only finitely many monomials less than any given monomial M.

9. Uniqueness of Reduced Gröbner Bases

In this section, K continues to denote a field, and we work with a fixed monomial ordering on $K[X_1, \ldots, X_n]$. Ideals in $K[X_1, \ldots, X_n]$ will always be specified by giving finite sets of generators. Our objective in this section is to show how any Gröbner basis can be "reduced" and that a "reduced" Gröbner basis for an ideal is unique. A by-product of the uniqueness argument will be a way of testing two ideals for equality.

Any finite set of generators of I that contains a Gröbner basis is again a Gröbner basis. Thus a constructed Gröbner basis will often be unnecessarily large. One simple kind of redundance is addressed by Lemma 8.27 below.

Lemma 8.27. If $\{g_1, \ldots, g_s\}$ is a Gröbner basis for a nonzero ideal I in $K[X_1, \ldots, X_n]$ and if $\text{LM}(g_1)$ lies in the ideal $\big(\text{LT}(g_2), \ldots, \text{LT}(g_s)\big)$, then $\{g_2, \ldots, g_s\}$ is a Gröbner basis of I.

REMARK. Lemma 8.17 shows how to check whether $\text{LM}(g_1)$ lies in the ideal $\big(\text{LT}(g_2), \ldots, \text{LT}(g_s)\big)$; all we have to do is see whether some $\text{LM}(g_j)$ for $j \geq 1$ divides $\text{LM}(g_1)$.

PROOF. By hypothesis, $\big(\text{LT}(g_2), \ldots, \text{LT}(g_s)\big) = \big(\text{LT}(g_1), \ldots, \text{LT}(g_s)\big) = \text{LT}(I)$. Therefore $\{g_2, \ldots, g_s\}$ is a Gröbner basis of I. (Recall that the definition of Gröbner basis does not assume that the set generates the ideal; Proposition 8.19 deduces that it generates.) □

A Gröbner basis $\{g_1, \ldots, g_s\}$ of a nonzero ideal I is said to be **minimal** if $\text{LC}(g_j) = 1$ for all j and if no $\text{LM}(g_i)$ is divisible by $\text{LM}(g_j)$ for some $j \neq i$. Lemma 8.27 shows that in trying to transform a Gröbner basis into a form for which a uniqueness result will apply, there is no loss of generality in assuming that the given Gröbner basis is minimal.

EXAMPLE. As in the example following Corollary 8.24, let I be the ideal in $K[X, Y]$ given by $I = (f_1, f_2)$ with $f_1(X, Y) = X^2 + 2XY^2$ and $f_2(X, Y) = XY + 2Y^3 - 1$. Then we saw that $\{f_1, f_2, f_3, f_4\}$ is a Gröbner basis of I in the lexicographic ordering, where $f_3(X, Y) = X$ and $f_4(X, Y) = 2Y^3 - 1$. The leading monomials are $\text{LM}(f_1) = X^2$, $\text{LM}(f_2) = XY$, $\text{LM}(f_3) = X$, and $\text{LM}(f_4) = Y^3$. The first two are divisible by the third. Therefore $\{X, Y^3 - \frac{1}{2}\}$ is the corresponding minimal Gröbner basis.

9. Uniqueness of Reduced Gröbner Bases

Unfortunately an ideal can have more than one minimal Gröbner basis, as is shown in Problem 17 at the end of the chapter. A Gröbner basis $\{g_1, \ldots, g_s\}$ of an ideal I is said to be **reduced** if it is minimal and if for each i, no monomial appearing in g_i with nonzero coefficient is divisible by $\text{LM}(g_j)$ for some $j \neq i$.

Theorem 8.28 (uniqueness of reduced Gröbner basis). If I is a nonzero ideal in $K[X_1, \ldots, X_n]$, then I has a unique reduced Gröbner basis, and this can be obtained algorithmically starting from any minimal Gröbner basis.

PROOF OF UNIQUENESS. Let $\{g_1, \ldots, g_s\}$ be any Gröbner basis. Since $\text{LT}(I) = \big(\text{LT}(g_1), \ldots, \text{LT}(g_s)\big)$, Lemma 8.17 shows that any $\text{LM}(f)$ for $f \in I$ is divisible by $\text{LM}(g_j)$ for some j. If $\{h_1, \ldots, h_t\}$ is a second Gröbner basis, then this argument shows that each $\text{LM}(h_i)$ is divisible by some $\text{LM}(g_j)$. Turned around, the argument shows that $\text{LM}(g_j)$ is divisible by some $\text{LM}(h_k)$. Since $\{h_1, \ldots, h_t\}$ is assumed minimal, $\text{LM}(h_k)$ cannot be divisible by $\text{LM}(h_i)$ if $i \neq k$. Thus $\text{LM}(h_i) = \text{LM}(h_k)$, and these equal $\text{LM}(g_j)$. Then it follows that $s = t$ and that we may enumerate any two minimal Gröbner bases in such a way that the leading monomial of the ith member of each basis is the same for each i with $1 \leq i \leq s$.

With this normalization in place, let us show that $g_i = h_i$. To do so, we expand $g_i - h_i$ as $g_i - h_i = \sum_{j=1}^{s} a_j h_j$ with $\text{LM}(g_i - h_i) = \max_j \text{LM}(a_j h_j)$ in accordance with (b) of Theorem 8.23. Choose k such that the maximum on the right side is attained at k, i.e., such that

$$\text{LM}(a_k) \text{LM}(h_k) = \text{LM}(g_i - h_i). \qquad (*)$$

Arguing by contradiction, suppose that the right side of $(*)$ is nonzero. Then it must be a monomial occurring in either g_i or h_i. Since the two Gröbner bases are reduced, no monomial occurring in g_i is divisible by $\text{LM}(g_k) = \text{LM}(h_k)$ if $k \neq i$, and similarly for monomials occurring in h_i. We conclude that $k = i$ and that $\text{LM}(h_i) = \text{LM}(g_i - h_i)$. But this is impossible by Proposition 8.18 if $g_i - h_i \neq 0$, since $\text{LM}(g_i) = \text{LM}(h_i)$ and $\text{LC}(g_i) = \text{LC}(h_i) = 1$. Therefore the right side of $(*)$ is 0, and $g_i = h_i$. □

PROOF OF EXISTENCE. Let $\{g_1, \ldots, g_s\}$ be a minimal Gröbner basis of I. As was shown in the proof of uniqueness, the leading monomials $\text{LM}(g_1), \ldots, \text{LM}(g_s)$ are independent of the choice of the actual minimal basis. Looking at the definition of "reduced," we see therefore that the property of being reduced is a property of each member g_i of the basis separately. That is, it is meaningful to say that g_i is reduced if no monomial appearing in g_i with nonzero coefficient is divisible by $\text{LM}(g_j)$ for some $j \neq i$. We shall show how to replace g_i by an element g'_i with the same leading monomial in such a way that the new set is still a Gröbner basis and g'_i is reduced, and then the proof will be complete. There is no loss of generality in taking $i = 1$.

Applying the generalized division algorithm (Proposition 8.20), we write

$$g_1 = \sum_{j=2}^{s} a_j g_j + r \qquad (**)$$

in such a way that

$$\text{LM}(g_1) = \max_{2 \leq j \leq s} \text{LM}(a_j g_j) \qquad (\dagger)$$

and that no monomial appearing in r with nonzero coefficient is divisible by $\text{LM}(g_j)$ for any $j \geq 2$. If we define g_1' to be this element r, then the element g_1' is reduced in the above sense, and the only question is whether $\{g_1', g_2, \ldots, g_s\}$ is a Gröbner basis. Since $\{g_1, \ldots, g_s\}$ is minimal, $\text{LM}(g_1)$ is not divisible by any $\text{LM}(g_j)$ for $j \geq 2$. Consequently $\text{LM}(g_1)$ appears with nonzero coefficient on the left side of $(**)$, and it does not appear in any of the terms $a_j g_j$ with nonzero coefficient on the right side. Consequently it appears in $r = g_1'$, and $\text{LM}(g_1) \leq \text{LM}(g_1')$. On the other hand, the equality $(\dagger)$ implies that $\text{LM}(g_1') \leq \text{LM}(g_1)$. Therefore $\text{LM}(g_1) = \text{LM}(g_1')$, and $\text{LT}(I) = \big(\text{LT}(g_1), \text{LT}(g_2) \ldots, \text{LT}(g_s)\big) = \big(\text{LT}(g_1'), \text{LT}(g_2) \ldots, \text{LT}(g_s)\big)$. Consequently $\{g_1', g_2, \ldots, g_s\}$ is a Gröbner basis by definition. □

Corollary 8.29 (solution of the ideal-equality problem). Let I and J be two nonzero ideals in $K[X_1, \ldots, X_n]$ specified in terms of finite sets of generators. Then $I = J$ if and only if the reduced Gröbner bases of I and J relative to a single monomial ordering are the same.

REMARK. As with the solution of problems listed in Corollaries 8.25 and 8.26, the desired end is independent of the monomial ordering, and in practice one might just as well start from a monomial ordering for which the computation of Gröbner bases is relatively easy.

PROOF. This result is immediate from Corollary 8.24 (constructive existence of Gröbner bases) and Theorem 8.28. □

10. Simultaneous Systems of Polynomial Equations

In this section we combine our techniques concerning the resultant and Gröbner bases to attack the original problem discussed in Section 1, that of solving systems of simultaneous polynomial equations in several variables. Our interest ultimately will be in the case that the underlying field is algebraically closed.

Corollary 8.26 and the Nullstellensatz already combine to give a criterion for such a system to have no solutions: We regard the system as the zero locus of an ideal, and we calculate a Gröbner basis for the ideal. Then the system has no solutions if and only if the Gröbner basis contains a constant polynomial, i.e., if and only if the reduced Gröbner basis is $\{1\}$.

Let us now consider the problem of finding the solutions when solutions exist. We begin with the case of two equations in two unknowns over the field $\mathbb{C}$, recalling what we know from the theory of the resultant. Consider the system

$$X^2Y + Y^2 = 5,$$
$$XY = 2.$$

Set $f(X, Y) = X^2Y + Y^2 - 5$ and $g(X, Y) = XY - 2$. To find points (x, y) with $f(x, y) = g(x, y) = 0$, using the style of Sections 1–3, we compute the resultant of f and g in the X variable, say, and obtain the polynomial $Y^4 - 5Y^2 + 4Y$. Setting this equal to 0 gives us $y = 0$, $y = 1$, and $y = \frac{1}{2}(-1 \pm \sqrt{17})$. We can then substitute each such y into $x^2y + y^2 = 5$ and get candidates (x, y). Doing so for $y = 0$ gives us no candidates, and doing so for each of the other three values of y gives us two values of x, differing only in a sign. So we get six pairs (x, y). However, only three of these satisfy the second given equation, $xy = 2$, one for each nonzero value of y. Thus the resultant gives us a handle on the problem of finding solutions, but it has two shortcomings: it produced a value of y yielding no solution pairs (x, y), and it produced extraneous x values.

To find points (x, y) with $f(x, y) = g(x, y) = 0$, using the style of Sections 7–10, we consider (f, g) as an ideal in $\mathbb{C}[X, Y]$, and we are interested in the locus of common zeros $V_{\mathbb{C}}((f, g))$ of the ideal. We start by finding a reduced Gröbner basis with respect to a suitable ordering. The usual lexicographic ordering will do fine here, and the result is $\{X + \frac{1}{2}Y^2 - \frac{5}{2}, Y^3 - 5Y + 4\}$. By what may seem to be good fortune, the second element depends on Y alone, and the roots are $y = 1$ and $y = \frac{1}{2}(-1 \pm \sqrt{17})$. If we substitute these values into the equation $x + \frac{1}{2}y^2 - \frac{5}{2} = 0$, we get one value of x for each y. We can solve because the coefficient 1 of x is nonzero for each y in question. No pair (x, y) that we obtain is superfluous because the locus of common zeros of f and g is identical with the locus of common zeros of the members of the Gröbner basis.

This approach raises several questions about a possible generalization:

(i) Under what conditions can we expect that a Gröbner basis for an ideal I in $K[X, Y]$ will contain a member that depends just on Y?
(ii) If the Gröbner basis contains no element that depends just on Y, then what can we expect?
(iii) If we are able to solve for values of y, under what conditions can we use the remaining member(s) of the Gröbner basis to solve for x?

Part of the answer to (i) is contained in the Elimination Theorem proved as Theorem 8.30 below. This theorem says for the lexicographic ordering that the members of a Gröbner basis that depend just on Y generate $I \cap K[Y]$; in fact, they form a Gröbner basis of this ideal of $K[Y]$. For the case that $I = (f, g)$, the resultant is a member of $I \cap K[Y]$. Thus a nonzero resultant ensures that some member of the Gröbner basis will depend just on Y; on the other hand, $I \cap K[Y]$ has to be a principal ideal in $K[Y]$, and any Gröbner basis of that principal ideal has to contain the ideal's generator (up to a scalar factor). By contrast, a zero resultant leads us to question (ii) because it says, by Theorem 8.1, that f and g have a common factor $h(X, Y)$ of positive degree in X as long as both f and g have positive degree in X. The largest power of X in h has as coefficient a polynomial in Y that has only finitely many roots, and if K is algebraically closed, then every y unequal to one of these roots will produce an x such that $h(x, y) = 0$ and therefore such that $f(x, y) = g(x, y) = 0$. In other words, except in degenerate cases a zero resultant implies that there cannot be a member of the Gröbner basis that depends just on Y. Finally the answer to (iii) lies deeper and is contained in the Extension Theorem, which is proved as Theorem 8.31 below.

Let I be a nonzero ideal in $K[X_1, \ldots, X_n]$, K being any field for now. If $0 \leq k \leq n - 1$, then the kth **elimination ideal** of I is the ideal $I \cap K[X_{k+1}, \ldots, X_n]$ in $K[X_{k+1}, \ldots, X_n]$. A monomial ordering on $K[X_1, \ldots, X_n]$ will be said to be of k-**elimination type** if any monomial containing any of $X_1, \ldots, X_k$ to a positive power is greater than any monomial in $X_{k+1}, \ldots, X_n$ alone. The usual lexicographic ordering is of k-elimination type for every k. An example of a monomial ordering of k-elimination type that is of great interest in applications is the one of Bayer–Stillman described in Example 4 of monomial orderings in Section 7.

Theorem 8.30 (Elimination Theorem). Let K be any field, let I be a nonzero ideal in $K[X_1, \ldots, X_n]$, let $0 \leq k \leq n$, and fix a monomial ordering of k-elimination type. If $\{g_1, \ldots, g_s\}$ is a Gröbner basis of I, then the subset of members of $\{g_1, \ldots, g_s\}$ depending only on $X_{k+1}, \ldots, X_n$ is a Gröbner basis of the kth elimination ideal $J = I \cap K[X_{k+1}, \ldots, X_n]$.

PROOF. Relabeling the members of $\{g_1, \ldots, g_s\}$, we may assume that the g_j's lying in J are $g_1, \ldots, g_t$. The first step is to show that $J = (g_1, \ldots, g_t)$. If $f \in J$ is given, we apply the generalized division algorithm (Proposition 8.20) and write $f = \sum_{i=1}^{s} a_i g_i + r$ with $\text{LM}(a_i g_i) \leq \text{LM}(f)$ for all i and with no monomial appearing in r with nonzero coefficient divisible by $\text{LM}(g_j)$ for any j. Corollary 8.21 shows that $r = 0$. If $a_i \neq 0$ and i is not $\leq t$, then $\text{LM}(a_i g_i)$ involves at least one of $X_1, \ldots, X_k$, and the definition of monomial ordering of k-elimination type implies that $\text{LM}(a_i f_i) > \text{LM}(f)$. It follows that $a_i = 0$ for $i > t$, and thus $J = (g_1, \ldots, g_t)$.

To see that $\{g_1, \ldots, g_t\}$ is a Gröbner basis of J, we apply Theorem 8.23. We are to show for each pair (g_j, g_k) with $S(g_j, g_k) \neq 0$ and $\{j, k\} \subseteq \{1, \ldots, t\}$ that there is an expansion $S(g_j, g_k) = \sum_{i=1}^{t} a_i g_i$ with $\text{LM}(a_i g_i) \leq \text{LM}(S(g_j, g_k))$. In view of the argument with f in the previous paragraph, it is enough to show that $S(g_j, g_k)$ lies in J. The formula is

$$S(g_j, g_k) = \frac{\text{LCM}(\text{LM}(g_j), \text{LM}(g_k))}{\text{LT}(g_j)} g_j - \frac{\text{LCM}(\text{LM}(g_k), \text{LM}(g_k))}{\text{LT}(g_k)} g_k.$$

The coefficient fractions are members of $K[X_{k+1}, \ldots, X_n]$, since the monomial ordering is of k-elimination type, and thus $S(g_j, g_k)$ is indeed in J. □

EXAMPLE. Formula for discriminant of a polynomial in one variable. This example is one that we have addressed before by specialized methods. We include it anyway because the use of Gröbner bases allows one to solve many similar problems that the specialized methods do not address. By way of illustration, let $(X - r)(X - s)(X - t)$ be a cubic polynomial. The discriminant is $D = (r - s)^2 (s - t)^2 (r - t)^2$. This is a polynomial that is symmetric in r, s, t, and the general theory of symmetric polynomials (in the problems for Chapter VIII in *Basic Algebra*) shows that it has to be a polynomial in the elementary symmetric polynomials $a = r + s + t, b = rs + rt + st, c = rst$. We seek a formula for D in terms of a, b, c. We form the ideal I in $K[r, s, t, D, a, b, c]$ given by

$$I = \left(D - (r-s)^2 (s-t)^2 (r-t)^2, a - (r+s+t), b - (rs+rt+st), c - rst \right).$$

With the variables enumerated as r, s, t, D, a, b, c, we use any monomial ordering of 4-elimination type, the lexicographic ordering for example, and form the reduced Gröbner basis of I. Calculation best done with the aid of a computer gives $D - a^2 b^2 + 4b^3 + 4a^3 c - 18abc + 27c^2$ and three other members of I that involve r, s, or t. Theorem 8.30 shows that the 4th elimination ideal is principal with generator $D - a^2 b^2 + 4b^3 + 4a^3 c - 18abc + 27c^2$. Thus the desired formula is $D = a^2 b^2 - 4b^3 - 4a^3 c + 18abc - 27c^2$.

Let us come to the Extension Theorem. The statement and proof of this theorem do not make use of Gröbner bases, but they do refer to the kth elimination ideal, which is identified explicitly in Theorem 8.30 with the aid of a Gröbner basis. The intention is that the theorem be applied inductively in any application, taking into account one additional variable at each step of an induction.

Theorem 8.31 (Extension Theorem). Let K be an algebraically closed field, let $I = (f_1, \ldots, f_s)$ be an ideal in $K[X_1, \ldots, X_n]$, and let J be the first elimination ideal of I in $K[X_2, \ldots, X_n]$. For each f_i, expand f_i in powers of X_1 as

$$f_i(X_1, \ldots, X_n) = g_i(X_2, \ldots, X_n) X_1^{l_i} + \text{(lower powers of } X_1)$$

with g_i in $K[X_2, \ldots, X_n]$ and g_i nonzero unless $f_i = 0$. Suppose that $(c_2, \ldots, c_n)$ lies in the zero locus $V_K(J) \subseteq K^{n-1}$. If $g_i(c_2, \ldots, c_n) \neq 0$ for some i, then there exists c_1 in K such that $(c_1, \ldots, c_n)$ is in the zero locus $V_K(I) \subseteq K^n$.

Before giving the proof, we need to extend the theory of the resultant slightly in such a way that it applies to s polynomials $f_1, \ldots, f_s$ rather than just to two. To do so, we introduce new indeterminates $U_2, \ldots, U_s$ and regard

$$F = U_2 f_2 + \cdots + U_s f_s$$

as a member of $K[U_2, \ldots, U_s, X_1, \ldots, X_n]$ whose degree $\deg_1 F$ in X_1 is the maximum of the degrees of $f_2, \ldots, f_s$ in X_1. We can then view f_1 as a member of the same polynomial ring $K[U_2, \ldots, U_s, X_1, \ldots, X_n]$ of degree $\deg_1 f_1$ and form the resultant of f_1 and F in the X_1 variable. This is computed as the determinant of some square matrix of size $\deg_1 f_1 + \deg_1 F$, and we are interested only in the case that $\deg_1 f_1 \geq 1$ and $\deg_1 F \geq 1$. When expanded in monomials $U^\alpha = U_2^{\alpha_2} \cdots U_s^{\alpha_s}$, the determinant is of the form

$$R(f_1, F) = \sum_\alpha h_\alpha(X_2, \ldots, X_n) U^\alpha$$

with each h_α in $K[X_2, \ldots, X_n]$. The polynomials h_α will be called the **generalized resultants** in the X_1 variable of the ordered pair $(f_1, \{f_2, \ldots, f_s\})$.

PROOF OF THEOREM 8.31. Let us abbreviate $\overline{X} = (X_2, \ldots, X_n)$ and $\bar{c} = (c_2, \ldots, c_n)$; we shall write

$$(X_1, \overline{X}) = (X_1, \ldots, X_n) \quad \text{and} \quad (X_1, \bar{c}) = (X_1, c_2, \ldots, c_n).$$

We seek $c_1 \in K$ with $f_j(c_1, c) = 0$ for all j. The assumption is that $g_i(\bar{c}) \neq 0$ for some i, and we may as well assume that this i is $i = 1$. If $\deg_1 f_1 = 0$, then f_1 is in J, and the conditions that $f_1 = 0$ on $V_K(J)$ and that $g_1(\bar{c}) \neq 0$ contradict one another; hence $\deg_1 f_1 \geq 1$.

As in the paragraph before the proof, put $F = U_2 f_2 + \cdots + U_s f_s$. If $\deg_1 F = 0$, then f_j is independent of X_1 for all $j \geq 2$, and hence f_j is in J for $j \geq 2$. In this case it is enough to find c_1 with $f_1(c_1, \bar{c}) = 0$. Since $g_1(\bar{c}) \neq 0$, $f_1(X_1, \bar{c})$ is a one-variable polynomial of degree $l_1 \geq 1$, and it is 0 for some value c_1. Thus the proof is complete if $\deg_1 F = 0$.

We may therefore assume that $\deg_1 F \geq 1$. Form the resultant in X_1 given by

$$R(f_1, F) = \sum_\alpha h_\alpha(\overline{X}) U^\alpha,$$

where the h_α's are the generalized resultants mentioned above. The main step is to prove that each h_α lies in the first elimination ideal J. Since h_α depends only on $\overline{X}$, it is enough to prove that each h_α is in I. We have arranged that each of f_1

and F has positive degree and has nonzero leading coefficient in X_1, and hence Theorem 8.1 shows that

$$af_1 + bF = R(f_1, F)$$

for some nonzero polynomials a and b in $K[U_2, \ldots, U_s, X_1, \overline{X}]$. Let the monomial expansions of a and b in terms of the U^α's be $a = \sum_\alpha a_\alpha U^\alpha$ and $b = \sum_\alpha b_\alpha U^\alpha$. Then we have

$$\sum_\alpha a_\alpha f_1 U^\alpha + \left(\sum_\beta b_\beta U^\beta\right)\left(\sum_{i=2}^s f_i U_i\right) = \sum_\alpha h_\alpha U^\alpha. \qquad (*)$$

Let e_i be the multi-index that is 1 in the ith place and 0 elsewhere. This has the property that $U^{e_i} = U_i$ for $2 \leq i \leq s$. We can rewrite $(*)$ as

$$\sum_\alpha h_\alpha U^\alpha = \sum_\alpha a_\alpha f_1 U^\alpha + \sum_\alpha \left(\sum_{\substack{(\beta,i) \text{ with} \\ 2 \leq i \leq s, \\ \beta + e_i = \alpha}} b_\beta f_i\right) U^\alpha.$$

Equating the coefficients of U^α on both sides gives

$$h_\alpha = a_\alpha f_1 + \sum_{\substack{(\beta,i) \text{ with} \\ 2 \leq i \leq s, \\ \beta + e_i = \alpha}} b_\beta f_i$$

and exhibits h_α as in I. Therefore h_α is in the elimination ideal J.

Since $\bar{c}$ lies in $V_K(J)$, $h_\alpha(\bar{c}) = 0$ for all α. Consequently

$$R(f_1, F)(U_2, \ldots, U_s, \bar{c}) = 0.$$

Theorem 8.1 shows that $f_1(X_1, \bar{c})$ and $F(U_2, \ldots, U_s, X_1, \bar{c})$ have a common factor of positive degree in X_1 provided either or both of two specific coefficients are nonzero. These are the coefficients of $X_1^{\deg_1 f_1}$ in $f_1(X_1, \bar{c})$ and of $X_1^{\deg_1 F}$ in $F(U_2, \ldots, U_s, X_1, \bar{c})$. The coefficient of $X_1^{\deg_1 f_1}$ in $f_1(X_1, \overline{X})$ is $g_1(\overline{X})$; thus the coefficient of $X_1^{\deg_1 f_1}$ in $f_1(X_1, \bar{c})$ is $g_1(\bar{c})$ and is nonzero by assumption. Therefore Theorem 8.1 is applicable.

The common factor of $f_1(X_1, \bar{c})$ and $F(U_2, \ldots, U_s, X_1, \bar{c})$ may be taken to be prime, and then it has to be a nonzero scalar multiple of $X_1 - c_1$ for some $c_1 \in K$, since that is the only kind of prime factor that divides $f_1(X_1, \bar{c})$, K being algebraically closed. Thus the element c_1 of K satisfies

$$f_1(c_1, \bar{c}) = 0 \quad \text{and} \quad F(U_2, \ldots, U_s, c_1, \bar{c}) = 0. \qquad (**)$$

Writing out F, we have
$$0 = F(U_2, \ldots, U_s, c_1, \bar{c}) = U_2 f_2(c_1, \bar{c}) + \cdots + U_s f_s(c_1, \bar{c}).$$
This is an identity in $K[U_2, \ldots, U_s]$, and each coefficient must be 0 on the right side. Thus $0 = f_2(c_1, \bar{c}) = \cdots = f_s(c_1, \bar{c})$. Since (**) shows that $f_1(c_1, \bar{c}) = 0$, this proves the theorem. □

11. Problems

1. How many points are in $\mathbb{P}^n_K$ if K is a finite field with q elements?
2. Resolve Cramer's paradox as formulated in Section 1.
3. **(Euler's Theorem)** Prove that if $F(X_1, \ldots, X_n)$ is any homogeneous polynomial of degree d, then $\sum_{j=1}^n X_j \frac{\partial F}{\partial X_j} = dF$.
4. Let A and B be unique factorization domains, and let $\iota : A \to B$ be a one-one homomorphism of commutative rings with identity. For each $h(X)$ in $A[X]$, let $h^\iota(X)$ be the member of $B[X]$ obtained by applying the substitution homomorphism that acts by ι on the coefficients and fixes X. Using resultants, prove that if $f(X)$ and $g(X)$ are two members of $A[X]$ such that $f^\iota(X)$ and $g^\iota(X)$ have a common factor in $B[X]$ that is not in B, then f and g have a common factor in $A[X]$ that is not in A.
5. Theorem 8.1 assumes that at least one of the coefficients f_m and g_n is nonzero. Sometimes this theorem is phrased with the stronger hypothesis that f_m and g_n are both nonzero. By comparing the resultants that are involved, show that all parts of the theorem with at least one of f_m and g_n nonzero are consequences of the theorem with both f_m and g_n nonzero.
6. Let K be an algebraically closed field, let f and g be members of $K[X_1, \ldots, X_n]$ with f irreducible, and suppose that $g(a_1, \ldots, a_n) = 0$ whenever $f(a_1, \ldots, a_n) = 0$. Give two proofs, one using the Nullstellensatz and one using resultants, that f divides g.
7. Factor the member $Y^3 - 2XY^2 + 2X^2Y - 4X^3$ of $\mathbb{C}[X, Y]_3$ into first-degree factors.
8. Find the intersections in $\mathbb{P}^2_\mathbb{C}$ of the zero loci of the projective plane curves $F(X, Y, W) = X(Y^2 - XW)^2 - Y^5$ and $G(X, Y, W) = Y^4 + Y^3W - X^2W^2$.
9. Let A be a unique factorization domain, let $B = A[Y_1, \ldots, Y_m, Z_1, \ldots, Z_n]$, let F and G be the polynomials in $B[X]$ given by
$$F(X) = \prod_{i=1}^m (X - Y_i) \quad \text{and} \quad G(X) = \prod_{j=1}^n (X - Z_j),$$
and let $R(Y_1, \ldots, Y_m, Z_1, \ldots, Z_n)$ be the resultant $R(F, G)$ with respect to X.
   (a) Show that $R(Y_1, \ldots, Y_m, Z_1, \ldots, Z_n)$ equals 0 if Y_i is set equal to Z_j.

(b) Deduce from (a) that $Y_i - Z_j$ divides $R(Y_1, \ldots, Y_m, Z_1, \ldots, Z_n)$.

(c) Deduce from (b) that $R(Y_1, \ldots, Y_m, Z_1, \ldots, Z_n) = c \prod_{i,j} (Y_i - Z_j)$ for some $c \neq 0$ in A depending on m and n.

10. Let $f(X)$ be in $K[X]$, K being a field, and let $f'(X)$ be the derivative of $f(X)$. Using the result of the previous problem and the computation at the beginning of Section V.4, prove that $R(f, f')$ is a nonzero multiple of the discriminant of f, the multiple depending only on deg f.

11. Let F and G be the homogeneous polynomials given by $F(X, Y, W) = (X^2 + Y^2)^2 + 3X^2YW - Y^3W$ and $G(X, Y, W) = (X^2 + Y^2)^3 - 4X^2Y^2W^2$. Calculate $I(P, F \cap G)$ for $P = [0, 0, 1]$.

12. Let G be a nonconstant homogeneous polynomial in $K[X, Y, W]_d$ vanishing at a point P of $\mathbb{P}_K^2$, let $m = m_P(G)$ be the order of vanishing of G at P, and let L be a projective line through P. Show from the definitions that L is a tangent line to G at P in the sense of Section 5 if and only if $i(P, L \cap G) \geq m + 1$ in the sense of Section 4.

13. Deduce relative to an arbitrary monomial ordering the (nonconstructive) existence of a Gröbner basis for a nonzero ideal I in $K[X_1, \ldots, X_n]$ from the form of a set of generators of the ideal $\mathrm{LT}(I)$.

14. For $1 \leq i \leq n$, let $w^{(i)}$ be the weight vector $w^{(i)} = (w_1^{(i)}, \ldots, w_n^{(i)})$ in $\mathbb{R}^n$, and suppose that these vectors are linearly independent. Show that the $w^{(i)}$ define a monomial ordering as in Example 5 of Section 7 if and only if for each j, the first i with $w_j^{(i)} \neq 0$ has $w_j^{(i)} > 0$.

15. This problem shows for two variables that every monomial ordering arises from a system of two independent weight vectors satisfying the condition in the previous problem. Let a monomial ordering be imposed on $K[X, Y]$.

(a) If $X > Y^q$ for all $q > 0$, show that the ordering is lexicographic and is determined by the system of two weight vectors $\{(1, 0), (0, 1)\}$.

(b) If $X < Y^q$ for some $q > 0$, show that there exists a unique real number $r \geq 0$ such that for all ordered pairs of integers $u \geq 0$ and $v \geq 0$, $X^u > Y^v$ if $ru > v$ and $X^u < Y^v$ if $ru < v$.

(c) If $X < Y^q$ for some $q > 0$ and if r is defined as in (b), prove that the monomial ordering is determined by the system of two weight vectors $\{(r, 1), (s, t)\}$ for a suitable (s, t).

16. In $K[X, Y]$, define $f(X, Y) = X^2Y + XY^2 + Y^2$, $f_1(X, Y) = XY - 1$, and $f_2(X, Y) = Y^2 - 1$. Show that

$$f(X, Y) = (X + Y)f_1 + 1 f_2 + r_1 = Xf_1 + (X + 1)f_2 + r_2$$

with $r_1(X, Y) = X + Y + 1$ and $r_2 = 2X + 1$ gives two decompositions in the lexicographic ordering of f relative to $\{f_1, f_2\}$ satisfying the conditions of the generalized division algorithm of Proposition 8.20. Conclude that the remainder term need not be unique, nor need the coefficients of f_1 and f_2.

17. Observe for any scalar a that the ideal $I = (X^2 + cXY, XY)$ in $K[X, Y]$ is independent of c.
    (a) Verify that $\{X^2 + cXY, XY\}$ is a minimal Gröbner basis of I relative to the lexicographic ordering for any choice of c.
    (b) Show that $\{X^2, XY\}$ is the reduced Gröbner basis for I.

Problems 18–20 characterize ideals in $K[X_1, \ldots, X_n]$ whose locus of common zeros is a finite set under the assumption that K is an algebraically closed field. Thus let K be an algebraically closed field, and let I be a nonzero ideal in $K[X_1, \ldots, X_n]$.

18. Under the assumption for each j with $1 \leq j \leq n$ that I contains a nonconstant polynomial $P_j(X_j)$, prove that $V_K(I)$ is a finite set.

19. Conversely under the assumption that $V_K(I))$ is a finite set, use the Nullstellensatz to produce for each j, a nonconstant polynomial $P_j(X_j)$ lying in I.

20. Impose the usual lexicographic ordering on monomials. Prove that $\mathrm{LT}(I)$ contains some $X_j^{l_j}$ for each j with $1 \leq j \leq n$ if and only if $V_K(I)$ is a finite set. (Educational note: The advantage of this characterization over the one in Problems 18–19 is that checking this one is easy by inspection once a Gröbner basis of I has been computed.)

Problems 21–23 relate solutions of simultaneous systems of polynomial equations to the theory of the Brauer group in Chapter III. A field L is said to satisfy **condition (C1)** if every homogeneous polynomial of degree d in n variables with $d < n$ has a nontrivial zero. The significance of this condition was shown in Problem 20 at the end of Chapter III: the Brauer group $\mathcal{B}(L)$ of such a field is necessarily 0. The present set of problems establishes that a simple transcendental extension of an algebraically closed field satisfies condition (C1). No knowledge of Chapter III is needed for these problems, but Problem 23 will take for granted a certain theorem to be proved in Chapter X.

21. Let K be an algebraically closed field, and let $L = K(X)$ be a simple transcendental extension. It is to be shown that any member $F(T_1, \ldots, T_n)$ of $L[T_1, \ldots, T_n]_d$ of the form $F(T_1, \ldots, T_n) = \sum_{i_1, \ldots, i_n} a_{i_1 \cdots i_n} T_1^{i_1} \cdots T_n^{i_n}$ has a nontrivial zero if $d < n$ and each $a_{i_1, \ldots, i_n}$ lies in the field $L = K(X)$.
    (a) Why is it enough to consider such polynomials with each $a_{i_1, \ldots, i_n}$ in the polynomial ring $K[X]$?
    (b) With the simplification from (a) in place, let δ be the maximum degree in X of the coefficients $a_{i_1 \cdots i_n}$. Let N be a positive integer to be specified. By looking for a solution of the form $T_i = \sum_{j=0}^{N} b_{ij} X^j$ with each b_{ij} in K, show that substitution of this formula into the formula $F(T_1, \ldots, T_n) = 0$ leads to a system of homogeneous polynomial equations over K in the unknowns b_{ij}, one of each degree from 0 to $\delta + Nd$.

22. (a) In the setting of the previous problem, show that the number of unknowns is $(N + 1)n$ and that the number of equations is at most $Nd + \delta + 1$.
    (b) Show for N sufficiently large that the number of equations is less than the number of unknowns.

23. The following theorem will be discussed in Chapter X: if K is algebraically closed and if $m \leq n$, then the locus of common zeros in $\mathbb{P}_K^n$ of m nonconstant homogeneous polynomials in $K[X_1, \ldots, X_{n+1}]$ is nonempty. Assuming this theorem, deduce from the previous two problems the conclusion that the field $L = K(X)$ satisfies condition (C1) if K is algebraically closed.

CHAPTER IX

The Number Theory of Algebraic Curves

Abstract. This chapter investigates algebraic curves from the point of view of their function fields, using methods analogous to those used in studying algebraic number fields.

Section 1 gives an overview, explaining how Riemann's theory of Riemann surfaces of functions ties in with the notion of an algebraic curve and explaining how such curves can be investigated through the discrete valuations of their function fields. It is shown that what needs to be studied is arbitrary function fields in one variable over a base field. It is known that every compact Riemann surface can be viewed as an algebraic curve irreducible over $\mathbb{C}$, and thus the function fields of compact Riemann surfaces are to be viewed as informative examples of the theory in the chapter.

Section 2 introduces the notion of a divisor, which is any formal finite $\mathbb{Z}$ linear combination of the discrete valuations of the function field that are trivial on the base field, and the notion of the degree of a divisor, which is the sum of its coefficients weighted suitably. Each nonzero member x of the function field gives rise to a principal divisor (x), and the main result of the section is that the degree of every principal divisor is 0. This is an analog for function fields of the Artin product formula for number fields.

Section 3 contains the definition of the genus of the function field under study. The main object of study is the vector space $L(A)$ for a divisor A; this consists of 0 and all nonzero members x of the function field such that $(x) + A$ is a divisor ≥ 0. Roughly speaking, it may be viewed as the space of functions on the zero locus of the curve whose poles are limited to finitely many points and to a certain order depending on the point. The genus is defined in terms of $\dim L(A) - \deg A$ when A is a divisor that is a large multiple of the pole part of any fixed principal divisor. The main result of the section is Riemann's inequality, which says that $\dim L(A) \geq \deg A + 1 - g$ for all divisors A, where $g \geq 0$ is the genus, and that g is the smallest integer that works in this inequality for all divisors A.

Sections 4–5 concern the Riemann–Roch Theorem, which gives an interpretation of the difference of the two sides of Riemann's inequality as $\dim L(B)$ for a suitable divisor B that can be defined in terms of A. Section 4 gives the statement and proof of the theorem, and Section 5 gives a number of simple applications.

1. Historical Origins and Overview

As was mentioned in Chapter VIII, modern algebraic geometry grew out of early attempts to solve simultaneous polynomial equations in several variables and out of the theory of Riemann surfaces. Chapter VIII discussed the impact of the first of these sources, and the present chapter discusses the impact of the second.

1. Historical Origins and Overview

The theory of Riemann surfaces was begun by Riemann and continued by Liouville, Abel, Jacobi, Weierstrass, and others. This section discusses briefly the point of view in these studies, which began as an effort to solve a problem in real analysis, moved into complex analysis, and finally arrived at investigations of affine plane curves over $\mathbb{C}$, but from a point of view quite different from the one in Chapter VIII. The end result is a study of the curve through the functions on its zero locus, and the approach has something in common with the approach to algebraic number theory in Chapter VI. It is not necessary to understand the background in maximum generality, and we shall be content with suitable examples.

Riemann was interested in saying something useful about seemingly intractable integrals like the one arising from the arc length of an ellipse; let us take

$$y = y(x) = \int^x \frac{dt}{\sqrt{(t-a)(t-b)(t-c)}},$$

where a, b, c are distinct constants, as a specific example. The lower limit of integration is unimportant, since it affects the value of the integral only by an additive constant. We sketch an analysis of the integral,[1] proceeding formally for the moment. Although y as a function of x seems intractable, any sort of inverse function has nice properties. The formula for y gives us

$$dy = \frac{dx}{\sqrt{(x-a)(x-b)(x-c)}},$$

and an inverse function $x = x(y)$ thus has derivative

$$\frac{dx}{dy} = \sqrt{(x-a)(x-b)(x-c)}.$$

Consequently we should expect that

$$\left(\frac{dx}{dy}\right)^2 = (x-a)(x-b)(x-c).$$

Of course, the singularities at a, b, c are problematic, and the square root might have a negative argument, depending on the location of x.

Riemann's starting point for a rigorous investigation was to let x be complex, rather than real, and to let the integral be taken over paths in $\mathbb{C}$. The result is then not an ordinary function $y(x)$, since the square root in the integrand is not a well-defined function for t in $\mathbb{C} - \{a, b, c\}$. We can make a choice for which the square root is well defined, however, as long as we restrict attention to a small neighborhood of a particular t. Thus we can visualize small overlapping disks each centered at a point along an arbitrary path of integration with t in $\mathbb{C} - \{a, b, c\}$ with the property that the integrand is well defined on each such

---

[1] For more details one can consult the author's book *Elliptic Curves*, pp. 165–183.

disk. The interpretation of the square root may be assumed to match on the intersection of any two disks. When a path goes around one or more of the singularities and we return to the same t, we view the new disk as the same as the old one if the values of the square root match, but as different if the values do not match. The union of the disks with this convention becomes a new domain of interest, and the function $F(t) = \sqrt{(t-a)(t-b)(t-c)}$ on $\mathbb{C} - \{a, b, c\}$ becomes a well-defined function $F(\zeta)$ on this new domain. This new domain is a relatively simple example of a **Riemann surface**, i.e., a connected 1-dimensional complex manifold.

In more modern language the new domain is a twofold covering of the three-times punctured plane $\mathbb{C} - \{a, b, c\}$, obtained as follows. We fix a base point z_0 in $\mathbb{C} - \{a, b, c\}$ and define a winding number for each of the points a, b, c as usual. The subset of the fundamental group of $\mathbb{C} - \{a, b, c\}$ for which the sum of the three winding numbers is even is a subgroup and corresponds, via standard covering-space theory, to a certain twofold covering space $\mathcal{R}$ of $\mathbb{C} - \{a, b, c\}$, the covering map being called e. This covering space is a new domain on which the integrand is well defined. On each fiber of the covering, e is two-to-one. Let ζ_0 be one of the two preimages of z_0. Let us adjoin points a^*, b^*, c^*, ∞^* to the covering space $\mathcal{R}$ and extend e by the definitions $e(a^*) = a, e(b^*) = b, e(c^*) = c$, $e(\infty^*) = \infty$. One can show that the complex structure extends from $\mathcal{R}$ to the enlarged space $\mathcal{R}^*$ in such a way that the extended e is a holomorphic function from $\mathcal{R}^*$ onto $\mathbb{C} \cup \{\infty\}$. The enlarged space $\mathcal{R}^*$ becomes a *compact* Riemann surface, and the extended e is a branched covering of the Riemann sphere $\mathbb{C} \cup \{\infty\}$. Topologically $\mathcal{R}^*$ turns out to be a torus, as we shall see in a moment.

Riemann in his own investigations went on to study the function theory of compact Riemann surfaces. The interest is in deciding whether there is a globally defined meromorphic function with poles/zeros only at chosen points and with poles/zeros at most/least of some specified order. If there is such a function, one wants to know the dimension of the space of such functions. The basic tool for addressing this question is the Riemann–Roch Theorem. In the context of Riemann surfaces, the Riemann–Roch Theorem has both an analysis aspect and an algebraic aspect. The analysis aspect may be viewed as using the theory of elliptic differential operators to prove existence of enough nonconstant meromorphic functions for the Riemann surface to acquire an algebraic structure. For the purposes of this book, we can just accept this circumstance and not try to extend it in any way; however, we will sketch in a moment how the algebraic structure can be obtained concretely for our example. The algebraic aspect may be viewed as mining this algebraic structure to deduce as many dimensionality relations as possible among the function spaces of interest. This is the theory that we shall want to extend; we return to our method for carrying out this project after producing the algebraic structure for our example by elementary means.

To introduce the algebraic structure in our example, we use our knowledge of $\mathcal{R}^*$ to make sense out of the expression

$$w(C) = \int_C F(\zeta)^{-1} d\zeta$$

for any piecewise smooth curve C on $\mathcal{R}^*$ that starts from the base point ζ_0. If C is given by $C(t)$ for t in an interval I, then this integral is to be equal to $w(C) = \int_{t \in I} F(C(t))(e \circ C)'(t)\, dt$. Let Γ_a, Γ_b, Γ_c be small loops in $\mathbb{C} - \{a, b, c\}$ respectively about a, b, c based at z_0, each having winding number 1, and define $\Gamma_1 = \Gamma_a \Gamma_b$ and $\Gamma_2 = \Gamma_b \Gamma_c$. Lift Γ_1 and Γ_2 to curves $\widetilde{\Gamma}_1$ and $\widetilde{\Gamma}_2$ in $\mathcal{R}^*$ based at ζ_0, and define

$$\omega_1 = \int_{\widetilde{\Gamma}_1} F(\zeta)^{-1} d\zeta \quad \text{and} \quad \omega_2 = \int_{\widetilde{\Gamma}_2} F(\zeta)^{-1} d\zeta.$$

It turns out that $\Lambda = \mathbb{Z}\omega_1 + \mathbb{Z}\omega_2$ is a lattice in $\mathbb{C}$ and that there is a well-defined function $w : \mathcal{R}^* \to \mathbb{C}/\Lambda$ such that whenever ζ is in $\mathcal{R}^*$ and C is a piecewise smooth curve from ζ_0 to ζ, then $w(\zeta) \equiv w(C) \bmod \Lambda$. The function $w(\zeta)$ is one-one onto and is biholomorphic. In particular, $\mathcal{R}^*$ is exhibited as homeomorphic to a torus.

Let $w^{-1} : \mathbb{C}/\Lambda \to \mathcal{R}^*$ be the inverse function of w, and let $\mu : \mathbb{C} \to \mathbb{C}/\Lambda$ be the quotient map. Then the composition $P = e \circ w^{-1} \circ \mu$ carries $\mathbb{C}$ to $\mathbb{C} \cup \{\infty\}$ and can be seen to satisfy $P'^2 = (P - a)(P - b)(P - c)$. In other words, P has been constructed rigorously as an inverse function to the original integral. Except for small details, P is the Weierstrass $\wp$ function for the lattice Λ in $\mathbb{C}$. It is almost true that $z \mapsto (P(z), P'(z))$ is a parametrization of the zero locus of the affine plane curve $y^2 - (x - a)(x - b)(x - c)$ defined over $\mathbb{C}$. The sense in which this parametrization fails is that $P(z)$ takes on the value ∞ at certain points. What happens more precisely is that $z \mapsto [P(z), P'(z), 1]$ is a parametrization of the zero locus of the projective plane curve $Y^2 W - (X - aW)(X - bW)(X - cW)$.

Our initial focus in this chapter is in mining this kind of algebraic-curve structure over $\mathbb{C}$ to deduce as many dimensionality relations as possible among interesting finite-dimensional subspaces of scalar-valued functions on the zero locus of the curve. For instance in the example above, one can ask for the dimension of the space of meromorphic functions on $\mathcal{R}^*$ with at worst simple poles at two specified points and with no other poles. The main theorem of this chapter, the Riemann–Roch Theorem, gives quantitative information about the dimension of this space and of similar spaces. The goal for this introduction is to frame this question as an algebra question about the algebraic structure and to see that some basic tools introduced in Chapter VI in the context of algebraic number theory are the appropriate tools to use here.

The primary object of study is the "function field" of the curve in question. Let us construct this function field for our example. The ideal

$$I = \bigl(Y^2 - (X-a)(X-b)(X-c)\bigr)$$

in $\mathbb{C}[X, Y]$ is prime, and the restrictions of all polynomial functions to its zero locus $V(I)$ may be identified with the integral domain $R = \mathbb{C}[X, Y]/I$ by the Nullstellensatz. It takes a little argument, which we omit, to justify saying that the meromorphic functions on the zero locus may be viewed as the field of fractions $\mathbb{F}$ of $\mathbb{C}[X, Y]/I$; suffice it to say for the moment that we insist that the behavior at all points of the locus, including any points on the line at infinity in the projective plane, be limited to poles and zeros, and that is why nonrational functions of (X, Y) do not appear. At any rate, $\mathbb{F}$ is what is taken as the function field of the curve. To have obtained a field by this construction, we could have started with any affine plane curve $f(X, Y)$ over $\mathbb{C}$ as in Chapter VIII, except that the principal ideal $(f(X, Y))$ in $\mathbb{C}[X, Y]$ has to be assumed to be prime to yield an integral domain as quotient. That is, $f(X, Y)$ has to be an irreducible polynomial; we say that the affine plane curve $f(X, Y)$ has to be assumed to be **irreducible** over $\mathbb{C}$.

The study of members of the function field $\mathbb{F}$ from the point of view of their poles and zeros is analogous to the problem of studying factorizations in the number-theoretic setting. This point was already made in Section VIII.7 of *Basic Algebra*, where the case of the affine plane curve above in which $(a, b, c) = (0, +1, -1)$ was studied in detail. For this one choice of (a, b, c), the integral domain $R = \mathbb{C}[X, Y]/I$ was observed to be the integral closure of $\mathbb{C}[X]$ in a finite separable extension of $\mathbb{C}(X)$, and it is a Dedekind domain by Theorem 8.54 of *Basic Algebra*; in fact, the same argument works for any choice of (a, b, c) as long as a, b, c are distinct complex numbers.

Unique factorization of elements into prime elements fails in this R, but we saw that a geometrically meaningful factorization instead is the factorization of nonzero ideals into prime ideals. This latter factorization is unique because R is a Dedekind domain. Meanwhile, since nonzero prime ideals are maximal in R, the Nullstellensatz shows[2] that the nonzero prime ideals in R correspond exactly to the points of the zero locus $V(I)$. Consequently the unique factorization of nonzero ideals in R has the geometric interpretation of associating orders of zeros and poles to members of R. This all seems very tidy, but there are at least three awkward matters that we need to take into account:

---

[2]Let $\varphi : \mathbb{C}[X, Y] \to R$ be the quotient homomorphism. If M is a maximal ideal in R, then $\varphi^{-1}(M)$ is a maximal ideal in $\mathbb{C}[X, Y]$ and hence is the set of all polynomials vanishing at some (x_0, y_0). To show that (x_0, y_0) is in $V(I)$, assume the contrary. Then there exists $g \in I$ with $g(x_0, y_0) \neq 0$. This g is not in the maximal ideal $\varphi^{-1}(M)$, and thus there exist $f \in \varphi^{-1}(M)$ and $h \in \mathbb{C}[X, Y]$ with $f + gh = 1$. Applying φ, we obtain $\varphi(f) = 1$, in contradiction to the fact that $\varphi(f)$ lies in the proper ideal M of R.

(i) we have not included information about zeros and poles at the points at infinity when the curve is viewed projectively, and that information surely plays some role,
(ii) the analysis of the function field $\mathbb{F}$ seems to rely on a subfield $\mathbb{C}(X)$ for which there is surely no canonical description,
(iii) the ring R no longer need be integrally closed if a, b, c are not assumed distinct, if for example $(a, b, c) = (0, 0, 1)$.

Point (ii) turns out to be an advantage, allowing us to work with the given curve from multiple perspectives. The "key observation" at the end of this section will make clear how we can take advantage of (ii).

Point (iii) is quite significant. The trouble with the curve $Y^2 - X^2(X-1)$ is that the curve has a singularity at $(0, 0)$ in the sense of Section VII.5. The maximal ideals of the ring $\mathbb{C}[X, Y]/(Y^2 - X^2(X-1))$ correspond to points on the zero locus of the curve; but the ring is not a Dedekind domain, and we have few tools for working with it. To handle matters properly, we have to form the function field directly as $\mathbb{F} = \mathbb{C}(X)[Y]/(Y^2 - X^2(X-1))$ and define R to be the integral closure of $\mathbb{C}[X]$ in $\mathbb{F}$. This ring R is bigger than $\mathbb{C}[X, Y]/(Y^2 - X^2(X-1))$ and is a Dedekind domain. Unfortunately its nonzero prime ideals no longer correspond exactly to points of the zero locus. Example 1 below will illustrate. What happens is that $\mathbb{F}$ readily provides information about the behavior of nonsingular points of the zero locus but not about singular points. Problems 5–11 at the end of the chapter address this matter for nonsingular points for affine plane curves more generally. The tool for making the connection for curves in higher dimension is Zariski's Theorem (Theorem 7.23), and we shall carry out the details in Chapter X when we treat the *geometry* of curves, as opposed to the number theory.

Point (i) is relevant and is easily handled. When we form the function field of the curve and take R to be the integral closure of $\mathbb{C}[X]$ in it, we can associate $\mathbb{C}[X]$ with the polynomials of $\mathbb{C}$ and think of them as embedded in the field $\mathbb{C}(X)$ of rational functions. The rational functions are all meaningful on the Riemann sphere $\mathbb{C} \cup \{\infty\}$, and we study behavior of rational functions near ∞ by writing them in terms of X^{-1} and regarding X^{-1} as a new variable that is near 0. In studying our curve, the points in the projective plane that we miss by considering just the affine curve are the ones that lie over ∞ in the Riemann sphere. We study them by considering the integral closure R' of $\mathbb{C}[X^{-1}]$ in $\mathbb{F}$. If the curve is nonsingular at all points lying over ∞, then these points correspond to the prime ideals of R' whose intersection with $\mathbb{C}[X^{-1}]$ is the prime ideal $X^{-1}\mathbb{C}[X^{-1}]$ of $\mathbb{C}[X^{-1}]$.

EXAMPLES.

(1) Affine plane curve $f(X, Y) = Y^2 - X^2(X-1)$. This polynomial is irreducible over $\mathbb{C}$ but is singular at $(0, 0)$ in the sense that $\frac{\partial f}{\partial X}$ and $\frac{\partial f}{\partial X}$ both

vanish there. Let $\mathbb{F} = \mathbb{C}(X)[Y]/(f(X,Y))$, and let x and y be the images of X and Y in $\mathbb{F}$. These elements lie in the ring $S = \mathbb{C}[X,Y]/(f(X,Y))$, whose maximal ideals correspond to points on the zero locus by the Nullstellensatz. All members of S are of the form $a(x) + yb(x)$, where a and b are arbitrary polynomials in one variable. Any proper ideal in S containing x has to be of the form $(x, yc_1(x), \ldots, yc_n(x))$ for some polynomials $c_1, \ldots, c_n$. A little argument using the fact that $\mathbb{C}[x]$ is a principal ideal domain shows that the ideal is of the form $(x, yc(x))$. Using products of x and polynomials, we see that we can discard all terms of $c(x)$ but the constant term. Hence the ideal is either (x) itself or is (x, y). The ideal (x) is not prime, since $y \cdot y$ is in it and y is not in it. The ideal (x, y) is maximal and hence prime. Since $(x, y)^2 = (x^2, xy, y^2) = (x^2, xy)$ is properly contained in (x), (x) is not the product of prime ideals in S. Thus S is not a suitable ring for investigating poles and zeros of members of the field $\mathbb{F}$. By contrast, a little computation shows that the integral closure R of $\mathbb{C}[x]$ in $\mathbb{F}$ is generated as a $\mathbb{C}$ algebra by x and $x^{-1}y$. This is a Dedekind domain, and the decomposition of the ideal (x) in R as a product of prime ideals can be checked to be $(x) = (x, x^{-1}y + i)(x, x^{-1}y - i)$. A factor on the right does not consist of all functions vanishing at some $(0, y_0)$ lying on the zero locus. The only point $(0, y_0)$ on the zero locus is $(0, 0)$, and the two prime factors of (x) say something about derivatives at that point. This example will be considered further in Problems 21–22 at the end of the chapter.

(2) Affine plane curve $f(X, Y) = Y^2 - X^4 + 1$. This polynomial is irreducible over $\mathbb{C}$ and is nonsingular at every point of its zero locus in $\mathbb{C}^2$. Again we form the function field $\mathbb{F}$, the members x and y of it, and the ring $\mathbb{C}[X, Y]/(f(X, Y))$. Using the fact that $X^4 - 1$ is square free, we can check that this ring is the full integral closure R of $\mathbb{C}[x]$ in $\mathbb{F}$. The ring R is a Dedekind domain, and its elements are all expressions $a(x)$ and $yb(x)$, where $a(x)$ and $b(x)$ are polynomials. Moreover, we have $(y + x^2)(y - x^2) = y^2 - x^4 = (x^4 - 1) - x^4 = -1$. Consequently the elements $y \pm x^2$ are nonconstant units in R, and they cannot have zeros or poles on the zero locus of $f(X, Y)$ in $\mathbb{C}^2$. Thus knowledge of the orders of zeros and poles at every point of the zero locus of $f(X, Y)$ in $\mathbb{C}^2$ does not determine a member of R up to a constant factor. Instead, we have to take into account the behavior at any points at infinity on the zero locus in the projective plane $\mathbb{P}^2_{\mathbb{C}}$. To see what this set is, we convert $f(X, Y)$ into a homogeneous polynomial of degree 4, specifically into $F(X, Y, W) = Y^2W^2 - X^4 + W^4$, and then we look for points $[x, y, w]$ with $F(x, y, w) = 0$ and $w = 0$. These have $x = 0$ and thus come down to $[0, y, 0]$. In other words, there is only one point at infinity on the zero locus of the curve. It is singular because all three partial derivatives of F are 0 there. The fact that it is singular means that we should not expect the prime ideals lying over $x^{-1}\mathbb{C}[x^{-1}]$ in the integral closure R' of $\mathbb{C}[x^{-1}]$ in $\mathbb{F}$ to correspond to the points at infinity on the curve. We return to this example shortly.

All these matters begin to sound quite complicated to sort out, but magically there is a simple way of handling them: for an affine plane curve irreducible over $\mathbb{C}$, we work with the field $\mathbb{F}$ of rational functions for the curve, ignoring the geometry of the curve, and we consider all discrete valuations on this field that are 0 on $\mathbb{C}^\times$. Discrete valuations were discussed at length in Section VI.2. They depend only on $\mathbb{F}$, not on the choice of a subring for which $\mathbb{F}$ is the field of fractions. As will be seen in Chapter X, the full set of discrete valuations of $\mathbb{F}$ gives information about all potential nonsingular points for any affine curve with function field $\mathbb{F}$, not necessarily planar; there will even be such a curve whose extension to be defined projectively is everywhere nonsingular, and then the points on the zero locus of the curve in projective space will be in one-one correspondence with the discrete valuations of $\mathbb{F}$.

Let us review what Chapter VI tells us about discrete valuations in our setting. Let $f(X, Y)$ be an irreducible polynomial in $\mathbb{C}[X, Y]$, let $\mathbb{F}$ be the field $\mathbb{C}(X)[Y]/(f(X, Y))$, let x and y be the images of X and Y in $\mathbb{F}$, and let R be the integral closure of $\mathbb{C}[x]$ in $\mathbb{F}$. This is a Dedekind domain by Theorem 8.54 of *Basic Algebra*. Corollary 6.10 classifies the discrete valuations of $\mathbb{F}$ that are 0 on $\mathbb{C}^\times$. It shows that all but finitely many correspond to prime ideals in R. There are only finitely many others. Corollary 6.10 tells us that these other discrete valuations can be described in terms of the integral closure R' of $\mathbb{C}[x^{-1}]$ in $\mathbb{F}$; this is another Dedekind domain whose field of fractions is $\mathbb{F}$. The exceptional discrete valuations of $\mathbb{F}$ arise from those prime ideals of R' that occur in the decomposition of the ideal $x^{-1}R'$ into prime ideals of R'. Geometrically we may view these additional discrete valuations as associated in some way with points at infinity in a projective space, but we can proceed with algebraic manipulations of these discrete valuations without invoking the geometric interpretation or using projective space.

EXAMPLE 2, CONTINUED. We continue with the affine plane curve $Y^2 - X^4 + 1$, the prime ideal $I = (Y^2 - X^4 + 1)$, and the ring R given as the integral closure of $\mathbb{C}[X]$ in the field $\mathbb{F} = \mathbb{C}(X)[Y]/I$. Corollary 6.10 divides the discrete valuations of $\mathbb{F}$ that are 0 on $\mathbb{C}^\times$ into two kinds. The ones of the first kind are built from the nonzero prime ideals of R. Since $y \pm x^2$ are units in R, all of these valuations take the value 0 on $y \pm x^2$. The discrete valuations of the second kind are those appearing in the decomposition of the ideal $x^{-1}R'$ in the integral closure R' of $\mathbb{C}[x^{-1}]$ in $\mathbb{F}$. The element $x^{-2}y$ is in R' because it is a root of the polynomial $Y^2 - (1 - x^{-4})$ in $\mathbb{C}[x^{-1}][Y]$. Hence R' contains x^{-1} and $x^{-2}y$. On the other hand, the most general element of $\mathbb{F}$ is of the form $a(x^{-1})x^{-2}y + b(x^{-1})$, where a and b are rational expressions in one variable, and this is a root of the polynomial

$$Y^2 - 2b(x^{-1})Y + \bigl(b(x^{-1})^2 - a(x^{-1})^2(1 - x^{-4})\bigr).$$

For this element to be in R', the coefficients must be in $\mathbb{C}[x^{-1}]$. This means that $b(X)$ is a polynomial and that $a(X)^2(1-X^4)$ is a polynomial. Since $1-X^4$ has no repeated roots, the latter condition forces $a(X)$ to be a polynomial. Thus x^{-1} and $x^{-2}y$ generate R' as a $\mathbb{C}$ algebra. Define ideals in R' by

$$P_1 = (x^{-1}, x^{-2}y + 1) \quad \text{and} \quad P_2 = (x^{-1}, x^{-2}y - 1).$$

Then it is straightforward to check the decompositions

$$(x^{-1}) = P_1 P_2, \quad (x^{-2}y + 1) = P_1^4, \quad \text{and} \quad (x^{-2}y - 1) = P_2^4.$$

Since $[\mathbb{F} : \mathbb{C}(x^{-1})] = 2$ and since x^{-1} is prime in $\mathbb{C}[x^{-1}]$, the ideal (x^{-1}) in R' is the product of at most two prime ideals, and it follows that P_1 and P_2 are prime ideals in R'. They are distinct because the difference of the respective second generators is a nonzero scalar. In view of Corollary 6.10, there are exactly two discrete valuations of $\mathbb{F}$ that are 0 on $\mathbb{C}^\times$ other than the ones coming from prime ideals of R, and these are the ones coming from the prime ideals P_1 and P_2 of R'. Let us call them v_1 and v_2. The above decompositions of principal ideals give $v_1(y + x^2) = v_1(x^{-1})^{-2} + v_1(x^{-2}y + 1) = (-2) + (+4) = +2$, whereas $v_1(y - x^2) = (-2) + (0) = -2$. Thus v_1 takes the distinct values $0, +2$, and -2 on $1, y + x^2$, and $y - x^2$. Similarly v_2 takes the values $0, -2$, and $+2$ on these elements.

We shall work with those discrete valuations of the field of rational functions for the curve under study that are 0 on the base field. These are canonical, independent of our choice of some Dedekind domain whose field of fractions is the given field. However, making a choice of Dedekind domain is convenient for making calculations. Then we can consider the discrete valuations as of two kinds, and which discrete valuations are of which kind will depend on our choice of Dedekind domain.

Context for the study in this chapter. Having concluded that the object to investigate is the field of rational functions of our curve and that the tools include the discrete valuations, we can now consider the context in which we should work. Let $\Bbbk$ be any field, not necessarily algebraically closed. We want to work with the "function field" of a suitable kind of curve defined over $\Bbbk$. If I is an ideal in $\Bbbk[X_1, \ldots, X_n]$, then the ring $R = \Bbbk[X_1, \ldots, X_n]/I$ is an integral domain if and only if the ideal I is prime, and in this case the field of fractions $\mathbb{F}$ of R can be taken to be the associated function field. Thus we restrict attention to the case that I is prime. To bring in the notion that the curve is to be 1-dimensional, we recall from Theorem 7.22 that the integral domain R has Krull dimension 1 in the sense of Section VII.4 if and only if the field of fractions $\mathbb{F}$ has transcendence degree 1 over $\Bbbk$. In this case, $\mathbb{F}$ is finitely generated as a field over $\Bbbk$, with a finite set of generators consisting of the elements $x_j = X_j + I$ for $1 \leq j \leq n$. That is, $\mathbb{F}$ is a function field in one variable over $\Bbbk$.

Conversely if $\mathbb{F}$ is a function field in one variable over $\Bbbk$, then $\mathbb{F}$ is a finite algebraic extension of a simple transcendental extension $\Bbbk(x_1)$. Let us write it as $\mathbb{F} = \Bbbk(x_1)[x_2, \ldots, x_n]$ for some n. Form the polynomial ring $\Bbbk[X_1, \ldots, X_n]$ and the ring homomorphism of this ring into $\mathbb{F}$ that fixes $\Bbbk$ and sends X_j into x_j. The image of this homomorphism is an integral domain R whose field of fractions is $\mathbb{F}$, and the kernel is a prime ideal I such that $R \cong \Bbbk[X_1, \ldots, X_n]/I$. Theorem 7.22 tells us that R has Krull dimension 1.

We are led to the following definition. For any field $\Bbbk$ and any integer $n \geq 1$, an ideal I in $\Bbbk[X_1, \ldots, X_n]$ is called an **affine curve irreducible**[3] **over** $\Bbbk$ if I is prime and the integral domain $R = \Bbbk[X_1, \ldots, X_n]/I$ has Krull dimension 1. An affine plane curve $(f(X, Y))$ in the sense of Chapter VIII will be an object of this kind if $f(X, Y)$ is an irreducible polynomial.[4]

The geometry of the zero loci of the curves we study will not play a role in the mathematics of this chapter; only the field of fractions $\mathbb{F}$ and the base field $\Bbbk$ will. We postpone to Chapter X any discussion of the geometry.[5] For any function field $\mathbb{F}$ in one variable over an arbitrary field $\Bbbk$, we shall study in detail those discrete valuations of $\mathbb{F}$ that are 0 on $\Bbbk$. We refer to such discrete valuations as the **discrete valuations of** $\mathbb{F}$ **defined over** $\Bbbk$. It will be helpful as motivation to remember for the special case in which $\Bbbk$ is algebraically closed

- that the members of $\mathbb{F}$ may be viewed as all rational functions on the zero locus of an affine curve irreducible over $\Bbbk$,
- that the order-of-a-zero function at any nonsingular point of this zero locus gives an example of a discrete valuation of $\mathbb{F}$ defined over $\Bbbk$, and
- that all discrete valuations of $\mathbb{F}$ defined over $\Bbbk$ arise in this way if the zero locus is nonsingular at every point and we take into account points at infinity in projective space.

However, the formal development will not make use of these interpretations.

---

[3] Beware of assuming too much irreducibility about such a curve. Just because I is prime does not mean that I remains prime when we extend the scalars and work with an algebraic closure $\Bbbk_{\text{alg}}$ of $\Bbbk$. For example, $X^2 + Y^2$ is an affine curve irreducible over $\mathbb{R}$, but it factors as $(X + iY)(X - iY)$ over $\mathbb{C}$ and is therefore not irreducible over $\mathbb{C}$.

[4] This change of context for the word "curve" from the definition in Chapter VIII is appropriate because of a change of emphasis: we shall now be studying an associated function field rather than the defining ideal. The word "curve" will undergo a genuine change in meaning in Chapter X: because of the Nullstellensatz, classical algebraic geometry in the form to be discussed in much of Chapter X places emphasis on zero loci defined by prime ideals of polynomials over an algebraically closed field, and it will be convenient to define the curve to be the zero locus rather than the defining ideal.

[5] In Chapter X we shall introduce two distinct notions of sameness for the zero loci under the assumption that the field is algebraically closed, namely "isomorphism" and "birational equivalence." The first is a refinement of the second. Birational equivalence will turn out to mean that the function fields are isomorphic. An important theorem says that each birational equivalence class of irreducible curves contains one and only one isomorphism class of curves that are everywhere nonsingular in the sense of Section VII.5.

What to expect from the study. When $\Bbbk$ is not necessarily algebraically closed, these interpretations break down, at least to some extent. Yet the main theorem of the chapter, the Riemann–Roch Theorem, is still geared to the geometric interpretation of discrete valuations in terms of poles and zeros. One may reasonably ask why one goes to the trouble of working in such a general context that the theory no longer has its geometric interpretation. The answer is that the investigation is to be regarded as one in number theory, not in geometry. For example, studying an affine plane curve over a field $\mathbb{F}_p$ is the same as studying solutions of congruences in two variables modulo a prime. Studying such a curve over the p-adic field $\mathbb{Q}_p$ is the same as studying solutions of such congruences modulo arbitrary powers of p. The Riemann–Roch Theorem is actually the first serious aid in making this study. The present chapter therefore does not constitute such a study; it merely prepares one for such a study. In addition, there is a side benefit to understanding the number theory that arises this way: the methods and results of this subject and of algebraic number theory have enough in common that the methods and results for each suggest methods and results for the other.

An especially tantalizing example of this phenomenon concerns zeta functions. The zeros with $0 < \operatorname{Re} s \leq 1$ for the Riemann zeta function, which is the meromorphic continuation to $\mathbb{C}$ of $\zeta(s) = \sum_{n=1}^{\infty} n^{-s} = \prod_{p \text{ prime}} (1 - p^{-s})^{-1}$, influence the error term in the distribution of the primes as asserted by the Prime Number Theorem. The classical Riemann hypothesis is the statement that the only such zeros occur on the line $\operatorname{Re} s = \frac{1}{2}$; it implies a high level of control of this error term. There is a corresponding zeta function for any algebraic number field, and to it corresponds a version of the Riemann hypothesis appropriate for prime ideals for the number field. Proofs or counterexamples for these versions of the Riemann hypothesis have been sought for more than a century.

Meanwhile, one can formulate a Riemann hypothesis for any function field in one variable over any finite field, and again the statement has consequences for the distribution of prime ideals. This time, however, the Riemann hypothesis is a theorem, stated and proved by A. Weil in 1940. One might hope that the methods used for Weil's theorem could shed enough light on the classical Riemann hypothesis to lead to a proof, but to date this has not happened.

Key observation to be used during the study. In the next section we shall make systematic use of the following construction for any function field $\mathbb{F}$ in one variable over the field $\Bbbk$. If x is any element of $\mathbb{F}$ transcendental over $\Bbbk$, then the only discrete valuations of $\mathbb{F}$ defined over $\Bbbk$ that take a nonzero value on x may be described as follows. Let R be the integral closure of $\Bbbk[x]$ in $\mathbb{F}$, and let R' be the integral closure of $\Bbbk[x^{-1}]$ in $\mathbb{F}$. Then R and R' are Dedekind domains by Corollary 7.14, whether or not $\mathbb{F}$ is a separable extension of $\Bbbk(x)$. Both have $\mathbb{F}$ as field of fractions. Let the ideals xR of R and $x^{-1}R'$ of R' have prime decompositions $xR = P_1^{e_1} \cdots P_g^{e_g}$ and $x^{-1}R' = Q_1^{e'_1} \cdots Q_{g'}^{e'_{g'}}$. Then the valuations v_{P_i} for $1 \le i \le g$ and v_{Q_j} for $1 \le j \le g'$ defined by P_i and Q_j have $v_{P_i}(x) = e_i$ and $v_{Q_j}(x) = -e'_j$, and no other discrete valuation of $\mathbb{F}$ that is defined over $\Bbbk$ takes a nonzero value on x. This observation follows from Corollary 6.10 and the definition of the discrete valuation associated with a nonzero prime ideal in a Dedekind domain.

2. Divisors

Let $\Bbbk$ be a field, and let $\mathbb{F}$ be a function field in one variable over $\Bbbk$. The first step is one of normalization: there is no loss of generality in replacing $\Bbbk$ by the larger field $\Bbbk'$ of all elements $\mathbb{F}$ that are algebraic over $\Bbbk$.[6]

Proposition 9.1. Let $\mathbb{F}$ be a function field in one variable over $\Bbbk$, and let $\Bbbk'$ be the subfield of all elements in $\mathbb{F}$ algebraic over $\Bbbk$. If x is in $\mathbb{F}^\times$, then every discrete valuation of $\mathbb{F}$ defined over $\Bbbk$ vanishes on x if and only if x is in $\Bbbk'$. Consequently $\mathbb{F}$ is automatically a function field in one variable over $\Bbbk'$, and as such, its discrete valuations defined over $\Bbbk'$ coincide with its discrete valuations defined over $\Bbbk$.

PROOF. If $x \in \mathbb{F}$ is transcendental over $\Bbbk$, then the observation at the end of Section 1 produces discrete valuations of $\mathbb{F}$ defined over $\Bbbk$ that take nonzero values on x. Conversely if $x \in \mathbb{F}^\times$ is algebraic over $\Bbbk$, we argue by contradiction. We may assume that $x \ne 0$. Suppose that v is a discrete valuation of $\mathbb{F}$ defined over $\Bbbk$ such that $v(x) \ne 0$. Possibly replacing x by x^{-1}, we may assume that $v(x) > 0$. Being nonzero algebraic over $\Bbbk$, x satisfies a polynomial equation

$$a_m x^m + a_{m-1} x^{m-1} + \cdots + a_1 x + a_0 = 0$$

with all $a_j \in \Bbbk$ and with $a_0 \ne 0$. For each j with $a_j \ne 0$, we have $v(a_j x^j) = v(a_j) + j v(x) = j v(x) > 0$. If $a_j = 0$, then $v(a_j x^j) = \infty > 0$. Thus $v(a_m x^m + a_{m-1} x^{m-1} + \cdots + a_1 x) > 0$. Since $v(a_0) = 0$, property (vi) of discrete valuations in Section VI.2 shows that

$$v\bigl((a_m x^m + a_{m-1} x^{m-1} + \cdots + a_1 x) + a_0\bigr) = v(a_0) = 0 \ne \infty = v(0),$$

contradiction.

The conclusions in the last sentence of the proposition now follow: Since $\mathbb{F}$ is generated over $\mathbb{F}$ by finitely many elements $x_1, \ldots, x_n$, it is generated over $\Bbbk'$ by the same elements. Moreover, any element of $\mathbb{F}$ transcendental over $\Bbbk$ is transcendental over $\Bbbk'$, since $\Bbbk'$ is algebraic over $\Bbbk$. Thus $\mathbb{F}$ is a function field in one variable over $\Bbbk'$. The first paragraph of the proof shows that every discrete valuation of $\mathbb{F}$ defined over $\Bbbk$ is defined over $\Bbbk'$, and the converse statement is immediate from the definition. $\square$

---

[6]The field $\Bbbk'$ is called the **field of constants** by some authors.

In accordance with Proposition 9.1, there is no loss of generality in replacing $\Bbbk$ by $\Bbbk'$ throughout. Changing notation, *we assume henceforth* that $\mathbb{F}$ is a function field in one variable defined over $\Bbbk$ and *that every element of $\mathbb{F}$ not in $\Bbbk$ is transcendental over $\Bbbk$*. These hypotheses will not be repeated for each result.

Suppressing $\Bbbk$ in the notation, we denote by $\mathbb{V}_\mathbb{F}$ the set of all discrete valuations of $\mathbb{F}$ defined over $\Bbbk$. A **divisor** is any member of the free abelian group $D_\mathbb{F}$ on $\mathbb{V}_\mathbb{F}$. Elements of $D_\mathbb{F}$ will be written additively,[7] and thus a typical member of $D_\mathbb{F}$ is

$$A = \sum_{v \in \mathbb{V}_\mathbb{F}} n_v v$$

with only finitely many of the integers n_v nonzero. We write $\text{ord}_v A$ for the integer n_v, calling it the **order** of A at v. The identity element of $D_\mathbb{F}$ is called **zero** and is denoted by 0.

Each x in $\mathbb{F}^\times$ defines a **principal divisor** (x) by the formula

$$(x) = \sum_{v \in \mathbb{V}_\mathbb{F}} v(x) v.$$

We verify that (x) is indeed a divisor by showing that $v(x)$ is nonzero for only finitely many v in $\mathbb{V}_\mathbb{F}$. For x in $\Bbbk$, $v(x) = 0$ for all v. All other x are transcendental over $\Bbbk$, and the observation at the end of Section 1 shows that exactly $g + g'$ members of $\mathbb{V}_\mathbb{F}$ are nonzero on x, where g and g' are certain positive integers depending on x.

It is sometimes convenient to decompose (x) as a particular difference of two divisors, writing $(x) = (x)_0 - (x)_\infty$ with

$$(x)_0 = \sum_{\substack{v \in \mathbb{V}_\mathbb{F}, \\ v(x) > 0}} v(x) v \quad \text{and} \quad (x)_\infty = \sum_{\substack{v \in \mathbb{V}_\mathbb{F}, \\ v(x) < 0}} (-v(x)) v.$$

This notation is motivated by the interpretation of (x) for the case $\Bbbk = \mathbb{C}$, which is discussed in an example below.

Because of the formula $v(xy) = v(x) + v(y)$, the set of principal divisors is a subgroup $P_\mathbb{F}$ of $D_\mathbb{F}$, and the mapping $x \mapsto (x)$ is a group homomorphism of $\mathbb{F}^\times$ onto $P_\mathbb{F}$. The quotient $C_\mathbb{F} = D_\mathbb{F}/P_\mathbb{F}$ is called the group of **divisor classes** of $\mathbb{F}$ over $\Bbbk$.

EXAMPLE. $\Bbbk = \mathbb{C}$. This is the setting of a compact Riemann surface, provided we take for granted that every compact Riemann surface can be realized as a nonsingular projective curve over $\mathbb{C}$. The field $\mathbb{F}$ is the field of global meromorphic

---

[7] Some authors use a multiplicative notation.

functions on the surface. A principal divisor can be viewed as a compilation of the orders of the zeros and poles of a nonzero global meromorphic function: each member of $\mathbb{V}_\mathbb{F}$ corresponds to a point of the surface, and the order of a principal divisor (x) with $x \in \mathbb{F}^\times$ at a point is positive if the meromorphic function x has a zero at the point, negative if x has a pole there. It is known that the sum of the orders of all the zeros of a nonzero global meromorphic function equals the sum of the orders of all the poles. In the current framework the statement is that the sum over $v(x)$ is 0 for every $x \in \mathbb{F}^\times$ when $\Bbbk = \mathbb{C}$.

Theorem 9.3 will generalize the fact about compact Riemann surfaces that $\sum_{v \in \mathbb{V}_\mathbb{F}} v(x) = 0$ for every $x \in \mathbb{F}^\times$ when $\Bbbk = \mathbb{C}$. When $\mathbb{C}$ is replaced by a more general field that is not necessarily algebraically closed, Proposition 6.9 already shows that the terms $v(x)$ in the corresponding sum have to be weighted by certain integers in order to yield sum 0. These integers are dimensions that are shown to be finite in the next proposition.

Proposition 9.2. Let v be any discrete valuation of $\mathbb{F}$ defined over $\Bbbk$, let R_v be the valuation ring, and let P_v be the valuation ideal. Then R_v and P_v are $\Bbbk$ vector spaces, and $\dim_\Bbbk R_v/P_v$ is finite.

REMARKS. The integer $f_v = \dim_\Bbbk R_v/P_v$ is called the **residue class degree** of the valuation v. The proof gives a method for computing f_v, and we shall make use of this method shortly in proving Theorem 9.3.

PROOF. The fact that R_v and P_v are $\Bbbk$ vector spaces is immediate from Proposition 9.1. Since v is not identically zero, there exists some $x \in \mathbb{F}$ with $v(x) \neq 0$, and x is transcendental by Proposition 9.1. Possibly replacing x by x^{-1}, we may assume that $v(x) > 0$. The observation at the end of Section 1 classifies those members of $\mathbb{V}_\mathbb{F}$ taking positive values on x. In that notation we decompose $(x)R$ as $P_1^{e_1} \cdots P_g^{e_g}$, and v is the valuation defined by P_j for some j. Theorem 6.5e shows that $R_v/P_v \cong R/P_j$. Since x is prime in $\Bbbk[x]$, the general theory of extensions of Dedekind domains shows that $P_j \cap \Bbbk[x] = x\Bbbk[x]$ and that $f_j = \dim_{\Bbbk[x]/(x)}(R/P_j)$ is finite. The field $\Bbbk[x]/(x)$ is isomorphic to $\Bbbk$, and thus the dimension over $\Bbbk$ of $R_v/P_v \cong R/P_j$ is f_j. □

The **degree** of a divisor A is the integer $\deg A = \sum_{v \in \mathbb{V}_\mathbb{F}} f_v \operatorname{ord}_v(A)$, where f_v is the residue class degree of v as defined in the remarks with Proposition 9.2. Degree is a homomorphism of $D_\mathbb{F}$ into $\mathbb{Z}$. We shall prove in Theorem 9.3 that principal divisors have degree 0. This result extends Proposition 6.9, which handles the special case of the function field $\Bbbk(x)$. Theorem 9.3 may be regarded as a function-field analog of the Artin product formula (Theorem 6.51) for number fields, but the proof is much easier for function fields because we can take advantage of the observation at the end of Section 1.

Theorem 9.3. The degree of every principal divisor is 0. In more detail, if (x) is a principal divisor with x not in $\Bbbk$, then $\deg(x)_0 = \deg(x)_\infty = \dim_{\Bbbk(x)} \mathbb{F}$, and hence $\deg(x) = \deg(x)_0 - \deg(x)_\infty = 0$.

PROOF. If x is in $\Bbbk^\times$, then Proposition 9.1 shows that $v(x) = 0$ for every $v \in \mathbb{V}_\mathbb{F}$, and hence $\deg(x) = 0$. Thus we may assume that x is transcendental over $\Bbbk$. Applying the observation at the end of Section 1 and using the notation from there, we know that the only v's for which $v(x) \neq 0$ are the ones relative to the prime ideals P_i of R and the prime ideals Q_j of R' such that

$$xR = P_1^{e_1} \cdots P_g^{e_g} \quad \text{and} \quad x^{-1}R' = Q_1^{e'_1} \cdots Q_{g'}^{e'_{g'}}. \tag{$*$}$$

Moreover, $v_{P_i}(x) = e_i$ and $v_{Q_j}(x) = -e'_j$. In addition, the proof of Proposition 9.2 showed that the respective residue class degrees are the usual indices f_i and f'_j associated to the decompositions $(*)$. Thus

$$\deg(x)_0 = \sum_{i=1}^g f_i e_i \quad \text{and} \quad \deg(x)_\infty = \sum_{j=1}^{g'} f'_j e'_j.$$

Two applications of Theorem 9.60 of *Basic Algebra* show that

$$\sum_{i=1}^g f_i e_i = \dim_{\Bbbk(x)} \mathbb{F} \quad \text{and} \quad \sum_{j=1}^{g'} f'_j e'_j = \dim_{\Bbbk(x^{-1})} \mathbb{F}.$$

Thus $\deg(x)_0 = \dim_{\Bbbk(x)} \mathbb{F}$, and $\deg(x)_\infty = \dim_{\Bbbk(x^{-1})} \mathbb{F}$. The theorem therefore follows from the fact that $\Bbbk(x) = \Bbbk(x^{-1})$. □

Let $D_{\mathbb{F},0}$ be the subgroup of all divisors of degree 0. Theorem 9.3 shows that $P_\mathbb{F} \subseteq D_{\mathbb{F},0}$. The quotient $C_{\mathbb{F},0} = D_{\mathbb{F},0}/P_\mathbb{F}$ is therefore a subgroup of $C_\mathbb{F} = D_\mathbb{F}/P_\mathbb{F}$ and is the group of all divisor classes of degree 0. This is a function-field analog of the class group for an algebraic number field; it can be shown to be finite if $\Bbbk$ is a finite field but it not if $\Bbbk$ is an arbitrary field.

3. Genus

In this section, $\mathbb{F}$ denotes a function field in one variable over a field $\Bbbk$, and we assume that every element of $\mathbb{F}$ outside $\Bbbk$ is transcendental over $\Bbbk$. We continue with the notation $\mathbb{V}_\mathbb{F}, D_\mathbb{F}, f_v, \mathrm{ord}_v A, \deg A,$ and (x) for $x \in \mathbb{F}^\times$, all as in Section 2.

If we were studying only what happens with $\Bbbk = \mathbb{C}$, we would be interested in the vector space of all meromorphic functions whose poles are limited to a certain finite set of points and are limited to some particular order at each of those points. The underlying compact Riemann surface is an ordinary closed orientable 2-dimensional manifold, and the dimensions of these spaces of meromorphic functions turn out to control the genus of this manifold. For general $\Bbbk$, we study the natural generalization of this situation.[8] The vector spaces of interest are defined in terms of divisors, and we will be led to a natural definition of genus of the curve under study.

We introduce a partial ordering on $D_\mathbb{F}$ by saying that two divisors A and B have $A \leq B$ if $\text{ord}_v A \leq \text{ord}_v B$ for all $v \in \mathbb{V}_\mathbb{F}$. The inequality $B \geq A$ is to mean the same thing as $A \leq B$. If $A \leq B$ and $A' \leq B'$, then $A + A' \leq B + B'$ because $\text{ord}_v(A + A') = \text{ord}_v A + \text{ord}_v A' \leq \text{ord}_v B + \text{ord}_v B' = \text{ord}_v(B + B')$. If $A \leq B$, then $-A \geq -B$.

For each divisor A, we shall study the $\Bbbk$ vector space

$$L(A) = \{0\} \cup \{x \in \mathbb{F}^\times \mid (x) \geq -A\} = \{x \in \mathbb{F} \mid v(x) \geq -\text{ord}_v A\}.$$

For $x \neq 0$, we can think of $v(x)$ as telling the order of the zero of x at a point corresponding to v. In that spirit, if $A \geq 0$, then $L(A)$ consists of all functions whose poles are limited to the set of v's for which $\text{ord}_v A \neq 0$, with the order of the pole bounded above by the number $\text{ord}_v A$. For general A, a similar interpretation is valid, except that the members of $L(A)$ are required also to vanish at certain points at least to certain orders.

We shall suppress any name for the function that embeds $\mathbb{V}_\mathbb{F}$ in $D_\mathbb{F}$. Thus for example if v_0 is in $\mathbb{V}_\mathbb{F}$, then $L(v_0)$ refers to $L(A)$ for the divisor A such that $\text{ord}_{v_0} A = 1$ and $\text{ord}_v A = 0$ when $v \neq v_0$.

Corollary 9.4. $L(0) = \Bbbk$, and $L(A) = 0$ if A is a nonzero divisor with $A \leq 0$.

PROOF. If $A \leq 0$ is nontrivial and if $x \in \mathbb{F}^\times$ were to have $(x) \geq -A$, then we would have $\deg(x) \geq -\deg A > 0$, in contradiction to the conclusion $\deg(x) = 0$ of Theorem 9.3. Thus $L(A) = 0$. Next, we have

$$L(0) = \{x \in \mathbb{F}^\times \mid v(x) = 0 \text{ for all } x\} \cup \bigcup_{v \in \mathbb{V}_\mathbb{F}} L(-v).$$

The first term on the right side is $\Bbbk^\times$, and the second term gives 0 by what we have just proved. Hence $L(0) = \Bbbk$. □

If $A \leq B$, then it follows from the definition that $L(A) \subseteq L(B)$. We shall be interested in how much $L(B)$ increases when B increases. This change is measured by what happens to the quotient space $L(B)/L(A)$. The key case is that $B = A + v_0$ for some $v_0 \in \mathbb{V}_\mathbb{F}$, and we treat that in the following lemma.

---

[8]In doing so, we follow the approach in the book by Villa Salvador, Chapter 3, but with different notation.

Lemma 9.5. If A is a divisor and v_0 is in $\mathbb{V}_\mathbb{F}$, then
$$\dim_\mathbb{k} L(A + v_0)/L(A) \leq f_{v_0} = \deg v_0.$$

PROOF. Put $f = f_{v_0}$, let R_{v_0} be the valuation ring of v_0, and let P_{v_0} be the valuation ideal of v_0. Since v_0 carries $\mathbb{F}^\times$ onto $\mathbb{Z}$, we can choose an element $y \in \mathbb{F}^\times$ with $v_0(y) = \text{ord}_{v_0}(A + v_0)$.

Let $f + 1$ members $x_1, \ldots, x_{f+1}$ of $L(A + v_0)$ be given. We shall produce an equation of linear dependence among the cosets $x_i + L(A)$, and this will prove the lemma. Computation gives
$$v_0(x_i y) = v_0(x_i) + v_0(y) = v_0(x_i) + \text{ord}_{v_0}(A + v_0) \geq 0$$
for $1 \leq i \leq f + 1$, since x_i is in $L(A + v_0)$. Hence $x_i y$ is in R_{v_0}. Since $\dim_\mathbb{k}(R_{v_0}/P_{v_0}) = f$, there exist members $c_1, \ldots, c_{f+1}$ of $\mathbb{k}$ not all 0 such that $\sum_{i=1}^{f+1} c_i(x_i y + P_{v_0}) = P_{v_0}$, i.e., such that $\sum_{i=1}^{f+1} c_i x_i y$ lies in P_{v_0}. Then $\sum_{i=1}^{f+1} c_i x_i$ lies in $y^{-1} P_{v_0}$, and
$$v_0\Big(\sum_{i=1}^{f+1} c_i x_i\Big) \geq -v_0(y) + 1 = -\text{ord}_{v_0}(A + v_0) + 1 = -\text{ord}_{v_0} A. \qquad (*)$$

Since each x_i is in $L(A + v_0)$, so is $\sum_{i=1}^{f+1} c_i x_i$. This fact and $(*)$ together show that $\sum_{i=1}^{f+1} c_i x_i$ is in $L(A)$, i.e., that $\sum_{i=1}^{f+1} c_i x_i + L(A)$ is the 0 coset. This proves the desired linear dependence and shows that $\dim_\mathbb{k} L(A + v_0)/L(A) \leq f$. $\square$

Theorem 9.6. If A and B are divisors such that $A \leq B$, then $L(B)/L(A)$ is finite-dimensional over $\mathbb{k}$ with
$$\dim_\mathbb{k} L(B)/L(A) \leq \deg B - \deg A.$$
Moreover, $L(A)$ and $L(B)$ are separately finite-dimensional over $\mathbb{k}$, and consequently
$$\dim_\mathbb{k} L(B) - \deg B \leq \dim_\mathbb{k} L(A) - \deg A.$$

REMARKS. We define $\ell(A) = \dim_\mathbb{k} L(A)$. This is finite by the theorem, and the resulting inequality of the theorem is that
$$\ell(B) - \deg B \leq \ell(A) - \deg A.$$

PROOF. The first conclusion is immediate from Lemma 9.5 by induction on $\sum_v (\text{ord}_v B - \text{ord}_v A)$. Fixing a reference point v_0 in $\mathbb{V}_\mathbb{F}$ and taking $A =$

$\sum_{\mathrm{ord}_v B \leq 0} (\mathrm{ord}_v B) v - v_0$ and applying Corollary 9.4 to A, we see that $L(A) = 0$. Therefore the first conclusion specializes to

$$\dim_\Bbbk L(B) - \deg B \leq -\deg A.$$

Since $\dim_\Bbbk L(B)$ is certainly nonnegative, this inequality implies that $L(B)$ is finite-dimensional. Then we can expand the left side of the first conclusion of the theorem to obtain

$$\dim_\Bbbk L(B) - \dim_\Bbbk L(A) = \deg B - \deg A,$$

and the proof is complete. □

The theorem identifies $\ell(B) - \deg B$ as a quantity of interest when we are trying to understand a divisor B. We shall undertake a study of this quantity, beginning first with the case of a divisor B equal to a multiple of the pole part $(x)_\infty$ of a principal divisor (x). Recall that the signs are arranged to have $(x)_\infty \geq 0$.

Lemma 9.7. For each x in $\mathbb{F}$ that is not in $\Bbbk$, there exists a constant C_x such that the multiple $p(x)_\infty$ of $(x)_\infty$ satisfies

$$\ell(p(x)_\infty) - \deg(p(x)_\infty) \geq C_x$$

for every integer p.

PROOF. Applying the observation at the end of Section 1, we form the integral closure R of $\Bbbk[x]$ in $\mathbb{F}$ and the integral closure R' of $\Bbbk[x^{-1}]$ in $\mathbb{F}$. The discrete valuations v for which $v(x) < 0$ are exactly those arising from prime ideals in the prime decomposition of $x^{-1}\Bbbk[x^{-1}]$, according to Corollary 6.10. Specifically the ideal $x^{-1}\Bbbk[x^{-1}]$ in R' decomposes as a product $Q_1^{e'_1} \cdots Q_{g'}^{e'_{g'}}$, and the corresponding discrete valuations have $v_{Q_k}(x^{-1}) = e'_k$. Theorem 9.3 shows that $\deg(x)_\infty = \dim_{\Bbbk(x)} \mathbb{F}$.

Let $n = \dim_{\Bbbk(x)} \mathbb{F}$. Choose a basis $y_1, \ldots, y_n$ of $\mathbb{F}$ over $\Bbbk(x)$ consisting of members of R. Each v arising from a prime ideal of R has $v(y_j) \geq 0$ for $1 \leq j \leq n$ by Proposition 6.7. The remaining v's all have $v(x) < 0$, and therefore there exists an integer $k \geq 0$ such that $v(y_j) \geq kv(x)$ for $1 \leq j \leq n$ and for all these remaining v's. For this value of the integer k, the elements $y_1, \ldots, y_n$ all lie in $L(k(x)_\infty)$.

Let $m \geq 0$ be arbitrary. The v's coming from some Q_k, i.e., those with $v(x) < 0$, have $v(x^i) \geq v(x^m)$ whenever $0 \leq i \leq m$, and the remaining v's, i.e., those with $v(x) \geq 0$, all have $v(x^i) \geq 0$ for $0 \leq i \leq m$. Therefore $1, x, x^2, \ldots, x^m$ all lie in $L((x^m)_\infty) = L(m(x)_\infty)$.

Multiplying, we see that $x^i y_j$ lies in $L\big((k+m)(x)_\infty\big)$ for $0 \leq i \leq m$ and $1 \leq j \leq n$. These elements $x^i y_j$ are linearly independent over $\mathbb{k}$, and therefore

$$\ell\big((k+m)(x)_\infty\big) \geq (m+1)n = (m+1)\deg(x)_\infty.$$

Since deg is a homomorphism from $D_\mathbb{F}$ into $\mathbb{Z}$,

$$\deg\big((k+m)(x)_\infty\big) = (k+m)\deg(x)_\infty.$$

Therefore each $m \geq 0$ has

$$\ell\big((k+m)(x)_\infty\big) - \deg\big((k+m)(x)_\infty\big) \geq (m+1-k-m)\deg(x)_\infty$$
$$= (1-k)\deg(x)_\infty.$$

We have therefore proved that

$$\ell(q(x)_\infty) - \deg(q(x)_\infty) \geq (1-k)\deg(x)_\infty$$

for all integers q that are sufficiently positive. If p is any integer, we can find q as above with $p \leq q$. Then $p(x)_\infty \leq q(x)_\infty$, and Theorem 9.6 shows that

$$(1-k)\deg(x)_\infty \leq \ell(q(x)_\infty) - \deg(q(x)_\infty) \leq \ell(p(x)_\infty) - \deg(p(x)_\infty).$$

This proves the lemma with $C_x = (1-k)\deg(x)_\infty$. $\square$

Lemma 9.8. If A is any divisor and x is any member of $\mathbb{F}^\times$, then $L((x)+A) \cong L(A)$ canonically. Therefore $\ell((x)+A) = \ell(A)$. In addition, $\deg((x)+A) = \deg A$.

PROOF. Define a $\mathbb{k}$ linear mapping $\varphi : L(A) \to \mathbb{F}$ by $\varphi(y) = x^{-1}y$. This is certainly one-one, and its image is contained in $L((x)+A)$ because any nonzero z in $L(A)$ has $(z) \geq -A$ and then also $(x^{-1}z) = -(x)+(z) \geq -(x)-A$. Similarly $\psi(y) = xy$ is one-one and carries $L((x)+A)$ into $L(A)$. By inspection, $\psi\varphi = 1$ and $\varphi\psi = 1$. Therefore $L((x)+A)$ and $L(A)$ are canonically isomorphic and have the same dimension over $\mathbb{k}$. For the last conclusion, $\deg((x)+A) = \deg(x) + \deg A = \deg A$ by Theorem 9.3. $\square$

Theorem 9.9 (Riemann's inequality). For each x in $\mathbb{F}$ that is not in $\mathbb{k}$, let g_x be the integer such that $1 - g_x$ is the largest possible C_x with

$$\ell\big(p(x)_\infty\big) - \deg\big(p(x)_\infty\big) \geq C_x$$

for every integer p. Then

(a) the integer $g = g_x$ is independent of x,
(b) g is ≥ 0,
(c) $\ell(A) - \deg A \geq 1 - g$ for every divisor A.

3. Genus

REMARKS. The integer g_x in the theorem exists by Lemma 9.7. Once it has been proved to be an integer g independent of x, it is called the **genus** of the function field $\mathbb{F}$ over $\mathbb{k}$.

PROOF. We begin by proving (c) with g replaced by g_x. Let C_x be any integer with the property that $\ell(p(x)_\infty) - \deg(p(x)_\infty) \geq C_x$ for all p. If a divisor A is given, we can write $A = A_0 - A_\infty$, where $A_0 = \sum_{\text{ord}_v A > 0} (\text{ord}_v A)v$ and $A_\infty = \sum_{\text{ord}_v A < 0} (-\text{ord}_v A)v$. Then $A \leq A_0$, and Theorem 9.6 shows that $\ell(A) - \deg A \geq \ell(A_0) - \deg A_0$. Thus it is enough to prove (c) for A_0. Let p be any integer ≥ 0. Since $A_0 \geq 0$, we have $p(x)_\infty - A_0 \leq p(x)_\infty$. Hence a second application of Theorem 9.6 shows that

$$\ell(p(x)_\infty - A_0) - \deg(p(x)_\infty - A_0) \geq \ell(p(x)_\infty) - \deg(p(x)_\infty) \geq C_x.$$

Since deg is a homomorphism, this inequality implies that

$$\ell(p(x)_\infty - A_0) \geq C_x + p \deg(x)_\infty - \deg A_0.$$

Fix an integer p large enough for the right side to be positive. For this p, the vector space $L(p(x)_\infty - A_0)$ is nonzero; let y be a nonzero member of it. This y has $(y) \geq -(p(x)_\infty - A_0)$, and hence $p(x)_\infty \geq A_0 - (y)$. A third application of Theorem 9.6, in combination with Lemma 9.8, shows that

$$\ell(p(x)_\infty) - \deg(p(x)_\infty) \leq \ell(A_0 - (y)) - \deg(A_0 - (y))$$
$$= \ell(A_0) - \deg A_0.$$

The left side is $\geq C_x$, and hence $\ell(A_0) - \deg A_0 \geq C_x$. Therefore

$$\ell(A) - \deg A \geq C_x \tag{$*$}$$

for every divisor A. Since one choice of C_x is $1 - g_x$, this proves (c).

Taking $A = p(y)_\infty$, we see that the best C_y has $C_y \geq C_x$. Since the roles of x and y can be interchanged, this proves (a). Finally if we take $A = 0$ in (c) and apply Corollary 9.4, we see that $1 - 0 \geq 1 - g$. Thus $g \geq 0$. This proves (b). □

EXAMPLES OF GENUS.

(1) $\mathbb{F} = \mathbb{k}(x)$ for a transcendental x. In the proof of Lemma 9.7, we have $n = 1$ and can take $y_1 = 1$. Then $k = 0$, and the proof of the lemma shows that the inequality of the lemma holds with $C_x = (1 - 0) \deg(x)_\infty = 1$. Therefore $1 - g \geq C_x = 1$, and $g \leq 0$. So $g = 0$ by Theorem 9.9b.

(2) $\mathbb{F} = \mathbb{C}[x, y]/(y^2 - x^4 + 1)$. This example was discussed in Section 1, and we have $x^{-1}R' = P_1P_2$ with $P_1 = (x^{-1}, x^{-2}y + 1)$ and $P_2 = (x^{-1}, x^{-2}y - 1)$. The corresponding valuations therefore have $v_{P_1}(x) = v_{P_2}(x) = -1$. Meanwhile, the elements 1 and y form a basis of $\mathbb{F}$ over $\Bbbk(x)$. The element 1 has $v_{P_1}(1) = v_{P_2}(1) = 0$; so 1 is in $L(p(x)_\infty)$ for every $p \geq 0$. Since $x^{-2}y$ is the sum of a generator of P_1 and a generator of P_2, $x^{-2}y$ lies in R'. Write $(x^{-2}y) = I_1 \cdots I_l$, where each I_j is a prime ideal in R'. Since $x^{-2}y$ and P_1 together generate 1, P_1 is not one of the ideals I_j. Similarly P_2 is not one of the I_j's. Thus $(y) = (x^{-1})^{-2}(x^{-2}y) = (P_1P_2)^{-2}I_1 \cdots I_l$, and we obtain $v_{P_1}(y) = v_{P_2}(y) = -2$. Hence y lies in $L(2(x)_\infty)$, and we can take $k = 2$ in the proof of Lemma 9.7. For this k, we have $C_x = (1 - 2)\deg(x)_\infty = -2$. Therefore $1 - g \geq C_x = -2$, and $g \leq 3$. In fact, $g = 1$ here, as a special case of the next example. Thus a routine use of the estimate from Lemma 9.7 has its limitations.

(3) $\mathbb{F} = \Bbbk[x, y]/(y^2 - p(x))$, where $p(x)$ is a square-free polynomial of degree m and $\Bbbk$ has characteristic $\neq 2$. Then $g = \frac{1}{2}m - 1$ if m is even and $g = \frac{1}{2}(m - 1)$ if m is odd. This computation will be carried out in Problems 12–20 at the end of the chapter.

Theorem 9.9 gives the lower bound of $1 - g$ for $\ell(A) - \deg A$ for all divisors A. There is also an upper bound, with the proviso that $L(A) \neq 0$.

Proposition 9.10. If A is any divisor such that $L(A) \neq 0$, then

$$\ell(A) - \deg A \leq 1.$$

Hence any divisor A with $\deg A \leq -1$ has $\ell(A) = 0$.

PROOF. Let y be a member of $\mathbb{F}^\times$ that lies in $L(A)$. Then every $v \in \mathbb{V}_\mathbb{F}$ has $v(y) \geq -\operatorname{ord}_v A$ and hence $0 \geq -\operatorname{ord}_v A - v(y) = -\operatorname{ord}_v(A + (y))$. This inequality says that $A + (y) \geq 0$. Then Corollary 9.4 and Theorem 9.6 together give

$$1 = \ell(0) - \deg 0 \geq \ell(A + (y)) - \deg(A + (y)),$$

and the right side equals $\ell(A) - \deg A$ by Lemma 9.8. Then $1 - \deg A \leq \ell(A) - \deg A \leq 1$, and we must have $\deg A \geq 0$ whenever $\ell(A) \geq 1$. □

4. Riemann–Roch Theorem

Riemann's inequality, proved in Section 3, shows that every divisor A satisfies $\ell(A) - \deg A \geq 1 - g$, where g is the genus of the curve in question. The Riemann–Roch Theorem, to be proved in the present section, gives an interpretation for the difference between the two sides of the inequality.

4. Riemann–Roch Theorem

In the classical setting of compact Riemann surfaces, the proof of the Riemann–Roch Theorem makes use of meromorphic differential forms, sometimes called abelian differentials by complex analysts. Meromorphic differential forms are objects that locally look like $f(z)\,dz$, where z is a local coordinate and $f(z)$ is a meromorphic function, and that fit together to be globally defined on the complex manifold. What the formula $f(z)\,dz = g(w)\,dw$ for fitting together means that in the overlap of the regions for two local coordinates z and w, $f(z)\,dz = g(w(z))\frac{dw}{dz}\,dz$ holds and hence $f(z) = g(w(z))\frac{dw}{dz}$. In the language of differential geometry, a meromorphic differential form is a meromorphic section of the cotangent bundle of the complex manifold. An important step that has to be carried out to make these differential forms useful is to prove a version of the Residue Theorem. This theorem says that the sum over all points of the manifold of the residues of the differential form is 0, the residue of $f(z)\,dz$ at the point corresponding to $z = 0$ being the coefficient of z^{-1} in the Laurent expansion[9] of $f(z)$ about 0. Once this theorem is in hand, one can begin to prove the Riemann–Roch Theorem.

In our present setting with the function field $\mathbb{F}$ in one variable over $\Bbbk$, it is not too hard to define an analog of meromorphic differential forms and to establish that they behave the way one would expect from differential calculus. In order to make use of these forms, one has to prove an analog of the Residue Theorem, and doing so requires some hard work. A. Weil discovered that this construction could be bypassed and that one could prove the theorem directly. The idea is to introduce the tool that differential forms make available and to skip the differential forms themselves.

It is worth understanding this background in a little more detail because otherwise the proof below may seem very strange indeed. To fix the ideas for this background only, suppose that the base field $\Bbbk$ is algebraically closed. Let us recall that elements of $\mathbb{V}_\mathbb{F}$ are meant to correspond to points of a zero locus in projective space, at least when the curve is everywhere nonsingular. We write this correspondence as $v \mapsto p(v)$. A local coordinate about $p(v)$ is denoted by a symbol like z classically, and in the setup with valuations, it is simply a member of the valuation ideal of v with $v(z) = 1$. A differential form that is given locally by classical expressions like $f(z)\,dz$ attaches to each v in $\mathbb{V}_\mathbb{F}$ the function $g_v \mapsto \mathrm{Residue}_{p(v)}(g_v f\,dz)$, where g_v is any Laurent expansion about $p(v)$.

Classically this Laurent expansion is to be convergent in some deleted neighborhood of $p(v)$, and it involves only finitely many negative powers of the local coordinate. The assumption that it converges is not important because if $v(f) = n$, then the only powers of z whose coefficients in g_v affect the residue at $p(v)$ are the kth powers for $k + n \leq -1$. Thus the assumption on g_v is that it is

---
[9] One has to show that this coefficient is independent of the choice of the local coordinate.

a member of the Laurent series field $\mathbb{k}((z))$. To compute the residue for $g_v f\, dz$, we need to know how to interpret $f(z)$ as a Laurent series about $p(v)$. Let R_v be the valuation ring of v, and let P_v be the valuation ideal. The field R_v/P_v is a finite extension of $\mathbb{k}$ and must be isomorphic to $\mathbb{k}$ because $\mathbb{k}$ is algebraically closed. For each $c \in \mathbb{k}$, choose a member $a_c \in R_v$ such that the coset $a + P_v$ corresponds to c; we may assume that $a_0 = 0$. Denote the set of these elements a_c by $R_\mathbb{k}$. If $v(f) = n$, then $h = z^{-n} f$ is in R_v, and thus some unique a_0 in $R_\mathbb{k}$ has the property that $h - a_0$ is in P_v. Hence $z^{-1}(h - a_0)$ is in R_v, and some unique a_1 in $R_\mathbb{k}$ has the property that $z^{-1}(h - a_0) - a_1$ is in P_v. From this, $z^{-1}(z^{-1}(h - a_0) - a_1)$ is in R_v, and we can continue to subtract members of $R_\mathbb{k}$ and divide by z in this way. The result is that $h = a_0 + a_1 z + a_2 z^2 + \cdots$ in the sense that $v(h - a_0 - a_1 z - \cdots - a_k z^k) \geq k + 1$ for every k. Therefore $f = z^n h = z^n(a_0 + a_1 z + a_2 z^2 + \cdots)$. If we replace each a_k by the corresponding member c_k of $\mathbb{k}$, then $z^n(c_0 + c_1 z + c_2 z^2 + \cdots)$ is the member of $\mathbb{k}((z))$ that we associate to f.

With this identification in place, we can regard the given differential form as yielding a $\mathbb{k}$ linear function

$$\text{Residue} : \prod_{v \in \mathbb{V}_\mathbb{F}} \mathbb{k}((z)) \to \prod_{v \in \mathbb{V}_\mathbb{F}} \mathbb{k}.$$

We want to cut down the domain of this mapping so the sum of the residues is meaningful for every member of the image. The local expressions $f(z)\, dz$ involve only finitely many poles in a neighborhood of each point, and compactness implies that there are only finitely many such points globally. Except at these points the residue of $g_v f\, dz$ can be nonzero only if g_v has a pole at $p(v)$. Thus we can ensure that the sum of the residues is meaningful if we assume that $v(g_v) \geq 0$ except for finitely many v.

For algebraic purposes the domain is still unnecessarily large. Since each local coordinate in the algebraic realization is actually a member of $\mathbb{F}$, the only members of $\mathbb{k}((z))$ that we need to handle at each point are the members of $\mathbb{F}$. So let $\mathcal{A}_\mathbb{F}^* = \prod_{v \in \mathbb{V}_\mathbb{F}} \mathbb{F}$, and let $\mathcal{A}_\mathbb{F}$ be the $\mathbb{k}$ subspace of all members $\{g_v\}$ of the product such that $v(g_v) < 0$ only finitely often. Then the differential form gives us a $\mathbb{k}$ linear functional

$$\text{Sum of Residues} : \mathcal{A}_\mathbb{F} \to \mathbb{k}.$$

We have seen that if the differential form is given by $f(z)\, dz$ locally near $p(v)$ and if $v(g_v) \geq -v(f)$, then the residue is 0 at $p(v)$. Hence there is some divisor A, depending on the differential form, such that if $v(g_v) \geq -\operatorname{ord}_v A$ for all $v \in \mathbb{V}_\mathbb{F}$, then all residues are 0 and the sum of the residues is 0. Consequently the kernel of the sum-of-residues map associated to the differential form contains all tuples $\{g_v\}$ of $\mathcal{A}_\mathbb{F}$ such that $v(g_v) \geq -\operatorname{ord}_v A$ for this divisor A and all v.

4. Riemann–Roch Theorem

Finally there is one more classical fact to bring into play. This is the Residue Theorem itself, saying that the sum of the residues is zero for any meromorphic differential form. If $\{g_v\}$ is actually a constant tuple with $g_v = h$ for some $h \in \mathbb{F}$, then the sum-of-residues map as defined above is giving us the classical sum of residues for the product of h and the given differential form. This sum is zero. In other words, every member of the diagonally embedded $\mathbb{F}$ in $\mathcal{A}_\mathbb{F}$ lies in the kernel of the sum-of-residues map associated to the differential form.

Weil's idea in a nutshell is that instead of developing differential forms, working with residues, and proving the consequence of the Residue Theorem, one should just start with any abstract linear functional on $\mathcal{A}_\mathbb{F}$ that satisfies the conditions that we noted above. Then the Riemann–Roch Theorem drops out fairly easily. This is the approach we shall follow. The abstract kind of linear functional on $\mathcal{A}_\mathbb{F}$ will be called a "differential" in what follows, as a reminder of the classical object that lies behind it.[10]

Without further ado, we proceed with the Riemann–Roch Theorem. In this section, $\mathbb{F}$ denotes a function field in one variable over a field $\Bbbk$, and we assume that every element of $\mathbb{F}$ outside $\Bbbk$ is transcendental over $\Bbbk$. We continue with the notation $\mathbb{V}_\mathbb{F}$, $D_\mathbb{F}$, f_v, $\mathrm{ord}_v A$, $\deg A$, and (x) for $x \in \mathbb{F}^\times$, all as in Sections 2–3, and with the notation $L(A)$ and $\ell(A)$ as in Section 3. If A is a divisor, we let

$$\delta(A) = \ell(A) - \deg A - (1 - g).$$

Riemann's inequality (Theorem 9.9) implies that $\delta(A) \geq 0$ for all A's and that $\delta(A) = 0$ for some A's. We seek an interpretation of $\delta(A)$.

Let $\mathcal{A}_\mathbb{F}^*$ be the ring of all functions from $\mathbb{V}_\mathbb{F}$ into $\mathbb{F}$, with the operations taken pointwise. It is customary to write such a function ξ as $v \mapsto \xi_v$ rather than as $v \mapsto \xi(v)$. Let $\mathcal{A}_\mathbb{F}$ be the subring[11] of all members ξ of $\mathcal{A}_\mathbb{F}^*$ such that $v(\xi_v) < 0$ for only finitely many v in $\mathbb{V}_\mathbb{F}$. We shall treat $\mathcal{A}_\mathbb{F}$ as an infinite-dimensional associative $\Bbbk$ algebra with identity.

Consider the diagonal map $\Delta : \mathbb{F} \to \mathcal{A}_\mathbb{F}$ defined by the formula $\Delta(x)_v = x$ for all $x \in \mathbb{F}$. Under this map, the member x of $\mathbb{F}$ goes to the function whose value at each v is x. The reason that $\Delta(x)$ is in $\mathcal{A}_\mathbb{F}$ and not just $\mathcal{A}_\mathbb{F}^*$ is that $v(x) < 0$ for only finitely many $v \in \mathbb{V}_\mathbb{F}$. The map Δ is a one-one $\Bbbk$ algebra homomorphism.

---

[10]Weil's argument dates to 1935. It appears in book form in Weil's *Basic Number Theory*, where the details are carried out when $\Bbbk$ is a finite field and where comments are made for general $\Bbbk$. Lang simplified Weil's argument and wrote it down for algebraically closed fields $\Bbbk$ in his *Introduction to Algebraic and Abelian Functions*. A version of this argument for general $\Bbbk$ appears in Villa Salvador's book. The present exposition benefits from all three of these books.

[11]For readers familiar with Section VI.10, the notation is intended to hint at "adeles" of $\mathbb{F}$. However, completions and topologies will play no role in the construction.

For each divisor A, define
$$\mathcal{L}(A) = \{\xi \in \mathcal{A}_{\mathbb{F}} \mid v(\xi_v) \geq -\mathrm{ord}_v(A)\}.$$
It is immediate from the definitions that
$$\mathcal{L}(A) \cap \Delta(\mathbb{F}) = \Delta(L(A)).$$
Let us see that
$$A \leq B \quad \text{if and only if} \quad \mathcal{L}(A) \subseteq \mathcal{L}(B).$$
In fact, the "only if" part of the statement is evident. Conversely suppose that $\mathcal{L}(A) \subseteq \mathcal{L}(B)$. Choose for each $v \in \mathbb{V}_{\mathbb{F}}$ an element π_v in $\mathbb{F}$ with $v(\pi_v) = 1$. The function $\xi_A : \mathbb{V}_{\mathbb{F}} \to \mathbb{F}$ defined by $(\xi_A)_v = \pi^{-\mathrm{ord}_v A}$ has $v((\xi_A)_v) = -\mathrm{ord}_v A$ and lies in $\mathcal{A}_{\mathbb{F}}$, since $\mathrm{ord}_v A$ is nonzero for only finitely many v. The definitions show that ξ_A lies in $\mathcal{L}(A)$, hence in $\mathcal{L}(B)$. Thus $-\mathrm{ord}_v(A) = v((\xi_A)_v) \geq -\mathrm{ord}_v B$, $\mathrm{ord}_v A \leq \mathrm{ord}_v B$, and $A \leq B$. This proves the "if" part of the displayed equivalence. If we apply the equivalence twice, we see that
$$A = B \quad \text{if and only if} \quad \mathcal{L}(A) = \mathcal{L}(B).$$

Let us take note of two operations on divisors A and the effect of these operations on the spaces $\mathcal{L}(A)$. If A and B are divisors, we define $C = \min(A, B)$ pointwise by the formula $\mathrm{ord}_v C = \min(\mathrm{ord}_v A, \mathrm{ord}_v B)$. Then C is a divisor with $C \leq A$ and $C \leq B$. Thus $\mathcal{L}(C) \subseteq \mathcal{L}(A)$ and $\mathcal{L}(C) \subseteq \mathcal{L}(B)$, and we consequently obtain
$$\mathcal{L}(\min(A, B)) \subseteq \mathcal{L}(A) \cap \mathcal{L}(B).$$
Still with A and B as divisors, we define $C = \max(A, B)$ pointwise by the formula $\mathrm{ord}_v C = \max(\mathrm{ord}_v A, \mathrm{ord}_v B)$. Then $A \subseteq C$ and $B \subseteq C$, from which we obtain $\mathcal{L}(A) \subseteq \mathcal{L}(C)$ and $\mathcal{L}(B) \subseteq \mathcal{L}(C)$. This proves the inclusion $\subseteq$ in the identity
$$\mathcal{L}(A) + \mathcal{L}(B) = \mathcal{L}(\max(A, B)).$$
To prove $\supseteq$, let ξ be in $\mathcal{L}(\max(A, B))$. We shall decompose ξ as a sum $\eta + \zeta$ in $\mathcal{L}(A) + \mathcal{L}(B)$ with one of η_v and ζ_v equal to 0 for each v. Let v be given. Since ξ is in $\mathcal{L}(\max(A, B))$, $v(\xi_v) \geq -\mathrm{ord}_v(\max(A, B)) = -\max(\mathrm{ord}_v A, \mathrm{ord}_v B)$. That is, $-v(\xi_v) \leq \max(\mathrm{ord}_v A, \mathrm{ord}_v B)$. If $-v(\xi_v) \leq \mathrm{ord}_v A$, then define $\eta_v = \xi_v$ and $\zeta_v = 0$; otherwise, we have $-v(\xi_v) \leq \mathrm{ord}_v B$, and we define $\eta_v = 0$ and $\zeta_v = \xi_v$. Then $v(\eta_v) \geq -\mathrm{ord}_v A$ for all v, and $v(\zeta_v) \geq -\mathrm{ord}_v B$ for all v. This proves $\supseteq$ in the displayed formula.

Lemma 9.11. If A and B are divisors with $A \leq B$, then
$$\dim_{\Bbbk} \left(\mathcal{L}(B)/\mathcal{L}(A) \right) = \deg B - \deg A.$$

PROOF. Proceeding inductively, we see that it is enough to handle the case that $B = A + v_0$, where v_0 is in $\mathbb{V}_{\mathbb{F}}$. Thus we are to show that

$$\dim_{\Bbbk} \left(\mathcal{L}(A + v_0)/\mathcal{L}(A) \right) = f_{v_0} = \deg(v_0). \qquad (*)$$

Put $f = f_{v_0}$, let R_{v_0} be the valuation ring of v_0, and let P_{v_0} be the valuation ideal of v_0. To prove $\leq$ in $(*)$, we argue as in the proof of Lemma 9.5. Since v_0 carries $\mathbb{F}^\times$ onto $\mathbb{Z}$, we can choose an element $y \in \mathbb{F}^\times$ with $v_0(y) = \text{ord}_{v_0}(A + v_0)$.

Let $f + 1$ members $\xi^{(1)}, \ldots, \xi^{(f+1)}$ of $\mathcal{L}(A + v_0)$ be given. We shall produce an equation of linear dependence among the cosets $\xi^{(i)} + \mathcal{L}(A)$, and this will prove $\leq$ in $(*)$. Computation gives

$$v_0(\xi^{(i)}_{v_0} y) = v_0(\xi^{(i)}_{v_0}) + v_0(y) = v_0(\xi^{(i)}_{v_0}) + \text{ord}_{v_0}(A + v_0) \geq 0$$

for $1 \leq i \leq f + 1$, with the inequality at the right holding because $\xi^{(i)}$ is in $\mathcal{L}(A + v_0)$. Hence $\xi^{(i)}_{v_0} y$ is in R_{v_0}. Since $\dim_{\Bbbk}(R_{v_0}/P_{v_0}) = f$, there exist members $c_1, \ldots, c_{f+1}$ of $\Bbbk$ not all 0 such that $\sum_{i=1}^{f+1} c_i(\xi^{(i)}_{v_0} y + P_{v_0}) = P_{v_0}$, i.e., such that $\sum_{i=1}^{f+1} c_i \xi^{(i)}_{v_0} y$ lies in P_{v_0}. Then $\sum_{i=1}^{f+1} c_i \xi^{(i)}_{v_0}$ lies in $y^{-1} P_{v_0}$, and

$$v_0 \left(\sum_{i=1}^{f+1} c_i \xi^{(i)}_{v_0} \right) \geq -v_0(y) + 1 = -\text{ord}_{v_0}(A + v_0) + 1 = -\text{ord}_{v_0} A. \qquad (**)$$

Since each $\xi^{(i)}$ is in $\mathcal{L}(A + v_0)$, so is $\sum_{i=1}^{f+1} c_i \xi^{(i)}_{v_0}$. This fact and $(**)$ together show that $\sum_{i=1}^{f+1} c_i \xi^{(i)}_{v_0}$ is in $\mathcal{L}(A)$, i.e., that $\sum_{i=1}^{f+1} c_i \xi^{(i)} + \mathcal{L}(A)$ is the 0 coset. This proves the desired linear dependence and shows that $\dim_{\Bbbk} \mathcal{L}(A + v_0)/\mathcal{L}(A) \leq f$.

To prove $\geq$ in $(*)$, we shall produce f members $\xi^{(j)}$ of $\mathcal{L}(A + v_0)$ that are linearly independent modulo $\mathcal{L}(A)$. We begin by choosing η in $\mathcal{L}(A)$ with $v_0(\eta_{v_0}) = -\text{ord}_{v_0} A$. (For example take any member η' of $\mathcal{L}(A)$, change η'_{v_0} to a new value on which v_0 takes the value $-\text{ord}_{v_0} A$, and leave η' unchanged at all other v.) Let $x_1, \ldots, x_f$ be a set of representatives in R_{v_0} of the f members of a $\Bbbk$ basis of the quotient R_{v_0}/P_{v_0}, and let π_{v_0} be a member of $\mathbb{F}$ with $v_0(\pi_{v_0}) = 1$. Define $\xi^{(j)}$ for $1 \leq j \leq f$ by

$$\xi^{(j)}_v = \begin{cases} \eta_v & \text{for } v \neq v_0, \\ \eta_{v_0} x_j \pi_{v_0}^{-1} & \text{for } v = v_0. \end{cases}$$

For each j, we have

$$v_0(\eta_{v_0} x_j \pi^{-1}) = v_0(\eta_{v_0}) + v(x_j) - v_0(\pi_{v_0})$$
$$= -\text{ord}_{v_0} A + v(x_j) - 1 \geq -\text{ord}_{v_0} A - 1,$$

and thus $\xi^{(j)}$ is in $\mathcal{L}(A + v_0)$. To prove the linear independence modulo $\mathcal{L}(A)$, suppose that $c_1, \ldots, c_f$ are members of $\Bbbk$ such that $\sum_{j=1}^{f} c_j \xi^{(j)}$ is in $\mathcal{L}(A)$. In this case we have an inequality $v_0\big(\sum_{j=1}^{f} c_j \xi^{(j)}\big) \geq -\text{ord}_{v_0} A$, which expands out as

$$v_0\Big(\sum_{j=1}^{f} c_j \eta_{v_0} x_j \pi_{v_0}^{-1}\Big) \geq v_0(\eta_{v_0}).$$

Since $v_0(\pi_{v_0}^{-1}) = -1$, subtraction of $v_0(\eta_{v_0})$ from both sides yields $v_0\big(\sum_{j=1}^{f} c_j x_j\big) \geq 1$. Therefore $\sum_{j=1}^{f} c_j x_j$ lies in P_{v_0}. By the assumed linear independence over $\Bbbk$ of the x_j's modulo P_{v_0}, all the c_j's are 0. Therefore the elements $\xi^{(j)}$ are linearly independent modulo $\mathcal{L}(A)$, and the proof of $\geq$ in (∗) is complete. □

Lemma 9.12. If A and B are divisors with $A \leq B$, then there is an exact sequence in the category of $\Bbbk$ vector spaces given by

$$0 \longrightarrow L(B)/L(A) \xrightarrow{\psi} \mathcal{L}(B)/\mathcal{L}(A)$$
$$\xrightarrow{\varphi} (\mathcal{L}(B) + \Delta(\mathbb{F}))/(\mathcal{L}(A) + \Delta(\mathbb{F})) \longrightarrow 0.$$

Consequently

$$\dim_{\Bbbk}(\mathcal{L}(B) + \Delta(\mathbb{F}))/(\mathcal{L}(A) + \Delta(\mathbb{F})) = (\ell(A) - \deg A) - (\ell(B) - \deg B)$$
$$= \delta(A) - \delta(B).$$

PROOF. The map ψ is induced by the map $\Delta : L(B) \to \mathcal{L}(B)$ followed by passage to the quotient. It descends to $L(B)/L(A)$ because $\Delta(L(A)) \subseteq \mathcal{L}(A)$, and it is one-one because $\Delta(L(B)) \cap \mathcal{L}(A) \subseteq L(A)$. The map φ is induced by the map $x \mapsto x + \Delta(\mathbb{F})$ followed by passage to the quotient. It descends to $\mathcal{L}(B)/\mathcal{L}(A)$ because $\mathcal{L}(A)$ maps into $\mathcal{L}(A) + \Delta(\mathbb{F})$, and it is onto because $x \mapsto x + \Delta(\mathbb{F})$ carries $\mathcal{L}(B)$ onto $\mathcal{L}(B) + \Delta(\mathbb{F})$. The composition $\varphi\psi$ is 0 because $L(B)$ maps under Δ into $\Delta(\mathbb{F})$, which lies in the 0 coset.

To prove the exactness, let $\xi + \mathcal{L}(A)$ be in $\ker \varphi$. This condition means that ξ is in $\mathcal{L}(B)$ and has $\xi + \Delta(\mathbb{F})$ in $\mathcal{L}(A) + \Delta(\mathbb{F})$. Thus there exists η in $\mathcal{L}(A)$ with $\xi - \eta$ in $\Delta(\mathbb{F})$. Since ξ and η are in $\mathcal{L}(B)$, $\xi - \eta$ is in $\mathcal{L}(B) \cap \Delta(\mathbb{F}) \subseteq \Delta(L(B))$. Hence $\xi + \mathcal{L}(A) = (\xi - \eta) + \mathcal{L}(A)$ lies in $\Delta(L(B)) + \mathcal{L}(A) = \text{image } \psi$, and exactness is proved.

From the exactness we obtain

$$\dim_{\Bbbk} \mathcal{L}(B)/\mathcal{L}(A) = \dim_{\mathbb{K}} L(B)/L(A) + \dim_{\Bbbk}(\mathcal{L}(B) + \Delta(\mathbb{F}))/(\mathcal{L}(A) + \Delta(\mathbb{F})).$$

The left side equals $\deg B - \deg A$ by Lemma 9.11, and the first term on the right side equals $\ell(B) - \ell(A)$ by the finite dimensionality of $L(B)$ and $L(A)$, which was proved as part of Theorem 9.6. The result follows. □

Theorem 9.13. There exists a divisor C such that $\mathcal{A}_{\mathbb{F}} = \mathcal{L}(C) + \Delta(\mathbb{F})$. For each divisor A,
$$\delta(A) = \dim_{\mathbb{k}}\left(\mathcal{A}_{\mathbb{F}}/(\mathcal{L}(A) + \Delta(\mathbb{F}))\right).$$

PROOF. Riemann's inequality produces a divisor C, specifically any sufficiently large positive power of a divisor $(x)_\infty$, such that $\delta(C) = 0$. If we can show that $\mathcal{A}_{\mathbb{F}} = \mathcal{L}(C) + \Delta(\mathbb{F})$, then the dimensional equality in Lemma 9.12 with $B = C$ will complete the proof of the present theorem.

Suppose that there exists a member ξ of $\mathcal{A}_{\mathbb{F}}$ that is not in $\mathcal{L}(C) + \Delta(\mathbb{F})$. For each $v \in \mathbb{V}_{\mathbb{F}}$, let $a_v = \min(v(\xi_v), -\mathrm{ord}_v C)$, and define $C' = -\sum_{v \in \mathbb{V}_{\mathbb{F}}} a_v v$. Since ξ is in $\mathcal{A}_{\mathbb{F}}$, only finitely many integers $v(\xi_v)$ are negative. This fact and the fact that C is a divisor together imply that only finitely many a_v are negative. Since C is a divisor, only finitely many integers $-\mathrm{ord}_v C$ can be positive, and thus only finitely many a_v can be positive. Therefore C' is a divisor.

The definition of C' is arranged in such a way that $C \leq C'$. Also, every v has $v(\xi_v) \geq a_v = -\mathrm{ord}_v C'$, and hence ξ lies in $\mathcal{L}(C')$. Consequently
$$\dim_{\mathbb{k}}(\mathcal{L}(C') + \Delta(\mathbb{F}))/(\mathcal{L}(C) + \Delta(\mathbb{F})) \geq 1.$$

By Lemma 9.12, $\delta(C) - \delta(C') \geq 1$. Since C was assumed to have $\delta(C) = 0$, we obtain $-\delta(C') \geq 1$, in contradiction to the fact that $\delta(A) \geq 0$ for every divisor A. We conclude that every ξ in $\mathcal{A}_{\mathbb{F}}$ lies in $\mathcal{L}(C) + \Delta(\mathbb{F})$. □

Theorem 9.13 gives a first interpretation of the difference $\delta(A)$ between the two sides of Riemann's inequality (Theorem 9.9). We shall now apply Theorem 9.13 and reinterpret $\delta(A)$ as the dimension $\ell(B)$ of a suitable divisor B obtained from A, and then we will have obtained the Riemann–Roch Theorem.

A **differential** of $\mathbb{F}$ is a $\mathbb{k}$ linear functional ω on $\mathcal{A}_{\mathbb{F}}$ with the property that ω vanishes on $\mathcal{L}(A)$ for some divisor A and ω vanishes also on $\Delta(\mathbb{F})$. The set of all differentials of $\mathbb{F}$ will be denoted by $\mathrm{Diff}(\mathbb{F})$. Let us observe that $\mathrm{Diff}(\mathbb{F})$ is a vector subspace of $\mathbb{k}$ linear functionals on $\mathcal{A}_{\mathbb{F}}$. Scalar multiplication by $\mathbb{k}$ is not an issue. To see that $\mathrm{Diff}(\mathbb{F})$ is closed under pointwise addition, let ω and ω' be differentials vanishing on $\mathcal{L}(A)$ and $\mathcal{L}(B)$, respectively. We have seen that $\mathcal{L}(\min(A, B)) \subseteq \mathcal{L}(A) \cap \mathcal{L}(B)$. Thus $\omega + \omega'$ vanishes on $\mathcal{L}(\min(A, B))$. Since $\omega + \omega'$ vanishes also on $\Delta(\mathbb{F})$, $\omega + \omega'$ is a differential.

The $\mathbb{k}$ vector space of differentials vanishing on $\mathcal{L}(A) + \Delta(K)$ may be identified with the vector space of $\mathbb{k}$ linear functionals on the quotient $\mathcal{A}_{\mathbb{F}}/(\mathcal{L}(A) + \Delta(\mathbb{F}))$, and the latter space is finite-dimensional of dimension $\delta(A)$ by Theorem 9.13. Since a finite-dimensional vector space and its dual have the same dimension, the $\mathbb{k}$ vector space of differentials vanishing on $\mathcal{L}(A) + \Delta(K)$ has $\mathbb{k}$ dimension $\delta(A)$.

In addition, $\mathrm{Diff}(\mathbb{F})$ carries a scalar multiplication by $\mathbb{F}$ that makes it into an $\mathbb{F}$ vector space. What is required to verify this statement is a definition, and then the verification of the properties of an $\mathbb{F}$ vector space is routine. If y is in $\mathbb{F}$ and ω is a differential, we define $y\omega$ on $\mathcal{A}_\mathbb{F}$ by $(y\omega)(\xi) = \omega(\Delta(y)\xi)$. The linear functional $y\omega$ vanishes on $\Delta(\mathbb{F})$ because Δ is a homomorphism. It is enough to check for $y \neq 0$ that

if ω vanishes on $\mathcal{L}(A)$, then $y\omega$ vanishes on $\mathcal{L}(A + (y))$,

where (y) is the principal divisor corresponding to y. To prove this vanishing, let ξ be in $\mathcal{L}(A + (y))$. Then $v(\xi_v) \geq -\mathrm{ord}_v(A + (y)) = -\mathrm{ord}_v A - \mathrm{ord}_v(y) = -\mathrm{ord}_v A - v(y)$, which implies that $v(\xi_v y) \geq -\mathrm{ord}_v A$, which implies that $\xi \Delta(y)$ lies in $\mathcal{L}(A)$, which implies that $\omega(\xi \Delta(y)) = 0$, which implies that $(y\omega)(\xi) = 0$. This proves the asserted vanishing, and it follows that $\mathrm{Diff}(\mathbb{F})$ carries a well-defined scalar multiplication by $\mathbb{F}$.

Each set $\mathcal{L}(A)$, where A is a divisor, will be called a **parallelotope** of $\mathcal{A}_\mathbb{F}$. These sets are large subsets of $\mathcal{A}_\mathbb{F}$, since $\dim_\mathbb{k} \mathcal{A}_\mathbb{F}/(\mathcal{L}(A) + \Delta(\mathbb{F}))$ is finite and $\dim_\mathbb{k} \mathcal{A}_\mathbb{F}/\Delta(\mathbb{F})$ is infinite. We are going to associate a particular parallelotope to each nonzero differential. Since we have seen that distinct parallelotopes correspond to distinct divisors, we shall obtain a way of associating a divisor to each nonzero differential.

Corollary 9.14. If ω is a nonzero differential and $\mathcal{L}(A)$ is a parallelotope in its kernel, then

$$\ell(A) \leq \delta(0) \quad \text{and} \quad \deg A \leq \delta(0) + g - 1.$$

Consequently there exists a unique maximum parallelotope on which ω vanishes.

REMARKS. In view of the remarks before the corollary, we therefore obtain a function $\omega \mapsto \mathrm{Div}(\omega)$ from the set $\mathrm{Diff}(\mathbb{F}) - \{0\}$ of nonzero differentials into the set $D_\mathbb{F}$ of divisors.

PROOF. If we know that $\ell(A) \leq \delta(0)$, then addition to this inequality of Riemann's inequality $\deg A - \ell(A) \leq g - 1$ as given in Theorem 9.9 shows that

$$\deg A \leq \delta(0) + g - 1$$

and proves the second inequality. The inequality $\ell(A) \leq \delta(0)$ is trivial if $L(A) = 0$.

Therefore we may assume in the two inequalities that $L(A) \neq 0$. Let y be any nonzero member of $L(A)$. Since the kernel of ω contains $\mathcal{L}(A)$, the kernel of $y\omega$ contains $\mathcal{L}(A + (y))$, by a computation made above. Meanwhile, the element y, being in $L(A)$, has $(y) \geq -A$ and hence $0 \leq A + (y)$. Therefore $\mathcal{L}(0) \subseteq \mathcal{L}(A + (y))$, and the kernel of $y\omega$ contains $\mathcal{L}(0)$. Since the kernel of $y\omega$ contains $\Delta(\mathbb{F})$, $y\omega$ is well defined on the quotient space $\mathcal{A}_\mathbb{F}/(\mathcal{L}(0) + \Delta(\mathbb{F}))$.

Now suppose that $y_1, \ldots, y_n$ is a $\Bbbk$ basis of $L(A)$. Let us use the fact that $\omega \neq 0$ to prove that $y_1\omega, \ldots, y_n\omega$ are linearly independent when viewed on $\mathcal{A}_\mathbb{F}/(\mathcal{L}(0) + \Delta(\mathbb{F}))$: If $c_1, \ldots, c_n$ are members of $\Bbbk$ not all 0, then $z = \sum_{j=1}^n c_j y_j$ is a nonzero member of $L(A)$, and we have just seen that $z\omega$ is well defined on $\mathcal{A}_\mathbb{F}/(\mathcal{L}(0) + \Delta(\mathbb{F}))$. Then we have $\sum_{j=1}^n c_j(y_j\omega) = \left(\sum_{j=1}^n c_j y_j\right)\omega = z\omega$, and this cannot act as 0 on $\mathcal{A}_\mathbb{F}/(\mathcal{L}(0) + \Delta(\mathbb{F}))$ without being identically 0 on $\mathcal{A}_\mathbb{F}$. Since any ξ_0 such that $\omega(\xi_0) \neq 0$ has the property that $z\omega(\Delta(z)^{-1}\xi_0) \neq 0$, the linear functionals $y_1\omega, \ldots, y_n\omega$ on $\mathcal{A}_\mathbb{F}/(\mathcal{L}(0) + \Delta(\mathbb{F}))$ are linearly independent.

We know that $\delta(0) = \dim_\Bbbk \mathcal{A}_\mathbb{F}/(\mathcal{L}(0) + \Delta(\mathbb{F}))$ by Theorem 9.13, and hence

$$n = \ell(A) \leq \delta(0).$$

This completes the proof of the two inequalities.

We turn to the existence and uniqueness of the maximum parallelotope on which ω vanishes. We continue to assume that $\omega \neq 0$. Now suppose that A is a divisor such that ω vanishes on $\mathcal{L}(A)$. Suppose that B is a divisor for which $B \leq A$ fails and for which $\omega(\mathcal{L}(B)) = 0$. We know that the divisor $\max(A, B)$ has the property that $\mathcal{L}(\max(A, B)) = \mathcal{L}(A) + \mathcal{L}(B)$. Since ω vanishes on $\mathcal{L}(A)$ and $\mathcal{L}(B)$, it follows that it vanishes on $\mathcal{L}(\max(A, B))$. Since $B \leq A$ fails, there exists some $v_0 \in \mathbb{V}_\mathbb{F}$ with $\text{ord}_{v_0} B > \text{ord}_{v_0} A$, and this v_0 has $\text{ord}_{v_0} \max(A, B) > \text{ord}_{v_0} A$. Thus $\deg \max(A, B) > \deg A$.

The second inequality proved above shows that the degree is bounded on all divisors whose parallelotopes are in $\ker \omega$. In finitely many steps we consequently arrive at a divisor C with $\mathcal{L}(C) \subseteq \ker \omega$ such that any divisor B with $\mathcal{L}(B) \subseteq \ker \omega$ has $B \leq C$. Then C is the unique maximum divisor on whose parallelotope ω vanishes. The parallelotope determines the divisor, and the proof of the corollary is complete. $\square$

Recall from Section 2 that the additive subgroup $P_\mathbb{F}$ of principal divisors within the group $D_\mathbb{F}$ of all divisors breaks $D_\mathbb{F}$ into equivalence classes known as **divisor classes**. The group $C_\mathbb{F} = D_\mathbb{F}/P_\mathbb{F}$ is the group of all divisor classes. The operation of a principal divisor (y), for $y \in \mathbb{F}^\times$, on a divisor A is $A \mapsto A + (y)$. On the other hand, we have seen that if a nonzero differential ω vanishes on $\mathcal{L}(A)$, then $y\omega$ vanishes on $\mathcal{L}(A + (y))$. In the notation of the remarks with Corollary 9.14, we therefore have

$$\text{Div}(y\omega) = \text{Div}(\omega) + (y).$$

A single orbit of nonzero differentials under the scalar-multiplication action on $\text{Diff}(\mathbb{F})$ by $\mathbb{F}^\times$ thus yields a single divisor class within $D_\mathbb{F}$. We shall show that $\text{Diff}(\mathbb{F})$ is 1-dimensional as an $\mathbb{F}$ vector space. Then the nonzero differentials form a single orbit under $\mathbb{F}^\times$, and the divisors that arise as $\text{Div}(\omega)$ for some nonzero differential ω form a single divisor class.

Lemma 9.15. As a vector space over $\mathbb{F}$, the space Diff($\mathbb{F}$) of differentials is 1-dimensional.

PROOF. First we prove that Diff($\mathbb{F}$) is nonzero. Referring to Theorem 9.13, we know that $\delta(A) = \dim_\mathbb{k} \left(\mathcal{A}_\mathbb{F} / (\mathcal{L}(A) + \Delta(\mathbb{F})) \right)$. If $\delta(A) > 0$, then there exist nonzero linear functionals on $\mathcal{A}_\mathbb{F}/(\mathcal{L}(A) + \Delta(\mathbb{F}))$, and the lift of such a nonzero linear functional to $\mathcal{A}_\mathbb{F}$ is a nonzero differential. Thus it is enough to produce a divisor A with $\delta(A) > 0$. Fix v_0 in $\mathbb{V}_\mathbb{F}$, and let $A = -2v_0$. Proposition 9.10 shows that $\ell(A) = 0$. Therefore

$$\delta(A) = \ell(A) - \deg A - (1-g) = 2 + g - 1 = g + 1 > 0,$$

and this A has $\delta(A) > 0$.

Now we shall prove that the $\mathbb{F}$ dimension of Diff($\mathbb{F}$) is at most 1. Arguing by contradiction, suppose that ω and ω' are differentials that are linearly independent over $\mathbb{F}$. If ω vanishes on $\mathcal{L}(A)$ and ω' vanishes on $\mathcal{L}(A')$, then $\omega + \omega'$ vanishes on $\mathcal{L}(A) \cap \mathcal{L}(A') \supseteq \mathcal{L}(C)$, where $C = \min(A, A')$. Let B be an arbitrary divisor. Suppose for the moment that $L(B) \neq 0$. If $y \neq 0$ is in $L(B)$, then $(y) \geq -B$, and $C + (y) \geq C - B$. So $\mathcal{L}(C + (y)) \supseteq \mathcal{L}(C - B)$. We have seen that the vanishing of ω on $\mathcal{L}(C)$ implies the vanishing of $y\omega$ on $\mathcal{L}(C + (y))$. Therefore $y\omega$ vanishes on $\mathcal{L}(C - B)$. Similarly $y\omega'$ vanishes on $\mathcal{L}(C - B)$.

Still with $L(B) \neq 0$, let $n = \ell(B)$, and let $x_1, \ldots, x_n$ and $y_1, \ldots, y_n$ be bases of $L(B)$ over $\mathbb{k}$. Then $x_1\omega, \ldots, x_n\omega, y_1\omega', \ldots, y_n\omega'$ are linearly independent over $\mathbb{k}$ because a relation

$$\sum_{i=1}^n a_i x_i \omega + \sum_{j=1}^n b_j y_j \omega' = 0$$

would mean that the members $x = \sum_{i=1}^n a_i x_i$ and $y = \sum_{j=1}^n b_j y_j$ of $\mathbb{F}$ have $x\omega + y\omega' = 0$. Since ω and ω' are assumed to be linearly independent over $\mathbb{F}$, $x = y = 0$. But then $a_i = 0$ for all i and $b_j = 0$ for all j. Consequently we can generate $2n$ linearly independent differentials that all vanish on $\mathcal{L}(C - B)$. These differentials may be regarded as linear functionals on the $\mathbb{k}$ vector space $\mathcal{A}_\mathbb{F}/(\mathcal{L}(C - B) + \Delta(\mathbb{F}))$, whose $\mathbb{k}$ dimension is $\delta(C - B)$ by Theorem 9.13.

Consequently

$$\delta(C - B) \geq 2\ell(B),$$

and this inequality is true also if $L(B) = 0$, by Riemann's inequality. Substituting from the formula for $\delta(\cdot)$, we obtain

$$\ell(C - B) - \deg(C - B) - 1 + g \geq 2\ell(B)$$
$$= 2\big(\deg B + 1 - g) + \delta(B)\big)$$
$$\geq 2 \deg B + 2 - 2g$$

because Riemann's inequality shows that $\delta(B) \geq 0$. Replacing $\deg(C - B)$ by $\deg C - \deg B$ gives

$$\deg B \leq \ell(C - B) - \deg C - 3 + 3g. \tag{$*$}$$

Proposition 9.10 shows that $\ell(C - B) \leq 1 + \deg(C - B)$ if $\ell(C - B) \neq 0$. In this case the two inequalities together give

$$2 \deg B \leq -2 + 3g;$$

hence $\ell(C - B) = 0$ if $\deg B$ is positive and sufficiently large. Choosing then a divisor B with $\deg B$ positive and sufficiently large, we have $\ell(C - B) = 0$, and $(*)$ gives

$$\deg B \leq -\deg C - 3 + 3g.$$

Since the right side is fixed and the left side can be made arbitrarily large, we have arrived at a contradiction. $\square$

As a result of Lemma 9.15, the divisors of the form $\text{Div}(\omega)$ for some nonzero differential ω constitute a single class in the group $C_{\mathbb{F}} = D_{\mathbb{F}}/P_{\mathbb{F}}$ of divisor classes. This class is called the **canonical class** of $\mathbb{F}$, and any divisor in the class is called a **canonical divisor**.

Theorem 9.16 (Riemann–Roch Theorem). Let $\mathbb{F}$ be a function field in one variable over a field $\Bbbk$, and suppose that every member of $\mathbb{F}$ not in $\Bbbk$ is transcendental over $\Bbbk$. If A is any divisor of $\mathbb{F}$ and C is any canonical divisor, then

$$\ell(A) = \deg A + (1 - g) + \ell(C - A),$$

where g is the genus of $\mathbb{F}$.

PROOF. Lemma 9.15 shows that there exists a nonzero differential ω_0. Let $C_0 = \text{Div}(\omega_0)$. Lemma 9.15 shows that $C = C_0 + (y_0)$ for some $y_0 \in \mathbb{F}^\times$. Then $\omega = y_0 \omega_0$ has

$$\text{Div}(\omega) = \text{Div}(y_0 \omega_0) = \text{Div}(\omega_0) + (y_0) = C_0 + (y_0) = C.$$

Let B be a divisor to be specified, and consider $C - B$. Any nonzero differential ω' vanishing on $\mathcal{L}(C - B)$ is of the form $\omega' = z\omega$ for some $z \in \mathbb{F}^\times$ by Lemma 9.15, and $\text{Div}(\omega') = \text{Div}(z\omega) = C + (z)$. Therefore $\mathcal{L}(C + (z)) \supseteq \mathcal{L}(C - B)$, $C + (z) \geq C - B$, and $(z) \geq -B$. This inequality means that z is in $L(B)$. Conversely if y is any nonzero element in $L(B)$, then $(y) \geq -B$ and $C + (y) \geq C - B$. So $\mathcal{L}(C + (y)) \supseteq \mathcal{L}(C - B)$. We know that $y\omega$ vanishes on $\mathcal{L}(C + (y))$, and hence $y\omega$ vanishes on $\mathcal{L}(C - B)$.

Consequently the differentials vanishing on $\mathcal{L}(C - B)$ are exactly the differentials $y\omega$ with y in $L(B)$. Such differentials vanish on $\Delta(\mathbb{F})$ by definition, and the space of them is $\mathbb{k}$ isomorphic to the space of $\mathbb{k}$ linear functionals on $\mathcal{A}_{\mathbb{F}}/\bigl(\mathcal{L}(C - B) + \Delta(\mathbb{F})\bigr)$. By Theorem 9.13 the latter space has $\mathbb{k}$ dimension $\delta(C - B)$, and hence the space of differentials in question has $\mathbb{k}$ dimension $\delta(C - B)$. In short,
$$\delta(C - B) = \ell(B).$$
Since B is arbitrary, we can specialize it to $B = C - A$. Then we obtain
$$\ell(C - A) = \delta(A) = \ell(A) - \deg A - (1 - g),$$
and the theorem follows. □

5. Applications of the Riemann–Roch Theorem

We begin with some immediate applications of the Riemann–Roch Theorem, and then we obtain some applications that require arguments that are a bit more subtle. Another application appears in the problems at the end of Chapter X.

Corollary 9.17. If C is any canonical divisor, then $\ell(C) = g$.

PROOF. Put $A = 0$ in Theorem 9.16, and use the fact given in Corollary 9.4 that $\ell(0) = 1$. □

Corollary 9.18. If C is any canonical divisor, then $\deg C = 2g - 2$.

PROOF. Put $A = C$ in Theorem 9.16, and apply Corollary 9.17 and Corollary 9.4. □

Corollary 9.19. Any divisor A with $\deg A > 2g - 2$ has $\delta(A) = 0$, i.e., $\ell(A) = \deg A + (1 - g)$.

PROOF. If $\deg A > 2g - 2$, then it follows from Corollary 9.18 that $\deg(C - A) < 0$. By Proposition 9.10, $\ell(C - A) = 0$. Then the corollary is immediate from Theorem 9.16. □

Corollary 9.20. If A is a divisor with $\deg A = 2g - 2$, then either A is a canonical divisor and $\ell(A) = g$, or A is not a canonical divisor and $\ell(A) = g - 1$.

PROOF. If A is a canonical divisor, then $\ell(A) = g$ by Corollary 9.17. Otherwise, the divisor $C - A$, which has degree 0 by Corollary 9.18, is not a principal divisor. Any nonzero y in $L(C - A)$ then would have $(y) \geq -(C - A)$ and $0 = \deg(y) \geq -\deg(C - A) = 0$; hence $v(y) = -\operatorname{ord}_v(C - A)$ for all v, and $(y) = C - A$, contradiction. Consequently $L(C - A) = 0$ and $\ell(C - A) = 0$. Theorem 9.16 now gives $\ell(A) = \deg A + (1 - g) = (2g - 2) + (1 - g) = g - 1$. □

5. Applications of the Riemann–Roch Theorem

EXAMPLES OF CANONICAL DIVISORS.

(1) Genus $g = 0$. In Corollary 9.20 with $g = 0$, the alternative $\ell(A) = g - 1 = -1$ is impossible, and therefore every divisor with degree -2 is a canonical divisor.

(2) Genus $g = 1$. In Corollary 9.20 with $g = 1$, take $A = 0$. Then $\ell(A) = 1 = g$ by Corollary 9.4. So Corollary 9.20 says that the divisor 0 is a canonical divisor.

Corollary 9.21. If v_0 is in $\mathbb{V}_\mathbb{F}$ and $n > \max(2g - 1, 0)$, then there exists a nonscalar x in $\mathbb{F}^\times$ with $(x)_\infty \leq nv_0$.

PROOF. Let $A = nv_0$, and let f_{v_0} be the residue class degree of v_0. Then $\deg A = nf_{v_0} \geq n > \max(2g - 1, 0)$, and Corollary 9.19 gives

$$\ell(A) = \deg A + (1 - g) = nf_{v_0} + (1 - g)$$
$$> \max(2g - 1, 0) + (1 - g) = \max(g, 1 - g) \geq 1.$$

Hence $\ell(A) \geq 2$, and $L(A)$ contains a nonscalar element x. This x has

$$-n = -\operatorname{ord}_{v_0} A \leq \operatorname{ord}_{v_0}(x) = \operatorname{ord}_{v_0}(x)_0 - \operatorname{ord}_{v_0}(x)_\infty = -\operatorname{ord}_{v_0}(x)_\infty,$$

and thus $(x)_\infty \leq nv_0$. □

Doubly periodic meromorphic functions on $\mathbb{C}$ in the subject of complex analysis may be viewed as meromorphic functions on some torus,[12] which is a compact Riemann surface of genus 1. The Weierstrass $\wp$ function for the torus in question has a double pole at one point, two zeros, and no other poles or zeros. It is therefore a function x with $(x)_\infty = 2v_0$ if v_0 is the discrete valuation corresponding to the location of the pole. Hence this x provides an example with equality holding in Corollary 9.21 when $g = 1$. A theorem of Liouville in this terminology says that there is no meromorphic function on the torus having just one simple pole and no other poles. The final corollaries abstract this result to our setting, but they need an additional hypothesis to ensure that $f_{v_0} = 1$. Certainly f_{v_0} will equal 1 if $\Bbbk$ is algebraically closed. We consider $g = 1$ and $g > 1$ separately. These corollaries will be generalized in Problems 23–25 at the end of the chapter.

Corollary 9.22. If $\Bbbk$ is algebraically closed, if v_0 is in $\mathbb{V}_\mathbb{F}$, and if $g = 1$, then every x in $\mathbb{F}$ with $(x)_\infty \leq v_0$ is a scalar multiple of the identity.

PROOF. Put $A = v_0$. We seek $x \in \mathbb{F}$ with $v_0(x) \geq -1 = -\operatorname{ord}_{v_0} A$ and with $v(x) \geq 0 = -\operatorname{ord}_v A$ for all other v. Thus we seek x in $L(A)$. This A has $\deg A = 1 = g = 2g - 1$. By Corollary 9.19, $\ell(A) = \deg A + (1 - g) = 1 + (1 - 1) = 1$. Since $L(A)$ already contains the multiples of the identity, it contains nothing else. □

---

[12] The particular torus is $\mathbb{C}/\Lambda$, where Λ is the lattice of periods.

Corollary 9.23. If $\Bbbk$ is algebraically closed, if v_0 is in $\mathbb{V}_\mathbb{F}$, and if $g > 1$, then every x in $\mathbb{F}$ with $(x)_\infty \leq v_0$ is a scalar multiple of the identity.

PROOF. We argue by contradiction. Suppose that x is a nonscalar element in $L(v_0)$. Take $r = 2g - 1$, and let $c_1, \ldots, c_r$ be distinct members of $\Bbbk$. For each j with $1 \leq j \leq r$, $x - c_j$ is in $L(v_0)$. Since $\deg(x - c_j) = 0$, there exists a unique $v_j \in \mathbb{V}_\mathbb{F}$ with $v_j(x - c_j) = 1$. The divisor of the element $(x - c_j)^{-1}$ is then $v_0 - v_j$. It follows that every $\Bbbk$ linear combination of the elements $(x - c_j)^{-1}$ lies in $L(A)$ for $A = v_1 + \cdots + v_r$. On the other hand, these elements are linearly independent because $v_j\bigl(\sum_{i=1}^r a_i(x - c_j)^{-1}\bigr) < 0$ if and only if $a_j \neq 0$. Thus $\ell(A) \geq 2g - 1$ and $\deg A = 2g - 1$. Since $\deg A > 2g - 2$, Corollary 9.19 is applicable and gives $\ell(A) = \deg A + 1 - g$. Thus $2g - 1 \leq \ell(A) = \deg A + 1 - g = 2g - 1 + 1 - g = g$, and we obtain the contradiction $g \leq 1$. □

6. Problems

1. Let $\mathbb{F}$ be a function field in one variable over the field $\Bbbk$, and let $\Bbbk'$ be the subfield of all members of $\mathbb{F}$ that are algebraic over $\Bbbk$.
   (a) Suppose that $t_1, \ldots, t_n$ are members of $\Bbbk'$ that are linearly independent over $\Bbbk$, and suppose that $x \in \mathbb{F}$ is transcendental over $\Bbbk$. Prove that $t_1, \ldots, t_n$ are linearly independent over $\Bbbk(x)$.
   (b) Deduce from (a) that $[\Bbbk' : \Bbbk] \leq [\Bbbk'(x) : \Bbbk(x)]$.
   (c) Deduce that $[\Bbbk' : \Bbbk] < \infty$.

Problems 2–4 concern perfect fields, which were defined in Section VII.3. The field $\Bbbk$ is perfect if either it has characteristic 0 or else it has characteristic p and the field map $x \mapsto x^p$ of $\Bbbk$ into itself is onto.

2. Prove that an algebraic extension of a perfect field is perfect.

3. When $\Bbbk$ is perfect, refine an argument in Section 1 by making use of Theorems 7.18, 7.20, 7.22, and the Theorem of the Primitive Element, and show that any function field in one variable is the function field of some affine plane curve irreducible over $\Bbbk$.

4. Let $\Bbbk$ be a perfect field. An affine plane curve $f(X, Y)$ irreducible over $\Bbbk$ is nonsingular at a point (a, b) of its zero locus if at least one of $\frac{\partial f}{\partial X}(a, b)$ and $\frac{\partial f}{\partial Y}(a, b)$ is nonzero. Using Bezout's Theorem and taking a cue from the proof of Theorem 7.20, prove that the curve can be singular at only finitely many points of its zero locus.

Problems 5–11 seek to attach a discrete valuation of the function field of an irreducible affine plane curve to each point of the zero locus at which the curve is nonsingular. Let $\Bbbk$ be a base field, let $f(X, Y)$ be an irreducible polynomial in $\Bbbk[X, Y]$, let $R =$

$\mathbb{k}[X, Y]/(f(X, Y))$, let x and y be the images of X and Y in R, and let $\mathbb{F}$ be the field of fractions of R. Suppose that $(a, b) \in \mathbb{k}^2$ has the property that $f(a, b) = 0$. The condition of nonsingularity of f at (a, b) is that one of $\frac{\partial f}{\partial X}$ and $\frac{\partial f}{\partial Y}$ be nonvanishing at (a, b), and it will be assumed that $\frac{\partial f}{\partial X}(a, b) \neq 0$. Observe from Lemma 7.16 that if S is any integral domain, if s is in S, and if $c(X)$ is in $S[X]$, then $c(X) - c(s) = (X-s)d(X)$ for some $d(X)$ in $S[X]$.

5. Let $f_1(X)$ be the member of $\mathbb{k}[X]$ defined as above to make $f(X, b) = (X - a)f_1(X)$. Using the fact that $\frac{\partial f}{\partial X}(a, b) \neq 0$, prove that $f_1(a) \neq 0$ and therefore also that $f_1(x) \neq 0$.

6. Let $g(X, Y)$ be a member of $\mathbb{k}[X, Y]$ with $g(x, y) \neq 0$. Prove that if $g(a, b) = 0$, then there exist $g_1(X)$ in $\mathbb{k}[X]$ and $h_1(X, Y)$ in $\mathbb{k}[X, Y]$ with

$$g(X, Y)f_1(X) - f(X, Y)g_1(X) = (Y - b)h_1(X, Y),$$

and deduce that $g(x, y) = (y - b)h_1(x, y)/f_1(x)$.

7. Show that there is a discrete valuation v_1 of $\mathbb{F}$ over $\mathbb{k}$ with $v_1(y - b) > 0$.

8. If $h(a, b) = 0$ in Problem 6, then the process can be repeated to give

$$g(x, y) = (y - b)^2 h_2(x, y)/f_1(x)^2.$$

It can be repeated again if $h_2(a, b) = 0$, and so on. By applying the valuation v_1 of the previous problem to $g(a, y)$, show that there is an upper bound to the integers $k \geq 0$ such that a nonzero member $g(x, y)$ in R can be written in the form $g(x, y) = (y - b)^k h_k(x, y)/f_1(x)^k$ for some $h_k(x, y)$ in R.

9. (a) Deduce that each nonzero $g(x, y)$ in R is of the form

$$g(x, y) = (y - b)^n h(x, y)/f_1(x)^n$$

with $n \geq 0$, $h(x, y)$ in R, and $h(a, b) \neq 0$, and that the integer n and the member $h(x, y)$ of R are uniquely determined by $g(x, y)$.

(b) Conclude that every nonzero member $g(x, y)$ of the field of fractions $\mathbb{F}$ is of the form $(y - b)^n h_1(x, y)/h_2(x, y)$ with n in $\mathbb{Z}$, $h_1(x, y)$ and $h_2(x, y)$ nonzero in R, $h_1(a, b) \neq 0$, and $h_2(a, b) \neq 0$.

(c) Prove in (b) that $g(x, y)$ uniquely determines n.

10. Write each nonzero $g(x, y)$ in $\mathbb{F}$ as in (b) of the previous problem, and put $v(g) = n$. Also, define $v(0) = \infty$. Show that the resulting function v is a well-defined valuation of $\mathbb{F}$ having R in its valuation ring, taking the value 0 on all members of R that are nonvanishing at (a, b), and having all members of R vanishing at (a, b) in its valuation ideal.

11. Prove that there is only one valuation of $\mathbb{F}$ over $\mathbb{k}$ taking the value 0 on all members of R that are nonvanishing at (a, b) and having all members of R vanishing at (a, b) in its valuation ideal.

Problems 12–20 compute the genus of certain function fields in one variable. Let $\mathbb{k}$ be a field of characteristic $\neq 2$, let $f(X)$ be a square-free nonconstant polynomial in $\mathbb{k}[X]$, let $\mathbb{F} = \mathbb{k}(X)[Y]/(Y^2 - f(X))$, and let x and y be the images of X and Y in $\mathbb{F}$. In these problems, p denotes a positive integer.

12. Verify that
    (a) the element x is transcendental over $\mathbb{k}$, y is algebraic over $\mathbb{k}(x)$ with $y^2 = f(x)$, and $\mathbb{F}$ is a function field in one variable over $\mathbb{k}$,
    (b) every member of $\mathbb{F}$ is uniquely of the form $a(x) + yb(x)$ with $a(x)$ and $b(x)$ in $\mathbb{k}(x)$,
    (c) every member of $\mathbb{F}$ not in $\mathbb{k}$ is transcendental over $\mathbb{k}$,
    (d) $\mathbb{F}/\mathbb{k}(x)$ is a Galois extension of degree 2, and the nontrivial element σ of $\text{Gal}(\mathbb{F}/\mathbb{k}(x))$ satisfies $\sigma(a(x) + yb(x)) = a(x) - yb(x)$ for $a(x)$ and $b(x)$ in $\mathbb{k}(x)$.

13. Prove that the integral closure of $\mathbb{k}[x]$ in $\mathbb{F}$ is the ring R of all elements $a(x) + yb(x)$ such that $a(x)$ and $b(x)$ are in $\mathbb{k}(x)$.

14. (a) Deduce from the previous problem that R is the set of all members z of $\mathbb{F}$ such that $v(z) \geq 0$ for all v in $D_\mathbb{F}$ that satisfy $v(x) \geq 0$.
    (b) Deduce from (a) that $L(p(x)_\infty) \subseteq R$.

15. Let v be any member of $D_\mathbb{F}$ with $v(x) < 0$.
    (a) Prove that every nonzero $c(x)$ in $\mathbb{k}[x]$ has $v(c(x)) = (\deg c)v(x)$.
    (b) Prove that $v(y) = \frac{1}{2}(\deg f)v(x)$.
    (c) Prove that if $a(x)$ and $b(x)$ are in $\mathbb{k}[x]$ with $\deg b + \frac{1}{2}\deg f \leq p$ and $\deg a \leq p$, then $v(a(x) + yb(x)) \geq pv(x)$.

16. Prove that if $a(x)$ and $b(x)$ are in $\mathbb{k}[x]$ with $\deg b + \frac{1}{2}\deg f \leq p$ and $\deg a \leq p$, then $a(x) + yb(x)$ lies in $L(p(x)_\infty)$.

17. (a) Prove that if v is in $D_\mathbb{F}$ and if σ is in $\text{Gal}(\mathbb{F}/\mathbb{k}(x))$, then the function v^σ defined by $v^\sigma(z) = v(\sigma(z))$ for $z \in \mathbb{F}$ is in $D_\mathbb{F}$.
    (b) Why is $v(x) < 0$ if and only if $v^\sigma(x) < 0$?
    (c) Deduce that if z is in $L(p(x)_\infty)$, then so is $\sigma(z)$.

18. (a) Using the previous problem, show that if $a(x)$ and $b(x)$ are in $\mathbb{k}[x]$ with $a(x) + yb(x)$ in $L(p(x)_\infty)$ and if v is a member of $D_\mathbb{F}$ with $v(x) < 0$, then $v(a(x)) \geq pv(x)$ and $v(a(x)^2 - f(x)b(x)^2) \geq 2pv(x)$. Conclude that $\deg a \leq p$ and $\deg(a^2 - fb^2) \leq 2p$.
    (b) Deduce that $L(p(x)_\infty)$ consists of all members $a(x) + yb(x)$ of R such that $\deg a \leq p$ and $\deg b + \frac{1}{2}\deg f \leq p$.

19. Calculate that $\ell(p(x)_\infty) = 2p + 2 - [\frac{1}{2}(1 + \deg f)]$ if $p \geq [\frac{1}{2}(1 + \deg f)]$. Here $[\,\cdot\,]$ denotes the greatest integer function.

20. (a) Why is $\deg(x)_\infty = 2$?
    (b) Using Corollary 9.19 with $A = p(x)_\infty$ for a suitable p, prove that the genus of $\mathbb{F}$ is $g = [\frac{1}{2}(1 + \deg f)] - 1$.

Problems 21–22 compute the genus of certain further function fields in one variable. The notation is as in Problems 12–20 except that $f(X)$ is allowed to have repeated factors. Suppose that $f(X) = g(X)^2 h(X)$, where $h(X)$ is a square-free nonconstant polynomial and $g(X)$ is in $\mathbb{k}[X]$. Let $\mathbb{F} = \mathbb{k}(X)[Y]/(Y^2 - f(X))$.

21. With $\mathbb{F}' = \mathbb{k}(X)[Z]/(Z^2 - h(X))$, exhibit a field isomorphism $\mathbb{F} \to \mathbb{F}'$ fixing $\mathbb{k}$.

22. Suppose that $f(X)$ has degree 3.
    (a) Prove that $\mathbb{F}$ has genus 1 if $f(X)$ has no repeated root in $\mathbb{k}$ and that $\mathbb{F}$ has genus 0 otherwise.
    (b) Prove that the affine plane curve $Y^2 - f(X)$ over $\mathbb{k}$ has a singularity in $\mathbb{k}_{\text{alg}}^2$ if and only if $f(X)$ has a repeated root in $\mathbb{k}_{\text{alg}}^2$. Here $\mathbb{k}_{\text{alg}}$ denotes an algebraic closure of $\mathbb{k}$.

Problems 23–25 introduce Weierstrass points. Let $\mathbb{k}$ be an algebraically closed field, and let $\mathbb{F}$ be a function field in one variable over $\mathbb{k}$ of genus g. Fix a discrete valuation v in $D_\mathbb{F}$.

23. Why is it true that $\ell(0v) = 1$, $\ell(1v) = 1$ if $g \geq 1$, $\ell((2g-1)v) = g$, $\ell(2gv) = g+1$, and $\ell(nv) \leq \ell((n+1)v) \leq \ell(nv) + 1$ for all integers $n \geq 0$?

24. Deduce from the previous problem that there exist exactly g integers $0 < n_1 < n_2 < \cdots < n_g < 2g$ such that there is no x in $\mathbb{F}$ with $(x)_\infty = n_i v$. (Educational note: The integers n_i are called the **Weierstrass gaps** of v, and $(n_1, \ldots, n_g)$ is the **gap sequence** for v. Classically when $\mathbb{F}$ is viewed as the function field of an everywhere nonsingular projective curve, then the points of the zero locus in projective space are in one-one correspondence with the members of $D_\mathbb{F}$; with this understanding, the point corresponding to v is called a **Weierstrass point** if the gap sequence for v is anything but $(1, 2, \ldots, g)$. Accordingly let us call v a **Weierstrass valuation** in this case.)

25. Prove that
    (a) v is a Weierstrass valuation if and only if $\ell(gv) > 1$.
    (b) 1 is a Weierstrass gap if $g > 0$.
    (c) v is not a Weierstrass valuation if $g = 0$ or $g = 1$.
    (d) if r and s are positive integers with sum $< 2g$ that are not Weierstrass gaps at v, then $r + s$ is not a Weierstrass gap at v.
    (e) if 2 is not a Weierstrass gap at v, then the gap sequence is $(1, 3, 5, \ldots, 2g-1)$.

CHAPTER X

Methods of Algebraic Geometry

Abstract. This chapter investigates the objects and mappings of algebraic geometry from a geometric point of view, making use especially of the algebraic tools of Chapter VII and of Sections 7–10 of Chapter VIII. In Sections 1–12, $\Bbbk$ denotes a fixed algebraically closed field.

Sections 1–6 establish the definitions and elementary properties of varieties, maps between varieties, and dimension, all over $\Bbbk$. Sections 1–3 concern varieties and dimension. Affine algebraic sets, affine varieties, and the Zariski topology on affine space are introduced in Section 1, and projective algebraic sets and projective varieties are introduced in Section 3. Section 2 defines the geometric dimension of an affine algebraic set, relating the notion to Krull dimension and transcendence degree. The actual context of Section 2 is a Noetherian topological space, the Zariski topology on affine space being an example. In such a space every closed subset is the finite union of irreducible closed subsets, and the union can be written in a certain way that makes the decomposition unique. Every nonempty closed set has a meaningful geometric dimension. In affine space the irreducible closed sets are the varieties, and each variety acquires a geometric dimension. The discussion in Section 2 applies in the context of projective space as well, and thus each projective variety acquires a geometric dimension. Moreover, any nonempty open subset of a Noetherian space is Noetherian. A nonempty open subset of an affine variety is called quasi-affine, and a nonempty open subset of a projective variety is called quasiprojective. Each quasi-affine variety or quasiprojective variety has a dimension equal to that of its closure, which is a variety.

Sections 4–6 take up maps between varieties. Section 4 introduces spaces of scalar-valued functions on quasiprojective varieties—rational functions, functions regular at a point, and functions regular on an open set. The section goes on to relate these notions for the different kinds of varieties. Section 5 introduces morphisms, which are a restricted kind of function between varieties. The tools of Sections 4–5 together show that for many purposes all the different kinds of varieties can be treated as quasiprojective varieties. Section 6 introduces rational maps between varieties; these are not everywhere-defined functions, but each can be restricted to an open dense subset on which it is a morphism. Rational maps with dense image correspond to field mappings of the fields of rational functions, with the order of the mappings reversed.

Section 7 concerns singularities at points of varieties, still over the field $\Bbbk$. Zariski's Theorem was stated in Chapter VII for affine varieties and partly proved at that time. In the current context it has a meaning for any point of any quasiprojective variety. The section proves the full theorem, which characterizes singular points in a way that shows they remain singular under isomorphisms of varieties.

Section 8 concerns classification questions over $\Bbbk$ for irreducible curves, i.e., quasiprojective varieties of dimension 1. From Section 6 it is known that two irreducible curves are equivalent under rational maps if and only if their fields of rational functions are isomorphic. The main theorem of Section 8 is that each such equivalence class of irreducible curves contains an everywhere nonsingular projective curve, and this curve is unique up to isomorphism of varieties. The points of this curve are parametrized by those discrete valuations of the underlying function field that are defined over $\Bbbk$.

Sections 9–12 relate the general theory of Sections 1–6 to the topic of simultaneous solutions of polynomial equations, as treated at length in Chapter VIII. Section 9 treats monomial ideals in $\Bbbk[X_1, \ldots, X_n]$, identifying their zero loci concretely and computing their dimension. The section goes on to introduce the affine Hilbert function of this ideal, which measures the proportion of polynomials of degree $\leq s$ not in the ideal. In the way that this function is defined, it is a polynomial for large s called the affine Hilbert polynomial of the ideal. Its degree equals the dimension of the zero locus of the ideal. Section 10 extends this theory from monomial ideals to all ideals, again concretely computing the dimension of the zero loci, obtaining an affine Hilbert polynomial, and showing that its degree equals the dimension of the zero locus of the ideal. Section 11 adapts the theory to homogeneous ideals and projective algebraic sets by making use of the cone in affine space over the set in projective space. Section 12 applies the theory of Section 11 to address the question how the dimension of a projective algebraic set is cut down when the set is intersected with a projective hypersurface. A consequence of the theory is the result that a homogeneous system of polynomial equations over an algebraically closed field with more unknowns than equations has a nonzero solution.

Section 13 is a brief introduction to the theory of schemes, which extends the theory of varieties by replacing the underlying algebraically closed field by an arbitrary commutative ring with identity.

1. Affine Algebraic Sets and Affine Varieties

We come now to the more geometric side of algebraic geometry. At least initially this means that we are interested in the set of simultaneous solutions of a system of polynomial equations in several variables. Because of the Nullstellensatz the natural starting point for the investigation is the case that the underlying field of coefficients is algebraically closed.

Accordingly, throughout Sections 1–6 of this chapter, $\Bbbk$ will denote an algebraically closed field.[1] We fix a positive integer n and denote by A the polynomial ring $A = \Bbbk[X_1, \ldots, X_n]$. Typical ideals of A will be denoted by $\mathfrak{a}, \mathfrak{b}, \ldots$. We begin by expanding on some definitions made in Section VIII.2. The set

$$\mathbb{A}^n = \{(x_1, \ldots, x_n) \in \Bbbk^n\}$$

is called **affine n-space**. Members of $\mathbb{A}^n$ are called **points** in affine n-space, and the functions $P \mapsto x_j(P)$ give the **coordinates** of the points.

To each subset S of polynomials in A, we associate the **locus of common zeros**, or **zero locus** of the members of S:

$$V(S) = \{P \in \mathbb{A}^n \mid f(P) = 0 \text{ for all } f \in S\}.$$

Any such set $V(S)$ is called an **affine algebraic set** in $\mathbb{A}^n$. If S is a finite set $\{f_1, \ldots, f_k\}$ of polynomials, we allow ourselves to abbreviate $V(\{f_1, \ldots, f_k\})$

---

[1]The exposition in these sections is based in part on Chapters 2, 4, and 6 of Fulton's book, Chapter I of Hartshorne's book, and Chapter I of Volume 1 of Shafarevich's books.

as $V(f_1, \ldots, f_k)$. It is immediate from the definitions that $V(S)$ is the same as $V(\mathfrak{a})$ if $\mathfrak{a}$ is the ideal in A generated by S. The Hilbert Basis Theorem shows that every ideal of A is finitely generated, and it follows that every affine algebraic set is of the form $V(f_1, \ldots, f_k)$ for some k and some polynomials $f_1, \ldots, f_k$.

In Chapter VIII we worked extensively with examples of ideals of A and their corresponding affine algebraic sets, and it will not be necessary to give further examples of that kind now.

Observe from the definition that $V(S) = \bigcap_{f \in S} V(f)$ for any subset S of A. It follows immediately that $S \mapsto V(S)$, as a function carrying each subset S of A to a subset $V(S)$ of $\mathbb{A}^n$, is inclusion reversing: $S_1 \subseteq S_2$ implies $V(S_1) \supseteq V(S_2)$. Using this same identity, we obtain the following further properties of V.

Proposition 10.1. Affine algebraic sets in $\mathbb{A}^n$ have the following properties:
  (a) $V(\varnothing) = V(0) = \mathbb{A}^n$ and $V(A) = \varnothing$,
  (b) $V\big(\bigcup_\alpha S_\alpha\big) = \bigcap_\alpha V(S_\alpha)$ if the S_α's are arbitrary subsets of A,
  (c) $V(S) = V(S_1) \cup V(S_2)$ if S_1 and S_2 are subsets of A and if S is defined as the set of all products $f_1 f_2$ with $f_1 \in S_1$ and $f_2 \in S_2$.

PROOF. Property (a) is immediate. For (b), we have
$$V\Big(\bigcup_\alpha S_\alpha\Big) = \bigcap_{f \in \bigcup_\alpha S_\alpha} V(f) = \bigcap_\alpha \bigcap_{f \in S_\alpha} V(f) = \bigcap_\alpha V(S_\alpha).$$

For (c), we observe first that $V(f_1 f_2) = V(f_1) \cup V(f_2)$ for any f_1 and f_2 in A. Then
$$V(S) = \bigcap_{\substack{f_1 \in S_1, \\ f_2 \in S_2}} V(f_1 f_2) = \bigcap_{f_1 \in S_1} \bigcap_{f_2 \in S_2} \big(V(f_1) \cup V(f_2)\big)$$
$$= \Big(\bigcap_{f_1 \in S_1} V(f_1)\Big) \cup \Big(\bigcap_{f_2 \in S_2} V(f_2)\Big) = V(S_1) \cup V(S_2). \qquad \square$$

Properties (a), (b), and (c) in the proposition are the axioms for the closed sets in a topology on $\mathbb{A}^n$. This topology is called the **Zariski topology** on affine n-space. Every one-point set is closed. The Zariski topology on $\mathbb{A}^n$ is never Hausdorff; for example, if $n = 1$, then it is the topology on $\mathbb{k}^1 = \mathbb{k}$ in which the nonempty open sets are the complements of the finite sets. Since one-point sets are closed and the topology is not Hausdorff, the Zariski topology on $\mathbb{A}^n$ is never regular. At first glance it looks like a useless topology, but we shall see already in Proposition 10.3b and again in Section 2 that it is quite helpful for handling the bookkeeping used in passing back and forth between algebra and geometry.

Next we introduce a function $E \mapsto I(E)$, carrying each subset E of $\mathbb{A}^n$ to an ideal $I(E)$ in A, by the definition
$$I(E) = \{f \in A \mid f(P) = 0 \text{ for all } P \in E\}.$$

Then $I(E) = \bigcap_{P \in E} I(\{P\})$. It follows immediately that $E \mapsto I(E)$ is inclusion reversing: $E_1 \subseteq E_2$ implies $I(E_1) \supseteq I(E_2)$. The result for $I(\cdot)$ that parallels Proposition 10.1 is as follows.

Proposition 10.2. For fixed n, the function $I(\cdot)$ has the following properties:
(a) $I(\varnothing) = A$ and $I(A) = 0$,
(b) $I(E_1 \cup E_2) = I(E_1) \cap I(E_2)$ if E_1 and E_2 are subsets of $\mathbb{A}^n$,
(c) $I(E_1 \cap E_2) \supseteq I(E_1) + I(E_2)$ if E_1 and E_2 are subsets of $\mathbb{A}^n$.

REMARKS. Equality can fail in (c). For example, if E_1 is the one-point set $\{0\}$ and E_2 is its complement, then $I(E_1 \cap E_2) = I(\varnothing) = A$, while $I(E_2) = 0$ and $I(E_1)$ consists of all members of A with 0 constant term.

PROOF. Property (a) is immediate. For (b), we have
$$I(E_1 \cup E_2) = \bigcap_{P \in E_1 \cup E_2} I(\{P\}) = \Big(\bigcap_{P \in E_1} I(\{P\})\Big) \cap \Big(\bigcap_{P \in E_2} I(\{P\})\Big) = I(E_1) \cap I(E_2).$$
In (c), the fact that $I(\cdot)$ is inclusion reversing implies that $I(E_1 \cap E_2) \supseteq I(E_1)$ and that $I(E_1 \cap E_2) \supseteq I(E_2)$. Since $I(E_1 \cap E_2)$ is closed under addition, (c) follows. □

This is all quite elementary. The less trivial question is the extent to which $V(\cdot)$ and $I(\cdot)$ are inverse to one another. Proposition 10.3 gives the answer.

Proposition 10.3. For fixed n,
(a) $I(V(\mathfrak{a})) = \sqrt{\mathfrak{a}}$ for each ideal $\mathfrak{a}$ in A,
(b) $V(I(E)) = \overline{E}$ for each subset E of $\mathbb{A}^n$, where $\overline{E}$ is the Zariski closure of E,
(c) $V(\mathfrak{a}) = V(\sqrt{\mathfrak{a}})$ for each ideal $\mathfrak{a}$ in A,
(d) any two ideals $\mathfrak{a}$ and $\mathfrak{b}$ in A have $\mathfrak{a}\mathfrak{b} \subseteq \mathfrak{a} \cap \mathfrak{b} \subseteq \sqrt{\mathfrak{a}\mathfrak{b}}$ and consequently have $V(\mathfrak{a} \cap \mathfrak{b}) = V(\mathfrak{a}\mathfrak{b}) = V(\mathfrak{a}) \cup V(\mathfrak{b})$.

REMARKS. Recall from Section VII.1 that $\sqrt{\mathfrak{a}}$ denotes the radical of $\mathfrak{a}$, consisting of all f in A such that f^k is in $\mathfrak{a}$ for some integer $k \geq 1$. The radical of $\mathfrak{a}$ equals $\mathfrak{a}$ itself if $\mathfrak{a}$ is prime.

PROOF. Conclusion (a) is the Nullstellensatz as formulated in Theorem 7.1b.

For (b), the definitions show that $V(I(E)) \supseteq E$. Since any set $V(S)$ is Zariski closed, we must have $V(I(E)) \supseteq \overline{E}$. On the other hand, the fact that $\overline{E}$ is closed means that $\overline{E} = V(S)$ for some S. Thus $V(S) = \overline{E} \supseteq E$, and the inclusion-reversing property of $I(\cdot)$ gives $I(V(S)) \subseteq I(E)$. Since the definitions imply that $S \subseteq I(V(S))$, we obtain $S \subseteq I(E)$. From the inclusion-reversing property of $V(\cdot)$, we conclude that $\overline{E} = V(S) \supseteq V(I(E))$.

For (c), (a) and (b) give $V(\sqrt{\mathfrak{a}}) = V(I(V(\mathfrak{a}))) = \overline{V(\mathfrak{a})} = V(\mathfrak{a})$ because $V(\mathfrak{a})$ is closed.

For (d), the inclusion $\mathfrak{ab} \subseteq \mathfrak{a} \cap \mathfrak{b}$ is immediate. If f is in $\mathfrak{a} \cap \mathfrak{b}$, then f is in $\mathfrak{a}$ and in $\mathfrak{b}$, and hence f^2 is in $\mathfrak{ab}$. Thus f is in $\sqrt{\mathfrak{ab}}$. Applying $V(\cdot)$ gives $V(\mathfrak{ab}) \supseteq V(\mathfrak{a} \cap \mathfrak{b}) \supseteq V(\sqrt{\mathfrak{ab}})$. Since $V(\mathfrak{ab}) = V(\sqrt{\mathfrak{ab}})$ by (c), $V(\mathfrak{a} \cap \mathfrak{b}) = V(\mathfrak{ab})$. Finally $V(\mathfrak{ab}) = V(\mathfrak{a}) \cup V(\mathfrak{b})$ by Proposition 10.1c. □

An **affine variety** is any affine algebraic set of the form $V(\mathfrak{p})$, where $\mathfrak{p}$ is a prime ideal[2] of A. That is, an affine variety is the locus of common zeros of any prime ideal of A.

For example, if f is an irreducible polynomial in A, then f is prime because A is a unique factorization domain, and consequently the principal ideal (f) is prime. Thus the zero locus in $\mathbb{A}^2$ of an irreducible polynomial f in $\mathbb{k}[X, Y]$ is an example of an affine variety. This particular kind of affine variety is called an **irreducible affine plane curve**.[3,4] More generally, if f is irreducible in $A = \mathbb{k}[X_1, \ldots, X_n]$ with $n \geq 2$, then the zero locus of f in $\mathbb{A}^n$ is called an **irreducible affine hypersurface**.[5] Another example of an affine variety is any translate of any vector subspace of $\mathbb{A}^n$. Examples of affine varieties other than irreducible hypersurfaces, translates of vector subspaces, and varieties built from other varieties in simple ways often take some work to establish. The reason is that it is usually not easy to show that a particular nonprincipal ideal is prime. Here is one example that is manageable.

EXAMPLE. The **twisted cubic** in $\mathbb{A}^3$ is the zero locus $V(\mathfrak{p})$ of the ideal $\mathfrak{p}$ in $\mathbb{k}[X, Y, Z]$ given by $\mathfrak{p} = (Y - X^2, Z - X^3)$; that is, $V(\mathfrak{p}) = \{(x, x^2, x^3) \mid x \in \mathbb{k}\}$. The substitution homomorphism φ that fixes $\mathbb{k}$ and sends X to X, Y to X^2, and Z to X^3 carries $\mathbb{k}[X, Y, Z]$ into $\mathbb{k}[X]$. It is onto $\mathbb{k}[X]$ because any polynomial in X alone is sent to itself by φ. The kernel of φ manifestly contains $\mathfrak{p}$. To see that it equals $\mathfrak{p}$, we argue by contradiction. Choose a polynomial f in $\ker \varphi$ not in $\mathfrak{p}$ whose degree in Z is as small as possible and whose degree in Y is as small as possible among those of minimal degree in Z. If Z occurs somewhere in f, then by replacing all occurrences of Z in f with X^3, we replace f by another member of $f + \mathfrak{p}$ of lower degree in Z, contradiction. Thus f has no Z in it. Arguing

---

[2]*Warning:* The books by Fulton and Hartshorne in the Selected References use the narrow definition of variety that is reproduced here. Some books by other authors allow all affine algebraic sets to be called varieties. Volume 1 of Shafarevich's books does not use the word "variety."

[3]*Warning:* This definition represents a change from Chapters VIII and IX, corresponding to a change in point of view. Previously the word "curve" referred to the ideal, and now it is to refer to the zero locus. From a mathematical standpoint Proposition 10.3 shows that this distinction is not important in the presence of the irreducibility and the fact that $\mathbb{k}$ is algebraically closed. The change thus represents only a matter of convenience for the exposition.

[4]Some authors build the condition of irreducibility into the definition of "curve," but this book does not.

[5]Some authors build the condition of irreducibility into the definition of "hypersurface," but this book does not.

similarly, we see that f has no Y in it. So f is a polynomial in X. Since φ acts as the identity on polynomials in X alone, $f = 0$. This contradiction shows that $\ker \varphi = \mathfrak{p}$. Since image $\varphi = \mathbb{k}[X]$ is an integral domain, $\mathfrak{p}$ is prime. By the Nullstellensatz, $\mathfrak{p}$ may be described alternatively as the ideal of all polynomials vanishing on $V(\mathfrak{p})$.

Every affine variety is nonempty, as a consequence of the Nullstellensatz. In fact, any prime ideal $\mathfrak{p}$ of A is contained in a maximal ideal $\mathfrak{m}$, whose zero locus is identified as some point P of $\mathbb{A}^n$. The inclusion $\mathfrak{p} \subseteq \mathfrak{m}$ implies that $V(\mathfrak{p}) \supseteq V(\mathfrak{m}) = \{P\}$. Affine varieties are characterized by a geometric irreducibility property that is stated in Corollary 10.4.

Corollary 10.4. The affine varieties in $\mathbb{A}^n$ are characterized as those nonempty Zariski closed sets that cannot be written as the union of two proper closed subsets.

REMARKS. One says that the affine varieties are those affine algebraic sets that are **irreducible**. Irreducible sets are nonempty by definition.

PROOF. Let $V(\mathfrak{p})$ be an affine variety with $\mathfrak{p}$ prime, and suppose that $V(\mathfrak{p}) = E_1 \cup E_2$ with E_1 and E_2 both closed and properly contained in $V(\mathfrak{p})$. Application of $I(\cdot)$ and use of Proposition 10.2b gives $I(V(\mathfrak{p})) = I(E_1) \cap I(E_2)$. Proposition 10.3a allows us to rewrite this conclusion as $\mathfrak{p} = \mathfrak{b}_1 \cap \mathfrak{b}_2$ with $\mathfrak{b}_1 = I(E_1)$ and $\mathfrak{b}_2 = I(E_2)$. By Problem 10a at the end of Chapter VII, $\mathfrak{p} = \mathfrak{b}_1$ or $\mathfrak{p} = \mathfrak{b}_2$. If $\mathfrak{p} = \mathfrak{b}_1$, then $V(\mathfrak{p}) = V(\mathfrak{b}_1) = V(I(E_1))$, and this equals E_1 by Proposition 10.3b because E_1 is closed. Similarly if $\mathfrak{p} = \mathfrak{b}_2$, then $V(\mathfrak{p}) = E_2$. Thus E_1 and E_2 cannot both be proper subsets of $V(\mathfrak{p})$.

Conversely suppose that E is an irreducible closed subset of $\mathbb{A}^n$. Let f and g be members of A with fg in $I(E)$. Then Propositions 10.3b and 10.1c give $E = V(I(E)) \subseteq V(fg) = V(f) \cup V(g)$. Therefore
$$E = \bigl(E \cap V(f)\bigr) \cup \bigl(E \cap V(g)\bigr)$$
exhibits E as the union of two closed sets. By irreducibility one of the two closed sets equals E. If $E = E \cap V(f)$, then $E \subseteq V(f)$ and $I(E) \supseteq I(V(f)) \supseteq (f)$. If $E = E \cap V(g)$, then similarly $I(E) \supseteq (g)$. Either way, one of f and g lies in $I(E)$. Since E is assumed nonempty, $I(E)$ is proper. Therefore $I(E)$ is prime. $\square$

2. Geometric Dimension

We continue to assume that $\mathbb{k}$ is an algebraically closed field and to write A for $\mathbb{k}[X_1, \ldots, X_n]$. If $\mathfrak{p}$ is a prime ideal in A, then the **dimension** of the affine variety $V(\mathfrak{p})$ was defined in Section VII.2 to be the transcendence degree of the field of fractions of the integral domain $A/\mathfrak{p}$ over $\mathbb{k}$. This quantity depends only on $V(\mathfrak{p})$ because $\mathfrak{p}$ can be recovered from $V(\mathfrak{p})$ by the formula $\mathfrak{p} = I(V(\mathfrak{p}))$ given in Proposition 10.3a. The integral domain $A/\mathfrak{p}$ is finitely generated as a $\mathbb{k}$ algebra with generators $X_1 + \mathfrak{p}, \ldots, X_n + \mathfrak{p}$, and Theorem 7.22 shows that this transcendence degree equals the Krull dimension of the ring $A/\mathfrak{p}$, which is denoted by $\dim A/\mathfrak{p}$. The latter quantity is the supremum of the indices d of all strictly increasing chains $\mathfrak{p}_0 \subsetneq \mathfrak{p}_1 \subsetneq \cdots \subsetneq \mathfrak{p}_d$ of prime ideals in $A/\mathfrak{p}$.

Because of this equality, it is natural to use the notion of Krull dimension in order to generalize the definition of dimension from varieties to all nonempty affine algebraic sets.[6] If $\mathfrak{a}$ is an any proper ideal in A, not necessarily prime, and $V(\mathfrak{a})$ is its locus of common zeros, we might first try defining $\dim V(\mathfrak{a})$ to be the Krull dimension of $A/\mathfrak{a}$. This approach is a bit cumbersome because two distinct ideals $\mathfrak{a}$ and $\mathfrak{a}'$ can have $V(\mathfrak{a}) = V(\mathfrak{a}')$; thus some argument would be needed to see that $\dim V(\mathfrak{a})$ is well defined before it would be possible to proceed.

Instead, we shall give a direct geometric definition of dimension in terms of the Zariski topology on $\mathbb{A}^n$. Theorem 10.7 later in this section will show that the geometric quantity $\dim V(\mathfrak{a})$ equals the Krull dimension of $A\big/\sqrt{\mathfrak{a}}$, thus that the dimension of an affine algebraic set has an algebraic formulation. From this result we shall deduce that $\dim V(\mathfrak{a})$ equals the Krull dimension of $A/\mathfrak{a}$ itself. This algebraic formulation of a definition will not yet allow us to compute dimensions concretely, but we shall introduce in Sections 9–11 an equivalent combinatorial definition of dimension that is computable in terms of Gröbner bases.

A topological space X will be said to be **Noetherian** if every strictly decreasing sequence of closed subsets is finite in length. An example is affine n-space $\mathbb{A}^n$. In fact, if $E_1, E_2, \ldots$ are closed sets in $\mathbb{A}^n$ with $E_1 \supseteq E_2 \supseteq \cdots$, then the corresponding ideals have $I(E_1) \subseteq I(E_2) \subseteq \cdots$. Since A is Noetherian, there exists some integer k with $I(E_k) = I(E_{k+1}) = \cdots$. Applying $V(\cdot)$ and using Proposition 10.3b, we obtain $E_k = E_{k+1} = \cdots$.

We can generalize the definition of irreducibility for closed sets from $\mathbb{A}^n$ to an arbitrary Noetherian topological space. Namely a nonempty closed set E is **irreducible** if it is not the union of two proper closed subsets. An important observation about any Noetherian topological space is that any nonempty relatively open subset U of an irreducible closed set V is dense in V; in fact, if $\overline{U}$ denotes the closure of U, then $V = \overline{U} \cup (V - U)$ exhibits V as the union of two closed subsets, and the irreducibility forces $\overline{U} = V$ since $V - U \neq V$.

Proposition 10.5. If X is a Noetherian topological space, then any closed subset is the finite union of irreducible closed subsets. This decomposition of a closed set as such a union may be chosen in such a way that none of the closed sets in the union contains another set in the union, and in this case the decomposition is unique.

---

[6]We shall leave the dimension of the empty set as undefined for now.

2. Geometric Dimension

PROOF. For existence of some decomposition of each closed set as a finite union of irreducible closed subsets, we argue by contradiction. Assuming that there exists some closed subset E of X that is not the finite union of irreducible closed subsets, we may assume by the Noetherian condition on X that E is minimal among all such counterexamples. Since E cannot itself be irreducible, we can write $E = E_1 \cup E_2$ with E_1 and E_2 closed and properly contained in E. Since E is minimal among all closed subsets that are not the finite union of irreducible closed subsets, E_1 and E_2 can be expressed as finite unions of irreducible closed subsets. Substituting these expressions into the equality $E = E_1 \cup E_2$ gives a contradiction to the fact that E is a counterexample.

This proves existence of a decomposition. By going through the sets in the decomposition one at a time and by discarding any set that is contained in another set, we obtain a decomposition as in the second sentence of the proposition.

For uniqueness, suppose that $E = E_1 \cup \cdots \cup E_k = F_1 \cup \cdots \cup F_l$ gives two decompositions of the asserted kind. Say that $k \geq l$. Since $F_i \subseteq E_1 \cup \cdots \cup \cdots \cup E_k$, we obtain $F_i = (F_i \cap E_1) \cup \cdots \cup (F_i \cap E_k)$. Irreducibility of F_i implies that $F_i = F_i \cap E_{j(i)}$ for some $j = j(i)$. Hence $F_i \subseteq E_{j(i)}$ for some function $j(i)$ from $\{1, \ldots, l\}$ to $\{1, \ldots, k\}$. Reversing the roles of the E_i's and the F_j's yields a function $i(j)$ such that $E_j \subseteq F_{i(j)}$. Then $F_i \subseteq E_{j(i)} \subseteq F_{i(j(i))}$. Since no F_i contains some $F_{i'}$ with $i' \neq i$, we conclude that $i(j(i)) = i$ for all i. Therefore $k = l$, and $i(\cdot)$ and $j(\cdot)$ are inverse to each other. □

Corollary 10.6. Every affine algebraic set in $\mathbb{A}^n$ can be expressed uniquely as the finite (possibly empty) union of affine varieties in such a way that none of the varieties contains another of the varieties.

REMARKS. For example,

$$V(X^2 - Y^2) = V(X+Y) \cup V(X-Y)$$

by Proposition 10.1c, and the affine algebraic set on the left side is expressed as the union of the affine varieties on the right.

PROOF. We saw before Proposition 10.5 that $\mathbb{A}^n$ is a Noetherian topological space, and Corollary 10.4 shows that the irreducible subsets are the affine varieties. The closed sets are the affine algebraic sets by definition, and hence the result is a special case of Proposition 10.5. □

The **geometric dimension** of a nonempty closed subset E of a Noetherian topological space X is the supremum of the integers $d \geq 0$ such that there exists a strictly increasing chain $E_0 \subsetneq E_1 \subsetneq \cdots \subsetneq E_d$ of irreducible closed subsets of E. This definition makes sense because a chain with $d = 0$ can always be formed with E_0 equal to one of the irreducible closed sets from Proposition 10.5;

however, there is no guarantee in this generality that the geometric dimension will be finite. In any event, it is clear from the definition that if two closed sets E and E' have $E \subseteq E'$, then the geometric dimension of E is $\leq$ the geometric dimension of E'.

In the case of a nonempty affine algebraic set $V(S)$, the geometric dimension of $V(S)$ is to refer to this kind of dimension relative to the Zariski topology.

EXAMPLES OF GEOMETRIC DIMENSION IN $\mathbb{A}^n$.

(1) Any one-point set in $\mathbb{A}^n$ is closed and plainly has geometric dimension 0. Any affine variety V with more than one point has geometric dimension ≥ 1, since $\{P\} \subsetneq V$ is a strictly increasing chain of irreducible closed sets if P is chosen as a point in V.

(2) $\mathbb{A}^n$ has geometric dimension n. This fact will follow from Theorem 10.7 below because A has Krull dimension n as a consequence of Theorem 7.22.

(3) Twisted cubic in $\mathbb{A}^3$, namely $\{(x, x^2, x^3) \mid x \in \mathbb{k}\}$. According to the example in Section 1, this is $V(\mathfrak{p})$ for the prime ideal $\mathfrak{p} = (Y - X^2, Z - X^3) \subseteq \mathbb{k}[X, Y, Z]$. The inclusions of prime ideals $(X, Y, Z) \supsetneq (Y - X^2, Z - X^3) \supsetneq (Y - X^2) \supsetneq 0$ give the strictly increasing chain $\{0\} \subsetneq V(\mathfrak{p}) \subsetneq \{(x, x^2, z)\} \subsetneq \mathbb{A}^3$, which is of the kind described for $\mathbb{A}^3$. If another term could be included between $\{0\}$ and $V(\mathfrak{p})$, then we would obtain a sequence showing that $\mathbb{A}^3$ has geometric dimension ≥ 4, in contradiction to Example 2. So $V(\mathfrak{p})$ has geometric dimension ≤ 1. In view of Example 1, $V(\mathfrak{p})$ has geometric dimension equal to 1.

Theorem 10.7. If $\mathfrak{a}$ is any proper ideal of A, then the following four quantities are equal:
  (a) the geometric dimension of $V(\mathfrak{a})$,
  (b) the Krull dimension of $A/\sqrt{\mathfrak{a}}$,
  (c) the maximum of the geometric dimension of V_j over all affine varieties V_j contained in $V(\mathfrak{a})$,
  (d) the Krull dimension of $A/\mathfrak{a}$.

REMARKS. We take these equal quantities as the definition of the **dimension** $\dim V(\mathfrak{a})$ of the affine algebraic set $V(\mathfrak{a})$. Because of Theorem 7.22, these quantities equal the transcendence degree over $\mathbb{k}$ of the field of fractions of $A/\mathfrak{a}$ in the case that $\mathfrak{a}$ is a prime ideal. For $\mathfrak{a} = 0$, we know that $\dim A = n$; hence the equal quantities in the theorem are $\leq n$.

PROOF. Let
$$E_0 \subseteq E_1 \subseteq \cdots \subseteq E_d \qquad (*)$$
be an increasing chain of irreducible closed subsets of $V(\mathfrak{a})$, and define $\mathfrak{p}_j$ to be the ideal $\mathfrak{p}_j = I(E_j)$. Then each $\mathfrak{p}_j$ is a prime ideal by Corollary 10.4, and also
$$\mathfrak{p}_d \subseteq \cdots \subseteq \mathfrak{p}_1 \subseteq \mathfrak{p}_0 \qquad (**)$$

2. Geometric Dimension

because $I(\cdot)$ is inclusion reversing. If $(*)$ is strictly increasing, then so is $(**)$; in fact, if $\mathfrak{p}_j$ were to equal $\mathfrak{p}_{j-1}$ for some j, then we would have $E_j = V(I(E_j)) = V(\mathfrak{p}_j) = V(\mathfrak{p}_{j-1}) = V(I(E_{j-1})) = E_{j-1}$, contradiction. In $(*)$, we have $E_d \subseteq V(\mathfrak{a})$, and thus Proposition 10.3a gives $\sqrt{\mathfrak{a}} = I(V(\mathfrak{a})) \subseteq I(E_d) = \mathfrak{p}_d$. In other words, any strictly increasing sequence $(*)$ of irreducible closed subsets of $V(\mathfrak{a})$ yields a strictly increasing sequence $(**)$ of prime ideals of A that contain $\sqrt{\mathfrak{a}}$.

Conversely if $(**)$ is a strictly increasing sequence of prime ideals of A containing $\sqrt{\mathfrak{a}}$, and if we define $E_j = V(\mathfrak{p}_j)$ for $0 \leq j \leq d$, then we obtain the sequence $(*)$ of irreducible closed subsets of $V(\sqrt{\mathfrak{a}}) = V(\mathfrak{a})$, and $(*)$ is strictly increasing, since an equality $E_j = E_{j-1}$ would imply that $\mathfrak{p}_j = I(V(\mathfrak{p}_j)) = I(E_j) = I(E_{j-1}) = I(V(\mathfrak{p}_{j-1})) = \mathfrak{p}_{j-1}$ because of Proposition 10.3a.

Thus the strictly increasing sequences $(*)$ of irreducible closed subsets of $V(\mathfrak{a})$ are in one-one correspondence with the strictly increasing sequences $(**)$ of prime ideals of A containing $\sqrt{\mathfrak{a}}$. Let $\varphi : A \to A/\sqrt{\mathfrak{a}}$ be the quotient homomorphism. Application of φ to $(**)$ yields a strictly increasing sequence of ideals of $A/\sqrt{\mathfrak{a}}$ by the First Isomorphism Theorem, and prime ideals map to prime ideals under this correspondence. Thus the existence of a strictly increasing sequence as in $(**)$ implies that the Krull dimension of $A/\sqrt{\mathfrak{a}}$ is $\geq d$. Meanwhile, the existence of a strictly increasing sequence as in $(*)$ implies that the geometric dimension of $V(\mathfrak{a})$ is $\geq d$. We have seen that these sequences are in one-one correspondence, and therefore the equality of (a) and (b) in the theorem follows.

In (c) certainly the geometric dimension of any V_j is $\leq$ the geometric dimension of $V(\mathfrak{a})$. If d_0 denotes the geometric dimension of $V(\mathfrak{a})$, then we can find a strictly increasing chain as in $(*)$ with $d = d_0$ and with all the sets contained in $V(\mathfrak{a})$. Corollary 10.4 shows that E_{d_0} is an affine variety contained in $V(\mathfrak{a})$, and the sequence $(*)$ shows that the geometric dimension of E_{d_0} is at least d_0. Thus $V_j = E_{d_0}$ is an affine variety contained in $V(\mathfrak{a})$ whose geometric dimension equals that of $V(\mathfrak{a})$.

To complete the proof, we show the equality of (b) and (d), i.e., we show that $A/\mathfrak{a}$ and $A/\sqrt{\mathfrak{a}}$ have the same Krull dimension. Since $\mathfrak{a} \subseteq \sqrt{\mathfrak{a}}$, it is enough to show that in any strictly increasing sequence of prime ideals as in $(**)$ such that all the ideals contain $\mathfrak{a}$, all the ideals actually contain $\sqrt{\mathfrak{a}}$. (Then the sequences $(**)$ for $\mathfrak{a}$ will be in one-one correspondence with the sequences for $\sqrt{\mathfrak{a}}$, and we can argue using the First Isomorphism Theorem as in the third paragraph of the proof.) Thus let x be in $\sqrt{\mathfrak{a}}$. By definition of radical, x^k lies in $\mathfrak{a}$ for some k. Since $\mathfrak{a} \subseteq \mathfrak{p}_d$, x^k lies in $\mathfrak{p}_d$. But $\mathfrak{p}_d$ is prime, and therefore x lies in $\mathfrak{p}_d$. Thus every ideal in the sequence $(**)$ for $\mathfrak{a}$ occurs in the sequence $(**)$ for $\sqrt{\mathfrak{a}}$, and the theorem follows. □

The dimension of an irreducible hypersurface in $A = \Bbbk[X_1, \ldots, X_n]$ is $n - 1$, as was observed in Section VII.5. Proposition 10.9 below will prove a converse.

Lemma 10.8. Every minimal nonzero prime ideal in A is principal.

PROOF. Let $\mathfrak{p}$ be a minimal nonzero prime ideal, let $f \neq 0$ be a nonzero member, and write f as the product of irreducible elements. Since $\mathfrak{p}$ is prime, one of the irreducible elements, say g, lies in $\mathfrak{p}$. Since A is a unique factorization domain, g is prime. Consequently (g) is a prime ideal of A lying in $\mathfrak{p}$. By minimality of $\mathfrak{p}$, $\mathfrak{p} = (g)$. $\square$

Proposition 10.9. Suppose that $\mathfrak{p}$ is a prime ideal of A and $V(\mathfrak{p})$ is the corresponding affine variety. If $\dim V(\mathfrak{p}) = n - 1$, then $\mathfrak{p}$ is principal, and hence $V(\mathfrak{p})$ is an irreducible hypersurface.

PROOF. For any $n \geq 1$, $\dim V(\mathfrak{p}) = n - 1 < n = \dim V(0)$ implies $\mathfrak{p} \neq 0$. Since $\dim V(\mathfrak{p}) = n - 1$, there exists a chain

$$0 = \mathfrak{q}_0 \subsetneq \mathfrak{q}_1 \subsetneq \cdots \subsetneq \mathfrak{q}_{n-1}$$

of prime ideals in $A/\mathfrak{p}$. If $\varphi : A \to A/\mathfrak{p}$ denotes the quotient homomorphism, then this chain lifts to A as

$$0 \subsetneq \mathfrak{p} \subsetneq \varphi^{-1}(\mathfrak{q}_1) \subsetneq \cdots \subsetneq \varphi^{-1}(\mathfrak{q}_{n-1}).$$

This chain has n members after the 0 at the left, and A has Krull dimension n. Consequently the first nonzero element, which is $\mathfrak{p}$, is a minimal nonzero prime ideal of A. By Lemma 10.8, $\mathfrak{p}$ is principal. $\square$

A **quasi-affine variety** is any nonempty Zariski open subset of an affine variety. These sets and their projective analogs, which will be defined in Section 3, will be the main objects of interest geometrically in Sections 1–6. If Y is a quasi-affine variety, then the closure $\overline{Y}$ is the affine variety in question because any nonempty relatively open subset of an affine variety is dense in the variety.[7]

Let us see that the relative Zariski topology on a quasi-affine variety Y makes Y into a Noetherian topological space. In fact, if X is a Noetherian topological space and Y is a topological subspace, then Y is Noetherian. To see this, we argue by contradiction, letting $E_1 \supseteq E_2 \supseteq \cdots$ be a strictly decreasing sequence of relatively closed sets in Y. Then the sequence of closures in X forms a decreasing sequence of closed sets in X with the property that $E_j = Y \cap \overline{E_j}$ for each j because E_j is assumed to be relatively closed in Y. It follows that the sequence of closures is strictly decreasing, contradiction.

Consequently any quasi-affine variety Y is Noetherian in the relative Zariski topology and has a meaningful geometric dimension. We write $\dim Y$ for this dimension.

---

[7]This important observation was made just before Proposition 10.5.

2. Geometric Dimension

Lemma 10.10. If Y is a quasi-affine variety in $\mathbb{A}^n$ and if E is a nonempty relatively closed subset of Y, then E is irreducible[8] for Y if and only if $\overline{E}$ is irreducible for $\mathbb{A}^n$.

REMARKS. We shall actually prove the stronger result that if Y is a nonempty open subset of a Noetherian topological space X (such as $\mathbb{A}^n$) and if E is a nonempty relatively closed subset of Y, then E is irreducible for Y if and only if $\overline{E}$ is irreducible for X. This stronger result will be used in Section 3.

PROOF. First we check that E reducible implies $\overline{E}$ reducible. If E is reducible, say is a union $E = E_1 \cup E_2$ with E_1 and E_2 relatively closed proper subsets of E, then $\overline{E} = \overline{E_1} \cup \overline{E_2}$. Each of $\overline{E_1}$ and $\overline{E_2}$ is a closed subset of $\overline{E}$. To see that $\overline{E_1}$ is proper, we argue by contradiction. If $\overline{E_1} = \overline{E}$, then intersecting both sides with Y gives the contradiction $E_1 = Y \cap \overline{E_1} = Y \cap \overline{E} = E$ because E_1 and E are both relatively closed. Similarly $\overline{E_2}$ is proper, and thus $\overline{E}$ is reducible.

Conversely suppose that $\overline{E}$ is reducible, say is a union $\overline{E} = F_1 \cup F_2$ with F_1 and F_2 closed in X and properly contained in $\overline{E}$. Intersecting both sides with Y gives $E = Y \cap \overline{E} = Y \cap (F_1 \cup F_2) = (Y \cap F_1) \cup (Y \cap F_2)$ because E is relatively closed. The sets $Y \cap F_1$ and $Y \cap F_2$ are relatively closed, and their union is E. To see that E is reducible, we argue by contradiction. If $Y \cap F_1 = E$, then $E \subseteq F_1$. Since F_1 is closed in X, $\overline{E} \subseteq F_1$. Thus F_1 is not a proper subset of $\overline{E}$, contradiction. Similarly we cannot have $Y \cap F_2 = E$, and therefore E is exhibited as the union of the two proper relatively closed subsets $Y \cap F_1$ and $Y \cap F_2$. □

Proposition 10.11. If Y is a quasi-affine variety in $\mathbb{A}^n$, then $\dim Y = \dim \overline{Y}$. Here $\dim \overline{Y}$ refers to the dimension of the affine variety $\overline{Y}$ in any of the senses of Theorem 10.7.

REMARKS. This proposition is a formal consequence of Lemma 10.10. The stronger statement that we actually prove is that if Y is a nonempty open subset of a Noetherian topological space X, then the geometric dimension of Y as a Noetherian space equals the geometric dimension of X as a Noetherian space.

PROOF. Let $E_0 \subseteq E_1 \subseteq \cdots \subseteq E_d$ be a strictly increasing sequence of relatively closed irreducible subsets of Y. Then $\overline{E_0} \subseteq \overline{E_1} \subseteq \cdots \subseteq \overline{E_d}$ is an increasing sequence of closed subsets of $\mathbb{A}^n$, each of which is irreducible by Lemma 10.10. Since $E_j = Y \cap \overline{E_j}$ for each j, the sets $\overline{E_j}$ are strictly increasing. Since the given sequence of sets E_j is arbitrary, it follows that $\dim Y \leq \dim \overline{Y}$.

For the reverse inequality, let $F_0 \subseteq F_1 \subseteq \cdots \subseteq F_d$ be a strictly increasing sequence of irreducible closed subsets of $\overline{Y}$. If E_j denotes $F_j \cap Y$, then $E_0 \subseteq$

---

[8] ... in the sense of not being the union of two relatively closed proper subsets.

$E_1 \subseteq \cdots \subseteq E_d$ is an increasing sequence of relatively closed subsets of Y, each of which is irreducible by Lemma 10.10. Since $F_j = \overline{E_j}$, the sets E_j are strictly increasing. Since the given sequence of sets F_j is arbitrary, it follows that $\dim \overline{Y} \leq \dim Y$. □

3. Projective Algebraic Sets and Projective Varieties

We continue to assume that $\mathbb{k}$ is an algebraically closed field and to write A for $\mathbb{k}[X_1, \ldots, X_n]$. In Section VIII.3 we studied the projective analogs of affine plane curves, and the task for the present section is to study similarly the projective analogs of general affine algebraic sets, affine varieties, and quasi-affine varieties.

As in Section VIII.3, **projective n-space** over $\mathbb{k}$ is defined set theoretically as the quotient

$$\mathbb{P}^n = \{(x_0, \ldots, x_n) \in \mathbb{k}^{n+1} - \{0\}\} \big/ \sim,$$

where $(x'_0, \ldots, x'_n) \sim (x_0, \ldots, x_n)$ if $(x'_0, \ldots, x'_n) = \lambda(x_0, \ldots, x_n)$ for some $\lambda \in \mathbb{k}^\times$. We write $[x_0, \ldots, x_n]$ for the class of $(x_0, \ldots, x_n)$ in $\mathbb{P}^n$.

Put $\widetilde{A} = \mathbb{k}[X_0, \ldots, X_n]$. The polynomials of interest for algebraic geometry relative to $\mathbb{P}^n$ are the homogeneous polynomials in $\widetilde{A}$. The definitions of "monomial," "total degree" of a monomial, "homogeneous polynomial," and "degree" of a homogeneous polynomial all appear in Section VIII.3; monomials are defined so as to have coefficient 1. By convention the 0 polynomial is homogeneous of every degree. We write $\widetilde{A}_d = \mathbb{k}[X_0, \ldots, X_n]_d$ for the $\mathbb{k}$ vector space of homogeneous polynomials of degree d. Each member F of $\widetilde{A}_d$ satisfies

$$F(\lambda x_0, \ldots, \lambda x_n) = \lambda^d F(x_0, \ldots, x_n)$$

for all $(x_0, \ldots, x_n) \in \mathbb{k}^{n+1}$ and $\lambda \in \mathbb{k}^\times$. Conversely the fact that the mapping of polynomials into polynomial functions is one-one for an infinite field implies that a member F of $\widetilde{A}$ is homogeneous of degree d if it satisfies the above displayed property. Four further properties of $\widetilde{A}_d$ from Section VIII.3 are that

- the zero locus of a member of $\widetilde{A}_d$ is well defined as a subset of $\mathbb{P}^n$,
- the monomials of total degree d form a $\mathbb{k}$ basis of the vector space $\widetilde{A}_d$,
- $\dim_\mathbb{k} \widetilde{A}_d = \binom{d+n}{n}$,
- any polynomial factor of a homogeneous polynomial over a field $\mathbb{k}$ is homogeneous.

An ideal $\mathfrak{a}$ in $\widetilde{A}$ is called a **homogeneous ideal** if it is the vector-space sum over $d \geq 0$ of its intersections with $\widetilde{A}_d$: $\mathfrak{a} = \bigoplus_{d=0}^\infty (\mathfrak{a} \cap \widetilde{A}_d)$. Any ideal in $\widetilde{A}$ that is generated by homogeneous polynomials is a homogeneous ideal. A special case of this fact is that if a $\mathbb{k}$ vector subspace $\mathfrak{a}_d$ of $\widetilde{A}_d$ is specified for each

3. Projective Algebraic Sets and Projective Varieties

integer $d \geq 0$, then $\mathfrak{a} = \bigoplus_{d=0}^{\infty} \mathfrak{a}_d$ is a homogeneous ideal if and only if for each $d \geq 0$ and $e \geq 0$, the inclusion $F\widetilde{A}_d \subseteq \widetilde{A}_{d+e}$ holds for each F in $\widetilde{A}_e$.

We can now imitate some of the development of Sections 1 and 2 for the present context as long as we stick to homogeneous polynomials in $\widetilde{A}$ and to homogeneous ideals. For any homogeneous polynomial F in $\widetilde{A}$, the set

$$V(F) = \{P = [x_0, \ldots, x_n] \in \mathbb{P}^n \mid F(x_0, \ldots, x_n) = 0\}$$

is well defined by the first bulleted property above. Thus if S is any set of homogeneous elements in $\widetilde{A}$, we can associate the **locus of common zeros** in $\mathbb{P}^n$, or **zero locus**, of the members of S by the formula

$$V(S) = \bigcap_{F \in S} V(F).$$

If $\mathfrak{a}$ is a homogeneous ideal, then $V(\mathfrak{a})$ by convention means $V(S)$, where S is the subset of all homogeneous members of $\mathfrak{a}$. Any such set $V(S)$ is called a **projective algebraic set** in $\mathbb{P}^n$. The function $S \mapsto V(S)$ is inclusion reversing. The analog of Proposition 10.1 in the present context is that projective algebraic sets have the following properties:

(i) $V(\varnothing) = V(0) = \mathbb{P}^n$ and $V(\widetilde{A}) = \varnothing$,
(ii) $V\big(\bigcup_\alpha S_\alpha\big) = \bigcap_\alpha V(S_\alpha)$ if the S_α's are arbitrary sets of homogeneous elements in $\widetilde{A}$,
(iii) $V(S) = V(S_1) \cup V(S_2)$ if S_1 and S_2 are sets of homogeneous elements in $\widetilde{A}$ and if S is defined as the set of all products $F_1 F_2$ with $F_1 \in S_1$ and $F_2 \in S_2$.

Consequently the projective algebraic sets in $\mathbb{P}^n$ form the closed sets for a topology on $\mathbb{P}^n$ called the **Zariski topology** on $\mathbb{P}^n$.

Next we associate to each point P of $\mathbb{P}^n$ a homogeneous ideal $I(P)$ in $\widetilde{A}$ by the definition

$$I(P) = \{F \in \widetilde{A} \mid F(x_0, \ldots, x_n) = 0 \text{ whenever } [x_0, \ldots, x_n] = P\}.$$

Problem 1 at the end of the chapter shows that $I(P)$ is indeed a homogeneous ideal. In terms of the ideals $I(P)$, we define $I(E) = \bigcap_{P \in E} I(P)$ for each subset E of $\mathbb{P}^n$. The result $E \mapsto I(E)$ is a function carrying subsets E of $\mathbb{P}^n$ to homogeneous ideals $I(E)$ in $\widetilde{A}^n$. The function $E \mapsto I(E)$ is inclusion reversing, and the same argument as for Proposition 10.2 shows that for each n it satisfies

(i) $I(\varnothing) = \widetilde{A}$ and $I(\mathbb{P}^n) = 0$,
(ii) $I(E_1 \cup E_2) = I(E_1) \cap I(E_2)$ if E_1 and E_2 are subsets of $\mathbb{P}^n$,
(iii) $I(E_1 \cap E_2) \supseteq I(E_1) + I(E_2)$ if E_1 and E_2 are subsets of $\mathbb{P}^n$.

If S is any set of homogeneous elements in $\widetilde{A}$ and if $V = V(S)$ is the corresponding projective algebraic set in $\mathbb{P}^n$, then we define the **cone** over V to be the subset of $\mathbb{A}^{n+1}$ given by

$$C(V) = (0, \ldots, 0) \cup \{(x_0, \ldots, x_n) \in \mathbb{A}^{n+1} \mid [x_0, \ldots, x_n] \in V\}.$$

This kind of set has the following two properties:
- (i) V nonempty implies that the ideals $I(C(V))$ and $I(V)$ in $\widetilde{A}$ are equal,
- (ii) any homogeneous ideal $\mathfrak{a}$ in $\widetilde{A}$ with $V(\mathfrak{a})$ nonempty in $\mathbb{P}^n$ has $C(V(\mathfrak{a}))$ equal to the subset $V(\mathfrak{a})$ in affine $(n+1)$-space.

Use of this device reduces a number of questions about $\mathbb{P}^n$ to questions about $\mathbb{A}^{n+1}$. An example is a projective analog of Proposition 10.3, which appears as the next proposition.

Proposition 10.12. For fixed n,

- (a) (**homogeneous Nullstellensatz**) a homogeneous ideal $\mathfrak{a}$ in $\widetilde{A}$ has $V(\mathfrak{a})$ empty in $\mathbb{P}^n$ if and only if there is an integer N such that $\mathfrak{a}$ contains $\widetilde{A}_k$ for $k \geq N$,
- (b) $I(V(\mathfrak{a})) = \sqrt{\mathfrak{a}}$ for each homogeneous ideal $\mathfrak{a}$ in $\widetilde{A}$ for which $V(\mathfrak{a})$ is nonempty in $\mathbb{P}^n$,
- (c) $V(I(E)) = \overline{E}$ for each subset E of $\mathbb{P}^n$, where $\overline{E}$ is the Zariski closure of E in $\mathbb{P}^n$.

REMARK. For clarity in the proof, let us write $V_a(\cdot)$ and $V_p(\cdot)$ to distinguish zero loci in $\mathbb{A}^{n+1}$ from zero loci in $\mathbb{P}^n$.

PROOF. For (a), $V_p(\mathfrak{a})$ is empty in $\mathbb{P}^n$ if and only if $V_a(\mathfrak{a})$ is contained in $\{0\}$ in $\mathbb{A}^{n+1}$, if and only if $\sqrt{\mathfrak{a}} = I(V_a(\mathfrak{a}))$ contains $(X_0, \ldots, X_n)$ by the affine Nullstellensatz. In this case if $f_1, \ldots, f_r$ are generators of $\sqrt{\mathfrak{a}}$, then the elements $f_1^m, \ldots, f_r^m$ are in $\mathfrak{a}$ for some m, and it follows that $\left(\sum_{j=1}^r c_r f_j\right)^k$ lies in $\mathfrak{a}$ for all scalars c_j whenever $k \geq rm$; hence $\widetilde{A}_k \subseteq \mathfrak{a}$ for $k \geq rm$. Conversely if $\sqrt{\mathfrak{a}}$ fails to contain some X_j, then X_j^k is not in $\mathfrak{a}$ for any $k \geq 1$, and $\widetilde{A}_k$ cannot be contained in $\mathfrak{a}$.

For (b), $I_p(V_p(\mathfrak{a})) = I_a(C(V_p(\mathfrak{a}))) = I_a(V_a(\mathfrak{a})) = \sqrt{\mathfrak{a}}$ by (i) of cones, (ii) of cones, and the affine Nullstellensatz.

Conclusion (c) is proved by the same argument as for Proposition 10.3b. $\square$

A **projective variety** is any *nonempty*[9] projective algebraic set of the form $V(\mathfrak{p})$, where $\mathfrak{p}$ is a prime homogeneous ideal in $\widetilde{A}$. If the ideal $\mathfrak{p}$ is the principal

---

[9]The prime homogeneous ideal $\mathfrak{p} = (X_0, \ldots, X_n)$ has $V(\mathfrak{p}) = \varnothing$, but no other prime homogeneous ideal $\mathfrak{q}$ has $V(\mathfrak{q}) = \varnothing$. In order to avoid trivial counterexamples to some results, we shall often want to exclude this particular prime ideal $\mathfrak{p}$ from consideration.

ideal generated by an irreducible homogeneous polynomial, then the ideal or the variety is called an **irreducible projective hypersurface**.[10]

Corollary 10.13. The projective varieties in $\mathbb{P}^n$ are characterized as those nonempty Zariski closed sets that cannot be written as the union of two proper closed subsets.

REMARK. Such a subset of $\mathbb{P}^n$ is said to be **irreducible**. As in the affine case, irreducible sets are understood to be nonempty.

PROOF. If $V(\mathfrak{p})$ is a projective variety, then the union of $\{0\}$ and the subset of $\mathbb{k}^{n+1}$ whose equivalence classes are in $V(\mathfrak{p})$ is an affine variety in $\mathbb{A}^{n+1}$. It is irreducible in $\mathbb{A}^{n+1}$, and this irreducibility in $\mathbb{A}^{n+1}$ implies irreducibility within $\mathbb{P}^n$.
Conversely if E is an irreducible closed subset of $\mathbb{P}^n$ and if F and G are *homogeneous* members of $\widetilde{A}$ with FG in $I(E)$, then we can argue as in the proof of Corollary 10.4 to see that one of F and G lies in $I(E)$ and that $I(E)$ is proper. Since $I(E)$ is a homogeneous ideal, this fact implies that $I(E)$ is prime. □

Since $\widetilde{A}$ is a Noetherian ring, it follows that $\mathbb{P}^n$ is a Noetherian topological space in the sense of Section 2. Consequently Proposition 10.5 is applicable. Combining this result with Corollary 10.13, we obtain the following corollary.

Corollary 10.14. Every projective algebraic set in $\mathbb{P}^n$ can be expressed uniquely as the finite (possibly empty) union of projective varieties in such a way that none of the varieties contains another of the varieties.

Geometric dimension is therefore meaningful for nonempty projective algebraic sets, and each such set in $\mathbb{P}^n$ has geometric dimension $\leq n$.

A **quasiprojective variety** is any nonempty Zariski open subset of a projective variety. Quasi-affine varieties and quasiprojective varieties will be the main objects of interest geometrically in Sections 1–7. If Y is a quasiprojective variety, then the relative Zariski topology on Y makes Y into a Noetherian topological space, just as in the quasi-affine case. Consequently Y has a meaningful geometric dimension. The arguments in Lemma 10.10 and Proposition 10.11 concerning quasi-affine varieties are arguments in point-set topology and valid proofs of facts about quasiprojective varieties. Therefore we obtain the following result.

Proposition 10.15. If Y is a quasiprojective variety in $\mathbb{P}^n$, then the closure $\overline{Y}$ in the Zariski topology of $\mathbb{P}^n$ is a projective variety, and the geometric dimensions of Y and $\overline{Y}$ are equal.

---

[10] As in the affine case, as long as the assumption of irreducibility is in force, the distinction between the ideal and the variety is unimportant.

We can identify $\mathbb{A}^n$ as a subset of $\mathbb{P}^n$ by the formula

$$\beta_0(x_1, \ldots, x_n) = [1, x_1, \ldots, x_n]$$

for $(x_1, \ldots, x_n)$ in $\mathbb{A}^n$. The complement of $\beta_0(\mathbb{A}^n)$ in $\mathbb{P}^n$ is the zero locus of the homogeneous polynomial X_0, and consequently $\beta_0(\mathbb{A}^n)$ is open in $\mathbb{P}^n$. Since the equality $\mathbb{P}^n = V(0)$ exhibits $\mathbb{P}^n$ as a projective variety, $\beta_0(\mathbb{A}^n)$ is a quasiprojective variety. We are going to show that β_0 respects topologies in that the Zariski topology of $\mathbb{A}^n$ is carried to the Zariski topology of the quasiprojective variety $\beta_0(\mathbb{A}^n)$. To do so, we make use of the corresponding transpose mapping $\beta_0^t : \widetilde{A} \to A$ on polynomials given by $\beta_0^t F = f$ with

$$f(X_1, \ldots, X_n) = F(\beta_0(X_1, \ldots, X_n)) = F(1, X_1, \ldots, X_n).$$

This is the substitution homomorphism that fixes $\mathbb{k}$, fixes $X_1, \ldots, X_n$, and carries X_0 to 1. Being an algebra homomorphism onto, β_0^t carries ideals of $\widetilde{A}$ to ideals of A. In particular, it carries homogeneous ideals of $\widetilde{A}$ to ideals of A.

Lemma 10.16. If $\mathfrak{a}$ is a homogeneous ideal in $\widetilde{A}$ and $\mathfrak{b} = \beta_0^t(\mathfrak{a})$ is its image under β_0^t, then β_0^t carries the set of homogeneous elements of $\mathfrak{a}$ onto $\mathfrak{b}$.

PROOF. Every member of $\mathfrak{b}$ is the sum of the images under β_0^t of finitely many homogeneous members of $\mathfrak{a}$. If $F_1, \ldots, F_k$ are these homogeneous members, then it is enough to produce $G_1, \ldots, G_k$ in $\mathfrak{a}$ all homogeneous of the same degree such that $\beta_0^t(F_j) = \beta_0^t(G_j)$ for all j. If $d_1, \ldots, d_k$ are the respective degrees of $F_1, \ldots, F_k$ and if $d = \max(d_1, \ldots, d_k)$, then the elements $G_j = X_0^{d-d_j} F_j$ have the required properties. □

Lemma 10.17. Let $\mathfrak{a}$ be a homogeneous ideal of $\widetilde{A}$, and let $\mathfrak{b}$ be the ideal of A given by $\mathfrak{b} = \beta_0^t(\mathfrak{a})$. Then $\beta_0(V(\mathfrak{b})) = V(\mathfrak{a}) \cap \beta_0(\mathbb{A}^n)$.

PROOF. If $(x_1, \ldots, x_n)$ is in $V(\mathfrak{b})$ and if F is a homogeneous member of $\mathfrak{a}$, then $f = \beta_0^t(F)$ is in $\mathfrak{b}$ with $0 = f(x_1, \ldots, x_n) = F(\beta_0(x_1, \ldots, x_n))$. Since F is arbitrary, $\beta_0(x_1, \ldots, x_n)$ is in $V(\mathfrak{a})$. Thus $\beta_0(V(\mathfrak{b})) \subseteq V(\mathfrak{a}) \cap \beta_0(\mathbb{A}^n)$.

For the reverse inclusion, let $[1, x_1, \ldots, x_n]$ be in $V(\mathfrak{a}) \cap \beta_0(\mathbb{A}^n)$. If f is in $\mathfrak{b}$, find by Lemma 10.16 a homogeneous F in $\mathfrak{a}$ with $\beta_0^t F = f$. Since $[1, x_1, \ldots, x_n]$ is in $V(\mathfrak{a})$, $F(1, x_1, \ldots, x_n) = 0$. Therefore $f(x_1, \ldots, x_n) = F(\beta_0(x_1, \ldots, x_n)) = F(1, x_1, \ldots, x_n) = 0$. Since f is arbitrary in $\mathfrak{b}$, the point $(x_1, \ldots, x_n)$ is in $V(\mathfrak{b})$, and $\beta_0(V(\mathfrak{b})) \supseteq V(\mathfrak{a}) \cap \beta_0(\mathbb{A}^n)$. □

Proposition 10.18. Under the inclusion $\beta_0 : \mathbb{A}^n \to \mathbb{P}^n$, the Zariski topology of affine n-space $\mathbb{A}^n$ coincides with the relative topology from $\mathbb{P}^n$.

PROOF. If we start from an affine algebraic set $V(\mathfrak{b})$ in $\mathbb{A}^n$, then Lemma 10.17 shows that $\beta_0(V(\mathfrak{b})) = V(\mathfrak{a}) \cap \beta_0(\mathbb{A}^n)$ for the homogeneous ideal $\mathfrak{a} = (\beta_0^t)^{-1}(\mathfrak{b})$ in $\widetilde{A}$. Since $V(\mathfrak{a})$ is Zariski closed in $\mathbb{P}^n$, $\beta_0(V(\mathfrak{b}))$ is exhibited as closed in the relative topology on $\beta_0(\mathbb{A}^n)$.

Conversely suppose that C is closed in the relative topology on $\beta_0(\mathbb{A}^n)$. Then it is of the form $\widetilde{C} \cap \beta_0(\mathbb{A}^n)$ for some projective algebraic set $\widetilde{C}$. The set $\widetilde{C}$ is of the form $V(\mathfrak{a})$ for some homogeneous ideal $\mathfrak{a}$. If $\mathfrak{b} = \beta_0^t(\mathfrak{a})$, then Lemma 10.17 shows that

$$\beta_0(V(\mathfrak{b})) = V(\mathfrak{a}) \cap \beta_0(\mathbb{A}^n) = \widetilde{C} \cap \beta_0(\mathbb{A}^n) = C,$$

and C is exhibited as β_0^t of an affine algebraic set in $\mathbb{A}^n$. □

Corollary 10.19. If V is a quasi-affine variety in $\mathbb{A}^n$, then $\beta_0(V)$ is a quasiprojective variety in $\mathbb{P}^n$. Moreover, the geometric dimension of V as a quasi-affine variety equals the geometric dimension of $\beta_0(V)$ as a quasiprojective variety.

REMARKS. In other words, the closure $\overline{\beta_0(V)}$ is a projective variety. It is called the **projective closure** of the quasi-affine variety V. If V is actually an affine variety, then it has an associated prime ideal in A, and the projective variety $\overline{\beta_0(V)}$ has an associated homogeneous prime ideal in $\widetilde{A}$. The correspondence between the prime ideal in A and the homogeneous prime ideal in $\widetilde{A}$ will be examined shortly.

PROOF. Because of the homeomorphism given by Proposition 10.18, Lemma 10.10 as restated in the lemma's remarks applies with $Y = \beta_0(\mathbb{A}^n)$, $X = \mathbb{P}^n$, and E equal to the closure of V in $\mathbb{A}^n$. The conclusion is that the closure of E in $\mathbb{P}^n$ is a projective variety, and the first conclusion of the corollary is proved. The second conclusion is immediate from the version of Proposition 10.11 mentioned in the remarks with that proposition. □

To each index i with $0 \leq i \leq n$, we can associate in a similar way a function $\beta_i : \mathbb{A}^n \to \mathbb{P}^n$. The formula for β_i is $\beta_i(x_1, \ldots, x_n) = [y_0, \ldots, y_n]$, where $y_j = x_{j+1}$ for $j < i$, $y_i = 1$, and $y_j = x_j$ for $j > i$. Just as in Proposition 10.18, under each β_i, the Zariski topology of affine n-space $\mathbb{A}^n$ coincides with the relative topology from $\mathbb{P}^n$. One consequence is that the notion of **projective closure** is meaningful if formed relative to any β_i in place of β_0. Another consequence is that $\mathbb{P}^n$ has a covering by $n+1$ open sets $\beta_i(\mathbb{A}^n)$ that are each Zariski homeomorphic to $\mathbb{A}^n$. The functions β_i may be viewed as playing a role similar to the inverses of charts in the definition of a smooth manifold.

Having used β_0 to associate a projective variety in $\mathbb{P}^n$ to each affine variety in $\mathbb{A}^n$ by passage to the topological closure, we turn to what happens with ideals.

Distinct homogeneous ideals in $\widetilde{A}$ can map under β_0^t to the same ideal in A; for example the principal ideals (1) and (X_0) in $\widetilde{A}$ both map to (1) in A. Theorem 10.20 will show that we can associate a particularly nice ideal of $\widetilde{A}$ to each ideal of A in such a way that prime ideals of A correspond to those nice ideals of $\widetilde{A}$ that are prime. Under this correspondence the ideals for an affine variety and its projective closure will match. It will be apparent from the construction in the proof that the ideal of $\widetilde{A}$ is generated by all homogeneous polynomials $F = F(f)$ of the form

$$F(X_0, \ldots, X_n) = X_0^d f(X_1/X_0, \ldots, X_n/X_0)$$

whenever $f \neq 0$ is in the ideal of A and $\deg f = d$.

Theorem 10.20. As a mapping of ideals in $\widetilde{A}$ to ideals in A, β_0^t is one-one from the set $\widetilde{\mathcal{I}}$ of all homogeneous ideals $\mathfrak{a}$ of $\widetilde{A}$ such that $X_0 F \in \mathfrak{a}$ implies $F \in \mathfrak{a}$ onto the set $\mathcal{I}$ of all ideals of A. Under this one-one correspondence prime ideals correspond to prime ideals.

PROOF. We are going to construct a two-sided inverse to the mapping induced by β_0^t from ideals in $\widetilde{\mathcal{I}}$ to ideals in $\mathcal{I}$.

Let $A_{\leq d}$ be the $\Bbbk$ vector space of all members of A, including the 0 polynomial, of degree $\leq d$. The homomorphism β_0^t carries $\widetilde{A}_d$ linearly into $A_{\leq d}$, and it carries the basis of homogeneous monomials in $\widetilde{A}$ of total degree d onto the basis of all monomials in A of total degree $\leq d$. Thus $\beta_0^t : \widetilde{A}_d \to A_{\leq d}$ is one-one onto. Observe for any f in $A_{\leq d}$ that the formula

$$F(X_0, \ldots, X_n) = X_0^d f(X_1/X_0, \ldots, X_n/X_0)$$

defines a member of $\widetilde{A}_d$. If we write $F = \varphi_d(f)$ when f and F are related in this way, then the function φ_d is a one-one $\Bbbk$ linear map from $A_{\leq d}$ into $\widetilde{A}_d$ such that $\varphi_d \beta_0^t$ is the identity on $\widetilde{A}_d$. Because of finite dimensionality, $\beta_0^t : \widetilde{A}_d \to A_{\leq d}$ and $\varphi_d : A_{\leq d} \to \widetilde{A}_d$ are two-sided inverses of one another.

Suppose that an ideal $\mathfrak{b}$ in A is given. Define $\mathfrak{a}_d = \varphi_d(\mathfrak{b} \cap A_{\leq d})$, and put $\mathfrak{a} = \bigoplus_{d=0}^{\infty} \mathfrak{a}_d$. According to remarks in the paragraph with the definition of homogeneous ideal, $\mathfrak{a}$ is a homogeneous ideal if $G \mathfrak{a}_d \subseteq \mathfrak{a}_{d+e}$ whenever G is in $\widetilde{A}_e$. Define $g = \beta_0^t(G)$. This polynomial has $\deg g \leq e$ and $\varphi_e(g) = G$, since $\varphi_e : A_{\leq e} \to \widetilde{A}_e$ is a two-sided inverse of $\beta_0^t : \widetilde{A}_e \to A_{\leq e}$. If f is in $\mathfrak{b} \cap A_{\leq d}$, then gf is in $\mathfrak{b} \cap A_{\leq (d+e)}$, and thus $G\varphi_d(f) = \varphi_e(g)\varphi_d(f) = \varphi_{d+e}(gf)$ is in $\mathfrak{a}_{d+e}$. This proves that $\mathfrak{a}$ is a homogeneous ideal in $\widetilde{A}$.

Under the construction $\mathfrak{b} \mapsto \mathfrak{a}$, let us see that $\mathfrak{a}$ is in $\widetilde{\mathcal{I}}$. If $X_0 F$ is in $\mathfrak{a}_{d+1}$, then we can write $X_0 F = \varphi_{d+1}(g)$ for some g in $\mathfrak{b} \cap A_{\leq d+1}$. That is, $X_0 F(X_0, \ldots, X_n) = X_0^{d+1} g(X_1/X_0, \ldots, X_n/X_0)$. Then $F(X_0, \ldots, X_n) = X_0^d g(X_1/X_0, \ldots, X_n/X_0)$. This formula shows that g is in $A_{\leq d}$ and that $F = $

3. Projective Algebraic Sets and Projective Varieties

$\varphi_d(g)$. Hence F is in $\mathfrak{a}_d$. In other words, the construction $\mathfrak{b} \mapsto \mathfrak{a}$ carries members of $\mathcal{I}$ to members of $\widetilde{\mathcal{I}}$.

Under the construction $\mathfrak{b} \mapsto \mathfrak{a}$, the homogeneous ideal $\mathfrak{a}$ has the property that

$$\beta_0^t(\mathfrak{a}) = \beta_0^t\Big(\bigoplus_{d=0}^{\infty} \mathfrak{a}_d\Big) = \sum_{d=0}^{\infty} \beta_0^t(\mathfrak{a}_d) = \sum_{d=0}^{\infty} (\mathfrak{b} \cap A_{\leq d}) = \mathfrak{b}.$$

Thus our construction starting from an ideal of A, passing to an ideal in the set $\widetilde{\mathcal{I}}$, and passing back to an ideal of A recovers the original ideal of A.

Now suppose that $\mathfrak{a}$ is in $\widetilde{\mathcal{I}}$. Put $\mathfrak{b} = \beta_0^t(\mathfrak{a})$. To see that the above passage to a member of $\widetilde{\mathcal{I}}$ recovers $\mathfrak{a}$ from $\mathfrak{b}$, we are to show that

$$\mathfrak{a} \cap \widetilde{A}_d = \varphi_d(\mathfrak{b} \cap A_{\leq d}). \qquad (*)$$

First we establish that

$$\beta_0^t(\mathfrak{a} \cap \widetilde{A}_d) = \beta_0^t(\mathfrak{a}) \cap A_{\leq d}. \qquad (**)$$

The inclusion $\subseteq$ in $(**)$ is easy because $\beta_0^t(\mathfrak{a} \cap \widetilde{A}_d) \subseteq \beta_0^t(\mathfrak{a})$ and $\beta_0^t(\widetilde{A}_d) \subseteq A_{\leq d}$. For the reverse inclusion, let f be in $\beta_0^t(\mathfrak{a} \cap \widetilde{A}_k) \cap A_{\leq d}$ for some k. This means that $\deg f \leq d$ and that $f = \beta_0^t(G)$ with $G \in \mathfrak{a} \cap \widetilde{A}_k$. Without loss of generality, we may assume that $k \geq d$. Let F be the element $F = \varphi_{\deg f}(f)$ of $\widetilde{A}_{\deg f}$. Then $X_0^{k-\deg f} F = \varphi_k(f)$, and $\beta_0^t(X_0^{k-\deg f} F) = \beta_0^t \varphi_k(f) = f = \beta_0^t(G)$. Hence $X_0^{k-\deg f} F$ and G are members of $\widetilde{A}_k$ with the same value under β_0^t. Since β_0^t is one-one on $\widetilde{A}_k$, $G = X_0^{k-\deg f} F$. Since G is in $\mathfrak{a}$ and since the ideal $\mathfrak{a}$ is in $\widetilde{\mathcal{I}}$, F is in $\mathfrak{a}$. Hence the element $X_0^{d-\deg f} F$ is in $\mathfrak{a} \cap \widetilde{A}_d$, and it has $\beta_0^t(X_0^{d-\deg f} F) = f$. This proves the inclusion $\supseteq$ in $(**)$. Application of φ_d to both sides of $(**)$ proves $(*)$ and completes the proof of the first statement of the theorem.

We are to show that prime ideals correspond to prime ideals. Let $\mathfrak{b}$ in $\mathcal{I}$ be prime, and let $\mathfrak{a}$ be the ideal in $\widetilde{\mathcal{I}}$ with $\beta_0^t(\mathfrak{a}) = \mathfrak{b}$. Let F and G be homogeneous elements in $\widetilde{A}$ of respective degrees d and e with FG in $\mathfrak{a}$. Then fg lies in $\mathfrak{b}$, where $f = \beta_0^t(F)$ and $g = \beta_0^t(G)$, and one of f and g lies in $\mathfrak{b}$ because $\mathfrak{b}$ is prime. Say f is in $\mathfrak{b}$. Then $F = \varphi_d(f)$ lies in the right side of $(*)$ and hence lies in the left side. Consequently F is in $\mathfrak{a}$, and $\mathfrak{a}$ is prime.

Conversely let $\mathfrak{a}$ in $\widetilde{\mathcal{I}}$ be prime, and let $\mathfrak{b} = \beta_0^t(\mathfrak{a})$. Suppose that f and g are members of A with fg in $\mathfrak{b}$. Put $d = \deg f$ and $e = \deg g$, and define $F = \varphi_d(f)$ and $G = \varphi_e(g)$. Then $FG = \varphi_{d+e}(fg)$ is in $\varphi_{d+e}(\mathfrak{b} \cap A_{\leq d+e})$, and $(*)$ shows that FG is in $\mathfrak{a} \cap \widetilde{A}_{d+e}$. Since $\mathfrak{a}$ is prime, one of F and G is in $\mathfrak{a}$. Say that F is in $\mathfrak{a}$. Then $f = \beta_0^t(F)$ is in $\mathfrak{b}$, and $\mathfrak{b}$ is prime. $\square$

Corollary 10.21. The inclusion $\beta_0 : \mathbb{A}^n \to \mathbb{P}^n$ sets up a one-one correspondence between the prime ideals in A and those prime homogeneous ideals in $\widetilde{A}$ that do not contain X_0.

PROOF. If $\mathfrak{a}$ is a prime homogeneous ideal in $\widetilde{A}$ and $X_0 F$ is in $\mathfrak{a}$, then either X_0 or F is in $\mathfrak{a}$. If we can always exclude X_0 from being in $\mathfrak{a}$, then F is in $\mathfrak{a}$, and the condition in the proposition for $\mathfrak{a}$ to be in $\widetilde{\mathcal{I}}$ is satisfied. The rest follows from Theorem 10.20. □

Corollary 10.22. Let $\mathfrak{a}$ be a prime homogeneous ideal of $\widetilde{A}$ not containing X_0, and let $\mathfrak{b} = \beta_0^t(\mathfrak{a})$ be the corresponding prime ideal of A. Then the Zariski closure in $\mathbb{P}^n$ of $\beta_0(V(\mathfrak{b}))$ is $V(\mathfrak{a})$.

REMARKS. In other words, if an affine variety V has $\mathfrak{b}$ as its ideal in A, then the projective closure of V has the corresponding $\mathfrak{a}$ from Theorem 10.20 as its ideal in $\widetilde{A}$.

PROOF. Corollary 10.19 shows that $\overline{\beta_0(V(\mathfrak{b}))} = V(\mathfrak{a}')$ for some prime homogeneous ideal of $\widetilde{A}$. Since $\beta_0(V(\mathfrak{b})) \subseteq V(\mathfrak{a})$ by Lemma 10.17 and since $V(\mathfrak{a})$ is closed in $\mathbb{P}^n$, $V(\mathfrak{a}') \subseteq V(\mathfrak{a})$. Arguing by contradiction, suppose that the inclusion is strict. Applying $I(\cdot)$ and using Proposition 10.12b, we obtain $\mathfrak{a}' \supseteq \mathfrak{a}$. Since application of $V(\cdot)$ to both sides of $\mathfrak{a}' \supseteq \mathfrak{a}$ has to yield a strict inclusion, we must have $\mathfrak{a}' \supsetneq \mathfrak{a}$. Choose G homogeneous in $\mathfrak{a}'$ that is not in $\mathfrak{a}$, and put $f = \beta_0^t G$. If $(x_1, \ldots, x_n)$ is in $V(\mathfrak{b})$, then $[1, x_1, \ldots, x_n]$ is in $\beta_0(V(\mathfrak{b})) \subseteq V(\mathfrak{a}')$, and hence $f(x_1, \ldots, x_n) = G(1, x_1, \ldots, x_n) = 0$. Thus f is in $I(V(\mathfrak{b})) = \mathfrak{b}$. Since $\deg f \leq \deg G$, the construction of $\mathfrak{a}$ from $\mathfrak{b}$ in the proof of Theorem 10.20 shows that $F = \varphi_{\deg G}(f)$ is in $\mathfrak{a}$. Then G and F are members of $\widetilde{A}_{\deg G}$ with $\beta_0^t(G) = f = \beta_0^t(F)$, and we obtain $G = F$, contradiction. □

EXAMPLE. Twisted cubic from the example in Section 1 and Example 2 in Section 2. The prime ideal $\mathfrak{b} \subseteq \mathbb{k}[X, Y, Z]$ is $(Y - X^2, Z - X^3)$, and we want to find the corresponding ideal $\mathfrak{a}$ given by Corollary 10.21. Let the additional indeterminate in $\widetilde{A}$ be W. Applying φ_2 and φ_3 to the respective generators $Y - X^2$ and $Z - X^3$ yields $WY - X^2$ and $W^2 Z - X^3$. These must be in $\mathfrak{a}$. So must

$$(W^2 Z - X^3) - X(WY - X^2) = W(WZ - XY)$$

and $\quad X(W^2 Z - X^3) - (WY + X^2)(WY - X^2) = W^2(XZ - Y^2).$

Since we seek a prime ideal for $\mathfrak{a}$ and W is not to be in $\mathfrak{a}$, $WZ - XY$ and $XZ - Y^2$ are in $\mathfrak{a}$. Thus $\mathfrak{a} \supseteq (WY - X^2, WZ - XY, XZ - Y^2)$. If $\mathfrak{c}$ denotes the ideal on the right, then $\mathfrak{a} \supseteq \mathfrak{c}$ and

$$\beta_0^t(\mathfrak{c}) = (Y - X^2, Z - XY, XZ - Y^2)$$
$$= (Y - X^2, Z - X^3, XZ - X^4) = (Y - X^2, Z - X^3) = \beta_0^t(\mathfrak{a}).$$

To show that $\mathfrak{a} = \mathfrak{c}$, it is enough according to Theorem 10.20 to show that if F is homogeneous and WF is in $\mathfrak{c}$, then F is in $\mathfrak{c}$. The three generators of $\mathfrak{c}$ are all in $\widetilde{A}_2$, and thus $\mathfrak{c} \cap \widetilde{A}_d = \widetilde{A}_{d-2}(\mathfrak{c} \cap \widetilde{A}_2)$. Hence it is enough to show that $\mathfrak{c} \cap \widetilde{A}_2$ contains no nonzero element divisible by W. Since $\mathfrak{c} \cap \widetilde{A}_2$ consists of all linear combinations of the three generators, we can check this fact by inspection. The result is that $\mathfrak{a} = \mathfrak{c}$. Once we know $\mathfrak{a}$, we can compute the projective closure of the twisted cubic from Corollary 10.22. We find that it consists of all $[w, x, y, z]$ of the form $[1, x, x^2, x^3]$ together with $[0, 0, 0, 1]$. We might have guessed this form for the projective closure from the parametric realization of the twisted cubic in $\mathbb{A}^3$ and from a passage to the limit, but proceeding in that fashion requires operations that we have certainly not justified.

4. Rational Functions and Regular Functions

We continue to assume that $\Bbbk$ is an algebraically closed field and to write A for $\Bbbk[X_1, \ldots, X_n]$ and $\widetilde{A}$ for $\Bbbk[X_0, \ldots, X_n]$. In this section we investigate certain classes of $\Bbbk$-valued functions on quasiprojective varieties, specifically the "rational" functions, the "regular" functions, and the local ring of functions regular at a particular point. For each kind of variety that we have introduced (affine, quasi-affine, projective, and quasiprojective), there are simple global definitions and there are complicated but *equivalent* local definitions for these notions. The complicated definitions have three advantages over the simple ones: they are virtually the same for all four kinds of varieties and therefore make it possible to work with all kinds of varieties uniformly, they make it possible in practice to construct a function by constructing only a local part of it, and they prepare the way better for a definition of isomorphism of varieties that does not insist on a particular dimension for the ambient affine or projective space.

In this section we shall first give the simple definitions in the affine and quasi-affine cases and then prove results saying that certain more complicated local-sounding versions of these definitions amount to the same thing as the simple definitions. Then we shall give the simple definitions in the projective and quasiprojective cases. Finally we shall relate the quasi-affine and quasiprojective cases and show that certain more complicated local-sounding definitions in the quasiprojective case amount to the same thing as the simple definitions.

We begin with affine varieties. Suppose that $V = V(\mathfrak{p})$ is an affine variety in $\mathbb{A}^n$, $\mathfrak{p}$ being a prime ideal in A. The **affine coordinate ring** of V is $A(V) = A/\mathfrak{p}$, which is an integral domain. Let us write the quotient homomorphism $A \to A(V)$ as $a \mapsto \bar{a}$. Because of the Nullstellensatz, $A(V)$ can be identified with the ring of all restrictions of polynomials to V; in particular, $\bar{a}(P)$ is meaningful for every $\bar{a} \in A(V)$ and $P \in V$.

Proposition 10.23. If V is an affine variety in $\mathbb{A}^n$, then the points P of V are in one-one correspondence with the maximal ideals $\mathfrak{m}_P$ of the affine coordinate ring $A(V)$, the correspondence being that $\mathfrak{m}_P$ is the maximal ideal of all members $\bar{a}$ of $A(V)$ with $\bar{a}(P) = 0$.

PROOF. Each $\mathfrak{m}_P$ is a maximal ideal, being the kernel of a multiplicative linear functional. In the reverse direction, if $\mathfrak{m}$ is a maximal ideal of $A(V)$, then its inverse image in A under the homomorphism $A \to A/\mathfrak{p} = A(V)$ is a maximal ideal M of A containing $\mathfrak{p}$, by the First Isomorphism Theorem. The Nullstellensatz shows that M consists of all polynomials vanishing at some point P. Applying $V(\cdot)$ to the inclusion $M \supseteq \mathfrak{p}$ gives $\{P\} = V(M) \subseteq V(\mathfrak{p}) = V$. Thus P is in V. □

Members of the field of fractions $\mathbb{k}(V)$ of $A(V)$ are called **rational functions** on V, and $\mathbb{k}(V)$ is called the **function field** on V. Rational functions on V are not really functions on V in the traditional sense, since their denominators can vanish here and there. By way of compensation, an allowable denominator never vanishes identically; the reason is that the construction of a field of fractions of an integral domain does not involve using the zero element of the integral domain in a denominator. If f is a rational function on V and P is in V, one says that f is **regular** at P, or **defined** at P, if there exist $\bar{a}$ and $\bar{b}$ in $A(V)$ with $\bar{b}(P) \neq 0$ such that $f = \bar{a}/\bar{b}$. In this case, an equality $\bar{a}/\bar{b} = \bar{a}'/\bar{b}'$ with $\bar{b}(P) \neq 0$ and $\bar{b}'(P) \neq 0$ implies that $\bar{a}\bar{b}' = \bar{a}'\bar{b}$, from which we see that $\bar{a}(P)\bar{b}'(P) = \bar{a}'(P)\bar{b}(P)$ and that $\bar{a}(P)/\bar{b}(P) = \bar{a}'(P)/\bar{b}'(P)$. Hence $f(P)$ can be defined unambiguously as $f(P) = \bar{a}(P)/\bar{b}(P)$. For P in V, the set of rational functions on V that are regular at P is a $\mathbb{k}$ algebra, as we see by carrying out the usual manipulations to add or multiply fractions. This $\mathbb{k}$ algebra is denoted by $\mathcal{O}_P(V)$. It has $A(V) \subseteq \mathcal{O}_P(V) \subseteq \mathbb{k}(V)$.

As in Proposition 10.23, let $\mathfrak{m}_P$ be the maximal ideal of all members $\bar{a}$ of $A(V)$ with $\bar{a}(P) = 0$. The localization of $A(V)$ with respect to this maximal ideal is exactly $\mathcal{O}_P(V)$. In fact, the localization is a subring of $\mathbb{k}(V)$ because $A(V)$ is an integral domain. The members of $\mathcal{O}_P(V)$ are exactly the quotients $f = \bar{a}/\bar{b}$ with $\bar{a}$ and $\bar{b}$ in $A(V)$ and with $\bar{b}$ not in $\mathfrak{m}_P$. Hence $\mathcal{O}_P(V) = S^{-1}A(V)$, where S is the set-theoretic complement of $\mathfrak{m}_P$. Thus $\mathcal{O}_P(V)$ is the asserted localization. It has a unique maximal ideal and is called the **local ring** of V at P.

A rational function is said to be **regular** on an open subset U of V if it is regular at every point of U. The regular functions on U form a $\mathbb{k}$ algebra denoted by $\mathcal{O}(U)$. In symbols the definition of $\mathcal{O}(U)$ is $\mathcal{O}(U) = \bigcap_{P \in V} \mathcal{O}_P(V)$.

When $A(V)$ is a unique factorization domain, the definition of regular at a point is simple enough to implement globally: we write $f = \bar{a}/\bar{b}$ in some fashion, reduce the fraction to lowest terms, and then read off all the points P for which f is defined from the single expression of f as a quotient. Ordinarily, however, $A(V)$ is not a unique factorization domain, and then the definition is more subtle, as the following example shows.

EXAMPLE. $V = V(\mathfrak{p})$ with $\mathfrak{p} = (XW - YZ)$ and $n = 4$. The polynomial $XW - YZ$ is irreducible, and thus V is an affine variety in $\mathbb{A}^4$. The affine coordinate ring is $A(V) = \mathbb{k}[W, X, Y, Z]/(XW - YZ)$. The quotient $f = \overline{X}/\overline{Y}$ is a rational function on V, since $\overline{Y}$ is not the 0 element of $A(V)$, and the definition shows that f is regular at all points (w, x, y, z) of V having $y \neq 0$. From $\overline{X}\,\overline{W} - \overline{Y}\,\overline{Z} = 0$, we have $\overline{X}/\overline{Y} = \overline{Z}/\overline{W}$, and thus f is defined also at all points (w, x, y, z) of V having $w \neq 0$. For example it is defined at the additional point $(w, x, y, z) = (1, 0, 0, 0)$. Actually, there exist no members $\bar{a}$ and $\bar{b}$ of $A(V)$ with $f = \bar{a}/\bar{b}$ and $\bar{b}(w, x, y, z) \neq 0$ whenever $xw = yz$ and one or both of w and y are nonzero. The details are carried out in Problem 8 at the end of the chapter.

The set of points P in the affine variety V at which a rational function f on V fails to be regular is called the **pole set** of f.

Proposition 10.24. If f is a rational function on the affine variety $V = V(\mathfrak{p})$, then the pole set of f is the affine algebraic set $V(\mathfrak{a}) \subseteq V(\mathfrak{p})$ corresponding to the ideal $\mathfrak{a} \supseteq \mathfrak{p}$ of all $b \in A$ such that $\bar{b}f$ is in $A(V)$.

PROOF. The set $\mathfrak{a}$ in the statement is an ideal in A that contains $\mathfrak{p}$. Hence $V(\mathfrak{a}) \subseteq V(\mathfrak{p})$. If P is in $V(\mathfrak{p})$ and f is defined at P, then there are members $\bar{a}$ and $\bar{b}$ of $A(V)$ with $\bar{b}(P) \neq 0$ such that $\bar{b}f = \bar{a}$; any representative of this $\bar{b}$ in A lies in $\mathfrak{a}$, and consequently P is not in $V(\mathfrak{a})$. Conversely if f is not defined at P, then no $\bar{b}$ such that $\bar{b}f$ is in $A(V)$ has $\bar{b}(P) \neq 0$. That is, no member b of $\mathfrak{a}$ has $\bar{b}(P) \neq 0$. So P is in $V(\mathfrak{a})$. This proves that the pole set of f is exactly $V(\mathfrak{a})$. □

Corollary 10.25. If $V = V(\mathfrak{p})$ is an affine variety, then
$$A(V) = \bigcap_{P \in V} \mathcal{O}_P(V).$$

REMARKS. In the notation introduced above, the corollary says that $A(V) = \mathcal{O}(V)$.

PROOF. The inclusion $\subseteq$ follows from the fact that $A(V) \subseteq \mathcal{O}_P(V)$ for each P. For the reverse inclusion, suppose that f lies in $\bigcap_{P \in V} \mathcal{O}_P(V)$. Then the pole set of f in V is empty. The pole set for f is the set $V(\mathfrak{a})$ for the ideal $\mathfrak{a}$ in Proposition 10.24, and it follows from the Nullstellensatz that $\mathfrak{a} = A$. Then 1 is in $\mathfrak{a}$, and the definition of $\mathfrak{a}$ shows that f is in $A(V)$. □

If we consider the complement of the pole set of f, then we see from Proposition 10.24 that the subset of V at which f is regular is (relatively) open in V. Hence it is empty or dense in V. On the set where f is regular, f is continuous into $\mathbb{A}^1$, according to the following proposition.

Proposition 10.26. If a rational function f on the affine variety V is regular on the nonempty open set U of V, then it is continuous from U into $\mathbb{A}^1$ with the Zariski topology (in which the proper closed sets are the finite sets).

PROOF. It is to be proved that f^{-1} of any finite subset of $\mathbb{A}^1$ is relatively closed in U. Since the finite union of closed sets is closed, it is enough to consider $f^{-1}(\{c\})$ for an element c of $\mathbb{k}$. This is the intersection with U of the pole set of $1/(f-c)$, which is relatively closed in U by Proposition 10.24. □

Now we can give the simple definitions in the quasi-affine case. Let the quasi-affine variety U in $\mathbb{A}^n$ have closure the affine variety V. If f is a rational function on V, then Proposition 10.24 shows that f is regular on a nonempty open subset of V. Since the intersection of any two nonempty open subsets is nonempty, f is regular on a nonempty open subset of U. Therefore it is meaningful to view f as a rational function on U. We define the **function field** of **rational functions** on U to be the same as the function field of V: $\mathbb{k}(U) = \mathbb{k}(V)$. The definition of **regular function** at P is the same for the quasi-affine variety U as for its Zariski closure V, and thus the **local ring** of U at P is given by $\mathcal{O}_P(U) = \mathcal{O}_P(V)$. A rational function is said to be **regular** on the quasi-affine variety U if it is regular at every point of U. Since $\mathbb{k}(U) = \mathbb{k}(V)$, the set of regular functions on U is the $\mathbb{k}$ algebra $\mathcal{O}(U) = \bigcap_{P \in U} \mathcal{O}_P(U)$.

The next step is to prove results saying that certain more complicated local-sounding definitions of the above notions amount to the same thing.

Lemma 10.27. If V is an affine variety, then any two members of the affine coordinate ring $A(V)$ that are equal on a nonempty open subset of V are the same.

PROOF. Subtracting, we may suppose that $\bar{a} \in A(V)$ is 0 on the nonempty open subset U of V. By Proposition 10.26, $\bar{a}$ is continuous from V into $\mathbb{A}^1$. The complement of $\bar{a}^{-1}(\{0\})$ has to be open in V and disjoint from U, and therefore it is empty. So $\bar{a}$ is everywhere 0 and is the 0 element of $A(V)$. □

Proposition 10.28. Let U be a nonempty open subset of the affine variety V in $\mathbb{A}^n$. Suppose that $f_0 : U \to \mathbb{k}$ is a function with the following property: for each P in U, there exist an open subset W of U containing P and polynomials a and b in A such that b is nowhere vanishing on W and $f_0 = a/b$ on W. Then there exists one and only one member f of $\mathbb{k}(V)$ such that f is regular on U and agrees with f_0 at every point of U.

REMARKS. For the quasi-affine case the more complicated local-sounding definition of "regular function" on U, mentioned in the first paragraph of this section, is what is assumed of f_0 in the statement of this proposition. The proposition says that such an f_0 necessarily comes from a global rational function on V that is regular on U in the sense just above.

PROOF OF UNIQUENESS. If there are two such members of $\Bbbk(V)$, then subtracting them gives a member g of $\Bbbk(V)$ that is 0 on U. By definition of $\Bbbk(V)$, $g = \bar{a}/\bar{b}$ with $\bar{a}$ and $\bar{b}$ in $A(V)$ with with $\bar{b} \neq 0$. Then $\bar{a} = g\bar{b}$ is a member of $A(V)$ that is 0 on U. By Lemma 10.27, $\bar{a} = 0$ in $A(V)$. Thus $g\bar{b} = 0$ in $\Bbbk(V)$. Since $\Bbbk(V)$ is a field and $\bar{b} \neq 0$, $g = 0$. □

PROOF OF EXISTENCE. If P is in U, then the hypothesis supplies some open subset W of U containing P and members a and b of A with b nowhere 0 on W and with $f_0 = a/b$ on W. Let $\bar{a}$ and $\bar{b}$ be the images of a and b in $A(V)$. Since b is not identically 0 on U, $\bar{b}$ is not the 0 element of $A(V)$. Therefore $f = \bar{a}/\bar{b}$ is a well-defined member of $\Bbbk(V)$, and it is regular on W and agrees with f_0 there. If we start with another point P' and an open subset W' of U containing P', then we similarly obtain $f' = \bar{a}'/\bar{b}'$ in $\Bbbk(V)$ that is regular on W' and agrees with f_0 there. The open subset $W \cap W'$ is nonempty, and $\bar{a}/\bar{b} = \bar{a}'/\bar{b}'$ on $W \cap W'$. Therefore $\bar{b}'\bar{a} = \bar{b}\bar{a}'$ on $W \cap W'$. By Lemma 10.27, $\bar{b}'\bar{a} = \bar{b}\bar{a}'$ as members of $A(V)$. Dividing, we obtain $f = f'$. Since the member f of $\Bbbk(V)$ is regular on an open neighborhood of each point of U, it is regular on U. □

Proposition 10.28 allows us also to give a local-sounding definition of rational function and see that it reduces to the original definition. Specifically we consider pairs (U_0, f_0) with U_0 nonempty open in the quasi-affine variety U and with f_0 satisfying the regularity condition on U_0 in the proposition.[11] Say that the pair (U_0, f_0) is equivalent to the pair (U_1, f_1) if $f_0 = f_1$ on $U_0 \cap U_1$. This relation is reflexive and symmetric. Let us see from the proposition why it is transitive. If (U_0, f_0) is equivalent to (U_1, f_1), then the existence part of the proposition yields three members of $\Bbbk(V)$—one for (U_0, f_0), one for $(U_0 \cap U_1, f_0) = (U_0 \cap U_1, f_1)$, and one for (U_1, f_1). The uniqueness part shows that the first two members of $\Bbbk(V)$ are equal and the last two are equal. Hence they are all equal. Now if (U_0, f_0) is equivalent to (U_1, f_1) and (U_1, f_1) is equivalent to (U_2, f_2), then we routinely find that $(U_0 \cap U_1, f_0)$ is equivalent to $(U_1 \cap U_2, f_2)$. From what we have just seen, (U_0, f_0) is equivalent to (U_2, f_2), and the relation is therefore transitive. We could take the union of all the sets U_0 appearing in the pairs within an equivalence class and obtain the **largest domain** within U on which the rational function in question is regular. This notion for a rational function will not be too useful for us, but an analogous notion for rational maps in Section 6 will be quite handy.

In similar fashion the local ring $\mathcal{O}_P(U)$ can be formulated in terms of "germs" of regular functions as follows. Fix P in U, and consider all pairs (U_0, f_0) such that U_0 is an open subset of U containing P and f_0 is a scalar-valued function on

---

[11]That is, for each P in U_0, there exist an open subset W of U_0 containing P and polynomials a and b in A such that b is nowhere vanishing on W and $f_0 = a/b$ on W.

U_0 satisfying the regularity condition on U_0 in the proposition.[12] Say that (U_0, f_0) is equivalent to (U_1, f_1) if $f_0 = f_1$ on some open neighborhood of U containing P. It is easy to see that the result is an equivalence relation. An equivalence class is called a **germ** of regular functions at P. Germs inherit a natural addition, scalar multiplication, and multiplication, and the set of germs at P is therefore a $\mathbb{k}$ algebra. The use of germs is the traditional device in mathematics for isolating local behavior of functions in arbitrarily small neighborhoods of points.

Corollary 10.29. Let U be a nonempty open subset of the affine variety V in $\mathbb{A}^n$, and let P be in U. To each germ $\{(U_0, f_0)\}$ of regular functions at P corresponds one and only one member f of $\mathbb{k}(V)$ that is associated via Proposition 10.28 to each pair (U_0, f_0). Moreover, this correspondence is a $\mathbb{k}$ algebra isomorphism of the ring of germs onto the local ring $\mathcal{O}_P(U)$.

PROOF. If (U_0, f_0) and (U_0', f_0') are two pairs in a germ at P, then the definition of germ gives a pair (W, g_0) such that W is a neighborhood of P contained in $U_0 \cap U_0'$ and g agrees with f_0 and f_0' on W. Proposition 10.28 supplies unique members f, f', and g of $\mathbb{k}(V)$ such that f is regular on U_0 and agrees with f_0 there, such that f' is regular on U_0' and agrees with f_0' there, and such that g is regular on W and agrees with g_0 there. The uniqueness in the proposition shows that $f = g$ and that $g = f'$. Therefore $f = f'$. So we have a well-defined map of germs into $\mathbb{k}(V)$.

The image f of the pair (U_0, f_0) is a member of $\mathbb{k}(V)$ that is regular on U_0, hence is defined at P. Thus the map on germs is into $\mathcal{O}_P(U)$. It is a $\mathbb{k}$ algebra homomorphism because of the definitions of the operations on germs. If the germ of (U_0, f_0) maps to 0, then f_0 is the 0 function on U_0, and any representative (W, g_0) of the germ with $W \subseteq U_0$ has g_0 equal to the 0 function on W. Thus the germ is the 0 germ, and the $\mathbb{k}$ algebra homomorphism is one-one. Finally if f is a member of $\mathcal{O}_P(U)$, then $f = \bar{a}/\bar{b}$ with $\bar{a}$ and $\bar{b}$ in $A(V)$ and with $\bar{b}$ nonvanishing at P. By Proposition 10.26, $\bar{b}$ is nonvanishing on some open neighborhood U_0 of P. Then the germ of (U_0, f_0) maps to f if f_0 is defined as the restriction of $\bar{a}/\bar{b}$ to U_0. Therefore the $\mathbb{k}$ algebra homomorphism is onto $\mathcal{O}_P(U)$. □

This completes the discussion of the definitions in the cases of affine and quasi-affine varieties. Next we consider projective varieties, beginning with the simple definitions. Let $V = V(\mathfrak{p})$ be a projective variety, $\mathfrak{p}$ being a prime homogeneous ideal in $\widetilde{A}$ different from $\bigoplus_{d \geq 1} \widetilde{A}_d$. The integral domain $\widetilde{A}(V) = \widetilde{A}/\mathfrak{p}$ is called the **homogeneous coordinate ring** of V. Since $\mathfrak{p}$ is homogeneous, we can write $\widetilde{A}(V)$ as

$$\widetilde{A}(V) = \bigoplus_{d=0}^{\infty} \widetilde{A}_d / (\widetilde{A}_d \cap \mathfrak{p}) = \bigoplus_{d=0}^{\infty} \widetilde{A}(V)_d.$$

---

[12]See the previous footnote.

Let us write the quotient homomorphism $\widetilde{A} \to \widetilde{A}(V)$ as $F \mapsto \overline{F}$. We say that $\overline{F}$ is **homogeneous** of degree d if it lies in $\widetilde{A}(V)_d = \widetilde{A}_d/(\widetilde{A}_d \cap \mathfrak{p})$.

Despite Proposition 10.12, homogeneous members of $\widetilde{A}(V)$ do not yield well-defined functions on V, and we cannot simply imitate the affine case in defining the function field of V. The **function field** $\Bbbk(V)$ of V is a certain *proper* subfield of the field of fractions of $\widetilde{A}(V)$, namely the set of all quotients $\overline{F}/\overline{G}$ with $\overline{F}$ and $\overline{G}$ homogeneous of the same degree and with $\overline{G} \neq 0$. If the common degree of $\overline{F}$ and $\overline{G}$ is d, then the quotient $\overline{F}/\overline{G}$ is homogeneous of degree 0 in $(x_0, \ldots, x_n)$ and is therefore well-defined on the equivalence class $[x_0, \ldots, x_n]$ in $\mathbb{P}^n$. Such quotients form a field because if $\overline{F}_1$ and $\overline{G}_1$ are homogeneous of degree d and $\overline{F}_2$ and $\overline{G}_2$ are homogeneous of degree e, then $\overline{F}_1/\overline{G}_1 + \overline{F}_2/\overline{G}_2 = (\overline{F}_1\,\overline{G}_2 + \overline{G}_1\,\overline{F}_2)/(\overline{G}_1\,\overline{G}_2)$ and $(\overline{F}_1\,\overline{F}_2)/(\overline{G}_1\,\overline{G}_2)$ are each the quotient of two members of $\widetilde{A}(V)$ that are homogeneous of degree $d + e$, the denominator not being the zero element, and because the inverse of $\overline{F}/\overline{G}$ is $\overline{G}/\overline{F}$. Elements of $\Bbbk(V)$ are called **rational functions** on V.

Although the values of homogeneous members of $\widetilde{A}$ are not meaningful on $\mathbb{P}^n$, the zero locus of such a polynomial *is* well defined. If $\overline{F}$ is a member of the quotient $\widetilde{A}(V)$ homogeneous of degree d, then its set of preimages in $\widetilde{A}_d$ is $F + (\widetilde{A}_d \cap \mathfrak{p})$. The members of $\widetilde{A}_d \cap \mathfrak{p}$ all vanish at every point of V, and therefore whether F vanishes at a point P of V depends only on the coset of F in $\widetilde{A}(V)$. Accordingly, a member h of $\Bbbk(V)$ is said to be **regular** at the point $P = [x_0, \ldots, x_n]$ of V, or **defined** at P, if h can be written as a quotient $h = \overline{F}/\overline{G}$ of homogeneous members of $\widetilde{A}(V)$ of the same degree in such a way that $\overline{G}(P) \neq 0$. In this case, $h(P)$ is well defined as the quotient $F(x_0, \ldots, x_n)/G(x_0, \ldots, x_n)$ for any $(x_0, \ldots, x_n)$ representing the point $P = [x_0, \ldots, x_n]$.

The set of points P in the projective variety V at which a rational function h on V fails to be regular is called the **pole set** of h. The proof of the following result is similar to the proof of Proposition 10.24 and is therefore omitted.

Proposition 10.30. If h is a rational function on the projective variety $V = V(\mathfrak{p})$, then the pole set of h is the projective algebraic set $V(\mathfrak{a}) \subseteq V(\mathfrak{p})$ corresponding to the homogeneous ideal $\mathfrak{a} \supseteq \mathfrak{p}$ generated by all homogeneous $G \in \widetilde{A}$ such that $\overline{G}h$ is in $\widetilde{A}(V)$.

As in the case of affine varieties, the set of members of $\Bbbk(V)$ regular at P in V is a $\Bbbk$ subalgebra of $\Bbbk(V)$ called the **local ring** of V at P and denoted by $\mathcal{O}_P(V)$.

Corollary 10.31. If $V = V(\mathfrak{p})$ is a projective variety, then

$$\Bbbk = \bigcap_{P \in V} \mathcal{O}_P(V).$$

REMARKS. The classical prototype of this corollary is that a rational function without poles on the Riemann sphere is constant. A direct proof of this fact for the Riemann sphere in the style of this book follows by applying Proposition 6.9 to the sum of the given rational function and any constant function. A generalization appears as Corollary 9.4.

PROOF. The inclusion $\subseteq$ is automatic. For the reverse inclusion, suppose that the rational function h on V lies in $\bigcap_{P \in V} \mathcal{O}_P(V)$. Then the pole set of h in V is empty. The pole set for h is the set $V(\mathfrak{a})$ for the ideal $\mathfrak{a}$ in Proposition 10.30, and it follows from the homogeneous Nullstellensatz (Proposition 10.12a) that $\widetilde{A}_N \subseteq \mathfrak{a}$ for all N sufficiently large. For any such N, $\widetilde{A}(V)_N h$ lies in $\widetilde{A}(V)$. It is homogeneous of degree N and hence is in $\widetilde{A}(V)_N$. Iterating this inclusion gives

$$\widetilde{A}(V)_N h^k \subseteq \widetilde{A}(V)_N \quad \text{for all } k \geq 0. \tag{$*$}$$

Since V is nonempty, some X_j is not in $\mathfrak{p}$; to fix the notation, let us suppose that X_0 is not in $\mathfrak{p}$. Then $\overline{X_0} \neq 0$. Inclusion $(*)$ shows that $\overline{X_0}^N h^k$ lies in $\widetilde{A}(V)$ for all $k \geq 0$. Thus h^k lies in the subset $\overline{X_0}^{-N} \widetilde{A}(V)$ of the field of fractions of $\widetilde{A}(V)$, and the ring $\widetilde{A}(V)[h]$, given by the substitution homomorphism $X \mapsto h$ applied to the polynomial ring $\widetilde{A}(V)[X]$, is exhibited as an $\widetilde{A}(V)$ submodule of the finitely generated $\widetilde{A}(V)$ module $\overline{X_0}^{-N} \widetilde{A}(V)$ of the field of fractions of $\widetilde{A}(V)$. Since $\widetilde{A}(V)$ is Noetherian as a homomorphic image of $\widetilde{A}$, $\widetilde{A}(V)[h]$ is a finitely generated $\widetilde{A}(V)$ module. By Proposition 8.35 of *Basic Algebra*, h is a root of some monic polynomial in $\widetilde{A}(V)[X]$. Say that h satisfies

$$h^l + c_{l-1} h^{l-1} + \cdots + c_1 h + c_0 = 0$$

with each c_j in $\widetilde{A}(V)$. Decomposing each term into homogeneous parts and equating to 0 the sum of the terms homogeneous of degree 0 shows that we can assume each c_j to be in $\widetilde{A}(V)_0 = \Bbbk$. That is, we may assume that h is algebraic over $\Bbbk$. Since $\Bbbk$ is algebraically closed, h is in $\Bbbk$. $\square$

If we consider the complement of the pole set of h, then we see from Proposition 10.30 that the subset of V at which h is regular is open in V. Hence it is empty or dense in V. On the set where h is regular, h is continuous into $\mathbb{A}^1$, according to the following proposition, whose proof is the same as for Proposition 10.26.

Proposition 10.32. If a rational function h on the projective variety V is regular on the nonempty open set U of V, then it is continuous from U into $\mathbb{A}^1$ with the Zariski topology (in which the proper closed sets are the finite sets).

The procedure for extending the above remarks from projective varieties to quasiprojective varieties is the same as for extending the earlier remarks from affine varieties to quasi-affine varieties. Let the quasiprojective variety U in $\mathbb{P}^n$ have closure the projective variety V. If h is a rational function on V, then Proposition 10.32 shows that h is regular on a nonempty open subset of V. Since the intersection of any two nonempty open subsets is nonempty, h is regular on a nonempty open subset of U. Therefore it is meaningful to view h as a rational function on U. Thus we define the **function field** of U to be the same as the function field of V: $\mathbb{k}(U) = \mathbb{k}(V)$. The definition of **regular function** at P is the same for the quasiprojective variety U as for its Zariski closure V, and thus the **local ring** of U at P is given by $\mathcal{O}_P(U) = \mathcal{O}_P(V)$. A rational function is said to be **regular** on the quasiprojective variety U if it is regular at every point of U. The set of regular functions on U is a $\mathbb{k}$ algebra denoted by $\mathcal{O}(U)$. Thus

$$\mathcal{O}(U) = \bigcap_{P \in U} \mathcal{O}_P(U).$$

For the special case that $U = V$, Corollary 10.31 shows that $\mathcal{O}(V)$ reduces to the constants.

The next step is to check that the simple definitions in this section in the affine and quasi-affine cases are consistent with the simple definitions in the projective and quasi-projective cases. Proposition 10.18 and Corollary 10.19 tell us the extent of the overlap—that any of the mappings $\beta_j : \mathbb{A}^n \to \mathbb{P}^n$ with $0 \leq j \leq n$ allows us to identify any quasi-affine variety with a quasiprojective variety. Thus what we need to show is that the definitions of function field, functions regular at a point, and functions regular on a variety amount to the same thing for a quasi-affine variety U and for the quasiprojective variety $\beta_j(U)$. For concreteness we shall take $j = 0$.

Corollaries 10.21 and 10.22 tell us exactly what we are to compare. The prime ideals $\mathfrak{a}$ of $\widetilde{A}$ not containing X_0 are in one-one correspondence with the prime ideals $\mathfrak{b}$ of A, the correspondence being $\mathfrak{b} = \beta_0^t(\mathfrak{a})$, and the Zariski closure of $V(\beta_0^t(\mathfrak{b}))$ in $\mathbb{P}^n$ is $V(\mathfrak{a})$. The correspondence does not yield a natural map of $\mathfrak{b}$ into $\mathfrak{a}$. Instead, the system of linear mappings $\varphi_d : A_{\leq d} \to \widetilde{A}_d$ given by

$$F(X_0, \ldots, X_n) = \varphi_d(f)(X_0, \ldots, X_n) = X_0^d f(X_1/X_0, \ldots, X_n/X_0)$$

is a system of inverses to the system of restrictions $\beta_0^t\big|_{\widetilde{A}_d} : \widetilde{A}_d \to A_{\leq d}$ of the homomorphism $\beta_0^t : \widetilde{A} \to A$ given by

$$f(X_1, \ldots, X_n) = \beta_0^t(F)(X_1, \ldots, X_n) = F(1, X_0, \ldots, X_n),$$

and these systems have the properties that

$$\mathfrak{a} \cap \widetilde{A}_d = \varphi_d(\mathfrak{b} \cap A_{\leq d}) \quad \text{and} \quad \beta_0^t(\mathfrak{a} \cap \widetilde{A}_d) = \mathfrak{b} \cap A_{\leq d}.$$

Proposition 10.33. Let a prime ideal $\mathfrak{a}$ of $\widetilde{A}$ not containing X_0 correspond to the prime ideal $\mathfrak{b}$ of A under the formula $\mathfrak{b} = \beta_0^t(\mathfrak{a})$ as in Theorem 10.20, and let $U = V(\mathfrak{b})$ and $V = V(\mathfrak{a})$ be the respective affine and projective varieties for $\mathfrak{b}$ and $\mathfrak{a}$, V being the Zariski closure of $\beta_0(U)$ in $\mathbb{P}^n$. Then β_0^t descends to a ring homomorphism ψ of $\widetilde{A}(V)$ onto $A(U)$, and ψ in turn induces a canonical field isomorphism $\Psi : \Bbbk(V) \to \Bbbk(U)$. Under the field isomorphism Ψ, the image of the local ring $\mathcal{O}_{\beta_0(P)}(V)$ is $\mathcal{O}_P(U)$ for each P in U.

PROOF. Since β_0^t carries $\widetilde{A}$ onto A and carries $\mathfrak{a}$ into $\mathfrak{b}$, β_0^t descends to a homomorphism ψ of $\widetilde{A}/\mathfrak{a} = \widetilde{A}(V)$ onto $A/\mathfrak{b} = A(U)$. If $\overline{F}$ and $\overline{G}$ are in the same homogeneous summand $\widetilde{A}(V)_d$ of $\widetilde{A}(V)$, then we define $\Psi(\overline{F}/\overline{G}) = \psi(\overline{F})/\psi(\overline{G})$ as a member of the field of fractions $\Bbbk(U)$ of $A(U)$. If $\overline{F}/\overline{G} = \overline{F}'/\overline{G}'$, then $\overline{F}\,\overline{G}' = \overline{F}'\,\overline{G}$. Applying ψ, using that ψ is a homomorphism, and reinterpreting matters in $\Bbbk(U)$, we see that $\Psi(\overline{F}/\overline{G}) = \Psi(\overline{F}'/\overline{G}')$, i.e., that Ψ is well defined. A similar argument that involves clearing fractions and applying ψ shows that Ψ respects addition and multiplication. Therefore Ψ is a field mapping of $\Bbbk(V)$ into $\Bbbk(U)$.

Let $A(U)_{\leq d}$ be the image of $A_{\leq d}$ in $A/\mathfrak{b} = A(U)$. Since β_0^t carries $\widetilde{A}_d$ onto $A_{\leq d}$ and carries $\mathfrak{a} \cap \widetilde{A}_d$ onto $\mathfrak{b} \cap A_{\leq d}$, ψ carries $\widetilde{A}(V)_d$ onto $A(U)_{\leq d}$. Any member of $\Bbbk(U)$ is the quotient of two members of $A(U)_{\leq d}$ for some d, and it is consequently Ψ of the quotient of the corresponding members of $\widetilde{A}(V)_d$. Therefore Ψ carries $\Bbbk(V)$ onto $\Bbbk(U)$ and is a field isomorphism.

Let $\overline{F}$ and $\overline{G}$ in $\widetilde{A}(V)$ be the cosets $F + \mathfrak{a}$ and $G + \mathfrak{a}$, let $f = \beta_0^t(F)$ and $g = \beta_0^t(G)$, and let $\bar{f}$ and $\bar{g}$ in $A(U)$ be the cosets of $f + \mathfrak{b}$ and $g + \mathfrak{b}$. Then $\psi(\overline{F}) = \bar{f}$ and $\psi(\overline{G}) = \bar{g}$, and hence $\Psi(\overline{F}/\overline{G}) = \bar{f}/\bar{g}$. Let $P = (x_1, \ldots, x_n)$ be in U, so that $\beta_0(P) = [1, x_1, \ldots, x_n]$ is in $\beta_0(U)$. Define $\beta_0^\#(P) = (1, x_1, \ldots, x_n)$ in $\mathbb{A}^{n+1}$, so that the class of $\beta_0^\#(P)$ in $\mathbb{P}^n$ is $\beta_0(P)$. Then $\bar{g}(P) = g(P) = (\beta_0^t G)(P) = G(\beta_0^\#(P)) = \overline{G}(\beta_0^\#(P))$. Therefore $\bar{f}/\bar{g}$ lies in $\mathcal{O}_P(U)$ if and only if $\overline{F}/\overline{G}$ lies in $\mathcal{O}_{\beta_0(P)}(V)$. So Ψ carries $\mathcal{O}_{\beta_0(P)}(V)$ onto $\mathcal{O}_P(U)$. $\square$

Corollary 10.34. Let V be a projective variety, and let U be a nonempty open subset of V. Then each member of $\mathcal{O}(U) \subseteq \Bbbk(V)$ is determined as an element in $\Bbbk(V)$ by its restriction to U.

PROOF. Subtracting two such members, we may assume that their difference h is 0 on U. We are to prove that $h = 0$ in $\Bbbk(V)$. For some j with $0 \leq j \leq n$, $\beta_j(\mathbb{A}^n) \cap V$ is nonempty, and we may assume that this is the case for $j = 0$. The subset $V_0 = \beta_0^{-1}(V)$ of $\mathbb{A}^n$ is an affine variety. Since U and $\beta_0(\mathbb{A}^n) \cap V$ are nonempty open subsets of V, their intersection is nonempty, and $U_0 = \beta_0^{-1}(U)$ is a nonempty open subset of V_0. Let $\Psi : \Bbbk(V) \to \Bbbk(V_0)$ be the field isomorphism in Proposition 10.33. By assumption, h is in $\mathcal{O}_{\beta_0(P)}(V)$ for every P in U_0. Since the value of h at P is 0, h is actually in the maximal ideal of $\mathcal{O}_{\beta_0(P)}(V)$ for P in U_0. Proposition 10.33 shows that $\Psi(h)$ is in the maximal ideal of $\mathcal{O}_P(V_0)$ for all P in U_0. Fix P_0 in U_0. Then we can write the member $\Psi(h)$ of $\Bbbk(V_0)$ as $\Psi(h) = \bar{a}/\bar{b}$ with $\bar{b}(P_0) \neq 0$. Since $\bar{b}$ is continuous on V_0 by Proposition 10.26, $\bar{b}(P)$ is nonzero for all P in some neighborhood W of P_0 contained in U_0. Then the formula $\Psi(h) = \bar{a}/\bar{b}$ shows explicitly that $\Psi(h)$ is defined at such points P and satisfies $\Psi(h)(P) = \bar{a}(P)/\bar{b}(P)$. Since $\Psi(h)$ is in the maximal ideal of $\mathcal{O}_P(V_0)$ for all P in U_0, $\Psi(h)(P) = 0$ for P in W. Hence $\bar{a}(P) = 0$ for P in W. Consequently $\bar{a}$ and 0 are two members of $A(V)$ that are equal on W, and Lemma 10.27 allows us to conclude that $\bar{a} = 0$. Therefore $h = 0$. □

Proposition 10.35. Let U be a nonempty open subset of the projective variety V in $\mathbb{P}^n$. Suppose that $h_0 : U \to \Bbbk$ is a function with the following property: for each P in U, there exist an open subset W of U containing P and homogeneous polynomials F and G in $\widetilde{A}$ of the same degree such that G is nowhere vanishing on W and $h_0 = F/G$ on W. Then there exists one and only one member h of $\Bbbk(V)$ such that h is regular on U and agrees with h_0 at every point of U.

REMARKS. For the quasiprojective case the more complicated local-sounding definition of "regular function" on U, mentioned in the first paragraph of this section, is what is assumed of h_0 in the statement of this proposition. The proposition says that such an h_0 necessarily comes from a global rational function on V that is regular on U in the sense just above.

PROOF. For each j with $0 \leq j \leq n$ such that $V_j = \beta_j(\mathbb{A}^n) \cap V$ is nonempty, $\beta_j^{-1}(V_j)$ is an affine variety, and $U_j = U \cap V_j$ is a nonempty open subset such that $h_{j,0} = h_0|_{U_j}$ is a function on U_j with the following property: for each P in U_j, there exist an open subset W of U_j containing P and homogeneous polynomials F and G in $\widetilde{A}$ of the same degree such that G is nowhere vanishing on W and $h_{j,0} = F/G$ on W. We pull back this situation by β_j^{-1}, writing $\beta_j^t h_{j,0}$ for the function on $\beta_j^{-1}(W)$ given by $(\beta_j^t h_{j,0})(Q) = h_{j,0}(\beta_j(Q))$. The set $\beta_j^{-1}(V_j)$ is an affine variety, and the Zariski closure of V_j in $\mathbb{P}^n$ is V. The homomorphism β_j^t on $\widetilde{A}$ descends to a ring homomorphism $\psi_j : \widetilde{A}(V) \to A(\beta_j^{-1}(V_j))$, and ψ_j induces a field isomorphism $\Psi_j : \Bbbk(V) \to \Bbbk(\beta_j^{-1}(V_j))$, according to Proposition 10.33.

The set $\beta_j^{-1}(U_j)$ is a nonempty open subset of the affine variety $\beta_j^{-1}(V_j)$, and $\beta_j^t h_{j,0}$ is a function on $\beta_j^{-1}(U_j)$ with the following property: for each P in $\beta_j^{-1}(U_j)$, there exist an open subset W of $\beta_j^{-1}(U_j)$ containing P and homogeneous polynomials F and G in $\tilde{A}$ of the same degree such that their images $\overline{F}$ and $\overline{G}$ in $\tilde{A}(V)$ have $\overline{G}$ nowhere vanishing on W and have $\beta_j^t h_{j,0} = \psi_j(\overline{F})/\psi_j(\overline{G}) = \Psi(\overline{F/G})$ on W. Proposition 10.33 says that $\psi_j(\overline{F}) = \bar{a}$ and $\psi_j(\overline{G}) = \bar{b}$ for members $\bar{a}$ and $\bar{b}$ of $A(\beta_j^{-1}(V_j))$. We are in the situation of Proposition 10.28 with $f_0 = \beta_j^t h_{j,0}$, and that proposition produces a unique member h_j of $\Bbbk(\beta_j^{-1}(V_j))$ that is regular on $\beta_j^{-1}(U_j)$ and agrees with $\beta_j^t h_{j,0}$ at every point of $\beta_j^{-1}(U_j)$.

The member h of $\Bbbk(V)$ that we seek is $h = \Psi_j^{-1}(h_j)$. To verify this assertion, we are to show that $\Psi_j^{-1}(h_j)$ is independent of j. Thus suppose that $V_i \cap V_j \neq \varnothing$. Fix P in $U_i \cap U_j = U \cap V_i \cap V_j$, and choose the above open neighborhood W of P small enough for the above construction to apply for both indices i and j. By the uniqueness in Proposition 10.28, h_j is the unique member of $\Bbbk(\beta_j^{-1}(V_j))$ that is regular on $\beta_j^{-1}(W)$ and agrees with $\beta_j^t h_{j,0} = \beta_j^t(h_0|_{U_j})$ at every point of $\beta_j^{-1}(W)$. Thus $\Psi_j^{-1}(h_j) = \overline{F/G}$ on W, where $\overline{F}$ and $\overline{G}$ are as in the previous paragraph. By the same uniqueness argument, $\Psi_i^{-1}(h_i) = \overline{F/G}$ on W. The difference $\Psi_i^{-1}(h_i) - \Psi_j^{-1}(h_j)$ is a member of $\Bbbk(V)$ that is regular on W and vanishes there. By Corollary 10.34, the difference is 0 as an element of $\Bbbk(V)$. Therefore $\Psi_j^{-1}(h_j)$ is independent of j, and we can take h to be this member of $\Bbbk(V)$. □

Just as in the quasi-affine case, it is possible in the quasiprojective case to give a local-sounding definition of rational function and a formulation of $\mathcal{O}_P(U)$ in terms of germs. We shall not use these notions, and we omit any further discussion of them.

5. Morphisms

The goal of this section and the next is to introduce maps that make the collection of all quasiprojective varieties over an algebraically closed field $\Bbbk$ into the objects of a category in a way that does not depend on the ambient space $\mathbb{A}^n$ or $\mathbb{P}^n$ of the variety. These maps will all be algebraic in nature, and there will be two choices of which class of maps to use, one involving good denominators and one allowing occasional bad denominators. The first kind of map will be called a "morphism," and the second kind of map will be called a "dominant rational map." The relationships between these two kinds of maps and the interpretation of these maps in terms of function fields will be of great importance in applying this theory.

A **variety** over the algebraically closed field $\Bbbk$ henceforth will be any affine, quasi-affine, projective, or quasiprojective variety as in the previous sections. To each such variety V, Section 4 associates a function field $\Bbbk(V)$, a local ring $\mathcal{O}_P(V) \subseteq \Bbbk(V)$ of regular functions at each point P, and a ring $\mathcal{O}(E) = \bigcap_{P \in E} \mathcal{O}_P(V) \subseteq \Bbbk(V)$ of regular functions on each nonempty open subset E of V. We have observed that each rational function on a variety V is regular on some nonempty open subset of V, namely the complement of the pole set. One further fact that we shall use about rational functions is the following.

Proposition 10.36. If P and Q are distinct points of a variety V, then there exists a rational function $h \in \Bbbk(V)$ such that h is defined at both P and Q, has $h(P) = 0$, and has $h(Q) \neq 0$.

PROOF. Without loss of generality, we may assume that V is projective. Say that $V \subseteq \mathbb{P}^n$. Let $\mathfrak{p}$ be the prime homogeneous ideal in $\widetilde{A} = \Bbbk[X_0, \ldots, X_n]$ such that $\widetilde{A}(V) = \widetilde{A}/\mathfrak{p}$, and let $F \mapsto \overline{F}$ be the quotient homomorphism $\widetilde{A} \mapsto \widetilde{A}(V)$. Let $P = [x_0, \ldots, x_n]$ and $Q = [y_0, \ldots, y_n]$. Choose a homogeneous polynomial F in $\widetilde{A}$ such that $F(x_0, \ldots, x_n) = 0$ and $F(y_0, \ldots, y_n) \neq 0$, and choose a homogeneous polynomial G with $\deg G = \deg F$ such that $G(x_0, \ldots, x_n) \neq 0$ and $G(y_0, \ldots, y_n) \neq 0$. Then $\overline{G}$ is not 0, and $h = \overline{F}/\overline{G}$ has the required properties. □

If U and V are varieties, then a continuous function $\varphi : U \to V$ relative to the Zariski topology is called a **morphism** if for each nonempty open subset E of V and each regular function f on E, the composition $f \circ \varphi$ is a regular function on the open subset $\varphi^{-1}(E)$ of U. Thus φ is to be continuous and is to induce by composition a function from $\mathcal{O}(E)$ into $\mathcal{O}(\varphi^{-1}(E))$ for each open subset E of V. An **isomorphism** of varieties is a morphism having an inverse function that is a morphism.

It is immediate that the composition of two morphisms is a morphism and that the identity function is a morphism. Thus the varieties over $\Bbbk$ form a category if morphisms are used as the maps.

EXAMPLES OF MORPHISMS. Suppose that $\Bbbk$ has characteristic different from 2. Let U be $\mathbb{P}^1$, written as

$$\mathbb{P}^1 = \{[s, t] \mid (s, t) \neq (0, 0)\},$$

and let V be the projective variety in $\mathbb{P}^2$ defined by the irreducible homogeneous polynomial $X^2 + Y^2 - Z^2$, i.e.,

$$V = \{[x, y, z] \mid x^2 + y^2 = z^2 \text{ and } (x, y, z) \neq (0, 0, 0)\}.$$

Let $\varphi : U \to V$ be the function given by

$$\varphi([s, t]) = [s^2 - t^2, 2st, s^2 + t^2].$$

This is well defined, and it is continuous because the Zariski closed proper subsets of V are the finite sets, whose inverse images are finite sets. If F and G are two homogeneous members of $\Bbbk[X, Y, Z]$ and if $\overline{F}$ and $\overline{G}$ are the images in $\widetilde{A}(V) = \Bbbk[X, Y, Z]/(X^2 + Y^2 - Z^2)$, we are to assume that $\overline{G}$ is not 0, i.e., that G is not divisible by $X^2 + Y^2 - Z^2$, and then $h = \overline{F}/\overline{G}$ is a typical rational function on V. We are to show that if h is regular on an open subset E of V, then $h \circ \varphi$ is regular on $\varphi^{-1}(E) \subseteq \mathbb{P}^1$. The expression $h = \overline{F}/\overline{G}$ exhibits h as regular on the open set E of points $[x, y, z]$ of V with $G(x, y, z) \neq 0$. The set $\varphi^{-1}(E)$ is the set of points $[s, t]$ in $\mathbb{P}^1$ with $G(s^2 - t^2, 2st, s^2 + t^2) \neq 0$. At such points the function $h \circ \varphi$ is given by

$$(h \circ \varphi)(s, t) = F(s^2 - t^2, 2st, s^2 + t^2)/G(s^2 - t^2, 2st, s^2 + t^2),$$

and it is given by a rational expression with nonvanishing denominator. Thus φ is a morphism.

Let us see that $\psi : V \to \mathbb{P}^1$ given by

$$\psi[x, y, z] = \begin{cases} [x + z, y] & \text{if } [x, y, z] \neq [1, 0, -1], \\ [-y, x - z] & \text{if } [x, y, z] \neq [1, 0, 1] \end{cases}$$

consistently defines another morphism. For the consistency we observe that $x^2 + y^2 = z^2$ implies that $(x + z)(x - z) = -y^2$; hence on the common domain of the two expressions, $[x + z, y] = [-y^2/(x - z), y] = [-y/(x - z), 1] = [-y, x - z]$. Continuity of ψ follows because the inverse image of any finite set is a finite set. For the regularity we observe that if F and G are homogeneous members of the same degree in $\widetilde{A}(\mathbb{P}^1) = \Bbbk[S, T]$ with $G \neq 0$ and if $h = F/G$, then the expression $h = F/G$ exhibits h as regular on the open set E of points $[s, t]$ in $\mathbb{P}^1$ with $G(s, t) \neq 0$. The set $\psi^{-1}(E)$ is the set of points $[x, y, z]$ on V with $G(x + z, y) \neq 0$. At such points the function $h \circ \psi$ is given by

$$(h \circ \psi)[x, y, z] = F(x + z, y)/G(x + z, y),$$

and it is given by a rational expression with a nonvanishing denominator. Thus ψ is a morphism. In other words, φ is an isomorphism.

Proposition 10.37. Let $\beta_0 : \mathbb{A}^n \to \mathbb{P}^n$ be the usual inclusion. If U is a quasi-affine variety in $\mathbb{A}^n$, then β_0 is an isomorphism of the quasi-affine variety U onto the quasiprojective variety $\beta_0(U)$.

PROOF. Proposition 10.18 shows that β_0 is a homeomorphism of U onto its image. The last conclusion of Proposition 10.33 implies that the regular functions for U match those for $\beta_0(U)$ under β_0, and the result follows. □

5. Morphisms

Theorem 10.38. Let U be any variety, let V be any affine variety, and let $A(V)$ be the affine coordinate ring of V. Then the morphisms $\varphi : U \to V$ are in one-one correspondence with the $\Bbbk$ algebra homomorphisms $\widetilde{\varphi} : A(V) \to \mathcal{O}(U)$ via the formula

$$\widetilde{\varphi}(f) = f \circ \varphi \qquad \text{for } f \in A(V).$$

REMARKS. Members f of $A(V)$ lie in $\mathcal{O}(V)$. The $\Bbbk$ algebra homomorphism $\widetilde{\varphi}$ is meaningful because the fact that φ is a morphism implies that $f \circ \varphi$ is in $\mathcal{O}(\varphi^{-1}(E))$ for every open E in V; here we take $E = V$ and $\varphi^{-1}(E) = U$. The proof of Theorem 10.38 will be preceded by a lemma.

Lemma 10.39. If U is a variety and V is an affine variety in $\mathbb{A}^n$, then a function $\psi : U \to V$ is a morphism if and only if $\overline{X_i} \circ \psi$ is a regular function on U for the image $\overline{X_i}$ in $A(V)$ of each coordinate function X_i with $1 \leq i \leq n$.

PROOF. If ψ is a morphism, then the definition of morphism forces $\overline{X_i} \circ \psi$ to be a regular function.

Conversely suppose ψ has the property that each $\overline{X_i} \circ \psi$ is a regular function. Then $f \circ \psi$ is a regular function on U for each f in $A(V)$, since every member of $A(V)$ is a polynomial in the elements $\overline{X_i}$. If E is a closed set in V, then E is the locus of common zeros of some set $\{f_\alpha\}$ of polynomials, and $\psi^{-1}(E)$ is the set of points P such that $f_\alpha(\psi(P)) = 0$ for all α. Hence $\psi^{-1}(E)$ is the locus of common zeros of a subset $\{f_\alpha \circ \psi\}$ of regular functions on U and is relatively closed in U. Thus ψ is continuous.

If E is nonempty open in V, then $\Bbbk(E) = \Bbbk(V)$ shows that each regular function h on E is locally the quotient of members of $A(V)$ with nonvanishing denominator. Let us write $h = f/g$ with g nonvanishing near a point of interest. Then $h \circ \psi = (f \circ \psi)/(g \circ \psi)$ is exhibited locally as a rational function with nonvanishing denominator. $\square$

PROOF OF THEOREM 10.38. Suppose that $\alpha : A(V) \to \mathcal{O}(U)$ is a $\Bbbk$ algebra homomorphism. Define $\psi : U \to V$ by $\psi(P) = (\alpha(\overline{X_1})(P), \ldots, \alpha(\overline{X_n})(P))$. Then $\overline{X_i} \circ \psi = \alpha(\overline{X_i})$ is in $\mathcal{O}(U)$ by definition of α, and Lemma 10.39 shows that ψ is a morphism.

The $\Bbbk$ algebra homomorphism $\widetilde{\psi}$ defined by $\widetilde{\psi}(f) = f \circ \psi$ has $\widetilde{\psi}(\overline{X_i}) = \overline{X_i} \circ \psi = \alpha(\overline{X_i})$. Since the elements $\overline{X_i}$ generate $A(V)$, $\widetilde{\psi} = \alpha$. Thus starting from α, forming ψ, and obtaining $\widetilde{\psi}$ recovers α. In the reverse direction if we start from φ, form $\widetilde{\varphi}$, and use the construction of the previous paragraph to obtain ψ, then $\psi(P) = \big(\widetilde{\varphi}(\overline{X_1})(P), \ldots, \widetilde{\varphi}(\overline{X_n})(P)\big) = \big(\overline{X_1}(\varphi(P)), \ldots \overline{X_n}(\varphi(P))\big) = \varphi(P)$ for P in U. Hence $\psi = \varphi$. Thus the function $\alpha \mapsto \psi$ is a two-sided inverse of the function $\varphi \mapsto \widetilde{\varphi}$. $\square$

Corollary 10.40. If U and V are affine varieties, then the morphisms $\varphi : U \to V$ are in one-one correspondence with the $\mathbb{k}$ algebra homomorphisms $\widetilde{\varphi} : A(V) \to A(U)$ via the formula

$$\widetilde{\varphi}(f) = f \circ \varphi \qquad \text{for } f \in A(V).$$

PROOF. This is immediate from Theorem 10.38, since Corollary 10.25 shows that $\mathcal{O}(U) = A(U)$. $\qquad\square$

Proposition 10.41. If U and V are varieties and if $\varphi : U \to V$ and $\psi : U \to V$ are morphisms such that $\varphi\big|_E = \psi\big|_E$ for some nonempty open set E in U, then $\varphi = \psi$.

PROOF. Let h be a rational function on V, and let E' be the nonempty open subset of V on which h is regular. Since φ and ψ are morphisms, $h \circ \varphi$ and $h \circ \psi$ are regular on the respective nonempty open subsets $\varphi^{-1}(E')$ and $\psi^{-1}(E')$ of U. The equality $\varphi\big|_E = \psi\big|_E$ shows that $h \circ \varphi$ and $h \circ \psi$ are equal on the nonempty open subset $E \cap \varphi^{-1}(E') \cap \psi^{-1}(E')$ of U. The function $h \circ \varphi - h \circ \psi$ is therefore a rational extension from $E \cap \varphi^{-1}(E') \cap \psi^{-1}(E')$ to U of the 0 function, and Proposition 10.34 shows that $h \circ \varphi - h \circ \psi = 0$ on U. Therefore $h \circ \varphi = h \circ \psi$ as elements of $\mathbb{k}(U)$ for every h in $\mathbb{k}(V)$.

Arguing by contradiction, suppose that P is a point in U for which $\varphi(P) \neq \psi(P)$. Then Proposition 10.36 produces h in $\mathbb{k}(U)$ such that h is regular on an open subset F of V containing $\varphi(P)$ and $\psi(P)$ and has $h(\varphi(P)) = 0$ and $h(\psi(P)) \neq 0$. Since φ and ψ are morphisms, $h \circ \varphi$ and $h \circ \psi$ are regular on the open set $\varphi^{-1}(F) \cap \psi^{-1}(F)$. Their respective values at P are $h(\varphi(P)) = 0$ and $h(\psi(P)) \neq 0$. Since $h \circ \varphi = h \circ \psi$ as rational functions, this is a contradiction. $\qquad\square$

Proposition 10.42. Suppose that U and V are varieties and that $\varphi : U \to V$ is a morphism. If P is in U, then φ induces a $\mathbb{k}$ algebra homomorphism $\varphi_P^* : \mathcal{O}_{\varphi(P)}(V) \to \mathcal{O}_P(U)$. Composition of morphisms goes to composition of these homomorphisms in the reverse order.

Proof. Propositions 10.33 and 10.37 together imply that we may assume U and V to be quasi-affine. Let f in $\mathbb{k}(V)$ be defined at $\varphi(P)$. Proposition 10.24 shows that the set E on which f is regular is open in V. Since φ is a morphism and f is regular on E, $f \circ \varphi$ is regular on the open subset $\varphi^{-1}(E)$ of U. Proposition 10.28, applied to $\varphi^{-1}(E) \subseteq U$, shows that there exists a unique member F of $\mathbb{k}(U)$ that is regular on $\varphi^{-1}(E)$ and agrees with $f \circ \varphi$ on $\varphi^{-1}(E)$. We put $\varphi_P^*(f) = F$. It is a routine matter to check that φ_P^* is a $\mathbb{k}$ algebra homomorphism and that compositions go to compositions in the reverse order. $\qquad\square$

6. Rational Maps

This section will introduce a second kind of map that makes the collection of all (quasiprojective) varieties over the algebraically closed field $\mathbb{k}$ into a category. These maps will not be ordinary functions, and the definition requires some care.

If U and V are varieties over the algebraically closed field $\mathbb{k}$, then a **rational map** $\varphi : U \to V$ is an equivalence class of pairs (E, φ_E), where E is a nonempty open set of U and φ_E is a morphism of E into V. The equivalence relation on two such pairs is that $(E, \varphi_E) \sim (E', \varphi_{E'})$ if $\varphi_E\big|_{E \cap E'} = \varphi_{E'}\big|_{E \cap E'}$. This is meaningful, since the intersection of any two nonempty open sets is nonempty. The relation $\sim$ is certainly reflexive and symmetric, and Proposition 10.41 shows that it is transitive. We can therefore take the union of the open subsets E such that some pair (E, φ_E) is in the equivalence class, and φ will be definable as a morphism on this union. This union is called the **largest domain** on which φ is a morphism.

A morphism from U to V defines a rational map. But a rational map need not be an everywhere-defined function, and forming the composition of two rational maps is problematic. For example, if E is the open subset of U on which a rational map $\varphi : U \to V$ is defined and F is the open subset of V on which a rational map $\psi : V \to W$ is defined, then it may happen that $\varphi(E)$ is disjoint from F. In this case the composition $\psi \circ \varphi$ makes no sense.

A rational map $\varphi : U \to V$ is said to be **dominant** if φ_E has dense image in V for some (and hence every) pair (E, φ_E) in the equivalence class. It is evident that the composition of two *dominant* rational maps makes sense as a rational map. The identity mapping is a dominant rational map, and thus the collection of all varieties over $\mathbb{k}$ becomes a category if the dominant rational maps are used as the maps of the category.

A **birational map** is a dominant rational map $\varphi : U \to V$ that has a dominant rational map $\psi : V \to U$ as a two-sided inverse. Two varieties admitting a birational map from the one to the other are said to be **birationally equivalent** varieties, or to be **birational**.

EXAMPLE. The irreducible affine plane curves defined by $T^2 - (S^4 + 1)$ and $Y^2 - (X^3 - 4X)$ are birationally equivalent if $\mathbb{k}$ has characteristic different from 2. Birational mappings in the two directions are given by

$$\left. \begin{array}{l} S = \dfrac{Y}{2X} \\[1em] T = \dfrac{Y^2 + 8X}{4X^2} \end{array} \right\} \quad \text{and} \quad \left\{ \begin{array}{l} X = \dfrac{2}{T - S^2} \\[1em] Y = \dfrac{4S}{T - S^2}. \end{array} \right.$$

The rational map from (X, Y) to (S, T) is a morphism on the complement of $(0, 0)$ in the locus $y^2 = x^3 - 4x$ in $\mathbb{A}^2$. The rational map from (S, T) to (X, Y) is a morphism on the entire locus $t^2 = s^4 + 1$ in $\mathbb{A}^2$.

Let $\varphi : U \to V$ be a dominant rational map, and let (E, φ_E) be any pair in the equivalence class φ. If $f \in \Bbbk(V)$ is a rational function on V, then the subset F of V on which f is defined is open and nonempty. So $f\big|_F$ is a regular function on F. Since φ_E is continuous and has dense image, $E' = \varphi_E^{-1}(F)$ is a nonempty open set in $E \subseteq U$. The function $\varphi_{E'}$ is a morphism from E' into F, and thus $f\big|_F \circ \varphi_{E'}$ is a regular function on E'. We can therefore regard it as a rational function on U, i.e., a member of $\Bbbk(U)$. Consequently the dominant rational map $\varphi : U \to V$ induces a function $\widetilde{\varphi} : \Bbbk(V) \to \Bbbk(U)$ that is easily seen to be a field mapping respecting $\Bbbk$. Compositions of dominant rational maps lead to compositions of such field mappings in the reverse order.

EXAMPLE, CONTINUED. The two irreducible affine plane curves in the example earlier in this section have been observed to be birationally equivalent. In view of the previous paragraph, their function fields must be isomorphic. Taking into account that the genus of a curve, as defined in Section IX.3, depends only on the function field, we see that the two curves must have the same genus. This equality is confirmed by Example 3 of genus in Section IX.3, which shows that the genus of $\Bbbk[x, y]/(y^2 - p(x))$, where $p(x)$ is a square-free polynomial of degree m in characteristic different from 2, is $\frac{1}{2}m - 1$ if m is even and is $\frac{1}{2}(m - 1)$ if m is odd. The two curves under study have $m = 4$ and $m = 3$, and the genus is 1 in both cases.

The main result of this section will be a converse to the construction just made, showing how to pass from a $\Bbbk$ algebra homomorphism between function fields to a dominant rational map in the reverse order. We require two lemmas.

Lemma 10.43. Let $V = V(f)$ be the hypersurface[13] in $\mathbb{A}^n$ defined by a nonconstant polynomial f in $\Bbbk[X_1, \ldots, X_n]$. Then the open set $\mathbb{A}^n - V$ is isomorphic to an affine variety, specifically to the hypersurface in $\mathbb{A}^{n+1}$ corresponding to the irreducible polynomial $X_{n+1} f(X_1, \ldots, X_n) - 1$ in $\Bbbk[X_1, \ldots, X_{n+1}]$.

REMARKS. Even though f is not assumed irreducible, $X_{n+1} f - 1$ is irreducible. In fact, consideration of the degree in X_{n+1} shows that the only possible nontrivial factorization is of the form $(X_{n+1} a - b)(c)$ with a, b, c in $\Bbbk[X_1, \ldots, X_n]$. Then $bc = 1$, and c has to be scalar. The open set $\mathbb{A}^n - V$ is a quasi-affine variety (having closure $\mathbb{A}^n$), and the lemma therefore asserts that this quasi-affine variety is isomorphic to a certain affine variety in $\mathbb{A}^{n+1}$.

PROOF. Let $W = V(X_{n+1} f - 1)$. Let $\varphi : W \to \mathbb{A}^n$ be the map defined by $\varphi(x_1, \ldots, x_{n+1}) = (x_1, \ldots, x_n)$ for $(x_1, \ldots, x_{n+1})$ in W. Then $X_j \circ \varphi$ is projection

---

[13]In the application of Lemma 10.43 to Lemma 10.44, it is important that the polynomial f is allowed to be reducible.

to the jth coordinate for $1 \leq j \leq n$, which is a regular function on W. Lemma 10.39 shows that φ is a morphism, and φ is one-one onto by inspection. The inverse function is given by $\varphi^{-1}(x_1, \ldots, x_n) = \big(x_1, \ldots, x_n, 1/f(x_1, \ldots, x_n)\big)$. Let $\overline{X_j}$ be the image of X_j in $\Bbbk[X_1, \ldots, X_{n+1}]/(X_{n+1}f - 1)$ for $1 \leq j \leq n+1$. Then $(\overline{X_j} \circ \varphi^{-1})(x_1, \ldots, x_n)$ equals x_j for $j \leq n$ and equals $1/f(x_1, \ldots, x_n)$ for $j = n+1$, and these are regular functions on the complement of $V(f)$ in $\mathbb{A}^n$. By Lemma 10.39, φ^{-1} is a morphism. □

Lemma 10.44. If V is a variety, then there is a base for the Zariski topology on V consisting of open sets that are isomorphic to affine varieties.

PROOF. Let P be in V, and let U be an open subset of V containing P. We are to produce an open subset W of U containing P that is isomorphic to an affine variety. Since any nonempty open set of a quasiprojective variety is a quasiprojective variety, U is a variety. Thus we may assume that $U = V$. Since any projective variety in $\mathbb{P}^n$ is covered by the affine varieties isomorphic via Proposition 10.37 to nonempty intersections with $\beta_j(\mathbb{A}^n)$, any quasiprojective variety is covered by quasi-affine varieties. Thus we may assume that $U = V$ is quasi-affine in $\mathbb{A}^n$. Let X be the closed subset $X = \overline{V} - V$ in $\mathbb{A}^n$, and let $\mathfrak{a} = I(X)$. Since P is in V, it is not in X, and there exists some f in $\mathfrak{a}$ with $f(P) \neq 0$. Let $Y = V(f)$. The point P is not in Y, and thus $W = V - V(f)$ is relatively open in V and contains P.

Being relatively open in V, W is a quasi-affine variety. Since f vanishes on X, $V(f)$ contains $X = \overline{V} - V$. Thus the equality $W = \overline{V} - V(f)$ exhibits W as a relatively closed subset of $\mathbb{A}^n - V(f)$, which Lemma 10.43 shows is isomorphic to an affine variety. Hence W itself is isomorphic to a quasi-affine variety that is closed in an affine variety. That is, W is isomorphic to an affine variety. □

Theorem 10.45. Let U and V be varieties, and let $\varphi \mapsto \widetilde{\varphi}$ be the function carrying dominant rational maps $\varphi : U \to V$ to field mappings $\widetilde{\varphi} : \Bbbk(V) \to \Bbbk(U)$ respecting the operations by $\Bbbk$ and given by

$$\widetilde{\varphi}(f) = (\text{class of } f\big|_F \circ \varphi_{E'}),$$

where f is in $\Bbbk(V)$, f is regular on F, (E, φ_E) is a pair in the class φ, and $E' = \varphi_E^{-1}(F)$. Then $\varphi \mapsto \widetilde{\varphi}$ is one-one onto the set of all field mappings from $\Bbbk(V)$ into $\Bbbk(U)$ respecting $\Bbbk$. Furthermore, if $P \in U$ and $Q \in V$ are points, then the maximal ideal of $\widetilde{\varphi}(\mathcal{O}_Q(V))$ is contained in the maximal ideal of $\mathcal{O}_P(U)$ if and only if P is in the largest domain on which φ is a morphism and has $\varphi(P) = Q$.

REMARK. The ring $\mathcal{O}_P(U)$ is the $\Bbbk$ vector space sum of its maximal ideal and the constants, since evaluation at P is a well-defined multiplicative linear functional on $\mathcal{O}_P(U)$, and a similar comment applies to $\mathcal{O}_Q(V)$. Whatever $\widetilde{\varphi}$

does, it certainly carries 1 to 1, and hence if $\widetilde{\varphi}$ carries the maximal ideal of $\mathcal{O}_Q(V)$ to the maximal ideal of $\mathcal{O}_P(U)$, then it carries $\mathcal{O}_Q(V)$ to $\mathcal{O}_P(U)$ also.

PROOF. We begin by inverting $\varphi \mapsto \widetilde{\varphi}$. Lemma 10.44 shows that any variety is covered by open subvarieties isomorphic to affine varieties, and the function fields of the variety and the subvarieties may all be identified with one another. Thus there is no loss in generality in assuming that V is an affine variety in $\mathbb{A}^n$. Let $\overline{X}_1, \ldots, \overline{X}_n$ be the images in $A(V)$ of $X_1, \ldots, X_n$, and suppose that a $\mathbb{k}$ algebra homomorphism $\gamma : \mathbb{k}(V) \to \mathbb{k}(U)$ is given. Then $\gamma(\overline{X}_1), \ldots, \gamma(\overline{X}_n)$ are rational functions on U, and we can find a nonempty open subset E of U on which all these functions are regular. Since γ is a homomorphism, γ yields by restriction of the images a homomorphism $\gamma : A(V) \to \mathcal{O}(E)$. Moreover, this version of γ is one-one on $A(V)$ because γ as a field mapping is one-one and because Proposition 10.34 shows that each member of $\mathcal{O}(E)$ extends in only one way to a member of $\mathbb{k}(U)$. Theorem 10.38 produces a morphism $\psi : E \to V$ such that $\widetilde{\psi} = \gamma$ for this restricted version of γ. Then the equivalence class φ of the pair (ψ, E) is a rational map of U into V.

To see that φ is dominant, suppose on the contrary that $\overline{\psi(E)}$ is a proper closed subset of V. Then we can find a polynomial f that is 0 on $\overline{\psi(E)}$ but is not identically 0 on V. The image $\bar{f}$ of f in $A(V)$ is nonzero. Since the restricted version of γ is one-one, $\gamma(\bar{f})$ is nonzero in $\mathcal{O}(E)$. However, $\gamma(\bar{f}) = \widetilde{\psi}(\bar{f}) = \bar{f} \circ \psi$, and the right side is 0 on E, contradiction.

The construction is arranged in such a way that if we start from φ, form $\widetilde{\varphi}$, and go through the construction to produce a rational map of U into V, then the resulting rational map is φ. In the reverse direction, suppose that we start from γ, produce φ, and then form $\widetilde{\varphi}$, and suppose that $\bar{f}$ in $\mathbb{k}(V)$ is in $A(V)$. If $E \subseteq U$ is as in the first paragraph of the proof, then a representative of φ is the pair (E, φ_E), where φ_E is the morphism such that $(\varphi_E)^{\sim} = \gamma$. Then $\gamma_\varphi(\bar{f})$ is the class of $\bar{f} \circ \varphi_E$, which equals $\widetilde{\varphi}(f)$ and hence $\gamma(\bar{f})$. In other words, γ and $\widetilde{\varphi}$ agree on $A(V)$; being field mappings, they agree on $\mathbb{k}(V)$. This completes the proof of the first conclusion of the theorem.

Now suppose that φ is a dominant rational map from U to V and that $\widetilde{\varphi}$ is the corresponding field map of $\mathbb{k}(V)$ to $\mathbb{k}(U)$. Let $P \in U$ and $Q \in V$ be points, suppose that there is an open neighborhood E of P such that (E, φ_E) is in the equivalence class φ, and suppose that $\varphi_E(P) = Q$. Lemma 10.44 shows that there is a base of open neighborhoods of Q in V consisting of open sets that are isomorphic to affine varieties. Since φ_E is by assumption continuous, we can select any such open neighborhood and assume that φ_E carries E into it. Thus there is no loss of generality in assuming that V is isomorphic to an affine variety. We associate to φ_E the $\mathbb{k}$ algebra homomorphism $(\varphi_E)^{\sim} : \mathcal{O}(V) \to \mathcal{O}(E)$ given by $(\varphi_E)^{\sim}(f) = f \circ \varphi_E$ for $f \in \mathcal{O}(V)$. This formula shows that the members f of $\mathcal{O}(V)$ that vanish at Q are carried to members of $\mathcal{O}(E)$ that vanish at P and that members of $\mathcal{O}(V)$ that do not vanish at Q go to members of $\mathcal{O}(E)$ that do not vanish at P. Therefore $(\varphi_E)^\sim$ carries $\mathcal{O}_Q(V)$ into $\mathcal{O}_P(E) = \mathcal{O}_P(U)$.

Conversely suppose that the field map $\widetilde{\varphi}$ has the property that the maximal ideal of $\widetilde{\varphi}(\mathcal{O}_Q(V))$ is contained in the maximal ideal of $\mathcal{O}_P(U)$. Possibly by passing to an open subneighborhood from the outset, we may assume by Lemma 10.44 that U and V are isomorphic to affine varieties. Dropping the isomorphism from the notation, we can write $\mathcal{O}(V) = A(V) = \Bbbk[y_1, \ldots, y_m]$ by Corollary 10.25. Each $\widetilde{\varphi}(y_j)$ is a rational function on U, which we can write as $\widetilde{\varphi}(y_j) = a_j/b_j$ with a_j and b_j in $\mathcal{O}(U) = A(U)$. The hypothesis on $\widetilde{\varphi}$ implies that $\widetilde{\varphi}(\mathcal{O}_Q(V)) \subseteq \mathcal{O}_P(U)$, hence that each $\widetilde{\varphi}(y_j)$ is regular at P. Thus we may take each denominator b_j to have $b_j(P) \neq 0$. Choose an open neighborhood of P on which all b_j are nonvanishing and an open subneighborhood E that is isomorphic to an affine variety. Since $\widetilde{\varphi}$ respects the field operations, it carries any polynomial in $y_1, \ldots, y_m$ to a quotient c/d with c and d in $\mathcal{O}(E)$ and with d nowhere 0 on E. Therefore c/d is in $\bigcap_{P' \in E} \mathcal{O}_{P'}(E) = \mathcal{O}(E)$. That is, $\widetilde{\varphi}$ carries $\mathcal{O}(V)$ into $\mathcal{O}(E)$. Since V is isomorphic to an affine variety, Corollary 10.25 and Theorem 10.38 show that $\widetilde{\varphi} : \mathcal{O}(V) \to \mathcal{O}(E)$ is given by the formula

$$\widetilde{\varphi}(h)(u) = h(\varphi_E(u)) \qquad (*)$$

for some morphism $\varphi_E : E \to V$ and all $h \in \mathcal{O}(V)$ and $u \in E$. The first part of the proof shows that the pair (E, φ_E) is in the equivalence class φ. Hence P is in the largest domain on which φ is a morphism. Arguing by contradiction, suppose that $\varphi_E(P) = Q' \neq Q$. Choose by Proposition 10.36 a rational function h on V that is defined at both Q and Q' and has $h(Q) = 0$ and $h(Q') \neq 0$. Then $\widetilde{\varphi}$ carries $\mathcal{O}_Q(V)$ and its maximal ideal into $\mathcal{O}_P(U)$ and its maximal ideal, and we obtain $0 = \widetilde{\varphi}(h)(P) = h(\varphi_E(P)) = h(Q') \neq 0$, contradiction. We therefore conclude that $\varphi_E(P) = Q$, and the proof of the second conclusion of the theorem is complete. $\square$

Corollary 10.46. If U and V are varieties, then the following conditions are equivalent:
  (a) U and V are birationally equivalent,
  (b) $\Bbbk(U)$ and $\Bbbk(V)$ are isomorphic as $\Bbbk$ algebras,
  (c) there are nonempty open subsets E of U and F of V such that E and F are isomorphic as varieties.

PROOF. The equivalence of (a) and (b) follows from Theorem 10.45 and the fact that composition of dominant rational maps corresponds to composition of homomorphisms of $\Bbbk$ algebras in the reverse order.

Let us check that (c) implies (a). If (c) holds, let $\varphi : E \to F$ and $\psi : F \to E$ be morphisms that are inverse to each other. Then the equivalence classes of

(E, φ) and (F, ψ) are rational maps from U to V and from V to U, respectively. The equivalence class of $(E, \psi \circ \varphi) = (E, 1_E)$ is the identity rational map on U, and the equivalence class of $(F, \varphi \circ \psi) = (F, 1_F)$ is the identity rational map on V. Hence the rational maps are inverses of one another. This proves (a).

Finally let us check that (a) implies (c). If (a) holds, let $\varphi : U \to V$ and $\psi : V \to U$ be rational maps that are inverse to each other. Let (E_1, φ) and (F_1, ψ) be pairs representing φ and ψ. Then a pair representing $\psi \circ \varphi$ is $(\varphi^{-1}(F_1), \psi \circ \varphi)$ because φ is a morphism on the open subset $\varphi^{-1}(F_1)$ of E_1 and ψ is a morphism on the open set F_1 containing $\varphi(\varphi^{-1}(F_1))$. Since $\psi \circ \varphi$ is the identity on U as a rational map, $\psi \circ \varphi$ is the identity morphism on $\varphi^{-1}(F_1)$. Put $E = \varphi^{-1}(F_1) \subseteq E_1$. Similarly $\varphi \circ \psi$ is the identity morphism on $\psi^{-1}(E_1)$, and we put $F = \psi^{-1}(E_1) \subseteq F_1$. Let us see that $\varphi(E) \subseteq F$. If e is in E, we are to exhibit some $e_1 \in E_1$ with $\psi(\varphi(e))$ in E_1, and then $\varphi(e)$ will be in $F = \psi^{-1}(E_1)$; for this purpose we can take $e_1 = e$, since $\psi \circ \varphi$ is the identity morphism on E. Similarly $\psi(F) \subseteq E$. Thus φ and ψ exhibit E and F as isomorphic varieties. This proves (c). $\square$

7. Zariski's Theorem about Nonsingular Points

Sections 1–6 have established the definitions and elementary properties of varieties, maps between varieties, and dimension. The present section concerns singularities, which are a fundamental topic of interest in algebraic geometry.[14] This topic was introduced in Section VII.5 in a context that we now recognize as affine varieties.

The definition of "nonsingular" was motivated by the classical Implicit Function Theorem. Let $\Bbbk$ be an algebraically closed field, let the affine space in question be $\mathbb{A}^n$, and let $\mathfrak{p}$ be the prime ideal such that the affine variety to study in $\mathbb{A}^n$ is $V(\mathfrak{p})$. If $\{f_i\}$ is a finite set of generators of $\mathfrak{p}$ and if P is in $V(\mathfrak{p})$, then P is said to be a **nonsingular** point of $V(\mathfrak{p})$ if rank $\left[\frac{\partial f_i}{\partial X_j}(P)\right] = n - \dim V(\mathfrak{p})$, and otherwise it is **singular**. Zariski's Theorem, which was formulated as Theorem 7.23 but only partially proved in Chapter VII, addressed this situation. In order to rephrase the theorem in our current notation, let $A(V)$ be the affine coordinate ring of V, and let $\Bbbk(V)$ be the field of fractions of $A(V)$, i.e., the function field of V. Let $\mathfrak{m}_P$ be the maximal ideal of all members of $A(V)$ vanishing at P, and let $\mathcal{O}_P(V)$ be the local ring at P; this is the localization of $A(V)$ with respect to the maximal ideal $\mathfrak{m}_P$ and is a subring of $\Bbbk(V)$. The maximal ideal of $\mathcal{O}_P(V)$, consisting of all members of $\Bbbk(V)$ defined and vanishing at P, will be denoted by M_P. Theorem 7.23, translated into this notation, is as follows.

---

[14]The exposition in this section is based in part on Chapter I of Hartshorne's book, Chapter III of Reid's book, and Chapter II of Volume 1 of Shafarevich's books.

Theorem 10.47 (Zariski's Theorem, rephrased). In the above notation,

$$\dim_{\mathbb{k}}(M_P/M_P^2) = \dim_{\mathbb{k}}(\mathfrak{m}_P/\mathfrak{m}_P^2) \geq \dim V(\mathfrak{p}),$$

and P is nonsingular if and only if equality holds. The set of nonsingular points of $V(\mathfrak{p})$ is nonempty and open.

Toward the proof of this theorem, we showed in Section VII.5 for all $P \in V(\mathfrak{p})$ that

(a) $\qquad \dim_{\mathbb{k}}(M_P/M_P^2) = \dim_{\mathbb{k}}(\mathfrak{m}_P/\mathfrak{m}_P^2),$

(b) $\qquad \dim_{\mathbb{k}}(\mathfrak{m}_P/\mathfrak{m}_P^2) + \operatorname{rank}\left[\frac{\partial f_i}{\partial X_j}(P)\right] = n,$

(c) $\qquad P$ is a nonsingular point if and only if $\dim_{\mathbb{k}}(\mathfrak{m}_P/\mathfrak{m}_P^2) = \dim V(\mathfrak{p}).$

In addition, we completed most of the proof in the special case that $V(\mathfrak{p})$ is an irreducible affine hypersurface by showing that

(d) $\qquad \dim_{\mathbb{k}}(\mathfrak{m}_P/\mathfrak{m}_P^2) \geq \dim V(\mathfrak{p}) \qquad$ for all $P \in V(\mathfrak{p}),$

(e) $\qquad \dim_{\mathbb{k}}(\mathfrak{m}_P/\mathfrak{m}_P^2) = \dim V(\mathfrak{p}) \qquad$ for some $P \in V(\mathfrak{p}).$

Our goal in this section is to complete the proof of Zariski's Theorem in the general case as stated by reducing (d) and (e) for the general case to what has already been proved for the special case that $V(\mathfrak{p})$ is an irreducible affine hypersurface. We need also to see in all cases that the set of nonsingular points is Zariski open.

Before proceeding, let us mention the significance of Theorem 10.47. The definition above of **nonsingular** and **singular** points extends immediately to quasi-affine varieties, using the same defining polynomials, and the theorem is then applicable because the open set of nonsingular points in an affine variety meets any nonempty open subset of the variety. In the projective case we can pull matters back to affine space by means of one of the maps $\beta_i : \mathbb{A}^n \to \mathbb{P}^n$. In this way we obtain definitions of **nonsingular** and **singular** point for quasiprojective varieties, and the theorem remains valid.[15] What is far from obvious with such a definition is that the decision nonsingular vs. singular for a point is unaffected by isomorphisms of varieties. On the other hand, the equivalent condition on M_P/M_P^2 as stated in Zariski's Theorem is manifestly unaffected by isomorphisms of varieties because of Proposition 10.42.

---

[15]Problems 13–16 at the end of the chapter show that the rank computation can alternatively be made directly with the homogeneous polynomials defining the projective variety in question.

Proposition 10.48. Any m-dimensional variety is birationally equivalent to an irreducible affine hypersurface H in $\mathbb{A}^{m+1}$.

PROOF. Let V be the variety in question. By definition of dim V, the function field $\Bbbk(V)$ is a finitely generated extension field of $\Bbbk$ of transcendence degree m over $\Bbbk$. Since algebraically closed fields are perfect, Theorem 7.20 shows that $\Bbbk(V)$ is "separably generated" over $\Bbbk$, and Theorem 7.18 shows as a consequence that $\Bbbk(V)$ has a "separating transcendence basis," i.e., a transcendence basis $\{x_1, \ldots, x_m\}$ such that $\Bbbk(V)$ is a finite separable algebraic extension of $\Bbbk(x_1, \ldots, x_m)$. By the Theorem of the Primitive Element, there exists an element x_{m+1} of $\Bbbk(V)$ such that $\Bbbk(V) = \Bbbk(x_1, \ldots, x_m)[x_{m+1}]$. Let $P(X_{m+1})$ be the minimal polynomial of x_{m+1} over $\Bbbk(x_1, \ldots, x_m)$. Writing out the equation $P(x_{m+1}) = 0$ and clearing fractions, we see that x_{m+1} satisfies a polynomial equation

$$a_r(x_1, \ldots, x_m)x_{m+1}^r + \cdots + a_1(x_1, \ldots, x_m)x_{m+1} + a_0(x_1, \ldots, x_m) = 0$$

in which the coefficient polynomials $a_j(X_1, \ldots, X_m) \in \Bbbk[X_1, \ldots, X_m]$ have no nontrivial common factor. In this case the polynomial $f(X_1, \ldots, X_{m+1})$ equal to

$$a_r(X_1, \ldots, X_m)X_{m+1}^r + \cdots + a_1(X_1, \ldots, X_m)X_{m+1} + a_0(X_1, \ldots, X_m)$$

is irreducible in $\Bbbk[X_1, \ldots, X_{m+1}]$. Thus the principal ideal (f) defines an irreducible affine hypersurface $H = V(f)$ in $\mathbb{A}^{m+1}$ whose affine coordinate ring is $\Bbbk[X_1, \ldots, X_{m+1}]/(f)$. The field of fractions $\Bbbk(H)$ is isomorphic to $\Bbbk(V)$, and H is birationally equivalent to V by the equivalence of (a) and (b) in Corollary 10.46. $\square$

Lemma 10.49. Every point P in $V(\mathfrak{p})$ has $0 \leq \dim_\Bbbk(M_P/M_P^2) \leq n$, and the set of points P in $V(\mathfrak{p})$ with $\dim_\Bbbk(M_P/M_P^2) \geq r$ is a Zariski closed subset for each integer r.

PROOF. The entries of the matrix $\left[\frac{\partial f_i}{\partial X_j}\right]$ are polynomials, and the set of points P of $V(\mathfrak{p})$ for which the matrix $\left[\frac{\partial f_i}{\partial X_j}(P)\right]$ has rank $\leq s$ is a Zariski closed subset, being the set on which all $(s+1)$-by-$(s+1)$ minors of the matrix vanish. By display formula (b) above, the set of points P for which $\dim_\Bbbk(\mathfrak{m}_P/\mathfrak{m}_P^2) \geq n - s$ is closed, and (a) therefore shows that the set with $\dim_\Bbbk(M_P/M_P^2) \geq n - s$ is closed. $\square$

PROOF OF THEOREM 10.47. Let $m = \dim V(\mathfrak{p})$, and let a birational mapping of $V(\mathfrak{p})$ to an affine hypersurface H of $\mathbb{A}^{m+1}$ be given. By the equivalence of (a)

7. Zariski's Theorem about Nonsingular Points

and (c) in Corollary 10.46, there exist nonempty open subsets E of $V(\mathfrak{p})$ and F of H that are isomorphic as varieties, say by an isomorphism $\varphi : E \to F$. Since $m = \dim V(\mathfrak{p}) = \dim H$, Proposition 10.11 shows that $m = \dim E = \dim F$ also. For each integer $r \geq 0$, let

$$S_r = \{P \in V(\mathfrak{p}) \mid \dim_{\mathbb{k}}(M_P/M_P^2) \leq r\},$$
$$T_r = \{P \in E \mid \dim_{\mathbb{k}}(M_P/M_P^2) \leq r\},$$
$$U_r = \{P \in F \subseteq H \mid \dim_{\mathbb{k}}(M_P/M_P^2) \leq r\}.$$

Lemma 10.49 shows that

S_r, T_r, U_r are relatively open in $V(\mathfrak{p}), E, F$, respectively, for each r. $\quad(*)$

Application of Proposition 10.42 to φ and φ^{-1} gives

$$\varphi(T_r) = U_r \qquad \text{for all } r \geq 0, \tag{**}$$

and the special case of Theorem 10.47 proved in Section VII.5 shows that

$$U_m \neq \varnothing \quad \text{and} \quad U_{m-1} = \varnothing. \tag{†}$$

Combining $(**)$ and $(†)$ yields

$$T_m \neq \varnothing \quad \text{and} \quad T_{m-1} = \varnothing. \tag{††}$$

Since $S_r \supseteq T_r$, the first of these shows that

$$S_m \neq \varnothing. \tag{‡}$$

If $S_{m-1} \neq \varnothing$, then $E \cap S_{m-1} \neq \varnothing$ because any two nonempty open subsets of $V(\mathfrak{p})$ have nonempty intersection; but $T_{m-1} = E \cap S_{m-1}$ would then be nonempty, in contradiction to $(††)$. Thus

$$S_{m-1} = \varnothing. \tag{‡‡}$$

In view of (a), (‡) proves (e) for $V(\mathfrak{p})$, and (‡‡) proves (d) for $V(\mathfrak{p})$. Because of (‡‡), Lemma 10.49 implies that S_m is Zariski open; thus the set of nonsingular points is open. □

8. Classification Questions about Irreducible Curves

Sections 1–7 give the fundamentals concerning (quasiprojective) varieties over the algebraically closed field $\mathbb{k}$. The remainder of the chapter will address aspects of three problems:

  (i) What are all varieties, or in what senses can varieties be classified?
  (ii) To what extent can one make computations in the subject?
  (iii) What can be said when the algebraically closed field $\mathbb{k}$ is replaced by a general commutative ring with identity?

Algebraic geometry is an enormous subject, going well beyond these problems. For example the investigation of the nature of singularities is in itself a large subject, with striking applications to topology and differential equations. The use of homological methods ties algebraic geometry closely to topology and to number theory, and these methods have bearing on the extent to which compact complex manifolds admit the structure of projective varieties. Algebraic geometry is an ingredient in the subject of invariant theory, which studies classical varieties using representation theory. It is an ingredient also in the subject of algebraic groups, which concerns varieties with a group structure in which multiplication and inversion are morphisms.

The present section concerns the first of the three problems listed above, and we limit our discussion to **irreducible curves**, i.e., to varieties of dimension 1. We say that an irreducible curve is **nonsingular** if it is nonsingular at every point. We are going to show in this section that each birational equivalence class of irreducible curves over $\mathbb{k}$ contains a nonsingular projective curve and that any two nonsingular projective curves in the birational equivalence class are isomorphic as projective varieties.[16] We also will get some information about how this nonsingular curve in the class is related to the other curves in the class. To a great extent the classification of irreducible curves will therefore have been reduced to the classification of the birational equivalence classes, which Corollary 10.46 says is the same thing as a classification of the function fields in one variable over $\mathbb{k}$. We will not have anything to say about classifying the function fields in one variable except to say that each class has a genus, according to Section IX.3, and that every nonnegative integer can arise as a genus, according to Example 3 of genus in Section IX.3.[17]

Chapter IX already contains clues about where to begin. Section IX.1 mentioned the relevance of Dedekind domains to the study, and Problems 5–11 at the end of that chapter attached a discrete valuation to each nonsingular point of any irreducible affine plane curve. The notions of Dedekind domains, discrete

---

[16]The exposition in this section is based in part on Chapter 7 of Fulton's book, Chapter I of Hartshorne's book, Chapter II of Reid's book, and Volume I by Zariski–Samuel.

[17]The subject of Teichmüller theory in effect addresses this question when $\mathbb{k} = \mathbb{C}$.

8. Classification Questions about Irreducible Curves   605

valuations, and nonsingular points are very closely related, and we begin with some equivalences concerning them. Recall from Sections 2 and 4 that the affine coordinate ring $A(C)$ of any irreducible affine curve C has Krull dimension 1. That is, the Noetherian domain $A(C)$ has the property that every nonzero prime ideal is maximal. We have seen that the local ring $\mathcal{O}_P(C)$ at any point is a localization of $A(C)$, namely the localization of $A(C)$ with respect to the maximal ideal $\mathfrak{m}_P$ of functions vanishing at P. Furthermore, the proper ideals of such a localization are exactly the sets $S^{-1}\mathfrak{a}$ with $\mathfrak{a}$ equal to an ideal disjoint from the set-theoretic complement of $\mathfrak{m}_P$ in $A(C)$. It follows that every nonzero prime ideal in $\mathcal{O}_P(C)$ is maximal. This conclusion extends to the quasiprojective case as a consequence of Proposition 10.33. Zariski's Theorem in Section 7 shows that nonsingularity of the point P of C can be detected from $\mathcal{O}_P(C)$. Consequently the following proposition is relevant.

Proposition 10.50. Let R be a Noetherian local ring that is an integral domain with the property that the only nonzero prime ideal is the maximal ideal. Let M be the unique maximal ideal of R, let K be the field of fractions of R, and let $F = R/M$ be the quotient field. Under the assumption that $M \neq 0$ and therefore that $R \neq K$, the following conditions on R are equivalent:

(a) R is integrally closed,
(b) R is a Dedekind domain,
(c) R is a principal ideal domain,
(d) R is the valuation ring relative to some discrete valuation of K,
(e) M is a principal ideal,
(f) $\dim_F M/M^2 = 1$.

REMARKS. Consider (f). To see how M/M^2 becomes an F vector space in a natural way, let $r + M$ be a member of F, and let $m + M^2$ be a member of M/M^2. Then $(r + M)(m + M^2) = rm + M^2$ is a well-defined scalar multiplication of F on M/M^2, and M/M^2 becomes a vector space over F. Nakayama's Lemma (Lemma 8.51 of *Basic Algebra*, restated in the present book on page xxiii) shows that an equality $MN = N$ for a finitely generated R module N is possible only if $N = 0$; since M itself is a finitely generated R module, being an ideal in a Noetherian ring, and since $M \neq 0$ by assumption, $M^2 = M$ is not possible. Therefore $\dim_F M/M^2 \geq 1$.

PROOF. If (a) holds, then R satisfies the three conditions (Noetherian, integrally closed, every nonzero prime ideal maximal) in the definition of Dedekind domain. Thus (a) implies (b). A Dedekind domain with only finitely many maximal ideals is a principal ideal domain by Corollary 8.62 of *Basic Algebra*, and thus (b) implies (c). A principal ideal domain is a unique factorization domain by Theorem 8.15 of *Basic Algebra*, and thus (c) implies (a) by Proposition 8.41 of *Basic Algebra*.

To see that (a) through (c) are equivalent to (d), first suppose that (a) through (c) hold. Then every fractional ideal in K relative to R is of the form M^k for some integer k. If $x \neq 0$ is in K, then the principal fractional ideal xR is of the form $xR = M^k$ for some k. Section VI.2 shows that the formula $v(x) = k$ (with $v(0) = \infty$) defines a discrete valuation on K, and the definition of v shows that the valuation ring of v is R. Hence (d) holds. Conversely if (d) holds, then R is a principal ideal domain by Proposition 6.2; thus (c) and necessarily (a) and (b) hold.

Let us prove that (e) and (f) are equivalent. If (e) holds, then we can write $M = (\pi)$ for some π in R. If $m + M^2$ is a given element of M/M^2, then m is of the form $m = r\pi$ for some r in R. Hence $(r + M)(\pi + M^2) = r\pi + M^2 = m + M^2$, and $\dim_F M/M^2 \leq 1$. Since the remarks before the proof show that $\dim_F M/M^2 \geq 1$, (f) holds.

If (f) holds, let $\{\pi + M^2\}$ be an F basis of M/M^2. If $m \in M$ is given, then $m + M^2 = (r + M)(\pi + M^2)$ for some $r \in R$. Therefore $m = r\pi + m'$ with $m' \in M^2$, and we see that $(\pi) + M^2 = M$. We shall apply Nakayama's Lemma in the local ring $R/(\pi)$ with maximal ideal $M/(\pi)$ and with module $N = M/(\pi)$: Given $m \in M$, we expand $m = r\pi + m'$ with $m' \in M^2$ as $m = r\pi + \sum_{i,j} m_i m_j$. Then the equality $m + (\pi) = \sum_{i,j} m_i m_j$ in $M/(\pi)$ shows that $m \equiv \sum_i m_i \sum_j m_j$, hence that the coset $m + (\pi)$ lies in $\sum_i (m_i + (\pi))(M/(\pi))$. In other words, $M/(\pi) = (M/(\pi))^2$. Nakayama's Lemma shows that $M/(\pi) = 0$, and therefore $M = (\pi)$. Thus (e) holds.

Finally let us prove that (c) and (e) are equivalent. If (c) holds, then M has to be principal, and hence (e) holds. Suppose that (e) holds, i.e., that $M = (\pi)$. Let I be a nonzero proper ideal in R. The ideal $N = \bigcap_{k=1}^{\infty} M^k$ is a finitely generated R module because R is Noetherian, and it has $MN = N$. By Nakayama's Lemma, $N = 0$. Since $I \subseteq M$ and since $I \neq 0$, there exists a largest integer $k \geq 1$ such that $I \subseteq M^k$. Choose $y \neq 0$ in I with y in $M^k = (\pi^k)$ but not in $M^{k+1} = (\pi^{k+1})$. Let us write $y = a\pi^k$ for some $a \in R$. Since y is not in M^{k+1} and since R is local, a is a unit in R. Hence $a^{-1}y = \pi^k$ is in I, and therefore $M^k = (\pi^k) \subseteq I$. Since we arranged that $I \subseteq M^k$, we obtain $I = M^k = (\pi^k)$. Thus (c) holds. □

Corollary 10.51. Let C be an irreducible quasiprojective curve over $\Bbbk$, and let $\Bbbk(C)$ be its function field. If P is a point of C, then the following conditions are equivalent:

(a) P is a nonsingular point,
(b) $\mathcal{O}_P(C)$ is the valuation ring of some discrete valuation of $\Bbbk(C)$ defined over $\Bbbk$,
(c) $\mathcal{O}_P(C)$ is integrally closed.

PROOF. Let M_P be the unique maximal ideal of $\mathcal{O}_P(C)$. Zariski's Theorem (Theorem 10.47) shows that (a) holds if and only if $\dim_{\Bbbk} M_P/M_P^2 = 1$. The corollary therefore follows from the equivalence of (f), (d), and (a) in Proposition 10.50, along with the observation that any discrete valuation produced by (d) has to be 0 on $\Bbbk^\times$. □

Corollary 10.52. If C is an irreducible affine curve over $\Bbbk$ with affine coordinate ring $A(C)$, then the following conditions on C are equivalent:
  (a) $A(C)$ is integrally closed,
  (b) $\mathcal{O}_P(C)$ is integrally closed for each point P of the curve,
  (c) C is nonsingular.

PROOF. If $A(C)$ is integrally closed, then Corollary 8.48c of *Basic Algebra* shows that each localization $\mathcal{O}_P(C)$ is integrally closed. Conversely if each $\mathcal{O}_P(C)$ is integrally closed and if a member f of the function field $\Bbbk(C)$ is given that is a root of a monic polynomial with coefficients in $A(C)$, then f is a root of the same polynomial with coefficients in $\mathcal{O}_P(C)$ and is in $\mathcal{O}_P(C)$ because $\mathcal{O}_P(C)$ is integrally closed. Corollary 10.25 shows that $A(C) = \bigcap_P \mathcal{O}_P(C)$. Therefore f lies in $A(C)$, and $A(C)$ is integrally closed. This proves that (a) and (b) are equivalent. The equivalence of (b) and (c) follows from Corollary 10.51. □

We turn our attention to constructing a nonsingular irreducible projective curve whose field of rational functions is a given function field $\mathbb{K}$ in one variable over $\Bbbk$. If C is any irreducible quasiprojective curve with $\Bbbk(C) = \mathbb{K}$, then Corollary 10.51 associates a discrete valuation of $\mathbb{K}$ over $\Bbbk$ to each nonsingular point of C. To get an idea what C must be like if it is to be nonsingular at every point, we now prove a theorem in the converse direction, associating a point of the curve to each discrete valuation of $\mathbb{K}$ over $\Bbbk$.

Theorem 10.53. Let C be an irreducible projective curve with function field $\Bbbk(C)$ equal to $\mathbb{K}$, and let v be a discrete valuation of $\mathbb{K}$ defined over $\Bbbk$. If R_v is the valuation ring of v and $\mathfrak{p}_v$ is the valuation ideal, then there exists a unique point P on the curve for which the maximal ideal M_P of $\mathcal{O}_P(C)$ has $M_P \subseteq \mathfrak{p}_v$.

PROOF OF UNIQUENESS. Assume the contrary. If P and Q are distinct points with $M_P \subseteq \mathfrak{p}_v$ and $M_Q \subseteq \mathfrak{p}_v$, then Proposition 10.36 constructs a function h in $\Bbbk(C)$ with h defined at P and Q, $h(P) = 0$, and $h(Q) \neq 0$. This function h is in M_P, and $h - h(Q)$ is in M_Q. The assumed inclusions of maximal ideals imply that $v(h) \geq 1$ and that $v(h - h(Q)) \geq 1$. On the other hand, $h(Q) \neq 0$ implies that $v(h(Q)) = 0$. Thus $0 = v(h(Q)) \geq \min\bigl(v(h(Q) - h), v(h)\bigr) \geq 1$, contradiction. □

PROOF OF EXISTENCE. It is shown in Problem 12 at the end of the chapter that any projective variety in $\mathbb{P}^r$ is isomorphic to a projective variety V in some $\mathbb{P}^n$ with $n \leq r$ such that V is not contained in any subvariety $\{[x_0, \ldots, x_n] \mid x_j = 0\}$

with $0 \leq j \leq n$. That being so, we may assume that C is a projective variety in $\mathbb{P}^n$ and that $C \cap \beta_j(\mathbb{A}^n) \neq \varnothing$ for $0 \leq j \leq n$, where $\beta_j : \mathbb{A}^n \to \mathbb{P}^n$ is the embedding defined after Proposition 10.18. Let $\widetilde{A}(C) = \mathbb{k}[X_0, \ldots, X_n]/I(C)$ be the homogeneous coordinate ring of C, and for each j, let x_j be the image of X_j in $\widetilde{A}(C)$. Since $I(C)$ does not contain X_j, x_j is not the 0 element of $\widetilde{A}(C)$. Since X_i and X_j are homogeneous of the same degree, each function x_i/x_j is a well-defined member of the function field $\mathbb{k}(C)$.

Let $N = \max_{i,j} v(x_i/x_j)$. Possibly by renaming some coordinate x_{j_0} as x_0, we may assume that $v(x_{i_0}/x_0) = N$ for some i_0. Then we have $v(x_i/x_0) = v(x_{i_0}/x_0) + v(x_i/x_{i_0}) = N - v(x_{i_0}/x_i) \geq 0$ for all i. Consequently each function x_i/x_0 lies in the subring R_v of $\mathbb{k}(C)$.

Theorem 10.20 and Corollary 10.22 show that $C_0 = \beta_0^{-1}(C)$ is an irreducible affine curve and that its prime ideal is $I(C_0) = \beta_0^t(I(C))$. Consequently the substitution homomorphism $\beta_0^t : \mathbb{k}[X_0, \ldots, X_n] \to \mathbb{k}[X_1, \ldots, X_n]$ descends to a homomorphism of $\widetilde{A}(C) = \mathbb{k}[X_0, \ldots, X_n]/I(C)$ onto $A(C_0) = \mathbb{k}[X_1, \ldots, X_n]/I(C_0)$ that carries x_0 in $\widetilde{A}(C)$ to 1 and carries the members $x_1, \ldots, x_n$ of $\widetilde{A}(C)$ to the generators of $A(C_0)$. The members x_i/x_0 of $\mathbb{k}(C)$ therefore get identified with the generators of $A(C_0)$, and we conclude that $A(C_0) \subseteq R_v$.

Define $\mathfrak{q} = \mathfrak{p}_v \cap A(C_0)$. This is a prime ideal of $A(C_0)$, and it pulls back under the quotient homomorphism $\mathbb{k}[X_1, \ldots, X_n] \to A(C_0)$ to a prime ideal $\widetilde{\mathfrak{q}}$ containing $I(C_0)$. Then $V(\widetilde{\mathfrak{q}})$ is an affine subvariety of C_0. Since $\dim C_0 = 1$, there are only two possibilities. One is that $\dim V(\widetilde{\mathfrak{q}}) = 1$, in which case $V(\widetilde{\mathfrak{q}}) = C_0$, $\widetilde{\mathfrak{q}} = I(C_0)$, and $\mathfrak{q} = 0$. The other is that $\dim V(\widetilde{\mathfrak{q}}) = 0$, in which case $V(\widetilde{\mathfrak{q}}) = \{P\}$ for some point P that necessarily lies on C_0. In the first case, v is 0 on every nonzero member of $A(C)$ and hence is 0 on $\mathbb{k}(C)^\times$, contradiction. Thus we are in the second case. Then $\widetilde{\mathfrak{q}}$ is maximal in $\mathbb{k}[X_1, \ldots, X_n]$, $\mathfrak{q}$ is maximal in $A(C_0)$, $\mathfrak{q}$ is the ideal $\mathfrak{m}_P$ of all members of $A(C_0)$ vanishing at P, and $A(C_0)/\mathfrak{q} \cong \mathbb{k}$. If S denotes the set-theoretic complement of $\mathfrak{q}$ in $A(C_0)$, then no member of S can be in $\mathfrak{p}_v$ because then $\mathfrak{q} + \mathbb{k}1 = A(C_0)$ would be in $\mathfrak{p}_v$, contradiction. Thus $v(s) = 0$ for all $s \in S$, and $M_P = S^{-1}\mathfrak{m}_P \subseteq \mathfrak{p}_v$. $\square$

Corollary 10.54. If φ is a rational map from an irreducible curve C' to an irreducible projective curve C, then the largest domain on which φ is a morphism contains every nonsingular point of C'. If C' is nonsingular, then φ is a morphism from C' into C.

PROOF. If φ is not dominant, then Problem 6 at the end of the chapter shows that φ is constant. Certainly the largest domain on which a constant φ is a morphism is C'.

Thus suppose that φ is dominant. Using the notation introduced early in Section 6, let $\widetilde{\varphi} : \mathbb{k}(C) \to \mathbb{k}(C')$ be the associated field map of function fields.

8. Classification Questions about Irreducible Curves   609

Since $\Bbbk(C)$ and $\Bbbk(C')$ both have transcendence degree 1 over $\Bbbk$ and since $\Bbbk(C)$ is finitely generated as a field over $\Bbbk$, the field $\Bbbk(C')$ is a finite algebraic extension of the field $\widetilde{\varphi}(\Bbbk(C))$. If v is any discrete valuation of $\Bbbk(C')$, then it follows from the finiteness of this extension that v cannot be identically 0 on $\widetilde{\varphi}(\Bbbk(C))^\times$; in fact, if it were identically 0, then the expansion $x = \sum_{j=1}^m c_j x_j$ of a general element x of $\Bbbk(C')$ in terms of a vector-space basis $\{x_1, \ldots, x_m\}$ of $\Bbbk(C')$ over $\widetilde{\varphi}(\Bbbk(C))$ would yield the inequality $v'(x) \geq \min_j v(x_j)$, which cannot be true for all x.

Meanwhile, if P is a nonsingular point of C', then Corollary 10.51 shows that $\mathcal{O}_P(C')$ is the valuation ring R_v for some valuation v of $\Bbbk(C')$ over $\Bbbk$. The maximal ideal M_P of $\mathcal{O}_P(C')$ equals the valuation ideal $\mathfrak{p}_v$ of v. Since the restriction of v to $\widetilde{\varphi}(\Bbbk(C))^\times$ is not identically 0, the restriction comes from some positive multiple e of a discrete valuation on $\widetilde{\varphi}(\Bbbk(C))$. Let v_0 be the corresponding discrete valuation of $\Bbbk(C)$; this is given by $v_0(f) = e^{-1} v(\widetilde{\varphi}(f))$. Let R_0 be its valuation ring and $\mathfrak{p}_0$ be its valuation ideal in $\Bbbk(C)$; the latter is given by $\mathfrak{p}_0 = \widetilde{\varphi}^{-1}(\mathfrak{p}_v)$. Theorem 10.53 shows that there exists a unique point Q on the curve C such that the maximal ideal M_Q of $\mathcal{O}_Q(C)$ is contained in $\mathfrak{p}_0$. That is, $M_Q \subseteq \mathfrak{p}_0 = \widetilde{\varphi}^{-1}(\mathfrak{p}_v)$. Application of $\widetilde{\varphi}$ gives $\widetilde{\varphi}(M_Q) \subseteq \widetilde{\varphi}\widetilde{\varphi}^{-1}(\mathfrak{p}_v) \subseteq \mathfrak{p}_v = M_P$. Theorem 10.45 shows that consequently P is in the largest domain on which φ is a morphism and that $\varphi(P) = Q$. □

Corollary 10.55. If two nonsingular irreducible projective curves are birationally equivalent, then they are isomorphic as varieties.

PROOF. This follows by applying Corollary 10.54 twice. □

Corollary 10.56. If C is a nonsingular irreducible projective curve with function field $\mathbb{K} = \Bbbk(C)$, then the points of C are in one-one correspondence with the discrete valuations of $\mathbb{K}$ defined over $\Bbbk$.

PROOF. This is the correspondence given in one direction by Corollary 10.51 and in the reverse direction by Theorem 10.53. □

Corollary 10.56 has a remarkable conclusion, but the corollary assumes the existence of a nonsingular projective curve, which we have not yet proved. In more detail we now know that a nonsingular point P of any irreducible projective curve C picks out a unique discrete valuation v of the function field $\mathbb{K} = \Bbbk(C)$, namely the one whose valuation ring is given by $R_v = \mathcal{O}_P(C)$, and that conversely when C is projective, any discrete valuation v' defined over $\Bbbk$ picks out a certain point P' of C with the property that $\mathcal{O}_{P'}(C) \subseteq R_{v'}$. If P is nonsingular and we go through the first step and then the second, using $v' = v$, we obtain $\mathcal{O}_{P'}(C) \subseteq \mathcal{O}_P(C)$. Proposition 10.36 shows that $P' = P$, and hence the second process inverts the first. That is what Corollary 10.56 says. Also, we know from Theorem 10.47 that many discrete valuations are involved in this process, since the set of nonsingular points of a variety is Zariski open. What we do not know is that any given discrete valuation over k ever yields a nonsingular point for *any* curve with the function field $\mathbb{K}$. This missing piece of information will be supplied in Corollary 10.58 below. To prove Corollary 10.58, we shall make use of the following theorem, which we need only in the case that the field k is our algebraically closed field $\Bbbk$. We postpone the proof of the theorem for a moment, and when we give the proof, we shall give it only for the case that the field k in the statement is algebraically closed.

Theorem 10.57. Let k be a field, let $R = k[x_1, \ldots, x_n]$ be a finitely generated integral domain over k, let K be the field of fractions of R, and let L be a finite algebraic extension of K. Then the integral closure T of R in L is a finitely generated R module.

Corollary 10.58. Let C be an irreducible projective curve with function field $\mathbb{K} = \Bbbk(C)$, let P be a point of C, and let M_P be the maximal ideal of $\mathcal{O}_P(C)$. Then there exists a discrete valuation v of $\mathbb{K}$ defined over $\Bbbk$ whose valuation ideal $\mathfrak{p}_v$ has $M_P \subseteq \mathfrak{p}_v$.

REMARKS. This result is a supplement to Theorem 10.53. It says that the map of that theorem, carrying discrete valuations of $\mathbb{K}$ defined over $\Bbbk$ to points of C, is onto.

PROOF. Without loss of generality, we may assume that C is affine. Let $\mathfrak{m}_P$ be the maximal ideal in the affine coordinate ring $A(C)$ consisting of all functions vanishing at P, and let S be the set-theoretic complement of $\mathfrak{m}_P$ in $A(C)$, so that $M_P = S^{-1}\mathfrak{m}_P$. Evaluation at P is a linear functional on $A(C)$ with kernel $\mathfrak{m}_P$, and therefore $A(C) = \mathfrak{m}_P + \Bbbk 1$. In other words, $\mathfrak{m}_P$ and any element of S together generate $A(C)$ as a $\Bbbk$ vector space.

If T denotes the integral closure of $A(C)$ in $\mathbb{K}$, then Theorem 10.57 implies that T is Noetherian, and Proposition 8.45 of *Basic Algebra* shows that every nonzero prime ideal of T is maximal. Hence T is a Dedekind domain. Proposition 8.53 of *Basic Algebra* shows that there exists a maximal ideal $\mathfrak{q}$ of T such that $\mathfrak{m}_P = A(C) \cap \mathfrak{q}$. Since T is a Dedekind domain, $\mathfrak{q}$ is contained in the valuation ideal $\mathfrak{p}_v$ of a unique discrete valuation v of $\mathbb{K}$, and T is contained in the valuation ring T_v of v. Thus $\mathfrak{m}_P \subseteq \mathfrak{p}_v$, and $S \subseteq T$ implies that $v(s) \geq 0$ for all $s \in S$. On the other hand, 1 lies in $\mathfrak{m}_P + \Bbbk s$ for any s in S, and hence $0 = v(1) \geq \min(1, v(s))$. Therefore $v(s) = 0$ for all $s \in S$, and $M_P = S^{-1}\mathfrak{m}_P \subseteq \mathfrak{p}_v$. □

Corollary 10.59. If $\mathbb{K}$ is a function field in one variable over $\Bbbk$ and if v is a discrete valuation of $\mathbb{K}$ defined over $\Bbbk$ with valuation ring R_v, then there exists an irreducible nonsingular *affine* curve C over $\Bbbk$ with function field $\mathbb{K}$ and with a point P such that $\mathcal{O}_P(C) = R_v$.

PROOF. Choose an element x of $\mathbb{K}$ such that $v(x) > 0$. Define $R = \Bbbk[x]$. Since $v(x) \neq 0$, x is transcendental over $\Bbbk$, and $\mathbb{K}$ is a finite algebraic extension of the field of fractions $\Bbbk(x)$ of R. Corollary 7.14 shows that the integral closure T of R in $\mathbb{K}$ is a Dedekind domain, and Theorem 10.57 shows that T is a finitely generated R module. Thus we can write T as $T = \Bbbk[x_1, \ldots, x_n]$ with $x_1 = x$. The substitution homomorphism with $X_j \mapsto x_j$ for all j carries $\Bbbk[X_1, \ldots, X_n]$ onto T and has a prime ideal $\mathfrak{p}$ as kernel, since T is an integral domain. Thus $V(\mathfrak{p})$ is an affine variety with T as its affine coordinate ring. The dimension of $V(\mathfrak{p})$ is the transcendence degree of $\mathbb{K}$ over $\Bbbk$, which is 1 by assumption. Thus $C = V(\mathfrak{p})$ is an irreducible curve. Since T is integrally closed by construction, Corollary 10.52 shows that C is nonsingular.

Let $R_v \subseteq \mathbb{K}$ be the valuation ring of v, and let $\mathfrak{p}_v$ be the valuation ideal. The inequality $v(x) > 0$ shows that v is ≥ 0 on $R = \Bbbk[x]$, and Proposition 6.7 says that v is consequently ≥ 0 on the integral closure T of R in $\mathbb{K}$. In other words, T is contained in R_v. Since T is a Dedekind domain and $\mathbb{K}$ is its field of fractions, Theorem 6.5 shows that $\mathfrak{q} = \mathfrak{p}_v \cap T$ is a nonzero prime (= maximal) ideal of T and that the discrete valuation $v_\mathfrak{q}$ of $\mathbb{K}$ over $\Bbbk$ determined by $\mathfrak{q}$ coincides with v. The maximal ideals of the affine coordinate ring of an affine variety correspond to the points of the variety by Proposition 10.23, and thus there exists a point P of C such that $\mathfrak{q}$ is the maximal ideal of T consisting of all functions vanishing at P. The localization of T with respect to $\mathfrak{q}$ is $\mathcal{O}_P(C)$ by definition and is R_v by Proposition 6.4. Therefore $\mathcal{O}_P(C) = R_v$. □

Corollary 10.60. Let C be the irreducible nonsingular affine curve constructed in Corollary 10.59 and having function field $\mathbb{K} = \Bbbk(C)$, and regard C as a subvariety of its projective closure $\overline{C}$. Then there are only finitely many discrete valuations v' of $\mathbb{K}$ defined over $\Bbbk$ such that the unique point P of $\overline{C}$ with $M_P \subseteq \mathfrak{p}_{v'}$, where M_P is the maximal ideal of $\mathcal{O}_P(\overline{C})$ and $\mathfrak{p}_{v'}$ is the valuation ideal of v', lies outside C.

PROOF. We go over the argument in Corollary 10.59 with the same element x and with any discrete valuation v' defined over $\Bbbk$ such that $v'(x) \geq 0$. This inequality implies that v' is ≥ 0 on $\Bbbk[x]$, and Proposition 6.7 then shows that v' is ≥ 0 on $T = A(C)$. Thus $A(C)$ is contained in the valuation ring $R_{v'}$ of v'. Define $\mathfrak{q} = \mathfrak{p}_{v'} \cap A(C)$. Arguing as in the existence proof for Theorem 10.53, we find that $\mathfrak{q}$ equals the ideal $\mathfrak{m}_P$ of all members of $A(C)$ vanishing at a certain point P of C, and that proof then shows that $M_P \subseteq \mathfrak{p}_{v'}$. By uniqueness in Theorem 10.53, this P is the one and only point produced by that theorem.

In other words, the only discrete valuations v' of $\mathbb{K}$ defined over $\Bbbk$ for which the point P lies outside C are those with $v'(x) < 0$. Corollary 6.10 shows that there are only finitely many of these. □

We come to the proof of Theorem 10.57, but only under the assumption that k is algebraically closed. The proof is rather technical, and the reader is encouraged to skip it on first reading. To underscore this point, the proof appears in small print. We need two lemmas.

Lemma 10.61. Let R be a Noetherian integrally closed domain with field of fractions F, let K be a finite *separable* extension of F, and let T be the integral closure of R in K. Then T is Noetherian and is finitely generated as an R module.

PROOF. In effect, this result was proved in *Basic Algebra*. In more detail: With the above assumptions and also the assumption that every nonzero prime ideal of R is maximal (i.e., that R is a Dedekind domain), the proof of Theorem 8.54 of *Basic Algebra* showed that T is a Dedekind domain. The hard part of that proof appeared in Section IX.15; it showed from the separability that T is finitely generated as an R module, and it did not make use of the assumption that every nonzero prime ideal of R is maximal. Since T is finitely generated and R is Noetherian, every R submodule of T is a finitely generated R module, by Proposition 8.34 of *Basic Algebra*. In particular, every ideal of T is finitely generated as an R module and therefore is finitely generated as a T module. Consequently T is Noetherian. $\square$

Lemma 10.62 (Noether Normalization Lemma). Let k be an infinite field, let $R = k[x_1, \ldots, x_n]$ be a finitely generated integral domain over k, and let $K = k(x_1, \ldots, x_n)$ be the field of fractions of k. Then for a suitable d with $0 \leq d \leq n$, there exist d linear combinations $y_1, \ldots, y_d$ of $x_1, \ldots, x_n$ with coefficients in k such that $y_1, \ldots, y_d$ are algebraically independent over k and such that every element of R is integral over $k[y_1, \ldots, y_d]$. If K is separably generated over k, then the y_i may be chosen in such a way that K is a separable extension of $k(y_1, \ldots, y_d)$.

REMARKS. It is immediate from the conclusion that d is the transcendence degree of K over k. The lemma is a result about the extension of rings that improves upon Theorem 7.7 for fields; the latter says that every field extension can be accomplished by a transcendental extension followed by an algebraic extension. The present lemma says that the passage from a field to a finitely generated integral domain can be accomplished by a full polynomial extension followed by an extension in which each generator is not merely algebraic but actually is a root of a monic polynomial with coefficients in the full polynomial ring.

PROOF. Let I be the kernel of the quotient homomorphism $k[X_1, \ldots, X_n] \to k[x_1, \ldots, x_n]$. The core of the proof involves a single nonzero f in I. The idea is to replace $X_1, \ldots, X_{n-1}$ by new indeterminates $X_1', \ldots, X_{n-1}'$ to make the equation $f(x_1, \ldots, x_n) = 0$ become a monic polynomial equation satisfied by x_n over $R' = k[X_1', \ldots, X_{n-1}']$. With $c_1, \ldots, c_{n-1}$ equal to members of k to be specified later, define $x_j' = x_j - c_j x_n$ for $1 \leq j \leq n - 1$. The equation $f(x_1, \ldots, x_n) = 0$ becomes

$$f(x_1' + c_1 x_n, \ldots, x_{n-1}' + c_{n-1} x_n, x_n) = 0. \qquad (*)$$

For a suitable choice of $c_1, \ldots, c_{n-1}$, we shall show in a moment that

$$\text{the polynomial} \quad f(X_1' + c_1 X_n, \ldots, X_{n-1}' + c_{n-1} X_n, X_n) \quad \text{is monic in } X_n \qquad (**)$$

after multiplication by a member of $k^\times$.

Assuming $(**)$, let us see how the first conclusion of the lemma follows by induction on n. For $n = 1$, there are two cases. One case is that K is a simple algebraic extension field of k, and then every element of the extension field $R = K$ is a root of its minimal polynomial over k. This is the case $d = 0$. The other case is that K is a simple transcendental extension, and then we can take $y_1 = x_1$. This is the case $d = 1$.

For the inductive step, assume the first conclusion of the lemma for $n - 1 \geq 1$, d being an integer with $0 \leq d \leq n - 1$. If $I = 0$, there is nothing to prove, since $x_1, \ldots, x_n$ are then algebraically independent and the lemma follows with $d = n$ and with $y_j = x_j$ for $1 \leq j \leq n$. If $I \neq 0$, fix $f \neq 0$ in I, and choose $c_1, \ldots, c_{n-1}$ in k to make (∗∗) hold. Then (∗) shows that x_n is a root of a monic polynomial with coefficients in $R' = k[x'_1, \ldots, x'_{n-1}]$. By the inductive hypothesis we can choose members $y'_1, \ldots, y'_d$ of R' with $0 \leq d \leq n - 1$ such that $y'_1, \ldots, y'_d$ are algebraically independent over k and such that every element of R' is integral over $k[y'_1, \ldots, y'_d]$. By transitivity of integral dependence, every element of $R'[x_n]$ is integral over $k[y'_1, \ldots, y'_d]$. Since the definition of x'_j in terms of x_j shows that $R'[x_n] = k[x'_1, \ldots, x'_{n-1}, x_n] = k[x_1, \ldots, x_{n-1}, x_n] = R$, every element of R is integral over $k[y'_1, \ldots, y'_d]$. This completes the induction, and the first sentence of conclusions of the lemma is proved except for (∗∗).

To prove (∗∗), let $r = \deg f$, and write $f = h_r + g$ with h_r nonzero and homogeneous of degree r and with $\deg g \leq r - 1$ (or $g = 0$). Then

$$f(X_1, \ldots, X_n) = f(X'_1 + c_1 X_n, \ldots, X'_{n-1} + c_{n-1} X_n, X_n)$$
$$= h_r(c_1 X_n, \ldots, c_{n-1} X_n) + \text{(terms involving } 1, X_n, X_n^2, \ldots, X_n^{r-1})$$
$$= h_r(c_1, \ldots, c_{n-1}, 1) X_n^r + \text{(terms involving } 1, X_n, X_n^2, \ldots, X_n^{r-1}).$$

Thus (∗∗) is proved if $c_1, \ldots, c_{n-1}$ can be chosen with the scalar $h_r(c_1, \ldots, c_{n-1}, 1)$ not 0. Here the fact that h_r is nonzero and homogeneous implies that $h_r(X_1, \ldots, X_{n-1}, 1)$ is not the 0 polynomial in $k[X_1, \ldots, X_{n-1}]$. Since k is an infinite field, Corollary 4.32 of *Basic Algebra* shows that the evaluation mapping of $k[X_1, \ldots, X_{n-1}]$ into the algebra of functions from k^{n-1} into k is one-one, and therefore there exist $c_1, \ldots, c_{n-1}$ with $h_r(c_1, \ldots, c_{n-1}, 1) \neq 0$. This proves (∗∗).

We are left with proving that if K is separably generated over k, then the y_i may be chosen with K separable over $k(y_1, \ldots, y_d)$. We proceed as above but with an amended version of (∗∗) that we mention in a moment. In the induction the extra hypothesis for $n = 1$ is that either x_1 is separable algebraic over k or x_1 is transcendental, and in both cases K is a separable extension of $k(y_1)$. For the inductive step when $I \neq 0$, Theorem 7.18 shows that $\{x_1, \ldots, x_n\}$ contains a separating transcendence basis; possibly by renumbering the variables, we may assume that this transcendence basis is a subset of $\{x_1, \ldots, x_{n-1}\}$. In particular, x_n is separable algebraic over $k(x_1, \ldots, x_{n-1})$. For the polynomial f, we start from the minimal polynomial of x_n over $k(x_1, \ldots, x_{n-1})$, next multiply by a common denominator to get all coefficients of powers of X_n to be in $k[x_1, \ldots, x_{n-1}]$, and then replace the occurrences of $x_1, \ldots, x_{n-1}$ by $X_1, \ldots, X_{n-1}$. The result is f. We choose $y'_1, \ldots, y'_d$ as above, and the inductive hypothesis shows that $k(x'_1, \ldots, x'_{n-1})$ is separable over $k(y'_1, \ldots, y'_d)$. If we can show that x_n is separable over $k(x'_1, \ldots, x'_{n-1})$, then we will have proved that K is a separable extension of $k(y'_1, \ldots, y'_d)$ because of the transitivity of separability. So the induction will be complete.

To get that x_n is separable over $k(x'_1, \ldots, x'_{n-1})$, it is enough to prove that we can arrange for

$$x_n \text{ to be a simple root of } f(x'_1 + c_1 X_n, \ldots, x'_{n-1} + c_{n-1} X_n, X_n) \quad (\dagger)$$

in addition to (∗∗). Indeed, then x_n is a root of a separable polynomial over $k(x'_1, \ldots, x'_{n-1})$ and hence is a separable element over $k(x'_1, \ldots, x'_{n-1})$. The condition (†) is the same as the condition that the derivative of (†) with respect to X_n, when evaluated at x_n, be nonzero. Thus we want to arrange that

$$f_n(x_1, \ldots, x_{n-1}, x_n) + c_1 f_1(x_1, \ldots, x_{n-1}, x_n) + \cdots + c_{n-1} f_{n-1}(x_1, \ldots, x_{n-1}, x_n) \neq 0, \quad (\dagger\dagger)$$

where the subscripts on f indicate first partial derivatives in the indicated variables. The left side of (††) is the sum of a constant and a linear functional on the vector space of all $(c_1, \ldots, c_{n-1})$ in k^{n-1}. The constant term is $f_n(x_1, \ldots, x_{n-1}, x_n)$, which is nonzero because x_n is separable over $k(x_1, \ldots, x_{n-1})$ and is therefore a simple root of its minimal polynomial over $k(x_1, \ldots, x_{n-1})$. Thus the left side of (††) is the value of a nonzero polynomial $p(X_1, \ldots, X_{n-1}) = a_n + \sum_{j=1}^{n-1} a_j X_j$ at $(c_1, \ldots, c_{n-1})$. Consequently (∗∗) and (††) will hold simultaneously if we choose a point $(c_1, \ldots, c_{n-1})$ in k^{n-1} at which the nonzero polynomial $p(X_1, \ldots, X_{n-1}) h_r(X_1, \ldots, X_{n-1}, 1)$ is not zero. $\square$

PROOF OF THEOREM 10.57 UNDER THE ASSUMPTION THAT k IS ALGEBRAICALLY CLOSED. The first step is to reduce to the case that $L = K$, i.e., that the field of fractions of R coincides with L. To do so, choose a vector-space basis $\{z_1, \ldots, z_r\}$ of L over K consisting of elements integral over R; this is possible by Proposition 8.42 of *Basic Algebra*. Put $S = R[z_1, \ldots, z_r]$. This is a finitely generated integral domain over k, all of its elements are integral over k, and it has L as field of fractions. The integral closure of R in L equals the integral closure of S in L.

Thus we may assume that $R = k[x_1, \ldots, x_n]$ is an integral domain with field of fractions K and that we are to prove that the integral closure T of R in K is a finitely generated R module. Let d be the transcendence degree of K over k. Since algebraically closed fields are perfect, Theorem 7.20 shows that K is separably generated over k. Lemma 10.62 is therefore applicable, and it produces d linear combinations $y_1, \ldots, y_d$ of $x_1, \ldots, x_n$ over k such that the subring $S = k[y_1, \ldots, y_d]$ of R is a full polynomial ring, every element of R is integral over S, and K is a separable extension of the field $k(y_1, \ldots, y_d)$. Since every element of T is integral over R, the transitivity of integral dependence implies that every element of T is integral over S. Therefore T is the integral closure of S in K. Being a full polynomial ring, S is Noetherian and is a unique factorization domain; the latter property implies that S is integrally closed, according to Proposition 8.41 of *Basic Algebra*. Taking S to be the Noetherian integrally closed domain in Lemma 10.61, we see that T is finitely generated as an S module. Since $S \subseteq R$, T is certainly finitely generated as an R module. $\square$

Now we come to the main theorem of this section.

Theorem 10.63. Every birational equivalence class of irreducible projective curves contains a nonsingular such curve, and this curve is unique within the equivalence class up to isomorphism of varieties. Any irreducible nonsingular quasiprojective curve is isomorphic to an open subvariety of some irreducible nonsingular projective curve.

REMARKS. The new content of the theorem is the existence of the nonsingular projective curve. The uniqueness is immediate from Corollary 10.55. The statement about nonsingular quasiprojective curves is a formality: Such a curve C_0 is birational to the nonsingular projective curve C produced by the theorem and also to the projective closure $\overline{C_0}$ of C_0. The birational maps from $\overline{C_0}$ into C and from C into $\overline{C_0}$ yield morphisms from C_0 into C and from C into $\overline{C_0}$ by Corollary 10.54; sorting out these morphisms shows that C_0 is isomorphic to an open subvariety of C.

The idea for proving the existence of the projective curve in the theorem is to start with any function field $\mathbb{K}$ in one variable over $\mathbb{k}$, take any discrete valuation v of $\mathbb{K}$ defined over $\mathbb{k}$ (these exist as a consequence of Section VI.2), and use Corollary 10.59 to obtain some irreducible nonsingular affine curve having $\mathbb{K}$ as function field and having its local ring at some point equal to the valuation ring of v. Corollary 10.60 shows that except for finitely many discrete valuations, we have associated a nonsingular point on some irreducible affine curve in the birational equivalence class to each discrete valuation of $\mathbb{K}$ defined over $\mathbb{k}$. Applying Corollary 10.59 to each of these exceptional discrete valuations, we end up with a finite set of irreducible nonsingular affine curves such that each discrete valuation of $\mathbb{K}$ over $\mathbb{k}$ corresponds to some point of at least one of the curves. We shall glue together these irreducible nonsingular affine curves in a suitable fashion to obtain the desired irreducible nonsingular projective curve.

The proof makes use of the fact that the product of two projective varieties is a projective variety and that morphisms behave as one might expect. Let us postpone the details of establishing a rigorous theory of product varieties, going right to the proof of Theorem 10.63.

PROOF OF THEOREM 10.63. Let $\mathbb{K}$ be the given function field, and let $C_1, \ldots, C_m$ be the irreducible nonsingular affine curves described two paragraphs before this paragraph. In each case the function field of the curve is isomorphic to $\mathbb{K}$ by some fixed isomorphism, but we shall treat this fixed isomorphism as if it were the identity in order to avoid unnecessary complications in the notation. Let $\mathbb{V}_\mathbb{K}$ be the set of discrete valuations of $\mathbb{K}$ defined over $\mathbb{k}$. For $v \in \mathbb{V}_\mathbb{K}$, we write $R_v \subseteq \mathbb{K}$ for the valuation ring of v and $\mathfrak{p}_v$ for the valuation ideal of v.

For definiteness let C_j be an affine variety in $\mathbb{A}^{k_j}$, and let $\overline{C}_1, \ldots, \overline{C}_n$ be the respective projective closures of $C_1, \ldots, C_m$ in $\mathbb{P}^{k_j}$. For any point P in $\overline{C}_j$, let M_P be the maximal ideal of the local ring $\mathcal{O}_P(\overline{C}_j)$.

Theorem 10.53 gives us for each j a well-defined function $\gamma_j : \mathbb{V}_\mathbb{K} \to \overline{C}_j$, and Corollary 10.58 says that γ_j is onto $\overline{C}_j$. The defining property of $\gamma_j(v)$ is that $M_{\gamma_j(v)} \subseteq \mathfrak{p}_v$, and it follows that $\mathcal{O}_{\gamma_j(v)}(\overline{C}_j) \subseteq R_v$. Corollary 10.51 shows that the inverse image under γ_j of any point in C_j is a singleton set, and Corollary 10.60 shows that the inverse image of any point of the complementary set $\overline{C}_j - C_j$ is a finite set. Let F be the finite subset $F = \bigcup_{j=1}^m \gamma_j^{-1}(\overline{C}_j - C_j)$ of $\mathbb{V}_\mathbb{K}$. For $v \notin F$, $\gamma_j(v)$ is a nonsingular point of C_j, and Corollary 10.51 shows that $\mathcal{O}_{\gamma_j(v)}(C_j) = R_v$. Hence also $M_{\gamma_j(v)} = \mathfrak{p}_v$. The construction of the curves $C_1, \ldots, C_m$ was arranged in such a way that

$$\text{each } v \in \mathbb{V}_\mathbb{K} \text{ has } \gamma_j(v) \text{ in } C_j \text{ for some } j. \qquad (*)$$

Let U_j be the open set of C_j given by $U_j = \gamma_j(\mathbb{V}_\mathbb{K} - F)$. The curves $\overline{C}_j$ are birationally equivalent because they all have $\mathbb{K}$ as function field, and Corollary 10.54 shows that the largest domain on which the birational map from $\overline{C}_j$ to $\overline{C}_1$ is a morphism includes all the nonsingular points of $\overline{C}_j$. In particular, it contains $U_j = \gamma_j(\mathbb{V}_\mathbb{K} - F)$. If φ_j is the morphism from U_j into $\overline{C}_1$, then Proposition 10.42 shows that φ_j induces a homomorphism $\varphi_{j,P}^* : \mathcal{O}_{\varphi_j(P)}(\overline{C}_1) \to \mathcal{O}_P(C_j)$ for $P \in U_j$. By assumption, the isomorphism $\widetilde{\varphi}_j : \mathbb{k}(C_1) \to \mathbb{k}(C_j)$ is normalized to be the identity. Since $\widetilde{\varphi}_j$ is the field mapping corresponding to the birational map φ_j, $\widetilde{\varphi}_j$ is an extension of $\varphi_{j,P}^*$. Thus $\varphi_{j,P}^*$ is the identity under our identifications: $\mathcal{O}_{\varphi_j(P)}(\overline{C}_1) = \mathcal{O}_P(C_j)$ for $P \in U_j$. Let $P = \gamma_j(v)$ with v in $\mathbb{V}_\mathbb{K} - F$, and let $\varphi_j(P) = \gamma_1(v')$ with v' in $\mathbb{V}_\mathbb{K}$. Then $R_v = \mathcal{O}_{\gamma_j(v)}(C_j) = \mathcal{O}_{\varphi_j(P)}(\overline{C}_1) \subseteq R_{v'}$, and it follows that $v' = v$. In particular, v' is in $\mathbb{V}_\mathbb{K} - F$, and $\gamma_1(v) = \varphi_j(\gamma_j(v))$. Hence

$$\varphi_j \circ \gamma_j : \mathbb{V}_\mathbb{K} - F \to U_1 \quad \text{is independent of } j,$$

and
$$\varphi_j : U_j \to U_1 \quad \text{is an isomorphism.}$$

The product $W = \overline{C}_1 \times \cdots \times \overline{C}_m$ is an m-dimensional closed subvariety of $\mathbb{P}^{k_1} \times \cdots \times \mathbb{P}^{k_m}$, which in turn is a projective variety in $\mathbb{P}^N$ for a suitably large N. For $1 \leq j \leq m$, let $\pi_j : W \to \overline{C}_j$ be the jth projection map; this is a morphism. The set $U_1 \times \cdots \times U_m$ is an open subvariety of W, and the "diagonal"

$$\Delta = \left\{ \delta(P) = \left(P, \varphi_2^{-1}(P), \ldots, \varphi_m^{-1}(P) \right) \mid P \in U_1 \right\}$$

of $U_1 \times \cdots \times U_m$ is an irreducible curve isomorphic to U_1. The closure $C = \overline{\Delta}$ is an irreducible projective curve. It is a closed subvariety of W, and it has Δ as an open subvariety. The curve Δ may be identified with U_1 via the projection π_1, and we may therefore identify the function field of Δ, which is the same as the function field of C, with $\mathbb{K}$.

We shall show that C is nonsingular. For each j, the restriction $\pi_j : C \to \overline{C}_j$ is a morphism, and the image contains all points $\pi_j(\delta(P)) = \varphi_j^{-1}(P)$ with $P \in U_1$. Hence it contains U_j, which is an open subset of $\overline{C}_j$. In other words, $\pi_j : C \to \overline{C}_j$ is a dominant morphism. For $P \in U_1$, we have $\pi_j(\delta(P)) = \varphi_j^{-1}(P)$. If $Q = \delta(P)$, this says that $\pi_j(Q) = \varphi_j^{-1}\delta^{-1}(Q)$, from which it follows that $\delta \circ \varphi_j$ is a two-sided inverse of π_j on Δ. Consequently the dominant morphism $\pi_j : C \to \overline{C}_j$ is a birational map. Let (V_j, ψ_j) be a pair in the class of the rational map π_j^{-1}; we may assume that V_j is the largest domain in $\overline{C}_j$ on which π_j^{-1} is a morphism.

Let P be any point of C, and let M_P be the maximal ideal of $\mathcal{O}_P(C)$. Corollary 10.58 shows that there is a member v of $\mathbb{V}_\mathbb{K}$ such that $M_P \subseteq \mathfrak{p}_v$. Choose $j = j(P)$ with $1 \leq j \leq m$ such that $\gamma_j(v)$ is in C_j. Since every point of C_j is a nonsingular point by construction, Corollary 10.54 shows that every point of C_j lies in the domain V_j on which ψ_j is defined as a morphism inverting π_j. Consequently the open subvariety $\pi_j^{-1}(C_j)$ of C is isomorphic to the nonsingular irreducible affine curve C_j, and the point P of C has an open neighborhood of nonsingular points. Since P is arbitrary, C is nonsingular. □

The remainder of this section develops a small theory of products of varieties in projective spaces. Most of the proofs are left to the problems at the end of the chapter. It is enough to handle the product of two varieties because general finite products of varieties can then be treated by induction.

We begin with the product of two projective spaces. Let $m \geq 1$ and $n \geq 1$ be integers, and put $N = (m + 1)(n + 1) - 1 = mn + m + n$. We shall exhibit

$\mathbb{P}^m \times \mathbb{P}^n$ as a projective variety in $\mathbb{P}^N$. To do so, we coordinatize $\mathbb{P}^m$, $\mathbb{P}^n$, and $\mathbb{P}^N$ by using x_i, y_j, and w_{ij} for $0 \leq i \leq m$ and $0 \leq j \leq n$. Then

$$\mathbb{P}^m = \{[x_0, \ldots, x_m]\}, \qquad \mathbb{P}^n = \{[y_0, \ldots, y_n]\},$$

and

$$\mathbb{P}^N = \{[w_{00}, w_{01}, \ldots, w_{m,n-1}, w_{mn}]\}.$$

The **Segre embedding** is the function

$$\sigma([x_0, \ldots, x_m], [y_0, \ldots, y_n]) = [x_0 y_0, x_0 y_1, \ldots, x_m y_{n-1}, x_m y_n],$$

i.e., $w_{ij} = x_i y_j$. Define $\mathfrak{a} \subseteq \mathbb{k}[W_{00}, \ldots, W_{mn}]$ to be the homogeneous ideal generated by all $W_{ij} W_{kl} - W_{il} W_{kj}$. Problems 17–19 at the end of the chapter show that σ is well defined and one-one, that the image of σ is $V(\mathfrak{a})$, and that $V(\mathfrak{a})$ is irreducible. Thus the Segre embedding exhibits $\mathbb{P}^m \times \mathbb{P}^n$ as a projective variety in $\mathbb{P}^N$. This variety is known as a **Segre variety**.[18]

Let $U \subseteq \mathbb{P}^m$ and $V \subseteq \mathbb{P}^n$ be projective algebraic sets. Then the Segre embedding σ carries $U \times V$ to a subset of $\mathbb{P}^N$, and we wish to see that $\sigma(U \times V)$ is a projective algebraic set in $\mathbb{P}^N$. Let us use the abbreviation $X = (X_0, \ldots, X_m)$. If $\alpha = (\alpha_0, \ldots, \alpha_m)$ is an $(m+1)$-tuple of nonnegative integers, we define $|\alpha| = \alpha_0 + \cdots + \alpha_m$ and $X^\alpha = X_0^{\alpha_0} \cdots X_m^{\alpha_m}$. We define $Y, \beta, |\beta|$, and Y^β similarly. Any monomial $X^\alpha Y^\beta$ with $|\alpha| = d$ and $|\beta| = e$ is said to be **bihomogeneous** of **bidegree** (d, e). A **bihomogeneous polynomial** of bidegree (d, e) is any linear combination of bihomogeneous monomials of bidegree (d, e).

The first observation is that any projective algebraic set S in $\mathbb{P}^m$ can be described as the locus of common zeros of a vector space of homogeneous polynomials in X of a fixed degree. In fact, we know that S is given by the locus of common zeros of a finite set of homogeneous polynomials $F_1(X), \ldots, F_r(X)$ of various degrees $d_1, \ldots, d_r$. Let us say that $d = \max_j d_j$. The point is that S is given by the locus of common zeros of a finite set of homogeneous polynomials all of degree d. The reason is that the locus of common zeros of $F_j(X)$ is the same as the locus of common zeros of $X_0^{d-d_j} F_j(X), \ldots, X_m^{d-d_j} F_j(X)$. The assertion about describing S follows.

Now let $U \subseteq \mathbb{P}^m$ be the locus of common zeros of homogeneous polynomials $F_1(X), \ldots, F_r(X)$ all of degree d, and let $V \subseteq \mathbb{P}^n$ be the locus of common zeros of homogeneous polynomials $G_1(Y), \ldots, G_r(Y)$ all of degree e. Then $U \times V$ is the locus of common zeros of the bihomogeneous polynomials $F_a(X) G_b(Y)$, all of bidegree (d, e). These cannot immediately be expressed in terms of the polynomials W_{ij} of the Segre embedding. However, if we use the same trick again, we can substitute the W_{ij}'s. Specifically suppose that $d \leq e$. Replace

---

[18] If we form the $(m+1)$-by-$(n+1)$ matrix whose $(i, j)^{\text{th}}$ entry is W_{ij}, then an equivalent description of the Segre variety is as the locus of common zeros of all 2-by-2 minors of this matrix.

$F_1(X), \ldots, F_r(X)$ by a family of $r(m+1)$ polynomials $F'_1(X), \ldots, F'_{r(m+1)}(X)$ homogeneous of degree e. Then the polynomials $F'_a(X)G_b(Y)$ are bihomogeneous of bidegree (e, e). When such a polynomial is expanded as a linear combination of monomials, each monomial has e factors from among $X_0, \ldots, X_m$ and e factors from among $Y_0, \ldots, Y_n$. We can pair the factors in whatever fashion we want and replace $X_i Y_j$ by W_{ij}. In this way our system of bihomogeneous polynomials can be rewritten as a system of polynomials $H_{ab}(W)$, together with the convention that $W_{ij} = X_i Y_j$. Then $\sigma(U \times V)$ is the locus of common zeros in $\mathbb{P}^N$ of the polynomials $H_{ab}(W)$ and the defining polynomials of the Segre variety.

Conversely if we have a projective algebraic set in $\mathbb{P}^N$, then its intersection with the Segre variety can be described as the locus of common zeros in $\mathbb{P}^m \times \mathbb{P}^n$ of a family of bihomogeneous polynomials in (X, Y). We have only to take the defining homogeneous polynomials $H(W)$ and substitute the definition $W_{ij} = X_i Y_j$ for W_{ij}. If $H(W)$ is homogeneous of degree e, then the result of the substitution is a polynomial bihomogeneous of bidegree (e, e).

Problems 20–21 at the end of the chapter show that if U and V are irreducible closed sets in $\mathbb{P}^m$ and $\mathbb{P}^n$, respectively, then $\sigma(U \times V)$ is irreducible in $\mathbb{P}^N$. Thus we can meaningfully speak of projective varieties in $\mathbb{P}^m \times \mathbb{P}^n$. The same pair of problems addresses what happens for quasiprojective varieties, showing that σ of any relatively open subset of a projective variety in $\mathbb{P}^m \times \mathbb{P}^n$ is a quasiprojective variety in $\mathbb{P}^N$.

Now that the notion of variety is meaningful in $\mathbb{P}^m \times \mathbb{P}^n$, with an interpretation in $\mathbb{P}^N$, we can similarly translate definitions and facts about morphisms to make them apply in $\mathbb{P}^m \times \mathbb{P}^n$. In particular, the projection of a variety to either factor $\mathbb{P}^m$ or $\mathbb{P}^n$ is a morphism on the variety. If U is a quasiprojective variety and if $\varphi_1 : U \to \mathbb{P}^m$ and $\varphi_2 : U \to \mathbb{P}^n$ are isomorphisms of U onto quasiprojective varieties in $\mathbb{P}^m$ and $\mathbb{P}^n$, then the diagonal $\Delta = \{(\varphi_1(u), \varphi_2(u)) \mid u \in U\}$ is a quasiprojective variety in $\mathbb{P}^m \times \mathbb{P}^n$, and the pair (φ_1, φ_2) is an isomorphism of varieties. These matters are discussed in Problem 22 at the end of the chapter.

9. Affine Algebraic Sets for Monomial Ideals

Sections 9–12 in part address aspects of the question of how much one can make explicit computations with affine and projective varieties. As a general rule, the tool for such computations is the theory of Gröbner bases, which were introduced in Sections VIII.7–VIII.10. The topic is an active area of continuing research.[19] One can think of immediate problems—such as finding the dimension of an algebraic set, determining the radical of an ideal when the ideal is given,

---

[19]The book edited by Buchberger and Winkler contains a number of expository "tutorials" that give an idea of the breadth of applications of the theory. The book contains also a certain number of research papers.

and deciding whether an ideal is prime. We shall concentrate on just one such problem, that of finding the dimension.[20]

Part of the abstract theory in this case dates back to Hilbert, but in combination with the theory of Gröbner bases it becomes easier to establish and relatively easy to implement computationally.[21] We shall prove in Section 12 as a consequence of this investigation the deep theorem that a system of simultaneous homogeneous polynomial equations having more equations than variables always has a nonzero solution.[22]

Hilbert associated a polynomial in one variable, now known as the "Hilbert polynomial," to each ideal of polynomials over an algebraically closed field. This polynomial encodes certain algebraic information about the ideal, and some features of this polynomial depend only on the geometry of the zero locus. In particular, the degree of the polynomial turns out to equal the geometric dimension of the zero locus, and that will be what interests us.

The theory behind Gröbner bases enables one to reduce the theory of the Hilbert polynomial to the case of a monomial ideal, for which it is relatively easy to understand.[23] We begin with that case in this section.

Let $\Bbbk$ be an algebraically closed field, consider affine space $\mathbb{A}^n$, and let $\mathfrak{a}$ be an ideal in $A = \Bbbk[X_1, \ldots, X_n]$. In this section we shall be interested in the case that $\mathfrak{a}$ is generated by monomials, in which case it is called a **monomial ideal**. The structure of monomial ideals is captured by Lemma 8.17, which says about such an ideal $\mathfrak{a}$ that

- for any polynomial $f \neq 0$ in $\mathfrak{a}$, each monomial term contributing to f lies in $\mathfrak{a}$,
- $\mathfrak{a}$ has a finite set of monomials as generators,
- if $\{M_1, \ldots, M_k\}$ is a set of monomials that generate $\mathfrak{a}$ and if M is any monomial in $\mathfrak{a}$, then some M_j divides M.

Let $e_1, \ldots, e_n$ be the standard basis of $\mathbb{A}^n$, and let $\langle e_{j_1}, \ldots, e_{j_k} \rangle$ be the linear span of $e_{j_1}, \ldots, e_{j_k}$. The vector space $\langle e_{j_1}, \ldots, e_{j_k} \rangle$ is called a **coordinate subspace** of $\mathbb{A}^n$. The ideal $\mathfrak{p}_k = (X_1, \ldots, X_k)$ in A is prime, and its variety is $V(\mathfrak{p}_k) = \langle e_{k+1}, \ldots, e_n \rangle$. Since $\mathfrak{p}_0 \subseteq \mathfrak{p}_1 \subseteq \cdots \subseteq \mathfrak{p}_n$ is a strictly increasing sequence of prime ideals in A and since A has Krull dimension n,

---

[20] Solutions to the other two problems are known as well. References may be found in Cox–Little–O'Shea. For determining the radical, see p. 177. For deciding whether an ideal is prime, see p. 207.

[21] The exposition in Sections 9–12 is based in part on Chapter 9 of the book by Cox–Little–O'Shea and in part on Chapter I of Hartshorne's book.

[22] For one equation with two variables, this amounts to the Fundamental Theorem of Algebra. For two equations with three variables, it amounts to the existence part of Bezout's Theorem as formulated in Theorem 8.5.

[23] Similarly the computations associated with Gröbner bases make it possible to reduce the computation of the Hilbert polynomial of a general ideal to the computation of the Hilbert polynomial of a monomial ideal.

no strictly increasing sequence of prime ideals containing $\mathfrak{p}_k$ can be longer than $\mathfrak{p}_k \subseteq \mathfrak{p}_{k+1} \subseteq \cdots \subseteq \mathfrak{p}_n$. It follows that the images of these ideals in $A/\mathfrak{p}$ give a strictly increasing sequence of prime ideals of maximal length and that $A/\mathfrak{p}$ has Krull dimension $n - k$. By Theorem 10.7 the geometric dimension of $V(\mathfrak{p}_k) = \langle e_{k+1}, \ldots, e_n \rangle$ is $n - k$. In other words, the geometric dimension of the vector subspace $\langle e_{k+1}, \ldots, e_n \rangle$ is the same as the vector-space dimension. Relabeling indices in this computation, we see that the geometric dimension of $\langle e_{j_1}, \ldots, e_{j_k} \rangle$ is k if the indices $j_1, \ldots, j_k$ are distinct.

Let us compute the geometric dimension of the zero locus of a general proper monomial ideal $(M_1, \ldots, M_k)$. If $\alpha = (\alpha_1, \ldots, \alpha_n)$ is a tuple of integers ≥ 0, we write X^α for $X_1^{\alpha_1} \cdots X_n^{\alpha_n}$ and $|\alpha|$ for $\alpha_1 + \cdots + \alpha_n$. Let $H_j = V(X_j)$ be the **coordinate hyperplane** of points in $\mathbb{A}^n$ with j^{th} coordinate 0. This is the linear span of all e_i for $i \neq j$, and it has geometric dimension $n - 1$. If a monomial X^α is given, then Proposition 10.1 shows that

$$V(X^\alpha) = \bigcup_{\alpha_j > 0} V(X_j) = \bigcup_{\alpha_j > 0} H_j$$

and then that

$$V(X^\alpha, X^\beta) = \left(\bigcup_{\alpha_i > 0} H_i \right) \cap \left(\bigcup_{\beta_j > 0} H_j \right) = \bigcup_{\alpha_i > 0, \, \beta_j > 0} (H_i \cap H_j).$$

Similarly $V(M_1, \ldots, M_k)$ is a finite union of k-fold intersections of coordinate hyperplanes. By Theorem 10.7 the geometric dimension of $V(M_1, \ldots, M_k)$ is the maximum dimension of the subspaces $H_i \cap H_j \cap \cdots$ appearing in the appropriate union for $M_1, \ldots, M_k$. To get the maximum dimension, we want as few distinct indices to appear in an intersection $H_i \cap H_j \cap \cdots$. If the smallest possible number of distinct indices is m, then we see that $V(M_1, \ldots, M_k)$ has geometric dimension $n - m$.

The insight is that to study $V(\mathfrak{a})$, one studies $A/\mathfrak{a}$, and that to study the latter, one considers what happens as a function of s to the part of $A/\mathfrak{a}$ that corresponds to degree at most s. In the case of a monomial ideal, this means that one is to study the monomials outside the ideal in question, particularly how the number of these monomials grows with s. Let $\mathcal{M}$ be the set of all monomials in $\Bbbk[X_1, \ldots, X_n]$. For our monomial ideal $\mathfrak{a}$, let $\mathcal{C}(\mathfrak{a})$ be the complementary subset to $\mathfrak{a}$ in $\mathcal{M}$ given by

$$\mathcal{C}(\mathfrak{a}) = \{ X^\alpha \mid X^\alpha \notin \mathfrak{a} \}.$$

Proposition 10.64. If $\mathfrak{a}$ is a proper monomial ideal in $\Bbbk[X_1, \ldots, X_n]$, then
(a) the vector subspace $V(\{X_i \mid i \notin \{j_1, \ldots, j_k\}\})$ is contained in $V(\mathfrak{a})$ if and only if $\{X^\alpha \in \mathcal{M} \mid \alpha \in \langle e_{j_1}, \ldots, e_{j_k} \rangle\}$ is contained in $\mathcal{C}(\mathfrak{a})$,
(b) the geometric dimension of $V(\mathfrak{a})$ equals the largest vector-space dimension of a coordinate subspace that lies in $\mathcal{C}(\mathfrak{a})$.

REMARK. The hypothesis "proper" is needed for (b), not for (a).

PROOF. For (a), first suppose that $V(\{X_i \mid i \notin \{j_1, \ldots, j_k\}\})$ is contained in $V(\mathfrak{a})$, and suppose that α is in $\langle e_{j_1}, \ldots, e_{j_k}\rangle$. Let $P = (x_1, \ldots, x_n)$ be the point with
$$x_i = \begin{cases} 1 & \text{for } i \in \{j_1, \ldots, j_k\}, \\ 0 & \text{for } i \notin \{j_1, \ldots, j_k\}. \end{cases} \quad (*)$$
Then P is on the zero locus of each X_i for $i \notin \{j_1, \ldots, j_k\}$, and hence P is in $V(\mathfrak{a})$. On the other hand, the value of the monomial X^α at P is 1. Since the value of every member of $\mathfrak{a}$ at P is 0, X^α cannot be in $\mathfrak{a}$. Thus X^α is in $\mathcal{C}(\mathfrak{a})$.

Next suppose that $E = V(\{X_i \mid i \notin \{j_1, \ldots, j_k\}\})$ is not contained in $V(\mathfrak{a})$. Say that $P = (x_1, \ldots, x_n)$ is in E but not $V(\mathfrak{a})$. The condition for P to be in E is that $x_i = 0$ for all $i \notin \{j_1, \ldots, j_k\}$. Since P is not in $V(\mathfrak{a})$, some member of $\mathfrak{a}$ is nonzero at P. The ideal is generated by monomials, and thus some monomial X^{α_0} in $\mathfrak{a}$ is nonzero at P. Let $\alpha_0 = (\alpha_1, \ldots, \alpha_n)$. The (nonzero) value of α_0 on P is $\prod_{i \text{ with } \alpha_i > 0} x_i^{\alpha_i}$. Now $x_i = 0$ for all $i \notin \{j_1, \ldots, j_k\}$, and consequently no i outside $\{j_1, \ldots, j_k\}$ can have $\alpha_i > 0$. Thus α_0 is in $\langle e_{j_1}, \ldots, e_{j_k}\rangle$, and α_0 exhibits $\{X^\alpha \in \mathcal{M} \mid \alpha \in \langle e_{j_1}, \ldots, e_{j_k}\rangle\}$ as failing to be contained in $\mathcal{C}(\mathfrak{a})$.

For (b), we saw before the proof that $V(\mathfrak{a})$ is the union of finitely many vector subspaces and that each vector subspace is an affine variety whose geometric dimension equals its vector-space dimension. By Theorem 10.7 the geometric dimension of $V(\mathfrak{a})$, $\mathfrak{a}$ being proper, is the maximum of the dimensions of these subspaces. Taking (a) into account, we conclude that (b) holds. $\square$

We seek a formula for the number of monomials in $\mathcal{C}(\mathfrak{a})$ of total degree $\leq s$ when s is large and positive. We begin with a lemma. For a monomial ideal $\mathfrak{a}$, the function carrying each integer $s \geq 0$ to the number of X^α in $\mathcal{C}(\mathfrak{a})$ with $|\alpha| \leq s$ is called the **affine Hilbert function** of $\mathfrak{a}$ and is denoted by $\mathcal{H}_a(s, \mathfrak{a})$. For $\mathfrak{a} = \Bbbk[X_1, \ldots, X_n]$, the affine Hilbert function is identically 0, and we shall usually not be interested in this case.

EXAMPLE. For $n = 1$ with one indeterminate X, the proper ideals of $\Bbbk[X]$ are 0 and (X^k) with $k > 0$. The monomials X^α with $|\alpha| \leq s$ are $1, X, X^2, \ldots, X^s$. By inspection, none of these is in $\mathfrak{a}$ if $\mathfrak{a} = 0$, and thus $\mathcal{H}_a(s, 0) = s + 1$. In the case of (X^k) with $k > 0$, the monomials X^α in $\mathcal{C}((X)^k)$ are $1, X, \ldots, X^{k-1}$, and thus $\mathcal{H}_a(s, (X^k))$ is $s + 1$ for $s \leq k - 1$ and is k for $s \geq k - 1$.

Theorem 10.65. If $\mathfrak{a}$ is a proper monomial ideal in $\Bbbk[X_1, \ldots, X_n]$, then the complementary set $\mathcal{C}(\mathfrak{a})$ of monomials is a disjoint union
$$\mathcal{C}(\mathfrak{a}) = C_0 \cup \cdots \cup C_n,$$
where C_k is a finite union of subsets of the form
$$E = \Big\{X^\alpha \in \mathcal{M} \mid \alpha \in \langle e_{j_1}, \ldots, e_{j_k}\rangle + \sum_{i \notin \{j_1, \ldots, j_k\}} a_i e_i\Big\}.$$

Here it is assumed that $\langle e_{j_1}, \ldots, e_{j_k} \rangle$ is a k-dimensional coordinate subspace and the coefficients a_i are particular integers ≥ 0.

REMARKS. The subsets of $\mathcal{M}$ of which the above set E is an example will be called **standard subsets** of $\mathcal{M}$ with k parameters. The member $\sum_{i \notin \{j_1, \ldots, j_k\}} a_i e_i$ of $\mathcal{M}$ is called the **associated translation** of E, and $\langle e_{j_1}, \ldots, e_{j_k} \rangle$ is called the **associated vector subspace** of E. Standard subsets of $\mathcal{M}$ with 0 parameters are singleton sets $\{X^\alpha\}$. An example of a standard subset of $\mathcal{M}$ with 1 parameter when $n = 2$ is $\{X_1^{\alpha_1} X_2^{\alpha_2} \mid \alpha_1 \geq 0, \alpha_2 = 2\} = \{X^\alpha \mid \alpha \in \langle e_1 \rangle + 2e_2\}$. It is apparent that the one and only circumstance in which C_n is nonempty is that $\mathcal{C}(\mathfrak{a}) = \mathcal{M}$, in which case $\mathfrak{a} = 0$.

PROOF. We proceed by induction on n, and we may assume that $\mathfrak{a} \neq 0$. The example above shows for $n = 1$ that $\mathcal{C}(\mathfrak{a})$ is a finite set if $\mathfrak{a}$ is a nonzero proper ideal. Thus $\mathcal{C}(\mathfrak{a}) = C_0$ in this case, and the base case of the induction is settled.

Assume inductively that the theorem has been proved for $n - 1$ indeterminates, and let $\mathfrak{a}$ be a nonzero ideal in $\Bbbk[X_1, \ldots, X_n]$. Let $\mathcal{M}_{n-1}$ and $\mathcal{M}_n$ denote the sets of monomials in $X_1, \ldots, X_{n-1}$ and $X_1, \ldots, X_n$, respectively. For $j \geq 0$, let $\mathfrak{a}_j$ be the ideal in $\Bbbk[X_1, \ldots, X_{n-1}]$ of all polynomials $f(X_1, \ldots, X_{n-1})$ such that $X_n^j f(X_1, \ldots, X_{n-1})$ is in $\mathfrak{a}$. The ideals $\mathfrak{a}_j$ are monomial ideals because $\mathfrak{a}$ is a monomial ideal, and $\mathfrak{a}_j \subseteq \mathfrak{a}_{j+1}$ for all j. Since $\Bbbk[X_1, \ldots, X_{n-1}]$ is Noetherian, there is some index l such that $\mathfrak{a}_j = \mathfrak{a}_l$ for all $j \geq l$. We apply the inductive hypothesis to $\mathfrak{a}_0, \mathfrak{a}_1, \ldots, \mathfrak{a}_l$, writing

$$\mathcal{C}(\mathfrak{a}_j) = C_{0,j} \cup \cdots \cup C_{n-1,j} \quad \text{for } 0 \leq j \leq l.$$

Here each $C_{k,j}$ is a finite union of standard subsets with k parameters in the $n-1$ indeterminates $X_1, \ldots, X_{n-1}$.

Let $C_{k,j} X_n^j$ be the set of all products of members of $C_{k,j}$ with X_n^j. We shall show that

$$\mathcal{C}(\mathfrak{a}) = C_0 \cup \cdots \cup C_n, \qquad (*)$$

where $C_0, \ldots, C_n$ are defined by

$$C_{k+1} = \bigcup_{j=0}^{\infty} C_{k,l} X_n^j \cup \bigcup_{j=0}^{l-1} C_{k+1,j} X_n^j \quad \text{for } 0 \leq k \leq n-1$$

and
$$C_0 = \mathcal{C}(\mathfrak{a}) - \bigcup_{k=1}^{n} C_k.$$

But first let us see that each C_{k+1} for $0 \leq k \leq n-1$ is a finite union of standard subsets of $\mathcal{M}_n$ with $k+1$ parameters. Each $C_{k+1,j}$ is a finite union of standard subsets of $\mathcal{M}_{n-1}$ with some associated translation γ such that $\gamma_n = 0$ and with an associated vector subspace $\langle e_{j_1}, \ldots, e_{j_{k+1}} \rangle$ such that $j_1 < \cdots < j_{k+1} < n$. Then each $C_{k+1,j} X_n^j$ is a finite union of standard subsets of $\mathcal{M}$ of the form X^α

with associated translation $\gamma + je_n$ and with the same associated vector space $\langle e_{j_1}, \ldots, e_{j_{k+1}} \rangle$. Similarly the set $\bigcup_{j=0}^{\infty} C_{k,l} X_n^j$ is a finite union of standard subsets of $\mathcal{M}$ with associated translation $\gamma + 0e_n$ and with associated vector space of the form $\langle e_{j_1}, \ldots, e_{j_k}, e_n \rangle$. Thus C_{k+1} is a finite union of standard subsets of $\mathcal{M}_n$ with $k+1$ parameters.

Let us verify (∗). The most general monomial in $\Bbbk[X_1, \ldots, X_n]$ is $X^\beta X_n^j$ with X^β in $\Bbbk[X_1, \ldots, X_{n-1}]$, and this monomial is in $\mathfrak{a}$ if and only if X^β is in $\mathfrak{a}_j$. Hence $X^\beta X_n^j$ is in $\mathcal{C}(\mathfrak{a})$ if and only if X^β is in $\mathcal{C}(\mathfrak{a}_j)$. Since $\mathfrak{a}_j = \mathfrak{a}_l$ for $j \geq l$, $\mathcal{C}(\mathfrak{a}_j) = \mathcal{C}(\mathfrak{a}_l)$ for $j \geq l$. Thus

$$\mathcal{C}(\mathfrak{a}) = \Big(\bigcup_{j=l}^{\infty} \mathcal{C}(\mathfrak{a}_l) X_n^j \Big) \cup \Big(\bigcup_{j=0}^{l-1} \mathcal{C}(\mathfrak{a}_j) X_n^j \Big). \tag{**}$$

If $j \leq l$, then $X^\beta X_n^l \in \mathcal{C}(\mathfrak{a})$ implies $X^\beta X_n^j \in \mathcal{C}(\mathfrak{a})$, since $X_n^{l-j} \mathfrak{a} \subseteq \mathfrak{a}$. Therefore $\mathcal{C}(\mathfrak{a}_j) \supseteq \mathcal{C}(\mathfrak{a}_l)$ for all $j \leq l$, and we see that $j \leq l$ implies that $\mathcal{C}(\mathfrak{a}_j) = \mathcal{C}(\mathfrak{a}_j) \cup \mathcal{C}(\mathfrak{a}_l)$. Substituting into (∗∗) and rearranging terms gives

$$\mathcal{C}(\mathfrak{a}) = \Big(\bigcup_{j=0}^{\infty} \mathcal{C}(\mathfrak{a}_l) X_n^j \Big) \cup \Big(\bigcup_{j=0}^{l-1} \mathcal{C}(\mathfrak{a}_j) X_n^j \Big). \tag{†}$$

For $j \leq l$, X^β is in $\mathcal{C}(\mathfrak{a}_j)$ if and only if X^β is in one of $C_{0,j}, \ldots, C_{n-1,j}$. Thus we can rewrite (†) as

$$\mathcal{C}(\mathfrak{a}) = \Big(\bigcup_{j=l}^{\infty} \bigcup_{k=0}^{n-1} C_{k,l} X_n^j \Big) \cup \Big(\bigcup_{j=0}^{l-1} \bigcup_{k=0}^{n-1} C_{k,j} X_n^j \Big)$$

$$= \Big(\bigcup_{j=l}^{\infty} \bigcup_{k=0}^{n-1} C_{k,l} X_n^j \Big) \cup \Big(\bigcup_{j=0}^{l-1} \bigcup_{k=0}^{n-2} C_{k+1,j} X_n^j \Big) \cup \Big(\bigcup_{j=0}^{l-1} C_{0,j} X_n^j \Big).$$

The first term on the right side contributes to C_{k+1}, with e_n to be adjoined to the basis vectors of the associated vector subspace $\langle e_{j_1}, \ldots, e_{j_k} \rangle$. Equating the terms on the two sides that contribute to C_{k+1} therefore yields (∗). The set C_0 is the last term on the right side. This is finite because each $C_{0,j}$ is finite, and therefore C_0 has the correct form. □

Lemma 10.66. Let E be a standard subset of $\mathcal{M}$ with k parameters, and let γ be its associated translation. Then the number of monomials X^α with $|\gamma| \leq s$ such that α is in E is equal to the binomial coefficient

$$\binom{k + s - |\gamma|}{s - |\gamma|}$$

if $s > |\gamma|$. This expression is a polynomial function of s of degree k, and the coefficient of s^k is $1/k!$.

PROOF. Let $\langle e_{j_1}, \ldots, e_{j_k}\rangle$ be the associated vector subspace for E. The associated translation γ is assumed to have $\gamma_i = 0$ for i in $\{j_1, \ldots, j_k\}$. We are to count monomials $X^\alpha = X^\gamma X^\beta$ with β in $\langle e_{j_1}, \ldots, e_{j_k}\rangle$ and with $|\gamma + \beta| \leq s$. Since $|\gamma| + |\beta| = |\gamma + \beta| \leq s$, the latter condition on β is that $|\beta| \leq s - |\gamma|$, which by assumption is ≥ 0. The entries of β are allowed to be arbitrary nonzero integers in the k entries $j_1, \ldots, j_k$, subject only to the limitation that the sum of the entries is to be $\leq s - |\gamma|$. The number of such β's equals the number of homogeneous monomials in $k + 1$ variables of total degree equal to $s - |\gamma|$. This number is recalled in a bulleted list in Section 3 and is $\binom{s-|\gamma|+k}{k} = \binom{s-|\gamma|+k}{s-|\gamma|}$. When expanded out, this binomial coefficient equals

$$\tfrac{1}{k!}(s+k-|\gamma|)(s+k-1-|\gamma|)\cdots(s+1-|\gamma|),$$

which is a polynomial function of s of degree k with leading coefficient $1/k!$. $\square$

Lemma 10.67. Let E and F be standard subsets of $\mathcal{M}$ with k and l parameters, respectively. Then $E \cap F$ either is empty or is a standard subset of $\mathcal{M}$ with m parameters, where $m \leq \min(k, l)$. Moreover, the only way that m can equal $\max(k, l)$ is for E to equal F.

PROOF. Denote the respective associated translations for E and F by γ_E and γ_F, and let S_E and S_F be the subsets of $\{1, \ldots, n\}$ such that $\langle e_i \mid i \in S_E\rangle$ and $\langle e_i \mid i \in S_F\rangle$ are the associated vector spaces for E and F, respectively. Let T_E be the subset of indices

$$T_E = \{i \in \{1, \ldots, n\} \mid (\gamma_E)_i > 0\},$$

and define T_F similarly. We are given that $|S_E| = k$ and $|S_F| = l$. Also, we are given that $S_E \cap T_E = \varnothing$ and $S_F \cap T_F = \varnothing$, i.e., that $T_E \subseteq S_E^c$ and $T_F \subseteq S_F^c$. If $E \cap F \neq \varnothing$, then there exist x and y with

$$\gamma_E + x = \gamma_F + y \quad \text{such that } x_i = 0 \text{ for } i \notin S_E \text{ and } y_j = 0 \text{ for } j \notin S_F. \quad (*)$$

Then $x_i = y_i = 0$ for $i \in S_E^c \cap S_F^c$, and we see that a necessary condition to have $E \cap F \neq \varnothing$ is that $(\gamma_E)_i = (\gamma_F)_i$ for $i \in S_E^c \cap S_F^c$. In this case the x and y in $(*)$ must have $x_i = (\gamma_F)_i$ for $i \in S_E \cap S_F^c$ and $y_i = (\gamma_E)_i$ for $i \in S_E^c \cap S_F$.

Conversely if $(\gamma_E)_i = (\gamma_F)_i$ for $i \in S_E^c \cap S_F^c$, then we can define $x_i = (\gamma_F)_i$ for $i \in S_E \cap S_F^c$, $y_i = (\gamma_E)_i$ for $i \in S_E^c \cap S_F$, and $x_i = y_i$ to be arbitrary for $i \in S_E \cap S_F$, and we obtain solutions of $(*)$. It is evident that all solutions of $(*)$ are obtained this way. Consequently $E \cap F$ is the standard subset of $\mathcal{M}$ with $|S_E \cap S_F|$ parameters; with associated translation γ having γ_i equal to γ_E on S_E^c, equal to γ_F on S_F^c, and equal to 0 on $S_E \cap S_F$; and with associated vector space $\langle e_i \mid i \in S\rangle$, where $S = S_E \cap S_F$.

The inequality $\dim_\Bbbk(S_E \cap S_F) \leq \min(\dim_\Bbbk S_E, \dim_\Bbbk S_F)$ is the inequality $m \leq \min(k, l)$ of the lemma. If $m = \max(k, l)$, then we must have $S = S_E = S_F$ and an equality $(\gamma_E)_i = (\gamma_F)_i$ for $i \in S_E^c \cap S_F^c$, i.e., for $i \notin S$. The latter equality implies that $\gamma_E = \gamma_F$. Hence $E = F$. $\square$

Theorem 10.68. If $\mathfrak{a}$ is a monomial ideal in $\Bbbk[X_1, \ldots, X_n]$ such that $V(\mathfrak{a})$ has geometric dimension d, then there exists a polynomial $H_a(s, \mathfrak{a})$ in one variable of degree d such that the affine Hilbert function $\mathcal{H}_a(s, \mathfrak{a})$ is equal to $H_a(s, \mathfrak{a})$ for all positive s sufficiently large. The leading coefficient of $H_a(s, \mathfrak{a})$ is positive.

REMARK. The polynomial $H_a(s, \mathfrak{a})$ is called the **affine Hilbert polynomial** of the monomial ideal $\mathfrak{a}$. It is of course uniquely determined.

PROOF. For s sufficiently large, we are to count the number of monomials X^α with $|\alpha| \leq s$ lying in the complementary set $\mathcal{C}(\mathfrak{a})$ to $\mathfrak{a}$. Proposition 10.64b and Theorem 10.65 together show that $\mathcal{C}(\mathfrak{a}) = C_0 \cup \cdots \cup C_d$ disjointly, with C_k equal to a finite union of standard subsets of $\mathcal{M}$ with k parameters and with C_d nonempty. The sets C_k being disjoint, it is enough to show that the number of such monomials in C_k is a function equal for large s to a polynomial of degree k, provided C_k is nonempty.

According to Lemma 10.66, if E is a standard subset of $\mathcal{M}$ with k parameters, if $s > 0$ is sufficiently large, and if γ is the translation parameter, then the number of monomials X^α in E with $|\alpha| \leq s$ is $\binom{k+s-|\gamma|}{s-|\gamma|}$ if $s > |\gamma|$, which is a polynomial of degree k with positive leading coefficient.

Because the sets E of this kind whose finite union is C_k may not be disjoint and because we seek an exact answer for the cardinality $|C_k|$ when s is large, we cannot simply add finitely many such expressions to obtain a value for $|C_k|$. We have to take into account the overlaps of the various sets E. Thus suppose that $C_k = E_1 \cup \cdots \cup E_r$ for standard subsets $E_1, \ldots, E_r$ of $\mathcal{M}$ with k parameters. Without loss of generality, we may assume that no two of the sets $E_1, \ldots, E_r$ are equal to one another. Let $E_1(s), \ldots, E_r(s)$ be the respective subsets of elements α with $|\alpha| \leq s$. We use the inclusion–exclusion formula, namely

$$\left| \bigcup_{i=1}^r E_i(s) \right| = \sum_i |E_i(s)| - \sum_{i_1 < i_2} |E_{i_1}(s) \cap E_{i_2}(s)| + \sum_{l=3}^r (-1)^{l+1} \sum_{i_1 < \cdots < i_l} \left| \bigcap_{j=1}^l E_{i_j}(s) \right|;$$

this is a formula in Boolean algebra that is readily proved by induction on r starting from the formula $|E \cup F| = |E| + |F| - |E \cap F|$.

Lemma 10.66 shows that $\sum_i |E_i(s)|$ is a sum of functions equal for large $s > 0$ to polynomials of degree d with positive leading coefficient. The leading coefficients cannot cancel, and thus the sum is for large $s > 0$ equal to a polynomial of degree d with positive leading coefficient. Each of the remaining terms on the right side of the inclusion–exclusion formula, according to Lemma 10.67, is plus or minus the number of monomials α with $|\alpha| \leq s$ in some standard subset E of $\mathcal{M}$ whose number of parameters is $< d$. Hence the sum of all those terms is a function equal for large s to a polynomial that is 0 or has degree $< d$. The theorem follows. $\square$

Proposition 10.69. A polynomial $P(s)$ in one variable of degree d takes integer values for s sufficiently large and positive if and only if it is an integer linear combination of the polynomials $s \mapsto \binom{s}{j}$ for $0 \leq j \leq d$.

PROOF. The sufficiency is immediate because $\binom{s}{j}$ is an integer for each j and s. For necessity, suppose that $P(s)$ is integer-valued and has degree d. Since $s \mapsto \binom{s}{j}$ is integer-valued of degree j with leading coefficient $1/j!$, $P(s)$ is certainly a rational linear combination of the polynomials $s \mapsto \binom{s}{j}$. We prove by induction on d that the coefficients are integers. For deg $P(s) = 0$, we have $\binom{s}{0} = 1$, and there is nothing to prove. Given an integer-valued $P(s)$ of degree d, write $P(s) = \sum_{j=0}^{d} a_j \binom{s}{j}$. Form

$$\Delta P(s) = P(s+1) - P(s) = \sum_{j=0}^{d} a_j \left[\binom{s+1}{j} - \binom{s}{j}\right] = \sum_{j=1}^{d} a_j \binom{s}{j-1} = \sum_{j=0}^{d-1} a_{j+1} \binom{s}{j},$$

the third equality holding by Pascal's triangle. Since $\Delta P(s)$ is integer-valued and has degree $d - 1$, the inductive hypothesis shows that a_{j+1} is an integer for $0 \leq j \leq d-1$; i.e., a_j is an integer for $1 \leq j \leq d$. Therefore $Q(s) = \sum_{j=1}^{d} a_j \binom{s}{j}$ is integer-valued. Since $P(s) - Q(s) = a_0$ is integer-valued and constant, a_0 is an integer. □

Corollary 10.70. If $\mathfrak{a}$ is a monomial ideal in $\Bbbk[X_1, \ldots, X_n]$ such that $V(\mathfrak{a})$ has geometric dimension d, then the affine Hilbert polynomial $H_a(s, \mathfrak{a})$ of $\mathfrak{a}$ is of the form $H_a(s, \mathfrak{a}) = \sum_{j=0}^{d} a_j \binom{s}{d-j}$ with integer coefficients a_j and with $a_0 > 0$.

PROOF. This follows by combining Theorem 10.68 and Proposition 10.69. □

10. Hilbert Polynomial in the Affine Case

We continue with an algebraically closed field $\Bbbk$ and with the polynomial ring $A = \Bbbk[X_1, \ldots, X_n]$. Let $\mathfrak{a}$ be an ideal in A. For each integer $s \geq 0$, let $A_{\leq s}$ be the vector subspace of A consisting of 0 and all elements of degree at most s, and put $\mathfrak{a}_{\leq s} = \mathfrak{a} \cap A_{\leq s}$. The inclusion of $A_{\leq s}$ into A descends to a $\Bbbk$ linear mapping $A_{\leq s}/\mathfrak{a}_{\leq s} \to A/\mathfrak{a}$, and this is one-one because $A_{\leq s} \cap \mathfrak{a} \subseteq \mathfrak{a}_{\leq s}$. Thus we can regard $A_{\leq s}/\mathfrak{a}_{\leq s}$, as s varies, as a sequence of successively better approximations to $A/\mathfrak{a}$. We define the **affine Hilbert function** $\mathcal{H}_a(s, \mathfrak{a})$ of $\mathfrak{a}$ by

$$\mathcal{H}_a(s, \mathfrak{a}) = \dim_{\Bbbk} A_{\leq s}/\mathfrak{a}_{\leq s} \qquad \text{for } s \geq 0.$$

When $\mathfrak{a}$ is a monomial ideal, this function is the one that was investigated in the previous section. In fact, the monomials of degree $\leq s$ form a vector-space basis of $A_{\leq s}$, and the monomials in $\mathfrak{a}$ of degree $\leq s$ form a basis of $\mathfrak{a}_{\leq s}$ because $\mathfrak{a}$ is spanned by monomials. If $\mathcal{C}(\mathfrak{a})$ denotes the set of monomials not in $\mathfrak{a}$, then the monomials of degree $\leq s$ within $\mathcal{C}(\mathfrak{a})$ descend to a basis of $A_{\leq s}/\mathfrak{a}_{\leq s}$. The number of such monomials gives the value of the affine Hilbert function as defined in the previous section, and thus the new definition is consistent with the old one in the case of monomial ideals.

When $\mathfrak{a}$ is a proper monomial ideal, we found in Theorem 10.68 that $\mathcal{H}_a(s, \mathfrak{a})$ equals a polynomial function of s for s sufficiently large and that the degree of this polynomial function equals the geometric dimension of the zero locus $V(\mathfrak{a})$ in the affine space $\mathbb{A}^n$. Our goal in this section is to show that these conclusions remain valid for all proper ideals $\mathfrak{a}$. The polynomial function that results for such an $\mathfrak{a}$ will be called the affine Hilbert polynomial of $\mathfrak{a}$.

We shall make the connection between general ideals $\mathfrak{a}$ and monomial ideals by means of the theory of Sections VIII.7–VIII.10. We recall the notion of a monomial ordering as defined in Section VIII.7. A monomial ordering $\leq$ is said to be a **graded monomial ordering** if $|\beta| < |\alpha|$ implies $X^\beta \leq X^\alpha$. The graded lexicographic ordering and the graded reverse lexicographic ordering (Examples 2 and 3 in Section VIII.7) are examples of graded monomial orderings, but the lexicographic ordering in Example 1 in that section is *not* a graded monomial ordering.

Fix a graded monomial ordering. As in Section VIII.7, $\text{LT}(f)$ denotes the leading monomial term of the polynomial f. By convention, $\text{LT}(0) = 0$. For our ideal $\mathfrak{a}$, we let $\text{LT}(\mathfrak{a})$ be the vector space of all linear combinations of polynomials $\text{LT}(f)$ for $f \in \alpha$. This is an ideal in A, and it is a monomial ideal. The connection between the goal of this section and the results of the previous section rests on the following remarkable theorem.

Theorem 10.71 (Macaulay). Let a graded monomial ordering be imposed on $\Bbbk[X_1, \ldots, X_n]$. If $\mathfrak{a}$ is any ideal in $\Bbbk[X_1, \ldots, X_n]$, then the affine Hilbert functions of $\mathfrak{a}$ and $\text{LT}(\mathfrak{a})$ coincide: $\mathcal{H}_a(s, \mathfrak{a}) = \mathcal{H}_a(s, \text{LT}(\mathfrak{a}))$.

PROOF. Fix $s \geq 0$. It is enough to prove that $\mathfrak{a}_{\leq s}$ and $\text{LT}(\mathfrak{a})_{\leq s}$ have the same $\Bbbk$ dimension. Since there are only finitely many monomials of degree $\leq s$, we can choose $f_1, \ldots, f_m$ in $\mathfrak{a}$ such that their leading monomials $\text{LM}(f_1), \ldots, \text{LM}(f_k)$ are distinct and form a vector-space basis of $\text{LT}(\mathfrak{a})_{\leq s}$. Without loss of generality, we may assume that $\text{LM}(f_1) > \cdots > \text{LM}(f_k)$. Certainly $\dim \text{LT}(\mathfrak{a})_{\leq s} = k$, and thus it is enough to show that $f_1, \ldots, f_k$ lie in $\mathfrak{a}_{\leq s}$ and form a vector-space basis of $\mathfrak{a}_{\leq s}$.

For each j, $\text{LM}(f_j - \text{LT}(f_j)) < \text{LM}(f_j)$. Since the monomial ordering is graded, this inequality implies that $\deg(f_j - \text{LT}(f_j)) \leq s$. But we know that $\deg(\text{LT}(f_j)) \leq s$, and therefore $\deg f_j \leq s$. Consequently f_j lies in $\mathfrak{a}_{\leq s}$.

To prove that $\{f_1, \ldots, f_k\}$ is linearly independent, suppose that $\sum_{j=1}^{k} c_j f_j = 0$ with all c_j in $\Bbbk$. Arguing by contradiction, suppose that not all c_j are 0. Let i be the least index j for which $c_j \neq 0$; then $\text{LM}(f_i) = \text{LM}(c_i f_i) = \text{LM}\left(-\sum_{j>i} c_j f_j\right) \leq \max_{j>i} \text{LM}(f_j)$, and we arrive at a contradiction. We conclude that $\{f_1, \ldots, f_k\}$ is linearly independent.

To prove that $\{f_1, \ldots, f_k\}$ spans $\mathfrak{a}_{\leq s}$, we again argue by contradiction. Among all g in $\mathfrak{a}_{\leq s}$ with g not in the linear span of $\{f_1, \ldots, f_k\}$, choose one for which $\text{LM}(g)$ is the smallest. Certainly $\text{LM}(g)$ is one of $\text{LM}(f_1), \ldots, \text{LM}(f_k)$. Say that $\text{LM}(g) = \text{LM}(f_i)$. For some scalar $c \neq 0$, we must have $\text{LT}(g) = \text{LT}(cf_i)$. Then $\text{LM}(g - cf_i) < \text{LM}(g)$, and the minimality of $\text{LM}(g)$ forces $g - cf_i$ to be in the linear span of $\{f_1, \ldots, f_k\}$. Since cf_i is in the linear span, so is g, contradiction. Thus $\{f_1, \ldots, f_k\}$ is a spanning set of $\mathfrak{a}_{\leq s}$. □

Corollary 10.72. If $\mathfrak{a}$ is an ideal in $\Bbbk[X_1, \ldots, X_n]$, then for all s sufficiently large, the affine Hilbert function $\mathcal{H}_a(s, \mathfrak{a})$ of $\mathfrak{a}$ equals a polynomial in s of the form $\sum_{j=0}^{d} a_j \binom{s}{d-j}$ with integer coefficients a_j and with $a_0 > 0$.

REMARKS. The polynomial in the statement of the corollary is called the **affine Hilbert polynomial** of $\mathfrak{a}$ and is denoted by $H_a(s, \mathfrak{a})$. It is the 0 polynomial if and only if $\mathfrak{a} = \Bbbk[X_1, \ldots, X_n]$.

PROOF. Theorem 10.71 says that $\mathcal{H}_a(s, \mathfrak{a}) = \mathcal{H}_a(s, \text{LT}(\mathfrak{a}))$. Consequently the result follows immediately by applying Corollary 10.70 to $\text{LT}(\mathfrak{a})$. □

Corollary 10.73. If a graded monomial ordering is imposed on $\Bbbk[X_1, \ldots, X_n]$ and if $\mathfrak{a}$ is any ideal in $\Bbbk[X_1, \ldots, X_n]$, then the affine Hilbert polynomials of $\mathfrak{a}$ and $\text{LT}(\mathfrak{a})$ coincide: $H_a(s, \mathfrak{a}) = H_a(s, \text{LT}(\mathfrak{a}))$.

PROOF. This is immediate from Theorem 10.71 and the definition of the affine Hilbert polynomial given in the remarks with Corollary 10.72. □

Corollary 10.74. If $\mathfrak{a}$ and $\mathfrak{b}$ are proper ideals of $\Bbbk[X_1, \ldots, X_n]$ such that $\mathfrak{a} \subseteq \mathfrak{b}$, then $\deg H_a(s, \mathfrak{a}) \geq \deg H_a(s, \mathfrak{b})$.

PROOF. Introduce a graded monomial ordering. The inclusion $\mathfrak{a} \subseteq \mathfrak{b}$ implies that $\text{LT}(\mathfrak{a}) \subseteq \text{LT}(\mathfrak{b})$. Therefore $\mathcal{C}(\text{LT}(\mathfrak{a})) \supseteq \mathcal{C}(\text{LT}(\mathfrak{b}))$. Proposition 10.64b shows that the geometric dimension of $V(\text{LT}(\mathfrak{a}))$ is the largest vector-space dimension of a coordinate subspace that lies in $\mathcal{C}(\text{LT}(\mathfrak{a}))$, and the same thing is true for $\text{LT}(\mathfrak{b})$. Thus the geometric dimension of $V(\text{LT}(\mathfrak{a}))$ is $\geq$ the geometric dimension of $V(\text{LT}(\mathfrak{b}))$. By Theorem 10.68, $\deg H_a(s, \text{LT}(\mathfrak{a})) \geq \deg H_a(s, \text{LT}(\mathfrak{b}))$. The result now follows immediately from Corollary 10.73. □

The affine Hilbert polynomial $H_a(s, \mathfrak{a})$ of $\mathfrak{a}$ depends on $\mathfrak{a}$, not just $V(\mathfrak{a})$, but we shall be interested mainly in the degree of $H_a(s, \mathfrak{a})$. Proposition 10.76, as amplified in Corollary 10.77, implies that the degree depends only on $V(\mathfrak{a})$. It requires a lemma.

Lemma 10.75. If $\mathfrak{a}$ is a monomial ideal in $\Bbbk[X_1, \ldots, X_n]$, then so is $\sqrt{\mathfrak{a}}$.

PROOF. The preliminary remarks in Section 9 show that $V(\mathfrak{a})$ is a finite union of coordinate subspaces. Let us write $V(\mathfrak{a}) = \bigcup_j E_j$ accordingly. By Proposition 10.2b, $\sqrt{\mathfrak{a}} = I(V(\mathfrak{a})) = I(\bigcup_j E_j) = \bigcap_j I(E_j)$. Since E_j is an affine variety and is equal to $V(X_{i_1}, \ldots, X_{i_k})$ for suitable $X_{i_1}, \ldots, X_{i_k}$, the Nullstellensatz shows that $I(E_j)$ is an ideal of the form $I(E) = (X_{i_1}, \ldots, X_{i_k})$. This is a monomial ideal, and it is therefore enough to show that the finite intersection of monomial ideals is a monomial ideal. By induction it is enough to show that $\mathfrak{b} \cap \mathfrak{c}$ is a monomial ideal if $\mathfrak{b}$ and $\mathfrak{c}$ are monomial ideals. If an element of $\mathfrak{b} \cap \mathfrak{c}$ is given, then that element is a linear combination of the monomials in $\mathfrak{b}$ and is also a linear combination of the monomials in $\mathfrak{c}$. Since $\mathcal{M}$ is linearly independent, the element is a linear combination of monomials lying in $\mathfrak{b} \cap \mathfrak{c}$. Therefore $\mathfrak{b} \cap \mathfrak{c}$ is a monomial ideal. $\square$

Proposition 10.76. If $\mathfrak{a}$ is a proper ideal in $\Bbbk[X_1, \ldots, X_n]$, then the degrees of the affine Hilbert polynomials $H_a(s, \mathfrak{a})$ and $H_a(s, \sqrt{\mathfrak{a}})$ are equal.

PROOF. Fix a graded monomial ordering. We begin by proving that

$$\mathrm{LT}(\mathfrak{a}) \subseteq \mathrm{LT}(\sqrt{\mathfrak{a}}) \subseteq \sqrt{\mathrm{LT}(\mathfrak{a})}. \qquad (*)$$

The left-hand inclusion is immediate because $\mathfrak{a} \subseteq \sqrt{\mathfrak{a}}$. For the right-hand inclusion, let $f \neq 0$ be in $\sqrt{\mathfrak{a}}$, and let $X^\alpha = \mathrm{LM}(f)$ be the leading monomial of f. Since f is in $\sqrt{\mathfrak{a}}$, f^r is in $\mathfrak{a}$ for some $r > 0$. Since the leading monomial of a product is the product of the leading monomials, $\mathrm{LM}(f^r) = X^{r\alpha}$. Thus a power of X^α is exhibited as in $\mathrm{LT}(\mathfrak{a})$, and X^α is in $\sqrt{\mathrm{LT}(\mathfrak{a})}$. This proves $(*)$.

Applying Corollary 10.74 to $(*)$, we obtain

$$\deg H_a(s, \mathrm{LT}(\mathfrak{a})) \geq \deg H_a(s, \mathrm{LT}(\sqrt{\mathfrak{a}})) \geq \deg H_a(s, \sqrt{\mathrm{LT}(\mathfrak{a})}). \qquad (**)$$

The ideal $\mathrm{LT}(\mathfrak{a})$ is a monomial ideal, and Lemma 10.75 shows that $\sqrt{\mathrm{LT}(\mathfrak{a})}$ is a monomial ideal. Then $\mathrm{LT}(\mathfrak{a})$ and $\sqrt{\mathrm{LT}(\mathfrak{a})}$ are monomial ideals with $V(\mathrm{LT}(\mathfrak{a})) = V(\sqrt{\mathrm{LT}(\mathfrak{a})})$, and Theorem 10.68 shows that

$$\deg H_a(s, \mathrm{LT}(\mathfrak{a})) = \deg H_a(s, \sqrt{\mathrm{LT}(\mathfrak{a})}).$$

Comparing this conclusion with $(**)$, we see that

$$\deg H_a(s, \mathrm{LT}(\mathfrak{a})) = \deg H_a(s, \mathrm{LT}(\sqrt{\mathfrak{a}})). \qquad (\dagger)$$

In combination with the equalities $H_a(s, \mathfrak{a}) = H_a(s, \mathrm{LT}(\mathfrak{a}))$ and $H_a(s, \sqrt{\mathfrak{a}}) = H_a(s, \mathrm{LT}(\sqrt{\mathfrak{a}}))$ given by Corollary 10.73, $(\dagger)$ completes the proof. $\square$

Corollary 10.77. If $\mathfrak{a}$ and $\mathfrak{b}$ are proper ideals in $\Bbbk[X_1, \ldots, X_n]$ with $V(\mathfrak{a}) \subseteq V(\mathfrak{b})$, then $\deg H_a(s, \mathfrak{a}) \leq \deg H_a(s, \mathfrak{b})$.

PROOF. Application of $I(\cdot)$ to the inclusion $V(\mathfrak{a}) \subseteq V(\mathfrak{b})$ gives $\sqrt{\mathfrak{a}} = I(V(\mathfrak{a})) \supseteq I(V(\mathfrak{b})) = \sqrt{\mathfrak{b}}$. Then Corollary 10.74 and Proposition 10.76 together yield $\deg H_a(s, \mathfrak{a}) = \deg H_a(s, \sqrt{\mathfrak{a}}) \leq \deg H_a(s, \sqrt{\mathfrak{b}}) = \deg H_a(s, \mathfrak{b})$. $\square$

Theorem 10.78. If $\mathfrak{a}$ is a prime ideal in $\Bbbk[X_1, \ldots, X_n]$, then the degree of the affine Hilbert polynomial $H_a(s, \mathfrak{a})$ equals the geometric dimension of the affine variety $V(\mathfrak{a})$.

PROOF. Define $d = \deg H_a(s, \mathfrak{a})$ and $V = V(\mathfrak{a})$, and let $A(V)$ be the affine coordinate ring $A(V) = \Bbbk[X_1, \ldots, X_n]/\mathfrak{a}$. Theorem 10.7 shows that $\dim V$ equals the Krull dimension of $A(V)$, and Theorem 7.22 shows that the latter equals the transcendence degree over $\Bbbk$ of the field of fractions $\Bbbk(V)$ of $A(V)$. Thus the theorem will follow if we show that $\Bbbk(V)$ has transcendence degree d over $\Bbbk$.

Let $\varphi : \Bbbk[X_1, \ldots, X_n] \to A(V)$ be the quotient homomorphism, and put $x_i = \varphi(X_i)$ for $1 \leq i \leq n$. Introduce a graded monomial ordering on $\mathcal{M}$. Corollary 10.73 shows that $H_a(s, \mathfrak{a}) = H_a(s, \mathrm{LT}(\mathfrak{a}))$, and Theorem 10.68 shows that $V(\mathrm{LT}(\mathfrak{a}))$ has geometric dimension d. We saw in Section 9 that the zero locus of a monomial ideal is the finite union of coordinate subspaces, and it follows that $V(\mathrm{LT}(\mathfrak{a})) \subseteq \mathbb{A}^n$ contains a coordinate subspace E of dimension d. Let E have as basis the standard vectors $e_{j_1}, \ldots, e_{j_d}$, so that

$$E = V(\{X_i \mid i \notin \{j_1, \ldots, j_d\}\}).$$

The set E is a variety, and thus $I(E) = (\{X_i \mid i \notin \{j_1, \ldots, j_d\}\})$. Also, $E \subseteq V(\mathrm{LT}(\mathfrak{a}))$, and hence $I(E) \supseteq I(V(\mathrm{LT}(\mathfrak{a}))) \supseteq \mathrm{LT}(\mathfrak{a})$. If X^α is a monomial in $\mathrm{LT}(\mathfrak{a})$, then it follows that X^α lies in the ideal generated by the X_i for $i \notin \{j_1, \ldots, j_d\}$. We can summarize this fact as follows: if we write $\Bbbk[X_{j_1}, \ldots, X_{j_d}]$ for the subring of $\Bbbk[X_1, \ldots, X_n]$ of polynomials involving only $X_{j_1}, \ldots, X_{j_d}$, then

$$\mathrm{LT}(\mathfrak{a}) \cap \Bbbk[X_{j_1}, \ldots, X_{j_d}] = 0. \tag{*}$$

If f is any nonzero member of $\Bbbk[X_{j_1}, \ldots, X_{j_d}]$, then its leading monomial $\mathrm{LM}(f)$ has to lie in $\Bbbk[X_{j_1}, \ldots, X_{j_d}]$, and thus $(*)$ implies that

$$\mathfrak{a} \cap \Bbbk[X_{j_1}, \ldots, X_{j_d}] = 0. \tag{**}$$

Using $(**)$ and notation introduced at the beginning of Section VII.4, we shall show that $x_{j_1}, \ldots, x_{j_d}$ are algebraically independent over $\Bbbk$, and then it follows that $d \leq \mathrm{tr.\,deg}\, A(V)$. Thus suppose that $g(Y_1, \ldots, Y_d)$ is a polynomial in $\Bbbk[Y_1, \ldots, Y_d]$ such that $g(x_{j_1}, \ldots, x_{j_d}) = 0$. We can identify $\Bbbk[Y_1, \ldots, Y_d]$

with $\Bbbk[X_{j_1}, \ldots, X_{j_d}] \subseteq \Bbbk[X_1, \ldots, X_n]$, and then the equality $g(x_{j_1}, \ldots, x_{j_d}) = 0$ means that $\varphi(g) = 0$, i.e., g is in $\mathfrak{a}$. Hence g is a member of $\mathfrak{a} \cap \Bbbk[X_{j_1}, \ldots, X_{j_d}]$, and $g = 0$ by (**). Therefore $x_{j_1}, \ldots, x_{j_d}$ are algebraically independent over $\Bbbk$.

For the reverse inequality, we are to prove that $d \geq \operatorname{tr.deg} A(V)$. Let $r = \operatorname{tr.deg} A(V)$. The elements $x_j = \varphi(X_j)$ generate $A(V)$ as a $\Bbbk$ algebra, and therefore they generate $\Bbbk(V)$ over $\Bbbk$ as a field. By Lemma 7.6b some subset $\{x_{j_1}, \ldots, x_{j_d}\}$ of $\{x_1, \ldots, x_n\}$ is algebraically independent. Consider the substitution homomorphism

$$\psi(h) = h(x_{j_1}, \ldots, x_{j_r})$$

of $\Bbbk[Y_1, \ldots, Y_r]$ into $A(V)$. This is one-one because the elements $x_{j_1}, \ldots, x_{j_d}$ by assumption are algebraically independent. Fix $s \geq 0$, and consider the restriction of ψ to $\Bbbk[Y_1, \ldots, Y_r]_{\leq s}$. If $h(Y_1, \ldots, Y_r)$ is a monomial Y^α in $\Bbbk[Y_1, \ldots, Y_r]_{\leq s}$ with $\alpha = (\alpha_1, \ldots, \alpha_r)$ and $|\alpha| \leq s$, then we see that

$$\psi(Y^\alpha) = \prod_{i=1}^r x_{j_i}^{\alpha_i} = \varphi\Big(\prod_{i=1}^r X_{j_i}^{\alpha_i}\Big).$$

In other words, $\psi(Y^\alpha)$ is the image under φ of a member of $\Bbbk[X_1, \ldots, X_n]$ of degree $\leq s$. Taking linear combinations of such monomials, we see that $\psi(h)$ is a one-one $\Bbbk$ linear mapping

$$\psi : \Bbbk[Y_1, \ldots, Y_r]_{\leq s} \to \Bbbk[X_1, \ldots, X_n]_{\leq s}/\mathfrak{a}_{\leq s} \subseteq A(V).$$

Therefore

$$H_\mathfrak{a}(s, \mathfrak{a}) = \dim_\Bbbk \big(\Bbbk[X_1, \ldots, X_n]_{\leq s}/\mathfrak{a}_{\leq s}\big) \geq \dim_\Bbbk \Bbbk[Y_1, \ldots, Y_r]_{\leq s} = \binom{r+s}{r}.$$

The binomial coefficient on the right side is a polynomial of degree r in s with positive leading coefficient. The left side is a polynomial in s of degree d. The inequality forces $d \geq r$, and the proof is complete. □

Proposition 10.79. If $\mathfrak{a}$ and $\mathfrak{b}$ are proper ideals in $\Bbbk[X_1, \ldots, X_n]$, then $\deg H_\mathfrak{a}(s, \mathfrak{a}\mathfrak{b}) = \max\big(\deg H_\mathfrak{a}(s, \mathfrak{a}), \deg H_\mathfrak{a}(s, \mathfrak{b})\big)$.

REMARKS. Proposition 10.1 points out that $V(\mathfrak{a}\mathfrak{b}) = V(\mathfrak{a}) \cup V(\mathfrak{b})$. Since the degree of the affine Hilbert polynomial of $\mathfrak{a}$ depends only on $V(\mathfrak{a})$, this proposition says that the degree associated with the union of two affine algebraic sets is the larger of the degrees associated with each of the sets.

PROOF. Impose a graded monomial ordering on $\mathcal{M}$. Let us check that

$$\big(\operatorname{LT}(\mathfrak{a})\big)\big(\operatorname{LT}(\mathfrak{b})\big) \subseteq \operatorname{LT}(\mathfrak{a}\mathfrak{b}) \subseteq \operatorname{LT}(\mathfrak{a} \cap \mathfrak{b}) \subseteq \sqrt{\big(\operatorname{LT}(\mathfrak{a})\big)\big(\operatorname{LT}(\mathfrak{b})\big)}. \quad (*)$$

In fact, let f be in $\mathfrak{a}$ and g be in $\mathfrak{b}$, and define $X^\alpha = \operatorname{LM}(f)$ and $X^\beta = \operatorname{LM}(g)$ to be the leading monomials of f and g. Then $X^{\alpha+\beta} = \operatorname{LM}(fg)$, and hence the product of any generator of $\operatorname{LT}(\mathfrak{a})$ and any generator of $\operatorname{LT}(\mathfrak{b})$ lies in $\operatorname{LT}(\mathfrak{a}\mathfrak{b})$.

This proves the first inclusion of $(*)$. The second inclusion is immediate because $\mathfrak{a}\mathfrak{b} \subseteq \mathfrak{a} \cap \mathfrak{b}$. If $X^\alpha = \mathrm{LM}(f)$ with $f \in \mathfrak{a} \cap \mathfrak{b}$, then $(X^\alpha)^2 = \mathrm{LM}(f)\,\mathrm{LM}(f)$ is in $\mathrm{LT}(\mathfrak{a})\,\mathrm{LT}(\mathfrak{b})$. Hence X^α is in $\sqrt{\bigl(\mathrm{LT}(\mathfrak{a})\bigr)\bigl(\mathrm{LT}(\mathfrak{b})\bigr)}$. Thus a generating set of $\mathrm{LT}(\mathfrak{a} \cap \mathfrak{b})$ lies in $\sqrt{\bigl(\mathrm{LT}(\mathfrak{a})\bigr)\bigl(\mathrm{LT}(\mathfrak{b})\bigr)}$, and the third inclusion of $(*)$ follows.

In $(*)$, the values of $V(\,\cdot\,)$ on the end two members are the same, according to Proposition 10.3c, and therefore

$$V\bigl(\mathrm{LT}(\mathfrak{a})\,\mathrm{LT}(\mathfrak{b})\bigr) = V(\mathrm{LT}(\mathfrak{a}\mathfrak{b})). \qquad (**)$$

The proposition now follows from the computation

$\max(\deg H_a(s, \mathfrak{a}),\, \deg H_a(s, \mathfrak{b}))$

$\qquad = \max(\deg H_a(s, \mathrm{LT}(\mathfrak{a})),\, \deg H_a(s, \mathrm{LT}(\mathfrak{b})))$ by Corollary 10.73

$\qquad = \max(\dim V(\mathrm{LT}(\mathfrak{a})),\, \dim V(\mathrm{LT}(\mathfrak{b})))$ by Theorem 10.68

$\qquad = \dim\bigl(V(\mathrm{LT}(\mathfrak{a})) \cup V(\mathrm{LT}(\mathfrak{b}))\bigr)$ by Theorem 10.7

$\qquad = \dim(V(\mathrm{LT}(\mathfrak{a})\,\mathrm{LT}(\mathfrak{b})))$ by Proposition 10.1c

$\qquad = \dim V(\mathrm{LT}(\mathfrak{a}\mathfrak{b}))$ by $(**)$

$\qquad = \deg H_a(s, \mathrm{LT}(\mathfrak{a}\mathfrak{b}))$ by Theorem 10.68

$\qquad = \deg H_a(s, \mathfrak{a}\mathfrak{b})$ by Corollary 10.73. $\square$

Corollary 10.80. If $\mathfrak{a}$ is any ideal in $\Bbbk[X_1, \ldots, X_n]$, then the geometric dimension of the affine algebraic set $V(\mathfrak{a})$ equals the degree of the affine Hilbert polynomial $H_a(s, \mathfrak{a})$.

PROOF. Write $V(\mathfrak{a}) = \bigcup_{j=1}^{k} V_j$ as a finite union of affine varieties V_j, and define $\mathfrak{p}_j = I(V_j)$. Since V_j is irreducible, $\mathfrak{p}_j$ is prime. Moreover, $V_j = V(I(V_j)) = V(\mathfrak{p}_j)$. Then Proposition 10.1c shows that $V(\mathfrak{p}_1\mathfrak{p}_2 \cdots \mathfrak{p}_k) = \bigcup_{j=1}^{k} V(\mathfrak{p}_j) = \bigcup_{j=1}^{k} V_j = V(\mathfrak{a})$. Proposition 10.79 and induction give

$$\deg H_a(s, \mathfrak{p}_1\mathfrak{p}_2 \cdots \mathfrak{p}_k) = \max_{1 \leq j \leq n} \deg H_a(s, \mathfrak{p}_j),$$

and Theorem 10.78 shows that the right side equals $\max_{1 \leq j \leq k} \dim V(\mathfrak{p}_j) = \max_{1 \leq j \leq k} \dim V_j$, which equals $\dim V(\mathfrak{a})$ by Theorem 10.7. $\square$

As a consequence of Corollary 10.80, we obtain an algorithm for computing the dimension of an affine algebraic set V when given an ideal $\mathfrak{a}$ whose locus of common zeros $V(\mathfrak{a})$ is V: We introduce any graded monomial ordering and compute $\mathrm{LT}(\mathfrak{a})$, using a Gröbner basis. Corollaries 10.73 and 10.80 together say that $\dim V(\mathfrak{a}) = \dim V(\mathrm{LT}(\mathfrak{a}))$. The remarks before Proposition 10.64 show how to compute $\dim V(\mathrm{LT}(\mathfrak{a}))$, and Proposition 10.64b gives an alternative method of computation.

11. Hilbert Polynomial in the Projective Case

In this section we consider the analog for projective space of the theory of Section 10. We continue with $\mathbb{k}$ as an algebraically closed field, and we let $\widetilde{A} = \mathbb{k}[X_0, \ldots, X_n]$. Our interest is in the zero locus $V(\mathfrak{a})$ in $\mathbb{P}^n$, as defined in Section 3, of a homogeneous ideal $\mathfrak{a}$ in $\widetilde{A}$. To relate matters to Section 10, we shall make use of the **cone** $C(V(\mathfrak{a}))$ over $V(\mathfrak{a})$, which was defined in Section 3 as

$$C(V(\mathfrak{a})) = (0, \ldots, 0) \cup \{(x_0, \ldots, x_n) \in \mathbb{A}^{n+1} \mid [x_0, \ldots, x_n] \in V(\mathfrak{a})\}.$$

The homogeneous ideal $\mathfrak{a}$ is in particular an ideal in $n+1$ variables, and its associated affine algebraic set is the subset $C(V(\mathfrak{a}))$ of $\mathbb{A}^{n+1}$. An affine Hilbert polynomial $H_a(s, \mathfrak{a})$ is therefore associated to $C(V(\mathfrak{a}))$, and its degree matches the geometric dimension of $C(V(\mathfrak{a}))$.

To get something directly related to the projective algebraic set $V(\mathfrak{a})$ in projective space $\mathbb{P}^n$, we make a new definition of Hilbert function. Let $\widetilde{A}_s = \mathbb{k}[X_0, \ldots, X_n]_s$ be the subspace $\widetilde{A}$ of all polynomials homogeneous of degree s. If $\mathfrak{a}$ is a homogeneous ideal in $\widetilde{A}$, let $\mathfrak{a}_s = \mathfrak{a} \cap \widetilde{A}_s$. The **Hilbert function**[24] of $\mathfrak{a}$ is the integer-valued function of $s \geq 0$ defined by

$$\mathcal{H}(s, \mathfrak{a}) = \dim_{\mathbb{k}} \widetilde{A}_s / \mathfrak{a}_s \qquad \text{for } s \geq 0.$$

We have $\widetilde{A}_{\leq s} = \widetilde{A}_s \oplus \widetilde{A}_{\leq s-1}$, and the fact that $\mathfrak{a}$ is homogeneous implies that $\mathfrak{a}_{\leq s} = \mathfrak{a}_s \oplus \mathfrak{a}_{\leq s-1}$. Consequently $\widetilde{A}_{\leq s}/\mathfrak{a}_{\leq s} \cong \widetilde{A}_s/\mathfrak{a}_s \oplus \widetilde{A}_{\leq s-1}/\mathfrak{a}_{\leq s-1}$. Therefore

$$\mathcal{H}(s, \mathfrak{a}) = \mathcal{H}_a(s, \mathfrak{a}) - \mathcal{H}_a(s-1, \mathfrak{a}).$$

This is the fundamental formula by which the algebraic part of the theory of the Hilbert function in the projective case can be reduced to the corresponding theory in the affine case.

We know that the affine Hilbert function is a polynomial for large s. Since

$$s^d - (s-1)^d = s^{d-1} - s^{d-2} + s^{d-3} - \cdots + (-1)^{d+1}$$

is a polynomial of one lower degree and with positive leading coefficient, it follows that the Hilbert function of $\mathfrak{a}$ is a polynomial for large s, that its degree is $\dim C(V(\mathfrak{a})) - 1$, and that its leading coefficient is positive. This polynomial is called the **Hilbert polynomial** of $\mathfrak{a}$ and is denoted by $H(s, \mathfrak{a})$. To connect the geometric part of the theory of the Hilbert function in the projective case to the corresponding theory in the affine case, we use the following proposition.

---

[24]It is traditional not to include the word "projective" or any subscript, even though the terminology is meant to refer to the projective case.

Proposition 10.81. If $\mathfrak{a}$ is a homogeneous ideal in $\Bbbk[X_0, \ldots, X_n]$ and if the corresponding projective algebraic set $V(\mathfrak{a})$ is nonempty, then

$$\dim C(V(\mathfrak{a})) = \dim V(\mathfrak{a}) + 1.$$

PROOF. The proof of Corollary 10.13 shows that $C(V(\mathfrak{a}))$ is irreducible in $\mathbb{A}^{n+1}$ if and only if $V(\mathfrak{a})$ is irreducible in $\mathbb{P}^n$. Since the dimension in both cases for a general $\mathfrak{a}$ is the maximum of the dimensions of irreducible closed subsets, it is enough to prove the dimensional equality in the irreducible case.

If we have a strictly increasing sequence of irreducible closed subsets $E_0 \subsetneqq E_1 \subsetneqq \cdots \subsetneqq E_d$ in $\mathbb{P}^n$, then each $C(E_j)$ is irreducible in $\mathbb{A}^{n+1}$, and the sequence $C(E_0) \subsetneqq C(E_1) \subsetneqq \cdots \subsetneqq C(E_d)$ in $\mathbb{A}^{n+1}$ consists of Zariski closed sets that are irreducible. Since the subset $\{0\}$ of $\mathbb{A}^{n+1}$ is irreducible and can be adjoined at the beginning of the latter sequence, we conclude that $\dim C(V(\mathfrak{a})) \geq \dim V(\mathfrak{a}) + 1$.

We need to prove the reverse inequality in the irreducible case. Since $V(\mathfrak{a})$ is assumed irreducible (and hence nonempty), we may assume that $\mathfrak{a}$ is prime and omits at least one of $X_0, \ldots, X_n$. To fix the notation, say that X_0 is not in $\mathfrak{a}$. Recall from Section 3 the substitution homomorphism $\beta_0^t : \Bbbk[X_0, \ldots, X_n] \to \Bbbk[X_1, \ldots, X_n]$ formed by setting $X_0 = 1$. Let $\mathfrak{b} = \beta_0^t(\mathfrak{a})$. This is a prime ideal in $\Bbbk[X_1, \ldots, X_n]$, according to Theorem 10.20. Let $A(C(V(\mathfrak{a}))) = \Bbbk[X_0, \ldots, X_n]/\mathfrak{a}$ and $A(V(\mathfrak{b})) = \Bbbk[X_1, \ldots, X_n]/\mathfrak{b}$. The homomorphism β_0^t descends to a homomorphism of $A(C(V(\mathfrak{a})))$ onto $A(V(\mathfrak{b}))$, which we denote by $\bar{\beta}_0^t$.

Let $x_0, \ldots, x_n$ be the images of $X_0, \ldots, X_n$ in $A(C(V(\mathfrak{a})))$. The element x_0 is transcendental over $\Bbbk$. In fact, the only alternative is that it is a scalar c, since $\Bbbk$ is algebraically closed; the equality $x_0 = c$ would imply that $X_0 - c$ is in $\mathfrak{a}$, and the fact that $\mathfrak{a}$ is homogeneous would imply that X_0 and c are separately in $\mathfrak{a}$, in contradiction to our choice of X_0. Consequently $\Bbbk(x_0)(x_1, \ldots, x_n)$ has transcendence degree $r = \dim C(V(\mathfrak{a})) - 1$ over $\Bbbk(x_0)$. Since $x_1, \ldots, x_n$ generate $\Bbbk(x_0)(x_1, \ldots, x_n)$ as a field over $\Bbbk(x_0)$, some subset $\{x_{j_1}, \ldots, x_{j_r}\}$ of $\{x_1, \ldots, x_n\}$ is a transcendence basis of $\Bbbk(x_0)(x_1, \ldots, x_n)$ as a field over $\Bbbk(x_0)$. Thus $\{x_0, x_{j_1}, \ldots, x_{j_r}\}$ is a transcendence basis of $\Bbbk(x_0, \ldots, x_n)$ over $\Bbbk$.

The elements $x_0, x_{j_1}, \ldots, x_{j_r}$ all lie in $A(C(V(\mathfrak{a})))$, and we consider their images $1, \bar{\beta}_0^t(x_{j_1}), \ldots, \bar{\beta}_0^t(x_{j_r})$ in $A(V(\mathfrak{b}))$. Suppose that $h(Y_1, \ldots, Y_r)$ is a polynomial in r variables exhibiting the last r of these images as algebraically dependent. That is, suppose that

$$h\bigl(\bar{\beta}_0^t(x_{j_1}), \ldots, \bar{\beta}_0^t(x_{j_r})\bigr) = 0. \tag{$*$}$$

Let h have degree d. We regard h as a member of $\Bbbk[X_1, \ldots, X_n]_{\leq d}$ that depends only on $X_{j_1}, \ldots, X_{j_n}$. With this notational change, $(*)$ reads

$$h(X_1, \ldots, X_n) \quad \text{is in } \mathfrak{b}. \tag{$**$}$$

We now refer to the details of the proof of Theorem 10.20 that are summarized before Proposition 10.33. The linear mapping φ_d with $\varphi_d(f)(X_0, \ldots, X_n) = X_0^d f(X_1/X_0, \ldots, X_n/X_0)$ is a two-sided inverse to $\beta_0^t : \mathbb{k}[X_0, \ldots, X_n]_d \to \mathbb{k}[X_1, \ldots, X_n]_{\leq d}$. Put $H = \varphi_d(h)$, so that $h = \beta_0^t(H)$. The detail in question is that

$$\mathfrak{a} \cap \mathbb{k}[X_0, \ldots, X_n]_d = \varphi_d\bigl(\mathfrak{b} \cap \mathbb{k}[X_1, \ldots, X_n]_{\leq d}\bigr). \tag{$\dagger$}$$

By $(**)$, $\varphi_d(h)$ is in the right side of $(\dagger)$. Since $(\dagger)$ is a valid identity, $\varphi_d(h)$ is in the left side. So H is in $\mathfrak{a}$. This means that $H(x_0, \ldots, x_n) = 0$. Remembering that H depends only on $X_0, X_{j_1}, \ldots, X_{j_r}$ and that $\{x_0, x_{j_1}, \ldots, x_{j_r}\}$ is a transcendence set, we see that $H = 0$. Therefore $h = 0$, and $\{\bar{\beta}_0^t(x_{j_1}), \ldots, \bar{\beta}_0^t(x_{j_r})\}$ is a transcendence set in $A(V(\mathfrak{b}))$. Thus

$$\dim V(\mathfrak{b}) = \mathrm{tr.\,deg}\, A(V(\mathfrak{b})) \geq r = \mathrm{tr.\,deg}\, A(C(V(\mathfrak{a})) - 1 = \dim C(V(\mathfrak{a})) - 1.$$

By Corollary 10.19, $\dim V(\mathfrak{b}) = \dim V(\mathfrak{a})$. Hence $\dim C(V(\mathfrak{a})) \leq \dim V(\mathfrak{a}) + 1$, and the proof is complete. $\square$

Corollary 10.82. If $\mathfrak{a}$ is a homogeneous ideal in $\mathbb{k}[X_0, \ldots, X_n]$ and if the corresponding projective algebraic set $V(\mathfrak{a})$ is nonempty, then $\dim V(\mathfrak{a})$ equals the degree of the Hilbert polynomial $H(s, \mathfrak{a})$.

PROOF. This is immediate from Proposition 10.81 because $\dim C(V(\mathfrak{a})) = \dim H_a(s, \mathfrak{a})$ and because $\deg H(s, \mathfrak{a}) = \deg H_a(s, \mathfrak{a}) - 1$. $\square$

We could also obtain a corollary relating $H(s, V(\mathfrak{a}))$ and $H(s, V(\mathrm{LT}(\mathfrak{a})))$ when a graded monomial ordering is imposed, and we could then give a geometric way of visualizing the dimension in terms of the projective case. But we shall not need these details, and we omit them.

12. Intersections in Projective Space

Hilbert polynomials are an appropriate tool for dealing with how a projective algebraic set intersects a lower-dimensional projective space. In this section we consider such intersections, and we obtain as a corollary the deep result that a system of homogeneous polynomial equations over an algebraically closed field $\mathbb{k}$ always has a nonzero solution if there are more variables than equations.

It will be convenient in this section to adopt the convention that the empty projective algebraic set has dimension -1 and that the 0 Hilbert polynomial has degree -1. To make use of this convention, we recall from the homogeneous Nullstellensatz (Proposition 10.12a) that a homogeneous ideal $\mathfrak{a}$ in $\mathbb{k}[X_0, \ldots, X_n]$ has $V(\mathfrak{a})$ empty in $\mathbb{P}^n$ if and only if there is an integer N such that $\mathfrak{a}$ contains $\mathbb{k}[X_0, \ldots, X_n]_k$ for $k \geq N$. In this case our definition makes $C(V(\mathfrak{a}))$ consist of $\{0\}$ alone.[25] With the convention that such ideals have $\dim V(\mathfrak{a}) = -1$ and $C(V(\mathfrak{a})) = \{0\}$, the formula of Proposition 10.81 remains valid, and we can therefore drop the assumption that $V(\mathfrak{a})$ is nonempty. As to Corollary 10.82, the definition of the Hilbert function when $\mathfrak{a}$ contains $\Bbbk[X_0, \ldots, X_n]_k$ for all sufficiently large k makes $\mathcal{H}(k, \mathfrak{a}) = 0$ for such k; therefore the Hilbert polynomial in this case is the 0 polynomial, and Corollary 10.82 continues to be valid even when $V(\mathfrak{a})$ is empty.

Theorem 10.83. If $\mathfrak{a}$ is any homogeneous ideal in $\Bbbk[X_0, \ldots, X_n]$ and if F is a homogeneous polynomial, then

$$\dim V(\mathfrak{a}) \geq \dim V(\mathfrak{a} + (F)) \geq \dim V(\mathfrak{a}) - 1.$$

In particular, $V(\mathfrak{a} + (F))$ is nonempty if $\dim V(\mathfrak{a}) \geq 1$.

PROOF. Since $\mathfrak{a} \subseteq \mathfrak{a} + (F)$ and since $V(\cdot)$ is inclusion reversing, we know that

$$\dim V(\mathfrak{a}) \geq \dim V(\mathfrak{a} + (F)).$$

To obtain the second inequality of the theorem, we shall compare the Hilbert polynomials $H(s, \mathfrak{a})$ and $H(s, \mathfrak{a} + (F))$, taking advantage of Corollary 10.82. Let $d = \deg F$, and suppose that $s > d$. The identity mapping on $\Bbbk[X_0, \ldots, X_n]_s$ descends to a $\Bbbk$ linear mapping

$$\varphi : \Bbbk[X_0, \ldots, X_n]_s / \mathfrak{a}_s \to \Bbbk[X_0, \ldots, X_n]_s \big/ (\mathfrak{a} + (F))_s,$$

and φ is onto, being formed from an onto map. To understand $\ker \varphi$, we shall use the $\Bbbk$ linear map

$$\psi : \Bbbk[X_0, \ldots, X_n]_{s-d} / \mathfrak{a}_{s-d} \to \Bbbk[X_0, \ldots, X_n]_s / \mathfrak{a}_s$$

induced by multiplication by F, which we view as carrying $\Bbbk[X_0, \ldots, X_n]_{s-d}$ into $\Bbbk[X_0, \ldots, X_n]_s / \mathfrak{a}_s$. Observe that if G is in $\Bbbk[X_0, \ldots, X_n]_{s-d}$, then FG is in $(\mathfrak{a} + (F))_s$, and therefore $\varphi \circ \psi = 0$, i.e., image $\psi \subseteq \ker \varphi$.

We shall prove that equality holds. Thus suppose that G is a member of $\Bbbk[X_0, \ldots, X_n]_s$ such that $G + \mathfrak{a}_s$ is in $\ker \varphi$, i.e., that G is in $(\mathfrak{a} + (F))_s$. Then we can write $G = G_1 + HF$ with G_1 in $\mathfrak{a}_s$ and H in $\Bbbk[X_0, \ldots, X_n]_{s-d}$. So $G - G_1 = HF$, and the coset $G + \mathfrak{a}_s = G - G_1 + \mathfrak{a}_s$ is ψ of $H + \mathfrak{a}_{s-d}$. We conclude that image $\psi = \ker \varphi$.

Now we compute

$$\dim_{\Bbbk} \Bbbk[X_0, \ldots, X_n]_s / \mathfrak{a}_s$$
$$= \dim_{\Bbbk}(\text{domain } \varphi) = \dim_{\Bbbk}(\ker \varphi) + \dim_{\Bbbk}(\text{image } \varphi)$$
$$= \dim_{\Bbbk}(\text{image } \psi) + \dim_{\Bbbk} \Bbbk[X_0, \ldots, X_n]_s \big/ (\mathfrak{a} + (F))_s$$
$$\leq \dim_{\Bbbk} \Bbbk[X_0, \ldots, X_n]_{s-d} / \mathfrak{a}_{s-d} + \dim_{\Bbbk} \Bbbk[X_0, \ldots, X_n]_s \big/ (\mathfrak{a} + (F))_s.$$

---

[25] Admittedly the inclusion of $\{0\}$ in the cone might seem unnatural if $\mathfrak{a} = \Bbbk[X_0, \ldots, X_n]$, but that is the definition that makes this particular $\mathfrak{a}$ behave like all other ideals.

In terms of Hilbert functions, this says that

$$\mathcal{H}(s, \mathfrak{a}) \leq \mathcal{H}(s - d, \mathfrak{a}) + \mathcal{H}(s, \mathfrak{a} + (F)).$$

For large s, this is an inequality of polynomials:

$$H(s, \mathfrak{a}) \leq H(s - d, \mathfrak{a}) + H(s, \mathfrak{a} + (F)).$$

Since $H(s, \mathfrak{a}) - H(s - d, \mathfrak{a})$ is a polynomial of one lower degree than $H(s, \mathfrak{a})$ with leading coefficient positive, we obtain

$$\deg H(s, \mathfrak{a}) - 1 \leq \deg H(s, \mathfrak{a} + (F)).$$

The second inequality of the theorem now follows from Corollary 10.82. The final assertion in the theorem takes into account the remarks in the paragraph preceding the statement of the theorem. □

Corollary 10.84. If $\mathfrak{a}$ is any homogeneous ideal in $\mathbb{k}[X_0, \ldots, X_n]$ and if $F_1, \ldots, F_r$ are homogeneous polynomials, then

$$\dim V(\mathfrak{a}) \geq \dim V(\mathfrak{a} + (F_1, \ldots, F_r)) \geq \dim V(\mathfrak{a}) - r.$$

In particular, $V(\mathfrak{a} + (F_1, \ldots, F_r))$ is nonempty if $\dim V(\mathfrak{a}) \geq r$.

PROOF. We use Theorem 10.83 inductively, first applying it to the ideal $\mathfrak{a}$ with $F = F_1$, then applying it to the ideal $\mathfrak{a} + (F_1)$ with $F = F_2$, and so on. This proves the first conclusion, and the second conclusion follows because of the convention that the empty set has dimension -1. □

Corollary 10.85. Over an algebraically closed field any system of homogeneous polynomial equations with more variables than equations has a nonzero solution.

PROOF. Let there be r equations and $n + 1$ variables with $n + 1 > r$, the equations being $F_1 = 0, \ldots, F_r = 0$. The zero locus for each equation is a subset of $\mathbb{P}^n$. Applying Corollary 10.84 with $\mathfrak{a} = 0$ shows that $\dim V(F_1, \ldots, F_r) \geq n - r \geq 0$ and that $V(F_1, \ldots, F_r)$ is not empty as long as $n \geq r$. □

Corollary 10.85 is the result in the present chapter that was anticipated in Problem 23 at the end of Chapter VIII.

13. Schemes

We conclude with some commentary about "schemes." The subject of algebraic geometry studied along the lines of Sections 1–12 suffers from at least two shortcomings. One concerns the coefficients that are involved. The original impetus for the subject came from systems of polynomial equations in several variables. These equations involve addition, subtraction, and multiplication, and the requirement that division be allowable is unnatural and cuts down the scope of the subject. It immediately cuts out Diophantine equations, for example, to say nothing of congruences modulo prime powers. It would be more natural to allow the coefficients to lie in any commutative ring with identity. The other shortcoming is that the definition of variety depends on an embedding whose chief role is to get past the stage of making definitions; soon the embedding is stripped away, and the interest is in varieties up to isomorphism. The situation is similar to the historical treatment of groups and of manifolds. Groups were for the most part originally conceived in terms of group actions, but eventually the groups were separated from the actions. Manifolds at first were defined as certain subsets of Euclidean space, but eventually they were given an intrinsic definition. It would be more in keeping with the wisdom gained from other areas of mathematics if varieties could be defined intrinsically right away.

Schemes, introduced and developed by A. Grothendieck in the late 1950s and early 1960s, accomplish both these objectives. The theory of schemes borrows ideas and techniques from many areas of mathematics, as will be apparent shortly. This section will briefly present some of the definitions, offer some examples, and show the sense in which varieties may be regarded as schemes.[26] The interested reader may want to read more, and this section will therefore conclude with some bibliographical remarks.

1. Spectrum. One preliminary remark is necessary. To isolate an affine variety from its ambient space $\mathbb{A}^n$, we can take advantage of Proposition 10.23, which says that the points of the variety correspond exactly to the maximal ideals of the affine coordinate ring.[27] The set of maximal ideals in a ring, however, is usually not an object that lends itself to use with mappings. For example the canonical inclusion of $\mathbb{Z}$ into $\mathbb{Q}$ is not reflected in any of the mappings of the singleton set $\{(0)\}$ of maximal ideals of $\mathbb{Q}$ into the set of maximal ideals of $\mathbb{Z}$. Instead, the theory of schemes works with *prime* ideals. These behave nicely in that the inverse image of a prime ideal under a homomorphism of rings with identity is a prime ideal.

---

[26]The material in this section is based in part on lectures by V. Schechtman given in 1991–92 and in part on the books by Gunning, Hartshorne, and Shafarevich in the Selected References.

[27]Readers familiar with some functional analysis will recognize that a similar thing happens with compact Hausdorff spaces; by a theorem of M. Stone, the points of the space correspond exactly to the maximal ideals of the algebra of continuous complex-valued functions on the space.

Thus we work with the category of commutative rings with identity, the motivating example being the affine coordinate ring of an affine variety over an algebraically closed field. If A is a ring in this category, the **spectrum** of A is the set Spec A of prime ideals of A. For example the spectrum of a field consists of the one element (0), that of a discrete valuation ring consists of 0 and the unique maximal ideal, that of a principal ideal domain consists of 0 and the principal ideals (f) such that f is an irreducible element, and that of $\mathbb{C}[X, Y]$ consists of the ideal (0), the maximal ideals corresponding to one-point sets in $\mathbb{C}^2$, and all prime ideals $(f(X, Y))$ of irreducible affine plane curves over $\mathbb{C}$.

The spectrum of A is understood to carry along with it two additional pieces of structure. The first piece of structure is an analog for Spec A of the Zariski topology.[28] To each ideal $\mathfrak{a}$ of A, we associate the subset $V(\mathfrak{a}) \subseteq$ Spec A of all prime ideals $\mathfrak{p}$ with $\mathfrak{a} \subseteq \mathfrak{p}$. The sets $V(\mathfrak{a})$ are easily seen to have the defining properties of the closed sets of a topology, and this topology will always be understood to be in place. It is immediate from the definition that $V(\mathfrak{a}) = V(\sqrt{\mathfrak{a}})$ for every ideal $\mathfrak{a}$. One checks for any prime ideal $\mathfrak{p}$ that $V(\mathfrak{p}) = \overline{\{\mathfrak{p}\}}$; consequently the one-point set $\{\mathfrak{p}\}$ is closed if and only if $\mathfrak{p}$ is a maximal ideal.

At least when A is Noetherian, Spec A is a Noetherian space, and a notion of dimension (not necessarily finite) is defined for each closed set in the usual way[29] as in Section 2; for A itself this coincides with the Krull dimension of A. In this situation the irreducible closed sets are the sets $V(\mathfrak{p})$ with $\mathfrak{p}$ prime. The fact that such a set is irreducible follows from the identity $V(\mathfrak{p}) = \overline{\{\mathfrak{p}\}}$; the converse assertion follows from the identity $V(\mathfrak{a}) = V(\sqrt{\mathfrak{a}})$ and the Lasker–Noether Decomposition Theorem (Problem 14 at the end of Chapter VII). By Proposition 10.5 every closed set is a finite union of irreducible closed sets, and thus we have a complete description of the closed sets. For example, in a principal ideal domain the closed sets consist of the finite sets of nonzero prime ideals, as well as the set of all prime ideals. For the ring $A = \mathbb{C}[X, Y]$, every proper closed set of Spec A is a finite union of singleton sets $\{(X - x_0, Y - y_0)\}$ and of sets

$$\{(f(X, Y))\} \cup \bigcup_{f(x_0, y_0) = 0} \{(X - x_0, Y - y_0)\}$$

with $f(X, Y)$ irreducible.

If $\varphi : A \to B$ is a homomorphism in our category of rings (always assumed to carry 1 to 1) and if $\mathfrak{p}$ is a prime ideal in B, then $\varphi^{-1}(\mathfrak{p})$ is a prime ideal in A. Thus the definition $^a\varphi(\mathfrak{p}) = \varphi^{-1}(\mathfrak{p})$ gives us a function $^a\varphi :$ Spec $B \to$ Spec A. If E is a subset of A, then we readily check that

$$(^a\varphi)^{-1}(V(E)) = (^a\varphi)^{-1}(\{\mathfrak{p} \mid \mathfrak{p} \supseteq E\}) = \{\mathfrak{q} \mid {}^a\varphi(\mathfrak{q}) \supseteq E\} = V(\varphi(E)),$$

---

[28] A little care is needed with the definitions when A is the 0 ring, which has an identity but no prime ideals. Then Spec A is empty, but we will want to allow it as part of the theory. So we need to allow the empty set as a topological space.

[29] The general theory treats dimension as defined even when A is not Noetherian, but it will be enough in this section to consider only the Noetherian case.

from which it follows that $^a\varphi$ is continuous. The function $^a\varphi$ can be fairly subtle. For example, if φ is the inclusion of $\mathbb{Z}$ into the ring R of algebraic integers in a number field and if P is a nonzero prime ideal in R, then $^a\varphi(P) = P \cap \mathbb{Z}$ is the corresponding prime ideal (p) in $\mathbb{Z}$; the continuity of $^a\varphi$ implies that each nonzero prime ideal (p) of $\mathbb{Z}$ arises in this way from only finitely many ideals P in R.

2. Structure sheaf. The second piece of additional structure carried by the spectrum of A is its "structure sheaf," which is a certain specific sheaf with base space Spec A. Sheaves were introduced by J. Leray in 1946 in connection with partial differential equations and by K. Oka and H. Cartan about 1950 in connection with the theory of several complex variables. As with vector bundles, sheaves may be viewed as having a base space carrying some topological information and fibers carrying some algebraic information; local sections will be of great interest. The initial example of a sheaf in several complex variables is the "sheaf of germs of holomorphic functions" on an open set in $\mathbb{C}^n$, germs being defined for holomorphic functions on an open set in the same way as they were defined in Section 4 for rational functions on a quasi-affine variety.

We shall define two general notions, "sheaf" and "presheaf," and compare them. The prototype of a presheaf in several complex variables is the collection of vector spaces of holomorphic functions on each nonempty open subset of the given open set; the prototype in classical algebraic geometry is the collection of regular functions on each nonempty open subset of a quasiprojective variety. In the general case, fix a category to describe the allowable structure on each fiber; common choices for the objects in this category are abelian groups, commutative rings with identity (called "rings" hereafter in this section), and unital R modules for some ring. In defining sheaves and presheaves, we shall write the definitions using abelian groups, since it is a simple matter to adjoin the additional structure when the fibers are rings or modules.

Let X be a topological space. A **presheaf** of abelian groups on the **base space** X is a collection $\{\mathcal{O}(U), \rho_{VU}\}$, parametrized by the open subsets U of X and the open subsets V of U, such that each $\mathcal{O}(U)$ is an abelian group, $\mathcal{O}(\varnothing)$ is the 0 group, each $\rho_{VU} : \mathcal{O}(U) \to \mathcal{O}(V)$ is a group homomorphism, each ρ_{UU} is the identity, and $\rho_{WV}\rho_{VU} = \rho_{WU}$ whenever $W \subseteq V \subseteq U$. We are to think of $\mathcal{O}(U)$ as a space of sections of some kind over U and ρ_{VU} as a restriction map carrying sections over U to sections over V. A **sheaf** of abelian groups on the **base space** X is a topological space $\mathcal{O}$ with a mapping $\pi : \mathcal{O} \to X$ such that π is a local homeomorphism onto, $\pi^{-1}(P)$ is an abelian group for each $P \in X$, and the group operations on each $\pi^{-1}(P)$ are continuous in the relative topology from $\mathcal{O}$. We are to think of the elements of a sheaf as germs obtained starting from a presheaf. The individual fibers $\pi^{-1}(P)$ of a sheaf are called **stalks**. One writes $(X, \mathcal{O})$ for the sheaf, sometimes abbreviating the notation to $\mathcal{O}$.

It is possible to construct a presheaf from a sheaf, and vice versa. If we are

13. Schemes

given a sheaf $\mathcal{O}$, we define a **section** s of $\mathcal{O}$ over U to be a continuous function $s : U \to \mathcal{O}$ such that $\pi \circ s = 1_U$. If $\mathcal{O}(U)$ denotes the abelian group of sections of $\mathcal{O}$ over U and if ρ_{VU} is the restriction map for sections, then $\{\mathcal{O}(U), \rho_{VU}\}$ is a presheaf. In the reverse direction if we start from a presheaf $\{\mathcal{O}(U), \rho_{VU}\}$ and form the kind of direct limit of abelian groups at each point that is suggested by the passage to germs, then it is possible to topologize the disjoint union of the abelian groups of germs so as to produce a sheaf. Passing from a sheaf to a presheaf and then back to a sheaf reproduces the original sheaf. But passing from a presheaf to a sheaf and then back to a presheaf does not necessarily reproduce the original presheaf. A necessary and sufficient condition on the presheaf $\{\mathcal{O}(U), \rho_{VU}\}$ for $\{\mathcal{O}(U), \rho_{VU}\}$ to result from passing to a sheaf and then back to a presheaf is that the presheaf be **complete** in the sense that both the following conditions hold:

(i) Whenever $\{U_j\}$ is an open covering of an open subset U of X and $f \in \mathcal{O}(U)$ is an element such that $\rho_{U_j,U}(f) = 0$ for all j, then $f = 0$.
(ii) Whenever $\{U_j\}$ is an open covering of an open subset U of X and f_j is given in $\mathcal{O}(U_j)$ for each j in such a way that $\rho_{U_j \cap U_k, U_j}(f_j) = \rho_{U_j \cap U_k, U_k}(f_k)$ for all j and k, then there exists $f \in \mathcal{O}(U)$ such that $\rho_{U_j,U}(f) = f_j$ for all j.

The **structure sheaf** of the spectrum of A is a certain sheaf of rings (Spec A, $\mathcal{O}$) with base space Spec A. Just as in the case of regular (= polynomial) functions on an affine variety, this sheaf will have the property that the ring of global sections is isomorphic to the original ring (cf. Corollary 10.25). We shall describe $\mathcal{O}$ by describing the presheaf. For each prime ideal $\mathfrak{p}$ of A, let $A_\mathfrak{p}$ be the localization of A at $\mathfrak{p}$, i.e., the localization of A relative to the multiplicative system consisting of the set-theoretic complement of $\mathfrak{p}$. This kind of localization is always a local ring. The idea is to define a ring $\mathcal{O}(U)$ of regular functions for each open subset U of Spec A in such a way that the stalk $\mathcal{O}_\mathfrak{p}$ at the point $\mathfrak{p}$ ends up being $A_\mathfrak{p}$ for each $\mathfrak{p}$. With affine varieties we were able to make the definition directly in terms of the function field of the variety, i.e., the field of fractions of A; both $\mathcal{O}(U)$ and the stalk $\mathcal{O}_P(U)$ at each point P ended up being subrings of this function field. The complication for general A is that we do not have a convenient analog of the function field available in which all the localizations are subrings. Thus we proceed by imitating the messier equivalent definition of **regular function** given in Proposition 10.28. Namely, for U open in Spec A, let $\mathcal{O}(U)$ be the set of functions s from U into the product $\prod_{\mathfrak{p} \in U} A_\mathfrak{p}$ such that $s(\mathfrak{p})$ is in the $\mathfrak{p}^{\text{th}}$ factor $A_\mathfrak{p}$ for each $\mathfrak{p}$ and such that s is locally a quotient of members of A in the following sense: for each $\mathfrak{p}$ in U, there is to be an open neighborhood V of $\mathfrak{p}$ within U and there are to be elements a and f in A such that for each $\mathfrak{q}$ in V, the element f is not in $\mathfrak{q}$ and $s(\mathfrak{q})$ equals a/f in $A_\mathfrak{q}$. (Recall that any element of A not in $\mathfrak{q}$ defines an element in the multiplicative system leading to $A_\mathfrak{q}$; f is to be such an element for each $\mathfrak{q}$ in V.) The mappings ρ_{VU} are taken as ordinary restriction mappings, and the result is a presheaf. This presheaf is complete, and the associated sheaf is the structure sheaf (Spec A, $\mathcal{O}$). An **affine scheme** is any sheaf of rings that is isomorphic in a suitable sense to the structure sheaf of some ring.

3. Scheme. To define "scheme" and the notion that a scheme is defined over some ring or some field, we need to back up and say a few more words about mappings in connection with sheaves. A **ringed space** is a sheaf of rings, (Spec A, $\mathcal{O}$) being an example. Let $(X, \mathcal{O}_X)$ and $(Y, \mathcal{O}_Y)$ be two ringed spaces, and let $\{\rho_{V^*U^*}\}$ and $\{\rho'_{VU}\}$ be their respective systems of restriction maps. A **morphism** $(\sigma, \psi) : (X, \mathcal{O}_X) \to (Y, \mathcal{O}_Y)$ **of ringed spaces** consists of a continuous function $\sigma : X \to Y$ and a collection ψ of homomorphisms $\psi_U : \mathcal{O}_Y(U) \to \mathcal{O}_X(\sigma^{-1}(U))$ such that

$$\psi_V \circ \rho_{\sigma^{-1}V, \sigma^{-1}U} = \rho'_{VU} \circ \psi_U$$

whenever U and V are open subsets of Y with $V \subseteq U$. The collection $\psi = \{\psi_U\}$ yields homomorphisms of stalks $\psi_P : \mathcal{O}_{Y, \sigma(P)} \to \mathcal{O}_{X, P}$ for each P in X.

One property of the definition is that if $\varphi : A \to B$ is a homomorphism of rings, then there is an associated morphism $(\sigma, \psi) : (\text{Spec } B, \mathcal{O}_B) \to (\text{Spec } A, \mathcal{O}_A)$ of ringed spaces. The continuous map $\sigma : \text{Spec } B \to \text{Spec } A$ is the map $\sigma = {}^a\varphi$ given by ${}^a\varphi(\mathfrak{p}) = \varphi^{-1}(\mathfrak{p})$ for any prime ideal $\mathfrak{p}$ of B. The mapping ψ on stalks carries $\mathcal{O}_{\text{Spec } A, \sigma(\mathfrak{p})} = \mathcal{O}_{\text{Spec } A, \varphi^{-1}(\mathfrak{p})}$ to $\mathcal{O}_{\text{Spec } B, \mathfrak{p}}$ and is what is induced on the stalk by composition with φ. It is not quite true that every morphism $(\sigma, \psi) : (\text{Spec } B, \mathcal{O}_B) \to (\text{Spec } A, \mathcal{O}_A)$ of ringed spaces arises from a ring homomorphism. The homomorphism (σ, ψ) of ringed spaces resulting from the ring homomorphism φ has the property that ψ carries the maximal ideal $M_{\varphi^{-1}(\mathfrak{p})}$ of the stalk $A_{\varphi^{-1}(\mathfrak{p})}$ into the maximal ideal $M_\mathfrak{p}$ of the stalk $B_\mathfrak{p}$. A morphism (σ, ψ) of ringed spaces whose stalks are local rings is called a **local morphism** if it has this property. With this definition one can show that every local morphism of ringed spaces $(\sigma, \psi) : (\text{Spec } B, \mathcal{O}_B) \to (\text{Spec } A, \mathcal{O}_A)$ arises from some ring homomorphism $\varphi : A \to B$. This result is to be compared with Corollary 10.40 for affine varieties.

An isomorphism of ringed spaces is automatically local if all the stalks are local rings. The reason is that an isomorphism of one local ring onto another carries the maximal ideal of the first onto the maximal ideal of the second. Thus the earlier definition of **affine scheme** as a ringed space that is isomorphic to some (Spec A, $\mathcal{O}$) concealed only the rather natural definition of isomorphism of ringed spaces, not the more subtle condition "local."

A **morphism of affine schemes** is a local morphism of the affine schemes as ringed spaces. Then the classes of all affine schemes and morphisms of affine schemes together form a category. A **scheme** is a ringed space $(X, \mathcal{O})$ such that each point of X has an open neighborhood for which the restriction of the ringed space to that part of the base is isomorphic to an affine scheme. One can define a natural notion of morphism for schemes, and the classes of all schemes and morphisms of schemes together form a category.

4. **Variety as a scheme.** Let V be an affine variety over an algebraically closed field, and let $A(V)$ be the affine coordinate ring. We have just seen how Spec $A(V)$ has the natural structure of an affine scheme. Since Spec $A(V)$ includes all prime ideals of $A(V)$, not just the maximal ideals, the continuous inclusion $V \to$ Spec $A(V)$ is not onto. However, there is a natural relationship between the two, and there is a natural relationship between their rings of regular functions. The reason is that morphisms of affine varieties correspond exactly (in contravariant fashion) to homomorphisms of the affine coordinate rings, which in turn correspond exactly to morphisms of affine schemes. From the point of view of categories, therefore, the categories of affine varieties and affine schemes match perfectly. This description blurs what happens to the underlying algebraically closed field of scalars, and one wants to be able to say that the categories of affine varieties over $\Bbbk$ and affine schemes over $\Bbbk$ match perfectly. Making this statement requires an additional construction, which will be sketched in the next subsection.

This correspondence can be extended suitably from affine varieties to quasiprojective varieties, and the interested reader can find details on page 30 of Volume 2 of Shafarevich's books.

5. **Scheme defined over a ring.** If A is a ring and $(X, \mathcal{O}_X)$ is a scheme, then a morphism of schemes $(\sigma, \psi) : X \to$ Spec A defines a homomorphism $A \to \mathcal{O}_X(U)$ of rings for each open subset U of X. Specifically $\psi_{\text{Spec } A}$ carries $\mathcal{O}_{\text{Spec } A}(\text{Spec } A) = A$ into $\mathcal{O}_X(X)$, and hence $\rho_{UX} \circ \psi_{\text{Spec } A}$ carries A into $\mathcal{O}_X(U)$ if $\{\rho_{VU}\}$ is the system of restriction maps for $(X, \mathcal{O}_X)$. The result is that $\mathcal{O}_X$ becomes a sheaf of A algebras.

Conversely if $\mathcal{O}_X$ is a sheaf of A algebras, then one can construct a morphism of schemes $X \to$ Spec A. In this case one says that $(X, \mathcal{O}_X)$ is a **scheme over** A. Every sheaf of abelian groups is a sheaf of $\mathbb{Z}$ algebras, and thus every scheme is a scheme over $\mathbb{Z}$. Schemes over $\mathbb{Z}$ are of special interest in number-theoretic situations, among others. The schemes produced from varieties in the previous subsection are schemes over the underlying field $\Bbbk$. The notion of a scheme over a field that is not algebraically closed is one way of extending the theory of varieties to have it apply when the underlying field is not algebraically closed.

6. **Role of homological algebra.** The sheaves of abelian groups over a fixed topological space X, with a natural definition of morphism, form a category, and one can define kernels and cokernels in this category. The result turns out to be an abelian category with enough injectives, and the homological algebra of Chapter IV is applicable. If $(X, \mathcal{O})$ is a sheaf over X, then formation of global sections, given by $(X, \mathcal{O}) \mapsto \mathcal{O}(X)$, is a covariant left exact functor. Since there are enough injectives in the category, the derived functors make sense, and the kth derived functor gives what is called the kth **sheaf cohomology** group $H^k(X, \mathcal{O})$ with coefficients in $\mathcal{O}$. This kind of cohomology is easy to use abstractly and hard to use concretely, but it can be shown to be isomorphic to other more concrete kinds of cohomology. In this way the cohomology of sheaves leads to generalizations

of Euler characteristics and Betti numbers that have significance in number theory and geometry.

In applications, there tends to be a ringed space $(X, \mathcal{R})$ (maybe a scheme) in the picture, and the sheaves $(X, \mathcal{O})$ often have the property that each stalk of $\mathcal{O}$ is a module for the corresponding stalk of $\mathcal{R}$. Then the above kind of theory is applicable for sheaves that are $\mathcal{R}$ modules in this sense, not merely sheaves of abelian groups. The interested reader can find details in Chapter III of Hartshorne's book.

BIBLIOGRAPHICAL REMARKS. The topic of schemes assumes knowledge of a certain core of algebraic geometry and commutative algebra, and it builds on more commutative algebra as it goes along. Some books mentioned in the Selected References that include algebraic geometry at the beginning level are those of Hartshorne (Chapter I), Harris, Reid, and Shafarevich (Volume 1). All these books have many geometric examples; this is particularly so for the book by Harris. Some books on commutative algebra are the ones by Atiyah–Macdonald, Eisenbud, Matsumura, and Zariski–Samuel. These lists are by no means exhaustive. There are in fact hundreds of books on the two subjects. To get a list of many of the ones in commutative algebra, one can search in the Library of Congress catalog at http://catalog.loc.gov, using the call number QA251.3; a few additional ones are sprinkled in among books with call number QA251. For books on algebraic geometry, one can search using the call number QA564.

The book by Eisenbud–Harris on schemes is an introductory one written in a style that makes it comparatively easy for the reader to get an overview of the subject. Two older books on schemes are the ones by Macdonald and Mumford. Hartshorne's book introduces schemes in Chapter II, and Volume 2 of Shafarevich's books is on that topic. The end of Volume 2 of Shafarevich's books contains a 20-page historical sketch of algebraic geometry, including discussion of some of the precursors of the subject of schemes.

14. Problems

In all problems, $\Bbbk$ is understood to be an algebraically closed field.
1. If P is in $\mathbb{P}^n$, show that the ideal $I(P)$ of members of $\Bbbk[X_0, \ldots, X_n]$ vanishing at all points $(x_0, \ldots, x_n)$ in $\Bbbk^{n+1} - \{0\}$ with $[x_0, \ldots, x_n] = P$ is homogeneous.
2. Let X be a Noetherian topological space.
   (a) Prove that X is compact.
   (b) Prove that every irreducible closed subset of X is connected.
3. (a) Prove that the image of a quasiprojective variety V under a regular function $f : V \to \mathbb{A}^1$ is connected.
   (b) Prove that if V is a projective variety and $\varphi : V \to \mathbb{A}^n$ is a morphism, then $\varphi(V)$ is a one-point set.

4. Let U be the quasi-affine variety $U = \mathbb{A}^2 - \{(0,0)\}$ in $\mathbb{A}^2$. Prove that $\mathcal{O}(U) = \mathbb{k}[X, Y]$.

5. Deduce from the previous problem, Corollary 10.25, and Theorem 10.38 that U is not isomorphic to an affine variety.

6. Prove that a rational map of an irreducible curve into an irreducible curve is dominant or is constant.

7. Let $\varphi : U \to V$ be a dominant morphism between quasiprojective varieties. Prove that the induced mapping of local rings $\varphi_P^* : \mathcal{O}_{\varphi(P)}(V) \to \mathcal{O}_P(U)$ given in Proposition 10.42 is one-one.

8. Let V be the affine variety $V = V(WX - YZ)$ in $\mathbb{A}^4$, let $A(V)$ be the affine coordinate ring $\mathbb{k}[W, X, Y, Z]/(WX - YZ)$, let $\overline{X}$ and $\overline{Y}$ be the images of X and Y in $A(V)$, and let $f = \overline{X}/\overline{Y}$ in the field of fractions of $A(V)$. Prove that there exist no members $\bar{a}$ and $\bar{b}$ of $A(V)$ with $f = \bar{a}/\bar{b}$ and $\bar{b}(w, x, y, z) \neq 0$ whenever $wx = yz$ and one or both of w and y are nonzero.

9. Let U and V be quasiprojective varieties, and let $\varphi : U \to V$ be a function. Suppose that U and V are unions of nonempty open subsets $U = \bigcup_{\alpha \in I} U_\alpha$ and $V = \bigcup_{\alpha \in I} V_\alpha$ such that $\varphi(U_\alpha) \subseteq V_\alpha$ for all α. Prove that φ is a morphism if and only if each $\varphi_\alpha : U_\alpha \to V_\alpha$ is a morphism.

10. This problem concerns local extensions of regular functions from quasiprojective varieties to open sets in the ambient affine or projective space.
    (a) Let V be an affine variety in $\mathbb{A}^n$, let U be a nonempty open subset of V, let f be in $\mathcal{O}(U)$, and let P be a point in U. Prove that there exist an open neighborhood U_0 of U about P in V, an open set $\widetilde{U}_0$ in $\mathbb{A}^n$, and a function F in $\mathcal{O}(\widetilde{U}_0)$ such that $U_0 = V \cap \widetilde{U}_0$ and such that F is an extension of $f\big|_{U_0}$.
    (b) Extend the result of (a) to make it valid for any quasiprojective variety V in $\mathbb{P}^n$.

11. Suppose that X and Y are quasiprojective varieties, that U and V are irreducible closed subsets of X and Y, respectively, and that $\varphi : X \to Y$ is a morphism such that $\varphi(U) \subseteq V$. Prove that $\varphi : U \to V$ is a morphism.

12. Prove that
    (a) the mapping $\varphi : \mathbb{P}^{n-1} \to \mathbb{P}^n$ given by $\varphi([x_0, \ldots, x_{n-1}]) = [x_0, \ldots, x_{n-1}, 0]$ is an isomorphism of $\mathbb{P}^{n-1}$ onto the projective hyperplane H_n corresponding to the homogeneous ideal (X_n) of $\mathbb{k}[X_0, \ldots, X_n]$,
    (b) any projective variety V in $\mathbb{P}^n$ that lies in H_n is isomorphic to a projective variety in $\mathbb{P}^{n-1}$,
    (c) any projective variety V in $\mathbb{P}^n$ is isomorphic to a projective variety V' in some $\mathbb{P}^r$ with $r \leq n$ that is not contained in any projective hyperplane defined by a homogeneous ideal (X_j) of $\mathbb{k}[X_0, \ldots, X_r]$.

Problems 13–16 relate the classical condition for detecting a singularity in the affine case to the corresponding condition in the projective case. The key is an identity traditionally known as Euler's Theorem that is proved as Problem 3 at the end of Chapter VIII. In these problems it is assumed that $F_1, \ldots, F_r$ are homogeneous polynomials in $\Bbbk[X_0, \ldots, X_n]$, that $P = [x_0, \ldots, x_n]$ is a point in $\mathbb{P}^n$ in their common locus of zeros, and that P is in the image of $\mathbb{A}^n$ under β_0, i.e., that $x_0 \neq 0$. Define $f_1, \ldots, f_r$ in $\Bbbk[X_1, \ldots, X_n]$ by $f_i(X_1, \ldots, X_n) = F_i(1, X_1, \ldots, X_n)$.

13. Define $J(F)(x_0', \ldots, x_n')$ to be the r-by-$(n+1)$ matrix whose $(i, j)^{\text{th}}$ entry is $\frac{\partial F_i}{\partial X_j}(x_0', \ldots, x_n')$ for $1 \leq i \leq r$ and $0 \leq j \leq n$, and define $J(f)(x_1', \ldots, x_n')$ to be the r-by-n matrix whose $(i, j)^{\text{th}}$ entry is $\frac{\partial f_i}{\partial X_j}(x_1', \ldots, x_n')$ for $1 \leq i \leq r$ and $1 \leq j \leq n$. Prove that rank $J(F)(x_0', \ldots, x_n') =$ rank $J(F)(\lambda x_0', \ldots, \lambda x_n')$ for all $\lambda \in \Bbbk^\times$.

14. With notation as in Problem 13, prove that the r-by-n matrix $J(f)(x_1', \ldots, x_n')$ equals the r-by-n matrix obtained by deleting the 0th column of the r-by-$(n+1)$ matrix $J(F)(1, x_1', \ldots, x_n')$.

15. Using Euler's Theorem (Problem 3 at the end of Chapter VIII), prove concerning the point P on the locus of common zeros of $F_1, \ldots, F_r$ that the 0th column of the matrix $J(F)(x_0, \ldots, x_n)$ is a linear combination of the other columns of the matrix.

16. Deduce for the point P on the locus of common zeros of $F_1, \ldots, F_r$ that
$$\text{rank } J(F)(x_0, x_1, \ldots, x_n) = \text{rank } J(f)(x_1/x_0, \ldots, x_n/x_0).$$

Problems 17–22 concern products of quasiprojective varieties. The Segre mapping $\sigma : \mathbb{P}^m \times \mathbb{P}^n \to \mathbb{P}^N$ with $N = mn + m + n$ was defined in Section 8 by $\sigma([x_0, \ldots, x_m], [y_0, \ldots, y_n]) = [w_{00}, \ldots, w_{mn}]$ with $w_{ij} = x_i y_j$. Let us abbreviate $[w_{00}, \ldots, w_{mn}]$ as $[\{w_{ij}\}]$ and $\Bbbk[W_{00}, \ldots, W_{mn}]$ as $\Bbbk[\{W_{ij}\}]$.

17. Prove that σ is well defined and one-one.

18. Every member $[\{w_{ij}\}]$ of image σ has $w_{ij} w_{kl} = w_{il} w_{kj}$ for all i, j, k, l. Prove conversely that every member $[\{w_{ij}\}]$ of $\mathbb{P}^N$ with $w_{ij} w_{kl} = w_{il} w_{kj}$ for all i, j, k, l is in image σ, and deduce that image $\sigma = V(\mathfrak{a})$, where $\mathfrak{a}$ is the ideal in $\Bbbk[\{W_{ij}\}]$ generated by all $W_{ij} W_{kl} - W_{il} W_{kj}$.

19. This problem will prove that $\mathfrak{a}$ is a prime ideal, and in particular it will follow that $V(\mathfrak{a})$ is irreducible. Let $\varphi : \Bbbk[\{W_{ij}\}] \to \Bbbk[X_0, \ldots, X_m, Y_0, \ldots, Y_n]$ be the substitution homomorphism given by setting $W_{ij} = X_i Y_j$. Then $\ker \varphi$ is an ideal containing $\mathfrak{a}$.
    (a) By introducing a suitable monomial ordering in $\Bbbk[\{W_{ij}\}]$, show that any monomial in $\Bbbk[\{W_{ij}\}]$ of total degree d is congruent modulo $\mathfrak{a}$ to a monomial of total degree d of the form $M = \prod_{i,j} W_{ij}^{a_{ij}}$ having the property that $a_{ij} > 0$ implies that $a_{kl} = 0$ for all (k, l) with $l > j$ and $k > i$. Call a monomial of this form **reduced**.

(b) Suppose that $M = \prod_{i,j} W_{ij}^{a_{ij}}$ and $M' = \prod_{i,j} W_{ij}^{b_{ij}}$ are two distinct reduced monomials. By considering the first W_{ij} for which $a_{ij} \neq b_{ij}$, prove that $\varphi(M) \neq \varphi(M')$.

(c) Deduce that $\ker \varphi = \mathfrak{a}$, and show why it follows that $\mathfrak{a}$ is prime.

20. Let $\mathfrak{p}$ be a prime ideal in $\Bbbk[X_0, \ldots, X_m]$, and let $R = \Bbbk[X_0, \ldots, X_m]/\mathfrak{p}$ be the quotient.

    (a) Prove that the ideal $\mathfrak{p}\,\Bbbk[Y_0, \ldots, Y_n]$ in $\Bbbk[X_0, \ldots, X_m, Y_0, \ldots, Y_n]$ generated by all products of members of $\mathfrak{p}$ and polynomials in $Y_0, \ldots, Y_n$ is prime.

    (b) By following the substitution homomorphism

    $$\Bbbk[\{W_{ij}\}] \to \Bbbk[X_0, \ldots, X_m, Y_0, \ldots, Y_n]$$

    with a substitution homomorphism $\Bbbk[X_0, \ldots, X_m, Y_0, \ldots, Y_n] \to R[Z]$, prove that whenever U is a projective variety in $\mathbb{P}^m$ and P is a point in $\mathbb{P}^n$, then $\sigma(U \times \{P\})$ is a projective variety in $\mathbb{P}^N$.

21. Let U and V be projective varieties in $\mathbb{P}^m$ and $\mathbb{P}^n$, respectively. Problem 20 shows that $\sigma(U \times \{v\})$ is a projective variety in $\mathbb{P}^N$ for each $v \in V$. Suppose that $\sigma(U \times V)$ is a union $E_1 \cup E_2$ of two closed sets in $\mathbb{P}^N$.

    (a) For i equal to 1 or 2, define $V_i = \{v \in V \mid \sigma(U \times \{v\}) \not\subseteq E_i\}$. Why is $V_1 \cap V_2 = \varnothing$?

    (b) Prove that V_1 and V_2 are open by using bihomogeneous polynomials to exhibit each of V_1 and V_2 as a neighborhood of each of its points.

    (c) Deduce from (b) that $\sigma(U \times V)$ is a projective variety in $\mathbb{P}^N$.

    (d) Show how to deduce from (c) that if U and V are quasiprojective varieties in $\mathbb{P}^m$ and $\mathbb{P}^n$, respectively, then $\sigma(U \times V)$ is a quasiprojective variety in $\mathbb{P}^N$.

22. (a) Prove that if U and V are quasiprojective varieties, then the projections of $U \times V$ to U and V are morphisms. Here the projection of $U \times V$ to U is understood to be the map $\sigma(u, v) \mapsto u$ of $\sigma(U \times V)$ into U, and similarly for the projection to V.

    (b) If $\varphi : U \to X$ and $\psi : U \to Y$ are morphisms, prove that $(\varphi, \psi) : U \to X \times Y$ when defined by $(\varphi, \psi)(u) = (\varphi(u), \psi(u))$ is a morphism.

    (c) If $\varphi : U \to X$ and $\psi : V \to Y$ are morphisms, prove that $\varphi \times \psi : U \times V \to X \times Y$ when defined by $(\varphi \times \psi)(u, v) = (\varphi(u), \psi(v))$ is a morphism.

Problems 23–25 make some observations about prime ideals and irreducible polynomials.

23. Let $I = (f_1, \ldots, f_r)$ be an ideal in $\Bbbk[X, Y]$ such that the zero locus $V(I)$ is irreducible and such that $f_1, \ldots, f_r$ are irreducible polynomials.

    (a) Prove that I is prime if $\dim V(I) = 1$.

    (b) Give an example to show that I need not be prime if $\dim V(I) = 0$.

24. Fix a monomial ordering for $\Bbbk[X_1, \ldots, X_n]$, and let I be a nonzero ideal in $\Bbbk[X_1, \ldots, X_n]$. Prove that if I is prime, then the members of any minimal Gröbner basis of I are irreducible polynomials.

25. Suppose that char($\Bbbk$) $\neq 2$. Within $\Bbbk[X, Y, Z]$, let E be the homogeneous subspace $\Bbbk[X, Y, Z]_2$. The six monomials in E form a $\Bbbk$ basis of E and may be used to identify E with $\Bbbk^6$. Under this identification prove that the subset of reducible polynomials in E, including the 0 polynomial, is an affine hypersurface of $\Bbbk^6$.

Problems 26–35 concern elliptic curves. An **elliptic curve** over $\Bbbk$ is a pair (E, O) consisting of a nonsingular irreducible projective curve E of genus 1 and a distinguished point O. These problems use the Riemann–Roch Theorem and its associated notation in Chapter IX in order to exhibit a concrete realization of such a curve in $\mathbb{P}^2$ with O on the line at infinity and with all other points of E in $\mathbb{A}^2$. Such a curve has a remarkable structure; for further information, including further applications of the Riemann–Roch Theorem to these curves, see the book by Silverman. Corollary 10.56 identifies the points of E with the discrete valuations of the function field $\Bbbk(E)$ over E. Let v_O be the discrete valuation corresponding to O.

26. For $n > 0$, prove that $\ell(nv_O) = n$. Use this result to find members x and y of $\Bbbk(E)$ whose divisors satisfy $(x)_\infty = 2v_O$ and $(y)_\infty = 3v_O$.

27. Prove that $[\Bbbk(E) : \Bbbk(x)] = 2$ and $[\Bbbk(E) : \Bbbk(y)] = 3$.

28. Why does it follow from the previous problem that $\Bbbk(E) = \Bbbk(x, y)$?

29. From the fact that $\ell(6v_O) = 6$, deduce a nontrivial linear dependence over $\Bbbk$ among the members $1, x, y, x^2, xy, y^2, x^3$ of $\Bbbk(E)$. Show that the coefficients of y^2 and x^3 are necessarily nonzero, and then scale x and y appropriately to show that the image of the function $\varphi : E - \{0\} \to \mathbb{P}^2$ defined by $\varphi(P) = [x(P), y(P), 1]$ is contained in the projective closure C of the zero locus of the polynomial $f(X, Y) = (Y^2 + a_1XY + a_3Y) - (X^3 + a_2X^2 + a_4X + a_6)$.

30. Prove that $f(X, Y)$ is irreducible and that C is therefore a projective curve.

31. Why is $\varphi : E - \{0\} \to C$ a morphism? Why does it follow that φ extends to a morphism $\Phi : E \to C$?

32. Deduce from Problem 28 that Φ is birational.

33. Show that C is nonsingular at its point at infinity.

34. Show that if C is singular at (x_0, y_0) in $\mathbb{A}^2$, then the member of $\Bbbk(E)$ given by $z = (y - y_0)(x - x_0)^{-1}$ has $v_O(z) = -1$ and $v_P(z) \geq 0$ for all P in $E - \{O\}$.

35. Deduce from Problems 33 and 34 that C is nonsingular, and explain why it follows that $\Phi : E \to Z$ is an isomorphism.

HINTS FOR SOLUTIONS OF PROBLEMS

Chapter I

1. We are interested in odd p's such that $\left(\frac{m}{p}\right) = +1$. Factor m as $\prod_j p_j^{k_j}$. Then quadratic reciprocity gives $\left(\frac{m}{p}\right) = \prod_j \left(\frac{p_j}{p}\right)^{k_j} = \prod_{k_j \text{ odd}} \left(\frac{p_j}{p}\right) = \prod_{k_j \text{ odd}} (-1)^{\frac{1}{4}(p-1)(p_j-1)} \left(\frac{p}{p_j}\right)$. We consider $p \equiv 1 \bmod 4$ and $p \equiv 3 \bmod 4$ separately. For $p \equiv 1 \bmod 4$, the set in question consists of those p's for which $\left(\frac{p}{p_j}\right)$ is -1 for an even number of those k_j's that are odd. This is the union over all such systems of minus signs of the intersection over j of the finitely many arithmetic progressions for which the residue $\left(\frac{p}{p_j}\right)$ equals the jth sign. For a single system of minus signs, the result is an arithmetic progression of the form $k \prod_{k_j \text{ odd}} p_j + b$ by the Chinese Remainder Theorem. Each of these contains a nonempty set of primes by Dirichlet's Theorem, and hence P is nonempty.

For $p \equiv 3 \bmod 4$, if $\prod_{k_j \text{ odd}}(-1)^{\frac{1}{2}(p_j-1)}$ is $+1$, then the set in question is of the same form as above. If $\prod_{k_j \text{ odd}}(-1)^{\frac{1}{2}(p_j-1)}$ is -1, then the set in question consists of those p's for which $\left(\frac{p}{p_j}\right)$ is -1 for an odd number of those k_j's that are odd, and this again is the finite union of arithmetic progressions.

2. For (a), the proof of necessity of Theorem 1.6b remains valid when the prime p is replaced by the integer m. For (b), the first paragraph of the proof of the sufficiency of Theorem 1.6b handles matters if m is odd.

3. For $D = -56$, H has order 4, but H' has order 3 because $3x^2 \pm 2xy + 5y^2$ are improperly equivalent but not properly equivalent. A 3-element set has no group structure such that a 4-element group maps homomorphically onto it.

4. For (a), the product of any two integers representable as $ax^2 + bxy + cy^2$ is representable by the class of the square, which is the class of the inverse because the class is assumed to have order 3. The class of the inverse is the class of $(a, -b, c)$, and this represents the same integers as (a, b, c).

For (b), we seek reduced triples. These are (a, b, c) with $|b| \leq a \leq c$ and with $b^2 - 4ac = D = -23$, and we know that $3ac \leq |D|$ and that b has the same parity as D. Hence b is odd, and the inequalities $3b^2 \leq 3a^2 \leq 3ac \leq 23$ show that $|b| = 1$. For $|b| = 1$, we have $1 - 4ac = -23$ and $ac = 6$. Since $a \leq c$, the possibilities with $|b| = 1$ are $(1, \pm 1, 6)$ and $(2, \pm 1, 3)$. Since $(1, 1, 6)$ and $(1, -1, 6)$ are properly equivalent by Proposition 1.7, $|b| = 1$ leads to just the three possibilities $(1, 1, 6)$, $(2, 1, 3)$, and $(2, -1, 3)$. Proposition 1.7 shows that these lie in distinct proper equivalence classes, and thus $h(-23) = 3$.

For (c), the general theory shows that $(1, 1, 6)$ corresponds to the identity class, and therefore the other two reduced forms are in classes of order 3.

For (d), we first track down what happens to the forms. If we write $\sim$ for proper equivalence, then we have

$$(2, 1, 3)(2, 1, 3) \sim (2, 1, 3)(3, -1, 2) \sim (2, 5, 6)(3, 5, 4)$$
$$= (6, 5, 2) \sim (2, -5, 6) \sim (2, -1, 3),$$

and the last form is improperly equivalent to $(2, 1, 3)$. The next step is to interpret this chain with actual variables. If the initial variables are x_1, y_1, x_2, y_2, then the change at the first step from $(2, 1, 3)$ to $(3, -1, 2)$ comes from $x_2 = y_2'$, $y_2 = -x_2'$ while leaving x_1 and y_1 unchanged as $x_1 = x_1'$, $y_1 = y_1'$. The change at the second step from $(2, 1, 3)$ to $(2, 5, 6)$ and from $(3, -1, 2)$ to $(3, 5, 4)$ comes from the translations $x_1' = x_1'' + y_1''$, $y_1' = y_1''$, $x_2' = x_2'' + y_2''$, $y_2' = y_2''$. The multiplication step comes from Proposition 1.9 and is given by $x_3 = x_1'' x_2'' - 2 y_1'' y_2''$ and $y_3 = 2 x_1'' y_2'' + 3 x_2'' y_1'' + 5 y_1'' y_2''$. And so on. The final result is that

$$(2x_1^2 + x_1 y_1 + 3 y_1^2)(2 x_2^2 + x_2 y_2 + 3 y_2^2) = 2X^2 + XY + 3Y^2,$$

where $X = x_1(-x_2 + y_2) + y_1(x_2 + 2y_2)$ and $Y = y_1(x_2 - y_2) + x_1(x_2 + y_2)$.

5. The equality $\begin{pmatrix} 1 & 0 \\ -a^{-1}b & 1 \end{pmatrix} \begin{pmatrix} 2a & b \\ b & 2c \end{pmatrix} \begin{pmatrix} 1 & -a^{-1}b \\ 0 & 1 \end{pmatrix} = \begin{pmatrix} 2a & -b \\ -b & 2c \end{pmatrix}$ shows this.

6. For reduced forms we seek (a, b, c) with $a > 0, c > 0, |b| \leq a \leq c$. We know that $3ac \leq |D| = 67$, and D odd implies b odd. From $3b^2 \leq 3a^2 \leq 3ac \leq 67$, we obtain $3b^2 \leq 67$ and $|b| \leq 4$. So $|b|$ is 1 or 3. For $|b| = 1$, $\frac{1}{4}(b^2 - D) = \frac{1}{4}(b^2 + 67) = 17$; then $17 = ac$, and $a = 1$ and $c = 17$. Since $(1, 1, 17)$ is properly equivalent to $(1, -1, 17)$ by Proposition 1.7, we obtain only one proper equivalence class from this pair. For $|b| = 3$, $\frac{1}{4}(b^2 - D) = \frac{1}{4}(9 + 67) = 19$ forces $ac = 19$ and then $a = 1$ and $c = 19$. Then $|b| \leq a$ is not satisfied. So $|b| = 3$ gives no proper equivalence classes, and $h(-67) = 1$.

7. The 6 cycles are

$(1, 8, -15), (-15, 7, 2), (2, 7, -15), (-15, 8, 1);$
$(-1, 8, 15), (15, 7, -2), (-2, 7, 15), (15, 8, -1);$
$(3, 8, -5), (-5, 7, 6), (6, 5, -9), (-9, 4, 7), (7, 3, -10), (-10, 7, 3);$
$(-3, 8, 5), (5, 7, -6), (-6, 5, 9), (9, 4, -7), (-7, 3, 10), (10, 7, -3);$
$(5, 8, -3), (-3, 7, 10), (10, 3, -7), (-7, 4, 9), (9, 5, -6), (-6, 7, 5);$
$(-5, 8, 3), (3, 7, -10), (-10, 3, 7), (7, 4, -9), (-9, 5, 6), (6, 7, -5).$

8. The form $(1, 1, 12)$ corresponds to the identity class, the classes of $(2, \pm 1, 6)$ are inverses of one another, and the classes of $(3, \pm 1, 4)$ are inverses of one another. The

group structure has to be cyclic, and any element other than the identity can be taken as a generator. Let us take a to be the class of $(2, 1, 6)$. We are to identify a^2. The form $(2, 1, 6)$ is aligned with itself (having the same b component), it has $j = 6/2 = 3$, and the composition formula of Proposition 1.9 leads to $(2 \cdot 2, 1, j) = (4, 1, 3)$. This is properly equivalent to $(3, -1, 4)$, and we do not have to follow through the algorithm of Theorem 1.6a to identify the product in our list. The result is that $a \leftrightarrow (2, 1, 6), a^2 \leftrightarrow (3, -1, 4), a^3 = (a^2)^{-1} \leftrightarrow (3, 1, 4), a^4 = a^{-1} \leftrightarrow (2, -1, 6)$, and $a^5 = 1 \leftrightarrow (1, 1, 12)$.

10. For (a), the result is known for n prime by Theorem 1.2. By induction and the definition of the Jacobi symbol, it is enough to handle $n = ab$ when a and b can be handled. We have $\frac{1}{2}(n-1) = \frac{1}{2}(ab-1) = \frac{1}{2}b(a-1) + \frac{1}{2}(b-1)$ $\equiv \frac{1}{2}(a-1) + \frac{1}{2}(b-1) \bmod 2$, the last step following because b is odd. Therefore $(-1)^{\frac{1}{2}(n-1)} = (-1)^{\frac{1}{2}(a-1)+\frac{1}{2}(b-1)} = \left(\frac{-1}{a}\right)\left(\frac{-1}{b}\right) = \left(\frac{-1}{n}\right)$, the last step following by Problem 9a.

For (b), we argue similarly, and the key computation is $\frac{1}{8}(n^2-1) = \frac{1}{8}(a^2b^2-1) = \frac{1}{8}b^2(a^2-1) + \frac{1}{8}(b^2-1) \equiv \frac{1}{8}(a^2-1) + \frac{1}{8}(b^2-1) \bmod 2$, the last step following because b^2 is odd.

11. Allowing primes to appear more than once, write factorizations of m and n as $m = \prod_{i=1}^{r} p_i$ and $n = \prod_{j=1}^{s} q_j$. Then Theorem 1.2 gives $\left(\frac{m}{n}\right) = \prod_{j=1}^{s} \prod_{i=1}^{r} \left(\frac{p_i}{q_j}\right) = \prod_{j=1}^{s} \prod_{i=1}^{r} \left(\frac{q_j}{p_i}\right)(-1)^{\frac{1}{2}(p_i-1)\frac{1}{2}(q_j-1)} = \left(\frac{n}{m}\right)(-1)^{\sum_{j=1}^{s}\sum_{i=1}^{r} \frac{1}{2}(p_i-1)\frac{1}{2}(q_j-1)}$. Since

$$\sum_{j=1}^{s}\sum_{i=1}^{r} \tfrac{1}{2}(p_i-1)\tfrac{1}{2}(q_j-1) = \left[\sum_{j=1}^{s} \tfrac{1}{2}(q_j-1)\right]\left[\sum_{i=1}^{r} \tfrac{1}{2}(p_i-1)\right]$$

and since $\sum_{j=1}^{s} \frac{1}{2}(q_j-1) \equiv \frac{1}{2}(n-1) \bmod 2$ and $\sum_{i=1}^{r} \frac{1}{2}(p_i-1) \equiv \frac{1}{2}(m-1) \bmod 2$ by the same argument as in Problem 10a, the required formula follows.

12. For (a), choose by Dirichlet's Theorem a sufficiently large prime p that is $\equiv 3 \bmod 8$ and is in particular $\equiv 3 \bmod 4$. If 8 divides $|G|$, then the fact that $|G|$ divides $p + 1$ implies that 8 divides $p + 1$. So $p \equiv -1 \bmod 8$. Since p was chosen with $p \equiv 3 \bmod 8$, this is a contradiction. So 8 cannot divide $|G|$.

For (b), choose by Dirichlet's Theorem a sufficiently large prime p that is $\equiv 7 \bmod 12$ and is in particular $\equiv 3 \bmod 4$. If 3 divides $|G|$, then 3 divides $p + 1$. Thus $p \equiv -1 \bmod 3$. Since also $p \equiv 3 \bmod 4$, $p \equiv 11 \bmod 12$. But p was chosen with $p \equiv 7 \bmod 12$. This is a contradiction, and 3 cannot divide $|G|$.

For (c) with an odd prime $q > 3$ given, choose by Dirichlet's Theorem a sufficiently large prime p that is $\equiv 3 \bmod 4q$ and is in particular $\equiv 3 \bmod 4$. If q divides $|G|$, then q divides $p + 1$, and $p + 1 \equiv 0 \bmod q$. Meanwhile, $p \equiv 3 \bmod 4q$ implies that $p + 1 \equiv 4 \bmod 4q$ and $p + 1 \equiv 4 \bmod q$, contradiction. So q cannot divide $|G|$.

13. For (a), choose by Dirichlet's Theorem a sufficiently large prime p that is $\equiv 5 \bmod 12$ and is in particular $\equiv 2 \bmod 3$ and $\equiv 1 \bmod 4$. If 4 divides $|G|$, then 4 divides $p + 1$, which is $\equiv 2 \bmod 4$. So 4 cannot divide $|G|$.

For (b), choose by Dirichlet's Theorem a sufficiently large prime p that is $\equiv 2 \bmod 9$ and is in particular $\equiv 2 \bmod 3$. If 9 divides $|G|$, then 9 divides $p+1$, which is $\equiv 3 \bmod 9$. So 9 cannot divide $|G|$.

For (c) with an odd prime $q > 3$ given, choose by Dirichlet's Theorem a sufficiently large prime p that is $\equiv 2 \bmod 3q$ and is in particular $\equiv 2 \bmod 3$. If q divides $|G|$, then q divides $p+1$, which is $\equiv 3 \bmod 3q$ and hence is $\equiv 3 \bmod q$. So q cannot divide $|G|$.

14. The integers in $\langle a, r \rangle$ are exactly the multiples of a, since such an integer n has to be of the form $n = ca + dr$ for integers c and d. This equation says that $n = ca$ and $0 = dr$, since 1 and r are linearly independent over $\mathbb{Q}$. The integer $N(s) = s\sigma(s)$ is in I because s is in I and $\sigma(s)$ is in R, and thus $N(s)$ has to be a multiple of a.

15. Write $I = \langle a, r \rangle$ with $a > 0$ an integer and r in I by Lemma 1.19b. As in the previous problem, the integer a is characterized uniquely in terms of I as the least positive integer in I. Put $r = b + g\delta$ for suitable integers b and g. Without loss of generality, we may assume that $g > 0$. Using the division algorithm and possibly replacing b by $b - na$ for some integer n, we may assume that $0 \le b < a$.

With these conventions in place, let us see that g necessarily divides a. The fact that $a\delta$ has to be in I means that $a\delta$ has an expansion $a\delta = c_1 a + c_2(b + g\delta)$ with integer coefficients. Then $a\delta = c_2 g \delta$, and g must divide a.

In particular, $0 < g \le a$ is forced. To see that b and g are uniquely determined, let $\{a, b' + g'\delta\}$ be another such $\mathbb{Z}$ basis. Since $b' + g'\delta = c_1 a + c_2(b + g\delta)$ and since symmetrically we have $b + g\delta = c_1' a + c_2'(b' + g'\delta)$, we obtain $g' = c_2 g = c_2 c_2' g'$. Therefore $|c_2| = 1$. Meanwhile, we must have

$$c_1 a + c_2 b = b' \quad \text{and} \quad c_2 g \delta = g' \delta.$$

The second of these equations shows that $c_2 > 0$. Thus $c_2 = 1$. Finally $c_1 a = b' - b$ with $0 \le b < a$ and $0 \le b' < a$ forces $b' - b = 0$. Therefore $a, b,$ and g are uniquely determined.

To complete the proof, we need to see that g divides b and that ag divides $N(b+g\delta)$. Since $a\delta$ is in I, $a\delta = c_1'' a + c_2''(b + g\delta)$. Hence $c_2'' g = a$ and $c_1'' a + c_2'' b = 0$. Substituting the first of these equations into the second gives $c_1'' c_2'' g + c_2'' b = 0$. Since $c_2'' \ne 0$ from the equality $c_2'' g = a$, $c_1'' g + b = 0$. Thus g divides b.

To see that ag divides $N(b + g\delta)$, we use the fact that $g\sigma(\delta)(b + g\delta)$ is in I to write $bg\sigma(\delta) + \delta\sigma(\delta)g^2 = d_1 ag + d_2 g(b + g\delta)$ for some integers d_1 and d_2. Then $N(b + g\delta) = b^2 + bg(\delta + \sigma(\delta)) + \delta\sigma(\delta)g^2 = b^2 + bg\delta + d_1 ag + d_2 g(b + g\delta)$. Equating coefficients of δ and 1 gives

$$0 = bg + d_2 g^2 \quad \text{and} \quad N(b + g\delta) = b^2 + d_1 ag + d_2 bg.$$

Since $g > 0$, the first of these equations gives $d_2 = -bg^{-1}$. Substituting into the second equation gives

$$N(b + g\delta) = b^2 + d_1 ag - (bg^{-1})bg = d_1 ag,$$

and we see that ag divides $N(b + g\delta)$.

16. We are to show that $\mathbb{Z}a + \mathbb{Z}(b + g\delta)$ is closed under multiplication by arbitrary members of R. It is enough to treat multiplication by 1 and by δ. There is no problem for 1. Since $\delta + \sigma(\delta)$ is in $\mathbb{Z}$, it is enough to show that there exist integers c_1, c_2, d_1, d_2 with

$$\delta a = c_1 a + c_2(b + g\delta) \quad \text{and} \quad \sigma(\delta)(b + g\delta) = d_1 a + d_2(b + g\delta).$$

In view of the assumed divisibility, we can put $c_2 = ag^{-1}, c_1 = -bg^{-1}, d_2 = -bg^{-1}$, and $d_1 = N(b + g\delta)(ag)^{-1}$. Then the first equation is certainly satisfied, and the question concerning the second equation, once we have multiplied it by g, is whether we have an equality

$$g\sigma(\delta)(b + g\delta) \stackrel{?}{=} N(b + g\delta) - b^2 - bg\delta.$$

The left side is $N(b + g\delta) - b(b + g\delta)$, and thus equality indeed holds.

17. From Section 7 the relevant formula is $N(I) = |\sqrt{D}|^{-1}|r_1\sigma(r_2) - \sigma(r_1)r_2|$. Here we can take $r_1 = a$ and $r_2 = c + d\delta$. Substitution gives

$$N(I) = |\sqrt{D}|^{-1}|a||\sigma(c + d\delta) - (c + d\delta)|$$
$$= |\sqrt{D}|^{-1}|a||c + d\sigma(\delta) - c - d\delta| = |\sqrt{D}|^{-1}|ad||\sigma(\delta) - \delta|.$$

The expression $|\sqrt{D}|^{-1}|\sigma(\delta) - \delta|$ arose in Section 7 in the computation of $N(R)$ and was shown to be 1. Thus $N(I) = |ad|$.

18. For (a), the algorithm of Section IV.9 of *Basic Algebra* shows how to align matters so as to compute the quotient of a free abelian group by a subgroup when the subgroup is given by generators. The given relationship between the generators a and $b + g\delta$ of Problem 15 with the $\mathbb{Z}$ basis of R is

$$\begin{pmatrix} a \\ b + g\delta \end{pmatrix} = \begin{pmatrix} a & 0 \\ b & g \end{pmatrix} \begin{pmatrix} 1 \\ \delta \end{pmatrix}.$$

The procedure is to do row and column operations on the coefficient matrix to bring it into diagonal form. Since g divides b, a column operation replaces the b by 0. We obtain a diagonal matrix with diagonal entries a and g, and the quotient group is identified as $(\mathbb{Z}/a\mathbb{Z}) \oplus (\mathbb{Z}/g\mathbb{Z})$. Thus ag is identified as the number of elements in the quotient group R/I. Problem 17 identified ag as $N(I)$, and thus $N(I)$ is the number of elements in R/I.

For (b), the inclusion $I \subseteq J$ induces a quotient mapping of the finite group R/I onto R/J. As a homomorphic image of R/I, R/J must have an order that divides the order of R/I. In view of (a), $N(J)$ divides $N(I)$. The equality $I = J$ holds if and only if the quotient mapping is one-one, and this happens, because of the finite cardinalities, if and only if $N(J) = N(I)$.

19. The relevant arguments for the first three parts of this problem already appear in Chapters VIII and IX of *Basic Algebra*, and thus we can be brief. For (a), the Chinese Remainder Theorem (Theorem 8.27 of *Basic Algebra*) shows that $R/IJ \cong R/I \times R/J$, and then $N(IJ) = N(I)N(J)$ by Problem 18a. For (b), the inductive argument for (∗∗) in the proof of Theorem 9.60 of *Basic Algebra* shows that $\dim_{\mathbb{Z}/p\mathbb{Z}} R/P^e = ef$, and thus $|R/P^e| = p^{ef}$. For (c), Corollary 8.63 of *Basic Algebra* and Problem 18a above together show that $N(I) = \prod_{j=1}^{n} N(P_j^{k_j})$ if $I = \prod_{j=1}^{n} P_j^{k_j}$ is the unique factorization of the ideal I. Since $N(P_j^{k_k}) = N(P_j)^{k_j}$ by (b), $N(I) = \prod_{j=1}^{n} N(P_j)^{k_j}$, and (c) follows immediately.

For (d), we use Problem 15 to write $I = \langle a, b + g\delta \rangle$; then

$$I\sigma(I) = (a^2, a(b + g\delta), a(b + g\sigma(\delta)), N(b + g\delta)).$$

Each of the generators on the right side lies in the principal ideal (ag). In fact, a^2 is in (ag) because g divides a, $a(b + g\delta)$ and $a(b + g\sigma(\delta))$ are in (ag) because g divides b, and $N(b+g\delta)$ is in (ag) because ag divides $N(b+g\delta)$. Therefore $I\sigma(I) \subseteq (ag)$. Since $N(I) = ag$ by Problem 17, Problem 19c shows that $N(I\sigma(I)) = N((ag))$. Then $I\sigma(I) = (ag) = (N(I))$ by Problem 18b.

20. The only ideal I with $N(I) = 1$ is $I = R$. Problem 19c therefore shows that a nontrivial factorization of $(p)R$ leads to a nontrivial factorization of its norm, which is p^2. This factorization must be $p^2 = p \cdot p$, and thus I factors nontrivially at most into two factors, each with norm p.

21. For (a), we use Problem 15 to write a nontrivial factor I of $(2)R$ as $I = \langle a, b + g\delta \rangle$. Problem 17 shows that $2 = N(I) = ag$ with g dividing a. Therefore $a = 2$ and $g = 1$. So the only possible factors are of the form $I = \langle 2, b + \delta \rangle$ with $0 \le b < a = 2$. Thus $b = 0$ or $b = 1$. When D is odd, we have $\text{Tr}(\delta) = 1$ and $N(\delta) = \frac{1}{4}(1 - m)$. Then $N(b + \delta) = b^2 + b\,\text{Tr}(\delta) + N(\delta) = b^2 + b + \frac{1}{4}(1 - m) \equiv \frac{1}{4}(1 - m) \bmod 2$. If $m \equiv 5 \bmod 8$, then we see that 2 does not divide $N(b + \delta)$, and thus $(2)R$ cannot have a nontrivial factor.

For (b), we again have $N(b + \delta) = b^2 + b\,\text{Tr}(\delta) + N(\delta) = b^2 + b + \frac{1}{4}(1 - m) \equiv \frac{1}{4}(1 - m) \bmod 2$, and the condition $m \equiv 1 \bmod 8$ makes the right side 0. Thus 2 divides $N(b+\delta)$, and $\langle 2, \delta \rangle$ and $\langle 2, 1+\delta \rangle$ are both ideals by Problem 16. The product of these ideals is $\langle 2, \delta \rangle \langle 2, 1 + \delta \rangle = (4, 2\delta, 2(1 + \delta), \delta^2)$ and contains $(2)R$ because $2 = 2(1 + \delta) - 2\delta$. Moreover, the product has norm 4 by Problems 17 and 19c, and this matches the norm of $(2)R$. Thus Problem 18b shows that $\langle 2, \delta \rangle \langle 2, 1+\delta \rangle = (2)R$.

For (c) and (d), $\delta = -\sqrt{m}$. Thus $N(b + \delta) = b^2 + b\,\text{Tr}(\delta) + N(\delta) = b^2 - m = b - \frac{1}{4}D$. If $D/4 \equiv 3 \bmod 4$, then $b - \frac{1}{4}D$ is divisible by 2 for $b = 1$. If $D/4 \equiv 2 \bmod 4$, then $b - \frac{1}{4}D$ is divisible by 2 for $b = 0$. With b taking on the appropriate value in the two cases, $\langle 2, b + \delta \rangle$ is an ideal by Problem 16. The square of this ideal is $(4, 2(b + \delta), (b - \sqrt{m})^2) = (4, 2(b + \delta), b^2 + m - 2m\sqrt{b})$. The definition of b makes $b^2 + m$ even in every case, and hence $\langle 2, b + \delta \rangle^2 \supseteq (2)R$. Since the norms of the ideals on the two sides are both 4, the two ideals must be equal.

Chapter I   655

22. Arguing as in the previous problem, we see that any nontrivial factor of $(p)R$ must have norm p and therefore must be given by $\langle p, x + \delta \rangle$ for some x such that p divides $N(x + \delta) = x^2 + x\operatorname{Tr}(\delta) + N(\delta)$.

For (a), $\operatorname{Tr}(\delta) = 1$ and $N(\delta) = \frac{1}{4}(1 - m) = \frac{1}{4}(1 - D)$, and the condition is that p divide $x^2 + x + \frac{1}{4}(1 - D)$. This means that $x^2 + x + \frac{1}{4}(1 - D) \equiv 0 \bmod p$ is to have a solution. When this happens, Problem 16 ensures that $\langle p, x + \delta \rangle$ is an ideal. Then $\langle p, x + \sigma(\delta) \rangle$ is an ideal as well, and the product of the two is $(p^2, p(x + \delta), p(x + \sigma(\delta)), N(x + \delta))$. Since p divides $N(x + \delta)$, this product ideal is contained in $(p)R$. The product ideal and $(p)R$ both have norm p^2, and therefore they are equal.

For (b), $\operatorname{Tr}(\delta) = 0$ and $N(\delta) = -m = -D/4$, and the condition is that p divide $x^2 - D/4$. This means that $x^2 - D/4 \equiv 0 \bmod p$ is to have a solution. When this happens, Problem 16 ensures that $\langle p, x + \delta \rangle$ is an ideal. Then $\langle p, x + \sigma(\delta) \rangle$ is an ideal as well, and the product of the two is $(p^2, p(x + \delta), p(x + \sigma(\delta)), N(x + \delta))$. Since p divides $N(x + \delta)$, this product ideal is contained in $(p)R$. The product ideal and $(p)R$ both have norm p^2, and therefore they are equal.

For (c), the respective conditions for factorization in (a) and (b) are that $x^2 + x + \frac{1}{4}(1 - D) \equiv 0 \bmod p$ and $x^2 - D/4 \equiv 0 \bmod p$ be solvable. In both cases the quadratic expression on the left side has discriminant D. Hence factorization occurs if and only if D is a square modulo p.

23. In both cases we are assuming that $(p)R$ has a factor $I = \langle p, x + \delta \rangle$ with $0 \le x < p$. Using Problem 15, let us write $\sigma(I) = \langle p, x + \sigma(\delta) \rangle = \langle p, y + \delta \rangle$ with $0 \le y < p$. Choose integers c and d with $x + \sigma(\delta) = cp + d(y + \delta)$. Since $\sigma(\delta) = \operatorname{Tr}(\delta) - \delta$, the equation is $x + \operatorname{Tr}(\delta) - \delta = cp + dy + d\delta$, and we obtain $x + \operatorname{Tr}(\delta) = cp + dy$ and $-\delta = d\delta$. Thus $d = -1$, $x + \operatorname{Tr}(\delta) = cp - y$, and $cp = x + y + \operatorname{Tr}(\delta)$. From $0 \le x < p$ and $0 \le y < p$, we have $0 \le x + y + \operatorname{Tr}(\delta) \le 2(p - 1) + \operatorname{Tr}(\delta) \le 2p - 1$. So c in the equation $cp = x + y + \operatorname{Tr}(\delta)$ has to be 1 or 0, and the equation is $x + y = p - \operatorname{Tr}(\delta)$ or $x + y = -\operatorname{Tr}(\delta)$. The condition that $\sigma(I) = I$ is the condition that $x = y$, hence that $2x = p - \operatorname{Tr}(\delta)$ or $2x = -\operatorname{Tr}(\delta)$. When D is odd, this says that $x = \frac{1}{2}(p - 1)$; when D is even, it says that $x = 0$.

24. Since $\sigma(\langle p, x + \delta \rangle) = \langle p, x + \sigma(\delta) \rangle$, the two factors are the same if and only if $\sigma(I) = I$. Problem 23 says that the latter equality holds for D odd if and only if $x = \frac{1}{2}(p - 1)$ and that it holds for D even if and only if $x = 0$. In the two cases we know from Problem 14 that p divides $N(x + \delta) = x^2 + x\operatorname{Tr}(\delta) + N(\delta)$.

When D is odd, this result says that p divides $x^2 + x + \frac{1}{4}(1 - D)$, hence that it divides $4x^2 + 4x + (1 - D) = (2x + 1)^2 - D$. Then p divides D if and only if p divides $2x + 1$, if and only if $x = \frac{1}{2}(p - 1)$.

When D is even, we know from Problem 14 that p divides $x^2 - m$. Hence p divides $4(x^2 - m) = 4x^2 - D = (2x)^2 - D$. Then p divides D if and only if p divides $2x$, if and only if $x = 0$.

25. Theorem 1.14 shows that the genus group G is the quotient of the abelian group H modulo its subgroup of squares. The subgroup of squares consists of the elements

in the product of the cyclic subgroups of orders $2^{k_1-1}, \ldots, 2^{k_r-1}, q_1^{l_1}, \ldots, q_s^{l_s}$, and the quotient is the product of r copies of a cyclic group of order 2. Thus G has order 2^r. The subgroup of elements of H whose order divides 2 is the product of the 2-element subgroups of the cyclic groups of orders $2^{k_1}, \ldots, 2^{k_r}$. It is a product of r copies of a cyclic group of order 2 and hence is abstractly isomorphic to G.

26. If P is a nonzero prime ideal, then so is $\sigma(P)$. Since $\sigma^2 = 1$, the mapping $P \mapsto \sigma(P)$ is a permutation of order 2 on the nonzero prime ideals. Evidently the prime ideals of type (i) above are permuted in 2-cycles, and the prime ideals of types (ii) and (iii) are left fixed.

If a nonzero ideal I has prime factorization $I = \prod_i P_i^{k_i}$, then $\sigma(I) = \prod_i \sigma(P_i)^{k_i}$. When $\sigma(I) = I$, we can match the factors and their exponents. We conclude that the factorization of I is as

$$I = \Big(\prod_{\substack{\text{pairs } (P_i, \sigma(P_i)) \\ \text{of type (i)}}} (P_i \sigma(P_i))^{k_i}\Big)\Big(\prod_{\substack{\text{ideals } P_i \\ \text{of type (ii)}}} P_i^{k_i}\Big)\Big(\prod_{\substack{\text{ideals } P_i \\ \text{of type (iii)}}} P_i^{k_i}\Big).$$

Each factor in the first product is of the form $(N(P_i))^{k_i}$ by Problem 19d, each factor in the second product is of the form $(p)^{k_i}$ for some prime p not dividing D, and each P_i^2 contributing to the third factor is of the form (p) for some prime p dividing D. The result follows.

27. For (a), the only nontrivial step in the displayed formula is the third equality, which follows because $x\sigma(x) = N(x) = 1$ by hypothesis. If we take $y = (1+x)^{-1}$, then the displayed formula gives $x = (1+x)(1+\sigma(x))^{-1} = y^{-1}\sigma(y)$ as required.

For (b), the equality $\sigma(y)y^{-1} = x$ remains valid when y is replaced by ny with $n \in \mathbb{Z}$, and thus we may take y to be in R. Now let y and z be in R with $\sigma(z)z^{-1} = x = \sigma(y)y^{-1}$. Then $\sigma(zy^{-1}) = zy^{-1}$, and zy^{-1} is in $\mathbb{Q}$. Among all $y \in R$ with $\sigma(y)y^{-1} = x$, let y_0 be one with $|N(y)|$ as small as possible; y_0 exists because $|N(y)|$ is an integer in each case. If $\sigma(z)z^{-1} = x$, write $z = u + v\delta$, $y_0 = a + b\delta$, and $zy_0^{-1} = p/q$ with $\text{GCD}(p, q) = 1$. Then $qu + qv\delta = qz = py_0 = pa + pb\delta$, and we obtain $qu = pa$ and $qv = pb$. Therefore q divides a and b, and $q^{-1}y_0 = q^{-1}a + q^{-1}b\delta$ is in R. Then $y = q^{-1}y_0$ is another element in R with $\sigma(y)y^{-1} = x$, and it contradicts the minimal choice of $|N(y_0)|$ unless $|q| = 1$. We conclude that $z = \pm py_0$.

28. In (a), $N(I^2) = N((x))$ says that $N(I)^2 = |N(x)|N(R) = |N(x)|$. Therefore $N(x^{-1}N(I)) = |N(x)|^{-1}N((N(I)) = |N(x)|^{-1}N(I)^2 = 1$, and $xN(I)^{-1}$ has norm 1.

In (b), Problem 27b gives us $y_0 \in R$ with $\sigma(y_0)y_0^{-1} = xN(I)^{-1}$. Then we compute that $\sigma((y_0)I) = \sigma(y_0)\sigma(I) = y_0xN(I)^{-1}\sigma(I) = y_0N(I)^{-1}(x)\sigma(I) = y_0N(I)^{-1}I^2\sigma(I) = y_0N(I)^{-1}((N(I))I = y_0I$.

For (c), suppose $N(y_0) > 0$. Then Problem 26a shows that $(y_0)I = (a)J_S$ for some $a \in \mathbb{Z}$, and this gives the required strict equivalence. If $N(y_0) < 0$, then $N(y_0\sqrt{m}) > 0$, and $\sigma((y_0\sqrt{m})I) = (y_0\sqrt{m})I$; Problem 26a shows that $(y_0\sqrt{m})I = (a)J_S$ for some $a \in \mathbb{Z}$, and this gives the required strict equivalence.

29. For (a), since $m < 0$ and m is neither -1 nor -3, the possible units are $\varepsilon = \pm 1$. The equality $\sigma(x) = \varepsilon x$ says that x is in $\mathbb{Z}$ if $\varepsilon = +1$, and it says that x is in $\mathbb{Z}\sqrt{m}$ if $\varepsilon = -1$.

For (b), when $m = -1$ or $m = -3$, we have $D = -4$ or $D = -3$; thus $g = 0$, and there is nothing to prove. For other values of $m < 0$, consider J_S. Then $N(J_S) = \prod_{p \in S} p$, and this is some divisor D' of D with no repeated factors. Let us write $J_S = \langle a, b + g\delta \rangle$ by Problem 15. Then $ag = D'$ and g divides a. Since D' is square free, $a = D'$ and $g = 1$. If J_S is principal, then (a) shows that $J_S = (c)$ for an integer c or $J_S = (d\sqrt{m})$ for an integer d.

Suppose $J_S = (c)$. Then $b + \delta = rc$ for some $r \in R$. Write $r = x + y\delta$ for integers x and y. Then $b + \delta = cx + cy\delta$ shows that $1 = cy$ and hence that c divides 1. Thus $J_S = R$, and the set S is empty.

Suppose $J_S = (d\sqrt{m})$. Then $b + \delta = dx\sqrt{m} + dy\delta\sqrt{m}$ for some integers x and y. If D is odd, then the equation reads $b + \frac{1}{2}(1 - \sqrt{m}) = dx\sqrt{m} + dy\frac{1}{2}(1 - \sqrt{m})\sqrt{m}$. This implies that $-\frac{1}{2}\sqrt{m} = d(x + \frac{1}{2}dy)\sqrt{m}$, hence that $-1 = d(2x + 1)$. Therefore $d = 1$, $J_S = (\sqrt{m}) = (\sqrt{D})$, $N(J_S) = |D|$, and $S = E$. If D is even, then the equation reads $b - \sqrt{m} = dx\sqrt{m} - dym$, and we obtain $-1 = dx$. So $d = 1$, $J_S = (\sqrt{m})$, $N(J_S) = m = D/4 = D'$. This is the product of all prime divisors of D if $D/4 \equiv 2 \bmod 4$ and all of them but 2 if $D/4 \equiv 3 \bmod 4$.

For (c), let E' be a subset of g members of E, and assume that the element of E that is not in E' is not 2 unless $D = -4$. If S and S' are two subsets of E', then $J_S J_{S'} = (n) J_T$, where $n = \prod_{p \in S \cap S'} p$ and $T = (S - S') \cup (S' - S)$. If J_S and $J_{S'}$ represent the same genera, then $J_S J_{S'}$ is principal, and J_T must be principal. The set T can be empty only if $S = S'$, and it has to be a subset of E' and thus cannot be all of E. According to (b), the only way that J_T can be principal is thus that $S = S'$ or that all of the conditions D even, $D/4 \equiv 2 \bmod 4$, and $T = E' = E - \{2\}$ are satisfied. In the latter case the construction of E' shows that $D = -4$, T is empty, and $S = S'$. Thus the ideals J_S for $S \subseteq E'$ represent distinct genera in every case.

For (d), the roots of unity are $\pm \varepsilon_1^k$. Since $N(\varepsilon_1) = -1$, the roots of unity of norm 1 are the $\pm \varepsilon_1^{2n}$. So suppose that $\varepsilon = \pm \varepsilon_1^{2n}$. Put $\varepsilon_0 = \varepsilon_1^n$. Then $\varepsilon_0 \sigma(\varepsilon_0) = N(\varepsilon_0) = (-1)^n$, and $\sigma(\varepsilon_1^n x) = \sigma(\varepsilon_0) \sigma(x) = \sigma(\varepsilon_0) \varepsilon x = (-1)^n \varepsilon_0^{-1} \varepsilon x = \pm(-1)^n \varepsilon_1^{-n} \varepsilon_1^{2n} x = \pm(-1)^n \varepsilon_1^n x = s \varepsilon_1^n x$ with $s = \pm(-1)^n$. If $s = +1$, then $\varepsilon_1^n x$ is in $\mathbb{Z}$, while if $s = -1$, then $\varepsilon_1^n x$ is in $\mathbb{Z}\sqrt{m}$. Then the same steps as in (b) and (c) finish the argument.

For (e), the four mentioned ideals are principal, and we have $(1) = J_S$ for S empty and $(\sqrt{m}) = J_S$ for S equal to the set of prime divisors of m. For these two ideals, $N(1) > 0$ and $N(\sqrt{m}) < 0$. Consider (y_0^+) and (y_0^-). The ideal (y_0^+) has $\sigma((y_0^+)) = (\sigma(y_0^+)) = (y_0^+ \varepsilon_1) = (y_0^+)$, and hence it is of the form $(n) J_S$ for some S. Then $y_0^+ = nr$ for some $r \in R$, and it follows that $n^{-1} y_0^+$ is in R. This contradicts the minimality of $|N(y_0^+)|$ unless $|n| = 1$. Hence $(y_0^+) = J_S$ for some S. Similarly $(y_0^-) = J_S$ for some S. Thus all four principal ideals are of the form J_S.

Let us see that the four principal ideals are distinct. Neither ideal (y_0^+) nor (y_0^-) can equal (1). In fact, if (y_0^+) were to equal (1), then y_0^+ would be a unit ε, and we would have $\varepsilon_1 = \sigma(y_0^+)(y_0^+)^{-1} = \sigma(\varepsilon)\varepsilon^{-1} = \varepsilon^{-2}$, in contradiction to the fact that ε_1 is fundamental. Similarly (y_0^-) cannot equal (1).

Since $\sigma(y_0^+\sqrt{m})(y_0^+\sqrt{m})^{-1} = -\sigma(y_0^+)\sqrt{m}(y_0^+)^{-1}(\sqrt{m})^{-1} = -\sigma(y_0^+)(y_0^+)^{-1} = -\varepsilon_1$, the definition of y_0^- shows that $y_0^+\sqrt{m} = ny_0^-$ for some integer n. Passing to norms gives $-mN(y_0^+) = n^2 N(y_0^-)$. Therefore $N(y_0^+)$ and $N(y_0^-)$ have opposite sign.

We have seen that two of the four elements $1, y_0^+, y_0^-, \sqrt{m}$ have positive norm, two have negative norm, and the two of positive norm generate distinct principal ideals. To see that the two of negative norm generate distinct ideals, we consider separately the cases $N(y_0^-) < 0$ and $N(y_0^+) < 0$. If $N(y_0^-) < 0$, we use the equation $-mN(y_0^+) = n^2 N(y_0^-)$ proved in the previous paragraph. If $(y_0^-) = (\sqrt{m})$, then cancellation gives $N(y_0^+) = +1$; then y_0^+ is a unit, and we have seen that it cannot be. If $N(y_0^+) < 0$, we use the definition of y_0^+ in the same way as in the previous paragraph to obtain $-mN(y_0^-) = n^2 N(y_0^+)$ for some integer n. Cancellation shows that $N(y_0^-) = +1$; then y_0^- is a unit, and we have seen that it cannot be. Thus the four principal ideals are distinct.

Now suppose that (x) is any principal ideal fixed by σ. As in the statement of the problem, we have $\sigma(x) = \varepsilon x$ for some unit ε. The most general unit is of the form $\varepsilon = \pm\varepsilon_1^n$. We shall produce constructively the element of Problem 27 corresponding to ε. Put $y_{0,n} = \varepsilon_1^{n/2}$ if n is even and $y_{0,n} = \varepsilon_1^{(n+1)/2} y_0$ if n is odd. For n even we have
$$\sigma(y_{0,n}x) = \sigma(y_{0,n})\varepsilon x = \pm\sigma(\varepsilon_1^{n/2})\varepsilon_1^n x = \pm\varepsilon_1^{-n/2}\varepsilon_1^n x = \pm y_{0,n}x,$$
and for n odd we have
$$\sigma(y_{0,n}x) = \sigma(y_{0,n})\varepsilon x = \pm\sigma(\varepsilon_1^{(n+1)/2}y_0)\varepsilon_1^n x = \pm\varepsilon_1^{-(n+1)/2}\sigma(y_0)\varepsilon_1^n x$$
$$= \pm\varepsilon_1^{(n-1)/2}\sigma(y_0)x = \pm\varepsilon_1^{(n-1)/2}y_0\varepsilon_1 x = \pm y_{0,n}x.$$

Thus $\sigma(y_{0,n}x) = \pm y_{0,n}x$ for all n. Therefore $y_{0,n}x$ is in $\mathbb{Z}$ or in $\mathbb{Z}\sqrt{m}$, depending on the sign $\pm$. Depending on the sign, $|N(y_{0,n}x)| = |N(y_{0,n})||N(x)|$ thus is either the square of an integer or m times the square of an integer. If n is even, then $|N(y_{0,n})| = 1$, and $|N(x)|$ is therefore either the square of an integer or m times the square of an integer. Since $|N(x)|$ is the value of the norm of (x), there are only two possible S's for which this can happen. If n is odd, then $|N(y_{0,n})| = a$ for a certain square-free integer > 1, as we have seen. Therefore $|N(x)|$ has to be either a^{-1} times the square of an integer or ma^{-1} times the square of an integer. So there are only two possible S's in this case. Thus there are only four possible S's in all cases, and these have been accounted for. So the number of principal ideals among the J_S's is exactly four. To complete the proof, we now argue as in (c) but consider only possibilities for which the product of two J_S's is n^2 times one of the two J_S's given by a principal ideal with a generator of positive norm.

30. Since D is fundamental, (a_1, b_1, c_1) is automatically primitive. Then Lemma 1.10 produces a properly equivalent form that represents some integer a relatively

prime to D. The rest follows from the argument in the second paragraph of the proof of sufficiency in Theorem 6b.

31. For (a), choose an integer r such that $b + 2ar = kD$ for some integer k; this is possible because $\gcd(D, 2a) = 1$. Then the translation $x = x' + ry'$, $y = y'$ leads from $ax^2 + bxy + cy^2$ to $ax'^2 + kDx'y' + c'y'^2$ for some c'. The discriminant of the new form is still $D = k^2 D^2 - 4ac'$, and thus $4ac' \equiv 0 \bmod D$. Since $\gcd(4a, D) = 1$, $c' \equiv 0 \bmod D$.

For (b), b has to be even because $D = b^2 - 4ac$ is even. Write $b = 2\bar{b}$. Choose an integer s such that $\bar{b} + as = kD$ for some k; this is possible because $\gcd(a, D) = 1$. Then the translation $x = x' + sy'$, $y = y'$ leads from $ax^2 + bxy + cy^2$ to $ax'^2 + 2kDx'y' + c'y'^2$ for some c'. The discriminant of the new form is $D = 4k^2 D^2 - 4ac'$, where $c' = (4a)^{-1} D(4k^2 D - 1) = a^{-1}(D/4)(4k^2 D - 1)$. Modulo D, this expression is $-\bar{a}(D/4)$, where $\bar{a}$ is an integer with $\bar{a} a \equiv 1 \bmod D$. Here a is odd, and hence $a^2 \equiv 1 \bmod 8$. If 2^u is the exact power of 2 dividing D, then $\bar{a} a \equiv 1 \bmod 2^u$, and hence $\bar{a} \equiv a \bmod 2^u$. If p is any odd prime dividing D, then p divides $D/4$, and hence $\bar{a}(D/4) \equiv 0 \equiv a(D/4) \bmod p$. Therefore $\bar{a}(D/4) \equiv a(D/4) \bmod D$, and we conclude that $c' \equiv -a(D/4) \bmod D$.

32. For (a), clearing fractions in the expression $ax^2 + kDxy + lDy^2 = r$ yields $au^2 + kDuv + lDv^2 = rw^2$. Suppose a prime p divides $\gcd(w, D)$. Then p divides au^2. Since $\gcd(a, D) = 1$, p divides u. Referring back to the equation, we see that p^2 divides au^2 and $kDuv$, hence divides lDv^2. Thus p divides lv^2. The discriminant is $D = k^2 D^2 - 4alD$, and divisibility of l by p would force p^2 to divide the left side D. Hence p does not divide l, and p must divide v. Then p divides both u and v, in contradiction to the minimality of the common denominator w. We conclude that $\gcd(w, D) = 1$. Taking the equation $au^2 + kDuv + lDv^2 = rw^2$ modulo D gives $au^2 \equiv rw^2 \bmod D$. Since r and w are relatively prime to D, so is u. Thus we can rewrite this congruence as $a \equiv d^2 r \bmod D$ for some integer d relatively prime to D.

For (b), the same argument gives $a' \equiv d'^2 r \bmod D$. Since d is relatively prime to D, we can rewrite the congruence for a as $r \equiv d^{-2} a \bmod D$, and then $a' \equiv d'^2 r \equiv (d^{-1} d')^2 a \bmod D$.

For (c), the given forms are properly equivalent over $\mathbb{Z}$ to (a, kD, lD) and to $(a', k'D, l'D)$, respectively, by Problem 31a. Proper equivalence over $\mathbb{Q}$ means that the two forms take on the same rational values, one of which is the integer a'. Part (b) therefore shows that $a' = as^2 + nD$ for some integers s and n, necessarily with $\gcd(s, D) = 1$. Modulo D, the forms are given by ax^2 and $a'x'^2$, and the first can be transformed into the second by the substitution $x = sx'$, $y = s^{-1}y'$, where s^{-1} is the multiplicative inverse of s in $\mathbb{Z}/D\mathbb{Z}$. In fact, substitution into ax^2 gives $a(sx')^2 = (as^2)x'^2 \equiv a'x'^2 \bmod D$. This substitution is given by the matrix $\begin{pmatrix} s & 0 \\ 0 & s^{-1} \end{pmatrix}$ in $\mathrm{SL}(2, \mathbb{Z}/D\mathbb{Z})$.

33. Part (a) is almost the same as Problem 32a. Clearing fractions leads to $au^2 + kDuv + (lD - a(D/4))v^2 = rw^2$, and the argument that no odd prime p divides $\gcd(w, D)$ is the same. Suppose that 2 divides w. The equation modulo 4

is then $au^2 - a(D/4)v^2 \equiv 0 \bmod 4$ with $D/4$ congruent to 2 or 3 modulo 4. Since 2 divides w, at least one of u and v must be odd. If $D/4 \equiv 3 \bmod 4$, the congruence becomes $a(u^2 + v^2) \equiv 0 \bmod 4$, which is impossible with at least one of u and v odd. If $D/4 \equiv 2 \bmod 4$, the congruence becomes $a(u^2 + 2v^2) \equiv 0 \bmod 4$, which again is impossible with at least one of u and v odd. Thus $\gcd(w, D) = 1$. Taking the equation modulo D and using the invertibility of r and w modulo D, we have $ar^{-1}w^{-2}(u^2 - (D/4)v^2) \equiv 1 \bmod D$.

For (b), let p be an odd prime divisor of D. The above congruence then becomes $ar^{-1}w^{-2}u^2 \equiv 1 \bmod p$. Similarly with the second form, there is some w' prime to D such that $a'r^{-1}w'^{-2}u'^2 \equiv 1 \bmod p$. Comparing the two expressions, we see that a modulo p is the product of a' and an invertible square.

For (c), the above congruence becomes $ar^{-1}w^{-2}(u^2+v^2) \equiv 1 \bmod 4$. This forces $u^2 + v^2 \equiv 1 \bmod 4$. Since w has to be odd, $w^2 \equiv 1 \bmod 4$. Hence $ar^{-1} \equiv 1 \bmod 4$. Similarly $a'r^{-1} \equiv 1 \bmod 4$, and therefore $a \equiv a' \bmod 4$.

For (d), the above congruence becomes $ar^{-1}(u^2 - (D/4)v^2) \equiv 1 \bmod 8$, since w is odd. If $D/4 \equiv 2 \bmod 8$, we obtain $ar^{-1}(u^2 - 2v^2) \equiv 1 \bmod 8$. Here u has to be odd, and thus $ar^{-1}(1 - 2v^2) \equiv 1 \bmod 8$. If v is even, this says that $a \equiv r \bmod 8$; if v is odd, it says that $a \equiv -r \bmod 8$. Putting this conclusion together with a similar conclusion about the second form, we obtain $a' \equiv \pm a \bmod 8$.

If $D/4 \equiv 6 \bmod 8$, we obtain $ar^{-1}(u^2 + 2v^2) \equiv 1 \bmod 8$. Here u has to be odd, and thus $ar^{-1}(1 + 2v^2) \equiv 1 \bmod 8$. If v is even, this says that $a \equiv r \bmod 8$; if v is odd, it says that $a \equiv 3r \bmod 8$. Putting this conclusion together with a similar conclusion about the second form, we obtain $a' \equiv a \bmod 8$ or $a' \equiv 3a \bmod 8$.

For (e), we shall assemble a member of $\mathrm{SL}(2, \mathbb{Z}/D\mathbb{Z})$ one prime at a time and use the Chinese Remainder Theorem. For odd primes p dividing D, choose s_p with $a' \equiv s_p^2 a \bmod p$, and introduce the matrix $M_p = \begin{pmatrix} s_p & 0 \\ 0 & s_p^{-1} \end{pmatrix}$ in $\mathrm{SL}(2, \mathbb{Z}/p\mathbb{Z})$. If $D/4 \equiv 3 \bmod 4$, introduce the matrix $M_2 = \begin{pmatrix} 1 & 0 \\ 0 & 1 \end{pmatrix}$ in $\mathrm{SL}(2, \mathbb{Z}/4\mathbb{Z})$. If $D/4 \equiv 2 \bmod 4$, let $M_2 = \begin{pmatrix} 1 & 6 \\ 1 & 7 \end{pmatrix}$ in $\mathrm{SL}(2, \mathbb{Z}/8\mathbb{Z})$ if $D/4 \equiv 6 \bmod 8$, and let $M_2 = \begin{pmatrix} 1 & 2 \\ 1 & 3 \end{pmatrix}$ in $\mathrm{SL}(2, \mathbb{Z}/8\mathbb{Z})$ if $D/4 \equiv 2 \bmod 8$. The Chinese Remainder Theorem produces a unique matrix with entries in $\mathbb{Z}/D\mathbb{Z}$ that is congruent to M_p modulo each odd prime divisor of D and is congruent to M_2 modulo the power of 2 dividing D. Call this matrix $M = \begin{pmatrix} \alpha & \beta \\ \gamma & \delta \end{pmatrix}$. It has determinant 1 modulo D and hence lies in $\mathrm{SL}(2, \mathbb{Z}/D\mathbb{Z})$. Then substitution of $x = \alpha x' + \beta y'$ and $y = \gamma x' + \delta y'$ into the form $a(x^2 - (D/4)y^2)$ modulo D leads to the form $a'(x^2 - (D/4)y^2)$ modulo D.

34. These problems establish a function from the set of equivalence classes of binary quadratic forms over $\mathbb{Z}$ with discriminant D, the equivalence relation being proper equivalence over $\mathbb{Q}$, *onto* the set of equivalence classes of binary quadratic forms over $\mathbb{Z}$ with discriminant D, the equivalence relation being proper equivalence over $\mathbb{Z}/D\mathbb{Z}$. The number of elements in the domain has to be $\geq$ the number of elements in the range.

35. The steps in solving Problems 32 and 33 involve relating a to r modulo each prime power dividing D. These relationships are the same as the relationships between a and r' if the form modulo D represents r' and $\mathrm{GCD}(r', D) = 1$, and the relationships are transitive. Thus the genus characters take the same values at r as they do at r', and they take the same values at a as well.

36. Multiplication is the operation on proper equivalence classes of forms that corresponds to composition of aligned representatives of the classes, and composition is defined in such a way that the set of values of the composition is the set of products of a value of one form by a value of the other. The values are unaffected by proper equivalence over $\mathbb{Z}$.

37. For (a), $D/4$ has an odd number $2t + 1$ of prime factors $4k + 3$. Use of the Jacobi symbol with a odd and p varying over the prime divisors of $D/4$ gives

$$\prod_p \left(\tfrac{a}{p}\right) = \prod_{p=4k+1}\left(\tfrac{a}{p}\right) \prod_{p=4k+3}\left(\tfrac{a}{p}\right) = \xi(a)^{2t+1} \prod_{p=4k+1}\left(\tfrac{p}{a}\right) \prod_{p=4k+3}\left(\tfrac{p}{a}\right) = \xi(a)\left(\tfrac{D/4}{a}\right).$$

Therefore

$$\xi(a) \prod_p \left(\tfrac{a}{p}\right) = \left(\tfrac{D/4}{a}\right) = \left(\tfrac{2}{a}\right)^2 \left(\tfrac{D/4}{a}\right) = \left(\tfrac{D}{a}\right).$$

For (b) and (c), say that the number of prime factors $4k + 3$ of $D/8$ is t. With p varying over the odd prime divisors of D, the same computation as above gives $\prod_p \left(\tfrac{a}{p}\right) = \xi(a)^t \left(\tfrac{D/8}{a}\right)$. Then $\left(\tfrac{D}{a}\right) = \left(\tfrac{2}{a}\right)\left(\tfrac{D/8}{a}\right) = \eta(a)\xi(a)^t \prod_p \left(\tfrac{a}{p}\right)$. One easily checks that t is even if $D/4 \equiv 2 \bmod 8$ and is odd if $D/4 \equiv 6 \bmod 8$, and the result follows.

38. For each odd prime divisor p of D, choose a residue r_p modulo p such that $\left(\tfrac{r_p}{p}\right) = s_p$. If D is even, choose an odd residue r_2 modulo 8 such that $\alpha(r_2) = s_2$. The Chinese Remainder Theorem produces an integer b prime to D such that $b \equiv r_p \bmod p$ for the odd p's and $b \equiv r_2 \bmod 8$. For this integer b and every $k \geq 0$, we have $\left(\tfrac{b+kD}{p}\right) = r_p$ for each odd p and $\alpha(b + kD) = s_2$. Dirichlet's Theorem says that $b + kD$ is a prime q for a suitable choice of k, and this prime q has the required properties.

39. Problem 37 showed that the product of the genus characters for an odd integer a such that $\mathrm{GCD}(a, D) = 1$ is $\left(\tfrac{D}{a}\right)$. Using the genus characters at $a = q$, we see that $\left(\tfrac{D}{q}\right) = 1$. Theorem 1.6b shows that q is primitively representable by some form (q, b, c) of discriminant D. The values of the genus characters for this form are their values on q, and we have arranged that these values are the various numbers s_p. Since there are $g + 1$ genus characters and the first g of them can be specified arbitrarily and still give a similarity class modulo D, there are at least 2^g similarity classes modulo D.

40. Problem 29 shows that the number of classes of type (i) is exactly 2^g. Problems 30–33 show that equivalence of type (i) implies equivalence of type (ii), and they therefore give a mapping of the set of classes of type (i) onto the set of classes of

type (ii). The definition of "similar modulo D" immediately implies that equivalence of type (ii) implies equivalence of type (iii), and therefore we obtain a mapping of the set of classes of type (ii) onto the set of classes of type (iii). Finally Problem 39 shows that there are at least 2^g classes of type (iii). The result follows.

Chapter II

1. The unital left $\mathbb{C}G$ modules correspond (via the universal mapping property of a group algebra) to representations of G on complex vector spaces. The theory in Chapter VII of *Basic Algebra* shows that every representation splits as the direct sum of irreducible representations, which correspond to simple left $\mathbb{C}G$ modules. Hence every unital left $\mathbb{C}G$ module is semisimple. The left regular representation of G, which corresponds to the left $\mathbb{C}G$ module $\mathbb{C}G$, decomposes as the sum of irreducible representations, each irreducible representation occurring as many times as its degree. The sum of all the irreducible subspaces of a given isomorphism type gives one of the factors $M_n(\mathbb{C})$ of $\mathbb{C}G$, and every factor arises this way.

2. For (a), rad $A = (\mathbb{C} + \mathbb{C}X)(X^2 + 1)$, and S will be the sum of two copies of $\mathbb{C}$. Finding S requires some computation. We can identify $A/(\text{rad } A)$ with the quotient $\mathbb{C}[X]/(X^2 + 1)$, and direct computation shows that the two idempotents in this notation having sum 1 are $\frac{1}{2i}(X + i)$ and $-\frac{1}{2i}(X - i)$. The proof of Proposition 2.23 shows how to lift these to idempotents in A. For the first one, put $a = \frac{1}{2i}(X + i)$ and $b = 1 - a = -\frac{1}{2i}(X - i)$, and observe that $(ab)^2 = 0$. The proposition gives the formula $e = \sum_{k=0}^{2} \binom{4}{k} a^{4-k} b^k = a^4 + 4a^3 b$, the term for $k = 2$ being 0. Then $e = a^3(a + 4b) = \frac{1}{16}(X + i)^3(-3X + 5i)$. So one contribution to S comes from $\mathbb{C}e$; the other will come from the complex conjugate in the form of $\mathbb{C}f$, where $f = \frac{1}{16}(X - i)^3(-3X - 5i)$.

We can check directly that e is an idempotent. In fact,

$$e^2 - e = e\left[\tfrac{1}{16}(X + i)^3(-3X + 5i) - 1\right].$$

The polynomial in square brackets vanishes at $X = i$, and so does its derivative. Thus the polynomial is divisible by $(X - i)^2$, and $e^2 - e = (X + i)^3(-3X + 5i) \times [(X - i)^2 Q(X)]$ is divisible by $(X^2 + 1)^2$.

For (b), the answer is yes. This problem anticipates Problem 5 below. The algebra S is spanned linearly by its idempotents, and Problem 5 shows that the idempotents are determined uniquely in the commutative case.

For (c), rad $A = (\mathbb{R} + \mathbb{R}X)(X^2 + 1)$. Call the subalgebra S_0. This subalgebra will be a 2-dimensional real subalgebra isomorphic to $\mathbb{C}$. To find it, we can go through the proof of Theorem 2.17 or we can use the Galois group. The latter method is a good bit easier. Thus we seek those members of S as in (a) that are fixed by complex conjugation. Since $S = \mathbb{C}e + \mathbb{C}\bar{e}$, the result is that $S_0 = \mathbb{R}(e + \bar{e}) + i\mathbb{R}(e - \bar{e})$. This is unique; in fact, any choice of S_0 has the property that $S_0 \otimes_\mathbb{R} \mathbb{C}$ is an S for (a), and we know that the S for (a) is unique.

3. Since rad A is a nilpotent ideal of A, (rad A) $\otimes_F B$ is a nilpotent ideal of $A \otimes_F B$, and therefore (rad A) $\otimes_F B \subseteq \text{rad}(A \otimes_F B)$. For the reverse inclusion Proposition 2.31 shows that $\text{rad}(A \otimes_F B) = I \otimes_F B$ for some two-sided ideal of A. If $(\text{rad}(A \otimes_F B))^n = 0$ and $a_1, \ldots, a_n$ are in I, then $(a_1 \otimes 1) \cdots (a_n \otimes 1)$ must be 0, and hence $a_1 \cdots a_n = 0$. Therefore $I \subseteq \text{rad } A$, and $\text{rad}(A \otimes_F B) \subseteq (\text{rad } A) \otimes_F B$.

4. For (a), suppose on the contrary that there is an infinite sequence $M_1, M_2, \ldots$ of distinct maximal ideals. Then we obtain a decreasing sequence of ideals $R \supseteq M_1 \supseteq M_1 M_2 \supseteq M_1 M_2 M_3 \supseteq \cdots$, and the Artinian property shows that $M_1 \cdots M_n = M_1 \cdots M_n M_{n+1}$ for some n. Since M_{n+1} is prime and $M_{n+1} \supseteq M_1 \cdots M_n$, M_{n+1} contains M_j for some j with $1 \leq j \leq n$. By maximality, $M_n = M_j$, and we have a contradiction.

In (b), every element of rad R is nilpotent because rad R is nilpotent. Conversely if $x \in R$ is nilpotent with $x^n = 0$, then Rx is nilpotent with $(Rx)^n = 0$, since $a_1 x a_2 x \cdots a_n x = a_1 a_2 \cdots a_n x^n = 0$ for any $a_1, \ldots, a_n \in R$. Thus $Rx \subseteq \text{rad } R$, and the nilpotent element x lies in rad R. This proves (b), and (c) follows because R is semisimple if and only if rad $R = 0$.

For (d), R semisimple implies that R is a product of full matrix rings over division rings. Commutativity implies that the matrices are all of size 1-by-1 and the division rings are all fields.

5. If e' is a second representative, then $e' = e + r$ with $r \in \text{rad } R$. If n is an odd integer large enough to have $r^n = 0$, then

$$0 = r^n = (e' - e)^n = \sum_{k=0}^{n} (-1)^k \binom{n}{k} (e')^{n-k} e^k = e' + \sum_{k=1}^{n-1} (-1)^k \binom{n}{k} e' e - e$$

$$= e' + \left(\sum_{k=0}^{n} (-1)^k \binom{n}{k} \right) e' e - e' e + e' e - e = e' + 0 - e' e + e' e - e = e' - e.$$

6. Let $M_1, \ldots, M_n$ be the finitely many maximal ideals, and put $N = M_1 \cdots M_n$. Nakayama's Lemma says that if I is any ideal contained in all maximal ideals, then the only finitely generated unital R module M having the property that $IM = M$ is $M = 0$. The Artinian property shows that $N^{k+1} = N^k$ for some k. We take $I = N$ and $M = N^k$ in Nakayama's Lemma. The R module M is finitely generated because Artinian implies Noetherian (Theorem 2.15), and hence Nakayama's Lemma shows that $N^k = 0$.

7. Let the maximal ideals be $M_1, \ldots, M_n$, and let $(M_1 \cdots M_n)^k = 0$. If P is a prime ideal, then $P \supseteq 0 = (M_1 \cdots M_n)^k$. Since P is prime, P contains one of the factors. Thus $P \supseteq M_j$ for some j.

8. It helps to have a multiplication table available. If the rows index a factor on the left and the columns index a factor on the right, then the resulting products are given by $\begin{pmatrix} R & M & 0 \\ 0 & 0 & M \\ 0 & 0 & S \end{pmatrix}$.

If I_2 is a left ideal of S and I_1 is a left R submodule of $R \oplus M$ containing MI_2, then $RI_2 = 0$, $MI_2 \subseteq I_1$, and $SI_2 \subseteq I_2$. Also, $RI_1 \subseteq I_1$, $MI_1 = 0$, and $SI_1 = 0$. Thus $AI_1 \subseteq I_1$ and $AI_2 \subseteq I_1 \oplus I_2$. Consequently $I_1 \oplus I_2$ is a left ideal of A.

In the reverse direction if J is a left ideal in A, then $I_1 = \begin{pmatrix} 1 & 0 \\ 0 & 0 \end{pmatrix} J \subseteq R \oplus M$ and $I_2 = \begin{pmatrix} 0 & 0 \\ 0 & 1 \end{pmatrix} J \subseteq S$ are such that $J = I_1 \oplus I_2$. Also, $r \in R$ implies $\begin{pmatrix} r & 0 \\ 0 & 0 \end{pmatrix} \begin{pmatrix} 1 & 0 \\ 0 & 0 \end{pmatrix} J = \begin{pmatrix} 1 & 0 \\ 0 & 0 \end{pmatrix} rJ \subseteq I_1$, while $(M \oplus S)I_1 = 0$; and $s \in S$ implies $\begin{pmatrix} 0 & 0 \\ 0 & s \end{pmatrix} \begin{pmatrix} 0 & 0 \\ 0 & 1 \end{pmatrix} J = \begin{pmatrix} 0 & 0 \\ 0 & 1 \end{pmatrix} sJ \subseteq I_2$, while $RI_2 = 0$ and $m \in M$ implies $\begin{pmatrix} 0 & m \\ 0 & 0 \end{pmatrix} \begin{pmatrix} 0 & 0 \\ 0 & 1 \end{pmatrix} J = \begin{pmatrix} 0 & m \\ 0 & 0 \end{pmatrix} J \subseteq \begin{pmatrix} 1 & 0 \\ 0 & 0 \end{pmatrix} \begin{pmatrix} 0 & m \\ 0 & 0 \end{pmatrix} J \subseteq \begin{pmatrix} 1 & 0 \\ 0 & 0 \end{pmatrix} J = I_1$.

9. For (a), suppose A is left Noetherian. The table produced in the solution of Problem 8 shows that $M \oplus S$ and $R \oplus M$ are two-sided ideals of A, and the respective quotient rings are R and S. As quotients of a left Noetherian ring, R and S have to be left Noetherian. If $\{M_i\}$ is an ascending chain of R submodules of M, then $\left\{\begin{pmatrix} 0 & M_i \\ 0 & 0 \end{pmatrix}\right\}$ is an ascending chain of left ideals of A, by Problem 8. The latter must be constant from some point on, and then the same thing is true for $\{M_i\}$.

Conversely suppose that R and S are left Noetherian and that the left R module M satisfies the ascending chain condition. If $\{J_i\}$ is an ascending chain of left ideals of A, then the corresponding sequence $\{(I_2)_i\}$ is an ascending chain of left ideals in S, and $\{(I_1)_i\}$ is an ascending chain of left R submodules of $R \oplus M$ containing MI_2. Since S is left Noetherian, $\{(I_2)_i\}$ is constant from some point on. Since $R = (R \oplus M)/M$ and M satisfy the ascending chain condition for their left R submodules, so does $R \oplus M$, and therefore $\{(I_1)_i\}$ is constant from some point on.

10. In view of Problem 9a, showing that A is left Noetherian amounts to showing that R and S are (left) Noetherian and M satisfies the ascending chain condition for its left R submodules. The ring S is Noetherian by assumption, and R is a field, hence is Noetherian. The action of R on M is the action of a field on itself, and the R submodules are trivial. In view of Problem 9b, A fails to be right Noetherian if the ascending chain condition fails for the right S submodules of $M = R$. If the ascending chain condition were to hold, then R would be a finitely generated S module, and the only denominators needed for members of the full field R of fractions would be those dividing the product of the denominators of the generators; these fractions are already in S, and hence S would equal R, contradiction.

The analogs of the results of Problem 9 for the Artinian case show that A fails to be either left or right Artinian if S is not Artinian. If s is a nonunit in S, then the chain of principal ideals $\{(s^k)\}$ is properly descending, since $(s^k) = (s^{k+1})$ implies $\varepsilon s^k = s^{k+1}$ for some unit ε and since the hypothesis that S is an integral domain allows us to cancel and obtain $\varepsilon = s$, contradiction.

11. Since R and S are fields, they are left and right Noetherian and Artinian. In view of Problem 9, we are to show that $M = R$ satisfies both chain conditions for its left R modules and neither chain condition for its right S modules. Since R is a

field, $M = R$ has only trivial R submodules and satisfies both chain conditions. For the S action on R, we are to examine the S vector subspaces of S. Since $\dim_S R$ is infinite, there exist both a properly increasing sequence of such subspaces and a properly decreasing one. Hence neither chain condition is satisfied.

12. For (a), the vector-space dimension over $\mathbb{F}$ is certainly 4, and computation shows that A is closed under products. The choices $a = 1$ and $b = 0$ show that A has an identity.

For (b), let $x \neq 0$ be in a two-sided ideal I. If $x = \begin{pmatrix} a & 0 \\ 0 & \sigma(a) \end{pmatrix}$, then x is invertible, and hence $I = A$. Otherwise suppose that some matrix $x = \begin{pmatrix} a & b \\ r\sigma(b) & \sigma(a) \end{pmatrix}$ with $b \neq 0$ is in I. With c as in the statement of the problem, $cx - xc = \begin{pmatrix} 0 & 2b\sqrt{m} \\ -2r\sigma(b)\sqrt{m} & 0 \end{pmatrix}$ is in I; this matrix is invertible since $b \neq 0$, and thus $I = A$.

To see that A is central, let x be in the center. The computation $0 = cx - xc$ shows that $b = 0$. Thus x is of the form $\begin{pmatrix} a & 0 \\ 0 & \sigma(a) \end{pmatrix}$. Such an x does not commute with $\begin{pmatrix} 0 & 1 \\ r & 0 \end{pmatrix}$ unless $a = \sigma(a)$, in which case x is in F.

13. The determinant is $a\sigma(a) - rb\sigma(b) = N_{K/F}(a) - rN_{K/F}(b)$ and equals 0 for a given r if and only if some pair $(a, b) \neq (0, 0)$ has $N_{K/F}(a) = rN_{K/F}(b)$. Since $r \neq 0$, both a and b are nonzero, and this equality then holds if and only if $r = N_{K/F}(ab^{-1})$.

In other words, some nonzero member of A has determinant 0 if r is a norm, and then A cannot be a division algebra. Conversely if r is not a norm, then every nonzero member of A is invertible as a matrix. Computation of the inverse matrix shows that it has the correct form to be in A. Hence A is a division algebra.

When A is not a division algebra, it is anyway finite-dimensional and central simple and has to be of the form $M_n(D)$ for some n and some division algebra D over F such that $\dim M_n(D) = 4$. The dimensional formula says that $n^2 \dim_F D = 4$. Since $n \neq 1$, we must have $n = 2$ and $D = F$.

14. The isomorphism follows from the computation $\begin{pmatrix} c & 0 \\ 0 & 1 \end{pmatrix} \begin{pmatrix} a & b \\ r\sigma(b) & \sigma(a) \end{pmatrix} \begin{pmatrix} c & 0 \\ 0 & 1 \end{pmatrix}^{-1} = \begin{pmatrix} a & bc \\ rc^{-1}\sigma(b) & \sigma(a) \end{pmatrix} = \begin{pmatrix} a & bc \\ r'\sigma(c)\sigma(b) & \sigma(a) \end{pmatrix} = \begin{pmatrix} a & bc \\ r'\sigma(bc) & \sigma(a) \end{pmatrix}$.

15. Direct computation.

16. If K is a maximal subfield, then $\dim_F K = 2$. Since the characteristic is not 2, $K = F(\sqrt{m})$ for some nonsquare $m \in F$. Define $i \in K$ to be $\sqrt{m}$.

The map $f : K \to D$ given by $f(a + bi) = a - bi$ is an algebra homomorphism into the central simple algebra D. So the Skolem–Noether Theorem produces $j \in D$ with $j(a + bi)j^{-1} = a - bi$ for all $a + bi$ in K, necessarily with j invertible. As in the proof of Theorem 2.50, $j^2 = r$ lies in F. Define $k = ij$. Then $k^2 = ijij = i(jij^{-1})j^2 = i(-i)j^2 = -rm$, and $-rm = k^2 = ijk$ implies that $k = -rm(j^{-1})(i^{-1}) = -rm(r^{-1}j)(m^{-1}i) = -ji$.

Let us check the multiplication table for $\{1, i, j, k\}$. We know that $i^2 = m$, $j^2 = r$,

$k^2 = -rm, ij = k$, and $ji = -k$. In addition, we have

$$jk = jij = (jij^{-1})j^2 = (-i)r = -ri,$$
$$kj = ijj = i(j^2) = ri,$$
$$ki = iji = i(jij^{-1})j = i(-i)j = -mj,$$
$$ik = iij = (i^2)j = mj.$$

Hence the F linear map φ from A into the given central simple algebra is an algebra homomorphism sending 1 into 1. Since A is simple, φ is one-one. Since A and the given algebra both have dimension 4, φ is onto. Thus φ is an algebra isomorphism. (We did not have to check directly that $\{1, i, j, k\}$ is linearly independent over F.)

17. A is an algebra by routinely checking that it is closed under multiplication. Manifestly A has an identity and has dimension 9 over F. If I is a nonzero two-sided ideal in A, let $x = a + bj + cj^2$ be nonzero in I, and assume that x is chosen in I such that as few of the coefficients a, b, c are nonzero as possible. Possibly by multiplying x by j or j^2 on the right, we may assume that $a \neq 0$. Choose $d \in K$ with $d, \sigma(d)$, and $\sigma^2(d)$ distinct. Computation shows that $dx - xd$ has one fewer nonzero coefficient. By minimality we must have $dx - xd = 0$; hence x must have had just one nonzero coefficient. Such an x is invertible, and thus 1 is in I and $I = A$. Hence A is simple. To see that A has just F as center, we test a general element $x = a + bj + cj^2$ for commutativity with both $d \in K$ and the element j, and we find that $b = c = 0$ and $a = \sigma(a) = \sigma^2(a)$.

18. Since A is finite-dimensional central simple, $A \cong M_n(D)$ for some n and some central division algebra D over F. Then $9 = \dim A = n^2 \dim_F D$, and the only possibilities are that $n = 3$ and $D = F$, or that $n = 1$. In the first case, $A \cong M_3(F)$, and in the second case, A is a division algebra. In the first case any column of A (when viewed as $M_3(F)$) is a 3-dimensional left A module; in the second case A has no proper nonzero left A modules.

19. Left multiplication by K makes A into a K vector space, and the left K submodules of A are the K vector subspaces. The F dimension of such a subspace is 3 times the F dimension. Hence the left K submodules of A are the subspaces of K dimension 1, which consist of all left K multiples of any nonzero vector.

Let $x = a_0 + b_0 j + c_0 j^2$ be nonzero in A. Then Kx is a left A module if and only if jx lies in Kx. Here $jx = \sigma(a_0)j + \sigma(b_0)j^2 + \sigma(c_0)j^3 = r\sigma(c_0) + \sigma(a_0)j + \sigma(b_0)j^2$. This equals dx for some $d \in K$ if and only if

$$r\sigma(c_0) = da_0, \quad \sigma(a_0) = db_0, \quad \text{and} \quad \sigma(b_0) = dc_0. \qquad (*)$$

Combining the second and third equations gives the necessary condition that $\sigma^2(a_0) = \sigma(db_0) = \sigma(d)\sigma(b_0) = \sigma(d)dc_0$. Applying σ gives the necessary condition $a_0 = \sigma^3(a_0) = \sigma(\sigma(d)dc_0) = \sigma^2(d)\sigma(d)\sigma(c_0) = \sigma^2(d)\sigma(d)r^{-1}da_0 = N_{K/F}(d)r^{-1}a_0$.

Thus it is necessary that some $d \in K$ have $N_{K/F}(d) = r$. Conversely if $d \in K$ has $N_{K/F}(d) = r$, then $x_0 = 1 + d^{-1}j + d^{-1}\sigma(d)^{-1}j^2$ has $a_0 = 1$, $b_0 = d^{-1}$, and $c_0 = d^{-1}\sigma(d)^{-1}$, and we observe that the conditions (∗) are satisfied; thus Kx_0 is a left A submodule.

Chapter III

1. For (a), define $f : A \times K \to \operatorname{End}_{B^o} A$ by $f(a, c)(a') = aa'c$ just as in the proof of Theorem 3.3. The verification that the action of right multiplication by $b \in B$ commutes with $f(a, c)$, i.e., that $f(a, c)$ is in $\operatorname{End}_{B^o} A$, uses that B commutes with K, and the verification that the extended map $f : A \otimes_F K \to \operatorname{End}_{B^o} A$ respects multiplication uses that K is commutative; otherwise the argument is the same as with Theorem 3.3. The algebra $A \otimes_F K$ is central simple over K, and B is an algebra over K because B contains K. Since $A \otimes_F K$ is simple, f is one-one.

For (b), let V be the unique-up-to-isomorphism simple finite-dimensional left B module. If the left B module B is the direct sum of m copies of V, then the proof of Theorem 2.2 shows that $B^o \cong \operatorname{End}_B B \cong M_m(D^o)$, where D^o is the central division algebra over K given by $D^o = \operatorname{End}_B V$. Hence $B \cong M_m(D)$. If V^o denotes the unique-up-to-isomorphism simple finite-dimensional left B^o module and if $D'^o = \operatorname{End}_{B^o}(V^o)$, then we have $B \cong \operatorname{End}_{B^o}(B^o) \cong M_{m'}(D'^o)$, and it follows that $m = m'$ and $D' \cong D^o$.

Since $B \subseteq A$, A is a right B module, hence a left B^o module, and A has to be the direct sum of some number n of copies of V^o. Then the same argument gives an isomorphism $\operatorname{End}_{B^o} A \cong M_n(D'^o) \cong M_n(D)$. The Double Centralizer Theorem gives $\dim_F A = (\dim_F B)(\dim_F K)$, and thus $\dim_K A = \dim_F B = (\dim_F K)(\dim_K B) = (\dim_F K)(m \dim_K V)$. Meanwhile, $\dim_K A = n \dim_K V$ and thus $n \dim_K V = (\dim_F K)(m \dim_K V)$. So $n = m \dim_F K$. Consequently $\dim_F \operatorname{End}_{B^o} A = n^2 \dim_F D = m^2 (\dim_F D)(\dim_F K)^2 = (\dim_F B)(\dim_F K)^2 = (\dim_F A)(\dim_F K) = \dim_F(A \otimes_F K)$, and the map f in (a) is onto.

For (c), application of (b) and an isomorphism from above gives $A \otimes_F K \cong \operatorname{End}_{B^o}(A) \cong M_n(D)$, and we have seen that $B \cong M_m(D)$. Thus $A \otimes_F K$ and B lie in the same Brauer equivalence class in $\mathcal{B}(K)$.

2. Take the product over σ of the equality $\rho(a(\sigma, \tau))a(\rho, \sigma\tau) = a(\rho, \sigma)a(\rho\sigma, \tau)$, and get $\rho\left(\prod_\sigma a(\sigma, \tau)\right) \prod_\sigma a(\rho, \sigma) = \prod_\sigma a(\rho, \sigma) \prod_\sigma a(\sigma, \tau)$. Canceling gives $\rho\left(\prod_\sigma a(\sigma, \tau)\right) = \prod_\sigma a(\sigma, \tau)$. Thus $\prod_\sigma a(\sigma, \tau)$ is fixed by every member of the Galois group and is in $F^\times$.

3. Proposition 3.32 and Theorem 3.31 show that $H^{2k}(\operatorname{Gal}(K/F), K^\times) \cong H^2(\operatorname{Gal}(K/F), K^\times)$ for $k \geq 1$ and $H^{2k+1}(\operatorname{Gal}(K/F), K^\times) \cong H^1(\operatorname{Gal}(K/F), K^\times)$ for $k \geq 0$. Then Corollary 3.34 gives $H^{2k} \cong F^\times/N_{K/F}(K^\times)$ for all $k \geq 1$, and Theorem 3.17 gives $H^{2k+1} = 0$ for all $k \geq 0$. Finally H^0 is the subgroup of elements in $K^\times$ fixed by $\operatorname{Gal}(K/F)$, and this is $F^\times$.

4. For (a), it is shown in Chapter IX of *Basic Algebra* that $\mathbb{Q}(e^{2\pi i/p})$ is a Galois extension of $\mathbb{Q}$ with cyclic Galois group of order $p-1$ whenever p is prime. Here $p=7$. Complex conjugation is a member of the Galois group of order 2, and K is the subfield fixed by this subgroup. Hence K has degree $6/2=3$ over $\mathbb{Q}$, and its Galois group is the quotient of a cyclic group of order 6 by the subgroup of order 2, hence is cyclic of order 3. The powers $\zeta^1,\dots,\zeta^6$ form a basis of the $\mathbb{Q}$ vector space $\mathbb{Q}(\zeta)$, and the sums of them with their images under complex conjugation span K. These sums are τ_1, τ_2, τ_3. Since there are only 3 such sums, they must be linearly independent over $\mathbb{Q}$. Put $\tau_k = \zeta^k + \zeta^{-k}$. Then τ_k depends only on k mod 7, and $\tau_k = \tau_{-k}$. Hence the only τ_k's that are not any of τ_1, τ_2, τ_3 are the ones with $k \equiv 0 \mod 7$. The members of the Galois group of $\mathbb{Q}(\zeta)$ carry ζ to ζ^k for $1 \le k \le 6$ and therefore carry τ_1 to τ_k, τ_2 to τ_{2k}, and τ_3 to τ_{3k}. None of $k, 2k, 3k$ is divisible by 7, and the result follows.

For (b), let $\sigma \in \text{Gal}(K/\mathbb{Q})$ have $\sigma(\tau_1) = \tau_2, \sigma(\tau_2) = \tau_3$, and $\sigma(\tau_3) = \tau_1$. For $x \in K$, we have $N_{K/\mathbb{Q}}(x) = x\sigma(x)\sigma^2(x)$. With $x = a\tau_1 + b\tau_2 + \tau_3$, we get 27 terms when everything is expanded out, and they are the ones listed.

For (c), $\tau_1 + \tau_2 + \tau_3 = -1$ because $\sum_{j=-3}^{3} \zeta^j = 0$. Next, $\tau_1 \tau_2 = (\zeta^1 + \zeta^{-1}) \times (\zeta^2 + \zeta^{-2}) = \zeta^3 + \zeta^{-3} + \zeta^{-1} + \zeta^1 = \tau_1 + \tau_3$, and the other two identities on the second line are similar. Finally $\tau_1^2 = (\zeta^1 + \zeta^{-1})^2 = \zeta^2 + 2 + \zeta^{-2} = \tau_2 + 2$, and the other two identities are similar.

For (d), let $\alpha, \beta, \gamma, \delta$ be the expressions involving τ_1, τ_2, τ_3 on the right side in (b). First we have $\tau_1^3 = \tau_1^2 \tau_1 = (\tau_2 + 2)\tau_1 = \tau_1\tau_2 + 2\tau_1 = 3\tau_1 + \tau_3$. Summing this expression and similar expressions for τ_2^3 and τ_3^3 gives $\alpha = 4(\tau_1 + \tau_2 + \tau_3) = -4$. Second $\beta = \tau_1 \tau_2 \tau_3 = (\tau_1 + \tau_3)\tau_3 = \tau_2 + \tau_3 + \tau_1 + 2 = 1$. In (d), the coefficient of abc is $\alpha + 3\beta = -4 + 3 = -1$, and the coefficient of $a^3 + b^3 + c^3$ is $\beta = 1$. Third $\tau_1^2 \tau_2 = \tau_1(\tau_1 + \tau_3) = (\tau_2 + 2) + (\tau_2 + \tau_3) = \tau_3 + 2\tau_2 + 2$. Similarly $\tau_2^2 \tau_3 = \tau_1 + 2\tau_3 + 2$ and $\tau_3^2 \tau_1 = \tau_2 + 2\tau_1 + 2$. The sum is $\gamma = 3(\tau_1 + \tau_2 + \tau_3) + 6 = 3$. Fourth $\tau_1 \tau_2^2 = \tau_1(\tau_3 + 2) = \tau_2 + \tau_3 + 2\tau_1$. Similarly $\tau_2 \tau_3^2 = \tau_1 + \tau_3 + 2\tau_2$ and $\tau_3 \tau_1^2 = \tau_1 + \tau_2 + 2\tau_3$. The sum is $\delta = 4(\tau_1 + \tau_2 + \tau_3) = -4$.

For (e), the norm modulo 3 is $(a^3 + b^3 + c^3) - abc - (a^2c + ab^2 + bc^2)$, and this is $\equiv (a + b + c) - abc - (a^2c + ab^2 + bc^2) \mod 3$. Any nonzero square is $\equiv 1 \mod 3$, and we consider cases. If 3 does not divide abc, then $a^2 \equiv b^2 \equiv c^2 \equiv 1 \mod 3$, and the norm is $\equiv -abc \not\equiv 0 \mod 3$. If 3 divides a but not bc, then $b^2 \equiv c^2 \equiv 1 \mod 3$, and the norm is $\equiv (b + c) - b \equiv c \not\equiv 0 \mod 3$. If 3 divides a and b but not c, then the norm is $\equiv c \not\equiv 0 \mod 3$, while if 3 divides a and c but not b, then the norm is $\equiv b \not\equiv 0 \mod 3$. The case that 3 divides all of a, b, c is excluded by the condition that $\text{GCD}(a, b, c) = 1$, and all other cases are handled by symmetry. Thus in all cases the norm is not divisible by 3.

For (f), let x, y, z be members of $\mathbb{Q}$ not all 0. Choose integers a, b, c and relatively prime integers n and d such that $x = n^{-1}da$, $y = nd^{-1}db$, $z = nd^{-1}c$, and $\text{GCD}(a, b, c) = 1$. Then $N_{K/\mathbb{Q}}(x\tau_1 + y\tau_2 + z\tau_3) = d^{-3}n^3 N_{K/\mathbb{Q}}(a\tau_1 + b\tau_2 + c\tau_3)$. Applying (e) and supposing that 3 is a norm, we obtain $3 = d^{-3}n^3(3k + (1 \text{ or } 2))$

for some integer k. Thus $3d^3 = n^3(3k + (1 \text{ or } 2))$. This equality forces n to divide d, and we may therefore take $n = 1$. Thus $3d^3 = 3k + (1 \text{ or } 2)$. The left side is divisible by 3, and the right side is not. Hence 3 is not a norm.

5. For (a), Dirichlet's Theorem (Theorem 1.21) says that there are infinitely many primes of the form $p = kn + 1$. For any such p, n divides $p - 1$. For (b) with this p, the Galois group of $\mathbb{Q}(e^{2\pi i/p})/\mathbb{Q}$ is cyclic of order $p - 1$ and has a cyclic subgroup of order $(p - 1)/n$. The corresponding subfield is a Galois extension of $\mathbb{Q}$ of degree n with cyclic Galois group.

6. For $0 \leq k < n$ and $0 \leq l < n$, we have $x_{\sigma^k} x_{\sigma^l} = j^k j^l = j^{k+l}$. Meanwhile, $x_{\sigma^{k+l}}$ equals j^{k+l} if $k + l < n$ and equals j^{k+l-n} if $k + l \geq n$. So $x_{\sigma^k} x_{\sigma^l} = x_{\sigma^{k+l}}$ if $k + l < n$ and $x_{\sigma^k} x_{\sigma^l} = j^n x_{\sigma^{k+l-n}} = r x_{\sigma^{k+l-n}}$ if $k + l \geq n$. Thus $a(\sigma^k, \sigma^l)$ has the stated value.

7. It is just a question of checking that $c_{\sigma^k} \sigma^k(c_{\sigma^l}) = a(\sigma^k, \sigma^l) c_{\sigma^{k+l}}$ with $a(\sigma^k, \sigma^l)$ as in the previous problem.

8. We have $\partial_0(1, \sigma^k) = 1 - \sigma^k$ and thus

$$f_0 \partial_0(1, \sigma^k) = 1 - \sigma^k = (\sigma - 1)(-(1 + \sigma + \cdots + \sigma^{k-1})).$$

If we put $f_1(1, \sigma^k) = -(1 + \sigma + \cdots + \sigma^{k-1})$, then we have $T f_1(1, \sigma^k) = f_0 \partial_0(1, \sigma^k)$ for all k.

Next, for $k \leq l$, we have $\partial_1(1, \sigma^k, \sigma^l) = (\sigma^k, \sigma^l) - (1, \sigma^l) + (1, \sigma^k) = \sigma^k(1, \sigma^{l-k}) - (1, \sigma^l) + (1, \sigma^k)$. Then $f_1 \partial_1(1, \sigma^k, \sigma^l)$ equals

$$-\sigma^k(1 + \sigma + \cdots + \sigma^{l-k-1}) + (1 + \sigma + \cdots + \sigma^{l-1}) - (1 + \sigma + \cdots + \sigma^{k-1}) = 0.$$

For $k > l$, the term (σ^k, σ^l) is replaced by $\sigma^k(1, \sigma^{n+l-k})$. Thus $\partial_1(1, \sigma^k, \sigma^l) = \sigma^k(1, \sigma^{n+l-k}) - (1, \sigma^l) + (1, \sigma^k)$. Then $f_1 \partial_1(1, \sigma^k, \sigma^l)$ is

$$-\sigma^k(1 + \sigma + \cdots + \sigma^{n+l-k-1}) + (1 + \sigma + \cdots + \sigma^{l-1}) - (1 + \sigma + \cdots + \sigma^{k-1})$$
$$= -(1 + \sigma + \cdots + \sigma^{n+l-1}) + (1 + \sigma + \cdots + \sigma^{l-1})$$
$$= \sigma^l(-(1 + \sigma + \cdots + \sigma^{n-1})).$$

If we define f_2 as in the problem, then in the two cases we have

$$k \leq l: \quad N f_2(1, \sigma^k, \sigma^l) = (1 + \sigma + \cdots + \sigma^{n-1})(0) = 0 = f_1 \partial_1(1, \sigma^k, \sigma^l),$$
$$k > l: \quad N f_2(1, \sigma^k, \sigma^l) = (1 + \sigma + \cdots + \sigma^{n-1})(-\sigma^l) = f_1 \partial_1(1, \sigma^k, \sigma^l).$$

9. To ψ in $\text{Hom}_{\mathbb{Z}G}(\mathbb{Z}G, K^\times)$, the chain map of the previous problem associates $\psi \circ f_2$ in $\text{Hom}_{\mathbb{Z}G}(\mathbb{Z}G(\{(1, g_1, g_2)\}), K^\times)$, and then the corresponding member of $C^2(G, K^\times)$ is $\Phi_2(\psi f_2)$ whose value at (g_1, g_2) is $\psi f_2(1, g_2, g_1 g_2)$. That is, $\Phi_2(\psi f_2)(\sigma^k, \sigma^l) = \psi f_2(1, \sigma^k, \sigma^{k+l})$, and this by Problem 8 is $\psi(0)$ if $k + l < n$ and is $\psi(-\sigma^{k+l-n}) = \psi(\sigma^{k+l-n})^{-1}$ if $k + l \geq n$.

10. Taking Proposition 3.32 into account, we see that the mapping whose kernel gives the cocycles is $\text{Hom}(T, 1) : \text{Hom}_{\mathbb{Z}G}(\mathbb{Z}G, K^\times) \to \text{Hom}_{\mathbb{Z}G}(\mathbb{Z}G, K^\times)$. Here $\text{Hom}(T, 1)\psi = \psi \circ T$. We are identifying ψ with $\psi(1)$ and also $\psi \circ T$ with $\psi(T(1)) = \psi(\sigma - 1) = (\sigma - 1)\psi(1)$ in additive notation. Hence the effect of $\text{Hom}(T, 1)$ is to carry y to $\sigma(y)y^{-1}$ in multiplicative notation. A necessary and sufficient condition for $\sigma(y)y^{-1}$ to be 1 is that y be in $F^\times$, since the subgroup of $K^\times$ fixed by G is $F^\times$.

11. Since $\psi(0) = 1$ and $\psi(\sigma^{k+l-n}) = \sigma^{k+l-n}\psi(1) = \psi(1) = r^{-1}$, the member a of $C^2(G, K^\times)$ that corresponds to ψ has

$$a(\sigma^k, \sigma^l) = \begin{cases} 1 & \text{if } k+l < n, \\ r & \text{if } k+l \geq n, \end{cases}$$

and this is the 2-cocycle of Problem 6.

12. Corollary 3.34 and Theorem 3.14 combine to give us a group isomorphism $\mathcal{B}(K/F) \cong F^\times/N_{K/F}(K^\times)$, and the above problems show that the element r of $F^\times$ used in defining A corresponds under this isomorphism to the coset of r^{-1}. Hence the order of the Brauer equivalence class of A equals the order of the coset of r, as required.

If A is not a division algebra, then $A \cong M_m(D)$ for some central division algebra D over F and for some integer $m > 1$. Here $\dim_F D = (n/m)^2 < n^2$. Corollary 3.15 then gives the contradiction that the order of the Brauer equivalence class of D, which is the same as the order of the class of A, divides n/m, which in turn is $< n$.

13. The Skolem–Noether Theorem shows that the image matrices under two different isomorphisms φ and ψ have to be conjugate to one another, say with $\varphi = C^{-1}\psi C$. Then

$$\det(\varphi(X1 - a \otimes 1)) = \det(C^{-1}\psi(C(X1 - a \otimes 1)))$$
$$= (\det C)^{-1} \det(\psi(X1 - a \otimes 1))(\det C)$$
$$= \det(\psi(X1 - a \otimes 1)).$$

14. Let $B = A \otimes_F K$. The left B module B is semisimple and is the direct sum of n isomorphic simple modules of dimension n. On each the operation of $a \otimes 1$ has characteristic polynomial $\det(X1 - a \otimes 1)$, and the characteristic polynomial for the direct sum of the spaces is the product of the characteristic polynomials.

15. Arguing by contradiction, we may assume that the statement is false for some monic $P = P(X)$ and that P has the lowest possible degree among all monic polynomials for which the assertion is false. Factor P over K into powers of distinct irreducible polynomials as $P = P_1^{d_1} \cdots P_k^{d_k}$. The n-fold product of $P_1^{d_1} \cdots P_k^{d_k}$ with itself is in $F[X]$ by assumption and is therefore invariant under $\text{Gal}(K/F)$. Consequently for each $\sigma \in \text{Gal}(K/F)$ and each P_i, there exists some P_j such that $P_j = \sigma(P_i)$. It follows that if H is the subgroup of $G = \text{Gal}(K/F)$ fixing P_1, then

$Q = \prod_{\sigma H \in G/H} \sigma P_1$ is the product of distinct irreducible factors of P and hence divides P. The polynomial Q is fixed by every member of G and hence is monic in $F[X]$. Thus $Q \neq P$. Then Q^n is in $F[X]$, and hence $(P/Q)^n$ is in $F[X]$. The fact that P is not in $F[X]$ implies that $Q \neq P$. Therefore $\deg(P/Q) < \deg P$. By the minimal choice of $\deg P$, P/Q is in $F[X]$. Therefore $P = (P/Q)Q$ is in $F[X]$, contradiction.

16. For a matrix m with entries in a field, passing to a larger field does not change $\det(X1 - m)$. Suppose we start with two finite Galois extensions K_1 and K_2 of F that split A. Let K_1 be a splitting field for a polynomial $g_1 \in F[X]$, and let K_2 be a splitting field for $g_2 \in F[X]$. Define K to be a splitting field for $g_1 g_2$. Then K is a finite Galois extension of F, and we can regard it as containing both K_1 and K_2. Applying the first sentence of this paragraph first to K_1 and K and then to K_2 and K, we see that the reduced characteristic polynomial is the same over K_1 as it is over K_2.

17. The formulas for $\mathrm{Nrd}_{A/F}(ab)$ and $\mathrm{Nrd}_{A/F}(1)$ follow from properties of determinants. From Problem 14 we observe that $\det a = (-1)^{n^2} \det(-a)$ and $\det(-\varphi(a \otimes 1)) = (-1)^n \det(\varphi(a \otimes 1))$. Substituting $X = 0$ into the formula therefore gives us $N_{A/F}(a) = \det a = (-1)^{n^2} \det(-a) = (-1)^{n^2} \det(-\varphi(a \otimes 1))^n = (-1)^{n^2}((-1)^n)^n \det(\varphi(a \otimes 1))^n = \det(\varphi(a \otimes 1))^n = \mathrm{Nrd}_{A/F}(a)^n$. If a is invertible, then $1 = \mathrm{Nrd}_{A/F}(1) = \mathrm{Nrd}_{A/F}(aa^{-1}) = \mathrm{Nrd}_{A/F}(a)\mathrm{Nrd}_{A/F}(a^{-1})$ shows that $\mathrm{Nrd}_{A/F}(a)$ is nonzero. Conversely if $\mathrm{Nrd}_{A/F}(a) \neq 0$, then $\mathrm{Nrd}_{A/F}(a) \neq 0$ and hence $\det L(a) \neq 0$. If $P(X)$ is the algebra polynomial of $L(a)$, then the Cayley–Hamilton Theorem shows that $P(L(a)) = 0$. Since $\det L(a) \neq 0$, $P(X)$ has a nonzero constant term. Therefore we can separate the constant term in the equation $P(L(a)) = 0$ to exhibit an identity of the form $L(a)Q(L(a)) = 1$ for some polynomial $Q(X)$, and the element $Q(a)$ is a 2-sided inverse to a in A. This proves (a), and the conclusion about division algebras is immediate.

18. The definition gives

$$m(dx_\rho) = \sum_\mu \mu(d)a(\mu, \rho)E_{\mu,\mu\rho},$$
$$m(cx_\tau) = \sum_\sigma \sigma(c)a(\sigma, \tau)E_{\sigma,\sigma\tau},$$
$$m\big((dx_\rho)(cx_\tau)\big) = m\big(d\rho(c)a(\rho, \tau)x_{\rho\tau}\big) = \sum_\mu \mu\big(d\rho(c)a(\rho, \tau)\big)a(\mu, \rho\tau)E_{\mu,\mu\rho\tau}.$$

Also we have

$$m(dx_\rho)m(dx_\rho) = \sum_{\mu,\sigma} \mu(d)a(\mu, \rho)\sigma(c)a(\sigma, \tau)E_{\mu,\mu\rho}E_{\sigma,\sigma\tau}$$
$$= \sum_\mu \mu(d)\mu\rho(c)a(\mu, \rho)a(\mu\rho, \tau)E_{\mu,\mu\rho\tau}.$$

This matches $m\big((dx_\rho)(cx_\tau)\big)$ by the cocycle relation for a.

For the reduced norm we have two one-one F algebra homomorphisms of A into $M_n(K)$, one via the mapping m above and one by the embedding $A \to A \otimes_F 1 \subseteq A \otimes_F K \cong M_n(K)$, and these are conjugate by the Skolem–Noether Theorem. Hence the determinant gives the same result in the two cases. The determinant in the second case gives the reduced norm, and hence it must give the reduced norm in the first case.

19. The algebra $\mathbb{H}$ can be realized as all complex matrices $x = \begin{pmatrix} \alpha & \beta \\ -\bar\beta & \bar\alpha \end{pmatrix}$, and $\mathrm{Nrd}_{\mathbb{H}/\mathbb{R}}(x) = |\alpha|^2 + |\beta|^2$ and $N_{\mathbb{H}/\mathbb{R}}(x) = (|\alpha|^2 + |\beta|^2)^2$ as a special case of Problem 18.

20. Let D be a finite-dimensional central division algebra over F, say with $\dim_F D = n^2$. Choose a basis $\{x_k\}$ of D over F, and expand elements of D as $x = \sum_{j=1}^{n^2} c_j x_j$. The function $P(c_1, \ldots, c_{n^2}) = \mathrm{Nrd}_{D/F}\big(\sum_{j=1}^{n^2} c_j x_j\big)$ is easily checked to be a homogeneous polynomial of degree n in n^2 variables, and condition (C1) says that it has a nontrivial zero if $n < n^2$. In this case the corresponding member x of D would be a nonzero element of D that fails to be invertible, and there is no such element. We conclude that $n < n^2$ is false, and that means that $n = 1$. Therefore F is the only finite-dimensional central division algebra over F, and $\mathcal{B}(F) = 0$.

Chapter IV

1. For (a), every free abelian group of finite rank is in the category, and such groups provide enough projectives.

Let $I = F \oplus T$ be a decomposition of an injective I as the direct sum of a free abelian group F of rank k and a torsion group T. The sequence $0 \to F \oplus T \to 2F \oplus T \to (\mathbb{Z}/2\mathbb{Z})^k \to 0$ is exact but not split unless $k = 0$, and thus $F = 0$. Thus every injective in the category is a finite group, and no infinite group in the category embeds into an injective.

For (b), every abelian group and in particular every torsion abelian group is a subgroup of a divisible group. The torsion subgroup of the divisible group is still divisible and is still an injective, and thus every group in the category embeds in an injective in the category.

Let P be a projective in the category mapping onto $\mathbb{Z}/2\mathbb{Z} = \{0, 1\}$ by a homomorphism τ, and let x be an element of P with $\tau(x) = 1$. If g is a generator of a cyclic group G of order 2^k, then there is a homomorphism φ of G onto $\mathbb{Z}/2\mathbb{Z}$ with $\varphi(g) = \tau(x) = 1$. Since P is projective, there exists a homomorphism $\sigma : P \to G$ with $\varphi\sigma = \tau$, and then we have $1 = \tau(x) = \varphi\sigma(x)$. Then $\sigma(x) = g^m$ for some odd integer m, and this has order 2^k. Hence x has order at least 2^k. Since k is arbitrary, x must have infinite order. But all groups in the category are torsion groups, and P therefore cannot exist.

2. Let p be a prime, and let $\mathcal{C}$ be the category of all abelian groups that are the underlying additive group of a vector space over the field of p elements. This category coincides with the category of all direct sums of copies of $\mathbb{Z}/p\mathbb{Z}$. Every such abelian

group is projective and injective for the category.

3. Every unital left R module is the direct sum of simple R modules. Hence every short exact sequence splits, and every module is both projective and injective for $\mathcal{C}_R$.

4. For (a), let I be injective. Given $x \in I$ and $a \neq 0$ in R, let $B = C = R$, let $\tau : R \to I$ have $\tau(r) = rx$, and let $\varphi : R \to R$ have $\varphi(r) = ra$. Setting up Figure 4.4, we obtain $\sigma : R \to I$ with $\tau = \sigma\varphi$. If we put $y = \sigma(1)$ and evaluate both sides at 1, then we obtain $x = \tau(1) = \sigma(\varphi(1)) = \sigma(a) = a\sigma(1) = ay$, as required.

For (b), suppose that the unital left R module I is divisible. Suppose that J is an ideal of R, and write $J = (a)$. Let $\varphi : J \to I$ be an R homomorphism. Since I is divisible, there exists y in I with $ay = \varphi(a)$. Then φ extends to the R homomorphism Φ with $\Phi(1) = y$. By Proposition 4.15, I is injective.

5. Proposition 4.20 shows that there exists an injective I_0 containing an isomorphic copy $\overline{M}$ of M. Problem 4 shows that I_0 is divisible, and hence $I_1 = I_0/\overline{M}$ is divisible. By Problem 4, I_1 is injective. Then $0 \to M \to I_0 \to I_1 \to 0$ is an injective resolution of M.

6. If a module M in $\mathcal{C}$ is given, we form the appropriate kind of resolution X in $\mathcal{C}$ needed to compute the derived functors of G, and the same X will be appropriate for computing the derived functors of $F \circ G$. The derived functors of G come from the homology or cohomology of $G(X)$ with $G(M)$ removed, and the derived functors of $F \circ G$ come similarly from $F(G(X))$. Thus the result follows from Proposition 4.4.

7. If a module M in $\mathcal{C}$ is given, we form the appropriate kind of resolution X in $\mathcal{C}$ needed to compute the derived functors of $G \circ F$ on M. Then $F(X)$ is the appropriate kind of resolution for computing the derived functors of G on $F(M)$, and the result follows.

8. For n odd, $H^n(G, M)$ is the cohomology of the complex

$$\operatorname{Hom}_{\mathbb{Z}G}(\mathbb{Z}G, M) \xleftarrow{N} \operatorname{Hom}_{\mathbb{Z}G}(\mathbb{Z}G, M) \xleftarrow{T} \operatorname{Hom}_{\mathbb{Z}G}(\mathbb{Z}G, M),$$

while for n even, $H^n(G, M)$ is the cohomology of the complex

$$\operatorname{Hom}_{\mathbb{Z}G}(\mathbb{Z}G, M) \xleftarrow{T} \operatorname{Hom}_{\mathbb{Z}G}(\mathbb{Z}G, M) \xleftarrow{N} \operatorname{Hom}_{\mathbb{Z}G}(\mathbb{Z}G, M).$$

This proves the isomorphisms concerning cohomology. For n odd $H_n(G, M)$ is the homology of the complex

$$\mathbb{Z}G \otimes_{\mathbb{Z}G} M \xrightarrow{N} \mathbb{Z}G \otimes_{\mathbb{Z}G} M \xrightarrow{T} \mathbb{Z}G \otimes_{\mathbb{Z}G} M,$$

while for n even, $H_n(G, M)$ is the homology of the complex

$$\mathbb{Z}G \otimes_{\mathbb{Z}G} M \xrightarrow{T} \mathbb{Z}G \otimes_{\mathbb{Z}G} M \xrightarrow{N} \mathbb{Z}G \otimes_{\mathbb{Z}G} M.$$

This proves the isomorphisms concerning homology.

9. For (a), let $T_{AB} : \text{Hom}_\mathcal{D}(F(A), B) \to \text{Hom}_\mathcal{C}(A, G(B))$ be the natural isomorphism. Naturality in B says for any $\psi : B \to B'$ that we have

$$\text{Hom}_\mathcal{C}(1_A, G(\psi)) \circ T_{AB} = T_{AB'} \circ \text{Hom}_\mathcal{D}(1_{F(A)}, \psi)$$

on $\text{Hom}_\mathcal{D}(F(A), B)$. Let P be projective in $\mathcal{C}$. We are to prove that $F(P)$ is projective in $\mathcal{D}$, thus to prove that $\text{Hom}_\mathcal{D}(F(P), \cdot)$ is exact. We need to show that whenever $\psi : B \to B'$ is onto in $\mathcal{D}$, then $\text{Hom}_\mathcal{D}(1_{F(P)}, \psi)$ is onto. By hypothesis, $G(\psi) : G(B) \to G(B')$ is onto in $\mathcal{C}$. The displayed equation with $A = P$ has $\text{Hom}_\mathcal{C}(1_P, G(\psi))$ onto, and T_{PB} and $T_{PB'}$ are given as isomorphisms. Therefore $\text{Hom}_\mathcal{D}(1_{F(P)}, \psi)$ is onto, as we were to show. The proof of (b) is similar.

10. Conclusion (a) follows from the natural isomorphism $\text{Hom}_S(P_R^S A, B) = \text{Hom}_S(S \otimes_R A, B) \cong \text{Hom}_R(A, \mathcal{F}_S^R B)$. Conclusion (b) follows from Problem 9a with $F = P_R^S$ and $G = \mathcal{F}_S^R$, since $\mathcal{F}_S^R$ is exact and therefore carries onto maps to onto maps. For (c), $P_R^S A$ is given by the tensor product $S \otimes_R A$, and this tensor product is an exact functor of A if S is projective as a right R module, by Proposition 4.19a.

For (d), part (c) says that $M \mapsto P_R^S M$ is an exact functor. Taking it to be F in Problem 7a and G to be $\text{Hom}_S(\cdot, N)$, we have $\text{Ext}_S^k(P_R^S M, N) = G^k(F(M))$. Problem 7a says that this is equal to $(G \circ F)^k$. Since $(G \circ F)(M) = \text{Hom}_S(P_R^S M, N) \cong \text{Hom}_R(M, \mathcal{F}_S^R N)$ has $(G \circ F)^k(M) = \text{Ext}_R^k(M, \mathcal{F}_S^R N)$, we obtain $\text{Ext}_S^k(P_R^S M, N) \cong \text{Ext}_R^k(M, \mathcal{F}_S^R N)$.

For (e), (b) shows that the chain complex $P_R^S X$ is projective over $P_R^S M$, and we are assuming that Y is exact (and projective) over $P_R^S M$. Theorem 4.12 says that the identity map on $P_R^S M$ extends to a chain map $f : P_R^S X \to Y$ that is unique up to homotopy. Dropping the terms in degree -1 and applying the functor $\text{Hom}_S(\cdot, N)$ to the diagram gives us a cochain map from the complex $\text{Hom}_S(Y, N)$ to the complex $\text{Hom}_S(P_R^S X, N) \cong \text{Hom}_R(X, \mathcal{F}_S^R N)$. Thus we get homomorphisms on cohomology $\text{Ext}_S^*(P_R^S M, N) \to \text{Ext}_R^*(M, \mathcal{F}_S^R N)$.

11. Conclusion (a) follows from the natural isomorphisms $\text{Hom}_S(A, I_R^S B) = \text{Hom}_S(A, \text{Hom}_R(S, B)) \cong \text{Hom}_R(S \otimes_S A, B) \cong \text{Hom}_R(\mathcal{F}_S^R A, B)$. Conclusion (b) follows from Problem 9b because $\mathcal{F}_S^R$ is exact and therefore carries one-one maps to one-one maps. For (c), $I_R^S = \text{Hom}_R(S, \cdot)$ is exact if S is projective as a right R module, by Proposition 4.19a.

For (d), part (c) says that $M \mapsto I_R^S M$ is an exact functor. Taking it to be F in Problem 7b and G to be $\text{Hom}_S(M, \cdot)$, we have $\text{Ext}_S^k(M, I_R^S N) = G^k(F(N))$. Problem 7b says that this is equal to $(G \circ F)^k$. Since $(G \circ F)(N) = \text{Hom}_S(M, I_R^S N) \cong \text{Hom}_R(\mathcal{F}_S^R M, N)$ has $(G \circ F)^k(M) = \text{Ext}_R^k(\mathcal{F}_S^R M, N)$, we obtain $\text{Ext}_S^k(N, I_R^S N) \cong \text{Ext}_R^k(\mathcal{F}_S^R M, N)$.

For (e), (b) shows that the cochain complex $I_R^S X$ is injective over $I_R^S N$, and we are assuming that Y is exact (and injective) over $I_R^S N$. Theorem 4.16 says that the identity map on $I_R^S N$ extends to a cochain map $f : Y \to I_R^S X$ that is unique up to homotopy. Dropping the terms in degree -1 and applying the functor $\text{Hom}_S(M, \cdot)$

to the diagram gives us a cochain map from the complex $\text{Hom}_S(M, Y)$ to the complex $\text{Hom}_S(M, I_R^S X) \cong \text{Hom}_R(\mathcal{F}_S^R M, X)$. Thus we get homomorphisms on cohomology $\text{Ext}_S^*(M, I_S^R N) \to \text{Ext}_R^*(\mathcal{F}_S^R M, N)$.

12. For (a), the definition of Φ_q is

$$(\Phi_q \varphi)(g_1, \ldots, g_q) = \varphi(1, g_1, g_1 g_2, \ldots, g_1 \cdots g_q)$$

for $\varphi \in \text{Hom}_{\mathbb{Z}G}(F_q, M)$. Putting $f = \Phi_q \varphi$ gives $(\rho^* f)(g_1, \ldots, g_q) = \rho^*(\Phi_q \varphi) = \Phi_q(\varphi \circ \rho) = (\Phi_q \varphi) \circ \rho$, as asserted.

For inflation the groups are $(G, G') = (G, G/H)$, and the map ρ is the quotient map; the effect is given by $(\text{Inf} f)(g_1, \ldots, g_q) = f(g_1 H, \ldots, g_q H)$ for f in $C^q(G/H, M^H)$. For restriction the groups are $(G, G') = (H, G)$, and the map is the inclusion; the effect is given by $(\text{Res}\psi)(h_1, \ldots, h_q) = \psi(h_1, \ldots, h_q)$ for $\psi \in C^q(G, M)$.

For (b), let f be in $C^1(G/H, M^H)$. Then $\text{Res}(\text{Inf}(f))(h) = \text{Inf}(f)(h) = f(hH) = f(H)$. The condition for f to be a cocycle is that $\delta_1 f = 0$, i.e., that $f(uv) = f(u) + u(f(v))$ for u and v in G/H. Taking u and v to be the identity coset H shows that $f(H) = 0$.

For (c), let $f \in C^1(G/H, M^H)$ be a cocycle. Then $\text{Inf}(f)(g) = f(gH)$. If this is a coboundary in $C^1(G, M)$, then there exists $\psi \in M$ with $\delta_0 \psi = f$, i.e., with $f(gH) = g\psi - \psi$ for all g. The left side depends only on the coset gH, and hence so must the right side. Then it follows that $gh\psi = g\psi$ for all $h \in H$ and that ψ is in M^H. Then the formula $f(gH) = g\psi - \psi$ exhibits f as a coboundary in $C^1(G/H, M^H)$.

For (d), let f be a cocycle in $C^1(G, M)$ such that $\text{Res} f$ is a coboundary in $C^1(H, M)$. The formula is $(\text{Res} f)(h) = f(h)$, and the coboundary condition shows that there is some $\psi \in M^H$ with $f(h) = h\psi - \psi$ for $h \in H$. Since ψ is in M^H, $f(h) = 0$ for all $h \in H$. The cocycle condition on f is that $f(uv) = f(u) + u(f(v))$ for all u and v in G. Taking v to be in H shows that $f(gh) = f(g)$ for all $h \in H$. Taking instead u to be in H shows that $f(hg) = h(f(g))$ for all $h \in H$. Since H is normal, $h(f(g)) = f(g)$ for all $h \in H$. Therefore f takes values in M^H and is Inf of the cocycle $\bar{f}$ in $C^1(G/H, M^H)$ given by $\bar{f}(gH) = f(g)$.

13. For (a), we have $(g_0 \varphi_m)(g) = \varphi_m(gg_0) = gg_0 m = \varphi_{g_0 m}(g)$, and $m \mapsto \varphi_m$ is a $\mathbb{Z}G$ homomorphism. Suppose that $\varphi_m = 0$. Then $gm = 0$ for all g and in particular for $g = 1$. Therefore $m = 0$, and $m \mapsto \varphi_m$ is one-one. Then it follows that the sequence is exact.

For (b), we know that $\mathbb{Z}G$ as an abelian group is free abelian. Then Problem 11d shows that $H^k(G, B) = \text{Ext}_{\mathbb{Z}G}^k(\mathbb{Z}, B) = \text{Ext}_{\mathbb{Z}G}^k(\mathbb{Z}, I_{\mathbb{Z}}^{\mathbb{Z}G}(\mathcal{F}_{\mathbb{Z}G}^{\mathbb{Z}} M)) \cong \text{Ext}_{\mathbb{Z}}^k(\mathbb{Z}, \mathcal{F}_{\mathbb{Z}G}^{\mathbb{Z}} M)$. Since $\text{Hom}_{\mathbb{Z}}(\mathbb{Z}, \cdot)$ is exact from $\mathcal{C}_{\mathbb{Z}}$ to itself, $\text{Ext}_{\mathbb{Z}}^k(\mathbb{Z}, \mathcal{F}_{\mathbb{Z}G}^{\mathbb{Z}} M) = 0$ for $k \geq 1$.

For (c), a $\mathbb{Z}$ basis of $\mathbb{Z}G$ consists of all 1-tuples (g) with $g \in G$, and a $\mathbb{Z}$ basis of $\mathbb{Z}H$ consists of all (h) with $h \in H$. Let $\{v\}$ be a set of representatives of the cosets of G/H, and let A be the free abelian group on $\{v\}$. The $\mathbb{Z}$-bilinear map $(v, (h)) \mapsto (vh)$

extends to a homomorphism of $A \otimes_{\mathbb{Z}} \mathbb{Z}H$ into $\mathbb{Z}G$ that is manifestly onto, and it is one-one because $\sum n_i(v_i h_i) = 0$ implies $n_i = 0$ for all i. Thus it is an isomorphism.

For (d), use of (c) gives $\mathcal{F}_{\mathbb{Z}G}^{\mathbb{Z}H} B \cong \mathcal{F}_{\mathbb{Z}G}^{\mathbb{Z}H} \operatorname{Hom}_{\mathbb{Z}}(\mathbb{Z}G, M) \cong \operatorname{Hom}_{\mathbb{Z}}(\mathcal{F}_{\mathbb{Z}G}^{\mathbb{Z}H}(\mathbb{Z}G), M)$
$\cong \operatorname{Hom}_{\mathbb{Z}}(A \otimes_{\mathbb{Z}} \mathbb{Z}H, M) \cong \operatorname{Hom}_{\mathbb{Z}}(\mathbb{Z}H, \operatorname{Hom}_{\mathbb{Z}}(A, M))$, and then $H^k(H, \mathcal{F}_{\mathbb{Z}G}^{\mathbb{Z}H} B) = 0$ for $k \geq 1$ by the same argument as in (b).

For (e), the long exact sequence for $\operatorname{Ext}_H^*(\mathbb{Z}, \cdot)$ that comes from the short exact sequence in (a) shows that $0 \to H^0(H, M) \to H^0(H, B) \to H^0(H, N) \to H^1(H, M)$ is exact. The right member is assumed to be 0, and the three middle members are isomorphic to M^H, B^H, and N^H.

For (f), consider the $\mathbb{Z}$ bilinear map $(1, (g)) \mapsto (gH)$ of $\mathbb{Z} \times \mathbb{Z}G$ into $\mathbb{Z}(G/H)$, and extend it to a $\mathbb{Z}$ linear map of $\mathbb{Z} \otimes_{\mathbb{Z}} \mathbb{Z}G$ into $\mathbb{Z}(G/H)$. The group H acts trivially on $\mathbb{Z}$ on the right, and it acts on $\mathbb{Z}(G/H)$ by left translation. Let h be in H. The passage $\mathbb{Z} \times \mathbb{Z}G \to \mathbb{Z}(G/H)$ has $(1h, (g)) \mapsto (gH)$ and $(1, h(g)) \mapsto h(gH) = (gH)$; thus the group homomorphism $\mathbb{Z} \otimes_{\mathbb{Z}} \mathbb{Z}G \to \mathbb{Z}(G/H)$ descends to a homomorphism of $\mathbb{Z} \otimes_{\mathbb{Z}H} \mathbb{Z}G$ into $\mathbb{Z}(G/H)$. This is certainly onto. To see that it is one-one, let $\sum_i n_i 1 \otimes (g_i) \mapsto 0$. Then $\sum_i n_i(g_i H) = 0$, and for each coset representative v in G, $\sum_{g_i \in vH} n_i(g_i) = 0$. So $\sum_i n_i(h_i^{-1}v) = 0$, and $\left(\sum_i n_i(h_i^{-1})\right)(v) = 0$. Then $\sum_i n_i(h_i^{-1}) = 0$ in $\mathbb{Z}H$ because (v) is invertible in $\mathbb{Z}G$, and it follows that the map is one-one.

For (g), (f) gives $B^H = \operatorname{Hom}_{\mathbb{Z}H}(\mathbb{Z}, \operatorname{Hom}_{\mathbb{Z}}(\mathbb{Z}G, M)) \cong \operatorname{Hom}_{\mathbb{Z}}(\mathbb{Z} \otimes_{\mathbb{Z}H} \mathbb{Z}G, M) \cong \operatorname{Hom}_{\mathbb{Z}}(\mathbb{Z}(G/H), M)$, and the same argument as in (b) shows that $H^k(G/H, B^H) = 0$ for $k \geq 1$.

Conclusion (h) is immediate because $q \geq 2$ and because all the cohomology associated with B has been shown to be 0 in degrees ≥ 1.

The commutativity in conclusion (i) follows because the inflation and restriction mappings are clearly functorial. The vertical mappings have been shown to be isomorphisms in (h). To see via induction that the top row is exact, we have to verify that $H^k(H, N) = 0$ for $k \leq q - 2$; but $H^k(H, N) \cong H^{k+1}(H, M)$ for all $k \geq 1$, and $H^{k+1}(H, M)$ is assumed to be 0 for $k + 1 \leq q - 1$. Therefore the bottom row is exact, and the induction is complete.

14–16. These problems are routine verifications.

17. Part (a) follows because $R \otimes_R A$ is naturally isomorphic to A. For (b), $F \otimes_R A \cong \bigoplus_{s \in S} (F_s \otimes_R A)$ and $1_F \otimes f$ corresponds to $\bigoplus (1_{F_s} \otimes f)$. The values of the various R homomorphisms are in the various spaces $F_s \otimes_R B$, whose sum is direct, and thus the kernel of $1_F \otimes f$ is the direct sum of the kernels. Then (b) follows. For (c), we see from (a) and (b) that free R modules are flat. In $\mathcal{C}_R$, every projective is a direct summand of a free module, and thus (c) follows by a second application of (b).

18. Consider $1 \otimes f : M \otimes_R A \to M \otimes_R B$. Any element of $\ker(1 \otimes f)$ is a finite sum $\sum m_i \otimes a_i$, and this lies in $\ker((1 \otimes f)|_{M_F})$, where F is the finite set of indices in question. Thus $\ker(1 \otimes f) \neq 0$ implies $\ker((1 \otimes f)|_{M_F}) \neq 0$ for some F. The converse is immediate because $\ker((1 \otimes f)|_{M_F}) \subseteq \ker(1 \otimes f)$ for all F.

19. The long exact sequence for tensor product over R is of the form

$$\cdots \to \mathrm{Tor}_1^R(A, F) \to \mathrm{Tor}_1^R(A, B) \to A \otimes_R K \to A \otimes_R F \to A \otimes_R B \to 0,$$

and $\mathrm{Tor}_1^R(A, F) = 0$ because F is projective for $\mathcal{C}_R$. This establishes the exactness of the sequence in the problem. If A is flat, then

$$0 \to \mathrm{Tor}_1^R(A, B) \to A \otimes_R K \to A \otimes_R F \to A \otimes_R B \to 0$$

is exact for each B, and $\mathrm{Tor}_1^R(A, B)$ must be 0 for each B. Conversely if $\mathrm{Tor}_1^R(A, B)$ is 0 for each B, then $A \otimes_R (\cdot)$ is an exact functor by Proposition 4.3. Hence A is flat by definition.

20. On the one hand, the long exact sequence associated to tensoring the short exact sequence given in (a) by B is of the form

$$0 \to \mathrm{Tor}_1^R(M, B) \to \mathrm{Tor}_1^R(T(M), B) \to F \otimes_R B \to M \otimes_R B \to T(M) \otimes_R B \to 0,$$

since F free implies $\mathrm{Tor}_1^R(F, B) = 0$. On the other hand, the given short exact sequence splits, and tensoring it by B must directly produce a short exact sequence

$$0 \to F \otimes_R B \to M \otimes_R B \to T(M) \otimes_R B \to 0.$$

Thus $\ker(F \otimes_R B \to M \otimes_R B) = 0$, and we must therefore have

$$\mathrm{image}(\mathrm{Tor}_1^R(T(M), B) \to F \otimes_R B) = \ker(F \otimes_R B \to M \otimes_R B) = 0.$$

Consequently $0 \to \mathrm{Tor}_1^R(M, B) \to \mathrm{Tor}_1^R(T(M), B) \to 0$ is exact. This proves (a).

For (b), Problem 18 shows that M is flat if and only if each M_F is flat, and (a) in combination with Problem 19 shows that each M_F is flat if and only if each $T(M_F)$ is flat. Now suppose that M is flat, so that $T(M_F)$ is flat for each finite subset F of M. This is true in particular for each finite subset F' of $T(M)$, and $T(M_{F'}) = M_{F'} = (T(M))_{F'}$. Hence Problem 18 shows that $T(M)$ is flat. Conversely suppose that $T(M)$ is flat. Then $T(M)_{F'}$ is flat for each finite subset F' of $T(M)$. Let F be a finite subset of M. Then M_F is a finitely generated R submodule, and the structure theorem shows that $T(M_F)$ is finitely generated. Let F' be a set of generators for it. Then $T(M_F) = M_{F'} = T(M)_{F'}$. This is flat by Problem 18, since $T(M)$ is flat, and the first sentence of this paragraph allows us to conclude that M is flat.

For (c), $T(M) \neq 0$ means that $am = 0$ for some nonzero $a \in R$ and $m \in M$. Let $i : (a) \to R$ be the inclusion, which is one-one. Then $i \otimes 1 : (a) \otimes_R M \to R \otimes_R M \cong M$ has $(i \otimes 1)(a \otimes m) = am = 0$. Thus the one-one map i is carried to the map $i \otimes 1$ that is not one-one, and tensoring with M is not exact. So M is not flat.

For (d), if M is flat, then $T(M) = 0$ by (c). Conversely if $T(M) = 0$, then $T(M)$ is flat, and (b) shows that M is flat.

21. Since $\partial'_{p,q}$ and $\partial''_{p,q}$ both lower $p+q$ by 1, they both carry E_{p+q} to E_{p+q-1}. Also, the hypotheses give $(\partial'_{p,q} + \partial''_{p,q})^2 = \partial'_{p-1,q}\partial'_{p,q} + \partial'_{p,q-1}\partial''_{p,q} + \partial''_{p-1,q}\partial'_{p,q} + \partial''_{p,q-1}\partial''_{p,q} = 0$, and we have a chain complex.

22. We compute that $\partial'_{p-1,q}\partial'_{p,q} = (\alpha_{p-1} \otimes 1)(\alpha_p \otimes 1) = \alpha_{p-1}\alpha_p \otimes 1 = 0$, $\partial'_{p,q-1}\partial''_{p,q} + \partial''_{p-1,q}\partial'_{p,q} = (\alpha_{p-1} \otimes 1)(-1)^p(1 \otimes \beta_q) + (-1)^{p-1}(1 \otimes \beta_q) \times (\alpha_p \otimes 1) = (-1)^p(\alpha_p \otimes \beta_q) - (-1)^p(\alpha_p \otimes \beta_q) = 0$, and that $\partial''_{p,q-1}\partial''_{p,q} = (-1)^p(1 \otimes \beta_{q-1})(-1)^p(1 \otimes \beta_q) = 1 \otimes \beta_{q-1}\beta_q = 0$.

23. The formulas for $\partial'_{p,q}$ and $\partial''_{p,q}$ show that $\ker \partial'_{p,q} = \ker \alpha_p \otimes_R D_q$ and that $\ker \partial''_{p,q} = C_p \otimes_R \ker \beta_q$. Since $\partial'_{p,q}E_{p,q}$ and $\partial''_{p,q}E_{p,q}$ lie in independent spaces, $\ker(\partial'_{p,q} + \partial''_{p,q}) = \ker \partial'_{p,q} \cap \ker \partial''_{p,q} = \ker \alpha_p \otimes_R \ker \beta_q$. Similarly $\partial'_{p+1,q}(E_{p+1,q}) = \alpha_{p+1}(C_{p+1}) \otimes_R D_q$ and $\partial''_{p,q+1}(E_{p,q+1}) = C_p \otimes_R \beta_{q+1}(D_{q+1})$, and hence

$$\text{image}(\partial'_{p+1,q} + \partial''_{p,q+1}) = \alpha_{p+1}(C_{p+1}) \otimes_R D_q + C_p \otimes_R \beta_{q+1}(D_{q+1}).$$

Thus if c is in C_p, d is in D_q, c' is in $\alpha_{p+1}(C_{p+1})$, and d' is in $\beta_{q+1}(D_{q+1})$, then $(\partial'_{p,q} + \partial''_{p,q})((c + c') \otimes (d + d'))$ is the sum of $(\partial'_{p,q} + \partial''_{p,q})(c \otimes d)$ and three terms that are in $\text{image}(\partial'_{p+1,q} + \partial''_{p,q+1})$. Consequently we obtain a well-defined homomorphism of $H_p(C) \otimes_R H_q(D)$ into $H_{p+q}(E)$.

24. Let ∂' and ∂'' be the boundary operators; these satisfy $\partial'\partial'' = -\partial''\partial'$. Let a be a cycle in $E_{-1,k}$, i.e., let $\partial''a = 0$. Since $\partial'a = 0$, the exactness for ∂' produces $c_{0,k} \in E_{0,k}$ with $a = \partial'c_{0,k}$. Since $\partial''a = 0$, this has $\partial'\partial''c_{0,k} = -\partial''\partial'c_{0,k} = -\partial''a = 0$. Now suppose inductively on $i \geq 0$ that $j \geq 0$ is defined by $i + j = k$ and that $c_{i,j} \in E_{i,j}$ is given with $\partial'\partial''c_{i,j} = 0$. By the assumed exactness, $\partial'\partial''c_{i,j} = 0$ implies $\partial''c_{i,j} = \partial'c_{i+1,j-1}$ for some $c_{i+1,j-1} \in E_{i+1,j-1}$, and then $\partial'\partial''c_{i+1,j-1} = -\partial''\partial'c_{i+1,j-1} = -\partial''\partial''c_{i,j} = 0$. The induction leads us nonuniquely to $c_{k,0} \in E_{k,0}$ such that $\partial'\partial''c_{k,0} = 0$. Define $b \in E_{k,-1}$ by $b = \partial''c_{k,0}$, and then $\partial'b = 0$. The result of the construction is therefore that we pass nonuniquely from the cocycle $a \in E_{-1,k}$ for ∂'' to a cocycle $b \in E_{k,-1}$ for ∂'.

Inverting the steps and the choices, we see that we can pass from b back to a. Thus if we can address the nonuniqueness, then the isomorphism in homology will have been established. We are to show that if $a \in E_{-1,k}$ at the start is a boundary relative to ∂'', then any system of choices leads to a result $b \in E_{k,-1}$ that is a boundary for ∂'. Since a is assumed to be a boundary for ∂'', $a = \partial''a'$ with $a' \in E_{-1,k+1}$. The element a' has $\partial'a' = 0$, and thus $a' = -\partial'a_{0,k+1}$ for some $a_{0,k+1} \in E_{0,k+1}$. Meanwhile, the above construction makes $a = \partial'c_{0,k}$. So $\partial'\partial''a_{0,k+1} = -\partial''\partial'a_{0,k+1} = \partial''a' = a = \partial'c_{0,k}$. By exactness, $c_{0,k} - \partial''a_{0,k+1} = \partial'b_{1,k}$ for some $b_{1,k} \in E_{1,k}$. This proves that $c_{0,k}$ is of the form $c_{0,k} = \partial''a_{0,k+1} + \partial'b_{1,k}$ with $a_{0,k+1} \in E_{0,k+1}$ and $b_{1,k} \in E_{1,k}$. (Note that this form for $c_{0,k}$ already implies that $\partial'\partial''c_{0,k} = 0$.)

Now suppose inductively on $i \geq 0$ that $j \geq 0$ is defined by $i + j = k$ and that $c_{i,j} \in E_{i,j}$ is given with $c_{i,j} = \partial''a_{i,j+1} + \partial'b_{i+1,j}$. The constructed element $c_{i+1,j-1} \in E_{i+1,j-1}$ has $\partial''c_{i,j} = \partial'c_{i+1,j-1}$ for some $c_{i+1,j-1} \in E_{i+1,j-1}$. Thus

$\partial' c_{i+1,j-1} = \partial''\partial' b_{i+1,j} = -\partial'\partial'' b_{i+1,j}$, and $c_{i+1,j-1} + \partial'' b_{i+1,j} = \partial' b_{i+2,j-1}$. If we put $a_{i+1,j} = -b_{i+1,j}$, then we have $c_{i+1,j-1} = \partial'' a_{i+1,j} + \partial' b_{i+2,j-1}$, and the induction goes through to $i = k$. Consequently any choice of $c_{k,0}$ obtained starting from the boundary a is of the form $c_{k,0} = \partial'' a_{k,1} + \partial' b_{k+1,0}$. The final step is to define $b = \partial'' c_{k,0}$, and then we have $b = \partial''\partial' b_{k+1,0} = -\partial'\partial'' b_{k+1,0}$, and b is exhibited as a boundary relative to ∂'.

25. Since each C_p is projective for $p \geq 0$, $C_p \otimes_R D$ is exact. Similarly $C \otimes_R D_q$ is exact for $q \geq 0$. The hypotheses of Problem 24 are satisfied, and the two homologies match.

26. $H_0(C) = H_0(C') = H_0(D) = \mathbb{Z}/2\mathbb{Z}$, and $H_p(C) = H_p(C') = H_p(D) = 0$ for $p \neq 0$. $H_0(C \otimes_\mathbb{Z} D) = H_0(C' \otimes_\mathbb{Z} D) = \mathbb{Z}/2\mathbb{Z}$, $H_1(C \otimes_\mathbb{Z} D) = 0$ and $H_1(C' \otimes_\mathbb{Z} D) = \mathbb{Z}/2\mathbb{Z}$, $H_p(C \otimes_\mathbb{Z} D) = H_p(C' \otimes_\mathbb{Z} D) = 0$ for $p \notin \{0, 1\}$.

27. Let $Z_p = \ker \partial'_p \subseteq C_p$, $B_p = \text{image } \partial'_{p+1} \subseteq C_p$, and $B'_p = B_{p-1}$. Since R is a principal ideal domain, Problem 20 shows that flat is equivalent to torsion free. Modules of the complex C are flat by assumption, hence torsion free. Modules of Z and B' are R submodules of these, hence are torsion free, hence are flat.

28. The long exact sequence in homology shows that

$$\text{Tor}_1^R(B', D) \to Z \otimes_R D \to C \otimes_R D \to B' \otimes_R D \to 0$$

is exact. Since B' is flat, Problem 19 shows that $\text{Tor}_1^R(B', D) = 0$.

29. For (a), the boundary map on $B'_p \otimes_R D_q$ in $B' \otimes_R D$ is $\partial' \otimes 1 + (-1)^p(1 \otimes \partial'')$, and $\partial' = 0$ on boundaries in B'_p.

For (b), tensoring with B' is an exact functor, since B' is flat. Therefore the exactness of $0 \to \overline{Z} \to D \xrightarrow{\partial''} \overline{B}' \to 0$ implies the exactness of

$$0 \to (B' \otimes_R \overline{Z})_n \to (B' \otimes_R D)_n \xrightarrow{(1 \otimes \partial'')_n} (B' \otimes_R \overline{B}')_n \to 0$$

for each n. From the exactness of this sequence, we can read off that $\ker(1 \otimes \partial'')_n$ within $(B' \otimes_R D)_n$ is $(B' \otimes_R \overline{Z})_n$ and that $\text{image}(1 \otimes \partial'')_n$ on $(B' \otimes_R D)_n$ is $(B' \otimes_R \overline{B}')_n$, which is the same thing as $(B' \otimes_R \overline{B})_{n-1}$.

For (c), the results of (b) show that

$$H_n(B' \otimes_R D) \cong \ker(1 \otimes \partial'')_n / \text{image}(1 \otimes \partial'')_{n+1} = (B' \otimes_R \overline{Z})_n / (B' \otimes \overline{B})_n.$$

Since tensoring with B' is exact, the exactness of $0 \to \overline{B} \to \overline{Z} \to H(D) \to 0$ implies the exactness of

$$0 \to B' \otimes_R \overline{B} \to B' \otimes_R \overline{Z} \to B' \otimes_R H(D) \to 0$$

in each degree. Thus $B' \otimes_R H(D) = (B' \otimes_R \overline{Z})/(B' \otimes_R \overline{B})$, and $H_n(B' \otimes_R D) \cong (B' \otimes H(D))_n = (B \otimes_R H(D))_{n-1}$.

Part (d) is handled in a fashion similar to (c).

30. For (a), $\operatorname{Tor}_1^R(\mathbb{Z}, H(D)) = 0$ because $\mathbb{Z}$ is flat.

In (b), comparison of the exact sequence with $\ker \omega_{n-1}$ with the exact sequence displayed before part (a) (but with n replaced by $n-1$) shows that $\ker \omega_{n-1}$ is isomorphic to $\operatorname{Tor}_1^R(H(C), H(D))_{n-1}$. Substituting for $\ker \omega_{n-1}$ and incorporating the isomorphism into the mapping into $H_n(B' \otimes_R D)$ leads to β'_{n-1} as the one-one mapping.

In (c), we have

$$\operatorname{coker}(\iota \otimes 1) = H_n(C \otimes_R D)/\operatorname{image}(\iota_n \otimes 1) = H_n(C \otimes_R D)/\ker(\partial'_n \otimes 1)$$
$$\cong \operatorname{image}(\partial'_n \otimes 1) = \ker \omega_{n-1} \cong \operatorname{Tor}_1^R(H(C), H(D))_{n-1}.$$

The composition of maps leading from $H_n(C \otimes_R D)$ to $H_n(B' \otimes_R D)$ has to be $\partial'_n \otimes 1$, and thus $\beta'_{n-1}\beta_{n-1} = \partial'_n \otimes 1$. The map β_{n-1}, apart from isomorphisms, is onto because q was constructed as onto.

Part (d) is completely analogous, and the resulting map α_n is one-one.

For (e), we know that α is one-one and that β is onto. Also, we have $\beta'_{n-1}\beta_{n-1}\alpha_n\alpha'_n = (\partial'_n \otimes 1)(\iota_n \otimes 1) = 0$. Since β'_{n-1} is one-one and α'_n is onto, $\beta_{n-1}\alpha_n = 0$. Finally suppose that x is in $\ker \beta_{n-1}$. Then x is in $\ker(\beta'_{n-1}\beta_{n-1}) = \ker(\partial'_n \otimes 1) = \operatorname{image}(\iota_n \otimes 1) = \operatorname{image}(\alpha_n\alpha'_n) = \operatorname{image}\alpha_n$. This completes the proof of exactness.

31. This is immediate.

32. Let $X = \{X_n\}$ and $Y = \{Y_n\}$. Then $\operatorname{Morph}(X, Y)$ is the subgroup of $\prod_{n=-\infty}^{\infty} \operatorname{Hom}(X_n, Y_n)$ consisting of those elements in the product satisfying the chain map conditions. A zero object is any tuple of 0's, and certainly product and coproduct make sense. One readily verifies that the tuple of kernels of a chain map furnishes a kernel for a chain map and that the tuple of cokernels furnishes a cokernel.

33. The additional objects and morphisms at the top of the extended diagram are $C_0 = 2\mathbb{Z}/8\mathbb{Z}$, $B_0 = \mathbb{Z}$, k given by 2 mod 8 $\mapsto$ 2 mod 8, $\widetilde{k}$ given by $\times 2$, $\widetilde{\psi}$ given by $1 \mapsto 2 \mod 8$, and $\widetilde{\varphi}$ given by $\times 4$. Since the composition of $\widetilde{k}$ followed by $\beta = \times 2$ is not 0, $(B_0, \widetilde{k})$ cannot be the kernel of β.

The additional objects and morphisms at the bottom of the extended diagram are $A'_0 = \mathbb{Z}/4\mathbb{Z}$, $B'_0 = \mathbb{Z}/16\mathbb{Z}$, p given by $1 \mapsto 1 \mod 4$, $\widetilde{p}$ given by $1 \mapsto 1 \mod 16$, $\widetilde{\varphi}'$ given by 1 mod 4 $\mapsto$ 4 mod 16, and $\widetilde{\psi}'$ given by 1 mod 16 $\mapsto$ 1 mod 4.

34. We give the argument only for $\operatorname{Hom}(M, \cdot)$. Let $0 \to A \xrightarrow{\varphi} B \xrightarrow{\psi} C \to 0$ be a given exact sequence, and form the sequence

$$0 \longrightarrow \operatorname{Hom}(M, A) \xrightarrow{\operatorname{Hom}(1,\varphi)} \operatorname{Hom}(M, B) \xrightarrow{\operatorname{Hom}(1,\psi)} \operatorname{Hom}(M, C).$$

We are to show that $\operatorname{Hom}(1, \varphi)$ is one-one and that exactness holds at $\operatorname{Hom}(M, B)$.

If σ is in $\operatorname{Hom}(M, A)$ with $\operatorname{Hom}(1, \varphi)(\sigma) = 0$, then $\varphi\sigma = 0$, and it follows that $\sigma = 0$ because φ is a monomorphism.

For the exactness at $\operatorname{Hom}(M, B)$, we use Theorem 4.42e. We know immediately that $\operatorname{Hom}(1, \psi)\operatorname{Hom}(1, \varphi) = \operatorname{Hom}(1, \psi\varphi) = \operatorname{Hom}(1, 0) = 0$. Thus suppose that

$\tau \in_m \text{Hom}(M, B)$ has $\text{Hom}(1, \psi)\tau \equiv 0$. This condition means that $\psi\tau \equiv 0$. Since the given sequence is exact, Theorem 4.42e produces some $\tau' \in_m A$ with $\varphi\tau' \equiv \tau$. In turn, this says that $\text{Hom}(1, \varphi)\tau' \equiv \tau$. By Theorem 4.42, we have exactness at $\text{Hom}(M, B)$.

35. We give the proof only that the splitting of exact sequences as indicated implies that P is projective. Thus suppose that a morphism $\tau \in \text{Hom}(P, B)$ and an epimorphism $\psi \in \text{Hom}(C, B)$ are given. We are to produce $\sigma \in \text{Hom}(P, C)$ with $\tau = \psi\sigma$. Let $(W, \widetilde{\psi}, \widetilde{\tau})$ be a pullback of (ψ, τ). Then $\tau\widetilde{\psi} = \psi\widetilde{\tau}$, and Proposition 4.40 shows that $\widetilde{\psi}$ is an epimorphism. Then it follows that

$$0 \to \text{domain}(\ker \widetilde{\psi}) \xrightarrow{\ker \widetilde{\psi}} W \xrightarrow{\widetilde{\psi}} P \to 0$$

is exact, and it must split by assumption. Thus there exists $\rho \in \text{Hom}(P, W)$ with $\widetilde{\psi}\rho = 1_P$. Put $\sigma = \widetilde{\tau}\rho$. Then $\psi\sigma = \psi\widetilde{\tau}\rho = \tau\widetilde{\psi}\rho = \tau 1_P = \tau$, as required.

Chapter V

1. If ξ is a root of $F(X)$, then the given formula shows that $D(\xi)$ is -23 and -31 in the two cases. These contain no square factor and therefore equal $D_\mathbb{K}$ in the two cases.

2. For (a), let $G(X) = F(X + \frac{2}{3}) = X^3 - \frac{4}{3}X + \frac{22}{27}$. Then $F(X)$ and $G(X)$ have the same discriminant, and the discriminant for $G(X)$ is given by the formula of Problem 1. It is -44.

For (b), let $x = a + b\xi + c\xi^2$ be given with a, b, c all in $\{0, 1\}$. The matrix of left-by-x in the ordered basis $(1, \xi, \xi^2)$ works out to be

$$\begin{pmatrix} a & -2c & -2b-4c \\ b & a & -2c \\ c & b+2c & a+2b+4c \end{pmatrix},$$

and the determinant of it is

$$a^3 + 2a^2(b + 4c) + 4c^3 - 2b(b + 2c)^2 + 4ac(b + 2c) + 2bc(a + 2b + 4c).$$

For x to be twice an algebraic integer, this determinant, which is the norm of x, has to be $\equiv 0 \mod 8$. All the terms are even except possibly the first, and thus a has to be even. That is, $a = 0$. The determinant then reduces to $4c^3 - 2b(b+2c)^2 + 4bc(b+2c)$. All terms here are divisible by 4 except possibly $-2b^2$. Thus b must be even. That is, $b = 0$. The determinant reduces in this case to $4c^3$. For this to be divisible by 8, c must be even. That is, $c = 0$. Proposition 5.2 consequently says that a further factor of 2^2 cannot be eliminated from the discriminant.

3. For (a), Theorem 5.21 and the remarks after it show that every equivalence class contains an ideal whose norm is $< (0.283) D_{\mathbb{K}}^{1/2}$. Proposition 5.8 shows that $D_{\mathbb{K}} = 3^5 = 243$. Thus every equivalence class contains an ideal with norm ≤ 4.

Conclusion (b) is immediate from Theorem 5.6 with $F(X) = X^3 - 3$. Conclusion (c) follows because $(\sqrt[3]{3} - 1)(\sqrt[3]{9} + \sqrt[3]{3} + 1) = (\sqrt[3]{3})^3 - 1 = 3 - 1 = 2$. Conclusion (d) is immediate from Proposition 5.10d.

For (e), any nonzero ideal is the product of powers of prime ideals associated with the various prime numbers. The ones corresponding to the prime numbers 2 and 3 are principal ideals by (b), (c), and (d). These are the only ones that need to be checked, according to (a). Thus every nonzero ideal is principal.

4. Conclusion (a) is immediate from Theorem 5.6, since $X^3 - 7$ factors modulo 2 as $(X + 1)(X^2 + X + 1)$. For (b), we show that no element $x = a + b\sqrt[3]{7} + c\sqrt[3]{49}$ has norm ± 2. Left multiplication by x carries 1 to $a + b\sqrt[3]{7} + c\sqrt[3]{49}$, carries $\sqrt[3]{7}$ to $7c + a\sqrt[3]{7} + b\sqrt[3]{49}$, and carries $\sqrt[3]{49}$ to $7b + 7c\sqrt[3]{7} + a\sqrt[3]{49}$. Thus its matrix is

$$\begin{pmatrix} a & 7c & 7b \\ b & a & 7c \\ c & b & a \end{pmatrix}.$$

The determinant is $a^3 + 49c^3 + 7b^3 - 21abc$, which is congruent modulo 7 to a^3. Modulo 7, the cubes are 0 and ± 1, and thus the congruence $a^3 \equiv \pm 2 \bmod 7$ has no solution.

5. Since the element $\sqrt{-1} + \sqrt{-5}$ has degree 4 over $\mathbb{Q}$, the minimal polynomial has degree 4. The product of $(X - (+\sqrt{-1} + \sqrt{-5}))$ and the Galois transforms $(X - (+\sqrt{-1} - \sqrt{-5}))$, $(X - (-\sqrt{-1} + \sqrt{-5}))$, and $(X - (-\sqrt{-1} - \sqrt{-5}))$ is $X^4 + 12X^2 + 16$, which is in $\mathbb{Z}[X]$.

6. The minimal polynomial of $\xi = \frac{1}{2}(\sqrt{-1} + \sqrt{-5})$ is $H(X) = X^4 + 2^{-2}12X^2 + 2^{-4}16 = X^4 + 3X^2 + 1$ with $|D(\xi)| = |N_{\mathbb{K}/\mathbb{Q}}(H'(\xi))|$. Here $H'(X) = 4X^3 + 6X = 2(2X^2 + 3)$. Since $\xi^4 + 3\xi^2 + 1 = 0$, we have $\xi^2 = -\frac{3}{2} \pm \frac{1}{2}\sqrt{5}$; thus $2\xi^2 + 3 = \pm\sqrt{5}$. So $|D(\xi)| = |N_{\mathbb{L}/\mathbb{Q}}(\pm 2\sqrt{5})|$. The four conjugates of $\sqrt{5}$ are $+\sqrt{5}$ twice and $-\sqrt{5}$ twice, and the norm is the product of the four conjugates. Thus $|D(\xi)| = |N_{\mathbb{L}/\mathbb{Q}}(\pm 2\sqrt{5})| = 2^4 5^2$.

7. These follow immediately by applying Theorem 5.6 to the indicated prime, 2 or 5, and the respective polynomials: $X^2 + 5$, $X^2 + X - 1$, and $X^2 + 1$.

8. With $\mathbb{Q} \subseteq \mathbb{K}' \subseteq \mathbb{L}$, the (e, f, g) for $\mathbb{L}/\mathbb{Q}$ has to be entry by entry $\geq$ the triple for $\mathbb{K}'/\mathbb{Q}$. The triple for $\mathbb{K}'/\mathbb{Q}$ is given in Problem 7b as $(1, 2, 1)$ for $p = 2$. Similarly from $\mathbb{Q} \subseteq \mathbb{K}'' \subseteq \mathbb{L}$, the (e, f, g) for $\mathbb{L}/\mathbb{Q}$ has to be $\geq (2, 1, 1)$. Thus $e \geq 2$, $f \geq 2$, and $g \geq 1$. Since $efg = 4$, equality must hold throughout: $(e, f, g) = (2, 2, 1)$.

This proves (a). Similarly for (b), we must have $(e, f, g) \geq (2, 1, 1)$ and $(e, f, g) \geq (1, 1, 2)$. Thus $(e, f, g) \geq (2, 1, 2)$. Since $efg = 4$, $(e, f, g) = (2, 1, 2)$.

9. In (a), Problem 8a shows that $(2)T = P^2$, and we know that $(2)R = \wp_2^2$. Then $P^2 = (2)T = (2)RT = \wp_2^2 T = (\wp_2 T)(\wp_2 T)$. Since P is prime, P divides $\wp_2 T$. For the equality $P^2 = (\wp_2 T)^2$ to hold, we must have $P = \wp_2 T$.

Similarly $(5)T = P_1^2 P_2^2$ and $(5)R = \wp_5^2$. Then $P_1^2 P_2^2 = (5)T = (5)RT = \wp_5^2 T = (\wp_5 T)^2$. Since P_1 and P_2 are prime, P_1 and P_2 must divide $\wp_5 T$. Therefore $P_1 P_2 = \wp_5 T$.

In (b), conclusion (a) shows that no prime ideal of R that divides $(2)R$ or $(5)R$ ramifies in T. Since $D(\xi)$ is divisible by no prime numbers other than 2 and 5, Theorem 5.6 shows that no prime ideal (p) of $\mathbb{Z}$ ramifies in T. Hence no prime ideal of R containing such a prime (p) of $\mathbb{Z}$ ramifies in T.

10. Roots of unity must map to roots of unity under the embedding, and there are only two roots of unity within $\mathbb{R}$. Hence there are no real-valued embeddings when $p > 2$. Thus the embeddings come in complex-conjugate pairs. The product $\sigma(x)\bar{\sigma}(x)$ is positive for $x > 0$, and $N_{\mathbb{K}/\mathbb{Q}}(x)$ is the product of these expressions over all such pairs.

11. For (a), $F(X)$ is the minimal polynomial of ζ^k when $\gcd(k, p) = 1$. Then $\zeta^k - 1$ is a root of $G(X) = F(X + 1)$ of the correct degree, and therefore $G(X)$ is the minimal polynomial of $\zeta^k - 1$. If $H(X)$ is the field polynomial of an element η, then $N_{\mathbb{K}/\mathbb{Q}}(\eta) = (-1)^{[\mathbb{K}:\mathbb{Q}]} H(0)$. In this instance $[\mathbb{K} : \mathbb{Q}] = p - 1$ is even. Taking $\eta = \zeta^k - 1$, we obtain $N_{\mathbb{K}/\mathbb{Q}}(\zeta^k - 1) = G(0) = F(1) = p$.

For (b), $\zeta - 1$ divides $\zeta^k - 1$, and hence the quotient is in R. If l is chosen with $lk \equiv 1 \bmod p$, then $\zeta - 1 = \zeta^{lk} - 1$, and $\zeta^k - 1$ divides $\zeta^{lk} - 1$. Therefore the reciprocal of $(\zeta^k - 1)/(\zeta - 1)$ is in R.

12. With $F(X)$ and $G(X)$ as in the previous problem, $F'(\zeta^k) = G'(\zeta^k - 1)$. Here $F(X) = (X^p - 1)/(X - 1)$ makes $G(X) = X^{-1}[(X + 1)^p - 1]$ and $G'(X) = X^{-2}[pX(X + 1)^{p-1} - (X + 1)^p + 1]$. Since $\zeta^{kp} = 1$,

$$F'(\zeta^k) = G'(\zeta^k - 1) = (\zeta^k - 1)^{-2}[p(\zeta^k - 1)\zeta^{k(p-1)} - \zeta^{kp} + 1] = (\zeta^k - 1)^{-1} p\zeta^{k(p-1)}.$$

The result now follows from the formula $\mathcal{D}(\zeta^k) = F'(\zeta^k)$.

13. Continuing from the previous problem gives

$$N_{\mathbb{K}/\mathbb{Q}}(F'(\zeta^k)) = N_{\mathbb{K}/\mathbb{Q}}(\zeta^k - 1)^{-1} p^{p-1} N_{\mathbb{K}/\mathbb{Q}}(\zeta^{k(p-1)}) = p^{p-2}.$$

The result follows from the computation $(-1)^{(p-1)(p-2)/2} D(\zeta^k) = N_{\mathbb{K}/\mathbb{Q}}(\mathcal{D}(\zeta^k)) = N_{\mathbb{K}/\mathbb{Q}}(F'(\zeta^k)) = p^{p-2}$.

14. For (a), we have $\lambda^k = (1 - \zeta)^k = \sum_{j=0}^{k} (-1)^j \binom{k}{j} \zeta^j$ and $\zeta^k = (1 - \lambda)^k = \sum_{j=0}^{k} (-1)^j \binom{k}{j} \lambda^j$. Conclusion (b) is a version of Problem 11b because the conjugates of ζ are the powers ζ^j for $1 \le j \le p - 1$. For (c), we have $p = \prod_{k=1}^{p-1} (1 - \zeta^k) = \prod_{k=1}^{p-1} (1 - \zeta) u_k = (1 - \zeta)^{p-1} \prod_{k=1}^{p-1} u_k$, where $u_k = (1 - \zeta^k)/(1 - \zeta)$. Each element u_k is a unit by Problem 11c, and (c) follows.

15. The identity $(p)R = (1 - \zeta)^{p-1}$ is immediate from Problem 14c. The extension $\mathbb{K}/\mathbb{Q}$ being Galois, we know that the prime decomposition of the ideal $(p)R$ is of the form $(p)R = P_1^e \cdots P_g^e$, where $p - 1 = efg$ and f is the common

value of all $\dim_{\mathbb{F}_p}(R/P_j)$. This latter fact says that no factorization of $(p)R$ into proper ideals can have more than $p-1$ factors, and $p-1$ factors occur only if all factors are prime. In this case, $(1-\zeta)$ is a proper ideal because $N_{\mathbb{K}/\mathbb{Q}}(1-\zeta) = p$. Thus each factor $(1-\zeta)$ is prime.

16. Following Proposition 5.2, suppose that a_j is an integer for each j with $s \leq j \leq k$ such that $0 \leq a_j \leq p-1$, $a_s \neq 0$, $a_k = 1$, and

$$a_s \lambda^s + a_1 \lambda^{s+1} + a_2 \lambda^{s+2} + \cdots + a_{k-1} \lambda^{k-1} + a_k \lambda^k = pr$$

with r in R. Subtracting all terms from the left side but the first and applying Problem 15 shows that $a_s \lambda^s$ lies in $(\lambda)^{s+1}$. Thus $(a_s)(\lambda)^s \subseteq (\lambda)^{s+1}$. Canceling gives $(a_s) \subseteq (\lambda)$, and this inclusion is a contradiction because $\text{GCD}(N((a_s)), N((\lambda))) = 1$.

17. Each step toward a $\mathbb{Z}$ basis multiplies a discriminant by a square, and it is enough to prove that a primitive element ξ for $\mathbb{K}/\mathbb{Q}$ lying in R has sgn $D(\xi) = (-1)^{r_2}$. We are thus to compute the sign of $\prod_{i<j}(\sigma_i(\xi) - \sigma_j(\xi))^2$. For a given pair (i, j), the factor $(\sigma_i(\xi) - \sigma_j(\xi))^2$ is matched by its complex conjugate elsewhere in the product unless σ_i and σ_j are both real or are complex conjugates of one another. The factor and its mate have a positive product, and pair with σ_i and σ_j both real contributes a positive square. If $\sigma_j = \overline{\sigma}_i$, then $\sigma_i(\xi) - \sigma_j(\xi)$ is purely imaginary, and its square is negative. Hence the sign is $(-1)^{r_2}$.

18. Let g be in $\text{Gal}(\mathbb{K}/\mathbb{Q}) = \{\sigma_1, \ldots, \sigma_n\}$. Replacing each σ_j by $g\sigma_j$ has the effect of permuting the columns of $[\sigma_j(\alpha_i)]$. If the permutation is even, then the terms contributing to P are the same before and after the permutation; otherwise they are interchanged. In either case, $P+N$ and PN are fixed. Since $P+N$ and PN are fixed by the Galois group, they are in $\mathbb{Q}$. The entries $\sigma_j(\alpha_i)$ of the matrix are in R, and thus P and N are in R. Consequently $P+N$ and PN are in $\mathbb{Z}$. The formula $D(\Gamma) = (P+N)^2 - 4PN$ shows that $D(\Gamma) \equiv (P+N)^2 \mod 4$. Any square of a member of $\mathbb{Z}$ is congruent to 0 or 1 modulo 4, and the result follows.

19. Let J be an ideal of $S^{-1}R$. Proposition 8.47 of *Basic Algebra* shows that $I = R \cap J$ is an ideal in R and that $J = S^{-1}I$. Since $I_1, \ldots, I_h$ is a complete set of representatives for the equivalence classes, $aI = bI_j$ for some j with $1 \leq j \leq h$. Let $(a)_S$ and $(b)_S$ be the principal ideals of $S^{-1}R$ generated by a and b. The fact that u is in $I_j \cap S$ means that $S^{-1}I_j = S^{-1}R$, and thus

$$(a)_S J = S^{-1}(a)S^{-1}I = S^{-1}(a)I = S^{-1}(b)I_j$$
$$= S^{-1}(b)S^{-1}I_j = S^{-1}(b)S^{-1}R = (b)_S. \qquad (*)$$

Hence J is principal. (In fact, the equality shows that $aj = b$ for some $j \in J$. Hence $ba^{-1} = j$ is an element of $J \subseteq S^{-1}R$, the principal ideal $(ba^{-1})_S$ of $S^{-1}R$ is meaningful, and $(ba^{-1})_S \subseteq J$. For the reverse inclusion let $j \in J$ be given, and use $(*)$ to write $aj = bx$ with $x \in S^{-1}R$. Then $j = (ba^{-1})x$ shows that j is in $(ba^{-1})_S$, and $J \subseteq (ba^{-1})_S$.)

20. For (a), write $ab = u^k$. Then $a^{-1} = u^{-k}b$ exhibits a^{-1} as in $S^{-1}R$. For (b), if $u^{-m}a$ is a unit in $S^{-1}R$, then $u^{-m}a^{-1} = u^{-l}c$ for some $c \in R$. Hence $ac = u^{l-m}$. Since ac is in R and u is not, $l - m = k$ with $k \geq 0$. Then a divides u^k.

21. For (a), write $(u) = P_1^{e_1} \cdots P_l^{e_l}$. Then $(u^h) = (P_1^h)^{e_1} \cdots (P_l^h)^{e_l} = (b_1^{e_1} \cdots b_l^{e_l})$. Thus $u^h = b_1^{e_1} \cdots b_l^{e_l}\varepsilon$ for some unit ε in R, each b_j divides u^h, and the conclusion follows from Problem 20a.

For (b), we have $(a)(b) = (u)^k = P_1^{ke_1} \cdots P_l^{ke_l}$. Since a and b are in R, this equality implies that $(a) = P_1^{r_1} \cdots P_l^{r_l}$. For each j, use the division algorithm to write $r_j = n_jh + t_j$ with $0 \leq t_j < h$. Then $P_j^{r_j} = (P_j^h)^{n_j}P_j^{t_j} = (b_j)^{n_j}P_j^{t_j}$, and consequently $(a) = (d)P_1^{t_1} \cdots P_l^{t_l}$ as required, where $d = \prod_{j=1}^l b_j^{n_j}$.

The argument for (c) was given in parentheses at the end of the solution of Problem 19.

22. Because of Problem 21d, we now have $(a) = (d)(c_i)$. Thus $a = dc_i\varepsilon$ for some unit ε in R. Since $u^k = ab = c_idb\varepsilon$, c_i divides u^k and is a unit in $S^{-1}R$ by Problem 20a.

23. Problem 22 shows that any unit of $S^{-1}R$ is a product of a power of u by a product $\prod_{j=1}^l b_j^{n_j}$, an element c_i, and a unit ε of R. Problem 21a shows that each b_j is a unit in $S^{-1}R$, and Problem 22 shows that each c_i is a unit in $S^{-1}R$. Thus $(S^{-1}R)^\times$ is generated by u, the finitely many elements b_j and c_i, and a finite set of generators of $R^\times$. (The group $R^\times$ is finitely generated by the Dirichlet Unit Theorem.)

24. $G(4/\xi) = (64\xi^{-3} - 16\xi^{-2} + 8\xi^{-1} + 8) = 8\xi^{-3}(\xi^3 + \xi^{-2} - 2\xi + 8) = 8\xi^{-3}F(\xi) = 0$. The element η is in $\mathbb{K}$, and it is exhibited as the root of a monic polynomial in $\mathbb{Z}[X]$; therefore it is in R.

25. For (a), $0 = F(\xi)/\xi = \xi^2 + \xi - 2 + 8\xi^{-1} = \xi^2 + \xi - 2 + 2\eta$. For (b), $0 = G(\eta)/\eta = \eta^2 - \eta + 2 + 8/\eta = \eta^2 - \eta + 2 + 2\xi$. Solving the first equation for ξ^2 gives the first formula in the table, and solving the second equation for η^2 gives the second formula in the table. The formula $\xi\eta = 4$ is immediate from the definition $\eta = 4/\xi$. The formulas in the table together show that any integer polynomial in ξ and η reduces to a $\mathbb{Z}$ combination of $1, \xi$, and η.

Conclusion (c) is clear. For (d), we have $\eta = 1 - \frac{1}{2}(\xi^2 + \xi)$, and this is not in $\mathbb{Z}(\{1, \xi, \xi^2\})$. For (e), we have $D((1, \xi, \xi^2)) = -2^2 \cdot 503$. Since the only square factor is 2^2, it follows that $\mathbb{Z}(\{1, \xi, \xi^2\})$ has index 2 in $\mathbb{Z}(\{1, \xi, \eta\})$ and that $D((1, \xi, \eta)) = -503$. This latter discriminant is square free and thus cannot be reduced further. Therefore $D_\mathbb{K} = -503$, and $\{1, \xi, \eta\}$ is a $\mathbb{Z}$ basis of R. Finally the formula $\eta = 1 - \frac{1}{2}(\xi^2 + \xi)$ shows that $\mathbb{Z}(\{1, \xi, \eta\}) = \mathbb{Z}(\{1, \xi, \frac{1}{2}(\xi^2 + \xi)\})$.

26. Application of φ to $\xi^2 = \xi + 2 - 2\eta$ gives $\overline{\xi}^2 = \overline{\xi}$. Similarly $\overline{\eta}^2 = \overline{\eta}$. The elements of a finite field of characteristic 2 fixed by the squaring map are 0 and 1. Hence $\overline{\xi}$ and $\overline{\eta}$ are in $\{0, 1\}$. Since $\mathbb{F} = \varphi(R)$ is generated by the values of φ on 1, ξ, and η, $\mathbb{F}$ has two elements. From $\xi\eta = 4$, it follows that $\overline{\xi}\overline{\eta} = 0$. Thus $\overline{\xi}$ and $\overline{\eta}$ cannot both be 1, and the only possibilities are the ones in the table.

27. Define $\varphi : R \to \mathbb{F}_2$ on ξ and η by one of the lines of the table of Problem 26, and set $\varphi(1) = 1$. Then φ extends to a well-defined additive homomorphism on $\mathbb{Z}(\{1, \xi, \eta\})$. We have to check that φ respects multiplication. It is enough to do so on additive generators. Thus we have to check that $\varphi(\xi^2) = (\varphi(\xi))^2$, that $\varphi(\eta^2) = (\varphi(\eta))^2$, and that $\varphi(\xi\eta) = (\varphi(\xi))(\varphi(\eta))$. Thus, for example, in the first one we want $-\varphi(\xi) + 2\varphi(1) - 2\varphi(\eta) = (\varphi(\xi))^2$. If we write the values of φ as triples corresponding to the three possible φ's, the left side is $-(0, 1, 0) + 2(1, 1, 1) - 2(0, 0, 1) \equiv (0, 1, 0) \bmod 2$, while the right side is $(0, 1, 0)^2 \equiv (0, 1, 0) \bmod 2$. These match, and this relation is verified. The other two relations are verified in similar fashion.

28. The norm of a kernel equals the number of elements in the image of the homomorphism, which is 2 in each case. Since each ideal has prime norm, the ideal is prime. Moreover, these ideals contain $(2)R$ and hence all figure into the prime factorization of $(2)R$. On the other hand, we must have $\sum e_i f_i = 3$ for the decomposition, and we have seen that there are at least three terms. So there are exactly three terms, and we must have $e_i = f_i = 1$ in each case. Therefore $(2)R = P_{0,0} P_{1,0} P_{0,1}$.

29. For (a), the elements listed are additive generators of the ideal in each case, and hence they are also ideal generators. For (b), $\eta = \eta(\xi + 1) - 2 \cdot 2$ shows that η is in the ideal $(2, \xi + 1)$. Thus $(2, \xi + 1, \eta) \subseteq (2, \xi + 1)$. The reverse inclusion is clear. In (c), the argument for $(2, \eta + 1)$ is completely symmetric. Let us see that $(2, \xi, \eta) = (2, \xi - \eta)$. The inclusion $\supseteq$ is clear. For the inclusion $\subseteq$, we use the two formulas

$$(-1 - \eta)2 + (-\xi)(\xi - \eta) = -2 - 2\eta - (-\xi + 2 - 2\eta) + 4 = \xi,$$
$$(3 + \xi)2 + (-\eta)(\xi - \eta) = 6 + 2\xi - 4 + (-2\xi - 2 + \eta) = \eta.$$

30. For (a), the field polynomial of $\theta - q$ is $H(X + q)$, and so the norm of $\theta - q$ is $-H(0+q)$, as required. In (b), the first two formulas come from the field polynomials $F(X)$ and $G(X)$ of ξ and η, and the other formulas follow from (a).

In (c), the fact that $N((\xi)) = |N_{\mathbb{L}/\mathbb{Q}}(\xi)| = 8$ shows that the prime factorization of (ξ) is into prime ideals whose norms are powers of two. Problem 28 shows that all such ideals have been identified, and thus $(\xi) = P_{0,0}^a P_{1,0}^b P_{0,1}^c$ for some exponents ≥ 0. Comparing norms shows that $a + b + c = 3$. Similar remarks apply to (η).

In (d), use of Problem 28 shows that $P_{0,0}^2 P_{1,0}^2 P_{0,1}^2 = ((2)R)^2 = (4)R = (\xi)(\eta) = P_{0,0}^{a+\alpha} P_{1,0}^{b+\beta} P_{0,1}^{c+\gamma}$. Then $a + \alpha = 2, b + \beta = 2$, and $c + \gamma = 2$ by unique factorization.

For (e), we observe from the kernels, or else we see from Problem 29a, that ξ is not in $P_{1,0}$ and that η is not in $P_{0,1}$. Hence $P_{1,0}$ does not appear in the prime factorization of (ξ), and $P_{0,1}$ does not appear in the prime factorization of (η). Therefore $b = \gamma = 0$.

For (f), the results of (e) and (d) combine to show that $a + \alpha = 2, \beta = 2$, and $c = 2$. Since $a + c = 3$ and $\alpha + \beta = 3, a = \alpha = 1$.

31. For (a), we see immediately from Problem 29a that $\xi + 1$ lies in $P_{1,0}$ but not in $P_{0,0}$ and not in $P_{0,1}$. For (b), the formula $|N_{\mathbb{K}/\mathbb{Q}}(\xi + 3)| = 2^2$ shows that

($\xi + 3$) is the product of exactly two of the prime ideals of norm 2; thus (a) implies that $(\xi + 3) = P_{1,0}^2$. Similarly $|N_{K/\mathbb{Q}}(\xi - 1)| = 2^3$, and (a) gives $(\xi - 1) = P_{1,0}^3$. Conclusion (c) is immediate from Problem 29a.

For (d), we have $(2)R \subseteq (2, \xi)$; thus $(2, \xi)$ is of the form $P_{0,0}^a P_{1,0}^b P_{0,1}^c$ with $a + b + c \leq 3$. Since ξ is not in $P_{1,0}$, $b = 0$. Since ξ is in $P_{0,0}$ and $P_{0,1}$, we must have $a > 0$ and $c > 0$. Since the inclusion $(2)R \subseteq (2, \xi)$ is proper (because ξ is not in $(2)R = 2\mathbb{Z}(\{1, \xi, \eta\})$), $N((2, \xi)) \leq 4$. Thus $a = c = 1$, and $(2, \xi) = P_{0,0} P_{0,1}$.

For (e), Problem 29a shows that $P_{0,1} = (2, \xi, \eta + 1)$. Thus $P_{0,1}^2$ contains 4 and $\xi(\eta + 1) = 4 + \xi$, hence ξ. If $P_{0,1}^2$ contains also $\xi + l$ with $l \equiv 2 \bmod 4$, then it contains $\xi + 2$, hence 2. This would mean that $P_{0,1}^2 \supseteq (2, \xi) = P_{0,0} P_{0,1}$. Since $P_{0,1}^2$ and $P_{0,0} P_{0,1}$ both have norm 4, they would have to be equal, and we would obtain $P_{0,1} = P_{0,0}$, contradiction.

For (f), Problem 30b gives $N((\xi + 2)) = 8$. In view of (c), $(\xi + 2) = P_{0,0}^a P_{0,1}^c$ with $a + c = 3$ and $c \geq 1$. Part (d) shows that $c \leq 1$. Thus $(\xi + 2) = P_{0,0}^2 P_{0,1}$. The argument for $(\xi - 2)$ is similar.

32. For (a), this kind of argument is done in a parenthetical remark at the end of the solution of Problem 19. For (b), we have $(\xi + 2) = r_{0,0}^2 P_{0,1}$ and $(\xi - 1) = P_{1,0}^3 = (\xi + 3) P_{1,0}$. Thus the same kind of argument shows that $P_{0,1}$ and $P_{1,0}$ are principal.

For (c), we factor $X^3 + X^2 - 2X + 8$ modulo 3; there is no root in $\mathbb{F}_3$, and hence the reduced polynomial is irreducible. By Theorem 5.6 the only prime ideal whose norm is a power of 3 has norm 3^3.

For (d), we factor $X^3 + X^2 - 2X + 8$ modulo 5 as $(X + 1)(X^2 - 2)$, and Theorem 5.6 gives us one prime ideal of norm 5 and one of norm 5^2. The one of norm 5, according to the theorem, is $(2, 1 + \xi)$. For (e), the technique of Problem 30a shows that $N((1 + \xi)) = 10$. Thus the only possibility for the prime factorization of $(1 + \xi)$ is as $(2, 1 + \xi)P$, where P is one of the three ideals of norm 2. For (f), since $(1 + \xi)$ and P are principal, $(2, 1 + \xi)$ is principal, by the same technique as in earlier parts.

For (g), the prime factorization of nonzero ideals allows us to conclude that every nonzero ideal of norm ≤ 6 is principal. Application of the technique after Theorem 5.21 shows that every ideal class has a representative with norm < 6.35, hence norm ≤ 6. All such ideals are principal, and therefore R is a principal ideal domain.

Chapter VI

1. Apply the Cauchy criterion. Since $|a_n + a_{n+1} + \cdots + a_m|_p \leq \max_{n \leq k \leq m} |a_k|_p$, the series is Cauchy, hence convergent, if and only if the terms tend to 0.

2. In (a), the equality $\text{GCD}(3, 2^n) = 1$ implies that there exist integers x_n and y_n such that $3x_n - 2^n y_n = 1$. Then $x_n - \frac{1}{3} = 2^n 3^{-1} y_n$. Applying the 2-adic absolute value gives $|x_n - \frac{1}{3}|_2 = 2^{-n} |y_n|_2 \leq 2^{-n}$, and this tends to 0. For example take $x_n = \frac{1}{3}(2^{2n-1} + 1)$. In (b), the argument with $\frac{a}{b}$ replacing $\frac{1}{3}$ is similar: to get

$|x - \frac{a}{b}|_2 \le 2^{-n}$, start by finding x and y with $bx - 2^n y = a$.

3. Write ideles as tuples indexed by $\infty, 2, 3, 5, \ldots$. If q is in $\mathbb{Q}$, then $\iota(q) = (q, q, q, q, \ldots)$. If this is to be in $\mathbb{R}^\times \times \prod_p \mathbb{Z}_p^\times$, then the only restriction on the first coordinate is that $q \ne 0$, but the other coordinates are restricted by $|q|_p = 1$ for all primes p. This means that q in lowest terms has no p in either the numerator or the denominator. So $q = \pm 1$. This proves (a).

In (b), let $(x_\infty, x_2, x_3, \ldots)$ be in $\mathbb{I}$. Since $|x_p|_p \ne 1$ for only finitely many p, there exists a unique positive rational q such that $|q|_p = |x_p|_p$ for all p. Define $z_p = x_p q^{-1}$ as a member of $\mathbb{Q}_p^\times$. Then $|z_p|_p = |x_p|_p |q|_p^{-1} = 1$ shows that $|z_p|_p = 1$ for all p. Finally define $r = x_\infty q^{-1}$ as a member of $\mathbb{R}^\times$. Then $(r, z_2, z_3, \ldots)$ is in $\mathbb{I}(S_\infty)$, and $(x_\infty, x_2, x_3, \ldots) = (q, q, q, \ldots)(r, z_2, z_3, \ldots)$.

4. In (a), the norm of the ideal divides the norm of any element, and if the norm of the ideal is prime, then the ideal is prime. With $K = \mathbb{Q}(\sqrt{-5})$, we have $N_{K/\mathbb{Q}}(1 \pm \sqrt{-5}) = 6$, $N_{K/\mathbb{Q}}(3) = 9$, and $N_{K/\mathbb{Q}}(2) = 4$. Therefore $N((1 \pm \sqrt{-5}, 3))$ divides $\gcd(6, 9) = 3$, and $N((1 \pm \sqrt{-5}, 2))$ divides $\gcd(6, 4) = 2$. One checks that these ideals are not all of R, and then the respective norms are 3 and 2. So the ideals are prime. In (b), $(1 + \sqrt{-5}) = (1 + \sqrt{-5}, 2)(1 + \sqrt{-5}, 3)$, and $(3) = (1 + \sqrt{-5}, 3)(1 - \sqrt{-5}, 3)$.

In (c), $\frac{1}{3}(1+\sqrt{-5})R = (1+\sqrt{-5}, 2)(1+\sqrt{-5}, 3)(1+\sqrt{-5}, 3)^{-1}(1-\sqrt{-5}, 3)^{-1}$
$= (1+\sqrt{-5}, 2)(1-\sqrt{-5}, 3)^{-1}$, and $(1+\sqrt{-5}, 3)$ does not appear.

In (d), $\frac{1+\sqrt{-5}}{3} = \frac{2(1+\sqrt{-5})}{2 \cdot 3} = \frac{2(1+\sqrt{-5})}{(1+\sqrt{-5})(1-\sqrt{-5})} = \frac{2}{1-\sqrt{-5}}$.

5. The mapping $\varphi : 1 + P_v^n \to P_v^n/P_v^{n+1}$ induced by $1 + x \mapsto x + P_v^{n+1}$ is a homomorphism from $1 + P_v^n$ under multiplication into P_v^n/P_v^{n+1} under addition because the equalities $\varphi(1 + x) = x + P_v^{n+1}$, $\varphi(1 + y) = y + P_v^{n+1}$, and

$$\varphi\bigl((1+x)(1+y)\bigr) = \varphi(1 + x + y + xy)$$
$$= x + y + xy + P_v^{n+1} = x + y + P_v^{n+1}$$

show that $\varphi\bigl((1 + x)(1 + y)\bigr) = \varphi(1 + x) + \varphi(1 + y)$. The kernel of φ is the set of all $1 + x$ with $x \in P_v^{n+1}$, i.e., $1 + P_v^{n+1}$, and the image is certainly all of P_v^n/P_v^{n+1}.

6. The composition $\mathbb{I}^1/\iota(K^\times) \to \mathbb{I}/\iota(K^\times) \to \mathcal{I}/\mathcal{P}$ induced by the inclusion $\mathbb{I}^1 \to \mathbb{I}$ and the passage from $\mathbb{I}$ to $\mathcal{I}$ discussed in Section 10 is onto $\mathcal{I}/\mathcal{P}$ because the composition is affected by only the nonarchimedean places and because any member of $\mathbb{I}$ can be adjusted at the archimedean places so as to be in $\mathbb{I}^1$. In addition, the composition is continuous if $\mathcal{I}/\mathcal{P}$ is given the discrete topology. Since $\mathbb{I}^1/\iota(K^\times)$ is compact, the discrete space $\mathcal{I}/\mathcal{P}$ has to be compact and must be finite.

7. Fix a finite subset S of places containing S_∞. Then the projection of $\prod_{w \in S} K_w^\times$ to $K_v^\times$ is continuous for each $v \in S$. Since also the inclusion $K_v^\times \to K_v$ is continuous, the composition $\prod_{w \in S} K_w^\times \to K_v$ is continuous. Thus the corresponding mapping $\prod_{w \in S} K_w^\times \to \prod_{w \in S} K_w$ is continuous. In similar fashion $\prod_{w \notin S} \mathbb{Z}_w^\times \to \mathbb{Z}_v$ is a

continuous function as a composition of continuous functions. Thus $\prod_{w \notin S} \mathbb{Z}_w^\times \to \prod_{w \notin S} \mathbb{Z}_w$ is continuous. Putting these two compositions together shows that $\mathbb{I}_K(S) \to \mathbb{A}_K(S)$ is continuous, and therefore $\mathbb{I}_K(S) \to \mathbb{A}_K$ is continuous. Since this is true for each S, it follows that $\mathbb{I}_K \to \mathbb{A}_K$ is continuous.

8. Each x_n lies in $\mathbb{A}_\mathbb{Q}(S_\infty)$, which is an open set in $\mathbb{A}_\mathbb{Q}$. For each prime p, $x_{n,p} = 1$ if n is large enough, and also $x_{n,\infty} = 1$ for all n. Since $\mathbb{A}_\mathbb{Q}(S_\infty)$ has the product topology, $\{x_n\}$ converges to (1). On the other hand, if $\{x_n\}$ were to converge to some limit x in $\mathbb{I}_\mathbb{Q}$, then x would have to lie in some $\mathbb{I}(S)$, and the ideles x_n would have to be in $\mathbb{I}(S)$ for large n. But $(x_{n,v})$ is not in $\mathbb{I}(S)$ as soon as v is outside S.

9. For fixed g in G, we have $d(\Phi(gx)) = d(\Phi(g)\Phi(x)) = d(\Phi(x))$, and hence $d(\Phi(\,\cdot\,))$ and $d(\,\cdot\,)$ are Haar measures on G. Any two Haar measures are proportional, and the result follows.

10. In (a) the equality is trivial if $c_1 c_2 = 0$. When $c_1 c_2 \neq 0$, we have $d(c_1 c_2 x) = |c_1 c_2|_F \, dx$ and also $d(c_1 c_2 x) = |c_1|_F d(c_2 x) = |c_1|_F |c_2|_F \, dx$, and it follows that $|c_1 c_2|_F = |c_1|_F |c_2|_F$ in this case as well.

The proof of continuity is harder (but is essential to make sense out of (b)). We first check continuity at each $c_0 \neq 0$. Let f be a continuous real-valued function vanishing off a compact set S, and let N be a compact neighborhood of c_0 not containing 0. If c is in N, then $f(c^{-1}x)$ is nonzero only for x in the compact set NS. Let $\epsilon > 0$ be given. Continuity of $(c, x) \mapsto f(c^{-1}x)$ allows us to find, for each x in NS, an open subneighborhood N_x of c_0 and an open neighborhood U_x of x such that $|f(c^{-1}y) - f(c_0^{-1}x)| < \epsilon$ for $c \in N_x$ and $y \in U_x$. Then $|f(c^{-1}y) - f(c_0^{-1}y)| < 2\epsilon$ for $c \in N_x$ and $y \in U_x$. The open sets U_x cover NS. Forming a finite subcover and intersecting the corresponding finitely many sets N_x, we obtain an open neighborhood N' of c_0 such that $|f(c^{-1}y) - f(c_0^{-1}y)| < 2\epsilon$ for $c \in N'$ whenever y is in NS. As a result, $c \mapsto \int_V f(c^{-1}x) \, dx$ is continuous at $c = c_0$. Therefore $c \mapsto |c|_V \int_V f(x) \, dx$ is continuous at c_0, and so is $c \mapsto |c|_V$.

To prove continuity at $c = 0$, we are to show that $\lim_{c \to 0} \int_V f(c^{-1}x) \, dx = 0$ for f as above. Let U be any compact neighborhood of 0 in V. Find a sufficiently small neighborhood N of 0 in V such that $c \in V$ implies that cS does not meet U^c. Then $c^{-1}U^c \cap S = \varnothing$. For such c's, we have $\left| \int_V f(c^{-1}x) \, dx \right| = \left| \int_U f(c^{-1}x) \, dx \right| \leq \|f\|_{\sup} (dx(U))$, and the desired limit relation follows.

For (b), we have $d(cx)/|cx|_F = (|c|_F \, dx)/(|c|_F |x|_F) = dx/|x|_F$. For (c), $|x|_F = |x|$ if $F = \mathbb{R}$, and $|x|_F = |x|^2$ if $F = \mathbb{C}$. For (d), $|x|_F = |x|_p$ if $F = \mathbb{Q}_p$. For (e), we have $I = p\mathbb{Z}_p$, and therefore the Haar measure of I is the product of $|p|_p = p^{-1}$ times the Haar measure of $\mathbb{Z}_p$. Hence the Haar measure of I is p^{-1}.

11. If F has characteristic $p' \neq 0$, then the sum $1 + \cdots + 1$ with p' terms is 0 in R, and it must be 0 in $R/\mathfrak{p}$. So $R/\mathfrak{p}$ must have characteristic p'. Thus any such $p' \neq 0$ must be p.

12. In (a), apply Corollary 6.29 with $f(X) = X^{q-1} - 1$ in $R[X]$. Every nonzero $\bar{a}$ is a simple root of the reduced polynomial $\overline{f}(X) = X^{q-1} - 1$ in $\mathbb{F}_q[X]$, simple

because $(q-1)(\bar{a})^{q-1} \neq 0$. The corollary produces a root a of $f(X)$ whose image in $R/\mathfrak{p}$ is $\bar{a}$. In this way we obtain $q-1$ distinct roots of 1 in R, each corresponding to a different coset in $R/\mathfrak{p}$. Together with 0, these exhaust the cosets of $R/\mathfrak{p}$.

In (b), if F has characteristic p, then raising to the pth power is a field mapping of F into itself. Since $q = p^m$, raising to the qth power is the m-fold iterate of a field map and is a field map. If a and b are two $(q-1)^{\text{st}}$ roots of 1 in R, then $(a \pm b)^q = a^q + (\pm b)^q = a + (\pm b)$, and so $a \pm b$ is a $(q-1)^{\text{st}}$ root of 1. Since the nonzero elements of E are closed under inverses, E is a subfield.

13. In (a) let x be in R. Problem 12 produces a unique $a_0 \in E$ with $x - a_0$ in $\mathfrak{p}$, i.e., with $v(x - a_0) \geq 1$. Then $v(t^{-1}(x - a_0)) \geq 0$, and Problem 12 produces a unique a_1 in E with $t^{-1}(x - a_0) - a_1$ in $\mathfrak{p}$. Continuing in this way, we obtain $a_0, \ldots, a_N$ in E with

$$t^{-1}(t^{-1}(\cdots(t^{-1}(x-a_0) - a_1) - \cdots) - a_{N-1}) - a_N$$

in $\mathfrak{p}$. Thus $v(x - \sum_{k=0}^{N} a_k t^k) \geq N + 1$. Since F is complete, $\sum_{k=0}^{\infty} a_k t^k$ converges with sum x. The statement about the value of v is clear.

In (b), the part about the series giving an element in R is immediate from Problem 1, since t^k has limit 0. The operations on R now match those on $\mathbb{F}_q[[t]]$, and the isomorphism follows. For (c), let x be given with $x \notin R$. Set $v(x) = -N$. Then $v(t^N x) = 0$, and we can apply (a) to write $t^N x = \sum_{k=0}^{\infty} a_k t^k$. Then $x = \sum_{k=0}^{\infty} a_k t^{k-N}$, as required.

14. In (a), the inclusion of the integers into R, followed by passage to the quotient $R/\mathfrak{p}$, is an additive homomorphism. Since $R/\mathfrak{p}$ has order q, q must map to the 0 coset, namely $\mathfrak{p}$.

Part (a) shows that $v(q) \geq 1$. Since $v(q) = v(p^m) = mv(p)$, $v(p)$ is positive, and (b) is proved. The same argument as in the proof of Ostrowski's Theorem shows that $v(p') = 0$ for all prime numbers other than p, and then (c) is immediate. For (d), it is enough to check equality of the absolute values in question on the element p, and for that we have $|p|_F^{1/(mv_0)} = q^{-v(p)/(mv_0)} = q^{-1/m} = p^{-1}$.

For (e), the map of $\mathbb{Q}'$ to $\mathbb{Q}$, when composed with the completion $\mathbb{Q} \to \mathbb{Q}_p$, is a homomorphism of valued fields into a complete field. It therefore extends uniquely as a homomorphism of the closure $\overline{\mathbb{Q}'}$ into $\mathbb{Q}_p$. The dense set $\mathbb{Q}'$ maps to the dense set $\mathbb{Q}$, and hence the extended map is an isomorphism.

Part (f) is just a repetition of the argument in Problems 13a and 13c. In (g), let $x = \sum_{k=0}^{\infty} a_k t^k$ be the expansion of f, and put $c_{j_0} = \sum_{k=0}^{v_0-1} a_k t_k$. Since $v(t) = 1$, we obtain $v(x - c_{j_0}) \geq v(t^{v_0}) = v_0 v(t) = v_0$. Therefore $v(p^{-1}(x - c_{j_0})) \geq 0$. Iterating this procedure as in Problem 13a, we obtain a convergent expansion $x = \sum_{k=0}^{\infty} c_{j_k} p^k$. For (h), we then have $x = \sum_{k=0}^{\infty} c_{j_k} p^k = \sum_{j=1}^{l} c_j \sum_{\{k \mid j_k = j\}} p^k$, and we see that x lies in $\sum_{j=1}^{l} \overline{\mathbb{Q}'} c_j$. Therefore $\dim[F : \overline{\mathbb{Q}'}] \leq l$.

15. Part (a) is immediate, and (b) follows from Theorem 6.33. For (c), $R/\mathfrak{p}$ corresponds to extracting the constant term from a power series in t, and thus $L/\wp \cong \mathbb{F}_{q^f}$ is of dimension f over $R/\mathfrak{p} \cong \mathbb{F}_q$. The computation $\wp T = tUT = tT = tRT = \mathfrak{p}T = P^e$ shows that K/L has ramification index e. For (d), each index (residue class

degree and ramification index) for K/F is the product of that index for K/L and that index for L/F. So e for L/F is 1, and f for K/L is 1.

16. For (a), the irreducible polynomial $\overline{g}(X)$ has to be separable, and therefore all of its roots in $\Bbbk_K$ are simple. Application of Hensel's Lemma in the form of Corollary 6.29 produces α. For (b), the polynomial $g(X)$ is monic with coefficients in R, and its root α is therefore a member of L integral over R. Thus α lies in U. The natural field map $U/\wp \to T/P$ takes $u + \wp$ to $u + P$, hence takes $\alpha + \wp$ to $\alpha + P = \overline{\alpha}$. Thus we can regard $\overline{\alpha}$ as a member of $\Bbbk_L$. Since $\Bbbk_F$ and $\overline{\alpha}$ generate $\Bbbk_K$ by construction of $\overline{\alpha}$, $\Bbbk_L = \Bbbk_F$.

For (d), let us use subscripts on the indices e and f to indicate the field extension in question. Then we have $e_{L/F} f_{L/F} = [L : F] = \deg g(X) = \deg \overline{g}(X) = [\Bbbk_K : \Bbbk_F] = f_{K/F}$ on the one hand and $f_{K/F} = [\Bbbk_K : \Bbbk_F] = [\Bbbk_L : \Bbbk_F] = f_{L/F}$ on the other hand. The two chains of equalities together show that $e_{L/F} = 1$, and the second one in combination with $f_{K/F} = f_{K/L} f_{L/F}$ shows that $f_{K/L} = 1$.

17. In (a), the element y_j exists and is unique because of the nondegeneracy of the trace form, which holds because K/F is separable (Theorem 8.54 and Section IX.15 of *Basic Algebra*).

In (b), the expression for the z_k's in terms of the y_j's shows that $\sum_{k=1}^{n} R z_k \subseteq \sum_{j=1}^{n} R y_j$. The assumption $\det A = \pm 1$ implies that $B = A^{-1}$ lies in $M_n(R)$. Since $y_j = \sum_k B_{kj} z_k$, we obtain $\sum_{j=1}^{n} R y_j \subseteq \sum_{k=1}^{n} R z_k$.

For (c), it is evident that the degree is at most $n-1$. Write $g(X) = \prod_j (X - \xi_j)$. The opening computations of Section V.4 show that $g'(\xi_i) = \prod_{j \neq i} (\xi_i - \xi_j)$. Therefore the value of the left side at ξ_k for the identity in question is

$$\sum_{i=1}^{n} \frac{\prod_{j \neq i} (\xi_k - \xi_j)}{\prod_{j \neq i} (\xi_i - \xi_j)}.$$

The numerator is 0 unless $i = k$. Thus only the ith term makes a contribution, and its value, namely 1, matches the value of the right side. Then (d) is a routine computation.

For (e), the rational expression $(1 + c_1 X + \cdots + c_n X^n)^{-1}$ on the left side is expanded in series using $(1 + Z)^{-1} = 1 - Z + Z^2 - Z^3 + \cdots$. Thus the left side is the sum of X^n and a series beginning with a multiple of X^{n+1}. The right side is $\sum_{k=0}^{\infty} \text{Tr}_{K/F} \left(g'(\xi)^{-1} \xi^k X^{k+1} \right)$, and the conclusion of the problem results by equating the indicated coefficients.

For (f), the result of (e) handles the entries with $i + j \leq n + 1$. For those with $n + 2 \leq i + j \leq 2n$, we write $\xi^{i+j-2} g'(\xi)^{-1}$ as $\xi^n \xi^{i+j-n-2} g'(\xi)^{-1}$, substitute for ξ^n recursively from the field polynomial, and check that the traces are in R by applying (e). Thus all A_{ij} are in R.

For (g), conclusion (f) shows that A is triangular with 1's on the off diagonal, and hence the determinant of A is ± 1. Put $z_k = \sum_j A_{jk} y_j$. Since $x_i = \xi^{i-1}$,

$$\text{Tr}_{K/F}(z_k x_i) = \sum_j A_{jk} \text{Tr}_{K/F}(y_j x_i) = A_{ik}$$
$$= \text{Tr}_{K/F}((g'(\xi)^{-1} \xi^{k-1}) \xi^{i-1}) = \text{Tr}_{K/F}((g'(\xi)^{-1} \xi^{k-1}) x_i).$$

Therefore $z_k = g'(\xi)^{-1}\xi^{k-1}$. Combining this equality with (b) shows that $\widehat{N} = \sum_j R y_j = \sum_k R z_k = \sum_k R g'(\xi)^{-1}\xi^{k-1} = g'(\xi)^{-1} N$.

18. For (a), the assumption $f = n$ makes $\dim_{\mathbb{k}_F}(\mathbb{k}_K) = n$. Thus $\deg g(X) = \deg \overline{g}(X) = n$. Since $\overline{g}(X)$ is irreducible, so is $g(X)$. The root α of $g(X)$ in K is such that $F(\alpha)$ is an n-dimensional subspace of K, hence equals K.

For (b), the conclusion $\widehat{N} \supseteq \widehat{T}$ follows from the definition. Since $\widehat{T} = \mathcal{D}(K/F)^{-1}$, we obtain $\mathcal{D}(K/F)^{-1} \subseteq \widehat{N} = g'(\alpha)^{-1}N \subseteq g'(\alpha)^{-1}T$.

For (c), the polynomial $\overline{g}(X)$ was constructed as irreducible, and $g(X)$ was constructed to reduce to $\overline{g}(X)$. Then $\overline{g}'(\overline{\alpha}) \neq 0$, and it follows that $g'(\alpha)$ is in T but not P. Thus $g'(\alpha)$ is a unit in T, and $g'(\alpha)^{-1}T = T$. Then $\mathcal{D}(K/F)^{-1} \subseteq T$. Since $\mathcal{D}(K/F)^{-1} \supseteq T$ also, $\mathcal{D}(K/F)^{-1} = T$, and $\mathcal{D}(K/F) = T$.

19. For (a), we may assume that $v(x_1) \leq v(x_j)$ for $j > 1$. If $v(x_1) < v(x_j)$ for all $j > 1$, then induction and use of property (vi) of discrete valuations shows inductively that $v(0) = v(x_1 + \cdots + x_m) = v(x_1)$, contradiction.

For (b), the element π is in T, and its minimal polynomial has coefficients in R because T is integral over R; in turn, the field polynomial is a power of the minimal polynomial. Since c_j is in R, we have $v_K(c_j) = nv_F(c_j)$, and therefore $v_K(c_j)$ is divisible by n.

For (c), apply (a) to the equality $c_0\pi^n + c_1\pi^{n-1} + \cdots + c_n = 0$ to produce indices $i < j$ with $v(c_i\pi^{n-i}) = v(c_j\pi^{n-j})$ and with $v(c_k\pi^{n-k}) \geq v(c_i\pi^{n-i})$ for all k. The equality involving i and j implies that $j - i = v_K(c_j) - v_K(c_i)$. From $i < j \leq n$, we have $n - i > 0$. Thus $v(c_i\pi^{n-i}) \geq v(c_i\pi) > 0$. By (b), $v(c_i\pi^{n-i}) \geq n$. So $v(c_k\pi^{n-k}) \geq n$.

In (d), the right side of the equality $j - i = v_K(c_j) - v_K(c_i)$ is divisible by n, by (b), and the left side is between 1 and n. Hence the two sides equal n, and we conclude that $i = 0$ and $j = n$. Thus the equality says that $n = v_K(c_n)$. Since c_n is in F and since $v_K = nv_F$, $v_F(c_n) = 1$. Therefore c_n is in $\mathfrak{p}$ but not $\mathfrak{p}^2$. The inequality $v_K(c_k\pi^{n-k}) \geq n$ implies that $v_K(c_k) \geq k$. For $1 \leq k \leq n$, this conclusion implies that $v_K(c_k) \geq 1$. Since c_K is in F and since $v_K = nv_F$, $v_F(c_k) > 0$ for $k \geq 1$. Thus c_k is in $\mathfrak{p}$ for $k \geq 1$.

In (e), the irreducibility is immediate from the Eisenstein irreducibility criterion, R being a principal ideal domain. Since the field polynomial is a power of the minimal polynomial, the field polynomial equals the minimal polynomial. Then the degree of $F(\pi)$ is n. Since $F(\pi)$ is an n-dimensional subfield of the n-dimensional field K, $K = F(\pi)$.

Part (f) is proved in the same way as Problem 14g. For (g), the expansion can be rewritten as $\sum_{k=0}^{\infty} a_k y_k = \sum_{i=0}^{\infty} \sum_{0 \leq j < e} a_{ei+j} y_{ei+j} = \sum_{0 \leq j < e} \pi^j \left(\sum_{i=0}^{\infty} a_{ei+j} \lambda^i \right)$. The term in parentheses is the most general member of R, and the left side is the most general member of T. Thus (g) follows.

In (h), conclusion (g) shows that $N = \sum_{k=0}^{n-1} R\pi^k$ equals T, and Problem 17 with $\xi = \pi$ shows that $\widehat{N} = g'(\pi)^{-1}N$. Thus $\mathcal{D}(K/F)^{-1} = \widehat{T} = g'(\pi)^{-1}T$. Multiplying by $(g'(\pi))\mathcal{D}(K/F)$, we obtain $\mathcal{D}(K/F) = (g'(\pi))$.

For (i), $g'(\pi) = e\pi^{e-1} + \sum_{k=1}^{n-1} c_{n-k} k \pi^{k-1} = e\pi^{e-1} + b$. In each term of b, $v_K(kc_{n-k}) \geq ev_F(c_{n-k}) \geq e$, and $v_K(\pi^{k-1}) = k - 1$. Thus $v_K(b) \geq e$. Meanwhile, $v_K(e\pi^{e-1}) = (e-1) + v_K(e)$. Thus $v_K(g'(\pi)) \geq \min\big((e-1) + v_K(e), v_K(b)\big)$, and property (vi) of discrete valuations shows that equality holds if the two members $(e-1) + v_K(e)$ and $v_K(b)$ of the minimum are unequal. If $v_K(e) = 0$, then the members are unequal, and we obtain $v_K(g'(\pi)) = e - 1$. Otherwise, we obtain $v_K(g'(\pi)) \geq e$. We know that $\mathcal{D}(K/F) = (g'(\pi)) = P^{v_k(g'(\pi))}$, and Lemma 6.47 follows.

Chapter VII

1. If x and y are members of L purely inseparable over K, then x^{p^e} and $y^{p^{e'}}$ are in K for suitable e and e'. Without loss of generality, let $e' \leq e$. Then x^{p^e} and y^{p^e} are in K, and hence $(x \pm y)^{p^e} = x^{p^e} \pm y^{p^e}$ are in K and so are $(xy)^{p^e} = x^{p^e} y^{p^e}$ and $(xy^{-1})^{p^e} = x^{p^e} y^{-p^e}$ if $y \neq 0$. So $x \pm y$, xy, and xy^{-1} are purely inseparable over K, the last of these if $y \neq 0$.

2. In view of Proposition 7.10, the given conditions imply that $[K(\alpha) : K] = p^e [K(\alpha^{p^e}) : K]$ and that $X^{p^\mu} - \alpha^{p^e}$ is irreducible over $K(\alpha^{p^e})$ for every $\mu \geq 0$. Since $\alpha^{p^{e-\mu}}$ is a root of this polynomial within $K(\alpha)$ for each $\mu \leq e$, $K(\alpha)$ has a chain of subfields
$$K(\alpha^{p^e}) \subsetneq K(\alpha^{p^{e-1}}) \subsetneq \cdots \subsetneq K(\alpha^p) \subsetneq K(\alpha)$$
in which the consecutive degrees of the extensions are all p. Let β be separable over K, and let $K(\alpha^{p^r})$ be the first of these fields to contain β. Arguing by contradiction, suppose that $r < e$. Then β and $\alpha^{p^{r+1}}$ generate $K(\alpha^{p^r})$ because $[K(\alpha^{p^r}) : K(\alpha^{p^{r+1}})]$ is prime. The separability of β over K implies that β is separable over $K(\alpha^{p^{r+1}})$, hence that $K(\alpha^{p^r})$ is separable over $K(\alpha^{p^{r+1}})$, hence that α^{p^r} is separable over $K(\alpha^{p^{r+1}})$. Since $(\alpha^{p^r})^p$ lies in $K(\alpha^{p^{r+1}})$, α^{p^r} is also purely inseparable over $K(\alpha^{p^{r+1}})$. By Corollary 7.12, α^{p^r} lies in $K(\alpha^{p^{r+1}})$. This contradicts the fact that the above chain of subfields is strictly increasing. We conclude that $r = e$. Hence all elements β separable over K lie in $K(\alpha^{p^e})$.

3. For suitable integers R_a, we form the tuple $z = (R_a + a\mathbb{Z})_{a \geq 1}$, using the realization of the inverse limit in Proposition 7.27. We have to specify the integers R_a. The condition for z to lie in $\widehat{\mathbb{Z}}$, coming from the condition $f_{ab} \circ f_b = f_a$ when a divides b, works out to be that $R_b - R_a$ is divisible by a whenever a divides b. After the integers R_a have been defined for all a, it is enough to check that $R_{pa} - R_a$ is divisible by a whenever p is prime.

For n odd, define $R_{2^c n} = nk + 1$, where k is the unique integer from 0 to $2^c - 1$ such that $nk + 1$ is divisible by 2^c. This k exists and is unique because $-n$ has an inverse modulo 2^c. One checks that $R_{2^{c+1} n} - R_{2^c n}$ is divisible by 2^c and by n, and that $R_{2^c pn} - R_{2^c n}$ is divisible by 2^c and by n if p is an odd prime. The definition makes $R_2 = 0$ and $R_q = 1$ for every odd prime q, and therefore z is not of the form z_c for any integer c.

4. The first part is immediate from Theorem 7.34. For the second part the group $\text{Gal}(\mathbb{R}/\mathbb{Q})$ is trivial. In fact, any member of $\text{Gal}(\mathbb{R}/\mathbb{Q})$ must fix $\mathbb{Q}$ and map squares in $\mathbb{R}$ to squares. It therefore respects the ordering. For any $r \in \mathbb{R}$, it fixes each rational less than r, and hence it fixes r.

5. Use $K_n = \mathbb{Q}(\sqrt{p_1}, \ldots, \sqrt{p_n})$, where p_n is the nth prime, and Proposition 7.30 to see that $\text{Gal}(K/\mathbb{Q})$ is an infinite product of groups of order 2. (A problem at the end of Chapter IX of *Basic Algebra* can help with this step.) The open subgroups of index 2 correspond to quadratic extensions of $\mathbb{Q}$, of which there are countably many. Since $\text{Gal}(K/\mathbb{Q})$ has uncountably many subgroups of index 2, such a subgroup H exists that is not open. The field extension $K/\mathbb{Q}$ is normal, and thus $\text{Gal}(K/\mathbb{Q})$ is a homomorphic image of $\text{Gal}(\mathbb{Q}_{\text{alg}}/\mathbb{Q})$, say by a homomorphism φ. Then $\varphi^{-1}(H)$ is the required subgroup of $\text{Gal}(\mathbb{Q}_{\text{alg}}/\mathbb{Q})$.

6. Suppose I is primary. If $b + I$ is a zero divisor in R/I, then ab is in I for some a not in I. Since I is primary, b^m is in I for some m. Thus $(b + I)^m = b^m + I = I$, and $b + I$ is nilpotent in R/I.

If every zero divisor in R/I is nilpotent, then the ideal 0 in R/I is primary because whenever $(a + I)(b + I) = I$ and $a + I \neq I$, then the nilpotence of $b + I$ implies that $b^m + I = I$ for some m. This says that the 0 ideal $0 + I$ in R/I is primary.

If the 0 ideal in R/I is primary and if ab is in I with a not in I, then $(a+I)(b+I) = I$ with $a + I \neq I$, and hence $(b + I)^m = I$ for some m, 0 being primary in R/I. This means that b^m is in I, and I is primary.

7. In (a), if xy is in $\sqrt{I}$, then $(xy)^m$ is in I for some m, and therefore either x^m is in I or y^{mn} is in I for some n, i.e., either x is in $\sqrt{I}$ or y is in $\sqrt{I}$.

In (b), let x be in $\sqrt{I}$, and choose n such that x^n is in I. Then x^n is in J because $I \subseteq J$. Since J is prime, some factor of x^n is in J, i.e., x is in J.

8. In (b), $R/I \cong \mathbb{C}[y]/(y^2)$. The zero divisors of R/I are cy with $c \in \mathbb{C}$, and $(cy)^2 = 0$ in R shows that cy is nilpotent in R. By Problem 6, I is primary. The radical $P = \sqrt{I}$ is (x, y) by inspection, and this is prime. Since $P^2 = (x^2, xy, y^2)$, we have $P^2 \subsetneq I \subsetneq P$. If $I = Q^n$ for some prime ideal Q, then $I \subseteq Q$, and Problem 7b shows that $\sqrt{I} \subseteq Q$. Since $\sqrt{I}$ is maximal in this case, Q has to be P.

In (c), $R/P \cong K[X, Y, Z]/(XY - Z^2, X, Z) \cong K[Y]$, and this is an integral domain. Hence P is prime. Next, $P^2 = (x^2, xz, z^2)$. Thus $xy = z^2$ lies in P^2. However, x is not in P^2, and y^m is not in P^2 for any $m > 0$. So P^2 is not primary.

9. Let a and b be in R with ab in I and a not in I. To show that I is primary, we are to show that b is in $\sqrt{I}$. We do this by showing that $(b) + I \subseteq \sqrt{I}$. The ideal $(b) + I$ is proper, since otherwise $1 = cb + x$ with $x \in I$, which implies that $a = cba + xa$ is in I, contradiction. Let J be a maximal ideal with $(b) + I \subseteq J$. It is enough to show that $\sqrt{I} \subseteq J$; in fact, then $\sqrt{I} = J$ because $\sqrt{I}$ is assumed maximal, and $(b) + I \subseteq \sqrt{I}$ as asserted. So let u be in $\sqrt{I}$. Then u^m is in $I \subseteq J$ for some m, and u is in J because J is prime.

This proves the first part. The second part follows from the observation that if J

is maximal, then $\sqrt{J^n} = J$. In fact, J^n contains all elements a^n for $a \in J$. So $\sqrt{J^n}$ has to contain all elements $a \in J$. Since J is maximal and $\sqrt{J^n}$ has to be proper, $\sqrt{J^n} = J$.

10. In (a), let P be a prime ideal, and suppose that $P = I \cap J$ nontrivially. If i is in I but not J and if j is in J but not I, then ij is in P, but i is not in P because i is not in J and similarly j is not in P because j is not in I.

In (b), $I^2 = (x^2, xy, y^2)$ is primary by Problem 9. The equality of I^2 with $(Rx + I^2) \cap (Ry + I^2)$ holds by inspection.

11. Arguing by contradiction, we can use the Noetherian property to obtain an ideal I maximal with respect to the property of not being a finite intersection of proper irreducible ideals. Since I is not irreducible, $I = A \cap B$ nontrivially. By maximality, A and B are intersections, and then so is I, contradiction.

12. Let Q be a proper irreducible ideal in R. Then 0 is a proper irreducible ideal in R/Q. We show that 0 is primary in R/Q, and then Problem 6 shows that Q is primary. Thus let $xy = 0$ in R/Q with $y \neq 0$ in R/Q. We want to see that some power of x is 0 in R/Q. In R/Q, we form the sequence of annihilators $\text{Ann}(x) \subseteq \text{Ann}(x^2) \subseteq \cdots$ and use the Noetherian property of R and its quotient R/Q to obtain $\text{Ann}(x^l) = \text{Ann}(x^{l+1})$ for some l. Let us see that the intersection $(x^l) \cap (y)$ is 0 in R/Q. In fact, if a is in (y), then $xy = 0$ implies $ax = 0$, and if a is in (x^l), then $a = bx^l$ and $0 = ax = bx^{l+1}$, from which we see that b is in $\text{Ann}(x^{l+1}) = \text{Ann}(x^l)$. Therefore $a = bx^l = 0$ in R/Q. Thus indeed $(x^l) \cap (y) = 0$. Since 0 is irreducible in R/Q and $(y) \neq 0$, we conclude that $(x^l) = 0$ and $x^l = 0$ in R/Q. This is what we were to show.

13. If ab is in Q and a is not in Q, then ab is in Q_i for all i and a is not in Q_{i_0} for some i_0. Since Q_{i_0} is primary, b^m is in Q_{i_0} for some m, i.e., b is in $\sqrt{Q_{i_0}} = P$. Since $\sqrt{Q_i} = P$ for all i, b^{k_i} is in Q_i for some k_i depending on i. Taking N to be the maximum of the integers k_i, we see that b^N is in each Q_i and hence is in their intersection Q. Thus Q is primary.

Problem 7b shows that $\sqrt{Q} \subseteq P$. On the other hand, if b is in P, we have just seen that some power b^N lies in Q. So b lies in $\sqrt{Q}$. Therefore $\sqrt{Q} = P$.

14. Problem 11 shows that every ideal is the finite intersection of proper irreducible ideals, and Problem 12 shows that these are primary. Thus if I is given, we have $I = \bigcap Q_i$ with each Q_i primary. Group all Q_i's whose associated prime ideal is the same P_j, and denote the intersection of these by Q'_j. The ideal Q'_j is primary by Problem 13. Then $I = \bigcap Q'_j$, and the Q'_j have distinct associated prime ideals. So condition (ii) is satisfied. Finally among all expressions for I as intersections satisfying (ii), choose one that involves the smallest number of primary ideals. This minimality forces (i) to hold.

Chapter VIII

1. $(q^{n+1} - 1)/(q - 1) = 1 + q + q^2 + \cdots + q^n$.

3. It is enough to consider a monomial $F(X_1, \ldots, X_n) = X^{\alpha_1} \cdots X^{\alpha_n}$ with $\sum_{j=1}^{n} \alpha_j = d$. Then $X_j \frac{\partial}{\partial X_j}(X^{\alpha_1} \cdots X^{\alpha_n}) = \alpha_j X^{\alpha_1} \cdots X^{\alpha_n}$, and the sum on j equals $dX^{\alpha_1} \cdots X^{\alpha_n}$.

4. If f^ι and g^ι have a nontrivial common factor in $B[X]$, then $0 = R(f^\iota, g^\iota) = \iota(R(f, g))$. Since ι is one-one, $R(f, g) = 0$. Therefore f and g have a nontrivial common factor in $A[X]$.

5. Let us show that if $g_n \neq 0$ and $f_m = 0$, then Theorem 8.1 for indices $(m-1, n)$ implies the theorem for indices (m, n), and vice versa. Assume for the moment that $m \geq 2$. Let $\mathcal{R}(f, g)$ be the resultant matrix of size $m + n$ that takes into account all coefficients $f_0, \ldots, f_m$ of f, and let $R(f, g)$ be its determinant. With $f_m = 0$, let $\mathcal{R}'(f, g)$ be the resultant matrix of size $m + n - 1$, and let $R'(f, g)$ be its determinant. The matrix $\mathcal{R}'(f, g)$ is obtained by erasing the mth row and last column of $\mathcal{R}(f, g)$. On the other hand, the only nonzero entry in the last column of $\mathcal{R}(f, g)$ is g_n. Expansion in cofactors therefore gives $R'(f, g) = g_n R(f, g)$. The hypotheses of Theorem 8.1 apply to f and g for either of these resultants, and we have just seen that the two conditions (c) are equivalent. Certainly the two conditions (a) are equivalent. For the two conditions (b), the resultant of size $m + n - 1$ tells us that $a' f + b' g = R'(f, g)$ with $\deg a' < n$ and $\deg b' < m - 1$. Certainly this implies that $af + bg = R(f, g)$ with $a = a' g_n$ and $b = b' g_n$. Conversely if $af + bg = R(f, g)$ with $\deg a < n$ and $\deg b < m$, we define $a' = a g_n^{-1}$ and $b' = b g_n^{-1}$. Then $a' f + b' g = R'(f, g)$ with $\deg a' < n$, and we need to see that $\deg b' = \deg b < m - 1$. Since $f_m = 0$, all the powers of X in af are $\leq (n-1) + (m-1)$, and the same must be true in bg. Since g has degree n, we must have $\deg b \leq m - 2 < m - 1$, as required.

Next we check what happens when $m = 1$ and we are comparing the resultant of size $n + 1$ and a degenerate resultant whose matrix is of size n and contains only the entries of g. The determinant formula is still valid, and we see that $R'(f, g) = g_0^n$, which is nonzero. Thus (a) and (c) are false for both sizes. For (b), we cannot have $af + bg = 0$ with $\deg b < 0$ and $b \neq 0$. We need to check that $af + bg = 0$ cannot happen with $\deg a < n$ and $\deg b < 1$; in fact, then $\deg bg = \deg g = n$, while $f_1 = 0$ implies that $\deg af < n + \deg f = n$. So we cannot have $af + bg = 0$ in this case either.

The result of these calculations is that Theorem 8.1 for (m, n) is equivalent to the theorem for $(m-1, n)$ if $g_n \neq 0$ and $f_m = 0$. Using induction, we see that the theorem for (m, n) is equivalent to the theorem for (k, n) if $g_n \neq 0$ and $f_{k+1} = \cdots = f_m = 0$. Taking $k = \deg f$ gives the desired result.

6. Proof via Nullstellensatz: Since f is irreducible and $K[X_1, \ldots, X_n]$ is a unique factorization domain, the principal ideal (f) is prime. Corollary 7.2 shows that g lies in (f): hence $g = hf$ for some h.

Proof via resultants: The idea is to arrange to have

$$af + bg = R(f, g), \qquad (*)$$

with the resultant taken with respect to X_n. Proposition 8.1 shows that this happens if f and g are of positive degree in X_n, and we shall show that either this is the case or else f divides g for easy reasons. Since f is nonconstant, it depends nontrivially on some X_j, and renumbering the variables allows us to assume that f depends nontrivially on X_n. Then f is of the form

$$f(X_1, \ldots, X_n)$$
$$= c_0(X_1, \ldots, X_{n-1}) + c_1(X_1, \ldots, X_{n-1})X_n + \cdots + c_r(X_1, \ldots, X_{n-1})X_n^r$$

with $r > 0$ and with c_r nonzero in $K[X_1, \ldots, X_{n-1}]$. If $g = 0$, then certainly f divides g. So we may assume that $g \neq 0$. Choose $a_1, \ldots, a_{n-1}$ in K such that

$$g(a_1, \ldots, a_{n-1}, X_n)c_r(a_1, \ldots, a_{n-1}) \neq 0. \qquad (**)$$

Then $f(a_1, \ldots, a_{n-1}, X_n)$ is a polynomial in X_n whose coefficient of X_n^r is nonzero. Since K is algebraically closed, this polynomial in X_n has a root, say a_n. Since $f(a_1, \ldots, a_n) = 0$, the hypothesis shows that $g(a_1, \ldots, a_{n-1}, a_n) = 0$, and $(**)$ allows us to conclude that $g = g(X_1, \ldots, X_n)$ depends nontrivially on X_n. This proves $(*)$.

To complete the proof, we show that $c_r R$ is 0 at every point $(b_1, \ldots, b_{r-1})$. Since K is infinite, it will follow that the polynomial $c_r R$ is 0; thus $R = 0$ because c_r is not the 0 polynomial. Then f and g will have a nontrivial common factor by Proposition 8.1, and f will have to divide g because f is prime. Thus suppose that $c_r(b_1, \ldots, b_{r-1}) \neq 0$. Then $f(b_1, \ldots, b_{r-1}, X_n)$ is a nonconstant polynomial in X_n and must have a root b_r, since K is algebraically closed. Hence $f(b_1, \ldots, b_r) = 0$, and the hypothesis on g shows that $g(b_1, \ldots, b_r) = 0$. By $(*)$, $R(b_1, \ldots, b_{r-1}) = 0$. This completes the proof.

7. $Y^3 - 2XY^2 + 2X^2Y - 4X^3 = (Y - 2X)(Y + i\sqrt{2}\,X)(Y - i\sqrt{2}\,X)$.

8. The resultant matrix in the W variable is

$$\begin{pmatrix} XY^4-Y^5 & -2X^2Y^2 & X^3 & 0 \\ 0 & XY^4-Y^5 & -2X^2Y^2 & X^3 \\ Y^4 & Y^3 & -X^2 & 0 \\ 0 & Y^4 & Y^3 & -X^2 \end{pmatrix},$$

and its determinant is $-X^3Y^9(Y - 2X)^2$. Substituting into either of the equations $F = 0$ and $G = 0$ gives the projective solutions (x, y, w) equal to $(1, 0, 0)$, $(0, 0, 1)$, and $(1, 2, 4 \pm 4\sqrt{2})$, up to nonzero scalar factors. (One has to check that both the equations $F = 0$ and $G = 0$ are satisfied.)

9. Introduce a new indeterminate $T = Y_i - Z_j$, and remove Y_i. Then $R(F, G) = R(Y_1, \ldots, T + Z_j, \ldots, Y_m, Z_1, \ldots, Z_n)$ is a polynomial in T, the Z_j's, and all the Y's except for Y_i. Also, $R(F, G) = 0$ when T is set equal to 0. Hence $R(F, G)$ is divisible by T. Then (a) and (b) follow. For (c), the polynomials $Y_i - Z_j$ are distinct primes. Since each divides $R(F, G)$, their product must divide. Their product has the same degree as $R(F, G)$, and the result follows.

10. We may assume that K is algebraically closed and that f is monic, say with $f(X) = \prod_{i=1}^{m}(X - \xi_i)$ and $f'(X) = m\prod_{j=1}^{m-1}(X - \eta_j)$. Then the previous problem gives $f'(\xi_i) = m\prod_{j=1}^{m-1}(\xi_i - \eta_j)$, and

$$R(f, f') = m^m c_{m,m-1} \prod_{i,j}(\xi_i - \eta_j) = m^m c_{m,m-1} \prod_{i=1}^{m} f'(\xi_i)$$

with $c_{m,m-1}$ equal to the constant c from Problem 9c when $n = m - 1$. According to Section V.4, the product is $(-1)^{n(n-1)/2}$ times the discriminant $D(f)$ of f. So the result follows.

11. Replace G by $G(X, Y, W) - (X^2 + Y^2)F(X, Y, W)$ to get $YWH(X, Y, W)$, where $H(X, Y, W) = (X^2 + Y^2)(X^2 - 3Y^2) - 4X^2YW$. Then

$$I(P, F \cap G) = I(P, F \cap YWH) = I(P, F \cap Y) + I(P, F \cap W) + I(P, F \cap H).$$

For $I(P, F \cap Y)$, we use the method of Section 4, looking at $F(t, 0, 1)$, which is t^4; thus $I(P, F \cap Y) = 4$. Since P is not on W, $I(P, F \cap W) = 0$.

For $I(P, F \cap H)$, replace H by $H(X, Y, W) - F(X, Y, W)$ to get $YJ(X, Y, W)$, where $J(X, Y, W) = -4X^2Y - 4Y^3 - 7X^2W + Y^2W$. Then

$$I(P, F \cap H) = I(P, F \cap YJ) = I(P, F \cap Y) + I(P, F \cap J),$$

and again $I(P, F \cap Y) = 4$. If the local expressions of F and J are denoted by f and j, then their lowest-order terms $f_3(x, y)$ and $j_2(x, y)$ are given by

$$f_3(x, y) = 3x^2y - y^3 = y(\sqrt{3}x + y)(\sqrt{3}x - y),$$
$$j_2(x, y) = -7x^2 + y^2 = -(\sqrt{7}x + y)(\sqrt{7}x - y).$$

Thus F and J have no tangent lines in common at P, and $I(P, F \cap J) = 3 \cdot 2 = 6$. Collecting the results, we find that $I(P, F \cap G) = 4 + 4 + 6 = 14$.

12. Let $P = [x_0, y_0, w_0]$, and choose $\Phi \in GL(3, K)$ with $\Phi(x_0, y_0, w_0) = (0, 0, 1)$. The local versions of G and L are $g(X, Y) = G(\Phi^{-1}(X, Y, 1))$ and $l(X, Y) = L(\Phi^{-1}(X, Y, 1))$. The expansion of g as a sum of homogeneous polynomials is $g = g_m + \cdots + g_d$ because $m = m_P(G) > 0$, and l is of the form $l(X, Y) = aX + bY$ because P lies on L. We can parametrize l by $\varphi(t) = (bt, -at)$, and then the definition of intersection multiplicity is that $I(P, L \cap G)$ is the least integer k such that the expression $g_k(\varphi(t)) = t^k g_k(b, -a)$ is nonzero. The definition of tangent line is any projective line L_i whose local version l_i is one of the factors of $g_m(X, Y) = c\prod_i(\alpha_i X + \beta_i Y)^{m_i}$. Then $g_m(\varphi(t)) = t^m g_m(b, -a) = c\prod_i(\alpha_i b - \beta_i a)^{m_i}$. If (a, b) is a multiple of some (α_i, β_i), then $g_m(\varphi(t)) = 0$; hence $I(P, L \cap G) \geq m + 1$. Otherwise $g_m(\varphi(t)) \neq 0$, and $I(P, L \cap G) = m$.

13. The linear span LT(I) of the members LT(f) for f in I is a monomial ideal and is of the form $(M_1, \ldots, M_k)$ for suitable monomials M_j each of the form LM(f_j) for some f_j in I. Then $\{f_1, \ldots, f_k\}$ is a subset of I such that $\big(\text{LT}(f_1), \ldots, \text{LT}(f_k)\big) = \text{LT}(I)$, and $\{f_1, \ldots, f_k\}$ is a Gröbner basis of I by definition.

14. If α, β, γ are vectors of exponents in monomials such that the first i with $w^{(i)} \cdot \alpha \neq w^{(i)} \cdot \beta$ has $w^{(i)} \cdot \alpha > w^{(i)} \cdot \beta$, then it equally true that the first i with $w^{(i)} \cdot (\alpha + \gamma) \neq w^{(i)} \cdot (\beta + \gamma)$ has $w^{(i)} \cdot (\alpha + \gamma) > w^{(i)} \cdot (\beta + \gamma)$. This proves that property (i) of monomial orderings holds with no further conditions on the weights. Property (ii) says for each vector α of nonnegative exponents not all 0 that the first i with $w^{(i)} \cdot \alpha \neq 0$ has $w^{(i)} \cdot \alpha > 0$. Applying this condition as a necessary condition to the jth standard basis vector $\alpha = e_j$, we see that the first i such that $w_j^{(i)} \neq 0$ must have $w_j^{(i)} > 0$ for (ii) to hold. On the other hand, if this condition holds for all j, then a suitable positive linear combination of these conditions gives (ii) for any α.

15. In (a), $a > a'$ implies that $X^{a-a'} \geq X > Y^{b'}$ for all $b' \geq 0$. Multiplying by $X^{a'}$ gives $X^a > X^{a'} Y^{b'}$. Since $Y^b \geq 1$ implies $X^a Y^b \geq X^a$, we conclude that $X^a Y^b > X^{a'} Y^{b'}$ for all b and b'. For $a = a'$, we observe that $b > b'$ implies that $Y^{b-b'} > 1$ and hence that $Y^b > Y^{b'}$. Multiplying by X^a gives $X^a Y^b > X^a Y^{b'}$. Hence the ordering is lexicographic.

In (b), we observe that an inequality between X^a and Y^b implies the same inequality between X^{na} and Y^{nb}. Consequently the particular inequality for X^a and Y^b depends only on the rational number a/b. The assumption for (b) is that $X < Y^q$, hence that $X^a \leq Y^{qa} \leq Y^b$ if $qa \leq b$, thus if $a/b \leq q^{-1}$. Thus the set S of rationals a/b such that $X^a > Y^b$ is bounded below by q^{-1}. Let r^{-1} be the greatest lower bound of S. We know then that $q^{-1} \leq r^{-1}$, hence that $r \leq q$. So $0 \leq r < \infty$, and r is a well-defined real number.

Suppose that $u/v < r^{-1}$. Then u/v is not in S, and so $X^u \leq Y^v$. In the reverse direction, suppose that $u/v > r^{-1}$. Then there is some rational c/d in S with $u/v > c/d \geq r^{-1}$; this has $X^c > Y^d$. Then $X^{ud} > X^{vc} > Y^{vd}$. Since $d > 0$, $X^u < Y^v$ would imply $X^{ud} < Y^{vd}$, which is false. Thus we must have $X^u > Y^v$. This proves (b).

For (c), the only rational u/v for which the inequality between X^u and Y^d is not decided is $u/v = r^{-1}$, and that only if r is rational. In this case a single weight vector will decide the correct inequality. All other inequalities between monomials follow from these. In fact, what needs deciding is the inequality between $X^a Y^b$ and $X^{a'} Y^{b'}$ when $a > a'$ and $b < b'$, and this is the same as the inequality between $X^{a-a'}$ and $Y^{b'-b}$.

16. The formulas for f are a matter of computation. Both satisfy the conditions of Proposition 8.20 because LM(f) $= X^2 Y$ is $\geq$ each of LM($(X + Y)f_1$) $= X^2 Y$, LM($1 f_2$) $= Y^2$, LM($X f_1$) $= X^2 Y$, and LM($(X + 1)f_2$) $= XY^2$ and because no term of r_1 or r_2 is divisible by LM(f_1) $= XY$ or LM(f_2) $= Y^2$.

17. In (a), we check that $\{X^2 + cXY, XY\}$ is a Gröbner basis using Theorem 8.23. The leading monomials of the two generators are X^2 and XY, and neither divides the

other. Since the leading coefficients are 1, this Gröbner basis is minimal.

In (b) when $c \neq 0$, $X^2 + cXY$ has a nonzero term whose monomial is divisible by the leading monomial of another generator; specifically the term cXY in $X^2 + cXY$ is divisible by the XY from the other generator. Following the procedure in Theorem 8.28, we find that $\{X^2, XY\}$ is the reduced Gröbner basis.

18. If $(c_1, \ldots, c_n)$ lies in $V_K(I)$, then c_j is one of finitely many roots of $P_j(X)$, for each j. Hence $|V_K(I)| \leq \prod_{j=1}^{n} \deg P_j$.

19. Fix j, and choose a polynomial Q_j in X that vanishes at the jth coordinate of every member of $V_K(I)$. Then $P_j(X_1, \ldots, X_n) = Q_j(X_j)$ is a polynomial vanishing on $V_K(I)$, and the Nullstellensatz shows that some power of it is in I. The result is a polynomial in X_j alone, as required.

20. If $V_K(I)$ is a finite set, then Problem 19 shows that I contains a nonconstant polynomial in X_j for each j. The leading monomial for the jth such polynomial has to be a power of X_j, and it lies in $\text{LT}(I)$. Conversely suppose that a power $X_j^{l_j}$ lies in $\text{LT}(I)$ for each j. Form a reduced Gröbner basis of I. Since the only monomials dividing $X_j^{l_j}$ are powers of X_j, there exist members g_j of the Gröbner basis for $1 \leq j \leq n$ such that

$$g_j(X_1, \ldots, X_n) = X_j^{m_j} + X_j^{m_j-1} a_{j,m_j-1} + \cdots + X_j a_{j,1} + a_{j,0}$$

for suitable polynomials $a_{j,m_j-1}, \ldots, a_{j,0}$ in $X_{j+1}, \ldots, X_n$. Then $V_K(I)$ is contained in $V_K((g_1, \ldots, g_n))$, and any member $(c_1, \ldots, c_n)$ of the latter has the property for each j that c_j is a root of a polynomial of degree m_j in one variable, once $(c_{j+1}, \ldots, c_n)$ is fixed. Thus $V_K(I)$ is contained in a finite set and has to be finite.

21. For (a), the coefficients $a_{i_1,\ldots,i_n}$ are given as in $K(X)$, and we look for solutions of $F(T_1, \ldots, T_n) = 0$. Clearing fractions in the coefficients, we see that it is enough to find a solution when each $a_{i_1,\ldots,i_n}$ has denominator 1.

For (b), substitution of $T_i = \sum_{j=1}^{N} b_{ij} X^j$, where each b_{ij} is an unknown in K, into the equation $F(T_1, \ldots, T_n) = 0$ gives

$$\sum_{i_1,\ldots,i_n} a_{i_1,\ldots,i_n} \Big(\sum_{j=1}^{N} b_{ij} X^j\Big)^{i_1} \cdots \Big(\sum_{j=1}^{N} b_{ij} X^j\Big)^{i_n} = 0.$$

We expand this out and set the coefficient of each power of X equal to 0. The largest possible power of X that can appear is the sum of the largest power of X in any $a_{i_1,\ldots,i_n}$, namely δ, and $\sum_{k=1}^{n} N i_k$. Since F is homogeneous of degree d, $\sum_{k=1}^{n} i_k = d$. Thus the largest possible power of X is $Nd + \delta$. We get one equation for each power of X that appears, and the unknowns are the various b_{ij}'s.

22. The number of equations is $\leq Nd + \delta + 1$, since the powers of X go from 0 to at most $Nd + \delta$. The number of unknowns is one for each index i with $1 \leq i \leq n$ and each possible power of X from 0 to N, hence exactly $(N+1)n$. For N sufficiently large we want to see that $Nd + \delta + 1 \leq (N+1)n$. Since $d < n$, the inequality in question is $\delta + 1 - n \leq N(n-d)$, and this is satisfied by taking N large enough.

23. In the context of Problem 22, we have a homogeneous system with more unknowns than equations (for large N). If the number of unknowns is $n+1$ and the number of equations is m, then we are looking for solutions in $\mathbb{P}^n_K$. Since the inequality $m \leq n$ is satisfied, the quoted theorem applies and produces a nonzero solution for the b_{ij}'s.

Chapter IX

1. For (a), we argue by contradiction. Suppose that $c_1(x), \ldots, c_n(x)$ are members of $\Bbbk(x)$, not all 0, such that $\sum_j c_j(x)t_j = 0$. Clearing fractions, we may assume that each $c_j(x)$ lies in $\Bbbk[x]$. If necessary, we can divide through by a power of x and arrange that some $c_j(x)$, say $c_{j_0}(x)$, has a nonzero constant term. The element x is by assumption transcendental over $\Bbbk$. Applying the substitution homomorphism of $\Bbbk[x]$ into $\Bbbk$ given by evaluation at 0 yields $\sum_j c_j(0)t_j = 0$. By the assumed linear independence of $t_1, \ldots, t_n$ over $\Bbbk$, $c_j(0) = 0$ for all j. This contradicts the fact that $c_{j_0}(0) \neq 0$. Then (b) is immediate. For (c), we know that $[\mathbb{F} : \Bbbk(x)] < \infty$, and therefore $[\Bbbk'(x) : \Bbbk(x)] < \infty$. By (b), $[\Bbbk' : \Bbbk] < \infty$.

2. This is immediate from Proposition 7.15. Alternatively, here is a direct proof. We may assume that the characteristic is p. It is enough to prove that if K is perfect and L is a finite extension, then L is perfect. Arguing by contradiction, we may assume that $[L : K]$ is as small as possible among all counterexamples. The image M of L under $x \mapsto x^p$ is a subfield of L, and M contains K because K is perfect. We cannot have $M = L$, since L is assumed not to be perfect. By construction of L, M is perfect. Composing $x \mapsto x^p$ from L into M with $x \mapsto x^{1/p}$ from M into itself, we obtain a field map of L onto M that fixes M. The result is a one-one M linear transformation of the finite-dimensional M vector space L onto a proper vector subspace, contradiction.

3. Let $\mathbb{F}$ be a function field in one variable over $\Bbbk$. Since $\Bbbk$ is perfect, Theorem 7.20 shows that $\mathbb{F}$ is separably generated. Let us write $\mathbb{F} = \Bbbk(x_1, \ldots, x_n)$. Theorem 7.18 shows that there is some x_j such that $\mathbb{F}$ is a separable extension of $\Bbbk(x_j)$. If we write x for x_j, then the Theorem of the Primitive Element shows that $\mathbb{F} = \Bbbk(x)[y]$ for some y algebraic over $\Bbbk(x)$. Put $R = \Bbbk[x][y] = \Bbbk[x, y]$; the field of fractions of R is $\mathbb{F}$. Let $g(x, Y)$ be the minimal polynomial of y over $\Bbbk(x)$. If $d(x)$ is a common denominator for the coefficients of $g(x, Y)$, then $d(x) \neq 0$ because x is transcendental over $\Bbbk$. If we set $f(X, Y) = d(X)g(X, Y)$, then $f(x, y) = 0$. Hence the substitution homomorphism $\Bbbk[X, Y] \to R$ given by replacing X by x and Y by y factors through to a homomorphism φ carrying $\Bbbk[X, Y]/(f(X, Y))$ onto R. The ring R is an integral domain; hence the ideal $(f(X, Y))$ is prime, and $f(X, Y)$ is irreducible. We can find an ideal I in $\Bbbk[X, Y]$ containing $(f(X, Y))$ such that φ descends to an isomorphism of $\Bbbk[X, Y]/I$ onto R. This ideal I has to be prime, and we let J be a maximal ideal

of $\Bbbk[X, Y]$ containing it. Then we have a chain of inclusions of prime ideals
$$0 \subsetneq (f(X, Y)) \subseteq I \subseteq J.$$
Theorem 7.22 shows that $\Bbbk[X, Y]$ has Krull dimension 2, and it follows that either $(f(X, Y)) = I$ in the above chain of inclusions, or $I = J$. The latter equality would mean that I is maximal and therefore that $R \cong \Bbbk[X, Y]/I$ is a field; this is not the case, and thus $(f(X, Y)) = I$. Hence $R \cong \Bbbk[X, Y]/(f(X, Y))$. Here $f(X, Y)$ is an affine plane curve irreducible over $\Bbbk$, and the field of fractions of R is by definition the function field of the curve; this field is $\mathbb{F}$, and the argument is complete.

4. The singular points are common zeros of f, $\frac{\partial f}{\partial X}$, and $\frac{\partial f}{\partial Y}$. If there are infinitely many, then Bezout's Theorem says that f and $\frac{\partial f}{\partial X}$ have a nontrivial common factor, and so do f and $\frac{\partial f}{\partial Y}$. Since f is irreducible and the partial derivatives reduce degrees in one or the other variable, we must have $\frac{\partial f}{\partial X} = \frac{\partial f}{\partial Y} = 0$ as polynomials. This is impossible in characteristic 0. In characteristic p, the first condition says that the only powers of X that appear in f are powers of X^p, and the second condition says that the only powers of Y that appear are powers of Y^p. The coefficients of f are powers of p because $\Bbbk$ is assumed perfect, and thus f is exhibited as a pth power, in contradiction to its assumed irreducibility.

5. Differentiate $f(X, b) = (X - a)f_1(X)$ and evaluate at (a, b) to obtain $\frac{\partial f}{\partial X}(a, b) = f_1(a) + (a - a)f_1'(a) = f_1(a)$.

6. Multiply the equation $g(X, b) = (X-a)g_1(X)$ by $f_1(X)$ and substitute to obtain $g(X, b)f_1(X) = f(X, b)g_1(X)$. Then the function $g(X, \cdot)f_1(X) - f(X, \cdot)g_1(X)$ is 0 at b and is of the form $g(X, Y)f_1(X) - f(X, Y)g_1(X) = (Y - b)h_1(X, Y)$, where $h_1(X, Y)$ for each X is a polynomial in Y. Since $(Y - b)h_1(X, Y)$ is equal to a polynomial in (X, Y), $h_1(X, Y)$ is a polynomial in (X, Y). To complete the problem, evaluate both sides at (x, y), and use the facts that $f(x, y) = 0$ and that $f_1(x) \neq 0$.

7. Since $\mathbb{F} = \Bbbk(x, y)$ is a function field in one variable, it is enough to see that y is transcendental over $\Bbbk$. Arguing by contradiction, suppose that there is some nonzero polynomial $c(Y)$ in $\Bbbk[Y]$ having y as a root. As a polynomial in $\Bbbk[X, Y]$, $c(Y)$ maps to $c(y) = 0$ when we pass to the quotient in $\Bbbk[X, Y]/(f(X, Y))$, and therefore $c(Y)$ is the product of $f(X, Y)$ by a polynomial. On the other hand, $\frac{\partial f}{\partial X}$ is not 0, and thus $f(X, Y)$ depends nontrivially on X. Hence the product of $f(X, Y)$ and any nonzero polynomial in (X, Y) depends nontrivially on X, contradiction. The result now follows from the observation at the end of Section 1.

8. Substituting a for x in the formula for $g(x, y)$ gives
$$g(a, y) = (y - b)^k h_k(a, y)/f_1(a)^k.$$
In this formula, $h_k(a, y)$ is a polynomial expression in y, hence also in $y - b$. Thus v_1 is ≥ 0 on it. The expression $f_1(a)^k$ is a nonzero member of $\Bbbk$, on which v_1 takes the value 0. Therefore
$$v_1(g(a, y)) = kv_1(y - b) + v_1(h_k(a, y)) \geq kv_1(y - b).$$

The left side is independent of k, and the right side is unbounded in k. Therefore there is some upper bound to the values of k for which $g(x, y)$ has an expansion of the kind in question.

9. For (a), we cannot have $h_k(a, b) = 0$ in Problem 8 for arbitrarily large k because of the bound found in Problem 8. If $k = n$ is the smallest k for which $h_k(a, b) \neq 0$, then the displayed formula holds with $h = h_n$. For uniqueness we substitute a for x and see that $g(a, y) = p_n(y)(y-b)^n$ for a polynomial p_n with $p_n(b) \neq 0$. We cannot have two such expressions involving distinct powers n because y is transcendental over $\Bbbk$.

For (b), we see from (a) that every nonzero member of R is of the required form with $n \geq 0$. Since $\mathbb{F}$ is the field of fractions of R, the same thing is true for $\mathbb{F}$ as long as we allow n to be arbitrary in $\mathbb{Z}$.

For (c), if we have two such expressions, we set them equal, clear fractions, and write the result as $(y-b)^k p(x, y) = q(x, y)$ for some $k \geq 0$ and for some polynomials p and q with $p(a, b) \neq 0$ and $q(a, b) \neq 0$. Substituting (a, b) for (x, y), we obtain 0 from $(y - b)^k p(x, y)$ unless $k = 0$, and we obtain something nonzero from $q(x, y)$. Therefore $k = 0$, and the required uniqueness follows.

10. From the definition we immediately have $v(g) = +\infty$ if and only if $g = 0$, as well as $v(gg') = v(g) + v(g')$ for all g and g'. We are to show that $v(g + g') \geq \min(v(g), v(g'))$. Thus write $g(x, y) = (y - b)^n h_1(x, y)/h_2(x, y)$ and $g'(x, y) = (y - b)^m h'_1(x, y)/h'_2(x, y)$ with $n \leq m$. Then $\min(v(g), v(g')) = \min(n, m) = n$. Also,
$$g + g' = (y - b)^n \frac{h_1 h'_2 + (y-b)^{m-n} h_2 h'_1}{h_2 h'_2}.$$

The numerator of the displayed fraction is a polynomial and can be written in the form of Problem 9a. Say that $(y - b)^k$ is the power of $(y - b)$ that appears in it, k being ≥ 0. Then $v(g + g') = n + k$, and this is $\geq n = \min(v(g), v(g'))$. The assertions about the valuation ring and the valuation ideal are clear.

11. Let v' be a second valuation having the stated properties. If $g(x, y)$ is given in $\mathbb{F}^\times$, decompose g as in Problem 9b, and apply v'. Then we obtain $v'(g(x, y)) = nv'(y - b) + v'(h_1(x, y)) - v'(h_2(x, y))$. The assumptions on v' show that $v'(h_1(x, y)) = v'(h_2(x, y)) = 0$. Therefore
$$v'(g(x, y)) = nv'(y - b) = v'(y - b)v(g(x, y)),$$
and $v' = v'(y - b)v$. By assumption, $v'(y - b)$ is positive. Since v' has to be onto $\mathbb{Z} \cup \{\infty\}$, we must have $v'(y - b) = 1$.

12. For (a), the argument is the same as with Problem 7 except that the roles of x and y are reversed. The partial derivative $\frac{\partial(y^2 - f(x))}{\partial y} = 2y$ is not the 0 element because the characteristic is not 2, and hence that earlier argument applies. Part (b) is elementary field theory, and (d) is a routine verification.

For (c), let $\Bbbk'$ be the subfield of elements of $\mathbb{F}$ algebraic over $\Bbbk$. Problem 1 shows that $[\Bbbk' : \Bbbk] \leq [\Bbbk'(x) : \Bbbk(x)] \leq [\mathbb{F} : \Bbbk] = 2$. Arguing by contradiction, suppose that $\{1, t\}$ is a basis of $\mathbb{k}'$ over $\mathbb{k}$. let $X^2 + uX + v$ be the minimal polynomial of t over $\mathbb{k}$; t satisfies $t^2 + ut + v = 0$. Problem 1a shows that $t = a(x) + yb(x)$ with $b(x) \neq 0$, and then t satisfies $t^2 - 2a(x)t + (a(x)^2 - f(x)b(x)^2) = 0$. Hence $ut + v = -2a(x)t + (a(x)^2 - f(x)b(x)^2)$. If $u \neq -2a(x)$, then we can solve for t and obtain the contradiction that t is in $\mathbb{k}(x)$. Thus $u = -2a(x)$, and also $v = a(x)^2 - f(x)b(x)^2$. Since x is transcendental over $\mathbb{k}$, the first of these shows that $a(x)$ does not involve x, i.e., $a(x)$ lies in $\mathbb{k}$. Then the second shows that $f(x)b(x)^2$ lies in $\mathbb{k}$, and unique factorization leads to the conclusion that $f(x)$ and $b(x)$ do not depend on x. This contradicts the assumption that $f(X)$ is nonconstant.

13. Let $z = a(x) + yb(x)$ be in the integral closure. Then so is the image of z under the nontrivial Galois group element σ, and so are $z + \sigma(z)$ and $z\sigma(z)$. The latter elements are $2a(x)$ and $a(x)^2 - f(x)b(x)^2$. Thus $a(x)$ is in the intersection of the integral closure with $\mathbb{k}(x)$, which is $\mathbb{k}[x]$ because $\mathbb{k}[x]$ is a principal ideal domain and is integrally closed. Then $f(x)b(x)^2$ is in $\mathbb{k}[x]$ by the same argument. Since $f(x)$ is square free, it follows that $b(x)$ is in $\mathbb{k}[x]$.

14. Part (a) is immediate from Corollary 6.6. Discrete valuations of $\mathbb{F}$ that are not in $D_\mathbb{F}$ play no role because of the inclusion $\mathbb{k} \subseteq R$: any discrete valuation that is ≥ 0 on R has to be 0 on $\mathbb{k}^\times$, since the image of $\mathbb{k}^\times$ under the valuation is a subgroup of $\mathbb{Z}$.

For (b), the condition for $z \neq 0$ to be in $p(x)_\infty$ is that $v(z) \geq -p\, \mathrm{ord}_v(x)_\infty$ for all $v \in D_\mathbb{F}$. If a particular v has $v(x) \geq 0$, then v does not contribute to $(x)_\infty$, and this condition says that $v(z) \geq 0$. By (a), z is in R.

15. For (a), let $c(x) = c_n x^n + \cdots + c_0 = x^n(c_n + c_{n-1}x^{-1} + \cdots + c_0 x^{-n})$ with $c_n \neq 0$. Then $v(c_n) = 0$, and $v(c_j x^{j-n}) > 0$ for $j < n$. Hence

$$v\big(x^n(c_n + c_{n-1}x^{-1} + \cdots + c_0 x^{-n})\big) = nv(x) + v(c_n + c_{n-1}x^{-1} + \cdots + c_0 x^{-n})$$
$$= nv(x) + v(c_n) = nv(x).$$

For (b), $2v(y) = v(y^2) = v(f(x)) = (\deg f)v(x)$, the latter equality holding by (a). In (c), we have

$$v(a(x) + yb(x)) \geq \min\big(v(a(x)), v(yb(x))\big)$$
$$= \min\big(v(a(x)), v(y) + v(b(x))\big)$$
$$= \min\big((\deg a)v(x), (\tfrac{1}{2}\deg f + \deg b)v(x)\big)$$
$$= v(x)\max\big(\deg a, \tfrac{1}{2}\deg f + \deg b\big) \geq pv(x).$$

16. Any $v \in D_\mathbb{F}$ with $v(x) \geq 0$ has $v(z) \geq 0 = -\mathrm{ord}_v(x)_\infty$ on all elements $z = a(x) + yb(x)$ with $a(x)$ and $b(x)$ in $\mathbb{k}[x]$, by Problems 13 and 14a. Suppose that $v(x) < 0$. Then Problem 15c and the assumptions on the degrees of $a(x)$ and $b(x)$ shows that $v(z) \geq pv(x) = -p\, \mathrm{ord}_v(x)_\infty$. Hence $(z) \geq -p(x)_\infty$, and z lies in $L(p(x)_\infty)$.

18. For (a), let σ be the nontrivial element of the Galois group. Problem 17c shows that if $z = a(x) + yb(x)$ is in $L(p(x)_\infty)$, then so is $\sigma(z) = a(x) - yb(x)$. Hence any $v \in D_\mathbb{F}$ with $v(x) < 0$ has $v(a(x) + yb(x)) \geq -p\,\mathrm{ord}_v(x)_\infty = pv(x)$ and $v(a(x) - yb(x)) \geq -p\,\mathrm{ord}_v(x)_\infty = pv(x)$. Consequently

$$v(a(x)) = v(2a(x)) \geq \min\left(v(a(x) + yb(x)), v(a(x) - yb(x))\right)$$
$$\geq \min\left(pv(x), pv(x)\right) = pv(x)$$

and

$$v\!\left(a(x)^2 - f(x)b(x)^2\right) = v(a(x) + yb(x)) + v(a(x) - yb(x)) \geq pv(x) + pv(x).$$

Using Problem 15a and the fact that $v(x) < 0$, we see from these two inequalities that $\deg a \leq p$ and $\deg(a^2 - fb^2) \leq 2p$.

For (b), Problem 14b shows that $L(p(x)_\infty) \subseteq R$, and Problem 13 shows that R consists of all $a(x) + yb(x)$ with $a(x)$ and $b(x)$ in $\mathbb{k}[x]$. Part (a) thus shows that $\deg a \leq p$ and $\deg(a^2 - fb^2) \leq 2p$. Since $\deg a \leq p$, the second of these inequalities shows that $\deg fb^2 \leq 2p$. Thus $\deg b + \frac{1}{2} \deg f \leq p$. In the reverse direction, if $a(x)$ and $b(x)$ are polynomials satisfying the degree relations, then Problem 16 shows that $a(x) + yb(x)$ is in $L(p(x)_\infty)$.

19. The polynomials $a(x)$ and $b(x)$ are limited only by the restrictions on their degrees. From $\deg a \leq p$, we get a space of dimension $p+1$. From $\deg b + \frac{1}{2} \deg f \leq p$, we have $\deg b \leq [p - \frac{1}{2} \deg f]$, and we get a space of dimension $[p - \frac{1}{2} \deg f] + 1$ if $[p - \frac{1}{2} \deg f] \geq 0$. Thus

$$\ell(p(x)_\infty) = (p+1) + [p - \tfrac{1}{2} \deg f] + 1$$
$$= 2p + 2 + [-\tfrac{1}{2} \deg f] = 2p + 2 - [\tfrac{1}{2}(1 + \deg f)]$$

if $p \geq -[-\frac{1}{2} \deg f] = +[\frac{1}{2}(1 + \deg f)]$.

20. Part (a) is immediate from Theorem 9.3, since $[\mathbb{F} : \mathbb{k}(x)] = 2$. For (b), Theorem 9.9 and Problem 19, in combination with the result of (a), show for sufficiently large positive p that

$$1 - g = \ell(p(x)_\infty) - p \deg(x)_\infty = 2p + 2 - [\tfrac{1}{2}(1 + \deg f)] - 2p.$$

Hence $g = [\frac{1}{2}(1 + \deg f)] - 1$.

21. Let $\Phi : \mathbb{k}(X)[Y] \to \mathbb{k}(X)[Z]$ be the substitution homomorphism that fixes $\mathbb{k}(X)$ and has $\Phi(Y) = g(X)Z$, and follow it with the quotient homomorphism to $\mathbb{k}(X)[Z]/(Z^2 - h(X))$. Then

$$\Phi(Y^2 - f(X)) = g(X)^2 Z^2 - f(X) = g(X)^2 (Z^2 - h(X)),$$

which goes to 0 in the quotient. Thus the composition of Φ followed by the quotient map descends to a field map $\varphi : \mathbb{k}(X)[Y]/(Y^2 - f(X)) \to \mathbb{k}(X)[Z]/(Z^2 - h(X))$. The inverse is constructed in the same way, starting from the formula $\Psi(Z) = g(X)^{-1} Y$.

22. For (a), the conclusion genus 1 when there are no repeated roots is immediate from Problem 20b with $\deg f = 3$. If there are repeated roots, then we can write $f(X) = g(X)^2 h(X)$ with $\deg g = \deg h = 1$. Applying Problem 21, we see that the genus is the same as for Problem 20b with $\deg f = 1$, i.e., the genus is 0.

For (b), a singularity occurs only at points (x, y) of the zero locus in $\mathbb{k}_{\text{alg}}^2$ at which both first partials are 0. Then $2Y = 0$, which says that $y = 0$ because the characteristic is not 2, and $f'(X) = 0$, which says that x is a root in $\mathbb{k}_{\text{alg}}$ of both $f(X)$ and $f'(X)$. This means that x is at least a double root in $\mathbb{k}_{\text{alg}}$ of $f(X)$.

23. The residue class degree f_v is 1, since $\mathbb{k}$ is algebraically closed. Thus $\deg nv = n$. Corollary 9.4 gives $\ell(0v) = 1$, Corollaries 9.22 and 9.23 together give $\ell(1v) = 1$ if $g \geq 1$, and Corollary 9.19 gives $\ell((2g-1)v) = \deg((2g-1)v) + (1-g) = (2g-1) + (1-g) = g$ and $\ell(2gv) = \deg(2gv) + (1-g) = g+1$. The inequality $\ell(nv) \leq \ell((n+1)v) \leq \ell(nv) + 1$ follows by combining Theorem 9.6, the fact that $A \leq B$ implies $L(A) \subseteq L(B)$, and the fact that $f_v = 1$.

24. For each $n \geq 0$,

$$L(nv) = \{0\} \cup \{x \in \mathbb{F}^\times \mid -(x)_\infty \geq -nv\} = \{0\} \cup \{x \in \mathbb{F}^\times \mid (x)_\infty \leq nv\}.$$

Thus $n \geq 1$ is a gap if and only if $\ell(nv) = \ell((n-1)v)$, and otherwise $\ell(nv) = \ell((n-1)v) + 1$ by the last fact in Problem 23.

Suppose that there are m gaps in passing from $\ell(0v)$ to $\ell(2gv)$. In the process we take $2g$ steps from $(n-1)v$ to nv, of which m are gaps and $2g - m$ are nongaps. (The gaps are certain of these integers n, $1 \leq n \leq 2g$.) Since $\ell(0v) = 1$ and $\ell(2gv) = g+1$ by Problem 23, the total number of nongaps is $(g+1) - 1 = g$. Solving $2g - m = g$ gives $m = g$. The formulas $\ell((2g-1)v) = g$ and $\ell(2gv) = g+1$ from Problem 23 show that $2g$ is not a gap.

25. For (a), if the gap sequence is $(1, 2, \ldots, g)$, then $1 = \ell(0v) = \ell(1v) = \ell(2v) = \cdots = \ell(gv)$. Conversely if the gap sequence is something else, let n with $1 \leq n \leq g$ be the first nongap; then $1 = \ell(0v) = \cdots = \ell((n-1)v) < \ell(nv) \leq \ell(gv)$.

For (b), Problem 23 gives $\ell(0v) = \ell(1v) = 1$ if $g \geq 1$, and thus 1 is a gap.

For (c), there are no integers strictly between 0 and $2g$ if $g = 1$, and the only such integer for $g = 1$ is 1. Part (b) shows that the gap sequence is indeed (1) if $g = 1$, and thus the gap sequence is always the standard one.

For (d), we have some x and y in $\mathbb{F}^\times$ with $(x)_\infty = rv$ and $(y)_\infty = sv$. Thus $(x) = (x)_0 - rv$ and $(y) = (y)_0 - sv$, and $(xy) = (x)_0 + (y)_0 - (r+s)v$. Since v does not contribute to $(x)_0$ and $(y)_0$, $(xy)_\infty = (r+s)v$, and thus $r+s$ is a nongap.

For (e), if 2 is a nongap, then iteration of (d) shows that $2, 4, 6, \ldots, 2g-2$ are nongaps. The only possible gaps are the remaining integers from 1 to $2g-1$, namely $1, 3, 5, \ldots, 2g-1$. There are g of these, and so all of them must be gaps.

Chapter X

1. If F is in $I(P)$, expand F as a sum of homogeneous terms $F = \sum_{d=0}^{\infty} F_d$. Then $0 = F(tx_0, \ldots, tx_n) = \sum_{d=0}^{\infty} F_d(tx_0, \ldots, tx_n) = \sum_{d=0}^{\infty} F_d(x_0, \ldots, x_n)t^d$ for all $t \in \Bbbk^{\times}$. Since $\Bbbk$ is infinite, every coefficient of this polynomial in t is 0. Thus each F_d is in $I(P)$, and $I(P)$ is generated by homogeneous elements.

2. In each part we argue by contradiction. For (a), if $\{X_\alpha\}$ is a system of nonempty closed subsets of X with the finite intersection property such that $\bigcap_\alpha X_\alpha = \varnothing$, then we can inductively define a strictly decreasing sequence of finite intersections of the X_α's, in contradiction to the Noetherian property. In (b), if E is a closed irreducible subset that is not connected, then $E = U \cup V$ with U and V nonempty, disjoint, and relatively open. Then $E = U^c \cup V^c$ contradicts the irreducibility of E.

3. For (a), the continuous image of a connected set is connected. Continuity is by Proposition 10.32, and connectedness is by Problem 2b applied to the Noetherian topological space V. For (b), if f is any polynomial function on $\mathbb{A}^n$, then $f \circ \varphi$ is in $\mathcal{O}(V)$ because φ is a morphism, and $f \circ \varphi$ is constant by Corollary 10.31. Then φ cannot have two distinct points in its image, since any two points in $\mathbb{A}^n$ can be distinguished by some polynomial.

4. Certainly $\mathcal{O}(U) \supseteq \Bbbk[X, Y]$. Also, the function field $\Bbbk(U)$ consists of all quotients of polynomials a/b with a and b in $\Bbbk[X, Y]$ and $b \neq 0$. Thus suppose that $f = a/b$ lies in $\mathcal{O}(U)$. By unique factorization in $\Bbbk[X, Y]$, we may assume that a and b are relatively prime. In the expression $f = a/b$, regularity at P implies that $b(P) \neq 0$ because an equality $a/b = c/d$ of two such expressions implies that $a = kc$ and $b = kd$ for some nonzero scalar k. Since f is regular everywhere in $\mathbb{A}^2$ except possibly at the origin, $b(X, Y)$ is nonvanishing away from the origin. However, if b is nonconstant, then $V(b)$ is a curve and has dimension 1, whereas the origin has dimension 0. We conclude that b is constant, and $f = a/b$ is in $\Bbbk[X, Y]$.

5. Arguing by contradiction, let $\varphi : W \to U$ be an isomorphism from an affine variety onto U. Then the map $\widetilde{\varphi} : \mathcal{O}(U) \to \mathcal{O}(W) = A(W)$ given by $\widetilde{\varphi}(f) = f \circ \varphi$ is an isomorphism. Let $\iota : U \to \mathbb{A}^2$ be the inclusion. The corresponding map on regular functions is $\widetilde{\iota} : A(\mathbb{A}^2) \to \mathcal{O}(U)$ given by $\widetilde{\iota}(h)(x, y) = h(x, y)$ for $(x, y) \neq (0, 0)$, and it is an isomorphism by Problem 4. Then $(\varphi \circ \iota)^{\sim} = \widetilde{\iota} \circ \widetilde{\varphi}$ is an isomorphism of $A(\mathbb{A}^2)$ onto $A(W)$. Its inverse has to be of the form $\widetilde{\psi}$ with $\widetilde{\psi}(g) = g \circ \psi$ for some isomorphism $\psi : \mathbb{A}^2 \to W$, according to Theorem 10.38. Since $\widetilde{\psi} \circ \widetilde{\varphi} \circ \widetilde{\iota}$ is the identity map on $A(\mathbb{A}^2)$, $\iota \circ \varphi \circ \psi$ is the identity map on $\mathbb{A}^2$. Using the definition of ι shows that $\varphi \circ \psi(x, y) = (x, y)$ for $(x, y) \neq (0, 0)$. Thus $\varphi \circ \psi$ is an isomorphism of $\mathbb{A}^2$ onto U that is the identity on U. This is a contradiction, since there is no possible image for $(0, 0)$ under $\varphi \circ \psi$ that makes $\varphi \circ \psi$ one-one.

6. Let φ be the rational map of the irreducible curve C into the irreducible curve C', and let (E, φ_E) be a morphism in the class φ. If φ is not dominant, then $\overline{\varphi_E(E)}$ is a proper closed subset of C' and must be finite. Hence $\varphi_E(E)$ is finite. The set E is connected by Problem 2b, and morphisms are continuous by definition. Therefore

$\varphi_E(E)$ is connected. Being connected and finite, it is a singleton set $\{y\}$. If φ_C is defined as everywhere equal to y on C, then (C, φ_C) is in the equivalence class φ. So φ is constant.

7. Suppose that f is a member of $\mathcal{O}_{\varphi(P)}(V)$ with $\varphi_P^*(f) = 0$. Since the set on which $f \in \Bbbk(V)$ is regular is open, there exists an open neighborhood E of $\varphi(P)$ on which f is defined. The morphism φ is continuous, and thus $\varphi^{-1}(E)$ is open in U. Since φ is a morphism and f is regular on E, $f \circ \varphi$ is regular on $\varphi^{-1}(E)$. According to the proof of Proposition 10.42, $\varphi_P^*(f)$ is defined to be the unique member of $\Bbbk(V)$ that agrees with $f \circ \varphi$ on $\varphi^{-1}(E)$. We are assuming $\varphi_P^*(f)$ to be 0, and thus $f \circ \varphi$ equals 0 on $\varphi^{-1}(E)$. By dominance of φ, $\varphi(\varphi^{-1}(E))$ is a dense subset of E. Thus the continuous function f is 0 on a dense subset of its domain E and is 0.

8. The inclusion $(WX - YZ) \subseteq (X, Z)$ yields a homomorphism φ of $A(V)$ onto $\Bbbk[W, X, Y, Z]/(X, Z) \cong \Bbbk[W, Y]$. Let $b' = \varphi(\bar{b})$. Then $b'(w, y) = \bar{b}(w, 0, y, 0)$ is a polynomial in (w, y) nonzero in the complement of the origin. The solution of Problem 4 shows that $b'(0, 0) \neq 0$. Thus $\bar{b}(0, 0, 0, 0) \neq 0$, and f is defined at $(0, 0, 0, 0)$. In view of the discussion of this example in Section 4, f is everywhere defined. Therefore it is in $\mathcal{O}(V)$, which equals $A(V)$ because V is an affine variety. Thus there is a polynomial g in $\Bbbk[W, X, Y, Z]$ whose image $\bar{g}$ in $A(V)$ equals $\overline{X/Y}$. Then $\overline{Y\bar{g}} = \overline{X}$, and $Yg = X + (WX - YZ)h$ for some polynomial h. So $Y(g + hZ) = X(1 + Wh)$. This implies that Y divides $1 + Wh$, which we see is impossible by evaluating at the origin.

9. The equivalence of continuity of φ and continuity of all φ_α will be taken as known. Suppose that $\varphi : U \to V$ is a morphism. Let an index α, an open set $E \subseteq V_\alpha$, and a member f of $\mathcal{O}(E)$ be given. We are to show that $f \circ \varphi_\alpha$ is in $\mathcal{O}(\varphi_\alpha^{-1}(E))$. Since φ is a morphism and E is open in V, we know that $f \circ \varphi$ is in $\mathcal{O}(\varphi^{-1}(E))$. By restriction, $f \circ \varphi_\alpha$ is in $\mathcal{O}(U_\alpha \cap \varphi^{-1}(E)) = \mathcal{O}(\varphi_\alpha^{-1}(E))$. Thus φ_α is a morphism.

In the reverse direction suppose that all $\varphi_\alpha : U_\alpha \to V_\alpha$ are morphisms. Let E be open in V, and let f be in $\mathcal{O}(E)$. We are to show that $f \circ \varphi$ is in $\mathcal{O}(\varphi^{-1}(E))$. Since $\varphi^{-1}(E) = \bigcup_\alpha (U_\alpha \cap \varphi^{-1}(E))$, it is enough to prove regularity of $f \circ \varphi$ on each $U_\alpha \cap \varphi^{-1}(E)$. On this open set, $f \circ \varphi$ equals $f \circ \varphi_\alpha$, which is regular because φ_α is a morphism. Thus φ is a morphism.

10. For (a), we use the equivalence of regularity with the condition in Proposition 10.28. Thus regularity at P in U means that there is a subneighborhood U_0 of U within V about P such that f equals a quotient $\bar{a}/\bar{b}$ on U_0 with $\bar{a}$ and $\bar{b}$ in $A(V)$ and with $\bar{b}$ nowhere vanishing on U_0. Choose polynomials a and b in $\Bbbk[X_1, \ldots, X_n]$ that restrict to $\bar{a}$ and $\bar{b}$ on V. Let U_0' be an open subset of $\mathbb{A}^n$ whose intersection with V is U_0. Since b is nowhere 0 on U_0 and is continuous on U_0', the subset $\widetilde{U}_0$ of U_1 on which b is nonvanishing is open and contains U_0. Then Proposition 10.28 shows that $F = a/b$ is a member of $\mathcal{O}(\widetilde{U}_0)$ whose restriction to U_0 equals f.

For (b), the result of (a) is local. Thus we can immediately allow V to be quasi-affine. Using Proposition 10.37, we can extend (a) to the case that V is quasiprojective.

11. Continuity is no problem. For the condition involving regularity, we use Problem 10. Let E be a relatively open set in V, and let f be in $\mathcal{O}(E)$. We are to show that $f \circ \varphi$ is in $\mathcal{O}(\varphi^{-1}(E))$. Thus let P be in $\varphi^{-1}(E) \subseteq U$; then $\varphi(P)$ is in $E \subseteq V$. Since f is in $\mathcal{O}(E)$, Problem 10 produces a relatively open neighborhood E_0 of $\varphi(P)$, an open subset $\widetilde{E}_0$ of Y with $\widetilde{E}_0 \cap V = E_0$, and a function F in $\mathcal{O}(\widetilde{E}_0)$ such that $F\big|_{E_0} = f\big|_{E_0}$. Since $\varphi : X \to Y$ is a morphism, $F \circ \varphi$ is in $\mathcal{O}(\varphi^{-1}(\widetilde{E}_0))$. Since $\varphi(\varphi^{-1}(\widetilde{E}_0) \cap U) \subseteq \widetilde{E}_0 \cap V = E_0$, $F \circ \varphi$ agrees with $f \circ \varphi$ on $\varphi^{-1}(\widetilde{E}_0) \cap U$. Thus $f \circ \varphi$ has an extension $F \circ \varphi$ from $\varphi^{-1}(\widetilde{E}_0) \cap U$ to $\varphi^{-1}(\widetilde{E}_0)$ that is in $\mathcal{O}(\widetilde{E}_0)$. The quotients that exhibit $F \circ \varphi$ as defined at points of $\varphi^{-1}(\widetilde{E}_0) \cap U$ exhibit $f \circ \varphi$ as defined there. The inclusion $\varphi^{-1}(E_0) = \varphi^{-1}(\widetilde{E}_0 \cap V) = \varphi^{-1}(\widetilde{E}_0) \cap \varphi^{-1}(V) \subseteq \varphi^{-1}(\widetilde{E}_0) \cap U$ shows that $f \circ \varphi$ is in $\mathcal{O}(\varphi^{-1}(E_0))$. This being true for all P in $\varphi^{-1}(E)$, $f \circ \varphi$ is in $\mathcal{O}(\varphi^{-1}(E))$.

12. Part (a) follows by applying instances of Problem 11 to φ and φ^{-1}. Then (b) follows by another application of Problem 11. Part (c) follows by inductive application of (b).

13. Let d_i be the degree of homogeneity of F_i. Then the ith row of the right-hand matrix is $\lambda^{d_i - 1}$ times the ith row of the left-hand matrix. Hence the dimension of the span of the rows is the same for the two matrices, and this number is the rank.

14. This comes down to the fact that differentiating with respect to X_j for $j > 0$ and then setting X_0 equal to 1 is the same as setting X_0 equal to 1 and then differentiating with respect to X_j.

15. For any of the functions F_i, the right side of the formula in Euler's Theorem is 0 at $(x_0, \ldots, x_n)$ by assumption. Hence Euler's Theorem gives $x_0 \frac{\partial F_i}{\partial X_0}(x_0, \ldots, x_n) = -\sum_{j=1}^n x_j \frac{\partial F_i}{\partial X_j}(x_0, \ldots, x_n)$. This says that

$$x_0 \times 0^{\text{th}} \text{ column of } J(F)(x_0, \ldots, x_n) = -\sum_{j=1}^n x_j \times j^{\text{th}} \text{ column of } J(F)(x_0, \ldots, x_n).$$

Since $x_0 \neq 0$, this is a relation of the required type.

16. Problem 13 shows that the left side equals rank $J(F)(1, x_1/x_0, \ldots, x_n/x_0)$, which Problem 15 shows to be equal to the rank of the matrix formed from the last n columns, which Problem 14 shows to be equal to the rank of $J(f)(x_1/x_0, \ldots, x_n/x_0)$.

18. Regard the elements w_{ij} as the entries of a matrix. The given condition is that every 2-by-2 subdeterminant of this matrix equals 0. The matrix is not 0, and consequently its rank is 1. Every matrix over $\Bbbk$ of rank 1 is of the form xy^t for column vectors x and y, and then $[\{w_{ij}\}]$ is exhibited as $\sigma\big([\{x_i\}], [\{y_j\}]\big)$.

19. For (a), one suitable monomial ordering is the lexicographic ordering that takes the elements W_{ij} in the order $W_{00}, W_{01}, \ldots, W_{mn}$ with W_{00} largest. Given a monomial M' of total degree d, choose among all monomials of total degree d the smallest one in the ordering that is congruent to M' modulo $\mathfrak{a}$. Write $M = \prod_{i,j} W_{ij}^{a_{ij}}$. If $a_{ij} > 0$ and if there exists (k, l) with $l > j$, $k > i$, and $a_{kl} > 0$, then $W_{ij} W_{kl}$ divides

M. Write $M_0 = M/W_{ij}W_{kl}$. Put $M'' = M_0 W_{il}W_{kj}$. Since $W_{ij}W_{kl} - W_{il}W_{kj}$ is in $\mathfrak{a}$, M'' is congruent to M modulo $\mathfrak{a}$. In the monomial ordering, all of the elements W_{kl}, W_{il}, W_{kj} are smaller than W_{ij}. Therefore $M'' < M$, in contradiction to the minimality of M.

In (b), let the largest W_{ij} whose exponents in M and M' are unequal be $W_{i_0 j_0}$. Let the products of the powers of the strictly larger monomials be N and N', respectively. It is enough to prove that $\varphi(M/N) \neq \varphi(M'/N')$. Then we have

$$M/N = \prod_{W_{ij} \leq W_{i_0 j_0}} W_{ij}^{a_{ij}} = W_{i_0 j_0}^{a_{i_0 j_0}} \prod_{\substack{(i,j) \text{ with} \\ i_0 < i \text{ or} \\ (i_0 = i \text{ and } j_0 < j)}} W_{ij}^{a_{ij}}$$

and a similar expression for M'/N'. The minimality condition says that $a_{ij} = 0$ if $i_0 < i$ and $j_0 < j$. Thus

$$M/N = \Big(\prod_{i_0 < i,\ j_0 \geq j} W_{ij}^{a_{ij}}\Big)\Big(\prod_{i_0 = i,\ j_0 \leq j} W_{ij}^{a_{ij}}\Big) = \Big(\prod_{k > i_0} \prod_{l \leq j_0} W_{kl}^{a_{kl}}\Big)\Big(\prod_{l \geq j_0} W_{i_0 l}^{a_{i_0 l}}\Big),$$

and $\quad \varphi(M/N) = \Big(\prod_{k > i_0} \prod_{l \leq j_0} X_k^{a_{kl}} Y_l^{a_{kl}}\Big)\Big(\prod_{l \geq j_0} X_{i_0}^{a_{i_0 l}} Y_l^{a_{i_0 l}}\Big).$

On the right side each pair of indices (k,l) occurs at most once. Thus an equality $\varphi(M/N) = \varphi(M'/N')$ would imply that $a_{kl} = b_{kl}$ for every (k,l). This proves (b).

In (c), we know that $\mathfrak{a} \subseteq \ker \varphi$. If equality fails, then there is a linear combination $\sum_r c_r M_r$ of monomials in $\ker \varphi$ that is not in $\mathfrak{a}$. Applying (a), we may assume that each M_r is reduced. Then $\sum_r c_r \varphi(M_r) = 0$. Each $\varphi(M_r)$ is a monomial, and (b) shows that the various monomials $\varphi(M_r)$ are distinct. Since the set of monomials is linearly independent, each c_r is 0. Therefore $\sum_r c_r M_r = 0$, contradiction.

20. For (a), compute the kernel of the natural substitution homomorphism of $\Bbbk[X_0, \ldots, X_m, Y_0, \ldots, Y_n]$ into $R[Y_0, \ldots, Y_n]$. For (b), let $P = [y_0, y_1, \ldots, y_n]$, $\mathfrak{p} = I(U) \subseteq \Bbbk[X_0, \ldots, X_m]$, and $\mathfrak{q} = I(\{P\}) \subseteq \Bbbk[Y_0, \ldots, Y_n]$. The inside homomorphism has kernel $\mathfrak{a}$ by Problem 19. The outside homomorphism takes $X_0, \ldots, X_m$ into R and takes each Y_j to $y_j Z$, where Z is an indeterminate; its kernel is isomorphic to $\mathfrak{pq}$. The kernel of the composition is $I(\sigma(U \times \{P\}))$, which is prime because $R[Z]$ is an integral domain.

21. See Fulton's book, page 145.

22. See Fulton's book, page 146.

23. For (a), Proposition 10.9 shows that $I(V(I)) = (h(X,Y))$ for an irreducible polynomial h if $\dim V(I) = 1$. The containment $I \subseteq I(V(I))$ shows that each f_j has to be of the form $f_j = a_j h$ for some a_j in $\Bbbk[X,Y]$. Since f_j and h are irreducible, a_j has to be a scalar. Thus $I = (h(X,Y))$, and I is prime. For (b), one can take $I = (Y + X^2, Y - X^2)$, which has $V(I) = \{(0,0)\}$ and which is not prime because it contains X^2 but not X.

24. Let $\{g_1, \ldots, g_s\}$ be a minimal Gröbner basis, and suppose that $g_j = ab$ is a nontrivial factorization of g_j in $\Bbbk[X_1, \ldots, X_n]$. Since I is prime, we may assume that a lies in I. Then $\mathrm{LM}(g_j) = \mathrm{LM}(a)\,\mathrm{LM}(b)$, and $\mathrm{LM}(a)$ lies in $\mathrm{LT}(I)$. Since $\{g_1, \ldots, g_s\}$ is a Gröbner basis, $\mathrm{LM}(a)$ lies in the monomial ideal $(\mathrm{LM}(g_1), \ldots, \mathrm{LM}(g_s))$. By Lemma 8.17, $\mathrm{LM}(g_i)$ divides $\mathrm{LM}(a)$ for some i. It follows that $\mathrm{LM}(g_i)$ divides $\mathrm{LM}(g_j)$. Since the Gröbner basis is minimal, $i = j$. That is, $\mathrm{LM}(g_i) = \mathrm{LM}(a) = \mathrm{LM}(g_j)$. Thus $\mathrm{LM}(b) = 1$, in contradiction to the assumption that the factorization of g_j is nontrivial.

25. Identify $a_{11}X^2 + 2a_{12}XY + a_{22}Y^2 + 2a_{13}XZ + 2a_{23}YZ + a_{33}Z^2$ with the symmetric matrix

$$A = \begin{pmatrix} a_{11} & a_{12} & a_{13} \\ a_{12} & a_{22} & a_{23} \\ a_{13} & a_{23} & a_{33} \end{pmatrix}.$$

By the Principal Axis Theorem choose an invertible matrix M such that $A' = M^t A M$ is diagonal. Put $\begin{pmatrix} X' \\ Y' \\ Z' \end{pmatrix} = M^{-1} \begin{pmatrix} X \\ Y \\ Z \end{pmatrix}$ and substitute. Then the given quadratic polynomial equals $\alpha X'^2 + \beta Y'^2 + \gamma Z'^2$, where α, β, γ are the diagonal entries of A'. If $\alpha\beta\gamma = 0$, this is reducible; it is readily checked to be irreducible if $\alpha\beta\gamma \neq 0$. Since $\alpha\beta\gamma = \det A' = (\det M)^2 \det A$, the reducible polynomials correspond to the affine hypersurface on which $\det A = 0$.

26. The first conclusion is a special case of Corollary 9.19. Then take x to be a nonconstant member of $L(2v_O)$, and take y to be a member of $L(3v_O)$ not in the linear span of $\{1, x\}$. Corollary 9.22 shows that $(x)_\infty = 2$, and then the equality $(y)_\infty = 3$ follows from the definitions.

27. These are special cases of Theorem 9.3.

28. Since $2 = [\Bbbk(E) : \Bbbk(x)] = [\Bbbk(E) : \Bbbk(x, y)][\Bbbk(x, y) : \Bbbk(x)]$, the integer $[\Bbbk(E) : \Bbbk(x, y)]$ divides 2. The corresponding equality with 3 and $\Bbbk(y)$ shows that $[\Bbbk(E) : \Bbbk(x, y)]$ divides 3. Therefore $[\Bbbk(E) : \Bbbk(x, y)] = 1$.

29. The values of v_O on the seven listed members of $\Bbbk(E)$ are $0, 2, 3, 4, 5, 6, 6$, respectively. The members are all in $L(6v_O)$, which has dimension 6 by Problem 28, and thus the listed members are linearly dependent. If y^2 or x^3 does not contribute to this dependence, then v_O takes distinct values on the remaining six members of $L(6v_O)$, and Problem 19a at the end of Chapter VI gives a contradiction. Hence the coefficients b and c of y^2 and x^3, respectively, are nonzero. If x and y are replaced by $-bcx$ and $bc^2 y$ and if the linear combination of terms is then divided by $b^3 c^4$, then the linear dependence takes the form $(y^2 + a_1 xy + a_3 x) - (x^3 + a_2 x^2 + a_4 x + a_6) = 0$, as required. Hence φ carries $E - \{0\}$ into $C \cap \mathbb{A}^2$.

30. Certainly $f(X, Y)$ is not divisible by any nonconstant polynomial in X. Thus the only possible reducibility is of the form $f(X, Y) = (Y + p(X))(Y + q(X))$. Expanding out the right side shows that

$$p(X) + q(X) = a_1 X + a_3,$$
$$p(X)q(X) = -(X^3 + a_2 X^2 + a_4 X + a_6).$$

The second equation shows that at least one of $p(X)$ and $q(X)$ has degree > 1, and then the first equation shows that $\deg p(X) = \deg q(X)$. But this equality would mean that $\deg p(X)q(X)$ is even, contradiction. Hence $f(X, Y)$ is irreducible.

31. The function φ is a morphism of $E - \{O\}$ into $C \cap \mathbb{A}^2$ by Lemma 10.39, and the composition with β_0 is a morphism into $\mathbb{P}^2$. Then φ is a morphism of $E - \{O\}$ into C by Problem 11. The class of $(E - \{O\}, \varphi)$ is therefore a rational map of E into C, and Corollary 10.54 shows that φ extends to a morphism $\Phi : E \to C$.

32. Let $\widetilde{\Phi} : \mathbb{k}(C) \to \mathbb{k}(E)$ be the field mapping that corresponds to Φ under Theorem 10.45. The field $\mathbb{k}(C)$ is generated by the functions x_0 and y_0 that pick out the coordinates of points of $C \cap \mathbb{A}^2$, and Theorem 10.45 shows that $\widetilde{\Phi}(x_0) =$ (class of $x_0 \circ \varphi$). For P in $E - \{O\}$, this has $\widetilde{\Phi}(x_0)(P) = x_0(\varphi(P)) = x(P)$, i.e., $\widetilde{\Phi}(x_0) = x$. Similarly $\widetilde{\Phi}(y_0) = y$. Therefore $\widetilde{\Phi}(\mathbb{k}(C)) = \mathbb{k}(x, y)$. By Problem 28, $\widetilde{\Phi}$ is onto $\mathbb{k}(E)$. By Corollary 10.46, Φ is birational.

33. The homogeneous polynomial of degree 3 from which $f(X, Y)$ arises is

$$F(X, Y, W) = (Y^2 W + a_1 XYW + a_3 YW^2) - (X^3 + a_2 X^2 W + a_4 XW^2 + a_6 W^3).$$

The points of C on the line at infinity arise by setting $W = 0$ and $F(X, Y, W) = 0$ simultaneously, and the only such point is $[0, 1, 0]$. Computation shows that $\frac{\partial F}{\partial W}(0, 1, 0) = 1$. Consequently $[0, 1, 0]$ is a nonsingular point of C.

34. A point (x_0, y_0) in $\mathbb{A}^2$ is a singular point of C if and only if $f(x_0, y_0) = \frac{\partial f}{\partial X}(x_0, y_0) = \frac{\partial f}{\partial Y}(x_0, y_0) = 0$. At (x_0, y_0), computation shows that

$$\frac{\partial^2 f}{\partial X^2} = -6X - 2a_2, \quad \frac{\partial^2 f}{\partial X \partial Y} = a_1, \quad \frac{\partial^2 f}{\partial Y^2} = 2, \quad \frac{\partial^3}{\partial X^3} = -6.$$

All higher-order derivatives are 0. Application of Taylor's formula about (x_0, y_0) therefore gives

$$f(X, Y) = (-3x_0 - a_2)(X - x_0)^2 + a_1(X - x_0)(Y - y_0) + (Y - y_0)^2 - (X - x_0)^3.$$

We put $X = x$ and $Y = y$, taking into account that $f(x, y) = 0$. After division by $(x - x_0)^2$, the result is that

$$((y - y_0)(x - x_0)^{-1})^2 + a_1(y - y_0)(x - x_0)^{-1} = (3x_0 + a_2) + (x - x_0).$$

That is, $z^2 + a_1 z = (3x_0 + a_2) + (x - x_0)$. Suppose that P is in $E - \{O\}$ and that $v_P(z) < 0$. Then we have $v_P(z + a_1) < 0$ and

$$0 \leq v_P\big((3x_0 + a_2) + (x - x_0)\big) = v_P(z^2 + a_1 z) = v_P(z) + v_P(z + a_1) < 0,$$

contradiction. Therefore $v_P(z) \geq 0$. Meanwhile, $v_O(x - x_0) = v_O(x) = -2$ and $v_O(y - y_0) = v_O(y) = -3$. Hence $v_O(z) = (-3) - (-2) = -1$.

35. Corollary 9.22 shows that no member of $\mathbb{k}(E)$ has the properties of z found in Problem 34. Thus C is nonsingular at every (x_0, y_0). In combination with Problem 33, this shows that C is everywhere nonsingular. By Corollary 10.55, Φ is an isomorphism.

SELECTED REFERENCES

Artin, E., *Theory of Algebraic Numbers*, notes by Gerhard Würges, George Striker, Göttingen, 1959.
Artin, E., C. J. Nesbitt, and R. M. Thrall, *Rings with Minimum Condition*, University of Michigan Press, Ann Arbor, 1944.
Atiyah, M. F., and I. G. Macdonald, *Introduction to Commutative Algebra*, Addison-Wesley Publishing Company, Reading, MA, 1969.
Borevich, Z. I., and I. R. Shafarevich, *Number Theory*, Academic Press, New York, 1966.
Brieskorn, E., and H. Knörrer, *Plane Algebraic Curves*, Birkhäuser, Basel, 1986.
Brown, K. S., *Cohmology of Groups*, Springer-Verlag, New York, 1982.
Buchberger, B. and F. Winkler (eds.), *Gröbner Bases and Applications*, Cambridge University Press, Cambridge, 1998.
Buell, D. A., *Binary Quadratic Forms*, Springer-Verlag, New York, 1989.
Cartan, H., and S. Eilenberg, *Homological Algebra*, Princeton University Press, Princeton, 1956.
Cassels, J. W. S., *Rational Quadratic Forms*, Academic Press, London, 1978.
Cassels, J. W. S., and A. Fröhlich (eds.), *Algebraic Number Theory*, Academic Press, London, 1967.
Cohn, H., *A Classical Invitation to Algebraic Numbers and Class Fields*, Springer-Verlag, New York, 1978.
Cox, D. A., *Primes of the Form $x^2 + ny^2$*, John Wiley & Sons, New York, 1989.
Cox, D., J. Little, and D. O'Shea, *Ideals, Varieties, and Algorithms*, Springer-Verlag, New York, 1992.
Dirichlet, P. G. L., *Lectures on Number Theory*, supplements by R. Dedekind, English translation of the original German, American Mathematical Society, Providence, 1999.
Dummit, D. S., and R. M. Foote, *Abstract Algebra*, Prentice Hall, Englewood Cliffs, NJ, 1991; second edition, Upper Saddle River, NJ, 1999; third edition, John Wiley & Sons, Hoboken, NJ, 2004.
Eisenbud, D., *Commutative Algebra with a View Toward Algebraic Geometry*, Springer-Verlag, New York, 1995.
Eisenbud, D., and J. Harris, *The Geometry of Schemes*, Springer-Verlag, New York, 2000.
Fabrizio, Tutorial on Groebner basis for a polynomial ideal, 2001, online as of 2007, `http://www.geocities.com/capecanaveral/hall/3131`.

Farb, B., and R. K. Dennis, *Noncommutative Algebra*, Springer-Verlag, New York, 1993.
Farkas, H. M., and I. Kra, *Riemann Surfaces*, Springer-Verlag, New York, 1980; second edition, 1992.
Freyd, P., *Abelian Categories: An Introduction to the Theory of Functors*, Harper and Row, New York, 1964.
Fröhlich, A., and M. J. Taylor, *Algebraic Number Theory*, Cambridge University Press, Cambridge, 1991.
Fulton, W., *Algebraic Curves: An Introduction to Algebraic Geometry*, Addison-Wesley Publishing Company, Redwood City, CA, 1989; originally published by W. A. Benjamin, Inc., New York, 1969, and reprinted in 1974.
Gauss, C. F., *Disquisitiones Arithmeticae*, English translation of the original Latin, Springer-Verlag, New York, 1986.
Griffiths, P. A., *Introduction to Algebraic Curves*, American Mathematical Society, Providence, 1989.
Gunning, R. C., *Introduction to Holomorphic Functions of Several Variables*, Volume III: Homological Theory, Wadsworth &Brooks/Cole, Pacific Grove, CA, 1990.
Hall, M., *The Theory of Groups*, The Macmillan Company, New York, 1959.
Hardy, G. H., and E. M. Wright, *An Introduction to the Theory of Numbers*, Clarendon Press, Oxford, 1938; second edition, 1945; third edition, 1954; fourth edition, 1960; fifth edition, 1979.
Harris, J., *Algebraic Geometry: A First Course*, Springer-Verlag, New York, 1992; reprinted with corrections, 1995.
Hartshorne, R., *Algebraic Geometry*, Springer-Verlag, New York, 1977.
Hasse, H., *Number Theory*, English translation of the original German, Springer-Verlag, Berlin, 1980; reprinted, 2002.
Hecke, E., *Lectures on the Theory of Algebraic Numbers*, English translation of the original German, Springer-Verlag, New York, 1981.
Hilton, P. J., and U. Stammbach, *A Course in Homological Algebra*, Springer-Verlag, New York, 1971; second edition, 1997.
Hua, L.-K., *Introduction to Number Theory*, English translation of the original Chinese, Springer-Verlag, Berlin, 1982.
Hungerford, T. W., *Algebra*, Holt, Rinehart and Winston, New York, 1974; reprinted, Springer-Verlag, New York, 1980; reprinted with corrections, 1996.
Ireland, K., and M. Rosen *A Classical Introduction to Modern Number Theory*, Springer-Verlag, New York, 1982; second edition, 1990.
Jacobson, N., *Basic Algebra*, Volume I, W. H. Freeman and Co., San Francisco, 1974; second edition, New York, 1985. Volume II, W. H. Freeman and Co., San Francisco, 1980; second edition, New York, 1989.
Jacobson, N., *Lectures in Abstract Algebra*, Volume I, D. Van Nostrand Company, Inc., Princeton, 1951; reprinted, Springer-Verlag, New York, 1975. Volume II, D. Van Nostrand Company, Inc., Princeton, 1953; reprinted, Springer-Verlag,

New York, 1975. Volume III, D. Van Nostrand Company, Inc., Princeton, 1964; reprinted, Springer-Verlag, New York, 1975.

Jacobson, N., *Structure of Rings*, American Mathematical Society, Providence, 1956; revised edition, 1964.

Jacobson, N., *The Theory of Rings*, American Mathematical Society, New York, 1943; multiple reprintings, American Mathematical Society, Providence.

Knapp, A. W., *Elliptic Curves*, Princeton University Press, Princeton, 1992.

Knapp, A. W., *Basic Real Analysis*, Birkhäuser, Boston, 2005.

Knapp, A. W., *Advanced Real Analysis*, Birkhäuser, Boston, 2005.

Knapp, A. W., *Basic Algebra*, Birkhäuser, Boston, 2006.

Knapp, A. W., and D. A. Vogan, *Cohomological Induction and Unitary Representations*, Princeton University Press, Princeton, 1995.

Lam, T. Y., *A First Course in Noncommutative Rings*, Springer-Verlag, New York, 1991; second edition, 2001.

Lang, S., *Algebra*, Addison-Wesley, Reading, MA, 1965; second edition, 1984; third edition, 1993; revised third edition, Springer, New York, 2002.

Lang, S., *Algebraic Number Theory*, Addison-Wesley, Reading, MA, 1970; reprinted, Springer-Verlag, New York, 1986; second edition, 1994.

Lang, S., *Introduction to Algebraic and Abelian Functions*, Addison-Wesley, Reading, MA, 1972; second edition, Springer-Verlag, New York, 1982; reprinted, 1995.

Mac Lane, S., *Categories for the Working Mathematician*, Springer, New York, 1971; second edition, 1998.

Macdonald, I. G., *Algebraic Geometry: Introduction to Schemes*, W. A. Benjamin, Inc., New York, 1968.

Massey, W. S., *Singular Homology Theory*, Springer-Verlag, New York, 1980.

Matsumura, H., *Commutative Algebra*, W. A. Benjamin, Inc., New York, 1970; second edition, Benjamin-Cummings, Reading, MA, 1980.

Matsumura, H., *Commutative Ring Theory*, Cambridge University Press, Cambridge, 1986; reprinted with corrections, 1989.

Muir, T., *The Theory of Determinants in the Historical Order of Development*, Volume I, Macmillan, London, 1906; Volume II, 1911; Volume III, 1920; Volume IV, 1923; Volume V, 1929. Reprint of Volumes I–II, Dover Publications, New York, 1960; reprint of Volumes III–IV, 1960.

Mumford, D., *The Red Book of Varieties and Schemes*, Lecture Notes in Mathematics, Volume 1358, Springer-Verlag, Berlin, 1988; second expanded edition, 1999.

Niven, I., and H. S. Zuckerman, *An Introduction to the Theory of Numbers*, John Wiley & Sons, New York, 1960; second edition, 1966; third edition, 1972; fourth edition, 1980; fifth edition, with H. L. Montgomery, 1991.

Reid, M., *Undergraduate Algebraic Geometry*, Cambridge University Press, Cambridge, 1988.

Rotman, J. J., *Notes on Homological Algebras*, Van Nostrand Reinhold Company, New York, 1970.

Rudin, W., *Principles of Mathematical Analysis*, McGraw–Hill Book Company, New York, 1953; second edition, 1964; third edition 1976.
St. Andrews, School of Mathematics and Statistics, University of St. Andrews, Scotland, *MacTutor History of Mathematics Archive, Biographies of Mathematicians*, http://www-groups.dcs.st-and.ac.uk for background, http://www-history.mcs.st-and.ac.uk/history for entry point, http://www-history.mcs.st-and.ac.uk/history/BiogIndex.html for indices of biographies of mathematicians, updated as of 2007,
Serre, J.-P., *Algèbre Locale, Multiplicités*, second edition, Lecture Notes in Mathematics, Volume 11, Springer-Verlag, Berlin, 1965.
Serre, J.-P., *A Course in Arithmetic*, Springer-Verlag, New York, 1973.
Serre, J.-P., *Local Fields*, Springer-Verlag, New York, 1979.
Shafarevich, I. R., *Basic Algebraic Geometry*, Springer-Verlag, Berlin, 1977; second edition, published as two volumes, 1994.
Silverman, J. H., *The Arithmetic of Elliptic Curves*, Springer-Verlag, New York, 1986.
Statistics and Computation Service, Information Technology Central Services, University of Michigan, *Ideal Theory and Gröbner Bases*, updated as of 2007, http://www.umich.edu/~gpcc/scs/magma/text833.htm.
Ueno, K., *An Introduction to Algebraic Geometry*, American Mathematical Society, Providence, 1997.
Van der Waerden, B. L., *Modern Algebra*, English translation of the original German, Volume I, Frederick Ungar Publishing Co., New York, 1949. Volume II, Frederick Ungar Publishing Co., New York, 1950.
Villa Salvador, G. D., *Topics in the Theory of Algebraic Function Fields*, Birkhäuser, Boston, 2006.
Walker, R. J., *Algebraic Curves*, Princeton University Press, Princeton, 1950; reprinted, Dover Publications, New York, 1962; reprinted, Springer-Verlag, New York, 1978.
Weil, A., *Basic Number Theory*, Springer-Verlag, New York, 1967; second edition, 1973; third edition, 1974; reprinted, 1995.
Weil, A., *Number Theory: An Approach through History: From Hammurapi to Legendre*, Birkhäuser, Boston, 1984; reprinted, 2007.
Zariski, O., and P. Samuel, *Commutative Algebra*, Volume I, D. Van Nostrand Co., Inc., Princeton, 1958; reprinted, Springer-Verlag, New York, 1975. Volume II, D. Van Nostrand Co., Inc., Princeton, 1960; reprinted, Springer-Verlag, New York, 1976.

INDEX OF NOTATION

This list indexes recurring symbols introduced in Chapters I through X (pages 1–648). For recurring symbols introduced in *Basic Algebra*, see the list of Notation and Terminology on pages xxi–xxiv. Some of the latter notation has been repeated here for the reader's convenience.

In the list below, each piece of notation is regarded as having a key symbol. The first group consists of those items for which the key symbol is a fixed Latin letter, and the items are arranged roughly alphabetically by that key symbol. The next group consists of those items for which the key symbol is a Greek letter. The final group consists of those items for which the key symbol is a variable or a nonletter, and these are arranged by type. To locate an item below, first proceed on the assumption that the key symbol is a Latin or Greek letter; if the item does not appear to be in the list, then treat it as if its key symbol is a variable or a letter.

$\mathbb{A}$, $\mathbb{A}_K$, 389, 559
$\mathbb{A}^n$, $\mathbb{A}_K^n$, 455, 559
$\mathcal{A}(K, \text{Gal}(K/F), a)$, 137
$\mathcal{A}_F$, 542, 543
$\mathcal{A}_F^*$, 542, 543
$A(V)$, 579
$\widetilde{A}$, 570
$\widetilde{A}_d$, 570
$\widetilde{A}(V)$, 584
$\widetilde{A}(V)_d$, 585
K_{alg}, 434
$^a\varphi$, 639
$\mathcal{B}(F)$, 126
$\mathcal{B}(K/F)$, 127
$\mathcal{C}$, 330
$\mathcal{C}(\mathfrak{a})$, 620
$\mathcal{C}_R$, 169
$C(V(\mathfrak{a}))$, 633
$C_\mathbb{F}$, 532, 549
$C_{\mathbb{F},0}$, 534
E^c, complement, xxi

coimage f, 240
coker f, 175
$\mathcal{D}(\xi)$, 279
$\mathcal{D}(K/F)$, 372
$D_\mathbb{F}$, 532, 549
$D_{\mathbb{F},0}$, 534
$D_\mathbb{K}$, 267
$D(\Gamma)$, 267
$\text{Diff}(\mathbb{F})$, 547
$\text{Div}(\omega)$, 548
d_{-1}, 194
d_n, 153
$X = \{(X_n, d_n)\}_{n=-\infty}^{\infty}$, 174
$\dim R$, 424
$\text{Ext}_R^n(A, B)$, 223
e_i, f_i, g, 275, 354
$\langle e_{j_1}, \ldots, e_{j_k} \rangle$, 619
$\text{ext}_R^n(A, B)$, 223
$\mathbb{F}_q[[X]]$, 347
$\mathbb{F}_q((X))$, 347
$F_\mathfrak{p}$, 346

Fr_q, 437
f_v, 533
G_P, 368
$\text{Gal}(F_2, F_1)$, 434
$\leq_{\text{GLEX}}$, $\leq_{\text{GREVLEX}}$, 494
g, 538
g_x, 538
$\mathcal{H}(s, \mathfrak{a})$, 633
$\mathcal{H}_a(s, \mathfrak{a})$, 621, 626
$H(s, \mathfrak{a})$, 633
$H_a(s, \mathfrak{a})$, 625, 628
H_j, 620
$H_n(X)$, 153, 172
$H^n(X)$, 153, 174
$H_*(X)$, 172
$H^*(X)$, 174
$H_n(G, M)$, 209
$H^n(G, M)$, 147
$\text{Hom}_R(A, B)$, 169
$h(D)$, 7, 14
$h_\mathbb{K}$, 299
$\mathbb{I}$, $\mathbb{I}_K$, 390
$\mathbb{I}^1$, 390
$\mathcal{I}$, 330, 393, 576
$\widetilde{\mathcal{I}}$, 576
$I = (r_1, r_2)$, 38
$I = \langle r_1, r_1 \rangle$, 38
$I(E)$, 560, 571
$I(P, F \cap G)$, 474
$I(P, L \cap F)$, 467
image f, 240
$J(\xi)$, 272
$\overline{K(S)}$, 409
$\overline{K(E)}$, 412
$\Bbbk$, 528, 559
$\Bbbk(V)$, 580, 585
$\Bbbk'$, 531
$\mathcal{L}(A)$, 544
$L(A)$, 535
$L(s, \chi)$, 63

$\text{LCM}(X^\alpha, X^\beta)$, 501
Log, 289
$\text{LM}(f)$, $\text{LC}(f)$, $\text{LT}(f)$, 496
$\text{LT}(I)$, 497
$\leq_{\text{LEX}}$, 493
$\ell(A)$, 536
$\lim_\leftarrow$, 439
$\mathcal{M}$, 493, 620
M_P, 600
M_x, 431
$\mathfrak{m}_P$, 600
$\mathfrak{m}_x$, 431
$m_P(F)$, 474
$N(I)$, 39, 273
$N_{A/F}(\cdot)$, 165
$N_{K/F}(\cdot)$, norm, xxiv
$\text{Nrd}_{A/F}(\cdot)$, 165
$\mathcal{O}(U)$, 580, 582, 587, 641
$\mathcal{O}_P(U)$, 582, 587
$\mathcal{O}_P(V)$, 580, 585
R^o, opposite ring, xxii
$\text{ord}_v(A)$, 532
$\mathbb{P}^2$, 456
$\mathbb{P}^n$, 457, 570
$\mathbb{P}^n_K$, 457
$\mathcal{P}$, 330, 393
$P_\mathbb{F}$, 532, 549
P_v, 322, 533
$\mathbb{Q}_p$, 316, 318
$\mathcal{R}(f, g)$, 451
$R(f, g)$, 451
$R(f_1, F)$, 514
$R_\mathfrak{p}$, 346
R_v, 322, 533
R_x, 431
Residue, 542
$\text{Residue}_{p(v)}$, 541
r_1, r_2, 348, 383
rad A, 79
$S(f_1, f_2)$, 502

Index of Notation 719

S_∞, 391
$S^{-1}R$, localization, xxiv
Spec A, 639
(Spec A, $\mathcal{O}$), 641
K_{sep}, 434
$\text{Tor}_n^R(A, B)$, 224
$\text{Tr}_{A/F}(\cdot)$, 165
$\text{Tr}_{K/F}(\cdot)$, trace, xxiv
$\text{Trrd}_{A/F}(\cdot)$, 165
A^t, transpose, xxi
$\text{tor}_n^R(A, B)$, 224
tr. deg R, 424
$V(C)$, $V_K(C)$, 455–456
$V(I)$, 429
$V(S)$, 559, 571
$V(f_1, \ldots, f_k)$, 559
$V_\mathbb{F}$, 532
$v_P(\cdot)$, 321
v_∞, 328
$X(S)$, 388
X^α, 494, 620
$x_j(P)$, 559
$\mathbb{Z}(\Gamma)$, 268
$\mathbb{Z}_p$, 318
$\widehat{\mathbb{Z}}$, 437
$\mathbb{Z}G$, integral group ring, xxiii

Greek

α_i, 369
β_0, 574
β_0^t, 574
β_i, 369, 575
$\delta(A)$, 543
δ_{ij}, Kronecker δ, xxi
ε, 149, 195
η_i, 369
ι, 390, 391
σ, 617, 646
$\sigma_1, \ldots, \sigma_n$, 283, 383
χ_0, 62

ω, 185, 547

Functors given by subscripts and superscripts

$R^\times$, units, xxiv
R_P, localization, xxiv
X^+, 194
M^G, invariants, 208
M_G, coinvariants, 208
$\widehat{M}$, dual fractional ideal, 372
$M_\mathfrak{p}$, 376
L^Φ, 460

Specific functions

$\alpha = (\alpha_1, \ldots, \alpha_n)$, multi-index, 494
$|\alpha|$, 620
$\left(\frac{a}{p}\right)$, Legendre symbol, 8
$\left(\frac{m}{n}\right)$, Jacobi sysmbol, 68
$[K : F]$, degree, xxiv
$|\cdot|_p$, 316
$|\cdot|$, absolute value, 331
$\|\cdot\|$, norm, 356
$(x)_0$, $(x)_\infty$, 532

Isolated symbols

$\sim$, Brauer equivalent, 124
$\simeq$, homotopic, 154
∂_n, 153, 172
∂_{-1}, 194
$\prod'$, restricted direct product, 388

Operations on sets and classes

RG, group algebra, xxiii
$\sqrt{I}$, radical, 405
$K[X_1, \ldots, X_{n+1}]_d$, 458
$A \xrightarrow{u} B$, morphism, 235

Miscellaneous

(x), principal divisor, 532
$(x_i)_{i \in I}$, 388

$I = (r_1, r_2)$, generated ideal, 38
$I = \langle r_1, r_2 \rangle$, 38
$[x, y, w]$, point in $\mathbb{P}^2$, 459
$[x_0, \ldots, x_n]$, point in $\mathbb{P}^n$, 570

$\varphi = \{(E, \varphi_E)\}$, rational map, 595
$X = \{(X_n, \partial_n)\}_{n=-\infty}^{\infty}$, 171
$(F, |\cdot|_F)$, valued field, 342
$\{\mathcal{O}(U), \rho_{VU}\}$, presheaf, 640

INDEX

Abel, 521
abelian category, 238
abelian group
    divisible, 196
    torsion, 169
abelian Lie algebra, 78
absolute discriminant, 35, 267
absolute norm of ideal, 39, 273
absolute value, 289, 331
    archimedean, 289
    discrete, 338
    of idele, 390
    nontrivial, 332
    normalized, 383, 384, 385, 386
    trivial, 331
acyclic resolution, 219
additive category, 233
additive functor, 170, 178
adele, 389
adjoint, 252
affine algebraic set, 559
    dimension of, 566
    irreducible, 563
affine coordinate ring, 579
affine curve, irreducible, 529
affine Hilbert function, 621, 626
affine Hilbert polynomial, 625, 628
affine hypersurface, irreducible, 430, 562
affine local coordinates, 461
affine n-space, 455, 559
affine plane curve, 455
    irreducible, 430, 524, 562
affine plane line, 455
affine scheme, 642
affine variety, 429, 562
algebra, xxiii
    abelian Lie, 78
    central, 111
    central simple, 111

crossed-product, 137
cyclic, 122, 162, 163
generalized quaternion, 121
Lie, 77
polynomial, 164
semisimple, 80
semisimple Lie, 78
simple, 80
simple Lie, 79
solvable Lie, 78
Weyl, 85
algebraic closure, separable, 434
algebraic set
    affine, 559
    irreducible affine, 563
    projective, 571
algebraically independent, 409
aligned primitive forms, 25
archimedean, 331, 333, 348
archimedean absolute value, 289
archimedean place, 383
archimedean valuation, 289
Artin product formula, 387, 390, 395
Artin reciprocity, 265
Artin's Theorem, 89
Artinian ring, 87
associated prime ideal, 446
associated translation, 622
associated vector subspace, 622
associative algebra
    semisimple, 80
    simple, 80
augmentation map, 149

Baer, 168
base field, 327
base space, 640
Bayer–Stillman ordering, 494
Bezout, 449

Bezout's Theorem, 453, 465, 471, 488
bidegree, 617
bifunctor, 223
bihomogeneous polynomial, 617
binary quadratic form, 3, 12
  similar, 74
birational, 595
birational map, 595
birationally equivalent, 595
Blichfeldt, 293
boundary, 172
boundary map, 172
boundary operator, 172
bounded sequence, 317
bracket, 78
Brauer equivalent, 124
Brauer group, 126
  relative, 127
Buchberger, 450
Buchberger's algorithm, 506

canonical class, 551
canonical divisor, 551
Cartan, E., 79
Cartan, H., 168
category
  abelian, 238
  additive, 233
  good, 169
Cauchy sequence, 317
Cayley, 77
central algebra, 111
central simple algebra, 111
centralizer, 114
chain complex, 171
  in abelian category, 240
  double, 257
  tensor product for, 258
chain map, 154, 155, 173
character
  Dirichlet, 62
  genus, 74
  multiplicative, 61
  principal Dirichlet, 62
Chase, 141
Chevalley, 165, 168
Chinese Remainder Theorem, xxiii, 30, 69,
    106, 314, 341, 367, 480, 483

class field, Hilbert, 265
class field theory, 265
class group
  form, 28
  ideal, 42, 265, 299, 330
  idele, 393
class number, 299, 393
  Dirichlet, 7, 14
co-invariant, 209
co-invariants functor, 209
coboundary, 174
coboundary map, 174
coboundary operator, 174
cochain complex, 173
cochain map, 154, 174
cocycle, 174
codomain of morphism, 232
cohomology, 153, 174
  sheaf, 168, 171, 218, 643
coimage in abelian category, 239
cokernel, 175
  of morphism, 236
    universal mapping property, 236
common discriminant divisor, 272
common index divisor, 272, 287, 310, 371
commutator ideal, 78
complete presheaf, 641
complete valued field, 343
  equal-characteristic case, 398
  unequal-characteristic case, 398
completion, 342
  universal mapping property of, 343
complex, 171
  in abelian category, 240
  chain, 171
  cochain, 173
  double, 257
  flat, 259
complex place, 383
composition formula, 24
condition (C1), 165, 518
cone, 572, 633
conic, 458
conjugate, 266, 288, 383
connecting homomorphism, 185, 187
connecting morphism in abelian category, 248
convergent infinite product, 51
convergent sequence, 317

coordinate, 455, 559
 affine local, 461
coordinate hyperplane, 620
coordinate ring
 affine, 579
 homogeneous, 584
coordinate subspace, 619
coproduct, xxiii
correspondence, one-one, xxi
countable, xxi
Cramer, 448
Cramer's paradox, 449
Cramer's rule, 448
crossed-product algebra, 137
cubic, 458
 twisted, 562
cubic extension, pure, 280
cubic number field, 279
cubical singular chain, 172
cubical singular homology, 172
cup product, 256
curve
 affine plane, 455
 elliptic, 648
 irreducible, 604
 irreducible affine, 529
 irreducible affine plane, 430, 524, 562
 projective plane, 458
cycle, 172
cyclic algebra, 122, 162, 163
cyclotomic field, 309

decomposition group, 368
Dedekind, 77
Dedekind Discriminant Theorem, 275, 371, 379, 381
Dedekind domain, xxiv, 266
 extension of, xxiv, 327, 417
Dedekind example, 287, 310
Dedekind's Theorem on Differents, 376
defined at a point, 580, 585
degenerate, 172
degree, 153
 of divisor, 533
 of inseparability, 415
 residue class, 275, 354, 533
 total, 457
 transcendence, 413

derived functor, 204
 formation of, 205
 long exact sequence, 209, 214
Dickson, 122
different, 279
 relative, 279, 372
differential, 543, 547
differential form, 541
dimension
 of affine algebraic set, 566
 of affine variety, 563
 geometric, 565
 Krull, 424, 528, 529, 564
 of zero locus, 423
Diophantus, 1
direct product, restricted, 388
direct sum in additive category, 233
directed set, 438
Dirichlet, 2, 24, 77
Dirichlet box principle, 297
Dirichlet character modulo m, 62
Dirichlet class number, 7, 14
Dirichlet L function, 63
Dirichlet pigeonhole principle, 297
Dirichlet series, 56
Dirichlet Unit Theorem, 290, 292, 384, 390, 395
Dirichlet's Theorem, 7, 50
discrete, 290
discrete absolute value, 338
discrete valuation, 322
 defined over $\Bbbk$, 529
discriminant, 12
 absolute, 35, 267
 of commutative semisimple algebra, 382
 field, 35, 264, 267
 fundamental, 33
 of ordered basis, 267
 relative, 275, 381
discriminant divisor, 272
divisible abelian group, 196
divisible module, 251
division algorithm, generalized, 499
divisor, 532
 principal, 532
divisor class, 532, 549
domain of morphism, 232
dominant rational map, 595

Double Centralizer Theorem, 115
double chain complex, 257
dual of fractional ideal, 372

Eckmann, 168
Eilenberg, 168
Eisenstein, 12
Eisenstein polynomial, 402
elimination ideal, 512
Elimination Theorem, 512
elimination type ordering, 494, 512
elliptic curve, 648
enough injectives, 202
enough projectives, 202
epi, 233
epimorphism, 233
equal-characteristic case, 398
equivalence class of forms
    ordinary, 13
    proper, 13
equivalence of
    absolute values, 333
    completions, 383
    forms, 13, 32
        improper, 13
        proper, 13, 32
    ideals, 40, 298
        narrow, 40
        strict, 40, 298
    morphisms, 242
Euler, 1, 3, 9, 50
Euler product, 50, 54, 60
    first-degree, 60
Euler's Theorem, 516, 646
exact complex, 175
exact functor, 179
    left, 182
    right, 183
exact on injectives, 222
exact on projectives, 222
exact sequence, 175
    in abelian category, 240
    long, 187, 188
    short, 175
    split, 200
Exchange Lemma, 412
Ext functor, 223
extension normal 435

extension
    of Dedekind domain, xxiv, 327, 417
    of integrally closed domain, 610
    purely transcendental, 409
    of valued field, 358
Extension Theorem, 512

factor set, 133
    trivial, 135
Fermat, 1, 3, 9
field discriminant, 35, 264, 267
field of formal Laurent series, 347
field of fractions, xxiv
field polynomial, 266
fine sheaf, 218
finiteness of class number, 390
first-degree Euler product, 60
flabby sheaf, 218
flat complex, 259
flat module, 256
form
    binary quadratic, 3, 12
    class group, 28
    negative definite, 14
    positive definite, 14
    reduced primitive, 18, 21
forms, primitive aligned, 25
Fourier inversion formula for finite abelian
    groups, 61
fractional ideal, 321
    principal, 321
    relative dual, 372
free resolution, 152, 195
Freudenthal, 168
Frobenius element, 437
Frobenius's Theorem about division algebras
    over the reals, 118, 160
function field, 419, 528, 580, 582, 585, 587
    in one variable, 326, 382, 528, 529
    in r variables, 419
functor
    additive, 170, 178
    co-invariants, 209
    derived, 204
    exact, 179
    Ext, 223
    global-sections, 218
    homology-of-groups, 209

invariants, 208
left exact, 182
right exact, 183
Tor, 224
functorial, 177
functoriality
   with long exact sequence, 191
   of long exact sequence of derived functors, 215, 218
   with snake diagram, 190
fundamental discriminant, 33
fundamental parallelotope, 293
Fundamental Theorem of Galois Theory, 443
fundamental unit, 36, 288

Galois, 77
Galois group, 434
gap sequence, 557
Gauss, 1, 3, 9, 24, 77
Gauss's group, 5, 28
Gelfand, 348
generalized division algorithm, 499
generalized quaternion algebra, 121
generalized resultant, 514
genus, 32, 539, 556, 557
   principal, 33
genus character, 74
genus group, 33, 70, 73
geometric dimension, 565
germ, 584
global field, 382
global-sections functor, 218
good category, 169
graded lexicographic ordering, 493
graded monomial ordering, 627
graded reverse lexicographic ordering, 494
Grothendieck, 638
Gröbner, 450
Gröbner basis, 450, 497, 564
   minimal, 508
   reduced, 509

Haar measure, 385
Halphen, 450
Hamilton, 77
Hensel, 279
Hensel's Lemma, 349, 351, 353, 399
Herstein, 130

Hilbert, 404
Hilbert Basis Theorem, xxiv, 491, 560
Hilbert class field, 265
Hilbert function, 633
   affine, 621, 626
Hilbert polynomial, 633
   affine, 625, 628
Hilbert's Theorem 90, 71, 145
homogeneous coordinate ring, 584
homogeneous ideal, 458
homogeneous member of homogeneous coordinate ring, 585
homogeneous Nullstellensatz, 572
homogeneous polynomial, 457
homology, 153, 172
   cubical singular, 172
   simplicial, 172
homology-of-groups functor, 209
homomorphism, 78
   connecting, 185, 187
   inflation, 254
   restriction, 254
   of valued field, 342
homotopic, 154, 173, 174, 193, 198
homotopy, 173, 174, 193, 198
Hopf, H., 167
Hopkins, 92
Hurewicz, 167
hyperplane coordinate, 620
hypersurface, irreducible affine, 430, 562
hypersurface, irreducible projective, 573

ideal
   fractional, 321
   in Lie algebra, 078
   principal fractional, 321
   valuation, 322
ideal class group, 42, 265, 299, 330
idele, 390
idele class group, 393
idempotent, 91, 369
   primitive, 369
image in abelian category, 239
Implicit Function Theorem, 428, 600
improper equivalence of forms, 13
independent, algebraically, 409
index, 272
   ramification, 275, 354

inertia group, 370
inertia subfield, 368
inflation homomorphism, 254
inflation-restriction sequence, 254
injective, 195
   in abelian category, 241
injective module, 195
injective resolution, 199, 205
inseparable element, 414
integral closure, xxiv, 610
integral domain, xxiii
integral element, xxiv
integrally closed, xxiv
intersection multiplicity, 467, 474
intersection number, 467
invariant, 208
invariants functor, 208
inverse limit, 439
   standard, 439
inverse system, 438
irreducible affine algebraic set, 563
irreducible affine curve, 529
irreducible affine hypersurface, 430, 562
irreducible affine plane curve, 430, 524, 562
irreducible closed set, 564, 573
irreducible curve, 604
irreducible element, xxiii
irreducible ideal, 446
irreducible projective hypersurface, 573
irredundant, 446
isomorphic idempotents, 97
isomorphism, 78
   of valued field, 342
   of varieties, 591

Jacobi, 521
Jacobi identity, 78
Jacobi symbol, 68
Jacobson radical, 89

kernel of morphism, 235
   universal mapping property, 235
Koszul, 168
Kronecker, 77
Krull dimension, 424, 528, 529, 564
Kummer, 77
Kummer's criterion, 275
Künneth Theorem, 258

Lagrange, 1, 4
Langlands reciprocity, 265
largest domain, 583, 595
Lasker–Noether Decomposition Theorem, 446, 639
lattice, 290
Law of Quadratic Reciprocity, 3, 8
least common multiple, 501
left adjoint, 252
left Artinian ring, 87
left exact functor, 182
left Noetherian ring, 87
left semisimple ring, 81
Legendre, 1, 4
Legendre symbol, 8
Leibniz, 7
Leray, 168
Levi, E. E., 79
lexicographic ordering, 493
Lie algebra, 77
   abelian, 78
   semisimple, 78
   simple, 79
   solvable, 78
Lie subalgebra, 78
line
   affine plane, 455
   at infinity, 459
   projective, 458
Liouville, 521
local expression, 462
local morphism, 642
local ring, xxiv
local ring at a point, 580, 582, 585, 587
local/global approach, 371
localization, xxiv
locus of common zeros, 429, 559, 571
long exact sequence, 187, 188
   functoriality with, 191
   of derived functors, 209, 214
      functoriality with, 215, 218

Mac Lane, 168, 420
Macaulay, 627
maps of a good category, 169
matrix units, 101
member in abelian category, 242
minimal Gröbner basis, 508

Minkowski, 301, 302
Minkowski Lattice-Point Theorem, 293, 384
modules of a good category, 169
monic, 232
mono, 232
monomial, 457
    reduced, 646
monomial ideal, 619
monomial ordering, 493
    graded, 627
monomorphism, 232
morphism, 169
    of affine scheme, 642
    local, 642
    of ringed space, 642
    of varieties, 591
multiplicative, 60
    strictly, 60
multiplicative character, 61
multiplicity of a tangent line, 478

Nakayama's Lemma, xxiii, 120, 605, 606
narrow equivalence of ideals, 40
natural, 177
negative, xxi
negatively oriented, 40
neighbor, 21
neighbor on the left, 21
neighbor on the right, 21
nil left ideal, 89
nilpotent element, 89, 446
nilpotent left ideal, 80, 90
Noether Normalization Lemma, 612
Noether–Jacobson Theorem, 130
Noetherian, xxiv
Noetherian ring, 87
Noetherian topological space, 564
nonarchimedean, 331, 335, 338
nonarchimedean place, 383
nonsingular curve, 604
nonsingular point, 429, 600, 601
nontrivial absolute value, 332
norm, 165, 356
    of ideal, 39
        absolute, 273
normal extension, 435
normalized absolute value, 383, 384, 385, 386
Nullstellensatz, 404, 561

    homogeneous, 572
number field, xxiv
    cubic, 279
    cyclotomic, 309
    quadratic, 35, 69, 263, 269

Oka, 168
one-one correspondence, xxi
order, 532
order of vanishing, 474
ordering
    Bayer–Stillman type, 494
    graded lexicographic, 493
    graded monomial, 627
    graded reverse lexicographic, 494
    k-elimination type, 494, 512
    lexicographic, 493
    monomial, 493
    total, 493
    from tuple of weight vectors, 494
ordinary equivalence class of forms, 13
oriented, 40
orthogonal idempotents, 97, 369
Ostrowski, 348
Ostrowski's Theorem, 336

p-adic absolute value, 316
p-adic integer, 279, 318
$\mathfrak{p}$-adic integer, 346
p-adic metric, 316
p-adic number, 279, 316, 318
$\mathfrak{p}$-adic number, 346
parallelotope, 548
    fundamental, 293
Peirce decomposition, 95
perfect field, 418, 554
place, 383
plane, projective, 456
plane curve
    affine, 455
    irreducible affine, 430, 524, 562
    projective, 458
plane line, affine, 455
Plücker, 450
point, 455, 456, 459, 559
points at infinity, 459
pole part, 537
pole set, 581, 585

positive, xxi
positively oriented, 40
presheaf, 640
　complete, 641
primary ideal, 445
prime element, xxiii
prime ideal, xxiii
　associated, 446
primitive, 12
primitive form, reduced, 18, 21
primitive forms, aligned, 25
primitive idempotent, 369
primitively represent, 14
principal Dirichlet character, 62
principal divisor, 532
principal fractional ideal, 321
principal genus, 33
problem
　ideal-equality, 510
　ideal-membership, 507
　proper-ideal, 507
product, xxiii
profinite group, 441
projective, 192
projective algebraic set, 571
projective closure, 575
projective hypersurface, irreducible, 573
projective in abelian category, 241
projective limit, 439
projective line, 458
projective module, 192
projective n-space, 457, 570
projective plane, 456
projective plane curve, 458
projective resolution, 195, 205
projective transformation, 460
projective variety, 572
proper equivalence of
　forms over $\mathbb{Q}$, 32
　forms over $\mathbb{Z}$, 13
proper equivalence class of forms, 13
pullback, 242
pure cubic extension, 280
　type of, 281
purely inseparable element, 415
purely inseparable extension, 416
purely transcendental extension, 409
pushout, 202, 243

quadratic form
　binary, 3, 12
　similar, 74
quadratic number field, 35, 69, 263, 269
quadratic reciprocity, 3, 8, 68
quartic, 458
quasi-affine variety, 568
quasiprojective variety, 573
quaternion algebra, 121

radical
　of algebra, 80
　of ideal, 405
　Jacobson, 89
　of Lie algebra, 78
　Wedderburn–Artin, 89, 91
ramification index, 275, 354
ramified, 367
ramify, 264, 275, 308
rational function, 580, 585
rational map, 595
　dominant, 595
rational point, 455, 456, 457, 459
real place, 383
reciprocity
　Artin, 265
　Langlands, 265
　quadratic, 3, 8, 68
reduced Gröbner basis, 509
reduced monomial, 646
reduced norm, 165
reduced polynomial, 165
reduced primitive form, 18, 21
reduced trace, 165
reducible ideal, 446
regular at a point, 580, 582, 585
regular function at a point, 587
regular function on an open set, 580, 582, 587, 641
regular point, 429
relative Brauer group, 127
relative different, 279, 372
relative discriminant, 275, 381
relative dual of fractional ideal, 372
represent, 14
　primitively, 14
residue class degree, 275, 354, 533, 322
Residue Theorem, 543

resolution, 194
　acyclic, 219
　free, 152, 195, 195
　injective, 199, 205
　projective, 205
　standard, 149
restricted direct product, 388
restriction homomorphism, 254
resultant, 449, 451
　generalized, 514
Riemann, 521
Riemann hypothesis, 530
Riemann sphere, 328
Riemann surface, 522
Riemann zeta function, 52, 58
Riemann's inequality, 538
Riemann–Roch Theorem, 522, 551, 648
right adjoint, 252
right Artinian ring, 87
right exact functor, 183
right Noetherian ring, 87
right semisimple ring, 81
ring of formal power series, 347
ringed space, 642

S-polynomial, 502
scheme, 642
　affine, 642
　defined over a ring, 643
Schmidt, 422
Schreier, 168
Schur's Lemma, 83
section, 640
Segre embedding, 617, 646
Segre variety, 617
semisimple associative algebra, 80
semisimple Lie algebra, 78
semisimple module, xxii
semisimple ring, 81, 84
separable algebraic closure, 434
separable element, 414
separable extension, 415
separable polynomial, 414
separable semisimple algebra over a field, 109
separably generated extension, 419
separating transcendence basis, 419
sheaf, 168, 640
　cohomology, 168, 171, 218, 643

fine, 218
flabby, 218
structure, 641
short exact sequence, 175
　in abelian category, 241
similar binary quadratic forms, 74
simple associative algebra, 80
simple Lie algebra, 79
simple module, xxii, 80
simple ring, 85
simplicial homology, 172
singular cube, 172
singular homology, 172
singular point, 429, 600, 601
Skolem–Noether Theorem, 113
snake diagram, 185, 261
　functoriality with, 190
Snake Lemma, 185, 248
solution of problem
　ideal-equality, 510
　ideal-membership, 507
　proper-ideal, 507
solvable Lie algebra, 78
spectral sequence, 171
spectrum, 639
split, 127
split exact sequence, 200
splitting field, 127
stalk, 640
standard inverse limit, 439
standard resolution, 149
standard subset, 622
Stickelberger's condition, 309
Stone, 638
strict equivalence of ideals, 40, 298
strictly multiplicative, 60
strong approximation property, 374
Strong Approximation Theorem, 372, 390, 391
structure sheaf, 641
summation by parts, 56

tangent lines, 478
tensor product of
　algebras, 104
　chain complexes, 258
　fields, 104
Theorem 90, Hilbert's, 71, 145
Tor functor, 224

Tornheim, 349
torsion abelian group, 169
torsion submodule, 257
total degree, 457
total ordering, 493
totally ramified, 367
trace, 165
transcendence basis, 409, 424
   existence, 411
   separating, 419
transcendence degree, 413
transcendence set, 409
translate of form, 26
triangular ring, 88
trivial absolute value, 331
trivial factor set, 135
twisted cubic, 562
type of pure cubic extension, 281

ultrametric inequality, 316, 331
unequal-characteristic case, 398
uniformizer, 323
uniformizing element, 323
unit, xxii, 36, 288
   fundamental, 36, 288
unital, xxii
Universal Coefficient Theorem, 261
universal mapping property of
   completion of valued field, 343
   cokernel, 236
   kernel, 235
unramified, 367

valuation, 322, 331
   archimedean, 289
   discrete, 322, 529
valuation ideal, 322
valuation ring, 322

valued field, 342
   complete, 343
   extension of, 358
   homomorphism of, 342
   isomorphism of, 342
variety, 590
   affine, 429, 562
   projective, 572
   quasi-affine, 568
   quasiprojective, 573
   as a scheme, 643
   Segre, 617

Weak Approximation Theorem, 340, 374
Wedderburn, 79, 86, 164
Wedderburn's Main Theorem, 94
Wedderburn's Theorem about finite division
   rings, 117, 160
Wedderburn's Theorem about semisimple
   rings, 83
Wedderburn–Artin radical, 89, 91
Weierstrass, 521
Weierstrass gap, 557
Weierstrass point, 557
Weierstrass valuation, 557
weight vectors, 494
Weil, 541, 543
Weyl algebra, 85

Zariski closure, 561, 578
Zariski topology, 560, 571
Zariski's Theorem, 431, 525, 601
zero locus, 455, 559, 571
zero member, 245
zero morphism, 233
zero object, 233
zeta function, 530
   Riemann, 52, 58